Monomers, Polymers and Composites from Renewable Resources

Edited by

Mohamed Naceur Belgacem

École Française de Papeterie et des Industries Graphiques (INPG)
Grenoble, France

Alessandro Gandini

CICECO, Chemistry Department
University of Aveiro, Portugal

Amsterdam • Boston • Heidelberg • London • New York • Oxford
Paris • San Diego • San Francisco • Singapore • Sydney • Tokyo

Elsevier
The Boulevard, Langford Lane, Kidlington, Oxford OX5 1GB, UK
Radarweg 29, PO Box 211, 1000 AE Amsterdam, The Netherlands

First edition 2008

British Library Cataloguing in Publication Data
A catalogue record for this book is available from the British Library

Library of Congress Cataloging-in-Publication Data
A catalog record for this book is available from the Library of Congress

ISBN: 978-0-08-045316-3

For information on all Elsevier publications
visit our web site at books.elsevier.com

Typeset by Charon Tec Ltd (A Macmillan Company), Chennai, India
www.charontec.com
Printed and bound in Great Britain

08 09 10 10 9 8 7 6 5 4 3 2 1

Contents

Foreword

Expert predictions about the future availability of fossil resources, viz. petrol, natural gas and coal, which are not renewable within a useful time scale, vary between one and three generations. The concern about the geopolitical situation related to these dwindling resources has therefore been increasing steadily in the last decade and has recently reached peak proportions with the apparently unstoppable soaring market prices of petrol.

The primary issue here is obviously energy, considering that more than 90 per cent of these resources are used as fuels, but the fact that the vast majority of organic chemicals and synthetic polymers are derived from them (as a result of the petrochemical 'revolution' of the post-Second World War boom), constitutes an equally serious challenge.

Whereas the energy issue is being vigorously debated through various alternative solutions (nuclear, biomass combustion, aeolian, geothermal, etc.), with the corresponding progressive implementations, the only alternative to fossil resources for the manufacture of commodity chemicals and polymers is the use of renewable vegetable and animal counterparts, that is, the biomass.

As a consequence, research initiatives to this effect are being implemented ubiquitously with a sense of increasing urgency, as witnessed by the growing investments assigned by the concerned ministries, supranational institutions (EU, UNIDO, etc.) and the private industrial sector. This is being accompanied by the dramatic increase in scientific publications, patents and international symposia covering the topic of the rational exploitation of renewable resources to produce commodities alternative to petrochemicals.

The incessant biological activities that the earth sustains thanks to solar energy provide not only the means of our survival, but also a variety of complementary substances and materials which have been exploited by mankind since its inception, albeit with a growing degree of sophistication. Suffice it to mention, as an example, wood as a source of shelter and, later, of paper. In modern times, the exploitation of renewable resources to prepare useful products and plastics was indeed quite prominent between about 1870 and 1940 (natural rubber for tyres, cellulose acetate and nitrate, plant-based dyes, drying oils, etc.). As already pointed out, however, a major shift in industrial chemistry took place, starting from the second quarter of the last century, which led to the supremacy of first coal and then petrol as the basis of its output in terms of most intermediates, commodities and polymers.

This trend witnessed its apex at the end of the last millennium, when competent assessments from various academic and industrial circles begun alerting the community about the need of returning to the exploitation of renewable resources, albeit, obviously, following a more rational and thorough strategy. The concept of the 'biomass refinery' puts forward an approach, similar to that of the classical petrol counterpart, in which each of the different components of a given natural resource is isolated by chemical or biochemical means with the aim of turning them into useful products. Thus, interesting chemicals and monomers for industry and medicine, compounds with a specific useful pristine structure, resins, natural fibres and oils (used as such or after adequate modification), as well as polymers produced by bacteria, could replace progressively petrol-based counterparts. To take wood again as an example, this strategy implies that its various components are separated and valorized, namely: (i) cellulose fibres, essentially for papermaking, but also as reinforcing elements in composite materials; (ii) lignin as a macromonomer for novel plastics, or as a source of valuable chemicals like vanillin; (iii) bark tannins for leather treatment and as components for resins and adhesives; (iv) specific minor chemicals present in knots for medical applications such as neutraceticals, etc.

In this way, the dependence on fossil resources steadily decreases and every country in the world profits equally from this radical change, since biomass is ubiquitous and the technologies associated with its refinery just have to be adapted to the local species.

Numerous countries have embarked on ambitious programmes devoted to these issues and some of them have already been turned into concrete achievements through collaboration between public and private sectors. This amply justifies the pursuit of further research and development activities.

The purpose of this book is to concentrate exclusively on one topic within this broad issue, namely the use of renewable resources as precursors or aids to novel macromolecular materials. It reflects the concerted effort of a number of specialists to propose a hopefully comprehensive scientific and technological appraisal of the state of the art and the perspectives related to the numerous facets of this realm. In the 15 years that have elapsed since the first, much more modest review on this topic was published by one of us*; progress has been astounding, both qualitatively and quantitatively. Hence, the initiative to prepare this collective volume, which we hope will constitute a useful working or consulting tool for all those involved, associated or simply interested in these important issues.

Given the global context, it does not seem preposterous to us to consider the materials discussed in this book as the *polymers of the future*.

The preparation of this book benefitted enormously from the unfailing editing advice of Joan Gandini, who devoted innumerable hours of her skills to help improve its quality. For this and for other precious heart-warming assistance, we are deeply grateful to her.

We wish to express our most sincere thanks to Vera Fernandes, who played a decisive role, with her secretarial competence and assiduity, in assuring the smooth progress of the different phases of the book construction.

* Gandini A., Polymers from renewable resources, in *Comprehensive Polymer Science, First Supplement*, Eds.: Aggarwal S.L. and Russo S., Pergamon Press, Oxford, 1992, pp. 528–573.

List of Contributors

Argüelles-Monal, Waldo	CIAD – Unidad Guaymas, Carret. al Varadero Nacional Km 6.6, Apdo. Postal 284, Col. Las Playitas, Guaymas, Sonora, 85480 Mexico. E-mail: waldo@cascabel.ciad.mx
Avérous, Luc	LIPHT-ECPM, University Louis Pasteur, 25 rue Becquerel-67087 Strasbourg Cedex 2, France. E-mail: AverousL@ecpm.u-strasbg.fr.
Belgacem, Mohamed Naceur	Laboratoire de Génie des Procédés Papetiers, UMR 5518, École Française de Papeterie et des Industries Graphiques (INPG), BP65, 38402 Saint Martin d'Hères, France. E-mail: naceur.belgacem@efpg.inpg.fr
Benvegnu, Thierry	Ecole nationale supérieure de chimie de Rennes, Campus de Beaulieu – avenue du Général Leclerc, 35700 Rennes, France. E-mail: Thierry.Benvegnu @ensc-rennes.fr
Borzani, Walter	Escola de Engenharia MauáCampus de São Caetano do Sul Praça Mauá 1 – São Caetano do Sul – SP CEP: 09580-900 Brasil. E-mail: borzani-@maua.br
Carvalho, Antonio José Felix de	Universidade Federal de São Carlos – Campus Sorocaba, P.B. 3031, 18043-970, Sorocaba, S.P. Brazil. E-mail: antonio.carvalho@if.sc.usp.br
Chodak, Ivan	Polymer Institute, Slovak Academy of Sciences, 842 36 Bratislava, Slovakia. E-mail: upolchiv@savba.sk
Dufresne, Alain	Laboratoire de Génie des Procédés Papetiers, UMR 5518, École Française de Papeterie et des Industries Graphiques (INPG), BP65, 38402 Saint Martin d'Hères, France. E-mail: alain.dufresne@efpg.inpg.fr
Galbis Pérez, Juan Antonio	Departamento de Química Orgánica y Farmacéutica. Facultad de Farmacia. Universidad de Sevilla, 41071 Sevilla (España). E-mail: jgalbis@us.es
Gandini, Alessandro	CICECO and Chemistry Department, University of Aveiro, 3810-193 Aveiro, Portugal. E-mail: agandini@dq.ua.pt
Garcia Martin, Maria de Gracia	Departamento de Química Orgánica y Farmacéutica. Facultad de Farmacia. Universidad de Sevilla, 41071 Sevilla (España). E-mail: graciagm@us.es
Gellerstedt, Göran L.F.	Department of Fibre and Polymer Technology, Royal Institute of Technology, KTH, 10044 Stockholm, Sweden. E-mail: ggell@pmt.kth.se
Goycoolea, F.M.	Centro de Investigación en Alimentación y Desarrollo, A.C., Apdo. Postal 1735, Hermosillo, Sonora 83000, Mexico. E-mail: fgoyco@ciad.mx
Heinze, Thomas	Centre of Excellence for Polysaccharide Research, Friedrich Schiller University of Jena, Humboldtstraße 10, D-07743 Jena. E-mail: thomas. heinze@uni-jena.de
Henriksson, E. Gunnar	Department of Fibre and Polymer Technology, Royal Institute of Technology, KTH, 10044 Stockholm, Sweden. E-mail: ghenrik@pmt.kth.se

Lemiégre, Löic Ecole nationale supérieure de chimie de Rennes, Campus de Beaulieu – avenue du Général Leclerc, 35700 Rennes, France. E-mail: loic.lemiegre @ensc-rennes.fr

Lora, Jairo H. GreenValue SA, 7 Camby Chase, Media PA 19063, USA. E-mail: jhlora @greenvalue-sa.com

Manzani, Danilo São Paulo State University, Institute of Chemistry, Dept. of General and Inorganic Chemistry, R. Prof. Francisco Degni, s/n – Araraquara, SP – Brazil. Zip – 14800-900. E-mail: daniloiq@bol.com.br

Messaddeq, Younes São Paulo State University, Institute of Chemistry, Dept. of General and Inorganic Chemistry, R. Prof. Francisco Degni, s/n – Araraquara, SP – Brazil. Zip – 14800-900. E-mail: younes@iq.unesp.br

Neto, Carlos Pascoal CICECO and Chemistry Department, University of Aveiro, 3810-193 Aveiro, Portugal. E-mail: cneto@dq.ua.pt

Pecoraro, Édison São Paulo State University, Institute of Chemistry, Dept. of General and Inorganic Chemistry, R. Prof. Francisco Degni, s/n – Araraquara, SP – Brazil. Zip – 14800-900. E-mail: pecoraro@iq.unesp.br

Peniche Covas, Carlos Andrés Centro de Biomateriales, Universidad de La Habana Ave. Universidad s/n e/ G y Ronda, A.P. 6120,10600 La Habana, Ciudad de La Habana, CUBA. E-mail: peniche@reduniv.edu.cu

Petzold, Katrin Centre of Excellence for Polysaccharide Research, Friedrich Schiller University of Jena, Humboldtstraße 10, D-07743 Jena. E-mail: katrin. petzold@uni-jena.de

Pizzi, Antonio ENSTIB, Université Henri Poincaré – Nancy 1, 27 Rue du Merle Blanc,BP 1041, 88051 EPINAL Cedex 9, France. E-mail: pizzi@enstib.uhp-nancy.fr

Plusquellec, Daniel Ecole nationale supérieure de chimie de Rennes, Campus de Beaulieu – avenue du Général Leclerc, 35700 Rennes, France. E-mail: Daniel. Plusquellec@ensc-rennes.fr

Popa, Valentin Technical University of Iasi, Faculty of Chemical Engineering, Iasi – 700050, Romania. E-mail: vipopa@ch.tuiasi.ro

Ribeiro, Sidney José Lima São Paulo State University, Institute of Chemistry, Dept. of General and Inorganic Chemistry, R. Prof. Francisco Degni, s/n – Araraquara, SP – Brazil. Zip – 14800-900. E-mail: sidney@iq.unesp.br

Rinaudo, Marguerite CERMAV – CNRS, BP53, 38041 Grenoble cedex 9, France. E-mail: Marguerite.Rinaudo@cermav.cnrs.fr

Silvestre, Armando Chemistry Department, University of Aveiro, 3810-193 Aveiro, Portugal. E-mail: armsil@dq.ua.pt

Spiridon, Iulana 'Petru Poni' Institute of Macromolecular Chemistry, Aleea 'Gr. Ghica Voda' 41A, Iasi-700487, Romania. E-mail: iulianaspiridon@yahoo.com

Zeng, Ming Faculty of Material Science and Chemical Engineering, China University of Geosciences (Wuhan), Wuhan 430074, China. E-mail: ming.zeng@nantes. inra.fr

Zhang, Lina Department of Chemistry, Wuhan University, Wuhan 430072, China. E-mail: lnzhang@public.wh.hb.cn

– 1 –

The State of the Art

Alessandro Gandini and Mohamed Naceur Belgacem

ABSTRACT

This chapter gives a general introduction to the book and describes briefly the context for which the editors established its contents and explains why certain topics were excluded from it. It covers the main raw materials based on vegetable resources, namely (i) wood and its main components: cellulose, lignin, hemicelluloses, tannins, rosins and terpenes, as well as species-specific constituents, like natural rubber and suberin; and (ii) annual plants as sources of starch, vegetable oils, hemicelluloses, mono and disaccharides and algae. Then, the main animal biomass constituents are briefly described, with particular emphasis on: chitin, chitosan, proteins and cellulose whiskers from molluscs. Finally, bacterial polymers such as poly(hydroxyalkanoates) and bacterial cellulose are evoked. For each relevant renewable source, this survey alerts the reader to the corresponding chapter in the book.

Keywords

Animal biomass, Vegetal biomass, Wood, Cellulose, Lignins, Hemicelluloses, Natural rubber, Suberin, Tannins, Rosins, Terpenes, Annual plants, Starch, Vegetable oils, Hemicelluloses, Mono and disaccharides, Polylactic acid, Algae, Chitin, Chitosan, Proteins, Cellulose whiskers, Bacterial polymers, Poly(hydroxyalkanoates), Bacterial cellulose

1.1 THE CONTEXT

The biosynthesis of macromolecules through enzymatic, bacterial and chemical polymerizations of specific molecular structures constitutes a key step in the evolution of living organisms. Natural polymers have therefore been around for a very long time and always constituted one of the essential ingredients of sustainability, first and foremost as food, but also as shelter, clothing and source of energy. These renewable resources have also played an increasingly important role as materials for humanity through their exploitation in a progressively more elaborated fashion. The ever improving technologies associated with papermaking, textile and wood processing, vegetable oils, starch and gelatin utilization, the manufacture of adhesives, etc. represent clear examples of the progressive sophistication with which man has made good use of these natural polymers throughout the millennia. Concurrently, natural monomers have been polymerized empirically for equally long periods for applications such as coatings, paint and ink setting, leather tanning, etc.

The progress of chemistry, associated with the industrial revolution, created a new scope for the preparation of novel polymeric materials based on renewable resources, first through the chemical modification of natural polymers from the mid-nineteenth century, which gave rise to the first commercial thermoplastic materials, like cellulose acetate and nitrate and the first elastomers, through the vulcanization of natural rubber. Later, these processes were complemented by approaches based on the controlled polymerization of a variety of natural monomers and oligomers, including terpenes, polyphenols and rosins. A further development called upon chemical technologies which transformed renewable resources to produce novel monomeric species like furfuryl alcohol.

The beginning of the twentieth century witnessed the birth of a novel class of materials, the synthetic polymers based on monomers derived from fossil resources, but the progress associated with them was relatively slow up to the Second World War and did not affect substantially the production and scope of the naturally based counterparts. Some hybrid materials, arising from the copolymerization between both types of monomers were also developed at this stage as in the case of the first alkyd resins. Interestingly, both monomers used in the first process to synthesize nylon in the late 1930s were prepared from furfural, an industrial commodity obtained from renewable resources, in a joint venture between Quaker Oats and DuPont.

The petrochemical boom of the second half of the last century produced a spectacular diversification in the structures available through industrial organic chemistry. Among these, monomers played a very significant role, as it transpires from the high percentage of such structures represented in the list of the most important chemical commodities in world production. The availability of a growing number of cheap chemicals, suitable for the production of macromolecular materials, gave birth to 'the plastic age', in which we still live today, with of course greatly enhanced quantitative and qualitative features.

This prodigious scientific and technical upsurge went to the detriment of any substantial progress in the realm of polymers from renewable resources. In other words, although these materials never ceased to exist, very modest investments were devoted to their development, compared with the astronomical sums invested in petrochemistry. As a consequence, although cotton, wool and silk textile are still plentiful, the competition of synthetic fibres has not stopped growing. Likewise, the incidence of natural rubber is very modest today, compared with its synthetic counterparts, not only in relative tonnage, but also in the continuously widening degree of sophistication associated with the properties of the latter materials. In virtually all other domains associated with polymeric materials, the present contribution of structures derived from renewable resources is very modest and has not played an appreciable role in terms of bringing about specific functional properties. On the other hand, paper has resisted all attempts to be replaced by synthetic polymers, although these have been playing a growing role as bulk and surface additives, albeit without modifying the essential constitution of this material, which still relies on the random assembly of cellulose fibres.

We are deeply convinced that this state of affairs has nothing to do with any consideration of relative merits associated with the different structures and chemical processes involved in either context. Its origin is instead to be found in a purely economic aspect (*i.e.* in the enormous difference in investment that favoured petrochemistry for the last half century) which was also the period when the chemical industry witnessed its fastest progress ever. The choice to finance R&D activities in polymers derived from fossil resources in a massive way was to the benefit of the corresponding materials as we know them today. This objective situation, however, does not prove anything against the potential interest of alternative counterparts made from renewable resources, simply because no such investments were ever made to that effect.

Should fossil resources be available for us to exploit for centuries to come, the above arguments would sound like a futile exercise in style. Their validity stems precisely and primarily from the very fact that fossil resources are dwindling and becoming progressively more expensive. Furthermore, they are not a commonly shared richness, since their global distribution is totally uneven, which implies that certain countries are heavily dependent on others in this respect. These problems are of course affecting the energy outlook in the first instance, since some 95 per cent of the fossil resources are used as fuel, but the looming crisis will inevitably affect the corresponding chemical industry as well.

The purpose of this book is to show, through its attempt to cover, as exhaustively as possible, the wide spectrum of materials already potentially available, that renewable resources are perfectly apt to provide as rich variety of monomers and polymers as that currently available from petrochemistry. Implicit in this statement is the condition that substantial investments should be placed in the future to carry out the required research.

All contributors to this volume provided stimulating evidence about the potentials in their own field, but they are also aware that their efforts will turn into industrial realities only with the joint intervention of adequately financed fundamental and applied research, viz. the implication of both public and private sectors. If, on the one hand, it is encouraging to see a very impressive increase in this type of activity worldwide, the situation relative to petrochemistry, on the other hand, is still at a very low level of competitiveness. Qualitatively, a change in awareness has indeed taken place, parallel to the preoccupation surrounding the energy issues. This book is intended to amplify these promising initial stirrings by providing very sound examples of what can be achieved thanks to this alternative strategy.

Renewable resources are intrinsically valuable in this realm because of their ubiquitous character, which gives any society precious elements of sustainability, including with respect to polymeric materials. In all the topics covered by this book, emphasis is made, explicitly or implicitly, to the essential fact that the specific sources utilized for the purpose of producing new polymers, are taken neither from food, nor from natural materials, but instead from *by-products* of agricultural, forestry, husbandry and marine activities. One of the best examples of this strategy is the production of furfural, since it can be carried out industrially virtually anywhere in the world (see Chapter 6), given the fact that any vegetable by-product containing pentoses represents an excellent raw material for its synthesis.

The term 'renewable resource', as used in this book, is defined as any animal or vegetable species which is exploited without endangering its survival and which is renewed by biological (short term) instead of geochemical (very long term) activities.

1.2 VEGETABLE RESOURCES

It is estimated that the world vegetable biomass amounts to about 10^{13} tons and that solar energy renews about 3 per cent of it per annum. Given its fundamental role in the maintenance of the oxygen level, the principle of sustainability limits its exploitation at most to that renewed percentage. With respect to the scope of this book, vegetable biomass can be divided into wood, annual plants and algae.

1.2.1 Wood

Wood is the most abundant representative of the vegetable realm and constitutes the paradigm of a composite material. It displays, on the one hand, a basic universal qualitative composition in terms of its major constituents (cellulose, lignin, hemicelluloses and polyphenols) and, on the other hand, species-specific components which can be polymeric, like poly-isoprene (natural rubber) and suberin, or small molecules, like terpenes, steroids, etc. An example of its morphology is shown in Fig. 1.1, which illustrates the role of the three basic components respectively as the matrix (lignin), the reinforcing elements (cellulose fibres) and the interfacial compatibilizer (hemicelluloses). The middle lamella (0.5–2 μm) is mainly composed of lignin (70 per cent), associated with small amounts of hemicelluloses, pectins and cellulose. The primary wall, often hard to distinguish from the middle lamella, is very thin (30–100 nm) and is composed of lignins (50 per cent), pectins and hemicelluloses. The secondary wall is the main part of the vegetal fibres. Its essential component is cellulose and it bears three layers, viz. the external, S_1 (100–200 nm), the central S_2 (the thickest layer of 0.5–8 μm) and the internal or tertiary layer, S_3 (70–100 nm) situated close to the lumen.

Wood is the structural aerial component of trees. The rest of their anatomy, namely roots, leaves, flowers and fruits, are not relevant to the aim of this book, and will therefore not be dealt with.

Cellulose dominates the wood composition, although its proportion with respect to the other main components can vary appreciably from species to species. Conversely, polyphenols are the least abundant components and moreover, can exhibit quite different structures. As for lignins and hemicelluloses, their relative abundance and their detailed structures are essentially determined by the wood family: softwoods are richer in lignins, whereas hardwoods are richer in hemicelluloses. These three basic polymeric components represent fundamental sources

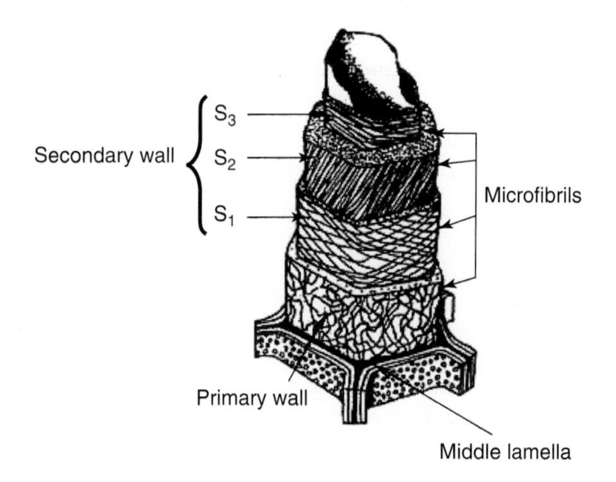

Figure 1.1 Typical morphology of a wood fibre.

of interesting materials and are thoroughly examined in this context, in specific chapters of this book, which focus their attention on the exploitation of these natural polymers after appropriate chemical treatment, with the aim of obtaining novel polymeric materials. The uses of wood itself as a structural material, as a source of furniture or flake boards and as a raw material in pulping, will not be treated here, because these applications call upon the exploitation of this fundamental natural resource through well-established technologies. Obviously, all these processes undergo regular improvement in their chemistry and engineering, but we deemed that their inclusion in the present treaty would have unduly overcharged its contents. The interested reader will find excellent monographs on each topic, going from introductory texts to highly specialized books [1–6].

The only area in which wood-related materials is witnessing important research contributions, concerns its physical and chemical modification, in order to protect it against degradation by various reagents or to obtain novel properties such as a thermoplastic behaviour. These aspects are treated in Chapter 20.

Similar considerations determined our decision not to include here chapters on the traditional uses of cellulose, like papermaking and cotton textiles, whose technologies are very thoroughly documented [7–10]. As for those cellulose derivatives which have been exploited for a very long time, like some cellulose esters and ethers, they will be dealt with in the appropriate chapter (Chapter 16), but again without a systematic treatment of the corresponding processes and properties [11–13].

1.2.1.1 Cellulose

Virtually all the natural manifestations of cellulose are in the form of semi-crystalline fibres whose morphology and aspect ratio can vary greatly from species to species, as shown in Fig. 1.2. The subunits of each individual fibre are the microfibrils which in turn are made up of highly regular macromolecular strands bearing the cellobiose monomer unit, as shown in Fig. 1.3.

The interest of cellulose as a source of novel materials is reflected in this book through four chapters dealing, respectively, with (i) the chemical bulk modification for the preparation of original macromolecular derivatives with specific functional properties (Chapter 16); (ii) the surface modification of cellulose fibres in view of their use as reinforcing elements in composite materials and as high-tech components (Chapter 18); (iii) the processing and characterization of these composites, including the use of nano fibres (Chapter 19) and (iv) the technology and applications associated with bacterial cellulose (Chapter 17). These contributions clearly show that cellulose, the most abundant and historically the most thoroughly exploited natural polymer, still provides new stimulating avenues of valorization in materials science and technology.

1.2.1.2 Lignins

Lignin, the amorphous matrix of wood, is characterized by a highly irregular structure compared with that of cellulose, and is moreover known to vary considerably as a function of wood family (*in situ*) and of the isolation process, which always involves a depolymerization mechanism. Figure 1.4 gives a typical example of the structure of a lignin macromolecule with its most characteristic building blocks. The pulping technology which calls upon a delignification mechanism based on the use of sulphites, yields lignin fragments bearing sulphonate moieties (*i.e.* polyelectrolytes).

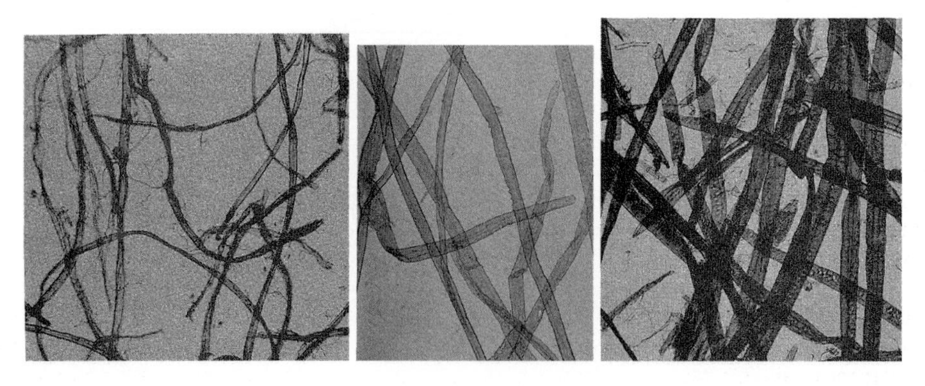

Figure 1.2 Cotton, pine and fir fibres.

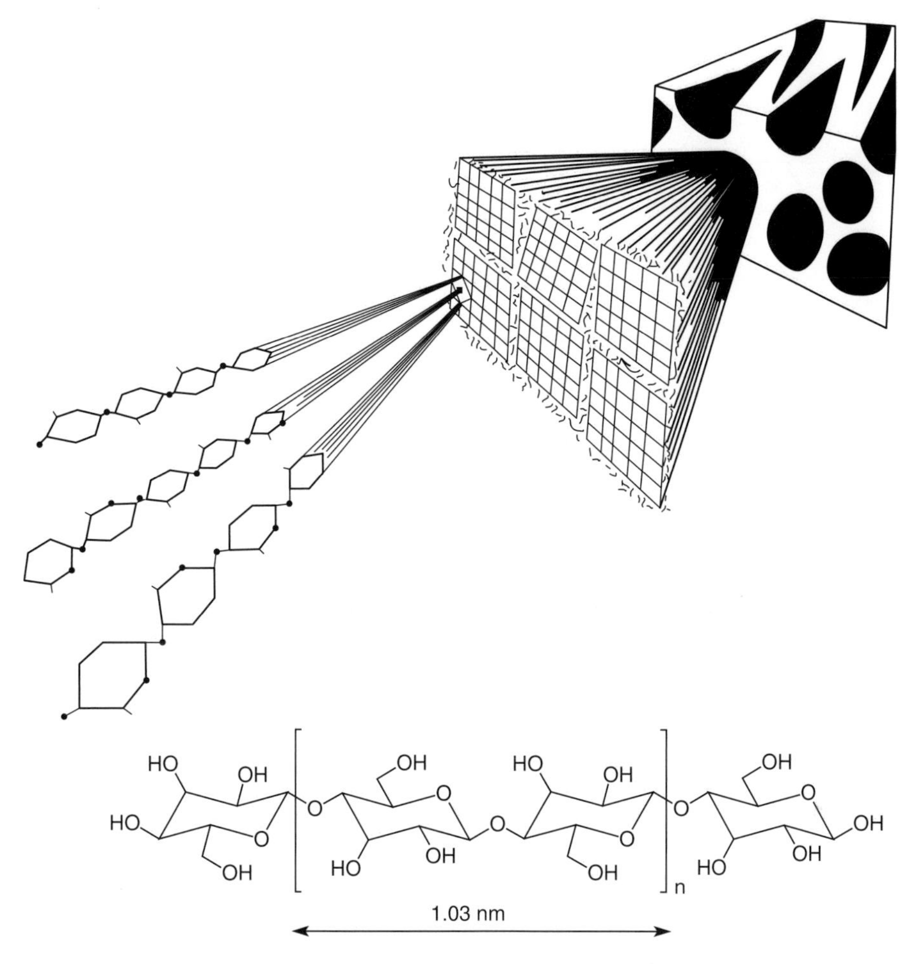

Figure 1.3 Schematic view of the components of cellulose fibre.

Traditionally, the aim of separating the wood components has been associated with papermaking, in which delignification isolates the cellulose fibres. In this context, the dissolved lignins have been utilized as fuel, which provides not only the energy required by the process, but also a convenient way of recovering its inorganic catalysts.

The idea of using these lignin fragments as macromonomers for the synthesis of polymers, by introducing them into formaldehyde-based wood resins, or by exploiting their ubiquitous aliphatic and phenolic hydroxyl groups, began to be explored only in the last quarter of the twentieth century. Given the fact that these industrial oligomers are produced in colossal amounts, it seems reasonable to envisage that a small proportion could be isolated for the purpose of producing new polymers, without affecting their basic use as fuel. Additionally, novel papermaking technologies, like the organosolv processes and biomass refinery approaches, like steam explosion, provide lignin fragments without the need of their use as a source of energy and with more accessible structures, in terms of lower molecular weights and higher solubility. Therefore, lignin macromonomers represent today a particularly promising source of novel materials based on renewable resources.

Three chapters are entirely devoted to lignin, covering (i) sources, structure and properties (Chapter 9), (ii) industrial processes and applications (Chapter 10) and (iii) their physical or chemical incorporation into novel macromolecular materials (Chapter 11). Moreover, its oxypropylation and the interest of the ensuing polyols are discussed in Chapter 12 and the properties of lignosulphonates as polyelectrolytes are dealt with in Chapter 24.

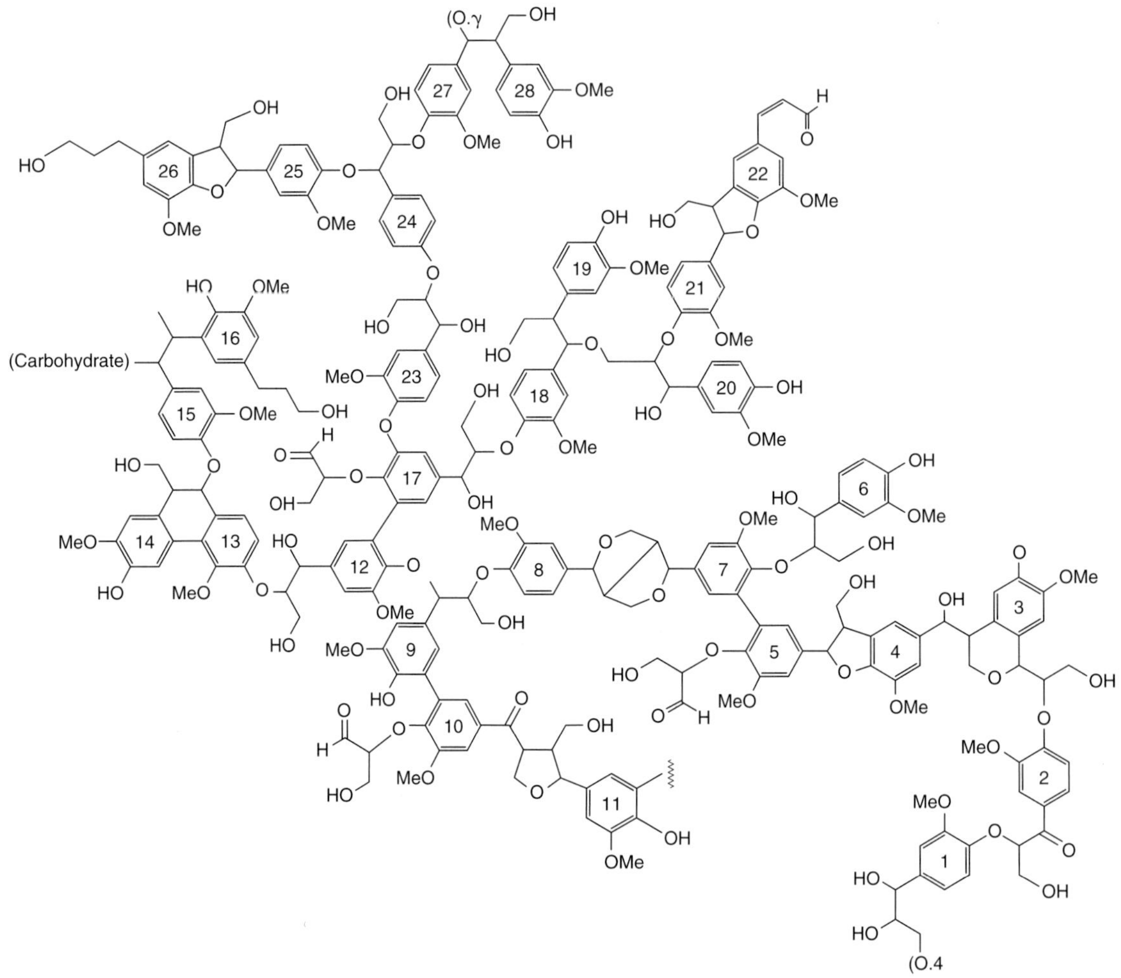

Figure 1.4 Lignin main moieties in a typical macromolecular assembly.

1.2.1.3 Hemicelluloses

Wood hemicelluloses are polysaccharides characterized by a relative macromolecular irregularity, compared with the structure of cellulose, both in terms of the presence of more than one monomer unit and by the possibility of chain branching. Figure 1.5 gives typical examples of such structures.

In papermaking processes, part of the wood hemicelluloses remain associated with the cellulose fibres, which results in the improvement of certain properties of the final material. The rest of these polysaccharides is dissolved together with lignin and in most processes it is burned with it. In some instances, however, particularly in the case of organosolv pulping or steam explosion technologies, the hemicelluloses can be recuperated as such. The utilization of wood hemicelluloses, but also of counterparts extracted from annual plants, has interested several industrial sectors for a long time, in particular that of food additives. In recent years, new possible outlets for hemicelluloses have been and are being explored, as discussed in Chapter 13.

1.2.1.4 Natural rubber

Turning now to more species-specific components, natural rubber is certainly one of the most important representatives. Different tropical trees produce different forms of poly(1,4-isoprene), which are exuded or extracted as an aqueous emulsion (latex) or as a sap-like dispersion, before coagulation. The *cis*-form of the polymer (Fig. 1.6(a)) tends to be amorphous and has a glass transition temperature of about −70°C, which makes it ideally suitable for application

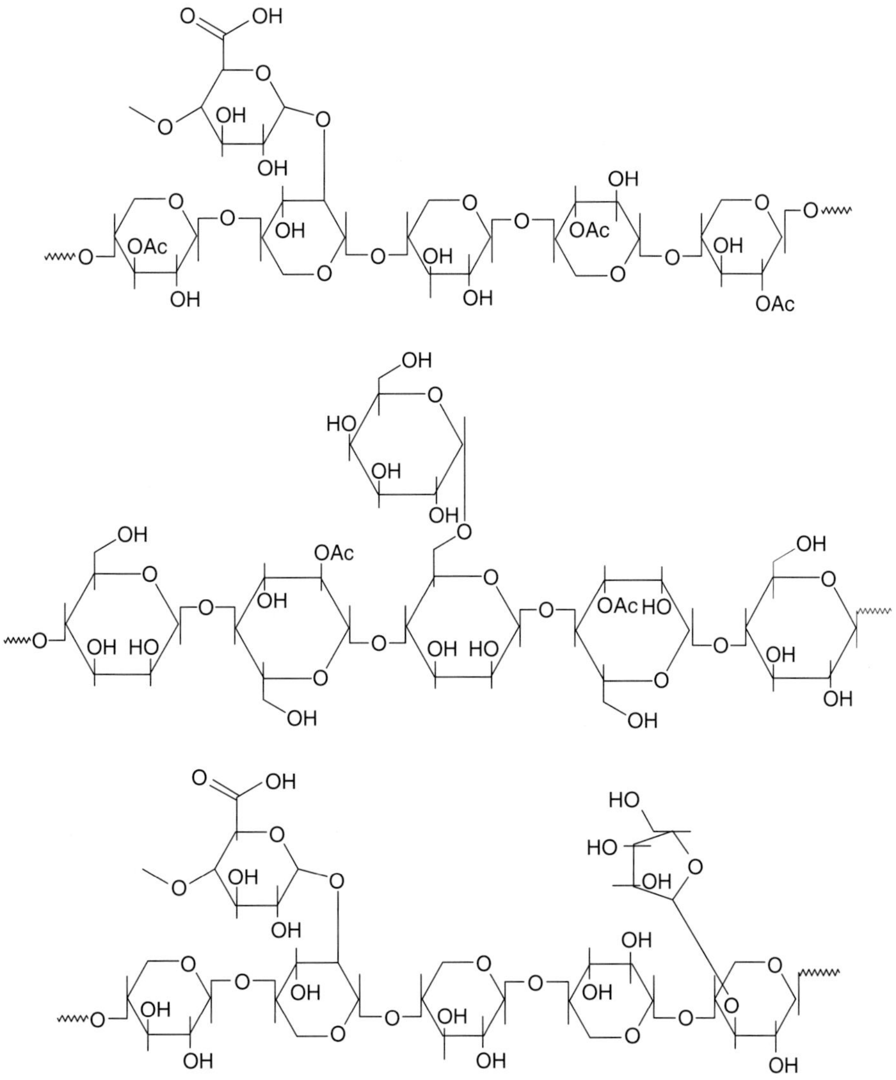

Figure 1.5 Three typical hemicellulose structures.

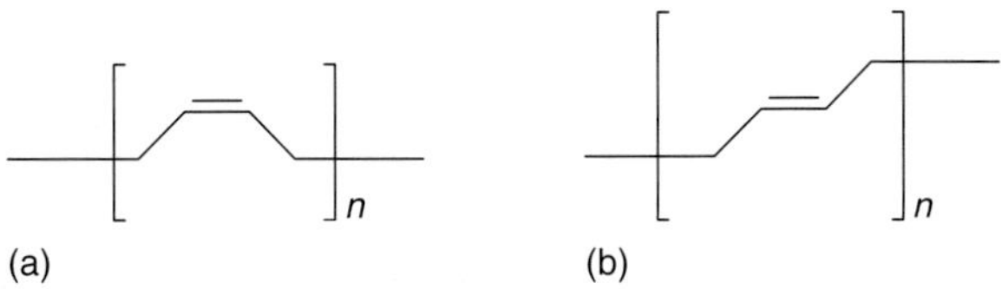

Figure 1.6 The two main structures of poly(1,4-isoprene) in natural rubber: (a) the *cis*-form and (b) the *trans*-form.

as elastomers, following chemical crosslinking (vulcanization) which involves some of its C=C insaturations. Its world production in 2004 was estimated at about 8 million tons. The *trans*-form (Fig. 1.6(b)), called *gutta percha* or balata, readily crystallizes forming rigid materials melting at about 70°C. As in the case of papermaking and cotton textile, the extraction and processing technology of these valuable natural polymers, as well as the preparation and optimization of their corresponding materials, represent a well-established and well-documented know-how [14] and is therefore not treated in this book. Examples of interesting recent contributions to the biosynthesis [15]

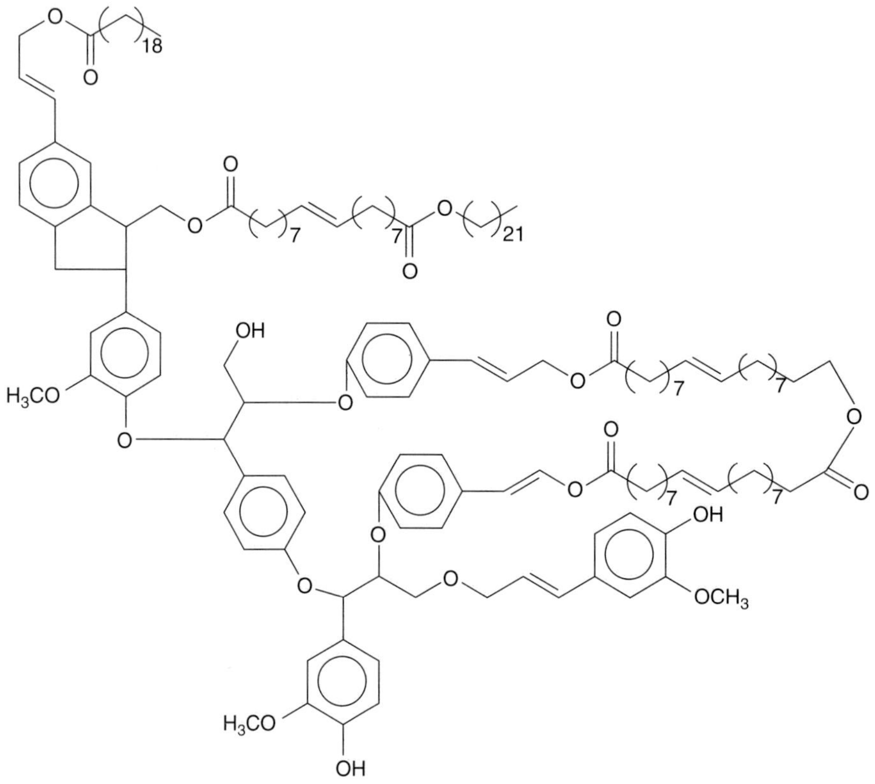

Figure 1.7 A schematic representation of the structure of suberin.

and chemical modification [16–18] of natural rubber are available. Moreover, the use of natural rubber in blends with other biopolymers like starch and lignins are discussed in the corresponding chapters.

1.2.1.5 Suberin

The other macromolecule found only in certain wood species is suberin. This non-linear polyester contains very long aliphatic moieties which impart a characteristic hydrophobic feature to the natural material that contains it. Figure 1.7 shows a schematic structure of suberin. By far the most representative species containing this polymer in its very thick bark (the well-known cork) is *Quercus suber*, which grows in the Mediterranean area, but Nordic woods like birch, also have a thin film of suberin coating their trunks. The sources of suberin, as well as the corresponding structure and composition are described in Chapter 14, together with the use of its monomeric components for the synthesis of novel macromolecular materials.

1.2.1.6 Tannins

Among the polyphenols present in the tree barks, tannins are by far the most interesting oligomers (molecular weights of 1 000–4 000) in terms of their utilization as macromonomers for the crosslinking of proteins in leather (tanning) and for macromolecular syntheses. The two representative structures of the flavonoid units in tannins are shown in Fig. 1.8. The most salient aspects related to the sources, structures and production of tannins and to their exploitation in polymer modification and manufacture are given in Chapter 8.

1.2.1.7 Wood resins

A number of resinous materials are secreted by trees. Their molecular weights are low (a few hundreds to a few thousands), hence a low melt viscosity, but their glass transition temperature can be as high as 100°C. Rosins (extracted

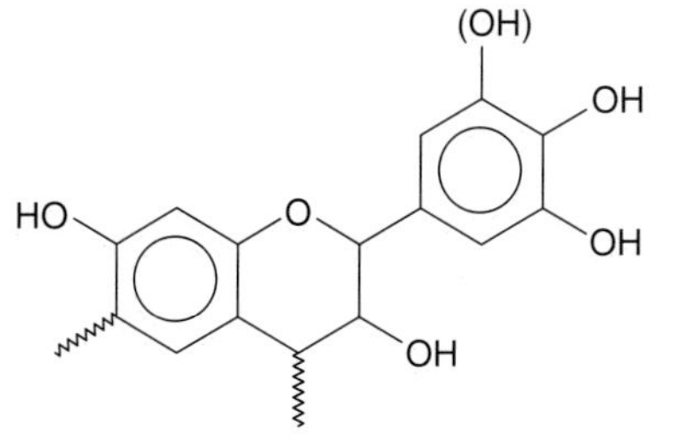

Figure 1.8 Two typical monomer units found in tannins.

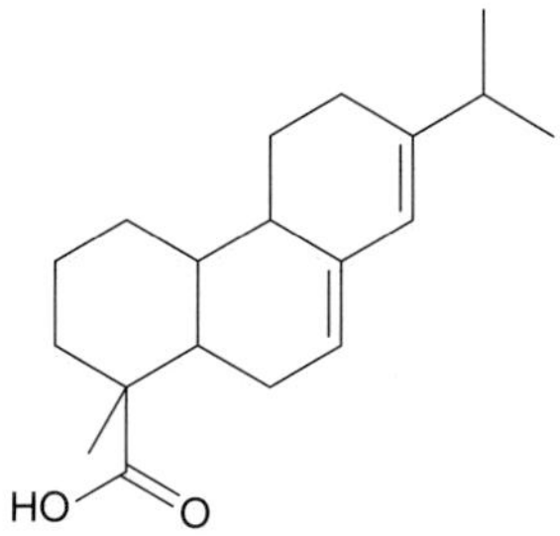

Figure 1.9 Abietic acid.

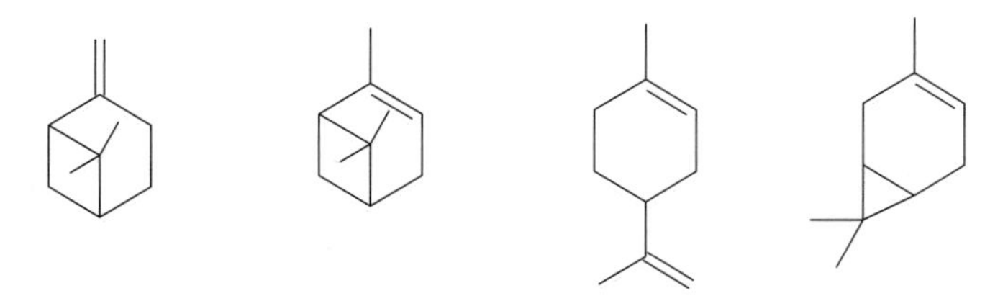

Figure 1.10 Four common terpenes.

from pine trees) are the most important representative of this family. They are made up of a mixture of unsaturated polycyclic carboxylic acids of which, abietic acid (Fig. 1.9) is the major representative. A detailed description of the sources, extraction, structures and chemical modification of these substances is given in Chapter 4, together with their use as sources of polymeric materials.

1.2.1.8 Terpenes

Apart from all these natural polymers and oligomers, either general or specific wood components, certain trees also produce monomers in the form of terpenes which are unsaturated cyclic hydrocarbons of general formula $(C_5H_8)_n$, with *n* mostly equal to 2. Figure 1.10 shows typical representatives of such compounds. Chapter 2 is devoted to the description of the sources of these monomers, their relative abundance and their polymerization.

Numerous other interesting molecules are found in the different elements of the tree anatomy, which find specific uses as a function of their structure (*e.g.* as medicines, cosmetics, dyestuffs, etc.), but which are not exploitable in polymer synthesis. These valuable compounds have been the object of much (still ongoing) research and development [19] which falls outside the scope of this book.

1.2.2 Annual plants

The term annual plant is used here to define plants and crops with a typical yearly turnover, but includes also species with shorter or longer cycles. The primary interest of annual plants, which have been optimized by human selection throughout the ages, is the production of food. Nevertheless, man has also exploited their residues for different purposes, including shelter, clothing, etc. In a complementary vein, annual plants have also been grown for the production of medicines, dyestuffs and cosmetics. In the framework of this book, attention will be focused on two different types of raw material, namely the fibrous morphologies of their basic structure and the substances they produce. Concerning the former, the lignocellulosic fibres of most plant stems fall into the same category as their wood counterpart and will therefore be discussed in the chapters devoted to cellulose. As for cotton (*i.e.* pure cellulose fibres annually produced by the corresponding plants) its traditional uses in textile, pharmaceutical aids, etc. reflect well-established and well-documented technologies, with little relevance to the primary scope of this book, as already pointed out in the case of papermaking and natural rubber technologies.

The relevant contribution of the output of annual plants to the realm of polymer synthesis and applications stems, instead, from some specific products, namely starch as a polymer, vegetable oils as triglyceride oligomers and hemicelluloses and monosaccharides as potential monomers or precursors to furan derivatives.

1.2.2.1 Starch

Starch is an extremely abundant edible polysaccharide present in a wide variety of tubers and cereal grains. In most of its manifestations, it is composed of two macromolecules bearing the same structural units, 1,4-*D*-glucopyranose, in linear (amylose, Fig. 11(a)) and highly branched architectures (amylopectin, Fig. 11(b)), present in different proportions according to the species that produces it. The utilization of starch or its derivatives for the production

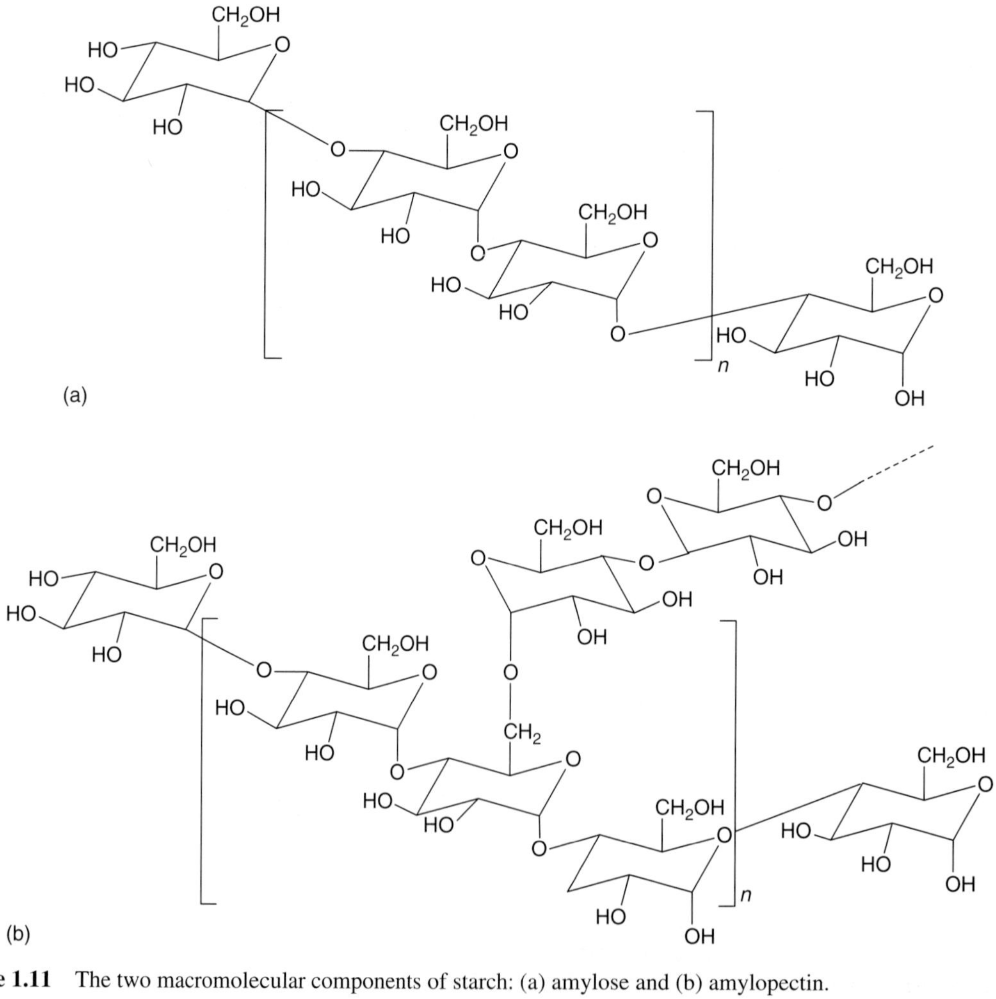

Figure 1.11 The two macromolecular components of starch: (a) amylose and (b) amylopectin.

of adhesives, or as wet-end additives in papermaking, constitute traditional applications to which a variety of novel materials, including plasticized starch, blends and composites, have been recently added, as discussed in Chapter 15.

1.2.2.2 Vegetable oils

Vegetable triglycerides are among the first renewable resources exploited by man primarily in coating applications ('drying oils'), because their unsaturated varieties polymerize as thin films in the presence of atmospheric oxygen. This property has been exploited empirically for millennia and has received much scientific and technological attention in the last few decades. These oils are extracted from the seeds or fruits of a variety of annual plants, mostly for human consumption. Within their general structure, consisting of glycerol esterified by three long-chain aliphatic acids (Fig. 1.12) bearing variable number of carbon atoms, the most relevant difference is undoubtedly the number of C=C insaturations borne by the chains, but other more peculiar features are also encountered (e.g. hydroxyl moieties). Their essential role as components of paints and inks constitute the most important application for the elaboration of materials. This traditional technology is presently being updated through research aimed at modifying the pristine structure of the oils in order to enhance their reactivity, particularly in the realm of photosensitive coatings, and thus render them competitive with respect to petroleum-based counterparts, like acrylic resins. Chapter 3 covers this reviving area of material science applied to renewable resources.

This chapter deals comprehensively with both triglycerides and fatty acids as sources of novel polymeric materials, but does not cover the growing interest in glycerol and its chemical transformation into other useful chemicals. This topic has been recently reviewed [20–22] and Chapter 7 discusses the interest of glycerol in the synthesis of surfactants derived from triglycerides.

1.2.2.3 Hemicelluloses

Annual plants produce a rich selection of hemicelluloses, often with quite different structures compared with those found in woods, although of course the basic chemical features are always those of polysaccharides. It follows that specific applications are associated with these different structural features, notably as food additives. The presence of charged monomer units is one of the most exploited characteristics, because of the ensuing rheological sensitivity to physical parameters. The properties and applications of plant and seaweed hemicelluloses are dealt with in Chapters 13 and 24, the latter dealing with the specific features related to polyelectrolytes.

In a different vein, plants rich in C5 hemicelluloses and more specifically xylans, are excellent raw materials for the production of furfural (Fig. 1.13). This simple technology, developed more than a century ago, has been applied to

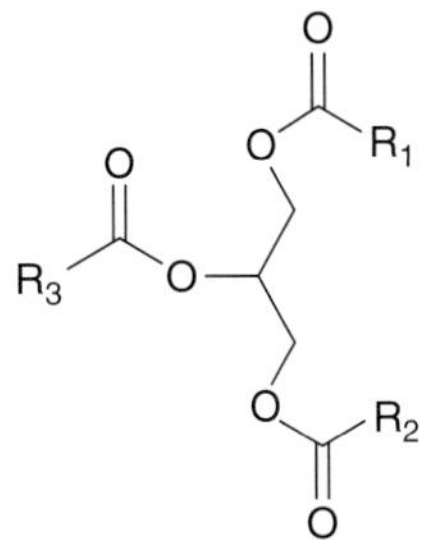

Where R_1, R_2 and R_3 are fatty acid chains.

Figure 1.12 The basic structure of triglycerides.

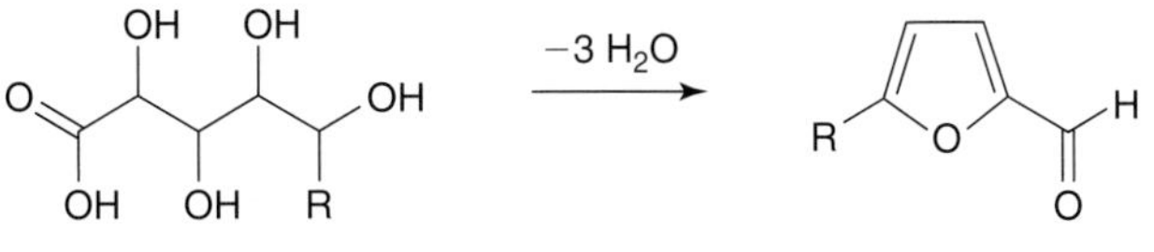

$-3\ H_2O$

R = H (furfural) or CH_3 (5-methyl furfural)

Figure 1.13 Schematic conversion of aldopentoses into furfural and 5-methyl furfural.

a whole host of annual plant residues, after the extraction of their food component, ranging from corn cobs, rice halls and sugar cane bagasse to olive husks. The industrial production of furfural is therefore possible in any country and indeed implemented in many of them, because of the wide variety of biomass containing its precursor and represents a beautiful example of the exploitation of renewable resources using a readily implemented and cheap process.

The use of furfural as such, as well as its transformation into a variety of furan monomers is discussed in Chapter 6 together with the synthesis and properties of the corresponding macromolecular materials.

Another possible exploitation of annual plant residues, after the separation of their foodstuff, is their conversion into polyols by oxypropylation. In this context, the whole of the residue is involved in the transformation, providing a convenient and ecological source of macromonomers. A typical example of this strategy is described in Chapter 12, in the case of sugar beet pulp, which is poor in xylans and cannot therefore be considered as a possible source of furfural.

1.2.2.4 Mono and disaccharides

The traditional use of some of the most important mono and disaccharides as sweeteners, whether energetic or not, is of course outside the scope of this book. The interest in using this family of compounds, produced by different annual plants, as precursors to novel materials, has increased considerably in recent years, mostly in three different directions, viz. (i) the conversion of fructose to hydroxymethyl furfural (Fig. 1.14), (ii) the synthesis of polycondensation materials using sugars as comonomers and (iii) the preparation of surfactants based on renewable resources.

Hydroxymethyl furfural is the other first generation furan derivative readily obtained from hexoses (C6 saccharides) [22]. Its role as a precursor to a large spectrum of monomers and the interest of the ensuing polymers constitute the topics of Chapter 6, complementary to those relative to furfural.

The synthesis of novel polymers, mostly polyesters and polyurethanes, based on mono and disaccharides (Fig. 1.15), together with their properties and possible applications, are described in Chapter 5.

A large section of Chapter 7 is devoted to the preparation of green surfactants based on sugars (Fig. 1.16). The inclusion of the topic of surfactants from renewable resources in this book stems from the fact that these compounds can play decisive roles in the synthesis, the processing and the application of polymers and must therefore be considered as an integral part of polymer science and technology.

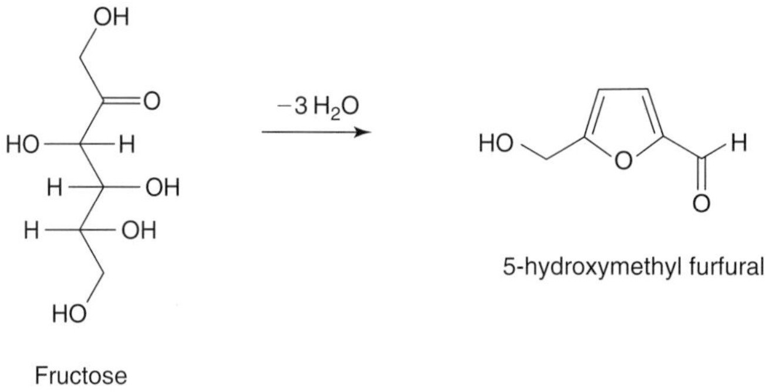

Fructose $\xrightarrow{-3\,H_2O}$ 5-hydroxymethyl furfural

Figure 1.14 Schematic conversion of fructose into 5-hydroxymethyl furfural.

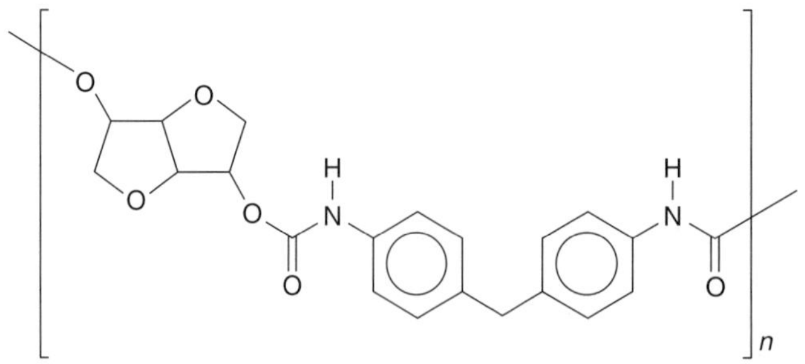

Figure 1.15 A sugar-based polyurethane.

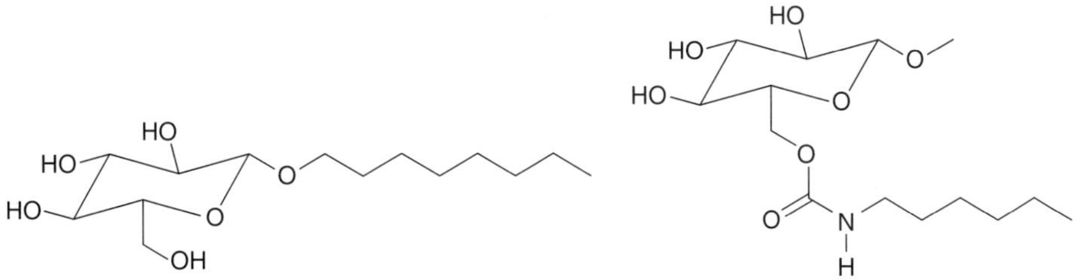

Figure 1.16 Two sugar-based non-ionic surfactants.

The fermentation of glucose has opened new avenues in the synthesis of polymers derived from renewable resources, with particular emphasis on the exploitation of lactic acid as a monomer. This topic is dealt with in Chapter 21.

1.2.3 Algae

Marine biomass is also a very interesting source of precursors to materials, both in terms of vegetable and animal resources. Polysaccharides derived from certain algae, like alginates, have been exploited for a long time as polyelectrolyte materials. This family of polymers are included in the contents of Chapter 24.

1.3 ANIMAL RESOURCES

As in the case of vegetable resources, all the traditional technologies of exploitation of materials derived from the animal realm will not receive a detailed treatment in this book. Thus, readers interested in leather [23], wool [24], silk [25], gelatin [26], animal fats and waxes [27] and carbon black [28], as well as animal-based resins like shellac [29], are invited to consult the corresponding monographs quoted here. The reason for these exclusions stems from the fact that the processes associated with the production of these materials have not been the object of any major qualitative improvement in recent times.

1.3.1 Chitin and chitosan

Chitin is undoubtedly the most abundant animal polysaccharide on earth. It constitutes the basic element of the *exo*-skeleton of insects and crustaceans, but it is also found in the outer skin of *fungi*. Chitin is a regular linear polymer whose structure differs from that of cellulose by the presence of *N*-methylamide moieties instead of the hydroxyl groups at C2 (Fig. 1.17). Given the susceptibility of this function to hydrolysis, chitin often bears a small fraction of monomer units in the form of primary amino groups resulting from that chemical modification.

Chitin is sparingly soluble even in very polar solvents, because of its high cohesive energy associated with strong intermolecular hydrogen bonds (NH—CO), which is also the cause of its lack of melting, because the temperature at which this phase change would occur is higher than that of the onset of its chemical degradation, just like with cellulose. It follows that the potential uses of chitin are strongly limited by these obstacles to processing. The possibility of exploiting chitin is therefore dependent on its transformation into its deacetylated derivatives through hydrolysis. As the proportion of the amide function converted into primary amino groups increases along the macromolecule, the

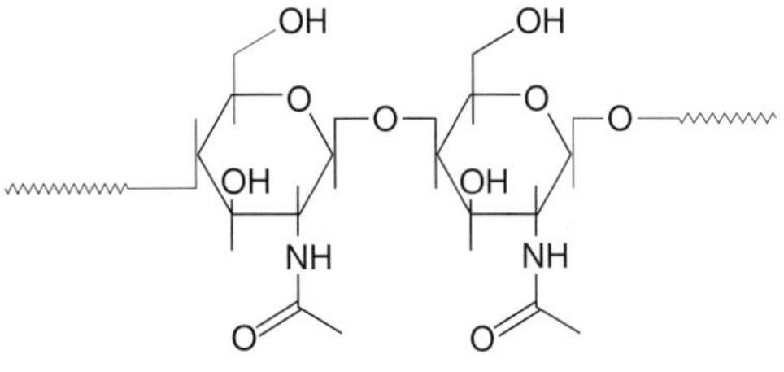

Figure 1.17 Chitin.

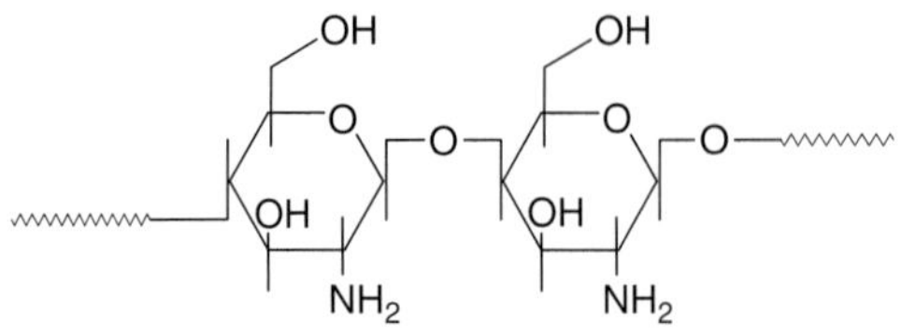

Figure 1.18 The main monomer units in chitosan.

corresponding material becomes progressively more soluble in such simple media as week aqueous acids or polar protic solvents. The ideal fully hydrolysed polymer takes the name of chitosan (Fig. 1.18), a term which is in fact also attributed to all random copolymers bearing more than about 50 per cent of amino monomer units.

Chitosan has become one of the most attractive polymers derived from renewable resources, because it possesses remarkable properties which find applications in many areas of material science and technology, particularly related to biomaterials and medical aids. It is not exaggerated to talk about the boom of chitosan-related research, considering the explosion of scientific and technical literature on this polymer, accompanied by the creation of learned societies and the frequent international meetings covering its progress. Industrial units devoted to the extraction of chitin followed by the production of chitosan are springing up throughout the world. The materials derived from chitin and chitosan are described in Chapter 25.

1.3.2 Proteins

Because of their highly polar and reactive macromolecular structure, proteins have attracted much attention in the last few decades, as possible sources of novel polymeric materials. The details of this activity are discussed in Chapter 23. A particularly interesting natural proteinic material is undoubtedly the spider dragline silk, because of its extraordinary mechanical properties. Given the obvious difficulties related to gathering viable amounts of this biopolymer, much research is being devoted to its bioengineering production [30].

1.3.3 Cellulose whiskers from molluscs

Although cellulose is the supreme example of a *predominantly vegetable* natural polymer, exotic animal species are known to produce this polysaccharide and, more particularly, some of its most regular manifestations. Thus, the tunicate mollusc has became the very symbol associated with cellulose whiskers (*i.e.* extremely regular nanorods with remarkable mechanical and rheo-modifying properties) as described in Chapter 19.

1.4 BACTERIAL POLYMERS

Although the polymerization induced by bacteria has been known and studied for a long time, the strategy based on using this biological activity to actually harvest commercial materials is a relatively recent endeavour. Two specific instances are prominent in this context, namely the production of poly(hydroxyalkanoates) and the synthesis of bacterial cellulose.

1.4.1 Poly(hydroxyalkanoates)

This family of polyesters and copolyesters (Fig. 1.19) has interested the polymer community both because of their remarkable physical properties and biodegradability. Efforts have been actively implemented to improve the economy of the biotechnological processes used to prepare these materials, so that they can become commercially competitive compared with petroleum-based polymers with similar properties. All these aspects are thoroughly tackled in Chapter 22.

Figure 1.19 The general formula of poly(hydroxyalkanoates).

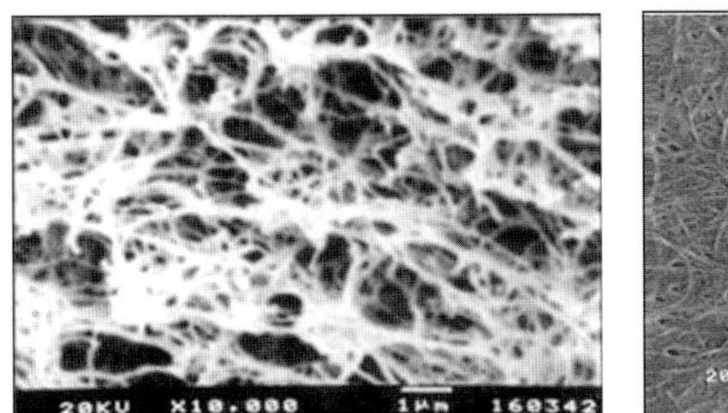

Figure 1.20 The unique morphology of bacterial cellulose.

1.4.2 Bacterial cellulose

Although the chemical structure of bacterial cellulose is identical to that of any other vegetable-based counterpart, its fibrous morphology (Fig. 1.20), as obtained directly in its biotechnological production, is unique and consequently the properties associated with this original material are also peculiar and promise very interesting applications. Details about this futuristic biopolymer are given in Chapter 17.

1.5 CONCLUSIONS

This book is not the first survey dealing with the interest of natural polymers or materials derived from renewable resources, since several monographs have been published in recent years [31–40], including two excellent reviews by Corma *et al.* covering the transformation of biomass for the production of fuel [41] and chemicals [21] and other recent contributions [42, 43]. The concept of the biorefinery [31, 44] is explicitly or implicitly at the basis of all these treaties, that is, the working hypothesis proposing a rational separation and exploitation of all the components of a given natural resource. The other common denominator to many of these collective overviews is the biodegradable character of the ensuing material.

We have attempted to gather in the present volume what we feel is a more comprehensive collection of monographs, with the materials science elements as the predominant feature. Of course, both the biorefinery and the biodegradability issues remain essential here, but within the primary focus spelled out in the title (*i.e.* ultimately the production of macromolecular materials from renewable resources).

REFERENCES

1. Fengel D., Wegener G., *Wood Chemistry Ultrastructure Reactions*, Walter de Gruyter, Berlin, 1989; Kennedy J.F., Philips, G.O., Williams P.A., *The Chemistry and Processing of Wood and Plant Fibrous Materials*, Woodhead, Cambridge, UK, 1996.
2. Hill C.A.S., *Wood Modification: Chemical, Thermal and Other Processes*, John Wiley & Sons, Ltd, Chichester, 2006.
3. Kettunen, P.O., Wood: Structure and properties, Mater. Sci. Found., 2006, Trans Tech Publications Ltd., Uetikon-Zuerich, 2006, pp. 29–30.
4. Rowell R.M., *Handbook of Wood Chemistry and Wood Composites*, CRC Press, Boca Raton, 2005.
5. Gullichsen J., Paulapuro H., *Papermaking Science and Technology*, Fapet Oy, Helsinki, 1999, Volumes 2 and 3.

6. Roberts J.C., *Paper Chemistry*, 2nd Edition, Chapman & Hall, United Kingdom, 1996; Eklund, D. Lindström, T., *Paper Chemistry: An Introduction*, DT Paper Science Publications, Grankulla, Finland, 1991.
7. Casey J.P., *Pulp and Paper Chemistry and Chemical Technology*, 3rd Edition, John Wiley and Sons, New York, 1980, Volumes 1–4.
8. Gullichsen J., Paulapuro H., *Papermaking Science and Technology*, Fapet Oy, Helsinki, 1999, Volumes 4–10.
9. Wakelyn P.J., Bertoniere N.R., French A.D., Thibodeaux D.P., Triplett B.A., Rousselle M.A., Goynes W.R. Jr., Edwards J.V., Hunter L., McAlister D.D., Gamble G.R., *Cotton Fibre Chemistry and Technology*, CRC Press, Boca Raton, 2007.
10. Ash M., Ash I., *Handbook of Textile Processing Chemicals*, Synapse Information Resources, Endicott, New York, 2001; Horrocks A.R., Anand S.C., *Handbook of Technical Textiles*, Woodhead Publ., Cambridge, UK, 2000.
11. Heinze T., Liebert T., Koschella A., *Esterification of Polysaccharides*, Springer, Berlin, 2006.
12. Hon D.N.S., Shiraishi N., *Wood and Cellulosic Chemistry*, Marcel Dekker Inc., New York, 1992.
13. Goetze K., *Synthetic Fibers by the Viscose Process* Springer-Verlag, 1967; Burt T.P., Heathwaite A.L., Trudgill T.S., *Nitrate: Processes, Patterns and Management*, Wiley, Chichester, UK, 1993.
14. Brendan R., *Rubber Compounding: Chemistry and Applications*, Marcel Dekker Inc., New York, 2004; Parry D.A.D., Squire J.M., Fibrous proteins: Coiled-coils, collagen and elastomers, *Adv. Protein Chem.*, **70**, 2005, Elsevier, San Diego, 2005.
15. Puskas J.E., Gautriaud E., Deffieux A., Kennedy J.P., *Prog. Polym. Sci.*, **31**, 2006, 533.
16. Kawahara S., Saito T., *J. Polym. Sci., Part A: Polym. Chem.*, **44**, 2006, 1561.
17. Kangwansupamonkon W., Gilbert R.G., Kiatkamjornwong S., *Macromol. Chem. Phys.*, **206**, 2005, 2450.
18. Jacob M., Francis B., Varughese K.T., Thomas S., *Macromol. Mater. Eng.*, **291**, 2006, 1119.
19. Murzin D.Y., Maki-Arvela P., Salmi T., Holmbom B., *Chem. Eng. Technol.*, **30**, 2007, 569.
20. Pagliaro M., Ciriminna R., Kimura H., Rossi M., Della Pina C., *Angew. Chem. Int. Ed.*, **46**, 2007, 4434.
21. Corma A., Iborra S., Velty A., *Chem. Rev.*, **107**, 2007, 2411.
22. Chheda J.N., Huber G.W., Dumesic J.A., *Angew. Chem. Int. Ed.*, **46**, 2007, 7164; Behr A., Eilting J., Irawadi K., Leschinski J., Lindner F., *Green Chem.*, **10**, 2008, 13.
23. Bienkiewicz K.J., *Physical Chemistry of Leather Making*, Krieger, Melbourne, FL, 1983.
24. Simpson W.S., Crawshaw G.H., *Wool: Science and Technology*, Woodhead Publ., Cambridge, UK, 2002.
25. Scheibel T., Silk, Special Issue of *Appl. Phys. A: Mater. Sci. Process.*, **82**, 2006, Springer, Heidelberg, 2006.
26. Schreiber R., Gareis H., *Gelatine Handbook*, Wiley-VCH, Weinheim, 2007.
27. Hamilton R.J., *Waxes: Chemistry, Molecular Biology and Functions*, The Oily Press, Dundee, UK 1995; Berger K.G., Animal Fats-BSE and After. *Proceedings of a Joint Meeting of the SCI Oils & Fats Group and the SCI Food Commodities & Ingredients Group*, London, UK, 10 June 1997, PJ Barnes & Associates, Bridgewater, UK, 1997.
28. Donnet J.B., Bansai R.C., Wang M.J., *Carbon Black: Science and Technology*, 2nd Edition, Marcel Dekker, New York, 1993.
29. Hicks E., *Shellac: Its Origin and Applications*, MacDonald, London, 1962.
30. Bini E., Po Foo C.W., Huang J., Karageorgiou V., Kitchel B., Kaplan D.L., *Biomacromolecules*, **7**, 2006, 3139.
31. Kamm B., Gruber P.R., Kamm M., *Biorefineries–Industrial Processes and Products*, Wiley VCH, Weinheim, 2006, Volumes 1 and 2.
32. Khemani K., C. Scholz, (Eds.), Degradable polymers and materials: Principles and practice, *ACS Symp. Ser.*, **939**, 2006, ACS, Washington DC.
33. Fingerman M., Nagabhushanam R., *Biomaterials from Aquatic and Terrestrial Organisms*, Science Publishers, Enfield, NH, 2006.
34. Im S.S., Kim Y.H., Yoon J.S., Chin I.J. (Eds), *Bio-Based Polymers: Recent Progress*, 2005, Wiley, New York.
35. Scott G., *Degradable Polymers, Principles and Applications*, 2nd Edition, Kluwer Academic, New York, 2003.
36. Steinbüchel A., Marchessault R.H., *Biopolymers for Medical and Pharmaceutical Applications*, Wiley, Weinheim, 2005.
37. Stevens C.V., Verhé R.G., *Renewable Bioresources: Scope and Modification for Non-Food Applications*, John Wiley & Sons, Ltd, Chichester, 2004.
38. Wool R.P., Sun X.S., *Bio-Based Polymers and Composites*, Elsevier, Amsterdam, 2005.
39. Wolf O., Techno-Economic Feasibility of Large-scale Production of Bio-Based Polymers in Europe, *Technical Report EUR 22103*, European Commission, 2005.
40. Argyropoulos D.S., (Ed.), Materials, chemicals and energy from forest biomass, *ACS Symp. Ser.*, **954**, 2007, ACS, Washingon DC.
41. Huber G.W., Iborra S., Corma A., *Chem. Rev.*, **106**, 2006, 4044.
42. Gallezot P., *Green Chem.*, **9**, 2007, 295.
43. Darder M., Aranda P., Ruiz-Hitzky E., *Adv. Mater.*, **19**, 2007, 1309.
44. Sanders J., Scott E., Weusthuis R., Mooibroek H., *Macromol. Biosci.*, **7**, 2007, 105.

– 2 –

Terpenes: Major Sources, Properties and Applications

Armando J.D. Silvestre and Alessandro Gandini

ABSTRACT

Turpentine, the volatile fraction of the so-called Naval Stores Industry is a complex mixture, mainly composed of monoterpenes. This chapter begins with a general overview of the major sources of turpentine, its composition and classical applications, including a systematic layout of the structure of terpenes. This is followed by a detailed discussion of the use of α- and β-pinene and other monoterpenes as monomers in conventional free radical and cationic homopolymerizations and copolymerizations, the latter involving other terpenes, monoterpene alcohols or synthetic monomers. Finally, the living cationic and living radical homo- and co-polymerization of β-pinene are thoroughly examined in terms of mechanisms and macromolecular structures. Throughout, the properties and possible applications of the materials ensuing from each type of system are critically examined.

Keywords

Turpentine, Monoterpenes, α-Pinene, β-Pinene, Cationic polymerization, Radical polymerization, Copolymerizations, Living polymerization, Mechanisms, Polymer properties

2.1 INTRODUCTION

The term 'terpenes' refers to one of the vastest families of naturally occurring compounds which bear a common biosynthetic pathway, but show enormous structural diversity depending on the number of carbon atoms (normally multiples of five), the variety of possible biosynthetic isomeric carbon skeletons and the stereochemical configurations.

Terpenes are secondary metabolites synthesized mainly by plants, but also by a limited number of insects, marine micro-organisms and fungi. These compounds were first considered as 'waste' products from plant metabolism with no specific biological role, but later, the involvement of some terpenes as intermediates in relevant biosynthetic processes was discovered [1]. Additionally, it has been well demonstrated that many terpenes play important ecological roles [1] as in plant defence, for example as insect repellents, and in symbiotic mechanisms, for example as attractants to specific insect species to stimulate cross pollination.

Most terpenes share isoprene (2-methyl-1,4-butadiene) as a common carbon skeleton building block. This structural relationship was identified by Wallach in 1887, who recognized that most terpenic structures result from the 'head-to-tail' condensation of isoprene units and this became known as the 'isoprene rule'. Based on this generic rule, terpenes can be classified according to the number of isoprene units (Table 2.1).

In the 1950s, Ruzicka proposed the 'biogenic isoprene rule' which states that all terpenes can be obtained by condensation, cyclization and/or rearrangement of a defined number of precursors sharing a common biosynthetic pathway [2]. The key precursor for monoterpenes is geranyl pyrophosphate, which, upon elimination of the pyrophosphate moiety, forms a *p*-menthane-type carbenium ion. Most common monoterpene skeletons are then derived from this carbocation through several rearrangement steps (Fig. 2.1).

Table 2.1

Major groups of terpene compounds according to the number of isoprenic units

Classification	Isoprene units
Hemiterpenes	1
Monoterpenes	2
Sesquiterpenes	3
Diterpenes	4
Sesterpenes	5
Triterpenes	6
Tetraterpenes	8
Polyterpenes	*n*

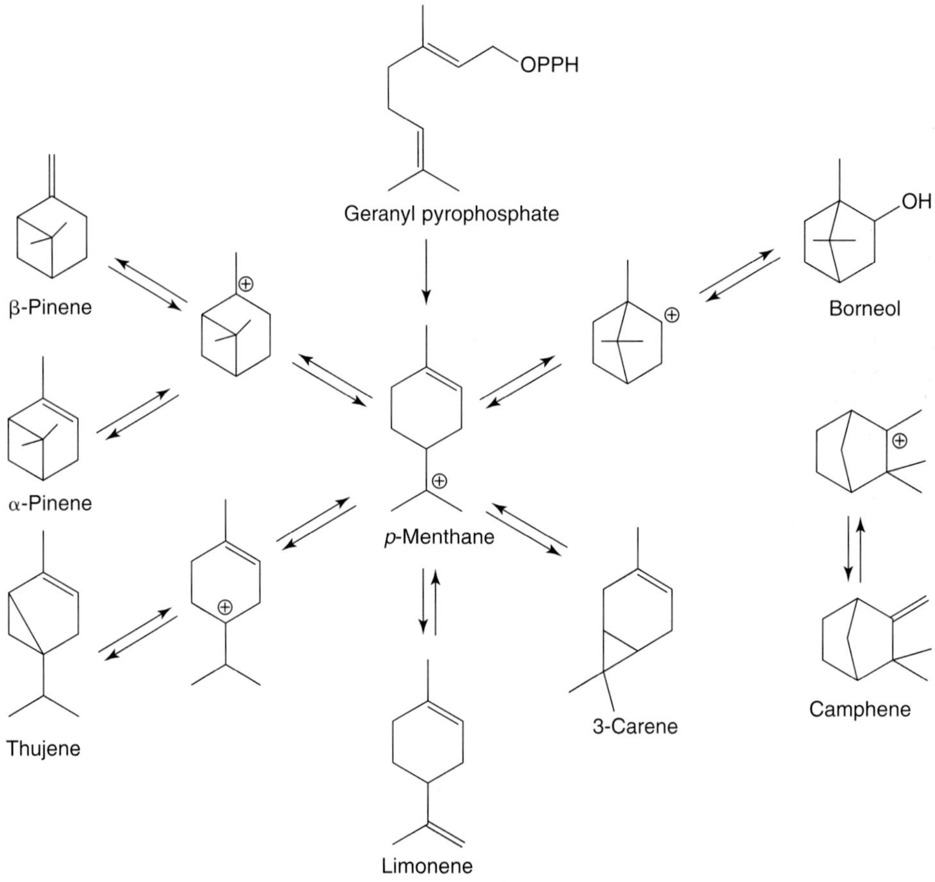

Figure 2.1 General outline of the biosynthetic pathway to the formation of main monoterpene skeletons.

Monoterpenes are extraordinarily diverse not only because of the variety of their basic skeletons, but also because a large number of stereoisomers are possible given the presence of stereogenic centres in every skeleton, and because of the wide variety of oxygenated derivatives (alcohols, aldehydes, ketones and carboxylic acids) that can be derived from those basic skeletons.

An important feature of monoterpenes (obvious from Fig. 2.1) is the ease with which the basic skeletons can interconvert to give rise to any monoterpene (or its derivatives) starting from any abundant precursor. This is one of the main reasons why pine turpentine has been used for decades as the most important source of terpenes in a vast number of industrial applications.

The systematic study of monoterpene chemistry started in the late 19th century, although their use as naturally occurring oils, known as 'essential oils', is a millenary activity. Thus for example, α-pinene rich oils were used as solvents by ancient Egyptians and Persians; menthol was used in Japanese medicine 2000 years ago and the medicinal use of camphor was reported as early as the 11th century [1]. The 'essential oil' designation finds its origin in the work of Paracelsus (15th century), who named the active principle of any medicinal substance as *quinta essentia*, and it is now used to designate the volatile terpenic fractions (mainly mono and sesquiterpenes) isolated from plants. Finally, these compounds were called terpenes by Kekulé, when he first identified terpene structures isolated from 'turpentine oil' extracted from *Pistacia therebinthus* L. Essential oils are predominantly composed of monoterpenes with a smaller fraction of sesquiterpenes. They find a very wide range of applications in cosmetics and pharmaceuticals [3, 4], insecticides [5], polymers [6] and other areas [1, 7].

Considering the aim of this book, only brief mentions will be given to applications other than polymeric materials and the discussion will be focused on turpentine, the main source of monoterpenes used as precursors for polymeric materials.

2.2 TURPENTINE

Turpentine is the common term given to the volatile fraction isolated from pine resin. For historical reasons, pine resin was called 'Naval Stores', due to its widespread use in the waterproofing of wooden ships. Depending on the way pine resin is isolated from wood, three products are distinguished viz. gum naval stores, obtained from tapping living trees; sulphate naval stores, recovered during the kraft pulping of pine wood (also known as tall oil resin) and wood naval stores, obtained from the solvent extraction of wood after harvesting. Details on the origin, extraction and purification of turpentine have been discussed in detail in the classic treatise Naval Stores by Zinkel and Russel [8].

Turpentine is by far the most widely produced essential oil in the world and consequently the privileged source for applications requiring large-scale supplies. The total world turpentine production showed a tendency to decline between the 1960s and the 1980s. However, statistics covering the next decade [9, 10] show that the turpentine production stabilized around 330 000 tons per year, with 70 per cent of the world production as sulphate turpentine, and the rest almost exclusively as gum turpentine [10, 11].

The turpentine chemical composition is strongly dependent on the tree species and age, geographic location and the overall procedure used to isolate it. In general, however, the major components are a few unsaturated hydrocarbon monoterpenes ($C_{10}H_{16}$) namely, α-pinene (45–97 per cent) and β-pinene (0.5–28 per cent) with smaller amounts of other monoterpenes (Fig. 2.2) [12].

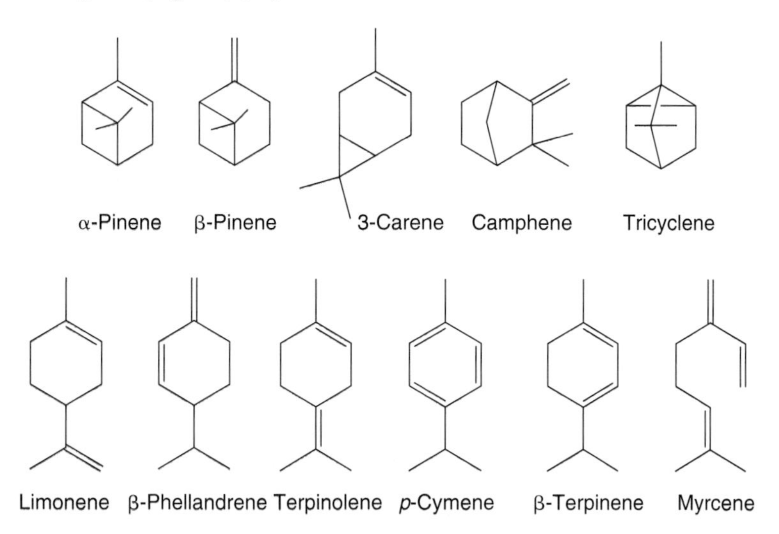

Figure 2.2 Chemical structures of the most common turpentine monoterpene components.

Turpentine was mainly used as a solvent in the 1940s (*ca.* 95 per cent) and only a minor fraction was exploited as a raw material for the chemical industry (*ca.* 5 per cent). Because of the progressive exploitation of the peculiar reactivity of its terpenic components [1, 12], these figures were completely reversed in the 1980s, with at least 95 per cent of the turpentine production being employed as Q reagent for the chemical industry [11, 13]. Historically, the most relevant uses of turpentine components as reagents include insecticides, although this application was progressively banned from the 1970s [5], pine oil production [7], flavours and fragrances and other fine chemicals [3, 4] as well as polyterpene resins [6, 13]. According to statistics related to the 1980–1990 decade, of the latter three applications, pine oil production dominated the turpentine use with 50 per cent, while the other two sectors used 25 per cent each.

Before dealing with the major topic of this chapter, that is, polymer synthesis, a brief overview of other turpentine applications is given in the following section.

2.3 TURPENTINE APPLICATIONS

2.3.1 Insecticides from turpentine

As mentioned above, the application of turpentine in the insecticide industrial production is no longer relevant [5], notably in the case of Thanite (isobornyl thiocianoacetate) and Toxaphene (a very complex mixture of chlorinated derivatives with an average empirical formula of $C_{10}H_{10}Cl_8$), both prepared from camphene (Fig. 2.3).

2.3.2 Synthetic pine oil from turpentine

Synthetic pine oil is prepared by the hydration of α-pinene with aqueous mineral acids (Fig. 2.4), and is mainly used in household cleaning and disinfection products [7]. The acidic conditions promote the formation of an intermediate carbocation which readily undergoes rearrangement to form predominantly the isomeric *p*-menthane carbocation, followed by water addition to generate essentially α-terpineol (Fig. 2.4) and minor amounts of fenchol and borneol.

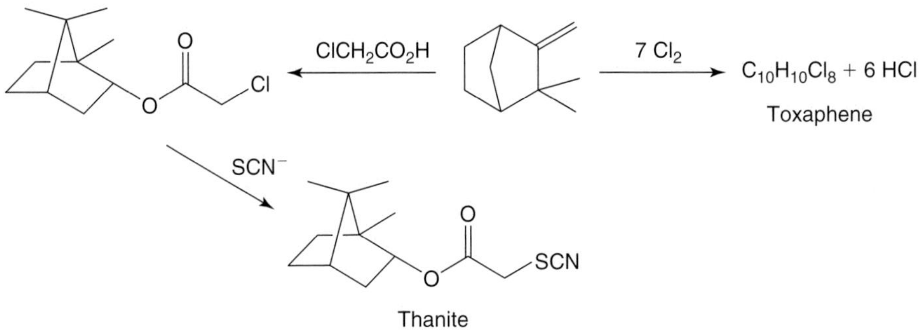

$ClCH_2CO_2H$ 7 Cl_2 $C_{10}H_{10}Cl_8$ + 6 HCl Toxaphene SCN^- SCN Thanite

Figure 2.3 Examples of insecticides prepared from turpentine components.

H^+ H OH

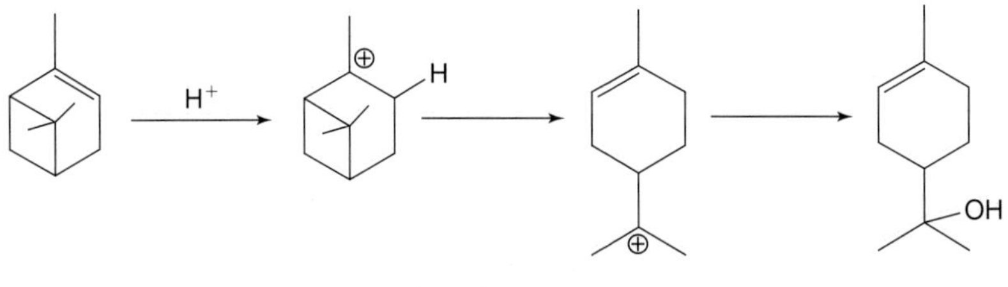

Figure 2.4 Conversion of α-pinene into synthetic pine oil.

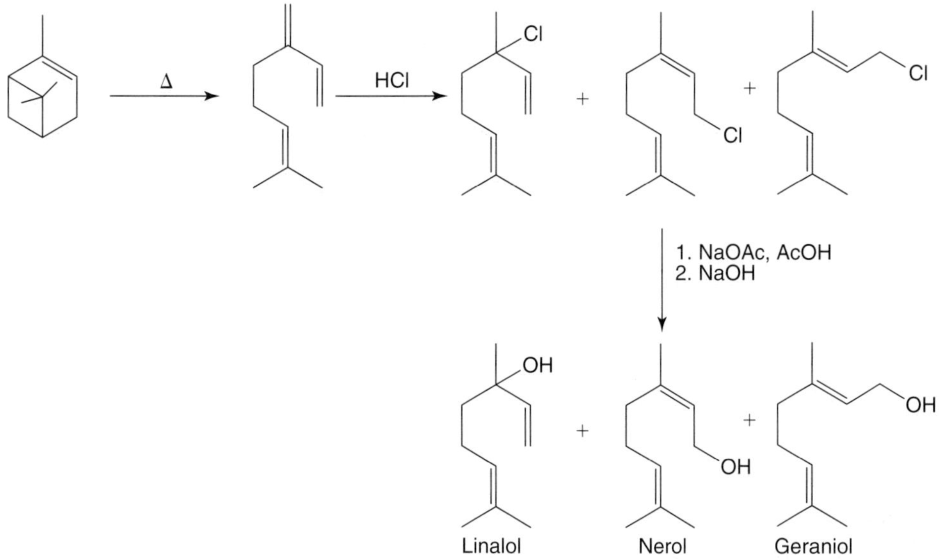

Figure 2.5 Synthesis of linalol, nerol and geraniol.

2.3.3 Flavours and fragrances from turpentine

The use of turpentine in the production of flavours and fragrances and other fine chemicals for bulk commodities mainly takes advantage of the thermal or acidic isomerization of the abundant α- and/or β-pinenes to produce the molecular skeletons of many naturally occurring fragrant chemicals (particularly when their natural sources are not abundant) and a wide variety of other fine chemicals [1, 3, 4].

Among the vast number of applications, an important example is the production of the so-called rose alcohols from β-pinene, viz. nerol, geraniol and linalool, key intermediates in the synthesis of many major fragrances. This process involves the thermal isomerization to myrcene, followed by addition of HCl, substitution by acetate and alkaline hydrolysis, as shown in Fig. 2.5 [3].

Apart from their applications as widespread fragrant raw materials, these terpenic alcohols, together with α-terpineol, have also attracted some interest in recent years as comonomers for radical copolymerization, as discussed later (Section 2.4.4.3).

Pinenes have also been used as feedstock to prepare several linear non-conjugated non-terpenic dienes (*e.g.* 3,7-dimethyl-1,6-octa-1,6-diene and 5,7-dimethylocta-1,6-diene) used in the synthesis of elastomeric ethylene–propylene–diene terpolymers [14].

2.4 POLYMERS FROM TERPENES

Abundant terpenes constitute the logical precursors to polyterpenic materials destined for bulk applications. In this context, pinenes, readily isolated on an industrial scale from turpentine, are obvious candidates. Although less abundant than α-pinene, β-pinene is by far the most studied homologue because of the higher reactivity of its exocyclic double bond in cationic polymerization. A considerable number of other monoterpenes have been tested as polymer precursors, using different polymerization mechanisms.

It is particularly relevant to this chapter and indeed to this book that the very first mention of the synthesis of a macromolecular material should have been through the use of a naturally occurring monomer mixture, namely turpentine. It was in fact Bishop Watson who reported in 1798 that the addition of a drop of sulphuric acid to turpentine produced a sticky resin, but it took another century for chemists to recognize that the properties of this material resembled those of natural rubber [15]. Another half century elapsed before the first systematic studies of the polymerization of terpenes, namely myrcene [16] and α- and β-pinene [17], were reported.

2.4.1 Conventional cationic polymerization

2.4.1.1 β-Pinene

The cationic polymerization of β-pinene (together with α-pinene) in toluene was first studied using several Lewis acid metal halides like $AlCl_3$, $AlBr_3$ and $ZrCl_4$, as initiators [17]. The high reactivity of the exomethylene double bond, the strain release resulting from the opening of the fused cyclobutane ring and the formation of a relatively stable tertiary carbenium ion are the driving forces of this reaction. The proposed polymerization mechanism (Fig. 2.6) involves the formation of a strong proton donor by the reaction of the Lewis acid with adventitious water (cocatalysis), followed by the protonation of the pinene molecule, whose ensuing carbenium ion rearranges to form a *p*-menthane type carbocation responsible for the propagation reaction [6, 17]. This process is extremely rapid and highly exothermic. In order to obtain polymers with viable molecular weights of a few thousand, the polymerization temperature must be particularly low, viz. around −40°C [6, 17], to reduce the relative kinetic contribution of the transfer reaction with respect to chain propagation (Fig. 2.6), a common feature in the cationic polymerization of aliphatic monomers.

Very recently, the cationic polymerization of β-pinene in toluene, using $AlCl_3$ as the initiator, was shown to proceed smoothly and rather rapidly under microwave irradiation [18].

Other Lewis acid initiators such as $SnCl_4$, $TiCl_4$, BF_3 and Et_2AlCl in various solvents and temperature ranges, were also studied [19], and polymers with Mn values in the range of 720–2 800 were obtained. In this study, it was shown that $TiCl_4$ and Et_2AlCl were the most efficient initiators, dichloromethane the best solvent and −80°C to 0°C the most suitable temperature range.

Of the numerous classical Lewis acids tested as initiators for the polymerization of β-pinene, $EtAlCl_2$ has proved by far the most efficient, particularly in terms of the possibility of preparing polymers with high molecular weights. Three studies have tackled this system over the last 15 years [20–22] using different strategies, particularly in the

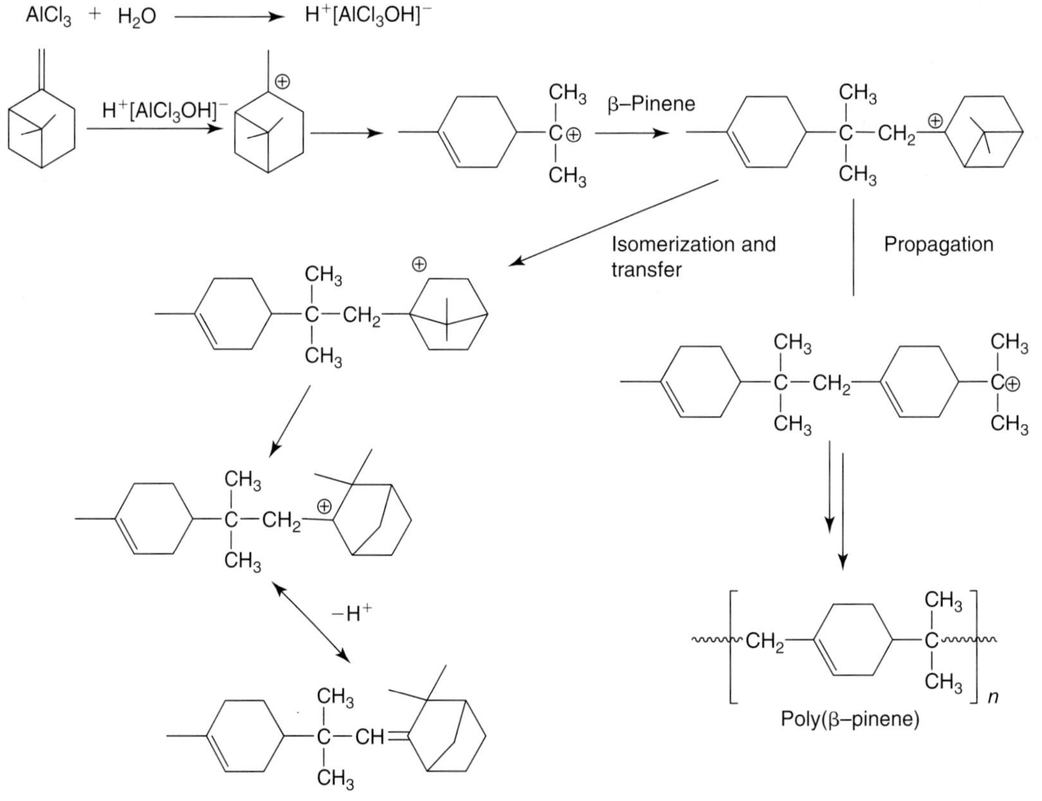

Figure 2.6 General mechanism of the cationic polymerization of β-pinene catalyzed by $AlCl_3$.

addition of specific bases, including 2,6-di-*tert*-butylpyridine, with its well-known property of proton trap, incapable of reacting with Lewis acids. The most interesting practical outcome of these investigations is associated with the attainment of Mn values which are particularly high for this monomer, ranging from 10 000 to 40 000, when working at low temperature [20, 22] and of a few thousand when working around room temperature [21]. These high-Mn poly(βPIN)s displayed a glass transition temperature of 90°C, which was further increased to 130°C after the polymer was hydrogenated [22].

Finally, the radiation-induced polymerization of β-pinene has also been studied [23]. Under extremely dry conditions, molecular weights in the range of 1 700–2 400 were obtained through a mechanism which was adequately demonstrated as being cationic in nature.

2.4.1.2 α-Pinene

Although α-pinene is readily protonated to form a tertiary carbocation, which can then rearrange to an unsaturated *p*-menthane isomer (Fig. 2.7), the attack of the endocyclic double bond to the isopropenyl cationic site is limited by steric hindrance.

The first reported study of the cationic polymerization of α-pinene [17] showed indeed that this isomer was vastly less reactive than its β-homologue because of the steric crowding associated with the polymerizable unsaturation. Experiments carried out at low temperature using various Lewis acid initiators gave extremely poor yields that were maximized only by working at 40°C, with the inevitable drawback of the drastic decrease in the molecular weight of the product, in fact mainly composed of dimeric structures [17]. An alternative α-pinene polymerization pathway involves the rearrangement of the tertiary pinane carbocation to a saturated secondary bornane isomer (Fig. 2.7). Both bornane and *p*-menthane repeating units (Fig. 2.7) have been detected in α-pinene homo- and co-polymers.

Despite this structural drawback, the fact that α-pinene is the most abundant turpentine component has spurred a considerable amount of research related to its cationic polymerization. Several laboratories have looked into the use of an initiating combination involving conventional Lewis acids and $SbCl_3$ as an activator [6, 24–28].

When the binary system $AlCl_3/SbCl_3$ was used in toluene at −15°C, poly(α-pinene)s with Mn of 1 140–1 600 were obtained [26, 27], with negligible proportions of dimers, as opposed to their high contribution in products prepared with only $AlCl_3$. Interestingly, whereas $SbCl_3$ seems essential for the optimization of α-pinene polymerization, its role in β-pinene polymerization is rather detrimental in terms of reaction rate and molecular weight distribution [27].

α-Pinene can also be polymerized using other aluminium halides ($AlBr_3$, $AlEtCl_2$) or other metal halides ($BF_3 \cdot OEt_2$, $SnCl_4$, $TiCl_4$, WCl_6) in conjunction with $SbCl_3$ [24]. All these aluminium halides tend to produce lower poly(αPIN) yields and Mn values (typically about 80 per cent and 700, respectively) when compared to $AlCl_3$. The other metal halides were shown to yield high amounts of dimers (20–40 per cent) along with small amounts of oligomers [24]. The ^{1}H-NMR characterization of the polymers obtained with aluminium-based initiators showed that the unsaturated *p*-menthane moiety is the predominant repeating unit (81–95 per cent) with much smaller contributions from saturated bornane counterparts [24].

The addition of nucleophilic additives, such as esters (ethyl benzoate or ethyl acetate), ethers (ethyl ether and 1,4-dioxane) and ammonium salts (nBu$_4$NCl) to the binary polymerization system $AlCl_3/SbCl_3$ allowed the α-pinene

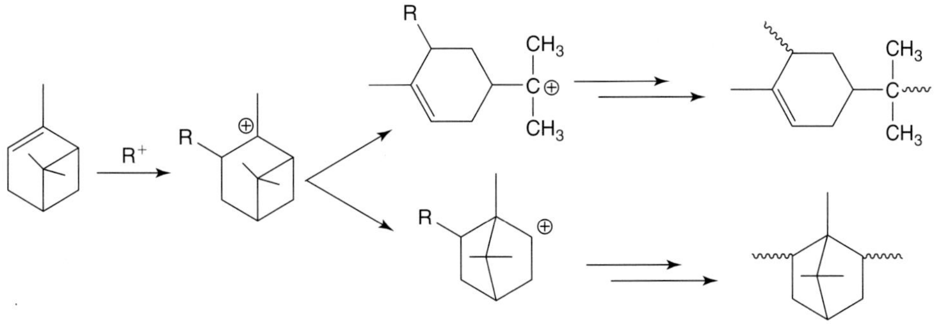

Figure 2.7 Alternative mechanisms of α-pinene cationic polymerization.

polymerization to be carried out at higher temperatures (0°C) without any significant decrease in molecular weights (Mn ~ 1 000) and with yields greater than 80 per cent with modest dimer contents (less than 6 per cent). However, these additives decreased the corresponding polymerization rates [25].

It has been reported that the cationic polymerization of α-pinene can be improved if, instead of using their simple mixture, $SbCl_3$ and $AlCl_3$ are previously melted together, thus favouring cross ionization by chloride ion transfer to give $SbCl_4^+$ as the direct initiating species, with $Al_2Cl_7^-$ as counter ion [29].

A recent addition to this topic deals with the hydrothermal deposit of $AlCl_3$ onto a zirconium mesoporous material and the use of the ensuing solid Lewis acid as a catalyst for both α-pinene polymerization and other conversions at room temperature [30].

2.4.1.3 Other terpenes and derivative

Several other terpenic components have been submitted to cationic polymerization. These include terpene oxides [31] and phellandrene [22]. The cationic polymerization of α- and β-pinene oxides, using BF_3 or PF_5 as initiators proceeded through the initial protonation of the oxyrane oxygen atom followed by the isomerization of the oxonium ion to give the tertiary exo-carbenium ion responsible for the chain growth, as shown in Fig. 2.8 [31]. This mechanism was corroborated by the ^{13}C-NMR of the ensuing polymers. However, this system was strongly affected by side reactions resulting from multiple rearrangements of the oxonium ion, which yielded a wide variety of secondary products and limited the polymer DP to 6–7 units [31].

The cationic polymerization of α-phellandrene, using $EtAlCl_2$, $SnCl_4$ or BF_3OEt_2 as initiators in various solvents produced polymers with Mn of 3 700–5 700 [22]. The cationic polymerization of limonene has also been reported [32] and its outcome found to be favoured by using the binary system $AlCl_3/SbCl_3$ instead of $AlCl_3$.

2.4.1.4 Copolymerization of β-pinene with synthetic monomers

The cationic random copolymerization of β-pinene with styrene and α-methylstyrene, initiated by $AlCl_3$ in dichloromethane, was investigated within a wide range of temperatures [33–35]. Regrettably, these studies were limited to a rather factual and empirical approach and left some major questions unanswered, such as the values of the reactivity ratios. The molecular weight of the copolymers, irrespective of the feed ratio and the reaction temperature, were very low in the first study [33] and increased somewhat in the subsequent ones [34, 35].

In the case of the β-pinene/isobutene combination initiated with $EtAlCl_2$, the ^{1}H-NMR spectrum of the copolymer showed an almost equal incorporation of the monomers and a high level of alternation [36].

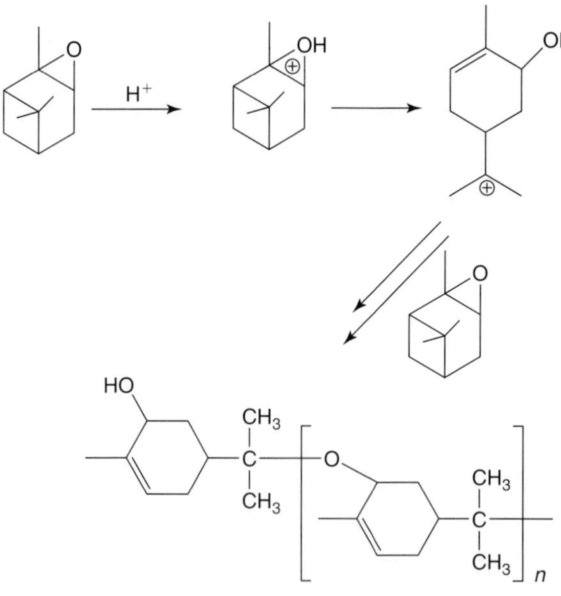

Figure 2.8 Cationic polymerization of α-pinene oxide.

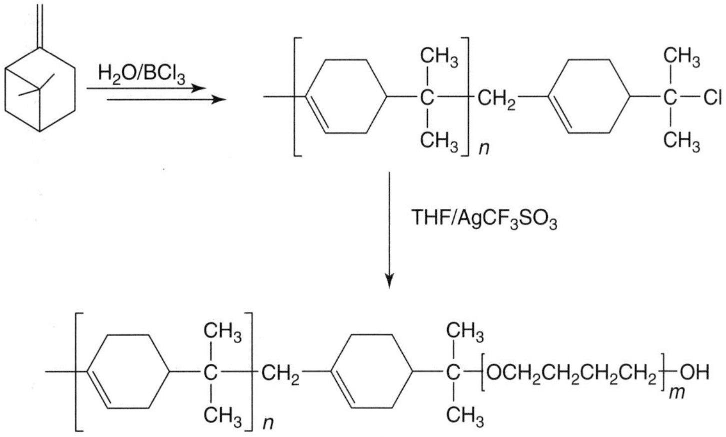

Figure 2.9 Synthesis of a di-block copolymers of β-pinene and THF through α-end chlorinated poly(βPIN).

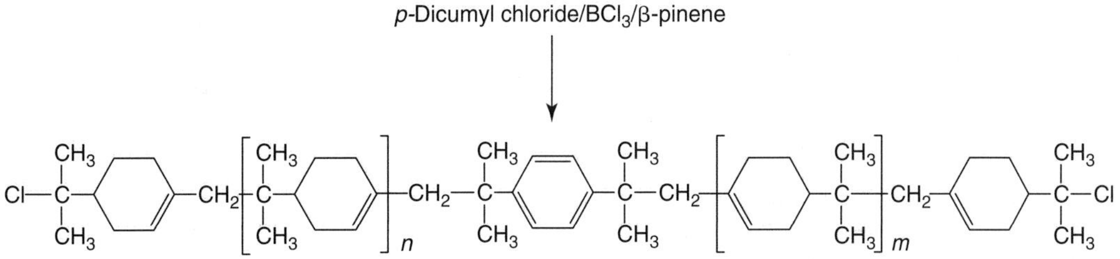

Figure 2.10 Synthesis of linear poly(βPIN) with two chlorine end groups.

The block copolymerization involving β-pinene has also been investigated [37]. This strategy involved the synthesis of a poly(βPIN) bearing tertiary chloride end groups and the attachment of poly(THF) blocks by their activation, as shown in Fig. 2.9 for chains with a single chlorinated site, thus giving a di-block structure.

When β-pinene was polymerized with BCl_3 in the presence of *p*-dicumylchloride or *p*-tricumylchloride, polymers with two (Fig. 2.10) and three (Fig. 2.11) chlorinated end groups were obtained as potential precursors for linear and star-shaped multi-block copolymers [37].

2.4.1.5 Copolymerization of α-pinene with synthetic monomers

The cationic copolymerization of α-pinene with several conventional monomers has been examined by various groups [38–41]. The reaction of α-pinene with isobutene using $EtAlCl_2$ in ethyl chloride at ~ −100°C yielded random copolymers with Mn values in the range of 3 200–29 000, increasing with decreasing α-pinene feed [38]. The unsaturated *p*-menthane moiety was the dominant repeating unit in these copolymers [38]. The copolymerization of α-pinene and styrene using $AlCl_3$ as the initiator gave copolymers with Mn of 2 320–3 080 [39]. However, as in the case of α-pinene homopolymerization, the use of the $SbCl_3/AlCl_3$ combination here reduced the difference in monomer reactivity and produced higher copolymer yields [40, 41].

2.4.2 Living cationic polymerization

2.4.2.1 Homopolymerization of β-pinene

The cationic polymerization of β-pinene with conventional systems has been shown above to give modest and, exceptionally, medium molecular weights associated with rather broad distribution. The development of novel initiating systems capable of providing controlled cationic polymerizations in terms of *quasi*-living conditions (LCP)

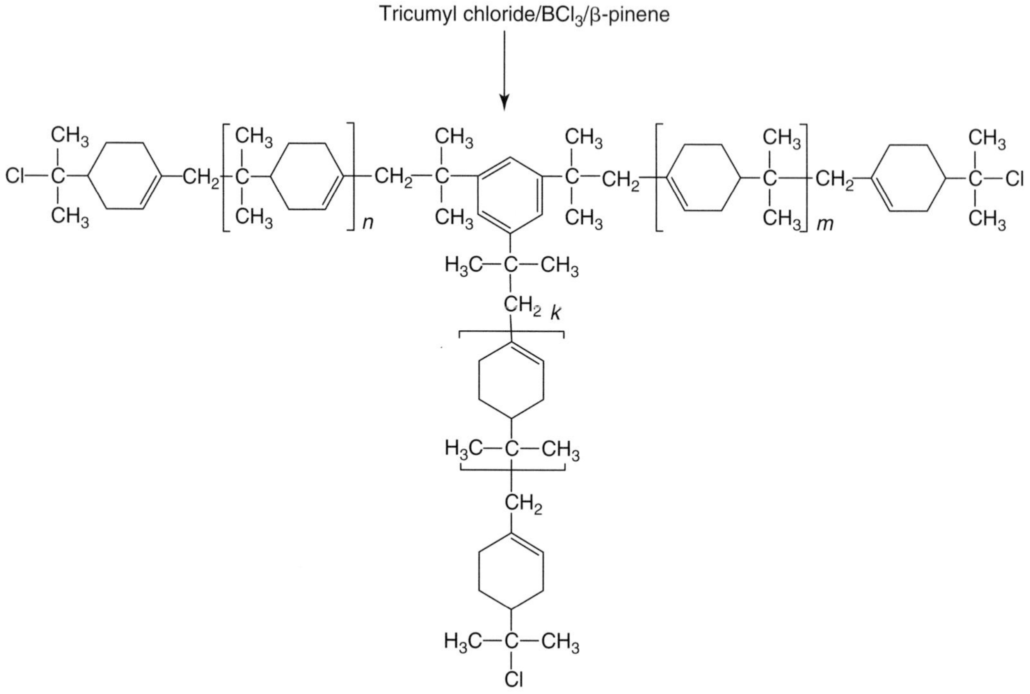

Tricumyl chloride/BCl_3/β-pinene

Figure 2.11 Synthesis of star-shaped poly(βPIN) with three chlorine end groups.

began some 20 years ago and was applied first to conventional monomers like styrene, isobutene and vinyl ethers, but this research was soon extended to terpenes after some failed attempts [20].

The first successful β-pinene LCP systems were reported by Lu *et al.* in 1997 [42–44]. The best conditions involved the use of the HCl adduct of 2-chloroethylvinyl ether, isopropoxytitanium chloride [$Ti(OiPr)Cl_3$] in the presence of tetra-*n*-butylammonium chloride [*n*BuNCl] in dichloromethane at −40°C to −78°C, viz. a system previously optimized for the LCP of styrene [45]. The living character of this system was clearly demonstrated by the fact that Mn increased linearly with monomer conversion, even after several monomer additions [43].

The proposed polymerization mechanism [43, 44], shown in Fig. 2.12, involves the rearrangement of the pinane skeleton with the formation of a *p*-menthane type repeating unit, as unambiguously demonstrated from the ^{1}H-NMR spectrum of the ensuing polymer. It follows that the strain release resulting from the cyclobutane ring opening is still a key driving force in this propagation mechanism.

Other initiating systems were also shown to give LCP conditions, namely through the use of other Lewis acids [43] and coinitiators [43, 46].

2.4.2.2 Copolymerization of β-pinene with synthetic monomers

The synthesis of block, as well as random copolymers of β-pinene with styrene and *p*-methylstyrene (*p*MeSt), was studied by living cationic polymerization, using both the styrene and vinyl ether adducts as initiators in the presence of $Ti(OiPr)Cl_3$ in methylene chloride at −40°C [44, 47]. For styrene (A) and β-pinene (B), both AB and BA block copolymers were obtained, as shown in Fig. 2.13, with Mn values of 4 000 and 3 600, respectively, and narrow Mw/Mn ratios (1.26 and 1.38, respectively). The efficiency of these block copolymerizations was attributed to the similar reactivity of the C—Cl bond derived from the two monomers [44].

The random living cationic polymerization of β-pinene with styrene and *p*MeSt was also inspected [44]. With the β-pinene/styrene mixtures, a faster consumption of the former monomer occurred, yielding tapered copolymers with Mn ~ 5 000 and MWD of 1.4 at 100/65 per cent consumption of β-pinene/styrene. In contrast, with the β-pinene/*p*MeSt mixtures, both monomers were consumed at nearly the same rate, yielding statistical copolymers with Mn ~ 5 000 and MWD of 1.5 at both monomer conversions higher than 90 per cent.

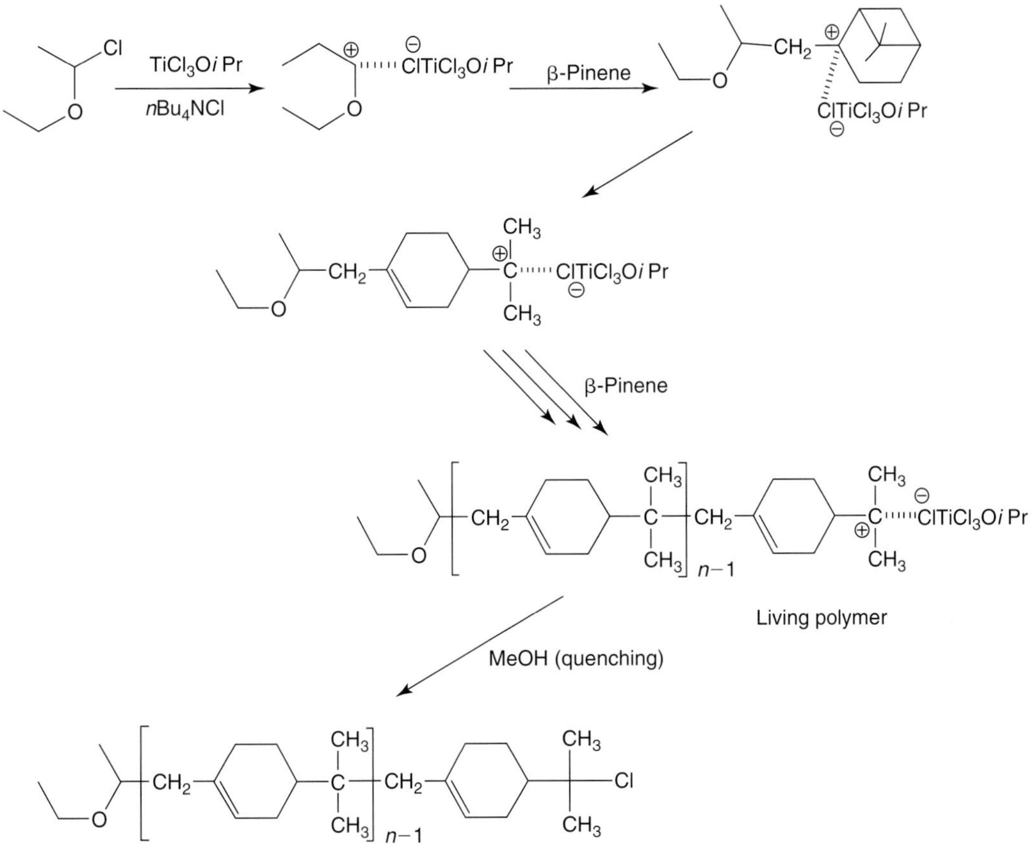

Figure 2.12 Mechanism of the LCP of β-pinene.

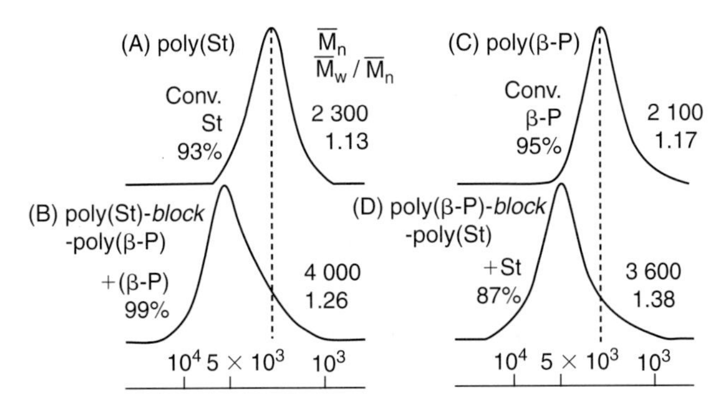

Figure 2.13 Molecular weight distribution curves for poly(St) A, poly(βPIN) B and the block copolymers obtained after addition of the other monomer (reprinted with permission from Reference [44]).

The random living cationic polymerization of β-pinene was also carried out with isobutene using 1-phenylethyl chloride/$TiCl_4$/$Ti(OiPr)_4$/$n$$Bu_4NCl$ [48]. Under these conditions, β-pinene and isobutene exhibited almost equal reactivities and were consumed at the same rate. This observation, together with proton NMR evidence, confirmed the random nature of the copolymer. Regardless of the monomer feed ratio, Mn increased with monomer conversion while preserving narrow MWDs (1.1–1.2), confirming the living character of the system [48].

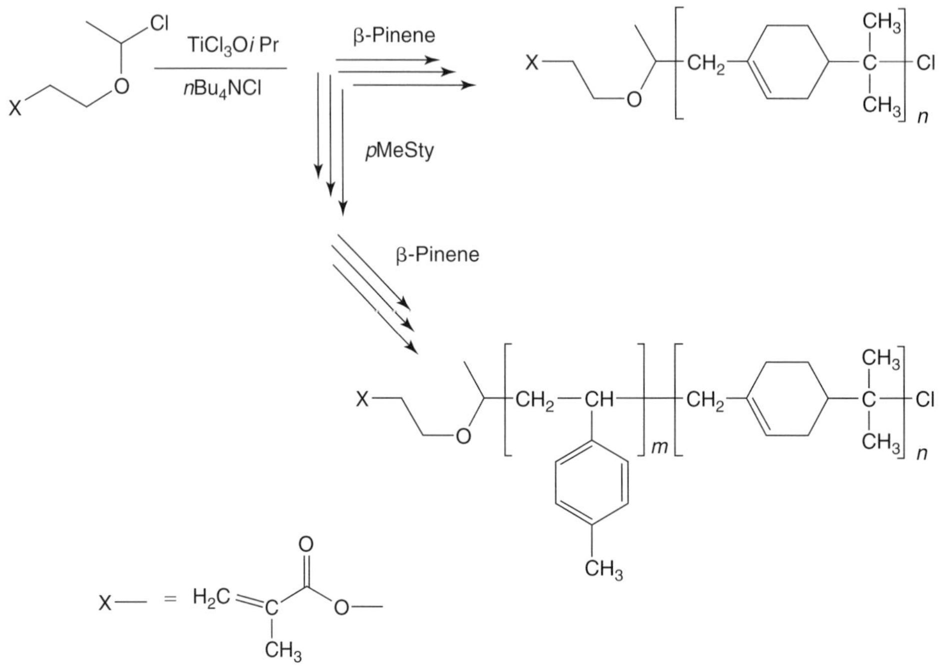

Figure 2.14 Synthesis of methacrylic end-functionalized poly(βPIN) and poly(*p*MeSt)-block-poly(βPIN) by LCP.

In a different vein, the LCP synthesis of poly(βPIN) bearing benzyl chloride end groups [49, 50] opened the way to the insertion of a poly(THF) block by the activation of this moiety with $AgClO_4$ or $AgSbF_6$. The ensuing copolymer was thus made up of a non-polar rigid segment of poly(βPIN) linked to a polar, soft and semi-crystalline poly(THF) segment [49].

Grafted copolymers have also been prepared from the brominated poly(βPIN), whose synthesis is shown in Fig. 2.15, by the activation of the C—Br bonds with Et_2AlCl leading to the cationic sites for the polymerization of styrene [51].

2.4.3 Living cationic-radical copolymerization of β-pinene with synthetic monomers

The possibility of controlling the cationic polymerizations of β-pinene using the systems described above, opened an additional avenue of macromolecular engineering which consists in introducing reactive functional groups at one or both ends of poly(βPIN) or its copolymers and even within their backbone. The interest of these moieties is in their possible exploitation as starting sites for the free radical polymerization of a suitable monomer [42, 49, 50, 52].

Figure 2.14 illustrates this strategy applied to the specific context in which the terminal reactive moiety for both homo- and co-polymers of β-pinene prepared by LCP is a methacrylic group [42], from which the free radical polymerization of methyl methacrylate (MMA) was carried out to give block copolymers with interesting properties [42].

Figure 2.15 shows a different and more sophisticated approach to the combination of cationic and radical mechanisms, since in this case both processes bear a living character. Allylic brominated poly(βPIN) with Br/β-pinene unit ratios of 1.0 and 0.5 (Mn of 2 810 and 2 420, respectively and a similar MWD of 1.3) were obtained by treatment with *N*-bromosuccinimide/AIBN and then used as macroinitiators in conjunction with CuBr and 2,2′-bipyridine for the atom transfer radical polymerization (ATRP) of acrylic monomers [52, 53].

These grafting processes clearly require further optimization since some side reactions were reported to provoke the formation of insoluble gels [52].

2.4.4 Radical polymerization of monoterpenes

Although it is generally assumed that monoterpenes do not undergo radical homopolymerization, in some experiments dealing with the thermal treatment of β-pinene [54], it was shown that the material obtained must have been

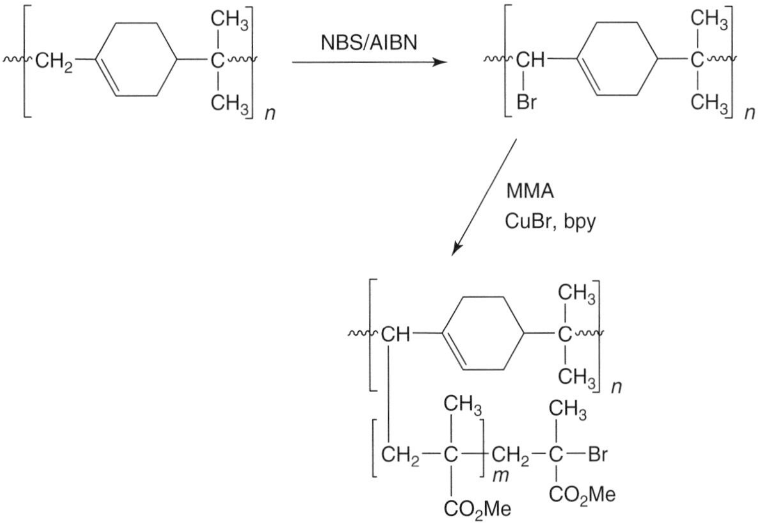

Figure 2.15 Synthesis of poly(βPIN)-*g*-poly(MMA).

produced by a free radical mechanism, since cationic and anionic inhibitors did not affect this process [55]. More recently, the radical homopolymerizations of α- and β-pinene have indeed been more clearly confirmed [39, 56]. However, these processes which called upon AIBN as initiator, only gave oligomers (Mn ~ 900) in modest yields (~30 per cent) [56]. It is, therefore, reasonable to conclude that the original assumption, although strictly speaking inaccurate, reflects the actual reality in the sense that there is no interest in using free radical initiation to prepare materials based on the homopolymerization of monoterpenes.

This unfavourable situation is however reversed when pinenes and other monoterpenes are used as comonomers in free radical copolymerizations with monomers bearing structures other than theirs. These systems are described below as a function of the specific terpene comonomer.

2.4.4.1 Copolymerizations using pinenes

A considerable amount of attention has been devoted to this topic in recent years, particularly in view of the advances related to controlled radical polymerization [56–60].

The conventional free radical copolymerization of β-pinene and MMA or St, initiated by AIBN, yielded copolymers with Mw of about 11 600 and 25 400, respectively and MWD of 1.5 and 1.7, respectively [56]. When 2,3,4,5,6-pentafluorostyrene (PFS) was used as comonomer, and benzoyl peroxide as the initiator, a PFS-rich copolymer incorporating isolated isomerized β-pinene units distributed between poly(PFS) segments was obtained [57]. This feature is related to the low reactivity of the β-pinene free radicalar towards its monomer, that is to a reactivity ratio close to zero. These PFS-β-pinene copolymers were shown to combine the typical high water contact angles of perflourinated polymers (hydrophobicity) with the optical activity of poly(βPIN) [57]. A recently published study indicated that the radical random copolymerization of both α- and β-pinene with styrene under microwave irradiation yields materials with Mn values considerably higher than those obtained under conventional conditions, but still with very low conversions [18].

The radical copolymerization of β-pinene with acrylonitrile (AN) using AIBN as the initiator produced very low product yields with Mn ~ 7 200 and MWD of 1.6 [59]. Again, the reactivity ratio relative to the terpenic monomer was close to zero and that of AN was 0.66, which produced a structure (Fig. 2.16) similar to that described above for PFS. This study had the added quality of kinetics and ^{1}H-NMR structural insights [59]. The addition of Lewis acids ($TiCl_4$, $SnCl_4$, $ZnCl_2$ and Et_nAlCl_{3-n}) produced an enhanced tendency to alternation, through the reduction of the reactivity ratio of AN to about 0.11, which resulted in an increase in the incorporation of β-pinene to levels as high as 43 per cent, that is to the formation of nearly alternated copolymer structures. Moreover, in the specific instance of $EtAlCl_2$ and especially of Et_2AlCl, higher copolymer yields were also obtained [59].

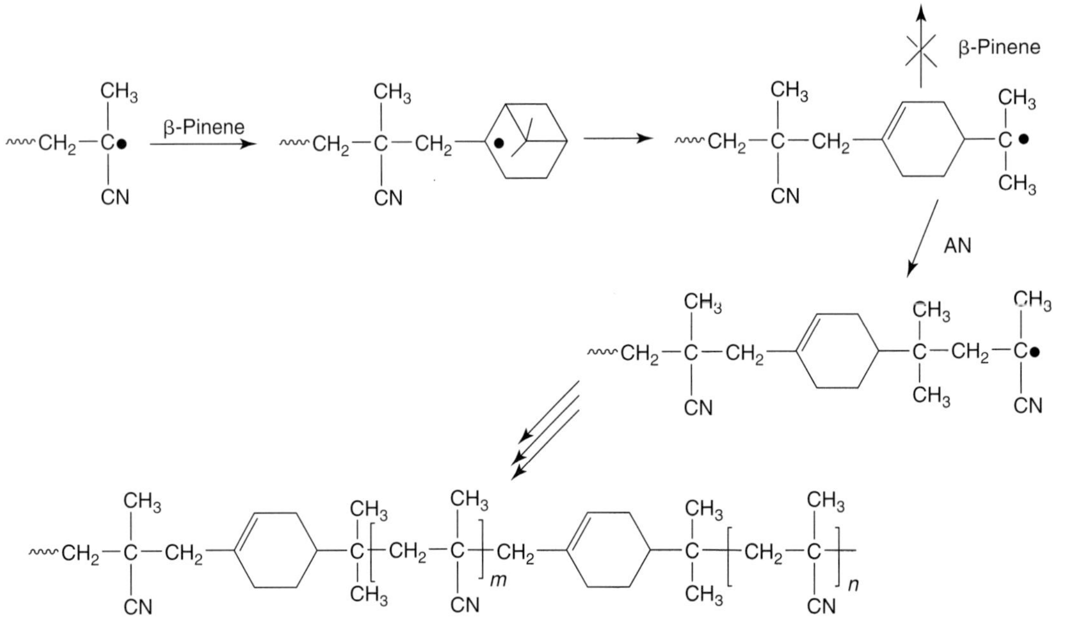

Figure 2.16 Proposed mechanism for the radical copolymerization of β-pinene and AN.

In the same study, the authors also investigated the use of the reversible addition-fragmentation transfer (RAFT) technique and showed that several such systems gave rise to a better controlled copolymerization, although the polydispersity of the ensuing copolymers depended on the nature of the actual agent used [59]. The most effective were cumyl dithiobenzoate (CDB) and 2-cyanopropyl-2-yl dithiobenzoate (CPDB) which gave polydispersities of 1.4 and 1.2, respectively. The living nature of these processes was confined to their first phase, since at high conversions the kinetics and Mn values strongly deviated from those predicted by a living system [59]. When Et_2AlCl was added to these optimized RAFT copolymerizations, both a narrow MWD (1.2) and a high β-pinene incorporation were achieved, suggesting a better reaction control and a higher tendency to alternation [59]. In spite of these improvements, the fact that the Mn of the copolymers did not increase with conversion, showed that even this system did not display a living character [59].

The AIBN-initiated radical copolymerization of β-pinene and methyl acrylate (MA) [58] with *n*-butyl acrylate (*n*BA) [60] produced results similar to those described above for AN [59] and thus the structure of the ensuing macromolecules were characterized by the presence of MA or *n*BA blocks alternated with single β-pinene units but the incorporation of β-pinene could again be enhanced by the addition of Et_2AlCl [58, 60]. When the authors attempted to turn this system into a controlled process by using RAFT reagents, the results were rather disappointing in the case of MA [58], whereas, when *n*BA was used [60], the results obtained were similar to those reported for AN [59].

To the best of our knowledge, only one publication describes the free radical copolymerization of α-pinene [56]. In this study, both MMA and St were studied as comonomers under standard conditions. Interestingly, it was shown that α-pinene is a more suitable comonomer than its β-isomer, since it gave higher rates, conversions and molecular weights under the same conditions [56]. Surprisingly, despite these more favourable features, α-pinene has not been investigated further in this context.

2.4.4.2 *Copolymerization of limonene*

The radical copolymerization of limonene with MMA [61, 62], AN [63], St [64] and *N*-vinylpirrolidone [65] gave alternating copolymers because both reactivity ratios were close to zero. Figure 2.17 shows the mechanism proposed for this process in the case of MMA.

More recently, liquid crystalline materials based on modified limonene–MMA copolymers have been described [66].

Finally, copolymers of limonene and maleic anhydride [67, 68] and terpolymers combining limonene with either MA and AN [69], or with St and MMA [70], have also been reported.

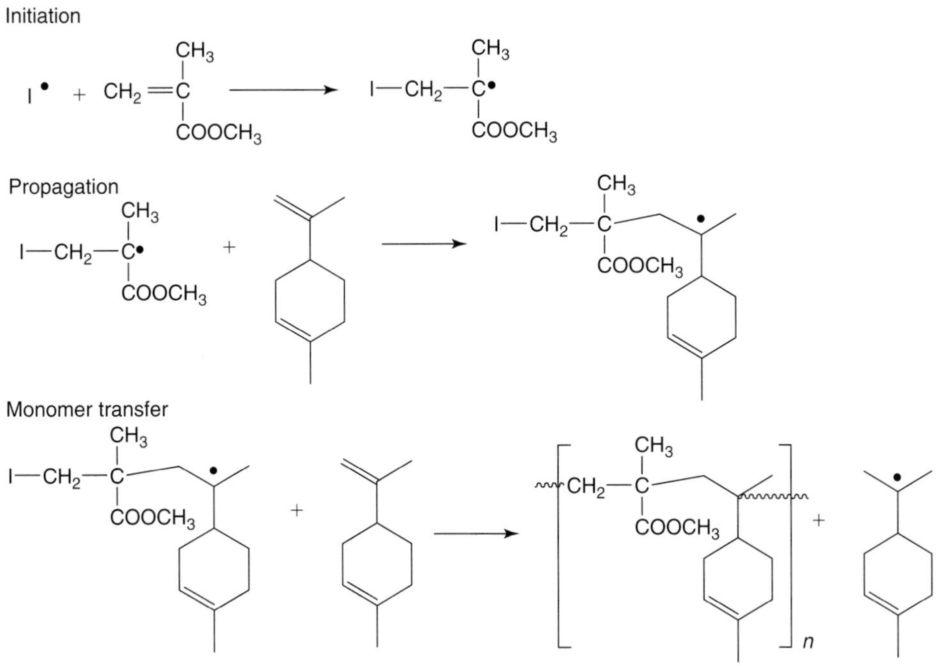

Initiation

Propagation

Monomer transfer

Figure 2.17 Proposed mechanism for the radical copolymerization of limonene and MMA.

2.4.4.3 Copolymerization of monoterpene alcohols

Like their hydrocarbon counterparts, most monoterpene alcohols do not undergo radical homopolymerization, but intervene in copolymerization with conventional monomers. The interest of the ensuing copolymers resides essentially in the fact that they bear a pendant hydroxyl group *per* terpene monomer unit.

α-Terpineol readily copolymerizes with MMA [71], St [72] and BMA [73]. In all these systems, the reactivity ratios were found to be very low, which favoured a high degree of alternation in the ensuing materials [71–73]. Fig. 2.18 depicts the mechanism proposed for the system involving MMA as comonomer [72].

Other authors have proposed mechanisms which differ from the one shown in Fig. 2.18, particularly in terms of the structure of the terpinyl free radical and hence of the structure of the corresponding monomer unit in the copolymers [71, 72, 74].

More recently, the radical copolymerization of α-terpineol with *N*-vinylpirrolidone was also studied. In this instance, the comonomers showed substantially different reactivity ratios resulting in the formation of random copolymers [75].

The radical copolymerization of linalol with AN [76], acrylamide [77], St [78] and BMA [79] yielded strongly alternating copolymers because of the very low reactivity ratios of both monomers (Fig. 2.19). The proposed polymer structures, as shown in Fig. 2.19 for the copolymer with AN, also involve the formation of the more stable tertiary radical intermediate [76, 77].

Likewise, the radical copolymerization of citronellol with vinyl acetate [80], St [81] and acrylamide [82] produced alternating copolymers whose structures evoke the same radical mechanism.

On the other hand, geraniol showed a substantially different reactivity in its copolymerization with St, leading to a lower alternating character [83].

Finally, it is worth mentioning that several radical terpolymerization studies using linalol [84, 85] and citronellol [86] have been carried out.

The obvious interest in interpenetrating polymer networks (IPN) based on very different macromolecular structures has spurred some studies involving terpenols. Thus the synthesis of IPN involving poly(St) and poly(citronellol-*alt*-MMA) [87] and of poly(α-terpineol-co-St) with ethyl acrylate crosslinked with divinylbenzene have been reported [74].

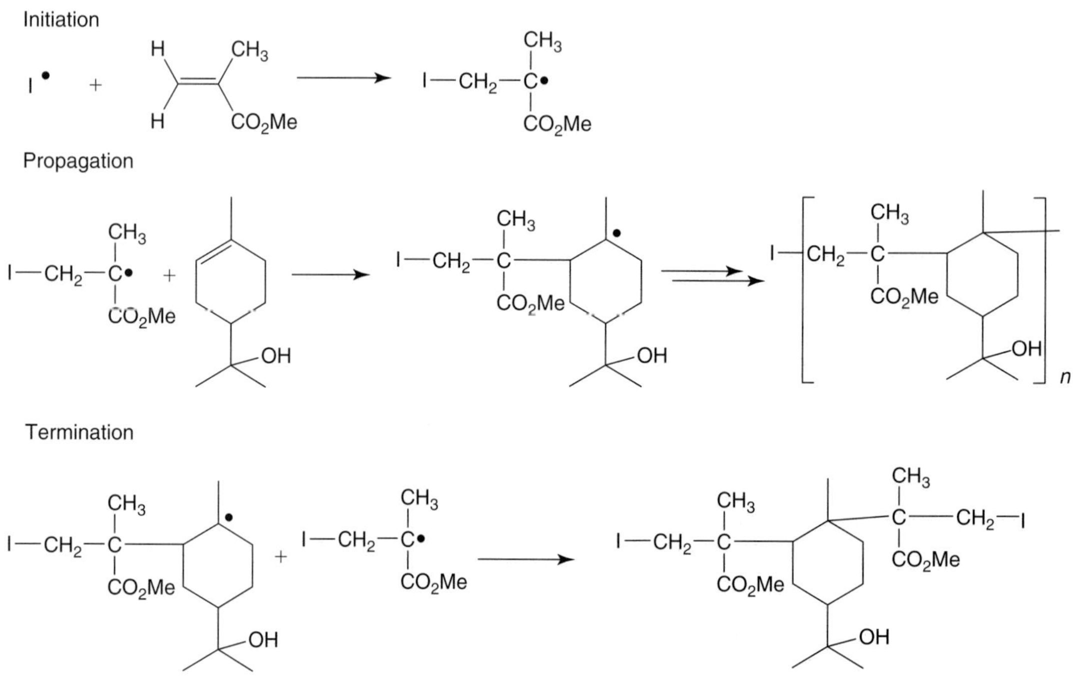

Figure 2.18 Proposed mechanism for the radical copolymerization of α-terpineol with MMA.

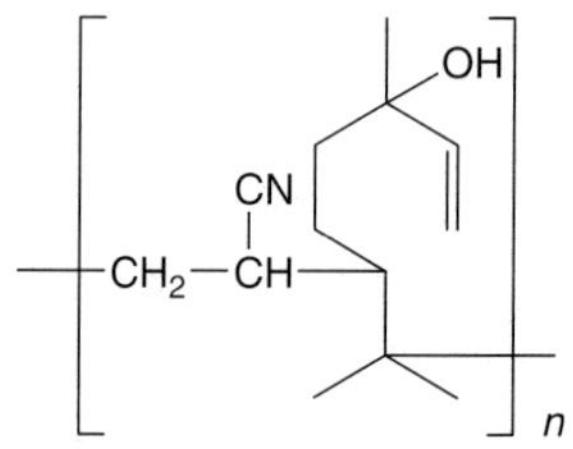

Figure 2.19 Structure of the linalol/AN copolymer.

An interesting exploitation of the OH side groups borne by these copolymers is provided by a recent study in which poly(α-terpineol-co-MMA) was modified by grafting phenyl benzoate mesogenic groups, bearing vinyl terminated polymethylene spacers, to give liquid crystalline materials [88].

2.4.5 Other monomers and polymerization systems

The emulsion polymerization of myrcene and its copolymerization with styrene and butadiene, followed by vulcanization of the ensuing polymers to obtain synthetic rubbers was reported as early as 1948 [16] but no other such system has been mentioned in the literature since then.

Only a few studies have been devoted to the Ziegler–Natta polymerization of terpene monomers. α- and β-pinene, limonene and camphene were investigated by various authors and the corresponding homopolymers were shown to be structurally analogous to those obtained by cationic mechanisms [89, 90]. More recently, this type of catalysis was successfully applied to copolymerization systems involving α-pinene and MMA or St [56].

Myrcene is, to our knowledge, the only monoterpene which has been the object of living anionic polymerization (LAP) and copolymerization studies [91]. These systems involved *N*-butyl lithium as the initiator, and either benzene or THF as solvent. When the reaction was carried out in the former, 1,4-addition was favoured (85–90 per cent),

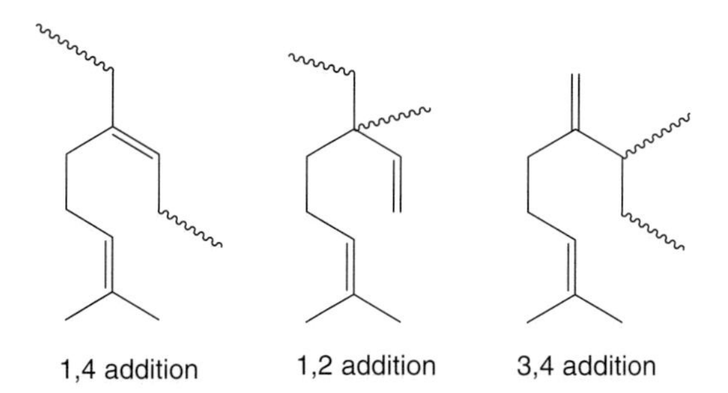

Figure 2.20 Structure of the 1,4- 1,2- and 3,4-type myrcene monomer units.

followed by 3,4-addition (11–15 per cent), whereas in THF, 1,4-addition accounted for only 39–44 per cent, followed by 3,4-addition, with 39–44 per cent, and by 1,2-addition, with 10–18 per cent (Fig. 2.20).

The corresponding copolymerization with styrene in benzene yielded a polymer rich in myrcene with a poly(St) block formed in the later polymerization stages. In THF, a styrene-rich copolymer was obtained in the early stages, with a polymyrcene block formed towards the end [91].

Alternating copolymers of limonene oxide and carbon dioxide have been prepared using β-diiminate zinc acetate complexes [92]. These polycarbonate-type copolymers had rather low Mn values (5 500–11 000), but narrow MDWs (close to 1.1). Cataldo and Keheyan [93] studied recently the polymerization of (—)-β-pinene with ψ-radiation, and found that the product was optically active, although such a feature had not been mentioned in earlier studies by other authors [94, 95].

2.5 POLYTERPENE APPLICATIONS

2.5.1 Tack and adhesion

Commercially available polyterpene resins, and particularly poly(PIN)s, are low molecular weight hydrocarbon-like polymers used as adhesive components to impart tack (to both solvent-based and hot-melt systems), provide high gloss, good moisture vapour transmission resistance and good flexibility for wax coating and ensure viscosity control and density increase to casting waxes. There is a considerable similarity in the context of these properties between rosin derivatives (see corresponding chapter in this book) and polyterpenes.

The ageing stability of an adhesive depends, to some extent, on the corresponding stability of its tackifing resin [96, 97]. Since poly(PIN)s contain unsaturations in their chemical structures, they are sensitive to atmospheric oxidation, which leads to the progressive loss of their tack. Photo-oxidation studies have shown that both poly(PIN)s behave similarly in this context, with α,β-unsaturated hydroperoxides as the primary products, which then undergo photolysis to give various low molecular weight degradation fragments [96]. Under thermal oxidation conditions, similar ageing processes are observed, but the reactivity of the two poly(PIN)s is now significantly different with the α-homologue being more fragile [97].

2.5.2 Other applications of polyterpenes and terpenes

Poly(βPIN) was used as a spray adjuvant in pesticide applications to crops to increase their deposition and to decrease their rate of decay, which resulted in increased efficiency [98]. This effect is achieved through the formation of an organic film covering the crop foliage, which protects the active component from rain and wind erosion as well as volatilization.

It was recently shown that doping poly(βPIN) with iodine produces a semi-conductive material, with conductivities of about 8×10^{-3} S cm^{-1}, typical of non-conjugated doped polymers [99]. This material also showed a large quadratic electro-optic effect which would make it useful for non-linear optics applications [100].

In recent years, some monoterpenes have attracted some interest as environment-friendly solvents for polymeric systems. Camphene, which is both harmless and readily sublimed, has been tested as solvent for polypropylene

[101, 102]. The thermally induced phase separation of these solutions was applied to the preparation of microporous polypropylene tubular membranes [102, 103]. Camphene has also been used in the preparation of gas-filled polymeric micro-bubbles by freeze drying polymer/camphene emulsions [104–106]. These spheres can be used as contrast agents for the medical ultrasonography of vital organs.

It has been shown that α-pinene oxide, in conjunction with $ZnEt_2$, constitutes an efficient catalytic system for the stereospecific polymerization of styrene oxide [107].

Limonene was tested as a solvent for recycling polystyrene [108], as well as a renewable polymerization solvent and chain transfer agent in ring-opening metathesis polymerizations [109].

Finally, poly(α-PIN) was used in the nonisothermal crystallization of isotactic polypropylene from blends containing up to 30 per cent of this polyterpene [110–112].

2.6 CONCLUDING REMARKS

Although the following observations might sound redundant in their similarity to the conclusions of many, if not all the chapters of this book, they reflect nonetheless what we consider as being very relevant points. The systematic literature survey that we carried out to ensure the comprehensive coverage of this chapter showed unambiguously that the last decade has witnessed a drastic increase in scientific and more applied research related to terpenes as source of materials. It seems obvious to us that the concept and the practical implementation of a biorefinery, that is the rational exploitation of renewable resources, find in terpenes an excellent additional justification.

REFERENCES

1. Erman W.F., *Chemistry of Monoterpenes*, Marcel Dekker, New York, 1985.
2. Ruzicka L., The isoprene rule and the biogenesis of terpenic compounds, *Experientia*, **9**(10), 1953, 357–367.
3. Albert R.M., Webb R.L., in *Fragrance and Flavour Chemicals* (Eds. Zinkel D.F. and Russel J.), Pulp Chemical Association, New York, 1989, pp. 479–509.
4. Breitmaier E., *Terpenes: Flavors, Fragrances, Pharmaca, Pheromones*, Wiley-VCH, Weinheim, 2006.
5. Buntin G.A., in *Insecticides* (Eds. Zinkel D.F. and Russel J.), Pulp Chemical Association, New York, 1989, pp. 531–559.
6. Ruckel E.R., Arlt H.J., in *Polyterpene Resins* (Eds. Zinkel D.F. and Russel J.), Pulp Chemical Association, New York, 1989, pp. 510–530.
7. Kelly M.J., Rohl A.E., in *Pine Oil and Miscellaneous Uses* (Eds. Zinkel D.F. and Russel J.), Pulp Chemical Association, New York, 1989, pp. 560–573.
8. Zinkel D.E., Russel J., *Naval Stores. Production, Chemistry, Utilization*, Pulp Chemical Association, New York, 1989.
9. Stauffer D.F., in *Production, Markets and Economics* (Eds. Zinkel D.F. and Russel J.), Pulp Chemical Association, New York, 1989, pp. 39–82.
10. Coppen J.J.W., Hone G.A., *Gum Naval Stores: Turpentine and Rosin from Pine Resin*, FAO, Rome, 1995.
11. Stauffer D.F., Production, markets and economics, in *Naval Stores. Production, Chemistry, Utilization* (Eds. Zinkel D.F. and Russel J.), Pulp Chemical Association, New York, 1989, pp. 39–82.
12. Derfer J.M., Traynor S.G., in *Chemistry of Turpentine* (Eds. Zinkel D.F. and Russel J.), Pulp Chemical Association, New York, 1989, pp. 225–260.
13. Cascaval C.N., Ciobanu C., Rosu L., Rosu D., Turpentine as the raw chemicals for the polymerization processes, *Mater. Plast.*, **42**(3), 2005, 229–232.
14. Dolatkhani M., Cramail H., Deffieux A., Linear nonconjugated dienes from biomass as termonomers in EPDM synthesis, 1. Study of their reactivity in homopolymerizations, copolymerizations and terpolymerizations, *Macromol. Chem. Phys.*, **196**(10), 1995, 3091–3105.
15. Ruckel E.R., Wojcik R.T., Arlt H.G., Cationic polymerization of alpha-pinene oxide and beta-pinene oxide by a unique oxonium ion-carbenium ion sequence, *Abstracts of Papers of the American Chemical Society*, **173**(MAR20), 1977, 49.
16. Johanson A.J., McKennon F.L., Goldblatt L.A., Emulsion polymerization of myrcene, *Ind. Eng. Chem.*, **40**(3), 1948, 500–501.
17. Roberts W.J., Day A.R., A study of the polymerization of alpha-pinene and beta-pinene with Friedel Crafts type catalysts, *J. Am. Chem. Soc.*, **72**(3), 1950, 1226–1230.
18. Barros M.T., Petrova K.T., Ramos A.M., Potentially biodegradable polymers based on alpha- or beta-pinene and sugar derivatives or styrene, obtained under normal conditions and on microwave irradiation, *Eur. J. Org. Chem.*(8), 2007, 1357–1363.

19. Martinez F., Cationic polymerization of beta-pinene, *J. Polym. Sci. A: Polym. Chem.*, **22**(3), 1984, 673–677.
20. Keszler B., Kennedy J.P., Synthesis of high-molecular-weight poly (beta-pinene), *Adv. Polym. Sci.*, **100**, 1992, 1–9.
21. Guine R.P.F., Castro J., Polymerization of beta-pinene with ethylaluminum dichloride ($C_2H_5AlCl_2$), *J. Appl. Polym. Sci.*, **82**(10), 2001, 2558–2565.
22. Satoh K., Sugiyama H., Kamigaito M., Biomass-derived heat-resistant alicyclic hydrocarbon polymers: poly(terpenes) and their hydrogenated derivatives, *Green Chem.*, **8**(10), 2006, 878–882.
23. Adur A.M., Williams F., Radiation-induced cationic polymerization of beta-pinene, *J. Polym. Sci. A: Polym. Chem.*, **19**(3), 1981, 669–678.
24. Higashimura T., Lu J., Kamigaito M., Sawamoto M., Deng Y.X., Cationic polymerization of alpha-pinene with aluminum-based binary catalysts. 2, *Makromol. Chem.-Macromol. Chem. Phys.*, **194**(12), 1993, 3441–3453.
25. Higashimura T., Lu J., Kamigaito M., Sawamoto M., Deng Y.X., Cationic polymerization of alpha-pinene with aluminum-based binary catalysts. 3. Effects of added base, *Makromol. Chem.-Macromol. Chem. Phys.*, **194**(12), 1993, 3455–3465.
26. Higashimura T., Lu J., Kamigaito M., Sawamoto M., Deng Y.X., Cationic polymerization of alpha-pinene with the binary catalyst Alcl3/Sbcl3, *Makromol. Chem.-Macromol. Chem. Phys.*, **193**(9), 1992, 2311–2321.
27. Lu J., Kamigaito M., Sawamoto M., Higashimura T., Deng Y.X., Cationic polymerization of beta-pinene with the AlC_3/$SbCl_3$ binary catalyst: Comparison with alpha-pinene polymerization, *J. Appl. Polym. Sci.*, **61**(6), 1996, 1011–1016.
28. Radbil A.B., Zhurinova T.A., Starostina E.B., Radbil B.A., Preparation of high-melting polyterpene resins from alpha-pinene, *Russ. J. Appl. Chem.*, **78**(7), 2005, 1126–1130.
29. Lu J., Liang H., Zhang R.J., Deng Y.X., Cationic polymerization of alpha-pinene initiated by $AlCl_3$/$SbCl_3$ – Studies on the nature of the initiating species, *Chem. J. Chin. Univ.*, **17**(8), 1996, 1314–1318.
30. Li L., Yu S.T., Liu F.S., Yang J.Z., Zhaug S.F., Reactions of turpentine using Zr-MCM-41 family mesoporous molecular sieves, *Catal. Lett.*, **100**(3–4), 2005, 227–233.
31. Ruckel E.R., Wojcik R.T., Arlt H.G., Cationic polymerization of alpha-pinene oxide and beta-pinene oxide by a unique oxonium ion carbenium ion sequence, *J. Macromol. Sci. Chem.*, **A10**(7), 1976, 1371–1390.
32. Lu J., Liang H., Zhang R.J., Deng Y.X., Comparison of cationic polymerization of alpha- and beta-pinenes and limonene, *Acta Polym. Sinica*(6), 1998, 698–703.
33. Pietila H., Sivola A., Sheffer H., Cationic polymerization of beta-pinene, styrene and alpha-methylstyrene, *J. Polym. Sci. A1 Polym. Chem.*, **8**(3), 1970, 727–737.
34. Snyder C., McIver W., Sheffer H., Cationic polymerization of beta-pinene and styrene, *J. Appl. Polym. Sci.*, **21**(1), 1977, 131–139.
35. Sheffer H., Greco G., Paik G., The characterization of styrene beta-pinene polymers, *J. Appl. Polym. Sci.*, **28**(5), 1983, 1701–1705.
36. Kennedy J.P., Chou T., Sequence distribution analysis of isobutylene–styrene and isobutylene–isoprene copolymers, *J. Macromol. Sci. Chem.*, **A10**(7), 1976, 1357–1369.
37. Kennedy J.P., Liao T.P., Guhaniyogi S., Chang V.S.C., New telechelic polymers and sequential copolymers by polyfunctional initiator-transfer agents (inifers). 19. Poly(beta-pinenes) carrying one, two, or three functional end groups. 1. The effect of reaction conditions on the polymerization of beta-pinene, *J. Polym. Sci. A: Polym. Chem.*, **20**(11), 1982, 3219–3227.
38. Kennedy J.P., Nakao M., Co-polymerization of isobutylene with alpha-pinene, *J. Macromol. Sci. Chem.*, **A11**(9), 1977, 1621–1636.
39. Khan A.R., Yousufzai A.H.K., Jeelani H.A., Akhter T., Copolymers from alpha-pinene. 2. Cationic copolymerization of styrene and alpha-pinene, *J. Macromol. Sci. Chem.*, **A22**(12), 1985, 1673–1678.
40. Deng Y.X., Peng C.P., Duan W., Zhang R.J., Zeng L.M., Studies on the cationic copolymerization of alpha-pinene and styrene with the $AlCl_3$/$SbCl_3$ catalyst, *Chem. J. Chin. Univ.*, **17**(9), 1996, 1467–1472.
41. Deng Y.X., Peng C.P., Liu P., Lu J., Zeng L.M., Wang L.C., Studies on the cationic copolymerization of alpha-pinene and styrene with complex $SbCl_3$/$AlCl_3$ catalyst systems. 1. Effects of the polymerization conditions on the copolymerization products, *J. Macromol. Sci.: Pure Appl. Chem.*, **A33**(8), 1996, 995–1004.
42. Lu J., Kamigaito M., Sawamoto M., Higashimura T., Deng Y.X., Living cationic isomerization polymerization of beta-pinene. 3. Synthesis of end-functionalized polymers and graft copolymers, *J. Polym. Sci. A: Polym. Chem.*, **35**(8), 1997, 1423–1430.
43. Lu J., Kamigaito M., Sawamoto M., Higashimura T., Deng Y.X., Living cationic isomerization polymerization of beta-pinene. Initiation with HCl-2-chloroethyl vinyl ether adduct $TiCl_3(OiPr)$ in conjunction with nBu(4)NCl, *Macromolecules*, **30**(1), 1997, 22–26.
44. Lu J., Kamigaito M., Sawamoto M., Higashimura T., Deng Y.X., Living cationic isomerization polymerization of beta-pinene. 2. Synthesis of block and random copolymers with styrene or *p*-methylstyrene, *Macromolecules*, **30**(1), 1997, 27–31.
45. Hasebe T., Kamigaito M., Sawamoto M., Living cationic polymerization of styrene with $TiCl_3(OiPr)$ as a Lewis acid activator, *Macromolecules*, **29**(19), 1996, 6100–6103.

46. Lu J., Liang H., Zhang R.F., Living cationic polymerization of beta-pinene with 1-phenylethyl chloride/$TiCl_4$/Ti(OiPr)(4), *Chem. J. Chin. Univ.*, **21**(12), 2000, 1932–1935.
47. Lu J., Liang H., Li B.E., Synthesis of block and graft copolymers of beta-pinene, *Acta Polym. Sinica*(6), 2001, 755–759.
48. Li A.L., Zhang W., Liang H., Lu J., Living cationic random copolymerization of beta-pinene and isobutylene with 1-phenylethyl chloride/$TiCl_4$/Ti(OiPr)(4)/nBu(4)NCl, *Polymer*, **45**(19), 2004, 6533–6537.
49. Lu J., Liang H., Zhang R.F., Li B., Synthesis of poly(beta-pinene)-b-polytetrahydrofuran from beta-pinene-based macroinitiator, *Polymer*, **42**(10), 2001, 4549–4553.
50. Lu L., Liang H., Li B., Zhang W., Synthesis of benzyl chloride capped beta-pinene macroinitiator and block copolymers, *Acta Polym. Sinica*(3), 2001, 357–360.
51. Hui L., Jiang L., Synthesis of poly(beta-pinene)-*g*-polystyrene from allylic brominated poly(beta-pinene), *J. Appl. Polym. Sci.*, **75**(5), 2000, 599–603.
52. Lu J., Liang H., Zhang W., Cheng Q., Synthesis of poly(beta-pinene)-*g*-poly(meth)acrylate by the combination of living cationic polymerization and atom transfer radical polymerization, *J. Polym. Sci. A: Polym. Chem.*, **41**(9), 2003, 1237–1242.
53. Li A.L., Liang H., Lu J., Synthesis of copolymers of beta-pinene by the combination of living cationic polymerization and ATRP, *Acta Polym. Sinica*(1), 2006, 151–155.
54. Kennedy J.P., Chou T., Crystalline poly(beta-pinene), *J. Polym. Sci. C: Polym. Lett.*, **14**(3), 1976, 147–149.
55. Adur A.M., Williams F., Effects of inhibitors on the thermal polymerization of beta-pinene, *J. Polym. Sci. C: Polym. Lett.*, **19**(2), 1981, 53–57.
56. Ramos A.M., Lobo L.S., Polymers from pine gum components: Radical and coordination homo and copolymerization of pinenes, *Macromol. Symp.*, **127**, 1998, 43–50.
57. Paz-Pazos M., Pugh C., Synthesis of optically active copolymers of 2,3,4,5,6-pentafluorostyrene and beta-pinene with low surface energies, *J. Polym. Sci. A: Polym. Chem.*, **44**(9), 2006, 3114–3124.
58. Wang Y., Li A.L., Liang H., Lu J., Reversible addition-fragmentation chain transfer radical copolymerization of beta-pinene and methyl acrylate, *Eur. Polym. J.*, **42**(10), 2006, 2695–2702.
59. Li A.L., Wang Y., Liang H., Lu J., Controlled radical copolymerization of beta-pinene and acrylonitrile, *J. Polym. Sci. A: Polym. Chem.*, **44**(8), 2006, 2376–2387.
60. Li A.L., Wang X.Y., Liang H., Lu J., Controlled radical copolymerization of β-pinene and *n*-butyl acrylate, *React. Funct. Polym.*, **67**, 2007, 481–488.
61. Sharma S., Srivastava A.K., Alternating copolymers of limonene with methyl methacrylate: Kinetics and mechanism, *J. Macromol. Sci.: Pure Appl. Chem.*, **A40**(6), 2003, 593–603.
62. Sharma S., Srivastava A.K., Free radical copolymerization of limonene with butyl methacrylate: Synthesis and characterization, *Indian J. Chem. Technol.*, **12**(1), 2005, 62–67.
63. Sharma S., Srivastava A.K., Radical copolymerization of limonene with acrylonitrile: Kinetics and mechanism, *Polym.-Plast. Technol. Eng.*, **42**(3), 2003, 485–502.
64. Sharma S., Srivastava A.K., Synthesis and characterization of copolymers of limonene with styrene initiated by azobisisobutyronitrile, *Eur. Polym. J.*, **40**(9), 2004, 2235–2240.
65. Sharma S., Srivastava A.K., Radical co-polymerization of limonene with *N*-vinyl pyrrolidone: Synthesis and characterization, *Designed Monom. Polym.*, **9**(5), 2006, 503–516.
66. Mishra G., Srivastava A.K., Side-chain liquid-crystalline polymers with a limonene-co-methyl methacrylate main chain: Synthesis and characterization of polymers with phenyl benzoate mesogenic groups, *J. Appl. Polym. Sci.*, **102**(5), 2006, 4595–4600.
67. Maslinskasolich J., Kupka T., Kluczka M., Solich A., Optically-active polymers. 2. Copolymerization of limonene with maleic-anhydride, *Macromol. Chem. Phys.*, **195**(5), 1994, 1843–1850.
68. Doiuchi T., Yamaguchi H., Minoura Y., Cyclo-copolymerization of D-limonene with maleic-anhydride, *Eur. Polym. J.*, **17**(9), 1981, 961–968.
69. Shukla P., Ali A., Srivastava A.K., Optically active terpolymer: Synthesis and characterization, *J. Indian Chem. Soc.*, **77**(1), 2000, 48–49.
70. Sharma S., Srivastava A.K., Synthesis and characterization of a terpolymer of limonene, styrene, and methyl methacrylate via a free-radical route, *J. Appl. Polym. Sci.*, **91**(4), 2004, 2343–2347.
71. Yadav S., Srivastava A.K., Kinetics and mechanism of copolymerization of alpha-terpineol with methylmethacrylate in presence of azobisisobutyronitrile as an initiator, *J. Polym. Res. Taiwan*, **9**(4), 2002, 265–270.
72. Yadav S., Srivastava A.K., Copolymerization of alpha-terpineol with styrene: Synthesis and characterization, *J. Polym. Sci. A: Polym. Chem.*, **41**(11), 2003, 1700–1707.
73. Yadav S., Srivastava A.K., Synthesis of functional and alternating copolymer of alpha-terpineol with butylmethacrylate, *Polym.-Plast. Technol. Eng.*, **43**(4), 2004, 1229–1243.
74. Prajapati K., Varshney A., Terpene-based semi-interpenetrating polymer networks initiated by *p*-nitrobenzyl triphenyl phosphonium ylide: Synthesis and characterization, *Designed Monom. Polym.*, **9**(5), 2006, 453–476.

75. Yadav S., Srivastava A.K., Kinetics and monomer reactivity ratios of *N*-vinylpirrolidone and alpha-terpineol, *J. Appl. Polym. Sci.*, **103**, 2007, 476–481.
76. Shukla A., Srivastava A.K., Kinetics and mechanism of copolymerization of linalool with acrylonitrile, *J. Macromol. Sci.: Pure Appl. Chem.*, **A40**(1), 2003, 61–80.
77. Shukla A., Srivastava A.K., Free radical copolymerization of acrylamide and linalool with functional group as a pendant, *High Perform. Polym.*, **15**(3), 2003, 243–257.
78. Shukla A., Srivastava A.K., Determination of reactivity ratios and kinetics of free radical copolymerization of linalool with styrene, *Polym. Adv. Technol.*, **15**(8), 2004, 445–452.
79. Pathak S., Srivastava A.K., Free radical co-polymerization of butylmethacrylate and linalool with a functional group as a pendant, *Designed Monom. Polym.*, **8**(5), 2005, 409–422.
80. Pandey P., Srivastava A.K., Benzoyl peroxide-initiated copolymerization of citronellol and vinyl acetate, *J. Polym. Sci. A: Polym. Chem.*, **40**(9), 2002, 1243–1252.
81. Srivastava A.K., Pandey P., Benzoyl peroxide-*p*-acetylbenzylidenetriphenyl arsoniumylide initiated copolymerization of citronellol and styrene, *Polym. Int.*, **50**(8), 2001, 937–945.
82. Srivastava A.K., Pandey P., Mishra G., Synthesis and characterization of functional copolymers of citronellol and acrylamide, *J. Appl. Polym. Sci.*, **102**(5), 2006, 4908–4914.
83. Srivastava A.K., Pandey P., Synthesis and characterisation of copolymers containing geraniol and styrene initiated by benzoyl peroxide, *Eur. Polym. J.*, **38**(8), 2002, 1709–1712.
84. Shukla A., Srivastava A.K., Terpolymerization of linalool, styrene, and methyl methacrylate: Synthesis, characterization, and a kinetic study, *Polym.-Plast. Technol. Eng.*, **41**(4), 2002, 777–793.
85. Srivastava A.K., Kamal M., Kaur M., Pandey S., Daniel N., Chaurasia A.K., Pandey P., Terpolymerization: A review, *J. Polym. Res. Taiwan*, **9**(3), 2002, 213–220.
86. Pandey P., Srivastava A.K., Synthesis and characterization of optically active and functional terpolymer of citronellol, styrene, and methyl methacrylate: A kinetic study, *Adv. Polym. Techol.*, **21**(1), 2002, 59–64.
87. Pandey P., Kamal M., Srivastava A.K., Interpenetrating polymer network of poly(styrene) and poly(citronellol-alt-methyl methacrylate). Synthesis and characterization, *Polym. J.*, **35**(2), 2003, 122–126.
88. Mishra G., Srivastava A.K., Side-chain liquid crystalline polymers with [alpha-terpineol-co-MMA] main chain: Synthesis and characterization of polymers with phenyl benzoate mesogenic group, *Polym. Bull.*, **58**(2), 2007, 351–358.
89. Marvel C.S., Hanley J.R., Longone D.T., Polymerization of beta-pinene with ziegler-type catalysts, *J. Polym. Sci.*, **40**(137), 1959, 551–555.
90. Modena M., Bates R.B., Marvel C.S., Some low molecular weight polymers of D-limonene and related terpenes obtained by Ziegler-type catalysts, *J. Polym. Sci. A: Gen. Papers*, **3**(3PA), 1965, 949–960.
91. Sivola A., *N*-butyllithium-initiated polymerization of myrcene and its copolymerization with styrene, *Acta Polytech. Scand.-Chem. Technol. Ser.*(134), 1977, 7–65.
92. Byrne C.M., Allen S.D., Lobkovsky E.B., Coates G.W., Alternating copolymerization of limonene oxide and carbon dioxide, *J. Am. Chem. Soc.*, **126**(37), 2004, 11404–11405.
93. Cataldo F., Keheyan Y., Radiopolymerization of beta(-)pinene: A case of chiral amplification, *Radiat. Phys. Chem.*, **75**(5), 2006, 572–582.
94. Bates T.H., Williams T.F., Radiolysis of terpene hydrocarbons – Isomerization and polymerization of alpha-pinene and beta-pinene, *Nature*, **187**(4738), 1960, 665–669.
95. Bates T.H., Best J.V.F., Williams T.F., Radiation chemistry of beta-pinene, *J. Chem. Soc.*(MAY), 1962, 1531–1540.
96. Binet M.L., Commereuc S., Chalchat J.C., Lacoste J., Oxidation of polyterpenes: A comparison of poly alpha, and poly beta, pinenes behaviours – Part I – Photo-oxidation, *J. Photochem. Photobiol. A: Chem.*, **120**(1), 1999, 45–53.
97. Binet M.L., Commereuc S., Verney V., Thermo-oxidation of polyterpenes: Influence of the physical state, *Eur. Polym. J.*, **36**(10), 2000, 2133–2142.
98. Blazquez C.H., Vidyarth A.D., Sheehan T.D., Bennett M.J., McGrew G.T., Effect of pinolene (beta-pinene polymer) on carbaryl foliar residues, *J. Agric. Food Chem.*, **18**(4), 1970, 681–684.
99. Vippa P., Rajagopalan H., Thakur M., Electrical and optical properties of a novel nonconjugated conductive polymer, poly(beta-pinene), *J. Polym. Sci. B: Phys.*, **43**(24), 2005, 3695–3698.
100. Rajagopalan H., Vippa P., Thakur M., Quadratic electro-optic effect in a nano-optical material based on the nonconjugated conductive polymer, poly(beta-pinene), *Appl. Phys. Lett.*, **88**(3), 2006.
101. Yang M.C., Perng J.S., Camphene as a novel solvent for polypropylene: Comparison study based on viscous behavior of solutions, *J. Appl. Polym. Sci.*, **76**(14), 2000, 2068–2074.
102. Yang M.C., Perng J.S., Comparison of solvent removal methods of microporous polypropylene tubular membranes via thermally induced phase separation using a novel solvent: Camphene, *J. Polym. Res. Taiwan*, **6**(4), 1999, 251–258.
103. Yang M.C., Perng J.S., Microporous polypropylene tubular membranes via thermally induced phase separation using a novel solvent – Camphene, *J. Membr. Sci.*, **187**(1–2), 2001, 13–22.

104. Bjerknes K., Sontum P.C., Smistad G., Agerkvist I., Preparation of polymeric microbubbles: Formulation studies and product characterisation, *Int. J. Pharm.*, **158**(2), 1997, 129–136.
105. Bjerknes K., Dyrstad K., Smistad G., Agerkvist I., Preparation of polymeric microcapsules: Formulation studies, *Drug Dev. Ind. Pharm.*, **26**(8), 2000, 847–856.
106. Bjerknes K., Braenden J.U., Smistad G., Agerkvist L., Evaluation of different formulation studies on air-filled polymeric microcapsules by multivariate analysis, *Int. J. Pharm.*, **257**(1–2), 2003, 1–14.
107. Huang Y.B., Gao L.X., Ding M.X., Polymerization of styrene oxide catalyzed by a diethylzinc/alpha-pinene oxide system, *J. Polym. Sci. A: Polym. Chem.*, **37**(24), 1999, 4640–4645.
108. Shin C.Y., Chase G.G., Nanofibers from recycle waste expanded polystyrene using natural solvent, *Polym. Bull.*, **55**(3), 2005, 209–215.
109. Mathers R.T., McMahon K.C., Damodaran K., Retarides C.J., Kelley D.J., Ring-opening metathesis polymerizations in D-limonene: A renewable polymerization solvent and chain transfer agent for the synthesis of alkene macromonomers, *Macromolecules*, **39**(26), 2006, 8982–8986.
110. Di Lorenzo M.L., Cimmino S., Silvestre C., Nonisothermal crystallization of isotactic polypropylene blended with poly(alpha-pinene). 2. Growth rates, *Macromolecules*, **33**(10), 2000, 3828–3832.
111. Di Lorenzo M.L., Cimmino S., Silvestre C., Nonisothermal crystallization of isotactic polypropylene blended with poly(alpha-pinene). I. Bulk crystallization, *J. Appl. Polym. Sci.*, **82**(2), 2001, 358–367.
112. Di Lorenzo M.L., Spherulite growth rates in binary polymer blends, *Prog. Polym. Sci.*, **28**(4), 2003, 663–689.

– 3 –

Materials from Vegetable Oils: Major Sources, Properties and Applications

Mohamed Naceur Belgacem and Alessandro Gandini

ABSTRACT

This chapter summarizes the most recent advances in the realm of the modification of vegetable oils aimed at their use in the preparation of polymeric materials. After a brief description of the most important industrial oils, their major sources, their worldwide production, some of their properties and isolation procedures are reviewed. The contents are then divided into different sections, including the crosslinking of oils by vinyl monomers and by metal-catalyzed reactions, the formation of interpenetrating networks, the modification of epoxidized and castor oils and the polymerization of the resulting products. Subsequently, different oil-based polyurethanes, polyamides, polyester-amides, alkyd resins and polyesters and poly(hydroxyalkanoates) are reviewed. With each family of oil-based polymers, the most promising materials and their relevant properties are singled out stressing their suitability to replace, at least partially, petroleum-based counterparts.

Keywords

Triglycerides, Soybean oil, Castor oil, Sunflower oil, Oil-based polymers, Fatty acids, Epoxidized oils, Interpenetrating networks, Crosslinking of oils, Oil-based polyurethanes, Oil-based polyamides, Oil-based polyester-amides, Oil-based alkyd resins, Oil-based polyesters, Oil-based poly(hydroxyalkanoates)

3.1 INTRODUCTION

Vegetal and animal oils and fats have been used for centuries in the production of coatings, inks, plasticizers, lubricants and agrochemicals, as reported in recent books and reviews [1–5]. As expected, this scientific literature is largely complemented by a wide range of patents [6–12]. In 2000, the production of oils and fats amounted to 17.4 kg per year per capita and it is estimated that these figures grow at a rate of 3.3 per cent per year. The total worldwide yearly production of these renewable materials was over 110 million tons in 2002. Vegetal oils constitute the main fraction of this production (*i.e.* 80 per cent, the remaining 20 per cent being animal fats). Only 15.6 million tons (around 15 per cent) of these raw materials are used as precursors to the synthesis of new chemical commodities and materials. The rest is used for animal (about 5 per cent) and human (*ca.* 80 per cent) consumption.

The main actors involved in oleochemistry are located in Asia, because of the climatic suitability for such agricultural activity: Malaysia, Indonesia and the Philippines being the main producers. Fats from animal biomass are produced mostly in the US and Europe. The respective production of fats and oils for the four main continents, Asia, North America, Europe and South America, is 44, 16, 15 and 14 per cent. The present chapter deals exclusively with the use of vegetable oils as sources of polymeric materials, because the structures of fats do not lend themselves meaningfully to that application.

Table 3.1

Average annual worldwide production of 17 commodity oils, for 2001–2005 period [1]

Oil	Average annual production (million tons)	Main producer
Soybean	26.52	US
Palm	23.53	Malaysia
Rapeseed/canola	15.29	Europe
Sunflower	10.77	Europe
Tallow	8.24	US
Lard	6.75	China
Butterfat	6.26	Europe
Groundnut	5.03	China
Cottonseed	4.49	China
Coconut	3.74	Philippines
Palm kernel	2.95	Malaysia
Olive	2.52	Europe
Corn	2.30	US
Fish	1.13	Peru
Linseed	0.83	Europe
Sesame	0.76	China
Castor	0.56	India
Total	**121.67**	

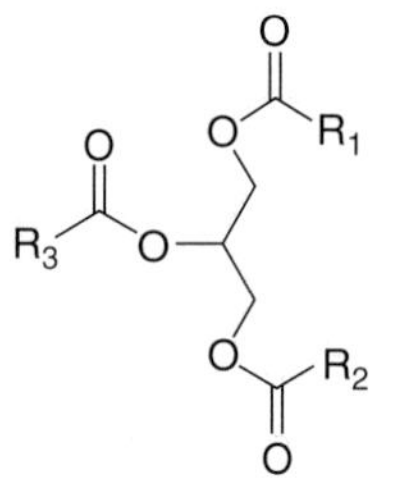

Where: R_1, R_2 and R_3 are fatty acid chains.

Scheme 3.1

Vegetal oils can be extracted from different species, as summarized in Table 3.1, which gives the average annual worldwide production of the 17 most important commodity oils. The major growth in the vegetable oil production is presently related to palm and rapeseed/canola oil [1].

As a general rule, the chemical composition of oils arises from the esterification of glycerol with three fatty acid molecules, as shown in Scheme 3.1. A given oil is always made up of a mixture of triglycerides bearing different fatty acid residues. The chain length of these fatty acids can vary from 14 to 22 carbons and contain 0–5 double bonds (DB) situated at different positions along the chain and in conjugated or unconjugated sequences. Table 3.2 collects the 16 most important oils with their specific chemical structures. Interestingly, none of them incorporated conjugated DB, which are however encountered in other less abundant oils.

In terms of chemical structure, few exceptions should be mentioned, since some fatty acids bear other types of functional groups, mainly epoxy rings and hydroxy moieties, triple bonds and ether functions [5], as discussed below. The fatty acid content in several common oils is given in Table 3.3 which, in addition, also provides the average number of DB per triglyceride unit. In the case of the more exotic castor, oiticica and tung oils, the main fatty acid residues are ricinoleic (87.5 per cent), licanic (74 per cent) of α-elaeostearic acids (84 per cent), respectively whose structures are provided in Scheme 3.2 below.

Alkyd resins are probably among the oldest polymers prepared from renewable resources through the esterification of polyhydroxy alcohols with polybasic and fatty acids. A considerable amount of technological research has

Table 3.2

Names and double bond positions of the most common fatty acids [1]

Chain length: Number of DB	Systematic name	Trivial name	Double bond position
12:0	Dodecanoic	Lauric	–
14:0	Tetradecanoic	Myristic	–
16:0	Hexadecanoic	Palmitic	–
18:0	Octadecenoic	Stearic	–
18:1	9-Octadecanoic	Oleic	9
18:2	9,12-Octadecadeinoic	Linoleic	9,12
18:3	6,9,12-Octadecatrienoic	γ-linolenic	6,9,12
18:3	9,12,15-Octadecatrienoic	α-linolenic	9,12,15
20:0	Eicosanoic	Arachidic	–
20:1	Eicosaenoic	–	9
20:4	Eicosatetraenoic	Arachidonic	5,8,11,14
20:5	Eicosapentaenoic	EPA	5,8,11,14,17
22:0	Docosanoic	–	–
22:1	Docosenoic	Erucic	13
22:5	Docosapentanoic	DPA	7,10,13,16,19
22:6	Docosahexanoic	DHA	4,7,10,13,16,19

Table 3.3

Main fatty acid contents in different oils [2, 13]

Oils	Fatty acid					Average number of DB per triglyceride
	Palmitic	Stearic	Oleic	Linoleic	Linolenic	
Canola	4.1	1.8	60.9	21.0	8.8	3.9
Corn	10.9	2.0	25.4	59.6	1.2	4.5
Cottonseed	21.6	2.6	18.6	54.4	0.7	3.9
Linseed	5.5	3.5	19.1	15.3	56.6	6.6
Olive	13.7	2.5	71.1	10.0	0.6	2.8
Soybean	11.0	4.0	23.4	53.3	7.8	4.6
Tung	–	4.0	8.0	4.0	–	7.5
Fish	–	–	18.20	1.10	0.99	3.6
Castor	1.5	0.5	5.0	4.0	0.5	3.0
Palm	39	5	45	9	–	–
Oiticica	6	4	8	8	–	–
Rapeseed	4	2	56	26	10	–
Refined tall	4	3	46	35	12	–
Sunflower	6	4	42	47	1	–

been devoted to the properties of these polyesters and the interested reader will find much detailed information in a monograph entirely devoted to them [14]. Here, we limit our coverage to the most relevant findings in the field of alkyd resins as polymeric materials.

3.2 PROPERTIES OF VEGETABLE OILS AND FATTY ACIDS

The physical and chemical properties of vegetable oils depend on their fatty acid distribution. The numbers of DB, as well as their positions within the aliphatic chain, affect strongly the oil properties. The actual number of carbon

Table 3.4

Some physical properties of triglyceride oils and fatty acids [15]

Name	Viscosity (mPa · s)	Specific gravity	Refractive index	Melting point (°C)
Castor oil	293.4 at 37.8°C	0.951 at 20°C	1.473–1.480 at 20°C	−20 to −10
Linseed oil	29.6 at 37.8°C	0.925 at 20°C	1.480–1.483 at 20°C	−20
Palm oil	30.92 at 37.8°C	0.890 at 20°C	1.453–1.456 at 20°C	33–40
Soybean oil	28.49 at 37.8°C	0.917 at 20°C	1.473–1.477 at 20°C	−23 to −20
Sunflower oil	33.31 at 37.8°C	0.916 at 20°C	1.473–1.477 at 20°C	−18 to −16
Myristic acid	2.78 at 110°C	0.844 at 80°C	1.4273 at 70°C	54.4
Palmitic acid	3.47 at 110°C	0.841 at 80°C	1.4209 at 70°C	62.9
Stearic acid	4.24 at 110°C	0.839 at 80°C	1.4337 at 70°C	69.6
Oleic acid	3.41 at 110°C	0.850 at 80°C	1.4449 at 60°C	16.3

Table 3.5

Iodine values of some unsaturated fatty acids and their triglycerides

Fatty acid (number of carbon atoms)	Iodine value of the	
	Acid	Triglyceride
Palmitoleic (C16)	99.8	95.0
Oleic (C18)	89.9	86.0
Linoleic (C18)	181.0	173.2
Linolenic (C18) and α-eleostearic	273.5	261.6
Ricinoleic (C18)	85.1	81.6
Licanic (C18)	261.0	258.6

atoms making up the aliphatic chains plays a very minor role, simply because most of these triglycerides have 18 and a few 16. Table 3.4 summarizes some relevant properties of common vegetable oils and fatty acids. The average degree of insaturations is measured by the iodine value. This parameter corresponds to the amount of iodine (mg) which reacts with the DB in 100 g of the oil under investigation. Vegetable oils are divided into three groups depending on their iodine values. Thus, oils are classified as 'drying', if their iodine value is higher than 130, 'semi-drying' if this parameter is comprised between 90 and 130 and 'non-drying' when it drops below 90. Iodine values of some common fatty acids and their triglycerides are given in Table 3.5.

3.3 ISOLATION OF VEGETABLE OILS

The isolation of vegetable oils can be achieved either mechanically or by solvent extraction [16]. The mechanical process consists of submitting the beans, cells and oil bodies to shearing, in order to liberate the oil. Heat is generated during this procedure and this can induce a negative effect on the proteins therein. The advantages of the mechanical isolation process reside in its low cost, low investment and safety in terms of environmental concerns, since it does not involve the use of solvents or hazardous substances. This process presents, nevertheless, the drawback of low yields of oil extraction, since the amount of oil left in the ensuing residues can reach values as high as 7 per cent.

The main principle of solvent extraction is that the solvent diffuses through the seeds, solubilizes and extracts the oil. Solvent extraction involves the use of organic solvents, the most common being hexane. The key parameter of this process is the diffusion rate of the solvent into the oil body. This process is more efficient than the mechanical counterpart, but has the drawback of using volatile organic solvents.

3.4 POLYMERS FROM VEGETABLE OILS

Soybean, fish, corn, tung, linseed and castor oils are the most common renewable resources used as precursors for the synthesis of vegetable oil-based polymers. Their composition in fatty acid is shown in Table 3.3 [3, 17]. The chemical structures of the most representative oils in these families are given in Scheme 3.2.

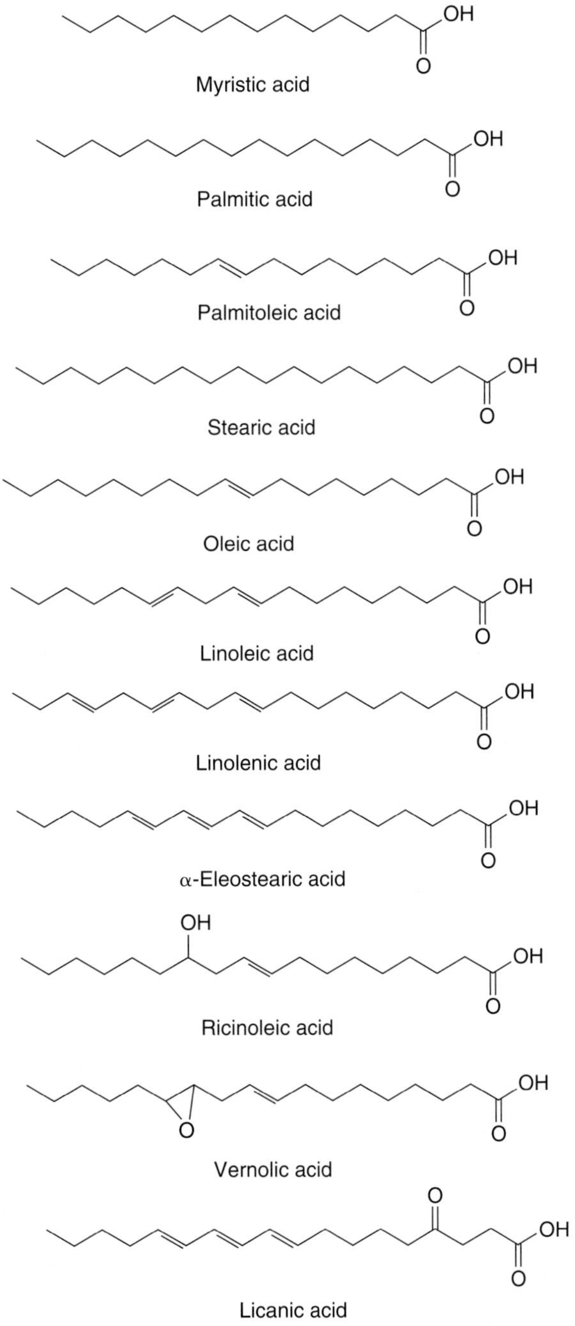

Myristic acid

Palmitic acid

Palmitoleic acid

Stearic acid

Oleic acid

Linoleic acid

Linolenic acid

α-Eleostearic acid

Ricinoleic acid

Vernolic acid

Licanic acid

Scheme 3.2

The free radical oxido-polymerization of unsaturated triglycerides in atmospheric conditions has been (and continues to be) exploited mostly in coating applications involving inks and paints. These systems and the mechanisms they involve represent well-established and well-documented traditional processes [15–17] and will not be discussed any further here, also because, in terms of bulk material properties, the ensuing networks display rather poor mechanical properties.

Other chemical exploitations of these molecules as macromonomers are instead becoming particularly promising, whether they involve reactions with the DB, the hydroxy groups or epoxy functions borne by some of their structures. Since this topic was covered by two very recent reviews [2, 3] and two books [1, 4], we will confine ourselves here to reviewing relevant studies in the field of macromolecular architecture with an emphasis on the potential viability of these novel materials derived from renewable resources. Of course, more recent contributions will also be assessed.

3.4.1 Crosslinking of oils by vinyl monomers

Linseed (**LO**), tung (**TO**), soybean (**SOYO**), fish (**FO**), sunflower (**SFO**), corn (**CORO**), oiticica (**OTO**) and dehydrated castor oils (**DCO**) are relatively DB rich molecules, which makes them suitable for copolymerization with vinyl monomers, such as styrene (**ST**), α-methyl styrene (**MST**), divinyl benzene (**DVB**) or cyclopentadiene (**CPD**). These oils have thus been used to produce polymeric networks through their copolymerization with vinyl monomers by the so-called grafting-through mechanism. The **SOYO** used for the preparation of polymeric materials is available in three grades, namely (i) regular, (ii) low saturated non-conjugated and (iii) conjugated saturated. The fatty acids which constitute **SOYO** are mainly C18 and the number of DB is about five per triglyceride molecule, as shown in Table 3.3. Two varieties of **SOYO** are available on the market, viz. non-modified and epoxidized (**ESOY**). **ESOY** can be submitted to further modifications, such as esterification or urethane formation, as described in the appropriate section below.

Regular **SOYO**, **LO**, **CORO**, **TO** and **FO** oils were extensively studied by Larock *et al.* [18–34], who submitted them to cationic copolymerization with vinyl monomers in the presence of boron trifluoride diethyl etherate (**BFE**), as initiator. Thus, native Norwegian and conjugated **FO**, as well as conjugated and non-conjugated **FO** ethyl esters, were submitted to copolymerizations with **ST**, **DVB**, **DCP** and **NBD**, in various combinations and proportions, and the ensuing thermosets characterized in terms of structures and mechanical properties [18, 19, 24, 26]. The glass transition temperatures of these materials varied from 11°C to 113°C, as a function of the monomer combination and composition. Generally, conjugated oils gave stiffer thermosets. The free oil, the gel fraction and the amount of incorporated oils varied from 15 to 32 per cent, 41 to 85 per cent and 30 to 50 per cent, respectively. As expected, the gel fraction of these materials exhibited a higher thermal stability. Some other comonomers were also tested with both conjugated and non-conjugated **FO** ethyl esters, namely furfural, *p*-benzoquinone, *p*-mentha-1,8-diene, furan, the Diels–Alder adduct between furfural and *p*-mentha-1,8-diene, maleic anhydride and vinyl acetate. The aim of these investigations was to modulate the glass transition temperature of the ensuing products.

SOYO alone, or in combination with other oils, was also studied in the context of cationic copolymerizations with the same comonomers mentioned above [20–23, 25, 27–29, 32, 33]. The **SOYOs** used were low saturated (**LSO**) or conjugated low saturated soybean (**CLS**) oils and the cationic initiator was blended with **FO**. The formulation of **TO** and **LSO** oils gave highly crosslinked networks with **ST** and **DVB** (95–96 per cent). **CORO** was also copolymerized with **ST** and **DVB** using **BFE**. The resulting copolymers were characterized by dynamic and thermal mechanical analysis (DMA and TMA), thermogravimetry (TGA), differential scanning calorimetry (DSC) and scanning electron microscopy (SEM). The gelation time in these very different systems varied from a few minutes to a few days. The Tgs and the moduli of the prepared copolymer varied from 0°C to 105°C and from 6 to 2000 MPa, respectively. These values are comparable to those of commercial rubbers synthesized from petroleum-based monomers. The free oil, the gel fraction and the amount of incorporated oils varied from, 12 to 27 per cent, 69 to 88 per cent and 29 to 56 per cent, respectively. These oil-based thermosets started to be degraded at temperatures higher than 400°C. Their toughness reached 4 MPa, for some combinations. The shape memory effect of these materials was also studied. It consisted in the establishment of three parameters, namely (i) their deformability (D) behaviour at temperatures higher than their Tg, (ii) the degree to which the deformation is subsequently fixed at ambient temperature (FD) and (iii) the final shape recovery (R) when the sample is reheated. Crosslinked samples with Tgs well above the ambient temperature exhibited good shape memory effects.

Very recently, novel thermosetting copolymers were prepared from **SOYO** and dicyclopentadiene (**DCP**) [32] using the catalytic system. The ensuing materials varied from tough and ductile to very soft rubbers. The gel fraction varied between 69 and 88 per cent and the Tgs were in the range of −20°C to 40°C. These materials lost 10 per cent of their weight at temperatures around 330°C. In the same vein, nanocomposites based on **SOYO-ST** and **SOYO-DVB** matrices reinforced with a reactive organo-modified montmorillonite clay were also prepared and showed, as expected, better mechanical properties and improved thermal resistance, compared with their respective matrices [33].

TO and **LO** were also tested in the same context [30, 31, 34] but the crosslinking was induced by a free radical mechanism using bezoyl- and *ter*-butylhydro-peroxides as well as azobisisobutyronitrile (AIBN), in the presence of acrylonitrile, **ST** and **DVB**. The onset of network formation occurred just above 140°C and fully crosslinked materials were obtained by post-curing at 160°C. The Tgs of these networks varied from 0°C to 110°C and the gel fraction reached 96 per cent for some combinations. The compressive modulus varied from 0.02 to 1.2 GPa, whereas the compressive strength ranged from 8 to 140 MPa. These materials were stable up to 300°C.

3.4.2 Crosslinking of modified oils by vinyl monomers

Castor oil (**CO**) and **LO** were reported to crosslink with **ST** in the presence of benzoyl peroxide, after being modified with acrylic acid (**AA**) through an inter-transesterification reaction [35].

The free radical crosslinking of bacterial polyesters obtained from **SOYO** has also been reported [36]. Thus, poly(-3-hydroxy octanoate) containing unsaturated side chains (**PHA–SOYO**) were produced from **SOYO** in the presence of *Pseudomonas oleovorans*. Subsequently, **PHA–SOYO** adducts were crosslinked, thermally or under UV irradiation, in the presence of a free radical initiator and ethylene glycol dimethacrylate as the crosslinker. The network yield reached 93 per cent and the Tg of the biopolyester increased from −60°C to −40°C after crosslinking.

SFO and **LO** were also styrenated after their transesterification with methyl methacrylate (**MMA**) in the presence of benzoyl peroxide [37] and the ensuing materials were characterized in terms of drying time, alkali and acid resistance and hardness. A more recent study reported the modification of **SOYO** and **SFO** by acrylamide derivatives [38] calling upon the use of Ritter's reaction. This rather fundamental study was limited to the determination of the reactivity ratios between **ST** and the different moieties of acrylamide derivatives.

Epoxynorborene derivatives of **LO** (see Scheme 3.3) were prepared and crosslinked by UV irradiation in the presence of tetraethylorthosiloxane (**TEOS**), in order to produce organic–inorganic hybrid films using (4-octyloxyphenyl) phenyl iodonium hexafluoroantimonate as the cationic photoinitiator [39]. Different formulations were studied, but in all cases the modified **LO** was in a large excess. The addition of 10 per cent of **TEOS** was found to be optimal in terms of the mechanical and adhesion properties of the composite films.

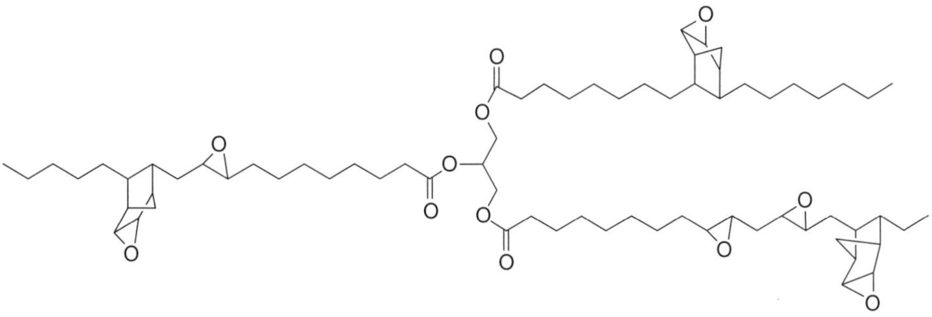

Scheme 3.3

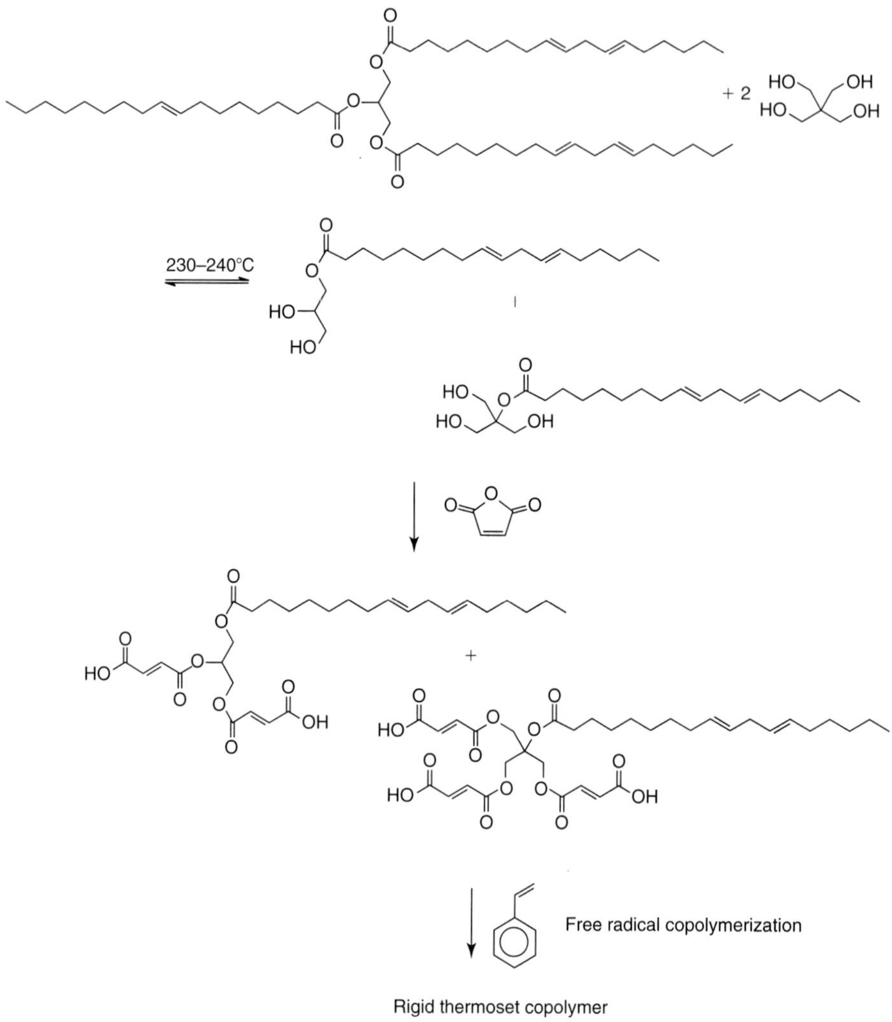

230–240°C

Free radical copolymerization

Rigid thermoset copolymer

Scheme 3.4

SOYO and **CO** were also submitted to alcoholysis with pentaerythritol or biphenol-A propoxylate and the ensuing mixtures reacted with maleic anhydride before crosslinking with **ST** [40, 41], as sketched in Scheme 3.4 for **SOYO**. For **CO,** the grafting of maleic anhydride is more favoured, since **CO** contains 3-OH functions per triglyceride molecule. The modified oils were carefully characterized by NMR spectroscopy, as shown in Fig. 3.1.

The crosslinking was followed by FTIR spectroscopy which showed that about 1 h was needed to reach a high gel fraction. Total conversion was attained after 3 h at 120°C. The **SOYO**-based polymers exhibited flexural moduli and strength in the range of 0.8–2.5 GPa and 32–112 MPa, respectively. Their glass transition temperatures varied from 72°C to 150°C. The surface hardness of the **SOYO**-crosslinked networks varied between 77 D and 90 D. **CO**-based thermosetting materials exhibited significantly improved moduli, strengths and Tg values, when compared with **SOYO**-based counterparts. These novel materials, based on renewable vegetable oils, show properties comparable to those of high performance unsaturated polyester resins which make them good candidates to replace those petroleum-based products.

3.4.3 Crosslinking of virgin oils by metal-catalyzed reactions

LO was submitted to auto-oxidative crosslinking in the presence of different metal-based catalysts [42–44], viz. titanium (IV) i-propoxide, titanium di-i-propoxide bis(acetylacetonate), cobalt, lead and zirconium-octonoates,

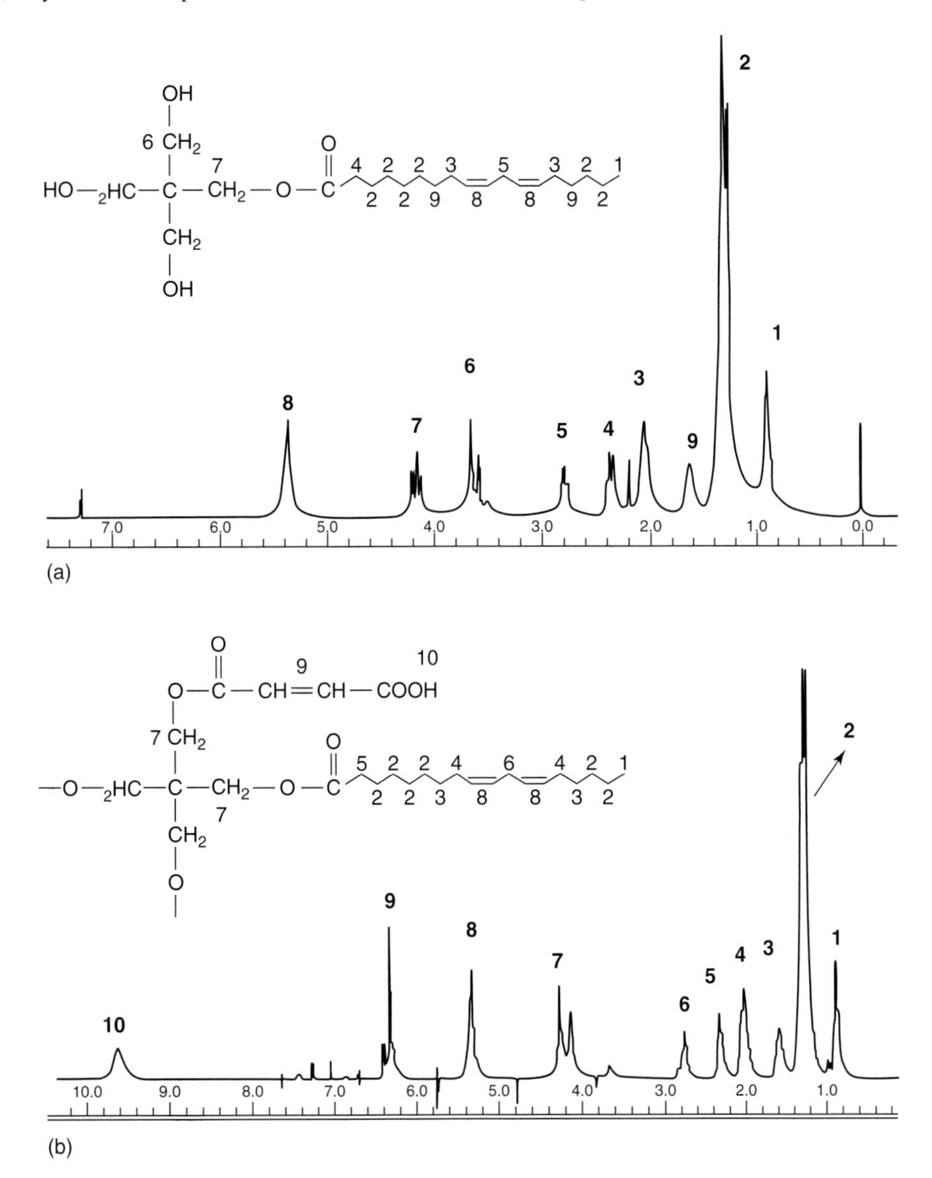

Figure 3.1 ^{1}H-NMR spectra of pentaerythritol-modified **SOYO** (a) and its maleic anhydride adduct (b). (Reproduced by permission of Wiley Periodicals, Inc. Copyright 2006. Reprinted from Reference [40].)

lead oxides and salts and iron oxides. The crosslinking reactions were followed by DSC. The increase in the amount of the catalyst induced a decrease in the heat of the reaction and increased the hardness of the crosslinked films. Cobalt salts were found to catalyse the oxidation step, whereas lead and zirconium homologues catalysed the polymerization reactions. The interest of these otherwise conventional systems lies solely in the purported possibility of replacing some toxic catalysts with environment-friendly counterparts, a claim which requires confirmation.

3.4.4 Crosslinking of modified oils forming interpenetrating networks

CO as such, after hydrogenation (**HCO**), or trans-esterification with glycerol, or modification with **LO**, or in combination with polyethylene glycol, were extensively used as precursors in the preparation of interpenetrating networks,

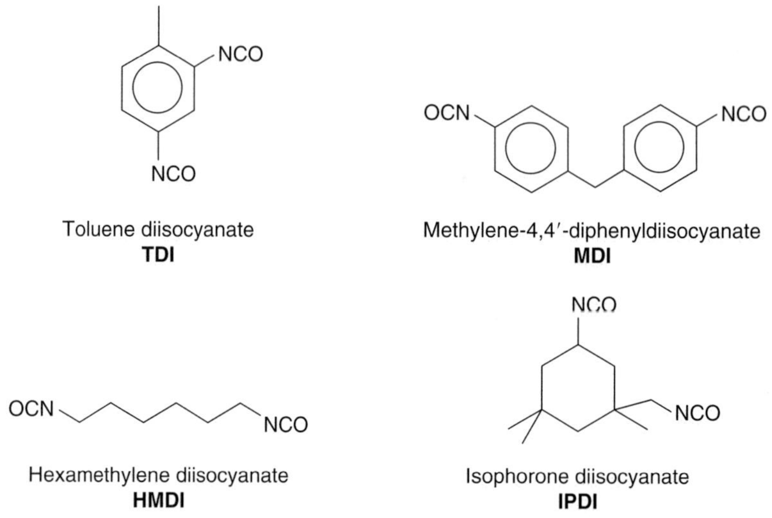

Toluene diisocyanate
TDI

Methylene-4,4′-diphenyldiisocyanate
MDI

Hexamethylene diisocyanate
HMDI

Isophorone diisocyanate
IPDI

Scheme 3.5

as recently reported [45–48] and reviewed [2]. The first step of this process consisted in the preparation of a polyurethane component which involved the use of different isocyanates, as summarized in Scheme 3.5. The most commonly used vinyl and acrylic monomers were **ST**, **DVB**, alkyl acrylates and methacrylates, **AA**, acrylonitrile, ethylene glycol dimethacrylate and 2-hydroxyethyl methacrylate. These systems will not be described at length here since the reader will readily find more details in the cited references [2, 45–48].

3.4.5 Modification of epoxidized oils and polymerization of the resulting products

Crivello's group has been involved in the cationic polymerization of epoxidized triglycerides like vernonia, lesquerella, crambe, rapeseed, canola, **SOYO**, **SFO** and meadowfoam [49–52]. Two routes were applied for the epoxidization, (i) glacial acetic acid in the presence of hydrogen peroxide and (ii) phase-transfer catalyzed by methyl octylammonium (diperoxotungsto)phosphate. In all cases, the optimal conditions yielding a high conversion were established. Glass-fibre reinforced composite materials, in which these epoxidized oils were used as matrix-precursors, were also prepared and characterized.

Polymer networks derived from the curing of epoxidized **LO** have also been prepared and characterized [53]. The crosslinking agents were phthalic (**PA**), tetrahydro phthalic (**TPA**), tetrahydromethyl phthalic (**TMPA**), hexahydromethyl phthalic (**HMPA**) and methyl endomethylene tetrahydrophthalic (**MTPA**) anhydrides and the reaction was catalyzed by different tertiary amines. The effect of the steric hindrance and stiffness of the hardeners was also studied.

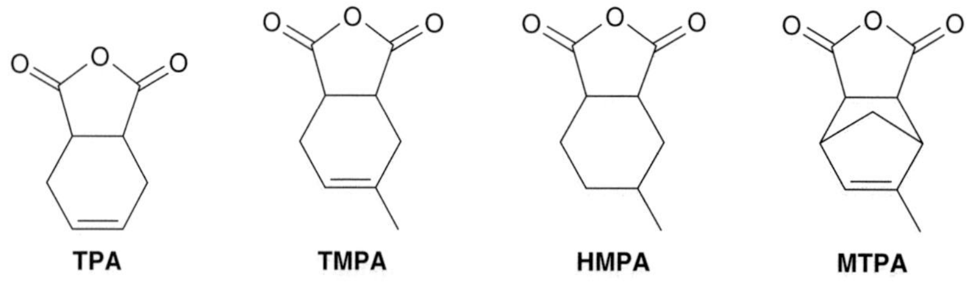

TPA **TMPA** **HMPA** **MTPA**

More recently, bromo-acrylated triglycerides were prepared from **SFO** and **SOYO** and the ensuing modified **SOYO** was homo-polymerized or co-polymerized with **ST** by free radical initiation [54]. This study was carried out with the aim of using the ensuing materials in the field of flame-resistant polymers. The FTIR, NMR and mass spectra of the bromo-acrylated methyl oleate were recorded and showed that the modification had been successful. **SOYO** gave a higher acrylate substitution than **SFO**. The glass transition temperatures of the polymers obtained were around 60 and 25°C, for **SFO**- and **SOYO**-based materials, respectively. These materials did indeed show good flame resistance [54].

CO was used in combination with a polyamide system in the preparation of coating adhesives [55] and some of these formulations showed excellent adhesion properties, satisfactory flexibility and good resistance to several chemicals. Epoxidized **CO** (**DECO**) were blended with **PMMA** in 80/20, 60/40 and 20/80 weight ratios, but these mixtures showed a certain degree of incompatibility [56, 57]. It follows that the possible application or copolymerization of these materials does not seem promising.

UV curable cycloalphatic epoxides were prepared from **LO** (**ELO**), as sketched in Scheme 3.6 and thoroughly characterized by FTIR, NMR and GPC. **ELO** was then efficiently UV cured and the conversion of the epoxy ring followed by monitoring the FTIR peak intensity at $831\,cm^{-1}$ [58]. Bio-based epoxy resins have been prepared by thermal cationic initiation using *N*-benzylpyrazinium hexafluoroantimonate (**BPH**) and *N*-benzylquinoxalinium hexafluoroantimonate (**BQH**) [59] and gave Tgs of 30–40°C. The mechanical properties of the films prepared by **BQH**-initiation were higher than those of the **BPH** counterpart, a fact which was attributed to the higher crosslink density reached with the former system.

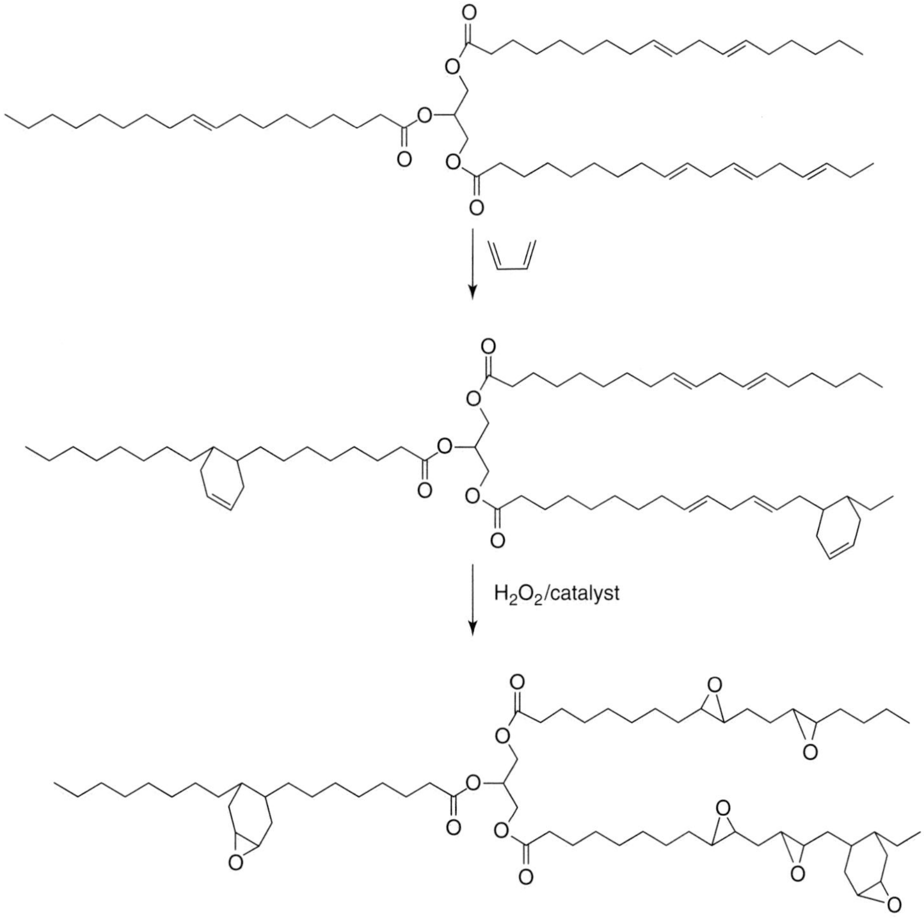

H_2O_2/catalyst

Scheme 3.6

More recently, a commercial sample of epoxidised **SOYO (ESOYO)** was thoroughly characterized [60–62], before being submitted to different modifications. The FTIR spectrum of **ESOYO** is given in Fig. 3.2, which shows the characteristic ester bands at 1743 cm^{-1} (carbonyl groups, C=O) and at 1159 cm^{-1} (ester moieties C—O), together with the oxirane peak at 823 cm^{-1}. The ^{1}H-NMR spectrum of **ESOYO** is presented in Fig. 3.3, and

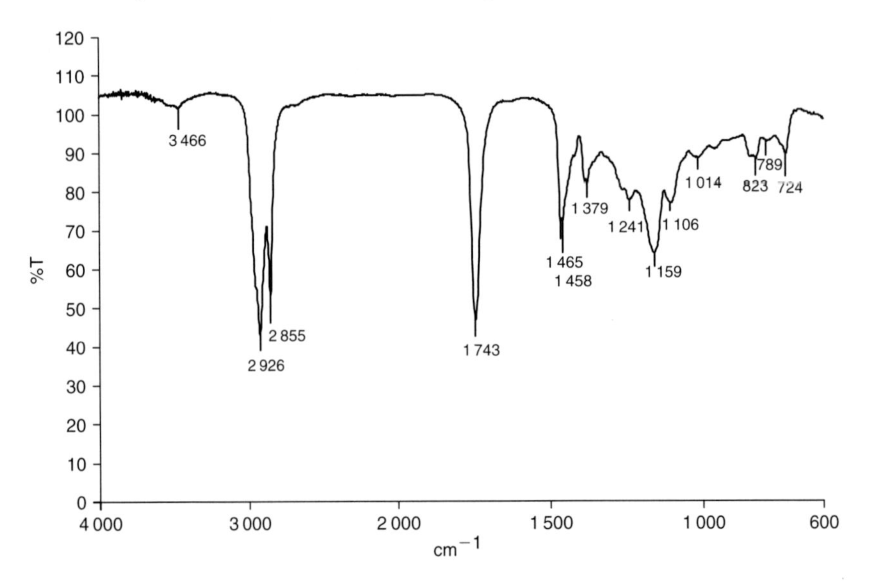

Figure 3.2 FTIR Spectrum of **ESOYO**, reported in reference [61]. (Reproduced by permission of the Wiley-VCH Verlag GmnH & Co. KGaA. Copyright 2006. Reprinted from Reference [61]).

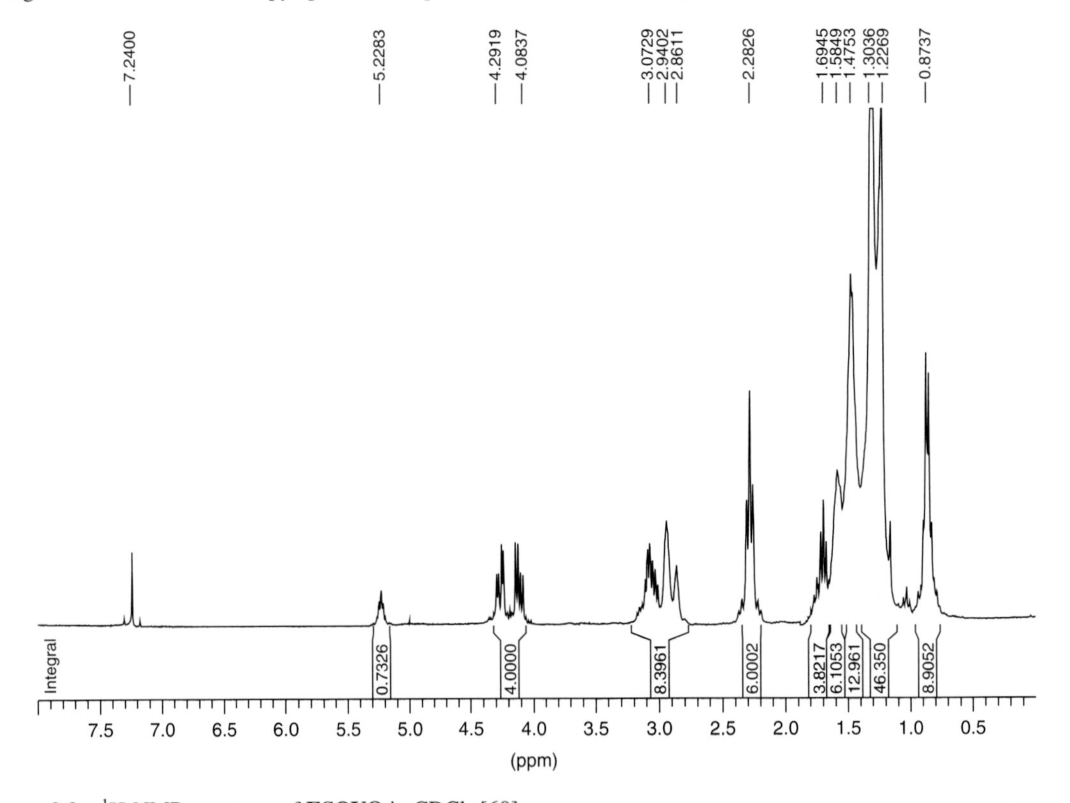

Figure 3.3 ^{1}H-NMR spectrum of **ESOYO** in $CDCl_3$ [60].

Table 3.6 summarizes its peak assignments. From this spectrum and the data presented in Table 3.6, the number of epoxy groups present in each triglyceride molecule was calculated. Thus, the intensity of the peaks between 2.9 ppm and 3.1 ppm, associated with epoxy functions, corresponded to 8.4 protons (*i.e.* about 4.5 oxirane moieties per molecule). This is in rather good agreement with the composition of **SOYO**, that is, on average, two linoleic and one oleic residues, as indicated in Table 3.3, which predicts the conversion of the five insaturations present in the initial triglyceride molecule into the corresponding oxirane function. The elemental analyses and the molecular weight determinations confirmed this structure.

ESOYO was then modified by reactions with **AA** or acryloyl chloride and the acrylated derivatives characterized by spectroscopic techniques before being submitted to free radical photopolymerization. The NMR spectrum of the acrylated **ESOYO** (**AESOYO**) is shown in Fig. 3.4. Table 3.6 summarizes the peak assignments and their

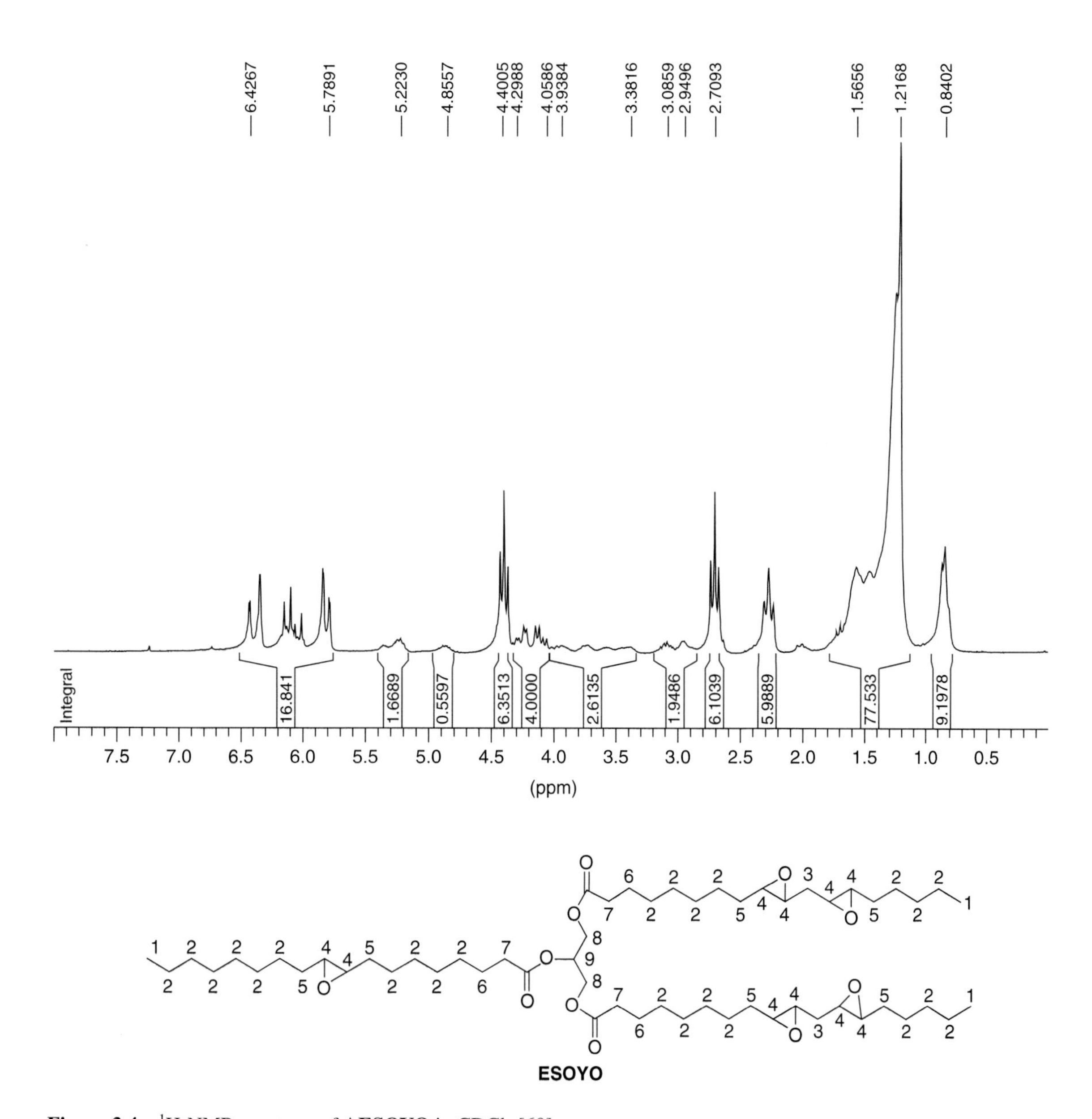

Figure 3.4 ^{1}H-NMR spectrum of **AESOYO** in $CDCl_3$ [60].

AESOYO

Figure 3.4 (*Continued*).

Table 3.6

^{1}H-NMR data relative to ESOYO [60]

ESOYO			AESOYO		
δ (ppm)	Intensity	Peak assignment	δ (ppm)	Intensity	Peak assignment
			5.8–6.4	16.8	H_4
5.2	1	H_9	5.2–5.3	1.6	H_9
4–4.3	4	H_8	4.3	6	H_{10}
2.9–3.1	8.4	H_4	4–4.3	4	H_8
2.3	6	H_7	3.4–4	2.6	H_5
1.7	4	H_3	2.9–3.1	2	H_{11}
1.6	6	H_6	2.7	6	H_3
1.5	13	H_5	2.3	6	H_7
1.3–1.4	46	H_2	1.3–1.6	77	H_6 and H_2
0.9	9	H_1	0.9	9	H_1

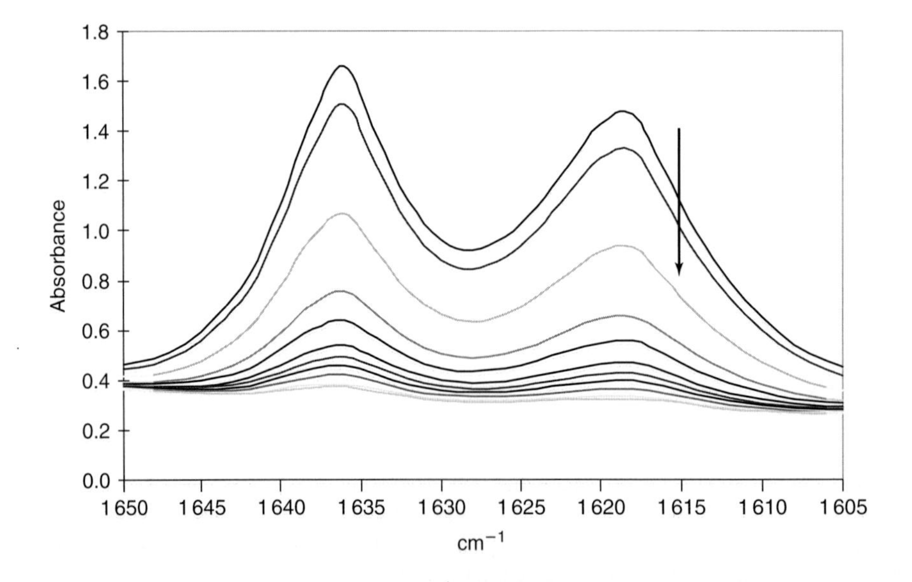

Figure 3.5 Evolution of the bands at 1 637 and 1 627 cm^{-1} in the FTIR spectra of **AESOYO** irradiated in solution in the presence of 4 per cent of Darocure 1 173 [60].

intensity. The irradiation of **AESOYO** was accompanied by FTIR spectroscopy in order to monitor the decrease of peak attributed to the C=C acrylic moieties, as shown in Fig. 3.5. This reaction rapidly generated a highly crosslinked network.

Of two very recent additions to the present context, one concerns the crosslinking of both **ELO** and epoxidized synthetic triglycerides obtained from ω-unsaturated fatty acids with either 4,4′-methylene dianiline or **PA** [63]. Films of these materials were glassy, since their Tgs varied from about 50°C to 130°C, and stiffer than those prepared from conventional bisphenol A-diglycidyl ether (**BADGE**)-based epoxy resins. However, the adhesion and the shear strength of these oil-based resins were better than those of classical **BADGE**-based counterparts. The other recent contribution deals with the free radical photocopolymerizations of **ESOYO** and maleinized **ESOYO** (**MESOYO**) with **ST**, vinyl acetate and **MMA** [64]. The authors concluded that these copolymers could be good candidates for liquid moulding resins.

3.4.6 Modification of CO and polymerization of the resulting products

The castor oil employed in this investigation was an industrial product whose FTIR spectrum, shown in Fig. 3.6, displays bands characteristic of OH functions at 3401 cm^{-1}, carbonyl groups at 1745 cm^{-1} and a weak peak at 3007 cm^{-1} attributed to insaturations.

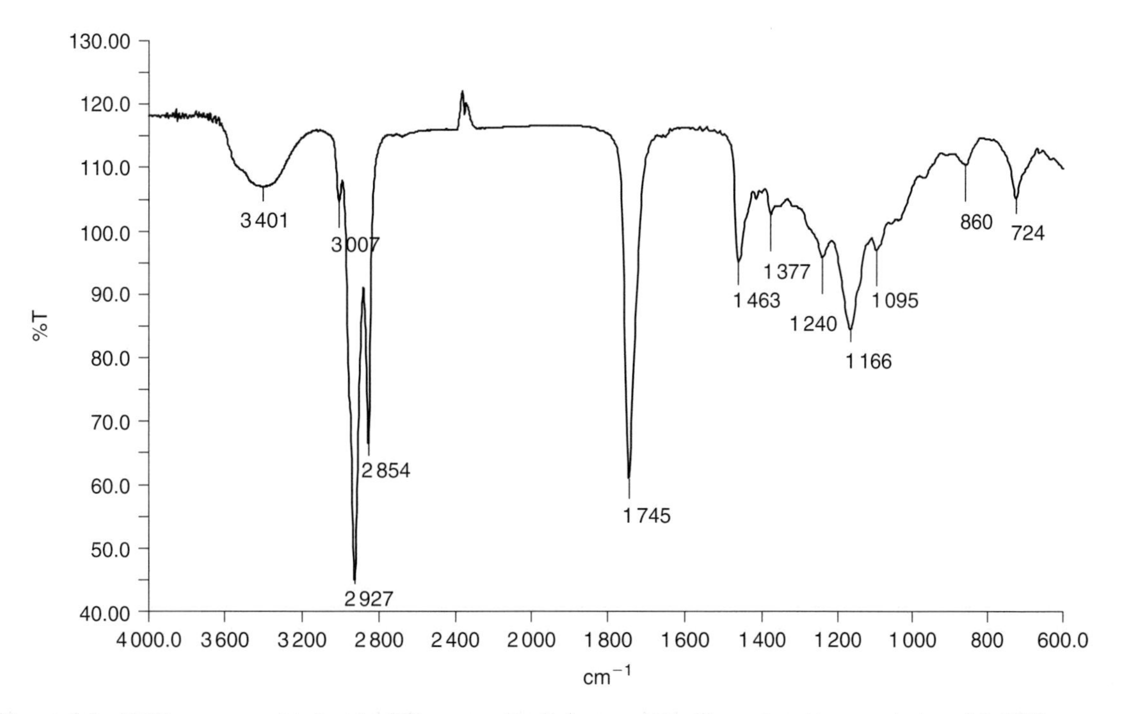

Figure 3.6 FTIR spectrum of industrial **CO**, reported in Reference [61]. (Reproduced by permission of the Wiley-VCH Verlag GmnH & Co. KGaA. Copyright 2006. Reprinted from Reference [61]).

The ^{1}H-RMN spectrum given in Fig. 3.7 had all the resonances expected for the **CO** structure, viz. 3-hydroxy groups and 3 insaturations per triglyceride molecule. Additionally, elemental analyses and molecular weight determination confirmed the proposed structure.

CO was modified by reacting it with **AA**, acryloyl chloride (**AC**) and α,α'-dimethylbenzylisocyanate (**TMI**). The latter reaction is depicted in Scheme 3.7 and the FTIR and NMR spectra of the **CO-TMI** adduct are given

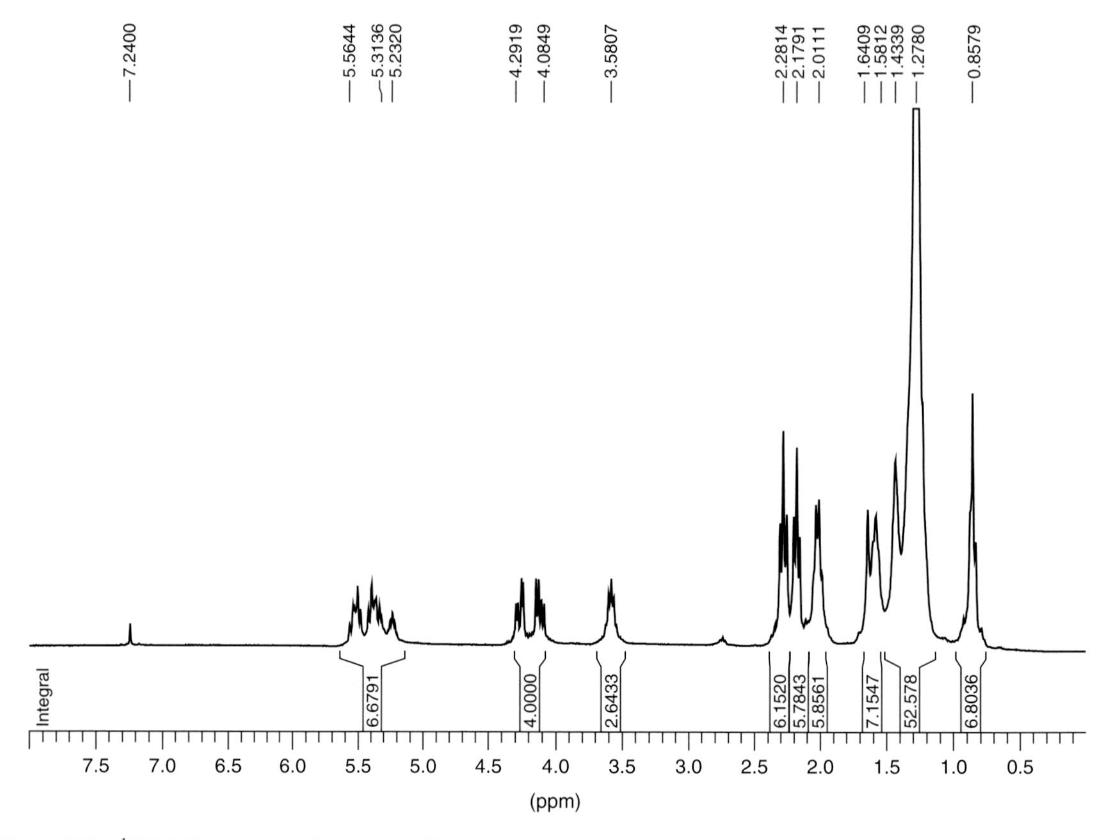

Figure 3.7 ^{1}H-NMR spectrum of industrial **CO** in $CDCl_3$ [60].

Scheme 3.7

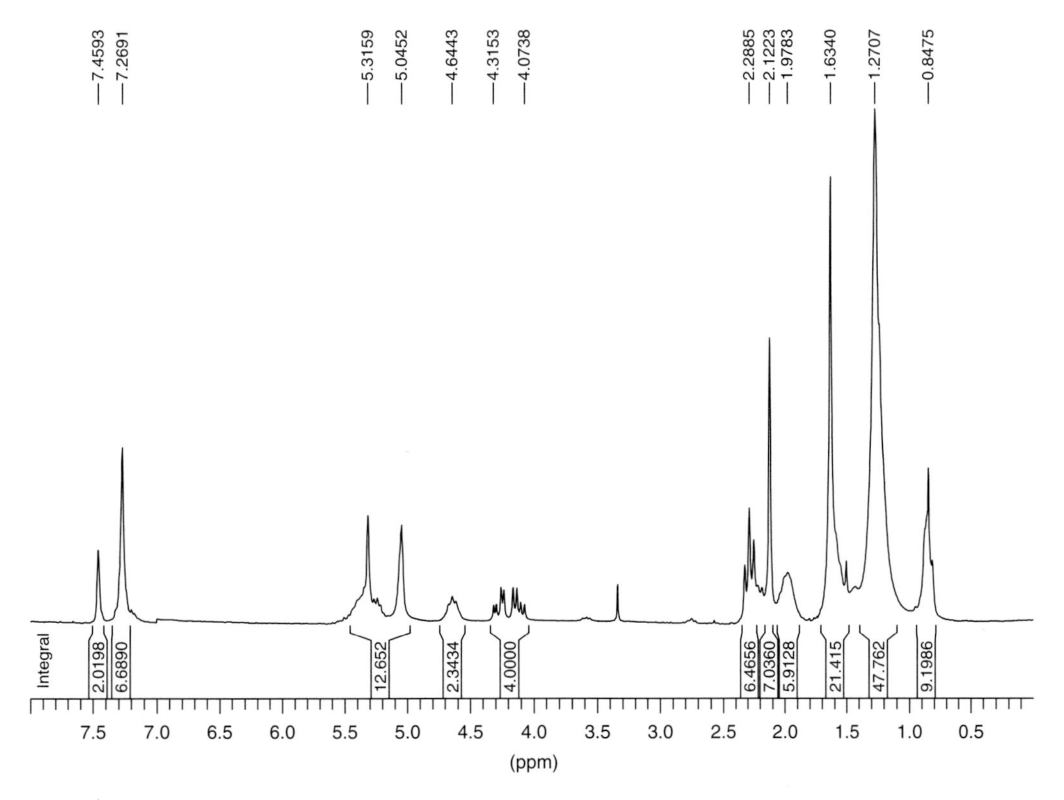

Figure 3.8 ^{1}H-NMR spectrum of **CO-TMI** in $CDCl_3$, reported in Reference [61]. (Reproduced by permission of the Wiley-VCH Verlag GmnH & Co. KGaA. Copyright 2006. Reprinted from Reference [61]).

in Figs 3.7 and 3.8, respectively, whereas Table 3.7 summarizes the NMR peak assignments for both **CO** and **CO-TMI**. The acrylated **CO** derivatives were found to crosslink via radical photopolymerization, whereas the **CO-TMI** adduct was crosslinked via cationic polymerization [60–62].

CO

CO–TMI

Table 3.7

^{1}H-NMR data relative to CO and CO–TMI [61]

δ (ppm)	Intensity	Peak assignment	δ (ppm)	Intensity	Peak assignment
7.24	–	The solvent	7.4	2	H_{14}
5.2–5.5	6.6	H_1 and H_7	7.3	7	H_{15}
4–4.3	4	H_2	5–5.3	13	H_{17}, H_1 and H_2
3.6	2.6	H_{10}	4.6	2	H_5
2.3	6	H_3	4–4.3	4	H_4
2.2	6	H_8	2.3	6	H_{10}
2	6	H_6	2.1	9	H_{16}, H_6
1.6	7	H_4	2	6	H_7
1.3–1.4	53	H_5	1.6	21	H_{11} and H_{13}
0.8	7	H_{11}	1.3	48	H_9, H_8
			0.8	9	H_3

3.4.7 Oil-based polyurethanes

Thanks to their good film-forming properties, **LO** and **SFO** were tested in paint formulations after their submission to chemical modification. Thus, initial triglycerides or their partially hydrolysed derivatives were reacted with **TDI**, **HMDI** or an NCO-terminated pre-polymer between 1,4-butanediol and **TDI** [65–67]. The viscosity of the **TDI-SFO**-based polymer solutions showed a non-Newtonian rheo-thinning behaviour, whereas those of the other polymers behaved as Newtonian fluids. **CO**-based polyurethanes gave polymers with Tgs varying from 25°C to 50°C and a gel fraction higher than 90 per cent. Water soluble polyurethanes were obtained from **SFO**-based maleinized fatty acids [68] and **TDI** in the presence of tertiary amines as catalysts. The films obtained displayed good resistance to abrasion and to mechanical impacts and good gloss.

Very recently, novel silicon-containing polyurethanes based on vegetable oils were prepared and characterized [69]. In particular, a silicon-fatty acid OH-terminated triglyceride (Si-oil-polyol) (Scheme 3.8) was prepared and reacted with methylenediphenyl diisocyanate (**MDI**). This polyol was characterized by its hydroxy number, functionality, molecular weight and NMR (Fig. 3.9). The Tgs of the corresponding polyurethanes decreased when the proportion of Si-oil-polyol was increased. It was found that higher silicon contents yielded materials which no longer burned in ambient air in the absence of complementary oxygen, indicating that these hybrids can be potential candidates for applications that require fire resistance.

3.4.8 Oil-based polyamides

The study of polyamides prepared from vegetable oils is an old field of research which has not reported any new data in the last three decades. Polyamides from **TO** and **SOYO** were used mainly in the paint industry because

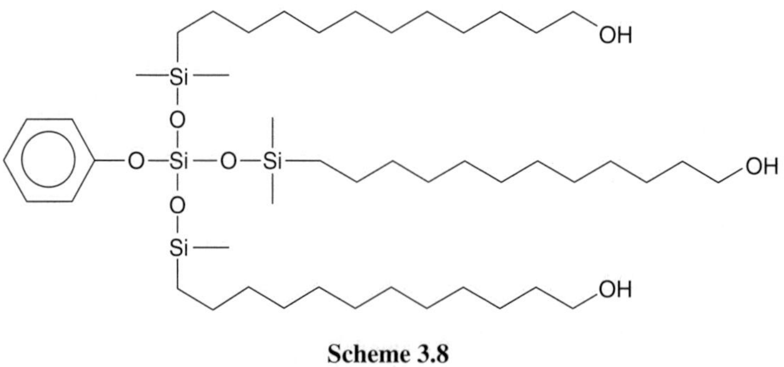

Scheme 3.8

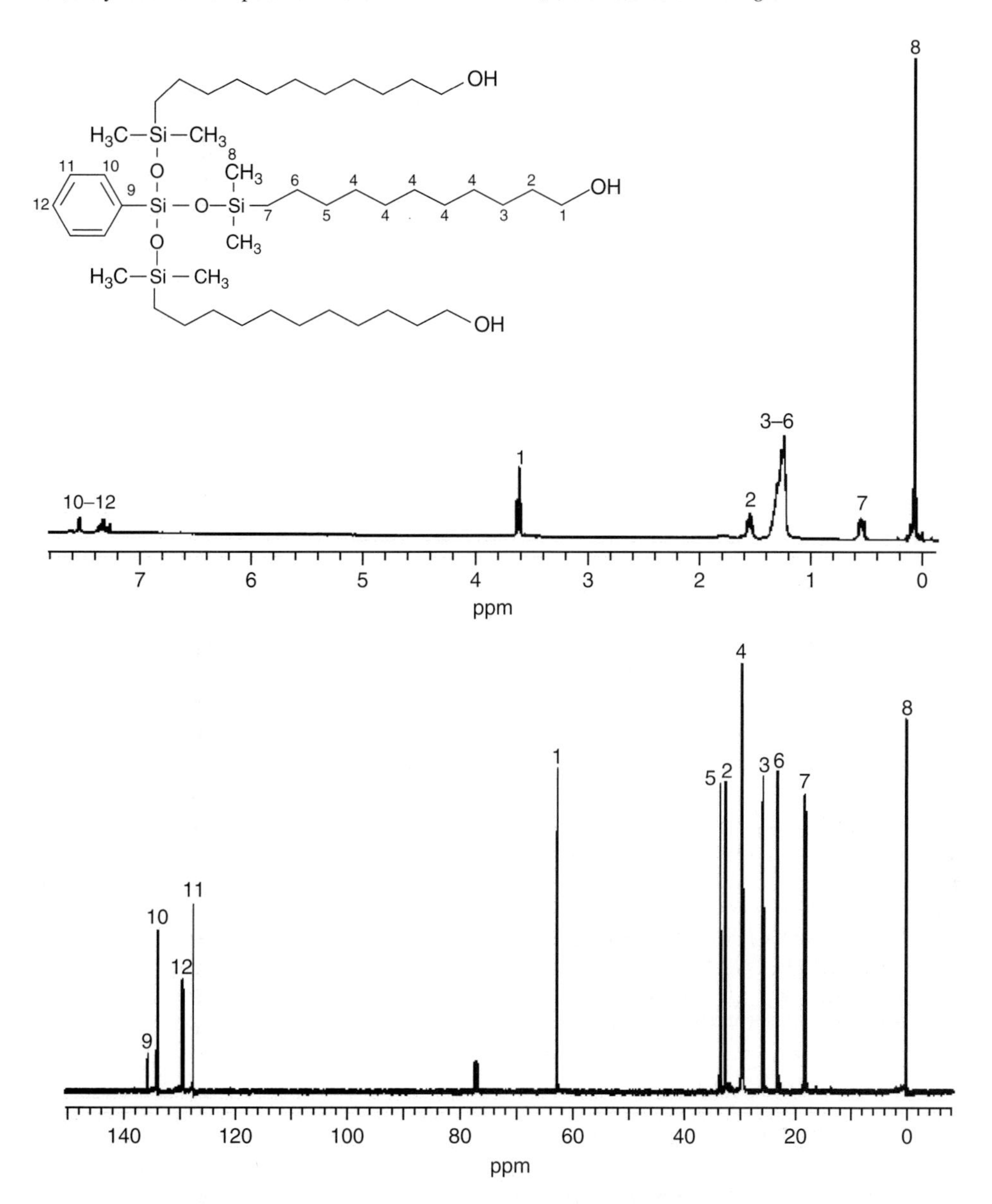

Figure 3.9 NMR spectra of the Si-oil-polyol reported in Reference [69]. (Reproduced by permission of the American Chemical Society, Copyright 2006. Reprinted from Reference [69]).

of their thixotropic character [70]. The best known polyamide in this context, Nylon 11, is now 30-years old and although this polymer has good properties like dimensional stability, chemical resistance and low cold brittleness temperature, research in this field has not been pursued [71, 72].

3.4.9 Oil-based polyester-amides

Polyester-amides, as such, or filled with alumina particles, were prepared from N,N′-bis(2 hydroxyethyl) **LO** (**HELA**) and phthalic acid, in the presence or absence of poly(styrene-co-maleic anhydride), with the aim of preparing novel surface coating materials [73–75]. These polymers were further modified with **TDI**, in order to

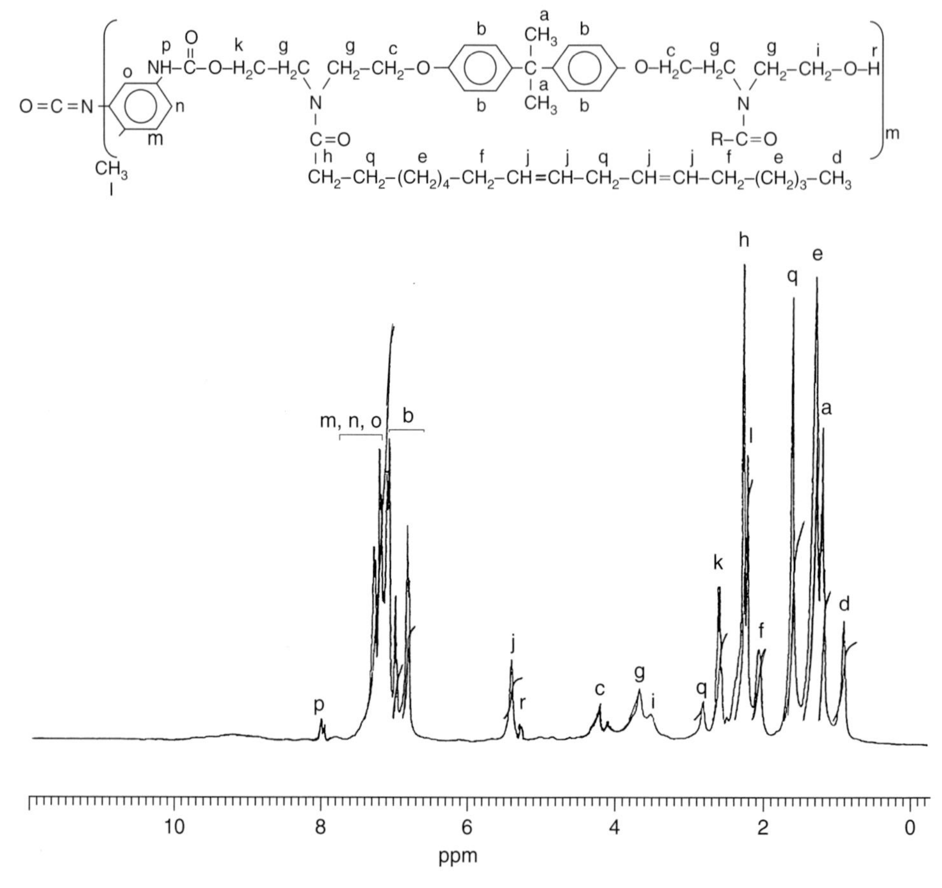

Figure 3.10 ^{1}H-NMR spectrum of PEA–TDI polyester-amide-urethane reported in Reference [76]. (Reproduced by permission of Elsevier. Copyright 2004. Reprinted from Reference [76]).

improve their adhesion, toughness and water and chemical resistances [76]. Figure 3.10 shows the NMR spectrum of these modified materials.

Other oils, specific to India like *Pongamia glabra* and Nahar (*Mesua ferrea*), as well as boron-based **SOYO,** have also been used in the preparation of polyester-amides [77–79]. These materials were used for coating at curing temperatures of 185°C–250°C. The toxicity evaluation and skin irritation tests showed that these polyester-amides were biologically safe for industrial applications. These materials also showed improved gloss, hardness adhesion and chemical resistance [77, 78]. The boron-based homologues were found to be good candidates for biomedical applications, because they exhibited anti-microbial properties [79].

3.4.10 Oil-based alkyd resins and polyesters

Alkyd resins can be prepared from either monoglycerides or fatty acids. The former approach is based on the alcoholysis of the oil by part of a polyol, followed by the esterification of the remaining hydroxy functions using a polyacid. The latter approach is instead based on a one-step process consisting of the direct reaction among the fatty acid, the polyol and a polyacid [14].

Traditionally, **SFO**, **SOYO, LO** and rapeseed oils have been used in the preparation of oil-modified polyesters, mainly in the field of offset printing inks, but new vegetable oils, such as rubber seed, karinatta, orange seed and melon seed oils were recently employed for this purpose [80]. Different anhydrides have been used to prepare alkyd resins, namely: maleic (**MAA**), succinic (**SA**), glutaric (**GA**), and phthalic (**PA**) [81]. Generally, in coating and printing, the drying time is one of the most relevant parameters. Resins prepared with **MAA** showed the best water resistance properties and the shortest drying time when **SFO** was used as the oil base.

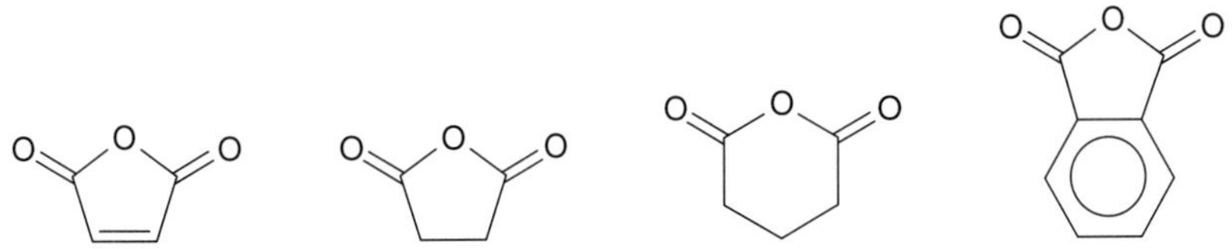

MAA SA GA PA

Liquid crystals from alkyd resins were reported 20 years ago [82, 83]. They were investigated with the aim of using them as coatings because they could be processed without a solvent and gave improved film properties. These materials were prepared by reacting *p*-hydroxybenzoic acid (**PHBA**) with hydroxy- or carboxy-terminated, or **SA**-modified alkydresins, as shown in Scheme 3.9.

Ecology-friendly water-based organic coatings have attracted much attention because of the absence of volatile organic compounds. For this purpose, ammonium salts of alkyd resins with high acid numbers have been used successfully [84–86]. Nakayama [87] prepared oil-based resin blends for water-borne paints by copolymerizing **ST**, **MMA** or **AA** with unsaturated fatty acid esterified glycidyl methacrylate. One of the resins reported in this study was an **ST**-allylalcohol copolymer, which was first esterified with **LO** fatty acids and then maleinized [87].

Recently reported highly branched oil-based alkyd resins were shown to be particularly interesting because their reduced viscosity allowed high solid contents coatings to be prepared [88].

Rubber seed oils (**RSO**) have also been used to prepare alkyd emulsions after their modification with **MAA** and **PA**, which produced water-soluble alkyd materials [86].

In another approach, **RSO**-based polyesters were prepared as shown in Scheme 3.10 and showed good chemical resistance [89]. Similar work based on the hydrosilylation of fatty acids and the preparation of hybrid organic–inorganic materials has also been reported [90].

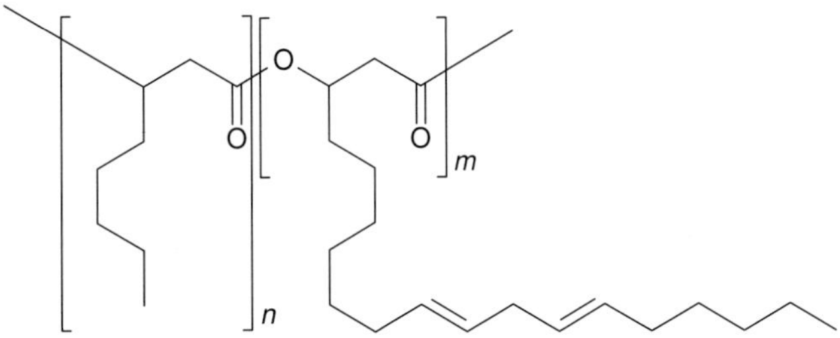

Scheme 3.9

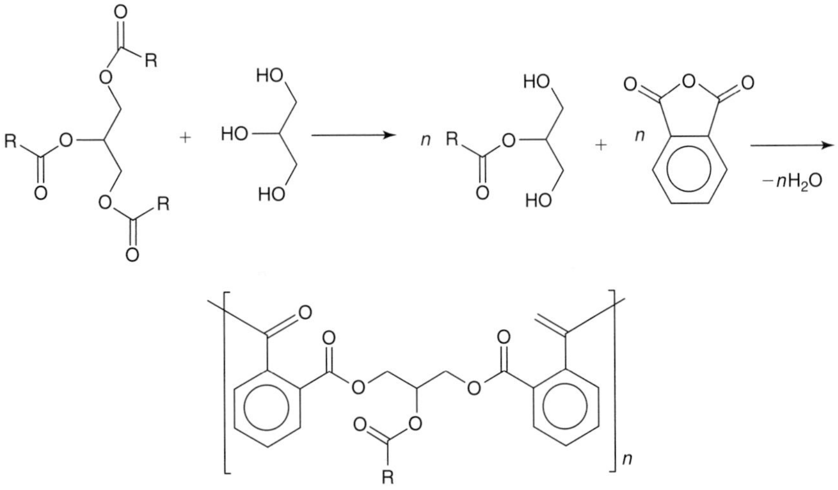

Scheme 3.10

A series of monooleate-polyethylene glycols with active carboxy functions arising from the condensation of succinic anhydride has been recently reported [91]. Their molecular weights, determined by GPC, varied from 760 to 10,340, depending on the reaction conditions and the monomers' stoichiometry. Partially bio-based unsaturated polyesters, containing epoxidized methyl linseedate, were prepared and showed that up to 25 per cent w/w of linseedate could be included into these formulations [92]. The incorporation of this vegetable oil derivative improved the thermal and mechanical properties of the ensuing polyesters.

3.4.11 Oil-based poly(hydroxyalcanoates)

Polyhydroxyalkanoates (PHAs) have attracted much attention because of their renewable character and good properties [93], as discussed in detail in Chapter 22. Vegetable oils are among the different feeding sources for the bacteria and only some features specific to this issue will be discussed in this chapter. Tallow oil has been used to prepare a semi-crystalline medium chain-length PHA, using the *Pseudomonas resinovorans* bacterium, with a Tg of −43°C and a melting temperature of 43°C [94]. The presence of insaturations within the side chains produced crosslinked materials when this PHA was irradiated with γ-rays. The radiation doses were found to increase the tensile properties of the films, but reduce their rate of biodegradation. Blends of tallow- and **LO**-based PHAs were also prepared and characterized [94]. Two years later, the same group prepared medium chain-length **LO**-based PHA (**PHA–L**), as a viscous liquid polymer [95]. The side chain insaturations were then converted into epoxy moieties (**PHA–LE**), using *m*-chloroperoxibenzoic acid, as shown in Scheme 3.11. Figure 3.11 shows the ^{13}C-NMR spectra of the **PHA–L** before and after epoxidation. As expected, the epoxidation increased the crosslinking rate and the tensile strengths of the corresponding networks.

The preparation of medium chain-length PHA using **LO** and *Pseudomonas putida* has also been reported [96]. These resins displayed a high ability to undergo oxidative drying (high amount of insaturations within the fatty acid side chains) and the ensuing paints had excellent coating properties with good adhesion to different substrates [96].

3.4.12 Miscellaneous

A wide variety of polymeric materials, other than those discussed previously, are described in the literature. Given once again the availability of recent monographs [1–5], we will limit our treatment of this rather marginal system, to some relevant examples. Polynaphthols were synthesized by combining artificial urushi and triglyceride [97, 98] and, in another vein, acrylic diene metathesis polymerization was applied to triglyceride [99–101]. Composites with oil-based matrices have been frequently described [4] and continue to arise interest as shown by recent contributions [102–104].

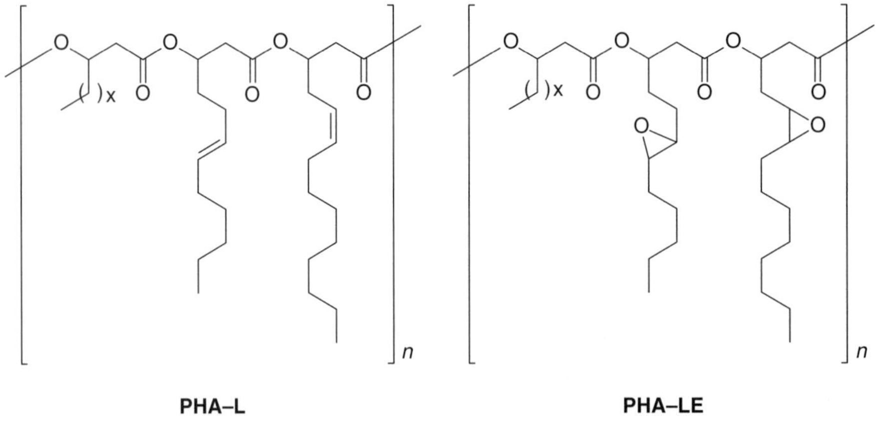

PHA–L **PHA–LE**

Scheme 3.11

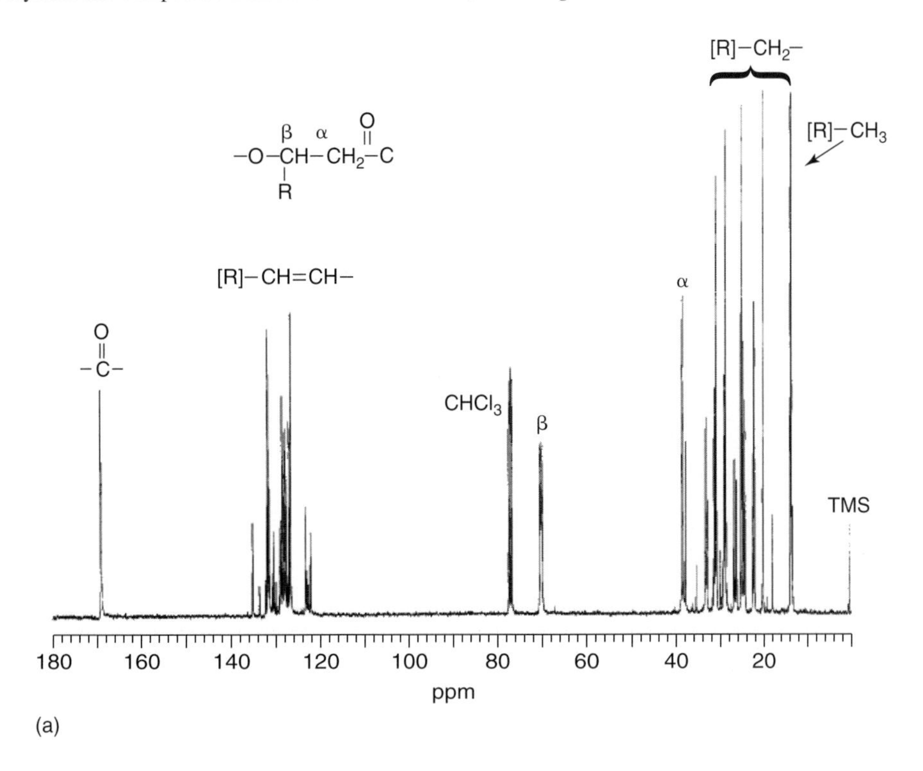

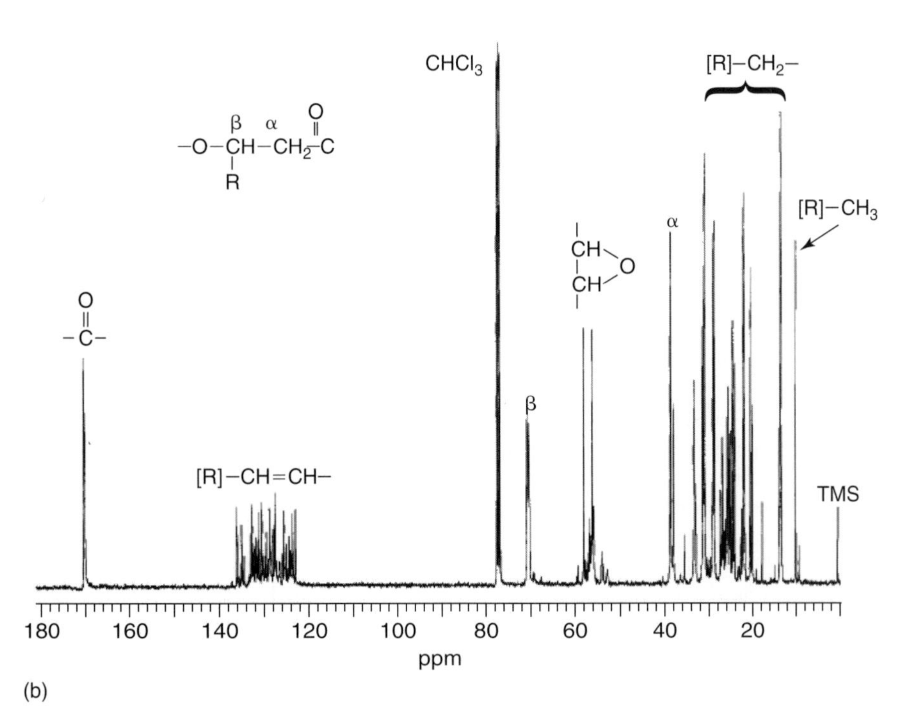

Figure 3.11 ^{13}C-NMR spectra of (a) **PHA–L** and (b) **PHA-LE**. (Reproduced by permission of Elsevier. Copyright 2000. Reprinted from Reference [95]).

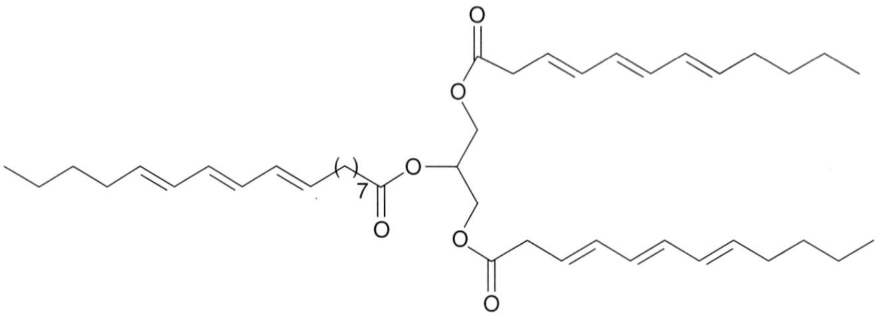

Scheme 3.12

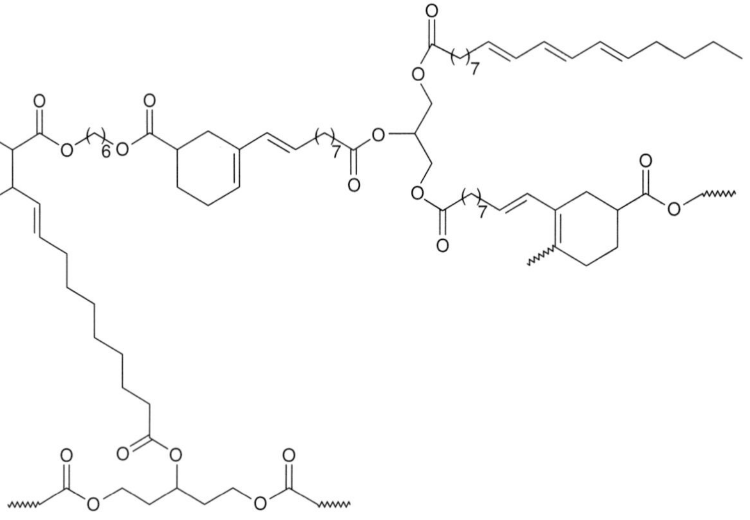

Scheme 3.13

Trumbo and Mote prepared crosslinked networks from **TO** (Scheme 3.12), *via* a Diels–Alder reaction [105], using 1,6-hexanediol- and 1,4-butanediol-diacrylates. They characterized their product and showed that some combinations gave films with 95 per cent of gel fraction. Scheme 3.13 shows the idealized network structure proposed by these authors. The corresponding films had high gloss and good solvent resistance.

Buriti oil (**BO**) was mixed with **PST** or **PMMA** and the ensuing blends characterized in terms of absorption and photoluminescence properties associated with the specific structure of this oil extracted from an Amazonian palm tree [106]. Liu *et al.* reported very recently that **SOYO** can be homo-polymerized in CO_2 supercritical conditions by cationic initiation with **BFE** without using a co-monomer [107].

Electrically conductive polyurethane membranes based on **CO** and doped with sulfonated polyaniline were prepared and characterized [108]. Their Tgs varied from -8 to 28°C, depending on the NCO/OH ratio.

SOYO's fatty acid was transformed into the corresponding terminal oxazoline derivative, which was then polymerized using a microwave assisted technology (Scheme 3.14) [109]. This polymer was found to crosslink when irradiated by UV light, because of the insaturations borne by the **SOYO** alkyl chains [109].

The inclusion of carbon black into polymers based on **ASOYO** was shown to display interesting modulated electrical resistivity, as a function of the filler content [110]. A percolation threshold was attained with only 1.2 per cent of carbon black and an associated resistivity of about $10^{11}\,\Omega$cm.

A recent original approach to the use of vegetable oils as precursors to polyurethanes describes the ozone-induced oxidation of the single insaturations borne by each chain of canola oil triglyceride followed by the cleavage and hydroxylation of the ensuing end groups (Scheme 3.15) [110]. This triol macromonomer was then mixed

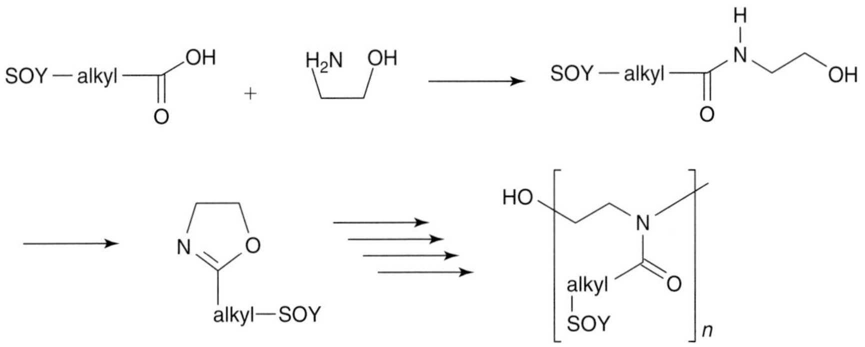

SOY — alkyl — OH O + H_2N OH → SOY — alkyl — H N OH O

→ N O alkyl—SOY → HO N alkyl SOY O n

Scheme 3.14

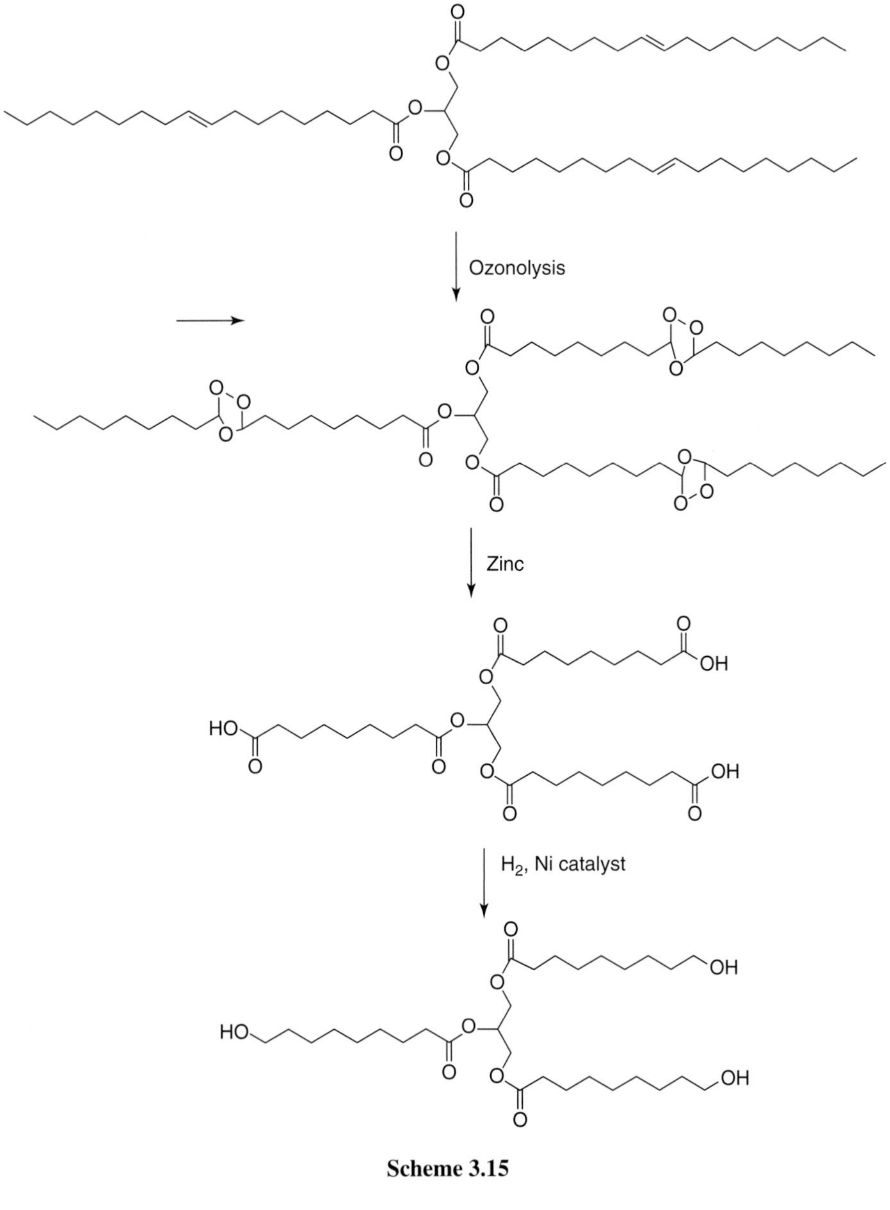

O O O O O O

Ozonolysis

O O O O O O O O O O O O O O O

Zinc

O O OH O HO O O O O O OH O O

H_2, Ni catalyst

O OH O HO O O O O OH O

Scheme 3.15

with **MDI** to prepare the corresponding polyurethanes which were thoroughly characterized. However, given the rather tortuous pathway leading to the triol and the fact that a half of the aliphatic carbon chain is lost in this process, the actual economic feasibility of this approach is doubtful.

3.5 CONCLUSIONS

The aim of this chapter was to highlight the potentiality of triglyceride oils in the field of the synthesis of polymeric materials from renewable resources and the main messages we tried to deliver are that:

(i) the presence of oil or oil derivatives (fatty acid chains) in different formulations for coating or printing improves the optical (gloss), physical (flexibility, adhesion) and chemical properties (resistances to water and chemicals) of the cured films;
(ii) in some instances, soybean oil-based polymeric materials have properties comparable with those of petroleum-based counterparts;
(iii) the modification of these oils can lead to macromonomers susceptible to polymerize via very fast light-induced mechanisms, a fact that again compares favourably with the behaviour of equivalent acrylic formulations.

The almost simultaneous publication of reviews and monographs on the topic covered by this chapter, together with the increasing flow of more specific studies covering a wide range of oils and related chemical modifications, represent indisputable proof about the relevance of this realm.

REFERENCES

1. Verhé R.G., Industrial products from lipids and proteins, in *Renewable Bioresources: Scope and Modification for Non-Food Applications* (Eds. Stevens C.V. and Verhé R.G.), John Wiley & Sons, Ltd, Chichester, 2004, pp. 208–250, Chapter 9.
2. Güner F.S., Yagci Y., Erciyes A.T., *Prog. Polym. Sci.*, **31**, 2006, 633.
3. Sharma V., Kundu P.P., *Prog. Polym. Sci.*, **31**, 2006, 983.
4. Wool R.P., Sun X.S., *Bio-Based Polymers and Composites*, Elsevier, Amesterdam, 2005.
5. Gunstone F., *Fatty acid and lipid chemistry*, Blackie Academic & Professional, New York, 1996.
6. Cunningham A., Yapp A., US Patent, 3,827,993, 1974.
7. Bussell G.W., US Patent, 3,855,163, 1974.
8. Hodakowski L.E., Osborn C.L., Harris E.B., US Patent, 4,119,640, 1975.
9. Trecker D.J., Borden G.W., Smith O.W., US Patent, 3,979,270, 1976.
10. Trecker D.J., Borden G.W., Smith O.W., US Patent, 3,931,075, 1976.
11. Salunkhe D.K., Chavan J.K., Adsule R.N., Kadam S.S., *World Oilseeds: Chemistry, Technology and Utilization*, Van Nostrand Reinhold, New York, 1992.
12. Force C.G., Starr F.S., US Patent, 4,740,367, 1988.
13. Khot S.N., Lascala J.J., Can E., Morye S.S., Williams G.I., Palmese G.R., Kusefoglu S.H., Wool R.P., *J. Appl. Polym. Sci.*, **82**, 2001, 703.
14. Deligny P.K.T., Tuck N., Alkyds and polyesters, in *Resins for Surface Coatings* (Ed. Oldring P.K.T.), Wiley, New York, 2000, Volume II.
15. Bailey A.E., *Bailey's Industrial Oil and Fat Products*, Wiley, New York, 1996.
16. Norris F.A., Extraction of fats and oils, in *Bailey' s Industrial Oil and Fat Products* (Ed. Swern D.), Wiley, New York, 1996, Volume II.
17. Johnson R.W., Fritz E E., *Fatty Acids in Industry*, Marcel Dekker, New York, 1989.
18. Li F., Larock R.C., Otaigbe J.U., *Polymer*, **41**, 2000, 4849.
19. Li F., Larock R.C., Marks D.W., Otaigbe J.U., *Polymer*, **41**, 2000, 7925.
20. Li F., Larock R.C., *J. Appl. Polym. Sci.*, **78**, 2000, 1044.
21. Li F., Larock R.C., *J. Appl. Polym. Sci.*, **78**, 2000, 2721.
22. Li F., Larock R.C., *J. Appl. Polym. Sci.*, **80**, 2001, 658.
23. Li F., Hanson M.V., Larock R.C., *Polymer*, **42**, 2001, 1567.
24. Li F., Perrenoud A., Larock R.C., *Polymer*, **42**, 2001, 10133.

25. Li F., Larock R.C., *J Polym Sci B: Polym Phys*, **39**, 2001, 60.
26. Marks D.W., Li F., Pacha C.M., Larock R.C., *J. Appl. Polym. Sci.*, **81**, 2001, 2001.
27. Li F., Larock R.C., *J. Appl. Polym. Sci.*, **84**, 2002, 1533.
28. Li F., Larock R.C., *Polym. Int.*, **52**, 2003, 126.
29. Li F., Hasjim J., Larock R.C., *J. Appl. Polym. Sci.*, **90**, 2003, 1830.
30. Li F., Larock R.C., *Biomacromolecules*, **4**, 2003, 1018.
31. Kundu P.P., Larock R.C., *Biomacromolecules*, **4**, 2005, 797.
32. Andjelkovic D.D., Larock R.C., *Biomacromolecules*, **4**, 2006, 927.
33. Lu Y., Larock R.C., *Biomacromolecules*, **7**, 2006, 2692.
34. Henna P.H., Andjelkovic D.D., Kundu P.P., Larock R.C., *J. Appl. Polym. Sci.*, **104**, 2007, 979.
35. Gultekin M., Beker U., Güner F.S., Erciyes A.T., Yagsi Y., *Macromol. Mater. Eng.*, **283**, 2000, 15.
36. Hazer B., Demirel S.I., Borcakli M., Eroglu M.S., Cakmak M., Erman B., *Polym. Bull.*, **46**, 2001, 389.
37. Akbas T., Beker U.G., Güner F.S., Erciyes A.T., Yagsi Y., *J. Appl. Polym. Sci.*, **88**, 2003, 2373.
38. Eren T., Küsefoglu S.H., *J. Appl. Polym. Sci.*, **97**, 2005, 2264.
39. Zong Z., He J., Soucek M.D., *Prog. Org. Coat.*, **53**, 2005, 83.
40. Can E., Wool R.P., Küsefoglu S., *J. Appl. Polym. Sci.*, **102**, 2006, 2433.
41. Can E., Wool R.P., Küsefoglu S., *J. Appl. Polym. Sci.*, **102**, 2006, 1497.
42. Tuman S.J., Chamberlain D., Scholsky K.M., Soucek M.C., *Prog. Org. Coat.*, **28**, 1996, 251.
43. Meneghetti S.M.P., de Souza R.F., Monteiro A.L., de Souza M.O., *Prog. Org. Coat.*, **33**, 1998, 219.
44. Turri B., Vinci S., Margutti S., Pedemonte E., *J. Therm. Anal. Calor.*, **66**, 2001, 343.
45. Fried R.J., *Polymer Science and Technology*, 2nd Edition, Pearson Education, New Jersey, 2003.
46. Sperling L.H., Manson J.A., Yenwo G.M., Devia-Manjarres N., Pulido J., Conde A., *Polym. Sci. Technol.*, **10**, 1977, 113.
47. Nayak P.L., *J. Macromol. Sci. Rev. Macromol. Chem. Phys.*, **40**, 2000, 1.
48. Yin Y., Yao S., Zhou X., *J. Appl. Polym. Sci.*, **88**, 2003, 1840.
49. Crivello J.V., Narayan R., *Chem. Mater.*, **4**, 1992, 692.
50. Crivello J.V., Carlson K.D., *Macromol. Reports*, **A33**(suppl 5 and 6), 1996, 251.
51. Crivello J.V., Narayan R., Sternstein S.S., *J. Appl. Polym. Sci.*, **64**, 1997, 2073.
52. Ortiz R.A., Lopez D.P., de L. Guillen Cosneros M., Valverde J.C.R., Crivello J.V., *Polymer*, **46**, 2005, 1535.
53. Boquillon N., Fringant C., *Polymer*, **41**, 2000, 8603.
54. Eren T., Küsefoglu S.H., *J. Appl. Polym. Sci.*, **91**, 2004, 2700.
55. Shukla V., Singh M., Singh D.K., Shukla R.I., *Surf. Coat. Int. Part B: Coat. Trans.*, **88**(B3), 2005, 157–230. 157
56. Ashraf S.M., Ahmad S., Riaz U., Alam M., Sharma H.O., *J. Macromol. Sci., Part A: Pure Appl. Chem*, **42**, 2005, 1409.
57. Ashraf S.M., Ahmad S., Riaz U., Sharma H.O., *J. Appl. Polym. Sci.*, **100**, 2006, 3094.
58. Zou K., Soucek M.D., *Macromol. Chem. Phys.*, **206**, 2005, 967.
59. Park S.J., Jin F.L., Lee J.R., Shin J.S., *Eur. Polym. J.*, **41**, 2005, 231.
60. H. Pelletier, Ph.D Thesis, Grenoble National Polytechnic Institute, 2005.
61. Pelletier H., Gandini A., *Eur. J. Lipid Sci. Technol.*, **108**, 2006, 411.
62. Pelletier H., Belgacem M.N., Gandini A., *J. Appl. Polym. Sci.*, **99**, 2006, 3218.
63. Earls J.D., White J.E., Lopez L., Lysenko Z., Dettloff M.L., Null M.J., *Polymer*, **48**, 2007, 712.
64. Esen H., Küsefoglu S.H., Wool R., *J. Appl. Polym. Sci.*, **103**, 2007, 626.
65. Guner F.S., Baranak M., Soytas S., Erciyes A.T., *Prog. Org. Coat*, **50**, 2004, 172.
66. Yeganeh H., Mehdizadeh M.R., *Eur. Polym. J.*, **40**, 2004, 1233.
67. Ozkaynak M.U., Atalay-Oral C., Tantekin-Ersolmaz S.B., Guner F.S., *Macromol. Symp.*, **228**, 2005, 177.
68. Gunduz G., Khalid A.H., Mecidoglu A., Aras L., Waterborne and air-drying oil-based resins, *Prog. Org. Coat.*, **49**, 2004, 259.
69. Lligadas G., Ronda J.C., Galia M., Cadiz V., *Biomacromolecules*, **7**, 2006, 2420.
70. Oldring P.K.T., Turk N., *Polyamides. Resins for Surface Coatings*, Wiley, New York, 2000, Volume III, pp. 131–197.
71. Nayak P.L., Natural oil-based polymers: Opportunities and challenges, *J. Macromol. Sci. Rev. Macromol. Chem. Phys.*, **40**, 2000, 1.
72. Koch R., Nylon 11, *Polym. News*, **3**, 1977, 302.
73. Ahmad S., Naqvi F., Verma K.L., Yadav S., *J. Appl. Polym. Sci.*, **72**, 1999, 1679.
74. Ahmad S., Ashraf S.M., Hasnat A., Yadav S., Jamal A., *J. Appl. Polym. Sci.*, **82**, 2001, 1855.
75. Zafar F., Ashraf S.M., Ahmad S., *Prog. Org. Coat.*, **50**, 2004, 250.
76. Alam M., Sharmin E., Ashraf S.M., Ahmad S., *Prog. Org. Coat.*, **50**, 2004, 224.
77. Ahmad S., Ashraf S.M., Naqvi F., Yadav S., Hasnat A., *Prog. Org. Coat.*, **47**, 2003, 95.
78. Mahapatra S.S., Karak N., *Prog. Org. Coat.*, **51**, 2004, 103.
79. Ahmad S., Haque M.M., Ashraf S.M., Ahmad S., *Eur. Polym. J.*, **40**, 2004, 2097.
80. Igwe I.O., Ogbobe O., *J. Appl. Polym. Sci.*, **75**, 2000, 1441.

81. Aydin S., Akçay H., Ozkan E., Guner F.S., Erciyes A.T., *Prog. Org. Coat.*, **51**, 2004, 273.
82. Chiang W.Y., Yan C.S., *J. Appl. Polym. Sci.*, **46**, 1992, 1279.
83. Chen D.S., Jones F.N., *J. Coat. Technol.*, **60**(756), 1988, 39.
84. Hofland A., *Surf. Coat. Int.*, **7**, 1994, 270.
85. Rødsrud G., Sutcliffe J.E., *Surf. Coat. Int.*, **1**, 1994, 7.
86. Aigbodion A.I., Okieimen F.E., Obazee E.O., Bakare I.O., *Prog. Org. Coat.*, **46**, 2003, 28.
87. Nakayama Y., *Prog. Org. Coat.*, **33**, 1988, 108.
88. Manczyk K., Szewzzyk P., *Prog. Org. Coat.*, **44**, 2002, 99.
89. Bakare I.O., Pavithran C., Okieimen F.E., Pillai C.K.S., *J. Appl. Polym. Sci.*, **100**, 2006, 3748.
90. Lliagadas G., Callau L., Ronda J.C., Cadiz V., *J. Polym. Sci., Part A: Polym. Chem.*, **43**, 2005, 6295.
91. Xiong F., Li J., Wang H., Chen Y., Cheng J., Zhu J., *Polymer*, **47**, 2006, 6636.
92. Miyagawa H., Mohanty A.K., Burgueno R., Drzal L.T., Misra M., *Ind. Eng. Chem. Res.*, **45**, 2006, 1014.
93. Braunegg G., Lefebvre G., Genser K.F., *J. Biotechnol.*, **65**, 1988, 127.
94. Ashby R.D., Cromwick A., Foglia T.A., *Int. J. Biol. Macromol.*, **23**, 1998, 61.
95. Ashby R.D., Foglia T.A., Solaiman D.K.Y., Liu C., Nunez A., Eggink G., *Int. J. Biol. Macromol.*, **27**, 2000, 355.
96. van der Walle G.A.M., Buisman G.J.H., Weusthuis R.A., Eggink G., *Int. J. Biol. Macromol.*, **25**, 1999, 123.
97. Kobayashi S., Uyama H., Ikeda R., Artificial urushi, *Chem. Eur. J.*, **7**, 2001, 4754.
98. Tsujimoto T., Uyama H., Kobayashi S., *Macromolecules*, **37**, 2004, 1777.
99. Tian Q., Larock R.C., *J. Am. Oil. Chem. Soc.*, **79**, 2002, 479.
100. Warwel S., Bruse F., Demes C., Kunz M., gen Klaas M.R., *Chemosphere*, **43**, 2001, 39.
101. Warwel S., Bruse F., Demes C., Kunz M., *Ind. Crops Prod.*, **20**, 2004, 301.
102. O'Donnell A., Dweib M.A., Wool R.P., *Comp. Sci. Technol.*, **64**, 2004, 1135.
103. Wang Y., Cao X., Zhang L., *Macromol. Biosci.*, **6**, 2006, 524.
104. Chen Y., Zhang L., Deng R., Liang H., *J. Appl. Polym. Sci.*, **101**, 2006, 953.
105. Trumbo D.L., Mote B.E., *J. Appl. Polym. Sci.*, **80**, 2001, 2369.
106. Duraes J.A., Drummont A.L., Piemental T.A.P.F., Murta M.M., Bicalho F., da S., Moreira S.G.C., Sales M.J.A., *Eur. Polym. J.*, **42**, 2006, 3324.
107. Liu Z., Sharma B.K., Erhan S.Z., *Biomacromolecules*, **8**, 2007, 233.
108. Amado F.D.R., Rodrigues L.F., Forte M.M.C., Ferreira C.A., *Polym. Eng. Sci.*, **46**, 2006, 1485.
109. Hoogenboom R., Schubert U.S., *Green Chem.*, **8**, 2006, 895.
110. Hernandez-Lopez S., Vegueras-Santiago E., Mercado-Posadas J., Snachez-Mendieta V., *Adv. Tech. Mater. Mater. Proc. J.*, **8**, 2006, 214.
111. Kong X., Narine S.S., *Biomacromolecules*, **8**, 2007, 2203.

– 4 –

Rosin: Major Sources, Properties and Applications

Armando J.D. Silvestre and Alessandro Gandini

ABSTRACT

Rosin exploitation, a part of the so-called Naval Stores Industry, is at least as old as the construction of wooden naval vessels. In recent years, rosin components have attracted a renewed attention, notably as sources of monomers for polymers synthesis. The purpose of the present chapter is to provide a general overview of the major sources and composition of rosin. It deals therefore with essential features such as the structure and chemical reactivity of its most important components, viz. the resin acids, and the synthesis of a variety of their derivatives. This chemical approach is then followed by a detailed discussion of the relevant applications, the resin acids and their derivatives, namely in polymer synthesis and processing, paper sizing, emulsion polymerization, adhesive tack and printing inks, among others.

Keywords

Rosin, Resin acids, Chemical modification, Paper sizing, Emulsion polymerization, Adhesive tack, Polymer chemistry and processing, Printing inks

4.1 INTRODUCTION

Rosin exploitation, a part of the so-called Naval Stores Industry, is at least as old as the construction of wooden naval vessels. It was however only during the first half of the twentieth century that the chemistry of this natural resource was studied in detail and new transformations and applications developed on a more scientific basis. A vast number of papers and patents are available on these topics and most of the relevant information were comprehensively reviewed in the classical book edited by Zinkel and Russell in 1989 [1], which covered all aspects, from the sources and processing, to the chemistry and applications of this versatile raw material. A less exhaustive survey, devoted to rosin applications for polymer synthesis, was also published in the same year by Maiti *et al.* [2].

Although the overall exploitation of rosin and its derivatives has been declining over the past few decades, it is the predicted depletion of fossil resources that is driving both the scientific community and industry to look for alternative renewable resources for the production of chemical commodities. It seems therefore logical that these abundant and cheap resins should witness an important comeback.

The purpose of the present chapter is to provide a general overview of the major sources, properties and both traditional and novel applications of rosin components, with particular emphasis on the most relevant contributions of the last couple of decades dealing with rosin as a source of monomers or additives for polymeric materials. A detailed analysis of the extensive bibliography on the various chemical transformations, as well as on other applications of rosin, falls outside the scope of this chapter, given the aim of this book.

Rosin, also known as colophony, is the common designation given to the non-volatile residue, mainly composed of resin acids, obtained after the distillation of the resin exuded by many conifer trees as a faintly aromatic, semi-transparent brittle solid. The most important source of rosin is pine trees (*Pinus* genus), not only because they are widespread in the Northern Hemisphere, but also because of their intensive use in the timber and pulp industries.

For historical reasons, pine resin was known as "Naval Stores", because of its use in the waterproofing of wooden ships. Depending on the way pine resin is isolated from wood, three products are distinguished, namely (i) *gum naval stores*, obtained by tapping living trees; (ii) *sulphate naval stores*, also known as tall oil rosin, recovered during the kraft pulping of pine wood and (iii) *wood naval stores*, also known as wood rosin, obtained from the solvent extraction of harvested wood.

The total world rosin production tended to decline between the 1960s and the 1980s. However, statistics published in the 1980s [3] and in 1994 [4], show that the rosin production stabilized at around 1.0–1.2 million tons per year. According to the 1994 survey [4], gum rosin amounted to 60 per cent of world production, tall oil rosin to around 35 per cent and wood rosin to only a few percent.

The evolution of gum rosin production has been affected by two distinct factors, viz. the intense competition of petroleum-based counterparts and the growing costs of man-power involved in tree tapping and resin recovery. Competition from petroleum-based products was quite strong in the second half of the twentieth century, given the ready availability, low cost and technological development of petrochemicals. As already pointed out, this trend might be reversed in view of the increasing costs of fossil resources which will probably stimulate novel fundamental and applied research on the exploitation of rosin as a source of chemicals and materials. High man-power costs have contributed to the decline of gum rosin production in the industrialized North-American and European countries with the result of shifting it to other areas of the world such as China and Indonesia, which in the mid-1990s were already responsible for 60 and 10 per cent of world production, respectively. Whereas production in China is unlikely to grow any further because of limiting forestry factors, Indonesia still has a significant growth potential [4]. Brazil, Russia and Portugal were once also important producers, but they too are witnessing a significant drop in production [4].

4.2 ROSIN CHEMICAL COMPOSITION

Regardless of its origin (gum, wood or tall oil), rosin is mainly composed (90–95 per cent) of diterpenic monocarboxylic acids, commonly known as resin acids whose generic formula is $C_{19}H_{29}COOH$. The remaining components are essentially made up of neutral compounds, the nature of which depends on the specific origin of the rosin [5]. The most common resin acids found in pine rosin are derived from the three basic tricyclic carbon skeletons abietane, pimarane and isopimarane and the less common bicyclic labdane skeleton (Fig. 4.1).

The abietane skeleton is shared by four resin acids (Fig. 4.2), namely abietic, neoabietic, palustric and levopimaric acids. The structures of these compounds, generally known as abietadienoic acids, differ only in the position of the conjugated double bond system, which is an important feature of this group of resin acids, because it

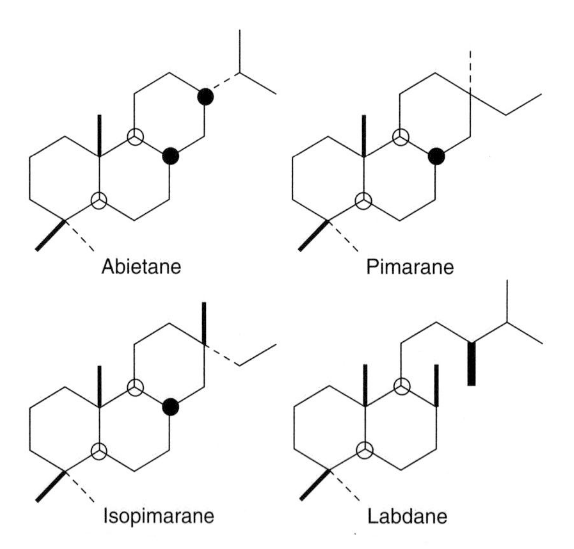

Figure 4.1 Diterpene carbon skeletons found in the most common resin acids.

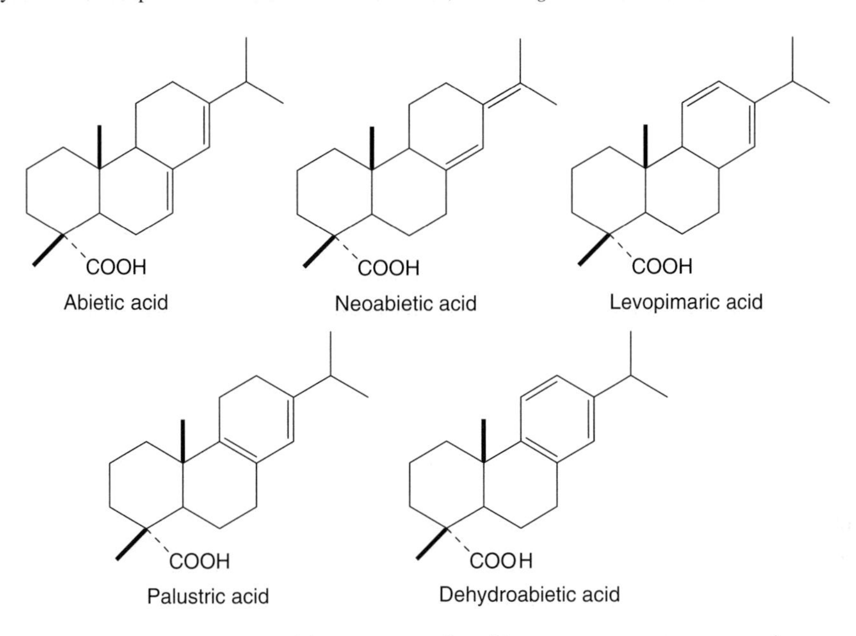

Figure 4.2 Structures of the most common abietane-type resin acids.

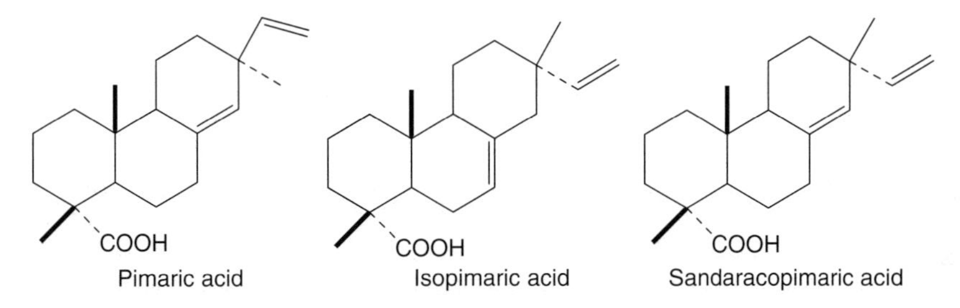

Figure 4.3 Structures of the most common pimarane-type resin acids.

influences their chemical reactivity and hence the applications of the ensuing products, as discussed below. The aromatic dehydroabietic acid is also found in small and variable amounts in various rosin species.

The most common pimarane-type acids are pimaric, isopimaric and sandaracopimaric acids (Fig. 4.3). Compared with the abietic counterparts, not only is the basic skeleton different but, more importantly, the double bond system is now not conjugated, a fact which reduces significantly the possible chemical exploitation of these compounds.

Although abietic-type acids are the dominant structures in most rosins, their relative abundance is quite variable, depending on the pine species and geographic origin. Furthermore, processing and handling conditions (*e.g.* temperature and pH) can induce the isomerization of the double bond system, leading to equilibrium mixtures, as discussed below.

From an industrial perspective, the quality of rosin and rosin derivatives is assessed on the basis of four basic parameters [4, 6], namely:

(1) The *acid number*, which is a measure of the amount of free carboxylic groups; a decrease in this value is indicative of decarboxylation and/or functionalization of the carboxylic moieties.
(2) The *saponification number*, which is a measure of the total amount of carboxylic groups; a decrease in this value is indicative of resin acid decarboxylation.
(3) The *colour*, whose intensity is a key detrimental factor in many applications, is a measure of rosin oxidation; an increase in colour intensity is therefore an indication of decreasing quality.
(4) The *softening point*, which is in fact a measure of the glass transition temperature associated with these complex mixtures of glassy materials; its value strongly influences the possible applications of these resins, as discussed below.

4.3 RESIN ACIDS CHEMICAL REACTIVITY

The chemical reactivity of resin acids is determined by the presence of both the double- bond system and the COOH group [5]. The carboxylic group is mainly involved in esterification, salt formation, decarboxylation, nitrile and anhydrides formation, etc. These reactions are obviously relevant to both abietic- and pimaric-type acids (Figs 4.1 and 4.3, respectively). The olefinic system can be involved in oxidation, reduction, hydrogenation and dehydrogenation reactions. Given the conjugated character of this system in the abietic-type acids, and the enhanced reactivity associated with it, much more attention has been devoted to these structures. In terms of industrial applications, salt formation, esterification, and Diels–Alder additions are the most relevant reactions of resin acids.

4.3.1 Reactions of the olefin system

The conjugated double bond system of abietadienoic acids is simultaneously a source of instability, detrimental for many rosin applications and, conversely, an interesting reactive centre for further modifications.

4.3.1.1 Oxidation, hydrogenation and dehydrogenation

The conjugated double bond is responsible for rosin yellowing arising from oxygen addition, isomerization and other reactions. In many applications, this colouring is associated with a loss of product quality. Oxygen addition is normally photochemically induced, leading to the formation of epoxides, hydroxylated derivatives and endoperoxides [7] (Fig. 4.4). These unwanted reactions can be suppressed by dehydrogenation or hydrogenation processes.

Rosin dehydrogenation leads to the removal of hydrogen, which, after double bond rearrangement, converts abietadienoic acids into dehydroabietic acid [5, 8–11]. This reaction is normally carried out at high temperatures (200–300°C), often in the presence of catalysts like metals (*e.g.* Pd, Ni), sulphur or iodine. The hydrogen thus released reacts *in situ* by adding to the pimaradienoic acids and, to a smaller extent, to the abietadienoic acids. Both reactions contribute to the rosin stabilization and these products are known as disproportionated rosins.

If dehydrogenation is carried out at extreme temperatures and/or for long reaction times, the reaction proceeds more extensively and is accompanied by decarboxylation to form the neutral aromatic compound retene (Fig. 4.5) [5].

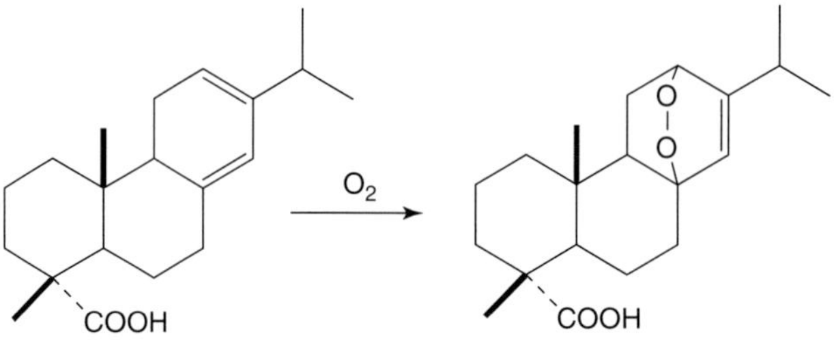

O_2

COOH

COOH

Figure 4.4 Oxidation of levopimaric acid with formation of an endoperoxide.

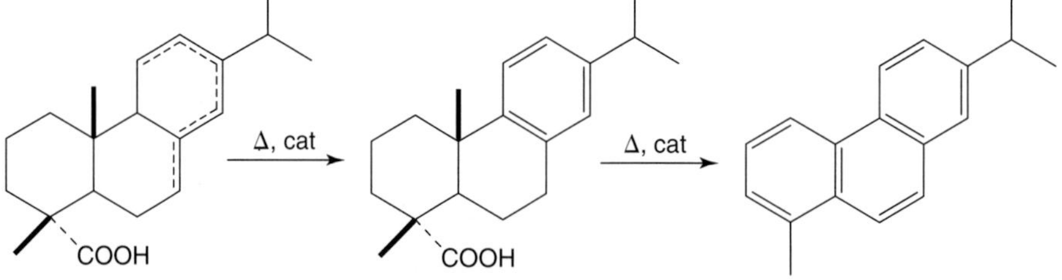

Δ, cat

Δ, cat

COOH

COOH

Figure 4.5 Conversion of abietadienoic acids into dehydroabietic acid and retene.

4.3.1.2 Functionalization of dehydroabietic acid aromatic ring

Dehydroabietic acid can undergo typical aromatic substitution reactions (*e.g.* acylation, chlorosulphonation, sulphonation and nitration) with preferential functionalization of the more reactive 12 position, followed in some cases by the 14 position. These reactions were first exploited in the 1930s–1940s [5, 12–14]. The nitration of dehydroabietic acid (Fig. 4.6) was one of the first reactions studied because of the synthetic versatility of the nitro group, namely as precursor of amino groups [14]. More recently, this reaction has been optimized using less harsh conditions [15, 16].

Hydrogenation is an efficient method for rosin stabilization. The reduction of the first (conjugated) double bond of abietadienoic acids with hydrogen in the presence of metal catalysts is relatively straightforward and frequently the ensuing dihydro-derivatives are stable enough for most applications. The reduction of the remaining double bond requires more severe conditions, but this process becomes necessary in the case of more stringent applications in terms of colour stability and oxygen resistance, such as with long-lifetime sealing agents, air-exposed varnishes, or even fresh-fruit film covering. These partially or fully hydrogenated rosins can then be used as precursors to other derivatives through the modification of the carboxylic functionality.

4.3.1.3 Isomerization

The chemical modifications of abietic-type acids through the conjugated double bond system must take into account a fundamental feature of these moieties, viz. their proneness to isomerization under heat and/or acidic conditions [17–22]. Thus, for example, abietadienoic acid mixtures undergo isomerization to give an equilibrium mixture of products in which abietic acid is the major component (~80 per cent), followed by palustric (~14 per cent) and neoabietic acids (~5 per cent), whereas levopimaric acid is only formed in trace amounts. Figure 4.7 shows the general mechanism of this isomerization process.

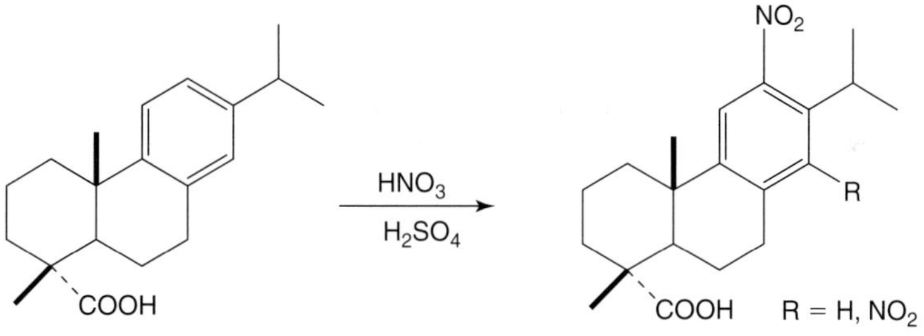

Figure 4.6 Nitration of dehydroabietic acid.

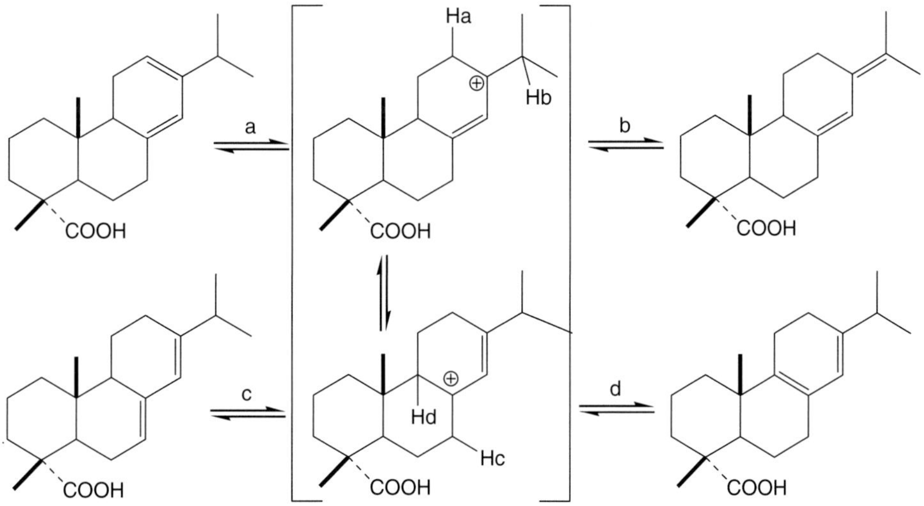

Figure 4.7 Mechanism of the acid-catalyzed isomerization of abietadienoic resin acids.

4.3.1.4 Diels–Alder reactions

When the Diels–Alder reaction with dienophiles is used to convert these acids into novel materials, its occurrence is optimized by the concomitant isomerization process giving levopimaric acid, which is the only homologue capable of producing an adduct. The equilibrium shown in Fig. 4.7 is then displaced by the consumption of levopimaric acid and the entire mixture is thus progressively consumed [5]. If, however, isomerization is minimized, this reaction can be used to isolate and quantify levopimaric acid [23]. Although the Diels–Alder reaction occurs essentially with levopimaric acid, it has been shown that in some instances it also takes place with palustric acid [24].

The Diels–Alder adduct between levopimaric acid and maleic anhydride, maleopimaric anhydride (Fig. 4.8) and the corresponding diacid are certainly the most important derivatives of this family mainly because of their applications in various domains. Levopimaric acid has also been used in the preparation of other adducts with a wide variety of dienophiles such as fumaric acid, acrylonitrile, acrylic acid, vinyl acetate and tetracyanoethylene [5].

The Diels–Alder reaction with maleic acid esters can be easily performed [25]. However, when it is carried out at high temperature (~250°C), partial decarboxylation takes place, with formation of ketone-type dimers [26], yielding polyfunctional derivatives (Fig. 4.9) that could be good monomeric units for polymer synthesis [27].

The functionalities of several Diels–Alder adducts, or their derivatives, have been thoroughly exploited in the synthesis of polymeric materials, as discussed below.

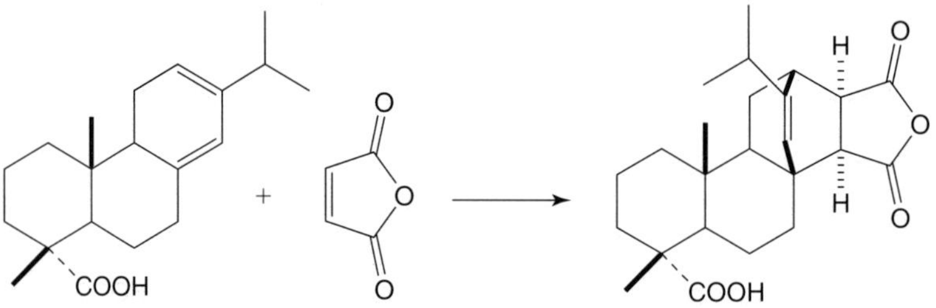

Figure 4.8 Diels–Alder reaction of levopimaric acid with maleic anhydride.

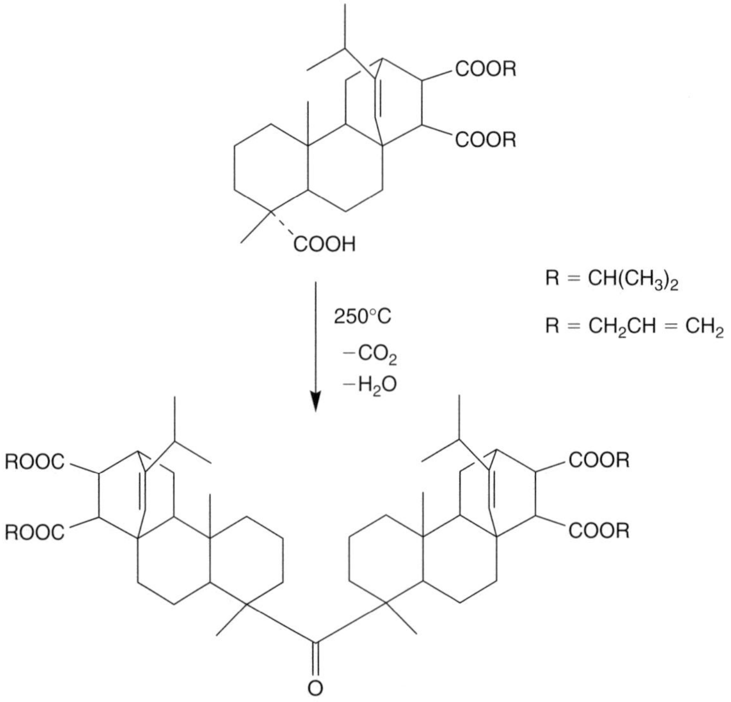

Figure 4.9 Formation of dimeric ketones of maleopimaric-type adducts [26].

4.3.1.5 Reactions with formaldehyde and phenol

Other relevant reactions of resin acids are those involving their addition to formaldehyde and/or phenols. Formaldehyde adds to abietic acid to form complex mixtures containing mono and dihydroxymethyl derivatives [28–31]. Under alkaline conditions, the formation of the mono derivative is favoured, whereas in acetic acid, the dihomologue (isolated as acetate) is the major product (Fig. 4.10). More recently, studies of the acid-catalyzed condensation of abietic acid with formaldehyde showed that, under such conditions the reactions mixtures were predominantly composed of trimeric structures [26, 32].

Since rosin can react with formaldehyde in a similar way as phenol, rosin components have also been used in the preparation of the corresponding rosin–phenol–formaldehyde resins with molecular weights and solubility that are adequate for their incorporation into printing inks. The incorporation of rosin components into phenol–formaldehyde prepolymers can occur through esterification or methylol condensation at one of the unsaturated carbons (Fig. 4.11).

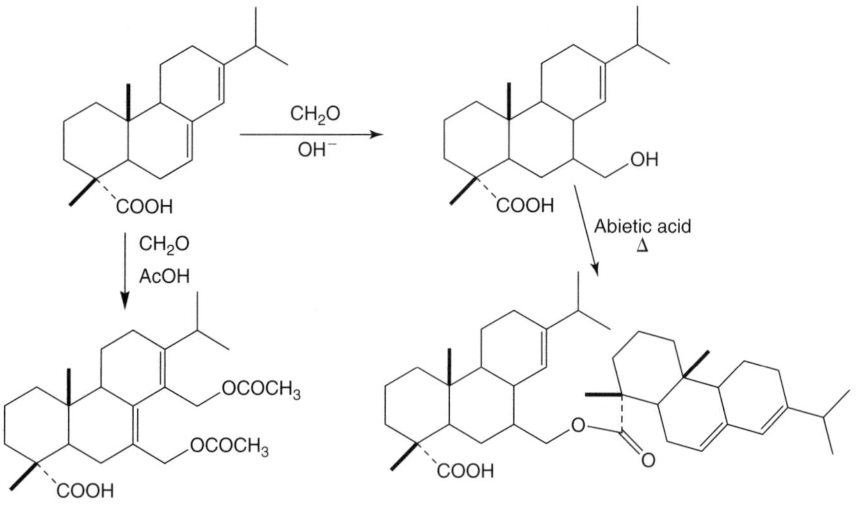

COOH
CH_2O
OH^-
OH
COOH
Abietic acid
Δ
CH_2O
AcOH
$OCOCH_3$
$OCOCH_3$
COOH
O
O
COOH

Figure 4.10 Addition of formaldehyde to abietic acid.

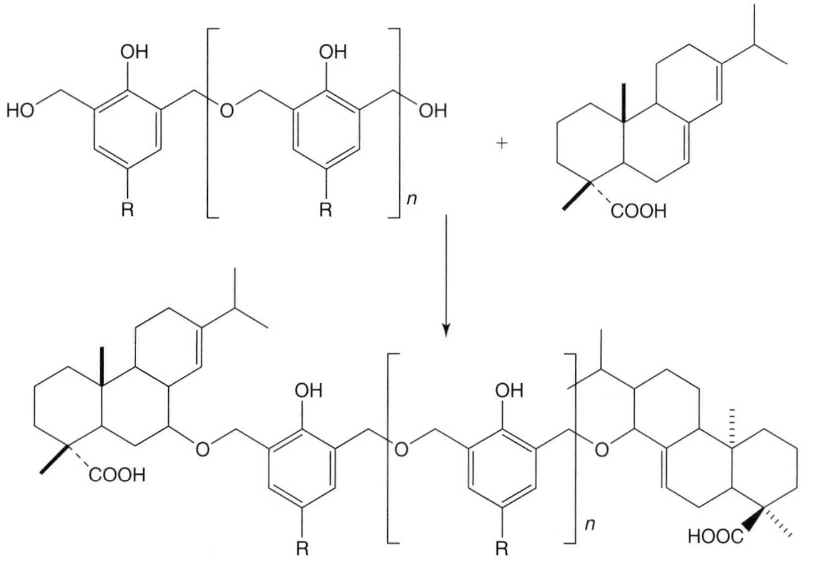

OH
OH
HO
O
OH
R
R
n
+
COOH
OH
OH
O
O
O
COOH
R
R
n
HOOC

Figure 4.11 Formation of rosin-modified phenol–formaldehyde resins.

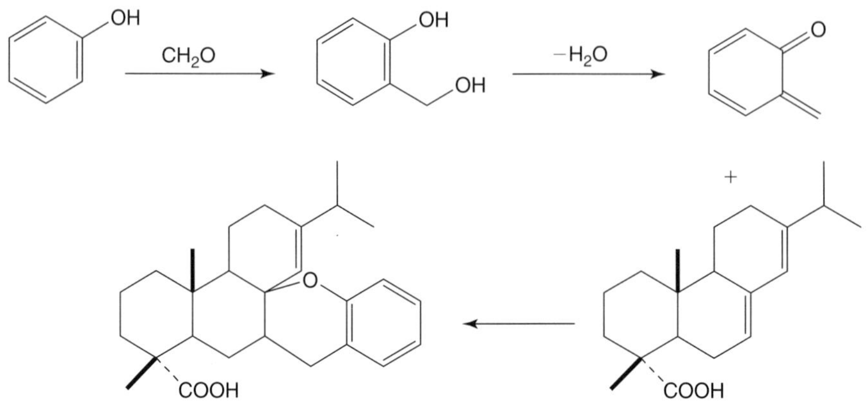

Figure 4.12 Formation of a chromane-type derivative of abietic acid through quinomemethide intermediate.

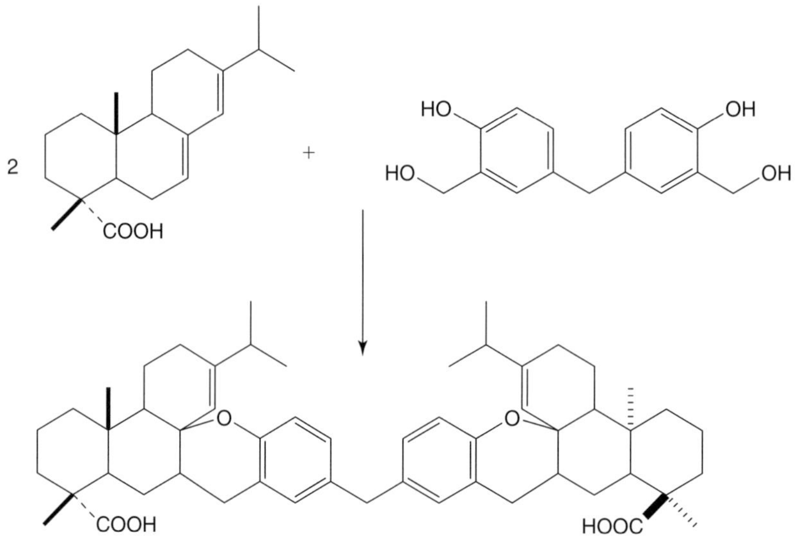

Figure 4.13 Formation of a chromane-type derivative of abietic acid by reaction with diphenylolpropane.

Another approach for the preparation of rosin-modified phenolics [33] involves the formation of quinonemethide by the reaction between phenol and formaldehyde, which then undergoes Diels–Alder addition with the less hindered abietic acid double bond forming a chromane-type derivative (Fig. 4.12).

Similar chromane structures (Fig. 4.13) can also be prepared by the reaction of abietic acid with diphenylolpropane [34].

Resins similar to those described in Fig. 4.11 were prepared by the reaction of abietic acid with aniline and formaldehyde, with application in pressure-sensitive adhesives [35].

Levopimaric acid can also form a Diels–Alder adduct with formaldehyde [5, 36, 37] that can easily undergo acid cleavage to yield a 12-hydroxymethylabietic acid, which, in turn, can be converted into the corresponding 12-hydroxymethylabiet-8-enoic acid and 12-hydroxymethylabietol derivatives (Fig. 4.14). These hydroxymethylated abietanols have been used as co-monomers in the synthesis of polyurethane [38, 39] and polyester [39] films and the corresponding alkoxylated derivatives in the preparation of polyurethane foams [40] and films [41].

Under strong acidic conditions and high temperatures, abietic-type acids can undergo dimerization [5, 42–44]. Although this transformation is frequently referred to as *polymerization*, the complex mixtures formed are predominantly composed of dimers (Fig. 4.15), with minor amounts of trimers [45]. These rosin derivatives find applications in printing inks.

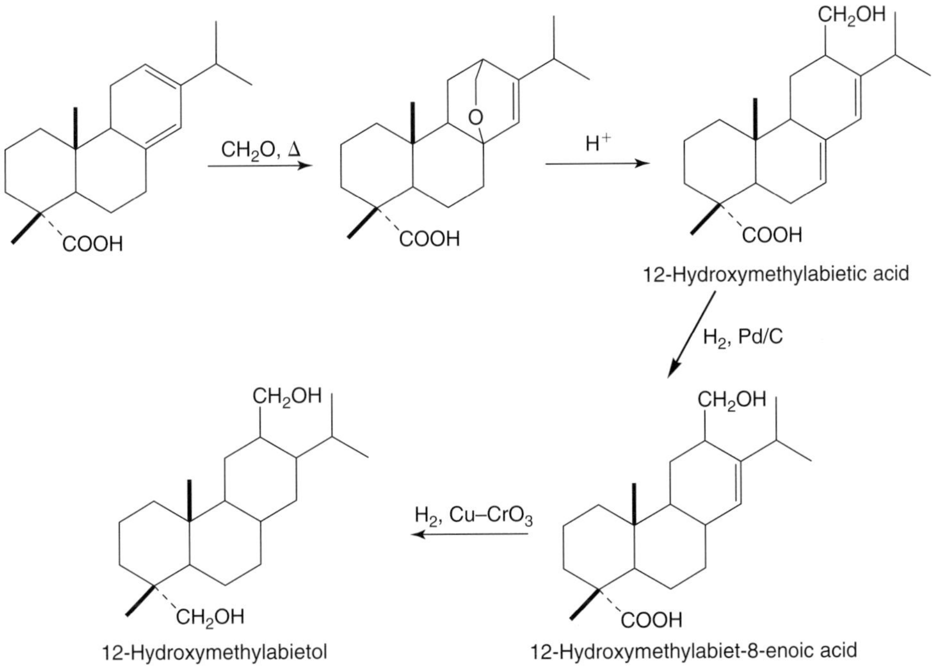

Figure 4.14 Formation of levopimaric adducts with formaldehyde and their conversion into 12-hydroxymethyl derivatives.

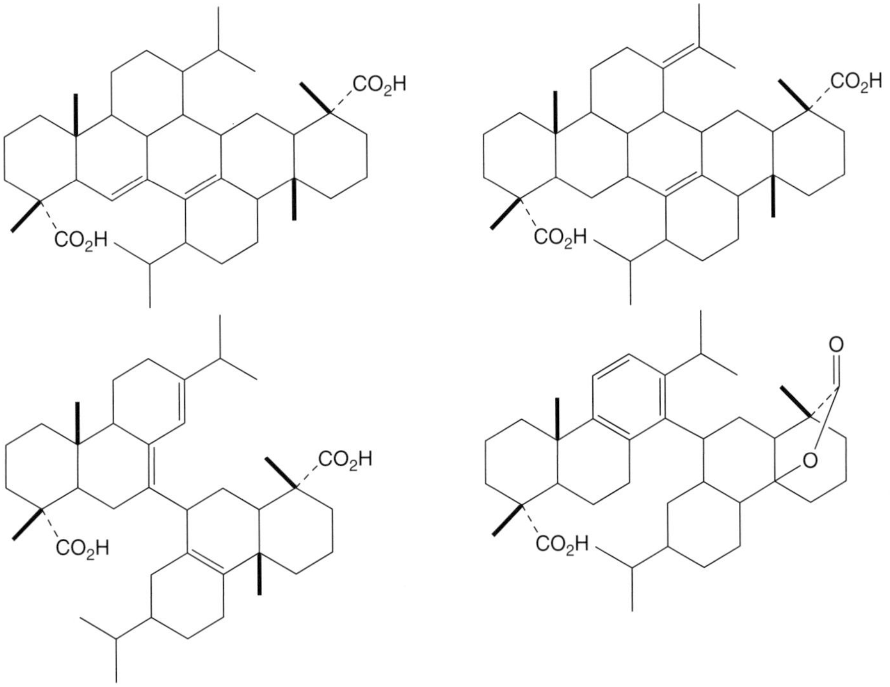

Figure 4.15 Typical dimeric structures of abietic-type acids.

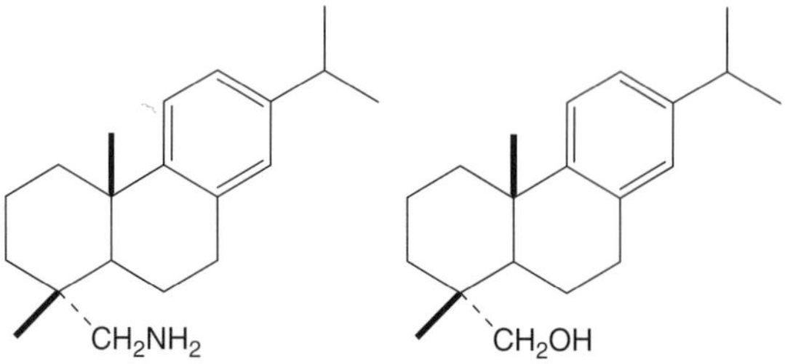

CH_2NH_2
Dehydroabietylamine

CH_2OH
Dehydroabietanol

Figure 4.16 Structures of dehydroabietylamine and dehydroabietanol.

4.3.2 Reactions of the carboxylic group

Na, Mg, Ca, Zn, Al and ammonium resin acid salts are produced industrially [5], albeit their relative importance has been shifting with time because of changes in their relevant applications. Na, K and ammonium salts are partially water soluble and were used as soaps in the past. Presently, resin acids sodium salts are mainly used as intermediates in paper sizing.

The esterification of resin acids is carried out industrially at high temperatures (260–300°C) using metal oxides as catalysts. The most common commercially available esters are those with methanol and polyols like ethylene glycol, diethylene glycol, glycerol and pentaerythritol. Glycerol esters were the first to be used in protective coatings, although pentaerythritol esters are harder and more durable for varnishes. The methyl esters are normally used as plasticizing agents, as described below.

The alkoxylation of free resin acids with reagents such as ethylene oxide, yields hydroxyl-terminated esters which can easily undergo polymerization with an excess of ethylene oxide to produce polyethers of variable chain lengths [46]. These rosin derivatives can be used as chain extenders in the manufacture of polyurethane foams [46].

4.3.3 Miscellaneous reactions

Other chemical transformations involving the carboxylic group have been explored, although their practical interest remains marginal. Anhydrides can be prepared by refluxing the corresponding acid in acetic anhydride [5]. These derivatives have found modest applications as such (*e.g.* in paper sizing), but they are also useful as precursors to nitrogen derivatives [5]. Rosin amines can be prepared from the corresponding nitriles (Eq. 4.1) by the reaction of molten rosin with ammonia at high temperature [5]:

$$RCOOH + NH_3 \rightarrow [RCONH_2] \rightarrow RCN \rightarrow RCH_2NH_2 \quad (4.1)$$

Dehydroabietylamine (Fig. 4.16) and the corresponding ammonium salts find a wide diversity of uses from cationic flotation agents to antioxidants, fungicides and anticorrosion materials [5].

The reduction of methyl dehydroabietate at high temperature leads to the formation of dehydroabietanol (Fig. 4.16), whose light colour and high stability make it a useful intermediate for the synthesis of a variety of esters used in protective coatings, adhesives and plasticizers [5].

4.4 MAJOR APPLICATIONS OF ROSIN AND DERIVATIVES

4.4.1 Paper sizing

Given the intrinsic hydrophylic character of cellulose fibres, specific additives are introduced into most papers to reduce this tendency to different extents according to their specific use, for example, to minimize the penetration of aqueous liquids, or the excessive wetting associated with certain printing processes [47]. This partial hydrophobization

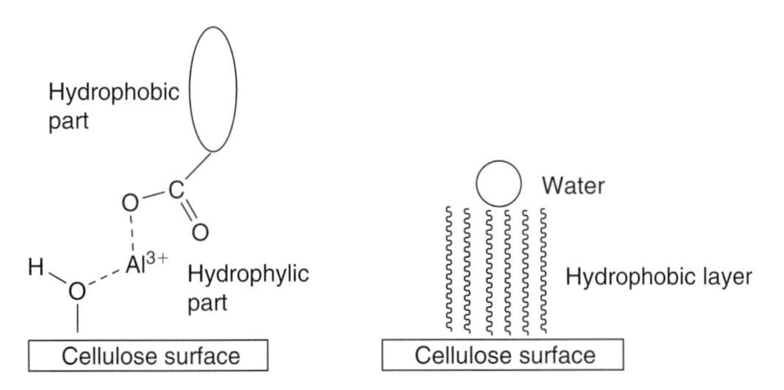

Figure 4.17 Interaction of aluminium resinates with cellulose surface.

can be achieved by bulk or surface treatment with adequate agents, which bind to the fibres' surface, either chemically or by other less strong interactions. This treatment is known as paper sizing. Among the numerous reagents tested for this purpose [48], three families dominate the actual industrial sizing namely, rosin and alum (aluminium sulphate), alkenyl succinic anhydrides (ASA) and alkyl ketene dimers (AKD) [47–50]. Sizing with rosin/alum was invented in 1807 [47, 49, 50] and had been the most important process up to the middle of the last century. However, the acidic conditions associated with its implementation was the main reason for its progressive abandonment because, on the one hand, cellulose degrades slowly by acid-catalyzed hydrolysis, thus inducing a progressive loss in the mechanical strength of the paper sheet and, on the other hand, acidic papers are incompatible with calcium carbonate, a very frequently used filler.

These problems provoked a steady decrease in the utilization of rosin alum in favour of the alternative use of AKD and ASA. However, since these sizing agents also have some drawbacks and are rather costly, research dealing with the implementation of rosin sizing under neutral conditions has been a continuous topic of investigation [48]. Rosin can be used in paper sizing in three forms, namely as sodium salts (known as neutral rosin), free rosin acids and modified rosin products. One of the most important rosin derivatives in this context is the maleopimaric acid adduct (Fig. 4.8), whose incorporation (8–15 per cent) into the sizing composition (a process referred to as "fortification") provides a marked improvement in efficiency compared with unmodified rosin [47]. Similar improvements can be achieved with fumaric or itaconic acid adducts, among others, although the maleopimaric acid adduct dominates the industrial applications of this family of agents.

Despite the increased hydrophilicity associated with the tricarboxylic functionality of the maleopimaric acid adduct, the corresponding increased anionicity reduces the tendency of the sizing agent to agglomerate and promotes a more uniform spreading over the sheet surface, while also increasing the reactivity with alum to form the corresponding aluminium rosinate, which is the actual sizing agent.

Several theories have been proposed to rationalize the mechanism of paper sizing with rosin [47, 49, 50]. An outline of the most relevant features involved in this process is provided here. Given that the cellulose surface is polar and negatively charged, the interaction with the sizing agent involves the *in situ* formation of an aluminium resinate, where the aluminium ions establish an electrostatic interaction between the negatively charged carboxylic groups of the resinate and the sheet surface (Fig. 4.17).

The uniform dispersion of the sizing agent over the cellulose surface, together with the alignment of the aliphatic triterpene skeletons away from the cellulose surface, creates a hydrophobic water-repellent layer (Fig. 4.17).

Effective sizing requires the presence of aluminium soluble complexes capable of exchanging one coordinating water molecule for a resinate anion, thus preserving the cationic nature of the ensuing species. An increase in the pH leads to a corresponding increase in the exchange between the coordinating water molecules and the hydroxy anions. Further water exchange with resinates can lead to the formation of neutral entities unable to interact with the cellulose surface, or, at even higher pH values, to the precipitation of aluminium hydroxide. This explains why sizing with rosin and alum is carried out normally in the pH range of 4–5.5 [49]. The inevitable requirements associated with rosin sizing at higher pH values are that, on the one hand, aluminium must remain in solution and, on the other hand, that its associated complexes are still cationic. The replacement of alum by other aluminium salts, such as polyaluminium chloride, was one of the first approaches in this context [48, 49]. Since then, many other additives have been tested in order to improve the cationicity of the aluminium species in neutral to basic conditions, so that rosin sizing can be applied to papermaking without the problems associated with acidity. Among them, nitrogen-containing polymers, namely polyamines [51–53] and polyimines [54–57], sometimes in conjunction with rosin esters [58–61], have attracted much attention because of their considerable cationic charge in neutral conditions. Despite these and other recent endeavours [62–67], some authors are still sceptical about the viability of the use of rosin in neutral/alkaline sizing [54, 55].

4.4.2 Emulsification

Because of their amphiphilic character, alkali resinates have been exploited both as polymer latex stabilizers and as surfactants in emulsion polymerization from the early development of these techniques, as in the pre-Second World War industrial example of the polymerization of 2-chloro-1,3-butadiene, to produce neoprene [68]. In the following decades, other emulsion polymerizations systems, like the synthesis of styrene–butadiene copolymers [68, 69], also called upon these surfactants, which are still being envisaged today, for example, for the polymerization of styrene [70] and chloroprene [71]. However, the reactivity of the conjugated double bond towards free radicals has made it more profitable to use hydrogenated or dehydrogenated rosins rather than their natural forms [68, 72].

4.4.3 Adhesive tack

As with most of the other sections of this chapter, the basic reference to the applications of rosin and its derivatives as components of adhesive formulations up to the mid-1980s, is the corresponding chapter by Kennedy *et al.* [73] of the classical Naval Store book. The major emphasis in that chapter was placed on the tackifying properties of rosin-based resins in pressure-sensitive, hot melt and sealant adhesives. Therefore, before proceeding to a critical bibliographic survey, it seems appropriate to define the concept of tack and its repercussion in the broad area of adhesion, as opposed to other contexts, like printing inks, paints and coatings.

Adhesive tack is the property that controls the instantaneous formation of a bond when an adhesive and a surface are brought into contact. In other words, tack is a measure of the "stickiness" of an adhesive while still in a liquid or viscoelastic state, or simply the resistance to separation which is rate and temperature dependent and involves viscoelastic deformation of the bulk adhesive. The appropriate measurement is the work expended in separation, rather than the force used. The critical features of tack are characterized by the "wetting stage", controlled by the surface energy, viscosity and thickness of the adhesive, and the debonding stage (peel strength), which is strongly influenced by the viscoelastic response of the adhesive. Although tack tests are generally simple to perform, this parameter lacks a strict physical definition or unit and the values obtained on a given apparatus are self-consistent, but not comparable with those from a different measuring system. Maximum tack occurs when the conditions of measurement and the properties of the tacky material combine to have a high energy loss within the adhesive material. This is the basis for the formulation and incorporation of "tackifiers" in adhesive formulations.

In the empirical world of tack experts, many professional terms are employed, like "green strength" (as initial tack is sometimes called), which relate to the resistance to separation before the adhesive has had a chance to vulcanize or crosslink. This characteristic may also be called "quick tack" or "aggressive tack". It may be one of the most important properties in determining the suitability of an adhesive, such as that placed on a pressure-sensitive tape, for a certain application. Associated with tack is "dry tack", which is a property of certain adhesives to stick to one another even though they seem to be dry to the touch. Autohesive tack (or autohesion) is the dry tack between materials having similar chemical compositions. "Tack range" is the time that an adhesive will remain in a tacky condition.

The relative decline in interest in using rosin as a tackifier in adhesive formulations, clearly reflected by the paucity of publications up to the mid-1990s, is counterbalanced by a dramatic increase in the number of papers published in the last decade in scientific and technical journals (more than 50), as predicted by an editorial in Chemical Week in 1995 [74]. A selection from this large output is given here to summarize some of the most salient features arising from the return to the utilization of this renewable resource as a key adhesive component.

A study of the surface properties of pentaerythritol rosin ester [75] showed that the most abundant moieties crowding this surface are the ester groups and that the surface energy was composed of a dispersive contribution of $\sim$22 mJ m^{-2} and a polar one of $\sim$10 mJ m^{-2}. This relatively low surface tension, compared with that of many polar polymers and indeed with most surfaces to be glued, explains in part the tackifying role of rosin derivatives added to polymer adhesive bases.

Rosin esters have been incorporated into a variety of macromolecules with potential adhesive properties in order to enhance their tack and in many of these studies the beneficial effect was compared with that of terpene and hydrocarbon resins. The polymer substrates include ethylene–vinyl acetate copolymers [76–78], natural rubber [79] and thermoplastic elastomers [80, 81], polyurethanes, where the rosin derivatives participate chemically in the polymer growth [82, 83] and acrylic resins in organic media [84] and as water emulsions [85]. Pressure-sensitive adhesives as well as hot melts were thus characterized in terms of tack and peel strength as a function of a variety of parameters, including the composition and average molecular weight of the base polymer, the amount of added tackifier and its miscibility in the polymer. Rosin esters gave satisfactory results with most polymers after the appropriate optimization of the adhesive compositions and compared favourably with other natural (terpenic) and synthetic (petroleum-based oligomers) tackifiers.

Some rosin derivatives have been found to provoke allergies [86], particularly if they are present in curative plasters.

4.4.4 Polymer chemistry and processing

Maiti *et al.* [2] thoroughly reviewed the technical literature up to the late 1980s related to the use of rosin and its derivatives as monomers or co-monomers in a variety of polymerization systems, as well as the properties of the ensuing polymers. Most of the documents cited in this monograph were patents unavailable to us, whose scientific content was therefore difficult to judge, but nonetheless, the interest of this survey lies in the fact that it emphasizes the extensive search of materials based on rosin. Some clear-cut investigations in this context are described here as typical examples of the exploitation of rosin and its derivatives as polymer precursors.

Maleopimaric anhydride can react with amines to form maleimides. If diamines are used, the corresponding maleimide-amines can be polymerized through a step process (Fig. 4.18), involving the condensation of the carboxylic moiety with the primary amine to form a poly(amide-imide), which was used in the formulation of gravure printing inks [87].

These poly(amide-imide) materials were extensively studied in the 1980s in various aspects (*e.g.* synthesis and preparation of blends with other polymeric materials), [2, 88–96] demonstrating the increasing interest in these materials. More recently, water soluble polyamides and poly(amide-imide)s [97] as well as polyhydroxyimides [98] structurally analogous to those described in Fig. 4.18 have been reported. Additionally, photoactive polymers were prepared by the condensation of a maleopimaric adduct with azo-dye type diamines [99]. The levopimaric

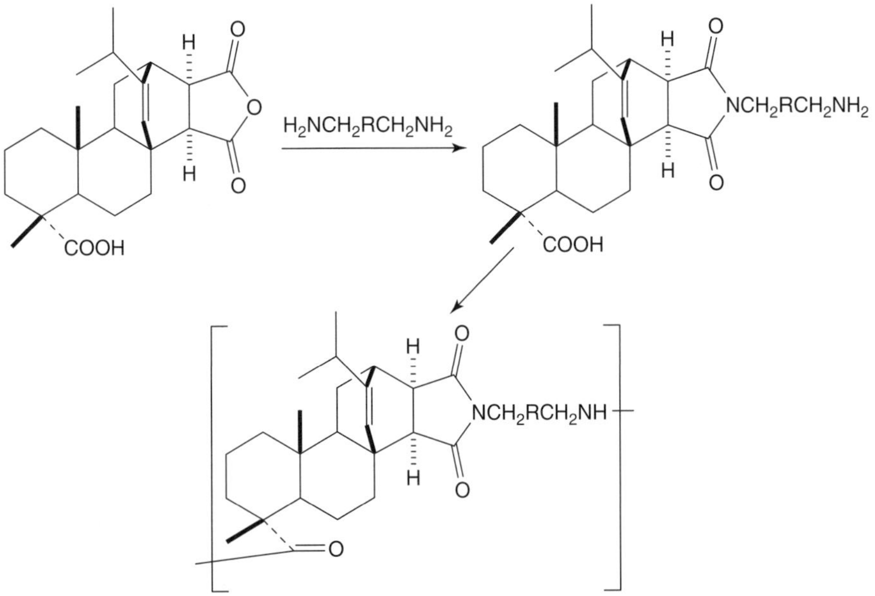

H O O H O COOH $H_2NCH_2RCH_2NH_2$ H O $NCH_2RCH_2NH_2$ H O COOH H O NCH_2RCH_2NH H O O

Figure 4.18 Synthesis and polycondensation of a rosin-based poly(amide-imide).

adduct with acrylic acid has been used as a polycondensation monomer, both as a diacid in the synthesis of polyesters [2, 100] and, after conversion of its two carboxylic groups into isocyanate functions, as a co-monomer in the preparation of polyurethanes [101].

Maleopimaric acid and its anhydride were converted into the corresponding peroxides with the aid of an organic peroxide and used as reactive compatibilizers in the extrusion of polyethylene and polyamide 6 [102]. The ensuing blends showed improved mechanical properties compared with those of physical mixtures of the two polymers, suggesting that the rosin-based peroxide had indeed played an active role in enhancing the quality of the interface between the two polymers.

The maleopimaric and acrylopimaric adducts, after a three-step synthesis with ethylene glycol catalyzed by *p*-toluene sulphonic acid (*p*TSA), followed by epychlorydrine and by acrylic or methacrylic acid, led to the formation of vinyl-type ester monomers (Fig. 4.19), which were then submitted to radical copolymerization with styrene and tested as metal coatings [103]. A similar approach to coating materials was recently applied to prepare unsaturated polyester resins based on resin acid adducts, glycols and maleic anhydride [104, 105].

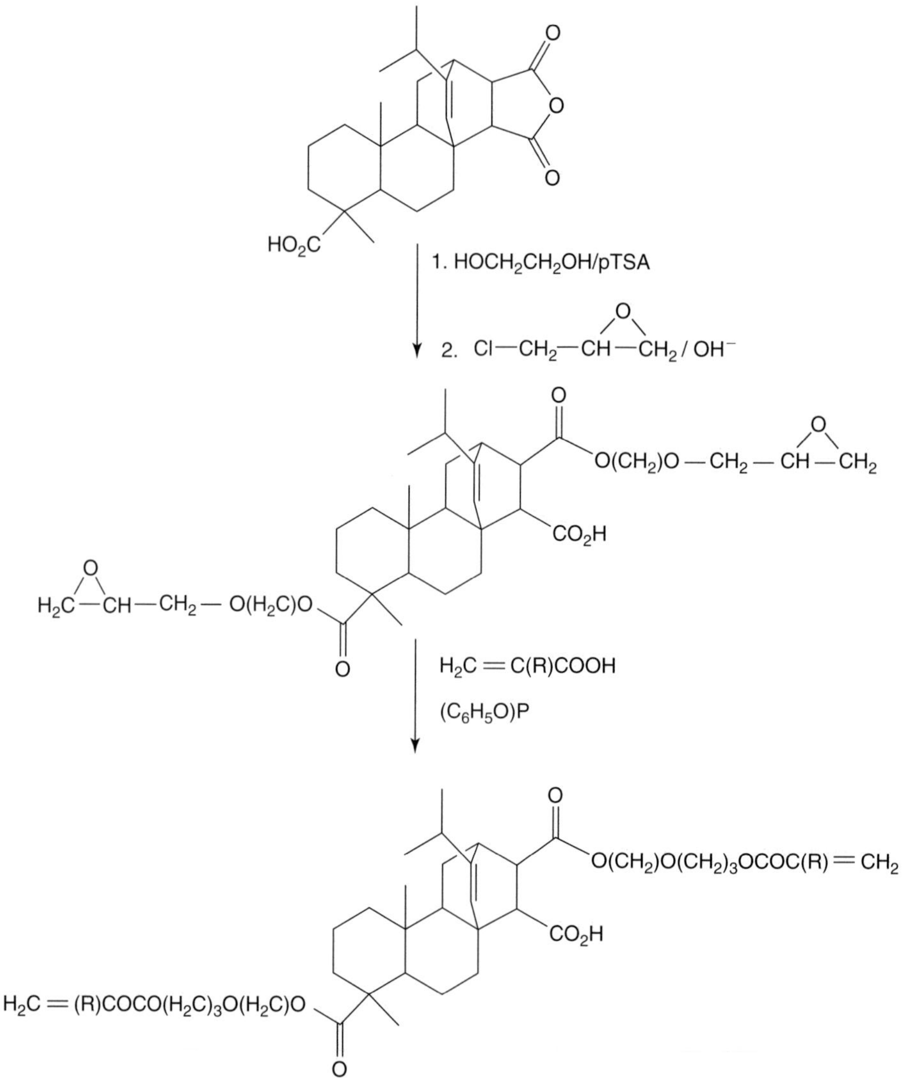

Figure 4.19 Synthesis of vinyl-type ester monomers from the maleopimaric adduct [103].

As shown above (Fig. 4.9), maleopimaric anhydride can be converted into diallyl derivatives [25], which can be further converted into dimeric-type ketones [27]. Both intermediates were used in polymerization tests that yielded crosslinked materials [25, 27].

When resin acid ketone dimers [26] (bisdienes) were submitted to the Diels–Alder reaction with an aromatic bismaleimide [106], high molecular weight thermally stable polyimides were obtained (Fig. 4.20). Similar materials were also prepared using resin acid dimers and aliphatic bismaleimides, as well as bisacrilamides [107]. Acrylonitrile adducts with resin acid dimers were also used as intermediates for the synthesis of polyamides with aliphatic diamines [108].

Resin acid dimer adducts with maleic anhydride and acrylic acid were also used to prepare epoxy resins [107, 109]. The epoxy precursors were prepared by the reaction of the Diels–Alder adducts with di(ethanol)amine, followed by treatment with epychlorydrine under alkaline conditions, as shown in Fig. 4.21 for the acrylic acid adduct [109]. After curing, the ensuing materials produce high stability coats [109].

On a different approach, poly(4-abietylmethylstyrene) was prepared in high yields by the reaction of poly (4-chloromethylstyrene) with sodium abietate. Near-UV irradiation of this novel polymer induced its crosslinking by photodimerization through the conjugated double bonds of the abietate moieties [110], presumably by the formation of dimeric structures similar to those described in Fig. 4.15. Abietic acid dimers were also used as diacids in the synthesis of polyamides by reaction with diamines [111].

Rosin and its derivatives are frequently used as polymer additives in different capacities. Their role as plasticizers is illustrated by recent work related to the paper coating for food packaging with a 3-hydroxybutirate/3-hydroxyvalerate copolymer (PHB/V) [112, 113]. The addition of tall oil rosin to the copolymer was found to improve its water vapour barrier properties as well as to reduce the pinhole density of the corresponding laminated papers.

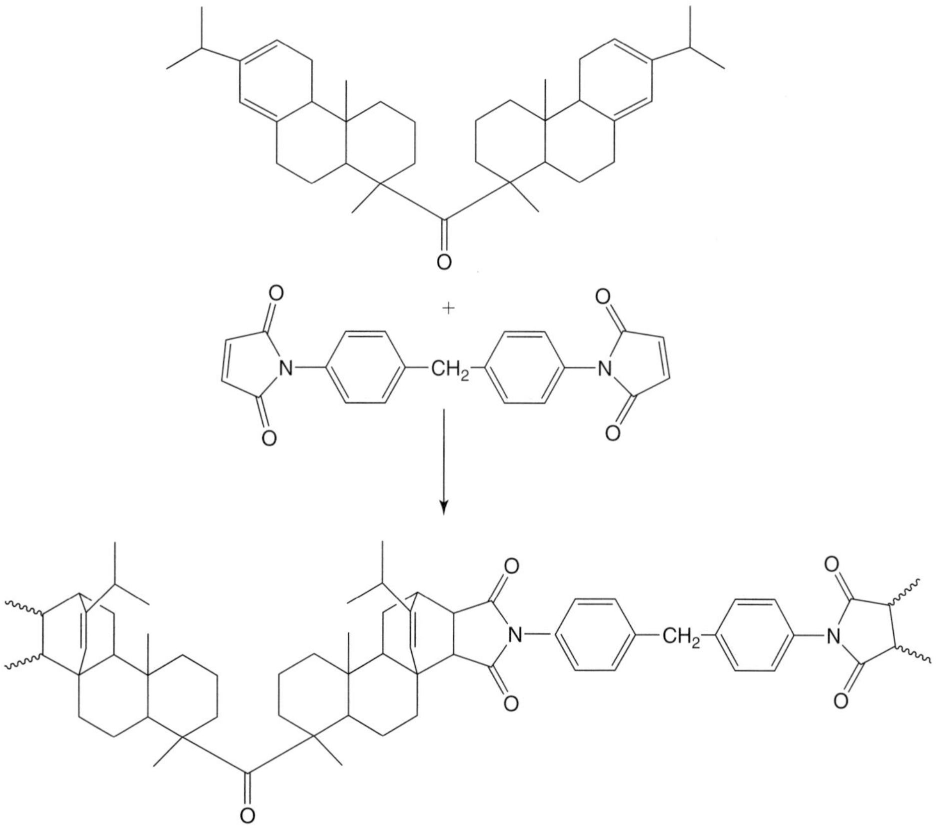

Figure 4.20 Synthesis of polyimides by Diels–Alder condensation of resin acid dimers with aromatic bismaleimides [106].

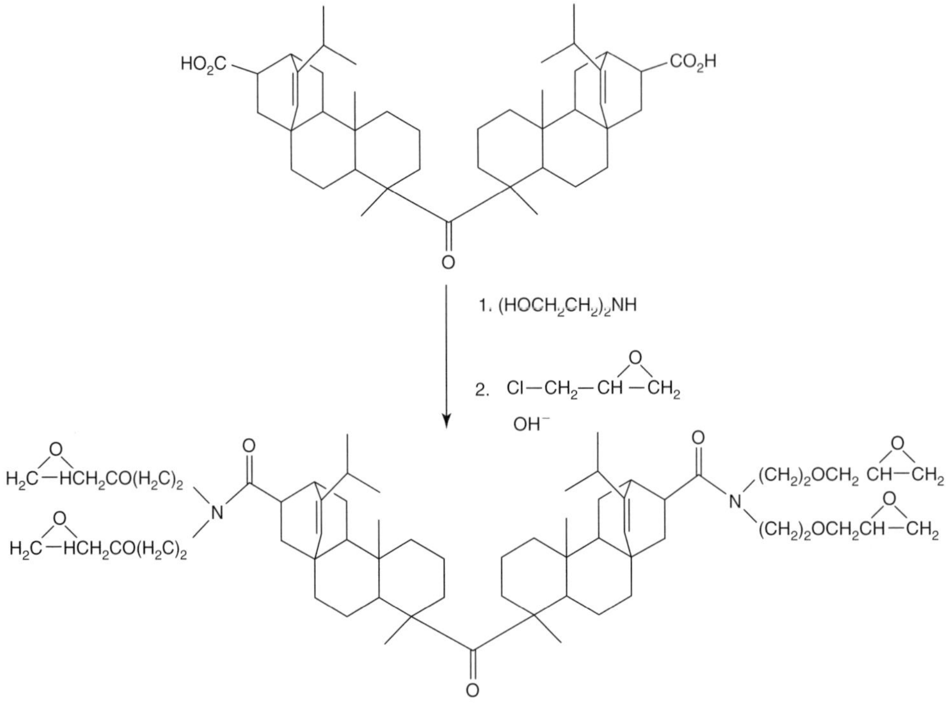

Figure 4.21 Synthesis of epoxy resins from resin acid dimer adduct with acrylic acid.

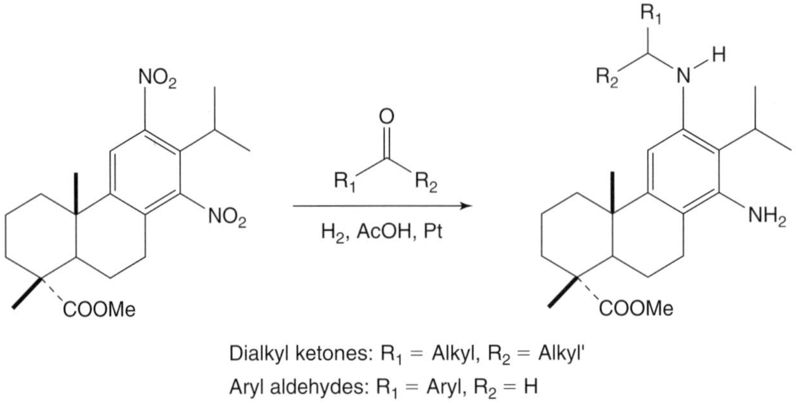

Figure 4.22 Synthesis of secondary amines of methyl dehydroabietate.

In another vein, secondary amines prepared by the reductive amination of appropriate carbonyl compounds with methyl 12,14-dinitrodehydroabietate (Fig. 4.22) gave promising results when used as stabilizers against thermal, photochemical and oxidative degradation of low-density polyethylene and ethylene–propylene elastomers [114–116]. This active role is not surprising given the well-known fact that aromatic amines are excellent free radical traps.

4.4.5 Printing inks

Four distinct processes dominate the printing technology [117], namely (i) offset lithography, by far the most important, (ii) letterpress, (iii) flexography and (iv) gravure. A description of the principles and peculiarities associated with each of them falls outside the scope of this chapter. In relation to their ink requirements, the major difference lies in the viscosity, which must be very high for the first two processes (paste inks) and low for the other two (liquid inks). Interestingly, although the properties of rosin itself do not make it very viable as a component of any of these inks, several of its derivatives have been, and are being, extensively used in the manufacture of all four types of printing inks. The reader is invited to consult Burke's chapter in the Naval Store book [34] for what is still the best available monograph on the use of rosin compounds in printing inks, even compared with the latest edition of the Printing Ink Manual [117]. It provides a clear technical introduction to the relationship between printing processes and the requirement of the corresponding inks, a thorough description of the types and roles of the rosin derivatives in each family of inks and a detailed literature survey up to the mid-1980s.

Any ink (varnish) must be transferred from the printing surface to the substrate to be printed (varnished) while still in a fluid state and thereafter it must dry through an appropriate physical or chemical process and turn into a solid film. The film-forming aptitude of an ink (varnish) is thus of paramount importance, as are the properties of the dried film. In the search for an optimal resin composition (the film precursor), these qualities represent therefore the first priority, but the solutions are not universal, for they must comply with the nature of the printing process. Since the only difference between an ink and a varnish is the presence of a suspended pigment in the former, the composition of the liquid phase is the same in both instances, for a given printing and drying process.

The most important rosin derivatives used in printing ink formulations are rosin oligomers and their esters, metal resinates, modified phenolic and alkyd resins, ester gums, maleic and fumaric acid adducts and their esters. Practically all types of printing inks can be manufactured with rosin-based components, because they provide good miscibility and compatibility with most film formers and other ink additives.

The following succinct account of the major utilizations of rosin derivatives in printing inks is aimed at giving a mere introduction to the art, which can be enriched by looking up the more comprehensive treatments already quoted [34, 117]. In contrast with the applications related to adhesion and tack, which are thoroughly documented by the numerous studies discussed in the preceding section, research dealing with printing ink formulations is much more confidential and the corresponding open documents are few and far between. It follows that the brief survey below is only based on general technical reports.

The high acid content of rosin and its oligomers may be lowered by reacting them with lime, calcium acetate or zinc oxide, or indeed a combination of these compounds. The ensuing metal resinates have high softening temperatures and are therefore inclined to give brittle films. When used in conjunction with a polymeric film former, like cellulose esters, in gravure inks, good rub and abrasion resistance are achieved, together with a reduction in green tack. Their incorporation in lithographic inks provides good pigment wetting and prevents gelation during the storage of drying oil varnishes.

Of the numerous rosin-based esters used in printing inks, the simplest are those formed by the direct condensation reaction of rosin or its oligomers with glycerol and pentaerythritol at high temperature. They find applications in letterpress and lithographic inks and are gaining use also in liquid ink compositions. The other ester family comprises the rosin adducts with maleic and fumaric acids and their esters. They are usually classified according to their solubility in different organic media. The lower their acid values (free COOH groups), the higher their solubility in non-polar solvents and vice versa. They are widely used in printing inks; in particular, high acid-value maleics are incorporated into water- and moisture-set letterpress inks and into water-based flexographic inks in conjunction with an amine or ammonia. Medium acid-value counterparts are added into alcohol-based gravure and flexographic inks in combination with polyamides or cellulose derivatives. Hydrocarbon soluble resins (low acid value) have the advantage of possessing a very pale coloration and are employed as additives to letterpress and lithographic gloss inks, heatset inks and overprint varnishes.

The third large group of rosin-based materials that are widely used in printing ink formulations are rosin-modified phenolic resins (*e.g.* Figs 4.11 and 4.13) and their esters. With their variety of properties, depending on the actual chemical composition, they represent an important family of materials in printing ink technology. The variables that allow such a diversity of properties to be obtained are (i) the actual phenol homologue, (ii) the classical phenol (P)/formaldehyde (F) ratio, (iii) the relative amount of PF resin with respect to rosin, (iv) the use and amount of a dicarboxylic acid structural modifier and (v) the final esterification (if any) in terms of both its extent and the type of polyol used. Typically, rosin is added to a PF prepolymer and the ensuing mixture is heated at 100–150°C in the presence of a basic catalyst. The temperature is then increased to ~200°C to complete the mutual condensation. If a high acid functionality is desired, the resin is then treated with a dicarboxylic acid. Finally, polyols like glycerol and pentaerythritol can be used in a further step to esterify some or most of the resin free COOH groups, in order to achieve different properties. Rosin-modified PF resins possess many specific properties, valuable in different printing ink types, as a function of their composition. They are compatible with alkyd resins and most modifiers used in oil-based inks and tend to give tough, glossy dry films with good resistance.

A recent interesting addition to this realm is the use of rosin-based resins in the manufacture of the new generation of ink jet inks, the phase-change inks [118].

4.4.6 Miscellaneous applications

The controlled release of pharmaceuticals and agrochemicals is a highly active research topic because, on the one hand, it contributes to maintain for long periods the constant levels required for effective disease treatment or pest control and, on the other hand, it sharply reduces the administered doses, with the concomitant health, environmental and economical benefits. Rosin and rosin esters (*e.g.* rosin and maleic rosin esters with glycerol, sorbitol, mannitol and pentaerythritol) have been investigated in the last couple of decades as coatings [119–123] and micro-encapsulating materials [83–91] for the controlled release of several pharmaceuticals, and as implant matrix for the delivery of drugs [124], given their established *in vitro* and *in vivo* degradation behaviour and biocompatibility [125]. Although rosin and hydrogenated rosin films tend to pose handling problems because of their brittle nature, this drawback can be easily overcome by adding adequate hydrophobic plasticizers (up to 20 per cent w/w), such as tributil sebacate and tributil citrate [125, 126].

Recent studies also indicated that rosin nanoparticles can be effective for the encapsulation and delivery of drugs [127]. Rosin formulations have also been used in the controlled release of carbofurane nematicides, thus allowing the applied doses to be reduced by half and to increase significantly their effectiveness and availability in the soil over longer periods of time [128].

Other applications of rosin derivatives and polymers include emulsifiers for pharmaceuticals and cosmetics [129, 130], flotation oils in the phosphate industry [131] and the formulation of polymer-concrete composites for the construction of temporary and semi-permanent military structures in arid zones [132].

4.5 CONCLUDING REMARKS

As in the case of other traditional renewable resources exploited by mankind since the inception of elementary technologies, for example, gelatin, soaps and vegetable oils, rosin and its derivatives suffered a decline in their utilization with the boom of petrochemistry in the second half of the last century. The renewed interest in these natural products is a growing reality because of the increasing awareness of the need to replace fossil resources with renewable counterparts. This inversion of tendency has already become visible in the case of rosin by the recent surge in publications devoted to the search both for more viable applications of known processes and commodities and for new materials with improved properties for razor-edge technologies.

REFERENCES

1. Zinkel D.E., Russel J., *Naval Stores. Production, Chemistry, Utilization*, Pulp Chemical Association, New York, 1989.
2. Maiti S., Ray S.S., Kundu A.K., Rosin – A renewable resource for polymers and polymer chemicals, *Prog. Polym. Sci.*, **14**(3), 1989, 297–338.
3. Stauffer D.F., Production, markets and economics, in *Naval Stores. Production, Chemistry, Utilization* (Eds. Zinkel D.F. and J. Russel , Pulp Chemical Association, New York, 1989, pp. 39–82.
4. Coppen J.J.W., Hone G.A., *Gum Naval Stores: Turpentine and Rosin from Pine Resin*, FAO, Rome, 1995.
5. Soltes J., Zinkel D.F., Chemistry of rosin, in *Naval Stores. Production, Chemistry, Utilization* (Eds. Zinkel D.F. and Russel J.), Pulp Chemical Association, New York, 1989, pp. 261–345.
6. Stump J.H., Quality control (tall oil, rosin, and fatty acids), in *Naval Stores. Production, Chemistry, Utilization* (Eds. Zinkel D.F. and Russel J.), Pulp Chemical Association, New York, 1989, pp. 846–868.
7. Schuller W.H., Lawrence R.V., Oxidation of levopimaric acid pine gum and various rosins with singlet oxygen, *Ind. Eng. Chem. Prod. Res. Develop.*, **6**(4), 1967, 266–268.
8. Fleck E.E., Palkin S., On the nature of pyroabietic acids, *J. Am. Chem. Soc.*, **60**, 1938, 921–925.
9. Fleck E.E., Palkin S., The dihydroabietic acids from so-called pyroabietic acids, *J. Am. Chem. Soc.*, **60**, 1938, 2621–2622.
10. Fleck E.E., Palkin S., The composition of so-called pyroabietic acid prepared without catalyst, *J. Am. Chem. Soc.*, **61**, 1939, 247–249.

11. Fleck E.E., Palkin S., The presence of dihydroabietic acid in pine oleoresin and rosin, *J. Am. Chem. Soc.*, **61**, 1939, 1330–1332.
12. Fieser L., Campbell W., Substitution reactions of dehydroabietic acid, *J. Am. Chem. Soc.*, **60**, 1938, 2631–2636.
13. Fieser L., Campbell W., Hydroxyl and amino derivatives of dehydroabietic acid and dehydroabietinol, *J. Am. Chem. Soc.*, **61**, 1939, 2528–2534.
14. Campbell W., Morgana M., Substitution reactions of dehydroabietic acid. II, *J. Am. Chem. Soc.*, **63**, 1941, 1838–1843.
15. Levinson A.S., Mononitration of methyl abieta-8,11,13-trien-18-oate, *J. Org. Chem.*, **36**(20), 1971, 3062–3064.
16. Gigante B., Prazeres A.O., Marcelocurto M.J., Mild and selective nitration by claycop, *J. Org. Chem.*, **60**(11), 1995, 3445–3447.
17. Schuller W.H., Lawrence R.V., Base-catalyzed isomerization of resin acids, *J. Org. Chem.*, **30**(6), 1965, 2080–2082.
18. Takeda H., Kanno H., Schuller W.H., Lawrence R.V., Effect of temperature on rosins and pine gum, *Ind. Eng. Chem. Prod. Res. Develop.*, **7**(3), 1968, 186–189.
19. Takeda H., Schuller W.H., Lawrence R.V., Thermal isomerization of methyl abietate, *J. Chem. Eng. Data*, **13**(4), 1968, 579–581.
20. Takeda H., Schuller W.H., Lawrence R.V., Thermal isomerization of abietic acid, *J. Org. Chem.*, **33**(4), 1968, 1683–1684.
21. Takeda H., Schuller W.H., Lawrence R.V., Thermal behavior of some resin acid esters, *J. Chem. Eng. Data*, **14**(1), 1969, 89–90.
22. Portugal I., Vital J., Lobo L.S., Resin acids isomerization – A kinetic-study, *Chem. Eng. Sci.*, **47**(9–11), 1992, 2671–2676.
23. Lloyd W.D., Hedrick G.W., Diels–Alder reaction of levopimaric acid and its use in quantitative determinations, *J. Org. Chem.*, **26**(6), 1961, 2029–2032.
24. Kotsuki H., Kataoka M., Matsuo K., Suetomo S., Shiro M., Nishizawa H., High-pressure organic-chemistry. 17. Diels–Alder reaction of methyl palustrate with maleic-anhydride and *N*-phenylmaleimide, *J. Chem. Soc. Perkin Trans. I*, **22**, 1993, 2773–2776.
25. Bicu I., Mustata F., Allylic polymers from resin acids, *Angew. Makromol. Chem.*, **246**, 1997, 11–22.
26. Bicu I., Mustata F., Study of the condensation products of abietic acid with formaldehyde at high-temperatures, *Angew. Makromol. Chem.*, **222**, 1994, 165–174.
27. Bicu I., Mustata F., Allylic polymers from resin acids – Monomer synthesis at high temperature, *Angew. Makromol. Chem.*, **255**, 1998, 45–51.
28. Minor J.C., Lawrence R.V., Esterification of methylolated rosin, *Ind. Eng. Chem.*, **50**(8), 1958, 1127–1130.
29. Royals E.E., Pyrolysis of esters. 1. Nonselectivity in the direction of elimination by pyrolysis, *J. Org. Chem.*, **23**(11), 1958, 1822.
30. Black D.K., Hedrick G.W., Condensation of paraformaldehyde with abietic acid and some of its derivatives, *J. Org. Chem.*, **32**(12), 1967, 3763–3767.
31. Stclair W.E., Minor J.C., Lawrence R.V., Fused metal resinates from aldehyde-modified rosin, *Ind. Eng. Chem.*, **46**(9), 1954, 1973–1976.
32. Bicu I., Mustata F., Study of the condensation products of abietic acid with formaldehyde, *Angew. Makromol. Chem.*, **213**, 1993, 169–179.
33. Hultzsch K., Uber Zusammenhange Und Abgrenzungen Bei Poly-Reaktionen, *Angew. Chem.*, **64**(13), 1952, 361.
34. Burke R.E., Rosin-based printing inks, in *Naval Stores. Production, Chemistry, Utilization* (Eds. Zinkel D.F. and J. Russel , Pulp Chemical Association, New York, 1989, pp. 667–700.
35. Mustata F., Bicu I., Epoxy aniline formaldehyde resins modified with resin acids, *Polimery*, **46**(7–8), 2001, 534–539.
36. Parkin B.A., Hedrick G.W., Chemistry of resin acids. I. Reaction of levopimaric acid with formaldehyde, *J. Org. Chem.*, **30**(7), 1965, 2356–2358.
37. Parkin B.A., Summers H.B., Settine R.L., Hedrick G.W., Production of levopimaric acid–formaldehyde adduct and hydroxymethylated materials from resin acids and rosin, *Ind. Eng. Chem. Prod. Res. Develop.*, **5**(3), 1966, 257–262.
38. Lewis J.B., Hedrick G.W., Hydroxymethylated derivatives of resin acids – Use in polyurethane films, *Ind. Eng. Chem. Prod. Res. Develop.*, **9**(3), 1970, 304–310.
39. Osman M.A., Marvel C.S., Polyesters from 12-hydroxymethyltetrahydroabietic acid and 12-hydroxymethyltetrahydroabietanol, *J. Polym. Sci. A1 Pol. Chem.*, **9**(5), 1971, 1213–1218.
40. Rohde W.A., Black D.K., Hedrick G.W., Resin–formaldehyde reaction – Separation of hydroxymethylated derivatives of resin acids, *Ind. Eng. Chem. Prod. Res. Develop.*, **12**(3), 1973, 241–246.
41. Bingham J.F., Marvel C.S., Polymers containing 12-hydroxymethyltetrahydroabietanol, *J. Polym. Sci. A1 Pol. Chem.*, **10**(2), 1972, 489–496.
42. Parkin B.A., Schuller W.H., Catalyst–solvent systems for dimerization of abietic acid and rosin, *Ind. Eng. Chem. Prod. Res. Develop.*, **11**(2), 1972, 156–158.
43. Parkin B.A., Schuller W.H., Polymerized rosin – Continuous method, *Ind. Eng. Chem. Prod. Res. Develop.*, **12**(3), 1973, 238–240.
44. Parkin B.A., Schuller W.H., Lawrence R.V., Thermal dimerization of rosin, *Ind. Eng. Chem. Prod. Res. Develop.*, **8**(3), 1969, 304–306.

45. Chang T.L., Analysis of tall oil by gel permeation chromatography, *Anal. Chem.*, **40**(6), 1968, 989–992.
46. Rohde W.A., Hedrick G.W., Esters of rosin acids and glycidyl ethers, *J. Am. Oil Chem. Soc.*, **47**(1), 1970, 3–4.
47. Niemo L., Internal sizing of paper, in *Papermaking Chemistry* (Ed. Niemo L.), Tappi Press, Helsinki, 1999.
48. Roberts J.C., Neutral and alkaline sizing, in *Paper Chemistry* (Ed. Roberts J.C.), Blackie Academic & Professional, London, 1996, pp. 140–160.
49. Gess J.M., The sizing of paper with rosin and alum at acid pHs, in *Paper Chemistry* (Ed. Roberts J.C.), Blackie Academic and Professional, Glasgow, 1996, pp. 120–139.
50. Strazdins E., Paper sizes and sizing, in *Naval Stores. Production, Chemistry, Utilization* (Eds. Zinkel D.F. and Russel J.), Pulp Chemical Association, New York, 1989, pp. 575–624.
51. Wu Z.H., Tanaka H., Behaviors of polyvinylamines in neutral rosin sizing, *Mokuzai Gakkaishi*, **41**(10), 1995, 911–916.
52. Wu Z.H., Chen S.P., Tanaka H., Effects of polyamine structure on rosin sizing under neutral papermaking conditions, *J. Appl. Polym. Sci.*, **65**(11), 1997, 2159–2163.
53. Wu Z.H., Chen S.P., Functions of polyamines in rosin sizing under neutral papermaking conditions, *Chin. J. Polym. Sci.*, **19**(2), 2001, 217–221.
54. Hartong B., Deng Y., Evidence of ester bond contribution to neutral to alkaline rosin sizing using polyethyleneimine–epichlorohydrin as a mordant, *J. Pulp Paper Sci.*, **30**(7), 2004, 203–209.
55. Xu Y., Hartong B., Deng Y., Neutral to alkaline rosin sizing using polyethyleneimine-epichlorohydrin (PEI-epi) as a mordant, *J. Pulp Pap. Sci.*, **28**(2), 2002, 39–44.
56. Zhuang J.F., Biermann C.J., Neutral to alkaline rosin soap sizing with metal-ions and polyethylenimine as mordants, *Tappi J.*, **78**(4), 1995, 155–162.
57. Biermann C.J., Rosin sizing with polyamine mordants from pH 3 to 10, *Tappi J.*, **75**(5), 1992, 166–171.
58. Li H.J., Ni Y.H., Sain M., Further understanding on polyethyleneimine-induced rosin ester sizing, *Nord. Pulp Pap. Res. J.*, **18**(1), 2003, 5–9.
59. Wang F., Kitaoka T., Tanaka H., Supramolecular structure and sizing performance of rosin-based emulsion size microparticles, *Coll. Surf. A Physicochem. Eng. Aspects*, **221**(1–3), 2003, 19–28.
60. Ito K., Isogai A., Onabe F., Rosin–ester sizing for alkaline papermaking, *J. Pulp Pap. Sci.*, **25**(6), 1999, 222–226.
61. Ito K., Isogai A., Onabe F., Retention behavior of size and aluminum components on handsheets in rosin–ester size alum systems, *J. Wood Sci.*, **45**(1), 1999, 46–52.
62. Kitaoka T., Isogai A., Onabe F., Endo T., Sizing mechanism of rosin emulsion size-alum systems. Part 4. Surface sizing by rosin emulsion size on alum-treated base paper, *Nord. Pulp Pap. Res. J.*, **16**(2), 2001, 96–102.
63. Wang F., Tanaka H., Mechanisms of neutral-alkaline paper sizing with usual rosin size using alum-polymer dual retention aid system, *J. Pulp Pap. Sci.*, **27**(1), 2001, 8–13.
64. Wang F., Wu Z.H., Tanaka H., Preparation and sizing mechanisms of neutral rosin size II: Functions of rosin derivatives on sizing efficiency, *J. Wood Sci.*, **45**(6), 1999, 475–480.
65. Wang T., Simonsen J., Biermann C.J., A new sizing agent: Styrene-maleic anhydride copolymer with alum or iron mordants, *Tappi J.*, **80**(1), 1997, 277–282.
66. Matsushita Y., Iwatsuki A., Yasuda S., Application of cationic polymer prepared from sulfuric acid lignin as a retention aid for usual rosin sizes to neutral papermaking, *J. Wood Sci.*, **50**(6), 2004, 540–544.
67. Wang F., Kitaoka T., Tanaka H., Vinylformamide-based cationic polymers as retention aids in alkaline papermaking, *Tappi J.*, **2**(12), 2003, 21–26.
68. Davis C.B., Rosin soaps as polymerization emulsifiers, in *Naval Stores. Production, Chemistry, Utilization* (Eds. Zinkel D.F. and Russel J.), Pulp Chemical Association, New York, 1989, pp. 625–644.
69. Carr C.W., Kolthoff I.M., Meehan E.J., Williams D.E., Studies on the rate of the emulsion polymerization of butadiene–styrene (75–25) as a function of the amount and kind of emulsifier used. 2. Polymerizations with fatty acid soaps, rosin soaps, and various synthetic emulsifiers, *J. Polym. Sci.*, **5**(2), 1950, 201–206.
70. Mayer M.J.J., Meuldijk J., Thoenes D., Emulsion polymerization of styrene with disproportionated rosin acid soap as emulsifier, *J. Appl. Polym. Sci.*, **59**(6), 1996, 1047.
71. Cochet F., Claverie J.P., Graillat C., Sauterey F., Guyot A., Emulsion polymerization of chloroprene in the presence of a maleic polymerizable surfactant: Control of gel formation at low conversion, *Macromolecules*, **37**(11), 2004, 4081–4086.
72. Zucolotto M., Darocha E.C., Vonholleben M.L.A., Disproportionation of rosin from pinus-elliotti – Obtention of emulsifier for polymerization, *Quim. Nova*, **18**(1), 1995, 78–79.
73. Kennedy N.L., Krajca K.E., Russel J., Rosin in adhesive tackifiers, in *Naval Stores. Production, Chemistry, Utilization* (Eds. Zinkel D.F. and Russel J.), Pulp Chemical Association, New York, 1989, pp. 645–664.
74. Fattah H., Adhesives and sealants – Rosin resin makers fight back, *Chem. Week*, **158**(45), 1996, 45.
75. Comyn J., Surface characterization of pentaerythritol rosin ester, *Int. J. Adhes. Adhes.*, **15**(1), 1995, 9–14.
76. Shih H.H., Hamed G.R., Peel adhesion and viscoelasticity of poly(ethylene-co-vinyl acetate)-based hot melt adhesives. 1. The effect of tackifier compatibility, *J. Appl. Polym. Sci.*, **63**(3), 1997, 323–331.

77. Takemoto M., Karasawa T., Mizumachi H., Kajiyama M., Miscibility between ethylene vinyl acetate copolymers and tackifier resins, *J. Adhesion*, **72**(1), 2000, 85–96.
78. Barrueso-Martinez M.L., Ferrandiz-Gomez T.D., Romero-Sanchez M.D., Martin-Martinez J.M., Characterization of EVA-based adhesives containing different amounts of rosin ester or polyterpene tackifier, *J. Adhesion*, **79**(8–9), 2003, 805–824.
79. Fujita M., Kajiyama M., Takemura A., Ono H., Mizumachi H., Hayashi S., Effects of miscibility on probe tack of natural-rubber-based pressure-sensitive adhesives, *J. Appl. Polym. Sci.*, **70**(4), 1998, 771–776.
80. Kim D.J., Kim H.J., Yoon G.H., Effects of blending and coating methods on the performance of SIS (styrene-isoprene-styrene)-based pressure-sensitive adhesives, *J. Adhes. Sci. Technol.*, **18**(15–16), 2004, 1783–1797.
81. Yang H., Sa U., Kang M., Ryu H.S., Ryu C.Y., Cho K., Near-surface morphology effect on tack behavior of poly(styrene-b-butadiene-b-styrene) triblock copolymer/rosin films, *Polymer*, **47**(11), 2006, 3889–3895.
82. Aran-Ais F., Torro-Pala A.M., Orgiles-Barcelo A.C., Martin-Martinez J.M., Addition of rosin acid during thermoplastic polyurethane synthesis to improve its immediate adhesion to PVC, *Int. J. Adhes. Adhes.*, **25**(1), 2005, 31–38.
83. Aran-Ais F., Torro-Palau A.M., Orgiles-Barcelo A.C., Martin-Martinez J.M., Characterization of thermoplastic polyurethane adhesives with different hard/soft segment ratios containing rosin as an internal tackifier, *J. Adhes. Sci. Technol.*, **16**(11), 2002, 1431–1448.
84. Hayashi S., Kim H.J., Kajiyama M., Ono H., Mizumachi H., Zufu Z., Miscibility and pressure-sensitive adhesive performances of acrylic copolymer and hydrogenated rosin systems, *J. Appl. Polym. Sci.*, **71**(4), 1999, 651–663.
85. Kim B.J., Kim S.E., Do H.S., Kim S., Kim H.J., Probe tack of tackified acrylic emulsion PSAs, *Int. J. Adhes. Adhes.*, **27**(2), 2007, 102–107.
86. Downs A.M.R., Sansom J.E., Colophony allergy: A review, *Contact Dermatitis*, **41**(6), 1999, 305–310.
87. Schuller W.H., Lawrence R.V., Culberts B.M., Some polyimide-amides from maleopimaric acid, *J. Polym. Sci. A1 Pol. Chem.*, **5**(8PA1), 1967, 2204–2207.
88. Maiti M., Maiti S., Polymers from renewable resources. 5. Synthesis and properties of a polyamideimide from rosin, *J. Macromol. Sci. Chem.*, **A20**(1), 1983, 109–121.
89. Ray S.S., Kundu A.K., Maiti M., Ghosh M., Maiti S., Polymers from renewable resources. 7. Synthesis and properties of polyamideimide from rosin–maleic anhydride adduct, *Angew. Makromol. Chem.*, **122**(JUN), 1984, 153–167.
90. Kundu A.K., Adhikari B., Ray S.S., Maiti M.M., Maiti S., Compatibility of polyblends of polyamideimide from rosin and novolac, *Abstr. Pap. Am. Chem. Soc.*, **188**(AUG), 1984, 49–50
91. Ray S.S., Kundu A.K., Ghosh M., Maiti S., A new route to synthesize polyamideimide from rosin, *Eur. Polym. J.*, **21**(2), 1985, 131–133.
92. Ray S.S., Kundu A.K., Maiti S., Polymers from renewable resources. 9. Synthesis and properties of a polyamideimide from rosin, *J. Macromol. Sci. Chem.*, **A23**(2), 1986, 271–283.
93. Kundu A.K., Ray S.S., Adhikari B., Maiti S., Polymer blends. 2. Compatibility and thermal-behavior of blends of novolac and polyamideimide from rosin, *Eur. Polym. J.*, **22**(5), 1986, 369–372.
94. Kundu A.K., Ray S.S., Maiti S., Polymer blends. 4. Blends of resol with polyamideimide synthesized from rosin, *Eur. Polym. J.*, **22**(10), 1986, 821–825.
95. Kundu A.K., Ray S.S., Mukherjea R.N., Maiti S., Polymer blends. 3. Preparation and evaluation of the blends of shellac with polyamideimide from rosin, *Angew. Makromol. Chem.*, **143**, 1986, 197–207.
96. Ray S.S., Kundu A.K., Maiti S., Polymers from renewable resources. 12. Structure property relation in polyamideimides from rosin, *J. Appl. Polym. Sci.*, **36**(6), 1988, 1283–1293.
97. Bicu I., Mustata F., Water soluble polymers from Diels–Alder adducts of abietic acid as paper additives, *Macromol. Mater. Eng.*, **280**(7–8), 2000, 47–53.
98. Mustata F., Bicu I., Polyhydroxyimides from resinic acids, *Polimery*, **45**(4), 2000, 258–263.
99. Kim S.J., Kim B.J., Jang D.W., Kim S.H., Park S.Y., Lee J.H., Lee S.D., Choi D.H., Photoactive polyamideimides synthesized by the polycondensation of azo-dye diamines and rosin derivative, *J. Appl. Polym. Sci.*, **79**(4), 2001, 687–695.
100. Roy S.S., Kundu A.K., Maiti S., Polymers from renewable resources. 13. Polymers from rosin acrylic-acid adduct, *Eur. Polym. J.*, **26**(4), 1990, 471–474.
101. Bingham J.F., Marvel C.S., Preparation and polymerization of a diisocyanate from Diels–Alder adduct of levopimaric acid and acrylic acid, *J. Polym. Sci. A1 Pol. Chem.*, **10**(3), 1972, 921–929.
102. Krivoguz Y.M., Yuvchenko A.P., Pesetskii S.S., Bei M.P., Functionalization of polyethylene with peroxy-containing formulations derived from resins modified with maleic anhydride. Properties of blends of functionalized polyethylene with polyamide 6, *Russ. J. Appl. Chem.*, **77**(6), 2004, 976–982.
103. Atta A.M., El-Saeed S.M., Farag R.K., New vinyl ester resins based on rosin for coating applications, *React. Funct. Polym.*, **66**(12), 2006, 1596–1608.
104. Atta A.M., El-Saeed S.M., Farag R.K., Synthesis of polyunsatuated polyester resins based on rosin acrylic acid adduct for coating applications, *React. Funct. Polym.*, **67**(9), 2007, 459–563.

105. Atta A.M., Nassar I.F., Bedawy H.M., Unsaturated polyester resins based on rosin maleic anhydride adduct as corrosion protections of steel, *React. Funct. Polym.*, **67**(7), 2007, 617–626.
106. Bicu I., Mustata F., Diels–Alder polymerization of some derivatives of abietic acid, *Angew. Makromol. Chem.*, **264**, 1999, 21–29.
107. Atta A.M., Mansour R., Abdou M.I., Sayed A.M., Epoxy resins from rosin acids: Synthesis and characterization, *Polym. Advan. Technol.*, **15**(9), 2004, 514–522.
108. Bicu A., Mustata F., Polymers from a levopimaric acid-acrylonitrile Diels–Alder adduct: Synthesis and characterization, *J. Polym. Sci. Polym. Chem.*, **43**(24), 2005, 6308–6322.
109. Atta A.M., Mansour R., Abdou M.I., El-Sayed A.M., Synthesis and characterization of tetra-functional epoxy resins from rosin, *J. Polym. Res.*, **12**(2), 2005, 127–138.
110. Kim W.S., Jang H.S., Hong K.H., Seo K.H., Synthesis and photocrosslinking of poly(vinylbenzyl abietate), *Macromol. Rapid Commun.*, **22**(11), 2001, 825–828.
111. Sinclair R.G., Berry D.A., Schuller W.H., Lawrence R.V., Polycondensation of resin-acid dimers with diamines, *J. Polym. Sci. A1 Pol. Chem.*, **9**(3), 1971, 801–804.
112. Kuusipalo J., PHB/V in extrusion coating of paper and paperboard: Part I: Study of functional properties, *J. Polym. Environ.*, **8**(1), 2000, 39–47.
113. Kuusipalo J., PHB/V in extrusion coating of paper and paperboard – Study of functional properties. Part II, *J. Polym. Environ.*, **8**(2), 2000, 49–57.
114. Jipa S., Zaharescu T., Setnescu R., Setnescu T., Brites M.J.S., Silva A.M.G., Marcelo-Curto M.J., Gigante B., Chemiluminescence study of thermal and photostability of polyethylene, *Polym. Int.*, **48**(5), 1999, 414–420.
115. Zaharescu T., Jipa S., Setnescu R., Wurm D., Brites M.J.S., Esteves M.A.F., Marcelo-Curto M.J., Gigante B., Effects of some secondary amines on the oxidation of ethylene–propylene elastomers, *Polym. Degrad. Stabil.*, **68**(1), 2000, 83–86.
116. Zaharescu T., Jipa S., Setnescu R., Brites J., Esteves M.A., Gigante B., Synergistic effects on thermal stability of ethylene–propylene elastomers stabilized with hindered phenols and secondary amines, *Polym. Test.*, **21**(2), 2002, 149–153.
117. Leach R.H., Pierce R.S., *The printing ink manual*, Dordrecht, Kluwer Academic Publishers, 1993.
118. Pekarovicova A., Bhide H., Fleming P.D., Pekarovic J., Phase-change inks, *J. Coating Technol.*, **75**(936), 2003, 65–72.
119. Ramani C.C., Puranik P.K., Dorle A.K., Study of diabietic acid as matrix forming material, *Int. J. Pharm.*, **137**(1), 1996, 11–19.
120. Pathak Y.V., Dorle A.K., Evaluation of pentaerythritol (rosin) estergum as coating materials, *Drug Dev. Ind. Pharm.*, **12**(11–13), 1986, 2217–2229.
121. Pathak Y.V., Dorle A.K., Evaluation of propylene–glycol rosin ester as microencapsulating material and study of dissolution kinetics, *Drug Dev. Ind. Pharm.*, **14**(15–17), 1988, 2557–2565.
122. Pathak Y.V., Nikore R.L., Dorle A.K., Study of rosin and rosin esters as coating materials, *Int. J. Pharm.*, **24**(2–3), 1985, 351–354.
123. Satturwar P.M., Fulzele S.V., Joshi S.B., Dorle A.K., Evaluation of the film-forming property of hydrogenated rosin, *Drug Dev. Ind. Pharm.*, **29**(8), 2003, 877–884.
124. Fulzele S.V., Satturwar P.M., Dorle A.K., Novel biopolymers as implant matrix for the delivery of ciprofloxacin: Biocompatibility, degradation, and *in vitro* antibiotic release, *J. Pharm. Sci.*, **96**(1), 2007, 132–144.
125. Fulzele S.V., Satturwar P.M., Dorle A.K., Study of the biodegradation and *in vivo* biocompatibility of novel biomaterials, *Eur. J. Pharma. Sci.*, **20**(1), 2003, 53–61.
126. Fulzele S.V., Satturwar P.M., Dorle A.K., Polymerized rosin: Novel film forming polymer for drug delivery, *Int. J. Pharm.*, **249**(1–2), 2002, 175–184.
127. Lee C.M., Lim S., Kim G.Y., Kim D., Kim D.W., Lee H.C., Lee K.Y., Rosin microparticles as drug carriers: Influence of various solvents on the formation of particles and sustained-release of indomethacin, *Biotechnol. Bioprocess Eng.*, **9**(6), 2004, 476–481.
128. Choudhary G., Kumar J., Walia S., Parsad R., Parmar B.S., Development of controlled release formulations of carbofuran and evaluation of their efficacy against Meloidogyne incognita, *J. Agric. Food Chem.*, **54**(13), 2006, 4727–4733.
129. Dhanorkar V.T., Gogte B.B., Dorle A.K., Formation and stability studies of multiple (w/o/w) emulsions prepared with newly synthesized rosin-based polymeric surfactants, *Drug Dev. Ind. Pharm.*, **27**(6), 2001, 591–598.
130. Dhanorkar V.T., Gawande R.S., Gogte B.B., Dorle A.K., Development and characterization of rosin-based polymer and its application as a cream base, *J. Cosmetic Sci.*, **53**(4), 2002, 199–208.
131. Snow R., Zhang J.P., Miller J.D., Froth modification for reduced fuel oil usage in phosphate flotation, *Int. J. Miner. Process.*, **74**(1–4), 2004, 91–99.
132. Gopal R., Polymer concrete composites for enhancement of mobility of troops in desert operations, *Mater. Sci. Eng. B Solid St. Mater. Adv. Technol.*, **132**(1–2), 2006, 129–133.

– 5 –

Sugars as Monomers

J.A. Galbis and M.G. García-Martín

ABSTRACT

Sustained efforts have been extensively devoted to prepare new polymers based on renewable resources and with higher degradability. Of the different natural sources, carbohydrates stand out as highly convenient raw materials because they are inexpensive, readily available, and provide great stereochemical diversity. This chapter describes the potential of sugar-based monomers as precursors to a wide variety of macromolecular materials, with particular emphasis on both the mechanisms of polymerization and the properties of the ensuing products.

Keywords

Sugar-based monomers, Carbohydrate monomers, Bifunctional sugar derivatives, Alditols, Aminosugars, Aldonic acids, Aldaric acids

5.1 INTRODUCTION

Biodegradable polymers obtained from renewable natural sources are currently receiving increasing attention because they are an alternative to the traditional petroleum-based plastics. Solid-waste management of plastics of agricultural or alimentary origin is demanding a rapid development of new biodegradable plastics. Another important application is in biomedicine, and in this field, biocompatibility is a required feature for resorbable biomedical devices.

Carbohydrates are an important natural source of building blocks for the synthesis of biodegradable polymers, especially for biomedical applications, because of their inherent properties of biocompatibility and biodegradability, and for the design of optically active polymers containing stereocentres in the repeating unit. There are various reasons for the interest in carbohydrates as polymer building blocks namely, (a) they are easily available, some even coming from agricultural wastes; (b) they are found in a very rich variety of chemical structures with great stereochemical diversity; and, above all, (c) they constitute a renewable source thanks to solar energy. Unfortunately, polycondensation of monomers derived from sugars is not straightforward. In the first place, their multifunctionality must usually be reduced, making use of appropriate protecting groups to prevent side reactions leading to undesirable products. Secondly, in order to obtain regio and stereoregular polymers, strict control of the stereochemical course of the polymerization is required. Otherwise, random orientation of the chiral unit will lead to atactic polymers.

Some reviews have recently been published on the synthetic carbohydrate-based polymers and glycopolymers [1–6]. However, they refer mainly to poly(vinylsaccharide)s and other conventional functionalized polymers having sugars as groups pendant from the main chain of the polymer. In this chapter, we shall describe those sugar-based monomers which lead to polymers having the sugar units incorporated into the main chain. The topic has also been excellently reviewed [7–10], but as the interest in this kind of carbohydrate-based polymer has been steadily increasing, a considerable number of papers have been published on the subject during the last few years.

The following sections report on the synthesis and polymerization of this type of sugar-derived monomers that have been published mainly during the past decade.

5.2 ALDITOLS

Acyclic alditols are sometimes referred to as glycitols or polyols. Polyols occur extensively in nature, but aldoses and ketoses can be reduced to alditols with the generation of new hydroxyl groups from the carbonyl functions; thus, D-glucose gives D-glucitol, trivially referred to as 'sorbitol'. The reduction of ketoses generates a new stereogenic centre, and gives two epimeric alditols; for example, D-fructose produces D-mannitol and D-glucitol.

5.2.1 Anhydroalditols

The thermally stable 1,4:3,6-dianhydrohexitols are the carbohydrate-based monomers most widely employed in the synthesis of chiral polymers because they are easily available. The three stereoisomeric dianhydroalditol monomers, having 1,4:3,6-dianhydro-D-*gluco*, -D-*manno*, and -L-*ido* configurations, have attracted increasing interest as building blocks for the synthesis of polycondensates, as reported by Kricheldorf [8]. These dianhydroalditols are referred to as isosorbide (DAS or DAG), isomannide (DAM), and isoidide (DAI). DAS is commercially produced in large quantities from starch by hydrolysis, subsequent reduction, and acid-catalyzed intramolecular dehydration. DAM is also readily available from D-mannose, by reduction and dehydration. Since L-idose is a rare sugar, and its synthesis from D-glucose requires several steps, DAI is obtained from DAS by a three-step procedure. DAS and DAM are currently commercially available. The DAS-based polymers are non-stereoregular, in contrast to those based on DAM and DAI, which are stereoregular.

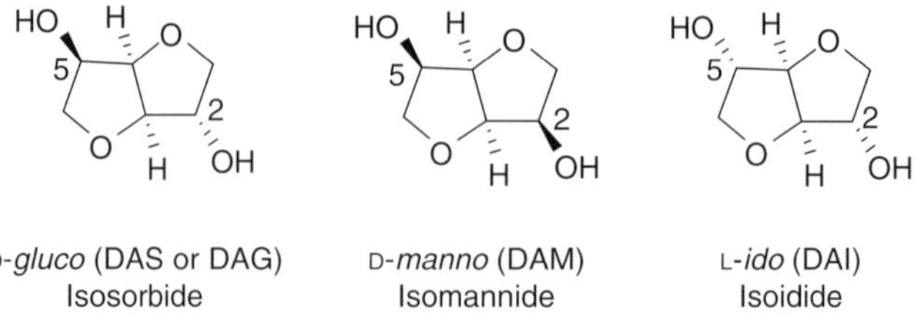

D-*gluco* (DAS or DAG) Isosorbide — D-*manno* (DAM) Isomannide — L-*ido* (DAI) Isoidide

The thermostability of these dianhydroalditol monomers may be higher than that of other aliphatic diols, which raises the T_g of the corresponding polymers. The most representative dianhydroalditol-based polymers that have been synthesized are polyesters, polycarbonates, polyethers, poly(ester carbonate)s, poly(ester anhydride)s, polyurethanes, polyureas, and poly(ester amide)s.

Thiem *et al.* prepared low-molecular weight polyterephthalates [11, 12] by melting condensation of terephthalic acid dichloride and 1,4:3,6-dianhydrohexitols. DAM turned out to be too labile under these conditions, giving cross-linked products. Storbeck *et al.* reinvestigated [13] the synthesis and characterization of these polyterephthalates; the reaction conditions were optimized, and pyridine was found to be the most useful acceptor of hydrogen chloride.

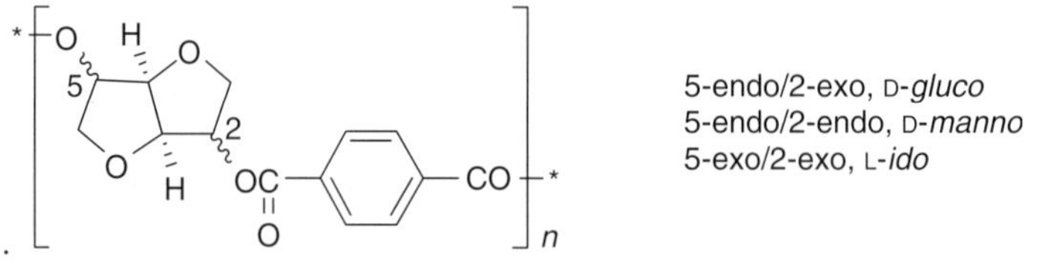

5-endo/2-exo, D-*gluco*
5-endo/2-endo, D-*manno*
5-exo/2-exo, L-*ido*

Poly(oxytetramethylene) is extensively used as a softening segment for polyurethane production. Thiem *et al.* studied, as an alternative, the ring-opening polymerization of 1,4-anhydro-2,3-di-*O*-alkyl-D-erythritols [14–16] using super-acids $HFSO_3$ and HCF_3SO_3, which led to novel functionalized poly(2,3-dialkoxytetramethylene)s. As previously demonstrated for ring-opening polymerization of carbohydrate acetals, as well as ethers, enhanced amounts of catalyst were required, in contrast to the case of non-carbohydrate starting materials.

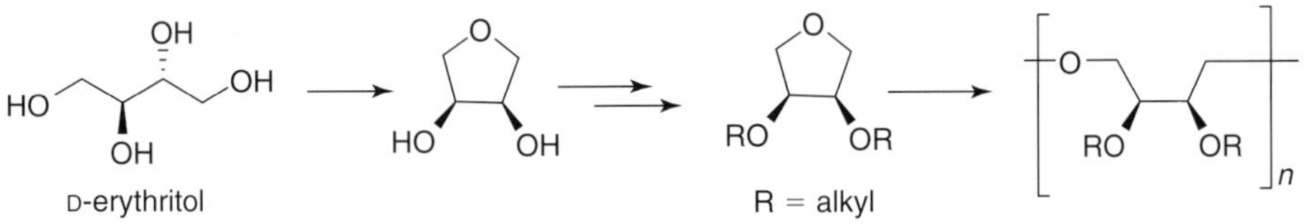

D-erythritol

R = alkyl

Base-catalyzed intramolecular-nucleophilic substitution of various DAS derivatives led to the formation of 1,4:2,5:3,6-trianhydro-D-mannitol, a tricyclic system of three interlinked oxolane rings, which afforded oligomeric soluble materials by ring-opening polymerization after treatment with CF_3SO_3H [17]. Ring-opening copolymerization of 1,4-anhydro-2,3-di-*O*-ethyl-D-erythritol or 1,4:2,5:3,6-trianhydro-D-mannitol with tetrahydrofuran (THF), using CF_3SO_3H as catalyst, afforded a series of copolyether polyols [18]. The polymers had medium molecular weights, with approximately 10 per cent of carbohydrate constituent.

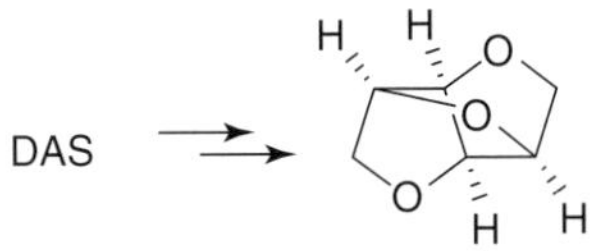

DAS

Braun and Bergmann [19, 20] used DAS and DAM for the syntheses of polyesters, polyethers, polyurethanes, and polycarbonates. They reported that the thermal and mechanical properties of the obtained polymers were similar to those of the respective counterparts obtained from usual petrochemical raw materials.

Okada *et al.* have worked extensively on biodegradable polymers based on dianhydroalditols as carbohydrate renewable resources. A series of polyesters were synthesized by the bulk polycondensation of the respective three stereoisomeric dianhydroalditols (DAS, DAM, and DAI) with aliphatic dicarboxylic acid dichlorides of 2–10 methylene groups [21, 22]. It was found that the biodegradability of the polyesters varied significantly depending on their molecular structures.

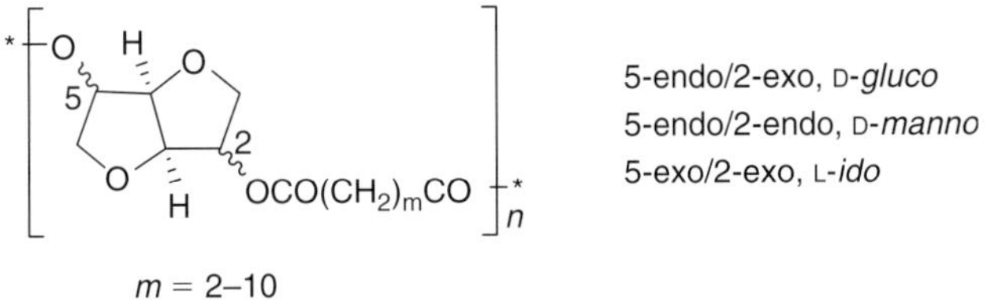

$OCO(CH_2)_mCO$

5-endo/2-exo, D-*gluco*
5-endo/2-endo, D-*manno*
5-exo/2-exo, L-*ido*

$m = 2–10$

DAS or DAM was also used as comonomer with dimethyl dialkanoates and monomers containing furan rings, for the synthesis of copolyesters by bulk polycondensations, in the presence of titanium isopropoxide or tetrabutyl-1,3-dichloro-distannoxane [23, 24]. In general, the biodegradability of the copolyesters of DAS decreased with increasing difuran dicarboxylate content, and copolymers containing sebacic acid units showed higher biodegradability. The enzymatic degradability of the polyesters based on isomeric 1,4:3,6-dianhydrohexitols and sebacic acid was found to decrease in the order DAS > DAM > DAI [25].

The synthesis and biodegradation behaviour of poly(ester amide)s composed of DAS, α-amino acid, and aliphatic dicarboxylic acid units were also described by Okada [26]. These products were synthesized by solution polycondensation of various combinations of *p*-toluenesulphonic acid salts of *O,O′*-bis-(α-aminoacyl)-1,4:3,6-dianhydro-D-glucitol and bis(*p*-nitrophenyl) esters of aliphatic dicarboxylic acids with methylene chain lengths of 4–10. The *p*-toluenesulphonic acid salts were obtained by the reactions of DAS with alanine, glycine, and glycylglycine, respectively, in the presence of *p*-toluenesulphonic acid. The poly(ester amide)s were, in general, degraded more slowly than the corresponding polyesters having the same aliphatic dicarboxylic acid units, both in composted soil and in an activated sludge.

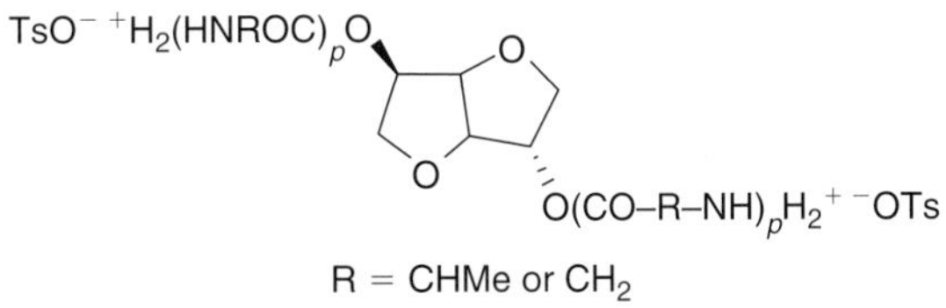

$TsO^- \ ^+H_2(HNROC)_pO$

$O(CO–R–NH)_pH_2^+ \ ^-OTs$

R = CHMe or CH_2

Poly(ester carbonate)s with different compositions were synthesized by bulk polycondensation of DAS with diphenyl sebacate and diphenyl carbonate in the presence of zinc acetate as a catalyst [27]. Biodegradability was found to be highest for the poly(ester carbonate)s with carbonate contents of 10–20 mol per cent, and to decrease markedly for those with the carbonate content above 50 mol per cent.

Okada published in 2002 a couple of reviews on biodegradable polymers from DAS and DAM and their related compounds [28] and chemical synthesis [29].

Novel random and alternating copolycarbonates from 1,4:3,6-dianhydrohexitols and aliphatic diols were synthesized in bulk or solution polycondensations [30]. Bulk polycondensations of 1,4:3,6-dianhydro-2,5-bis-*O*-(phenoxycarbonyl)-D-glucitol and -D-mannitol with α,ω-alkanediols of different number of methylene groups, afforded random copolycarbonates. However, solution polycondensations of the corresponding *p*-(nitrophenoxycarbonyl) monomers in solution gave well-defined copolycarbonates having regular structures consisting of alternating sugar carbonate and aliphatic carbonate moieties.

x = 4, 6, 8,10

Random copolycarbonates

Alternating copolycarbonates

The environmental and enzymatic degradability of copolycarbonates consisting of DAS and DAM and alkylene diols or oligo(ethylene glycols) was investigated [31].

Tartaric acid has also attracted a great deal of interest as a substrate for the synthesis of functional polymers based on carbohydrates; for example, L-tartaric acid is a natural product mainly obtained from a large variety of fruits. Okada has recently accomplished the study of novel polycarbonates with pendant functional groups, based on 1,4:3,6-dianhydrohexitols and L-tartaric acid derivatives [32]. Solution polycondensations of 1,4:3,6-dianhydro-bis-*O*-(*p*-nitrophenoxycarbonyl)hexitols and 2,3-di-*O*-methyl-L-erythritol or 2,3-*O*-isopropylidene-L-erythritol afforded polycarbonates having pendant methoxy or isopropylidene groups, respectively. Subsequent acid-catalyzed deprotection of the isopropylidene groups gave well-defined polycarbonates having pendant hydroxyl groups regularly distributed along the polymer chain. The degradation of these polymers was remarkably fast.

EDPA, DMAP

Sulpholane, 60°C, 24 h

Ar = *p*-nitrophenyl

R = Me
R = Isopropylidene

2-exo, D-*gluco*
2-endo D-*manno*

The 1,4:3,6-dianhydroalditols (mainly) and other saccharide derivatives were found to be useful chiral components of cholesteric materials with interesting optical properties, capable of forming Grandjean textures [8]. Chirality plays an important role in combination with the liquid crystalline (LC) character of both low- or high-molecular weight materials. A systematic study reported by Vill *et al.* [33] on the twisting power of DAS, DAM, and DAI derivatives contributed to increasing the interest in these monomers as building blocks of cholesteric polymers.

Appropriate combinations of monomers should be sought that allow the incorporation of the sugar component without loss of the LC character, because the stereochemistry of the sugar diols is highly unfavourable for an LC character of their polymers – in fact, all homopolymers derived from sugar diols are isotropic. Kricheldorf studied LC cholesteric copoly(ester-imide)s based on DAS or DAM [8]. The comonomers to obtain these chiral thermotropic polymers were *N*-(4-carboxyphenyl)trimellitimide, 4-aminobenzoic trimellitimide, 4-aminocinnamic acid trimellitimide, adipic acid, 1,6-hexanediol, and 1,6-bis(4-carboxyphenoxyl)hexane. Apparently, the poly(ester-imide) chains are so stiff that the twisting power of the sugar diol has little effect.

Cholesteric polyesters were prepared from silylated derivatives of 2,3-di-*O*-isopropylidene-D-threitol, DAS, or DAM with dicarboxylic acid dichlorides by polycondensation in solution [34]. Trifluoroacetic acid–water allowed an easy cleavage of the isopropylidene group without hydrolysis of the polyester. All these polyesters formed a broad cholesteric phase, and the polymers containing 5 or 10 mol per cent sugar diol displayed a blue Grandjean texture.

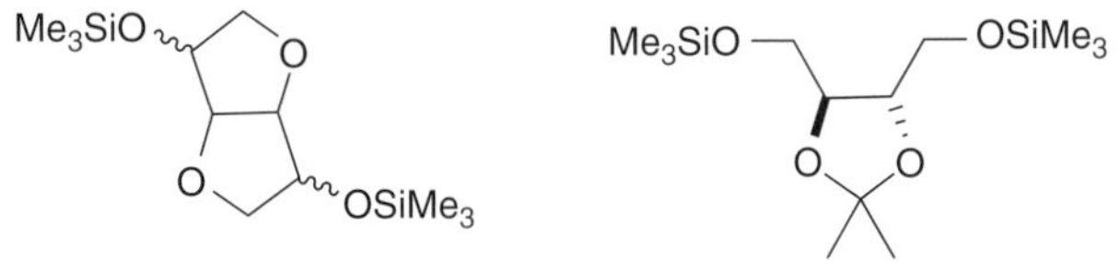

Me_3SiO $OSiMe_3$ Me_3SiO $OSiMe_3$

5-endo/2-exo, D-*gluco*
5-endo/2-endo, D-*manno*

Photosetting cholesteric polyesters derived from DAS and 4-hydroxycinnamic acid [35], and from 2,5-bis(dode cyloxy)terephthalic acid and 4,4′-dihydroxybiphenyl, were also described [36].

Cyclic aliphatic polyesters of DAS were prepared by polycondensation with aliphatic dicarboxylic acid dichlorides with pyridine as catalyst and HCl acceptor [37]. Cyclic polyesters were isolated by a selective extraction of the linear polymers.

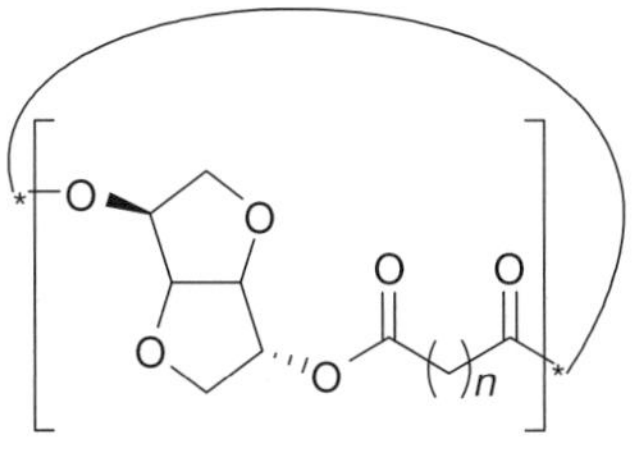

$n = 4, 6, 8, 10$

Recently, Kricheldorf has published the synthesis of copolyesters by ring-opening copolymerization of DAS, ϵ-caprolactone, and suberic acid in the presence of catalysts [38].

DAS- and DAM-homopolycarbonates have been prepared by different authors [20, 39]. The interfacial polycondensation, typical for the synthesis of aromatic polycarbonates, is not useful with alditols, including DAS, because they are water-soluble and less acidic than diphenols. The DAS-homopolycarbonate was prepared by phosgenation of the sugar diol, with phosgene or diphosgene in pyridine-containing solvent mixtures at low temperatures. The polycondensation of the DAS-bischloroformate in pyridine is an alternative approach.

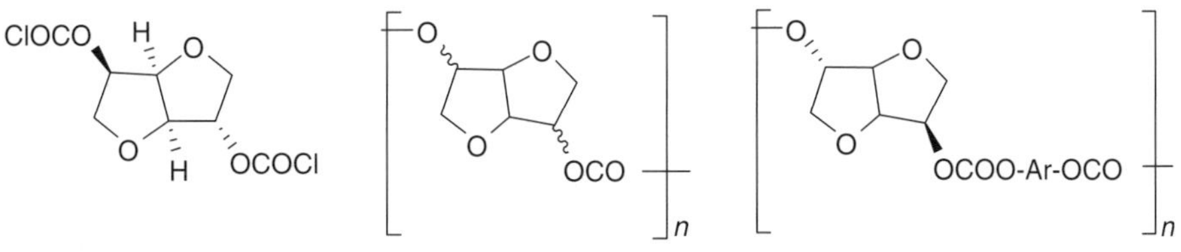

DAS-bischloroformate

Ar = Phenols or bisphenols

Cholesteric polycarbonates derived from DAS bis(phenyl carbonate), methyl hydroquinone, and 4,4′-dihydrobiphenyl were better obtained by polycondensation in pyridine-containing organic solvents, as reported by Kricheldorf *et al.* [39]. Trichloromethylchloroformate (TCF) was successfully used for all these polycondensations. The use of solid TCF is advantageous over phosgene because it is easy to apply in stoichiometric amounts, easy to store, and commercially available.

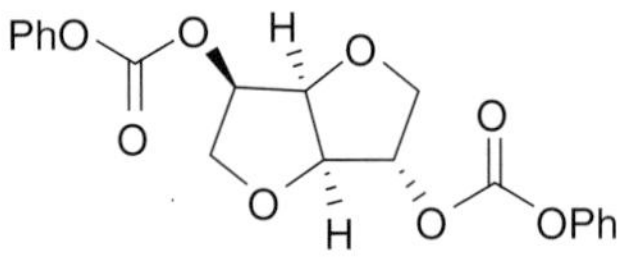

DAS bis(phenylcarbonate)

The same group also synthesized a series of ternary polycarbonates derived from DAS-chloroformate, hydroquinone 4-hydroxybenzoate (HQHB) as mesogenic diphenol, and 4,4′-dihydroxychalcone in pyridine [40]. Particularly noteworthy is the finding that the alternating copolycarbonate of HQHB formed a broad cholesteric phase, despite the unfavourable stereochemistry of DAS. The ternary copolycarbonates formed a cholesteric melt and a Grandjean texture upon shearing. It seems that the combination of DAS and carbonate groups is more favourable for the stabilization of a cholesteric molecular order than is the combination of DAS and ester groups.

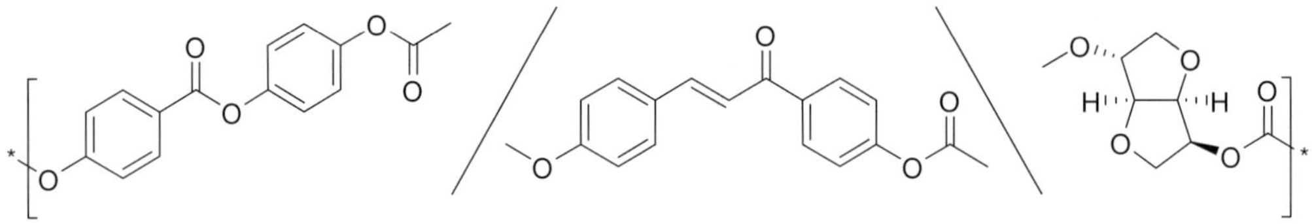

Cholesteric polycarbonates derived from DAS and 2,5-bis(4′-hydroxybenzylidene)cyclopentanone have also been described [41].

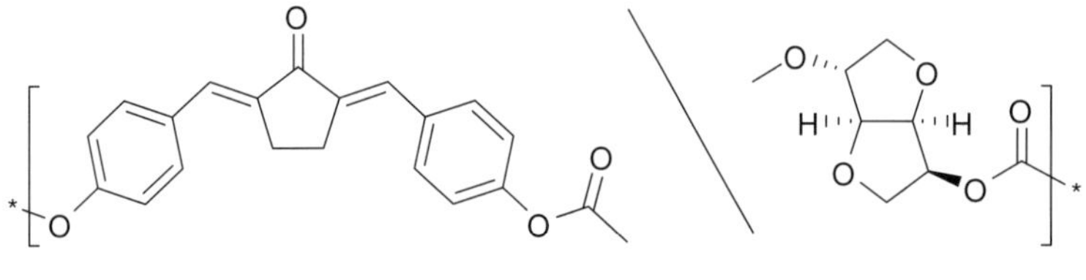

Recently, DAS and equimolar amounts of various diols were polycondensed with diphosgene in pyridine. Different bisphenols, 1,3-bis(4-hydroxybenzyloxy) propane, and 1,4-cyclohexanediol were used as comonomers [42]. In some cases, large amounts of cyclic oligo- and polycarbonates were formed.

Polyurethanes containing DAS have been prepared by several research groups, and complex polyurethanes with an elastomeric character and good mechanical properties were described in a few patents [8]. These polymers were obtained from DAS and diisocyanates in the presence of suitable catalysts (*e.g.* Braun and Bergmann used triethylamine in dimethylsulphoxide [20]).

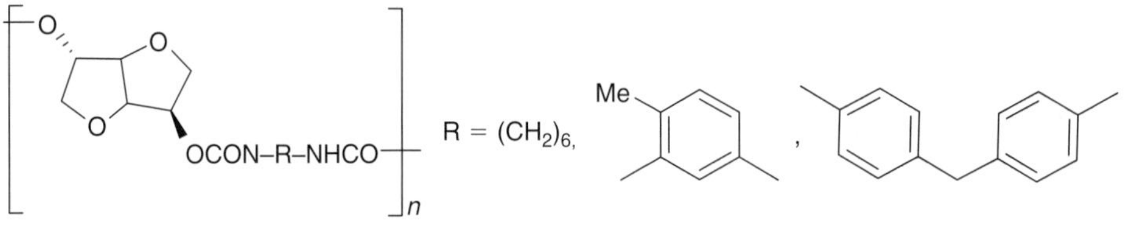

R = $(CH_2)_6$,

Polyethers derived from DAS were mentioned in a patent describing the synthesis and properties of epoxy resins, and linear polyethers were prepared from free DAS, which was reacted with α,ω-bischloroalkanes or 1,6-dibromohexane [8].

—$CH_2CH_2OCH_2CH_2$—
—$(CH_2CH_2)_2CH_2CH_2$—
—$(CH_2)_6$—
—H_2C—C_6H_4—CH_2—

A series of polyphosphites, polyphosphates, polythiophosphates, and other polymers containing sulphone functions, based on DAS, have also been described [8, 43].

Recently, an efficient synthesis of polyethers from DAS and 1,8-dibromo or dimesyl octane by microwave-assisted phase transfer catalysis has been reported [44].

5.2.2 *O*-protected alditols

The most common *O*-protecting groups of the secondary hydroxyl groups of the alditol monomers are acetal, ester, and ether groups. The ether group is of course the most resistant *O*-protecting group of alditol monomers under the polycondensation reaction conditions, but also the most difficult to remove from the resulting polymers.

Within the framework of systematic research to explore the potential of sugar-based polymers, Muñoz-Guerra along with Galbis *et al.* have employed various *O*-methyl alditol derivatives to prepare novel polymers, as well as chemical modification of other well-known materials. Poly(ethylene terephthalate) (PET), poly(ethylene isophthalate) (PEI), and poly(butylene terephthalate) (PBT) have been chemically modified by insertion of a series of *O*-methyl alditols. Among them, 2,3-di-*O*-methyl-L-threitol was inserted into PET [45] and the preparation of PET and PEI analogues by total replacement of the ethylene glycol (EG) units with 2,3,4,5-tetra-*O*-methyl-hexitols having D-*manno* and *galacto* configurations [46]. They also obtained analogues to PET, PEI, and PBT by using 2,3,4-tri-*O*-methyl-L-arabinitol or 2,3,4-tri-*O*-methyl-xylitol [47, 48]. The *O*-methyl-alditol monomers used are easily obtained from the respective commercially available diethyl L-tartrate, pentoses, and hexitols [46, 49, 50].

L-threitol

Sugar	R^1	R^2
L-arabinose	OH	H
D-xylose	H	OH

Sugar	R^1	R^2
L-arabinitol (**Ar**)	OMe	H
Xylitol (**Xy**)	H	OMe

D-mannitol, R^1 = OH, R^2 = H
D-galactitol, R^1 = H, R^2 = OH

D-mannitol, R^1 = OMe, R^2 = H
D-galactitol, R^1 = H, R^2 = OMe

Galbis *et al.* have also described the use of 2,3,4-tri-*O*-methyl-L-arabinitol and 2,3,4-tri-*O*-methyl-xylitol in the synthesis of polycarbonates and polyesters. These pentitols were polycondensed using a commercial solution of phosgene in toluene; whereby, homopolycarbonates (PsuC) and copolycarbonates with BPA, P(Su-*co*-BPAC), were obtained in high yields [51]. Both showed high resistance to chemical hydrolysis; however, they were enzymatically degraded in different degrees. The fastest degradation promoted by lipase B from *Candida antarctica* was observed for the fully xylitol-based polycarbonate, followed by copolycarbonates also based on xylitol, which revealed a marked stereospecificity of the enzyme towards this sugar.

PSuC

P(Su-*co*-BPAC)

Su	R¹	R²
Ar	OMe	H
Xy	H	OMe

A variety of carbohydrate-based linear polyesters [52] of the poly(alkylene dicarboxylate) type were obtained by polycondensation reactions of the alditols 2,3,4-tri-*O*-methyl-L-arabinitol and 2,3,4-tri-*O*-methyl-xylitol, and aldaric acids 2,3,4-tri-*O*-methyl-L-arabinaric acid and 2,3,4-tri-*O*-methyl-xylaric acid. Butanediol and adipic acid were also used as comonomers. Copolyesters of the poly(alkylene-*co*-arylene dicarboxylate) type were obtained using bisphenols as comonomers. Chemical polycondensation reactions were conducted in bulk or in solution. Enzymatic polycondensation reactions of adipic acid with the above-mentioned alditols were carried out successfully using Lipozyme® and Novozyme® 435. The hydrolytic degradations of some of these polyesters were also described.

Aldaric acid	R¹	R²
2Ar, L-arabinaric	OMe	H
2Xy, Xylaric	H	OMe

L-arabino
Xylo

L-arabino-L-arabino
Xylo-Xylo

Butanediol
Bisphenol-A

L-arabino
Xylo

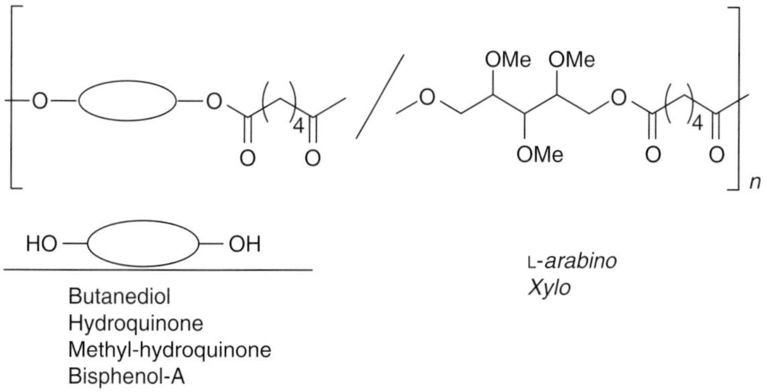

HO—⬭—OH

Butanediol
Hydroquinone
Methyl-hydroquinone
Bisphenol-A

L-*arabino*
Xylo

Bioerodible polymers having pendant functional groups are of particular interest, since they are capable of covalent pro-drug formation. Acemoglu *et al.* [53] prepared biodegradable poly(hydroxyalkylene carbonate)s from the optically active and racemic 2,3-*O*-isopropylidene-threitol and 2,4:3,5-di-*O*-isopropylidene-L-mannitol with diethyl carbonate in the presence of dibutyl tin oxide. The isopropylidene groups of the polycarbonates were hydrolyzed, and derivations were made on the hydroxy groups to obtain esters, orthoesters, and carbamates. The deprotected polycarbonates were water-soluble, and degraded in a few weeks by a mechanism in which hydroxy groups were shown to participate.

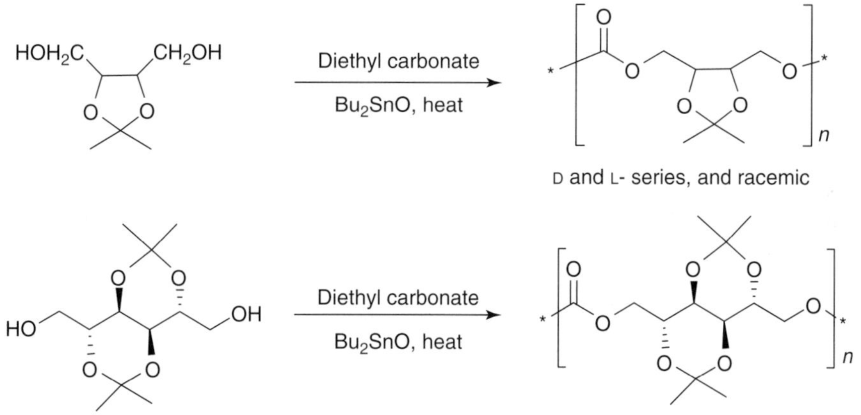

HOH_2C CH_2OH —Diethyl carbonate / Bu_2SnO, heat→

D and L- series, and racemic

HO OH —Diethyl carbonate / Bu_2SnO, heat→

The first apparently cholesteric polyester derived from an alditol such as 1,2:5,6-di-*O*-isopropylidene-D-mannitol was reported by Chiellini *et al.* [54]. Surprisingly, an acidolytic deprotection yielded a seemingly thermotropic polyester.

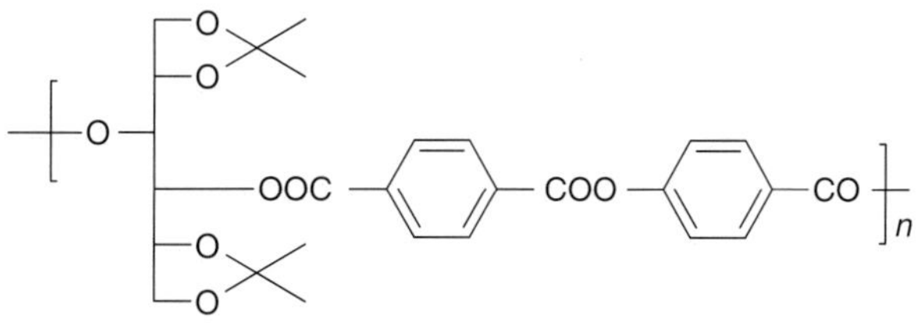

5.2.3 Unprotected alditols

For the last few years, Kobayashi's and Gross's groups have been independently working on the synthesis of alditol-based polyesters by lipases as catalysts for selective polyol polymerizations. Lipases and proteases are well known as providing regioselectivity during esterification reactions at mild temperatures. In addition to this, the use of enzymes in polyester synthesis is noteworthy from the standpoint of 'green chemistry'. The first example of a sugar alcohol used as monomer in an enzyme-catalyzed polymerization was reported by Kobayashi *et al.* [55]. In this case, the activation of carboxylic acids with electron-withdrawing groups was thought to be necessary. They proved the regioselective polymerization at positions 1 and 6 of D-glucitol (sorbitol) and divinyl sebacate, using lipase derived from *Candida antarctica* in acetonitrile.

Alditols are insoluble in non-polar organic media; instead, they are soluble in polar solvents such as pyridine, dimethyl sulphoxide, 2-pyrrolidone, and acetone. However, these solvents cause large reductions in enzyme activity. A simple and versatile strategy for performing selective lipase-catalyzed condensation polymerizations between dicarboxylic acids and polyols was explored by Gross's group. Instead of using organic solvents, the monomers adipic acid, glycerol, and sorbitol were solubilized within binary or ternary mixtures [56]. Thus, the polymerization reactions were performed without activation of adipic acid. Immobilized Lipase B from *Candida antarctica* (Novozyme 435) catalyzed bulk polycondensations of three monomers, forming a monophasic ternary mixture that resulted in hyperbranched polyesters with octanediol–adipate and glycerol–adipate repeating units [57]. The key to the success of the method is the use of a highly active and selective lipase as the catalyst, as well as adjusting the reaction mixture so that it is monophasic. Thus, polyesters containing many hydroxyl functionalities, such as poly(sorbityl adipate), were prepared without the need to use protection–deprotection chemistry.

Kobayashi *et al.* have worked extensively on enzymic polymerization for the synthesis of both novel and natural oligo- and polysaccharides [58–61]. They induced the regioselective polymerization of divinyl ester with triols by lipase catalysis, to give soluble polymers of relatively high-molecular weight. The 1,3-disubstituted glyceride unit was mainly formed from glycerol and immobilized lipase derived from *Candida antarctica*, and the microstructure depended on the lipase origin [62]. They also accomplished the enzymatic syntheses of a new class of crosslinkable polyesters by the polymerization of divinyl sebacate and glycerol, using *Candida antarctica* lipase in the presence of unsaturated higher fatty acids; thus, the polyesters had an unsaturated group in the side chain [63, 64]. Biodegradable epoxide-containing polyesters derived from glycerol and unsaturated fatty acids were also enzymatically synthesized [65]. Kobayashi has recently published a review on the enzymatic synthesis of polyesters via polycondensation [66].

Very recently, Gross *et al.* reported the use of various unprotected alditols, such as erythritol, xylitol, ribitol, D-mannitol, D-glucitol, and D-galactitol, in the preparation of 'sweet polyesters' obtained by lipase-catalyzed polycondensation with 1,8-octanediol and adipic acid [67]. Polymerizations were performed under conditions that allowed vacuum regulation and pressure for water removal, and the selected lipase was N435 (*Candida antarctica* Lipase B, physically immobilized on Lewatit beads). It was found that the polyol reactivity is determined by the stereochemical configuration of carbons closest to terminal hydroxyl groups. The (*R,R*) configuration at C2/C5 of D-mannitol is a key factor that led to rapid formation of the corresponding polyesters with high-molecular weights.

5.3 ALDONIC ACIDS AND LACTONES

Aliphatic polyesters may be classified into two groups, depending on the bond constitution of the monomers: poly(hydroxy acid)s (*i.e.* polyhydroxyalkanoates (PHAs)) and poly(alkylene dicarboxylate)s [68]. The former are polymers of hydroxy acids (α, β,…ω-hydroxy acids), obtained by ring-opening polymerization or polycondensation reactions. The latter are synthesized by the polycondensation reaction of diols with dicarboxylic acids. Aldonic and aldaric acids can be used to prepare both groups of polyesters.

The polycondensation reaction of aldonic acids to render polyesters has not been much explored. However, pioneering papers describing this kind of polycondensation are known. Drew and Haworth [69] described the polycondensation of 2,3,4-tri-*O*-methyl-L-arabinonic acid through its 1,5-lactone, to give an oligomeric material not completely characterized. Later, Mehltretter and Mellies [70] studied the polyesterification of the difunctional *O*-protected sugar 2,4;3,5-di-*O*-methylene-D-gluconic acid, obtaining low-molecular weight polyesters.

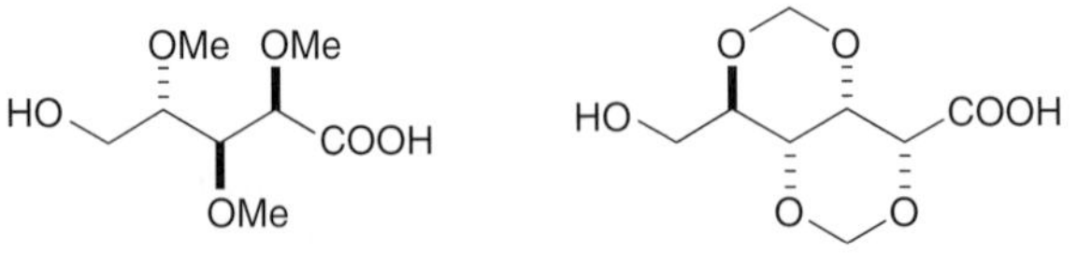

We have described an easy preparation of 2,3,4-tri-*O*-methyl-D-xylose and 2,3,4-tri-*O*-methyl-L-arabinose [50], which can be oxidized to give the corresponding lactones [71], susceptible to ring-opening polymerization. However, in our hands, all the attempts to polymerize these lactones were unsuccessful.

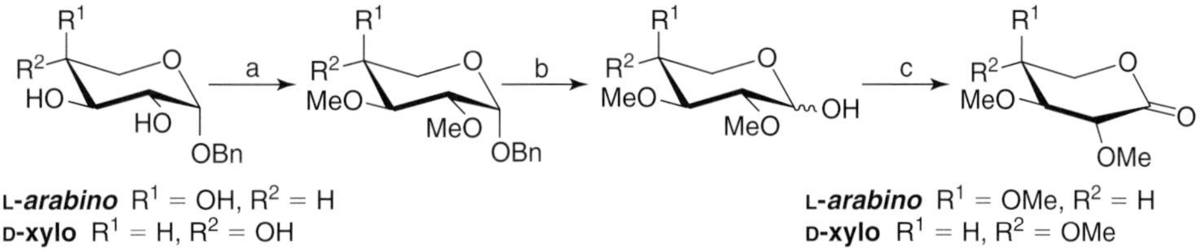

L-*arabino* R^1 = OH, R^2 = H
D-xylo R^1 = H, R^2 = OH

L-*arabino* R^1 = OMe, R^2 = H
D-xylo R^1 = H, R^2 = OMe

(a) MeI, KOH, Me_2SO; (b) H_2, Pd-C, MeOH; (c) PCC.

2,3,4,5-Tetra-*O*-methyl-D-glucono-1,6-lactone has been prepared from D-glucose as a crystalline compound in acceptable yields by two different routes [72].

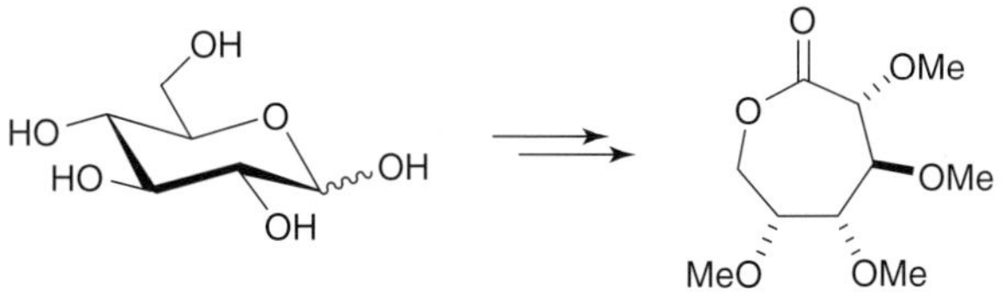

Although the assays of homopolymerization of this lactone were also unsuccessful, we carried out a copolymerization experiment by the bulk ring-opening polymerization of a mixture of L-lactide and 2,3,4,5-tetra-*O*-methyl-D-glucono-1,6-lactone, in a ratio of 5:1, using tin (II) 2-ethylhexanoate ($SnOct_2$) as initiator. From the copolymerization reaction mixture, we obtained two copolymers, containing 1.3 and 2.2 per cent of the carbohydrate monomer, respectively, as determined by NMR studies.

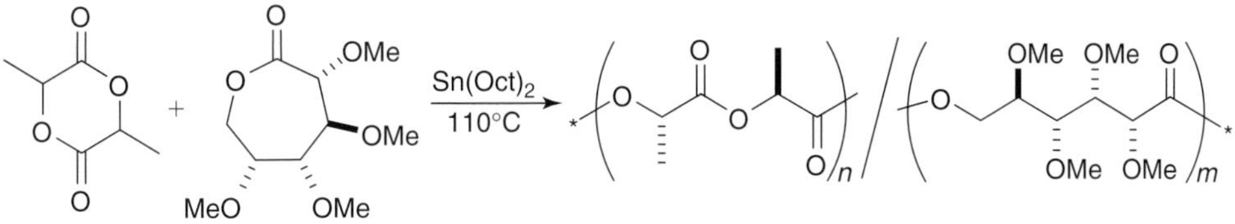

Recently, Varela *et al.* [73] obtained similar results from 2,3,4,5-tetra-*O*-methyl-D-galactono-1,6-lactone. The homopolymerization of this lactone, promoted by aluminium isopropoxide [$Al(O^iPr)_3$] or scandium triflate [$Sc(OTf)_3$], was attempted, but failed in both cases. The copolymerization with ε-caprolactone, using scandium triflate as initiator, gave a low-molecular weight copolymer (~3 000) with an incorporation of the sugar to the polymeric chain of about 10 per cent, as estimated by NMR.

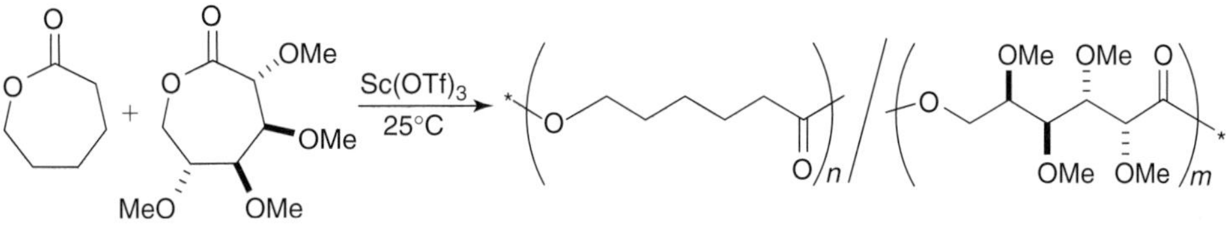

5.4 ALDARIC ACIDS

Unprotected and *O*-protected aldaric acids have been extensively used in polycondensation reactions, mainly in the preparation of polyamides and polyesters. A review has been published on potentially important aldaric acids as monomers for the preparation of polyamides [74].

Muñoz-Guerra *et al.* carried out a detailed study on polytartaramides (*i.e.* polyamides derived from tartaric acid). Various types of polytartaramide, differing either in the diamine used as comonomer, or in the nature of the *O*-protecting groups used to block the two secondary hydroxyl groups of the tartaric acid, have been synthesized, and their properties evaluated in connection with their structure.

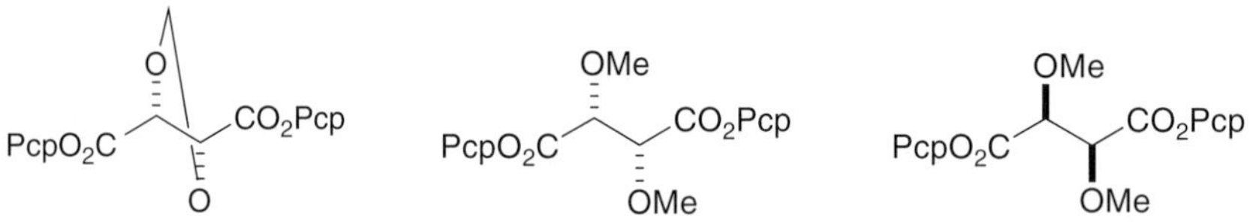

Thus, stereoregular polyamides were prepared from bis(pentachlorophenyl) 2,3-*O*-methylene-L-tartrate [75] and bis(pentachlorophenyl) 2,3-di-*O*-methyl-L-tartrate [76], and their crystal structure [77] and hydrolytic degradation [78] carefully investigated. Stereocopolyamides derived from 2,3-di-*O*-methyl-D- and -L-tartaric acids and hexamethylenediamine, with enantiomeric D/L ratios ranging from 1:9 to 1:1, were also obtained by the active ester polycondensation method [79]. The crystal structure of the ensuing racemic copolyamides obtained from the racemic mixture of 2,3-di-*O*-methyl-D- and -L-tartaric acids and hexamethylenediamine, or by equimolecular mixture of the two enantiomerically pure D- and L-polyamides, was investigated [80]. The same group also described the preparation of poly(ester amide)s derived from 2,3-di-*O*-methyl-L-tartaric acid, more sensitive to hydrolysis than the polyamides, and their mechanism of hydrolytic degradation has been investigated in relation with their microstructure [81]. Aregic [82] and regioregular [83] polytartaramides made from di-*O*-methyl-L-tartaric acid and ethyl L-lysine have been reported. Recently [84], the same authors prepared a series of polyamides from hexamethylenediamine and the pentachlorophenyl esters of 2,3-di-*O*-acyl-L-tartaric acid by polycondensation in solution. Both *O*-alkanoyl and *O*-benzoyl esters were used as hydroxyl-protecting groups. The controlled hydrolysis of the protecting groups yielded poly(hexamethylene-L-tartaramide)s with different content of free hydroxyl groups.

Kiely *et al.* initiated the preparation of a novel type of stereoregular and non-stereoregular hydroxylated nylons (polyhydroxypolyamides) by the direct polycondensation of alkyl esters of unprotected aldaric acids (D-glucaric, galactaric, xylaric, D-mannaric) and their lactones, with linear aliphatic and arylalkylenediamines [85–88], and with aza- and oxa-alkylenediamines [89, 90]. They obtained high yields of solid polymers with low-molecular weight, some of them being soluble in water. The mechanism of formation [91], conformational analysis [92], and adhesive properties [93] of these polyhydroxypolyamides were also studied.

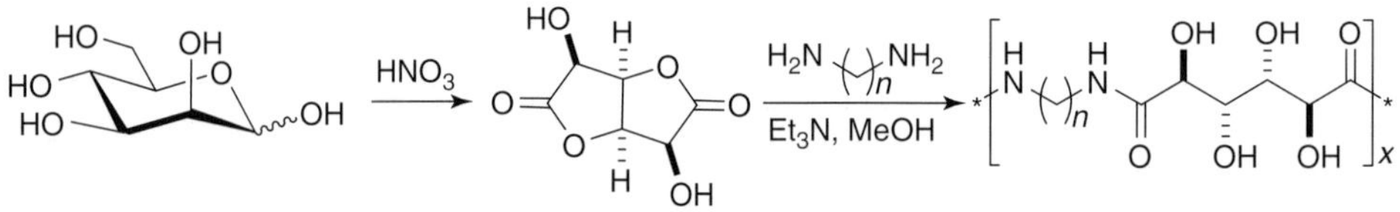

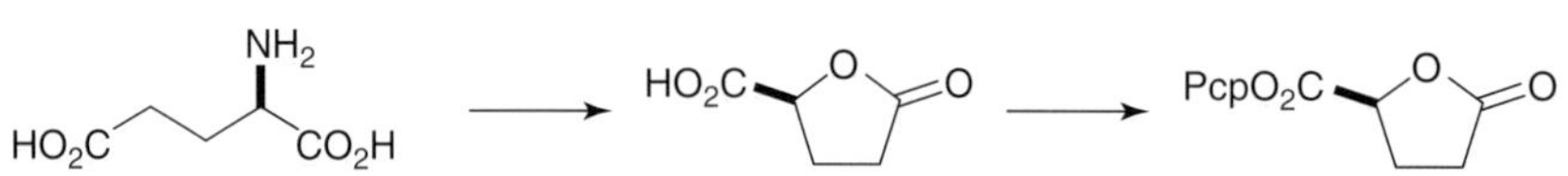

Varela *et al.* described some stereoregular hydroxylated polymannaramides [94] by the reaction of D-mannaro-1,4:6,3-dilactone with even-numbered alkylenediamines ($n = 2, 6, 8, 10, 12$). Hydroxylated stereoregular and non-stereoregular polyamides were also prepared by the same authors [95] from hexamethylene diamine and pentachlorophenyl (2*S*)-5-oxo-2-tetrahydrofurancarboxylate, synthesized [96] from the chiral (2*S*)-2-hydroxypentanedioic acid 5,2-lactone. The latter was obtained [97] by deamination of the easily available L-glutamic acid.

Galbis *et al.* [50] described the preparation of the pentachlorophenyl esters of 2,3,4-tri-*O*-methyl-L-arabinaric (and xylaric) acids as suitable bifunctional monomers for linear polycondensations.

L-*arabino* R^1 = OMe, R^2 = H
D-*xylo* R^1 = H, R^2 = OMe

García-Martín *et al.* reported [98] on the AABB polyamides based on these pentaric acids. Two different types of polycondensate were prepared: fully sugar-based polyamides (**PA–ArAr** and **PA–XyXy**), and polyamides derived from aldaric acids and aliphatic diamines (**PA–*n*Ar** and **PA–*n*Xy**, ***n*** = 6, 8, 12). In all these cases, *aregic* polymers were formed, since both sugar configurations lack the C_2 axis.

PA–ArAr R^1 = OMe, R^2 = H
PA–XyXy R^1 = H, R^2 = OMe

PA–*n*Ar R^1 = OMe, R^2 = H
PA–*n*Xy R^1 = H, R^2 = OMe

Recently [52], a variety of carbohydrate-based linear homo- and co-polyesters were obtained from these pentaric acids by polycondensation reactions (see Section 5.2.2).

2,3,4,5-Tetra-*O*-methyl-D-mannaric and galactaric acids and their bis(pentachlorophenyl) esters have been prepared [99] as crystalline compounds, in good yields, from D-mannitol and galactitol, respectively. A new stereoregular fully sugar-based polyamide, analogous to Nylon 66, has also been prepared [99] by polycondensation of bis(pentachlorophenyl) 2,3,4,5-tetra-*O*-methyl-D-mannarate with 1,6-diamino-1,6-dideoxy-2,3,4,5-tetra-*O*-methyl-D-mannitol dihydrochloride [100].

D-mannitol R^1 = OH, R^2 = H
Galactitol R^1 = H, R^2 = OH

D-*manno* R^1 = OMe, R^2 = H
Galacto R^1 = H, R^2 = OMe

EDPA, NMP

Pcp: pentaclhoropenyl; DCC: dicyclohexylcarbodiimide; NMP: *N*-methylpyrrolidinone

Some other fully sugar-based polyamides have been obtained from these activated hexaric acids and the 1,5-diamino-2,3,4-tri-*O*-methyl-pentitols having the L-*arabino* and *xylo* configuration [101].

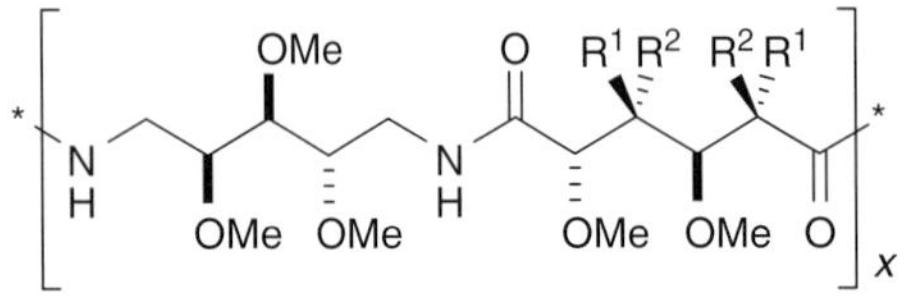

PA–ArMn R^1 = OMe, R^2 = H
PA–ArGa R^1 = H, R^2 = OMe

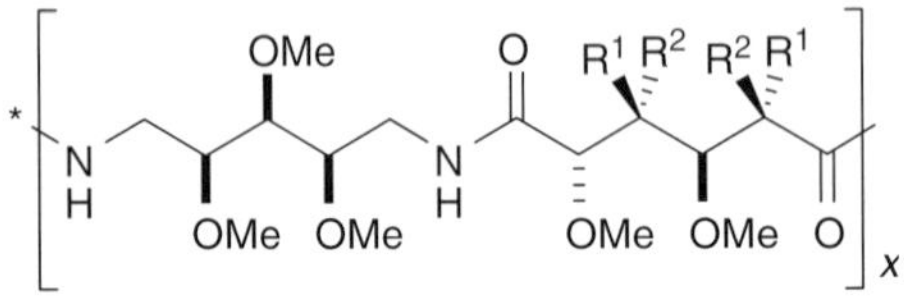

PA–XyMn R^1 = OMe, R^2 = H
PA–XyGa R^1 = H, R^2 = OMe

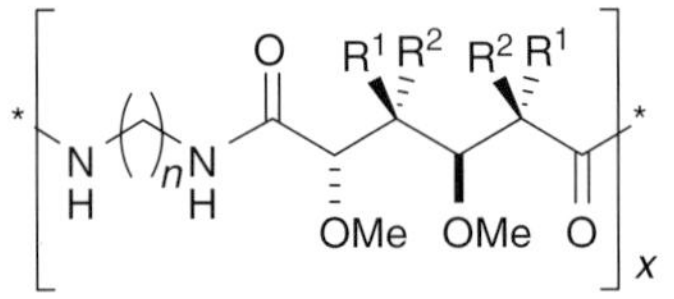

PA–*n*Mn R^1 = OMe, R^2 = H
PA–*n*Ga R^1 = H, R^2 = OMe

We also prepared [101] a series of polyamides (**PA–*n*Su**) derived from these hexaric acids and non-carbohydrate alkylenediamines with even number of methylenes (n = 4, 6, 8, 10, 12).

5.5 AMINOSUGARS

Polyamides and polyurethanes are the most widely studied carbohydrate-based polymers. The preparation of hydrophilic nylon-type polyamides containing sugars was a great challenge, as reported in two reviews by Thiem [7] and Varela [10]. In general, carbohydrate-based monomers present low-thermal stability; thus, initial approaches to the synthesis of polyamides by melt polycondensation led only to brittle fibers of low-molecular weights. The interfacial polycondensation technique made possible the synthesis of the first chiral nylon-type polyamide based on carbohydrates. The polymerizable sugar monomers required for the synthesis of polyamides should be diamino-sugars, aldaric acids, or aminoaldonic acids. As mentioned above, the multifunctionality of the sugars requires several protection and deprotection steps. The introduction of amino groups is generally accomplished by the preparation of a sulphonyl ester, nucleophilic displacement by azide, and subsequent hydrogenation. Alternative methods have also been described [7].

5.5.1 Diaminoanhydroalditols

Carbohydrates such as saccharose, D-glucitol, and diaminosaccharides such as trehalose, have been widely used for the syntheses of highly functionalized polyol-based polyurethanes. The preparation of polyurethanes by the reaction of dianhydrohexitols with aliphatic diisocyanates was explored in 1983, and technical applications of dianhydrohexitols were found as chain extenders in polyurethane syntheses, as patented.

The stable 2,5-diamino-2,5-dideoxy-1,4:3,6-dianhydrohexitol dihydrochlorides with D-*gluco*, L-*ido*, and D-*manno* configurations were prepared by Thiem *et al.* according to known procedures.

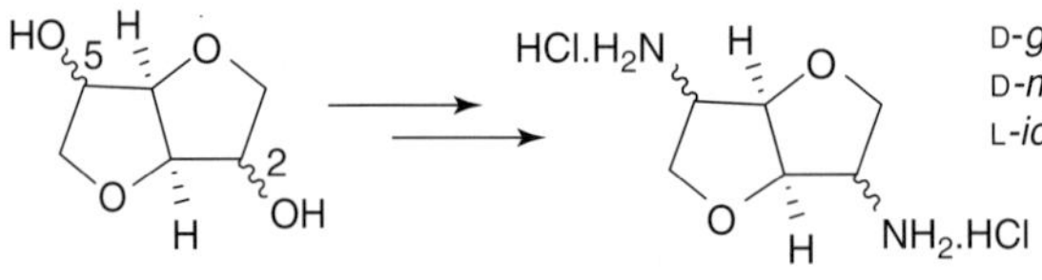

D-*gluco* (2-exo/5-endo)
D-*manno* (2-endo/5-endo)
L-*ido* (2-exo/5-exo)

They developed an initial approach to novel polyurethanes with these diamino monomers, aliphatic diamines, and the bischloroformate of DAS, by phase transfer polycondensation [102].

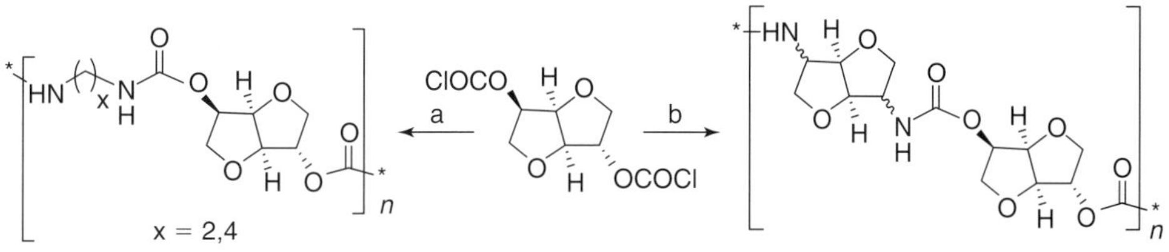

x = 2,4

a: $(CH_2)_n(\overset{+}{N}H_3)_2\,(C_2O_4^{2-})$ b: Diamino-DAS, diamino-DAM or diamino-DAI

The catalyzed polyaddition of the monomer 2-deoxy-1,4:3,6-dianhydro-2-isocyanato-L-iditol gave the corresponding polyurethane [103]. However, the intermediate aminochloroformyl derivative underwent spontaneous polycondensation by catalytic hydrogenation.

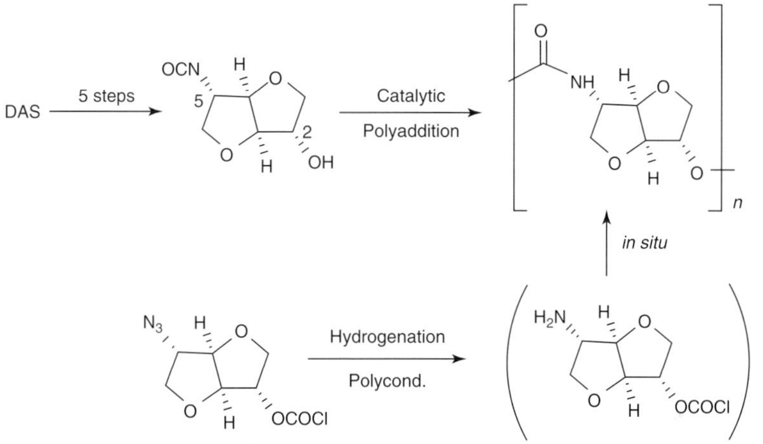

The stable diamino dihydrochlorides of the dianhydrohexitols were transformed into the corresponding diisocyanates by reaction with phosgene [104].

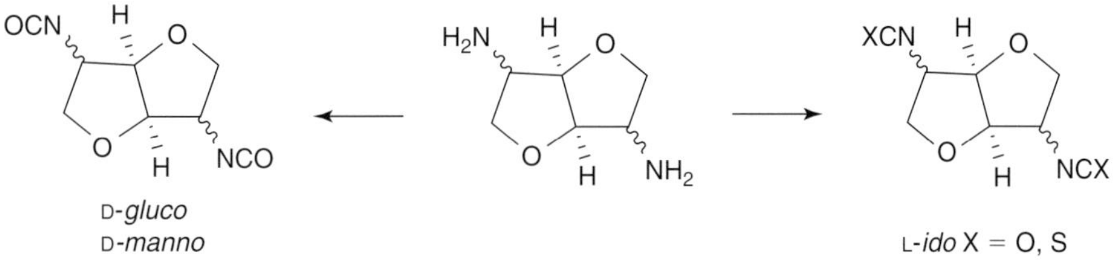

The addition of phosgene proceeded better when the amino groups were orientated out of the molecular plane (*exo* configuration); thus, the yields of the three isomer derivatives decreased in the order L-*ido*, D-*gluco*, and D-*manno*. The use of diphosgene gave unsuccessful results. The dithiodiisocyanato derivatives were prepared from the corresponding diamino-dianhydrohexitol and thiophosgene, and the same stereochemical reasons gave best yields for the L-*ido* compound. Polyurethanes and polyureas were synthesized from the D-*gluco*, D-*manno*, and L-*ido* monomers, and poly(thio)urethanes and poly(thio)ureas from the corresponding L-*ido* monomer.

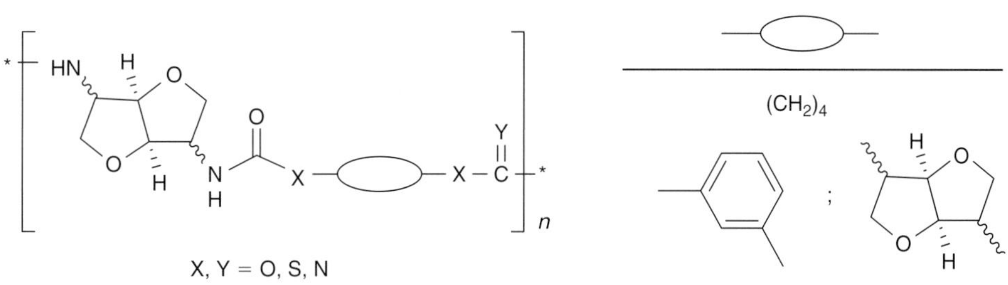

X, Y = O, S, N

The interfacial polycondensation of 2,3-diamino-1,4-anhydro-anhydroalditols with D,L-*threo* and *erythro* configurations, and the three 2,5-diamino-1,4:3,6-dianhydrohexitol derivatives with purchasable aromatic and aliphatic dicarboxylic acid chlorides, gave polyamides that were fully characterized [105].

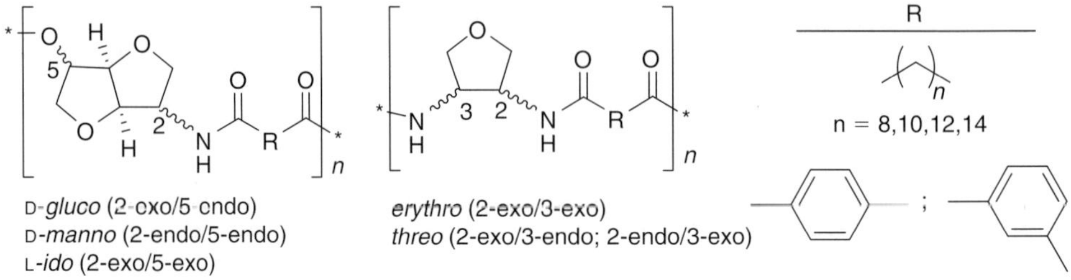

D-*gluco* (2-exo/5-endo)
D-*manno* (2-endo/5-endo)
L-*ido* (2-exo/5-exo)

erythro (2-exo/3-exo)
threo (2-exo/3-endo; 2-endo/3-exo)

R

n = 8,10,12,14

Hydrophilic polyamides were prepared by the polycondensation reaction of 2,3,4,5-tetra-*O*-acetyl-galactaroyl dichloride with diaminoanhydroalditols [106].

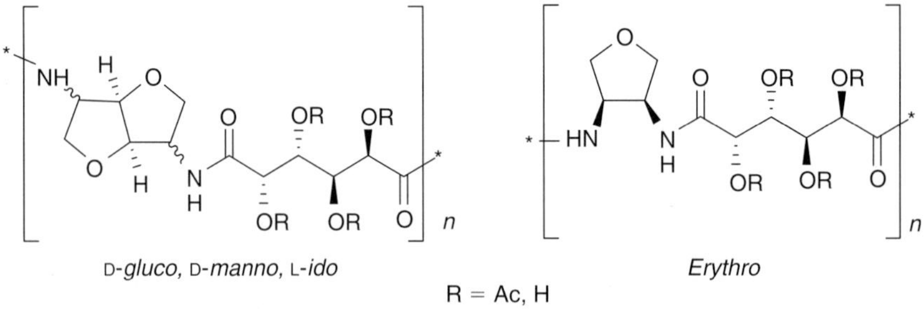

D-*gluco*, D-*manno*, L-*ido*

Erythro

R = Ac, H

Polyamides derived from D-glucose and D-glucosamine by interfacial and solution polycondensations of the sugar diamino derivatives with aromatic and aliphatic acyl chlorides have also been described [107]. The presence of an anomeric benzyl group did not decrease the reactivity of the 2-amino function. Similar chiral polyamides were synthesized from the 1,7-diamino derivative, which was prepared from D-glucal.

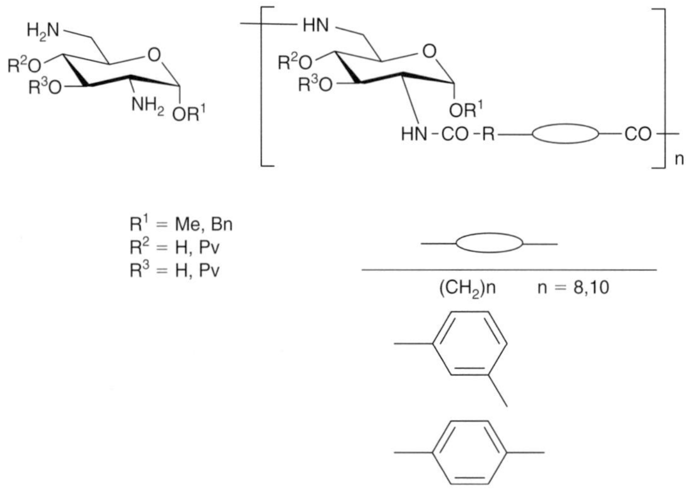

R^1 = Me, Bn
R^2 = H, Pv
R^3 = H, Pv

$(CH_2)n$ n = 8,10

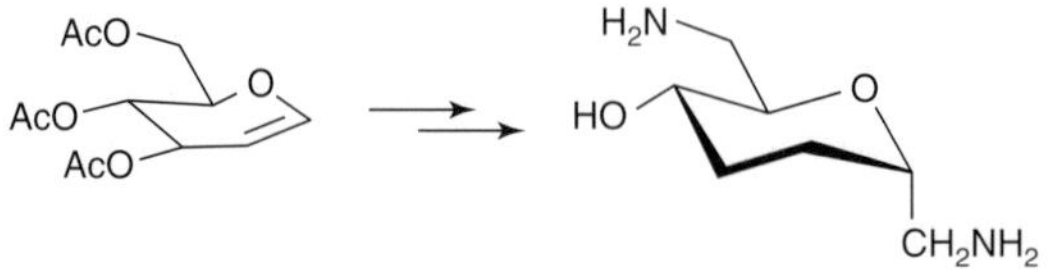

5.5.2 Amino- and diaminoalditols

Muñoz-Guerra *et al.* and Galbis *et al.* initiated in the past decade a systematic study of a series of sugar-based AABB linear polyamides. These polyamides are derived from appropriately *O*-protected aldaric acids and 1,ω-diamino-dideoxyalditols. The diaminoalditol monomers of L-*threo*, L-*arabino*, *xylo*, D-*manno*, and L-*ido* configurations were obtained from the corresponding alditols, having the secondary hydroxyl groups protected as methyl ether.

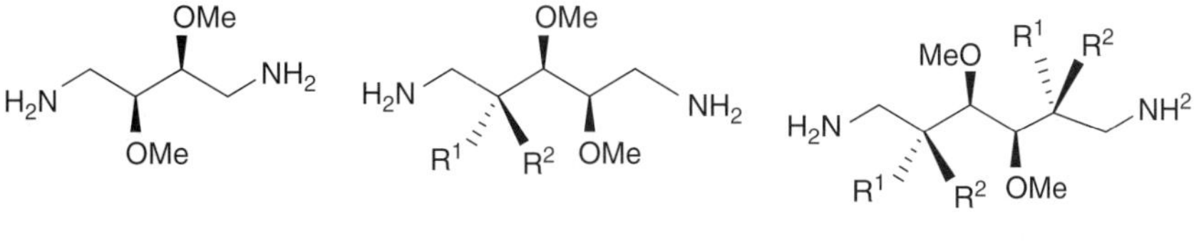

L-*threo* L-*arabino*, R^1 = OMe, R^2 = H D-*manno*, R^1 = OMe, R^2 = H
xylo R^1 = H, R^2 = OMe L-*ido*, R^1 = H, R^2 = OMe

The transformation of the primary hydroxyl groups into the amino function was carried out by the usual procedure (sulphonation-nucleophilic displacement by azide-reduction). Thus, Muñoz-Guerra *et al.* synthesized (2*S*,3*S*)-2,3-di-methoxy-1,4-butanediamine [49] from natural L-tartaric acid as a raw material, to prepare stereoregular polytartaramides (Nylon 4,*n* analogues). Since L-tartaric acid contains a C_2 symmetry axis, the resulting polyamides are regiochemically ordered. The polycondensation in a chloroform solution of this diamine, activated as the *N,N′*-bis(trimethylsilyl) derivative, with activated pentachlorophenyl esters of aliphatic dicarboxylic acids, afforded stereoregular polyamides. This method allows polycondensations under mild conditions, and the corresponding polyamides are obtained in good yields and with acceptable molecular weights. Because of their stereoregularity, these substituted polyamides present an interesting combination of properties since they are highly crystalline, have mechanical properties comparable to those of Nylons, and may undergo hydrolytic degradation under physiological conditions.

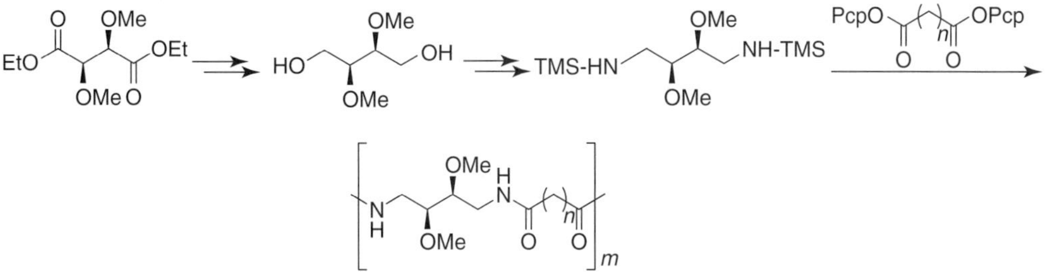

n = 4, 6, 8, 10, 12

Muñoz-Guerra *et al.* also studied stereoregular polyamides fully based on D- and L-tartaric acid [108]. The bispentachlorophenyl esters of both 2,3-di-*O*-methyl-tartaric acids were condensed with (2*S*,3*S*)-2,3-dimethoxy-1,4-butanediamine to obtain optically active (PTA–LL) and racemic (PTA–LD) polytartaramides. Fibre-oriented and powder X-ray studies of these polyamides demonstrated that PTA–LL crystallized in an orthorhombic lattice, whereas PTA–LD seemed to adopt a triclinic structure. In both cases, the polymeric chain appears to be in a folded conformation, more contracted than in the common γ-form of conventional Nylons.

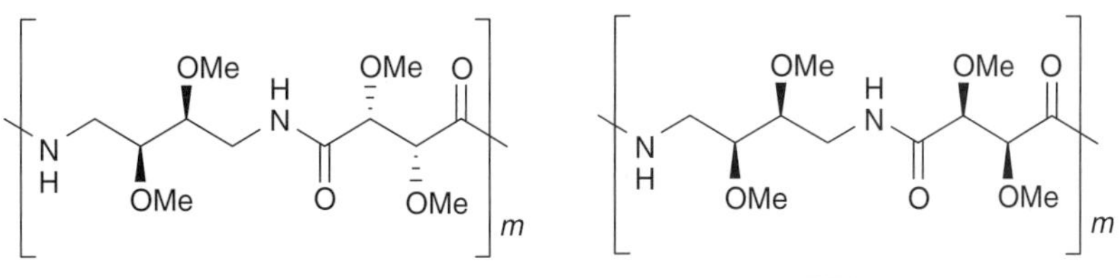

PTA-LL **PTA-LD**

The 1,5-diamino-1,5-dideoxy-2,3,4-tri-*O*-methylpentitols of L-*arabino* and *xylo* configurations [50] gave aregic polyamides when polycondensated with activated aliphatic dicarboxylic acids or aldaric acids [98]. Those polyamides, entirely based on xylose were, as expected, not optically active.

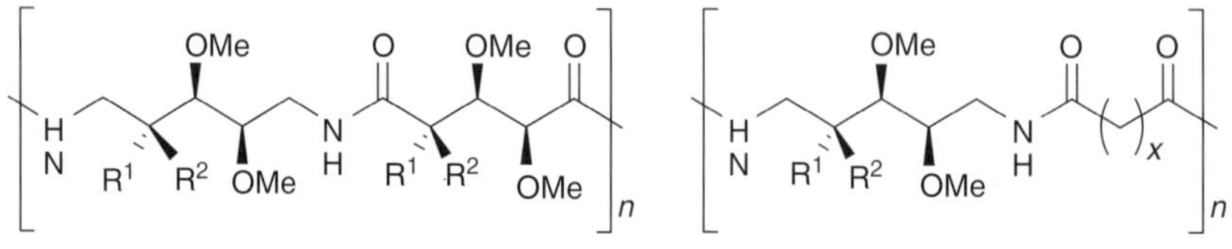

L-*arabino*, R^1 = OMe, R^2 = H x = 8, 10
xylo, R^1 = H, R_2 = OMe

The synthesis of 1,6-diamino-1,6-dideoxy-2,3,4,5-tetra-*O*-methyl-D-mannitol and its L-iditol analogue from D-mannitol has been described [100]. These diamines, containing a two-fold axis, gave stereoregular AABB polyamides on polycondensation with terephthaloyl dichloride and dipentachlorophenyl esters, or dichlorides of aliphatic dicarboxylic acids, in solution or under interfacial polycondensation conditions [99, 109]. In spite of the regioregularity present in the polymeric chains, these optically active polyamides could not be crystallized.

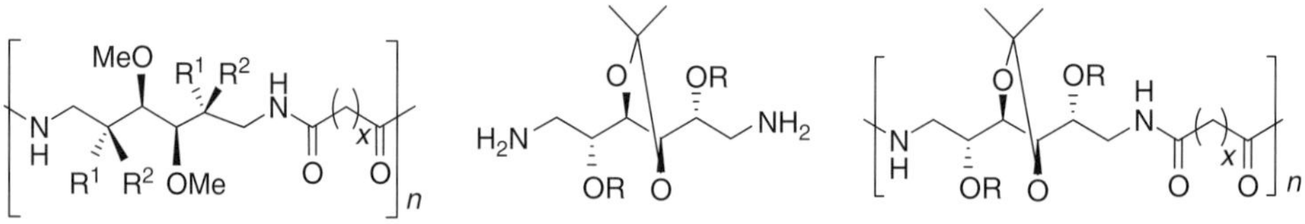

x = 0, 2, 4, 6, 8
PA-Mn*n*, R^1 = OMe, R^2 = H
PA-Id*n* R^1 = H, R^2 = OMe

R = H, Me x = 2, 4, 6, 8,10

1,6-Diamino-1,6-dideoxy-2,5-di-*O*-methyl-3,4-*O*-isopropylidene-D-mannitol was also prepared and polymerized with active esters of dicarboxylic acids to give the corresponding polyamides. Aregic polyamides of the AABB type were also obtained by the polycondensation of 1,5-diamino-1,5-dideoxypentitols of L-*arabino* and *xylo* configurations with activated and *O*-methyl D-mannaric and galactaric acids [101].

We have also prepared 1-amino-1-deoxy-2,3,4-tri-*O*-methyl-L-arabinitol and -D-xylitol, which were transformed into a series of poly(ester amide)s [110, 111] and copoly(ester amide)s [112], whose structure and properties were carefully studied. Their hydrolytic degradation was also investigated, and a mechanism proposed for this process [113–115].

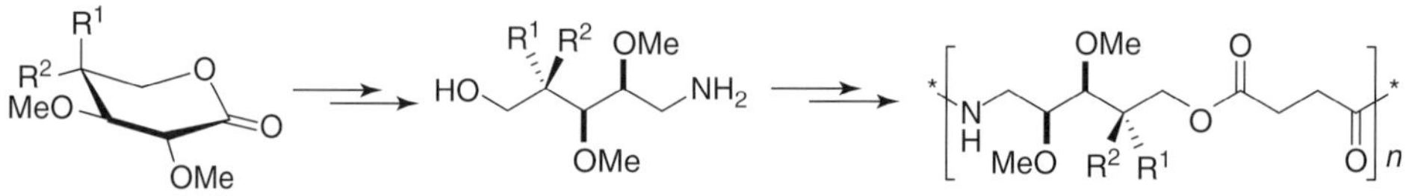

L-*arabino* R^1 = OMe, R^2 = H
Xylo R^1 = H, R^2 = OMe

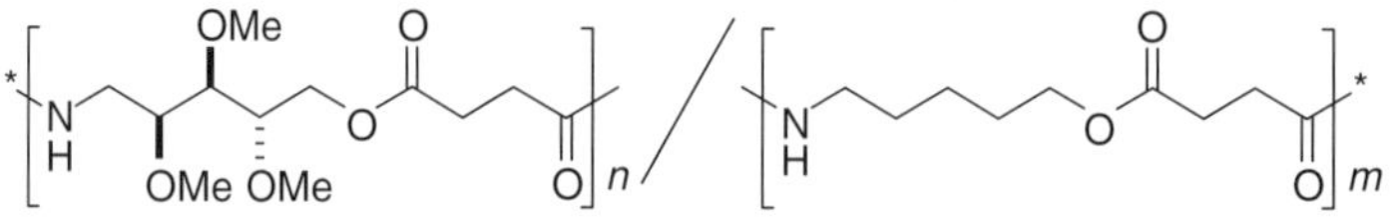

5.5.3 Aminoaldonic acids

Galbis *et al.* synthesized two stereoregular polygluconamides starting from D-glucosamine and D-glucose [116]. The synthesis of the polyamide of the polypeptide type [116, 117] was accomplished by the ring-opening polymerization of the *N*-carboxyanhydride obtained by treatment of 2-amino-2-deoxy-3,4,5,6-tetra-*O*-methyl-D-gluconic acid hydrochloride, in turn obtained from D-glucosamine, with TCF in THF.

OH HO HO O OH NH_2 → COOH NH_2.HCl MeO OMe OMe CH_2OMe $\xrightarrow{\text{TCF, THF}}$ O O HN O NH_2.HCl MeO OMe OMe CH_2OMe $\xrightarrow{\text{Et}_3\text{N, DMF}}$ H N O n MeO OMe OMe CH_2OMe

D-glucosamine

A methoxylated polyamide analogous to Nylon 6 was obtained in several steps from D-glucose [116, 118] through the preparation of a dimeric active ester of 6-amino-6-deoxy-2,3,4,5-tetra-*O*-methyl-D-gluconic acid. This polyamide was highly crystalline, and gave resistant films with a spherulitic texture.

D-glucose → C_6Cl_5OOC OMe OMe OMe OMe O N H OMe OMe OMe OMe NH_2.HCl $\xrightarrow{\text{EDIPA, DMF}}$ H N OMe OMe O OMe OMe n

The same group described the syntheses of some derivatives of 5-amino-5-deoxy-L-arabinonic acid, 5-amino-5-deoxy-D-xylonic acid, and (*S*)-5-amino-4-hydroxypentanoic acid, which were performed in several steps from L-arabinose, D-xylose, and (*S*)-(+)-glutamic acid, respectively [119]. These monomers were polymerized to give methoxylated homo and copolyamides analogous to Nylon 5 [120, 121].

R^1 R^2 MeO O O OMe → H_2N R^1 R^2 OMe OH OMe O → * H N R^1 R^2 OMe OMe O *

L-arabino R^1 = OMe, R^2 = H
Xylo R^1 = H, R^2 = OMe

HO O OMe OH O → H_2N OMe OH O → * H N OMe O *

Romero Zaliz and Varela [122] recently prepared, from the D- and L-galactono-1,4-lactones, the corresponding 6-amino-6-deoxy-2,3,4,5-tetra-*O*-methyl-D-(and L-)galactonic acids, which were polymerized by the procedure described by Galbis *et al.* [116] to give stereoregular sugar-based polyamides analogous to Nylon 6 [123].

OH HO O O H HO OH → H_2N OMe OMe OH OMe OMe O → * N H OMe OMe OMe OMe O *

Fleet *et al.* described some oligomers of *O*-protected-6-amino-6-deoxy-D-allonate [124] and D-galactonate [125], as intermediates for the preparation of polyhydroxylated Nylon 6 analogues.

A chiral β-polyamide of the Nylon 3 type was also synthesized [126, 127] by Galbis *et al.* by the ring-opening polymerization of the β-lactam derived from 3-amino-3-deoxy-2,4,5,6-tetra-*O*-methyl-D-altronic acid.

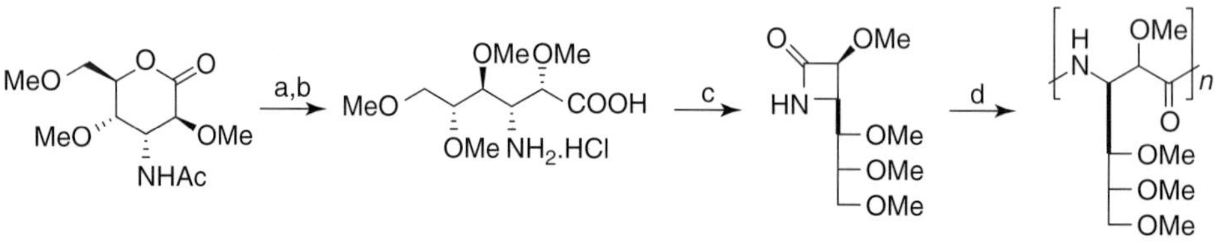

(a) MeI, HOH, THF; (b) 4M HCl; (c) 1. DIPEA, MeCN; 2. MsCl, $NaHCO_3$, MeCN; (d) K^tOBu, CH_2Cl_2.

Some other chiral Nylon 3 analogues have also been prepared by the same authors by the ring-opening polymerization of chiral β-lactams derived from D-glyceraldehyde [128, 129]. Both the enantiomerically pure (2*R*,3*R*) and the racemic (2*R*,3*R* and 2*S*,3*S*) β-polyamides were obtained, and their properties compared.

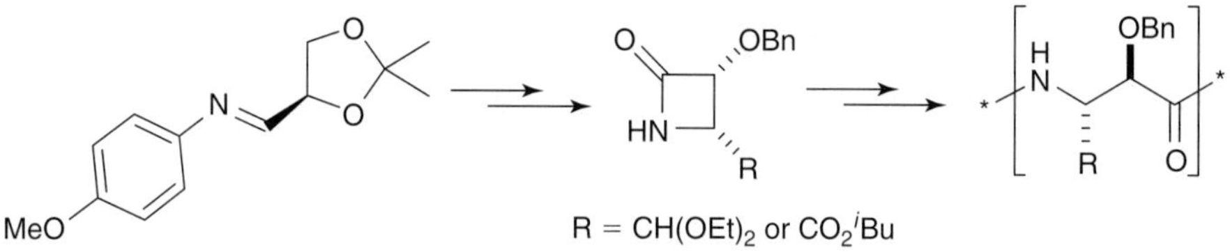

R = $CH(OEt)_2$ or $CO_2{}^iBu$

5.6 MISCELLANEOUS

Several modified D-glycosylamine and D-glucosamine monomers were synthesized by Thiem *et al.* to carry out catalytic polymerizations leading to polymers with urea, urethane, and amide linkages [130]. The reactivity of these monomers was studied, and it was found that the anomeric hydroxyl groups were more reactive than the amino groups. With selectively blocked D-glucosamine monomers, the reaction yielded preferentially polyurethanes incorporating the anomeric hydroxy function.

Very recently, Thiem *et al.* explored the synthesis and degradability of novel carbohydrate-segmented silicones polyamides, obtained from activated glucaric and galactaric acid derivatives and different α,ω-diaminoalkyl polydimethylsiloxanes [131]. Galactaric acid-segmented silicones display a higher T_g than their glucaric acid analogues. These materials can be partially degraded by enzymes, with the amide bond functioning as a predetermined breaking point.

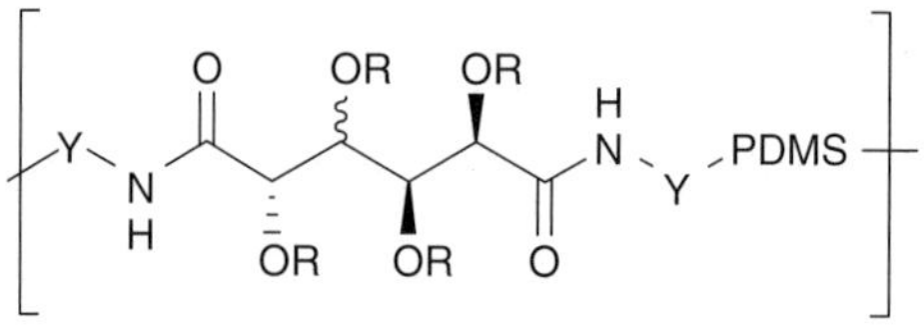

R = Ac or H
Y = $(CH_2)_2NH(CH_2)_3$ or $(CH_2)_3$
PDMS = $SiMe_2(SiMe_2O)_{70}$ or $SiMe_2(SiMe_2O)_{15}$

The same authors prepared a series of allyl-group-containing bifunctional carbohydrate derivatives that were reacted with hydrodimethylsilyl-terminated polysiloxane using Speier's catalyst [132].

Chen and Gross [133] synthesized high-molecular weight copolycarbonates derived from L-lactide and the 3,5-cyclic carbonate of 1,2-*O*-isopropylidene-D-xylofuranose. The best results were obtained using $Sn(Oct)_2$ as catalyst, with 83:17 mol/mol ratio of the comonomers. Even though the monomer reactivity ratio of L-lactide is much greater than that of the xylose monomer, very short sugar segment lengths were formed. This fact was explained by intramolecular exchange reactions or a xylose monomer insertion reaction during propagation, or by other, more-complex phenomena. Subsequent deprotection of the ketal structure of the copolymer was successfully carried out using CF_3COOH.

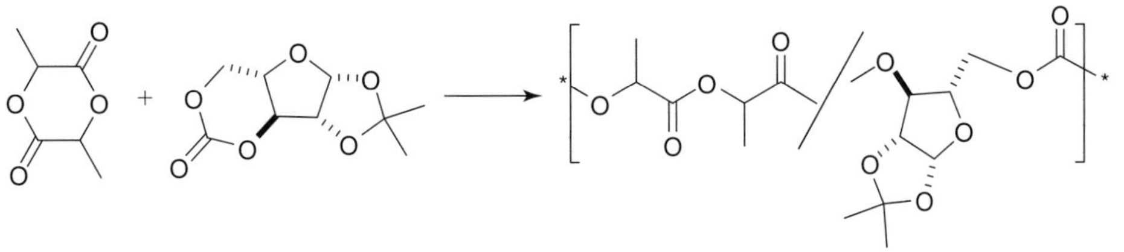

The same xylose monomer was homopolymerized by ring-opening polymerization [134] with different organometallic catalysts such as methylaluminoxane (MAO), $AlEt_3 \cdot 0.5\ H_2O$, ${}^tBuO^-K^+$, $ZnEt_2 \cdot 0.5\ H_2O$, and $Y(O^iPr)_3$. In another study, the same authors also polymerized the xylose monomer with trimethylene carbonate in the presence of catalysts; the best results were obtained using MAO and $ZnEt_2 \cdot 0.5\ H_2O$ [135]. Copolymers containing 8–83 per cent of xylose presented an alternating structure and were amorphous. The ketal deprotection was also carried out; and the resulting polymers were soluble in dimethylformamide (DMF), which suggested low chain crosslinking. The original polymers presented increased T_g as the xylose content increased, whereas the T_g of the unprotected copolymers decreased.

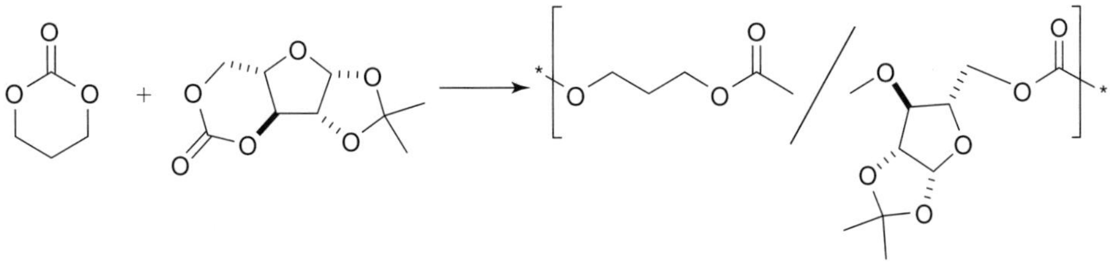

The new 4,4′-cyclic carbonate monomer of 1,2-isopropylidene-3-benzyloxy-D-pentofuranose was copolymerized with L-lactide in the presence of $Sn(Oct)_2$ at 130°C [136]. The benzyl ether and ketal groups were selectively removed, so that the units within the copolymers could have one, two, or three hydroxyl groups.

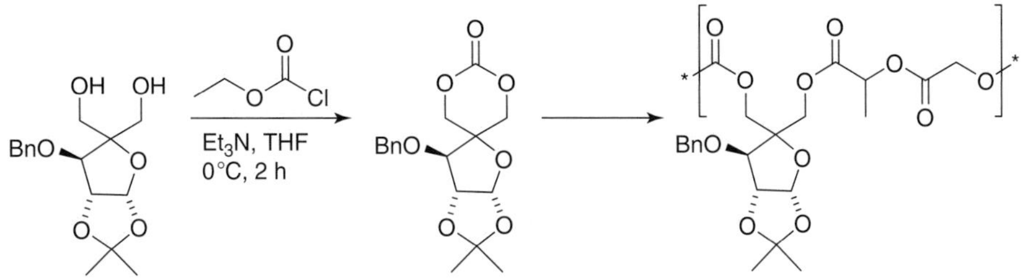

The ring-opening polymerization of the cyclic carbonate methyl 4,6-*O*-benzylidene-2,3-*O*-carbonyl-α-D-glucopyranoside has recently been described. This polymerizes at relatively low temperatures (60°C), without any elimination of carbon dioxide, to give polycarbonates in good yields and medium molecular weights [137].

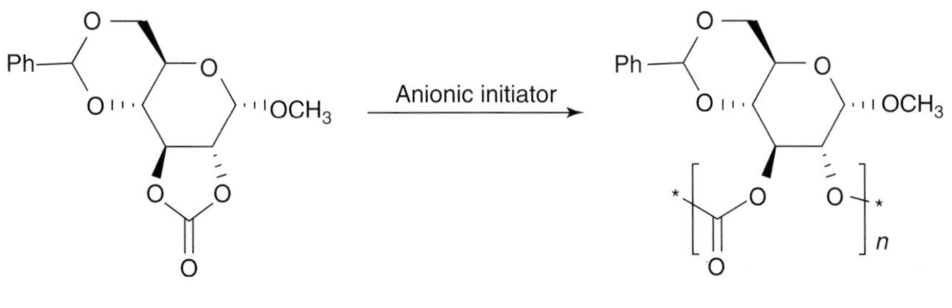

REFERENCES

1. Varma A.J., Kennedy J.F., Galgali P., Synthetic polymers functionalized by carbohydrates: A review, *Carbohyd. Polym.*, **56**, 2004, 429–445.
2. Cunlife D., Pennadam S., Alexander C., Synthetic and biological polymers-merging the interface, *Eur. Polym. J.*, **40**, 2004, 5–25.
3. Ladmiral V., Melia E., Haddleton D.M., Synthetic glycopolymers: An overview, *Eur. Polym. J.*, **40**, 2004, 431–449.
4. Narain R., Jhurry D., Wulf G., Synthesis and characterization of polymers containing linear sugar moieties as side groups, *Eur. Polym. J.*, **38**, 2005, 273–280.
5. Okada M., Molecular design and synthesis of glycopolymers, *Prog. Polym. Sci.*, **26**, 2001, 67–104.
6. Carneiro M.J., Fernandes A., Figueiredo C.M., Fortes A.G., Freitas A.M., Synthesis of carbohydrate based polymers, *Carbohyd. Polym.*, **45**, 2001, 135–138.
7. Thiem J., Bachmann F., Carbohydrate-derived polyamides, *Trends Polym. Sci.*, **2**, 1994, 425–432.
8. Kricheldorf H.R., "Sugar diols" as building blocks of polycondensates, *J. Macromol. Sci., Rev. Macromol. Chem. Phys.*, **C37**, 1997, 599–631.
9. Gonsalves K.E., Mungara P.M., Synthesis and properties of degradable polyamides and related polymers, *Trends Polym. Sci.*, **4**, 1996, 25–31.
10. Varela O., Orgueira H.A., Synthesis of chiral polyamides from carbohydrate-derived monomers, *Adv. Carbohyd. Chem. Biochem.*, **55**, 1999, 137–174.
11. Thiem J., Lüders H., Synthesis of polyterephtalates derived from dianhydrohexitols, *Polym. Bull.*, **11**, 1984, 365–369.
12. Thiem J., Lueders H., Synthesis and directed polycondensation of starch-derived anhydroalditol building blocks, *Starch/Staerke*, **36**, 1984, 170–176.
13. Storbeck R., Rehahn M., Balluff M., Synthesis and properties of high-molecular-weight polyesters based on 1,4:3,6-dianhydrohexitols and terephthalic acid, *Makromol. Chem.*, **194**, 1993, 53–64.
14. Thiem J., Häring T., Ring-opening polymerization of cis-3,4-dimethoxyoxolane, *Makromol. Chem.*, **188**, 1987, 711–718.
15. Thiem J., Strietholt W.A., Häring T., Ring-opening polymerization of carbohydrate-derived dialkoxyoxolanes. Approaches to functionalized poly(oxytetramethylens)s, *Makromol. Chem.*, **190**, 1989, 1737–1753.
16. Thiem J., Haering T., Strietholt W.A., Studies on the synthesis of polyethers from anhydropolyols, *Starch/Staerke*, **41**, 1989, 4–10.
17. Strietholt W.A., Thiem J., Höweler U.F.B., Synthesis and ring-opening polymerization of 1,4:2,5:3,6-trianhydro-D-mannitol and structure studies by MNDO calculations, *Makromol. Chem.*, **192**, 1991, 317–331.
18. Thiem J., Strietholt W.A., Ring-opening copolymerization of anhydroalditols with tetrahydrofuran, *Macromol. Chem. Phys.*, **196**, 1995, 1487–1493.
19. Braun D., Bergmann M., 1,4:3,6-Dianhydrohexite als Bausteine für Polymere, *J. Prakt. Chem.*, **334**, 1992, 298–310.
20. Braun D., Bergmann M., Polyesters with 1,4:3,6-dianhydrosorbitol as polymeric plasticizers for PVC, *Angew. Makromol. Chem.*, **199**, 1993, 191–205.
21. Okada M., Okada Y., Tao A., Aoi K., Synthesis and degradabilities of polyesters from 1,4:3,6-dianhydrohexytols and aliphatic dicarboxylic acids, *J. Polym. Sci. Part A: Polym. Chem.*, **33**, 1995, 2813–2820.
22. Okada M., Okada Y., Tao A., Aoi K., Biodegradable polymers based on renewable resources: Polysters composed of 1,4:3,6-dianhydrohexitol and aliphatic dicarboxylic units, *J. App. Polym. Sci.*, **62**, 1996, 2257–2265.
23. Okada M., Tachikawa K., Aoi K., Biodegradable polymers based on renewable resources. II. Synthesis and biodegradability of polyesters containing furan rings, *J. Poly. Sci. Part A: Polym. Chem.*, **35**, 1997, 2729–2737.
24. Okada M., Tachikawa K., Aoi K., Biodegradable polymers based on renewable resources. III. Copolyesters composed of 1,4:3,6-dianhydro-D-glucitol, 1,1-bis(5-carboxy-2-furyl)ethane and aliphatic dicarboxylic acid units, *J. App. Polym. Sci.*, **74**, 1999, 3342–3350.
25. Okada M., Tsunoda K., Tachikawa K., Aoi K., Biodegradable polymers based on renewable resources. IV. Enzymatic degradation of polyesters composed of 1,4:3,6-dianhydro-D-glucitol and aliphatic dicarboxylic acid moieties, *J. App. Polym. Sci.*, **77**, 2000, 338–346.
26. Okada M., Yamada M., Yokoe M., Aoi K., Biodegradable polymers based on renewable resources. V. Synthesis and biodegradation behavior of poly(ester amide)s composed of 1,4:3,6-dianhydro-D-glucitol, α-amino acid, and aliphatic dicarboxylic acid units, *J. App. Polym. Sci.*, **81**, 2001, 2721–2743.
27. Okada M., Yokoe M., Aoi K., Biodegradable polymers based on renewable resources.VI. Synthesis and biodegradability of poly(ester carbonate)s containing 1,4:3,6-dianhydro-D-glucitol and sebacic acid units, *J. App. Polym. Sci.*, **86**, 2002, 872–880.
28. Okada M., Aoi K., Biodegradable polymers from 1,4:3,6-dianhydro-D-glucitol (isosorbide) and its related compounds, *Current Trends Poly. Sci.*, **7**, 2002, 57–70.
29. Okada M., Chemical synthesis of biodegradable polymers, *Prog. Polym. Sci.*, **27**, 2002, 87–133.

30. Yokoe M., Aoi K., Okada M., Biodegradable polymers based on renewable resources. VII. Novel random and alternating copolycarbonates from 1,4:3,6-dianhydrohexitols and aliphatic diols, *J. Polym. Sci.: Part A: Polym. Chem.*, **41**, 2003, 2312–2321.
31. Yokoe M., Aoi K., Okada M., Biodegradable polymers based on renewable resources. VIII. Environmental and enzymatic degradability of copolycarbonates containing 1,4:3,6-dianhydrohexitols, *J. App. Polym. Sci.*, **98**, 2005, 1679–1687.
32. Yokoe M., Aoi M., Okada M., Biodegradable polymers based on renewable resources. IX. Synthesis and biodegradation behavior of polycarbonates based on 1,4:3,6-dianhydrohexitols and tartaric acid derivatives with pendant functional groups, *J. Polym. Sci.: Part A: Polym. Chem.*, **43**, 2005, 3909–3919.
33. Vill V., Fischer F., Thiem J., Helical twisting power of carbohydrate derivatives, *Z. Naturforsch*, **43**, 1988, 1119–1125.
34. Schwarz G., Kricheldorf H.R., New polymer synthesis. LXXXIII. Synthesis of chiral and cholesteric polyesters from silylated "sugar diols", *J. Polym. Sci. Part A: Polym. Chem.*, **34**, 1996, 603–611.
35. Sapich B., Stumpe J., Krawinkel T., Kricheldorf H.R., New polymer syntheses. 95. Photosetting cholesteric polyesters derived from 4-hydroxycinnamic acid and isosorbide, *Macromolecules*, **31**, 1998, 1016–1023.
36. Kricheldorf H.R., Wulff D.F., Layer structures. 11. Cholesteric polyesters derived from isosorbide, 2,5-bis(dodecyloxy)tere phthalic acid and 4,4′-dihydroxybiphenyl, *Polymer*, **39**, 1998, 6145–6151.
37. Kricheldorf R., Chatti S., Schwarz G., Krueger R.P., Macrocycles 27: Cyclic aliphatic polyesters of isosorbide, *J. Polym. Sci. Part A: Polym. Chem.*, **41**, 2003, 3414–3424.
38. Kricheldorf H.R., Copolyesters of ε-caprolactone, isosorbide and suberic acid by ring-opening copolymerization, *J. Macromol. Sci. Part A: Pure Appl. Sci.*, **43**, 2006, 967–975.
39. Kricheldorf H.R., Sun S.J., Gerken A., Ch. Chang T., Polymers of carbonic acid 22. Cholesteric polycarbonates derived from (S)-[(2-methylbutyl)thio]hydroquinone or isosorbide, *Macromolecules*, **29**, 1996, 8077–8082.
40. Kricheldorf H.R., Sun Sh.J., Chen Ch.P., Chang T.Ch., Polymers of carbonic acid. XXIV. Photoreactive, nematic or cholesteric polycarbonates derived from hydroquinone-4-hydroxybenzoate 4,4′-dihydroxychalcone and isosorbide, *J. Polym. Sci. Part A: Polym. Chem.*, **35**, 1997, 1611–1619.
41. Sun S.J., Schwarz G., Kricheldorf H.R., Ch. Chang T., New polymers of cabonic acid. XXV. Photoreactive cholesteric polycarbonates derived from 2,5-bis(4′-hydroxybenzylidene)cyclopentanone and isosorbide, *J. Polym. Sci. Part A: Polym. Chem.*, **37**, 1999, 1125–1133.
42. Chatti S., Kricheldorf H.R., Schwarz G., Copolycarbonates of isosorbide and various diols, *J. Polym. Sci. Part A: Polym. Chem.*, **44**, 2006, 3616–3628.
43. Kricheldorf H.R., Al Masri M., New polymer syntheses. LXXXII. Syntheses of poly(ether-sulfone)s from silylated aliphatic diols including chiral monomers, *J. Polym. Sci. Part A: Polym. Chem.*, **33**, 1995, 2667–2671.
44. Chatti S., Bortolussi M., Loupy A., Blais F.C., Bogdal D., Majdoub M., Efficient synthesis of polyethers from isosorbide by microwave-assisted phase transfer catalyst, *Eur. Polym. J.*, **38**, 2002, 1851–1861.
45. Kint D.P.R., Wingstrom E., Martínez de Ilarduya A., Alla A., Muñoz-Guerra S., Poly(ethylene terephthalate) copolyesters derived from (2S,3S)-2,3-dimethoxy-1,4-butanediol, *J. Polym. Sci. Part A: Polym. Chem.*, **39**, 2001, 3250–3262.
46. Zamora F., Mancera M., Rivas M., Hakkou K., Roffé I., Alla A., Muñoz-Guerra S., Galbis J., Aromatic polyesters from naturally occurring monosaccharides: Poly(ethylene terephthalate) and poly(ethylene isophthalate) analogs derived from D-mannitol and galactitol, *J. Polym. Sci. Part A: Polym. Chem.*, **43**, 2005, 4570–4577.
47. Zamora F., Hakkou K., Alla A., Espartero J.L., Muñoz-Guerra S., Galbis J., Aromatic homo- and copolyesters from naturally occurring monosaccharides: PET and PEI analogs derived from L-arabinitol and xylitol, *J. Polym. Sci. Part A: Polym. Chem.*, **43**, 2005, 6394–6410.
48. Alla A., Hakkou K., Zamora F., Martínez de Ilarduya A., Galbis J., Muñoz-Guerra S., Poly(butylene terephthalate) copolyesters derived from L-arabinitol and xylitol, *Macromolecules*, **39**, 2006, 1410–1416.
49. Bou J.J., Iribarren I., Muñoz-Guerra S., Synthesis and properties of stereoregular polyamides derived from L-tartaric acid: Poly[(2S,3S)-2,3-dimethoxybutylene alkanamide]s, *Macromolecules*, **27**, 1994, 5263–5270.
50. García-Martín M.G., Ruiz Perez R., Benito Hernández E., Galbis J.A., Synthesis of L-arabinitol and xylitol monomers for the preparation of polyamides. Preparation of an L-arabinitol-based polyamide, *Carbohyd. Res.*, **333**, 2001, 95–103.
51. García-Martín M.G., Ruiz Pérez R., Benito Hernández E., Espartero J.L., Muñoz-Guerra S., Galbis J.A., Carbohydrate-based polycarbonates. Synthesis, structure, and biodegradation studies, *Macromolecules*, **38**, 2005, 8664–8670.
52. García-Martín M.G., Ruiz Pérez R., Benito Hernández E., Galbis J.A., Linear polyesters of the Poly[alkylene (and co-arylene) dicarboxylate] type derived from carbohydrates, *Macromolecules*, **39**, 2006, 7941–7949.
53. Acemoglu M., Bantle S., Mindt T., Nimmerfall F., Novel bioerodible poly(hydroxyalkylene carbonate)s: A versatile class of polymers for medical and pharmaceutical applications, *Macromolecules*, **28**, 1995, 3030–3037.
54. Chiellini E., Galli G., Po R., Chiral liquid-crystalline polymers. XII. New chiral polyesters by chemical modification of a nonmesophasic polymer, *Polym. Bull.*, **23**, 1990, 397–402.
55. Uyama H., Klegraf E., Wada S., Kobayashi S., Regioselective polymerization of sorbitol and divinyl sebacate using lipase catalyst, *Org. Lett.*, 2000, 800–801.

56. Kumar A., Kulsrestha S., Gao W., Gross R.A., Versatile route to polyol polyesters by lipase catalysis, *Macromolecules*, **36**, 2003, 8219–8221.
57. Kulsrestha S., Gao W., Gross R.A., Glycerol copolyesters: Control of branching and molecular weight using a lipase catalyst, *Macromolecules*, **38**, 2005, 3193–3204.
58. Fujita M., Shoda S., Kobayashi S., Xylanase-catalyzed synthesis of a novel polysaccharide having a glucose–xylose repeating unit, a cellulose–xylan hybrid polymer, *J. Am. Chem. Soc.*, **120**, 1998, 6411–6412.
59. Kobyashi S., Makino A., Matsumoto H., Kunii S., Ohmae M., Kiyosada T., Makiguchi K., Matsumoto A., Horie M., Shoda S.I., Enzymatic polymerization to novel polysaccharides having a glucose-*N*-acetylglucosamine repeating unit, a cellulose–chitin hybrid polysaccharide, *Biomacromolecules*, **7**, 2006, 1644–1656.
60. Kobyashi S., Makino A., Tachibana N., Ohmae M., Chitinase-catalyzed synthesis of a chitin–xylan hybrid polymer: A novel water-soluble β(1 ← 4) polysaccharide having an *N*-acetylglucosamine-xylose repeating unit, *Macromol. Rapid Comm.*, **27**, 2006, 781–786.
61. Makino A., Ohmae M., Kobayashi S., Enzymatic synthesis of cellulose, chitin and chitosan, *Bio Industry*, **22**, 2005, 12–21.
62. Uyama H., Inada K., Kobayashi S., Regioselective polymerization of divinyl sebacate and triols using lipase catalyst, *Macromol. Rapid Comun.*, **20**, 1999, 171–174.
63. Tsujimoto T., Uyama H., Kobayashi S., Enzymatic synthesis of cross-linkable polyesters from renewable resources, *Biomacromolecules*, **2**, 2001, 29–31.
64. Tsujimoto T., Uyama H., Kobayashi S., Enzymatic synthesis and curing of biodegradable crosslinkable polyesters, *Macromol. Biosci.*, **2**, 2002, 329–335.
65. Uyama H., Kuwabara M., Tsujimoto T., Kobayashi S., Enzymatic synthesis and curing of biodegradable epoxide-containing polyesters from renewable resources, *Biomacromolecules*, **4**, 2003, 211–215.
66. Uyama H., Kobayashi S., Enzymatic synthesis of polyesters via polycondensation, *Adv. Polym. Sci.*, **194**, 2006, 133–158.
67. Hu J., Gao W., Kulshrestha A., Gross R.A., "Sweet polyesters": Lipase-catalyzed condensation-polymerizations of alditols, *Macromolecules*, **39**, 2006, 6789–6792.
68. Sudesh K., Doi Y., Polydroxyalkanoates, in *Handbook of Biodegradable Polymers* (Ed. Bastoli C.), Rapra Technology Ltd., Shrewsbury, UK, 2005.
69. Drew H.D.K., Haworth W.N., Studies in Polymerisation. Part I. 2:3:4-Trimethyl L-arabinolactone, *J. Chem. Soc.*, 1927, 775–779.
70. Mehltretter C.L., Mellies R.L., The polyesterification of 2,4;3,5-di-*O*-methylene-D-gluconic acid, *J. Am. Chem. Soc.*, **77**, 1955, 427–428.
71. Zamora F., Bueno M., Molina I., Orgueira H.A., Varela O., Galbis J.A., Synthesis of carbohydrate-based monomers that are precursors for the preparation of stereoregular polyamides, *Tetrahedron: Asymm.*, **7**, 1996, 1811–1818.
72. Molina Pinilla I., Bueno Martínez M., Galbis J.A., Synthesis of 2,3,4,5-tetra-*O*-methyl-D-glucono-1,6-lactone as a monomer for the preparation of copolyesters, *Carbohyd. Res.*, **338**, 2003, 549–555.
73. Romero Zaliz C.L., Varela O., Facile synthesis of a D-galactono-1,6-lactone derivative, a precursor of a copolyester, *Carbohyd. Res.*, **341**, 2006, 2973–2977.
74. Kiely D.E., Carbohydrate diacids: Potential as commercial chemicals and hydrophobic polyamide precursors, *ACS Symp. Ser.*, **784**, 2001, 64–80.
75. Rodríguez-Galán A., Bou J.J., Muñoz-Guerra S., Stereoregular polyamides derived from methylene-L-tartaric acid and aliphatic diamines, *J. Polym. Sci. Part A: Polym. Chem.*, **30**, 1992, 713–721.
76. Bou J.J., Rodríguez-Galán A., Muñoz-Guerra S., Optically active polyamides derived from L-tartaric acid, *Macromolecules*, **26**, 1993, 5664–5670.
77. Iribarren I., Alemán C., Bou J.J., Muñoz-Guerra S., Crystal structures of optically active polyamides derived from di-*O*-methyl-L-tartaric acid and 1,*n*-alkanediamines: A study combining energy calculations, diffraction analysis, and modeling simulations, *Macromolecules*, **29**, 1996, 4397–4405.
78. Ruiz-Donaire P., Bou J.J., Muñoz-Guerra S., Rodríguez-Galán A., Hydrolytic degradation of polyamides based on L-tartaric acid and diamines, *J. Appl. Polym. Sci.*, **58**, 1995, 41–54.
79. Regaño C., Martínez de Ilarduya A., Iribarren I., Rodríguez Galán A., Galbis J.A., Muñoz-Guerra S., Stereocopolyamides derived from 2,3-di-*O*-methyl-D- and -L-tartaric acids and hexamethylenediamine. 1. Synthesis, characterization, and compared properties, *Macromolecules*, **29**, 1996, 8404–8412.
80. Iribarren I., Alemán C., Regaño C., Martínez de Ilarduya A., Bou J.J., Muñoz-Guerra S., Stereocopolyamides derived from 2,3-di-*O*-methyl-D- and -L-tartaric acids and hexamethylenediamine. 2. Influence of the configurational composition on the crystal structure of optically compensated systems, *Macromolecules*, **29**, 1996, 8413–8424.
81. Villuendas I., Molina I., Regaño C., Bueno M., Martínez de Ilarduya A., Galbis J.A., Muñoz-Guerra S., Hydrolytic degradation of poly(ester amide)s made from tartaric and succinic acids. Influence of the chemical structure and microstructure on degradation rate, *Macromolecules*, **32**, 1999, 8033–8040.
82. Bou J.J., Muñoz-Guerra S., Synthesis and characterization of a polytartaramide based on L-lysine, *Polymer*, **36**, 1995, 181–186.

83. Majó M.A., Alla A., Bou J.J., Herranz C., Muñoz-Guerra S., Synthesis and characterization of polyamides obtained from tartaric acid and L-lysine, *Eur. Polym. J.*, **40**, 2004, 2699–2708.
84. Alla A., Oxelbark J., Rodríguez-Galán A., Muñoz-Guerra S., Acylated and hydroxylated polyamides derived from L-tartaric acid, *Polymer*, **46**, 2005, 2854–2861.
85. Kiely D.E., Chen. L., Lin T.-H., Hydroxylated nylons based on unprotected esterified D-glucaric acid by simple condensation reactions, *J. Am. Chem. Soc.*, **116**, 1994, 571–578.
86. Chen. L., Kiely D.E., Synthesis of stereoregular head, tail hydroxylated nylons derived from D-glucose, *J. Org. Chem.*, **61**, 1996, 5847–5851.
87. Kiely D.E., Chen. L., Lin T.-H., Synthetic polyhydroxypolyamides from galactaric, Xylaric, D-glucaric, and D-mannaric acids and alkylenediamine monomers–Some comparisons, *J. Polym. Sci. Part A: Polym. Chem.*, **38**, 2000, 594–603.
88. Styron S.D., Kiely D.E., Ponder G., Alternating stereoregular head, tail–tail, head–poly(alkylene D-glucaramides derived from a homogeneous series of symmetrical diamido-di-D-glucaric acid monomers, *J. Carbohyd. Chem.*, **22**, 2003, 123–142.
89. Carter A., Morton D.W., Kiely D.E., Synthesis of some poly(4-alkyl-4-azaheptamethylene-D-glucaramides), *J. Polym. Sci. Part A: Polym. Chem.*, **38**, 2000, 3892–3899.
90. Morton D.W., Kiely D.E., Synthesis of poly(azaalkylene aldaramide)s and poly(oxaalkylenealdaramide)s derived from D-glucaric and D-galactaric acids, *J. Polym. Sci. Part A: Polym. Chem.*, **38**, 2000, 604–613.
91. Viswanathan A., Kiely D.E., Mechanisms for the formation of diamides and polyamides by aminolysis of D-glucaric acid esters, *J. Carbohyd. Chem.*, **22**, 2003, 903–918.
92. Styron S.D., French A.D., Friedrich J.D., Lake C.H., Kiely D.E., MM3(96) Conformational analysis of D-glucaramide and X-ray crystal structures of three D-glucaric acid derivatives-Models for synthetic poly(alkylene D-glucaramides), *J. Carbohyd. Chem.*, **21**, 2002, 27–51.
93. Morton D.W., Kiely D.E., Evaluation of the film and adhesive properties of some block copolymerpolyhydroxypolyamides from esterified aldaric acids and diamines, *J. Appl. Polym. Sci.*, **77**, 2000, 3085–3092.
94. Orgueira H.A., Varela O., Synthesis and characterization of stereoregular AABB-type polymannaramides, *J. Polym. Sci. Part A: Polym. Chem.*, **39**, 2001, 1024–1030.
95. Orgueira H.A., Erra-Balsels R., Nonami H., Varela O., Synthesis of chiral polyhydroxy polyamides having chains of defined regio and stereoregularity, *Macromolecules*, **34**, 2001, 687–695.
96. Orgueira H.A., Varela O., Stereocontrolled synthesis of stereoregular, chiral analogs of nylon 5,5 and nylon 5,6, *Tetrahedron: Asymm.*, **8**, 1997, 1383–1389.
97. Ravid U., Silverstein R.M., Smith L.R., Synthesis of the enantiomers of 4-substituted-γ-lactones with known absolute configuration, *Tetrahedron*, **34**, 1978, 1449–1452.
98. García-Martín M.G., Benito Hernández E., Ruiz Pérez R., Alla A., Muñoz-Guerra S., Galbis J.A., Synthesis and characterization of linear polyamides derived from L-arabinitol and xylitol, *Macromolecules*, **37**, 2004, 5550–5556.
99. Mancera M., Roffé I., Rivas M., Galbis J.A., New derivatives of D-mannaric and galactaric acids. Synthesis of a new stereoregular Nylon 66 analog from carbohydrate-based monomers having the D-manno configuration, *Carbohyd. Res.*, **338**, 2003, 1115–1119.
100. Mancera M., Roffé I., Rivas M., Silva C., Galbis J.A., Synthesis of D-mannitol and L-iditol monomers for the preparation of polyamides. Preparation of new regioregular AABB-type polyamides, *Carbohyd. Res.*, **337**, 2002, 607–611.
101. Mancera M., Zamora F., Roffé I., Bermúdez M., Alla A., Muñoz-Guerra S., Galbis J.A., Synthesis and properties of poly(D-mannaramide)s and poly(galactaramide)s, *Macromolecules*, **37**, 2004, 2779–2783.
102. Thiem J., Lüders H., Synthesis and properties of polyurethanes derived from dianhydrohexytols, *Makromol. Chem.*, **187**, 1986, 2775–2785.
103. Bachmann F., Reimer J., Ruppenstein M., Thiem J., Synthesis of a novel starch-derived AB-type polyurethane, *Macromol. Rapid Comunn.*, **19**, 1998, 21–26.
104. Bachmann F., Reimer J., Ruppenstein M., Thiem J., Synthesis of novel polyurethanes and polyureas by polyaddition reactions of dianhydrohexytol configurated diisocyanates, *Macromol. Chem. Phys.*, **202**, 2001, 3410–3419.
105. Thiem J., Bachmann F., Synthesis and properties of polyamides from anhydro- and dianhydroalditols, *Makromol. Chem.*, **192**, 1991, 2163–2182.
106. Bachmann F., Thiem J., Syntheis of hydrophilic carbohydrate-derived polyamides, *J. Polym. Sci. Part A. Polym. Chem.*, **30**, 1992, 2059–2062.
107. Thiem J., Bachmann F., Synthesis and properties of polyamides from glucosamine and glucose derivatives, *Makromol. Chem.*, **194**, 1993, 1035–1057.
108. Bou J.J., Iribarren I., Martínez de Ilarduya A., Muñoz-Guerra S., Stereoregular polyamides entirely based on tartaric acid, *J. Polym. Sci. Part A: Polym. Chem.*, **37**, 1999, 983–993.
109. Mancera M., Roffe I., Al-Kass S.S.J., Rivas M., Galbis J.A., Synthesis and characterization of new stereoregular AABB-type polyamides from carbohydrate-based monomers having D-manno and L-ido configurations, *Macromolecules*, **36**, 2003, 1089–1097.

110. Molina I., Bueno M., Galbis J.A., Synthesis of a stereoregular polyesteramide from L-arabinose, *Macromolecules*, **28**, 1995, 3766–3770.
111. Molina I., Bueno M., Zamora F., Galbis J.A., Synthesis and properties of stereoregular poly(esteramide)s derived from carbohydrates, *J. Polym. Sci. Part A: Polym. Chem.*, **36**, 1998, 67–77.
112. Molina I., Bueno M., Galbis , Zamora F., Galbis J.A., Carbohydrate-based copolymers. Synthesis and characterization of copoly(ester amide)s containing L-arabinose units, *Macromolecules*, **35**, 2002, 2977–2984.
113. Molina I., Bueno M., Galbis , Zamora F., Galbis J.A., Hydrolytic degradation of poly(ester amide)s derived from carbohydrates, *Macromolecules*, **30**, 1997, 3197–3203.
114. Villuendas I., Molina I., Regaño C., Bueno M., Martínez de Ilarduya A., Galbis J.A., Muñoz-Guerra S., Hydrolytic degradation of poly(esteramide)s made from tartaric and succinic acids: Influence of the chemical structure and microstructure on degradation rate, *Macromolecules*, **32**, 1999, 8033–8040.
115. Molina I., Bueno M., Galbis J.A., Carbohydrate-based copolymers. Hydrolysis of copoly(ester amide)s containing L-arabinose units, *Macromolecules*, **35**, 2002, 2985–2992.
116. Bueno M., Galbis J.A., García-Martín M.G., De Paz M.V., Zamora F., Muñoz-Guerra S., Synthesis of stereoregular polygluconamides from D-glucose and D-glucosamine, *J. Polym. Sci.: Part A. Polym. Chem.*, **33**, 1995, 299–305.
117. García-Martín M.G., De Paz-Báñez M.V., Galbis J.A., Preparation of 2-amino-2-deoxy-3,4,5,6-tetra-*O*-methyl-D-gluconic acid hydrochloride. An intermediate in the preparation of polypeptide-type polyamides, *Carbohyd. Res.*, **240**, 1993, 301–305.
118. Bueno M., Zamora F., Ugalde M.T., Galbis J.A., Some derivatives of 6-amino-6-deoxy-D-gluconic acid that are precursors for the synthesis of polyamides, *Carbohyd. Res.*, **230**, 1992, 191–195.
119. Zamora F., Bueno M., Molina I., Orgueira H.A., Varela O., Galbis J.A., Synthesis of carbohydrate-based monomers that are precursors for the preparation of stereoregular polyamides, *Tetrahedron: Asymm.*, **7**, 1996, 1811–1818.
120. Bueno M., Zamora F., Molina I., Orgueira H.A., Varela O., Galbis J.A., Synthesis and characterization of optically active polyamides derived from carbohydrate-based monomers, *J. Polym. Sci. Part A: Polym. Chem.*, **35**, 1997, 3645–3653.
121. Orgueira H.A., Bueno M., Funes J.L., Varela O., Galbis J.A., Synthesis and characterization of chiral polyamides derived from glycine and (S)-5-amino-4-methoxypentanoic acid, *J. Polym. Sci. Part A: Polym. Chem.*, **36**, 1998, 2741–2748.
122. Romero Zaliz C.L., Varela O., Straightforward synthesis of derivatives of D- and L-galactonic acids as precursors of stereoregular polymers, *Tetrahedron: Asymm.*, **14**, 2003, 2579–2586.
123. Romero Zaliz C.L., Varela O., Synthesis of stereoregular poly-*O*-methyl-V- and L-polygalactonamides as nylon 6 analogues, *Tetrahedron: Asymm.*, **16**, 2005, 97–103.
124. Hunter D.F.A., Fleet G.W.J., Towards hydroxylated nylon 6: Olygomers from a protected 6-amino-6-deoxy-D-allonate, *Tetrahedron: Asymm.*, **14**, 2003, 3829–3831.
125. Mayes B.A., Stetz R.J.E., Watterson M.P., Edwards A.A., Ansell C.W.G., Tranter G.E., Fleet G.W.J., Towards hydroxylated nylon 6: Linear and cyclic olygomers from a protected 6-amino-6-deoxy-D-galactonate. A novel class of carbopeptoid-cyclodextrin (CPCD), *Tetrahedron: Asymm.*, **15**, 2004, 627–638.
126. García-Martín M.G., De Paz Báñez M.V., Galbis J.A., Preparation of 3-amino-3-deoxy-2,4,5,6-tetra-*O*-methyl-D-altronic acid hydrochloride, an intermediate in the preparation of a chiral β-polyamide (nylon 3 analog), *J. Carbohyd. Chem.*, **19**, 2000, 805–815.
127. García-Martín M.G., De Paz Báñez M.V., Galbis J.A., Preparation and reactivity of some 3-deoxy-D-altronic acid derivatives, *J. Carbohyd. Chem.*, **20**, 2001, 145–157.
128. García-Martín M.G., De Paz-Báñez M.V., Galbis J.A., Synthesis of poly[isobutyl (2S,3R)-3-benzyloxyaspartate], *Makromol. Chem. Phys.*, **198**, 1997, 219–227.
129. García-Martín M.G., De Paz-Báñez M.V., García-Alvárez M., Muñoz-Guerra S., Galbis J.A., Synthesis and structural studies of 2,3-disubstituted poly(β-peptide)s, *Macromolecules*, **34**, 2001, 5042–5047.
130. Bachmann F., Ruppenstein M., Thiem J., Synthesis of aminosaccharide-derived polymers with urea, urethane and amide linkages, *J. Polym. Sci. Part A. Polym. Chem.*, **39**, 2001, 2332–2341.
131. Henkensmeier D., Abele B.V., Candussio A., Thiem J., Synthesis, characterization and degradability of polyamides derived from aldaric acids and chain end functionalized polydimethylsiloxanes, *Polymer*, **45**, 2004, 7053–7059.
132. Henkensmeier D., Abele B.V., Candussio A., Thiem J., Syntheseis of carbohydrate-segmented polydimethylsiloxanes by hydrosilylation, *J. Polym. Sci., Part A: Polym. Chem.*, **43**, 2005, 3814–3822.
133. Chen X., Gross R.A., Versatile copolymers from L-lactide and D-xylofuranose, *Macromolecules*, **32**, 1999, 308–314.
134. Shen Y., Chen X., Gross R.A., Polycarbonates from sugars: Ring opening polymerization of 1,2-*O*-isopropylidene-D-xylofuranose-3,5-cyclic carbonate (IPXTC), *Macromolecules*, **32**, 1999, 2799–2802.
135. Shen Y., Chen X., Gross R.A., Aliphatic polycarbonates with controlled quantities of D-xylofuranose in the main chain, *Macromolecules*, **32**, 1999, 3891–3897.
136. Kumar R., Gao W., Gross R.A., Functionalized polylactides: Preparation and characterization of L-lactide-co-pentofuranose, *Macromolecules*, **35**, 2002, 6835–6844.
137. Haba O., Tomizuka H., Endo T., Anionic ring-opening polymerization of methyl 4,6-*O*-benzylidene-2,3-*O*-carbonyl-α-D-glucopyranoside: A first example of anionic ring-opening polymerisation of five-membered cyclic carbonate without elimination of CO_2, *Macromolecules*, **38**, 2005, 3562–3563.

– 6 –

Furan Derivatives and Furan Chemistry at the Service of Macromolecular Materials

Alessandro Gandini and Mohamed Naceur Belgacem

ABSTRACT

This chapter reports the most relevant aspects related to the preparation and the characterization of macromolecular structures incorporating furan heterocycles or moieties arising from them. It begins with a short discussion on the major chemical features of furans and on the synthesis of furfural and hydroxymethylfurfural, which constitute the basic precursors to monomers suitable for both chain- and step-growth polymerizations. It then deals with the free radical, cationic, anionic and stereospecific polymerization and copolymerizations and with step-growth systems of suitable furan monomers. Particular emphasis is placed on furan polyesters, polyamides, polyurethanes and conjugated oligomers as the most promising macromolecular architectures. The next section is devoted to the application of the Diels–Alder reaction to the synthesis and properties of linear, crosslinked and dendritic macromolecules, with a strong accent on the reversible nature of these constructions. Finally, the aging of furan polymers is briefly discussed.

Keywords

Furan, Furfural, Hydroxymethylfurfural, Furan polymers, Chain polymerizations, Step-growth polymerizations, Furan polyesters, Furan polyamides, Furan polyurethanes, Furan-containing conjugated oligomers, Diels–Alder reaction, Dendrimers, Reversible crosslinking

6.1 INTRODUCTION

Furan derivatives are ubiquitous in nature in a wide variety of structures, but all of them appear in very modest amounts within specific vegetable and animal (including human) species. The interest of these compounds is mostly relevant to phytochemists and other natural product practitioners [1]. The chemistry associated with the furan heterocycle has been the subject of extensive studies over the last century and is sill a very active field of research because of its important repercussions in areas such as synthons, pharmaceuticals and other fine chemicals [2], liquid crystals [3], as well as polymer science and technology [4]. The classic treatise by Dunlop and Peters [5], which placed furan chemistry in a highly visible perspective, was followed by a number of comprehensive reviews and monographs which progressively updated the state of the art [1–4].

The purpose of this chapter is to deal exclusively with the use of furan compounds and the exploitation of specific features related to furan chemistry with the aim of synthesizing polymeric materials. Implicit in this treatment is the fact that vegetable renewable resources, in the form of mono, oligo and polysaccharides, are excellent sources of two first generation furans which, in turn, represent sources of a variety of monomers and other derivatives relevant to polymer synthesis. Although this topic has been reviewed on previous occasions [4], important advances have enriched it in recent years. An attempt will therefore be made to provide here a balanced treatment covering both the most salient achievements reported in the past several decades and novel promising contributions and perspectives.

Two striking differences distinguish the essence of this chapter from most other chapters, namely: (i) the fact that the furan compounds relevant to polymer synthesis are not found as such in nature but are instead obtained from parent renewable resources and (ii) it is possible in principle to envisage a whole new realm of polymer materials based on furan monomers and furan chemistry, covering a very wide spectrum of macromolecular structures. Concerning the first point, the massive availability of saccharidic precursors and their relatively simple conversion into furan derivatives, eliminate in fact any apparent problem of absence of such natural structures. As for the second point, its unique relevance has to do with the potential perspective of a viable alternative to the present reality based on polymer chemistry derived from fossil resources. In other words, the biomass refinery concept would be applied here to the synthesis of different furan monomers, simulating the equivalent petroleum counterpart.

The impressive achievements of the petroleum-based synthesis of monomers, and the huge success of the ensuing polymeric materials, are ineluctable realities which combine pervasive technical performances with remarkable economy (at least for the time being). It follows that the challenges faced by the possibility of a novel branch of polymer chemistry based on furan monomers are daunting. We hope to show in this chapter that daunting does not mean impossible within a realistic framework of research efforts and time span.

6.2 THE FURAN HETEROCYCLE AND SOME OF ITS CHEMICAL FEATURES

Furan (**1**), a volatile colourless liquid with an aromatic scent, is one of the major representatives of the five-member unsaturated heterocycle family, which also includes pyrrole (**2**) and thiophene (**3**), as well as several other less widespread homologues.

(1) (2) (3)

Molecular orbital calculations provide the following set of resonance-contributing structures for the furan heterocycle:

(1) (1a) (1b) (1c)

In furan chemistry, substitution reactions, such as alkylation, halogenation, sulphonation and nitration, occur regioselectively at C2 and/or C5 when these positions are not substituted, suggesting that **1a** is the dominant resonance structure. Compared with its classical homologues **2** and **3**, furan displays the lowest aromatic and the highest dienic character, as illustrated in Scheme 6.1 which also gives the two limiting structures of benzene and cyclopentadiene, respectively.

This predominant dienic character represents a peculiar chemical feature that bears important mechanistic consequences in both organic chemistry and more especially here, in original approaches to macromolecular synthe-

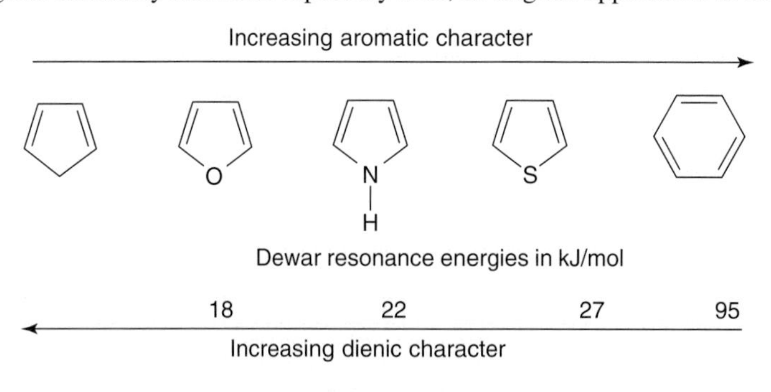

Scheme 6.1

sis. Probably the most exemplary of these consequences is the ease with which furan and some of its derivatives undergo the Diels–Alder (DA) reaction as dienes. Of all cycloadditions, this is perhaps the most emblematic because of its widespread use in organic syntheses involving a large combination of dienes and dienophiles [6]. It is not excessive to assert that the DA reaction between furan and maleic anhydride represents the very symbol of this reversible interaction, as found in many organic chemistry textbooks.

The following examples of DA cyclizations (Schemes 6.2–6.4) illustrate schematically the formation of the corresponding adducts using either a mono- or a bis-dienophile and emphasize the reversibility of these reactions which is promoted by an increase in temperature.

The extension of these mechanisms to DA reagents bearing multiple functionalities opens the way to various original approaches to polymer syntheses as discussed below.

The pronounced regioselectivity of the electrophilic substitution at the unsubstituted C2 and C5 positions (see Scheme 6.5), can also be exploited for building interesting macromolecular architectures by cationic polymerization as illustrated in Section 6.5.2.

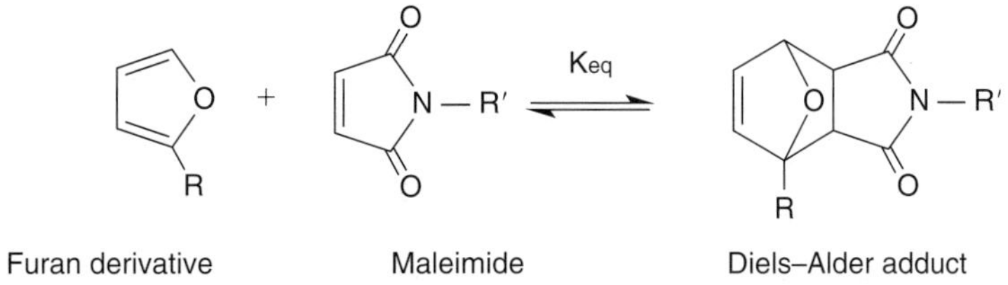

Furan derivative + Maleimide $\overset{K_{eq}}{\rightleftharpoons}$ Diels–Alder adduct

Scheme 6.2

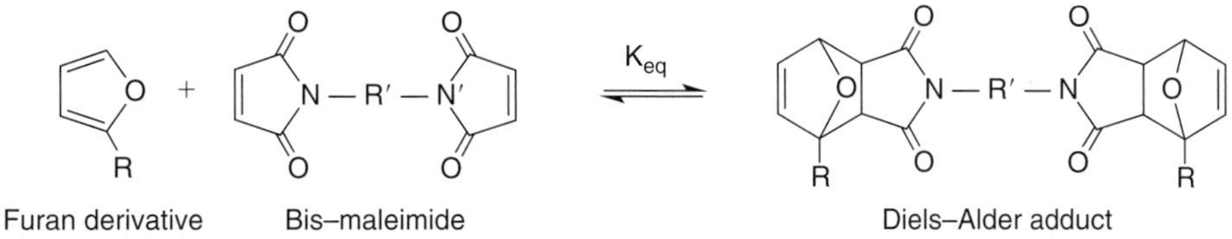

Furan derivative + Bis–maleimide $\overset{K_{eq}}{\rightleftharpoons}$ Diels–Alder adduct

Scheme 6.3

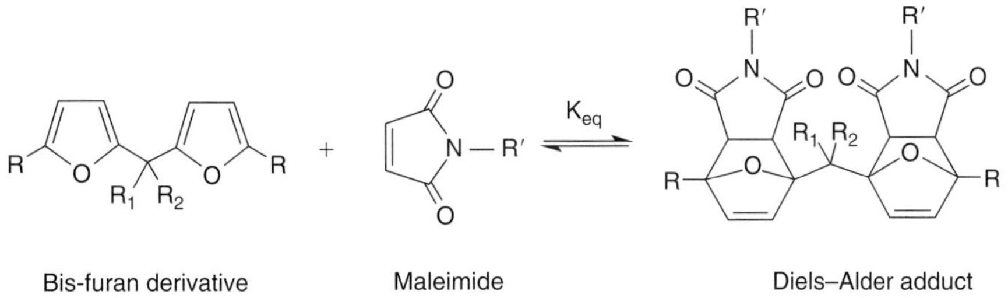

Bis-furan derivative + Maleimide $\overset{K_{eq}}{\rightleftharpoons}$ Diels–Alder adduct

Scheme 6.4

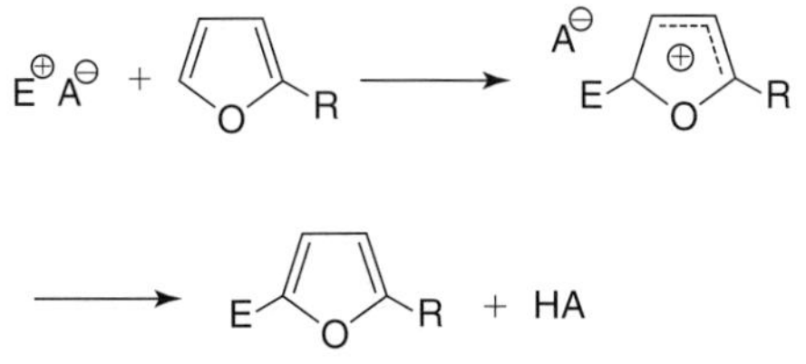

$E^{\oplus}A^{\ominus}$ + furan–R → $A^{\ominus}$ E–furan$^{\oplus}$–R → E–furan–R + HA

Scheme 6.5

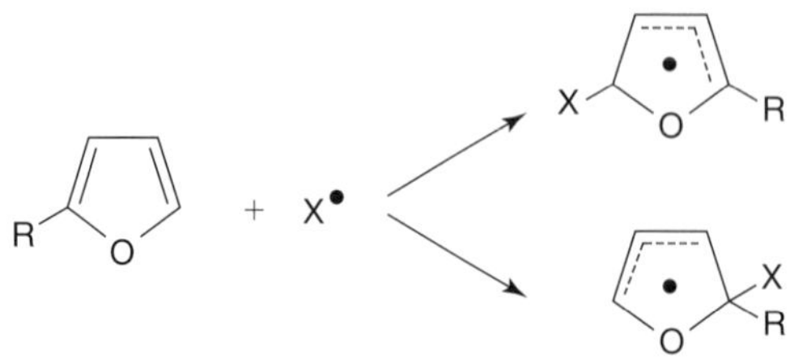

Scheme 6.6

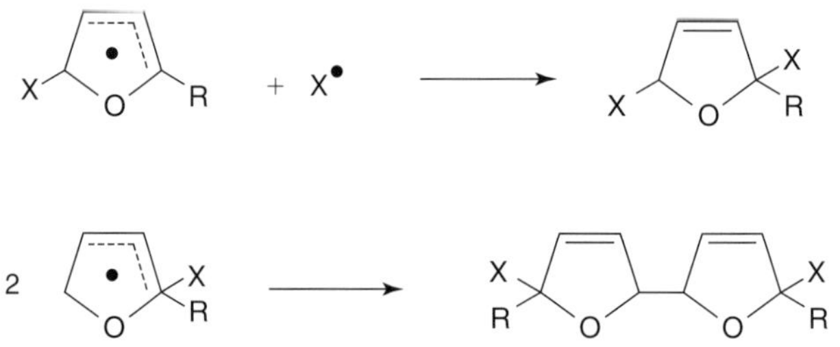

Scheme 6.7

Scheme 6.8

Another peculiar feature of the furan heterocycle has to do with its behaviour in the presence of free radicals. The addition of primary radicals occurs predominantly at C2 or C5, with a kinetic preference for the unsubstituted position. The resulting furan radicals differ in structure (Scheme 6.6), but display a common sluggishness towards propagation with further furanic substrate molecules because of their relatively good stabilization. The two major reaction pathways open to these intermediates are therefore either coupling with other primary radicals (or among themselves) to give various mono- and di-dihydrofurans (Scheme 6.7) or homolytic aromatic substitution resulting in the corresponding novel furan structure (Scheme 6.8) [4d, 7].

The fact that the typical furyl radicals in Scheme 6.6 are unable to propagate in the classical mode of free radical polymerization because of their resonance stabilization, has profound consequences in terms of the reactivity of furan monomers in this type of polymerization and the role of furan derivatives as possible perturbing agents in free radical chain reactions in general. These specific aspects will be illustrated below.

The sensitivity of the furan ring to acid-catalyzed hydrolysis must finally be mentioned as one of its typical chemical features. Its intervention in the context of this monograph must be seen as an undesired event to be avoided, or at least minimized, since its mechanism leads to the destruction of the heterocycle with the formation of aliphatic carbonyl compounds, as illustrated in the simplified Scheme 6.9. It is therefore clear that any polymerization system requiring the preservation of the furan or cognate structures in the final product would be marred by side reactions caused by the presence of moisture in an acidic medium. Curiously, the acid-catalyzed hydrolysis of 2,5-dimethyfuran in a water–ethanol medium shows self-oscillating features [8].

6.3 FURFURAL AND HYDROXYMETHYLFURFURAL

Furfural or 2-furancarboxyaldehyde (**F**) was first obtained in the early nineteenth century and became an industrial commodity about a century later, to reach an industrial production today of some 280 000 tons per year [9]. It can be readily and economically prepared from a vast array of agricultural and forestry wastes containing pentoses (see Chapter 13) in sufficient amounts to justify a commercial exploitation. Examples of these renewable resources are corn cobs, oat and rice hulls, sugarcane bagasse, cotton seeds, olive husks and stones, as well

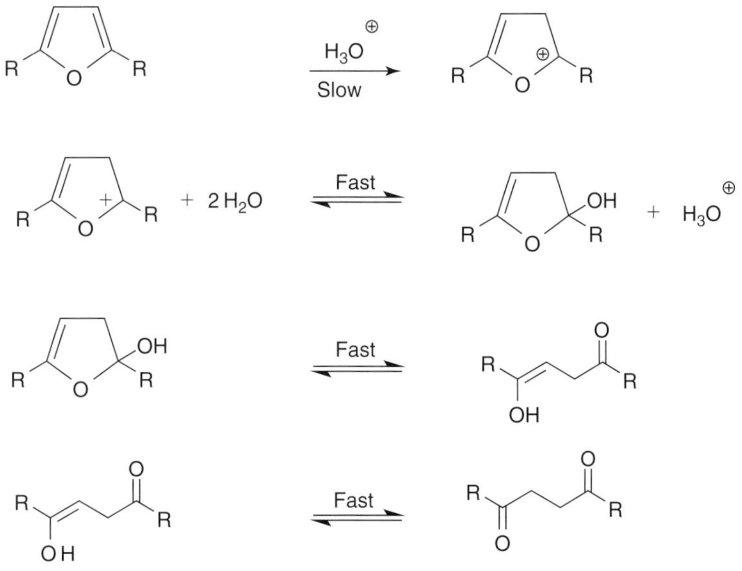

Scheme 6.9

as wood chips. This lack of specificity explains why **F** is produced industrially all over the globe, irrespective of the geo-economic context, simply because virtually every country cultivates a species containing pentoses. Yields up to about 15 per cent of **F** with respect to dry matter are obtained from processes which always involve an aqueous acidic medium and fairly high temperature [10]. The reaction sequence leading to **F** first goes through the hydrolysis of the polymeric pentoses down to the corresponding monosaccharide (aldopentose), which is then dehydrated progressively and finally cyclized to give **F** (from xylose), and in much smaller proportions, 5-methylfurfural (**MF**) (from rhamnose) as shown in Scheme 6.10. **F** and **MF** are then separated by distillation.

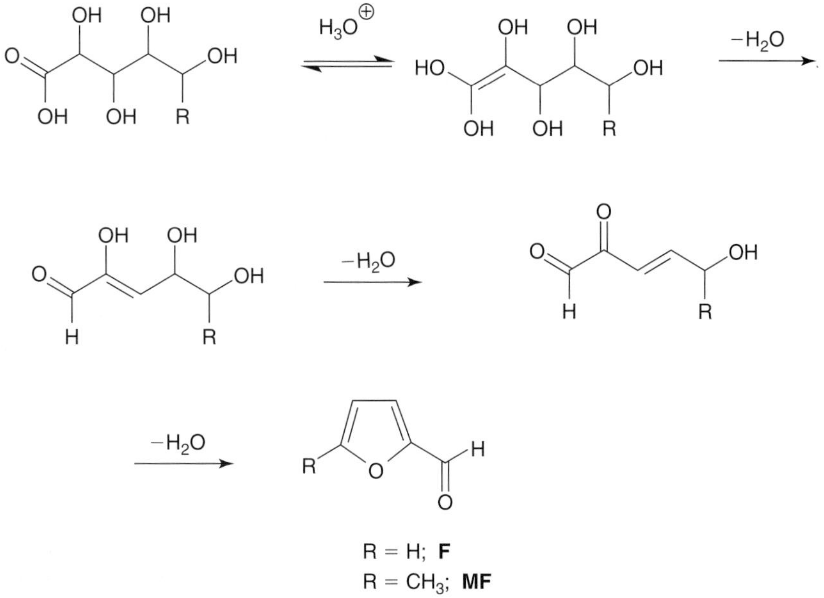

R = H; **F**
R = CH_3; **MF**

Scheme 6.10

At the current market price of about $1 per kg [9], **F** is a viable chemical commodity, both as such and as a precursor to other furan and non-furan derivatives.

Hydroxymethylfurfural or 2-hydoxymethyl-5-furancarboxyaldehyde (**HMF**) represents the other major first generation furan derivative whose synthesis is carried out from hexoses in the form of the corresponding mono, oligo and polysaccharides through a sequential mechanism entirely equivalent to that shown in Scheme 6.10, but ending with **HMF**. This process has been the object of much laboratory and technological research without reaching yet beyond the stage of pilot plants [4e, 11]. Recent promising novel approaches [12] suggest that the industrial production of **HMF** could soon become a reality. Given the high sensitivity of this compound to resinification induced by acidic impurities and other agents, it seems likely that its commercialization might involve its preliminary transformation into the corresponding dialdehyde (2,5-furancarboxydialdehyde, **FCDA**) or diacid (2,5-furandicarboxylic acid, **FDCA**).

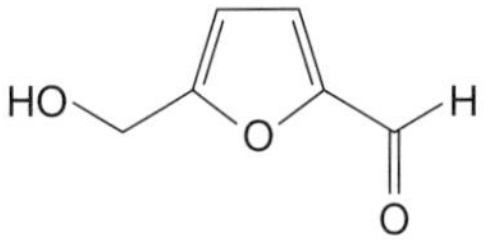

HO H O O

HMF

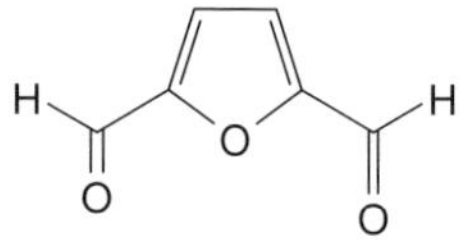

H H O O O

FCDA

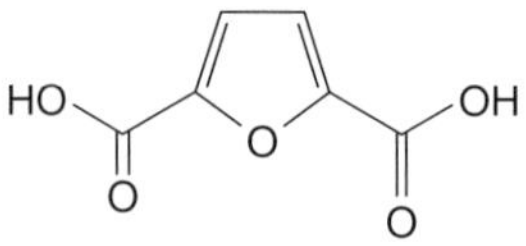

HO OH O O O

FDCA

These two basic furan derivatives obtained directly from cheap vegetable resources find important uses as chemical precursors in a variety of industrial and fine chemical processes which are well documented [4e, 11, 13]. Only those leading to furan monomers and to other compounds which can be exploited in polymer synthesis will be examined further hereafter.

6.4 FURAN MONOMERS

F and **MF** are typical precursors to furan monomers bearing a moiety which can be polymerized by chain-reaction mechanisms. Scheme 6.11 provides a non-exhaustive array of entries into such structures, which have all been synthesized, characterized and polymerized [4]. A major exception to this general postulate is constituted by furfuryl alcohol (2-hydroxymethylfuran, **FA**), which is in fact still today the most important commercially available furan compound, obtained by the catalytic reduction of **F** involving more than 80 per cent of its world production. **FA** is widely used as a polycondensation monomer and does not therefore belong to the class of compounds shown in Scheme 6.11.

HMF on the other hand, is ideally suited as a precursor to bifunctional furan monomers to be used in step growth reactions as shown in Scheme 6.12 which includes **FDCA** and **FCDA**. Again, all these compounds were synthesized and used to prepare the corresponding polycondensates, like polyesters, polyamides, polyurethanes, etc [4].

Another approach to the synthesis of bifunctional monomers which has also been widely explored, calls upon the acid-catalyzed condensation of the corresponding monofunctional furan derivatives with an aldehyde or a ketone, as shown in Scheme 6.13 [4].

These difuranic monomers were also the object of detailed polycondensation studies, as discussed below.

All the monomer structures shown in Schemes 6.11–6.13 are in fact the furan counterparts of the typical aliphatic and aromatic monomers that have been synthesized from petrochemical precursors and widely studied since the beginning of last century. In other words, the only original characters of these novel families is, on the one hand the fact that they all bear one or several furan heterocycles and, on the other hand, the fact that they all derive from precursors which are based on vegetable renewable resources.

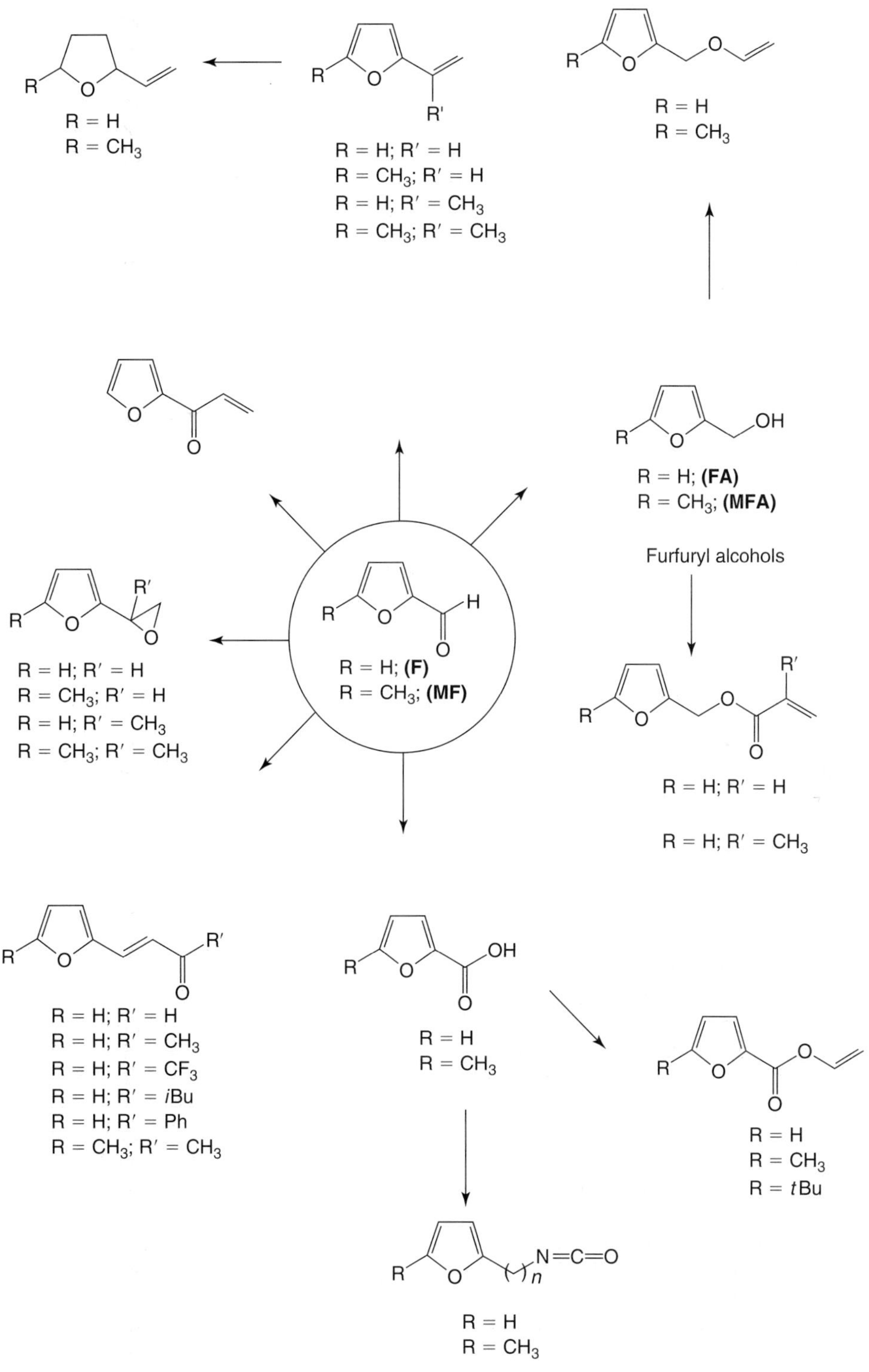

R = H
R = CH_3

R = H; R′ = H
R = CH_3; R′ = H
R = H; R′ = CH_3
R = CH_3; R′ = CH_3

R = H
R = CH_3

R = H; **(FA)**
R = CH_3; **(MFA)**

Furfuryl alcohols

R = H; R′ = H
R = CH_3; R′ = H
R = H; R′ = CH_3
R = CH_3; R′ = CH_3

R = H; **(F)**
R = CH_3; **(MF)**

R = H; R′ = H

R = H; R′ = CH_3

R = H; R′ = H
R = H; R′ = CH_3
R = H; R′ = CF_3
R = H; R′ = *i*Bu
R = H; R′ = Ph
R = CH_3; R′ = CH_3

R = H
R = CH_3

R = H
R = CH_3
R = *t*Bu

N=C=O

R = H
R = CH_3

Scheme 6.11

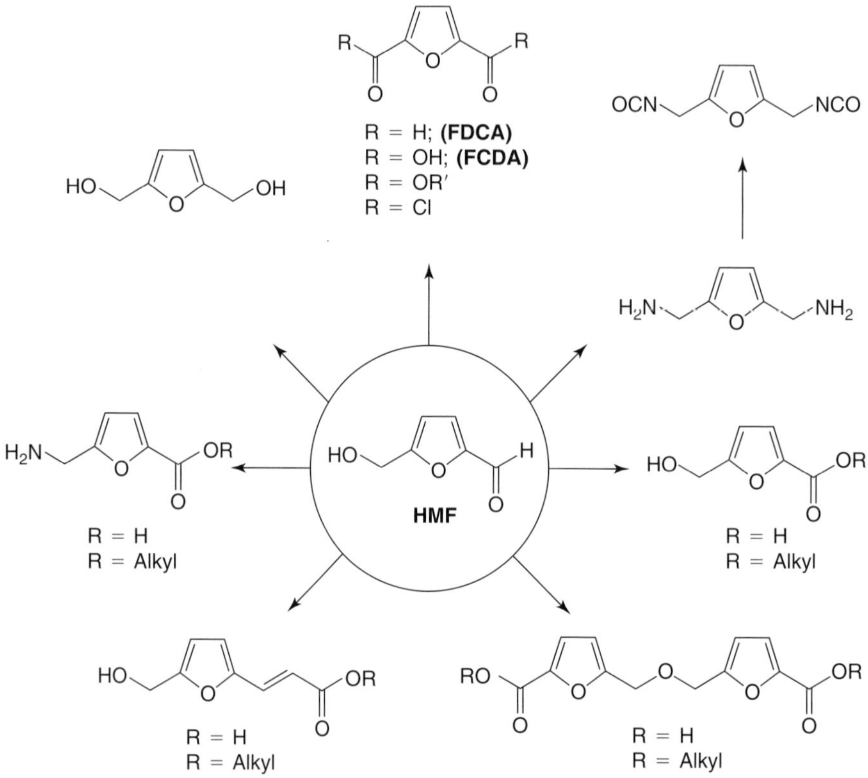

Scheme 6.12

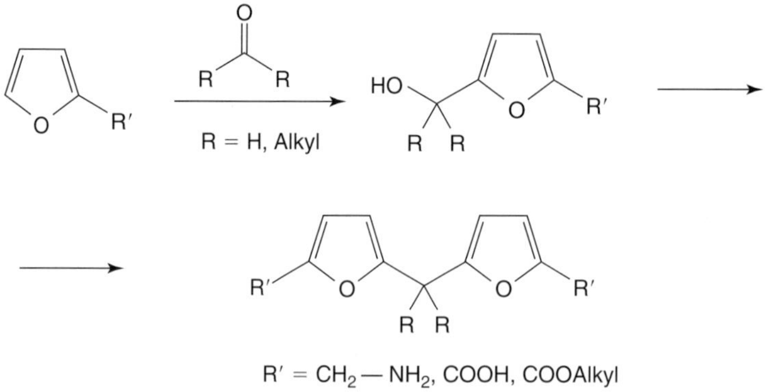

Scheme 6.13

Whereas the monomers shown in Scheme 6.11 are specifically suited for chain polymerization reactions through the moieties external to the ring and thus give rise to polymers or copolymers with the heterocycle pendant to their backbone, the monomers shown in Schemes 6.12 and 6.13 give rise to polycondensate architectures in which the heterocycle is an integral part of the backbone.

6.5 POLYMERS FROM CHAIN REACTIONS

The monomers appearing in Scheme 6.11 are all susceptible to polymerize via chain reactions but, according to their specific structure, their actual response to different types of initiation varies considerably. The organization of this section follows therefore the nature of the polymerization mechanism, viz. via free radical, cationic, anionic and stereospecific initiation.

6.5.1 Free radical polymerization

Given the mechanistic considerations related to the reaction of furans with free radicals (see Schemes 6.6–6.8), it follows that among the monomer structures shown in Scheme 6.11, two limiting behaviours were to be expected, as indeed confirmed experimentally. On the one hand, furan monomers bearing an alkenyl moiety deprived of any stabilizing environment after its transformation into a free radical are not expected to polymerize to any sensible extent because the preferred free radical structure would be that involving the much more stable furyl moiety. This was indeed the behaviour observed with 2-vinyl furoate [4d, 7]. On the other hand, a pronounced stabilization of the alkenyl moiety by an adequate structure, in fact higher than that of the alternative furyl radical, would favour the normal chain growth, which was exactly what happened with furfuryl acrylates and methacrylates [4d], although, even with these monomers, chain branching and crosslinking were found to occur in modest amounts at high conversions [14], indicating that some furyl radicals were indeed formed by the attack of primary or macromolecular radicals onto the rings pendant to the formed polymer chains. Intermediate situations were also studied, notably with 2-vinyl furan [4d].

Copolymerization systems involving the combination of a furan monomer and a conventional counterpart, confirmed this general state of affairs in that only the use of polymerizable furan monomers yielded the expected copolymers. Conversely, comonomers like 2-vinyl furoate either retarded the homopolymerization of well-stabilized standard monomers like methacrylates, or actually inhibited the polymerization of poorly stabilized conventional monomers like vinyl acetate [4d, 7].

In conclusion, although only a modest number of furan monomers responded adequately to free radical initiation, interesting fallout related to the peculiar behaviour of the furan heterocycle in this chemical environment was its exploitation in the synthesis of novel inhibitors based on the generic formula **4**. The trapping of a free radical $X^{\bullet}$ by such molecules produces the corresponding highly stabilized Structure **5**, in which the unpaired electron is delocalized over the entire conjugated moiety. This family of compounds displayed a very high efficiency in stopping free radical based reactions like polymerizations, photochemical processes and oxidations, indeed higher than that of most conventional inhibitors [4d, 7].

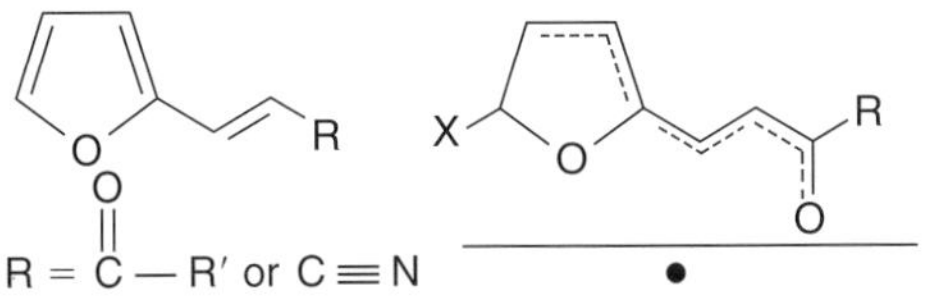

R = C(=O)—R′ or C≡N

(4) **(5)**

A recent extension of this inhibiting strategy [15] was applied in the context of the free radical functionalization of polypropylene during its processing in the melt. Furanacrylic derivatives bearing the basic Fu—CH═CH—C═O moiety, as in Structure **4**, were found to reduce considerably the extent of polymer degradation produced by the free radical chain reactions arising when the typical maleic anhydride/peroxide pair is used without an inhibitor.

Although furan and its methylated homologues do not polymerize by free radical initiation for the reasons discussed above, their DA adducts as well as their 1:1 charge transfer complexes with maleic anhydride are sensitive to this type of activation and give polymers with different structural units and molecular weights in the region of a few thousand [4d]. The properties of these materials have not been the subject of any published study, and hence their possible applications cannot be assessed. Recently, difurylmethane was also copolymerized with maleic anhydride to give a branched alternating copolymer [16].

6.5.2 Cationic polymerization

Furan and its methylated homologues undergo cationic polymerization initiated by both Lewis and Brønsted acids, but the fact that the ensuing branched products (DP = 50–300) bear complex and irreproducible structures, made up of both dihydro- and tetrahydro-furan rings [4a, 4d], casts serious doubts about their interest as materials. Claims related to the synthesis of regular macromolecular structures with these monomers were not substantiated by convincing spectroscopic evidence.

2-Alkenylfurans are readily activated by even the mildest cationic initiators because of their pronounced nucleophilic character. A comprehensive study of these systems involved the four homologues 2-vinylfuran (**VF**), 2-vinyl-5-methylfuran (**MVF**), 2-isopropenylfuran (**IPF**) and 2-isopropenyl-5-methylfuran (**MIPF**) and various Lewis and Brønsted acids (mostly trifluoroacetic acid) [4d].

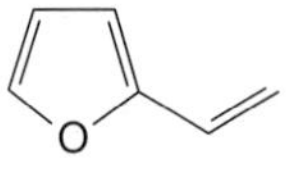

VF 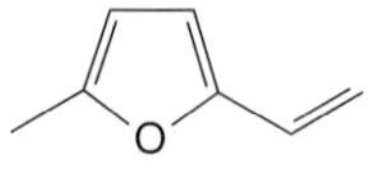 **MVF**

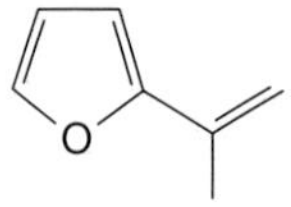

IPF 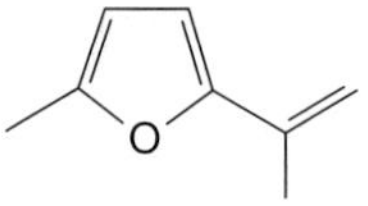 **MIPF**

The choice of these monomers was dictated by preliminary results which showed that **VF** gave polymers bearing different units, which suggested the occurrence of two types of side reactions accompanying the expected propagation through the vinyl unsaturation. On the one hand, an active species could add onto a monomer molecule through an alkylation reaction at C5 (occurring through both normal and anti-Markovnikoff substitutions), instead of incorporating it through the normal vinyl growth. This was to generate Structure **6**. To verify this possibility, **MVF** was tested and indeed the monomer sequence of the ensuing polymer followed the expected regular enchainment through the vinyl moiety.

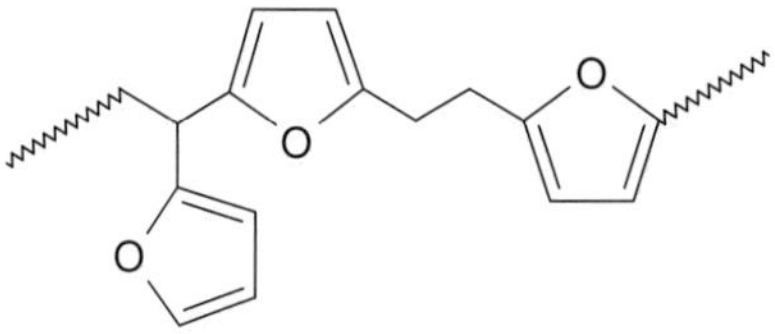

(6)

On the other hand, both poly(**VF**) and poly(**MVF**) were found to contain conjugated unsaturated sequences, formed during the polymerization, but also after complete conversion (giving rise to electronic spectra with bathochromically shifting maxima, all the way into the visible), thus suggesting that the monomer was not involved in the mechanism of their generation. The possibility of hydride/proton losses through reactions between terminally unsaturated polymer molecule and an active species (Scheme 6.14) was confirmed by the fact that **IPF** produced macromolecules deprived of polyunsaturations, although of course these polymers bore the units testifying to the occurrence of C5 alkylation.

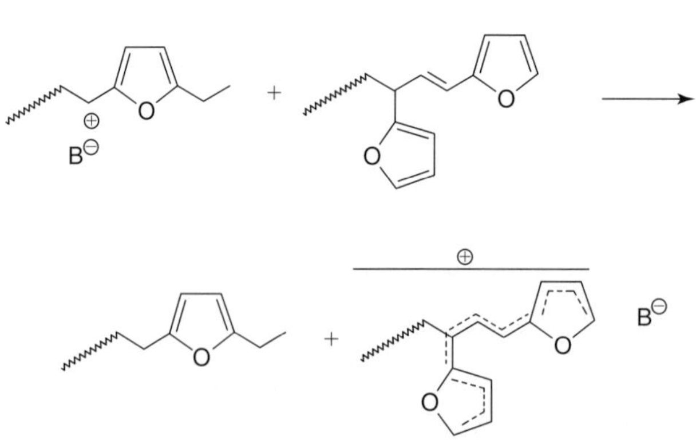

Scheme 6.14

Both side reactions could therefore be eliminated by calling upon **MIPF**, which gave the regular Structure **7**, with DPs close to 100, when polymerized at low temperature to reduce the impact of chain-transfer reactions.

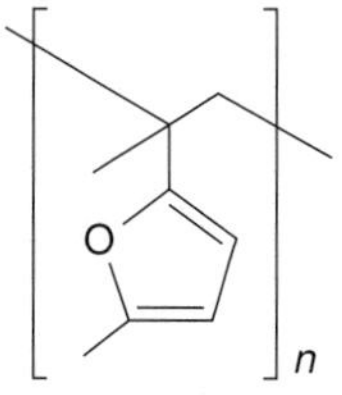

$n = 100$

(7)

An interesting exploitation of the complicated behaviour of **VF** in cationic polymerization consisted in pushing the anomalous reactions to the extreme by using very strong initiators like triflic acid in relatively high concentrations within the bulk monomer and leaving the mixture to stand at room temperature well beyond the complete monomer consumption. These conditions led to black insoluble products because the extent of conjugated sequences had been maximized and the branching associated with alkylation reactions had been allowed to form a polymer network. These materials displayed a very high proton affinity (up to 60 meq g^{-1}) in organic media, promoted by the strong nucleophilic character of the long conjugated sequences and could be used as proton exchange resins, since they could be easily regenerated by an alkali treatment [4d, 17]. This process has also been applied to prepare silica particles covered with the **VF** resin [18].

Other furan monomers which polymerize cationically include 2-furfuryl vinyl ether, 2-vinyl furoate (albeit through a polyalkylation mechanism giving a polyester incorporating the ring into the polymer backbone), **F** and **MF** as co-monomers in conjunction with substituted styrenes and vinyl ethers, as well as 2-furfurylidene methyl ketone (obtained by the base-catalyzed condensation of **F** with acetone) and its homologues [4d].

In a different vein, the high reactivity of the furan heterocycle towards electrophilic substitution at its C2 and C5 positions has been exploited in cationic polymerization to prepare functional oligomers and block copolymers. Thus, in the polymerization of isobutene, 2-methylfuran plays the role of dominant transfer agent with respect to the monomer transfer reaction and gives oligomers bearing a furan end group, which can be appropriately modified, for example by the DA reaction [4a]. An extension of this concept uses bifuran derivatives to play the role of joining moiety by allowing two successive alkylation reactions arising from the sequential cationic polymerization of two different monomers, thus producing the corresponding block copolymers [4a]. This original strategy, developed and patented some 20 years ago [4a, 19], was recently revived in a couple of laboratories, without any appreciable qualitative novelty [20, 21]. Of course, this strategy can also be applied to insert grafts onto a trunk polymer bearing pendant furan moieties with free C5 positions. In this case, the cationic polymerization of the future graft is conducted in the presence of the trunk so that grafting-onto takes place.

6.5.3 Anionic polymerization

The preparation of 2-furyl lithium as an intermediate in organic synthesis is a routine operation. However, sodium naphthalene does not give 2-furyl sodium with furan, but only a resinous product, and sodium was reported to react with furan at 4 K to give the radical anion **8**, which again generates a mixture of unidentified oligomeric products [4d].

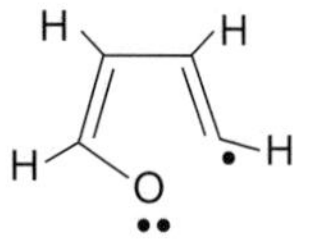

(8)

2-Alkenylfurans do not polymerize under the effect of conventional anionic initiators and the only system which produced a modest yield of polymer (regular structure, DP $\sim$ 100) involved **VF** with sodium biphenyl in HMTP at room temperature [4d]. Monomer conversions higher than about 10 per cent could not be attained, despite a thorough study of the relevant reaction conditions.

2-Furyloxirane (**FO**) is very prone to anionic initiation through the opening of the epoxide ring, even with such mild nucleophiles as alcohols, and gives the corresponding polyether [4d, 4e]. Whereas conventional initiators like tBuOK induced both the α and the β opening of the ring and gave therefore irregular polymer structures in terms of both monomer enchainment and stereospecificity, $Al(iPrO)_3$ acted regiospecifically at the α position and produced regular macromolecules with appreciable tacticity (**9**), as revealed by ^{13}C NMR spectroscopy [4d].

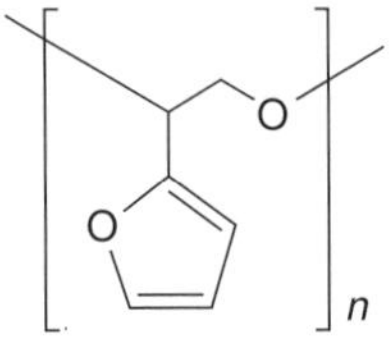

(**9**)

The fact that alcohols are sufficiently nucleophilic to initiate the anionic polymerization of **FO** (a unique feature, not shared by other epoxides, including styrene oxide), provides a useful way of preparing block, star and graft copolymers from, respectively, diols, polyols and OH-bearing polymers such as cellulose and poly(vinyl alcohol). The **FO** blocks and grafts however, are oligomeric in size, because transfer reactions limit their growth [4d].

Another interesting and unique peculiarity related to the behaviour of this monomer is its reactivity towards traces of moisture, without any catalyst. This spontaneous ring opening generates the corresponding diol, which in turn initiates the polymerization of the excess **FO**, as shown in Scheme 6.15. In other words, **FO** polymerizes in the presence of traces of water! [4d].

Furan isocyanates **FI**, **MFI**, **FMI** and **MFMI** were found to give mixtures of polymers (Scheme 6.16) and cyclic trimers (Scheme 6.17) when activated with classical anionic initiators like *n*BuLi and sodium naphthalene,

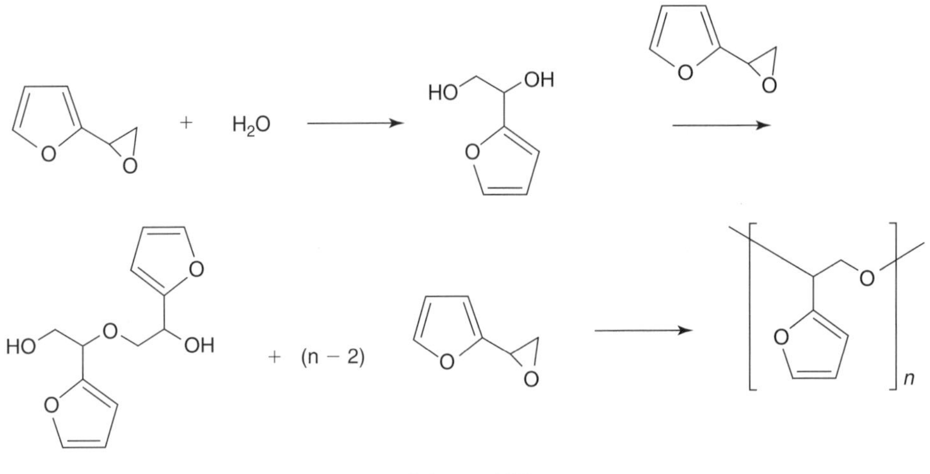

+ H_2O

HO OH

HO O OH + (n − 2)

Scheme 6.15

R O N=C=O *n*

FI: R = H, *n* = 0

MFI: R = CH_3, *n* = 0

FMI: R = H, *n* = 1

MFMI: R = CH_3, *n* = 1

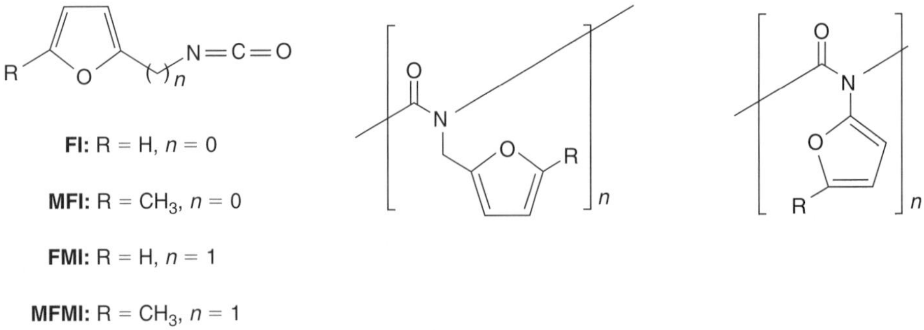

Scheme 6.16

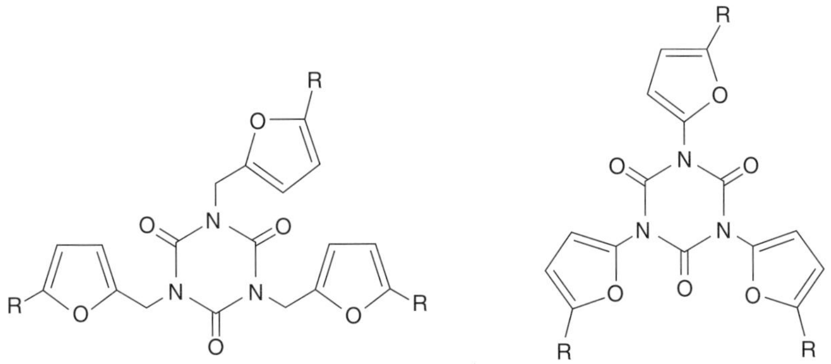

Scheme 6.17

whereas NaCN, the classical promoter of aliphatic isocyanate polymerization, failed to polymerize them, probably because the cyanide anion interacted preferentially with the furan heterocycle. Solvent of low polarity favoured the formation of polymers, as opposed to polar media, which induced predominantly the formation of the corresponding cyclic trimers [4a].

Other monomers which were found to respond positively to anionic polymerization were the 2-furfurylidene ketones, which gave phantom polymers following a propagation mechanism involving the isomerization of the active species before each addition step [4d, 4e].

6.5.4 Stereospecific polymerization

Apart from the partial tacticity of the poly**FO** prepared with $Al(iPrO)_3$ mentioned in the preceding section, no other report of the synthesis of a stereoregular furan polymer has appeared to the best of our knowledge. Numerous attempts to prepare such structures from 2-alkenylfurans using a variety of catalytic combinations failed [4d]. This aspect deserves more attention, because there are no obvious reasons that would exclude *a priori* the possibility of synthesizing stereoregular (and crystallizable) poly(alkenylfuran)s.

6.6 POLYMERS FROM STEP-GROWTH REACTIONS

Before examining the behaviour of the monomers in Schemes 6.12 and 6.13, it is important to summarize the major aspects related to the polycondensation of **F** and **FA**, since these processes provide interesting materials, despite their complexity and the irregular nature of the ensuing resins.

6.6.1 The resinification of furfural

F is seldom employed on its own as a monomer, mostly because the resinous products it produces under the effect of heat and/or acid catalysis are blackish crosslinked materials of little practical interest [4a, 4d]. Some of the mechanistic features of these systems have however been elucidated [4d, 22]. Interestingly, the liquid-phase photoirradiation of **F** [4d] produced linear oligomers whose basic structure suggested condensation reactions, but no intervention of free radicals, viz. Scheme 6.18.

The condensation products of **F** with acetone constitute interesting precursors to resins which have found applications as adhesives, corrosion-resistant coatings and floors for the chemical industry [4a, 4c, 4d]. Numerous other monomer combinations involving **F** have been exploited to prepare materials for different uses [4a, 4c, 4d] and recent additions to these studies include chelate polymers for the adsorption of metal ions [23], nanocomposites incorporating Fe_2O_3 [24], an investigation of the reductive electrochemical polymerization of **F** in acetonitrile [25] and the anticorrosion protection offered by the ensuing polymer [26].

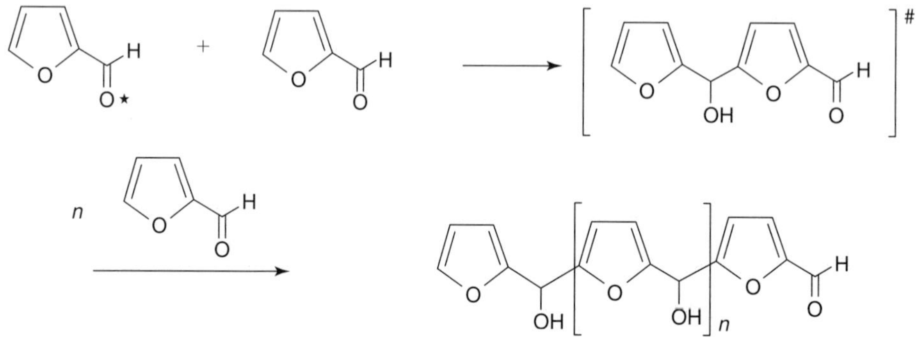

Scheme 6.18

Finally, a considerable amount of work has been devoted to the development of wood adhesives consisting of resins in which furfural is used in conjunction with phenol and formaldehyde or in which furfural-based diamines and diisocyanates were the basic components [27, 28].

Notwithstanding these and other contributions, the role of **F** as a monomer or comonomer is unlikely to acquire a major impact in polymer technology, as compared to **FA**, which has a decisive part to play, as discussed in the next section.

6.6.2 The self-condensation of furfuryl alcohol

FA is the dominant furan commodity on the market precisely because it has provided for decades, and is still providing today, a number of processes and resins, which fulfil precise and outstanding performances in various technological areas. These systems and materials represent as yet by far the most important industrial issues related to furan-based polymers.

First the chemistry will be discussed. The intricate set of reactions that compose the acid-catalyzed and the polycondensation of **FA** constitute the most attentively searched puzzle in the realm of furan polymers because of striking contradictions between an expectedly straightforward macromolecular growth and a surprisingly complex behaviour. Given the by-now well-argumented reactivity of the furan ring in terms of its regiospecific electrophilic substitution at C5, the predominant reaction characterizing the self-condensation of **FA** should lead to a linear poly(2,5-furylmethylene), that is a linear colourless thermoplastic material, following the overall Scheme 6.19, even allowing for some head-to-head condensations giving occasional —CH_2—O—CH_2— bridges.

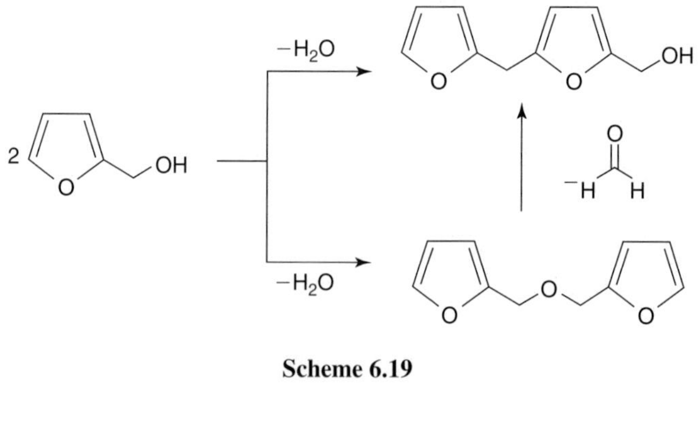

Scheme 6.19

The reality with all such systems, independent of the specific experimental conditions (strength and concentration of the acid, temperature, type of solvent used, if any), is qualitatively different to such a degree that it has puzzled chemists for decades. In fact, although the expected self-condensation reaction shown in Scheme 6.19 does indeed represent the basic growth mechanism, other important events intervene to modify drastically the polymer structure, to the point that the actual product is crosslinked and deeply coloured [4c, 4d].

It was only 20 years ago that the details of the two reactions responsible for these structural anomalies, which alter in a fundamental way the properties of polyFA, were unravelled, thanks to a systematic investigation involving the use of several model compounds. Whereas the details of this study can be found in the original publication, suffice it to give here a brief account of the essential issues. The colour build-up is mechanistically similar to that observed in the cationic polymerization of 2-vinylfuran and stems from the ease of hydride loss associated with a C—H bond in which the carbon atom is directly linked to the furan heterocycle (here to two of them, which makes much more likely). The driving force is of course the stabilization of the ensuing carbenium ion, which can subsequently readily induce a proton loss giving a neutral unsaturated moiety. The repetition of this cycle generates conjugated sequences, as shown in Scheme 6.20.

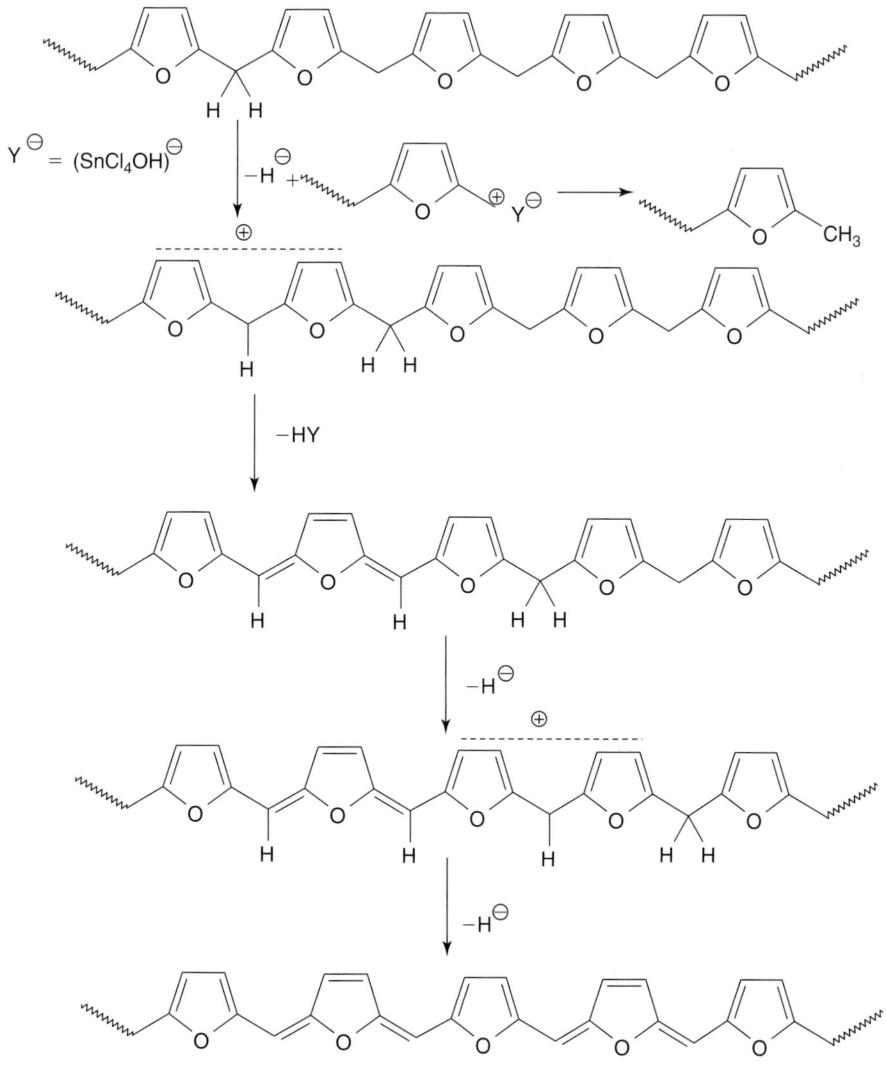

$Y^{\ominus} = (SnCl_4OH)^{\ominus}$

$-H^{\ominus}$

$-HY$

$-H^{\ominus}$

$-H^{\ominus}$

CH_3

Scheme 6.20

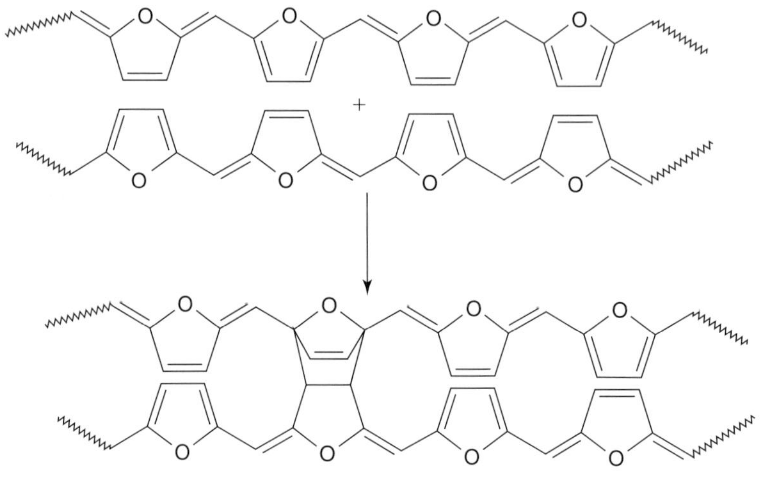

Scheme 6.21

It is only after the formation of some of these unsaturated moieties and of the accompanying =CH— sites linking a furan ring to an exounsaturated 2,5-dihydrofuran counterpart, that the polymer chains begin to couple through a DA mechanism (Scheme 6.21) and this leads to their eventual crosslinking.

Given the ineluctable fact that both 'side' reactions are intrinsic to the very nature of the -2,5-Fu—CH_2— unit formed in the initial **FA** self-condensation oligomers, no obvious way to avoid, or indeed minimize them, has yet been found.

Despite these inevitable drawbacks, **FA**-based resins have found numerous useful applications, which have not ceased to multiply. The classical uses of these resins [4c, 4d, 27], for some of which they are practically irreplaceable, include metal casting cores and moulds, corrosion-resistant coatings, polymer concretes, wood adhesives and binders, sand consolidation and well plugging, low flammability and low smoke release materials and carbonaceous products including graphitic electrodes and carbon micro-particles. Recent additions to this wide catalogue of applications relate to novel polymer concretes [29], wood preservation [30] and stabilization [31], lignocellulosic fibre surface treatment [32] and also high-tech materials like micro- and meso-porous carbon, carbon-silica [33] and PFA-silica materials [34], nanostructured and nanocomposite carbons [35] and Nafion-PFA [36] and sol–gel-based nanocomposites [37].

The future of **FA** as a cheap precursor to a variety of useful materials appears therefore to reflect a growing exploitation in traditional as well as state-of-the-art technologies. Additionally, as pointed out in various subsequent sections, **FA** is finding other novel utilizations as precursor to multifunctional furan monomers.

6.6.3 Polyesters

The pioneering work carried out in Moore's laboratory in the 1970s and 1980s represents the first systematic approach to furan polyesters (after numerous isolated and rather qualitative publications on the topic), which were prepared from 2,5-furandicarboxylic derivatives in conjunction with aliphatic, furanic and tetrahydrofuranic diols as well as with bisphenol A [4d, 4e]. In the same period, the furan hydroxyacid **10** was polymerized to give both linear and cyclic oligomers at relatively low temperatures, but its condensation with ethylene glycol produced black intractable resins [4d, 38].

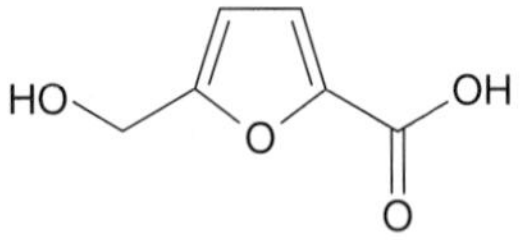

HO OH O O

(10)

The synthesis of polyesters based on various furan diacids derivatives and dianhydrosugar diols [4d, 39] were reported in the context of biomass-derived polymers with biodegradable properties, but regrettably these studies were not pursued.

A series of studies based on the combination of the difuran–dicarboxylic derivatives **11** and a wide choice of diols **12** were undertaken recently using all the classical procedures for the synthesis of polyesters **13**, namely transesterification, condensation between the acid chloride moiety and the OH function and direct condensation. This comprehensive investigation included the structural characterization of all the polymers obtained, the determination of their molecular weight and molecular weight distribution, as well as of the thermal transitions and stability. From this broad set of results, interesting structure properties relationships could be drawn [4e].

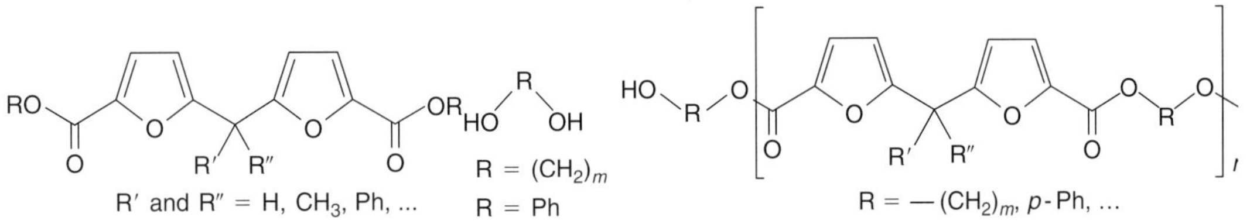

R′ and R″ = H, CH_3, Ph, ...

(11)

R = $(CH_2)_m$
R = Ph

(12)

R = —$(CH_2)_m$, *p*-Ph, ...

(13)

Thus, for example, the glass transition temperature varied considerably as a function of both the nature of the groups appended on the carbon atom joining the two furan rings in **11** and the structure of the diol **12,** in particular when going from aliphatic to aromatic homologues.

In another vein, the polytransesterification of monomer **14**, readily obtained from **HMF**, yielded the interesting material **15** because of the photosensitivity of the -2,5-Fu—CO—CH=CH— moiety (see Section 6.7.2) which reacted by coupling with another such moiety (statistically belonging to another polyester chain), thus ultimately providing a rigid crosslinked product. Copolymers based on **14** and aliphatic hydroxyesters were also prepared and the ensuing crosslinked materials displayed glass transition temperatures well below room temperature and were therefore elastomers [40].

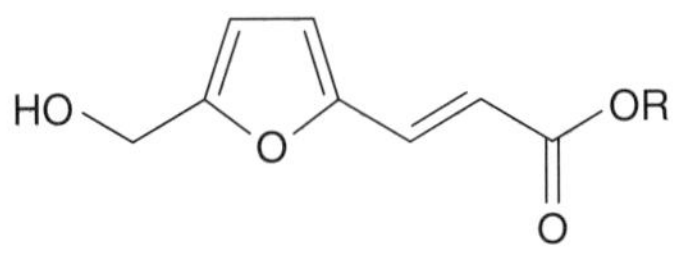

(14)

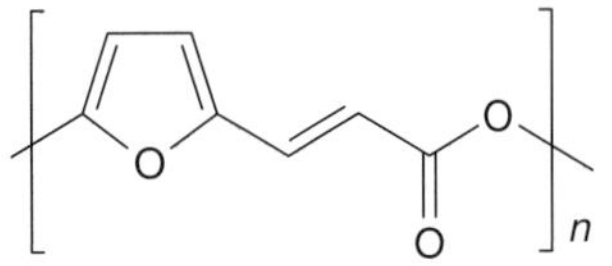

(15)

It is well known that fully aromatic polyesters are intractable materials in terms of processing because they are sparingly soluble even in very polar solvents and they start degrading before reaching their melting temperature. The replacement of up to 50 per cent of the aromatic dicarboxylic moieties by 2,5-furan counterparts was shown to reduce Tm correspondingly and make these copolyesters more apt to be processed, while keeping the required mechanical properties [4d].

Two recent additions to the topic of furan-based polyesters include macromolecular structures synthesized from 2,5-bis(hydroxymethyl)furan and various natural dicarboxylic acids such as succinic, fumaric and maleic acid [41] and the ring-closing depolymerization of polyesters **13** [42].

Curiously, to the best of our knowledge, no detailed study is available on the synthesis of the polyester arising from 2,5-furan dicarboxylic acid in combination with ethylene glycol, viz. the exact counterpart of PET which is by far the most widespread commercial polyester. The obvious interest of comparing the performances of the two polymers stems from the possibility of proposing a novel material, alternative to PET, based on a diacid readily available from the oxidation of **HMF** (a classical example of a compound derived from renewable resources as discussed above in Section 6.3). This void is presently being fulfilled in ongoing research carried out at Aveiro University.

6.6.4 Polyamides

As with furan polyesters, following a number of isolated and rather qualitative studies, a real scientific interest in furan-based polyamides began in the 1980s with a thorough study carried out in Moore's laboratory [43] which called upon the use of 2,5-furan dicarboxylic monomers in conjunction with aliphatic diamines. These new 'nylons' were however not inspected in terms of molecular weight, crystallinity or physical properties.

A series of furan-aromatic polyamides were prepared at a later date using the then novel technique called 'direct condensation', which was applied to the polycondensation of both 2,5- and 3,4-dicarboxylic acids with a number of aromatic diamines [4d, 4e]. All these polymers were thoroughly characterized and showed regular structures, high molecular weights and interesting crystallization and thermal properties, thus indicating that a novel class of material similar to aramides could be prepared using the furan diacid instead of the corresponding aromatic monomer. In particular, the polyamide **16**, arising from the simple combination of the 2,5-furan diacid (readily prepared by the oxidation of **HMF**) with 1,4-phenylene diamine, displayed properties entirely comparable to those of the homologous Kevlar. These included a remarkable thermal stability, up to 400°C, very pronounced mechanical properties and lyotropic liquid crystal behaviour.

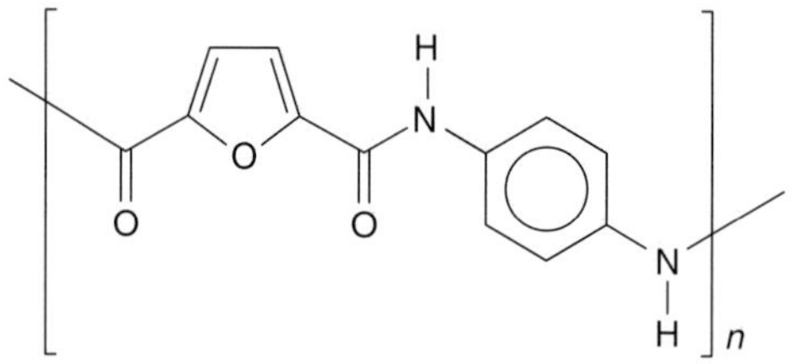

(16)

The corresponding all-furan polyamides **17–19**, obtained with 2,5- and 3,4-bis(furfuryl) amines, were also prepared and characterized but gave disappointing properties particularly in terms of thermal stability because of the lability of the hydrogen atoms in the methylene groups attached to the amine-derived furan rings.

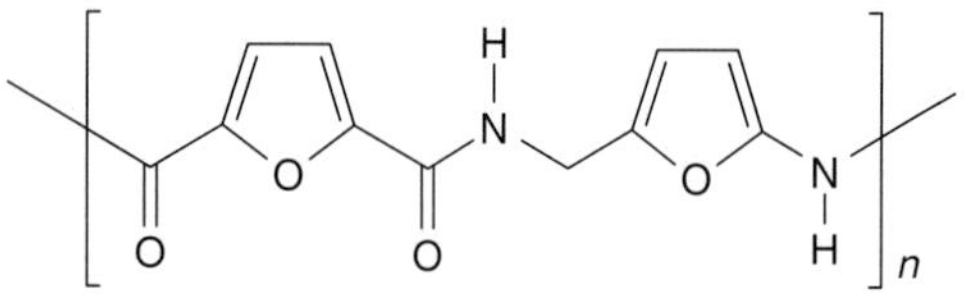

(17) 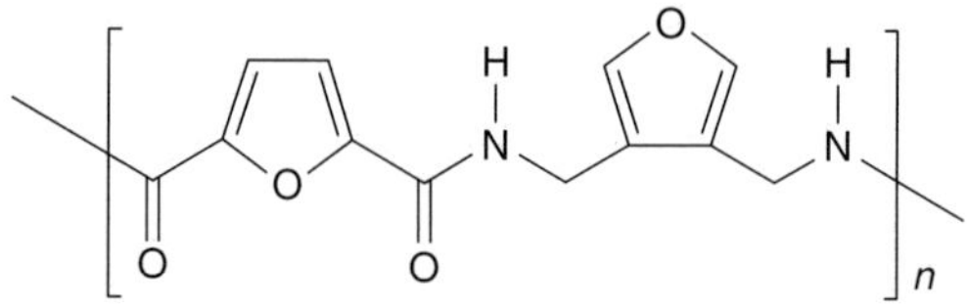 **(18)**

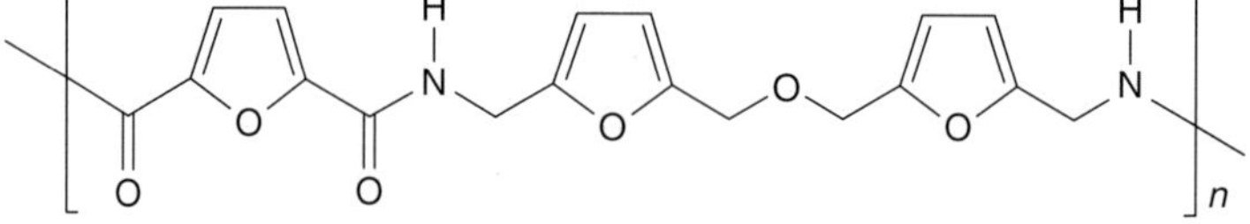

(19)

Recently, a systematic study has been undertaken dealing with the synthesis and characterization of a wide spectrum of furan polyamides **20** which simulate the structures of polyesters **13** since they were prepared using monomers **11** and a large choice of both aliphatic and aromatic diamines including furan homologues [4e, 44].

Again, different synthetic procedures were applied in order to optimize both polymer yields and molecular weights and very instructive relationships could be drawn from the multiplicity of the ensuing macromolecular structures. Figure 6.1 shows the ^{1}H-NMR spectrum of polyamide **20a** and Fig. 6.2 the DSC tracing of polyamide **20b**.

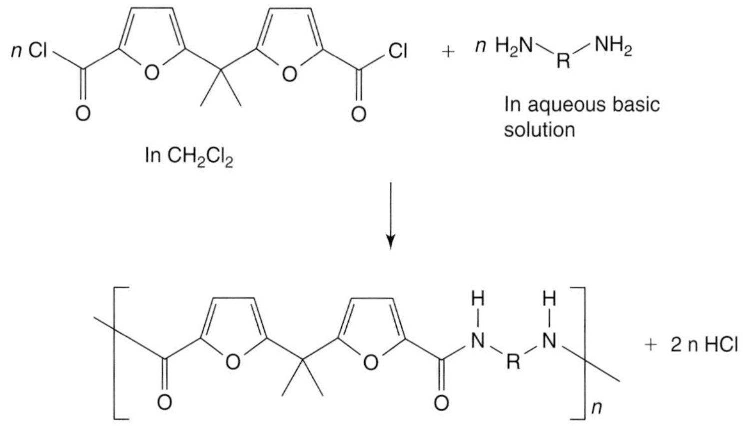

n Cl O O O Cl O + *n* H_2N R NH_2

In CH_2Cl_2

In aqueous basic solution

H H
O O N R N
O O
n

+ 2 n HCl

(20)

R = C_2H_6, C_4H_8, C_6H_{12}, C_8H_{16}, , , , O , S O O

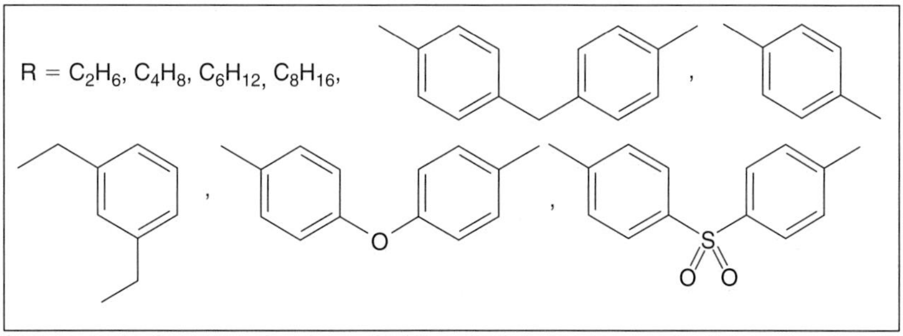

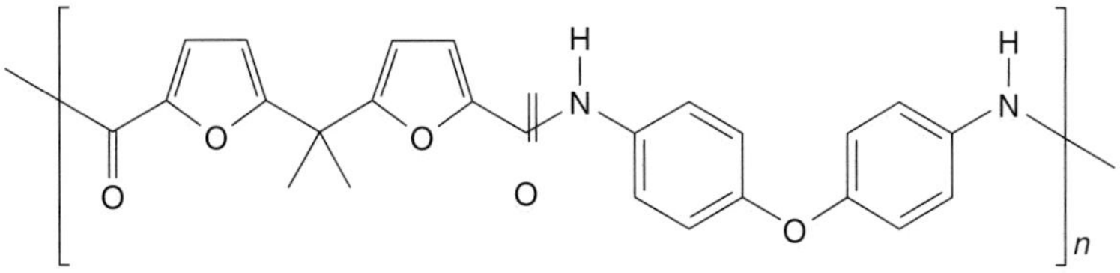

H H
O O N N
O O O
n

(**20a**)

H H
O O N N
O O S
O O
n

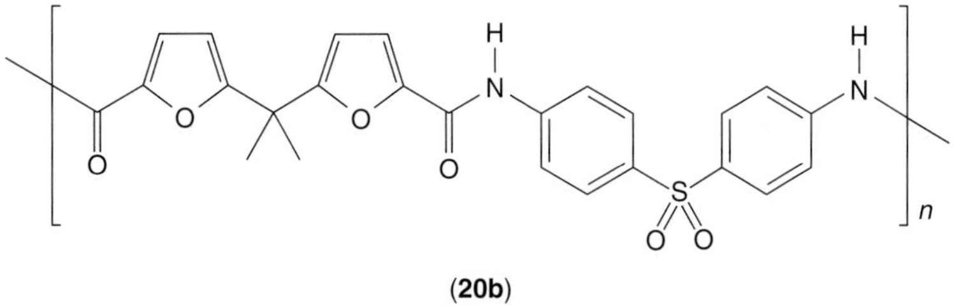

(**20b**)

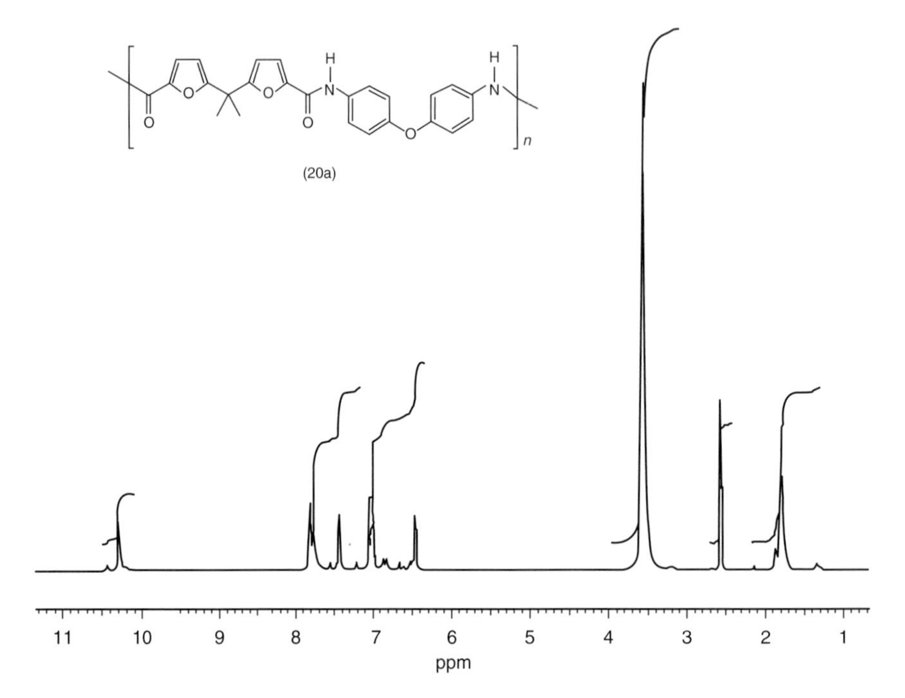

Figure 6.1 ^{1}H-NMR spectrum of polyamide **20a**, reported in Reference [44a]. (Reproduced by permission of the Elsevier. Copyright 2004. Reprinted from Reference [44a]).

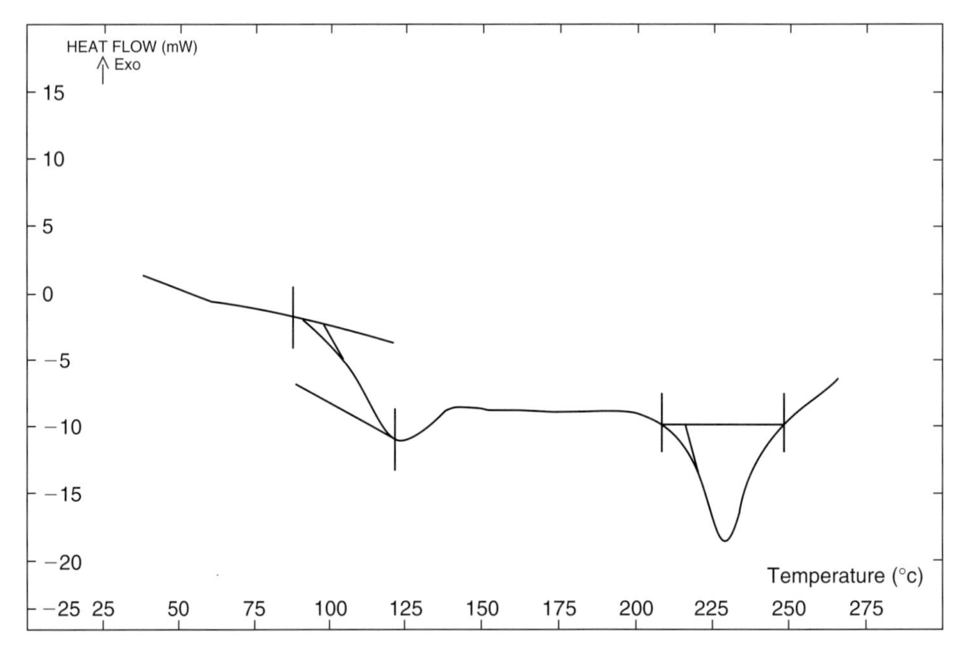

Figure 6.2 DSC tracing of polyamide **20b**, reported in Reference [44a]. (Reproduced by permission of Elsevier. Copyright 2004. Reprinted from Reference [44a]).

A peculiar route to a unique furan polyamide was applied to *N*-hydroxymethylfuramide prepared *in situ* from 2-furamide and an excess of formaldehyde, which gave rise to polymer **21** by an acid-catalyzed polycondensation mechanism involving the C5 electrophilic substitution of the heterocycle [4e].

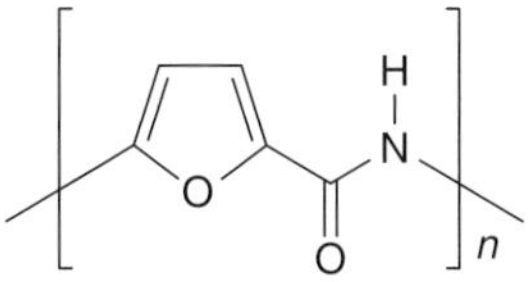

(21)

6.6.5 Polyurethanes

Once more, the interest in furan-based polyurethanes was scanty before a very thorough approach to the issue was carried out by us in the 1990s [4d, 4e]. This investigation involved the synthesis of both model compounds and new monomers bearing isocyanate functions, the study of the kinetics of their condensation reactions with alcohols and diols, the preparation and characterization of model mono- and di-urethanes and finally the actual systematic synthesis of a whole spectrum of polyurethanes. These polymers bore the furan ring in the main chain (*e.g.* **22**) and/or as pendant moieties (*e.g.* **23**). The heterocycles arose from either the diisocyanate monomer, or the diol, or from both. Other moieties incorporated in these linear polyurethanes included aliphatic (*e.g.* **24**, **25**) and aromatic (*e.g.* **23**) structures.

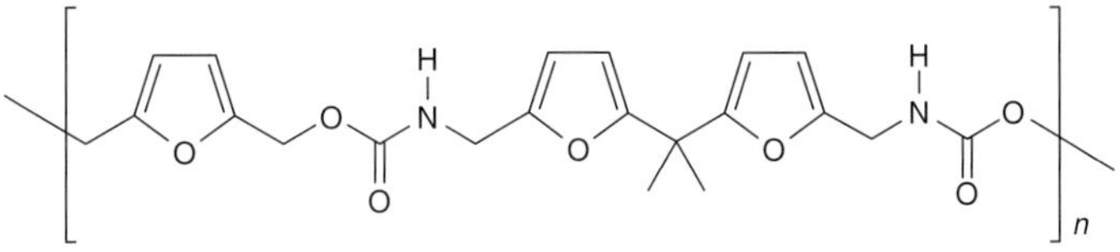

(22)

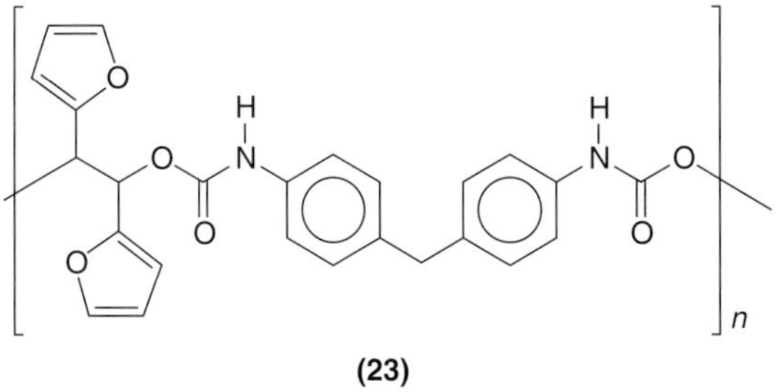

(23)

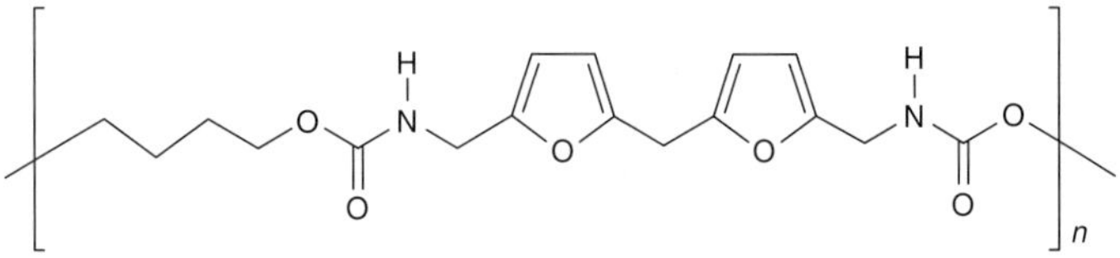

(24)

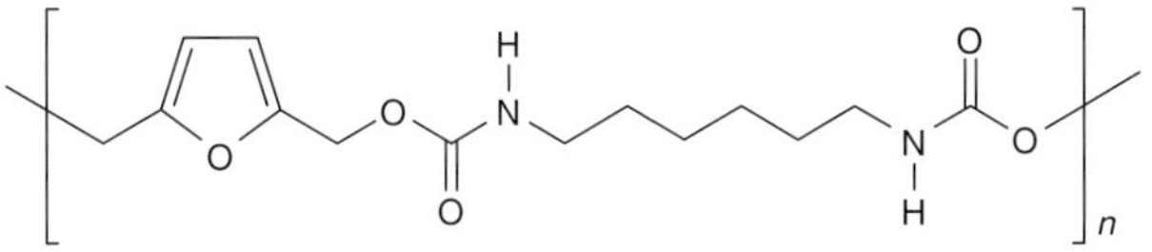

(25)

This comprehensive study showed first of all that the use of both types of furan monomers did not entail the occurrence of any unwanted side reaction, since all the polymers displayed a perfectly regular structure, as assessed by FTIR and NMR spectroscopy.

The second observation has to do with the very pronounced reactivity of furan isocyanates towards alcohols, higher than that of aromatic counterparts, even when the NCO group is separated from the heterocycle by a methylene spacer. This was so dramatic that in fact 2,5-furandiisocyanate could not be of any practical use because of its excessive nervousness.

The most relevant outcome here has to do of course with the structure–property relationships gathered from this investigation which gave clear-cut information about Tg, Tm and the corresponding extent of crystallization, thermal stability, mechanical properties and phase separation phenomena associated with a series of thermoplastic elastomers, like **26**, prepared together with all the other linear materials.

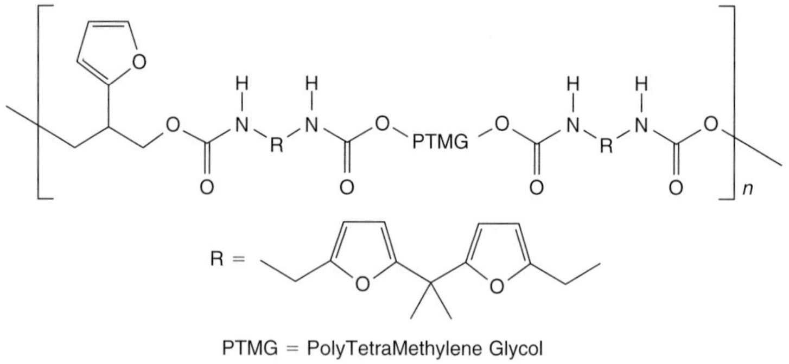

PTMG = PolyTetraMethylene Glycol

(26)

6.6.6 Other polymers

Among the numerous furan polymers derived from polycondensation involving systems other than polyesters, polyamides or polyurethanes, the most relevant materials include:

- polyschiff bases prepared from 2,5-furandicarboxaldehyde (readily obtained from the controlled oxidation of **HMF**) and aromatic diamines (*e.g.* **27**) and similar systems, including co-polyschiff bases incorporating flexible oligoether segments [4d, 4e];

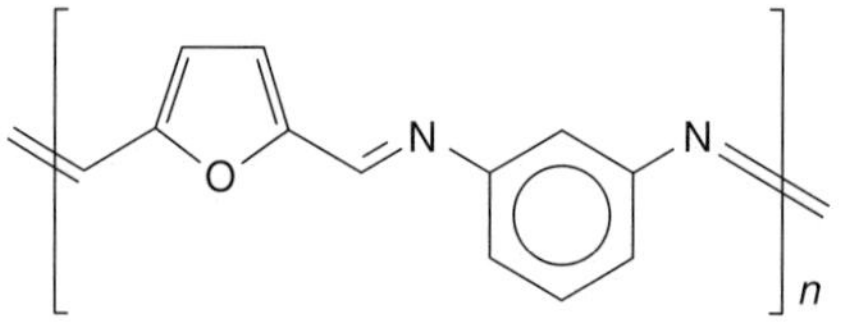

(27)

- polyethers following a single study of the phase-transfer condensation of 2,5-bis(hydroxymethyl) furan with various dibenzylic halides [4d];
- polyhydrazides formed from furan 2,5-dihrazides and diacid chlorides (*e.g.* **28**) and the corresponding poly-1,3,4-polyoxadiazoles arising from their dehydration (*e.g.* **29**) [4e, 45];

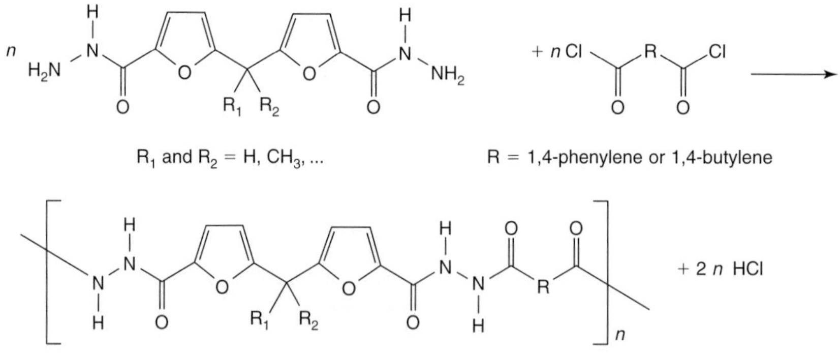

R_1 and R_2 = H, CH_3, ...   R = 1,4-phenylene or 1,4-butylene

+ 2 *n* HCl

(28)

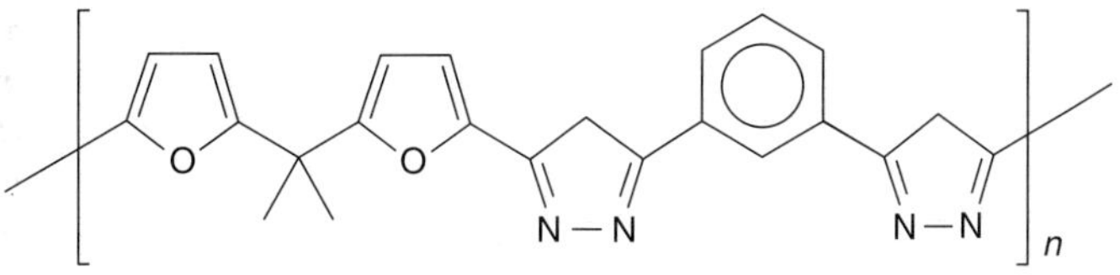

(29)

- polyamide-imides synthesized in solution from aromatic dianhydrides and 2-furoic acid dihydrazides (*e.g.* **30**) [46];

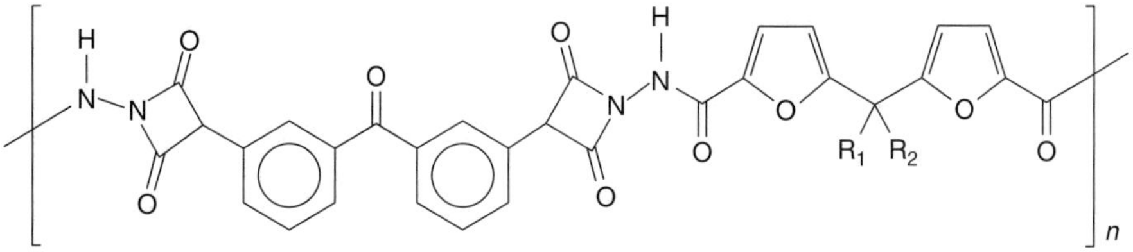

$R_1/R_2 = CH_3/CH_3;\ CH_3/C_6H_5;\ CH_3/C_2H_5;\ CH_3/C_5H_{11};\ CH_3/CF_3$

(30)

- polyureas (*e.g.* **31**) and poly(parabanic acid)s (*e.g.* **32**) obtained, respectively, from difuran diamines and aliphatic diisocyanates and the subsequent heterocyclization with oxalyl chloride [47];

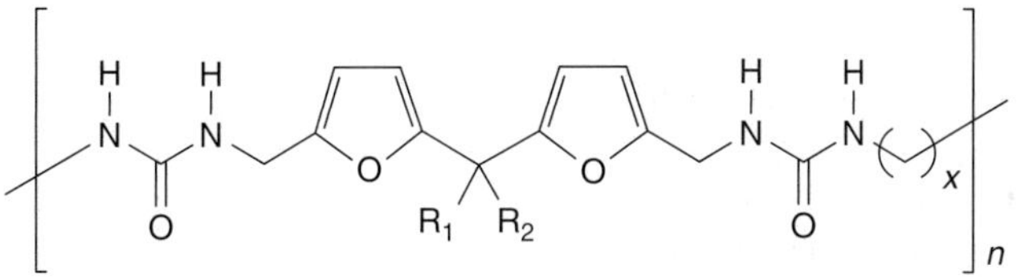

$R_1/R_2 = CH_3/CH_3;\ C_2H_5/C_2H_5;\ CH_3/H;\ CH_{13}/H$

$x = 4$, 6 and 8

(31)

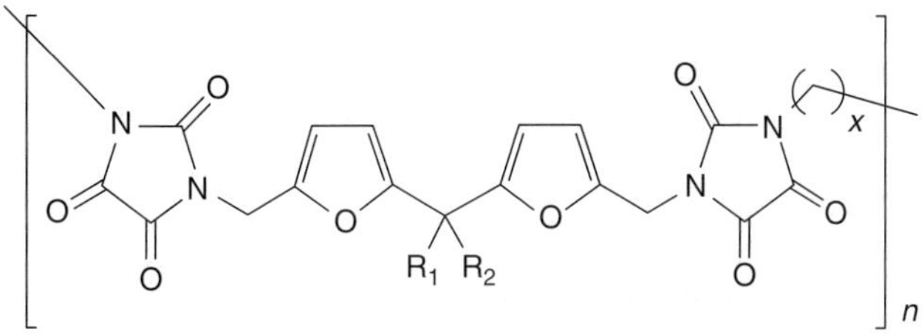

$R_1/R_2 = CH_3/CH_3;\ C_2H_5/C_2H_5;\ CH_3/H;\ CH_{13}/H$

$x = 4$, 6 and 8

(32)

- quasi-spiro polyacetals (*e.g.* **33**) arising from the polycondensation of 2,5-furandicarboxaldehyde with pentaerithritol [4d, 4e];

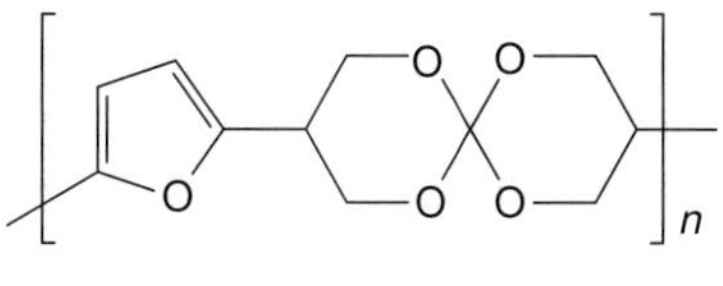

(33)

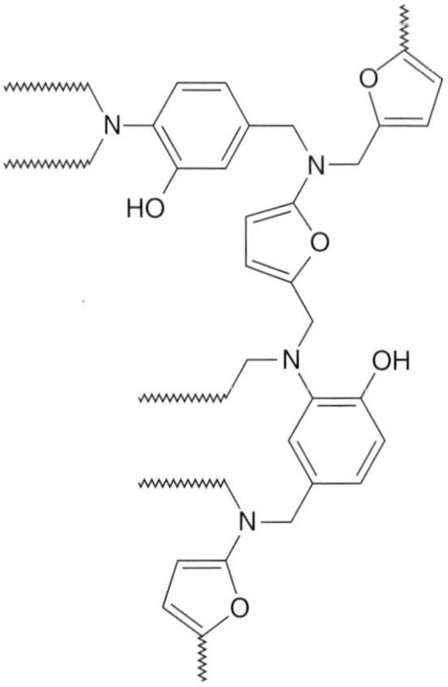

Scheme 6.22

- polybenzoxazines prepared from furan-containing benzoxazines by temperature-driven ring opening leading to crosslinked structures, as shown in Scheme 6.22 [48].

6.7 CONJUGATED OLIGOMERS AND POLYMERS

The growing interest in conjugated macromolecular structures, associated with their possible high-tech applications, as in opto-electronic devices, has spurred a large variety of materials based on aliphatic, aromatic and heterocyclic [49] moieties linked in various modes. The possibility of introducing the furan heterocycle in these polymers has obviously retained the attention of specialists, particularly in view of the success of related structures based on thiophene and pyrrole. These studies can be divided into two types of approaches, namely polymers and copolymers synthesized using furan itself (or 2-alkylfurans) and polymers bearing 2,5-furylvinylene moieties.

6.7.1 Polyfuran

Despite numerous attempts at synthesizing the ideal polymer **34** [4d, 49, 50] the preparation of polyfuran has not been a clear-cut issue in terms of structural assessment and material performance.

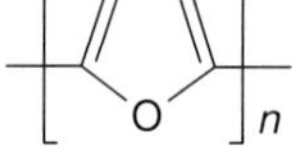

(34)

Whether by chemical or electro-chemical means, the expected structure has eluded chemists for some 25 years because what is in fact obtained is a material which suggests the existence of extensive conjugation by its blackish colour, but whose macromolecular architecture is much more irregular and complicated than the linear chain **34**, notwithstanding some claims related to a thorough structural elucidation. It follows that what all these authors happily call 'polyfuran' is true to its name only because it is obtained by polymerizing furan, albeit through insofar

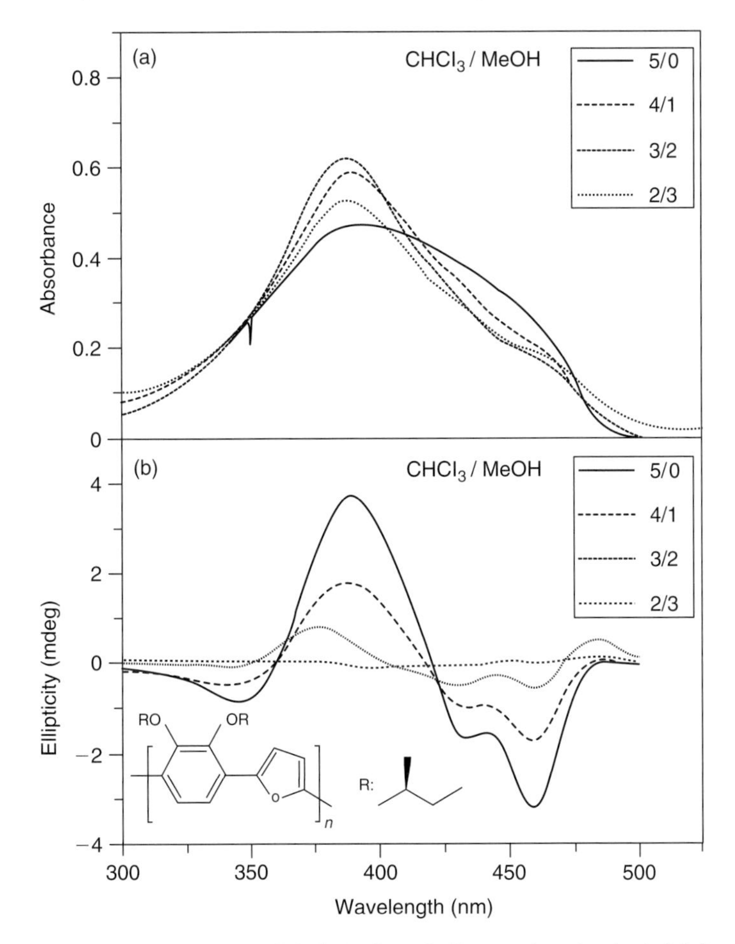

Figure 6.3 Solvatochromic studies of poly(1,4-(2,3-di-(*S*)-2-methylbutoxyphenylene)-*alt*-2,5-furan) in different solvents: (a) UV–vis measurements; (b) CD measurements, reported in Reference [53b]. (Reproduced by permission of the American Chemical Society. Copyright 2002. Reprinted from Reference [53b]).

obscure mechanisms. This state of affairs is reflected by the fact that these materials have not achieved any of the success that the homologous polypyrrole and polythiophene have enjoyed.

The insertion of furan moieties within the macromolecular structure composed mostly of other conjugated units, like homologous heterocycles [51], substituted anilines [52] or aromatic rings [53], has been much more successful in terms of structure regularity. Among these, mixed furan–thiophene oligomers have been prepared and thoroughly characterized [54]. Figure 6.3 shows the UV-vis and the circular dichroism features of one of these conjugated structures in solvent mixtures.

A recent addition to this rather modest field, consists in preparing polyfuran composites with both inorganic and carbon nanoparticles [55].

6.7.2 Poly(2,5-furylene vinylene)

The success of poly(1,4-phenylene vinylene) (**PPV**) and its derivatives was accompanied by studies aimed at preparing similar structures in which the aromatic ring was substituted by a heteroaromatic counterpart [56]. The furan homologue was first prepared using a synthetic process which simulated that of **PPV**, but in those studies more emphasis was placed on the characterization of the reactive intermediates rather than on the final polymer [4d, 56].

A radically different approach was developed in later years which gave the possibility of building stepwise the units of poly(2,5-furylene vinylene) (**PFV**) from the dimer **35** to polymers **36** with DPs reaching about 100 and to isolate well-defined pure oligomers [4e, 57]. This synthesis is based on the base-catalyzed self-condensation of **MF** through a mechanism which only allows the reaction between the terminal aldehyde group of the growing macromolecule with the methyl group of the *monomer*, as shown in Scheme 6.23.

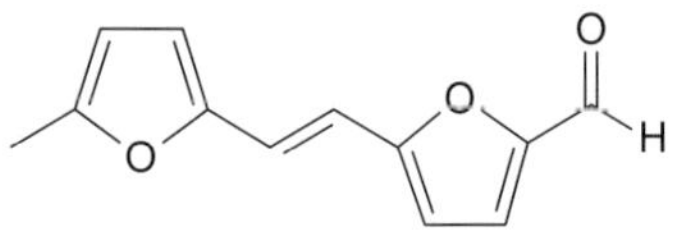

(35) 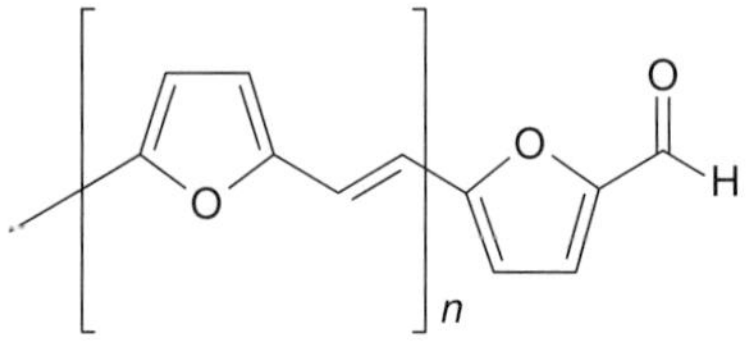 **(36)**

The electrical conductivity of **36** was close to $1\,S\,cm^{-1}$ at room temperature. Its brittle consistency made it difficult to process films from it but, thanks to the terminal aldehyde group, block copolymers could readily be prepared by reacting these monofunctional conjugated polymers with mono- or di-functional polyethers bearing terminal primary amino groups and possessing very low glass transition temperatures. The ensuing internally plastified materials were readily soluble in common solvents, and thus easily processable [4d].

In a more specific study, pure individual oligomers with DPs ranging from 2 to 5 and oligomeric mixtures with DPs of 6 to 10 were synthesized and thoroughly characterized in terms of their structure and electronic spectra before and after protonation [4d, 57]. This investigation also involved the preparation of the thiophene analogues as well as furan–thiophene mixed oligomers. All these compounds were found to display a pronounced photoluminescent behaviour [57] and the wavelength of the emitted radiation increased with their DP, enabling to cover the entire visible spectrum. This feature is illustrated in Fig. 6.4 for the specific case of the oligo(furylene vinylene) series.

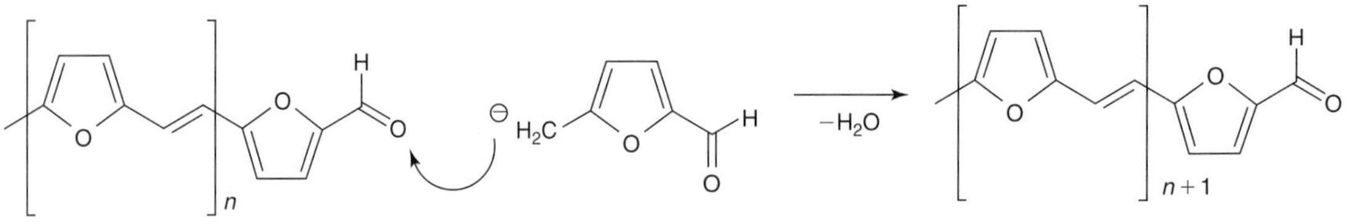

$-H_2O$

Scheme 6.23

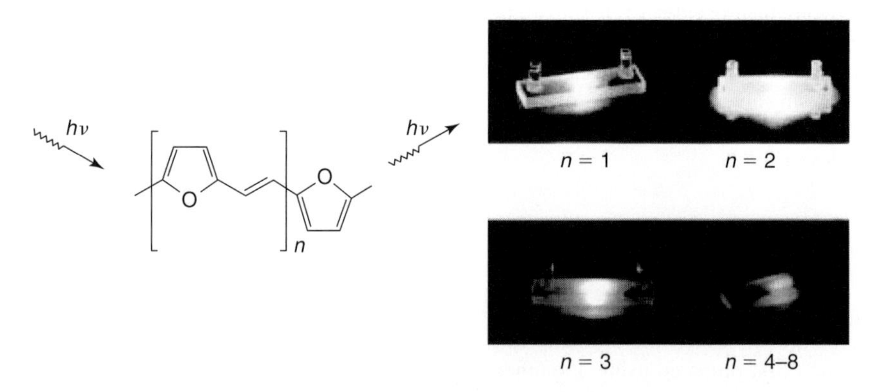

Figure 6.4 Photoluminescence of oligo(furylene vinylene)s of different lengths [57b] (see colour plate section).

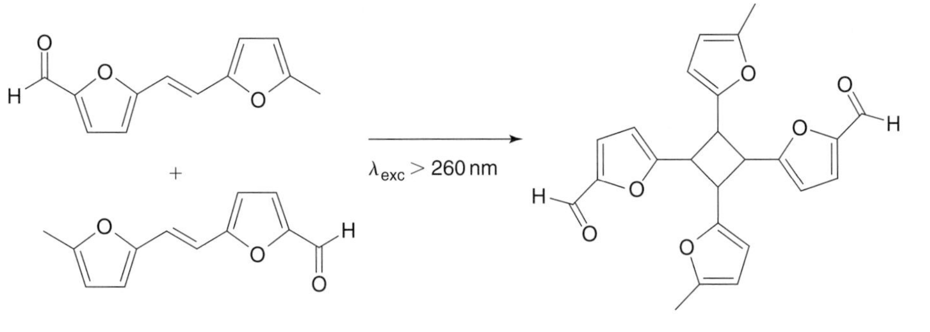

$\lambda_{exc} > 260\,nm$

Scheme 6.24

Dimer **35** was found to possess a straightforward single photo-chemical pathway which led to its dimerization when it was irradiated with near UV light, as shown in Scheme 6.24 [4d, 58]. Later studies showed that this behaviour was indeed common to molecules bearing the general Fu—CH=CH—C=O chromophore [58].

This photo-chemical feature was exploited in a number of investigations dealing with the possibility of preparing photocrosslinkable polymers. Thus poly(vinyl alcohol) was modified by condensing dimer **35** with two of its OH groups, which appended the furan chromophores by forming acetal moieties. The ensuing polymer **37**, with as little as 5 per cent photosensitive units, readily crosslinked in a few minutes under the irradiation of a standard medium-pressure mercury lamp [59]. In another application, Schiff bases were prepared between dimer **35** and mono-, di- and tri-functional Jeffamines. Following irradiation of these photosensitive oligomers, macrodimers, linear polymers and crosslinked structures were obtained, respectively, through the coupling of the terminal chromophores. These materials displayed interesting properties as polymer electrolytes [60]. A recent addition to this strategy called upon the introduction of the same furan chromophores as infrequent side groups to the chitosan backbone, together with grafted oligo(ethylene oxide) branches. The ensuing material had excellent film-forming aptitudes and was of course also a very good iron conductor because of the abundant polyether grafts and when irradiated, it crosslinked to give dimensionally stable thin films [61].

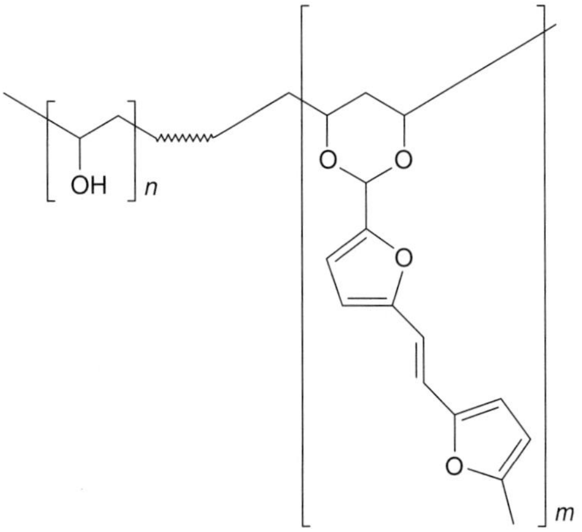

(37)

Whereas all the above materials bore the photosensitive moiety as side or end groups, the unsaturated polyester **15** and its copolymers were characterized by the presence of a similar chromophore as part of the main polymer chain. This internal placement did not hinder its photo reactivity [62], since these materials readily crosslinked upon irradiation [40].

It is worth noting that all these latter examples of materials whose key feature calls upon a specific behaviour of furan moieties, require modest contributions from these heterocycles in terms of its quantitative presence in the macromolecules. This situation is similar to that described in Section 6.5.2, in which furan derivatives were used to modify the end groups of some polymers or to synthesize block copolymers by cationic polymerization. In both instances, therefore, it is not the dominant presence of the heterocycle that determines the specific properties of the final materials (as in the case of polymers and copolymers bearing furan monomer units), but instead the fact that a small or even minute percentage of these structures introduces an original mechanistic feature, associated with the peculiar chemistry of the heterocycle, which transforms the behaviour and properties of the final material. This crucial point will be met again in the context of the application of the DA reaction in Section 6.8.

Cyclic oligomers bearing some resemblance to the open-ended **PVFs**, were recently reported, albeit in very small yields, from the reaction of furan with perfluorobenzaldehyde [63]. Their absorption spectra, centred between 500 and 600 nm, reflected a substantial degree of conjugation.

6.8 THE APPLICATION OF THE DA REACTION TO FURAN POLYMERS

The strategies which make use of the DA reaction to prepare original polymers are discussed here with specific reference to the diene–dienophile combination consisting of a furan and a maleimide ring, respectively. We first reviewed this topic a decade ago [4d] and returned to it very recently [64]. However, the growing interest in developing novel functional materials using this approach justifies a brief treatment here with up-to-date contributions. The other interesting application of DA adducts in macromolecular synthesis calls upon their ring opening metathesis polymerization (ROMP), which has been reviewed previously by us [4d] and, because it has not received further attention since that critical coverage, will not be dealt with here.

Before tackling the aspects related to reversible macromolecular structures, a recent study on surfactants deserves a brief mention here because of its originality, albeit outside the realm of polymer materials. The idea consists in preparing amphiphilic structures in which the hydrophilic moiety is separated from the hydrophobic one by a DC adduct [65]. The surfactant displays its expected activity in aqueous media until the temperature is raised close to their boiling point, where it decomposes by the retro-DA reaction, thus generating two surface-inactive fragments. Because of the very low concentration of these fragments, associated with the classical values of surfactant utilization, the kinetics of recombination of the two fragments to regenerate the original structure is far too low to envisage such an event. The application of this principle falls into the broader area, very topical today, of degradable surfactants. Thus, surfactant **38** was synthesized and its thermo-reversible character tested, as shown in Fig. 6.5, which represents the ^{1}H-NMR spectra of **38** after solid state heating at different temperatures.

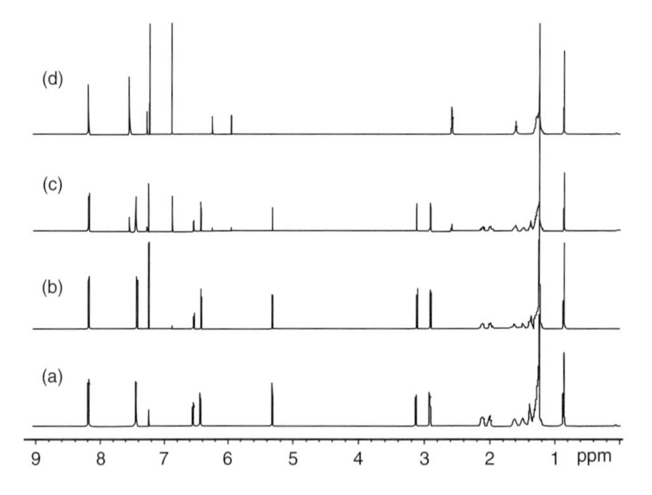

Figure 6.5 ^{1}H-NMR spectra of **38** after solid state heating for 30 min for (a) 25°C, (b) 105°C, (c) 120°C and (d) 125°C, reported in Reference [65] (Reproduced by permission of American Chemical Society. Copyright 2005. Reprinted from Reference [65]).

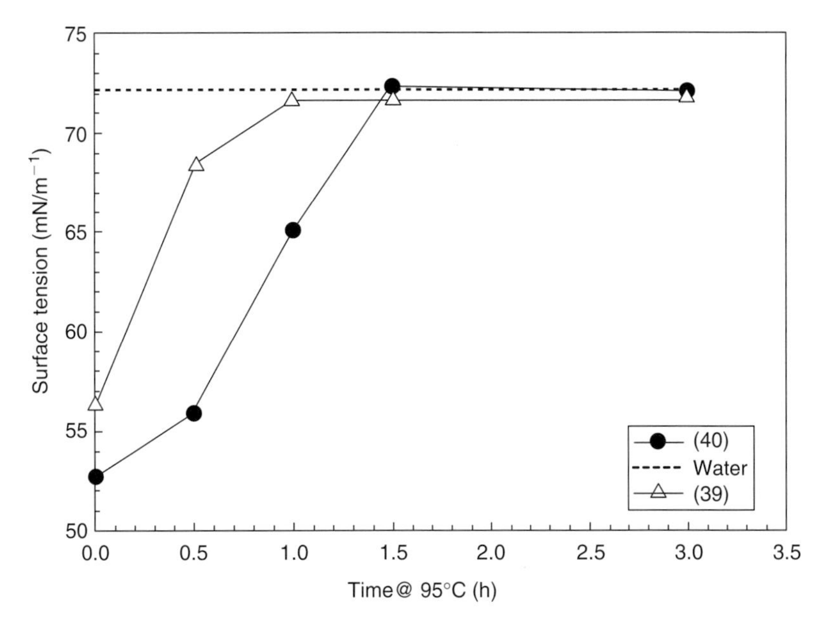

Figure 6.6 Changes in the dynamic surface tension of aqueous solutions of **39** and **40**, as a function of time at 95°C, reported in Reference [65] (Reproduced by permission of the American Chemical Society. Copyright 2005. Reprinted from Reference [65]).

Figure 6.6 represents the changes in dynamic surface tension (operated in an equilibrium mode) for 20 mM aqueous solutions of anionic surfactants **39** and **40**, as a function of time at 95°C and measured at 26°C. The dashed line represents the surface tension value of water at 25°C (72.1 mN m^{-1}).

O O N O OH O

(38)

O O N O $O^{\ominus}$ $K^{\oplus}$ O

(39)

O O N O O $O^{\ominus}$ $K^{\oplus}$ O

(40)

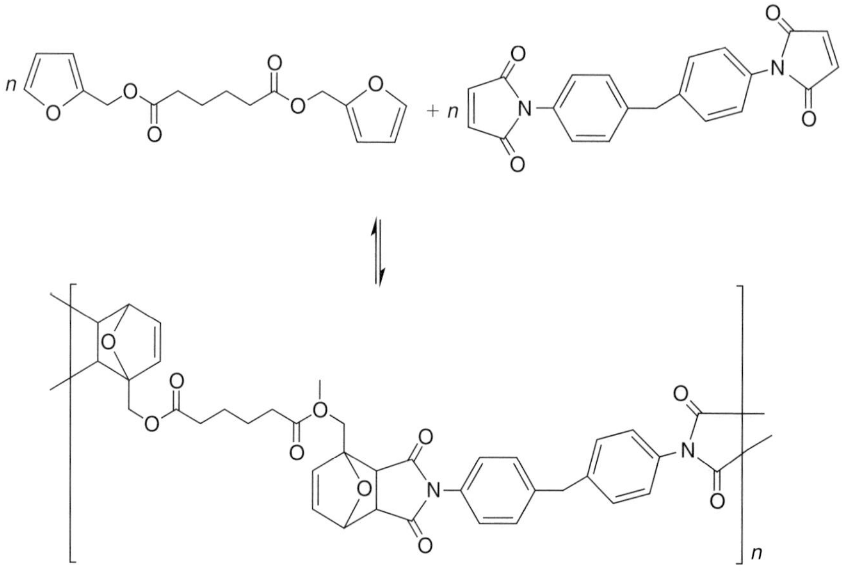

Scheme 6.25

6.8.1 Linear step-growth polymerization

The vast majority of systems falling within this category are based on the use of a bismaleimide and a difuran derivative, viz. a classical linear polycondensation of the A-A + B-B type. Since this synthetic challenge was first tackled some 20 years ago [4d, 64], numerous investigations have been published with different monomer combinations and with different approaches both in terms of the experimental conditions used for the polymer synthesis and the characterization and purpose of the ensuing materials [4d, 64].

Among the most thorough studies of these systems, Kuramoto *et al.* [66] were the first to carry out both the polymerization and the thermal depolymerization of the ensuing DA polyadduct as shown in Scheme 6.25.

A few years later, Brand and Klapper [67] paid closer attention to the reversibility of this linear polycondensation using NMR spectroscopy and viscosity measurements, also applied to model compounds. This fine piece of research represents, in our view, the best example of how to approach these complex systems, an example which has not been followed systematically in subsequent studies.

As a rule, the molecular weights obtained in these syntheses are not particularly high because, on the one hand, the use of relatively elevated temperatures shifts the equilibrium in favour of a growing contribution of the retro-DA reaction (viz. depolymerization) and on the other, polymers tend to lose their solubility as their DP increases, thus precipitating out of the reaction medium.

The interest in synthesizing thermoplastic polymers made up of DA adduct units, bridged by moieties bearing different flexible or rigid structures, can vary from one investigation to another. The reversible character associated with the DA reaction is of course one of the primary motives for these studies because it opens the way to preparing materials which can be thermally depolymerized (in the case of furan–maleimide adducts just above 100°C) in view of their ablation and/or recycling. Conversely, in other applications, the polymer structure obtained in these syntheses should be preserved even at high temperatures and, in that case, the chemical modification of the adduct to render it thermally stable, becomes indispensable. The aromatization of the DA adduct through a catalyzed dehydration reaction (Scheme 6.26) is a typical solution to this requirement.

The use of a single A-B monomer bearing both the furan and the maleimide ring represents an interesting alternative to this type of polymer synthesis because, as in all polycondensation reactions, this approach ensures the ideal initial stoichiometry. One such monomer (**41**) was prepared, characterized and polymerized, [68] but the single methylene bridge joining the two complementary DA rings made the structure susceptible to degradation, that

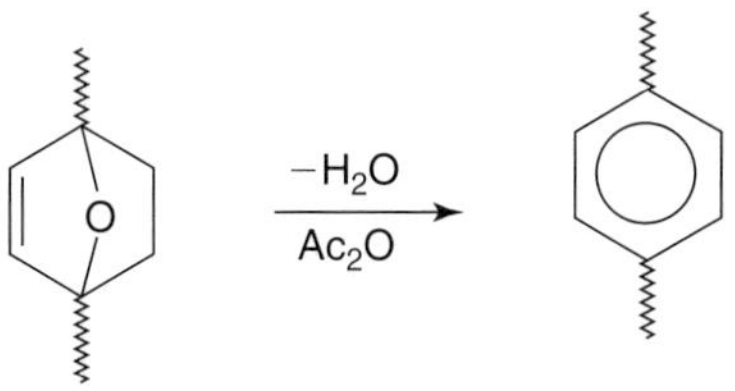

$-H_2O$
Ac_2O

Scheme 6.26

is **41** turned out to be difficult to handle. Work is in progress in the Aveiro laboratory to prepare more manageable structures of this type.

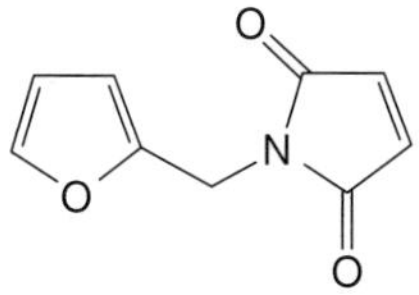

(41)

The only other study of this nature known to us describes the synthesis of two other A-B monomers [64] as well as their solution and bulk polymerization, but the proposed mechanism for their chain growth is unclear. The aim of this investigation was to prepare thermally stable materials and therefore the DA adducts in the polymers were aromatized by dehydration with acetic anhydride.

6.8.2 Networks and dendrimers

The extension of these DA-based polymerizations to multi-functional furan and/or maleimide monomers naturally leads to crosslinked materials, whose original feature is related to the fact that they can be readily decrosslinked by a simple thermal treatment. Since the first report of a mendable material based on this principle, but applied in cycles [69], several similar studies have appeared in the literature [64] including interesting applications in the realm of thermally removable foams [64] and adhesives [64]. It seems likely that such a simple and useful strategy will continue to draw attention for the preparation of novel intelligent materials.

Another approach leading to non-linear macromolecular structures incorporating DA adducts has been put forward recently in the form of thermally reversible dendrimers [64, 70]. This interesting piece of research describes aromatic ether architectures joined by DA adducts and reaching up to the third generation. The backward mechanism, based on the retro-DA reaction progressively reduced this structure to lower generations.

6.8.3 Reversible crosslinking of linear polymers

Perhaps the most active area within the broader scope discussed here, refers to the preparation of linear polymers bearing pendant furan or maleimide moieties and their DA-based crosslinking with the complementary bi-functional reagent, viz. a bismaleimide or difuran compound, respectively [64]. The obvious reason for the growing interest in these materials relates to the possibility of recycling them after a simple thermal treatment capable of reverting their structure to the original linear thermoplastic architecture. The first example of this strategy which immediately comes to mind, is the possibility of recycling tyres, a major industrial and ecological issue yet to be solved.

After the pioneering study that Stevens and Jenkins published in 1979 [71], a real interest in this topic started only more than 10 years later with a series of studies aimed at testing the feasibility of the reversible pathway [64].

The best approach to this type of system included an original concept based on the necessity of trapping the difunctional molecule liberated during the retro-DA undoing of the crosslinked polymer in order to avoid a subsequent reconstruction of the network upon cooling [64]. This was successfully achieved by introducing in the decrosslinking medium an excess of mono-functional DA reagent (*e.g.* 2-methylfuran in the case of the liberation

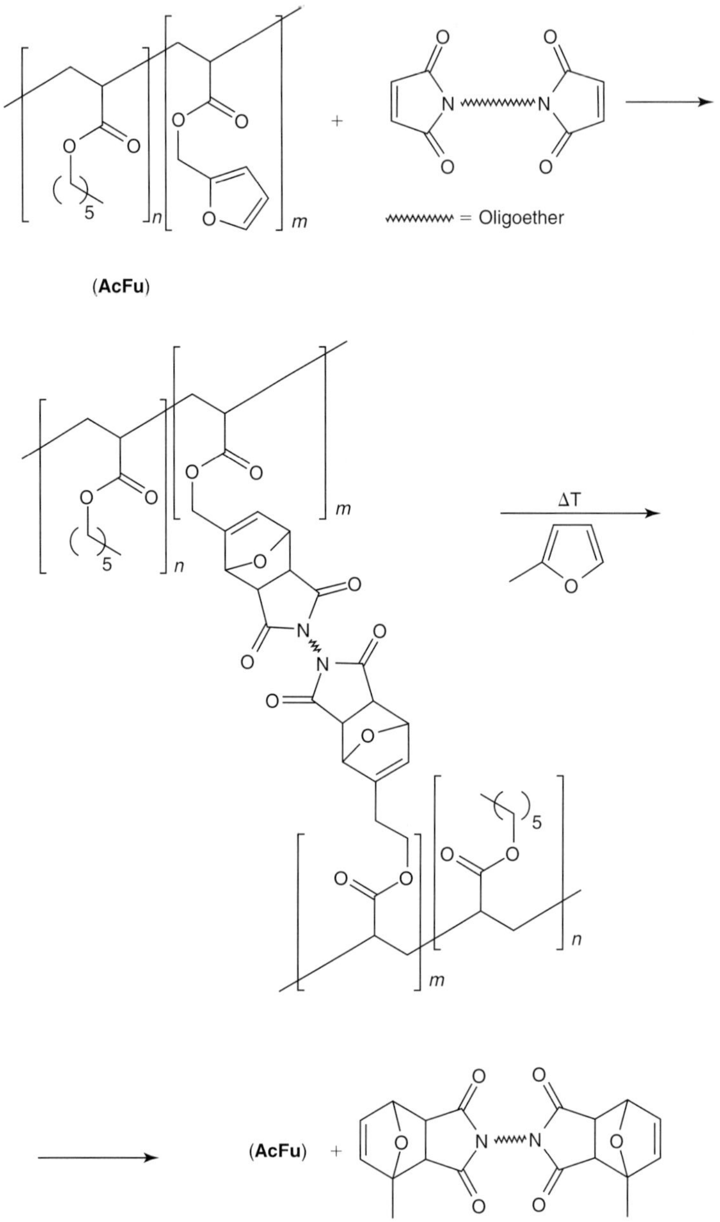

(AcFu)

=Oligoether

ΔT

(AcFu)

Scheme 6.27

of a bismaleimide and *N*-methylmaleimide when a difuran compound is liberated). In this way, styrene-type [72] and acrylic [73] copolymers with pendant furan moieties, as well as silicone structures bearing pendant maleimide groups [73], were first crosslinked with a bismaleimide and a difuran derivative, respectively. Subsequently, these networks were suspended in appropriate media containing the monofunctional trapping molecule and heated at reflux just above 100°C, whereby the original linear copolymers were progressively regenerated as shown by the corresponding solubilization of the material. Thus, the DA/retro-DA mechanism reached a full cycle with the complete recovery of the starting thermoplastic copolymers even after their solution was cooled down to room temperature. An example of this cycle is shown in Scheme 6.27 [73].

Several experimental observations [64] have shown that the retro-DA reaction becomes kinetically relevant in this context at about 100°C and virtually negligible below about 70°C.

In the past few years, the application of these principles has acquired an enhanced relevance, as reflected by the numerous publications, which have appeared since our latest review [64], both with linear [74] and crosslinked [75] structures.

6.9 MISCELLANEOUS DENDRIMERS

Two recent papers on furan-containing dendrimers have been added to the original work based on the DA reaction discussed in Section 6.8.2. The first describes the formation of star polymers induced by the cyclotrimerization of furan derivatives bearing aliphatic aldehyde functions attached at the 2 and 5 ring positions [76]. This hyper-branched structure was therefore generated by the formation of trifunctional acetal moieties, as shown in **42**, and should therefore be readily hydrolyzed making the whole process reversible, although this latter aspect was not pointed out by the authors.

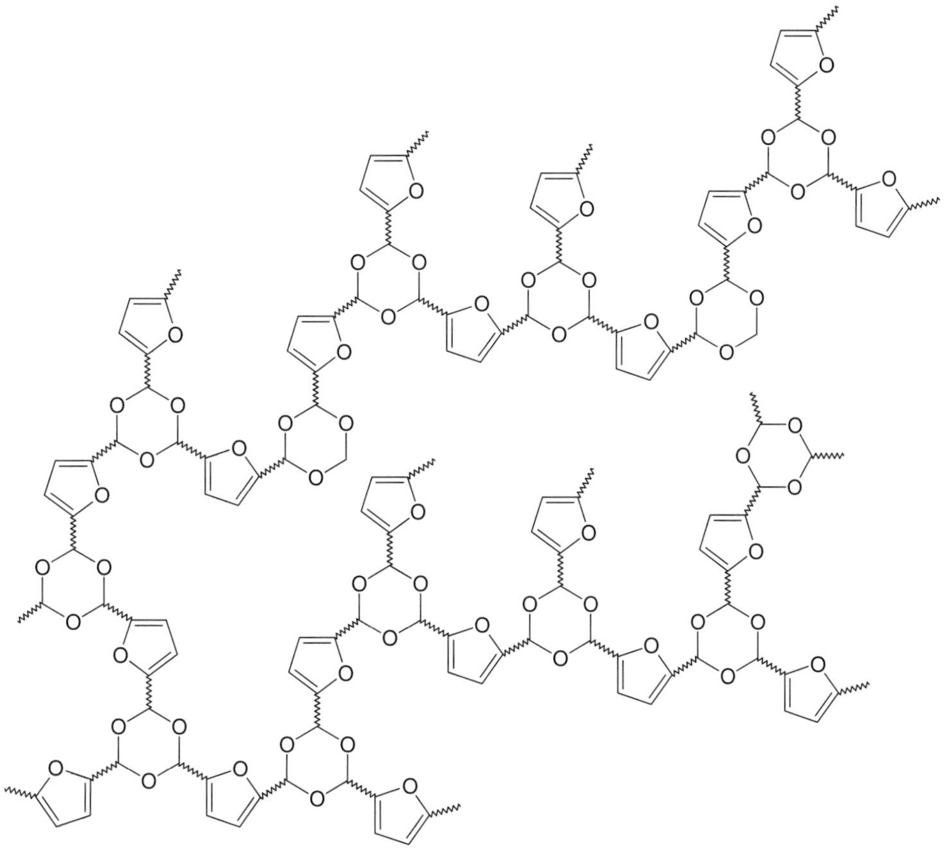

(42)

The other approach was based on the acid-catalyzed coupling between a phenolic group and the C5 position of the furan ring [77] which gave rise to the second generation dendrimer, **43**. This approach gave rise to the preparation of novel molecular resists based on first generation dendrimers incorporating furan moieties. Figure 6.7 shows the SEM micrograph of the resist film formulated with **44**.

A novel hybrid dendritic copolymer bearing conjugated photosensitive furan rings was also prepared with the aim of developing crosslinkable polymer electrolytes [78].

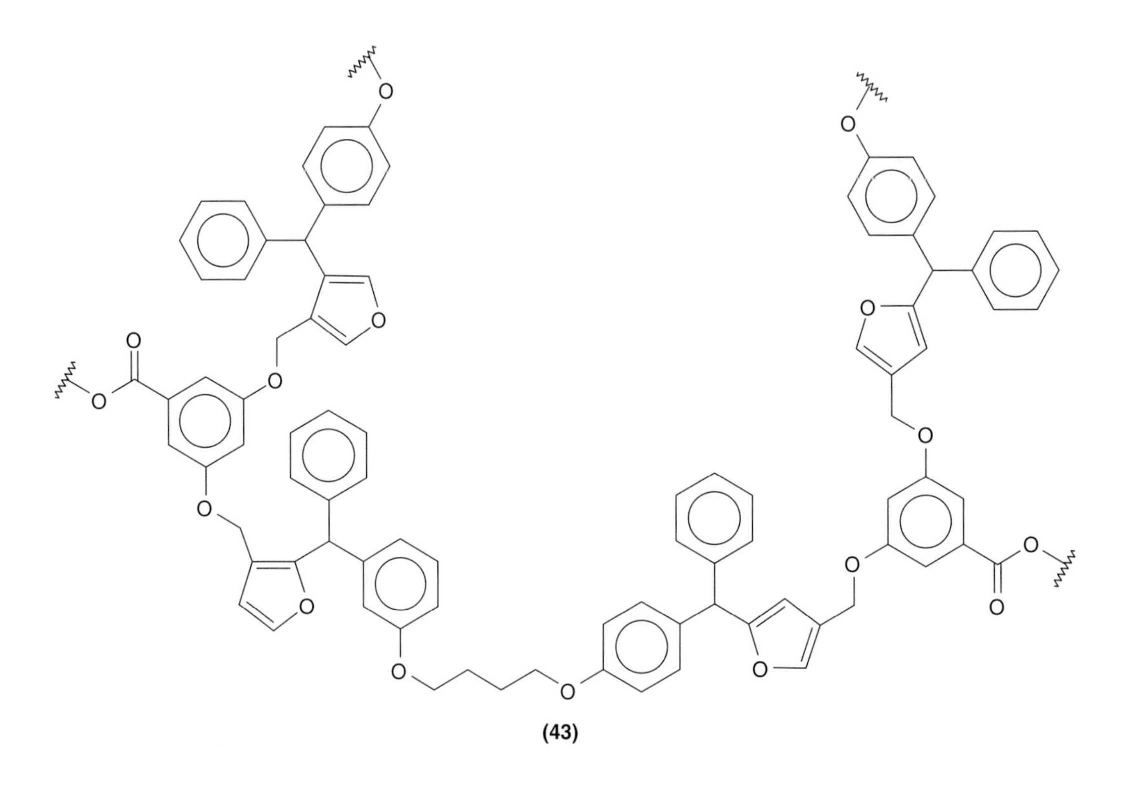

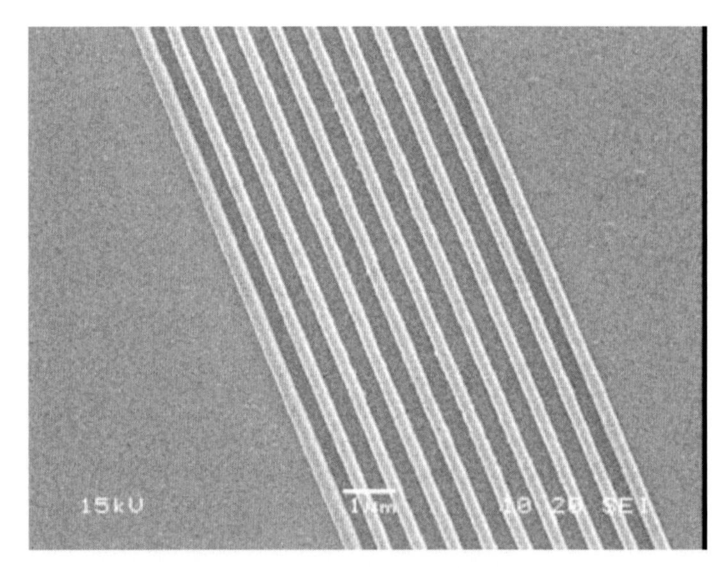

Figure 6.7 SEM image of negative-tone line and space patterns for the resist film formulated with **44**, reported in Reference [77] (Reproduced by permission of Wiley-VCH Verlag GmbH & Co. KGaA. Copyright 2006. Reprinted from Reference [77]).

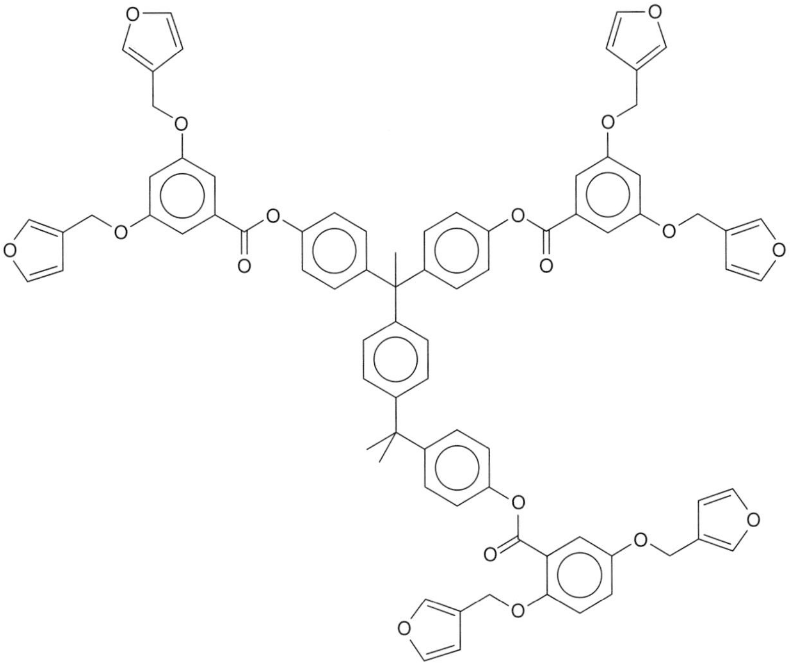

(44)

6.10 THE AGING OF FURAN POLYMERS

By far the most fragile structures in furan polymers are a tertiary C—H bond attached to one or more heterocycles and a secondary C—H bond attached to two rings. This high lability arises from the fact that, independent of the specific nature of the living H species, namely a proton, a hydride ion or a hydrogen atom, the intermediate structure that will be left is strongly stabilized by the adjacent heterocycle thanks to its dienic character. In other words, whether this intermediate is a carbanion, a carbenium ion or a free radical, the charge or the unpaired electron will be delocalized over the neighbouring furan ring(s).

The ease with which this mechanism occurs has already been emphasized when discussing the side reactions associated with the cationic polymerization of 2-vinylfurans (Section 6.5.2) and with the self-condensation of furfuryl alcohol (Section 6.6.2). However, its negative impact is also found, albeit with slower rates, in isolated furan polymers which possess the culprit C—H moiety as in the case of poly(2-vinylfuran) which slowly develops an insoluble fraction caused by the free radical mechanism shown in Scheme 6.28. These reactions are also responsible for the colour that is often associated with furan polymers.

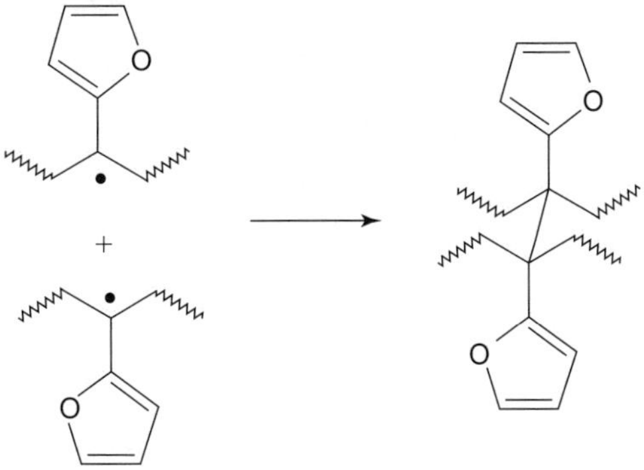

Scheme 6.28

Secondary C—H bonds attached to a single furan ring do not display such a pronounced lability, as shown by polymers like furfuryl acrylates which do not suffer substantial aging.

It must be emphasized that the presence of carbonyl groups attached to furan rings represents on the contrary an element of stabilization in furan polymers. Thus, for example, furan polyesters and polyamides are white and do not suffer any structural modification over very long periods of time.

6.11 CONCLUSIONS

This condensed survey of how furan monomers and furan chemistry can benefit polymer science hopefully provides two major reflections. The first has to do with the fact that it is possible in principle to build a comprehensive new family of macromolecular materials in which some, if not all, of the precursors are derived from renewable resources. In other words, there is no intrinsic limit to the variety of furan monomers that can be synthesized from **F**, **HMF** and their homologues and hence no limit to the polymers and copolymers that they can generate. This constitutes an encouraging potential in terms of the possibility of gradually replacing fossil-derived monomers and polymers by this alternative family.

The second reflection concerns a more qualitative aspect, namely the fact that some peculiarities associated with the chemistry of the furan heterocycle can be put to the advantage of preparing materials with specific functional properties, otherwise difficult to realize. Ample evidence of these unique features is provided in this chapter.

As with other domains dealt with in this book, the road ahead is much longer than what has been achieved up to now in the general endeavour to make polymers from renewable resources. Whether the competition with the existing materials derived from dwindling resources will be successful or not depends almost entirely on the very concrete issue of the extent of financial sponsoring these novel research initiatives will receive.

REFERENCES

1. For selected reviews, see (a) Donnely D.M.X., Meegan M.J., in *Comprehensive Heterocyclic Chemistry*, Eds.: Katritzky A.R., and Rees C.W., Vol. 4, Pergamon, Oxford, 1984, p. 657. (b) Keay B.A., Dibble P.W., in *Comprehensive Heterocyclic Chemistry II*, Eds.: Katritsky A.R., Rees C.W., and Scriven E.F.W., Vol. 2, Elsevier, Oxford, 1997, p. 395. (c) Hou X.L., Yang Z., Wong H.N.C., *Prog. Heterocycl. Chem.*, **15**, 2003, 167.
2. For selected reviews, see (a) Bird C.W., Cheeseman G.W., in *Comprehensive Heterocyclic Chemistry*, Eds.: Katritzky A.R., and Rees C.W., Vol. 4, Pergamon, Oxford, 1984, Chapters 1–3. (b) Dean F.M., Sargent M., in *Comprehensive Heterocyclic Chemistry*, Eds.: Katritzky A.R., and Rees C.W., Vol. 4, Pergamon, Oxford, 1984, Chapters 10–11. (c) Hou X.L., Cheung H.Y., Hon T.Y., Kwan P.L., Lo T.H., Tong S.Y., Wong H.N.C., *Tetrahedron*, **54**, 1998, 1955. (d) Wright D.L., *Prog. Heterocycl. Chem.*, **17**, 2005, 1. (e) Sperry J.B., Wright D.L., *Curr. Opin. Drug Discov. Develop.* **8**, 2005, 723. (f) Brown R.C.D., *Angew. Chem., Int. Ed.*, **44**, 2005, 850. (g) Kirsch S.F., *Org. Biomol. Chem.*, **4**, 2006, 2076.
3. Petrov V.F., Pavluchenko A.I., *Mol. Cryst. Liq. Cryst.*, **393**, 2003, 1.
4. (a) Gandini A., *Adv. Polym. Sci.*, **25**, 1977, 47. (b) Gandini A. in *Encyclopedia of Polymer Science and Engineering*, Eds.: Mark H.F., Bikales N.M., Overberger C.G., Menges G., Second Edition, Volume 7, John Wiley, New York, 1988, p. 454. (c) McKillip W.J., *ACS Symp. Ser.*, **385**, 1989, 408. (d) Gandini A., Belgacem M.N., *Progr. Polym. Sci.*, **22**, 1997, 1203. (e) Moreau C., Gandini A., Belgacem M.N., *Topics Catal.*, **27**, 2004, 11.
5. Dunlop A.P., Peters F.N., *The Furans*, Reinhold Publishing Co., New York, 1953.
6. (a) Fringuelli F., Taticchi A., *The Diels–Alder Reaction*, John Wiley and Sons, Chichester, UK, 2002. (b) Kappe C.O., Murphree S.S., Padwa A., *Tetrahedron*, **53**, 1997, 14179. (c) Rulisek L., Sebek P., Havlas Z., Hrabal R., Capek P., Svatos A., *J. Org. Chem*, **70**, 2005, 6295.
7. Gandini A., Rieumont J., *Tetrahedron Lett.*, **25**, 1976, 2101.
8. Gubina T.I., Pankratov A.N., Labunnskaya V.I., Rogacheva S.M., *Chem. Heterocycl. Comp.*, **40**, 2004, 1396.
9. Win D.T., *Au J. T.*, **8**, 2005, 185.
10. Theander O., Nelson D.A., *Adv. Carbohydr. Chem. Biochem.*, **46**, 1988, 273.
11. Lewkowski J., *ARKIVOC*, **Part 1**, 2001, 17.
12. (a) Asghari F.S., Yoshida H., *Carbohydr. Res.*, **341**, 2006, 2379. (b) Román-Leshkov Y., Chheda J.N., Dumesic J.A., *Science*, **312**, 2006, 1933. (c) Chheda J.N., Roman-Leshkov Y., Dumesic J.A., *Green Chem.*, **9**, 2007, 342.
13. (a) Lichtenthaler F.W., *Acc. Chem. Res.*, **35**, 2002, 728. (b) Lichtenthaler F.W., Peters S., *C. R. Chimie*, **7**, 2004, 65.
14. Lange J., Davidenko N., Rieumont J., Sastre R., *Polymer*, **43**, 2002, 1003.

15. Romani F., Corrieri R., Braga V., Ciardelli F., *Polymer* **42**, 2002, 1115; Coiai S., Passaglia E., Aglietto M., Ciardelli F., *Macromolecules* **37**, 2004, 8414.
16. Ddamba W.A.A., Ngila J.C., Mokoena T.T., Motlhagodi K., *S. Afr. J. Chem.*, **55**, 2002, 1.
17. Salon M.C., *Doctorate Thesis*, Grenoble National Polytechnic Institute, France, 1984.
18. Spange S., Hohne S., Franche V., Gunther H., *Macromol. Chem. Phys.*, **200**, 1999, 1054.
19. Gandini A., Salon M.C., Dutch Patent 8900137, 1989.
20. Hadjikyriacou S., Faust R., *Macromolecules*, **32**, 1999, 6393; Faust R., *Macromol. Symp.*, **157**, 2000, 101; Kwon Y., Faust R., *Adv. Polym. Sci.*, **167**, 2004, 107.
21. Lange A., Rath H.P., Lang G., *Macromol. Symp.*, **215**, 2004, 209.
22. Sánchez R., Hernández C., Rieumont J., *Polym. Degrad. Stabil.*, **61**, 1998, 513; Sánchez R., Rieumont J., Tavares M.I.B., *Polym. Test.*, **17**, 1998, 395.
23. Zhou L.C., Li Y.F., Zhang S.J., Chang X.J., Hou Y.F., *J. Appl. Polym. Sci.*, **99**, 2006, 1620.
24. Gang L., Li Y.F., Zhou L.C., Li B.N., Men X.H., Li X.Z., *High Perform. Polym.*, **17**, 2005, 469.
25. Lucho A.M.S., Gonçalves R.S., *J. Macromol. Sci. Pure Appl. Chem.*, **A42**, 2005, 791.
26. Grosser F.N., Gonçalves R.S., *Anti-corros. Method Mater.*, **52**, 2005, 78.
27. Belgacem M.N., Gandini A., Furan-based adhesives, in *Handbook of Adhesive Technology* (Eds. Pizzi A. and Mittal K.L.), 2nd Edition, Marcel Dekker, New York, 2003, Chapter 30.
28. Wu D., Fu R., *Micropor. Mesopor. Mater.*, **96**, 2006, 115.
29. Muthukumar M., Mohan D., *J. Polym. Res.*, **12**, 2005, 231.
30. Lande S., Westin M., Scheider M.H., *Manag. Environ. Quality*, **15**, 2004, 529; Lande S., Eikenes M., Westin M., *Scand. J. Forest Res.*, **19**(suppl. 5), 2004, 14; Lande S., Westin M., Schneider M., *Scand. J. Forest Res.*, **19**(suppl 5), 2004, 22.
31. Baysal E., Ozaki S.K., Yalinkilic M.K., *Wood Sci. Technol.*, **38**, 2004, 405.
32. Toriz G., Arvidsson R., Westin M., Gatenholm P., *J. Appl. Polym. Sci.*, **88**, 2003, 337; Trindade W.G., Hoareau W., Megliatto J.D., Razera A.T., Castellan A., Frollini E., *Biomacromolecules* **6**, 2005, 2485.
33. Müller H., Rehak P., Jäger C., Hartmann J., Meyer N., Spange S., *Adv. Mater.*, **12**, 2000, 1671; Kawashima D., Aihara T., Kobayashi Y., Kyotani T., Tomita A., *Chem. Mater.* **12**, 2000, 3397; Zarbin A.J., Bertholdo R., Oliveira M.A.F.C., *Carbon* **40**, 2002, 2413.
34. Yao J., Wang H., Chan K.Y., Zhang L., Xu N., *Micropor. Mesopor. Mater.*, **82**, 2005, 183. Príncipe M., Suárez H., Jimenez G.H., Martínez R., Spange S., *Polym. Bull.*, **58**, 2007, 619.
35. Yao J., Wang H., Liu J., Chan K.Y., Zhang L., Xu N., *Carbon* **43**, 2005, 1709; Yi B., Rajagopalan R., Foley H.C., Kim U.J., Liu X., Eklund P.C., *J. Am. Chem. Soc.*, **128**, 2006, 11307; Wang H., Yao J., *Ind. Eng. Chem. Res.*, **45**, 2006, 6393.
36. Liu J., Wang H., Cheng S., Chan K.Y., *Chem. Commun.*, **728**, 2004.
37. Grund S., Kempe P., Baumann G., Seifert A., Spange S., *Angew. Chem. Int. Ed.*, **46**, 2007, 628.
38. Hirai H., Naito K., Hamasaki T., Goto M., Koinuma H., *Macromol. Chem. Phys.*, **185**, 1984, 2347.
39. Okada M., Tachikawa K., Aoi K., *J. Polym. Sci. Polym. Chem. Ed.*, **35**, 1997, 2729; Okada M., Tachikawa K., Aoi K., *J. Appl. Polym. Sci.*, **74**, 1999, 3342.
40. Lasseuguette E., Gandini A., Belgacem M.N., Timpe H.J., *Polymer*, **46**, 2005, 5476.
41. Hatanaka K., Yoshida D., Okuyama K., Miyagawa A., Tamura K., Sato N., Hashimoto K., Sagehashi M., Sakoda A., *Kobunshi Ronbunshu*, **62**, 2005, 316.
42. Kamoun W., Salhi S., Abid S., Tessier M., El Gharbi R., Fradet A., *e-Polymer*, **36**, 2005, 1.
43. Moore J.A., Bunting W.W., *Adv. Polym. Synth.*, **31**, 1985, 51.
44. (a) Abid S., El Gharbi R., Gandini A., *Polymer*, **45**, 2004, 5793. (b) Gharbi S., Gandini A., *J. Soc. Chim. Tunis.*, **6**, 2004, 17.
45. Afli A., Gharbi S., El Gharbi R., Gandini A., *J. Soc. Chim. Tunis.*, **5**, 2003, 3.
46. Abid S., El Gharbi R., Gandini A., *Polymer*, **45**, 2004, 6469.
47. Abid S., Matoussi S., El Gharbi R., Gandini A., *Polym. Bull.*, **57**, 2006, 43.
48. Liu Y.L., Chou C.I., *J. Polym. Sci. Polym. Chem. Ed.*, **43**, 2005, 5267.
49. Chen W., Xue G., *Prog. Polym. Sci.*, **30**, 2005, 783.
50. Benvenuti F., Galletti A.M.R., Carlini C., Sbrana G., Nanninin A., Bruschi P., *Polymer*, 38, 1997, 4973; Demirboga B., Önal A.M., *Synth. Met.*, **99**, 1999, 237; Del Valle M.A., Ugalde L., Díaz F.R., Bodini M.E., Barnède J.C., *J. Appl. Polym. Sci.*, **92**, 204, 1346; Talu M., Kabasakaloglu M., Yildirim F., Sari B., *Appl. Surf. Sci.* **181**, 2001, 51; **218**, 2003, 84.
51. Shilabin A.G., Entezami A.A., *Eur. Polym. J.*, **36**, 2000, 2005.
52. Gok A., Sari B., Talu M., *J. Appl. Polym. Sci.*, **88**, 2003, 2924; **98**, 2005, 2048; *J. Polym. Sci. Polym. Phys. Ed.*, **42**, 2004, 3359.
53. (a) Tsuie B., Reddinger J.L., Sotzing G.A., Soloducho J., Katritzky A.R., Reynols J.R., *J. Mater. Chem.*, **9**, 1999, 2189. (b) Dubus S., Marceau V., Leclerc M., *Macromolecules*, **35**, 2002, 9296. (c) Teare G.C., Ratcliffe N.M., Ewen R.J., Smith J.R., Campbell S.A., *Smart Mater. Struct.*, **12**, 2003, 129.

54. (a) Miyata Y., Nishinaga T., Komatsu K., *J. Org. Chem.*, **70**, 2005, 1147. (b) Alakhras F., Holze R., *Synth. Metals*, **157**, 2007, 109.
55. Ballav N., Biswas M., *Polym. Int.*, **53**, 2004, 1467. 54, 2005, 725.
56. Cho B.R., *Progr. Polym. Sci.*, **27**, 2002, 307.
57. (a) Coutterez C., Gandini A., *Recent Advances in Environmentally Compatible Polymers*, Woodhead Publishing, Cambridge, 2001. p.17. (b) Coutterez C., Ph.D. Thesis, Institut National Polytechnique de Grenoble, 1998.
58. Baret V., Gandini A., Rousset E., *J. Photochem. Photobiol.*, **A103**, 1997, 169.
59. Waig Fang S., Timpe H.J., Gandini A., *Polymer*, **43**, 2002, 3505.
60. Albertin L., Stagnaro P., Coutterez C., Le Nest J.F., Gandini A., *Polymer*, **39**, 1998, 6187.
61. Gandini A., Hariri S., Le Nest J.F., *Polymer*, **44**, 2003, 7565.
62. Lasseuguette E., Gandini A., Timpe H.J., *J. Photochem. Photobiol.*, **A174**, 2005, 222.
63. Reddy J.S., Mandal S., Anand V.G., *Org. Lett.*, **8**, 2006, 5541.
64. Gandini A., Belgacem M.N., *ACS Symp. Ser.*, **954**, 2007, 280.
65. McElhanon J.R., Zifer T., Kline S.R., Wheeler D.R., Loy D.A., Jamison G.M., Long T.M., Rahimian K., Simmons B.A., *Langmuir*, **21**, 2005, 3259.
66. Kuramoto N., Hayashi K., Nagai K., *J. Poly. Sci. Polym Chem. Ed.*, **32**, 1994, 2501.
67. Brand T., Klapper M., *Des. Monomers Polym.*, **2**, 1999, 287.
68. Goussé C., Gandini A., *Polym. Bull.*, **40**, 1998, 389.
69. Chen X., Dam M.A., Ono K., Mal A., Shen H., Nutt S.R., Sheran K., Wudl F., *Science*, **295**, 2002, 1698.
70. McElhanon J.R., Wheeler D.R., *Org. Lett.*, **3**, 2001, 2681; Szalai M.L., McGrath D.V., Wheeler D.R., Zifer T., McElhanon J.R., *Macromolecules*, **40**, 2007, 818.
71. Stevens M.P., Jenkins A.D., *J. Polym. Sci. Polym. Chem. Ed.*, **17**, 1979, 3675.
72. Goussé C., Gandini A., Hodge P., *Macromolecules*, **31**, 1998, 314.
73. Gheneim R., Pérez-Berumen C., Gandini A., *Macromolecules*, **35**, 2002, 7246.
74. Teramoto N., Arai Y., Shibata M., *Carbohydr. Polym.*, **64**, 2006, 78; Watanabe M., Yoshie N., *Polymer*, **47**, 2006, 4946; Ahmad J., Ddamba W.A.A., Mathokgwane P.K., *Asian J. Chem.*, **18**, 2006, 1267; Costanzo P.J., Demaree J.D., Beyer F.L., *Langmuir*, **22**, 2006, 10251; He F., Tang Y., Yu M., Wang S., Li Y., Zhu D., Adv. Funct. Mater., **17**, 2007, 996.
75. Liu Y.L., Hsieh C.Y., Chen Y.W., *Polymer*, **47**, 206, 2581; Liu Y.L., Hsieh C.Y., *J. Polym. Sci. Polym. Chem. Ed.* **44**, 2006, 905; Liu Y.L., Chen Y.W., *Macromol. Chem. Phys.*, **208**, 2007, 224; Gotsmann B., Duerig U., Frommer J., Hawker C.J., *Adv. Funct. Mater.*, **16**, 2006, 1499.
76. Wang H.S., Yu S.J., *Tetrahedron Lett.*, **43**, 2002, 1051.
77. Mori H., Nomura E., Hosoda A., Miyake Y., Taniguchi H., *Macromol. Rapid Commun.*, **27**, 2006, 1792.
78. Froimowicz P., Paez J., Gandini A., Belgacem M.N., Strumia M., *Macromol. Symp.*, **245–246**, 2006, 51.

– 7 –

Surfactants from Renewable Sources: Synthesis and Applications

Thierry Benvegnu, Daniel Plusquellec and Loïc Lemiègre

ABSTRACT

The development of surfactants based on natural renewable resources is a concept that is gaining recognition from the detergent and cosmetic industries. This new class of biocompatible and biodegradable surfactants should be a response to the increasing consumer demand for products that are both greener and more powerful. The present chapter reviews the major contributions to the synthesis and characterization of amphiphilic molecules derived from natural oils and fats (lipophilic moiety) and incorporating sugar-, polyol- or ionic-type residues as the polar headgroups. It covers various aspects of their synthesis, industrial production, applications and regulations.

Keywords

Surfactants, Renewable resources, Carbohydrates, Polyols, Oils, Fats, Cationic emulsifiers, Gemini, Bolaamphiphiles synthesis, Surface properties, Applications

7.1 INTRODUCTION

Surfactants are one of the most representative chemical commodities, not only in terms of quantity, but also in view of the great variety of applications in household, industry and agriculture areas [1]. These 'surface-active agents' have in common the same basic molecular structure – a hydrophilic moiety (polar head group) attached to a hydrophobic backbone (alkyl chain) (Fig. 7.1) – and these two parts provide a compound with interfacial activity and give rise to a wide range of surface chemistry functions: wetting, emulsifying, softening, solubilizing, foaming/defoaming, rheology-modifying, detergency and surface conditioning.

The current world wide production of surfactants amounts to around 12.5 million tons with a growth rate of 500 000 t per year. About 60 per cent of the surfactant production is used in household detergents, 30 per cent in industrial and technical applications (as antistatic agents, lubricants and levelling agents, example for textile production, as flotation agents, for example in mining, oil production and wastewater treatment and as emulsifiers in the food industry, in road constructions and for the production of dies, coatings and plastics), 7 per cent in industrial and institutional cleaning and 6 per cent in personal care. The surfactant consumption is expected to grow at an average annual rate of 2 per cent in North America, to 3.7 million tons in 2010, and by 1.5 per cent in Western Europe, to about 2.5 million tons over the same period. Growth rates are expected to be higher in developing markets, with 3.3 per cent per year predicted for Latin America and 4.2 per cent per year for Asia over the 2000–2010 period [2].

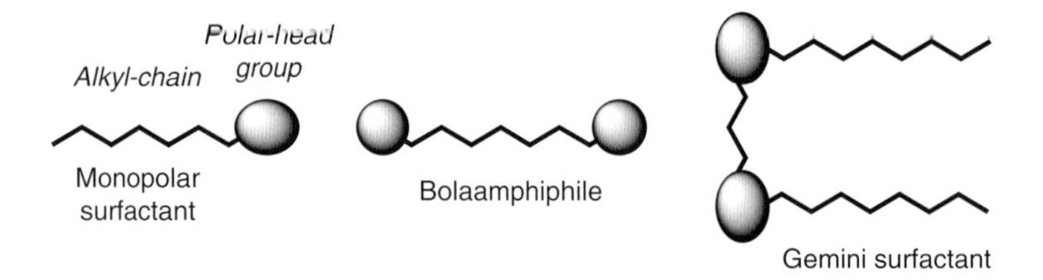

Figure 7.1 Schematic representation of mono and bipolar surfactants.

Surfactant molecules can be broadly divided into four groups, according to their electrical charge in water solution. These groups are anionic (negative charge), cationic (positive charge), non-ionic (no charge), or amphoteric (the molecules contain both positive and negative charges). The anionic surfactants are the 'traditional' ones to which belong, among others, common soaps and they are still the commercially most important. Non-ionic counterparts constitute a smaller group, but are the fastest growing, whereas the other two groups are comparatively small on the market. Properties of surfactants are additionally governed by the alkyl chain length of their hydrophobic part: wetting agents (C_8–C_{10}), detergents (C_{12}–C_{16}), emulsifiers and softeners (C_{18}–C_{22}).

Surfactants can be derived from both petrochemical feedstock and renewable resources (plant and animal oils, microorganisms). They were originally made from renewable resources like fats and oils, whereas today, the majority is of petrochemical origin. But things may slowly be beginning to turn back. The current pressure to move away from non-renewable petroleum feedstocks and towards plants as raw materials has led to considerable effort to develop surfactants from oleochemical feedstocks or other vegetable sources. Many recently developed surfactants are an attempt to satisfy the modern consumers' desire for products to be 'more natural'. Like most plant-based products, surfactants derived from renewable raw materials (RRMs) are characterized by their positive impact on the environment, biodegradability, low or non-toxicity and innocuousness for human health. In addition, the use of renewables in surfactant production can contribute to save fossil resources, such as crude oil and natural gas, and to the reduction of fossil carbon dioxide emissions (CO_2) and hence could be part of a strategy to mitigate the greenhouse effect. There is a good chance that the use of biomass could contribute to these goals, because the quantity of carbon dioxide originated from biomass is equivalent to the amount which was previously withdrawn from the atmosphere during its growth. Finally, the natural origin of these surfactant molecules is a significant communication and marketing asset.

These features, in conjunction with the rising consumer interest for agricultural products, have driven a strong progression in plant-based surfactants. The European market in surfactants represented in 2002 a volume of approximately 2.5 million tons, of which around 25 per cent came from plants. The stakes are thus high for a truly renewable alternative in this sector, and could represent several thousand hectares of crops, knowing that it takes around 60 000 hectares of cropland to produce 100 000 tons of vegetable-based surfactants.

Among surfactants based on renewable chemicals, some are made entirely with RRM (*e.g.* they are 100 per cent natural), whereas others possess only a hydrophobic portion based on RRM (hemipetrochemicals). Vegetable oils furnish fatty chains (acids, alcohols, amines, esters) with a carbon chain length comprised between either 8 and 18 (saturated caprylic, capric, lauric, myristic, palmitic and stearic chains from tropical coconut, palm or palm kernel oils) or 18 and 22 (unsaturated oleic, linoleic, and linolenic chains from European rapeseed and sunflower oils). In this chapter, we will focus on the synthesis and applications of surfactants from entirely natural and renewable resources. We will first study carbohydrate-based surfactants that are gaining more and more recognition from a number of industries ranging from agrochemicals to personal care products and food. We will also discuss (poly)glycerol ester-type surfactants derived entirely from oleochemicals that contribute to increase the value of the major secondary product of oleochemistry (*i.e.* glycerol). Following that, we will present recent research on biodegradable cationic emulsifiers for road industry or cosmetics. In all of these surfactants, the basic structure is that of a single (or multiple) lipophilic tail carrying a single hydrophilic head group as most surfactant development has been confined to this arrangement. However, a more radical approach is to use one or two lipophiles and two hydrophiles in the same molecule. These structures, known as bolaamphiphile (or bolas) and Gemini (or dimeric) surfactants (Fig. 7.1), will be described in the last part of the chapter. As a conclusion, we will finally analyse the perspectives for RRM surfactants in the context of new regulatory policies (notably REACH).

7.2 CARBOHYDRATE-BASED SURFACTANTS

Carbohydrate-based surfactants are the final result of a product concept that is based on the greatest possible use of renewable resources. Whereas the derivatization of fats and oils to produce a variety of different surfactants for a broad range of applications has a long tradition and is well established, the production of surfactants based on fats, oils, and carbohydrates on a larger industrial scale is relatively new. Considering the amphiphilic structure of a typical surfactant with a hydrophilic head group and a hydrophobic tail, it has always been a challenge to attach a carbohydrate molecule as a perfect polar head, due to its numerous hydroxyl groups, to a fat and oil derivative such as a fatty acid or a fatty alcohol. Sugar surfactants have several rather flexible properties. Furthermore, sugar esters and alkyl glycosides are generally non-toxic and non-cumulative. Their properties are temperature insensitive, in contrast to the likewise non-ionic alkyl-PEG surfactants [3]. Although scientists have reported numerous ways of making ester or glycosidic linkages and also described a large number of different carbohydrates used in such reactions, it is clear from an industrial perspective that only a few carbohydrates fulfil the criteria of price, quality, and availability to be interesting raw materials. These include sucrose from sugar beet or sugar-cane (0.3 €/kg; worldwide production 120 000 000 t per year), glucose derived from starches (0.6 €/kg; worldwide production 5 000 000 t per year), sorbitol as the hydrogenated glucose derivatives (0.7 € per kg; worldwide production 900 000 t per year), lactose from mammalian milk (0.6 € per kg; worldwide production 295 000 t per year), and possibly pentoses (wheat straw, wheat bran) and uronic acids (beet pulps, citrus peels or algae). The prices given are intended as a benchmark, rather than a basis of negotiations between producers and customers. Quotations for less pure products are, in part, sizably lower, and thus may readily be used for large-scale preparative purposes.

7.2.1 Sucrose esters

Of all the natural sugars used as renewable sources of raw materials, sucrose **1** is produced by almost every green plant and is therefore widespread in nature. With an annual production of 120 000 000 tons, it represents the world's most abundantly produced organic compound; moreover, sucrose is available at very low cost that is, 0.3 € per kg and at a very high level of purity. Large volume markets of surfactants represent therefore an obvious target for sucrose.

A prerequisite for transforming sucrose into non-ionic surfactants is the introduction of a long alkyl or acyl chain at one of its eight hydroxyls. Sucrose acid esters, more commonly known as sucrose esters or sucroesters, fit the requirements for green chemistry development because they are both biodegradable and can be produced from cheap, renewable and widely available resources: cane or beets sucrose and fat or oil triglycerides. Most of sucroesters are odourless and tasteless, or slightly bitter, allowing them to find applications both in food and personal care products.

Sucrose is a non-reducing disaccharide composed of a α-D-glucopyranosyl unit bonded to the anomeric carbon of a β-D-fructofuranoside (Fig. 7.2). The acid lability of the glycosidic linkage, coupled with the insolubility of sucrose in most common organic solvents, limits its chemical reactions.

The original preparation of sucrose esters of fatty acids involves the transesterification of a triglyceride molecule with sucrose in the presence of a basic catalyst at 90°C in *N,N*-dimethylformamide (DMF) as solvent (Fig. 7.3) [4].

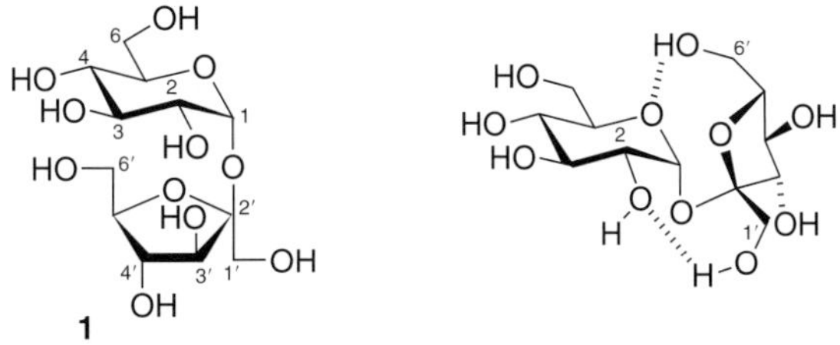

Figure 7.2 Common representation of the chemical formula of sucrose and its intramolecular hydrogen bonds in its crystallized form.

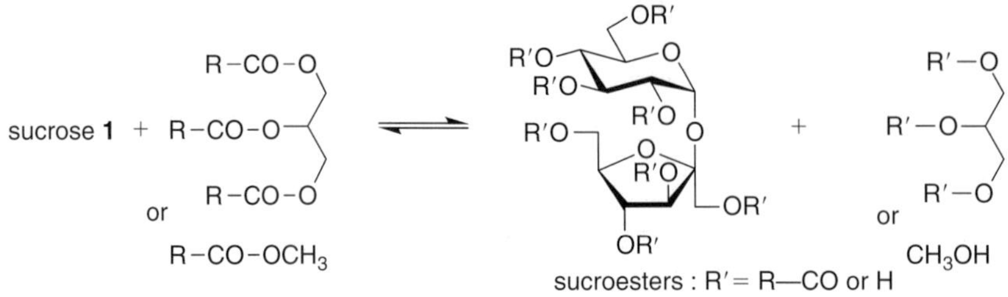

Figure 7.3 Commercial routes to sucroesters.

DMF was later replaced by dimethylsulphoxide (DMSO), a less expensive and less toxic solvent. The reaction product contains more than 50 per cent of monoesters, di- and higher esters, unreacted sucrose and di- and triglycerides.

Fatty acid methyl esters are currently used in transesterification reactions with sucrose. The formation of methanol, which can be distilled off, drives the esterification process in favour of the sucrose ester and improves yields. A solvent-less process using a slurry of sucrose and potassium carbonate in fatty acid methyl ester or triglyceride oil at 130°C has been developed more recently [5].

These procedures are commonly used to produce sucrose esters such as stearates, tallowates, oleates, palmitates, myristates and laurates that are actually complex product mixtures containing 70 per cent of monoester and 30 per cent of di-, tri- and poly-esters. Global production of sucroesters is estimated at 5 000 t per year.

The demand for such compounds and therefore their market value, could still increase substantially if the reaction processes, especially for the synthesis of well-defined products, can be further optimized.

The selective monoacylation of free sucrose encounters regioisomer problems owing to (i) rather similar reactivities of the eight hydroxyl groups and (ii) easy intramolecular migrations of acyl groups in the unprotected derivatives. The eight hydroxyl groups, numbered as shown in Fig. 7.2, include (i) three primary hydroxyls at carbons 6, 6′ and at neopentylic carbon 1′; (ii) five secondary hydroxyls at carbons 2,3,4, 3′ and 4′. Theoretically, the reaction of sucrose with one molar equivalent of acylating reagent may therefore provide eight possible regioisomeric monoesters. Nevertheless, regioselective chemical modifications of the free substrate have already been proposed [6].

In modifying sucrose for the preparation of sucrose esters, great attention must be focused on the structure, the conformation of sucrose in solution, the reaction conditions (solvent, electrophilic reagent, catalysis, temperature, etc.) and the purification procedures of the reaction product.

Indeed, the solution conformation of sucrose is surmized to depend on the nature of the solvent and therefore the disruption of one or both of the intramolecular hydrogen bonds that are present in the crystal structure. For aprotic polar solvents such as DMSO and DMF, the occurrence of two conformations with competitive intramolecular hydrogen bonds was suggested, namely 2-O•••HO-1′ and 2-O•••HO-3′, the equilibrium being in favour of the former (Fig. 7.4) [7].

The conformation of sucrose in water is more controversial, but it is now generally accepted that there are no direct intramolecular hydrogen bonds between the glucose and the fructose moieties. However, some data reveal the existence of indirect ones, an interresidue waterbridge linking the glucose-2-*O* with the fructosyl-1-*O*. The linkage geometry in aqueous solution closely resembles that found in the crystal and in polar aprotic solvents [7a]. Chemists have to keep in mind the possible connections between the structure of sucrose and the relative reactivities of its different hydroxyls.

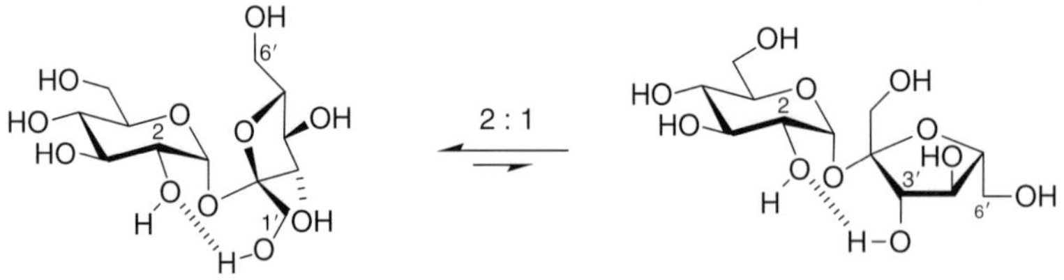

Figure 7.4 The conformation of sucrose in polar aprotic solvents such as DMSO and DMF.

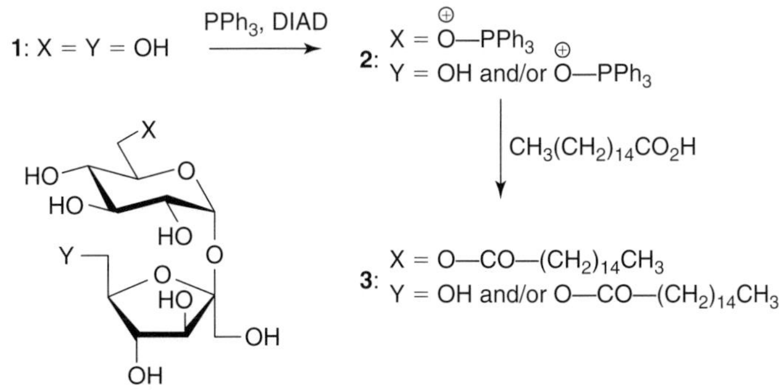

Figure 7.5 Synthesis of 6-*O* (6′-*O*)-(di)palmitoyl sucrose by a Mitsunobu approach.

Acylation of the least hindered 6 and 6′ hydroxyl groups can be achieved in low to moderate yields using exceptionally mild reaction conditions or the 6-position can be acylated with bulky acid chlorides that is, pivaloyl chloride [8]. In a different approach, the application of the Mitsunobu reaction allowed Bottle and Jenkins [9] to prepare bulky oxophosphonium intermediates at the primary 6 and/or 6′-positions. A substitution reaction by palmitate led to the 6-*O*-palmitoyl sucrose (Fig. 7.5). Interestingly, the 6-position appeared to be more reactive than the 6′; but, nevertheless, the 6-6′-dipalmitate could be synthesized by using a 2.6 molar equivalent of palmitic acid. However, the neopentylic-like 1′-position is considerably more hindered in such a substitution process.

An interesting approach to the regioselective synthesis of 6-*O*-acyl sucrose involves the use of sucrose dibutylstannyl-acetal intermediate **4** (Fig. 7.6). This acetal reacted with fatty anhydrides in DMF at room temperature giving, in all cases, the single product 6-*O*-acyl sucrose **5** [10]. In order to avoid the production of toxic tin by-products, our own contribution was based on a quite different one-pot strategy and made possible the acylation of free sucrose with *N*-acyl-thiazolidine-2-thiones **6** in DMF and in the presence of 1,8-diazabicyclo(5.4.0)-undec-7-ene (DBU). Long chain 6-*O*-acyl sucroses were isolated in 60 per cent yields [11]. Compounds **5** could also be easily obtained by the quantitative intramolecular isomerizations of 2-*O*-acyl sucroses (vide supra) in the presence of DBU in DMF at room temperature.

The preferential reaction of the 6-hydroxyl group of sucrose with *N*-acyl-thiazolidine-2-thiones **6** in the presence of DBU is noteworthy. Indeed, we have also shown that regioselectivity was enhanced towards 6′-*O*-acylsucroses **8** (*i.e.* acylation of the fructosyl moiety) when a weaker base, such as 1,4-diazabicyclo(2.2.2)octane (DABCO), was used. Best results are obtained with 3-acyl-5-methyl-1,3,4-thiadiazol-2(3 H)-thiones **7** in DMF at low temperature [12]. Such conditions give predominantly 6′-*O*-acylsucroses **8** along with 6-*O*- and 1′-*O*-acylates,

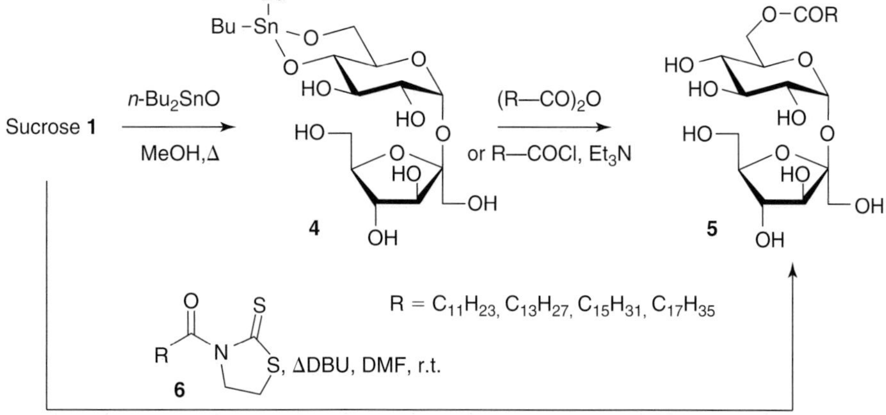

Figure 7.6 Synthesis of 6-*O*-acyl sucroses via a dibutylstannylene acetal approach(10) or a one-pot approach using *N*-acyl-thiazolidine-2-thiones (11).

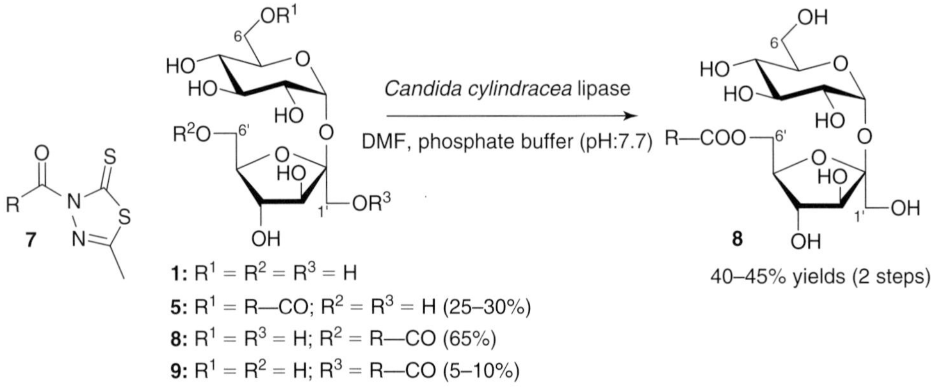

Candida cylindracea lipase

DMF, phosphate buffer (pH:7.7)

1: $R^1 = R^2 = R^3 = H$

5: R^1 = R—CO; $R^2 = R^3$ = H (25–30%)

8: $R^1 = R^3$ = H; R^2 = R—CO (65%)

9: $R^1 = R^2$ = H; R^3 = R—CO (5–10%)

8 40–45% yields (2 steps)

Figure 7.7 Regioselective synthesis of 6′-*O*-acylsucroses.

5 and **9** respectively (Fig. 7.7). To make the purification of esters **8** easier, we predicted that *Candida cylindracea* lipase (CCL) could deacylate the by-products as indeed verified for the elimination of esters **5** from the crude mixtures.

In order to rationalize these data, one can assume that DABCO acts as a weak base (when compared with DBU) as well as a nucleophilic catalyst. DABCO deprotonates therefore the more acidic hydroxyls of sucrose, the more nucleophilic alkoxides (*i.e.* C-2 and C-6′ hydroxyls) that are intramolecularly hydrogen bonded (Fig. 7.4) [6b]. On the other hand, reagents **7** could easily lead with DABCO to bulky loose ion pairs which preferably acylate the less hindered alkoxide, that is to say the C6′.

Regioselective base activation of the more acidic hydroxyl group to the more nucleophilic alkoxide may therefore be considered as a powerful entry into well-defined derivatives of sucrose. As indicated earlier, 2-oxygen of sucrose is involved in an intramolecular or an indirect hydrogen bond. The direct consequence on the acidity of the OH group, and therefore on the nucleophilicity of the resulting 2-oxyanion, may be expected. We anticipated that the gradual addition of a base into a solution containing unprotected sucrose and a suitable acylating reagent that did not react on the base, would activate the 2-OH group to a more nucleophilic alkoxide. Indeed, when NaH (catalytic amount) was added to a solution of sucrose and 3-lauroylthiazolidine-2-thione **6** ($R = C_{11}H_{23}$) in pyridine or DMF, the reaction mixture gave a 70 per cent yield of 2-*O*-lauroylsucrose [8]. This procedure was extended to a variety of fatty acid derivatives and was further successfully used to prepare 6-*O*-acylsucroses through the controlled intramolecular isomerization with DBU (Fig. 7.8) [11].

The use of hydrolytic enzymes in organic solvents represents an alternative approach for the regioselective synthesis of sucroesters. Of particular interest are vinyl esters of fatty acids which are used to drive the esterification process to completion through tautomerization of the enol by-product to the corresponding aldehyde (Fig. 7.9).

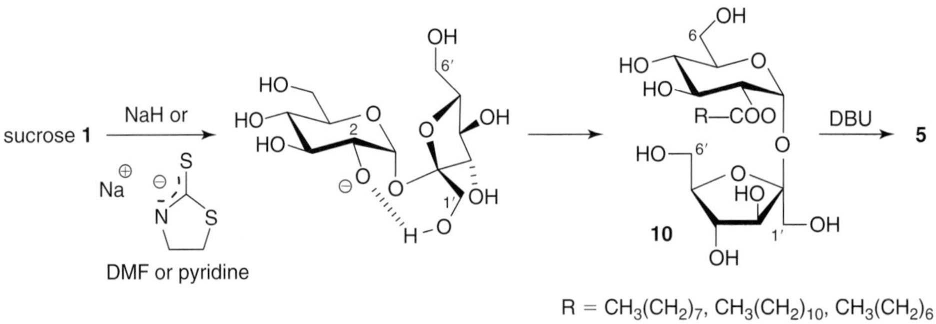

sucrose **1** NaH or DMF or pyridine DBU **5**

$R = CH_3(CH_2)_7, CH_3(CH_2)_{10}, CH_3(CH_2)_6$

Figure 7.8 Regioselective synthesis of 2-*O*-acylsucroses.

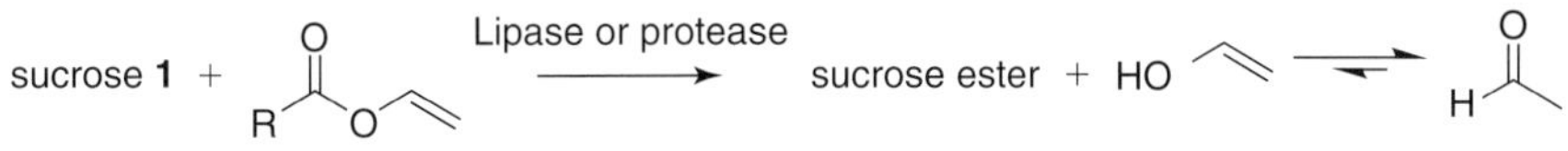

Figure 7.9 Enolesters in transesterification procedures.

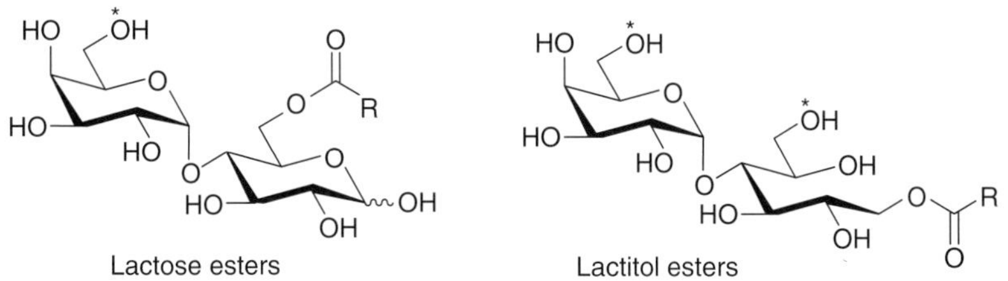

Figure 7.10 The structures of lactose and lactitol mono-ester surfactants. The asterisks denote alternative locations for the fatty acid substitution.

Thus, 6-*O*-lauroyl(pamitoyl)sucroses were obtained in 70–80 per cent yields by performing the transesterification in a mixture of *tert*-amyl alcohol and DMSO in the presence of lipase from *Humicola lanuginose* [13].

Proteases can also catalyse the esterification of carbohydrates. Thus 1′-*O*-butyl sucrose was prepared from sucrose and trichloroethyl butyrate with subtilisin in anhydrous DMF [14]. This strategy has been extended to commercially available insoluble crosslinked forms of subtilisin to prepare regiospecifically 1′-*O*-acylsucroses from fatty acids and vinyl esters in high yields [15].

A judicious choice of electrophilic reagent, catalyst and/or reaction medium is a requirement for a regioselective synthesis of well defined sucrose esters. Three main processes are observed: secondary 2-OH selective reactions based on electronic factors; primary OH groups, that is, 6-OH and/or 6′-OH selective reactions based on steric hindrance and 1′-OH selective reaction based on protease catalysis. These reactions open new routes to sucrose esters as products of commercial significance or as intermediates in sugar chemistry. As an example, the placement of a protective ester group at the 6-position of the glycopyranoyl moiety and the subsequent chlorination and cleavage of the ester group provide high intensity, non-metabolizable sweeteners [16].

Sucrose esters of fatty acids having 12 or more carbon atoms display surface active properties. Most of them are odourless and tasteless (or slightly bitter) allowing them to find applications both in food and personal care products [17]. Sucroesters were approved and freely permitted in Japan for use as food additives in 1959 for both their emulsifying ability and their heat stability. In addition, they are well known to protect food proteins from thermal denaturation and inhibit the growth of *Escherichia coli* and other bacteria.

Covering a wide hydrophilic – lipophilic – balance (HLB) range, monoesters can be used to stabilize 'oil-in-water'emulsions, whereas higher esters give more lipophilic surfactants which can stabilize 'water-in-oil' emulsions.

The surface activity of some well defined monoesters of sucrose [10] and their self-organizing properties [18] have been described recently revealing exceptionally low critical micellar concentration (CMC) values and original self-organizing arrangements, respectively. The physical, chemical and biological properties of most pure derivatives still need to be evaluated and their scale up to be developed for other important applications in the cosmetic and pharmaceutical industries that require pure compounds.

Additional disaccharidic esters such as lactose and lactitol esters with varying alkyl chain lengths were reported in the literature (Fig. 7.10) [19]. These derivatives behave similarly to sucrose esters with the same hydrocarbon chain length, although they have a tendency towards closer packing at the air–water interface for the open chain type.

7.2.2 Alkyl polyglycosides and analogues

The sugar surfactants of choice for industries such as agrochemicals and detergents are not the sucrose esters, which are often considered to be rather too expensive, but another set of sugar surfactants, the so-called alkyl glucosides

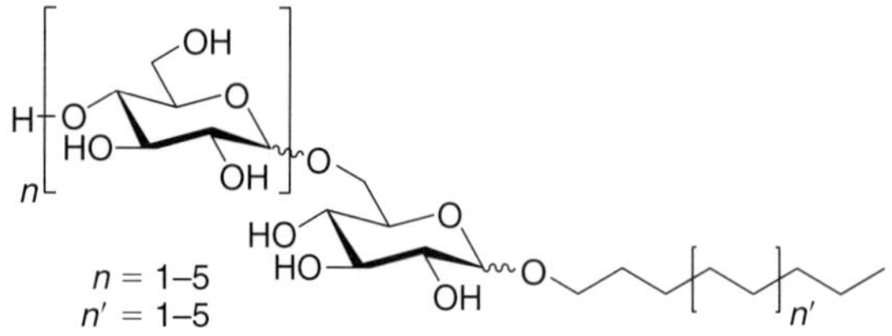

Figure 7.11 Alkyl polyglycosides (APG).

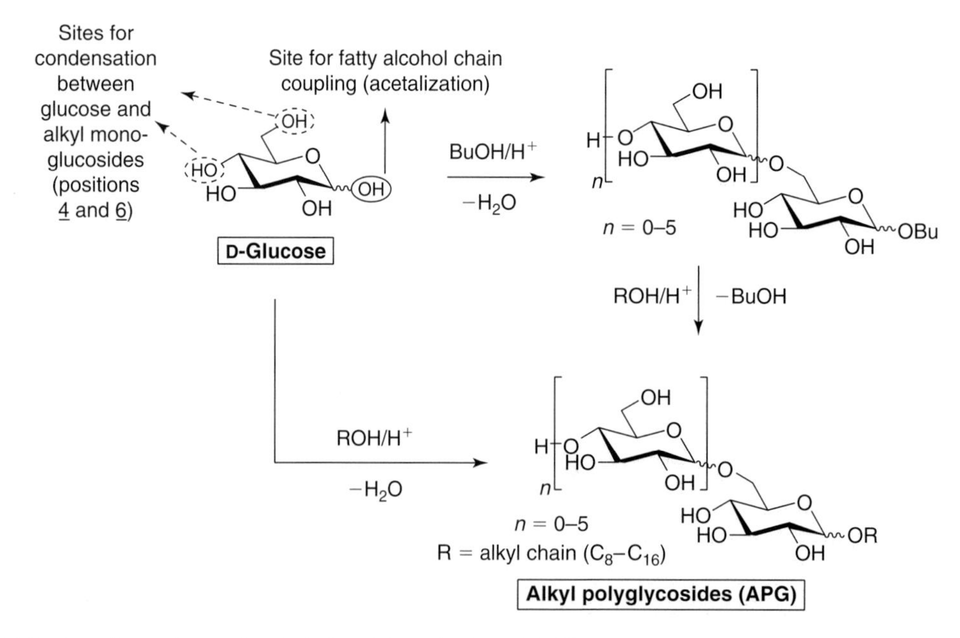

Figure 7.12 Reaction pathways for the industrial production of APG.

(AGs) or alkyl polyglycosides (APG – a registered trademark of Henkel). APG consumption represents a world market of 100 000 t per year, including 50 000 tons for Europe, and is growing fast. Marketed originally in the late 1970s, they are produced today mainly by Cognis (capacity of 50 000 t per year), SEPPIC, Kao, Zeneca, Union carbide and BASF. APGs consist of fatty alcohols mainly obtained from coconut or palm kernel oil (C_{12-14} fatty alcohols) and palm or rapeseed oil (C_{16-18} fatty alcohols) as the hydrophobic part (also from petrochemicals), and glucose obtained from corn starch, wheat or potatoes, as the hydrophilic part, their hydrophilicity varying through the degree of oligomerization (Fig. 7.11). APG are complex mixtures of isomers characterized by a polysaccharide unit varying between 1 and 6 (commercially available APG have an average degree of polymerization between 1.3 and 1.6).

The industrial production of APG requires at least three manufacturing operations. The first stage is the acid-catalyzed condensation (acetalization) between glucose and an aliphatic alcohol (Fig. 7.12).

There are currently two acetalization methods that are used industrially: (1) the direct acetalization consists of condensing glucose and an alcohol directly in a solvent-free reaction (Fischer glycosidation); (2) the transacetalization requires the preliminary synthesis of a butylpolyglucoside (carried out without solvent and catalyzed by an acid) which then reacts with an aliphatic alcohol. The transacetalization has the advantage of using cheap sources of glucose such as glucose syrup or starch, whereas direct acetalization, although simpler, requires more expensive anhydrous glucose. Whatever the method of acetalization used, the resulting APG is always obtained in the form of a complex mixture. The diversity of products formed can be explained partly because of the various possibilities of isomerization (furanoside, *e.g.* five-membered cycle or pyranoside, *e.g.* six-membered cycle, forms) and anomerization (α and β) known to occur with sugars under acidic conditions, the

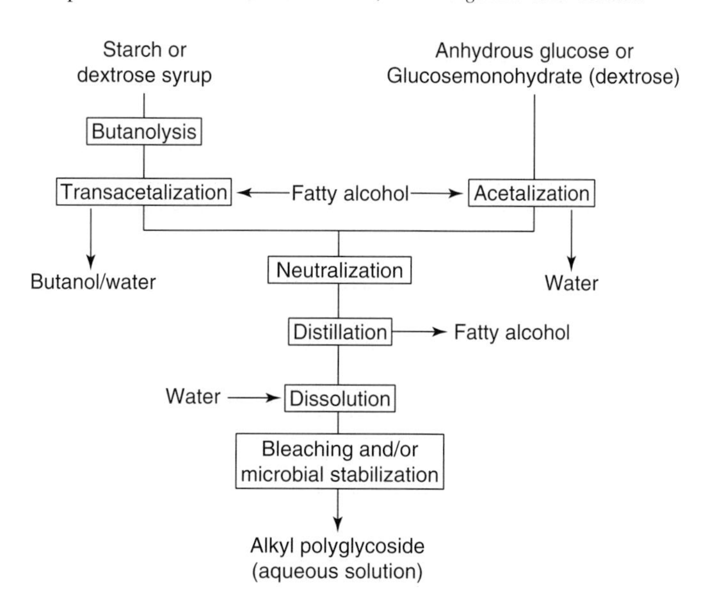

Figure 7.13 Flow diagram for the production of alkyl polyglycosides based on different carbohydrate sources.

pyranoside forms (>90 per cent) and alpha anomers (>65 per cent) being dominant. Next, the condensation between glucose and monoglucosides leads to polymerization giving alkyl polyglucosides. Monoglucosides have four hydroxyl groups that are liable to be acetalized, but for stereoisomeric reasons, the 1→4 and 1→6 bonds are favoured. A reduction of reaction products can be achieved by varying some parameters during the acetalization step, such as temperature, pressure, type of catalyst and alcohol quantity. The second manufacturing stage consists in stopping the acetalization reaction by neutralising the acid catalyst. As the acetalization reaction requires an excess of alcohol, the third stage consists in the elimination of this excess by distillation. In some cases, additional refining steps are envisaged comprising APG bleaching and microbial stabilization (Fig. 7.13).

Polyglycosides with a short alkyl chain (C_8–C_{10}) are readily water-soluble and applicable for a wide range of surfactants from mild furniture-care products to industrial cleaning agents; those with a medium chain length (C_{12}–C_{14}) show a strong synergistic effect in combination with other surfactants, and are suitable for detergents, all-purpose cleaners, shampoos and cleansing cosmetics. The long-chain homologues (C_{16}–C_{18}), insoluble in water, find applications in cosmetic creams because of their good oil/water emulsifying properties. The advantages of APG in comparison with other surfactants are numerous: (1) unlike most surfactants (such as non-ionic ethoxylates), they do not precipitate from solution at high temperatures (*i.e.* they do not exhibit a cloud point phenomenon) so they can be used without fear of phase separation. And because of their far greater tolerance to high-electrolyte concentration than any other non-ionic surfactants, APG can be added to highly ionic products (pesticides); (2) in regard to their ecological (complete and rapid biodegradation in the environment both aerobically and anaerobically, mainly resulting from the presence of an acetal linkage), toxicological, and dermatological properties, APG have extraordinary product safety characteristics. They have a good compatibility with eyes, skin and mucous membranes and even reduce irritant effects of surfactant combinations. The synergistic interactions between APG and most traditional primary surfactants commonly used in detergents, such as linear alkylbenzene sulphonate (LAS), mean that they make ideal cosurfactants; (3) in personal care products, they represent a new concept in compatibility, properties and care. One of Cognis's APG products, called Lamesoft P065, a lamellar dispersion of the monoglyceride GMO in an APG, is used in body washes and soap bars, and has also shown to be effective in two-in-one shampoos, making clear formulations possible [20].

APGs are not the only examples of sugar-based surfactants resulting from a glycosidation reaction (acetalization) of carbohydrates. Alkyl monoglucosides were prepared using standard glycosylation reactions (Koenigs–knorr syntheses) to prepare specific glycosides; for instance, β-1-*n*-octyl-D-glucopyranoside is widely used in biomembrane research for the extraction of water-insoluble membrane proteins without denaturation (Fig. 7.14). It is

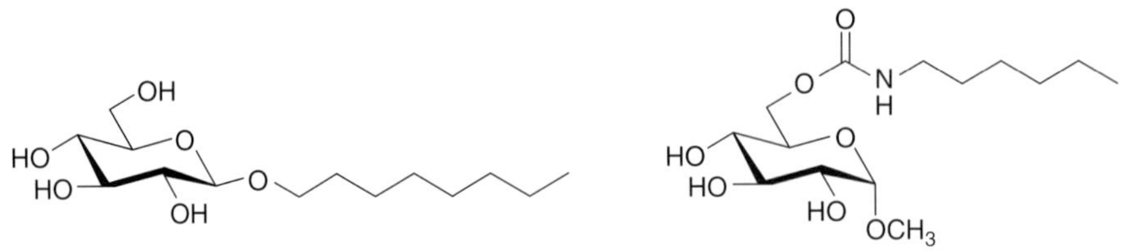

β-*n*-octyl-D-glycopyranoside

HECAMEG

Figure 7.14 Structure of glucoside surfactants for membrane studies.

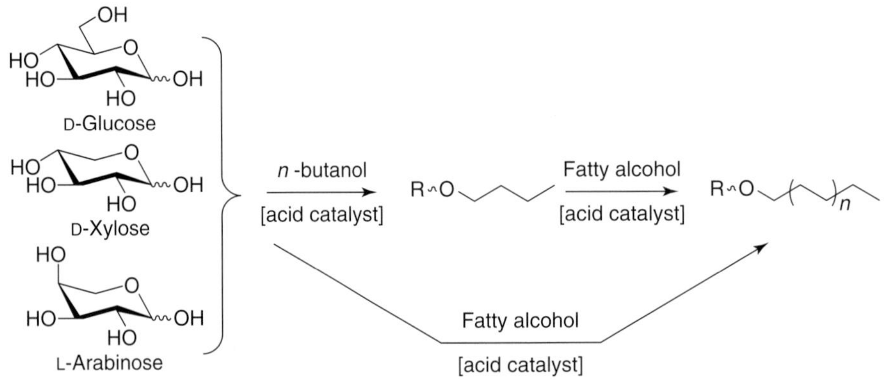

D-Glucose

D-Xylose

L-Arabinose

n-butanol [acid catalyst]

Fatty alcohol [acid catalyst]

Fatty alcohol [acid catalyst]

Figure 7.15 Reaction pathways for the industrial production of bran-based surfactants.

preferred over other non-ionic surfactants, such as polyethylene glycol alkyl ethers, because it has a very high CMC, and hence, it can easily be separated from the proteins by dialysis. The syntheses of *n*-alkyl-β-D-glucosides require expensive reagents (silver catalysts) and are rather time-consuming (several steps). Additional glucose-derivatives such as 6-*O*-(*N*-heptylcarbamoyl)-methyl-α-D-glucopyranoside (HECAMEG), a very mild surfactant useful for membrane protein studies, were synthesized by a simple and low cost procedure from methyl-α-D-glucopyranoside [21].

More recently, monosaccharide mixtures of pentoses, easily available from wheat straw and wheat bran, were used by Soliance, a subsidiary of ARD (Agro Industry R & D – owned by a consortium of French agricultural cooperatives specialized in wheat and sugar beet production) for the preparation of bran-based surfactants. Wheat straw provides mainly pentoses: L-arabinose and D-xylose; wheat bran gives a mixture of pentoses and glucose. All these monosaccharides, or mixtures thereof, can be used as starting materials in the acetalization step (Fig. 7.15), following the same strategies as for APG production (direct glycosidation or synthesis of butylglycosides followed by a transglycosylation reaction using fatty alcohols). The pentoses derived from hemicellulose have the advantage of being more reactive than glucose during the glycosidation reaction, allowing gentler reaction temperatures, and hence improving the energy efficiency of the process. During the reaction, an oligomerization of the saccharides occurs. Using various starting mixtures, a wide range of amphiphiles can be obtained with good foaming and detergency (C_8–C_{14} alkyl chain length) or emulsifying (C_{14}–C_{22} alkyl chain length) properties (Soliance already markets an emulsifier under the name of Emuliance®) [22].

Innovative research was also developed for the production of monosaccharidic surfactants through a direct synthesis of tautomerically and anomerically pure alkyl glycosides from unprotected sugars. The Fischer glycosylation of fatty alcohols is the most commonly used method for preparing alkyl glycosides (see APG and pentose derivatives), but the reaction invariably produces mixtures of α,β-glycopyranosides and corresponding furanosides owing to: (i) tautomeric equilibria and (ii) *in situ* anomerizations. Thus, it was found that, depending on the promoter and on the reaction conditions, either alkyl glycofuranosides from D-glucose (D-Glc*f*), D-galactose (D-Gal*f*) or D-mannose (D-Man*f*) or pyranosidic derivatives (D-Glc*p*), (D-Man*p*), (D-GlcNAc) and (D-Fruc*p*) from D-glucose, D-mannose, *N*-acetyl-D-glucosamine and a ketose of commercial significance, D-fructose, were obtained in

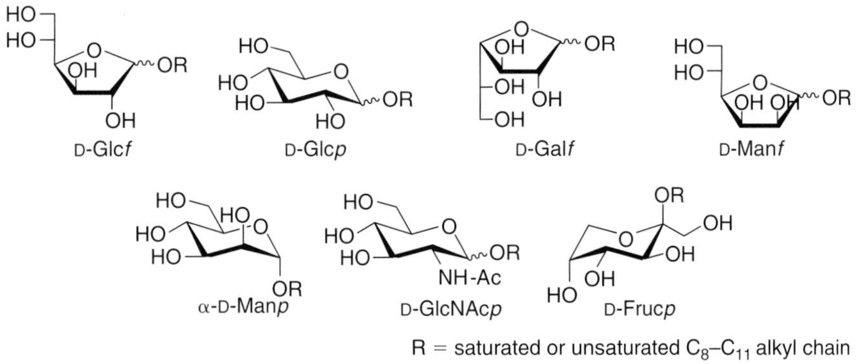

R = saturated or unsaturated C_8–C_{11} alkyl chain

Figure 7.16 Alkyl D-glycosides from *O*-unprotected sugars using $FeCl_3$ or $BF_3.OEt_2$ as catalysts.

moderate-to-good yields (Fig. 7.16). The synthesis of tautomerically and anomerically well-defined alkyl *O*-glycosides from unprotected carbohydrates requires: (i) a faster reaction of the alcohol with the sugar than the self-condensation of the monosaccharide and (ii) control of the equilibria involving substrates, intermediates and products, including the control of the final α,β stereoselectivity of the products. In order to avoid the self-condensation of monosaccharides, the reactions are performed in heterogeneous media, the unprotected carbohydrates being slightly soluble in solvents such as tetrahydrofuran (THF), 1,4-dioxane, acetonitrile and dichloromethane. After experimentation, it was found that the glycosylation reaction indeed occurred at room temperature and was best achieved in THF by the use of iron(III) chloride or boron trifluoride ethyl etherate Lewis acids as promoters. Interestingly, a dichotomy between $FeCl_3$ and $BF_3.OEt_2$ was noteworthy in terms of furanosides/pyranosides ratio since alkyl D-glycofuranosides could be obtained as the single products with the former promoter, whereas the thermodynamically more stable pyranosides were exclusively isolated in the presence of the latter [23]. The surface tension and critical micellar concentration of the alkyl glycosides were determined by tensiometric measurements. Furanosides generally possess very interesting surfactant properties. They are able to reduce the surface tension of pure water by more than 40 $mN\,m^{-1}$. Thus, the surface tension values above the cmc varied between 25 and 30 $mN\,m^{-1}$, whereas those obtained for the tautomeric pyranosic compounds are generally greater than 30 $mN\,m^{-1}$. Furanosides are therefore more efficient surfactants than the corresponding pyranosides. This result probably reflects a greater destructuring effect of the furanosides on the 3D structure of water. Consequently, the shape of the glucidic moiety has a great influence on the micellization for amphiphiles of similar chain lengths and polarities, and this effect could constitute a new factor to modulate surfactant properties [24]. In the case of D-fructopyranosides, atypical liquid crystalline self-organisation properties were found, probably resulting from the amphiphilicity of the fructopyranosides [25].

Uronic acids (*e.g.* alkyl glycosiduronic acids) were also employed as starting materials for the preparation of carboxylic acid-containing carbohydrate surfactants. Beet and citrus pulp supply D-galacturonic acid from pectins. D-Glucofuranurono-6,3-lactone ('D-glucurone') can be isolated from wheat bran. Finally, brown algae furnish D-mannuronic acid and L-guluronic acid that compose the heteropolysaccharide alginates. The direct synthesis of tautomerically and anomerically pure glycosiduronic acids from *O*-unprotected uronic acids remains particularly difficult owing to (i) the requirement of a higher activation at the anomeric position for uronic acids, in comparison with their neutral analogues; (ii) tautomeric equilibria of some sugars in solution; (iii) *in situ* anomerizations; and (iv) competitive *O*-glycosidation and esterification processes. Alkyl D-galactofuranosiduronic acids were synthesized by the direct glycosidation of totally *O*-unprotected D-galacturonic acid with fatty alcohols in heterogeneous media using THF as a solvent, ferric chloride as a promoter and calcium chloride as an additive [26]. The furanosiduronic acids were obtained in 50–80 per cent overall yields with a high β-stereoselectivity (typically α ≈ β 1:9). The β-anomers crystallized out of diethyl ether-light petroleum mixtures, thus affording anomerically pure compounds. Alkyl D-galactopyranosiduronic acids were prepared through a procedure involving (Fig. 7.17): (i) glycosidation of D-galacturonic acid with fatty alcohols in THF in the presence of $BF_3.OEt_2$ at 30°C; (ii) concentration of the reaction mixture (total evaporation of THF) at 30°C under reduced pressure for *in situ* isomerization (furanosides -> pyranosides) and anomerization (β > α); (iii) saponification (2.5 $mol\,dm^{-3}$ NaOH in water: acetone; 15 min at room temperature) and removal of fatty alcohol by extraction; (iv) acidification

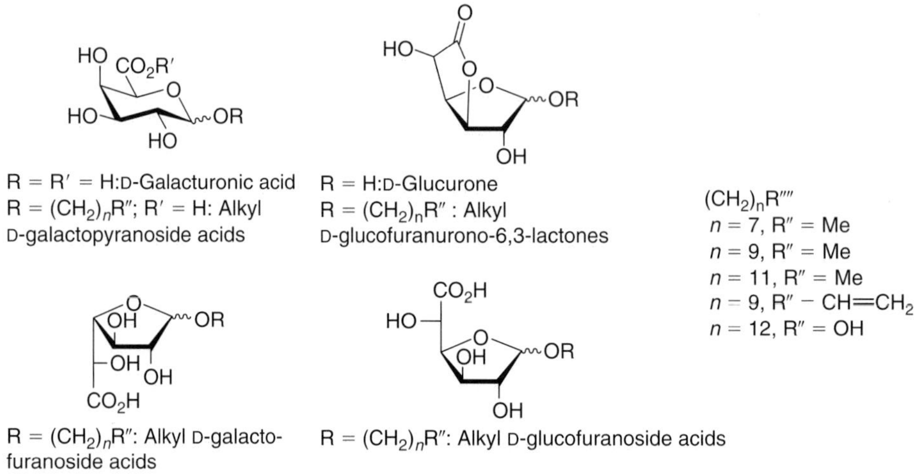

R = R′ = H:D-Galacturonic acid
R = $(CH_2)_nR''$; R′ = H: Alkyl D-galactopyranoside acids

R = H:D-Glucurone
R = $(CH_2)_nR''$: Alkyl D-glucofuranurono-6,3-lactones

$(CH_2)_nR''''$
$n = 7$, R″ = Me
$n = 9$, R″ = Me
$n = 11$, R″ = Me
$n = 9$, R″ − CH=CH$_2$
$n = 12$, R″ = OH

R = $(CH_2)_nR''$: Alkyl D-galactofuranoside acids

R = $(CH_2)_nR''$: Alkyl D-glucofuranoside acids

Figure 7.17 Alkyl D-glycosiduronates from O-unprotected D-uronic acids using $FeCl_3$ or $BF_3.OEt_2$ as catalysts.

of the aqueous phase and isolation of the α-galactopyranosiduronic acids by extraction. In the D-glucuronic series, treatment of D-glucurone with fatty alcohols and $BF_3.OEt_2$ in refluxing THF provided crystalline furanoside lactones in good-to-excellent yields and high stereoselectivity (β:α = 10:1) which could be saponified in high yields into crystalline β-D-glucofuranosidic acids. These results are of much interest since they represent one of the very few examples where *O*-glycosiduronic acids have been synthesized without protecting groups. These compounds should find applications as new surfactants [27] and liquid crystals [28].

More recently, novel biocompatible surfactants derived from alginate were developed for applications in detergents and cosmetics [29]. These surfactants are characterized by the presence of a monosaccharidic α-D-mannuronate residue attached to one or two fatty alcohols through ester and/or glycosidic linkages. Alginate, a heteropolysaccharide from brown algae, is composed of (1,4)-linked β-D-mannuronic acid (M) and α-L-guluronic acid (G) units in the form of homopolymeric (MM- or GG-blocks) and heteropolymeric sequences (MG- or GM-blocks). Controlled acid hydrolysis of commercially available alginate from *Laminaria digitata* gives saturated oligomannuronates on a multigram scale (Fig. 7.18). Subsequent one-pot acid glycosidic bond hydrolysis, esterification and stereocontrolled Fischer glycosidation in butanol with the related oligouronates, using methanesulphonic acid, efficiently provides *n*-butyl-(*n*-butyl α-D-mannopyranosiduronate). Double-tailed amphiphiles were next obtained from this key intermediate by transesterification/transglycosylation processes in fatty alcohols. Aqueous

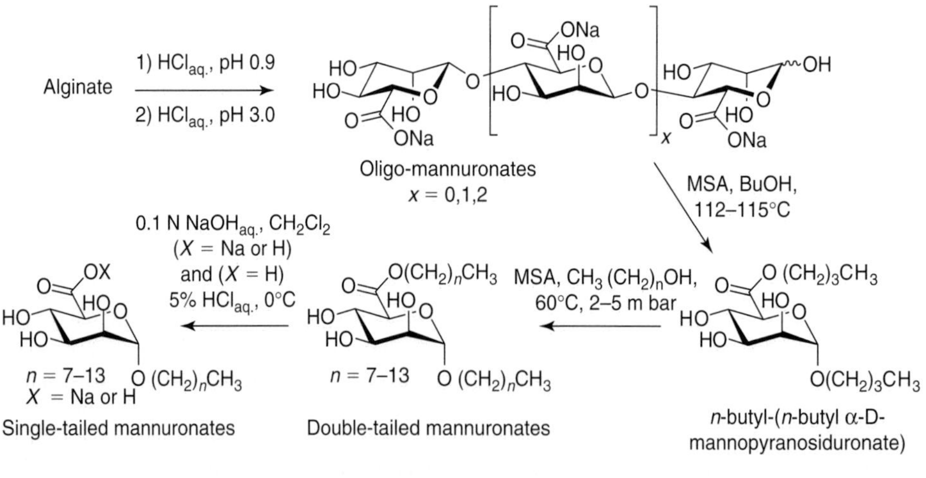

Alginate → 1) $HCl_{aq.}$, pH 0.9; 2) $HCl_{aq.}$, pH 3.0 → Oligo-mannuronates x = 0,1,2

MSA, BuOH, 112–115°C → *n*-butyl-(*n*-butyl α-D-mannopyranosiduronate)

MSA, $CH_3(CH_2)_nOH$, 60°C, 2–5 m bar → Double-tailed mannuronates (n = 7–13)

0.1 N $NaOH_{aq.}$, CH_2Cl_2 (X = Na or H) and (X = H) 5% $HCl_{aq.}$, 0°C → Single-tailed mannuronates (n = 7–13, X = Na or H)

Figure 7.18 Stereoselective synthesis of α-D-mannuronate surfactants.

basic and acid treatments furnished anionic or neutral single-tailed surfactants. These original alginate-derived amphiphiles exhibit attractive surface-tension and foaming properties. Anionic and neutral single tailed surfactants with C_{12}–C_{14} fatty chains reduce the surface tension to values (29–30 mN m^{-1}) comparable to those obtained with commercial non-ionic surfactants (Polyethylene glycol, APG). In all cases, carboxylate derivatives exhibited higher cmc values than their neutral counterparts (0.28–1.62 mmol L^{-1} for anionic compounds and 0.13–0.16 mmol L^{-1} for neutral counterparts). Mannuronic acid derivatives are high-foaming surfactants which can be compared with the ether sulphate derivative. In particular, similar values are observed for surfactant with a C_{12} aliphatic chain and SDS, both in regard to foam performance and foam stability. Finally, the variability of the double-tailed amphiphiles, in terms of hydrophilic-hydrophobic balance, also permits to envisage additional applications such as oil-in-water and water-in-oil emulsions. Preliminary results show that very stable emulsions can be obtained with vegetable (sunflower) and mineral (paraffin) oil and fatty esters (capric/caprylic) formulations, including these new amphiphiles. Furthermore, with the aim of reducing the cost of these surfactants (price of alginates: 11–12 € per kg), additional procedures were recently carried out directly on brown seaweeds (<0.2 € per kg), to provide oligomannuronates in quite satisfactory yields (12.5 per cent) [30]. These promising results may allow the production of these novel alkyl D-mannopyranosiduronate surfactants with an overall cost compatible with their potential application markets.

7.2.3 Fatty acid glucamides

An advantageous property of glycosides and amides over esters is that they are less sensitive to hydrolysis under alkaline conditions. Thus, fatty acid glucamides with a linear headgroup derived from D-glucitol coupled *via* an amide linkage to an alkyl chain were developed as additional glucose-derived surfactants. The synthesis to produce *N*-methyl-*N*-acyl glucamides involves the reductive amination of D-glucose with methylamine, which smoothly generates the corresponding aminoalditol (Fig. 7.19). In a subsequent reaction step, this intermediate is converted with fatty acid methyl esters to the corresponding fatty acid amide carrying a methyl group and a pentahydroxylated six-carbon chain at the amido nitrogen. To avoid significant amounts of unreacted *N*-methylglucamine, which could be considered as precursors for potentially carcinogenic nitrosamines, Procter & Gamble developed an optional reaction with acetic anhydride in the finished product. Free secondary amines are acetylated in this step, and the resulting acetates can remain in the final product [31]. Among these surfactants, oleoyl-*N*-methyl-glucamide was additionally synthesized enzymatically [32].

Like APG, glucamides show synergetic effects with other types of surfactants and due to their polyol structure, they have a low-irritation potential. Disadvantages of glucamides are their affinity towards calcium ions and their low solubility and therefore they always have to be formulated with sequestering agents in order to avoid precipitation. *N*-Methyllauroylglucamine enters in dish-washing compositions [33]. Today, the main producers are Pfizer, Hatco and Clariant with an estimated total production capacity of approximately 40 000 t per year. They are only used in detergency. Lactose-derived glycamides, namely *N*-acetyl-*N*-alkyllactosylamines (*N*-alkyl-aminolactitols)

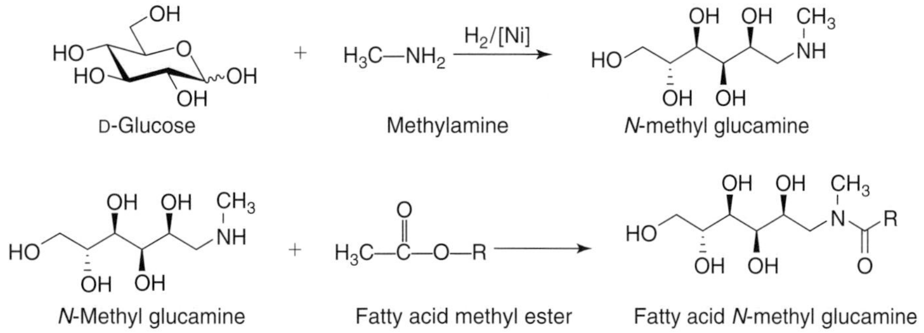

D-Glucose + Methylamine $\xrightarrow{H_2/[Ni]}$ *N*-methyl glucamine

N-Methyl glucamine + Fatty acid methyl ester → Fatty acid *N*-methyl glucamine

Figure 7.19 Two-step synthesis of fatty acid glucamides by reductive alkylation of methylamine with glucose using Raney nickel as the hydrogenation catalyst to obtain *N*-methyl glucamine, which is acylated by a base-catalyzed reaction with fatty acid methyl ester in a second step.

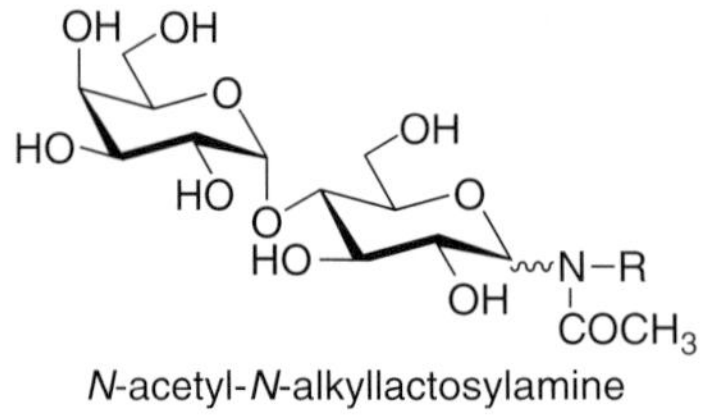

N-acetyl-*N*-alkyllactosylamine

Figure 7.20 Structure of *N*-acetyl-*N*-alkyllactosylamine

were recently synthesized and exhibited attractive interfacial properties, particularly suitable for the extraction of proteins from membranes (Fig. 7.20) [34].

7.2.4 Sorbitan esters

Another class of carbohydrate-derived surfactants, called sorbitan esters (anhydro sorbitol) is derived from sorbitol, a hexitol produced by the catalytic reduction of glucose. In acidic conditions, sorbitol is dehydrated to sorbitan and even further to bicyclic isosorbide. Sorbitan fatty acid esters, commercially available as 'Span' surfactants, can be produced by a direct or indirect industrial process (Fig. 7.21). The first industrial route is based on a two-step procedure, with acid-catalyzed sorbitol cyclization to sorbitan, followed by a high-temperature alkali-catalyzed transesterification [35]. Acid catalysts include H_3PO_2, H_3PO_3, *p*TosOH; [36] and bases are simply NaOH or KOH. In these procedures, pressure, temperature and catalysts are adjusted to minimize colour development in the product. Phosphoric acid of medium strength catalyses the conversion of sorbitol to sorbitan in high yields at 230°C, the subsequent dehydration to isosorbide proceeding at a significantly lower rate, so that phosphoric acid represents a selective catalyst. However, the dehydration products are mainly of a dark colour. The best colour can be allegedly achieved by a NaH_2PO_4 catalysis [37]. Partial esters of sorbitol can also be prepared by enzymatic esterification [38]. In the second process, sorbitan esters are produced by direct base- or acid-catalyzed reactions of sorbitol with fatty acids at elevated temperature, or by base-catalysed transesterifications of sorbitol with triglycerides or fatty acid methyl esters [36d, 39]. Homogeneous catalysts, either acids (H_2SO_4, H_3PO_2, *p*TosOH) or alkalis (NaOH, KOH, alkaline carbonates) are generally used in the absence of a solvent. Depending on the type and amount of fatty materials used, various sorbitan esters (e.g., laureates, oleats or stearates) are produced with

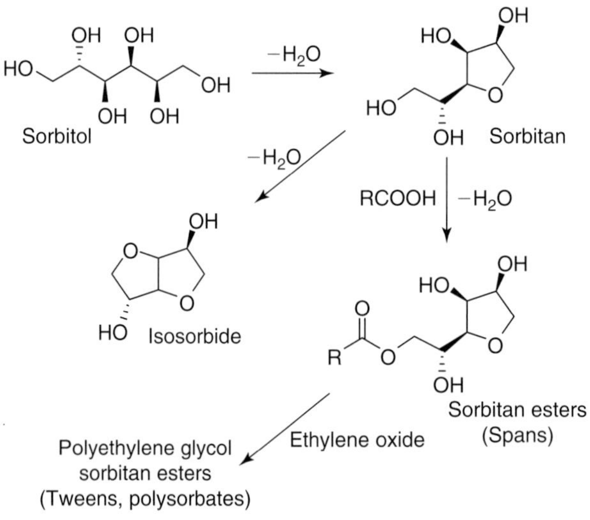

Sorbitol
$-H_2O$
Sorbitan
$-H_2O$
RCOOH $-H_2O$
Isosorbide
Sorbitan esters (Spans)
Ethylene oxide
Polyethylene glycol sorbitan esters (Tweens, polysorbates)

Figure 7.21 Synthesis of sorbitan esters by intramolecular dehydration of sorbitol in the presence of an acid at 150–200°C and subsequent base-catalyzed esterification with fatty acids at 200–250°C.

hydrophilic–lipophilic balance (HLB) values in a range of 1–8. The annual world production of sorbitan esters exceeds 25 000 tons (main manufacturers: Akcros, Dai-ichi Kogyo, Cognis, Kao and SEPPIC). These products are used as emulsifiers and solubilizers in food, cosmetic and pharmaceutical products. The Spans become more water-soluble when polyoxyethylene chains are grafted onto their cyclic moiety. These products are known as Tweens or polysorbates [40] and have found applications in the same fields.

7.3 SURFACTANTS BASED ON RRMs ENTIRELY FROM OLEOCHEMISTRY: POLYGLYCEROL ESTERS

The valorization of a raw material reaches an optimum when secondary products are also exploited. The main secondary product of oleochemistry is glycerol (coproduct of triglyceride hydrolysis and methanolysis processes), with a growing contribution from industrial vegetable oils. Within this context, the use of glycerol, as well as vegetable oils, as starting materials for the manufacture of surfactants represents a convenient strategy for the development of surface-active products entirely derived from oleochemistry. Glycerol itself is not suitable as a primary constituent of the hydrophilic part of the surfactant and polyglycerols are needed to increase the hydrophilicity and to adjust the hydrophilic–hydrophobic balance (HLB) of the products. Fatty acid esters of these polyols, called polyglycerol esters (PGE), have been developed leading to applications in cosmetic or food emulsifiers. Perhaps they will not substitute ethoxylates but rather play a role in several niche markets. New compounds could find applications in offshore oil drilling, in the decontamination of polluted soils, or in the protection of wood [41]. In general, the preparation of PGE involves two subsequent steps, viz. (1) the polymerization of glycerol in the presence of a small amount of alkali (base catalysis) and (2) esterification of the resulting polyglycerol. Conventional methods to polymerize glycerol require drastic process conditions, namely temperatures above 200°C in the presence of homogeneous alkaline catalysts (hydroxides, carbonates and oxides of several metals). The condensation reaction usually involves the more reactive terminal hydroxyl groups of the glycerol molecules, but some branched and even cyclic by-products, resulting from the reaction of the secondary hydroxyl groups, are always formed (Fig. 7.22). Carbon dioxide or nitrogen is bubbled through the reaction mixture to prevent dehydration of glycerol to the undesired acrolein. New catalyst systems have been recently proposed for the selective oligomerization of glycerol. Stable active and basic mesoporous catalysts were used for the chemo- and regioselective conversion of glycerol to linear oligoglycerols. For example, mesoporous solids modified by ceasium impregnation led, with the best selectivity and yield, to di- and tri-glycerol [42]. Additionally, the alkaline (KOH) polymerization of glycidol at a reaction temperature of ca. 120°C was found to afford selectively the linear polyglycerols [43].

The production of fatty acid esters of polyglycerols has been usually achieved through (1) direct esterification of the polyol using either alkali [44] or acid catalysts [45], (2) transesterification of the polyol with a triglyceride or a fatty acid methyl ester in the presence of a suitable alkaline catalyst [46] or (3) addition polymerization of glycidol to a fatty acid or to a fatty acid monoglyceride catalyzed by acids [47]. These processes generally lead to the formation of mixtures in which the fatty acid chains are distributed among all available hydroxyl groups, whose proportions mainly depend on temperature and reagent molar ratio. Moreover, the molecular composition

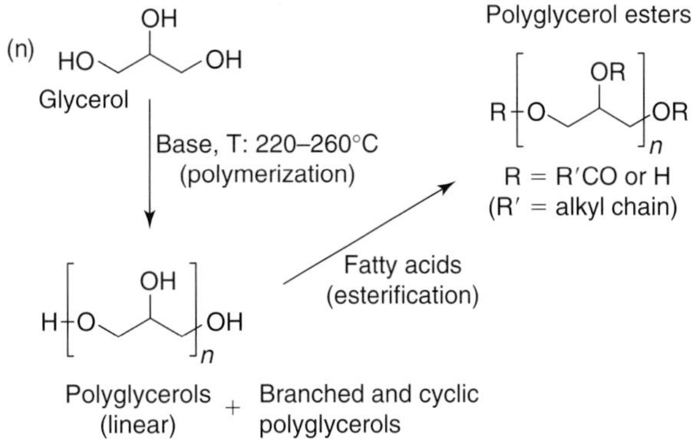

Figure 7.22 Synthesis of polyglycerol esters.

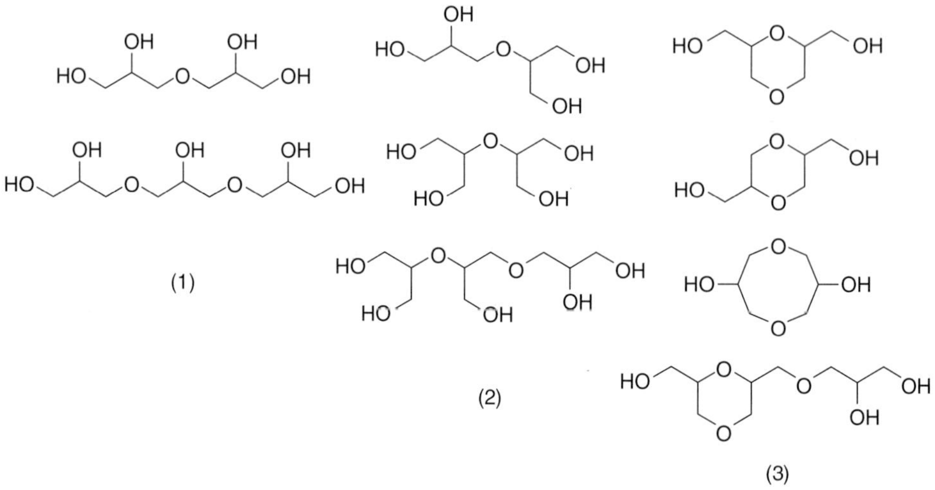

Figure 7.23 Some examples of di- and triglycerol isomers (1), (2) and (3) being respectively linear, branched and cyclic isomers.

of the product is very difficult to determine and correlations between structure and performances cannot be established precisely. Within this context, the project 'Polyglycerols chemistry ecology and applications of polyglycerol esters' (FAIR-CT96-1829) provided a complete description of polyglycerol-based surfactants to clarify the relationship between structure and environmental impact, on the one hand, and the physico-chemical properties, on the other hand, of PGEs [48]. The synthesis of selected linear, branched and cyclic polyglycerols (Fig. 7.23) with a defined degree of oligomerization (2–6) and their regiocontrolled monoesterification or monoetherification with C_{12} or $C_{18:1}$ fatty chains furnished important information relative to (1) their detergent properties (low cmc values and good surface activity; all polyglycerols have low-interfacial tension against *n*-decane and therefore good emulsification capability and a very pronounced foaming behaviour); (2) skin irritation (the position of the acyl/alkyl chain determines the toxicity of linear surfactants; the nature, number and location of neighbouring hydroxyl functions, primary or secondary, on the polyglycerol backbone influence the irritation score); (3) biodegradability (linear polyglycerols are more biodegradable than branched and cyclic structures); (4) ecotoxicity and bioaccumulation (increasing lipophilicity and decreasing polymerization lead to higher toxicity).

Polyglycerols have been known since the beginning of the twentieth century. However, due to the difficulties encountered for the industrial development of high-purity products, it is only in the last decade that their uses increased. Solvay is the world-leader in the production of high purity polyglycerols and manufactures diglycerol, a distilled product of 90 per cent minimum purity, and polyglycerol-3, a grade with a narrow oligomer distribution typically containing a minimum 80 per cent of di-, tri-, and tetraglycerol. As a result of their multifunctional properties and harmless nature, PGE are used in many applications in food [49] and cosmetic industries [50]. They notably function as emulsifiers, dispersants, thickeners, solubilizers minimum, spreading agents and emollients. More recently, new industrial applications based on PGEs have been developed. This includes their utilization as antifogging and antistatic additives, lubricants and plasticizers [51].

7.4 NOVEL BIODEGRADABLE PLANT-DERIVED CATIONIC EMULSIFIERS FOR ROAD CONSTRUCTION AND COSMETICS

Cationic surfactants fall into several categories depending on the nature of their cationic polar heads. Some of them have functional groups susceptible to protonation (*e.g.* amines) and thus display cationic properties particularly in acidic media, while others, such as quaternary ammonium salts, exhibit a permanent positive charge. In household products, cationic surfactants are primarily applied in fabric softeners and hair preparations. Other applications of cationic surfactants include disinfectants and biocides, emulsifiers, wetting agents and processing additives. By volume, the most important cationic surfactants in household products are the alkyl ester ammonium salts that

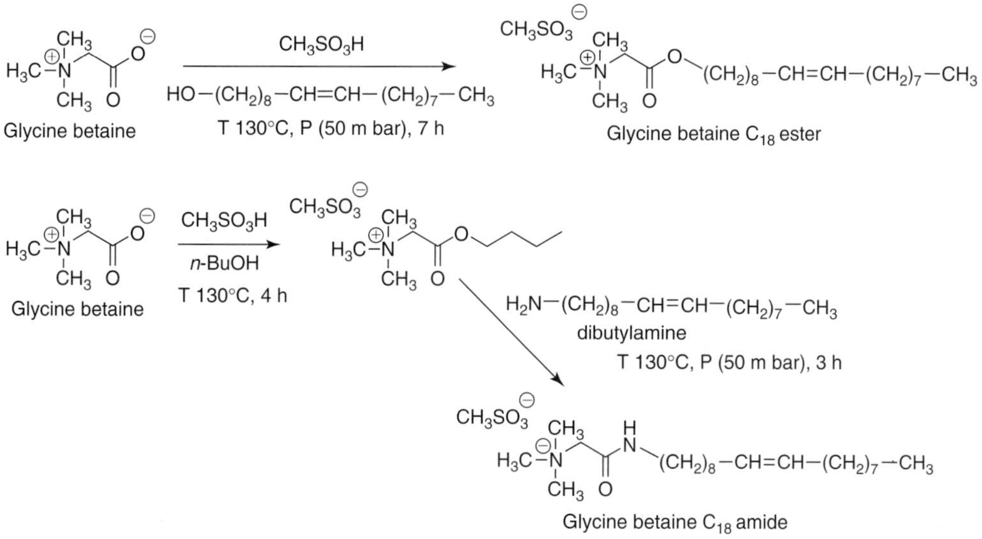

Figure 7.24 Synthesis of glycine betaine C_{18} ester and amide.

are used in fabric softeners (esterquats). These surfactants generally have an acute aquatic toxicity and a low-ultimate biodegradability and their use was recently reduced or even abandoned in certain European countries like Germany and Netherlands. These undesirable effects on the environment, coupled with the consumer demand for 'greener' products, are leading manufacturers to focus on the production of cationic surfactants from alternative, less harmful raw materials. Glycine betaine, a not very expensive natural substance possessing a quaternary trimethylalkylammonium moiety and a carboxylate function, constitutes a prime raw material for the preparation of biodegradable and biocompatible cationic surfactants (Fig. 7.24). Accounting for 27 per cent in weight of molasses of sugar beet and obtained after extraction of saccharose, it remains currently a little developed by-product of the sugar-industry. Within this context, novel glycine betaine esters and amides were recently produced from tropical oils (copra, palm kernel) or European oils (sunflower, rapeseed), conveniently, economically and with an environmentally acceptable process (no solvent, no waste) [52]. In particular, fatty chains with 18 carbon atoms derived from European vegetable oils have been used for their emulsifying properties compatible with numerous industrial and technical applications (road bitumen emulsifying applications, emulsions for uses in cosmetics).

The environment-friendly synthesis of glycine betaine esters was carried out through the direct esterification reaction between glycine betaine and fatty alcohols, preferably stearic and oleic alcohols, catalyzed by methanesulphonic acid without any solvent (Fig. 7.24). Glycine betaine amides were prepared in two steps: first, glycine betaine reacts with *n*-butanol in the presence of methanesulphonic acid as catalyst. In a second step, the short butyl chain is replaced by a longer chain in an aminolysis reaction with fatty amines, particularly C_{18} stearic and oleic amines. Purification procedures (precipitation in organic solvents like *n*-butanol or ethanol, or use of crude mixtures obtained without any additional treatment) furnished a large variety of emulsifying formulations characterized by various compositions (several controlled ratios of residual free glycine betaine, fatty alcohols, methanesulphonic acid). The scale-up of one oleic ester-based mixture obtained without any purification was performed on a 60 kg scale for road making application. In road construction, bitumen products are typically applied in conjunction with a mineral aggregate.

The good adhesion properties of bitumen to aggregates allows it to act as a binder, while the aggregate provides mechanical strength. At ambient temperature, bitumen is a highly viscous to almost solid substance that is extremely difficult to work with, but can, however, be changed into a workable form by applying heat, by blending it with petroleum solvents or by emulsification in water. Working with bitumen at high temperatures (150–180°C) is very dangerous, with a risk of serious burns and requires moreover costly equipment for heating, storage and application, which must be performed on-site. Bitumen that has been mixed with petroleum solvents is more workable, but the solvents, usually kerosene, are fire hazards and produce hydrocarbon emissions contributing to air pollution. In contrast, bitumen emulsions can be applied without heating (although they are still prepared at moderately high temperatures) and do not have the handling and environmental hazards associated with hot mixes. Bitumen emulsions also improve adhesion to aggregates, and can be applied in a good range of weather conditions [53]. The bitumen is dispersed throughout the continuous water phase as minute particles, typically 0.1–5 μm in diameter, as an oil-in-water emulsion. Bitumen–water emulsions must break in a controlled manner upon being laid with the mineral aggregate and, as a function of the speed of this breaking process, various technologies for road making are used: surface dressing with rapid breaking emulsions (the emulsion is sprayed onto the road surface and chippings are spread on top and normally rolled to ensure proper embedment and alignment) and cold mixes (this category covers several different technologies, among which open-graded storable cold asphalt for patch working, combined cement/bitumen emulsion mixes and dense-graded cold asphalt for wearing course) that require slower breaking emulsions. The emulsifier in a bitumen emulsion system has thus three functions: it reduces the interfacial tension between bitumen and water, stabilizes the emulsion and assists the adhesion of the bitumen. Cationic emulsifiers for bitumen emulsions are usually employed in Europe (2 000 000 tons of bitumen emulsion produced per year), particularly in France, which produces 50 per cent of the total European bitumen emulsion (0.2–2 per cent of surfactants). Commercially available cationic emulsifiers are mainly surfactants that are ionized in an acidic environment and include polyamines, amidoamines and imidazolines, which are the most widely used emulsifiers. These cationic surfactants, that come entirely or partially from petrochemicals, were found to exhibit a low biodegradability and a high aquatic toxicity. Within this context, glycine betaine C_{18} esters represent a new family of plant-derived emulsifiers that limit environmental pollution. These formulations were found to have very promising physico-chemical properties in terms of their use in surface dressing technologies, namely, chemical stability, viscosity of formulation, bitumen emulsifying properties (adhesivity, emulsion breaking) and a high biodegradability (ultimate biodegradability above 70 per cent, according to OECD 301) [54]. After an experimental phase (1 000 m road coating using a bitumen emulsion based on the glycine betaine oleic ester surfactant), the project arrives today at the stage of industrial development for the production, on a large scale, of this new vegetable-derived surfactant, which should make it possible to exploit important quantities of raw materials of agricultural origin (in particular European oils) in non-food fields. Additional applications of these novel biodegradable cationic surfactants in cosmetic formulations (shampoos, body washes, creams) are currently being developed [52].

7.5 GEMINI SURFACTANTS AND BOLAAMPHIPHILES

Changes in the molecular structure of surfactants characterized by a single hydrophobic tail connected to an ionic or non-ionic polar headgroup have attracted the attention of chemists. In particular, the covalent linking of two hydrophilic groups was envisaged to yield surfactant dimers such as Gemini surfactants and bolaamphiphiles. Even if several dimeric surfactants have been synthesized and patented for more than 50 years (especially cationic ones), only few examples related to renewable sources can be found. However, as these bipolar surfactants show interesting surface tension lowering, rheological and self-organizing properties, the number of publications and patents is expected to increase rapidly in the next few years.

7.5.1 Gemini surfactants

Gemini surfactants are defined as surfactants made up of two identical amphiphilic moieties connected at the level of the headgroups, or of the alkyl chains, but still very close to the head groups, by a spacer group which can be hydrophobic, flexible, or rigid (Fig. 7.25).

In the field of Gemini amphiphiles from renewable sources, sugar derivatives are of special interest. Two types of sugar gemini surfactants can be distinguished, viz. those involving pyranoses and those containing reduced carbohydrates. Engberts *et al.* described sugar-based Gemini surfactants **11** possessing both reduced glucoside and mannoside derivatives combined with oleyl, oleoyl or saturated chains through an oligoethylene glycol or

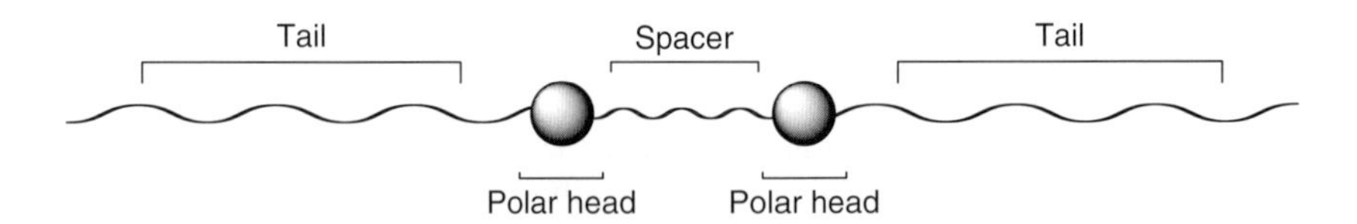

Figure 7.25 Schematic representation of gemini surfactants.

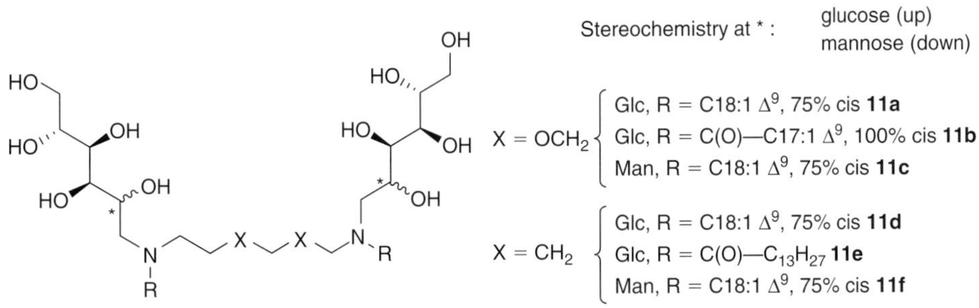

Stereochemistry at * : glucose (up) mannose (down)

X = OCH_2 { Glc, R = C18:1 Δ^9, 75% cis **11a**; Glc, R = C(O)—C17:1 Δ^9, 100% cis **11b**; Man, R = C18:1 Δ^9, 75% cis **11c**

X = CH_2 { Glc, R = C18:1 Δ^9, 75% cis **11d**; Glc, R = C(O)—$C_{13}H_{27}$ **11e**; Man, R = C18:1 Δ^9, 75% cis **11f**

Figure 7.26 Representative examples of sugar-based Gemini surfactants (reduced sugars).

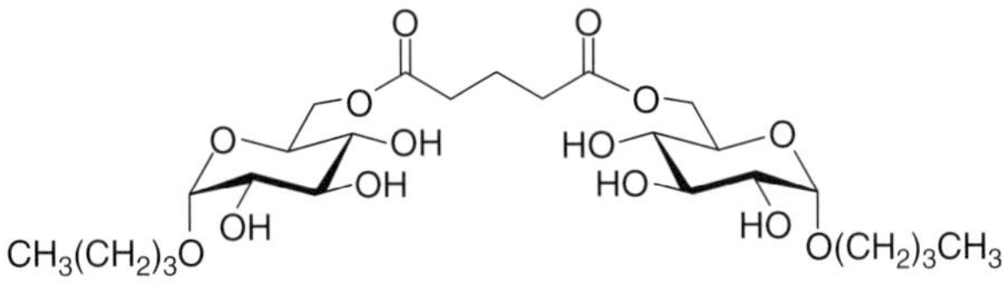

Figure 7.27 Sugar-based Gemini surfactant **12**.

oligomethylene spacer (Fig. 7.26) [55]. Their synthesis involves the preparation of a dimeric structure followed by a double reductive amination of D-glucose or D-mannose in the presence of an oligomethylene or an oligoethylene glycol diamine (spacer). Then, the acylation or alkylation reaction is conducted under standard conditions (respectively acylation with a long chain anhydride and reductive amination of long-chain aldehydes).

The incorporation of cyclic carbohydrates into Gemini surfactants was developed by Castro *et al.* (Fig. 7.27) [56]. These sugar geminis have been synthesized from AGs which can be easily prepared *via* a Fisher-type glycosylation reaction of free glucose by *n*-butanol. Convenient protection of the sugar moiety, followed by selective deprotection of the primary alcohol, permits the formation of gemini **12** by a double esterification with propanedicarboxylic acid.

Formally, Gemini surfactants possess a structure resembling a pair of conventional single-chain surfactants covalently connected by a spacer. It is therefore appropriate to compare Gemini and conventional surfactant properties and, in doing so, intriguing results appear. Indeed, Gemini show generally a low cms, good oil solubilization and the formation of threadlike micelles [57]. These properties, and especially their high-surface activity, are the reason for the multiple applications of Gemini in detergency, cosmetics, pharmaceuticals, food, metallurgy, paints, oilfields and polymers. In particular, carbohydrate-derived Gemini have been described in a number of patents by Procter & Gamble in which they are claimed as components in laundry, cleaning, fabric, and personal care compositions [58].

7.5.2 Bolaamphiphiles

The term 'bolaamphiphile' or 'bola' is related to structures composed of a hydrophobic core bearing two polar headgroups at opposite ends (Fig. 7.28) [59]. They are naturally found in archaebacteria microorganisms where they

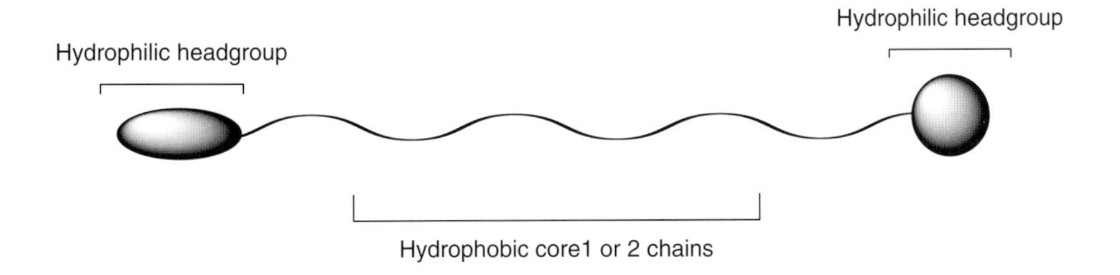

Figure 7.28 Schematic representation of a bolaamphiphile.

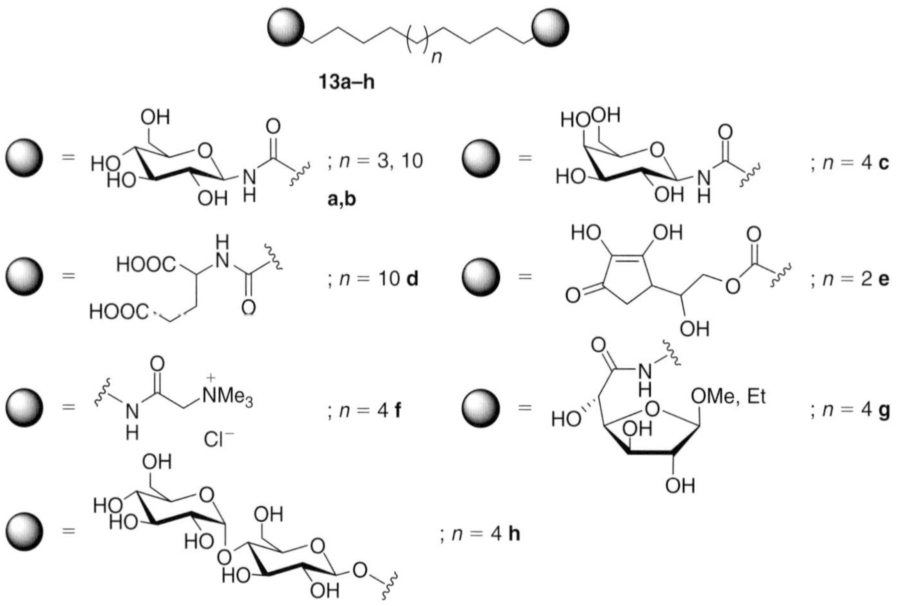

Figure 7.29 Symmetrical bolaamphiphiles possessing a single polymethylene bridging chain.

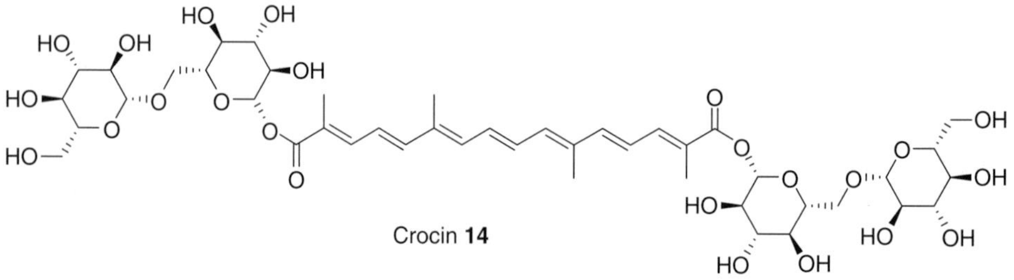

Figure 7.30 Natural carotenoid glycoside crocin **14**.

improve significantly the integrity of lipid membranes under extreme conditions (high or low temperatures, low pH, high salinity) [60]. Due to their particular structures and properties, synthetic bolas have been prepared, thus enlarging the family of surfactants and the possibilities of new applications. Here we will focus on bolas issued from renewable sources and characterized by the presence of a single bridging chain.

Among the synthetic bolaamphiphiles reported in the literature, only few are issued from renewable sources, mainly fatty acids and sugar derivatives (Fig. 7.29). Pyranose bolaamphiphiles **13a–c** were prepared by the condensation of β-1-aminoglucosides with the corresponding α,ω-dicarboxylic acid chlorides [61]. Both glucose **13a,b** and galactose **13c** derivatives were synthesized with three lengths of oligomethylene chains ($n = 3, 4, 10$). Furanose-type headgroups were introduced through an amide linkage between D-glucuronolactone and a saturated C_{12} α, ω-diamine, thus providing diamides **13g** [62]. Standard glycosylation methods were used to introduce two disaccharide polar headgroups into the terminal sites of 1,12-dodecanediol (**13h**) [63]. Additionally, glutamic acid [64], glycine betaine [65] or vitamin C [66] were also used as natural starting materials for the production of neutral or ionic bipolar surfactants.

Substituted and/or unsaturated lipophilic chains, such as carotenoids, are also available and can be combined with two sugars polar ends affording neutral surfactants like crocin **14,** a natural carotenoid disaccharide (*ca.* 25 per cent of saffron) (Fig. 7.30) [67].

Access to unsymmetrical derivatives was envisaged, especially to study symmetrical or dissymmetrical organization of bolas in lipid membranes. They are usually prepared by selective reactions of one functional group of the unprotected symmetrical hydrophobic core. Shimizu and Masuda [68] developed the synthesis of bolaamphiphiles **15a–g** containing an oligomethylene chain connected with a sugar at one end and a free carboxylic acid

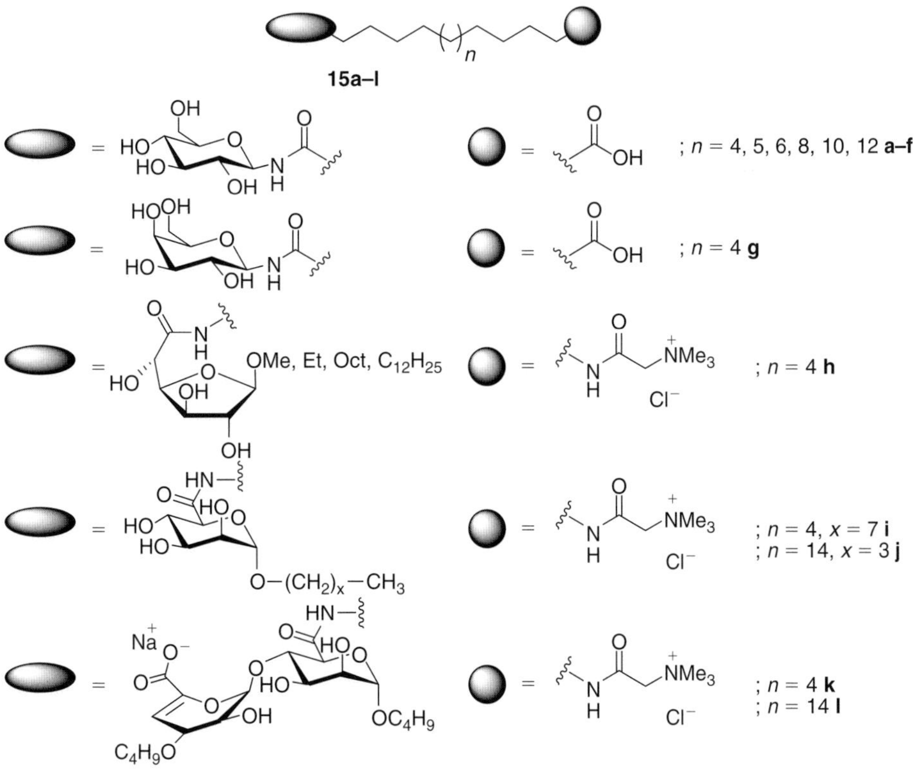

Figure 7.31 Unsymmetrical bolaamphiphiles possessing a single polymethylene bridging chain.

at the other. The preparation of these unsymmetrical surfactants involves the reaction of oligomethylene (14–22 methylenes) dicarboxylic acid chlorides with one equivalent of β-1-amino-glucoside or -galactoside (Fig. 7.31). Recently, Benvegnu *et al.* [69] described the synthesis of surfactants **15h–l** derived from D-glucurone or alginate oligosaccharides as sugar moieties and containing positively-charged glycine betaine. They are easily accessible through the reaction of monoester or lactone carbohydrates with an alkyl diamine chain, followed by a condensation of the second free primary amine with an activated glycine betaine derivative.

Bolaform surfactants are generally less effective than conventional surfactants, but they often display unusual self-assembling properties. Compared with monopolar surfactants, the additional polar headgroup of bolaamphiphiles generally induces both a higher water solubility increasing the cmc and the surface tension, and a decrease in the aggregation number. Bolaamphiphiles can span a membrane, having one headgroup on the outside of the membrane and one on the inside. Extremophile membranes are stabilized in this way [59a]. The results obtained for the synthetic bipolar surfactants possessing a polymethylene linker and natural polar headgroups clearly demonstrated the existence of complex relationships between chain composition, spacers and headgroups that influence the structure of their supramolecular aggregates (monolayered or bilayered vesicles, disks, fibers, ribbons, nanotubes and helices), the stretched or U-bent conformation of the lipids, as well as the kinetics of their interconversion (Fig. 7.32) [55, 59, 62, 63, 69]. Additionally, dissymmetrical arrangements in tubular assemblies [68] and in monocrystals [70] have been found in the case of some unsymmetrical structures.

When particular functions such as thiol, alcohol, carboxylic acid, nitrile or silyl derivatives, are present at one end of the bolas, they can easily stack on solid surfaces like gold, silver, copper or silica and form planar molecular monolayer (coated particles) (Fig. 7.33). The hydrophobic core may serve as a barrier or a solvent and the available headgroup allows molecular recognition in solution.

The potential applications of bolaamphiphiles include the formation of monolayer vesicles for drug/gene delivery, ultra thin monolayer membranes, inclusion of functionalities into membranes, and disruption of biological membranes [59a].

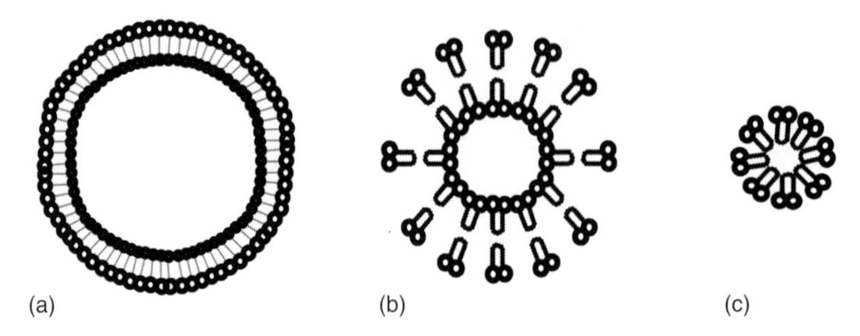

Figure 7.32 Organization of bolaamphiphiles in (a) a monolayer membrane, (b) a U-bent bilayer membrane and (c) a U-bent micelle.

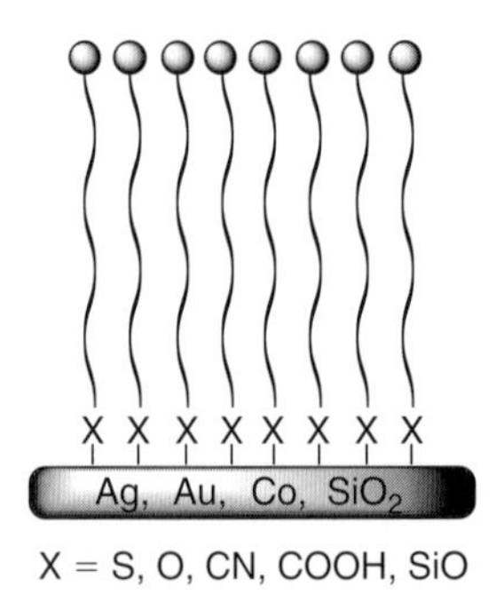

Figure 7.33 Schematic representation of monolayer coated particles.

7.6 CONCLUSIONS AND PERSPECTIVES

It seems clear that with the examples of recent product innovations from RRMs, the successful development of environmentally compatible and powerful surfactants in the sense of a sustained development has been demonstrated. RRMs are biocompatible, usually easily biodegradable and generally non-toxic, unless they have been chemically modified by undesired functional groups. These properties are particularly important in the context of the growing number of European laws, regulations and directives. The new European Detergent Regulation that entered into force in October 2005 is the most recent one. The main elements of this regulation are additional requirements with respect to ultimate biodegradability (60 per cent degradation within 28 days) of surfactants, special new requirements for the additional declarations of detergent ingredients on the packaging, including possible sensitizing ingredients, and additional compulsory information concerning detergent formulations to be provided for medical professionals and consumers (Commission Regulation (EC) No 907/2006 of 20 June 2006 amending Regulation (EC) No 648/2004 of the European Parliament and of the Council on detergents, Official Journal of the European Union). The proposed new biodegradation tests ensure a higher level of environmental protection, especially of the aquatic environment. While previous legislation was only applicable to anionic and non-ionic surfactants, this Regulation includes all four surfactant families: anionic, non-ionic, cationic and amphoteric. Consequently, innovative research and development in new biodegradable products, particularly for the introduction of new cationic surfactants, should be favoured in order to propose alternatives to non-compliant surfactants found in formulations impacted by the European Detergent Regulation.

Other EU initiatives include REACH (Registration, Evaluation and Authorization of Chemicals) to help regulate new and existing chemical substances. The surfactant industry will be largely sensitive to the profile and the requirements of the future REACH legislation. In general, surfactants are manufactured in quantities greater than 1 t per year (industry would be obliged to register all marketed chemicals above 1 ton annually within 11 years). Consequently, the major surface-active components will need registration under REACH. The positive impact of this new regulation on the development of RRM-based surfactants is not so obvious if we consider the high entrance barrier expected for the introduction of new substances [71].

To make more surfactants based on RRMs commercially viable, more innovation is needed, since technical constraints often limit the applicability of renewables. If the barriers to vegetable-derived chemicals inherent in the EINECS (European Inventory of Existing Commercial Chemical Substances)/ELINCS (European List of Notified Chemical Substances) system were removed, such innovation would be facilitated. Much applied research has to be encouraged and supported by national trade organizations (ACTIN in the UK, FNR in Germany, AGRICE in France, AIACE in Italy) and by the European Commission. Finally, the policy measures introduced to reduce greenhouse gas emissions might shift the balance of competition between RRM and fossil raw materials in favour of the former and thus act as drivers for an increased industrial use of renewable raw material [1c].

REFERENCES

1. (a) Ehrenberg J., Current situation and future prospects of EU industry using renewable raw materials, coordinated by the *European Renewable Resources & Materials Association, European Commission DG Enterprise Unit E.1: Environmental Aspects of Industry Policy*, Brussels, February 2002. (b) Deleu M., Paquot M., From renewable vegetable resources to microorganisms: New trends in surfactants, *C. R. Chim.*, **7**(6–7), 2004, 641–646. (c) Patel M., Surfactants based on renewable raw materials, *J. Ind. Ecol.*, **7**(3-4), 2004, 47–62.
2. Oils and Fats International, 22(1), 36–37, 39–40, January 2006.
3. Shinoda K., Carlsson A., Lindman B., On the importance of hydroxyl groups in the polar head-group of non-ionic surfactants and membrane lipids, *Adv. Colloid Interface Sci.*, **64**, 1996, 253–271.
4. Hass H.B., Snell F.D., York W.C., Osipow L.I., Process for producing sugar esters, Patent # US 2893990, 1990.
5. Parker W.J., Khan R.A., Mufti K.S., Process for the production of surface active agents comprising sucrose esters, Patent # GB 1399053, 1973.
6. (a) Polat T., Linhardt R.J., Synthesis and applications of sucrose-based esters, *J. Surfact. Deterg.*, **4**(4), 2001, 415–421. (b) Queneau Y., Fitrepeann J., Trombotto S., The chemistry of unprotected sucrose: the selectivity issue, *C. R. Chim.*, **7**(2), 2004, 177–188.
7. (a) Immel S., Lichtenthaler F.W., The conformation of sucrose in water: A molecular dynamics approach, *Liebigs Ann.*(11), 1995, 1925–1937. (b) Chauvin C., Baczko K., Plusquellec D., New highly regioselctive reactions of unprotected sucrose. Synthesis of 2-O-acylsucroses and 2-O-(N-alkylcarbamoyl)sucroses, *J. Org. Chem.*, **58**(8), 1993, 2291–2295.
8. Khan R., Chemistry and new uses of sucrose: How important?, *Pure Appl. Chem.*, **56**(7), 1984, 833–844.
9. Bottle S., Jenkins I.D., Improved synthesis of cord factor analogues, *J. Chem. Soc., Chem. Commun.*(6), 1984, 385. 385
10. Vhalov I.R., Vhalova P.I., Linhardt R.J., Regioselctive synthesis of sucrose monoesters as surfactants, *J. Carbohyd. Chem.*, **16**(1), 1997, 1–10.
11. Baczko K., Nugier-Chauvin C., Banoud J., Thibault P., Plusquellec D., A new synthesis of 6-O-acylsucroses and of mixed 6,6′-di-O-acylsucroses, *Carbohyd. Res.*, **269**(1), 1995, 79–88.
12. Chauvin C., Plusquellec D., A new chemoenzymatic synthesis of 6′-O-acylsucroses, *Tetrahedron Lett.*, **32**(29), 1991, 3495–3498.
13. Ferrer M., Cruces M.A., Bernabe M., Ballesteros A., Plou F.J., Lipase catalyzed regioselective acylation in two solvent mixtures, *Biotechnol. Bioeng.*, **65**(1), 1999, 10–16.
14. Riva S., Chopineau J., Kielboom A.P.G., Klibanov A.M., Protease catalysed regioselective esterification of sugars and related compounds in anhydrous dimethylformamide, *J. Am. Chem. Soc.*, **110**(2), 1988, 584–589.
15. Polat T., Bazin H.G., Linhardt R.J., Enzyme-catalysed regioselective synthesis of sucrose fatty acid ester surfactants, *J. Carbohyd. Chem.*, **16**(9), 1997, 1319–1325.
16. Hough L. in *Carbohydrates as Organic Raw Materials* (Ed. Lichtenthaler F.W.), VCH, Weinheim, 1991, pp. 33–55.
17. Davies E., Sugaring – The surfactant pill, *Chem. Brit.*, December 2000, 24–27. 2000
18. Molinier V., Kouwer P.J.J., Fitremann J., Bouchu A., Mackenzie G., Queneau Y., Goodby J.W., Self-organizing properties of monosubstituted sucrose fatty acid esters: The effect of chain length and unsaturation, *Chem. Eur. J.*, **12**(13), 2006, 3547–3557.
19. (a) Söderman O., Johansson I., Polyhydroxyl-based surfactants and their physico-chemical properties and applications, *Curr. Opin. Colloid Interface Sci.*, **4**(6), 2000, 391–401. (b) Drummond C.J., Wells D., Nonionic lactose and lactitol based surfactants: comparison of some physico-chemical properties, *Coll. Surf.*, **141**(1), 1998, 131–142.
20. (a) Hill K., von Rybinski W., Stoll G. (Eds), *Alkyl polyglycosides: Technology, properties and applications*, 1996, VCH, Weinheim. (b) von Rybinski W., Hill K., Alkyl K., Polyglycosides – Properties and applications of a new class of surfactants, *Angew. Chem. Int. Ed.*, **37**(10), 1998, 1328–1345.
21. Plusquellec D., Chevalier G., Talibart R., Wroblewski H., Synthesis and characterization of 6-O-(N-heptylcarbamoyl)-methyl-alpha-D-glucopyranoside, a new surfactant for membrane studies, *Anal. Biochem.*, **179**(1), 1989, 145–153.

22. Bertho J.-N., Mathaly P., Dubois V., De Baynast R, Process for the preparation of surface active agents using wheat by-products and their applications, Patent # US 5688930, 1997.
23. Ferrières V., Bertho J.-N., Plusquellec D., A new synthesis of O-glycosides from totally O-unprotected glycosyl donors, *Tetrahedron Lett.*, **36**(16), 1995, 2749–2752.
24. Goodby J.W., Haley J.A., Mackenzie G., Watson M.J., Plusquellec D., Ferrières V., Amphitropic liquid-crystalline properties of some novel alkyl furanosides, *J. Mater. Chem.*, **5**(12), 1995, 2209–2220.
25. Ferrières V., Benvegnu T., Lefeuvre M., Plusquellec D., Mackenzie G., Watson M.J., Haley J.A., Goodby J.W., Pindak R., Durbin M.K., Diastereospecific synthesis and amphiphilic properties of new alkyl β-D-fructopyranosides, *J. Chem. Soc., Perkin Trans.* (2), 1999, 951–959.
26. Bertho J.-N., Ferrières V., Plusquellec D., A new synthesis of D-glycosiduronates from unprotected D-uronic acids, *J. Chem. Soc., Chem. Commun.*(13), 1995, 1391–1393.
27. (a) sz Nijs M.P., Maat L., Kieboom A.P.G., Two-step chemo-enzymic synthesis of octyl 6-O-acyl-α-D-glucopyranoside surfactants from glucose, *Recl. Trav. Chim. Pays-Bas*, **109**(7–8), 1990, 429–433. (b) J.-N. Bertho, Ph.D. Thesis, Université de Rennes 1, 17 juin 1994.
28. Auvray X., Labulle B., Petipas C., Bertho J.-N., Benvegnu T., Plusquellec D., Thermotropic liquid-crystalline properties of some novel hexuronic acid derivatives bearing a single or two alkyl chains, *J. Mater. Chem.*, **7**(8), 1997, 1373–1376.
29. Roussel M., Benvegnu T., Lognoné V., Le Deit. H., Soutrel I., Laurent I., Plusquellec D., Synthesis and physico-chemical properties of novel biocompatible alkyl D-mannopyranosiduronate surfactants derived from alginate, *Eur. J. Org. Chem.*(14), 2005, 3085–3094.
30. Benvegnu T., Plusquellec D., Roussel M., Preparation, surface tension, micelle formation, and liquid crystal property, of (alkyl-D-mannopyranoside)uronic acid derivatives and their use in cosmetics, Patent # FR 2840306, WO 2003104248, 2005.
31. (a) Laughlin R.G., Fu Y.-C., Wireko F.C., Scheibel J.J., Munyon R.L., in *Novel surfactants, Preparations, Applications, and Biodegradability* (Ed. Holmberg K.), Marcel Dekker, New York, 1998, pp. 1–30. (b) Scheibel J.J., Connor D.S., Shumate R.E., St. Laurent J.C.T.R.B., Patent # EP-B 0558515, US 598462, 1990.
32. Maugard T., Remaud-Simeon M., Petre D., Monsan P., *Enzymatic synthesis of glycamide surfactants by amidification reaction, Tetrahedron*, **53**(14), 1997, 5185–5194.
33. Ruback W., Schmidt S., van Beklum H., Roper H., Voragen F., Eds *Carbohydrates as Organic Raw Materials III*, VCH, Weinheim, 1996, p. 231.
34. Lattes A., Rico-Lattes I., Perez E., Blanzat M., Reactions and synthesis in surfactant systems New glycolipids having biological activities: key role of their organization, in Surfactant Science Series, Volume 100, Marcel Dekker Inc., 2001.
35. Hoydonckx H.E., De Vos D.E., Chavan S.A., Jacobs P.A., Esterification and transesterification of renewable chemicals, *Top. Catal.*, **27**(1–4), 2004, 83–96.
36. (a) Sawada H., Nishikawa N., Kurosaki T., Process for the preparation of a sorbitan ester, Patent # DE 3729335, 1988. (b) Stockburger G.J., Process for preparing sorbitan esters, Patent # US 4297290, 1981. c) Ellis J.M.H., Lewis J.J., Beattie J., Manufacture of fatty acid esters of sorbitan as surfactants, Patent # US 6362353, 2002. d) Milstein N., Improved esterification of oxyhydrocarbon polyols and ethers thereof, and products therefrom, Patent # WO 9200947, 1992.
37. Smidrkal J., Cervenkova R., Filip V., Two-stage synthesis of sorbitan esters, and physical properties of the products, *Eur. J. Lipid Sci. Technol.*, **106**(12), 2004, 851–855.
38. Otero C., Arcos J.A., Berrendero M.A., Torres C., Emulsifiers from solid and liquid polyols: Different strategies for obtaining optimum conversions and selectivities, *J. Mol. Catal. B: Enzym.*, **11**(4–6), 2001, 883–892.
39. (a) Gruetzmacher G.O., Raggon J.W., Wlodecki B., Low calorie fat substitute, Patent # US 5612080, 1997. (b) Ellis J.M.H., Lewis J.J., Beattie J., Manufacture of fatty acid esters of sorbitan as surfactants, Patent # WO 9804540, 1998.
40. Biermann M., Lange F., Piorr R., Ploog U., Rutzen H., Schindler J., *Surfactants in Consumer Products*, Springer-Verlag, Berlin, 1987. pp. 23–132
41. Claude S., Heming M., Hill K., Development of glycerol markets. Part 2. Research, developments and conclusions, *Lipid Technol. Newsletter*, October 2000, 108–113.
42. Barrault J., Clacens J.-M., Pouilloux Y., Selective oligomerization of glycerol over mesoporous catalysts, *Top. Catal.*, **27**(1–4), 2004, 137–142.
43. Dobson K.S., Williams K.D., Boriack C.J., The preparation of polyglycerol esters suitable as low-caloric fat substitutes, *J. Am. Oil Chem. Soc.*, **70**(11), 1993, 1089–1092.
44. (a) Nakamura T., Yamashita M., Patent # US 0035238, 2002. (b) Lemke D., Processes for preparing linear polyglycerols and polyglycerol esters, Patent # WO 0236534, 2002. c) Ishitobi M., Patent # US 123439, 2002.
45. (a) Jakobson G., Siemanowski W., Uhlig K.H., Patent # US 5424469, 1995. (b) Jakobson G., Siemanowski W., Uhlig K.H., Polyglycerol fatty acid ester mixture, Patent # US 5466719, 1995.
46. Márquez-Alvarez C., Sastre E., Pérez-Pariente J., Solid catalysts for the synthesis of fatty esters of glycerol, polyglycerols and sorbitol from renewable resources, *Top. Catal.*, **27**(1–4), 2004, 105–117.

47. (a) Endo T., Maruo K., Production of polyglycerol mono-fatty acid ester, Patent # JP 2000143794, 2000. (b) Endo T., Ueno T., Production of polyglycerol monofatty acid ester, Patent # JP 2000143584, 2000.
48. (a) http://www.biomatnet.org/secure/Fair/R1829.htm. (b) Cassel S., Debaig C., Benvegnu T., Chaimbault P., Lafosse M., Plusquellec D., Rollin P., Original synthesis of linear, branched and cyclic oligoglycerol standards, *Eur. J. Org. Chem.*, **5**, 2001, 875–896. (c) Debaig C., Benvegnu T., Plusquellec D., Synthesis of linear and cyclic polyglycerols. Polyglycerylated surfactants: synthesis and characterization, *OCL.*, **9**(2–3), 2002, 155–161. (d) Lagarden M., Schroeder K.R., Hofmann R., Cassel S., Lafosse M., Rollin P., Benvegnu T., Debaig C., Plusquellec D., Hill K., New primary monoalkyl ethers of autocondensation products of glycerol and mixtures are used in production of nutrient and surfactant formulations, such as cosmetics, pharmaceuticals, laundry or other detergents, Patent # DE 19959000, 2001. (e) Lagarden M., Schmiedel P., Schroeder K.R., Cassel S., Debaig C., Hill K., Rollin P., Lafosse M., Benvegnu T., Plusquellec D., Novel primary mono-fatty acid esters of glycerol self-condensation products, useful as surfactants in cosmetics, pharmaceuticals, foods, detergents or cleaning compositions, comprises specific structure, Patent # DE 19949518, 2001.
49. Norn V., Polyglycerol Esters, in *Emulsifiers in Food Technology* (Ed. Whitehurst R.J.), Blackwell, Oxford, 2004, pp. 110–130. Chapter 5
50. Plasman V., Caulier T., Boulos N., Polyglycerols: Versatile ingredients for personal care, *Happi*, November 2004, 94–97. 2004.
51. Plasman V., Caulier T., Boulos N., Polyglycerol esters demonstrate superior antifogging properties for films, *Plastics Additives and Compounding*, March/April 2005, 30–33. 2005
52. Antoine J.-P., Marcilloux J., Lefeuvre M., Plusquellec D., Benvegnu T., Goursaud F., Parant B., Preparation of glycine betaine surfactants for use in cosmetics, Patent # FR 2869913, WO 2005121291, 2005.
53. Gutierrez X., Silva F., Chirinos M., Leiva J., Rivas H., Bitumen-in-water emulsions: An overview on formation, stability and rheological properties, *J. Dispersion Sci. Technol.*, **23**(1–3), 2002, 405–418.
54. Antoine J.-P., Marcilloux J., Plusquellec D., Benvegnu T., Goursaud F., Bituminous aqueous emulsion with glycine betaine ester/amide surfactant for roads coatings, Patent # FR 2869910, WO 2005121252, 2005.
55. (a) Johnsson M., Engberts J.B.F.N., Novel sugar-based gemini surfactants: Aggregation properties in aqueous solution, *J. Phys. Org. Chem.*, **17**(11), 2004, 934–944. (b) Johnsson M., Wagenaar A., Engberts J.B.F.N., Sugar-based gemini surfactant with a vesicle-to-micelle transition at acidic pH and a reversible vesicle flocculation near neutral pH, *J. Am. Chem. Soc.*, **125**(3), 2003, 757–760. (c) Fielden M.L., Perrin C., Kremer A., Bergsma M., Stuart M.C., Camilleri P., Engberts J. B.F.N., Sugar-based tertiary amino gemini surfactants with a vesicle-to-micelle transition in the endosomal pH range mediate efficient transfection *in vitro*, *Eur. J. Biochem.*, **268**(5), 2001, 1269–1279. (d) Johnsson M., Wagenaar A., Stuart M.C.A., Engberts J.B.F.N., Sugar-based gemini surfactants with pH-dependent aggregation behavior: Vesicle-to-micelle transition, *Critical Micelle Concentration, and Vesicle Surface Charge Reversal, Langmuir*, **19**(11), 2003, 4609–4618. (e) Pestman J. M., Terpstra K.R., Stuart M.C.A., van Doren H.A., Brisson A., Kellogg R.M., Engberts J.B.F.N., Nonionic bolaamphiphiles and gemini surfactants based on carbohydrates, *Langmuir*, **13**(25), 1997, 6857–6860.
56. Castro M.J.L., Kovensky J., Fernandez C.A., Gemini surfactants from alkyl glucosides, *Tetrahedron Lett.*, **38**(23), 1997, 3995–3998.
57. Zana R., Novel Surfactants, in *Preparation, Applications and biodegradability* (Ed. Holmberg K.), Marcel Dekker, New York, 1998, pp. 241–277.
58. (a) Scheibel J.J., Connor D.S., Fu Y.-C., Bodet J.–F., Brown L.A., Vinson P.K., Reilman R.T., Polyhydroxy diamines and their use in detergent compositions, PCT WO 9519951, 1995. (b) Scheibel J.J., Connor D.S., Fu Y.-C., Gemini polyhydroxy fatty acid amides, PCT WO 9519953, 1995. (c) Scheibel J.J., Connor D.S., Fu Y.-C., Poly polyhydroxy fatty acid amides and laundry, cleaning, fabric and personal care composition containing them, PCT WO 9519954, 1995. (d) Foley P.R., Clarke J.M., Fu Y.–C., Vinson P.K., Liquid dishwashing detergent compositions, PCT WO 9520026, 1995. (e) Scheibel J.J., Connor D.S., Fu Y.-C., Gemini polyhydroxy fatty acid amides, US 5534197, 1996.
59. (a) Fuhrhop J.-H., Wang T., Bolaamphiphiles, *Chem. Rev.*, **104**(6), 2004, 2901–2937. (b) Fuhrhop J.-H., Fritsch D., Bolaamphiphiles form ultrathin, porous and unsymmetric monolayer lipid membranes, *Acc. Chem. Res.*, **19**(5), 1986, 130–137.
60. (a) Patel G.B., Sprott G.D., *Crit. Rev. Biotechnol.*, **19**(4), 1999, 317–357. (b) De Rosa M., Gambacorta A., Nicolaus B., *J. Memb. Sci.*, **16**, 1983, 287–294. (c) Lo S.-L., Chang E.L., *Biochem. Biophys. Res. Commun.*, **167**(1), 1990, 238–243.
61. Shimizu T., Masuda M., Stereochemical effect of even-odd connecting links on supramolecular assemblies made of 1-glucosamide bolaamphiphiles, *J. Am. Chem. Soc.*, **119**(12), 1997, 2812–2818.
62. Benvegnu T., Lecollinet G., Guilbot J., Roussel M., Brard M., Plusquellec D., Novel bolaamphiphiles with saccharidic polar headgroups: Synthesis and supramolecular self-assemblies, *Polym. Int.*, **52**(4), 2003, 500–506.
63. (a) Garamus V.M., Milkereit G., Gerber S., Vill V., Micellar structure of a sugar based bolaamphiphile in pure solution and destabilizing effects in mixtures of glycolipids, *Chem. Phys. Lett.*, **392**(1–3), 2004, 105–109. (b) Gerber S., Garamus V. M., Milkereit G., Vill V., Mixed micelles formed by SDS and a bolaamphiphile with carbohydrate headgroups, *Langmuir*, **21**(15), 2005, 6707–6711.

64. Zhan C., Gao P., Liu M., Self-assembled helical spherical-nanotubes from an L-glutamic acid based bolaamphiphilic low molecular mass organogelator, *Chem. Commun.*, **4**, 2005, 462–464.
65. Legros N., Ph.D Thesis, Université de Rennes, 1997.
66. Ambrosi M., Fratini E., Alfredsson V., Ninham B.W., Giorgi R., Lo Nostro P., Bagliono P., Nanotubes from a vitamin C-based bolaamphiphile, *J. Am. Chem. Soc.*, **128**(22), 2006, 7209–7214.
67. Naess S.N., Elgsaeter A., Foss B.J., Li B., Sliwka H.-R., Partali V., Melo T.B., Naqvi K.R., Hydrophilic carotenoids: Surface properties and aggregation of crocin as a biosurfactant, *Helv. Chim. Acta,*, **89**(1), 2006, 45–53.
68. Masuda M., Shimizu T., Lipid nanotubes and microtubes: Experimental evidence for unsymmetrical monolayer membrane formation from unsymmetrical bolaamphiphiles, *Langmuir*, **20**(14), 2004, 5969–5977.
69. (a) Roussel M., Lognoné V., Plusquellec D., Benvegnu T., Monolayer lipid membrane-forming dissymmetrical bolaamphiphiles derived from alginate oligosaccharides, *Chem. Commun.*, **34**, 2006, 3622–3624. (b) Guilbot J., Benvegnu T., Legros N., Plusquellec D., Dedieu J.-C., Gulik A., Efficient synthesis of unsymmetrical bolaamphiphiles for spontaneous formation of vesicles and disks with a transmembrane organization, *Langmuir*, **17**(3), 2001, 613–618.
70. Masuda M., Shimizu T., Multilayer structure of an unsymmetrical monolayer lipid membrane with a 'head-to-tail' interface, *Chem. Commun.*, **23**, 2001, 2442–2443.
71. (a) Hauthal H.G., Dynamic surfactants and nanostructured surfaces for an innovative industry, CESIO 2004, *SÖFW-Journal*, **130**, October, 2004, 3–17, 2004. (b) Smith G., World surfactants congress, *CESIO 6th, J. Surfactants Deterg.*, **7**(4), October 2004, 337–341, 2004.

– 8 –

Tannins: Major Sources, Properties and Applications

Antonio Pizzi

ABSTRACT

After a brief historical introduction and the distinction between hydrolysable and condensed tannins, a description of their chemistry and a short historical review on their use in leather tanning, the more recent developments in tannins for adhesives with and without the use of any aldehyde-yielding compounds, even without the use of any hardeners, are described. Examples of the use of tannins for other industrial, nonleather, applications are reported. In particular, this chapter focuses briefly on their new intended use in the medical and pharmaceutical fields. New data on their antiviral effectiveness against a great number of different viruses compared to their higher, lower or absent cytotoxicity are also presented.

Keywords

Tannins, History, Sources, Adhesives, Wood, Resins, Antiviral activity, Cytotoxicity, Leather, Nonleather uses, Pharmaceutical, Medical

8.1 HISTORY OF TANNINS EXTRACTION

Tannins are a renewable resource that is 'coming of age' in several fields different from their usual, classical application, namely hide tanning to produce heavy duty leather. Leather tanning has been used for centuries, millennia in fact, by immersing hides in pits in which tree bark or wood rich in tannin, such as oak, had been left in. This method took up to one full year to produce good leather. However, the actual tannin extraction industry is relatively more recent. It started in Lyon, France and in northern Italy in the 1850s to satisfy the need for black dyes for silk clothes. As the fashion of women silk blouses waned towards the 1870s, the multitude of small chestnut tannin extraction factories that had sprung up underwent a dramatic change of fortune, many going into bankruptcy and closing down, others combining to build enough critical mass to find an alternative use for tannin extract [1].

The few surviving producers, managed to convince the leather manufacturers that, by dissolving tannin extract in the treatment pits, leather could be manufactured in just 1 month, instead of the 1 year it took on an average with the traditional bark bath technology. As a consequence, the tannin extraction industry started its second life for an application quite different from the original one. The advantages of tannin extract in leather making, and the timesaving involved by its use, were such that the industry underwent rapid expansion and prospered. The short availability of materials in Europe to satisfy the rapidly growing demand for tannin extracts for leather prompted the opening of factories in far-away countries and the use of new types of tannins. Thus, in the early 1900s, tannin factories using quebracho tannin from South America and Mimosa tannin from Southern and Central Africa started their extraction in industrial quantities and exported them to the main northern hemisphere markets.

The two World Wars gave a considerable impulse to the expansion of tannin extraction, considering that all the armies marched on shoes with leather soles. To give a typical quantitative example, in 1946, just after the end of World War II, a major producer such as South Africa manufactured about 110 000 metric tons of dry tannin extract solids. That year however, was the zenith of the use of vegetable tannins for leather making, because from then on rubber and neoprene soles in everyday shoes, cheaper and readily available, progressively replaced leather counterparts. By the beginning of the 1970s, the total tanning production was down to 72 000 tons per year. This was followed by a further decrease in a period in which hides supply dwindled, due to the decrease and difficulties of cattle farming in the 1960s and 1970s. Finally, the considerable shift in customers' tastes in shoes, which came about with the comfortable sport and leisure shoes 'boom', gave a final blow to the use of tannins in leather making. It is interesting that today the same country that produced 110 000 tons of tannin extract in 1946, now only produces 42 000, only half of which are still used for leather manufacture.

As a consequence of the steady dwindling in tannin sales for leather, in the 1960s and 1970s the industry started to desperately look for new applications for these natural products. After all, they had survived the collapse of the silk dyes market more than a century earlier, and a new lease of life could perhaps be found in other fields. Many applications were tried, from varnish primers for metals, which effectively were in use in Britain for some time during the 1960s and 1970s, to antipollution flocculating agents that were successful for about 15 years in the 1970s and 1980s, before being superseded by better synthetic materials [18]. About 600 tons tannin per year, are still in use as ore flotation agents, especially in a couple of feldspar mines in southern Africa [18]. Furthermore, fluidifying agents for drilling mud and superplasticizing additives for cement were developed [101]. To the best knowledge of the author, about 70–80 tons per year of tannin-based cement additives are still used. The main use found, however, was for tannin adhesives for wood panels and other wood products. Production started in 1973 and reached 4 500 dry tons/year of tannin by 1978, and this figure has now risen to around 25 000.

The application of tannins as wood adhesives contributed to save a few tannin extraction factories and to stabilize the situation in some major producing countries, but did not translate into an immediate and definitive rescue of that industry. Worldwide, leather still consumes more tannin than adhesive formulations. This is because oil-derived synthetic adhesives are cheaper than the tannin-based counterparts. The first oil crisis of 1974 at least convinced a few southern hemisphere industrialists (in Australia, South Africa and New Zealand) to use tannin adhesives to avoid the difficulties of supplying synthetic adhesives. The phenomenon persisted with a couple of still existing exceptions, but it did not take off in Europe or North America. Synthetic adhesives became cheap again and in abundant supply after 1974, being strongly 'pushed' by all the big chemical companies. Considerable renewed interest in tannins started again after the year 2000 as a result of two factors: (i) the recent marked increase in oil prices that raised disproportionately the market price of all synthetic adhesives, thus favouring natural raw materials. As a consequence, tannin has now become much cheaper than phenol and competes with the cheapest adhesive of them all, urea–formaldehyde resins; (ii) the recent severe tightening of formaldehyde emission regulations, mainly the introduction of the extremely severe Japanese standard, a regulation that is now starting to spill over into other countries [2, 3].

Although tannin adhesives are now fast becoming an interesting industrial proposition as an alternative to synthetic homologues, the use of tannins is expanding rapidly to an even more interesting field, namely to the pharmaceutical/medicine areas. Thus, a fourth market transformation has started and it appears set to overtake tannin transformation into wood adhesives, at least in terms of monetary added value. The therapeutic virtues of the addition of tannins in wine (consequence of the so-called French Paradox), the use of tannins to cure some gastrointestinal diseases, the increasing use of tannins as food supplements in North America, and the research on the beneficial effects of tannins in a multitude of diseases, even serious ones such as cancer and virus-induced sicknesses, are in full swing. This is because the value added to the base cost of tannins is considerable. Indeed, since tannins to be used for human consumption must obviously be thoroughly purified, their price is 40–50 times higher than that of tannins for industrial applications.

8.2 MAJOR SOURCES

The sources of tannins are very varied. There is a multitude of trees and shrubs which contain tannins. For both hydrolysable and condensed structures, the species rich in tannins are many. Notable for either their present or past economic and/or industrial importance are black wattle or black mimosa bark (*Acacia mearnsii*), quebracho wood (*Schinopsis balansae or lorentzii*), oak bark (*Quercus* spp.), chestnut wood (*Castanea sativa*), mangrove wood, *Acacia catechù*, *Uncaria gambir*, sumach, myrabolans (*Terminalia* and *Phyllantus* tree species), divi-divi (*Caesalpina coraria*), algarobilla chilena, tara, and the bark of several species of pines and firs, among them *Pinus radiata* and *Pinus nigra*, not counting even more plants with extractable tannins.

As regards the tannin origin by location, the areas of strong industrial production today are Brazil, South Africa, India, Zimbabwe, Tanzania for mimosa tannin; Argentina for quebracho tannin; Indonesia for mangrove and for cube Gambier tannins; and Italy and Slovenia for chestnut tannin. There are many other small to very small producers a bit everywhere, for example, small pine tannin factories in Turkey and Chile, an oak tannin factory in Poland and a grape pip tannin factory in France.

8.3 USES

The variety of uses of tannins has been illustrated in the introduction. Some, which were important in the past but are no longer now, will not be described, as the literature on the subject is extant [1, 14, 15, 18]. The major existing uses or past uses of tannins are listed below, and the more important will be developed later in more detail.

(1) Leather manufacture.
(2) Adhesives, in particular wood adhesives.
(3) Wine, beer and fruit juices additives.
(4) Ore flotation agents.
(5) Cement superplasticizers.
(6) Medical and pharmaceutical applications.

Of these, leather manufacture will only be briefly discussed as it is beyond the scope of this review and because extensive technical and scientific literature is available on the subject [1, 5, 7, 14, 15–19, 27]. Suffice it to say that it is the interaction between the phenolic hydroxy groups of the tannins and the polar groups of proteins that give rise to a very strongly associated whole.

8.4 TANNIN STRUCTURE

The term natural vegetable tannins is used loosely to define two broad classes of chemical compounds of mainly phenolic nature, namely condensed or polyflavonoid tannins and hydrolysable tannins. The recognized oligomeric nature of condensed tannins [4–7] contrasts with the allegedly nonpolymeric nature of hydrolysable tannins [5–7].

8.4.1 Hydrolysable tannins

Hydrolysable tannins, including chestnut (*Castanea sativa*), myrabolans (*Terminalia* and *Phyllantus* tree species), divi-divi (*Caesalpina coraria*), tara, algarobilla, valonea, oak and several other commercial tannin extracts are reputed to be mixtures of simple phenols such as gallic and ellagic acids and of esters of a sugar, mainly glucose, with gallic and digallic acids, and with more complex structures containing ellagic acid (Fig. 8.1).

Notwithstanding their alleged lack of a polymeric nature, they can form complex structures. It must be noted first that carbohydrates are intimately and covalently linked to the phenolic moieties in the structure of

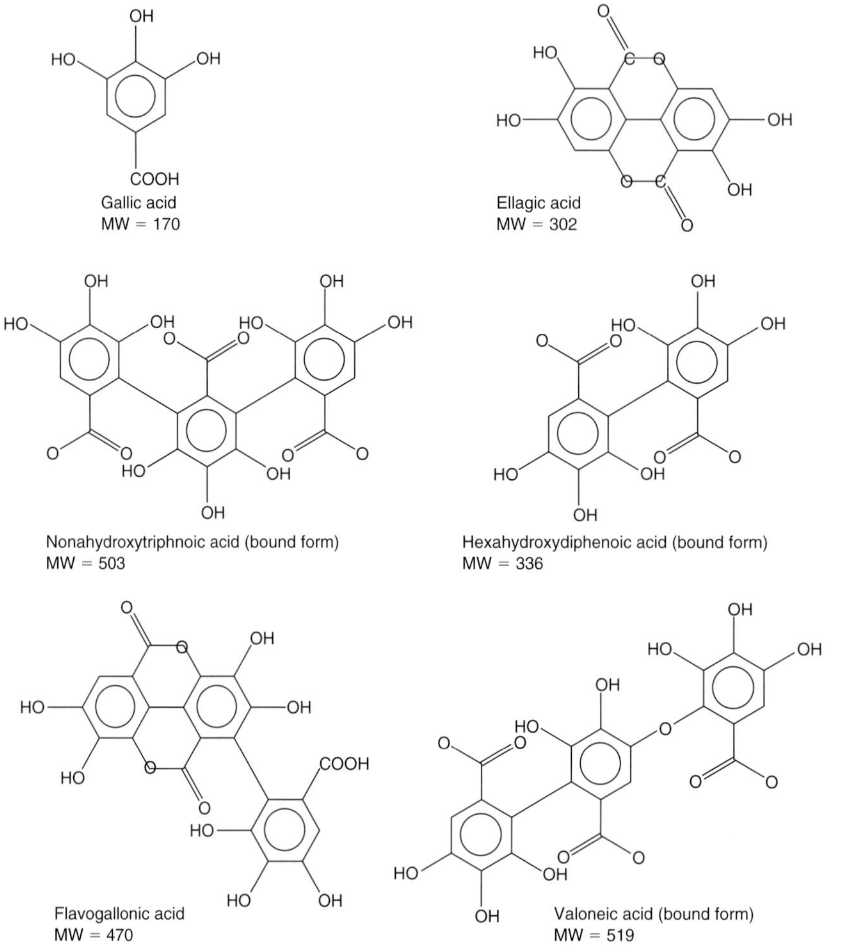

Gallic acid
MW = 170

Ellagic acid
MW = 302

Nonahydroxytriphnoic acid (bound form)
MW = 503

Hexahydroxydiphenoic acid (bound form)
MW = 336

Flavogallonic acid
MW = 470

Valoneic acid (bound form)
MW = 519

Figure 8.1 Chemical species characteristic of the low molecular weight fraction of hydrolysable tannins.

these tannins, and are therefore to be considered as part of the tannin itself. Indeed, several studies [4, 5] identified the major constituents of the main commercial hydrolysable tannin, chestnut tannin extract (an ellagitannin), as the positional isomers castalagin and vescalagin (**I**), present respectively in 14.2 and 16.2 per cent by mass.

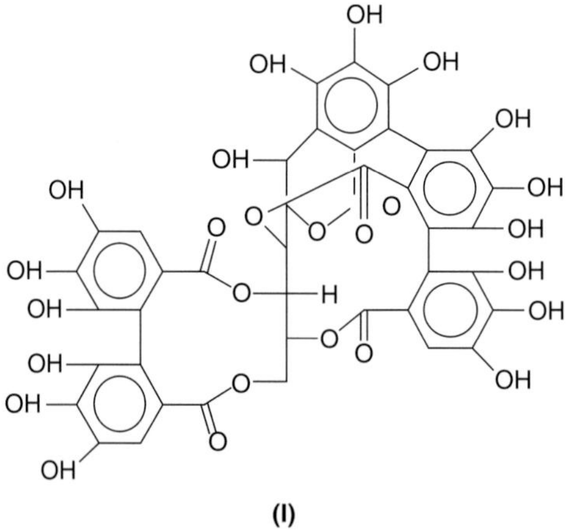

(I)

The rest of the tannin was found to contain 6.6 per cent of the positional isomers castalin and vescalin (**II**) [4, 5], 6 per cent of gallic acid, and 3 per cent of pentagalloyl glucose monomer.

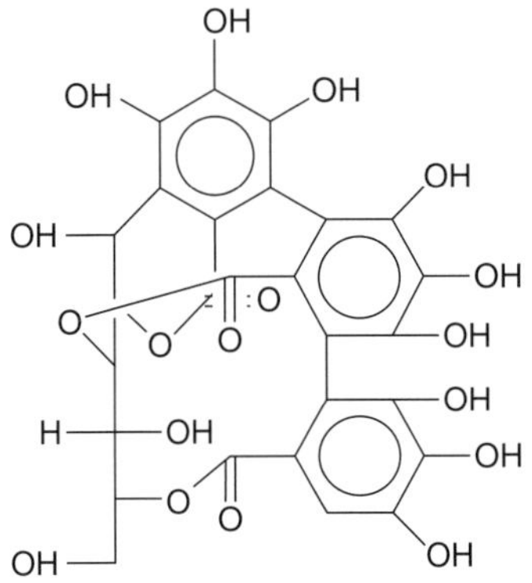

(II)

It must be pointed out however that the authors of the study advancing relative composition percentages [5] clearly state that the disadvantage of the chromatographic technique they used is its strong adsorption of several types of tannins, particularly tannins composed of large molecules. This limitation might well slant the percentages of lower to higher molecular weight components in the analysis of the chestnut extract [4]. Notwithstanding such a limitation, two classes of compounds have however mass predominance in chestnut tannins, namely 28.8 per cent of small molecules, the formula of which is shown in Fig. 1, and 25.4 per cent (or higher, see reasons above) of an unknown and difficult to isolate fraction of apparently a very much higher molecular weight [5] and a very low TLC Rf. This fraction appears to be composed of a number of closely related components giving a continuous TLC smudge of Rf values between 0 and 0.33, which has recently been identified as a mixture of oligomers of pentagalloyl glucose [8], leading to the hypothesis that castalagin, vescalagin, vescalin and castelin are nothing but the hydrolysis of the real structure of the tannin as present in nature [8]. Circumstantial evidence strongly suggests that their repeat unit is:

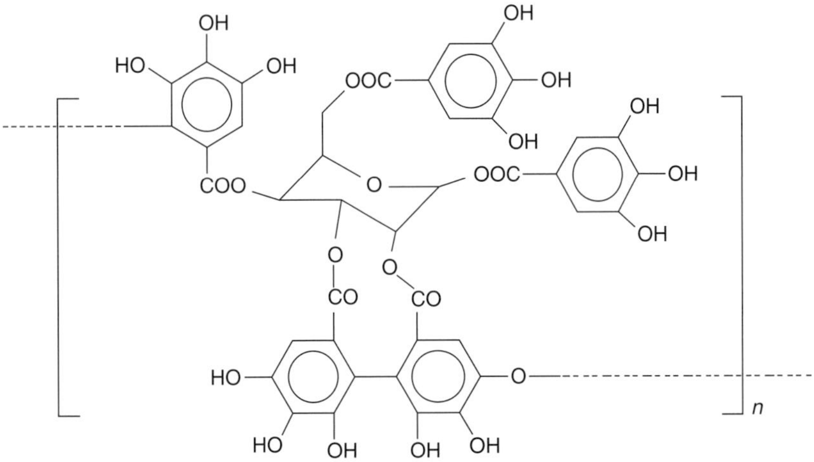

Although these tannins can be reacted with formaldehyde and other aldehydes, the rates of these interactions are low, and they are therefore not favoured for the preparation of resins. They have, however, been used successfully as partial substitutes (up to 50 per cent) of phenol in the manufacture of phenol–formaldehyde resins [9, 10]. Their chemical behaviour towards formaldehyde is analogous to that of simple phenols of low reactivity and their moderate use as phenol substitutes in the above-mentioned resins does not present difficulties. Their lack of macromolecular structure, the low level of phenol substitution they allow, their low nucleophilicity, limited worldwide production and relatively high price, somewhat decrease their chemical and economical interest for resin production. Consequently, their main use is for leather tanning where their performance, especially in terms of clarity of colour and light resistance, is truly excellent.

8.4.2 Condensed (polyflavonoid) tannins

Condensed tannins constitute more than 90 per cent of the total world production of commercial tannins (200 000 tons per year) [11]. Their high reactivity towards aldehydes and other reagents renders them both chemically and economically more interesting for the preparation of adhesives, resins and other applications apart from leather tanning. The main commercial species, such as mimosa and quebracho, also yield excellent heavy duty leather. Condensed tannins and their flavonoid precursors are known for their wide distribution in nature and particularly for their substantial concentration in the wood and bark of various trees. These include various *Acacia* (wattle or mimosa bark extract), *Schinopsis* (quebracho wood extract), *Tsuga* (hemlock bark extract), *Rhus* (sumach extract) species, and various *Pinus* bark extract species, from which commercial tannin extracts are manufactured.

The structure of the flavonoid constituting the main monomer of condensed tannins may be represented as follows:

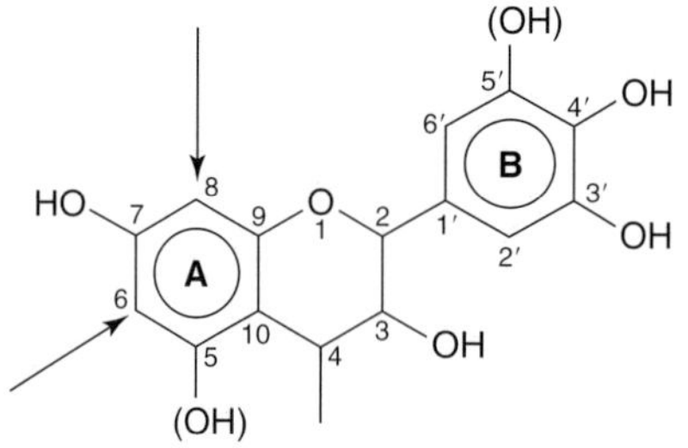

This flavonoid unit is repeated 2–11 times in mimosa and quebracho tannins [11, 12], with an average degree of polymerization of 4–5, and up to 30 times for pine tannins, with an average degree of polymerization of 6–7 for their soluble extract fraction [13]. The nucleophilic centres on the A-ring of a flavonoid unit tend to be more reactive in aromatic substitution, than those found on the B-ring. This is due to the vicinal hydroxyl substituents, which cause general activation in the B-ring, without any localized effects such as those found in the A-ring [11].

The following three-unit fragment illustrates two typical tannin structures:

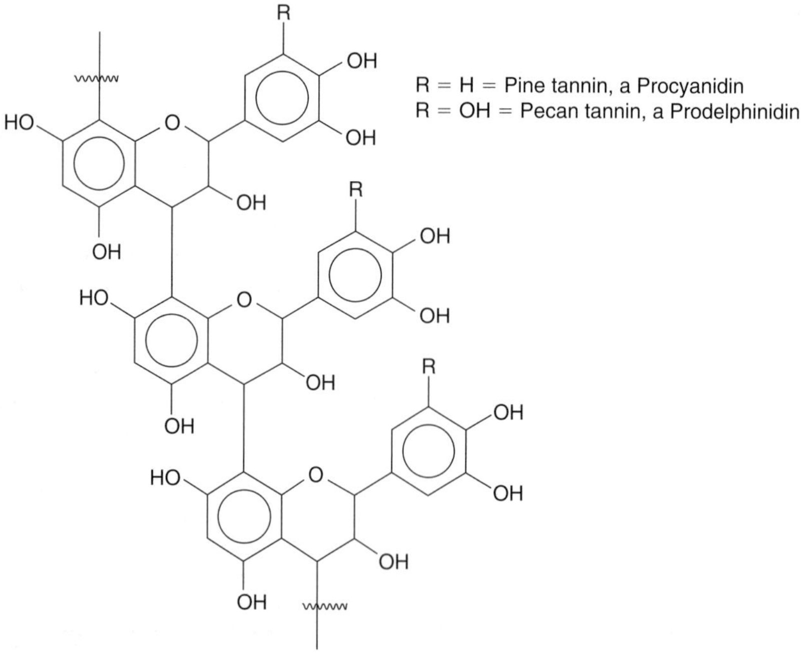

R = H = Pine tannin, a Procyanidin
R = OH = Pecan tannin, a Prodelphinidin

Formaldehyde and other aldehydes react with tannins to induce polymerization through methylene bridge linkages at reactive positions on the flavonoid molecules, mainly the A-rings. The reactive positions of the A-rings are the 6 *or* 8 locations (according to the type of tannin) of all the flavonoid units, and both of them for the upper terminal flavonoid units. The A-rings of mimosa and quebracho tannins show reactivity towards formaldehyde comparable to that of resorcinol [11, 14–17]. Assuming the reactivity of phenol to be 1 and that of resorcinol to be 10, the A-rings have a reactivity of 8–9. However, because of their size and shape, the tannin molecules lose their mobility and flexibility at a relatively low level of condensation with formaldehyde, so that the available reactive sites are too far apart for further methylene bridge formation. The result may be incomplete polymerization and therefore poor material properties. Bridging agents with longer molecules should be capable of joining the distances that are too long for methylene bridges. Alternatively, other techniques can be used to solve this problem [11, 100].

In condensed tannins from mimosa bark, the main polyphenolic pattern is represented by flavonoid analogues based on resorcinol A-rings and pyrogallol B-rings. These constitute about 70 per cent of the tannins. The secondary but parallel pattern is based on resorcinol A-rings and catechol B-rings [11, 14]. These tannins represent about 25 per cent of the total of mimosa bark tannin fraction. The remaining parts of the condensed tannin extract are the 'nontannins' [14]. They may be subdivided into carbohydrates, hydrocolloid gums and small amino and imino acid fractions [11, 14]. The hydrocolloid gums vary in concentration from 3 to 6 per cent and contribute significantly to the viscosity of the extract despite their low concentration [11, 14]. Similar flavonoid A- and B-ring patterns also exist in quebracho wood extract (*Schinopsis balansae*, and *Lorentzii*) [15–17], but no phloroglucinol A-ring pattern, or probably a much lower quantity of it, exists in the quebracho extract [17–19]. Similar patterns to wattle (mimosa) and quebracho are encountered in hemlock and Douglas fir bark extracts. Completely different patterns and relationships are found instead in pine tannins [20–22] which present only two main patterns: one represented by flavonoid analogues based on phloroglucinol A-rings and catechol B-rings [20, 22] and the other, present in a much lower proportion, represented by phloroglucinol A-rings and phenol B-rings [20, 22]. The A-rings of pine tannins then possess only the phloroglucinol type of structure, much more reactive towards formaldehyde than a resorcinol-type counterpart, with important consequences in the use of these tannins for adhesives.

8.5 ANALYSIS

Various methods of analysis are available for the determination of tannin content. These methods can generally be grouped into two broad classes:

(1) *Methods aimed at the determination of tannin material content in the extract*: The classical method of this type still used is the hide-powder method. These methods were devised to determine which percentage of the extract would participate in leather tanning. Their main drawback for their use in adhesives is their incapacity of detecting and determining the approximate 3–6 per cent of monoflavonoids and biflavonoids, or phenolic 'nontannins', present in the extract, which do not contribute to tanning capacity, but which do definitely react with formaldehyde and contribute to adhesive preparation.
(2) *Methods aimed at the determination of phenolic materials present in the extract that can be reacted with formaldehyde*: These methods were devised particularly for tanning extract used in adhesives and are all based on the determination of some of the products of the reaction of the flavonoids with formaldehyde.

The accepted methods of the first type comprise the hide-powder method [23], the refractometric method and various visible ultraviolet, and infrared spectroscopic methods. The accepted methods of the second type include comparative methods such as the Stiasny–Orth method [24, 25] and its modifications, all these being gravimetric methods today largely obsolete due to the lack of reliability consequent to the coprecipitation of some carbohydrates together with the phenolic material of the tannin extract and to the results being expressed in an absolute value, which is never convertible to a percentage of useful material in the extract. The Lemme sodium bisulphite backtitration method [26], the ultraviolet spectrophotometric molibdate ion method [27], and the infrared spectrophotometric methods [28] give instead correct percentage results.

8.6 A FEW CONSIDERATIONS ON LEATHER MANUFACTURE

The leather tanning industry is one of the most ancient processes still in operation. Extensive reviews and articles on the use of vegetable tannins in leather making exist [1, 5, 7, 14, 15–19, 27]. A brief general and comparative outline of vegetable versus other types of tanning is given here for the information of the reader who might not know this industry. Although the technology of leather manufacture has evolved over centuries, and even in recent years, the basic principles for the production of leather have remained essentially the same. Hide proteins, mainly collagen, are rendered insoluble and dimensionally more stable by treatment with chemical products able to join them and thus render them both more resistant to mechanical wear and less susceptible to biological degradation and other types of attack. The main products used today for leather tanning are (i) acid salts of trivalent chromium, mainly used for the manufacture of soft leathers for shoe uppers and for leather bags; (ii) forestry-derived, natural vegetable tannins, such as chestnut and flavonoid extracts, mainly used for the manufacture of heavy, rigid and hard leathers for shoe soles, saddles, belts and other implements subject to intensive wear; (iii) aldehydes, in particular formaldehyde and glutaraldehyde; (iv) sulphonated synthetic polymers, such as acid phenol–formaldehyde novolak-type resins and (v) a number of other synthetic resins and compounds (acrylics, oxazolidines, aminoplastic resin, etc.).

Each of the products mentioned above is more apt than the others to the manufacture of certain specific types of leather. The fact remains however, that the first two in the above list account for more than 90 per cent of all the leather manufactured today, and that the process based on trivalent chromium salts accounts by itself for about 70 per cent of the total. Chrome tanning is particularly suited for soft leather, as it does not affect hide flexibility and renders the leather very lightfast and very stable both chemically and physically. It produces leather of excellent antishrinkage ability, as indicated by its high shrinkage temperature in testing. The forestry-derived vegetable tannins have, instead, a strong astringent effect (they fix very effectively on the collagen structure) and give considerable 'body', hardness and toughness to the leather produced with them. However, they have the considerable disadvantage of exhibiting marked darkening problems when exposed to light and, even worse, to shrink at a much lower temperature. It is these two main disadvantages that have somewhat limited their application in relation to the ubiquitousness of chromium salts. Conversely, their combination with some synthetic resins such as MUF [29, 30], give light-coloured leathers presenting high resistance to light-induced degradation, as well as other advantages similar to those of chrome tanning.

Chromium salts are becoming less acceptable in many types of industry due to potential effluent pollution. Furthermore, well-defined quality standards as regards leather product skin-contact allergic reactions, have also been introduced for finished products, for instance in leather clothing and interior car linings. In this respect, two of the requirement limits to comply with are the amount of both leachable trivalent chromium, which generally does not constitute a problem, and of one of its tanning derivatives, namely the more dangerous, highly toxic hexavalent chromium salts. Recent norms [31, 32] limit severely the proportion of these compounds in leathers to be used in direct contact with human skin, such as watch straps, shoe uppers, etc.). Furthermore, the treatment of tanning waste waters represents one of the major problems in the leather industry, especially today that the relevant European Commission norms impose ever more stringent effluent limits. The waste waters are generally treated to abate (never eliminate) chromium salt residues. However it has proved difficult to find suitable alternatives to chromium salts up to now.

In the case of natural tannins, their sensitivity to photo-oxydation limits their use to applications where such a characteristic is of no consequence. It is the tannin phenolic structure itself which renders photo-oxydation possible [33], with transformation of tannin phenolic groups to coloured quinones, but it has been shown that this effect can be drastically limited if the tannins are condensed with sulphonated synthetic aminoplast resins such as MUF [29, 30]. Conversely, while the use of synthetic aminoplast resins is developing in the tanning industry, as they give leathers of a certain degree of softness and flexibility, and particularly suitable for colouring, their weak point is the excessive and inevitable presence of free formaldehyde [34] and their poor tanning capability due to their low astringency. Polyphenolic vegetable tannins are well known to act as powerful free formaldehyde scavengers, as they react rapidly and irreversibly with this compound [11, 35]. Their combination with aminoplast MUF resins thus reduces markedly the photo-oxydation of vegetable tannins through synergy with the synthetic resin, reduces formaldehyde emission to just about zero, yields relatively soft but also tough leather and eliminates the need for chromium salts. Furthermore, such a mix can achieve leather shrinkage temperatures that at least match the chromium salt performance on this parameter.

8.7 TANNIN-BASED ADHESIVES

8.7.1 Wood adhesives

In condensed polyflavonoid tannin molecules, the A-rings of the constituent flavonoid units retain only one highly reactive nucleophilic centre, the remainder accommodating the interflavonoid bonds. Resorcinolic A-rings (wattle) show reactivity towards formaldehyde comparable to, though slightly lower than, that of resorcinol [36]. Phloroglucinolic A-rings (pine) behave instead as phloroglucinol [37]. Pyrogallol or catechol B-rings are by comparison unreactive and may be activated by anion formation only at relatively high pH [38]. Hence the B-rings do not participate in the reaction except at high pH values (pH 10), where the reactivity towards formaldehyde of the A-rings is so high that the ensuing tannin–formaldehyde adhesives have unacceptably short pot lives [36]. In the usual tannin adhesive practice, only the A-rings are used to generate the network. With regard to the pH dependence of the reaction with formaldehyde, it is generally accepted that the reaction rate of wattle tannins with formaldehyde is the slowest in the pH range 4.0–4.5 [39]; for pine tannins, the range is between 3.3 and 3.9.

Formaldehyde is generally the aldehyde used in the preparation, setting and curing of tannin-based adhesives. It is normally added to the tannin extract solution at the required pH, preferably in its polymeric form of paraformaldehyde, which is capable of fairly rapid depolymerization under alkaline conditions, and as urea–formalin concentrates. Hexamethylenetetramine (hexamine) may also be added to these resins because of its potential formaldehyde releasing action under heat. Hexamine is, however, unstable in acid media [40], but becomes more stable with increasing pH values. Hence, under alkaline conditions, the liberation of formaldehyde might not be as rapid and as efficient as wanted. Also, it has been fairly widely reported, with a few notable exceptions [41], that bonds formed with hexamine as hardener are not as boil resistant [42] as those formed by paraformaldehyde. The reaction of formaldehyde with tannins may be controlled by the addition of alcohols to the system. Under these circumstances, some of the formaldehyde is stabilized by the formation of hemiacetals (*e.g.* $CH_2(OH)(OCH_3)$) if methanol is used [11, 37]. When the adhesive is cured at an elevated temperature, the alcohol is driven off at a fairly constant rate and formaldehyde is progressively released from the hermiacetal. This ensures that less formaldehyde is volatilized when the reactants reach curing the curing temperature and that the pot life of the adhesive is extended. Other aldehydes have also been substituted for formaldehyde [11, 36, 39, 41].

In the reaction of polyflavonoid tannins with formaldehyde two competitive mechanisms take place:

(1) The reaction of the aldehyde with tannin and with low molecular weight tannin–aldehyde condensates, which is responsible for the aldehyde consumption.
(2) The liberation of formaldehyde, which become available again for reaction. This release is probably due to the decomposition of the unstable $—CH_2—O—CH_2—$ ether bridges initially formed into $—CH_2—$ counterparts.
(3) In the case of some tannins (*e.g.* quebracho tannin) a third important reaction occurs, namely the simultaneous hydrolysis of some interflavonoid bonds, hence a depolymerization reaction, partly counteracting and thus slowing down the hardening process [12, 42, 43].

Considering the fact that the two major existing industrial polyflavonoid tannins, namely mimosa and quebracho tannins, are very similar and both composed of mixed prorobinetinidins and profisetinidins, it was difficult to rationalize this anomalous behaviour of quebracho tannin. It has now been possible to determine by both NMR [42] and particularly by laser desorption mass spectrometry (MALDI–TOF), applied to mimosa and quebracho tannins and some of their modified derivatives [12] that: (i) mimosa tannin is predominantly composed of prorobinetinidins, while quebracho is predominantly composed of profisetinidins; (ii) mimosa tannin is heavily branched due to the presence of considerable proportions of 'angular' units in its structure, while quebracho tannin is almost completely linear [12]. This latter structural difference is the one which contributes to the considerable differences in viscosity of water solutions of the two tannins and which induces the interflavonoid link of quebracho to be more easily hydrolysable, because of the linear structure of this tannin. This feature confirms the NMR findings [12, 42] showing that this tannin is subject to polymerization/depolymerization equilibria. The specificity of the quebracho structure also explains the decrease in viscosity arising from acid/base treatments to yield tannin adhesive intermediates after a certain level of hydrolysis of the tannin itself and not only of the carbohydrates present in the extract (see also Section 8.4). Such a tannin hydrolysis does not appear to occur with mimosa tannin in which the interflavonoid link is completely stable against this attach.

It is interesting to note that whereas $—CH_2—O—CH_2—$ ether bridged compounds have been isolated for the phenol–formaldehyde [40] reaction, their existence for fast-reacting phenols such as resorcinol and phloroglucinol has been postulated, but they have never been isolated, since these two phenols have always been considered too reactive towards formaldehyde. They are detected indirectly by the surge in the concentration of formaldehyde observed in kinetic studies, as a consequence of the methylene ether bridge decomposition [44].

When heated in the presence of strong mineral acids, condensed tannins are subject to two competitive reactions. One is degradative leading to lower molecular weight products, and the second is condensative, as a result of the hydrolysis of heterocyclic rings (*p*-hydroxybenzyl ether links) [38]. The generated *p*-hydroxybenzylcarbenium ions condense randomly with nucleophilic centres on other tannin units to form 'phlobaphenes' or 'tanner's red' [38, 45–47]. Other modes of condensation (*e.g.* free radical coupling of B-ring catechol units) cannot be excluded in the presence of atmospheric oxygen. In predominantly aqueous conditions, the formation of phlobaphene or insoluble condensates predominates. These reactions, characteristic of tannins and not of synthetic phenolic resins, must be taken into account when formulating tannin adhesives.

The sulphitation of tannins is one of the oldest and most useful reactions in flavonoid chemistry. Slightly sulphited water is sometimes used to increase tannin extraction from the bark containing it. In certain types of adhesives, the total effect of sulphitation has both negative and positive connotations. The latter aspects are related to both a higher concentration of tannin phenolics in adhesive applications, due to enhanced solubility and decreased viscosity, and to a higher moisture retention by the tannin resins, giving a slower adhesive film dry out and hence a longer assembly time [48]. As for the negative aspects, the presence of sulphonate groups promotes a higher sensitivity to moisture with a consequent adhesive deterioration and bad water resistance of the cured glue line even with adequate crosslinking [48–51].

In recent years, the importance of the marked colloidal nature of tannin extract solutions has come to the fore [42, 43, 52–60]. It is the presence of both polymeric carbohydrates in the extract, as well as of the higher molecular fraction of the polyphenolic tannins, which determines the colloidal state of tannin extract solutions in water [42, 52]. This feature affects many of the reactions which lead to the formation and curing of tannin adhesives, to the point where reactions not thought possible in solution become instead, not only possible, but, the favoured ones [42, 52]. Conversely, reactions mooted to be of determinant importance when found on models not in a colloidal state, have in reality been shown to be inconsequential to tannin adhesives and their applications [35, 59].

8.8 TECHNOLOGY OF INDUSTRIAL TANNIN ADHESIVES

8.8.1 Wood adhesives

The purity of vegetable tannin extracts varies considerably. Commercial wattle bark extracts normally contain 70–80 per cent active phenolic ingredients. The nontannin fraction, consisting mainly of simple sugars and high molecular weight hydrocolloid gums, does not participate in the resin formation with formaldehyde. Sugars reduce the strength and water resistance in direct proportion to the amount added. Their effect is a mere dilution effect of the adhesive resin solids, with the consequent proportional worsening of adhesive properties. The hydrocolloid gums, instead, have a much more marked effect on both original strength and water resistance of the adhesive [11, 37, 58]. If it is assumed that the nontannins in tannin extracts have a similar influence on adhesive properties, it can be expected that unfortified tannin–formaldehyde networks can achieve only 70–80 per cent of the performance shown by synthetic adhesives.

In many glued wood products, the demands on the glue line are so high that unmodified tannin adhesives are unsuitable. The possibility of refining extracts has proved fruitless largely because the intimate association between the various constituents makes industrial fractionation difficult. Fortification is in many cases the most practical approach to reducing the effect of impurities and generally consists of copolymerization of the tannin with phenolic or aminoplastic resins [36, 37, 58, 61]. It can be carried out during the manufacture of the adhesive resin, during gluemix assembly, just before use, or during adhesive use. If added in sufficient quantity, various synthetic resins have been found effective in reducing the nontannin fraction to below 20 per cent and in overcoming other structural problems [36, 37]. The main resins used are phenol–formaldehyde and urea–formaldehyde resols with a medium-to-high methylol group content. These resins can fulfil the functions of hardeners, fortifiers, or both. Generally, they are used as fortifiers in between 10 and 20 per cent of total adhesive solids, with paraformaldehyde used as a hardener. Such an approach is the favourite one for marine-grade plywood adhesives. These fortifiers are particularly suitable for the resorcinol types of condensed tannins, such as mimosa. They can be copolymerized with the tannins during resin manufacture, during use, or both [11, 35–37, 58]. Copolymerization and curing are based on the condensation of the tannin with the methylol groups carried by the synthetic resin. Since tannin molecules are generally large, the rate of molecular growth in relation to the rate of linkage is high, so that tannin adhesives generally tend to have fast gelling and curing times and shorter pot lives than those of synthetic phenolic adhesives. From the point of view of reactivity, phloroglucinol tannins, such as pine tannins, are much faster than mainly resorcinol tannins such as mimosa. The usual ways of slowing them down and, for instance, lengthening the adhesive pot life are:

(1) To add alcohols to the adhesive mix to form hemiacetals with formaldehyde which therefore act as retardants of the tannin–formaldehyde reaction.
(2) To adjust the adhesive pH to obtain the required pot life and rate of curing.
(3) To use hexamine as a hardener which, under the current conditions, gives a very long pot life at ambient temperature but still a fast curing time at higher temperatures.

The viscosity of bark extracts is strongly dependent on concentration and increased very rapidly above 50 per cent. Compared to synthetic resins, tannin extracts are more viscous at the concentrations normally required in adhesives. The high viscosity of aqueous solutions of condensed tannins is due to the following causes, in order of importance:

(1) *Presence of high molecular weight hydrocolloid gums in the tannin extract* [58, 60]: The viscosity is directly proportional to the amount of gums present in the extract [58, 60].
(2) *Tannin–tannin, tannin–gum and gum–gum hydrogen bonds*: Aqueous tannin extract solutions are not true solutions but, rather, colloidal suspensions in which water access to all parts of the molecules present is very slow. As a consequence, it is difficult to eliminate intermolecular hydrogen bonds by dilution only [58, 60].
(3) *Presence of high molecular weight tannins in the extract* [12, 58, 60].

The high viscosity of tannin extract solutions has also been correlated to the proportion of very high molecular weight tannins present in the extract. This effect is not well defined. In most adhesive applications, such as in plywood adhesives, the viscosity is not critical and can be manipulated by dilution.

In the case of particleboard adhesives, the decrease in viscosity is, instead, an important prerequisite. When reacted with formaldehyde, unmodified condensed tannins give adhesives having characteristics that do not suit particleboard manufacture, namely, high viscosity, low strength and poor water resistance. The most commonly used process to eliminate these disadvantages in the preparation of tannin-based particleboard adhesives consists of a series of subsequent acid and alkaline treatments of the tannin extract, causing hydrolysis of the gums to simple sugars and some tannin structural changes, thus improving the viscosity, strength and water resistance of the unfortified tannin–formaldehyde adhesive [11, 35]. Furthermore, such treatments may cause the partial rearrangement of the flavonoid molecules that causes liberation of some resorcinol *in situ* in the tannin, rendering it more reactive, allowing better crosslinking with formaldehyde and, ultimately, yielding an adhesive which, without addition of any fortifier resins, gives a truly excellent performance for exterior-grade particleboard [11, 35]. This modification cannot be carried out too extensively, in order to avoid the precipitation of the tannin from the solution by the formation of 'phlobaphenes'.

Typical results are shown in Table 8.1.

Table 8.1

Unfortified tannin–formaldehyde adhesives obtained by acid–alkali treatment, for exterior-grade particleboard: Example of industrial board results [11, 77, 78]

	Swelling after a 2 h boil				
Panel density (g/cm^3)	Measured wet (%)	Measured dry (irreversible swelling) (%)	Original internal bond (IB) tensile perpendicular (kg/cm^2)	IB after a 2 h boil (kg/cm^2)	Cyclic test after five cycles measured (%)
0.700	11.0	0.0	13.0	9.0	3.0

Particular gluing and pressing techniques have been developed for tannin particleboard adhesives [62, 63] to achieve pressing times much shorter than those traditionally obtained with synthetic phenol–formaldehyde adhesives, although recent advances in the latter materials have markedly limited such an advantage [64, 65]. Pressing times of 7 s mm^{-1} of panel thickness have been achieved and of 9 s mm^{-1} at 190–200°C pressing temperatures are in daily operation: these are pressing times that are becoming comparable to those obtainable with urea–formaldehyde or melamine–formaldehyde resins at the same pressing temperatures. The success of these simple types of particleboard adhesives relies heavily on industrial application technologies rather than just on the preparation technology of the adhesive itself [58, 62, 66]. A considerable advantage is the much higher moisture content of the resinated chips tolerable with these adhesives than with any of the synthetic phenolic and aminoresin counterparts. In the case of wood particleboard and of oriented strandboard (OSB) panels, the technology so developed allows hot-pressing at moisture contents of around 24 per cent, against values of 12 per cent for traditional synthetic adhesives, and presents other advantages as well [58, 66, 67].

The best adhesive formulation for phloroglucinolic tannins, such as pine tannin extracts is, instead, a comparatively new and is also capable of giving excellent results when using resorcinol tannins such as a wattle tannin extract [68–71]. The adhesive gluemix consists only of a mix of an unmodified tannin extract 50 per cent solution to which paraformaldehyde and polymeric nonemulsifiable 4,4′-diphenylmethane diisocyanate (commercial pMDI) are added [68–71]. The proportion of tannin extract solids to pMDI can be as high as 70/30 w/w, but can be much lower in pMDI content. This adhesive is based on the peculiar mechanism by which the pMDI in water, is hardly deactivated to polyureas because it reacts faster with the hydroxymethyl groups of a formaldehyde-based resin, be it a tannin or another resin [69, 71].

The properties of the particleboards manufactured with this system using pine tannin adhesives are listed in Table 8.2. The results obtained with this system are quite good and not too different from those produced by some of the other tannin adhesives already described. In the case of phloroglucinolic tannin extracts, no pH adjustment of the solution is needed. One point that was given close consideration is the deactivating effect of water on the isocyanate group of pMDI. It has been found that the amount of deactivation by water in a concentrated solution (50 per cent or over) of a phenol is much lower than previously thought [68–71]. This is the reason why aqueous tannin extract solutions and pMDI can be mixed without any substantial pMDI deactivation by the water present.

The quest to decrease or completely eliminate formaldehyde emission from wood panels bonded with adhesives, although not really necessary in tannin adhesives due to their very low emission (as with most phenolic adhesives), has nonetheless promoted here too some research to further reduce formaldehyde emission. This has centred into two lines of investigation: (i) tannin autocondensation (see paragraph D later on), and (ii) the use of a hardener not emitting at all, simply because no aldehyde has been added to the tannin [72–74]. Methylolated nitroparaffins and, in particular, the simpler and least expensive exponent of their class, namely trishydroxymethyl nitromethane [72, 73], function well as hardeners for a variety of tannin-based adhesives, while affording considerable side advantages to the adhesive and to the bonded wood joint. In panel products such as particleboard, medium density fibreboard and plywood, the joint performance which is obtained is of the exterior/marine-grade type, combined with a very advantageous and very considerable lengthening in gluemix pot life is obtained. Furthermore, the use of this hardener is coupled with such a marked reduction in formaldehyde emission from the bonded wood panel that it is limited exclusively to the formaldehyde released only by the heated wood, or even less, thus functioning as a mild depressant of emission from the wood itself. Moreover, trishydroxymethyl nitromethane can be mixed in any proportion with traditional formaldehyde-based hardeners for tannin adhesives, its proportional substitution of such hardeners inducing a correspondingly marked decrease in the formaldehyde emission of the wood panel, without affecting its exterior/marine-grade

Table 8.2

Properties of a particleboard manufactured using pine tannin adhesives

Panel density	Swelling after a 2 h boil: Measured wet (%)	Swelling after a 2 h boil: Measured dry (irreversible swelling) (%)	Original internal bond (IB), tensile perpendicular (kg/cm^2)	IB after a 2 h boil (kg/cm^2)	IB retention after a 2 h boil (%)
0.690	15.0	4.3	8.4	4.3	51

performance. Medium density fibreboard (MDF) industrial plant trials confirmed all the properties reported above [72, 73]. A cheaper but equally effective alternative to hydroxymethylated nitroparaffins is the use of hexamine as tannin hardener. This sometimes causes problems of an early agglomeration in some tannins [74] and a better solution which overcames this drawback was to use a mixture of formaldehyde and an ammonium salt as the hardener.

8.8.2 Corrugated cardboard adhesives

The adhesives developed for the manufacture of damp-ply-resistant corrugated cardboard are based on the addition of spray-dried wattle extract, urea–formaldehyde resin, and formaldehyde to a typical Stein–Hall starch formula with 18–22 per cent starch content [75, 76]. The wattle tannin–urea–formaldehyde copolymer formed *in situ* and any free formaldehyde left in the glue line are absorbed by the wattle tannin extract. The wattle extract powder should be added at level of 4–5 per cent of the total starch content of the mix (*i.e.* carrier plus slurry). Successful results can be achieved in the range of 2–12 per cent of the total starch content, but 4 per cent is the recommended starting level. The final level is determined by the degree of water hardness and desired bond quality. This wattle extract UF-fortifier system is highly flexible and can be adapted to damp-proof a multitude of basic starch formulations.

8.8.3 Cold-setting laminating and fingerjointing adhesives for wood

A series of different novolak-like materials are prepared by copolymerization of resorcinol with resorcinol A-rings of polyflavonoids, such as condensed tannins [77–79]. The copolymers formed have been used as cold-setting exterior-grade wood adhesives, complying with the relevant international specifications. Several formulations are used. The system most commonly employed commercially relies on the simultaneous copolymerization of resorcinol and of the tannin resorcinol A-ring, thanks to their comparable reactivity towards formaldehyde. The following scheme summarizes the principle of this system:

Resorcinol + HCHO + flavonoid tannin

HO, HO, CH_2, HO, O, OH, OH, OH

+ Uncondensed resorcinol and flavonoids

The final mixture is an adhesive that can be set and cured at ambient temperature by the addition of paraformaldehyde. Other cold-set systems exist and are described in the more specialized literature [11, 77–79]. Some typical results obtained with these adhesives are given in Table 8.3.

Table 8.3

Typical results of tannin–resorcinol–formaldehyde cold-setting adhesives used on beech strips according to British Standard BS 1204 [80]

	Dry	After 24 h cold-water soak	After 6 h boil
Tensile strength (N)	3 200–3 800	2 300–2 900	2 200–2 800
Wood failure (%)	90–100	75–100	80–100

A particularly interesting system now used extensively in several southern hemisphere countries is the so-called honeymoon' fast-setting, separate-application system [81, 82]. In this process, one of the surfaces to be mated in the joint is spread with a standard synthetic phenol–resorcinol–formaldehyde adhesive plus a paraformaldehyde hardner. The second surface is spread with a 50 per cent tannin solution at pH 12. When the two surfaces are jointed, fingerjoints develop enough strength to be installed within 30 min and laminated beams (glulam) need to be clamped for only 2.5–3 h instead of the traditional 16–24 h, with a consequent considerable increase in factory productivity. This adhesive system also provides full weather- and boil-proof capabilities.

8.8.4 TYRE CORD ADHESIVES

Another application of condensed tannin extracts that has proved technically successful is as a tyre cord adhesive. Both thermosetting tannin formulation [83] and tannin–resorcinol–formaldehyde formulations have been tested successfully.

8.9 NEW CONCEPTS AND PRINCIPLES

8.9.1 Surface catalysis

As in the case of other formaldehyde-based resins, the interaction energies of tannins with cellulose obtained by molecular mechanics calculations [59] tend to confirm the effect of surface catalysis induced by cellulose also on the curing and hardening reaction of tannin adhesives. The considerable energies of interactions obtained can effectively explain a weakening of the heterocycle ether bond leading to the accelerated and easier opening of the pyran ring in a flavonoid unit, as well as the ease with which hardening by self-condensation can occur. As in synthetic formaldehyde-based resins, the same effect explains the decrease in the activation energy of the condensation of polyflavonoids with formaldehyde leading to exterior wood adhesives curing and hardening [84].

8.9.2 Hardening by tannins self-condensation

The self-condensation reactions characteristic of polyflavonoid tannins have only recently been used to prepare adhesive polycondensates hardening in the absence of aldehydes [85]. This self-condensation reaction is based on the opening under alkaline and acid conditions of the O1—C2 bond of the flavonoid repeating unit and the subsequent condensation of the reactive centre formed at C2 with the free C6 or C8 sites of a flavonoid unit on another tannin chain [85–89]. Although this reaction may lead to considerable increases in viscosity, gelling does not generally occur. However, gelling does occur when the reaction takes place (i) in the presence of small amounts of dissolved silica (silicic acid or silicates) catalyst and some other catalysts [85–90], and (ii) on a lignocellulosic surface [89]. In the case of the more reactive procyanidin- and prodelphinidin-type tannins, such as pine tannin, cellulose catalysis is more than enough to cause hardening and to produce boards of strength satisfying the relevant standards for interior-grade panels [89]. In the case of the less reactive tannins, such as mimosa and quebracho, the presence of a dissolved silica or silicate catalyst is essential to achieve the panel strength required by the relevant standards. Self-condensation reactions have been shown to contribute considerably to the dry strength of wood panels bonded with tannins, but to be relatively inconsequential in contributing to the bonded-panel exterior-grade properties, which are instead attained by polycondensation reactions with aldehydes [89–91]. Combinations of tannin self-condensation and reactions with aldehydes, and combinations of radical and ionic reaction, have been used to decrease the proportion of aldehyde hardener needed, as well as to decrease considerably further the already low formaldehyde emission caused by the use of tannin adhesives [89–91].

8.10 CEMENT SUPERPLASTICIZERS

Plasticizers or dispersion agents are additives which are incorporated into concrete to improve its workability, reduce its water content needs, and enhance its strength development. Unfortunately, these properties are mutually exclusive. Thus, all plasticizers also cause retardation in the setting and in the early strength development of concrete. Superplasticizers are, instead, substances that in small amounts are able to strongly fluidify a cement mix *without* retarding its setting. Heavily sulphonated melamine–formaldehyde resins are the only other superplasticizers known in cement technology, other than tannins. This enhanced workability without any addition of extra water entails neither a loss of final strength, nor any gross retardation of concrete strength. Furthermore, no decrease in initial strength is observed. Their remarkable plasticizing action is demonstrated by slumps of 200 mm without increases in water content or by water reduction of up to 30 per cent [92].

The effect of a superplasticizer is due to its sulphonic groups being oriented towards water, but also adsorbing on the cement grain surface in sufficient numbers to form a monolayer around the grain. The combination of electrostatic repulsion and large ionic size brings about a rapid dispersion of the individual cement grains. In doing so, water trapped within the original flocks is released and can contribute to the mobility of the cement paste and, hence, to the workability of the concrete. Superplasticizers do not cause much reduction in the surface tension of water. The adsorption of the anions on the surface of the cement grain is also less tenacious than in the case of retarders and the course of the hydration reaction is not hindered at normal dosage levels. It follows that, for normal superplasticizers, there is no significant retardation of setting and hardening.

Polyflavonoid tannins have structures capable of complexing metallic ions such as Fe^{2+}/Fe^{3+} and aluminium ions through the ortho hydroxyl groups of the B-ring of the flavonoid units [11]. They can also be sulphonated, and often are, to improve their solubility in water, with the consequent opening of their etherocyclic pyran ring and the introduction of sulphonic groups at the C2 sites of some of the flavonoid units [11]. They also contain up to 20 per cent monomeric and polymeric carbohydrates. Notwithstanding the well-known retarding effect of carbohydrates on cement, this is not the case in the presence of the tannins in the tannin extract. These characteristics render polyflavonoid tannins an interesting material for use as dispersing/plasticizing agents for cements, which are materials mostly composed of calcium and iron silicates and aluminates.

Sulphonated mimosa, quebracho and pine tannin extracts all behave well as cement superplasticizers, with mimosa and pine being the better ones [93]. A dosage of 0.25–0.5 per cent in cement has a noticeable effect of fluidification. The tannin extract cement superplasticizing behaviour was ascribed to the balance of different effects, namely (1) their increase in molecular weight by self-condensation induced by the presence of silicate and aluminium components of cement [72, 73, 93]; (2) the decrease first and then the stabilization of the molecular weight and the improved solubility induced by the introduction of sulphonic groups in the tannin structure and (3) the stabilization of the molecular weight induced by the addition of urea, through its hindrance to tannin self-condensation and its decrease of the tannin extracts colloidal association in water.

8.11 MEDICAL/PHARMACEUTICAL APPLICATIONS

Tannins are well known to have an antimicrobial activity. This is logical as their capability to tan proteins means that they will complex irreversibly also with the proteins in bacterial membranes, thus inhibiting any activity they might have. It follows that, pharmaceuticals containing tannins and aimed at curing bacterial intestinal infections have been around already for some time. Some studies on their anticavity effectiveness have also been conducted [94]. Additionally, the use of tannins in other pharmaceutical medical applications, have been reported, particularly concerning their antitumour and anticancer activity [95–98]. More recently, work on their antiviral effectiveness has been conducted [99]. The data in Tables 4 to 11 present preliminary results obtained on the antiviral activity of 12 different flavonoid and hydrolysable tannins, obtained at the medical department of Leuven University [99]. These results evaluated both the effectiveness of 12 different tannins, as measured by the Minimum Inhibitory Concentration (MIC) required to reduce virus-induced cytopathogenicity by 50 per cent. The lower the MIC value, the better the compound as an antiviral substance. Equally important, the results in the tables report the Minimum Cytotoxic Concentration (MCC) required to cause a microscopically detectable alteration in the normal cell morphology. The higher the MCC, the less toxic is the compound to the patient's cells and the better it is as an antiviral substance. Thus, what is sought is the lowest possible MIC and the highest possible MCC. These results were obtained as *in vitro* screening tests, which implies that *in vivo* conformations are required. Nonetheless, their novelty and thoroughness justify their full report here. Different tannins have been shown to be very effective against different viruses, the nature of the different polyphenolic groups being the cause of this behaviour. Most likely, they tan the proteins and associate with the carbohydrates of the virus membrane, through an interaction similar to that associating them with hide proteins and carbohydrates to give leather.

Table 8.4

Tannin concentration required to protect CEM cells against the cytopathogenicity of HIV by 50% (MIC)

Anti-HIV-1 and -HIV-2 activity of the compounds in human T-lymphocyte (CEM) cells		
Compound	EC_{50} (μg/ml) HIV-1	EC_{50} (μg/ml) HIV-2
1. Mimosa tannin	6.0 ± 0.0	>20
2. Mimosa tanin intermediate [100]	5.0 ± 1.4	>20
3. Chestnut tannin	1.4 ± 0.5	>20
4. Tara + Chestnut mix	5.0 ± 1.4	>20
5. Quebracho standard	6.5 ± 0.7	>20
6. Quebracho highly purified	7.5 ± 0.7	>20
7. Quebracho highly sulphited	7.0 ± 1.4	>20
8. Pecan nut tannin	5.0 ± 1.4	>20
9. Cube Gambier	9.0 ± 1.4	>20
10. Radiata pine tannin	7.0 ± 1.4	>20
11. Maritime pine tannin	7.5 ± 0.7	>20
12. Sumach tannin	11.0 ± 1.4	>20
13. Spruce tannin	>100	>100

EC_{50} = effective concentration or concentration required to protect CEM cells against the cytopathogenicity of HIV by 50%

Table 8.5

Cytotoxicity and antiviral activity of compounds in HEL cell cultures, Herpes and vesicular stomatitis viruses. Tannins added prior to virus administration

Cytotoxicity and antiviral activity of compounds in E_6SM cell cultures						
Compound	Minimum cytotoxic concentration[a] (μg/ml)	Minimum inhibitory concentration[b] (μg/ml)				
		Herpes simplex virus-1 (KOS)	Herpes simplex virus-2 (G)	Vaccinia virus	Vesicular stomatitis virus	Herpes simplex virus^{-1} TK^{-} KOS ACV^{r}
1	200	40	16	16	>80	40
2	⩾40	40	16	16	>80	40
3	⩾40	40	16	16	>80	47
4	⩾40	40	48	16	>80	47
5	40	47	36	16	>80	47
6	40	80	36	16	>80	47
7	⩾40	40	40	16	>80	36
8	40	16	16	16	>80	36
9	8	>80	>80	>80	>80	>80
10	40	40	>80	16	>80	40
11	40	>80	>80	80	>80	47
12	40	36	36	16	>80	36
13	40	>16	>16	>16	>16	>16
Brivudin	>400	0.128	400	16	>400	>400
Ribavirin	>400	>400	>400	400	>400	>400
Acyclovir	>400	0.384	0.128	>400	>400	48
Ganciclovir	>100	0.0064	0.0064	100	>100	2.4

[a] Required to cause a microscopically detectable alteration of normal cell morphology.

[b] Required to reduce virus-induced cytopathogenicity by 50%.

Table 8.6

Cytotoxicity and antiviral activity of compounds in HEL cell cultures, vesicular stomatitis, Coxsackie and respiratory syncytial viruses

Cytotoxicity and antiviral activity of compounds in HeLa cell cultures				
Compound	Minimum cytotoxic concentration[a] (μg/ml)	Minimum inhibitory concentration[b] (μ/ml)		
		Vesicular stomatitis virus	Coxsackie virus B4	Respiratory syncytial virus
1	400	>80	>80	12 ± 5
2	400	>80	>80	12 ± 5
3	400	>80	>80	43 ± 4
4	400	43.4 ± 4	>80	35 ± 7
5	400	>80	>80	40 ± 0.2
6	400	>80	>80	40 ± 0.2
7	400	>80	>80	43 ± 5
8	400	>80	>80	9 ± 1
9	400	>80	>80	>80
10	400	>80	>80	40 ± 0.2
11	400	>80	>80	40 ± 0.2
12	80	>16	>16	>16
13	⩾80	>80	>80	>80
Brivudin	>400	>400	>400	>400
(S)-DHPA	>400	400	>400	>400
Ribavirin	>400	48	240	9.4

[a] Required to cause a microscopically detectable alteration of normal cell morphology.
[b] Required to reduce virus-induced cytopathogenicity by 50%.

Table 8.7

Inhibitory effects of tannins on the proliferation of murine leukemia cells (L1210/0), murine mammary carcinoma cells (FM3A) and human T-lymphocyte cells (Molt4/C8, CEM/0)

Compound	IC_{50} (μg/ml)[a]			
	L1210/0	FM3A/0	Molt4/C8	CEM/0
1	18 ± 0	153 ± 66	74 ± 18	58 ± 0
2	16 ± 1	148 ± 74	66 ± 27	61 ± 1
3	17 ± 0	141 ± 7	98 ± 22	65 ± 2
4	17 ± 0	114 ± 1	75 ± 57	56 ± 0
5	12 ± 4	76 ± 16	20 ± 1	51 ± 30
6	15 ± 2	79 ± 27	33 ± 21	45 ± 27
7	14 ± 2	82 ± 26	40 ± 27	55 ± 25
8	21 ± 6	⩾200	81 ± 7	66 ± 11
9	13 ± 4	80 ± 22	17 ± 2	18 ± 1
10	65 ± 4	⩾200	65 ± 28	71 ± 9
11	53 ± 23	⩾200	94 ± 1	111 ± 40
12	17 ± 0	18 ± 1	17 ± 0	18 ± 2
13	49 ± 16	>200	145 ± 78	83 ± 20

[a] 50% inhibitory concentration.

Table 8.8

Cytotoxicity and antiviral activity of compounds in HEL cell cultures, influenza viruses. Compounds added prior to virus administration

Cytotoxicity and antiviral activity of compounds in MDCK cell cultures				
Compound	Minimum cytotoxic concentration[a] (μg/ml)	EC_{50}		
		Influenza A H1N1	Influenza A H3N2	Influenza B
		MTS	MTS	MTS
1	100	3.3 ± 1.2	1.7 ± 1.3	2.3 ± 1.9
2	100	2.2 ± 0.1	1.7 ± 0.6	2.3 ± 1.5
3	100	4.0 ± 2.8	2.0 ± 0	1.4 ± 0.8
4	33.3	4.1 ± 2.8	2.2 ± 0.8	1.4 ± 0.9
5	100	1.7 ± 0.1	2.1 ± 0.4	3.5 ± 3.2
6	100	5.4 ± 3.8	3.7 ± 1.6	3.6 ± 3.0
7	100	4.4 ± 2.8	1.9 ± 0.4	3.4 ± 3.0
8	33.3	2.1 ± 0.1	3.0 ± 1.5	1.8 ± 1.1
9	100	5.5 ± 4.6	4.4 ± 3.6	2.7 ± 2.6
10	100	4.2 ± 3.1	2.7 ± 1.0	2.7 ± 2.7
11	100	2.9 ± 1.2	2.2 ± 0.3	1.5 ± 1.1
12	20	2.0 ± 1.6	0.9 ± 0.2	2.6 ± 1.9
13	100	9.9 ± 5.7	9.5 ± 4.8	1.9 ± 1.9
Oseltamivir carboxylate (μM)	>100	0.05	0.65	10.65
Ribavirin (μM)	60	4.55	6.32	9.07
Amantadin (μM)	>100	21.39	0.78	>100
Rimantadin (μM)	>100	18.45	0.05	>100

[a] Required to cause a microscopically detectable alteration of normal cell morphology.

Table 8.9

Cytotoxicity and antiviral activity of compounds in HEL cell cultures, Corona viruses

	Feline Corona virus (FIPV)		Human Corona (SARS) virus	
	EC50 (μg/ml)	CC50 (μg/ml)	EC50 (μg/ml)	CC50 (μg/ml)
1	52 ± 19	>100	>100	>100
2	67 ± 47	>100	>100	>100
3	49 ± 17	>100	>100	>100
4	43 ± 2	>100	>100	>100
5	49 ± 10	⩾100	44 ± 10	>100
6	55 ± 19	>100	49 ± 21	>100
7	32 ± 1	>100	40 ± 1	>100
8	72 ± 40	>100	>100	>100
9	⩾100	>100	>100	>100
10	20 ± 21	>100	>100	>100
11	44 ± 5	>100	56 ± 13	>100
12	7.8 ± 8.0	81 ± 13	>100	>100
13	63 ± 32	>100	>100	>100

Table 8.10

Cytotoxicity and antiviral activity of compounds in HEL cell cultures, Herpes and Vaccinia viruses. Tannins added after virus administration

Cytotoxicity and antiviral activity of compounds in HEL cell cultures						
Compound	Minimum cytotoxic concentration[a] (μg/ml)	Minimum inhibitory concentration[b] (μg/ml)				
		Herpes simplex virus-1 (KOS)	Herpes simplex virus-2 (G)	Vaccinia virus	Vesicular stomatitis virus	Herpes simplex virus^{-1} TK^{-} KOS ACV^{r}
1	200	6 ± 2	1.8 ± 0.2	>40	>40	6 ± 2
2	200	4 ± 0	2 ± 1	8 ± 0	>40	4 ± 0
3	200	15 ± 1	3 ± 1	24 ± 1	>40	17 ± 3
4	200	15 ± 1	8 ± 0	24 ± 1	>40	20 ± 2
5	⩾40	>40	>40	20 ± 2	>40	>40
6	⩾40	>40	8 ± 0	>40	>40	>40
7	⩾40	8 ± 0	>40	20 ± 2	>40	>40
8	40	4 ± 0	4 ± 0	8 ± 0	>8	4 ± 0
9	⩾40	>40	>40	>40	>40	>40
10	40	8 ± 0	4 ± 0	8 ± 0	>8	>8
11	40	>8	8 ± 0	>8	>8	>8
12	40	>8	8 ± 0	>8	>8	>8
13	200	>40	>40	>40	>40	>40
Brivudin (μM)[c]	>250	0.016	10	6	>250	50
Ribavirin (μ)[c]	>250	250	50	30	50	250
Acyclovir (μM)[c]	>250	0.08	0.08	>250	>250	50
Ganciclovir (μM)[c]	>100	0.0064	0.032	>100	>100	12
Compounds added prior to virus administration						

[a] Required to cause a microscopically detectable alteration of normal cell morphology.
[b] Required to reduce virus-induced cytopathogenicity by 50%.
[c] Controls.

Table 8.11

Cytotoxicity and antiviral activity of compounds in Vero cell cultures, influenza viruses

Cytotoxicity and antiviral activity of compounds in Vero cell cultures						
Compound	Minimum cytotoxic concentration[a] (μg/ml)	Minimum inhibitory concentration[b] (μg/ml)				
		Para-influenza-3 virus	Reovirus-1	Sindbis virus	Coxsackie virus B4	Punta Toro virus
1	400	>80	>80	>80	>80	>80
2	>400	>400	>400	149.5 ± 24	149.5 ± 24	169.5 ± 13
3	400	>80	>80	>80	>80	>80
4	400	>80	>80	>80	>80	>80
5	400	>80	>80	>80	>80	>80
6	400	>80	>80	>80	>80	>80
7	80	>40	>40	>40	>40	>40
8	80	>40	>40	>40	>40	>40
9	400	>80	>80	>80	>80	>80
10	400	>80	>80	>80	>80	>80
11	400	>80	>80	>80	>80	>80
12	400	>80	>80	80	>80	>80
13	80	>16	>16	>16	>16	>16
Brivudin[c]	>400	>400	>400	>400	>400	>400
(S)-DHPA[c]	>400	>400	>400	>400	>400	>400
Ribavirin[c]	>400	240	80	240	>400	240

[a] Required to cause a microscopically detectable alteration of normal cell morphology.
[b] Required to reduce virus-induced cytopathogenicity by 50%.
[c] Controls.

REFERENCES

1. Calleri L., Le Fabbriche Italiane de Estratto di Castagno, Silva, S.Michele Mondovi (CN), Italy 1989.
2. Japanese Standards Association (JIS). Particleboard. JIS A 5908, Tokyo, Japan 1994.
3. CARB – California Air Resources Board, Composite Wood Products Public Workshop, 20 June, 2006.
4. Fengel, D., Wegener G., Wood, Chemistry, Ultrastructure, Reactions, Walther DeGruyter, Berlin, 1984.
5. Tang H.R., Hancock R.A., Covington A.D, Studies on commercial tannin extracts, *XXI IULTCS (International Union of Leather Trades Chemists), Proceedings*, Barcelona, Spain, September 1991, pp. 1503–1527.
6. Haslam E., *The Chemistry of Vegetable Tannins*, Academic Press, London, 1966.
7. Roux D.G., Reflections on the chemistry and affinities of the major commercial condensed tannins, in *Plant Polyphenols 1* (Eds. Hemingway R.W. and Laks P.E.), Plenum, New York, 1992, pp. 27–39.
8. Pasch H., Pizzi A., *J. Appl. Polym. Sci.*, **85**(2), 2002, 429–437.
9. Kulvik E., *Adhes. Age*, **18**(3), 1975.
10. Kulvik E., *Adhes. Age*, **19**(3), 1976.
11. Pizzi A., Wood adhesives–Chemistry and technology, Volume I, Ed. Pizzi A.,) Marcel Dekker, New York, 1983, Chapter4; *J. Macromol. Sci. Chem.* Ed. **C18**(2), 1980, 247.
12. Pasch H., Pizzi A., Rode K., *Polymer*, **42**(18), 2001, 7531–7539.
13. Pizzi A., *Advanced Wood Adhesives Technology*, Dekker, New York, 1994.
14. Roux D.G., *Modern Applications of Mimosa Extract*, Leather Industries Research Institute, Grahamst6wn, South Africa, 1965. pp. 34–41
15. King H.G.C., White T., *J. Soc. Leather Traders' Chem.*, **41**, 1957, 368.
16. Roux D.G., Paulus E., *Biochem. J.*, **78**, 1961, 785.
17. King H.G.C., White T., Huges R.B., *J. Chem. Soc.*, 1961, 3234.
18. Roux D.G., Ferreira D., Hundt H.K.L., Malan E., *Appl. Polymer Symp.*, **28**, 1975, 335.
19. Clark-Lewis J.W., Roux D.G., *J. Chem. Soc.*, 1959, 1402.
20. Hemingway R.W., McGraw G.W., *Appl. Polymer Symp.*, **28**, 1976.
21. Rossouw D.duT., Pizzi A., McGillivray G., *J. Polymer Sci. Chem. Ed.*, **18**, 1990, 3323.
22. Porter L.J., *New Zeal. J. Sci.*, **17**, 1974, 213.
23. Roux D.G., *J. Soc. Leather Trades' Chem.*, **35**, 1951, 322.
24. Stiasny E., Orth F., *Collegium*, **24**(50), 1924, 88.
25. Wissing A., *Svensk Papperstid*, **20**, 1955, 745.
26. A.Lemme, US Patent 3,232,80 (1966).
27. Roux D.G., *LIRI report*, Leather Industries Research Institute, Grahamstown, South Africa, 1971.
28. Tanac, private comunication, Montenegro, Brazil (1999).
29. Simon C., Pizzi A., *J.Appl.Polymer Sci.*, **88**(8), 2003, 1889–1903.
30. Simon C., Pizzi A., *JALCA-J. Am. Leather Chem. Assoc.*, **98**(3), 2003, 83–96.
31. International Norm, International Leather Union I.U.C./18, Determination of the leather content of Cr^{VI}, 1995.
32. European Union Directives 91/271/Cee and 91/676/Cee on water effluents, 1999.
33. Noferi M., Masson E., Merlin A., Pizzi A., Deglise X., *J. Appl. Polymer Sci.*, **63**, 1997, 475–482.
34. Marutzky R., Formaldehyde emission, in Wood Adhesives Chemistry and Technology, Dekker, New York, 1989, Volume 2.
35. Pizzi A., *Advanced Wood Adhesives Technology*, Dekker, New York, 1994.
36. Pizzi A., Scharfetter H.O., *J. Appl. Polymer Sci.*, **22**, 1978, 1945.
37. Scharfetter O., Pizzi A., Rossouw D.duT., *IUFRO Conference on Wood Gluing*, Merida, Venezuela, October 1977, 16.
38. Roux D.G., Ferreira D., Hundt H.K.L., Malan E., *Appl. Polymer Symp.*, **28**, 1975, 335.
39. K.F. Plomley, Paper 39, Division of Australian Forest Products Technology, 1966.
40. Megson N.J.L., *Phenolic Resins Chemistry*, Butterworth, Sevenoaks, Kent, England, 1958.
41. Pizzi A., Scharfetter H.O., *CSIR Special Report HOUT 138*, Pretoria, South Africa, 1977.
42. Pizzi A., Stephanou A., *J. Appl. Polymer Sci.*, **51**, 1994, 2109.
43. Garnier S., Pizzi A., Vorster O.C., Halasz L., *J. Appl. Polymer Sci.*, **86**(4), 2002, 852–863. 86(4), 2002, 864–871.
44. Rossouw D.duT., Pizzi A., McGillivray G., *J. Polymer Sci. Chem. Ed.*, **18**, 1990, 3323.
45. Brown R., Cummings W., *J. Chem. Soc.*, 1959, 4302.
46. Freudenberg K., Alonso de Larna J.M., *Annalen*, **612**, 1958, 78.
47. Brown R., Cummings W., Newbould J., *J. Chem. Soc.*, 1961, 3677.
48. Pizzi A., *Colloid Polymer Sci.*, **257**, 1979, 37.
49. Dalton L.K., *Aust. J. Appl. Sci.*, **1**, 1950, 54.
50. Dalton L.K., *Aust. J. Appl. Sci.*, **4**, 1953, 54.
51. Parrish J.R., *J. S. Afr. Forest Ass.*, **32**, 1958, 26.

52. Pizzi A., Stephanou A., *J. Appl. Polymer Sci.*, **51**, 1994, 2125.
53. Merlin A., Pizzi A., *J. Appl. Polymer Sci.*, **59**, 1996, 945.
54. Masson E., Merlin A., Pizzi A., *J. Appl. Polymer Sci.*, **60**, 1996, 263.
55. Masson E., Pizzi A., Merlin A., *J. Appl. Polymer Sci.*, **60**, 1996, 1655.
56. Masson E., Pizzi A., Merlin A., *J. Appl. Polymer Sci.*, **64**, 1997, 243.
57. Garcia R., Pizzi A., Merlin A., *J. Appl. Polymer Sci.*, **65**, 1997, 2623.
58. Pizzi A., *Forest Products J.*, **28**(12), 1978, 42.
59. Pizzi A., Stephanou A., *J. Appl. Polymer Sci.*, **50**, 1993, 2105.
60. Garnier S., Pizzi A., Vorster O.C., Halasz L., *J. Appl. Polymer Sci.*, **81**(7), 2001, 1634–1642.
61. Specification for particleboard, Deutschen Normenausschuss V100 and V313, DIN 68761, Part 3, 1967.
62. Pizzi A., *Adhesives Age*, **21**(9), 1978, 32.
63. Pizzi A., *Holzforschung Holzverwertung*, **31**(4), 1979, 85.
64. Pizzi A., Stephanou A., *Holzforschung*, **48**(2), 1994, 150.
65. Zhao C., Pizzi A., Garnier S., *J. Appl. Polymer Sci.*, **74**, 1999, 359.
66. Pizzi A., Valenzuela J., Westermeyer C., *Holz Roh Werkstoff*, **52**, 1994, 311.
67. Pichelin F., Pizzi A., Frühwald A., Triboulot P., *Holz Roh Werkstoff*, **59**(4), 2001, 256–265.
68. Pizzi A., *J. Macromol. Sci. Chem. Ed.* **A16**(7), 1981, 1243; Holz Roh Werkstoff 40, 1982, 293.
69. Pizzi A., Von Leyser E.P., Valenzucla J., *Holzforschung*, **47**, 1993, 69.
70. Pizzi A., *J. Appl. Polymer Sci.*, **25**, 1980, 2123.
71. Pizzi A., Walton T., *Holzforschung*, **46**, 1992, 541.
72. Trosa A., Pizzi A., *Holz Roh Werkstoff*, **59**(4), 2001, 266–271.
73. Trosa, A., Doctoral Thesis, University Henri Poincaré – Nancy 1, Nancy, France 1999.
74. Pichelin F., Kamoun C., Pizzi A., *Holz Roh Werkstoff*, **57**(5), 1999, 305.
75. McKenzie A.E., Yuritta Y.P., *Appita*, **26**, 1974.
76. Custers P.A.J.L., Rushbrook R., Pizzi A., Knauff C.J., *Holzforschung Holzverwertung*, **31**(6), 1979, 131.
77. Pizzi A., Roux D.G., *J. Appl. Polymer Sci.*, **22**, 1978, 1945.
78. Pizzi A., Roux D.G., *J. Appl. Polymer Sci.*, **22**, 1978, 2717.
79. Pizzi A., *J. Appl. Polymer Sci.*, **23**, 1979, 2999.
80. Specification for synthetic adhesive resins for wood, BritishStandard BS 1204, Parts 1 and 2, 1965.
81. Pizzi A., Rossouw D.duT., Knuffel W., Singmin M., *Holzforschung Holzverwertung*, **32**(6), 1980, 140.
82. Pizzi A., Cameron F.A., *Forest Prod. J.*, **34**(9), 1984, 61.
83. Chung K.H., Hamed Q.R., in *Chemistry and Significance of Condensed Tannins,* (Eds. Hemingway R.W. and Karchesy J.J.), Plenum Press, New York, 1989.
84. Pizzi A., Mtsweni B., Parsons W., *J. Appl. Polymer Sci.*, **52**, 1994, 1847.
85. Meikleham N., Pizzi A., Stephanou A., *J. Appl. Polymer Sci.*, **54**, 1994, 1827.
86. Pizzi A., Stephanou A., *Holzforschung Holzverwertung*, **45**(2), 1993, 30.
87. Pizzi A., Meikleham N., *J. Appl. Polymer Sci.*, **55**, 1995, 1265.
88. Pizzi A., Meikleham N., Stephanou A., *J. Appl. Polymer Sci.*, **55**, 1995, 929.
89. Pizzi A., Meikleham N., Dombo B., Roll W., *Holz Roh Werkstoff*, **53**, 1995, 201.
90. Garcia R., Pizzi A., *J. Appl. Polymer Sci.*, **70**(6), 1998, 1083.
91. Garcia R., Pizzi A., *J. Appl. Polymer Sci.*, **70**(6), 1998, 1093.
92. Khalil S.M., *Cement Concrete Res.*, **3**, 1973, 677.
93. Kaspar H.R.E., Pizzi A., *J. Appl. Polymer Sci.*, **59**, 1996, 1181–1190.
94. Mitsunaga, T. in *Plant Polyphenols 2* Eds. Gross G.G., Hemingway R.W., Yoshida T., Kluwer Academic/Plenum Publishers, New York, pp. 555–574.
95. Yang L.-L., Wang C.-C., Yen K.-Y., Yoshida T., Hatano T., Okuda T. in *Plant Polyphenols* 2 Eds. Gross G.G., Hemingway R.W., Yoshida T., Kluwer Academic/Plenum Publishers, New York, pp. 615–628.
96. Nakamura Y., Matsuda M., Honma T., Tomita I., Shibata N. Warashina N., Noro T., Hara Y., in *Plant Polyphenols 2* Eds. Gross G.G., Hemingway R.W., Yoshida T., Kluwer Academic/Plenum Publishers, New York, pp. 629–642.
97. Miyamoto K., Murayama, T., Hatano T.,Yoshida T.,Okuda T., in *Plant Polyphenols 2* Eds. .Gross G.G., Hemingway R.W.,Yoshida T., Kluwer Academic/Plenum Publishers, New York, pp. 643–664.
98. Noro T., Ohki T., Noda Y., Warashina T., Noro K., Tomita I., Nakamura Y., in *Plant Polyphenols 2 –* Eds. Gross G.G., Hemingway R.W., Yoshida T., Kluwer Academic/Plenum Publishers, New York, pp. 665–674.
99. Balzarini J., Persoons L., Absillis A., Van Berckelaer L., Pizzi A., Rijk Universiteit Leuven, Belgium and University of Nancy 1, France, unpublished results 2006.
100. Pizzi A., *Forest Prod. J.*, **28**(12), 1978, 42–47.
101. Kaspar H.R.E., Pizzi A., *J. Appl. Polymer Sci.*, **59**, 1996, 1181–1190.

– 9 –

Lignins: Major Sources, Structure and Properties

Göran Gellerstedt and Gunnar Henriksson

ABSTRACT

Lignin is one of the most predominant biopolymers present in plants. Together with cellulose and hemicelluloses, lignin builds up the cell wall in an arrangement which is regulated on the nano-scale and results in lignin–carbohydrate network structures. The molecular complexity of lignin renders all isolation and identification processes difficult and, consequently, many structural questions still remain. In this chapter, our present knowledge about the formation of lignin in plants, its presence in different types of plants as well as several different approaches taken to reveal the chemical structure, is summarized. Furthermore, a brief discussion about the chemical changes introduced in lignin as the result of different types of delignification processes, such as kraft and sulphite pulping and steam explosion, is included.

Keywords

Analytical methods, Annual plants, Cellular structures, Dehydropolymerizate (DHP), Hardwoods, Kraft lignins, Lignification, Lignins, Lignosulphonates, Milled wood lignins (MWL), Softwoods, Steam explosion lignins

9.1 INTRODUCTION

Plants are eukaryotic organisms with an ability to utilize light for the fixation of carbon dioxide, having cell walls rich in cellulose and with an ability to use starch as nutrition storage inside the cells. In the evolution of plants, lignin (the term lignin is derived from the Latin word for wood, *lignum*) was introduced some 440 million years ago when the group of vascular plants started to develop. Thereby, a new cell type, the tracheid, was formed typified by its elongated feature and its thick hydrophobic cell wall. In addition to cellulose and lignin, all vascular plants also contain other polysaccharides, hemicelluloses and low molecular mass compounds, extractives. The presence of tracheids permitted an efficient transport of water in the plant and provided the strength necessary for the development of larger species. Simultaneously, the development of root systems took place, thus also allowing plants to grow in dryer environments.

The first vascular plants to develop were vascular cryptogams reproduced using spores and, today, such plants can still be found as the herb families of club mosses, ferns, horsetails and in fern trees. In the further evolution of plants, seed fertilization developed some 360 million years ago and paved the way for an expansion of novel types of trees which successively overruled the vascular cryptogam trees due to a more efficient reproduction system and a greater ability to colonize dryer areas of land. The first trees to develop all had naked seeds often organized in cones and are referred to as gymnosperms. About 150 million years ago, a dramatic new development occurred with plants-carrying flowers, the fertilization being done by insects and with seeds developed inside a fruit body. Such plants, the angiosperms, rapidly took over in importance and, today, more than 90 per cent of all land-living plant species are of this type. Simultaneously, the angiosperms developed more advanced types of leafs and the

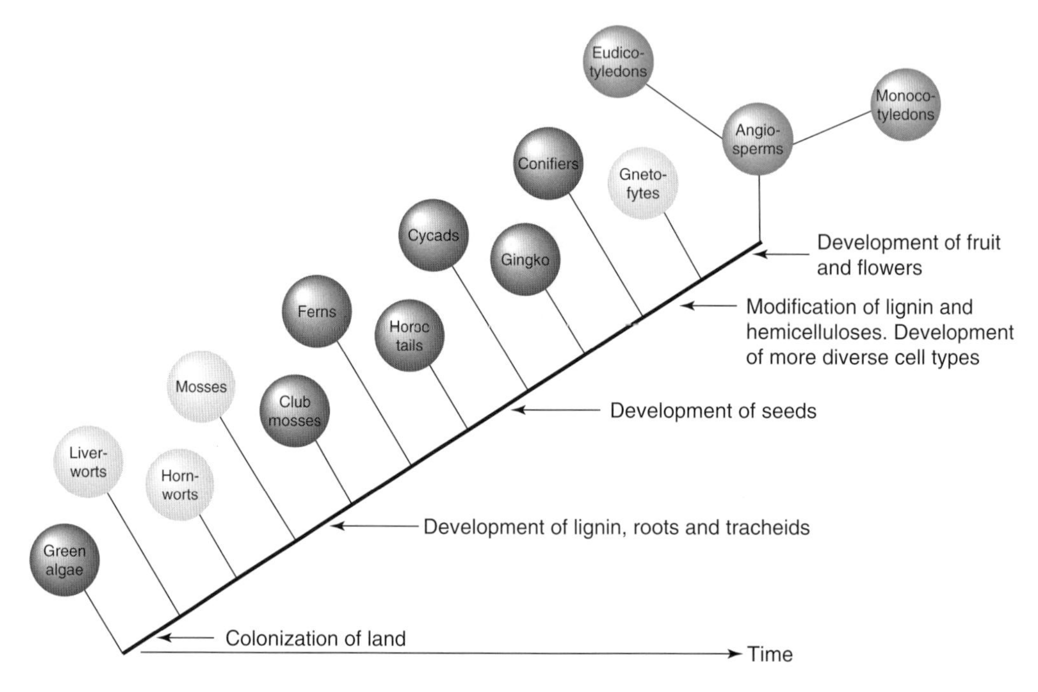

Figure 9.1 A schematic view of the evolution of plants.

cells became more specialized with vessel cells taking care of the water and nutrition transport. The structure of lignin and hemicelluloses was modified in comparison to the gymnosperms. Usually, the angiosperms are divided into two classes, *viz.* the monocotyledons, which include species such as grasses, bamboo and palm trees, and the eudicotyledons represented among others by leaf-carrying trees like birch, aspen and eucalyptus. The woody plants of the latter group are referred to as hardwoods. A simplified schematic representation of the evolution of plants is given in Fig. 9.1.

9.2 NOMENCLATURE OF LIGNIN

Lignins are formed by polymerization of cinnamyl alcohols (monolignols) which differ in structure depending on plant type. In coniferous wood, the lignin is built up almost exclusively by coniferyl alcohol (G-units) with minor amounts of coumaryl alcohol (H-units) present. The latter is, however, a major constituent in compression wood lignin. In hardwoods, on the other hand, both coniferyl alcohol and sinapyl alcohol (S-units) are used as building blocks and in monocotyledonous tissue, all three alcohols are used as lignin precursors [1–3]. Figure 9.2 shows the numbering system and the structure of the three major monolignols, together with some minor monolignols found as lignin end-groups or present in specific plants [4, 5]. Table 9.1 shows the degree of participation of the major monolignols in different types of plants.

9.3 BIOSYNTHESIS OF MONOLIGNOLS AND THE FORMATION OF LIGNIN

The three major monolignols (Fig. 9.2) used to synthesize lignin polymers are formed in the cytoplasm via the shikimate pathway which produces phenylalanine as key intermediate [7–9]. Through further enzyme mediated deamination, hydroxylation, reduction and methylation reactions, the final lignin precursors are formed as depicted in Fig. 9.3.

The further reactions of monolignols in the plant cell wall to form lignin may occur through an initial laccase or peroxidase oxidation of the monolignol giving rise to a resonance-stabilized phenoxy radical as depicted in Fig. 9.4 [10–12]. This step is followed by a coupling reaction and model experiments have shown that the first product, the

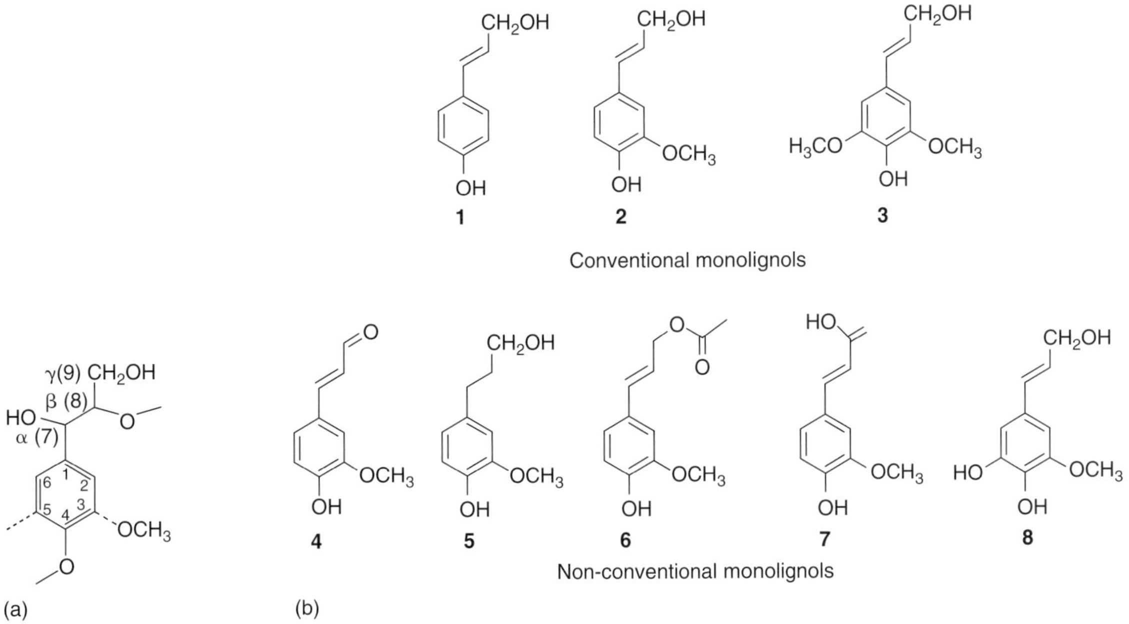

Figure 9.2 (a) Numbering system in monolignols. (b) Types of monolignols found as building blocks in lignin. 1 = *p*-coumaryl alcohol (H-unit), 2 = coniferyl alcohol (G-unit), 3 = sinapyl alcohol (S-unit), 4 = coniferaldehyde, 5 = dihydroconiferyl alcohol, 6 = coniferyl alcohol-9-acetate, 7 = ferulic acid, 8 = 5-hydroxyconiferyl alcohol.

Table 9.1

The participation of different monolignols in lignin from various plants

Plant type	*p*-Coumaryl alcohol	Coniferyl alcohol	Sinapyl alcohol
		(%)	
Coniferous; softwoods	<5[a]	>95	0[b]
Eudicotyledonous; hardwoods	0–8	25–50	45–75
Monocotyledonous; grasses	5–35	35–80	20–55

[a] Higher amount in compression wood.
[b] Some exceptions exist [6].

dimer, will be formed by involvement of the β-carbon in one unit and either the phenolic hydroxyl, the aromatic C5 (only in G- or H-units) or the β-carbon in the other [10–12]. It should be noted that the theoretically possible coupling products between two C5 centred radicals or between a phenoxy and a C5 radical are not formed to any noticeable extent. The stepwise further growth of the polymer is thought to involve an end-wise addition of a new monolignol radical to the growing polymer chain. In the latter, however, only the phenolic hydroxyl, the aromatic C5 or C1 positions are available for coupling. Thus, the overall frequency of inter-unit linkages involving the C–β carbon can be expected to be completely dominant in any lignin. Coupling in the aromatic C1 position takes place to a low extent and predominantly results in the formation of a spirodienone structure with a secondary lignin fragmentation reaction as an alternative [13]. Other possible modes of coupling may occur if two growing polymer chains are in close proximity, by formation of C5/C5 (in G-units) or phenolic hydroxyl/C5 linkages. Thereby, branching or crosslinking points in the lignin polymer are created. Branching points in the growing lignin polymer are also created by the formation of dibenzodioxocin structures. These are prevalent in guaiacyl lignins and formed by internal trapping after coupling of a C–β radical with a 5-5′ phenoxy radical [14, 15]. The various types of inter-unit linkages and their abbreviated nomenclature are depicted in Fig. 9.5.

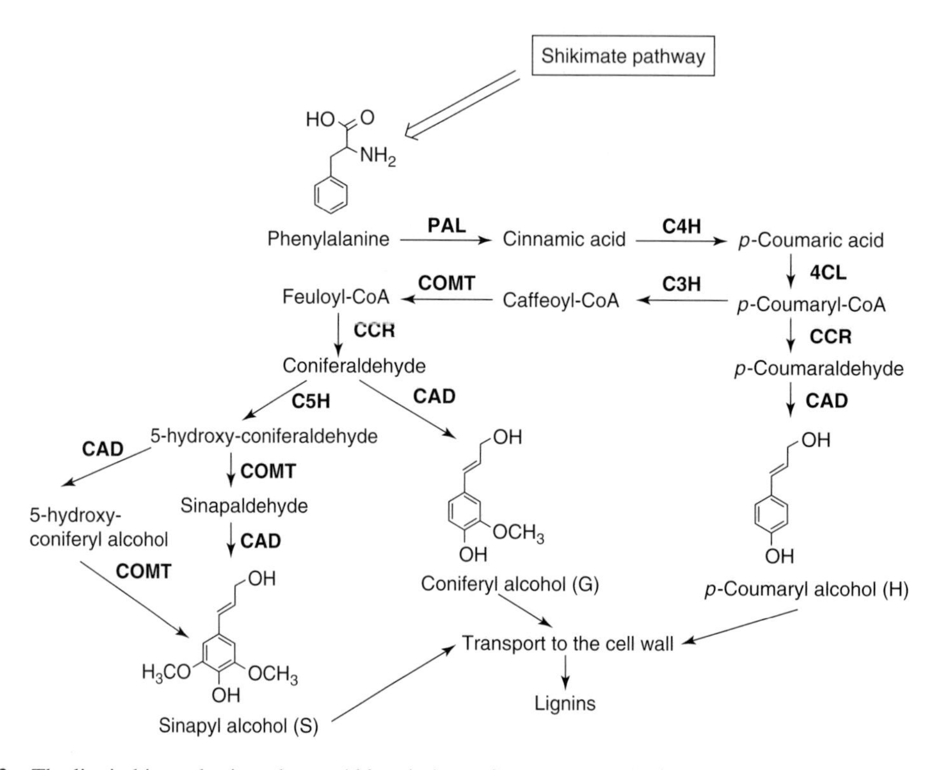

Figure 9.3 The lignin biosynthesis pathway. Abbreviations of enzymes are: PAL = phenylalanine–ammonia-lyase; C4H = cinnamate 4-hydroxylase; 4CL = hydroxycinnamate:CoA-ligase; C3H = 4-hydroxycinnamate 3-hydroxylase; COMT = S-adenosyl-methionine:caffeate/5-hydroxyferulate-O-methyltransferase; CAD = hydroxycinnamyl alcohol dehydrogenase; C5H = coniferaldehyde-5-hydroxylase.

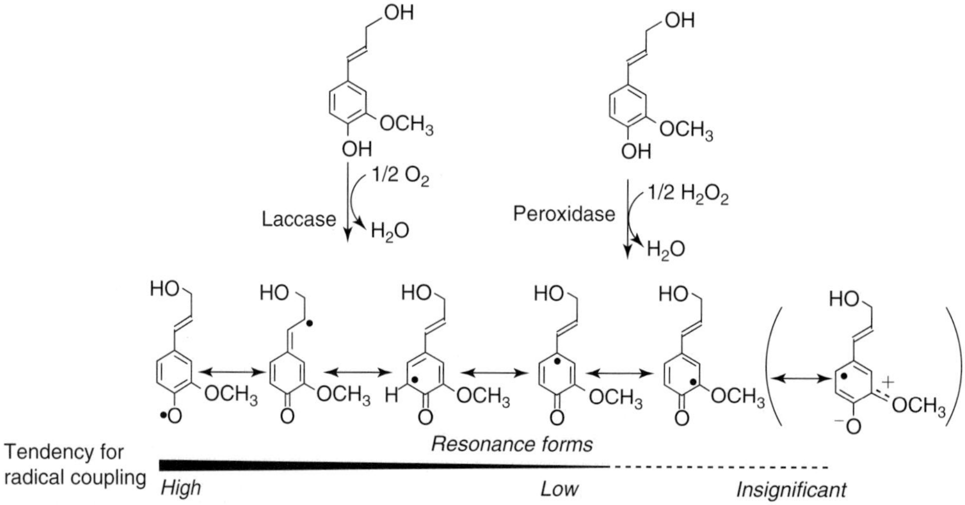

Figure 9.4 Enzymatic formation of resonance-stabilized monolignol radicals. The relative reactivity in coupling reactions is indicated by the intensity of the line.

Recently, it has been suggested that lignin is not built up by a combinatorial polymerization of monolignols, but has a more ordered structure with a repeating unit larger than that given by the phenylpropane units [16, 17]. Thus, by a template polymerization involving cell wall proteins, oligomeric lignin fragments should be synthesized and added together in a controlled manner. Until now, however, there has been no experimental data to support this theory.

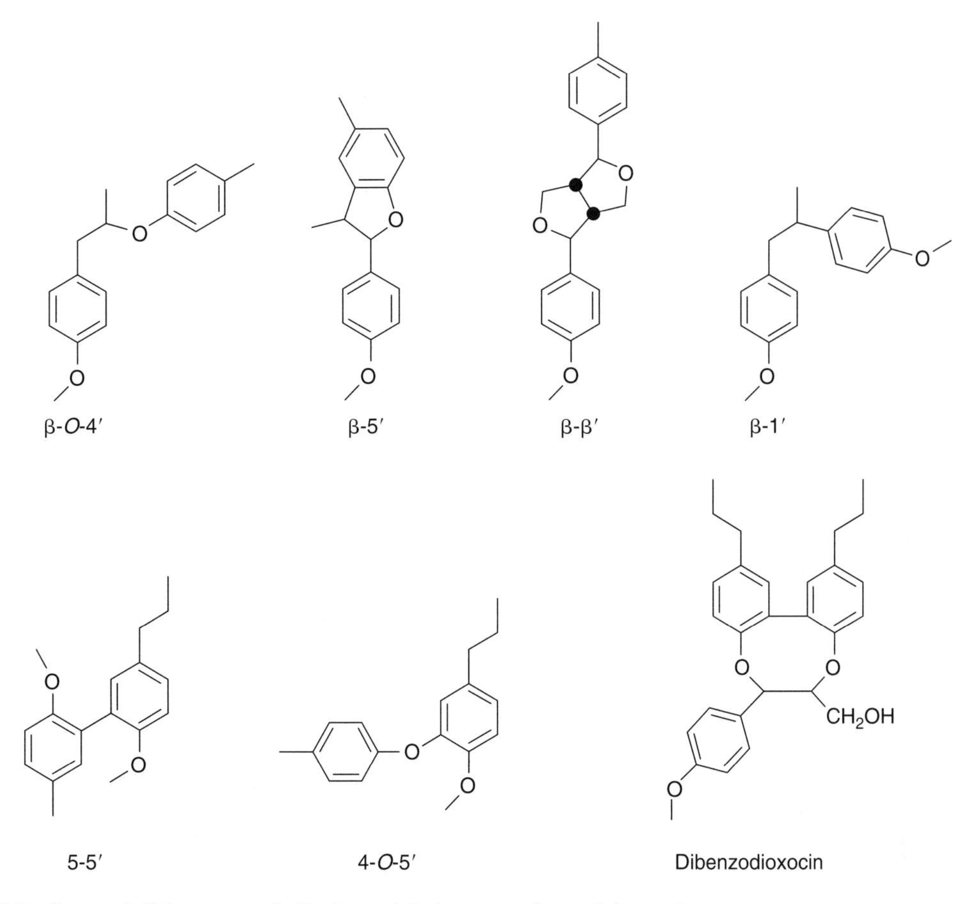

Figure 9.5 Inter-unit linkage types in lignins and their commonly used denotation.

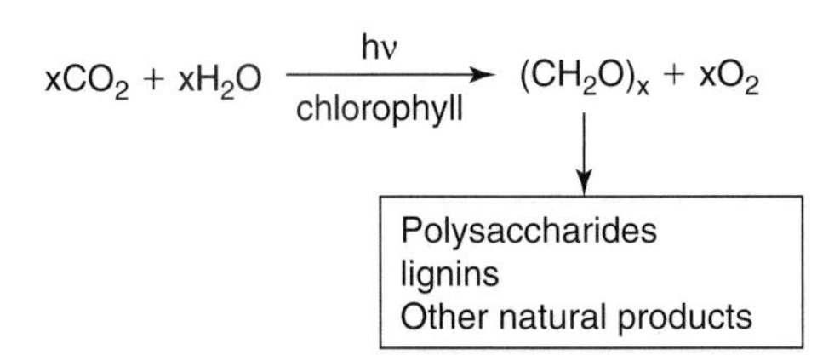

Figure 9.6 The fixation of carbon dioxide in nature through the photosynthesis reaction.

9.4 MAJOR SOURCES OF LIGNIN

Lignin is present in all vascular plants making it second to cellulose in abundance among polymers in nature. Since lignin, like many other biomass components, is formed via the photosynthesis reaction (Fig. 9.6), it is renewable and it has been estimated that the annual production of lignin on earth is in the range of $5–36 \times 10^8$ tons.

In woody plants from the gymnosperm and angiosperm phylum, the lignin content is in the order of 15–40 per cent [18] whereas in herbs, the lignin content is less than 15 per cent [19]. Low lignin content is usually also encountered in annual plants. In Table 9.2, some representative values for the content of lignin in various types of commercially important plants are given.

Many softwood and hardwood species, together with certain types of annual plants, have commercial interest as a source of cellulose fibres for the production of paper and board products. Thus, in technical fibre liberation processes, such as alkaline or sulphite pulping, huge quantities of lignin are dissolved as alkali lignin and lignosulphonates, respectively. With few exceptions (see Chapter 10), these lignins are, however, never isolated, but burnt

Table 9.2

Lignin content in various types of plants

Plant Scientific/Common name	Lignin content	Reference
Gymnosperms		
Picea abies, Norway spruce	28	[20]
Picea abies, Norway spruce (compression wood)	39	[21]
Pinus radiata, Monterey pine	27	[22]
Pinus sylvestris, Scots pine	28	[22]
Pseudotsuga menziesii, Douglas fir	29	[22]
Tsuga canadensis, Eastern hemlock	31	[22]
Angiosperms – Eudicotyledons		
Acasia mollissima, Black wattle	21	[22]
Betula verrucosa, Siver birch	20	[23]
Eucalyptus globulus, Blue gum eucalyptus	22	[22]
Eucalyptus grandis, Rose eucalyptus	25	[24]
Populus tremula, European aspen	19	[23]
Corchorus capsularis, Jute	13	[25]
Hibiscus cannabinus, Kenaf	12	[25]
Linun usitatissiumum, Flax	2.9	[25]
Angiosperms – Monocotyledons		
Oryza species, Rice straw	6.1	[26]
Saccharum species, Bagasse	14	[26]

together with other wood constituents liberated in the pulping liquor in order to produce the steam required for the process. In all commercial pulping processes, as well as in emerging processes, such as wood hydrolysis for the production of biofuels, the lignin is structurally altered in comparison to the native lignin. Broadly, these types of lignin can be described as being heterogeneous polyphenols with molecular masses in the range of 100–300 000, often with a high degree of polydispersity (see Section 9.6).

In growing plants, on the other hand, the lignin constitutes an integral part of the cell walls with chemical linkages to all types of polysaccharide constituents present. For spruce wood, it has been shown that the major portion of lignin is covalently linked to the hemicelluloses (*i.e.* xylan and glucomannan) with a minor amount being linked to cellulose [27]. The concentration of lignin is, however, not evenly distributed throughout the cell wall and, despite a high concentration of lignin in the middle lamella, the predominant portion is located in the S2 layer of the secondary wall due to its large relative volume [28]. A detailed analysis of the lignin distribution in two wood species, one softwood and one hardwood, has been published and is shown in Table 9.3 [29].

All attempts to isolate lignin from wood or other types of biomass must be preceded by some mechanical disintegration of the material. Usually, intensive milling of the material is employed whereby the structural integrity (*i.e.* cell types), cell layers and any inhomogeneities at the macromolecular level, is eliminated and, from such materials, only an average lignin structure can be obtained. Despite these drawbacks, almost all present knowledge about the structure of lignins is based on comprehensive milling of the plant material, followed by solvent extraction with dioxane and (sometimes) further purification to yield low to moderate yields of lignin. In most isolation procedures, the lignin contains minor impurities of carbohydrates.

9.4.1 Milled wood lignin

In 1956, the first description of the isolation of lignin from spruce wood according to these principles was published and the lignin was denoted as milled wood lignin (MWL) [30]. By employing extremely long extraction times (sequential extraction with dioxane–water, 96:4, for ~6 weeks), around 50 per cent of the lignin could be isolated. The author concluded that 'MWL is a very useful material for lignin chemists'. In a subsequent paper, several other softwood and hardwood species were used to produce the corresponding MWLs and the similarities

Table 9.3

Distribution of lignin in various cell wall layers of softwood tracheids and hardwood fibres

Wood cell	Cell wall layer	Tissue volume (%)	Lignin (% of total)	Lignin concentration (%)
Loblolly pine tracheids (softwood)				
Early wood	S1	13	12	25
	S2	60	44	20
	S3	9	9	28
	ML + P	12	21	49
	CC	6	14	64
Late wood	S1	6	6	23
	S2	80	63	18
	S3	5	6	25
	ML + P	6	14	51
	CC	3	11	78
White birch fibres (hardwood)				
	S	73	60	19
	ML + P	5	9	40
	CC	2	9	85

between various softwood lignins and the differences between hardwood lignins could be clearly demonstrated [31]. Furthermore, it was shown that a second solvent extraction could be used to provide a lignin material still containing substantial amounts of carbohydrates and this was denoted lignin carbohydrate complex (LCC) [32].

As an alternative to the isolation of lignin by solvent extraction, liquid–liquid partition of dissolved wood meal between aqueous sodium thiocyanate and benzyl alcohol has been suggested [33]. In this rather complicated procedure, it was possible to isolate about 40 per cent of the lignin in spruce wood. The lignin still contained some 10 per cent of carbohydrates. In a further alternative to the original method, the milled wood was treated with an enzyme cocktail containing cellulolytic and hemicellulolytic activities, thereby facilitating the subsequent solvent extraction. Fractions of lignin, having different amounts of remaining carbohydrates, could be isolated with a total yield of lignin of 57 per cent from spruce and 68 per cent from sweetgum respectively [34, 35]. An even higher total yield of lignin was obtained by first extracting milled spruce wood with water, followed by extraction with dioxane. After enzymatic hydrolysis of the polysaccharides present in the residue, a second extraction with dioxane was carried out affording a total yield of 84 per cent of the lignin [36]. Enzymatic hydrolysis of extensively milled wood has also been combined with acid hydrolysis employing 0.01 M hydrochloric acid. From both spruce and poplar, around 60–70 per cent of the lignin could be isolated with this method [37]. Table 9.4 gives some data on MWLs from various sources.

Almost all isolation procedures for lignin published till date have been preceded by an intensive milling of pre-extracted wood meal in a vibratory or rotatory mill. In the former type of mill, a milling time of 48 h has frequently been used [30, 34] while in the latter, about 1–2 weeks is required [36, 37]. During the milling, certain types of chemical changes take place in the lignin such as cleavage of β-*O*-4 linkages and introduction of carbonyl groups [34]. This reaction will result in the formation of new phenolic lignin end-groups with a simultaneous fragmentation of the lignin macromolecule as depicted in Fig. 9.7.

For extractives-free spruce wood, the concentration of free phenolic hydroxyl groups has been determined and values of 10–13 per 100 phenylpropane units have been obtained [38–40]. The corresponding value for spruce MWL is in the order of 20–30 (Table 9.4) depending on the intensity of milling and way of extracting the lignin. Consequently, all published analytical data on the structural composition of native lignin fail to give accurate values for the concentration of β-*O*-4 structures, since various amounts of this linkage have been mechano-chemically cleaved during the isolation procedure.

In an alternative way of isolating lignin from wood, pre-extracted spruce wood meal has been treated with a commercial mono-component endoglucanase [27, 41]. Thereby, the crystalline structure of cellulose is destroyed and the now amorphous wood can be homogeneously swollen in aqueous urea. By subsequent fractionation of the material in aqueous alkali, all lignin can be recovered as various LCCs (Table 9.5). Also in this case, however, the wood

Table 9.4

Representative data on MWL isolated from different wood species

Sample origin	OCH_3 (%)	Phenolic OH (%)	Carbohydrate (%)	Lignin yield (%)[a]	Reference
Picea abies, Norway spruce	15.45	30	1.9	19	[30]
Picea abies, Norway spruce	15.2	20	4.1	17	[34]
Picea abies, Norway spruce, CEL-96[b]	15.2	20	4.3	28	[34]
Picea mariana, Black spruce	15.41	28	n.a.[c]	n.a.	[31]
Picea mariana, Black spruce	15.3	23	<9.6	25	[37]
Picea mariana, Black spruce, EAL[d]	13.7	21	<7.7	69	[37]
Pinus sylvestris, Scots pine	15.74	27	n.a.	n.a.	[31]
Betula verrucosa, Silver birch	21.51	n.d.	7.5	n.a.	[31]
Liquidambar styraciflua, Sweetgum	21.4	14	3.6	17	[34]
Liquidambar styraciflua, Sweetgum, CEL-96[b]	21.5	13	3.8	43	[34]
Populus tremuloides, Trembling aspen	19.6	21	<8.0	24	[37]
Populus tremuloides, Trembling aspen, EAL[d]	20.9	21	<4.3	61	[37]

[a] Based on the lignin content in wood.

[b] Ball milling of wood followed by enzymatic hydrolysis of polysaccharides prior to extraction of the residue with dioxane–water (96:4).

[c] n.a. = not analysed.

[d] Ball milling of wood followed by enzymatic hydrolysis of polysaccharides and subsequent acid hydrolysis with 0.01 M HCl of the residue in dioxane–water (85:15).

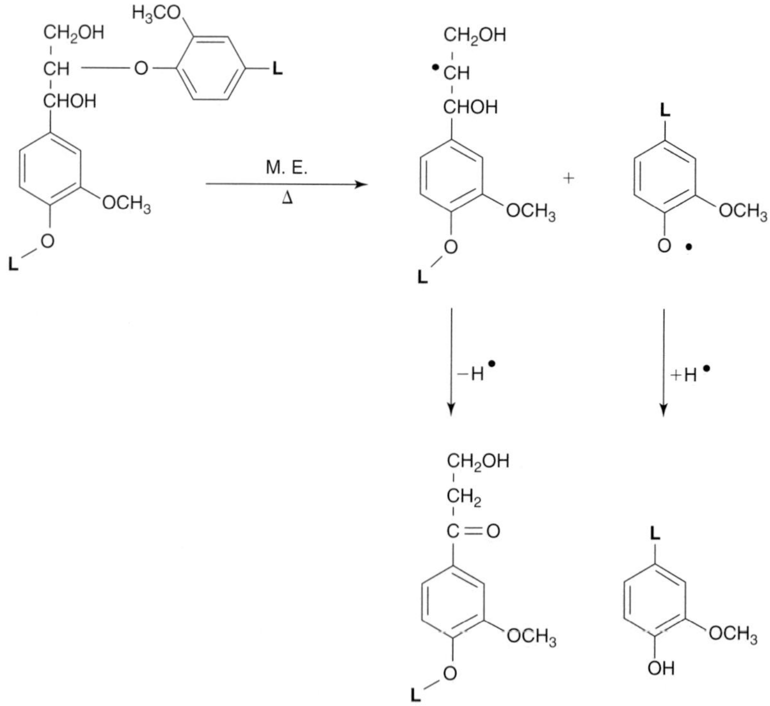

Figure 9.7 Lignin fragmentation reaction suggested to take place during the milling of wood. M.E. = mechanical energy.

Table 9.5

Types of LCC isolated from spruce wood meal after treatment with endoglucanase followed by swelling in urea and fractionation in aqueous alkali [27, 41]

Type of lignin–carbohydrate complex (LCC)	Lignin yield (%)
GalactoGlucoMannan – Lignin	8
Glucan – Lignin	4
GlucoMannan – Lignin[a]	48
Xylan – Lignin[a]	40

[a] Sum of two different fractions.

structure must be degraded to some extent prior to enzyme treatment by vibratory ball milling for 3h and, thus, neither this lignin can be regarded as completely native although the milling intensity has been greatly reduced.

9.4.2 Dehydropolymerizate

In the early work on lignin structure, enzymatic polymerization of coniferyl alcohol, employing either laccase or peroxidase together with oxygen or hydrogen peroxide respectively, has frequently been used to produce 'lignin-like' substances denoted dehydrogenation polymers (DHPs) (see Fig. 9.4) [10]. The results from this work were used to construct the first generation of statistical lignin structures in which the then identified relative amounts of different inter-unit linkages were included (Fig. 9.8).

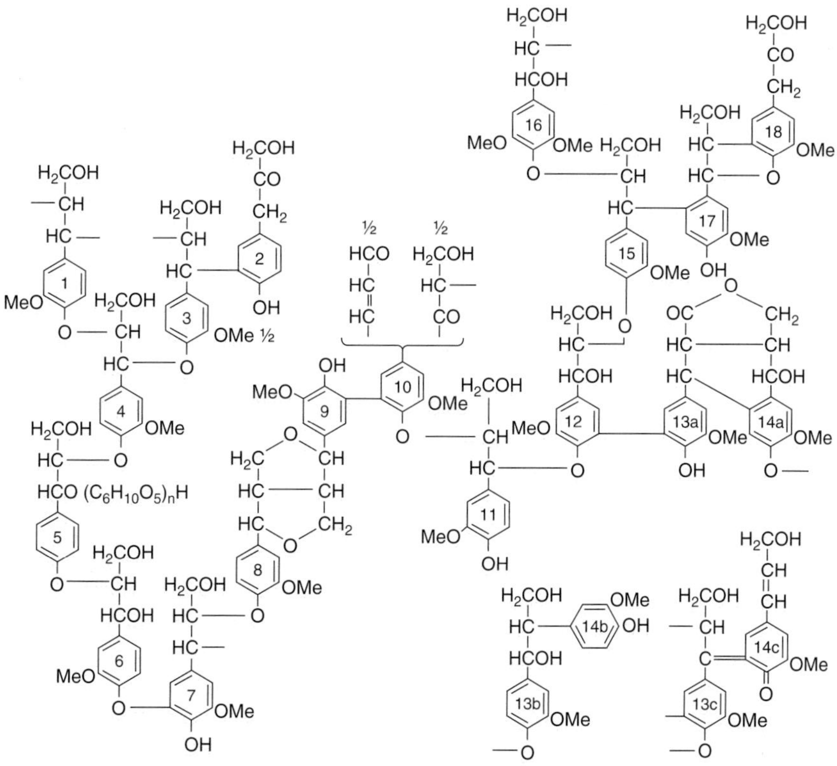

Figure 9.8 A first generation schematic formula of spruce lignin showing the most important inter-unit linkages between the phenylpropane units [42]. Reproduced by permission of Springer Science and Business Media.

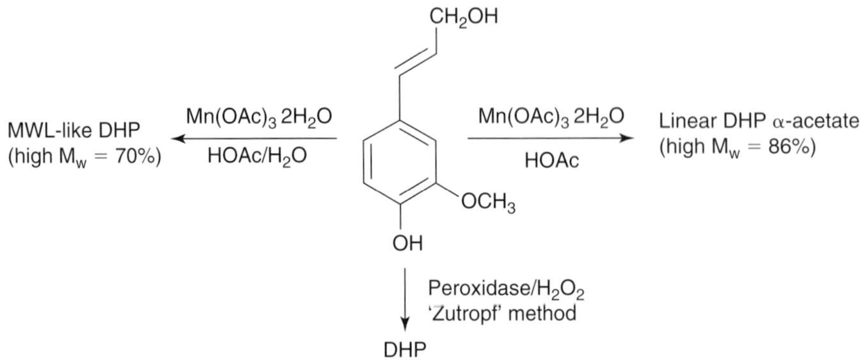

Figure 9.9 Synthesis of DHP using the 'Zutropf method' with peroxidase and hydrogen peroxide or manganese(III)-acetate in acetic acid with or without the presence of water. The biomimetic DHPs were fully acetylated and the high molecular weight fraction obtained after chromatography on polystyrene gel [46].

In the synthesis of DHP, two different modes have been employed, the 'Zulauf' and the 'Zutropf' methods. The former involves a stepwise addition of enzyme to a solution of coniferyl alcohol whereas in the latter, the reverse is used. From a structural point of view, the two methods give slightly different DHPs but the Zutropf–DHP is considered as being somewhat more 'lignin-like' [43]. For both types of DHPs, however, considerable structural differences are encountered in comparison with MWL.

Attempts to improve the synthesis of DHP in order to obtain a more lignin-like polymer have been made. In one such system, coniferin, together with β-glucosidase, peroxidase and hydrogen peroxide, prepared *in situ* from oxygen, glucose oxidase and liberated glucose, were used. Thereby, a very low concentration of hydrogen peroxide could be maintained throughout the oxidation. In different variants of the procedure, the reaction was performed with or without the presence of a polysaccharide and at different pH-values [44]. Based on this model system, it was found that pH, the concentration of monomer radicals and the carbohydrate matrix all affected the structure of the resulting DHP to some extent [45]. Despite these modifications, however, the DHPs were still found to be quite different from MWL.

In alternative ways of preparing DHP by employing biomimetic approaches, a much closer structural resemblance to MWL could be obtained. In one method, manganese(III) acetate was used to polymerize coniferyl and/or sinapyl alcohol [46]. The reaction conditions were found to exert a profound influence on the structure of the resulting polymer and lignin-like, as well as linear polymers, were produced as schematically illustrated in Fig. 9.9. In a similar approach, manganese(III) was produced *in situ* by the oxidation of manganese(II) with manganese peroxidase. Oxidation of coniferyl alcohol was carried out in a two-vessel reaction flask separated by a semi-permeable membrane allowing the manganese(III) to freely move between the vessels, while keeping the enzyme and coniferyl alcohol separated. Again, a lignin-like DHP could be formed as revealed by thioacidolysis in combination with size exclusion chromatography and ^{13}C NMR spectroscopy [47].

9.5 THE STRUCTURE OF LIGNIN

The quantification of lignin present in wood and other plants is usually done indirectly by weighing the solid residue that remains after the complete hydrolysis of all polysaccharides present in the material [48]. The insoluble material, denoted Klason lignin, is thought to be formed by a comprehensive condensation of the original lignin structure by the strongly acidic conditions used in the hydrolysis. In addition, a minor amount of acid-soluble lignin can be determined by measuring the UV absorbance of the hydrolysis liquor, usually at 205 nm [48]. Although the latter value is often added together with the value for Klason lignin to provide the total amount of lignin in a sample, its origin is not known with certainty and, for example, carbohydrates may well contribute degradation products with absorbance in the same region (Gellerstedt, unpublished).

For the quantification of lignin in a large number of similar samples, near-infrared spectroscopy (NIR) is well suited. Once a suitable set of reference samples has been determined using the Klason lignin method, the data can be used to calibrate the NIR signal. Subsequently, the lignin content in unknown samples can be registered in a convenient and rapid way by collecting the respective NIR spectra [49].

Among the early methods for the identification of lignin in a sample, colour reactions have been frequently used [50]. Thus, the treatment of wood or mechanical pulp with a mixture of phloroglucinol and hydrochloric acid results in the development of a reddish colour attributable to a condensation product between coniferaldehyde end-groups in lignin and phloroglucinol.

9.5.1 Wet chemistry methods

Today, a large number of wet chemistry methods exist by which lignin can be identified on the basis of its pattern of degradation products. Such methods have also been extensively used to get information about the structural details of different types of lignins. In Table 9.6, some commonly used methods have been collected. Each one of these methods will provide a piece of structural information but in most cases, the information is at best semi-quantitative.

Based on comprehensive work on spruce and birch MWL employing predominantly oxidation with permanganate–hydrogen peroxide and acidolysis respectively as analytical tools, a large number of degradation products has been identified and quantified [51, 56]. Based on these data, the concentration of individual inter-unit linkages present in the two lignins has been calculated and a structural scheme of spruce lignin, similar to that depicted in Fig. 9.8, was constructed [61]. An approach employing catalytic hydrogenation of spruce wood and subsequent identification of the degradation products was found to yield similar results in terms of functional groups and relative frequency of inter-unit linkages [54]. In a similar way, thioacetolysis (*i.e.* acid catalysed condensation of lignin with thioacetic acid), followed by alkaline hydrolysis and reduction with Raney nickel, has been applied on beech wood meal to degrade the lignin into a mixture of monomeric and oligomeric products. Several of these were identified and quantified providing data which formed the basis for a constitutional scheme of beech lignin [62, 63].

9.5.2 Nuclear magnetic resonance

Contrary to the wet chemical methods, analysis by nuclear magnetic resonance (NMR) does not degrade the sample and, consequently, all structural features in a lignin can be visualized. Despite the fact that analysis by either proton or carbon NMR results in crowded spectra with a severe overlap of signals, numerous spectra of lignins have been published and a large number of individual signals assigned [64–69]. By the use of spectral editing with, for example, the DEPT-sequence, permitting the separate analysis of CH-, CH_2- and CH_3-groups respectively, further details of the assignment of individual signals have been achieved [70]. Furthermore, the use of inverse gated decoupling combined with long delay times between pulses has permitted the quantitative analysis of ^{13}C NMR spectra thus providing reliable data on the amount of different types of carbon atoms in spruce MWL [71, 72].

Two-dimensional correlation spectra such as the HSQC-sequence can be used to separate the carbon and proton signals, thus permitting a much higher degree of accuracy in signal assignment. Furthermore, such spectra can be integrated in the *z*-direction to give quantitative information provided that a suitable reference signal has been chosen [73]. In Table 9.7, the result of such an integration is shown together with similar data taken from the literature and based on NMR, as well as on wet chemical methods.

Table 9.6

Wet chemistry methods for the analysis of lignin and lignin structures

Method	Reaction principle	Reference
Oxidation with $KMnO_4 - H_2O_2$	Oxidative elimination of side chains	[51, 52]
Thioacetolysis	Cleavage of alkyl-aryl ethers	[53]
Hydrogenolysis	Reductive cleavage of ethers	[54, 55]
Acidolysis	Hydrolytic cleavage of ethers	[56, 57]
Thioacidolysis	Hydrolytic cleavage of ethers	[4, 58]
Acetyl bromide – Zinc/Acetic anhydride	Reductive cleavage of ethers	[59, 60]

Table 9.7

Frequency of inter-unit linkages in MWL from spruce (*Picea abies*) based on integration of a 2D HSQC spectrum using the aromatic C2 as internal standard

Lignin structure	Number per C9-unit	Reference		
		[74][a]	[54][b]	[61][c]
β-*O*-4′	0.43	0.45	0.45	0.50
β-5′	0.12	0.09	0.14	0.12
Dibenzodioxocin	0.05	0.07		
β−β′	0.035	0.06	0.07	0.02
β-1′	0.02	0.01	0.09	0.07
Spirodienone	0.02	0.02		
Coniferyl alcohol	0.02	0.02	0.04	
Coniferaldehyde	0.03	0.04	0.04	
Vanillin	0.02	0.05	0.06	
Dihydroconiferyl alcohol	0.02	0.02	0.04	
5–5′		0.13	0.11	0.11
4-*O*-5′			0.035	0.04

[a] Based on quantitative ^{13}C NMR.
[b] Based on hydrogenolysis of ezomatsu (*Picea jezoensis*) wood.
[c] Based predominantly on permanganate–hydrogen peroxide oxidation.

9.5.3 Analysis of β-*O*-4 structures

The quantitatively most important inter-unit linkage in lignin, the β-*O*-4 linkage, can be cleaved in alkaline as well as in acidic conditions. For analytical purposes, alkaline cleavage is used as part of the thioacetolysis reaction sequence [53], whereas acidic conditions can be applied directly to the wood material [75]. In the latter method, hydrochloric acid has been used to afford monomeric and dimeric lignin degradation products which can be identified but, in addition, the reaction also gives rise to products of higher molecular mass, at least in part formed by condensation reactions [56]. Although the method can be used as an analytical tool to give a relative quantification of the number of β-*O*-4 structures, the presence of condensation reactions is a serious drawback [57]. The chemistry of lignin encountered in acid media, *viz.* the intermediate formation of a benzylic cation, followed by either a proton elimination reaction or an addition of a nucleophilic carbon atom, readily explains the observed behaviour. The presence of an external nucleophile may, however, in analogy to sulphite pulping or analysis by thioacetolysis, prevent any condensation reaction.

In thioacidolysis, a Lewis acid, boron trifluoride etherate, is used together with a strong nucleophile, ethanethiol, to hydrolyse all β-*O*-4 linkages present in a lignin, wood or pulp sample [20, 58]. The predominant product from the cleavage reaction (Fig. 9.10), a phenylpropane structure (two diastereoisomeric forms) having all side-chain carbons substituted with ethylthio groups, can be readily quantified. Furthermore, several other monomeric degradation products can also be identified, thus providing information about the variability of lignin end-groups [4].

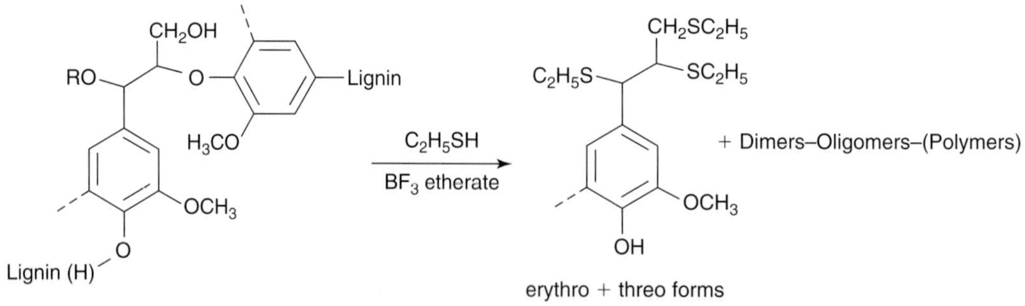

Figure 9.10 Thioacidolysis of lignin and formation of ethylthio-substituted degradation products.

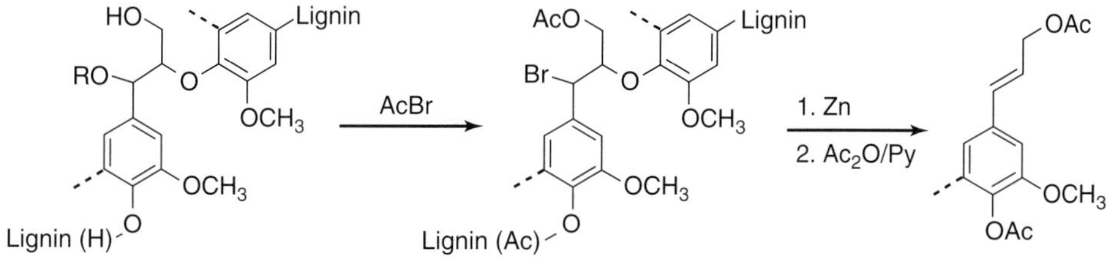

Figure 9.11 Analysis of β-*O*-4 structures in lignin according to the DFRC method (derivatization followed by reductive cleavage) resulting in the formation of cinnamyl alcohol derivatives suitable for quantification. AcBr = acetyl bromide, Ac_2O/Py = acetic anhydride/pyridine [59].

From wood, the reaction gives a high yield of low molecular mass degradation products (monomers to trimers, with traces of higher M_w products) and, after treatment, no lignin can be detected in the remaining wood material. Thioacidolysis, when applied to kraft pulps, on the other hand, in addition to monomers–trimers also results in the presence of polymeric material formed as a consequence of pulping reactions [20].

In a later extension of the thioacidolysis reaction, a second step, a Raney nickel reduction, was introduced, thereby also permitting the analysis of dimeric and trimeric degradation products by gas chromatography [20, 76]. By this method, more than 20 different dimeric and 16 trimeric structures, representing all the various inter-unit linkages depicted in Fig. 9.5, except the β-*O*-4 and dibenzodioxocin structures, have been qualitatively identified from spruce wood. Similar data from birch and aspen wood are also available [23].

A different approach to the analysis of β-*O*-4 structures involves a two-step procedure with acetyl bromide treatment followed by a reductive cleavage of the β-ether substituent as depicted in Fig. 9.11 [59, 60]. The reaction can be applied to isolated lignins, as well as on cell wall material, and results in the formation of acetylated cinnamyl alcohols which can be readily quantified by gas chromatography. The yield of the major compounds originating from G-, H- and S-units seems, however, to be inferior to the corresponding values obtained from thioacidolysis [20, 60, 76].

Thioacidolysis results in a high yield of the major monomeric degradation product (in two diastereoisomeric forms) from softwoods, as well as from hardwoods. In addition, it is possible to identify several other monomers in minor amounts (Fig. 9.2) [4]. The latter are assumed to constitute lignin end-groups attached to the rest of the polymer through β-*O*-4′ linkages. By pre-swelling of the wood in water prior to the thioacidolysis reaction, an apparent quantitative delignification can be achieved and the yield of the major monomer(s) substantially increased, as shown in Table 9.8. From this table, it is also obvious that the milling used to extract MWL from the wood samples

Table 9.8

Yield of the main monomer(s) on thioacidolysis of spruce, birch and aspen wood, spruce TMP and the corresponding MWLs together with the number of phenolic hydroxyl groups per 100 phenylpropane (C9) units [20, 23]

Sample	Yield of the main monomer(s) (μmol/g Klason lignin)	Content of phenolic OH[a], Number per 100 C9-units[b]
Spruce wood	1332	n.a.[c]
Spruce wood (pre-swollen)[d]	1682 (31%)[e]	13
Spruce MWL[f]	986	20
Spruce TMP[g] (pre-swollen)[d]	1498	14
Birch wood (pre-swollen)[d]	672 (G) + 2318 (S) = 2990 (63%)[e]	7.6
Birch MWL[f]	403 (G) + 809 (S) = 1212	n.a.[c]
Aspen wood (pre-swollen)[d]	866 (G) + 1942 (S) = 2808 (58%)[e]	10
Aspen MWL[f]	609 (G) + 863 (S) = 1472	n.a.[c]

[a] Determination by periodate oxidation according to [40].
[b] Assumed molecular mass ($g\,mol^{-1}$) for one C9-unit: spruce = 183; birch = 210; aspen = 206.
[c] n.a. = not analysed.
[d] Sample pre-swollen in water over-night.
[e] Percentage of the theoretical yield of C9-units per gram Klason lignin.
[f] Yield given as $\mu mol g^{-1}$ MWL.
[g] Thermomechanical pulp.

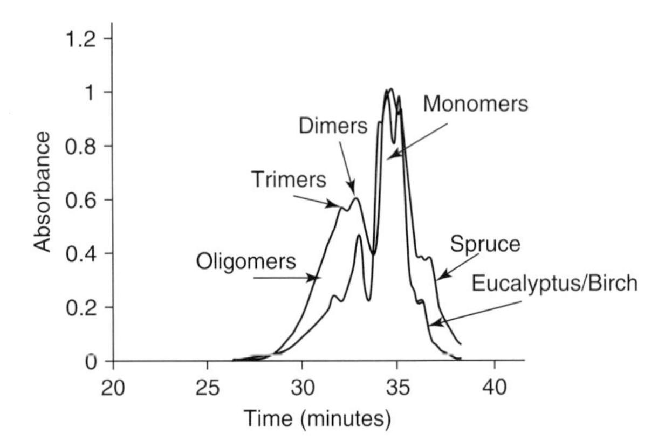

Figure 9.12 HPSEC on styragel of the acetylated lignin degradation products obtained after thioacidolysis of spruce, birch and eucalyptus wood respectively.

results in substantially reduced amounts of monomeric thioacidolysis products, as discussed above (see Figure 9.7). Simultaneously, the number of phenolic hydroxyl groups is strongly increased. Even the comparatively mild milling procedure used in the production of thermomechanical pulp (TMP), results in an increased number of phenolic hydroxyl groups and a reduction in monomer yield. In the two hardwood samples shown in Table 9.8, the yield of S-units is high and reflects the fact that β-*O*-4′ structures are completely dominant in syringyl type lignins, since the β-5′ and 5-5′ coupling modes cannot take place. Generally, the content of phenolic hydroxyl groups in hardwoods is lower as compared to softwoods; a result in line with a high abundance of linear β-*O*-4′-linked syringyl lignin. In guaiacyl lignins, on the other hand, β-*O*-4′ moieties may be present in both linear and branched/crosslinked structures, thus resulting in a considerably lower yield of the monomeric thioacidolysis product.

When applied to wood, the thioacidolysis reaction seems to result in a complete dissolution of lignin, since no Klason lignin can be detected in the residue. After acetylation, the degraded lignin fragments can be conveniently analysed by high performance size exclusion chromatography (HPSEC) in tetrahydrofuran [20]. From spruce, as well as from birch and eucalyptus wood, similar chromatograms can be obtained, although the relative intensity of the various peaks is different (Fig. 9.12) (Gellerstedt, unpublished). By the use of calibration substances, it can be shown that no polymeric material is present in the degradation mixture with monomers, dimers and trimers constituting the predominant types of products.

This result is completely in line with the view that the lignin polymer in both softwoods and hardwoods is built up with β-*O*-4′ linkages as the completely dominating type of inter-unit linkage. The number of such linkages has been estimated based on analytical data from MWL, and values of about 50 and 60 per 100 phenylpropane units for spruce and birch, respectively, have been obtained [77]. Obviously, these values must be much higher for the native lignin in wood since, as discussed above, the milling procedure used to produce MWL involves a comprehensive cleavage of β-*O*-4′ linkages (see Fig. 9.7 and Table 9.8).

9.5.4 Lignin heterogeneity

By a selective modification of the cellulose structure in spruce wood, by treatment with an endoglucanase, followed by swelling of the wood in aqueous urea and dissolution in alkali, a subsequent stepwise acidification will result in a complete recovery of all the lignin as different LCCs (see Table 9.5) [27,41]. On thioacidolysis followed by HPSEC of the various LCCs, large differences in the resulting lignin chromatograms were encountered, as demonstrated in Fig. 9.13.

Thus, the thioacidolysis mixture obtained from xylan-linked lignin showed a complete dominance of monomers with only a small amount of dimers present. The glucomannan lignin complex, on the other hand, gave a chromatogram similar to that obtained from wood (Fig. 9.12). A reasonable explanation for these results is the presence of two distinctly different types of lignin in spruce wood; one having a very strong dominance of β-*O*-4′ linkages, being linked to xylan and with a rather linear structure, the other being linked to glucomannan and having a more complex structure comprising the various linkages depicted in Fig. 9.5. The former type of lignin

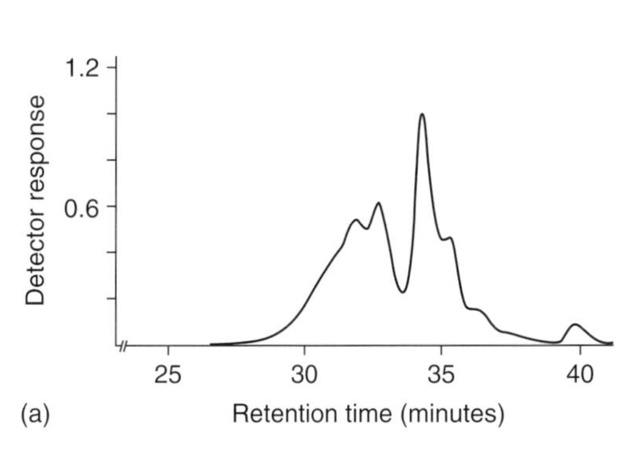

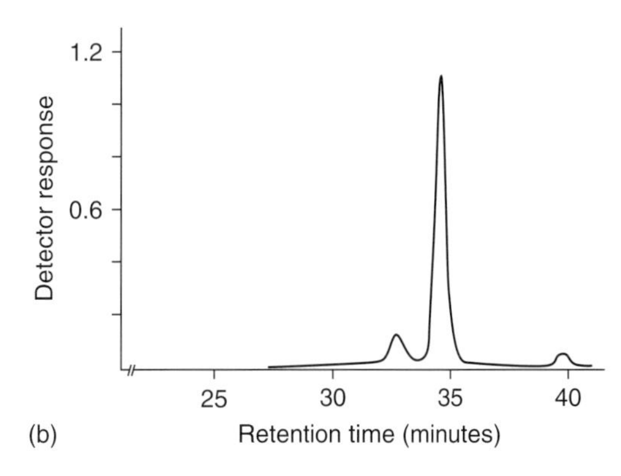

Figure 9.13 HPSEC of thioacidolysis products obtained from (a) glucomannan-linked lignin, and (b) xylan-linked lignin [41].

Table 9.9

Distribution of guaiacyl and syringyl lignin in various morphological regions of white birch (*Betula papyrifera*) [78]

Morphological differentiation	Guaiacyl/Syringyl
Fibre, S2-layer	12:88
Vessel, S2-layer	88:12
Ray parenchyma, S-layer	49:51
Middle lamella (fibre–fibre)	91:9
Middle lamella (fibre–vessel)	80:20
Middle lamella (fibre–ray)	100:0
Middle lamella (ray–ray)	88:12

should be expected to be particularly sensitive to milling, resulting in a low yield of thioacidolysis products with a corresponding high amount of free phenolic hydroxyl groups. This has been found to be the case, thus lending further support to the view that a linear β-*O*-4′-linked lignin is present in spruce wood [41].

For birch wood, it was shown a long time ago, by the use of UV microscopy, that the structure of lignin is not uniform throughout the fibre and vessel elements. Whereas the secondary wall in the fibre fraction was found to contain an almost pure syringyl lignin (*i.e.* a linear lignin with a predominance of β-*O*-4′ linkages) the lignin present in vessels was of the guaiacyl type. Guaiacyl lignin was also found in the middle lamella region [29]. In a further development of the analytical method, more precise data for the presence of G- and S-units throughout the cell walls were obtained (Table 9.9) [78].

By the use of an immunogold labelling technique with specific response for condensed guaiacyl and guaiacyl–syringyl lignin and for non-condensed guaiacyl–syringyl lignin respectively, the data in Table 9.9 have recently been qualitatively confirmed in transmission electron microscopy studies of aspen wood [79]. Further support for the presence of linear lignin structures of the β-*O*-4′ type present in hardwood has also been found by the mass spectrometric analysis of a low molecular mass fraction of eucalyptus lignin, showing the arrangement of S (and G)-units in pentamer and hexamer fragments (Fig. 9.14) [80].

9.5.5 The structure of native lignin

The growing awareness that the structure of lignin inside the fibre wall is not uniform, but probably adapted to its site of formation, chemical surrounding and other as yet unknown factors makes the presentation of statistical formulas based on MWL-data alone highly uncertain. The presence of a substantial amount of lignin in spruce, as well as in various hardwood species strongly enriched in β-*O*-4′ structures suggests, however, that linear lignin macromolecules are abundant in both softwood and hardwood lignin. These, as well as other types of lignin macromolecules,

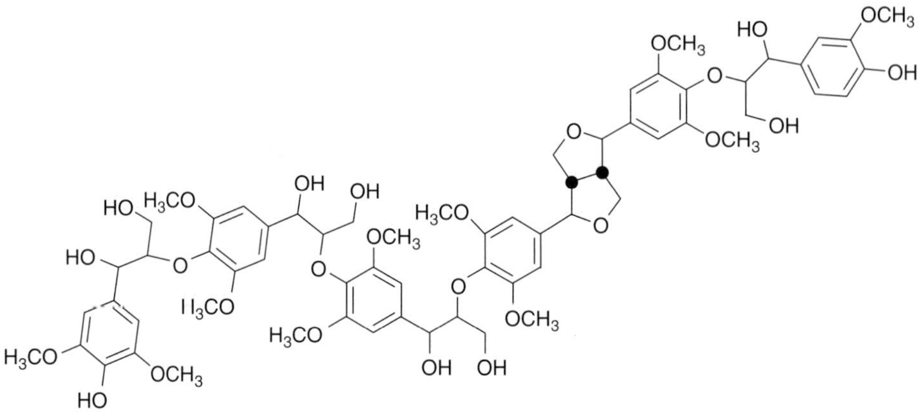

Figure 9.14 A hexameric lignin fragment from eucalyptus (*E. grandis*) identified by MS-analysis. Further pentamer and hexamer fragments were also found [80].

seem to be chemically linked to the polysaccharides thus forming network structures in the wood tissue. In technical processes, the different types of lignin macromolecules will show different reactivity, thus resulting in inhomogeneous dissolution at the nano-scale [81–83].

The knowledge collected till date on the structure of lignin distinguishes two different types of macromolecules; *viz.* a linear structure predominantly built up by β-*O*-4′ linkages in G- as well as in S-units, with minor amounts of β−β′, β-5′ and β-1′ linkages present. The number of 5-5′ and 4-*O*-5′ linkages in such lignin is assumed to be very low or zero. As a consequence, the number of free phenolic hydroxyl groups is also very low. In the second type of lignin, a much more branched structure should prevail, 5-5′ and 4-*O*-5′ linkages being present and the number of phenolic hydroxyl groups quite high. In spruce, this type of lignin should be linked predominantly to glucomannan. The presence of such a lignin requires the formation of chain-like fragments which subsequently are linked together by a secondary formation of 5-5′ and 4-*O*-5′ linkages (see Section 9.3). The variability of the chain-like fragments can be substantial, thus resulting in lignin macromolecules of considerable difference from each other. Still, the number of β-*O*-4′ linkages must be high, however, in order to fulfil all structural requirements concluded from present knowledge. A segment of such lignin is shown in Fig. 9.15.

9.6 TECHNICAL LIGNINS

For a more comprehensive discussion of technical lignins, the reader should refer to Chapter 10.

9.6.1 Kraft lignins

Delignification of wood and other biomass for the production of kraft pulp involves treatment at high temperature with an aqueous solution of sodium hydroxide and sodium sulphide. Under these conditions, most of the β-*O*-4′ structures in lignin are hydrolysed (>95 per cent) and the resulting lignin fragments dissolved in the alkaline solution [84]. Several other degradative lignin reactions also take place under the harsh conditions prevailing in the digester and most of the phenylpropane side-chains are in part eliminated, in part modified. The process results in the dissolution of around 90–95 per cent of all lignin present in the starting material. By acidification of the pulping liquor, the dissolved lignin can be recovered to a great extent as a complex mixture of phenolic structures with molecular mass in the range of ~150–200 000 [85–89].

By solvent fractionation, it has been shown that a predominant portion of softwood kraft lignin is of low molecular mass with a low degree of polydispersity as shown in Table 9.10. The remainder, on the other hand, has both high molecular mass and high polydispersity, presumably due to the presence of both polysaccharides and condensed lignin structures in these fractions [88]. A similar fractionation technique applied to a birch kraft lignin gave the results shown in Table 9.11, again with a high molecular mass tail resulting in a high degree of polydispersity for one of the fractions [89].

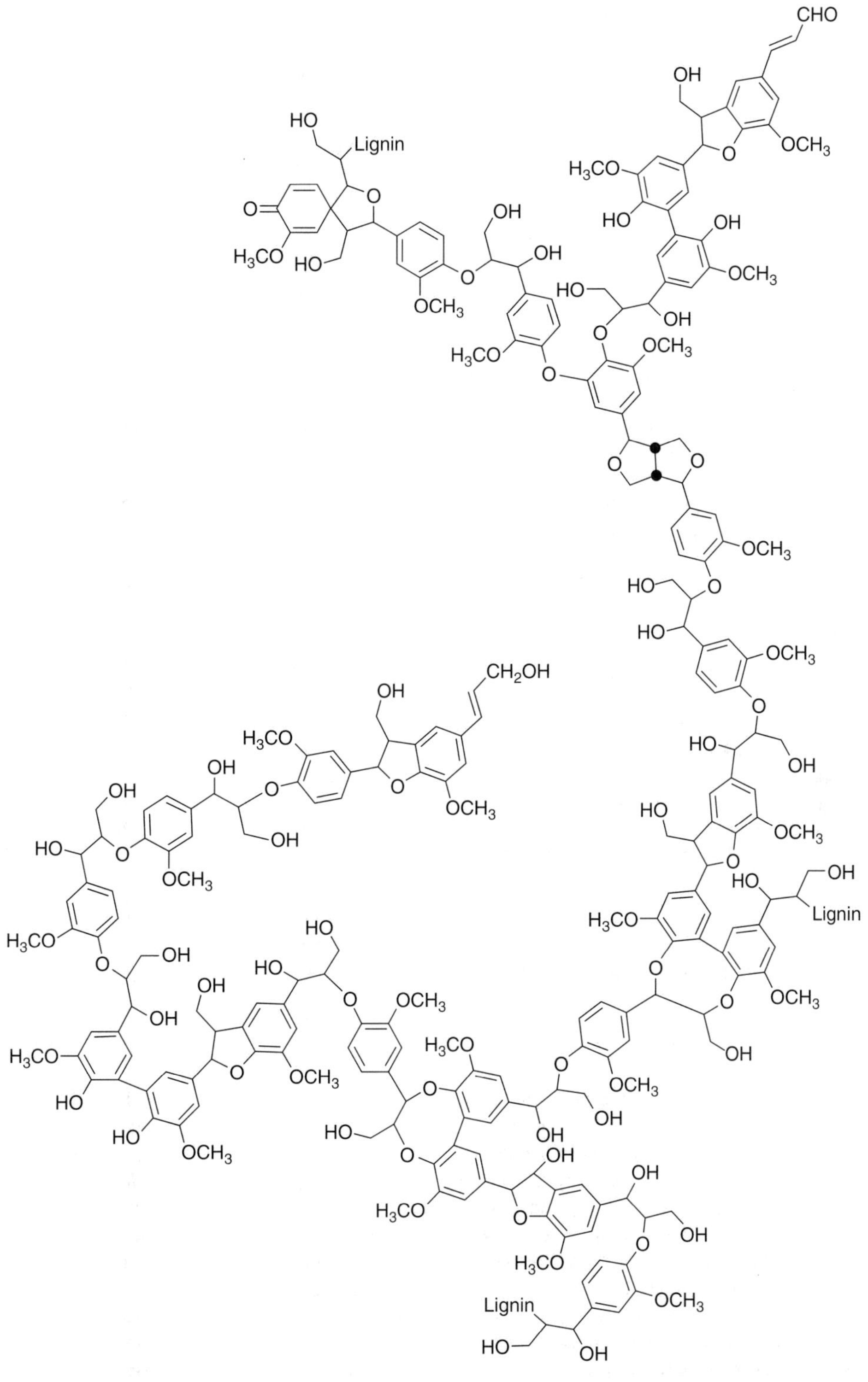

Figure 9.15 Statistical scheme of the glucomannan-linked lignin from spruce. The structure contains 25 phenylpropane units with a total of 4 free phenolic hydroxyl groups. The proportion of inter-unit lignin linkages is: 12 β-*O*-4′ (2 in dibenzodioxocin), 5 β-5′, 4 5-5′ (2 in dibenzodioxocin), 1 4-*O*-5′, 1 β−β′, 1 β-1′, 1 coniferaldehyde, 1 coniferyl alcohol.

Table 9.10

Analytical data on kraft lignins obtained after solvent fractionation of isolated industrial softwood black liquor lignin [88]

Fraction No[a]	Yield (%)	M_n^b	M_w^b	M_w/M_n	Phenolic OH (mmol/g)	Aliphatic OH (mmol/g)	Carboxyl (mmol/g)
1	9	450	620	1.4	5.1	1.0	2.3
2	22	900	1 290	1.4	5.0	2.3	1.1
3	26	1 710	2 890	1.7	4.3	2.1	0.8
4	28	3 800	82 000	22	3.9	2.4	0.4
5	14	5 800	180 000	31	3.0	2.4	0.3

[a] Fraction 1 = soluble in methylene chloride.
Fraction 2 = residue soluble in *n*-propanol.
Fraction 3 = residue soluble in methanol.
Fraction 4 = residue soluble in methanol/methylene chloride (7:3).
Fraction 5 = residue.
[b] Acetylated lignin fractions.

Table 9.11

Analytical data on kraft lignins obtained after solvent fractionation of isolated industrial birch black liquor lignin [89]

Fraction no[a]	Yield (%)	M_n^b	M_w^b	M_w/M_n	Phenolic OH (mmol/g)	Aliphatic OH (mmol/g)
1	32	650	910	1.4	5.0	1.0
2	38	1 320	2 110	1.6	4.4	2.2
3	30	3 760	87 080	23	2.9	3.7

[a] Fraction 1 = soluble in methylene chloride.
Fraction 2 = residue soluble in methanol.
Fraction 3 = residue.
[b] Acetylated lignin fractions.

Table 9.12

Elemental and methoxyl analysis data for softwood kraft lignin fractions from Table 9.10 together with data for spruce MWL and purified pine kraft lignin [88]

Fraction no	Carbon (%)	Hydrogen (%)	Oxygen (%)	Sulphur (%)	OCH_3 (%)	Sugars (%)
1	67.8	6.43	23.0	2.7	15.4	–
2	65.3	6.48	26.8	1.4	16.0	–
3	64.7	5.96	27.7	1.5	14.9	–
4	64.5	5.73	28.2	1.3	13.9	0.3
5	59.0	5.54	32.8	1.2	11.3	8.7
MWL, spruce[a]	63.8	6.0	29.7	–	15.8	
Kraft lignin[b]	64.3	6.0	27.9	1.8	13.8	0.9

[a] Reference [30].
[b] Reference [91].

NMR analysis of the various lignin fractions from pine/spruce and birch respectively [88, 89], as well as of unfractionated kraft lignin from pine [71], has shown that comprehensive changes of the side-chain structure take place during pulping with a strong reduction of oxygen-linked carbons and a concomitant increase of aliphatic methine, methylene and methyl groups. The content of aromatic rings with free phenolic hydroxyl groups is high, in particular in the low molecular mass fractions. The presence of carboxyl groups (in softwood) is substantial and can be explained by radical coupling reactions between certain fatty acids, present among the wood extractives, and phenolic end-groups in the lignin [90]. Moreover, all the lignin fractions shown in Table 9.10 also contain considerable amounts of sulphur as shown in Table 9.12. Similar data were also found for the birch lignin fractions.

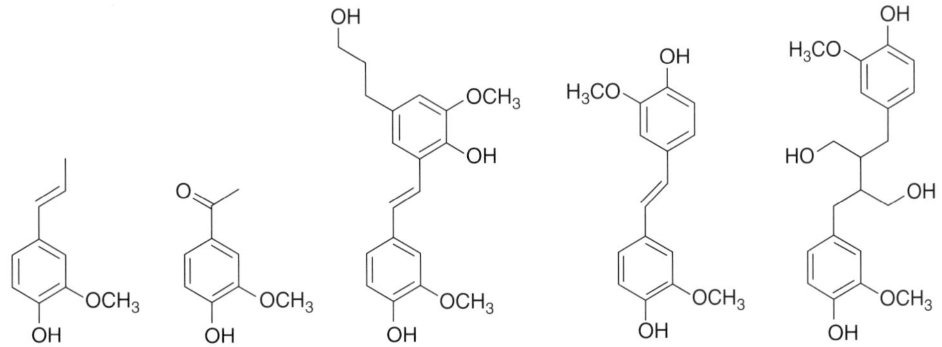

Figure 9.16 Some representative lignin fragments present in spruce kraft pulping liquor.

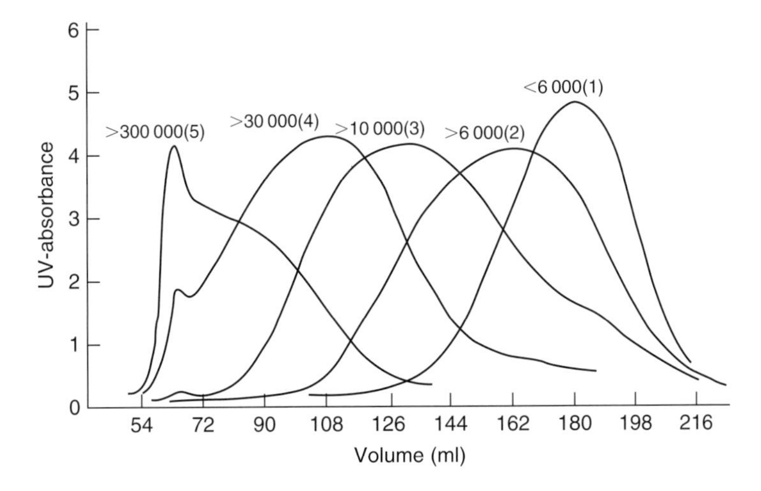

Figure 9.17 Size exclusion chromatography of lignosulphonate fractions after ultrafiltration. Fraction number and approximate molecular mass are shown in the figure [92].

Several hundred low molecular mass lignin-derived components containing one or two aromatic rings have been identified in kraft pulping liquors from spruce and birch respectively [86, 87]. Some prominent structures are shown in Fig. 9.16.

9.6.2 Sulphite lignin

The major reaction in sulphite pulping involves the introduction of sulphonate groups in the C_α-positions in lignin via the intermediate formation of carbo-cations. Thereby, any lignin–carbohydrate linkages present in these positions are cleaved and, together with some hydrolytic cleavage of β-*O*-4′ linkages, the lignin is solubilized. Condensation, predominantly between C_α and C_6 in adjacent lignin structures, is a major competing reaction, but can be suppressed by having a high concentration of bisulphite ions present in the pulping liquor. The highly water-soluble lignosulphonate can be purified from other substances present in the pulping liquor by ultrafiltration and by using membranes having different M_w cut-off values, a series of lignosulphonate fractions can be obtained as shown in Fig. 9.17 [92]. By elemental and methoxyl analysis on these fractions, a degree of sulphonation of about 0.4–0.5 per phenylpropane unit has been found. The liberation of phenolic hydroxyl groups, on the other hand, is limited but clearly visible in those lignosulphonate fractions having a low molecular mass distribution (Table 9.13). In comparison with kraft lignins, the side-chains in lignosulphonates are still intact to a high degree, as revealed by ^{13}C NMR spectroscopy, although some α-carbon atoms are linked in condensation reactions [93].

Table 9.13

Analytical data of lignosulphonate fractions after ultrafiltration of a technical softwood acid sulphite pulping liquor [92]

Fraction no	Yield (%)	M_w	Phenolic OH (mmol/g)	Sulphonate/Phenylpropane
1	20	590	2.01	0.48
2	29	1440	0.91	0.52
3	18	3690	0.65	0.43
4	4	6500	0.50	0.40
5	6	10880	0.49	0.39
6	22	21950	0.51	0.36

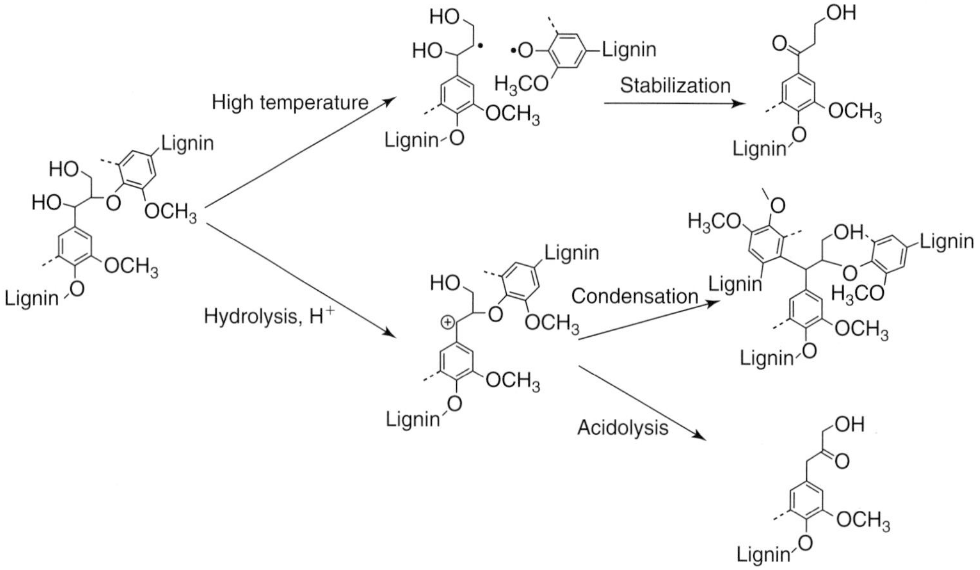

Figure 9.18 Types of lignin reactions encountered during steam explosion. The relative importance may differ depending on the steam treatment conditions.

9.6.3 Steam explosion lignin

By subjecting wood or other biomass to high temperature steam treatment, followed by a rapid pressure release, the fibrous mass is 'exploded' and liberated fibres together with fibre bundles are formed. By adjusting the time and temperature, different degrees of wood polymer modification and degradation can be achieved. Although the process is not commercial at present, it has gained much attention as a possible means for simple and cheap separation of wood polymers (*e.g.* for the production of micro-crystalline cellulose and bio-based ethanol). In particular, hardwood species such as aspen are suitable raw materials, since the lignin portion can be extracted, to a large extent, by either aqueous alkali or by organic solvents leaving a residue highly enriched in cellulose [94, 95].

In a pure steam explosion process without any added chemicals, the reaction conditions are weakly acidic, due to the release of acetic acid from the hemicellulose. Thus, the major reaction types are similar to those present in acidic sulphite pulping, *viz.* hydrolysis of polysaccharides and hydrolysis and condensation of lignin. In addition, due to the high temperature usually employed (~200°C), homolytic cleavage reactions of, for example, β-*O*-4′ linkages in lignin can be assumed to take place. The reaction types for lignin are summarized in Fig. 9.18. Altogether, these reactions result in a highly heterogeneous lignin structure containing both degraded lignin fragments and recombined fragments through condensation reactions [96]. Consequently, the number of β-*O*-4′ structures is much lower, as compared with that of the starting material and the content of phenolic end-groups higher. In addition, the number of carbonyl groups is considerably increased due to hydrolytic and/or homolytic cleavage reactions.

REFERENCES

1. Fengel D., Wegener G., *Wood. Chemistry, Ultrastructure, Reactions*, Walter de Gruyter, Berlin, 1984. pp. 132–181.
2. Sjöström E., *Wood Chemistry. Fundamentals and Applications*, 2nd Edition, Academic Press Inc., San Diego, USA, 1993. pp. 71–91.
3. Sakakibara A., Sano Y., Chemistry of lignin, in *Wood and Cellulosic Chemistry* (Eds. Hon D.N.-S. and Shiraishi N.), 2nd Edition, Marcel Dekker Inc., New York, USA, 2001, pp. 109–174.
4. Rolando C., Monties B., Lapierre C., Thioacidolysis, in *Methods in Lignin Chemistry* (Eds. Lin S.Y. and Dence C.W.), Springer-Verlag, Berlin Heidelberg, Germany, 1992, pp. 334–349.
5. Ralph J., Lu F., The DFRC method for lignin analysis. 6. A simple modification for identifying natural acetates on lignins, *J. Agr. Food Chem.*, **46**, 1998, 4616–4619.
6. Chen C.-L., Nitrobenzene and cupric oxide oxidations, in *Methods in Lignin Chemistry* (Eds. Lin S.Y. and Dence C.W.), Springer-Verlag, Berlin Heidelberg, Germany, 1992, p. 301.
7. Higuchi T., Lignin biochemistry: Biosynthesis and biodegradation, *Wood Sci. Technol.*, **24**, 1990, 23–63.
8. Boudet A.-M., Lignins and lignification: Selected issues, *Plant Physiol. Biochem.*, **38**, 2000, 81–96.
9. Donaldson L.A., Lignification and lignin topochemistry – An ultrastructural view, *Phytochemistry*, **57**, 2001, 859–873.
10. Freudenberg K., in *Constitution and Biosynthesis of Lignin* (Eds. Freudenberg K. and Neish A.C.), Springer-Verlag, Berlin-Heidelberg, 1968, pp. 47–122.
11. Katayama Y., Fukuzumi T., Enzymatic synthesis of three lignin-related dimers by an improved peroxidase–hydrogen peroxide system, *Mokuzai Gakkaishi*, **24**, 1978, 664–667.
12. Ralph J., Lundquist K., Brunow G., Lu F., Kim H., Schatz P.F., Marita J.M., Hatfield R.D., Ralph S.A., Christensen J.H., Boerjan W., Lignins: Natural polymers from oxidative coupling of 4-hydroxyphenylpropanoids, *Phytochemistry Rev.*, **3**, 2004, 29–60.
13. Zhang L., Gellerstedt G., Ralph J., Lu F., NMR studies on the occurrence of spirodienone structures in lignins, *J. Wood Chem. Technol.*, **26**, 2006, 65–79.
14. Karhunen P., Rummakko P., Sipilä J., Brunow G., Kilpeläinen I., Dibenzodioxocins; a novel type of linkage in softwood lignins, *Tetrahedron Lett.*, **36**, 1995, 169–170.
15. Argyropoulos D.S., Jurasek L., Kristofova L., Xia Z., Dun Y., Palus E., Abundance and reactivity of dibenzodioxocins in softwood lignin, *J. Agr. Food Chem.*, **50**, 2002, 658–666.
16. Lewis N.G., A 20th century roller coaster ride: A short account of lignification, *Current Opin. Plant Biol.*, **2**, 1999, 153–162.
17. Davin L.B., Lewis N.G., Lignin primary structure and dirigent sites, *Current Opin. Biotechnol.*, **16**, 2005, 407–415.
18. Fengel D., Wegener G., *Wood. Chemistry, Ultrastructure, Reactions*, Walter de Gruyter, Berlin, 1984. pp. 56–58
19. dos Santos Abreu H., Maria M.A., Reis J.L., Dual oxidation ways toward lignin evolution, *Floresta e Ambiente*, **8**, 2001, 207–210.
20. Önnerud H., Gellerstedt G., Inhomogeneities in the chemical structure of spruce lignin, *Holzforschung*, **57**, 2003, 165–170.
21. Önnerud H., Lignin structures in normal and compression wood. Evaluation by thioacidolysis using ethanethiol and methanethiol, *Holzforschung*, **57**, 2003, 377–384.
22. Sjöström E., *Wood Chemistry. Fundamentals and Applications*, 2nd Edition, Academic Press Inc., San Diego, USA, 1993, p. 249.
23. Önnerud H., Gellerstedt G., Inhomogeneities in the chemical structure of hardwood lignins, *Holzforschung*, **57**, 2003, 255–265.
24. Bassa A., Sacon V.M., da Silva Junior F.G. and Barrichelo L.E.G. Polpacao kraft convencional e modificada para madeiras de Eucalyptus grandis e hibrido (E. grandis x E. urophylla), *35° Congresso e Exposicao Anual de Celulose e Papel*, Sao Paulo, Brazil, October 14–17, 2002, Proceedings, p. 5.
25. del Rio J.C., Rodriguez I.M. and Gutierrez A. Chemical characterization of fibers from herbaceous plants commonly used for manufacturing of high quality paper pulps. *9th European Workshop on Lignocellulosics and Pulp*, Vienna, Austria, August, 27–30, 2006, Proceedings, pp. 109–112.
26. http://www.vcn.vnn.vn/sp_pape/spec_5_4_2001_13.htm.
27. Lawoko M., Henriksson G., Gellerstedt G., Characterization of lignin–carbohydrate complexes (LCCs) of spruce wood (*Picea abies* L.) isolated with two methods, *Holzforschung*, **60**, 2006, 156–161.
28. Fergus B.J., Procter A.R., Scott J.A.N., Goring D.A.I., The distribution of lignin in sprucewood as determined by ultraviolet microscopy, *Wood Sci. Technol.*, **3**, 1969, 117–138.
29. Fergus B.J., Goring D.A.I., The distribution of lignin in birch wood as determined by ultraviolet microscopy, *Holzforschung*, **24**, 1970, 118–124.
30. Björkman A., Studies on finely divided wood. Part 1. Extraction of lignin with neutral solvents, *Svensk Papperstidn.*, **59**, 1956, 477–485.
31. Björkman A., Person B., Studies on finely divided wood. Part 2. The properties of lignins extracted with neutral solvents from softwoods and hardwoods, *Svensk Papperstidn*, **60**, 1957, 158–169.

32. Björkman A., Studies on finely divided wood. Part 3. Extraction of lignin–carbohydrate complexes with neutral solvents, *Svensk Papperstidn*, **60**, 1957, 243–251.
33. Brownell H.H., Isolation of milled wood lignin and lignin–carbohydrate complex, *Tappi*, **48**(9), 1965, 513–519.
34. Chang H-m., Cowling E.B., Brown W., Adler E., Miksche G., Comparative studies on cellulolytic enzyme lignin and milled wood lignin of sweetgum and spruce, *Holzforschung*, **29**, 1975, 153–159.
35. Glasser W.G., Barnett C.A., The structure of lignins in pulps. II. A comparative evaluation of isolation methods, *Holzforschung*, **33**, 1979, 78–86.
36. Bezuch B., Polcin J., The structural study of milled wood and enzymatically isolated lignin fractions from spruce wood, *Cell. Chem. Technol.*, **12**, 1978, 473–482.
37. Wu S., Argyropoulos D.S., An improved method for isolating lignin in high yield and purity, *J. Pulp Pap. Sci.*, **29**, 2003, 235–240.
38. Yang J.-M., Goring D.A.I., The phenolic hydroxyl content of lignin in spruce wood, *Can. J. Chem.*, **58**, 1980, 2411–2414.
39. Gellerstedt G., Lindfors E.-L., Structural changes in lignin during kraft cooking. Part 4. Phenolic hydroxyl groups in wood and kraft pulps, *Svensk Papperstidn*, **87**, 1984, R115–R118.
40. Lai Y.-Z., Guo X.-P., Variation of the phenolic hydroxyl group content in wood lignins, *Wood Sci. Technol.*, **25**, 1991, 467–472.
41. Lawoko M., Henriksson G., Gellerstedt G., Structural differences between the lignin–carbohydrate complexes present in wood and in chemical pulps, *Biomacromolecules*, **6**, 2005, 3467–3473.
42. Freudenberg K., in *Constitution and Biosynthesis of Lignin* (Eds. Freudenberg K. and Neish A.C.), Springer-Verlag, Berlin-Heidelberg, 1968, p. 103.
43. Nimz H., Mogharab I., Ludemann H.-D., ^{13}C-Kernresonanzspektren von Ligninen, 3. Vergleich von Fichtenlignin mit kunstlichem Lignin nach Freudenberg, *Macromol. Chem.*, **175**, 1974, 2563–2575.
44. Terashima N., Atalla R.H., Ralph S.A., Landucci L.L., Lapierre C., Monties B., New preparations of lignin polymer models under conditions that approximate cell wall lignification. I. Synthesis of novel lignin polymer models and their structural characterization by ^{13}C NMR, *Holzforschung*, **49**, 1995, 521–527.
45. Terashima N., Atalla R.H., Ralph S.A., Landucci L.L., Lapierre C., Monties B., New preparations of lignin polymer models under conditions that approximate cell wall lignification. II. Structural characterization of the models by thioacidolysis, *Holzforschung*, **50**, 1996, 9–14.
46. Landucci L.L., Reaction of *p*-hydroxycinnamyl alcohols with transition metal salts. 3. Preparation and NMR characterization of improved DHPs, *J. Wood Chem. Technol.*, **20**, 2000, 243–264.
47. Önnerud H., Zhang L., Gellerstedt G., Henriksson G., Polymerization of monolignols by redox shuttle-mediated enzymatic oxidation: A new model in lignin biosynthesis, *Plant Cell*, **14**, 2002, 1953–1962.
48. Dence C.W., The determination of lignin, in *Methods in Lignin Chemistry* (Eds. Lin S.Y. and Dence C.W.), Springer-Verlag, Berlin Heidelberg, Germany, 1992, pp. 33–61.
49. Wallbäcks L., Edlund U., Norden B., Berglund I., Multivariate characterization of pulp using solid-state carbon-13 NMR, FTIR and NIR, *Tappi J.*, **74**(10), 1991, 201–206.
50. Nakano J., Meshitsuka G., The detection of lignin, in *Methods in Lignin Chemistry* (Eds. Lin S.Y. and Dence C.W.), Springer-Verlag, Berlin Heidelberg, Germany, 1992, pp. 23–32.
51. Erickson M., Larsson S., Miksche G.E., Gaschromatografische Analyse von Ligninoxydationsprodukten. VII. Ein verbessertes Verfahren zur Charakterisierung von Ligninen durch Methylierung und oxydativen Abbau, *Acta Chem. Scand.*, **27**, 1973, 127–140.
52. Gellerstedt G., Permanganate oxidation, in *Methods in Lignin Chemistry* (Eds. Lin S.Y. and Dence C.W.), Springer-Verlag, Berlin Heidelberg, Germany, 1992, pp. 322–333.
53. Nimz H., Uber ein neues Abbauverfahren des Lignins, *Chem. Ber.*, **102**, 1969, 799–810.
54. Sakakibara A., A structural model of softwood lignin, *Wood Sci. Technol.*, **14**, 1980, 89–100.
55. Sakakibara A., Hydrogenolysis, in *Methods in Lignin Chemistry* (Eds. Lin S.Y. and Dence C.W.), Springer-Verlag, Berlin Heidelberg, 1992, pp. 350–368.
56. Lundquist K., Low-molecular weight lignin hydrolysis products, *Appl. Polym. Symp.*, **28**, 1976, 1393–1407.
57. Lapierre C., Rolando C., Monties B., Characterization of poplar lignins acidolysis products: Capillary gas–liquid and liquid–liquid chromatography of monomeric compounds, *Holzforschung*, **37**, 1983, 189–198.
58. Lapierre C., Monties B., Rolando C., Thioacidolysis of lignin: Comparison with acidolysis, *J. Wood Chem. Technol.*, **5**, 1985, 277–292.
59. Lu F., Ralph J., Derivatization followed by reductive cleavage (DFRC method), a new method for lignin analysis: Protocol for analysis of DFRC monomers, *J. Agr. Food Chem.*, **45**, 1997, 2590–2592.
60. Lu F., Ralph J., The DFRC method for lignin analysis. 2. Monomers from isolated lignins, *J. Agr. Food Chem.*, **46**, 1998, 547–552.

61. Adler E., Lignin chemistry – Past, present and future, *Wood Sci. Technol.*, **11**, 1977, 169–218.
62. Nimz H., Chemistry of potential chromophoric groups in beech lignin, *Tappi*, **56**(5), 1973, 124–126.
63. Nimz H., Beech lignin – Proposal of a constitutional scheme, *Angew. Chem.*, **13**, 1974, 313–321.
64. Lundquist K., Proton (^{1}H) NMR spectroscopy, in *Methods in Lignin Chemistry* (Eds. Lin S.Y. and Dence C.W.), Springer-Verlag, Berlin Heidelberg, Germany, 1992, pp. 242–249.
65. Robert D., Carbon-13 nuclear magnetic resonance spectrometry, in *Methods in Lignin Chemistry* (Eds. Lin S.Y. and Dence C.W.), Springer-Verlag, Berlin Heidelberg, Germany, 1992, pp. 250–273.
66. Ludemann H.-D., Nimz H., Carbon-13 nuclear magnetic resonance spectra of lignins, *Biochem. Biophys. Res. Comm.*, **52**, 1973, 1162–1169.
67. Nimz H.H., Robert D., Faix O., Nemr M., Carbon-13 NMR spectra of lignins. 8. Structural differences between lignins of hardwoods, softwoods, grasses and compression wood, *Holzforschung*, **35**, 1981, 16–26.
68. Lundquist K., NMR studies of lignins. 2. Interpretation of the ^{1}H NMR spectrum of acetylated birch lignin, *Acta Chem. Scand.*, **B33**, 1979, 27–30.
69. Lundquist K., Paterson A., Ramsey L., NMR studies of lignins. 6. Interpretation of the ^{1}H NMR spectrum of acetylated spruce lignin in a deuterioacetone solution, *Acta Chem. Scand.*, **B37**, 1983, 734–736.
70. Bardet M., Foray M.-F., Robert D., Use of the DEPT pulse sequence to facilitate the ^{13}C NMR structural analysis of lignins, *Macromol. Chem.*, **186**, 1985, 1495–1504.
71. Gellerstedt G., Robert D., Quantitative ^{13}C NMR analysis of kraft lignins, *Acta Chem. Scand.*, **B41**, 1987, 541–546.
72. Robert D., Chen C.-L., Biodegradation of lignin in spruce wood by *Phanerochaete chrysosporium*: Quantitative analysis of biodegraded spruce lignins by ^{13}C NMR spectroscopy, *Holzforschung*, **43**, 1989, 323–332.
73. Zhang L. and Gellerstedt G. Quantitative 2D HSQC NMR determination of polymer structures by selecting suitable internal standard references. *Magn. Res. Chem.* **45**, 2007, 37–45.
74. Capanema E.A., Balakshin M.Y., Kadla J.F., A comprehensive approach for quantitative lignin characterization by NMR spectroscopy, *J. Agr. Food Chem.*, **52**, 2004, 1850–1860.
75. Gellerstedt G., Lindfors E.-L., Lapierre C., Monties B., Structural changes in lignin during kraft cooking. Part 2. Characterization by acidolysis, *Svensk Papperstidn*, **87**, 1984, R61–R67.
76. Lapierre C., Pollet B., Monties B., Rolando C., Thioacidolysis of spruce lignin: GC–MS analysis of the main dimers recovered after Raney nickel desulphuration, *Holzforschung*, **45**, 1991, 61–68.
77. Sjöström E., *Wood Chemistry. Fundamentals and Applications*, 2nd Edition, Academic Press Inc., San Diego, USA, 1993. p. 82.
78. Saka S., Goring D.A.I., The distribution of lignin in white birch wood as determined by TEM–EDXA, *Holzforschung*, **42**, 1988, 149–153.
79. Grunwald C., Ruel K., Kim Y.S., Schmitt U., On the cytochemistry of cell wall formation in poplar trees, *Plant biol.*, **4**, 2002, 13–21.
80. Evtuguin D.V., Amado F.M.L., Application of electrospray ionization mass spectrometry to the elucidation of the primary structure of lignin, *Macromol. Biosci.*, **3**, 2003, 339–343.
81. Lawoko M., Henriksson G., Gellerstedt G., New method for the quantitative preparation of lignin–carbohydrate complex from unbleached softwood kraft pulp: Lignin–polysaccharide networks I., *Holzforschung*, **57**, 2003, 69–74.
82. Lawoko M., Berggren R., Berthold F., Henriksson G., Gellerstedt G., Changes in the lignin carbohydrate complex in softwood during kraft and oxygen delignification: Lignin–polysaccharide networks II., *Holzforschung*, **58**, 2004, 603–610.
83. Lawoko M., Henriksson G., Gellerstedt G., Characterization of lignin–carbohydrate complexes from spruce sulfite pulp, *Holzforschung*, **60**, 2006, 162–165.
84. Gellerstedt G. and Zhang L. Chemistry of TCF-bleaching with oxygen and hydrogen peroxide, in *Oxidative Delignification Chemistry. Fundamentals and Catalysis*, Ed. Argyropoulos D.S., *ACS Symposium Series* **785**, 2001, pp. 61–72.
85. Goring D.A.I., Polymer properties of lignin and lignin derivatives, in *Lignins. Occurrence, Formation, Structure and Reactions* (Eds. Sarkanen K.V. and Ludwig C.H.), Wiley-Interscience, New York, USA, 1971, pp. 695–768.
86. Gierer J., Lindeberg O., Reactions of lignin during sulfate pulping. Part XIX. Isolation and identification of new dimers from a spent sulfate liquor, *Acta Chem. Scand.*, **B34**, 1980, 161–170.
87. Niemelä, K. "*Low-Molecular-Weight Organic Compounds in Birch Kraft Black Liquor*". Ph.D. Thesis, Helsinki University of Technology, 1990.
88. Mörck R., Yoshida H., Kringstad K., Hatakeyama H., Fractionation of kraft lignin by successive extraction with organic solvents. I. Functional groups, ^{13}C NMR-spectra and molecular weight distributions, *Holzforschung*, **40**(Suppl.), 1986, 51–60.
89. Mörck R., Reiman A., Kringstad K., Fractionation of kraft lignin by successive extraction with organic solvents. III. Fractionation of kraft lignin from birch, *Holzforschung*, **42**, 1988, 111–116.
90. Gellerstedt G., Majtnerova A., Zhang L., Towards a new concept of lignin condensation in kraft pulping. Initial results, *C. R. Biol.*, **327**, 2004, 817–826.

91. Gellerstedt G., Lindfors E.-L., Structural changes in lignin during kraft pulping, *Holzforschung*, **38**, 1984, 151–158.
92. Lin S.Y. and Detroit W.J. Chemical heterogeneity of technical lignins – Its significance in lignin utilization. *1st International Symposium on Wood and Pulping Chemistry*, Stockholm, Sweden, 1981, Proceedings, Volume 4, pp. 44–52.
93. Hassi H.Y., Chen C-L. and Gratzl J.S. Carbon-13 NMR spectroscopic characterization of lignosulfonates. *Tappi Research and Development Conference*, Appleton, USA 1984, Proceedings, pp. 249–260.
94. DeLong E.A. Method of rendering lignin separable from cellulose and hemicellulose in lignocellulosic material and the product so obtained. Can. Pat. #1,096,374, 1981.
95. Josefsson T., Lennholm H., Gellerstedt G., Steam explosion of aspen wood. Characterization of reaction products, *Holzforschung*, **56**, 2002, 289–297.
96. Robert D., Bardet M., Lapierre C., Gellerstedt G., Structural changes in aspen lignin during steam explosion treatment, *Cell. Chem. Technol.*, **22**, 1988, 221–230.

– 10 –

Industrial Commercial Lignins: Sources, Properties and Applications

Jairo Lora

ABSTRACT

This chapter is devoted to review the interest of industrial lignins as materials, rather than fuels. It examines the different industrial technologies related to pulp manufacture in terms of the delignification process and, more importantly, the properties of the corresponding isolated lignins. Sulphite, kraft and soda lignins are therefore reviewed with an emphasis on their structural differences and hence their different domains of applications, including the unique polyelectrolyte features associated with lignosulphonates. Novel emerging technologies related to biomass conversion are also critically discussed, in particular steam explosion and organosolv pulping, with the purpose of illustrating the types of lignins produced and their potential usage.

Keywords

Lignin, Industrial production, Sulphite pulping, Kraft pulping, Soda pulping, Lignosulphonates, Properties, Applications, Steam explosion, Organosolv technology

10.1 INTRODUCTION

Industrial lignins are currently obtained as co-products of the manufacture of cellulose pulp for paper and some chemical derivatives. Although the amount of lignin extracted in pulping operations around the world is estimated to be over 70 million tons per year, less than 2 per cent is actually recovered for utilization as a chemical product. The majority of the lignin extracted in pulp and paper operations is actually burned as part of the treatment of the spent pulping liquors for recovery of energy and regeneration of pulping chemicals.

In this chapter lignins that are available as commercial products in industrial quantities are discussed. The type of pulping process determines the type of lignin industrially available. There are three types of lignins, which correspond to the three dominant chemical pulping processes, namely sulphite, kraft and soda. During pulping the lignin structure is unavoidably modified from that in the original feedstock. For each type of lignin the main reactions and process involved, properties of the resulting lignin and its applications and main producers are reviewed. Chapter 9 of this book addresses the structure and properties of native lignins (*i.e.* as they exist in nature). The reader is also referred to that chapter for information on the type of repeating units normally found in the lignin structure and on the nomenclature used to designate the carbon atoms in the molecule.

In addition to currently available industrial lignins, this chapter includes a discussion on novel lignins that are potentially available in the near future from biorefinery operations, for instance, for the production of ethanol from lignocellulosic biomass.

10.2 SULPHITE LIGNIN

Sulphite lignin is produced in the sulphite pulping process and has historically been the most abundant type of industrial lignin commercially available. About 80 years ago, sulphite was the dominant type of pulping process [1]. It was initially developed based on the use of calcium bisulphite, an inexpensive pulping chemical. The lack of suitable technology for the regeneration of this chemical from the spent pulping liquor propitiated the development of uses for calcium lignosulphonate.

The sulphite pulping technology evolved to encompass the use of other sulphites and bisulphites, such as magnesium, ammonium and sodium. The resulting lignosulphonates found also applications, in some cases specially adapted to a given cation. For these pulping chemicals, recovery systems based on combustion of the organic components of the spent liquor were developed, facilitating environmental compliance without having to rely on lignosulphonate sales. With time, the sulphite pulping process started to lose ground relative to the kraft counterpart, which is more versatile and produces stronger pulps, and has a robust chemical recovery process. Thus, the days of sulphite pulping dominance are long gone and sulphite pulp production has declined from nearly 20 million tons in the 1980s [1] to about 7 million tons nowadays [2]. Current expectations are for continuing decline of sulphite pulp production, since no new sulphite mills have been built in over a quarter of a century and the shut down of antiquated inefficient units will negate small incremental increases that may happen at a few places. In spite of this trend, sulphite lignin production has managed to remain relatively stable, meeting a worldwide demand in which recent declines in consumption in Western Europe, North America and Japan have been offset by increases in China, India and other parts of the world.

10.2.1 Process

In the sulphite pulping process, wood is digested at 140–170°C with an aqueous solution of a sulphite or bisulphite salt of sodium, ammonium, magnesium or calcium. The type of salt and its solubility and dissociation characteristics determine the pH of the digestion. For instance, the pulping medium is highly acidic when calcium bisulphite is the pulping agent, but is practiced in highly alkaline environment when using sodium sulphite. During the digestion process, several chemical events take place, including the cleavage of linkages between the lignin and the carbohydrates, the scission of carbon–oxygen bonds that interconnect lignin units, and the sulphonation of the lignin aliphatic chain. The latter reaction is the most critical in sulphite pulping and gives sulphite lignin its main characteristics. This reaction is the result of the attack of the negatively charged sulphite or bisulphite ions on the lignin structure. The targets of such a nucleophilic attack depend on the pH of the sulphite delignification. Thus, at low pH, carbenium (benzylium) ions of phenolic or non-phenolic units are sulphonated at the α position of the side chain (Fig. 10.1), while at higher pH levels quinone methide structures of phenolic units are the targets [3, 4].

In the acid pH range, weakly nucleophilic sites in the aromatic ring of lignin compete with the sulphite and bisulphite ions for the carbonium ion intermediates, forming condensed structures. Such condensation reactions are detrimental to pulp production and can be minimized by a judicious choice of pulping conditions [4].

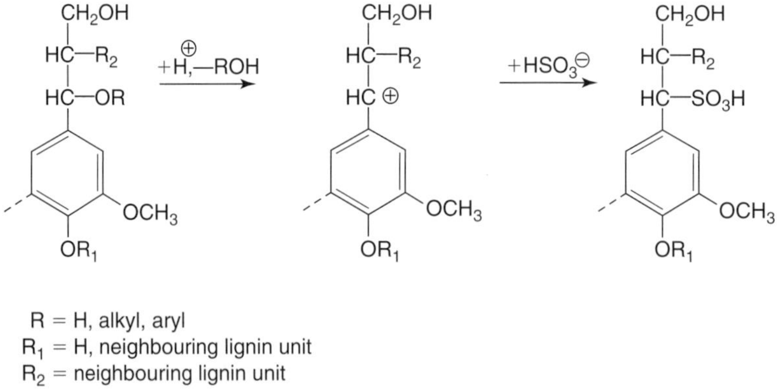

Figure 10.1 Main reaction scheme for lignosulphonate formation during acid sulphite pulping [3].

During pulping, about 4–8 per cent sulphur is incorporated into the lignin molecules [5, 6], mostly in the form of sulphonate groups. Such groups cause the lignin to become water-soluble and prevent its recondensation, a reaction that is detrimental to pulping as it results in the redeposition of lignin on the fibres. As lignin is removed, the cellulosic fibres are liberated and once the required degree of delignification has been attained, the resulting pulp is separated from the spent pulping liquor by filtration and washing. Most lignosulphonates available in the market today are co-produced with pulps of low lignin content, but some may be obtained in conjunction with the production of pulps with relatively high lignin content, such as pulps used for paper packaging grades. Such a difference in pulping process intensity probably causes variances in the lignin structure.

Sulphite pulping (or any other pulping process, for that matter) does not selectively remove lignin. The carbohydrate components (particularly the hemicelluloses) are solubilized to a considerable extent. As a result, the spent sulphite pulping liquor contains not only lignosulphonate (or sulphonated lignin), but also hemicellulose sugars, and the spent pulping chemicals. Furthermore, carbohydrates appear to be chemically combined with some fractions of the lignosulphonate [7]. All these components are highly soluble in water and are difficult to separate from one another by methods that rely on solubility differences.

Crude spent sulphite liquor in a concentrated form, or as a powder, is available as a product (Fig. 10.2), but purified forms of sulphite lignin have broader applications and higher value [8]. Such purified lignins are obtained by diverse techniques (sometimes involving the simultaneous recovery of co-products) such as:

- Alcoholic fermentation of the sugars in the sulphite liquor and subsequent distillation of the resulting beer, which leaves behind a low sugar content lignosulphonate.
- Sugar removal by chemical destruction.
- Ultrafiltration to remove a sugar-rich permeate, therefore increasing the lignosulphonate content of the retentate fraction.
- Precipitation.

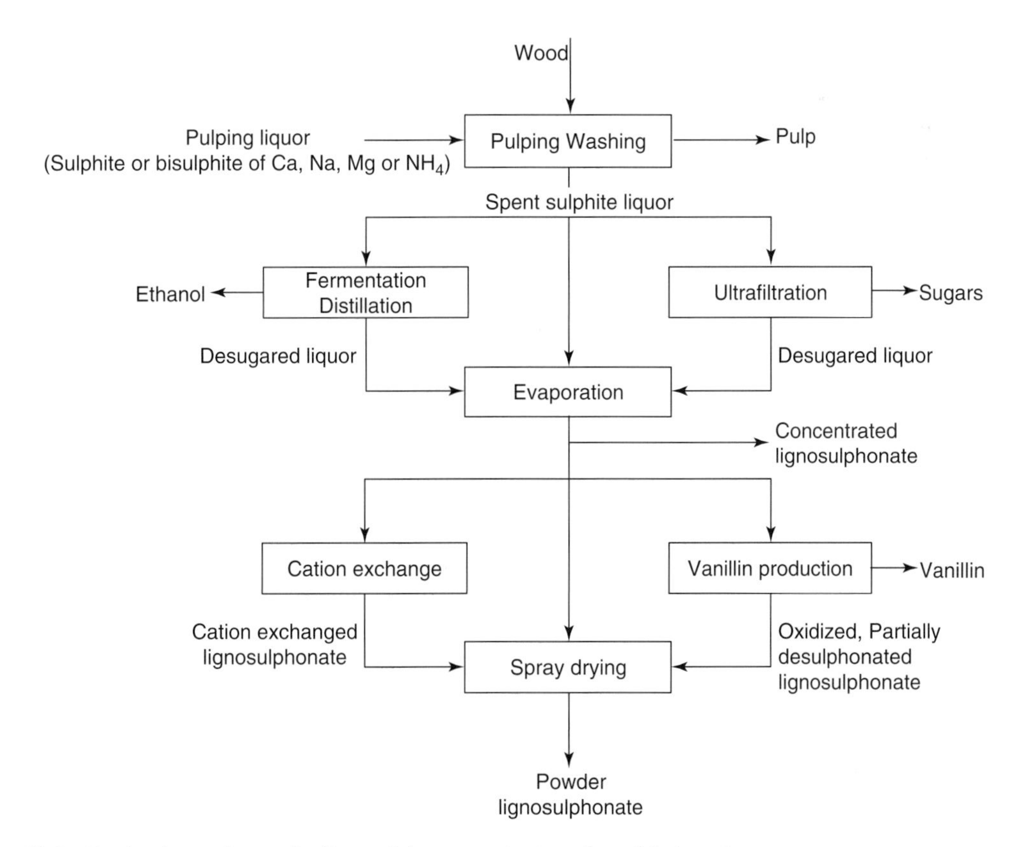

Figure 10.2 Production pathways for lignosulphonate and selected modified versions.

In addition to purified lignosulphonates, chemically modified lignosulphonates are also produced. The modifications currently in use include:

- Changing the cation (or base) from that originally present in the pulping liquor to a different one, as required for specific applications. This is normally done through the use of ion-exchange resins.
- Oxidation, which is normally done for the production of vanillin. After vanillin recovery, the remaining low sugar, partially desulphonated, oxidized lignosulphonate is recovered and used in specific applications.
- Various other reactions, such as carboxylation, amination, crosslinking, depolymerization and graft copolymerization [9]. Sometimes combinations of these modifications are practiced.

10.2.2 Properties

There is a dearth of information on fundamental analysis and characterization of industrial lignins in the published scientific and technical literature, causing many gaps in the detailed understanding of the structure and macromolecular properties of the products industrially available. The situation appears to be worse for sulphite lignins, in spite of the fact that these are the most widely used lignin products. While advancements in modern analytical techniques have made significant contributions to the elucidation of the properties of other lignins (such as native lignin, change in lignin structure during kraft pulping, and lignins from emerging processes, such as organosolv delignification), lignins from the sulphite pulping process have benefited much less from the latest methods. Several factors may explain this, including the declining importance of sulphite pulping, the greater complexity of lignosulphonates as compared to the lignin molecule, (due to the multiple reactions of lignin with the very active sulphite and bisulphite species present in the pulping medium and the possibility to have various counter ions associated with the molecule), the lack of widely accepted standard methods for the preparation of purified lignosulphonates, and the fact that the protocols developed for the lignins that have received more attention require adaptation before they can be applied to lignosulphonates. One exception, which illustrates the complexity of lignosulphonates, is the recent application of two-dimensional ^{1}H-^{13}C NMR spectroscopy to hardwood and softwood lignosulphonates, which has been used to show differences in functional groups for high and low molecular weight fractions, presence of sugar residues, and the nature of methylene moieties [10].

The sulphur content of lignosulphonates is reported in the literature at 4–8 per cent [6, 11, 12]. Most of the sulphur appears in the lignin as sulphonate, but other types of organic sulphur, sulphite and sulphate, are also present. Due to their sulphonate content, lignosulphonates are very soluble in water and are insoluble in organic solvents. According to technical literature from the company Lignotech Borregaard (a leading lignosulphonate producer), the degree of sulphonation in its industrial products ranges from 0.17 to 0.65 sulphonate groups per phenyl propane unit, or about 0.9–3.3 sulphonate groups per 1 000 grams of lignin. The high end is represented by de-sugared lignosulphonates extracted from pulping liquors and the low end by modified products, such as oxidized lignosulphonates obtained as co-products in vanillin production [13].

Other functional groups that are present in lignosulphonates include phenolic hydroxyl, aliphatic hydroxyl and carbonyl groups. Their quantity is very dependent on the degree of purity, molecular weight and the production process [14, 15].

Sugar content is an important property of industrial lignosulphonates. The presence of sugars has implications that may be negative in some applications, such as being a retardant in concrete cure. Crude spent sulphite liquor may have reducing sugar contents as high as 35 per cent (solids basis). Low sugar lignosulphonates are produced by fermentation of the sugars, chemical degradation and ultrafiltration [8, 12, 16, 17]. In addition, essentially sugar-free (but oxidized and partially desulphonated) lignosulphonates are obtained as co-products in the manufacture of vanillin from the spent sulphite liquor [13].

The range of values reported in the literature for the molecular weights of lignosulphonates is quite large, from less than 1 000 to 150 000 and even higher. Polydispersities as high as 7 have been reported [5, 18]. It has been recently proposed that the behaviour of the lignosulphonate molecule in an aqueous medium is better explained by assuming that the macromolecule follows a randomly branched polyelectrolyte conformation [19], rather than the microgel model proposed earlier [18]. In the randomly branched polyelectrolyte model, the backbone of the molecule is less sulphonated than the branches. The use of laser correlation spectroscopy suggests that lignosulphonates are nano-dimensional systems with two domains of particle sizes: one corresponding to individual macromolecules with size ~10 nm and the other to associated macromolecules with 100–200 nm size [20].

There is great potential for lignosulphonate characterization based on the application of techniques that rely on hydrophobic interactions. Thus, reversed phase chromatography has been applied to the analytical and preparative fractionation of lignosulphonates on the basis of polarity and has been found to be able to resolve sugar-free lignosulphonates from those covalently bound with carbohydrates [21]. Hydrophobic interaction chromatography, which is extensively used in protein purification and is based on the hydrophobic attraction between a target molecule and a stationary phase, has recently been shown to be useful to classify lignosulphonates according to their hydrophilicity [22, 23]. Inverse gas chromatography (where the stationary phase is the polymer being characterized, which is done by injecting volatile probes of known properties) has been successfully used to characterize the dispersive component of surface free energy (a parameter that for films is evaluated by contact angle measurements) of powder lignosulphonates and other lignins [24, 25].

Regarding the thermal properties of lignosulphonates, when examined by differential scanning calorimetry (DSC), they do not exhibit a glass transition temperature. In high temperature DSC decomposition runs they show two peaks, one at about 320°C and the other at about 450°C. The first peak is associated with the degradation of carbohydrates and the second with lignin itself [26].

10.2.3 Applications

The main applications for lignosulphonates are as dispersants and binders and are discussed in greater detail elsewhere [8]. They include:

- *Concrete water reducers*: This is the largest current application for lignins. The dispersant capacity of lignosulphonates (mainly calcium and sodium) is the basis for their use as additives that help increase the workability of concrete and allow a reduction in the amount of water required for proper mixing and handling. By using less water, it is possible to have stronger concrete. Lignosulphonates tend to cause some retardation of concrete setting, particularly when their sugar content is high, and they may also cause air entrainment. Lignosulphonates are also used in the manufacture of grouts and mortars and in cement manufacture by the wet process, where they reduce energy consumption. The use of lignosulphonate in concrete admixtures and other construction-related applications is expected to continue growing significantly in countries such as China and India, which have rapidly expanding economies that require multiple large infrastructure projects and other civil works.
- Other applications that benefit from the tensioactive properties of lignosulphonates include dispersants for gypsum wallboard (where the use of lignosulphonates permits a reduction in the amount of water used for forming the board and consequently a decrease of the drying energy), agrochemicals (for instance, to make water dispersible or wettable powders), dyestuffs (where they may serve as grinding aids as well), bitumen (specially when preparing cold-mix bitumen emulsions), pigments (particularly carbon black used in inks). The dispersing properties of lignosulphonates also make them components of drilling fluids for oil recovery, markets for which in the past chrome and ferrochrome lignosulphonates were extensively used. Currently, a more environmentally friendly alternative is lignosulphonate grafted with acrylates [8, 9]. Other dispersant applications of interest are formulations for scale formation inhibition in industrial waters for a variety of other industrial cleaning purposes. Lignosulphonates are also used as thinning agents for molasses in animal feed.
- *Animal feed binders*: On a volume basis, this is the second most important application of lignosulphonates. Ammonium or calcium lignosulphonate are preferred for the manufacture of pellets, which must be robust enough for handling in industrial equipment without generation of fines and dust. In addition to acting as a binder for the pellets, lignosulphonates provide lubrication in the pelletizing equipment.
- Lignosulphonates are used in other industrial binding applications, such as for green strength improvement of refractories, components in foundry resins, granulation of fertilizers and other agrochemicals, and for briquetting of mineral ores and coal dust. Lignosulphonates have been used in the formulation of thermoset phenol and urea formaldehyde resins used as binders for fibreglass insulation. Because of their high water solubility and low flow when heated, lignosulphonates are less suitable for use in exterior grade resins used as plywood and oriented strand board adhesives. Another major volume utilization (but of a rather low value) is the utilization of crude spent sulphite liquor for soil stabilization and dust control.

- Lignosulphonates are also used in miscellaneous applications, such as additives to stiffen packaging grade papers [27–29], as leather tanning agents [30], and as carriers for agricultural micronutrients [31], among others. Softwood lignosulphonates are catalytically oxidized for conversion to vanillin, a flavour ingredient and pharmaceutical intermediate [32]. The latter process was widely used in the past, but only one facility remains in operation in the world. The residual material that remains after vanillin production is a partially oxidized, low sugar lignin derivative that finds special use as a dispersant for dyes and as an expander for negative pastes of lead-acid storage batteries [13].

10.2.4 Main producers

Borregaard LignoTech is the largest producer of lignosulphonates worldwide. Its 11 manufacturing units in Europe, Asia, Africa and the Americas produce more than 500 000 metric tons (dry basis) of lignin products annually, obtained as co-products from the pulping of hardwoods and softwoods. The manufacture of lignin products by Borregaard LignoTech is integrated in some of their factories with ethanol and/or vanillin production. Their lignin product range consists of more than 200 different offerings, based principally on calcium and sodium lignosulphonates and including full sugar content, fermented, double fermented, ultrafiltered, oxidized and ion-exchanged products [33].

Tembec is the second largest lignosulphonate producer in the world. The company has three sulphite pulp mills in Europe and North America, two of which are major producers of lignosulphonates for sale. Tembec can potentially produce 570 000 metric tons of lignosulphonates per year, but sales are less than that quantity, since the company processes a major fraction of its liquor through its recovery boilers, depending on energy cost and requirements, market conditions and other considerations. The company offers sodium, ammonium and magnesium lignosulphonates from hardwood and softwood in liquid as well as powder form. Complementary processes used by Tembec include alcoholic fermentation (to produce low sugar products) and ion-exchange to produce sodium base lignosulphonate [34].

La Rochette Venizel has one facility in France, which offers sodium, ammonia, magnesium and calcium lignosulphonates. Unlike most other lignosulphonates, which are co-produced with low yield, bleachable pulps, La Rochette Venizel's products are co-produced with high yield, packaging grade pulps, using a milder cooking process [35].

Nippon Paper Chemicals is a Japanese producer of calcium, sodium and magnesium lignosulphonates encompassing a range of properties, including variations in molecular weights and degrees of sulphonation [36].

Cartiere Burgo is an Italian producer of liquid and powder calcium, ammonium and sodium lignosulphonates from red pine, with capacity for about 40 000 tons per year (solids basis) [37].

Domsjö Fabriker AB (formerly part of the MoDo group) has one facility in Sweden with capacity to produce about 25 000 tons of softwood lignosulphonate per year. The facility is integrated with ethanol production [38]. The company has a technology licensing agreement with Nippon Paper Chemicals for the manufacture of concrete additives [36].

10.3 KRAFT LIGNIN

Although the kraft process is the most predominant pulping process worldwide, the recovery of kraft lignin for chemical uses is not practiced broadly at this time. Kraft pulp mills have evolved into very large facilities that are integrated with a highly engineered system for recovery of pulping chemicals and energy which is based on the combustion of the spent pulping liquor (black liquor). The recovery system is essential to the economic and environmental performance of the kraft pulp mill and its use simplifies the business model of pulp and paper producers, allowing them to concentrate on their core paper business. Therefore, relative to sulphite lignin, the quantity of kraft lignin recovered for chemical use is rather small; worldwide only one company is currently practicing it on an industrial basis. This situation may change in the near future, since due to steady efficiency improvements (particularly in Scandinavia), a modern kraft pulp mill nowadays may generate an excess of energy relative to its needs. In addition, the ability to implement incremental kraft pulp production is a desirable goal. Thus, the extraction of lignin has been proposed as a strategy to allow pulp capacity expansion and keep energy production and consumption in balance within the mill. The extracted lignin can then be used inside the mill to generate exportable electricity or outside the pulp mill for energy or other uses.

10.3.1 Process

In the kraft process, the fibrous feedstock is digested with a mixture of sodium hydroxide and sodium sulphide at about 170°C. During the digestion several reactions take place, including cleavage of lignin–carbohydrate linkages, depolymerization of the lignin, its reaction with hydrosulphide ion and its recondensation. Lignin depolymerization during kraft pulping occurs principally by the cleavage of α and β aryl ether bonds, first in phenolic units and in a later phase of delignification in non-phenolic units. Such reactions generate moieties with free phenolic groups, which are soluble in the alkaline environment prevalent during the digestion.

Quinone methide structures are involved as intermediates in kraft delignification, as illustrated in Fig. 10.3. The α carbon of such intermediates reacts with the hydroxyl, hydrosulphide and sulphide ions present in the pulping medium. Since the sulphide and hydrosulphide ions are much stronger nucleophiles than the hydroxyl group, benzylthiol structures are preferentially formed [4, 39]. This strongly nucleophilic species displaces the aryl ether attached to the β carbon forming an episulphide and causing the depolymerization of the lignin molecule and the formation of a phenolic hydroxyl group in the displaced aryl group, hence increasing lignin solubility in alkali.

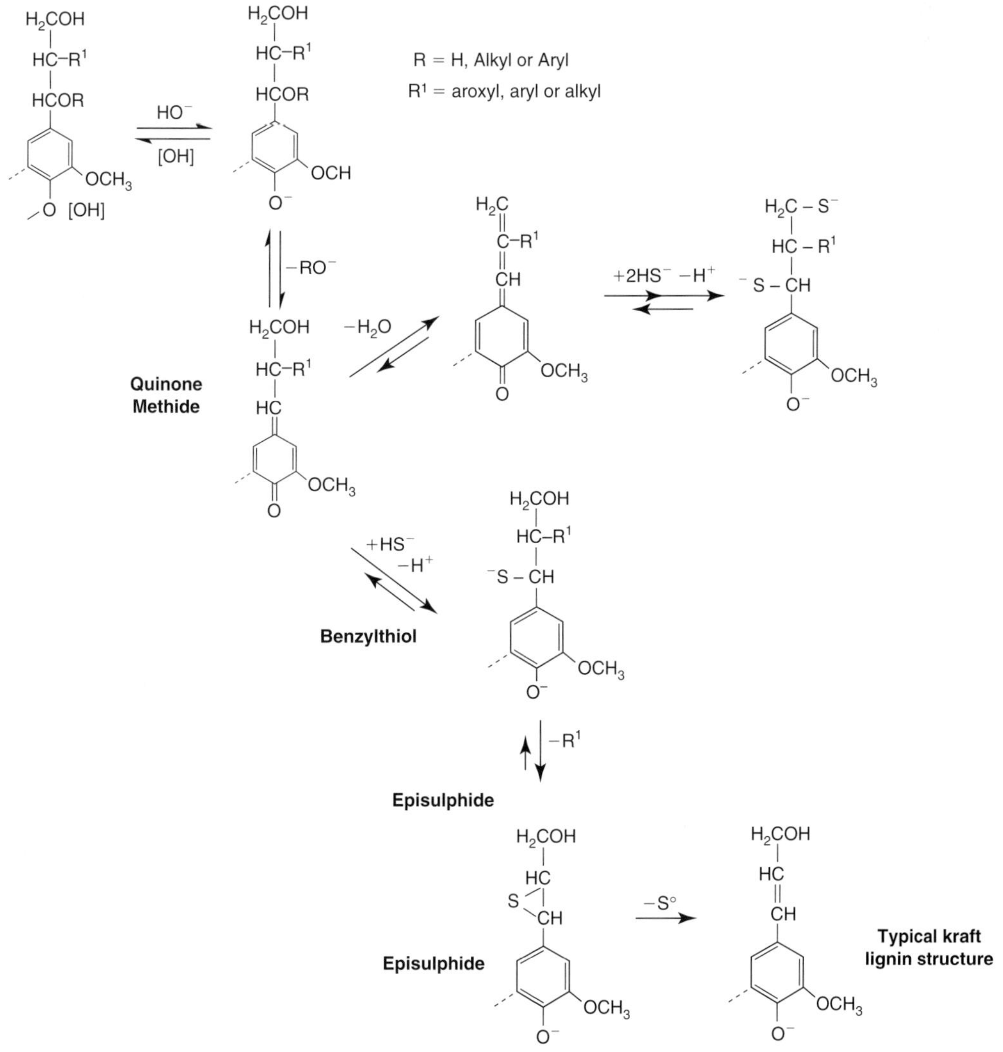

Figure 10.3 Main reaction scheme for the formation of kraft lignin during pulping. Adapted from [39].

The lignin episulphide intermediate in turn can lose elemental sulphur to give conjugated structures. The elimination of sulphur from the episulphide intermediate explains the relatively low sulphur content of kraft lignins. Unlike the sulphonation reaction in sulphite pulping, which produces a water-soluble lignin throughout the entire pH range, the reactions in kraft pulping do not lead to the formation of a lignin soluble in water under acid or neutral conditions.

Kraft pulping appears to be particularly efficient for cleaving dibenzodioxocins, the eight-member rings that have been proposed as a major branching point in wood lignins [40–42]. As a result, 5,5′-biphenyls are formed in high yield and with great selectivity. During alkaline pulping lignin–lignin condensation reactions also take place, therefore counteracting to some extent the lignin fragmentation processes [4]. In this case carbanions formed in the degraded lignin compete for reactive quinone methide sites with the nucleophiles supplied by the pulping liquor. In the case of kraft pulping, the intermediate benzylthiol and episulphide structures may protect the quinone methide intermediates and prevent their participation in such condensation reactions, which would explain the lower amount of condensed diphenyl methane structures observed in kraft pulping relative to soda pulping [42]. Another type of condensation that takes place may involve the formaldehyde released during pulping by the elimination of methylol groups in the lignin side chain. Such formaldehyde may form diarylmethane structures via reactions similar to phenol–formaldehyde (PF) alkaline condensation reactions.

The decreasing water solubility of kraft lignin as the pH is lowered, is the principle used for separation of lignin from the black liquor. As the pH is reduced, the ionization of the molecule is decreased and self-aggregation takes place. Since other components of the black liquor (such as inorganic constituents and the sugars and their degradation products) are soluble in water over a wide pH range, by lowering the pH and precipitating the lignin it is possible to recover it as a product relatively low in ash and carbohydrates.

In carrying out the precipitation, essentially any acidifying medium can be used. However, early on it was recognized that carbon dioxide gas (for instance, from flue gases) was a convenient and cost effective way to accomplish the acidification of alkaline black liquors to a pH of about 9.5 [43]. In the case of kraft wood black liquors at such a pH, the lignin exists as a colloidal suspension that can then be coagulated, that is, heated to about 95–100°C to increase the particle size from ~0.5 to 5 μm or greater and form a more easily filterable material. After filtration, washing with mineral acid of the resulting lignin cake may be applied to reduce the level of contaminants. This can be done by several means, for instance by resuspension in a mineral acid solution followed by filtration [44] or by displacement washing with an acid solution [45] (Fig. 10.4).

Most current commercial applications for kraft lignin involve its chemical derivatives, rather than the lignin as such. The most widely used kraft lignin derivatives are water-soluble sulphonated products, which are used as surfactants. Since the starting lignin material is low in sugars and ash, the resulting sulphonated lignin has high purity and may be used in high value-added applications. The introduction of sulphonate groups into kraft lignin is normally accomplished by sulphomethylation. This reaction can be carried out in a number of ways, such as by reacting the lignin either simultaneously or sequentially with formaldehyde and sodium sulphite [46, 47]. As a result, a methanesulphonate group is introduced at the lignin aromatic ring. Other methods have been proposed to introduce sulphonate groups in kraft lignin, including oxidative sulphonation and sulphonation in oleum, but such methods apparently are not practiced industrially.

Other important derivatives of kraft lignin are lignin amine derivatives prepared by the Mannich reaction [48–50]. Some sulphonated kraft lignins that are used as dispersants are also ethoxylated [51].

10.3.2 Properties

Industrial unmodified kraft lignins (such as Indulin AT produced by MeadWestvaco or Curan 100, formerly produced by Lignotech Borregaard) are characterized by a relatively high degree of purity. Since kraft lignin is not very soluble in water, it is possible to fractionate out water-soluble components, such as sugars and ash, as part of the precipitation process described above. Ash contents are usually below 3 per cent. A recent analytical round robin of lignin samples has reported acid insoluble and soluble lignin contents of about 90 and 3 per cent, respectively, and sugar contents below 2.3 per cent [52]. Carbohydrate moieties in kraft lignin seem to be more predominantly associated with the high molecular weight fraction [53, 54].

Kraft lignins contain 1.5–3.0 per cent sulphur, some of it organically bound and some as elemental sulphur [55]. For softwood kraft lignin, the sulphur content appears to be higher in low molecular weight fractions [53].

The literature has several reports on the molecular weight distribution of kraft lignins. Preparations isolated in the laboratory from industrial black liquors have been reported to have weight average molecular weights ranging from

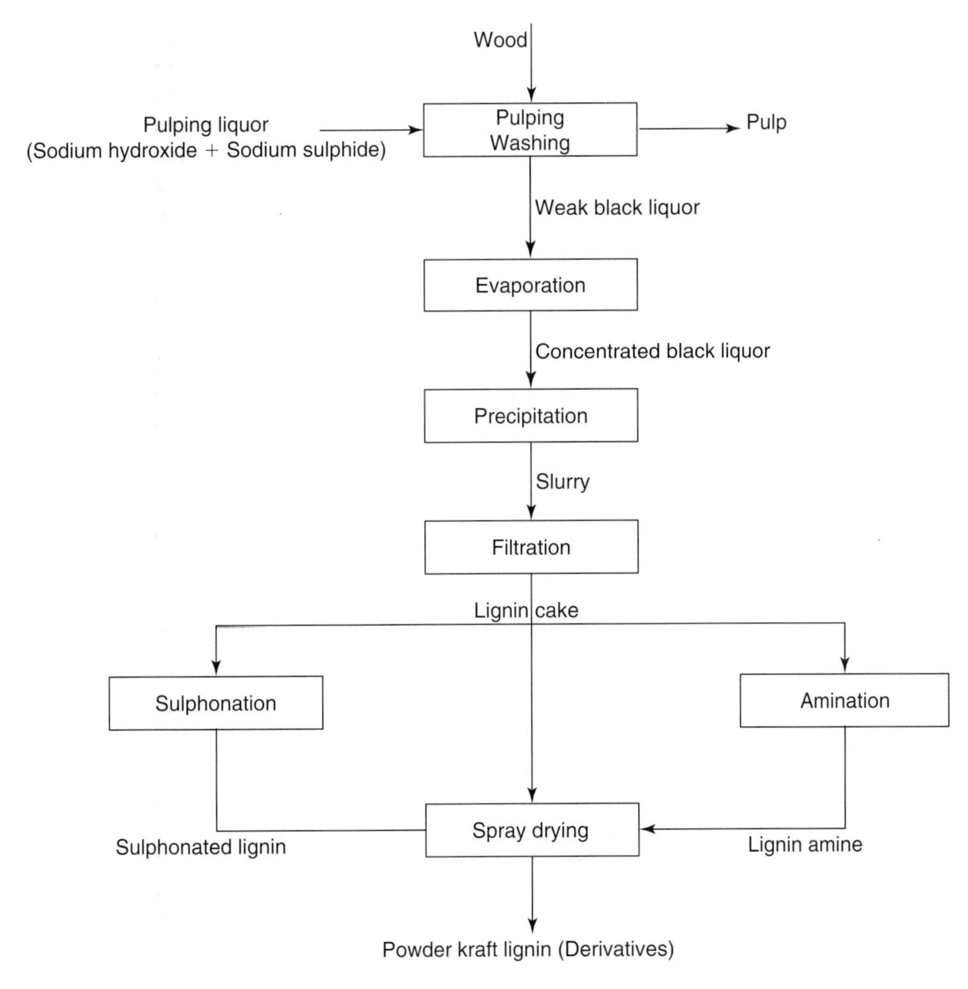

Figure 10.4 Diagram for production of kraft lignin and selected derivatives.

2500 to 39000 [53, 54, 55–57]. Measurements done on industrial products suggest a narrower range [5, 52, 58]. Standard methods for molecular weight determination of lignins have not been established as yet, in spite of significant ongoing efforts in that direction [59].

The total hydroxyl content of industrial kraft lignin has been reported at 1.2–1.27 groups per C9 unit, of which 56–60 per cent are phenolic in nature. Kraft lignin does not have aldehyde groups, since these are very labile under alkaline pulping conditions. About 0.15 conjugated and non-conjugated carbonyl groups are present. Other functional groups include benzylic alcohol, carboxylic acids and unsaturated structures. NMR studies indicate that pine kraft lignin has about 2.5 unsubstituted positions in its aromatic rings [60, 61].

Industrial kraft lignin exhibits a glass transition temperature at around 140°C. In high temperature decomposition runs, it exhibits a multiplet peaking at 452°C with two shoulders at 432°C and 482°C. This is in contrast to lignosulphonates, which display no glass transition temperature and have a low temperature degradation peak because of their carbohydrate content [26].

Concerning kraft lignin derivatives such as sulphonated kraft lignins (which have more extensive commercial application than the unmodified product), there is relatively little information available. According to their manufacturer, they are available with weight average molecular weights between about 3000 and 15000 and degrees of sulphonation in the range of 0.7 to 3.3 sulphonate group per 1000 g of lignin [62]. The latter is in the range of sulphonation levels exhibited by lignosulphonates produced in sulphite pulping.

Hydrophobic interaction chromatography appears to be promising to classify kraft and other lignins according to their degree of sulphonation [23].

10.3.3 Applications

The main applications for kraft lignin derivatives are as dispersants and emulsifiers. They include:

- *Dye dispersants*: Sulphonated kraft lignins are the benchmark for dispersed dyes. Such lignins are made by processes in which the molecular weight, degree of sulphonation or number of free phenolic groups are controlled to get products that meet the requirements of the dyestuff industry, such as grinding efficiency, rheology control, thermal stability, staining and counteracting strength reduction of diazo dyes. The design of kraft lignin based dispersants is a balancing act. For instance, increasing the degree of sulphonation, increases the dispersing efficiency, but reduces heat stability, that is, the adsorption of the dye to the dispersant at a higher temperature. To obtain good dispersion and heat stability, combinations of surfactants used in specific order of addition are sometimes recommended.
- *Agrochemical dispersants*: Sulphonated and ethoxylated sulphonated kraft lignins are used as dispersants for crop protection products formulated in tablets, microcapsule suspensions, wettable powders and produced by pan granulation, spray drying, fluid bed agglomeration and others. The molecular weight and degree of sulphonation can be tailored to the application.
- Sulphonated kraft lignins are also components of air entrainment formulations that give mortar and/or concrete a microstructure that results in better freeze/thaw performance. They are also used as battery expanders (similar to oxidized, partially desulphonated lignosulphonates) to increase the surface of the paste in lead-acid batteries.
- A high-tech application that is in its early stages of development is the production of conductive polymers, in which ethoxylated sulphonated kraft lignin has been used as dopant in the template-guided polymerization of aniline. These materials encounter use in electrostatic dissipative devices, electromagnetic interference shielding of sensitive electronic equipment, production of radar invisible coatings and corrosion protection, among others [63]. Unlike other conductive polymers, those based on sulphonated kraft lignin are more dispersible in water and other solvents and promise to be more versatile and economic.
- *Asphalt emulsifiers*: Unmodified kraft lignins are reported to be good stabilizers for oil-in-water emulsions [64]. Aminated kraft lignin products are particularly effective to prepare asphalt emulsions that are stable at temperatures below water freezing to above water boiling, regardless of dilution or presence of weak acids, bases or salts [30, 50].
- Unmodified kraft lignins have a few applications, principally as anti-oxidant for fat during meat rendering and as carriers, adsorbents and UV screens for active compounds in formulations of crop protection chemicals [62].
- *Production of low molecular weight chemicals*: Dimethylsulphoxide (DMSO) is produced from kraft black liquor by Gaylord Chemical, a US manufacturer [65]. The process involves adding elemental sulphur to concentrated black liquor and reacting at high temperature to form dimethyl sulphide (DMS) and methyl mercaptan. The DMS is oxidized with nitrogen oxide and oxygen to form DMSO [32].

10.3.4 Main producer

Mead-Westvaco (who recently acquired Lignotech Borregaard relatively minor kraft lignin business) in the US is the only industrial producer of kraft lignin and derivatives for chemical use worldwide. The company produces mostly low sugar content sulphomethylated lignins from wood at various degrees of sulphonation and molecular weights. Most products are offered in powder form and are used all over the world. The production capacity of the company has been previously reported at 35 000 ton per year (dry basis) [8].

10.4 SODA LIGNIN

Soda pulping was industrialized in 1853 [1] and has traditionally been used for non-wood fibres, such as straw, sugarcane bagasse, flax, etc. Such raw materials played a dominant role as sources of pulp up to about a century ago and they still remain an important fibre source for many types of papers in certain developing countries (noticeably China, India and other Asian and South American countries) and for specialty grades in developed countries. Soda pulping is also used to produce high yield hardwood pulps that are employed to make packaging papers and boards.

Non-wood fibre soda pulp mills are normally small in capacity, since the feedstock is bulky and often is not produced at a steady rate all year around. Technology developed for much larger wood pulping operations concerning the processing of the black liquor to recover energy and pulping chemicals is normally not affordable to the typical small non-wood based pulp mill. Lignin recovery is one of the alternatives proposed to handle non-woods soda black liquors. Removal of the lignin reduces the black liquor's chemical oxygen demand (COD) by about 50 per cent and generates a revenue stream from an effluent, effectively enhancing the economic performance of the mill. In recent years, the first two soda lignin recovery facilities were installed in France and India [66], and it is expected that additional facilities will be added in the near future, as this approach to handle non-wood spent liquors gets further established.

10.4.1 Process

In the soda pulping process the fibrous feedstock is digested with an aqueous solution of sodium hydroxide. Since non-wood fibres have a relatively open and more accessible structure and low lignin content, the pulping temperature can be 160°C or lower. There are some similarities between soda pulping and the kraft process and reactions such as cleavage of lignin–carbohydrate linkages and depolymerization of the lignin and its recondensation take place. Lignin depolymerization during soda pulping also occurs principally by the cleavage of α and β aryl ether bonds, first in phenolic units and, in a later phase of delignification, in non-phenolic units. The generation of free phenolic groups in such reactions results in lignin fragments that are soluble in the alkaline environment prevalent during the digestion.

Quinone methide structures originating from free phenolic groups are formed in soda pulping, but in this case their nucleophilic attack is not very important, given the relatively weak nucleophilic nature of the hydroxyl group. Instead, there is abstraction of the β proton or elimination of formaldehyde from a methylol group attached to the β carbon, leading to the formation of alkali stable stilbene or styryl aryl structures. In soda pulping, lignin depolymerization takes place mostly in non-phenolic β aryl ether units (Fig. 10.5). In this context, the formation of quinone methide structures is not possible; instead hydroxyl groups in the α or γ carbon assist in the alkaline elimination of the aryl group with the formation of an intermediate epoxide, which under alkaline conditions opens up to form a glycol [4]. Soda pulping is not as efficient as kraft at cleaving dibenzodioxocins. In the case of wood pulping, in addition to 5,5′ biphenyls (which are essentially the only product from the cleavage of such units in kraft pulping), enol ethers and vanillin are formed [40–42].

In soda pulping there are similar condensation reactions as those described above for kraft pulping. In this case, however, the black liquor does not have strong nucleophiles (such as sulphide or hydrosulphide) that might block condensation of the quinone methide intermediates with the carbanions. Hence, lignin condensation may be more prevalent for soda than for kraft pulping.

It should be mentioned that the above mechanisms have been studied in the literature principally from the point of view of pulping of wood, and they may not apply entirely to non-wood soda pulping because of the structural differences in non-wood lignins relative to wood lignins.

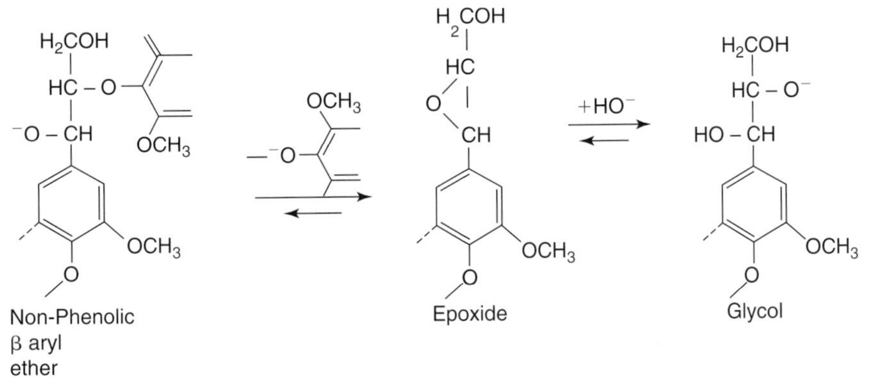

H_2COH | $HC-O$ | ^-O-CH OCH_3 OCH_3 O Non-Phenolic β aryl ether ⇌ OCH_3 ^-O ; H_2COH HC O CH OCH_3 O Epoxide $\underset{\leftarrow}{\xrightarrow{+HO^-}}$ H_2COH $HC-O^-$ $HO-CH$ OCH_3 O Glycol

Figure 10.5 Main reactions leading to the formation of soda lignin [39].

As in the case of kraft lignin recovery, the recovery of soda lignin is based on acid precipitation and adjustment of other process variables. An example of a process used industrially to recover non-wood lignins from soda black liquors is the Lignin Process System (LPS). In this case the pH of the black liquor is reduced, usually with a mineral acid, to form a lignin slurry, which is filtered; the lignin cake obtained is washed and dried to generate a high purity lignin powder with less than 5 per cent moisture [67].

10.4.2 Properties

Soda lignins are significantly different from lignosulphonates, as they are low molecular weight, insoluble in water, and obtained with low levels of sugar and ash contaminants. They have more in common with kraft lignins, resembling them in molecular weight and hydrophobicity. Among the commercially available lignins, soda lignins are unique in the sense that they are sulphur-free, and therefore can be considered as being closer to lignins as they exist in nature. Another important difference for current commercial products has to do with their genetic origin, which impacts functional groups. Currently available industrial soda lignins are mostly obtained from non-wood plants, while commercial kraft lignins are obtained from woods. In nature, non-wood lignins have structural differences relative to wood lignins. In addition to guaiacyl and syringyl moieties, non-woods have significant amounts of *p*-hydroxyl groups. They also contain etherified phenolic acids that serve as crosslinking bridge with carbohydrates [68, 69]. As discussed elsewhere [70, 71], native lignins from non-woods show structural diversities according to genetic origin, and one would expect such differences to carry on to the corresponding industrial versions.

Similar to unmodified kraft lignins, unmodified soda lignins are available as water insoluble products with relatively low content of ash and sugars. Sugar content for soda lignin recovered by the LPS process has been reported in the 2–3 per cent range. The type of sugars found, however, depends on the species of the pulping feedstock. For instance, straw lignin has predominance of pentoses, while hexoses are more abundant in hemp and flax. This is probably a reflection of the saccharide distribution of the hemicelluloses of the different species. Regarding ash, one of the main components are silicates, which naturally occur in non-wood species. In general, silica contents below 1 per cent have been reported for products offered by a leading soda lignin supplier [72]. Non-wood soda lignins have been reported to contain 0.8–1.6 per cent nitrogen, which is significantly higher than those observed in wood lignins. This nitrogen is assumed to be related, at least in part, to protein residues.

Industrial soda lignins are reported to have total hydroxyl contents in the range of 4.4–5 mmole g^{-1} [61, 73], about equally divided between aliphatic and aromatic. As expected for materials of non-wood origin, soda non-wood lignins are polycarboxylates, that is, polymeric materials that contain substantial amounts of carboxylic acid groups (about 2.1–2.3 mmole g^{-1}), making them somewhat less hydrophobic than kraft wood lignins.

Recent results on molecular weight distribution of soda lignins obtained using alkaline size exclusion chromatography mention weight average molecular weights in the range of 6 900 to 8 500 and polydispersities of about 3. These values are in a similar range as those found for kraft wood lignins [52] but, as discussed above, the lack of universally accepted standard protocols for molecular weight determination of a wide range of lignins, compromises their reliability.

Soda lignins do flow when heated and their glass transition temperatures have been reported in the range of 158–185°C [71]. The lack of sulphur gives them some advantage over kraft lignins for applications for which thermal flow is a requirement and volatile sulphurous emissions are undesirable.

10.4.3 Applications

Because of their lack of sulphur and closeness to lignin in nature, soda lignins have potential applications that are not available to kraft and sulphite lignins. For instance, they can be used more readily in thermoset binding applications, where use of heat without evolution of sulphurous volatiles is a must. The natural carboxylic acid functionality of non-wood soda lignins gives it the profile of a polycarboxylate-type dispersant. Soda lignins are also better suited in applications for which biological activity is important, such as in the animal feed and nutrition area.

The main applications for soda lignins include the following:

- *Component in PF formulations*: Soda lignins are currently used as partial replacement for phenol in the manufacture of PF resins used as binders in plywood [74], foundry sands and moulding compounds, among others.

With current trends towards reducing dependency on fossil raw materials, it is expected that the use of such high purity lignins as sustainable raw materials will increase in the near future, perhaps expanding to other applications beyond the PF domain [71].

- *Animal health and nutrition*: Lignin in nature is a component of dietary fibre. High purity sulphur-free soda lignins are the industrially available lignins that most closely resemble such native lignins. Products from such lignins have been proposed for the management of enteric disturbances in several animal species, including ruminants and monogastrics [75, 76]. Soda lignins are a natural alternative to the use of antibiotics, which is under increasing regulation in Europe and other places.
- *Dispersants*: Unmodified soda lignins from annual plants have some dispersant ability, given their carboxylic acid content and have been used as dispersants for biological slime control in industrial water circuits [77]. Similar to kraft lignins, soda lignins are converted to high purity, water-soluble derivatives via the introduction of sulphonate groups and the resulting dispersants may be used as concrete additives and other applications [78, 79].

10.4.4 Main producers

GreenValue SA is the main producer in the emerging field of high purity sulphur-free soda lignins. This group has two production facilities, including the largest sulphur-free lignin factory in the world, which is located in Asia. Total capacity is in excess of 10000 ton per year on a dry basis. The group offers lignins derived from non-wood plant sources, such as flax, wheat straw and other grasses. Products are offered in powder and solution form. In addition to unmodified soda lignin, the company also produces modified products, including high purity sulphonated lignins [73].

Northway Lignin Chemical is a small North American producer which offers 50 per cent solids soda liquor, containing soda lignin, ash and sugars and their derivatives. The product is obtained from wood pulping and is used as a low end binder [80].

10.5 OTHER FUTURE POTENTIAL SOURCES OF INDUSTRIAL LIGNINS

Lignocellulosic biomass sources are expected to become preferred feedstocks for ethanol biorefineries, since traditional sources of carbohydrates for alcohol conversion (such as grain or sucrose) will not be available in sufficient amounts and under economically competitive conditions to meet increasing demand for sustainable transportation fuels. Fast growth trees and annual non-wood plants will probably be the predominant feedstock for such biorefineries. The use of lignocellulosic biomass as feedstock for ethanol production will generate a lignin-rich stream amounting to more than 60 per cent by weight of the ethanol produced. This lignin fraction has potential use as an energy source and as a chemical intermediate and product. The calorific value of lignin is 93 per cent of the calorific value of alcohol, and therefore the energy content of the lignin co-produced will be more than half of the energy content of the alcohol generated.

Recent projections indicate that the use of alcohol as fuel in the US is going to increase by about 9.5 billion litres by the year 2012 [81]. If 25 per cent of the alcohol increase comes from non-woody biomass, at least 1.1 million tons of lignin could be potentially available. Similar scenarios are expected for Europe and other parts of the world. How this lignin resource is managed is expected to have a significant impact on the economics and ultimate success of the biorefinery.

Biomass conversion technologies typically involve a pre-treatment, which is designed to break down the linkages that exist between lignin and carbohydrate components in the feedstock. Most pre-treatments are either catalyzed by added mineral acid or autocatalyzed by biomass-derived organic acids, as occurs in steam explosion or autohydrolysis. During acid pre-treatments, lignin undergoes depolymerization, caused by scission of β-*O*-4 and other linkages. Electrophilic attack of side-chain carbenium ions to lignin aromatic rings may to some extent reverse the depolymerization reaction, forming condensed structures linked by carbon–carbon bonds. If conducted under the right conditions of time, temperature and acidity, the acid catalyzed pre-treatments may result in high yields of a partially depolymerized lignin that, although not soluble in water, can be readily extracted with organic solvents or dilute alkali [82–84].

The extraction of lignin from the pre-treated biomass ahead of hydrolysis has its advantages. Lignin binds with enzymes, effectively poisoning these biocatalysts. Thus, if less lignin is present during enzymatic treatment, enzyme requirements decrease. Furthermore, dewatering of the residue after acid or enzymatic hydrolysis is made more difficult by the presence of lignin.

In another category of pre-treatment processes, delignification is actually an integral part of the pre-treatment. This is exemplified by organosolv processes, in which the biomass is treated with an organic solvent at high temperature and pressure to obtain a cellulose solid residue with high carbohydrate content and an extract from which high purity lignin may be recovered. Alkaline delignification pre-treatments have also been proposed in connection with biorefinery operations.

As discussed elsewhere in this chapter, soda delignification is a well-known process widely practiced for the production of pulp and high purity lignin from non-wood fibres [85]. The cellulose fraction could be converted to alcohol by saccharification and fermentation, particularly if the fibre properties of the feedstock are not the most suitable for paper manufacture. In addition to soda delignification, ammonia explosion and lime are among other alkaline processes being considered.

Various biomass conversion technologies have advanced to demonstration and pilot scale, and some lignins originating from such efforts have been available for industrial development purposes. Among these were lignins from steam explosion available in the 1980s, for instance, Sucrolin, offered by a South African company, and Angiolin produced by the US University Virginia Tech in pilot quantities [86].

The Alcell organosolv delignification technology was demonstrated on mixed hardwoods pulping at the semi-commercial level in the latter part of the twentieth century [87] and has been taken over recently by the Canadian company Lignol Innovations, which has incorporated the process into a biorefinery concept. In the Lignol biorefinery process, lignin is recovered from the spent solvent by precipitation (which typically involves adjusting concentration, pH and temperature), filtration and drying [88]. Organosolv lignins are usually high purity, low molecular weight products with narrow molecular weight distributions [89]. These lignins show a low glass transition temperature and exhibit flow when heated. They have high solubility in organic solvents, and are very hydrophobic and practically insoluble in water [71]. Organosolv lignin from the Alcell process has been used industrially at a significant level in the past in various applications and is among the most studied lignins in academic and applied industrial circles in recent times. A very comprehensive review covering the properties, applications and potential of organosolv lignins has been published recently [90].

For a number of years, the Brazilian company Dedini [91] has operated an acid catalyzed organosolv delignification demonstration plant that starting with sugar cane bagasse simultaneously generates a solution of sugars, lignin and other co-products. After separation of the lignin, the sugar solution is co-fermented to alcohol with sucrose from cane juice.

Organic acids (particularly acetic and formic acid) have also received attention as organosolv delignification solvents. The processes of the French organization CIMV [92] and of the Finnish company Chempolis [93, 94] are examples of biorefinery technologies that are currently pursued at pre-commercial level and that claim to recover lignins of high purity with linear structure and very low softening temperatures.

REFERENCES

1. Ingruber O.V., Kocurek M.J., Wong A.W. (Eds), *Pulp and Paper Manufacture, Volume 4, Sulfite Science and Technology*, 1985, Tappi Press, Atlanta.
2. Sixta H. (Ed.), *Handbook of Pulp*, 2006, Wiley-VCH Verlag GmbH & Co., Weinheim.
3. Gellerstedt G., The reactions of lignin during sulfite pulping, *Svensk Papperstidning*, **79**(16), 1976, 537–543.
4. Gierer J., The reactions of lignin during pulping – A description and comparison of conventional pulping processes, *Svensk Papperstidning*, **73**(18), 1970, 571–596.
5. Glasser W.G., The potential role of lignin in tomorrow's wood utilization technologies, *Forest Prod. J.*, **31**(3), 1981, 24–29.
6. Freudenberg K., Maercker G., Nimz H., *Chem. Ber.*, **97**, 1964, 903.
7. Forss K., Sagfors P.-E., Kokkonen R., The composition of birch lignin. Fractionation of lignosulphonates and lignosulphonate–carbohydrate compounds, *Proceedings International Symposium on Wood and Pulping Chemistry – Poster Presentations*, Vancouver, Canada, 1985 pp. 23–26.
8. Gargulak J.D., Lebo S.E., Commercial use of lignin-based materials Chapter 15, in Lignin: Historical, Biological and Materials Perspectives, Eds. Glasser W.G., Northey R.A., Schultz T.P., *ACS Symposium Series*, American Chemical Society, Washington, DC, 1999.
9. Lin S.Y. Graft polymerization of lignosulphonate – An investigation of reaction mechanism and its relation to property modification, *Proceedings Seventh International Symposium on Wood and Pulping Chemistry Volume 1*, Beijing, 1993, pp. 16–23.
10. Lutnaes B.F., Lauten R.A., Myrvold B.O., Ovrebo H.H., Characterisation of lignosulphonates by NMR spectroscopy, *Euromar Conference*, July 16–21, New York, UK, 2006.

11. Shulga G., Soloddovniks P., Shakels V., Anisckevicha O., New semi-interpenetrating polymer networks incorporating soluble lignin, *Proceedings of the 7th ILI Forum—Barcelona*, International Lignin Institute, Lausanne, 2005, pp. 159–162.
12. Gasche U.P., Hüsler W., Utilization of spent sulfite liquor – processes now in use and future applications of new methods, *Proceedings International Sulfite Pulping Conference*, Toronto 1982.
13. Anonymous, Lignin chemicals presented by Borregaard – Technical bulletin 600 E, Sarpsborg, Norway, 1980.
14. Alonso M.V., Rodriguez J.J., Oliet M., Rodriguez F., Garcia J., Gilarranz M.A., Characterization and structural modification of ammonic lignosulfonate by methylolation, *J. Appl. Polym. Sci.*, **82**, 2001, 2661–2668.
15. Alonso M.V., Oliet M., Rodriguez F., Garcia J., Gilarranz M.A.R., odriguez J.J., Modification of ammonium lignosulfonate by phenolation for use in phenolic resins, *Bioresource Technol.*, **96**, 2005, 1013–1018.
16. Bansal I.K., Wiley A.J., Membrane processes for fractionation and concentration of spent sulphite liquors, *Tappi J.*, **58**(1), 1975, 125–130.
17. Bar-Sinai Y.L., Wayman M., Separation of sugars and lignin in spent sulphite liquor by hydrolysis and ultrafiltration, *Tappi J.*, **59**(3), 1976, 112–114.
18. Goring D.A.I., Polymer properties of lignin and lignin derivatives, Chapter 17, in *Lignins – Occurrence, Formation, Structure and Reactions* (Eds. Sarkanen K.V. and Ludwig C.H.), John Wiley & Sons, New York, 1971.
19. Myrvold B.O., A new model for the structure of lignosulphonates, *Proceedings of the 7th ILI Forum – Barcelona*, International Lignin Institute, Lausanne, 2005, pp. 39–42.
20. Parfenova L., Polyelectrolyte expansion and crossover area in solutions of lignosulphonates modified by membrane methods, *Proceedings of the 9th European Workshop on Lignocellulosics and Pulp*, Vienna, Austria, 27–30 August, 2006, pp. 425–427.
21. Forss K., Kokkonen K., Sagfors P.E., Reversed phase chromatography of lignin derivatives Chapter 13, in Lignin Properties and Materials, (Eds. Glasser W.G., Sarkanen S.), *ACS Symposium Series 397*, American Chemical Society, Washington, 1989.
22. Gretland K.S., Gustafsson J., Bräten S.M., Fredheim G.E., Ekeberg D., Characterization of lignosulphonates and sulphonated kraft lignin by hydrophobic interaction chromatography, *Proceedings of the 7th ILI Forum – Barcelona*, International Lignin Institute, Lausanne, 2005, pp. 125–127.
23. Ekeberg D., Gretland K.S., Gustafsson J., Braten S.M., Frdheim G.E., Characterisation of lignosulphonates and kraft lignin by hydrophobic interaction chromatography, *Anal. Chim. Acta*, **565**, 2006, 121–128.
24. Belgacem M.N., Blayo A., Gandini A., Surface characterization of polysaccharides, lignins, printing ink pigments, and ink fillers by inverse gas chromatography, *Journal of Colloid and Interface Sci.*, **182**, 1996, 431–436.
25. Belgacem M.N., Gandini A., Inverse gas chromatography as a tool to characterize dispersive and acid base properties of the surface of fibers and powders Chapter 2, in *Interfacial Phenomena in Chromatography* (Ed. Pefferkorn E.), Marcel Dekker, Inc., New York, 1999.
26. Lora J.H., Creamer A.W., Wu L.C.F. and Goyal G.C., Chemicals generated during alcohol pulping: Characteristics and applications, *Proceedings Volume 2, 6th International Symposium on Wood and Pulping Chemistry*, Melbourne, Australia, 1991, pp. 431–438.
27. Web site of BIOTECH Lignosulfonate Handels-GesmbH http://www.biot.org/english/index_produkte.html Visited on October 15, 2006.
28. Web site of La Rochette Venizel Novibond http://www.novibond.com/novibond/paper.htm Visited on November 03, 2006.
29. Jopson R.N., Saturation technology for corrugated containers, *Tappi J.*, **76**(4), 1993, 207–214.
30. Hoyt C.H., Goheen D.W., Polymeric products Chapter 20, in *Lignins – Occurrence, Formation, Structure and Reactions* (Eds. Sarkanen K.V. and Ludwig C.H.), John Wiley & Sons, New York, 1971.
31. Adolphson C. and Simmons R.W., Compositions having available trace elements and process of making same and providing for nutrition of plants, shrubs and trees, US Patent 3,244,505, 1966.
32. Goheen D.W., Low molecular weight chemicals Chapter 19, in *Lignins – Occurrence, Formation, Structure and Reactions* (Eds. Sarkanen K.V. and Ludwig C.H.), John Wiley & Sons, New York, 1971.
33. Web site of Lignotech Borregaard http://www.lignotech.com/ Visited on October 15, 2006.
34. Web site of Tembec Lignosulphonate products http://www.arbo.ca/pages/en/aboutus.html Visited on October 15, 2006.
35. Web site of La Rochette Venizel – Novibond products http://www.novibond.com/novibond/ Visited on October 16, 2006.
36. Web site of Nippon Paper Chemicals http://www.npchem.co.jp/english/product/lignin/index.html Visited on October 16, 2006.
37. Web site of Cartiere Burgo http://www.burgo.com/burgo/Ligninsolfonati_en/index.htmVisited on November 16, 2006.
38. Web site of Domsjo Fabriker http://www.domsjoe.com/ Visited on November 12, 2006.
39. Gierer J., Chemistry of delignification, Part 1: General concept and reactions during pulping, *Wood Sci. Technol.*, **1**, 1985, 289–312.
40. Karhunen P., Rummakko P., Sipila J., Brunow G., Kilpelainen I., Dibenzodioxocin: A novel type of linkage in softwood lignin, *Tetrahedron Lett.*, **36**, 1995, 169–170.

41. Argyropoulos D.S., Jurasaek L., Kristopova L., Xia Z., Sun Y., Palus E., Abundance and reactivity of dibenzodioxocins in softwood lignin, *J. Agric. Food Chem.*, **50**(4), 2002, 658–666.
42. Argyropoulos D.S., Salient reactions in lignin during pulping and oxygen bleaching: An overview, *J. Pulp Pap. Sci.*, **29**(9), 2003, 308–313.
43. Rinman E.L., Process of treating waste liquors of soda pulp mills, US Patent 1,005,882 1911.
44. Keillen J.J., Ball F.J., Gressang R.W., Method of coagulating colloidal lignates in aqueous dispersions, US Patent 2,623,040 1952.
45. Loutfi H., Blackwell B., Uloth V., Lignin recovery from kraft black liquor: preliminary process design, *Tappi J.*, **74**(1), 1991, 203–210.
46. Adler E. and Hagglund E.K.M., Method of producing water soluble products from black liquor lignin, US Patent 2,680,113 1954.
47. Dillig P., Method for producing low electrolyte lignosulfonates, US Patent 4,740,590 1988.
48. Merton B.J., Cationic bituminous emulsions, US Patent 3,126,350 1964.
49. Ball J.C., Inhibition of corrosion of iron in acids, US Patent 2,863,870 1958.
50. Borgfeldt M.J., Anionic bituminous emulsions, US Patent 3,123,569 1964.
51. MeadWestvaco web site http://www.meadwestvaco.com/c_dir/chempdb.nsf/PDBProductName/REAX%20825E/$File/REAX_825E.pdf?OpenElement Visited on November 15, 2006.
52. Abaecherli A., Gosselink R.J.A., de Jong E., Baumberger S., Hortling B., Bonini C.D., Auria M., Zimbardi F., Barisano D., Duarte J.C., Sena-Martins G., Ribeiro B., Koukios E., Koullas D., Avgerinos E., Vasile C., Cazacu G., Mathey R., Ghidoni D., Gellerstedt G., Li J., Quintus-Leino P., Piepponen S., Laine A., Koskinen P., Gravitis J., Suren J., Fasching M., Intermediary status of the round robins in the Eurolignin network, *Proceedings of the 7th ILI Forum – Barcelona*, International Lignin Institute, Lausanne, 2005, pp. 119–124.
53. Morck R., Yoshida H., and Kringstad K., Fractionation of kraft lignin by successive extraction with organic solvents. 1. Functional groups, ^{13}C-NMR-Spectra and molecular weight distributions, *Holzforschung*, **40**(Suppl.), 1986, 51–60.
54. Morck R., Reimann R., Kringstad K., Fractionation of kraft lignin by successive extraction with organic solvents. 3. Fractionation of kraft lignin from birch, *Holzforschung*, **42**, 1988, 111–116.
55. Gellerstedt G., Lindfors E.-L., Structural changes in lignin during kraft pulping, *Holzforschung*, **38**, 1984, 151–158.
56. Shulga G., The coil-to-globule transition of softwood kraft lignin in very dilute aqueous solutions, *Proceedings 9th European Workshop in Lignocellulosics and Pulping*, Vienna, 2006, pp. 510–513.
57. Dutta S., Garvere T.M. and Sarkanen S., Modes of association between kraft lignin components Chapter 12 in Lignin – Properties and Materials, (Eds. Glasser, W. and Sarkanen, S.), *ACS Symposium series 397*, Washington, DC, 1989.
58. Gosselink R.J.A., Snijder M.H.B., Kranenbarg A., Keijsers E.R.P., de Jong E., Stigsson L.L., Characterisation and application of NovaFiber lignin, *Indus. Crops and Products*, **20**, 2004, 191–203.
59. Gosselink R.J.A., de Jong E., Abächerli A. and Guran B., Activities and results of the thematic network Eurolignin, *Proceedings of the 7th ILI Forum – Barcelona*, International Lignin Institute, Lausanne, 2005, pp. 25–30.
60. Marton J., Reactions in alkaline pulping, Chapter 16, in *Lignins – Occurrence, Formation, Structure and Reactions* (Eds. Sarkanen K.V. and Ludwig C.H.), John Wiley & Sons, New York, 1971.
61. Cateto C.A., Barreiro M.F., Rodrigues A.E., Brochier-Salon M.C., Thielemans W. and Belgacem M.N., FTIR and NMR studies of lignin acetylation, *Proceedings of the 9th European Workshop on Lignocellulosics and Pulp*, Vienna, Austria, August 27–30, 2006, pp. 192–195.
62. Mead-Westvaco Web Site http://www.meadwestvaco.com/c_dir/chempdb.nsf/search#SearchResults Visited on November 15, 2006.
63. Berry B.C., Viswanathan T., Lignosulfonic acid – doped polyaniline (Ligno-Pani) – A versatile conducting polymer, in *Chemical Modification, Properties and Usage of Lignin* (Ed. Hu T.Q.), Kluwer Academic/Plenum Publishers, New York, 2002, pp. 21–40.
64. Rojas O., Bullon J., Yaambertt F., Forgiarini A., Argyropoulos D.S., Gaspar A.R., Phase behavior of lignins and formulations of oil-in-water emulsions, *Proceedings 9th European Workshop in Lignocellulosics and Pulping*, Vienna, 2006, pp. 471–475.
65. Gaylord Chemical Inc. web site http://www.gaylordchemical.com/products/dmso.htm Vsited on November 30, 2006.
66. Lora J.H., Biorefinery non-wood lignins: Potential commercial impact, *92nd Annual Meeting Preprints – Book C*, Pulp and Paper Technical Association of Canada, Montreal, 2006, pp. C3–C6.
67. Abächerli A. and Doppenberg F., PCT WO 98/42912, 1998.
68. Jacquet J., Pollet B., Lapierre C., Mhamdi F., Rolando C., New ether-linked ferulic acid-coniferyl alcohol dimmers identified in grass straws, *J. Agric. Food Chem.*, **43**, 1995, 2746–2751.
69. Lam T.B.T., Iiyama K., Stone B.A., Quantitative determination of lignin polysaccharides associations in wheat internode cell walls: Ferulic acid bridges, in *Appita Proceedings, 6th International Symposium n Wood and Pulping Chemistry, Volume 1*, Melbourne, 1991, pp. 29–33.

70. Lora J.H., Characteristics, industrial sources and utilization of lignins from non-wood plants, in *Chemical Modification, Properties and Usage of Lignin* (Ed. Hu T.Q.), Kluwer Academic/Plenum Publishers, New York, 2002, pp. 267–282.
71. Lora J.H., Glasser W.G., Recent industrial applications of lignin: A sustainable alternative to nonrenewable materials, *J. Polym. Environ.*, **10**(1/2), 2002, 39–48.
72. Gosselink R.J.A., Abächerli A., Semke H., Malherbe R., Käuper P., Nadif A., van Dam J., Analytical protocols for characterization of sulphur-free lignin, *Indus. Crops and Products*, **19**(3), 2004, 271–281.
73. Asian Lignin Manufacturing web site http:www.asianligninn.com Visited on 20 November, 2006.
74. Khan M.A., Lora J.H., *Ply Gazette*, August, 2006, 68–77.
75. Montserrat M., Garcia L., Marrero E., Vinardell P., Does lignin affect intestinal morphometry, *6th Internet World Congress for Biomedical Sciences*, Web site http://www.uclm.es/inabis2000/posters/files/012/session.htm#1, Visited on 15 November, 2006.
76. Cruz R., Dopico D., Figueredo J., Rodriguez M., Martinez G., Uso de la lignina de bagazo con fines medicinales, *Rev. Med. Exp. INS*, **14**(1), 1997, 67–71.
77. Oberkofler J., Replacement of biocides by special low molecular weight lignin based compounds for bacterial and deposit control of industrial water cycles, *Proceedings of the 7th ILI Forum – Barcelona*, International Lignin Institute, Lausanne, 2005, pp. 79–81.
78. Chang D.Y., Chan S.Y.N., Straw pulp waste liquor as a water reducing admixture, *Mag. Concr. Res.*, **47**(171), 1995, 113–118.
79. He W., Tai D., and Lin S., A comparative study of the reactivities of straw and wood alkali lignin in sulfonation reactions, *Appita Proceedings 6th International Symposium on Wood and Pulping Chemistry 1*, Atlanta, USA, 1991, pp. 509–515.
80. Northway Lignin Chemical web site http://www.duenorth.net/northway/polybind300.htm Visited on November 15, 2006.
81. Anonymous, 2003 World Ethanol Production, 2003, *F.O. Lichts World Ethanol and Biofuels Report*, **2**(4), October, 28.
82. Wayman M., Lora J.H., Process for depolymerization and extraction of lignin, Canadian Patent No. 1,147,105, 1983.
83. Lora J.H., Wayman M., Delignification of hardwoods by autohydrolysis and extraction, *Tappi*, **61**(6), 1978, 47–50.
84. Wayman M., Lora J.H. and Gulbinas E., Material and energy balances in the production of ethanol from wood, Chapter 13, in Chemistry for Energy, (Ed. Tomlinson M.), *American Chemical Society Symposium Series No. 90*, Washington, 1979.
85. Lora J.H., Options for black liquor processing in non-wood pulping, Paper 20, in *Proceedings of International Conference on Cost Effectively Manufacturing Paper and Paperboard from Non-wood Fibres and Crop Residues*, Pira International, Surrey, UK, 2001.
86. Glasser W.G., Davé V., Frazier C.E., Molecular weight distribution of (semi-) commercial lignin derivatives, *J. Wood Chem. Technol.*, **13**(4), 1993, 545–559.
87. Pye E.K., Lora J.H., The Alcell Process: A proven alternative to kraft pulping, *Tappi J.*, **74**(3), 1991, 113–118.
88. Lora J.H., Wu C.F., Cronlund M., Katzen R., Lignin recovery process, US Patent No. 4,764,596, 1988.
89. Lora J.H., Creamer A.W., Wu C.F., Goyal G.C. et al, Industrial scale production of organosolv lignins; characteristics and applications, in *Cellulosics: Chemical, Biochemical and Material Aspects* (Ed. Kennedy J.F.), Ellis Horwood Ltd, 1993, pp. 251–256.
90. Pye E.K., Industrial Lignin Production and Applications, Chapter 5, in *Biorefineries – Industrial Processes and Products* (Eds. Kamm B., Gruber P.R. and Kamm M.), Wiley-VCH GmbH & Co. KGaA, Weinheim Germany, 2006, pp. 165–200.
91. Lahr Filho D., Vaz C.E., Lamonica H.M., Integration of DHR Process (Fast acid Hydrolysis Dedini) with a Typical Sugar and Ethanol Factory, *Abstracts of the ISSCT Co-products Workshop*, Copersucar, Piracicaba, Brazil, 2003.
92. Delmas M., Biomass refining and furanic chemistry, Book of Abstracts, Biomass Derived Pentoses Conference, Reims, 2006.
93. Rousu P., Rousu P., Anttila J., Tanskanen P., A novel biorefinery – production of pulp, bioenergy and green chemicals from nonwood materials. *Procedings Tappi Engineering Pulping and Environmental Conference*, Atlanta, 2006.
94. Anttila J., Rousu P., Tanskanen J., Chemical recovery in a non-wood pulping process based on formic acid, *Procedings Tappi Engineering Pulping and Environmental Conference*, Atlanta, 2006.

– 11 –

Lignins as Components of Macromolecular Materials

Alessandro Gandini and Mohamed Naceur Belgacem

ABSTRACT

This chapter surveys the use of lignin, as such or after modification, as an additive to other polymers, a blend component and a macromonomer. After a brief description of the use of lignin as a filler in different polymeric vehicles, such as inks and paints, the use of this renewable macromolecule as a co-reactant in phenol–formaldehyde and urea–formaldehyde resins, is described. Then, different systems aiming at preparing poly(urethane)s, polyesters, epoxy resins and resins based on lignins and furans are reviewed, and those giving the most interesting materials discussed in some depth. The modification of lignin with alkenyl groups and the polymerization of the ensuing macromonomers are also critically assessed. The following topic is the preparation of lignin-based carbon fibres and activated carbons and their use as reinforcing elements in composites and selective adsorbents, respectively. The preparation of aromatic monomers from lignin and the interest of their polymers conclude the chapter.

Keywords

Lignins as a filler, Lignins as a macromonomer, Lignins in blends, Resins based on lignins and furans, Chemically modified lignins, Lignin-based polyurethanes, Lignin-based polyesters, Lignin-based epoxy resins, Lignin-containing formaldehyde-type resin, Alkenyl lignins, Lignin-based activated carbon, Lignin-based carbon fibres, Aromatic monomers.

11.1 INTRODUCTION

Following the exhaustive preceding monographs on the structure and technology related to the different forms of lignins, their intervention as physical and/or chemical actors in the manufacture of polymers and composites is the subject of this third chapter devoted to that natural polymer. Whereas the role of lignin in wood and annual plant anatomies is always that of the crosslinked matrix surrounding the cellulose fibres, its exploitation as a constituent of synthetic macromolecular materials can only arise from the use of lignin fragments or oligomers, as obtained from the various delignification processes discussed in the preceding chapters. These fragments will simply be called lignins in the course of this chapter, since they represent the only actual sources which intervene in the elaboration of the materials discussed herein. They represent an annual industrial output of some 50 million tons and their use is still today restricted essentially to energy production by combustion in the paper mills. Only a very limited percentage is recovered for other applications. Yet a number of sound arguments have been put forward to show that lignins are very promising sources of materials and chemicals, namely, among others (i) their ready availability in huge amounts at modest prices; (ii) the presence of reactive moieties in their structure which enables chemical modifications, chain extension and polymerizations to be conducted; (iii) the possibility of their controlled chemical or enzymatic splicing into monomeric species or value-added chemicals; (iv) their potential use as components of polymer blends or as fillers in other polymers. These are indeed the topics of the present chapter since they all deal with viable prospects related to the preparation of novel macromolecular materials.

Given the specific vegetable source and the delignification technology used, a large variety of structures and molecular weights are inevitably obtained; hence the chemical and physical properties of lignins can vary considerably. These differences are likely to play a role when they are used as merely physical additives (*e.g.* in composite materials). When lignins, as obtained, or after suitable chemical modification, are instead used as macromonomers participating in the construction of novel polymer architectures, the differences in their structure and molecular weight have no significant impact on the mechanisms associated with these chain growths. Their impact will however be relevant to some properties of the ensuing materials.

All lignins, irrespective of how they are obtained and from which vegetable source, are brown amorphous solids with glass transition temperatures ranging between 70°C and 170°C, depending on their specific structure and molecular weight (and molecular weight distribution). Obviously, a higher degree of aromatic contribution, particularly with inter-aromatic bridges and multiple substitutions, as well as a high molecular weight, will contribute to increase the T_g and vice versa.

All these lignins possess nevertheless, one universal set of features which is most relevant to the present context and that is the fact that, regardless of the specific structure and molecular weight, they all incorporate aliphatic and phenolic OH groups and unsubstituted aromatic sites. Inevitably, the relative proportions of these three moieties can vary considerably, mostly as a consequence of the precursor species, but their presence is ubiquitous simply because they are totally representative of the general structure of the natural polymer. Other moieties can also be referred to as universal in lignins, such as the typical propyl aliphatic sequence and the methoxy groups attached to the aromatic ring, but they play no role in terms of chemical reactivity for polymerization. Conversely, certain potentially reactive moieties like carboxylic groups and C=C unsaturations, are not ubiquitous in lignins because their presence depends on both the structure of their precursor and the delignification process adopted to fragment it. It follows that if general procedures are to be investigated in order to give lignins a viable status as macromonomers, the universal functional groups will be the only logical choice. In other words, whether lignins are used as macromonomers, as obtained, or after chemical modification, the OH groups and/or the unsubstituted aromatic sites must be considered the only exploitable moieties.

The above considerations are not limited to concepts of chemical modification because even when lignins are considered as additives in the preparation of polymer blends or composites, their hydroxyl functions represent a key structural element in terms of polar contributions and sources of hydrogen bonds which will affect the quality of the interfacial interactions of the ensuing materials, just as the less-polar (ether groups, aromatic rings, etc.) and non-polar (aliphatic sequences) moieties will, in terms of hydrophobic interactions.

The subdivision of this chapter follows the primary criterion based on whether lignins intervene only physically or also chemically in the manufacturing of synthetic polymers. The secondary criterion adopted here has to do with whether the lignins are used as such, that is, as recovered from industrial delignification processes, or after suitable chemical modifications.

Previous monographs on the use of lignins in macromolecular materials testify to the growing interest in this lively scientific and technological realm since the 1990s, [1] which includes international meetings devoted to it. The establishment in that period of the International Lignin Institute, [2] whose activities are steadily expanding, as well as the increasing number of laboratories involved in this field has generated some exciting achievements and it is the purpose of this chapter to assess them within the framework of previous major contributions.

11.2 LIGNINS AS PHYSICAL COMPONENTS

The introduction of lignins, as such or after suitable chemical modifications, as fillers or blend components into natural and synthetic polymers has been the subject of a vast number of studies which cover an impressive selection of materials and vary considerably in both purpose and depth. Feldman is among the most active contributors to this field and has reviewed it thoroughly twice, the second time covering work done up to the end of the second millennium [1c,1f]. One of us carried out similar surveys at the same time [1a,1e]. These monographs should be considered as background bibliography by the interested readers. Since then, numerous original contributions have appeared as discussed below.

11.2.1 Unmodified lignins

Notwithstanding their structural fine details, lignins must be considered as relatively polar macromolecules because of the substantial impact of the two types of OH groups borne by all of them, which give rise to inter- and

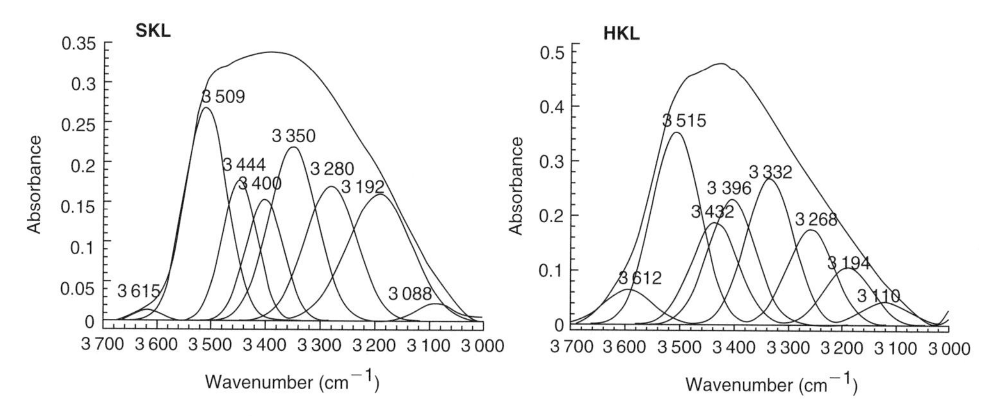

Figure 11.1 Deconvoluted FTIR spectra of the *ν*OH region of two KLs. (Reproduced by permission of American Chemical Society. Copyright 2005. Reprinted from Reference [3]).

intramolecular associations through hydrogen bonding [3] and thus enhance the material cohesive energy, as reflected by the relatively high T_g of lignins. Figure 11.1 shows the deconvoluted hydroxyl stretching region of the softwood kraft lignin (SKL) and hardwood kraft lignin (HKL). The number of bands contained in the broad OH peak for both SKL and HKL do not vary significantly, except for a change in the relative intensity of several absorptions.

In a more prosaic, but inevitably relevant tone, lignins isolated from industrial or pilot delignification processes are cheap commodities and their direct use in the manufacture of novel materials is therefore particularly attractive, if the operation leads to a viable exploitation.

Before examining the behaviour of macromolecular materials incorporating unmodified lignins, it is worth mentioning a recent interesting study on the plasticization of kraft lignin (KL) [4], which highlighted the existence of two distinct phenomena, one related to small amounts of additives capable of establishing hydrogen bonds with the lignin's OH groups, like water and poly(ethylene glycol) (PEG), and the other involving the conventional interactions based on solubility parameters and associated with high concentrations of plasticizer. It was shown that quantities of water up to about twice the number of lignin OH groups had the same softening effect (T_g decrease) as that of acetylating them through the progressive reduction in hydrogen bonding. However, further water addition played a minor plasticizing role, whereas additives with a structure similar to that of the lignin monomer units (*e.g.* vanillin and ferulic acid) as well as low degree of polymerization (DP) PEGs, were found to display a high efficiency.

A systematic investigation of materials made up of homogeneous mixtures of kraft or organosolv lignins and high molecular weight poly(ethylene oxide) (PEO) [5] showed that the solubility of lignin in all proportions, as already reported in a different study, [6] was mostly caused by intermolecular hydrogen bonding between the lignin phenolic OH and the PEO ether oxygen atoms, but that other weaker interactions must also occur. The differential scanning calorimetry (DSC) tracings of these blends are displayed in Fig. 11.2.

In the specific instance of organosolv lignin mixtures, [5c] continuous fibres were readily spun from the melt, and although these materials were excessively sensitive to moisture, they found applications as precursors to carbon fibres, as discussed below. These systems are the only example of blends of unmodified lignins with another polymer which are homogeneous over the entire composition range. KL, mixed with poly(vinyl acetate) and two plasticizers, gave homogeneous thermoplastic materials with up to 85 per cent of KL [7]. However, their fragility in aqueous media was a serious handicap towards possible applications.

The surface energy of non-ionic lignins is close to 70 mJ m^{-2}, with a dispersive contribution of about 45 mJ m^{-2} [8, 9]. It follows that, as in the case of cellulose fibres (whose composites are discussed in Chapter 19), the quality of an interface involving lignin and another polymer will depend on the structure of the latter. Essentially non-polar macromolecules, like polyolefins, will give rise to a relatively high interfacial energy and thus poor adhesion, whereas polymers bearing polar groups, like polyesters and polysaccharides, will establish a much more compatible interface and thus strong adhesion. The degree of dispersion of lignin in heterogeneous polymer blends will therefore depend on both the technique adopted for their preparation and the structural interactions of the two polymers.

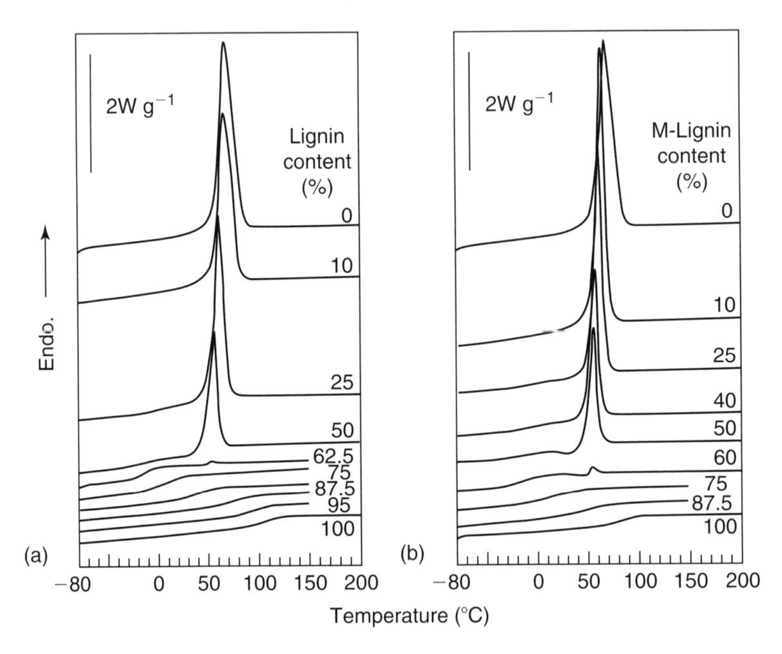

Figure 11.2 DSC tracing of blends of PEO-hardwood kraft lignin (a) before, and (b) after methylation. (Reproduced by permission of American Chemical Society. Copyright 2003. Reprinted from Reference [5a]).

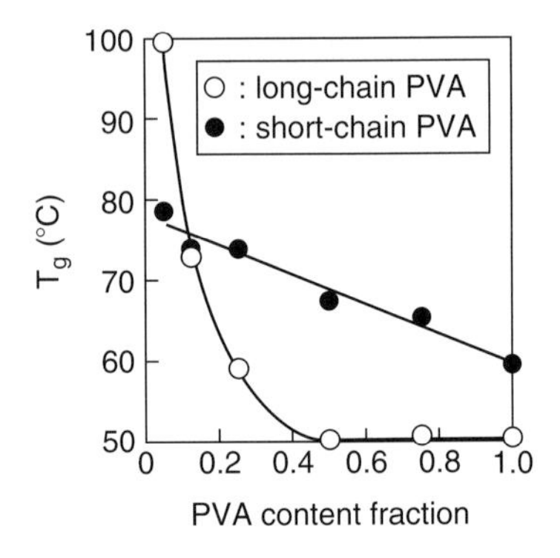

Figure 11.3 T_g versus composition curves for the long-chain and short-chain PVA lignin/PVA blends. (Reproduced by permission of American Chemical Society. Copyright 2003. Reprinted from Reference [10]).

Despite the polar character of poly(vinyl alcohol) (PVA) (or perhaps because this feature is too pronounced!), its blends with KL are heterogeneous, but specific intermolecular interactions were shown to exist in these materials, which were melt spun and thoroughly characterized [10]. Figure 11.3 shows the evolution of the glass transition temperature as a function of the PVA/lignin ratios, for two types of PVA.

With another polar polymer, namely a polyurethane elastomer, less than about 10 wt% of lignin apparently give homogeneous blends with interesting mechanical and thermal properties, which become however progressively worse as the lignin content is increased and phase separation enhanced [11]. The intermolecular interactions of lignin blended with various synthetic polymers of different polarity have been examined and discussed in terms of miscibility and the quality of extruded fibres [12]. The addition of a few per cent of lignosulphonic acid into polysulphone gave homogeneous membranes with good proton conductivities (*i.e.* potentially useful in the assembly of fuel cells [13]).

The incorporation of lignins in poly(olefins) and rubbers has attracted much attention within the very important context of polymer stabilization against degradation mechanisms involving free radicals, *viz.* oxidation, photolysis

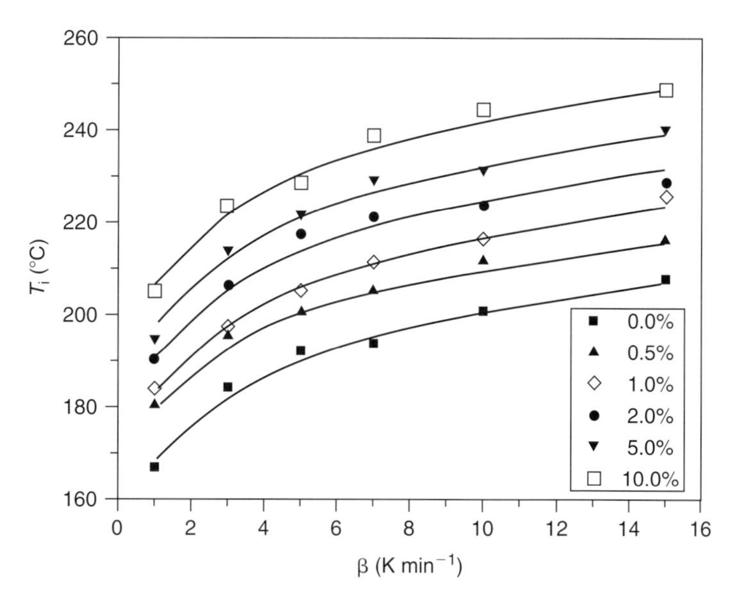

Figure 11.4 Experimental and fitted dependence of the onset of oxidation temperatures, as a function of the heating rate for polypropylene stabilized with different amounts of lignin. (Reproduced by permission of Elsevier. Copyright 2005. Reprinted from Reference [14]).

and thermolysis. This approach is entirely plausible, given the abundant phenolic content of all lignins and the well-known role of these structures as radical traps. The contribution of Košiková's group to the study of this topic is fundamental, although other laboratories have occasionally provided additional evidence. The most extensively examined polymer is poly(propylene) (PP), both as pristine and recycled material, for which the stabilization role of small percentages of lignins against photo-oxidation processes has been clearly established [14]. Thus, the onset of the thermal degradation of PP–lignin blends was found to increase with increasing lignin content, as shown in Fig. 11.4.

Higher lignin contents have recently been shown to protect this polymer against thermal degradation, [15] as shown in Fig. 11.5, which illustrates the thermal degradation of PP–lignin blends as a function of the amount of added lignin.

Poly(ethylene) has been recently protected from UV degradation with both a conventional [16] and a novel sulphur-free lignin [17]. When added to natural rubber (1–5 wt%), lignin plays as good a role as a commercial antioxidant [18] and when introduced in styrene–butadiene rubber in significant amounts, it also plays the role of a reinforcing filler [19]. The diffusion of aromatic hydrocarbons through natural rubber was found to be reduced by the incorporation of KL in relatively high proportions [20].

A recent addition to the field called upon the study of the effect of Alcell lignin on the rheological properties of printing inks, varnishes and paints [21]. The lignin particles ($\Phi < 0.2\,\mu m$) were incorporated into a series of commercial viscous products and found to be adequately wetted and dispersed. The changes in viscosity and tack of the ensuing suspensions were recorded, on the one hand, as a function of composition and ink source and, on the other hand, as a function of the measuring parameters, *viz.* temperature, tack roller speed and shear rate. It was found that this organosolv lignin (OL) brought about noticeable improvements in the properties of the viscous media used for offset inks and paints, particularly in terms of tack and misting reduction. These positive aspects were not marred by any detrimental effects on their other physical and chemical (drying kinetics) properties. The remarkable suppression of misting is highlighted in Table 11.1 for a variety of commercial inks (alkyd resins), paints and varnishes.

Blends of lignins with biopolymers have also attracted a good deal of attention, from Glasser's pioneering studies with cellulose and some of its derivatives to more recent investigations involving other polysaccharides, proteins and bacterial polyesters [1e,22]. Recent work on starch–lignin films illustrates this aspect adequately; given the growing impact of starch-related materials, as discussed in Chapter 15. Baumberger's group has devoted much effort to the study of the interactions of these two polymers and to the properties of the ensuing blends [23]. Figure 11.6 shows the effect of electron-beam irradiation on the mechanical properties of lignin–starch blends.

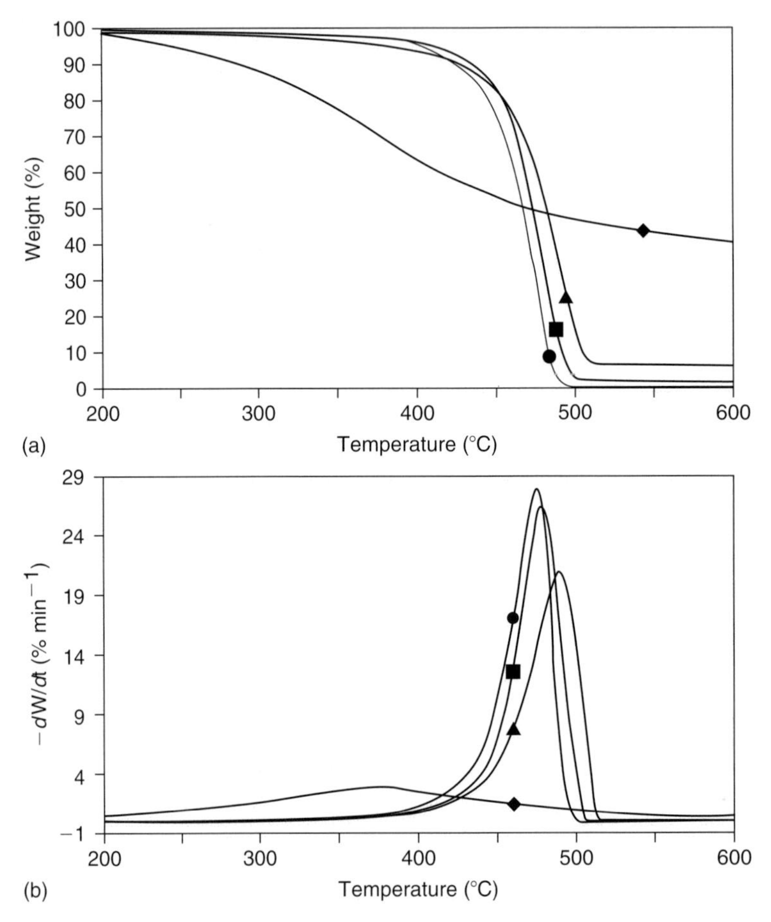

Figure 11.5 (a) TGA, and (b) DTG thermograms under nitrogen atmosphere for (●) pure PP; (■) PP/lignin 95/5; (▲) PP/lignin 85/15; (◆) pure lignin. (Reproduced by permission of Elsevier. Copyright 2006. Reprinted from Reference [15]).

Table 11.1

Misting of various commercial resins containing variable amounts of suspended **OL** [21]

	Temperature	% of misting as a function of **OL** content (% w/w)				
Suspension	(°C)	0	2	10	20	40
Alkyd resin B	30	35	33	25	0	0
Alkyd resin C	30	66	63	46	13	0
Paint F*	30	25	24	13	0	0
Paint G*	30	8	0	0	0	0
Varnish E (at 30°C)	30	10	0	0	0	0
Varnish E (at 50°C)	50	63	60	8	0	0
		% of misting as a function of temperature (°C)				
		0	9.2	16.0	20.3	42.0
Alkyd resin A	20	9	0	–	0	0
	30	23	0	0	0	0
	40	45	23	0	0	0
	50	65	–	27	17	0
	60	70	43	33	22	0

* Paints F and G were commercial paints presenting misting problems.

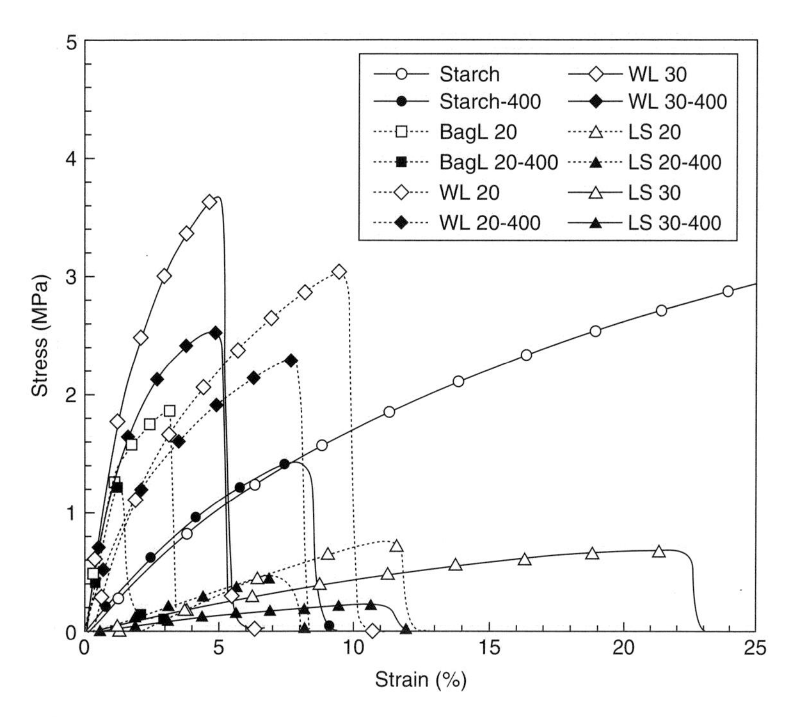

Figure 11.6 Effect of the 400 kGy-irradiation on the stress–strain curves (1 mm/min stretching speed) relative to the control and starch/lignins films after storage at 75 per cent RH. (Reproduced by permission of the American Chemical Society. Copyright 2006. Reprinted from Reference [23c]).

The lignins used were from different sources, namely: bagasse alkali (BagL), wheat straw (WL) and lignosulphonate (LS) lignins, incorporated in the starch matrix.

The primary aim of this research was to develop thermoplastic materials in which the hydrophilic character of starch would be reduced by the presence of lignin, but of course the thermomechanical properties of these composites also represented an important issue. All the blends were heterogeneous and the presence of lignin (whatever its source) did not affect the morphology of the amylose and amylopectin phases, although the latter starch component displayed a higher interfacial compatibility with lignin. Moreover, the overall compatibility between the two natural polymers was enhanced by moisture content and lignin lower DPs. LS gave the most intimate interpolymer interactions, but the other lignins, with their more hydrophobic character, improved the water resistance, particularly in cast films. Electron-beam irradiation of these films at high doses produced contrasting results in that their surface water resistance was enhanced to the detriment of their mechanical properties, both properties being affected by the occurrence of crosslinking and degradation reactions involving starch and lignin. This energetic post-treatment, applied in a more controlled fashion, represents a promising tool for the optimization of the functional properties of these materials.

Lignosulphonic acid has found a very appropriate application as a dopant for polyaniline, [24] as already pointed out in the previous chapter of this book. This high-tech material (LIGNO-PANI™) is an excellent illustration of a currently held working hypothesis which purports that lignins can be rationally exploited to give very original value-added commodities. The recently reported composite based on lignin and natural fibres, registered as ARBOFORM®, [25] seems to comply with these considerations, although little has been disclosed about its detailed chemistry (composition) and physics (processing).

11.2.2 Chemically modified lignins

The properties of lignin-based blends and composites can be improved by appropriate chemical modifications aimed at optimizing the interactions between or among the components of these materials, essentially by reducing the corresponding interfacial energy and thus enhancing their compatibility. This logical strategy however, has the inevitable drawback associated with the cost of the modification, which in the case of lignins can alter dramatically the economic feasibility related to the commercial exploitation of the materials in question. Only those realizations capable of fulfilling a technological requirement of high priority, whether because there is no known alternative, or because the improvements associated with the novel product are particularly noteworthy, will justify such an added cost. These considerations seem entirely obvious to us, but some of the approaches proposed both in this specific context and in the field of lignins as macromonomers discussed below, appear to ignore them or minimize the impact of economic considerations.

Three publications on the use of modified lignins claimed a major breakthrough in developing thermoplastic materials with very high lignin contents. The first [26] reported some properties of '100 per cent KL-based polymeric materials' which were in fact methylated and/or ethylated lignins (*i.e.* lignin ethers prepared by treatments with diethyl sulphate and diazomethane). These materials were not inspected further as such and attention was focused instead on blending them with aliphatic polyesters [27] and to prepare other modified lignins bearing both ether and ester moieties, whose blends with aliphatic esters and PEO were in turn characterized [28]. The mechanical and thermal properties of the blends were thoroughly assessed as a function of composition, plasticizing polymer and modification and were shown to reflect typical features associated with a viscoelastic behaviour in the near-T_g region, often including a semicrystalline component arising from the polyester. These materials are certainly viable in terms of such features, but the basic question is whether they would be economically and ecologically competitive with a wide range of available counterparts, considering the separation procedures applied to obtain the desired lignin fraction, the requirements and reagents associated with the alkylation processes and the cost of the plasticizing polyesters. In other words, the obvious interest of using lignins as the major precursors seems to be negatively counterbalanced by the operations involved in their modification and blending. Figure 11.7 shows the effect of poly(1,4-butylene adipate) on the progressive plasticization of etherified KL samples [27]. A very recent addition to this saga [29] does not provide any results that would alter the above considerations.

The reaction of lignins, including LS, with epichloridrin [30] gives products which have been tested both in blends and in reactive compositions. Blends of these modified lignins with a poly(ethylene)–poly(propylene) mixture, [31] hydroxypropyl cellulose [32] and polyalkanoates [33] have been reported and their properties do not seem to warrant the additional requirements associated with the modification.

Homogeneous blends of PVA with 5–15 per cent of a KL modified by reaction with a maleimide-substituted aromatic carboxylic acid provided a means to improve the thermal and photochemical stability of the vinyl polymer [34]. Thus, the thermal resistance of the PVA-modified lignin blends increased, as illustrated in Fig. 11.8 [34]. Spectroscopic

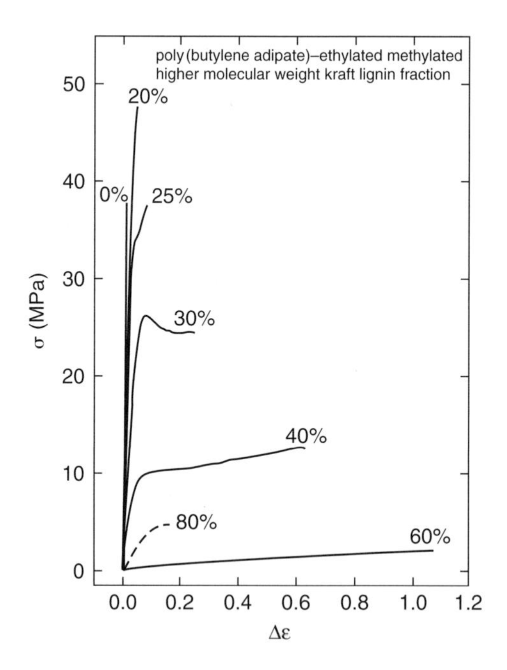

Figure 11.7 Progressive plasticization of ethylated methylated KL-based polymeric material by poly(1,4-butylene adipate). (Reproduced by permission of American Chemical Society. Copyright 2002. Reprinted from Reference [27]).

evidence of hydrogen bonding interactions between the two polymers was provided in this study, as shown in Fig. 11.9, which shows the chemical shifts of the PVA protons in NMR spectra in the region of 4–5 ppm [34].

Hydroxypropyl lignin, prepared with propylene oxide (PO) in a basic aqueous medium, was mixed with soy protein to give nano-heterogeneous blends with better mechanical properties and higher glass transition

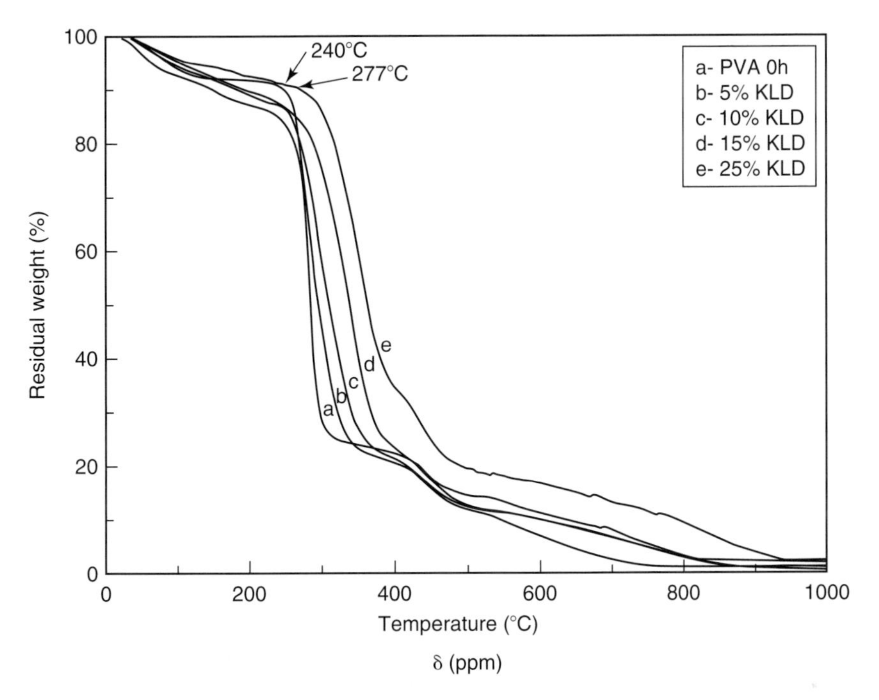

Figure 11.8 TGA curves of PVA and blends containing 5–25 per cent kraft lignin derivative (KLD). (Reproduced by permission of Elsevier. Copyright 2006. Reprinted from Reference [34]).

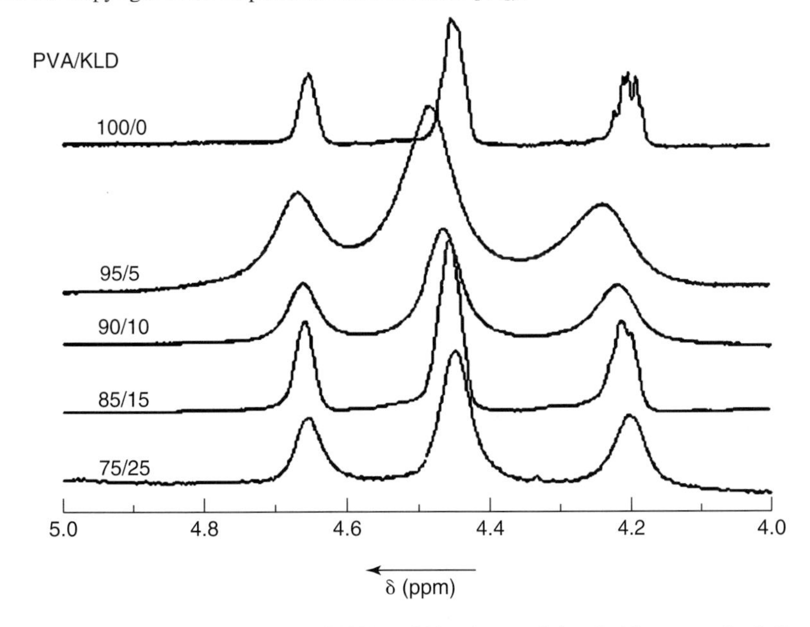

Figure 11.9 ^{1}H NMR spectra (4–5 ppm region) of PVA and blends containing 5–25 per cent kraft lignin derivative (KLD) in DMSO-d6, at room temperature. (Reproduced by permission of Elsevier. Copyright 2006. Reprinted from Reference [34]).

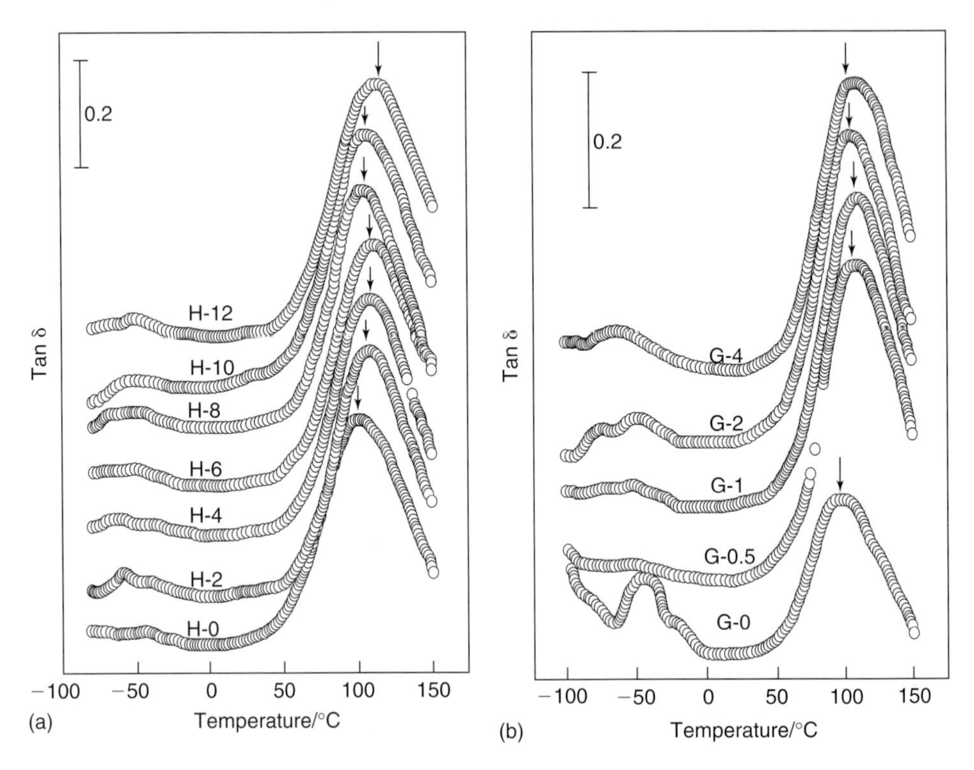

Figure 11.10 DMTA spectra of the (a) H-series, and (b) G-series sheets plotted in the form of tan δ versus temperature. The arrows indicate $T\alpha$. (Reproduced by permission of Wiley Periodicals. Copyright 2006. Reprinted from Reference [35]).

temperatures than those of the pristine protein [35]. Figure 11.10 shows the shifts observed for tan δ in the dynamical mechanical curves demonstrating the improved interface quality. The H-series means a content of hydroxypropyl alkaline lignin corresponding to 0, 2, 4, 6, 8, 10 and 12 wt%. The G-series indicates that 0, 0.5, 1, 2 and 4 mL, corresponding to 0, 1.7, 3.3, 6.6 and 13.2 wt%, respectively, of glutaraldehyde (GA) were added.

Another interesting family of materials involving modified lignins was described by Glasser's group [22] who prepared blends of biodegradable thermoplastics including cellulose esters, polyhydroxybutyrate and a starch–caprolactone copolymer with different lignin esters. The justification for using these lignin derivatives was that the corresponding unmodified organosolv lignin gave unsatisfactory materials. A thorough characterization of all these combinations showed very different levels of interaction between each polymer couple.

In a different vein, the modification of lignins by plasmas generated with different gases was successfully carried out in a rotating reactor [36]. As with other solid substrates, the modifications were confined to the surface of the lignin particles and were therefore suitable for the improvement of the interfacial compatibility in the processing of blends or composites.

Finally, the recent discovery of an accidentally prepared lignin sulphate [37] has opened potentially valuable avenues of medicinal applications.

11.3 LIGNINS AS MACROMONOMERS

As with blends and composites, the use of lignins for the synthesis of macromolecular materials has been based either on pristine reagents, as obtained in the delignification processes, or on appropriately modified homologues. The present survey of recent progress covers both approaches separately, as in Section 11.2. Previous reviews on the exploitation of lignins as macromonomers covered the topic more or less systematically up to the pertinent chapters in the book on lignin edited by Hu in 2002 [1].

11.3.1 Unmodified lignins

As already emphasized above in the case of physical mixtures of lignins with other polymers, it seems important to us to reiterate that, particularly for polymeric materials which have to compete with low priced existing petroleum-based counterparts, the use of lignins as macromonomers should not be accompanied by their costly modifications. In other words, the ideal situation is obviously that in which lignins can be introduced in the polymerizing medium as obtained from the delignification process. If that optimum approach cannot be applied because of reactivity or other constraints, then the modifications introduced into the lignin structure to solve these problems, should be carefully considered in terms of their economic and ecologic viability.

11.3.1.1 Phenol–formaldehyde and urea–formaldehyde resins

The partial replacement of phenol in the formaldehyde-based (PF) resins appears as one of the most obvious utilization of lignins and indeed much work has been devoted to the synthesis and assessment of these materials [1b,1e]. Given the huge market associated with particleboard, plywood and fibreboard adhesives, the possible incorporation of lignins in their formulation would represent a very profitable operation. Pizzi [38] has analysed the issue from an industrial standpoint and highlighted the advantages and difficulties associated with these systems. The major obstacle to such an industrial development is the limited reactivity of the aromatic sites of lignin compared with that of phenol, because of their higher degree of substitution, which introduces considerable steric hindrance. An interesting way to overcome this relatively slow step is to start the process by methylolating the lignin before mixing it with PF resins and even employing oligomeric aromatic diisocyanates as co-reagents. These process modifications have largely eliminated the problem of the longer pressing times induced by the presence of lignins, making it possible to use more than 50 per cent of methylolated lignins in the resin formulations.

More recent work on this topic include the use of sulphur-free lignins replacing up to 30 per cent of phenol [17] and of 5 per cent of sodium LS into PF–resol resins, which was found to promote condensation reactions [39].

The surface properties of these resins have been determined using both contact angle measurements and inverse gas chromatography [40] and compared with those of a standard PF counterpart. The dispersive contribution to the surface energy increased with the introduction of KL, whereas the polar contribution was not significantly affected. With sodium LS, the major change took now the form of a considerable increase in the donor properties of the resin surface, attributed to the presence of the —SO_3^- moieties. A recent interesting addition to this topic deals with lignin-based wood panel adhesives prepared without formaldehyde [41].

The introduction of lignins in urea–formaldehyde (UF) resins has also been extensively investigated [38] and successful industrial applications for particleboard have been operating with about 10 per cent of spent sulphite liquor in the UF formulation. Higher lignin contents have been applied to improve water resistance. In these resins, the lignin does not replace another reagent, as in the case of the PF counterparts, but simply intervenes as a third component, whose reactivity is lower, but whose interests lie in its low price and hydrophobic contribution to the final crosslinked material.

11.3.1.2 Poly(urethane)s

Although the Glasser group's pioneering work on lignin-based poly(urethane)s (PU) was mostly based on modified lignins, and will therefore be discussed in detail below, some of the materials prepared and reported in that systematic series of studies were indeed made up with unmodified lignins, but only for the sake of comparison. The reasons for the modifications were attributed to the modest accessibility of the lignins' hydroxyl groups for their reaction with isocyanate moieties.

Feldman's group was the first to tackle the possibility of preparing poly(urethane–urea)s to be used as sealing materials in which the oligomeric multiple isocyanate, that is normally the single precursor which crosslinks upon addition of the required amount of water, was mixed with 5–20 wt% of unmodified KL [42]. The characterization of these materials showed unambiguously that the lignin had indeed participated chemically to the construction of the network through some of its OH groups.

Two laboratories have been involved more recently in the synthesis and characterization of PUs based on unmodified lignins. In both approaches, the trick aimed at enhancing the lignins' reactivity consisted in using a diol as third monomer, preferably capable of dissolving the lignin.

Our contribution to this field started with a study of PU foams prepared from KL and hexamethylene isocyanate in the presence of an oligo(caprolactone) macrodiol [43]. In this system, the lignin mass contribution gave viable

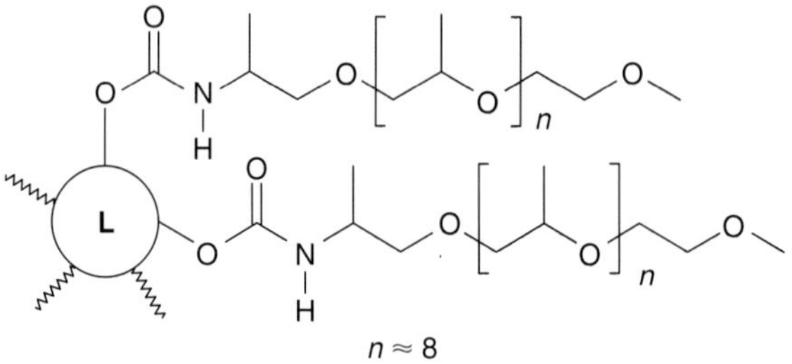

$n \approx 8$

Scheme 11.1

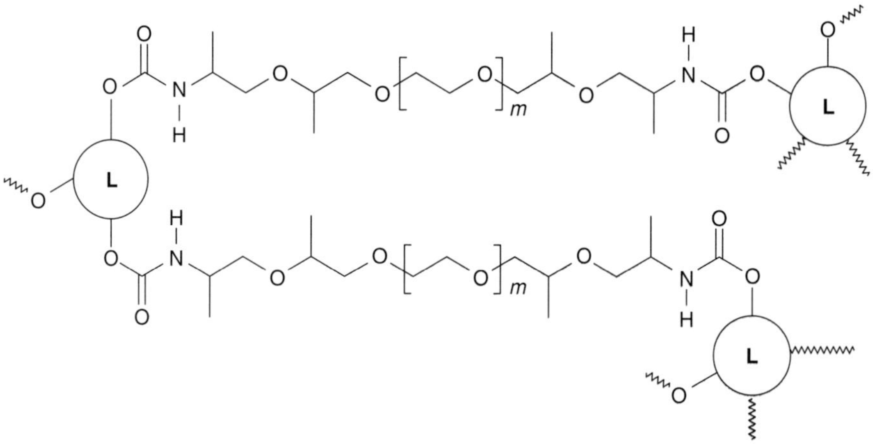

Scheme 11.2

materials up to about 40 per cent. Subsequent studies [44] called upon the use of Alcell lignin (*i.e.* a mixture of oligomers with Mn ~ 700 and Mw ~ 1 700 and a T_g of 64°C) whose reactivity was considerably higher than that of KL because of the reduction of steric crowding around the hydroxyl groups. In order to prepare elastomeric PUs we synthesized oligoether isocyanates bearing very low glass transition temperatures. Before carrying out the actual polymer syntheses, various model reactions were studied with the aim of establishing comparative reactivity criteria related to both the difference between phenolic and aliphatic OH groups and the role of steric hindrance around each type. Thus, various lignin models, as well as a model monoisocyanate, were employed together with the actual monomers, *viz.* Alcell lignin, an oligoether monoisocyanate and the corresponding diisocyanate. FTIR and NMR spectroscopy were used to assess these reactivity features and the conclusion was that, although differences existed in terms of the nature of the OH group and its steric availability, all the reactions went to completion, indicating that it was possible in this context to involve every single hydroxyl function. The two types of materials which were then prepared consisted respectively of thermoplastic elastomers or pastes, when the monoisocyanate was used, and partly or fully crosslinked elastomers, when the comonomer was the diisocyanate. The structure of both these low T_g PUs are shown in Schemes 11.1 and 11.2.

This work was then extended to oxygen-organosolv lignins isolated from spent liquors after delignification of aspen and spruce in different acidic water/organic solvent media [45]. It was found that the nature of the organic solvent used in the delignification process had an effect on the reactivity of the ensuing lignins because it affected their OH content. On the whole however, the properties of the polyurethane networks, prepared with the same diisocyanate shown in Scheme 11.2, were similar to those of the corresponding materials based on Alcell lignin discussed above.

The broad investigation carried out in Hatakeyama's laboratory [46] encompassed various unmodified lignins including Kraft, organosolv and sulphonate varieties which were dissolved into oligoether diols before mixing the ensuing solution with a multi-functional aromatic isocyanate. The strategy here for improving the reactivity of the lignin OH groups consisted in using the macrodiol as both the solvent for lignin and a comonomer. This second

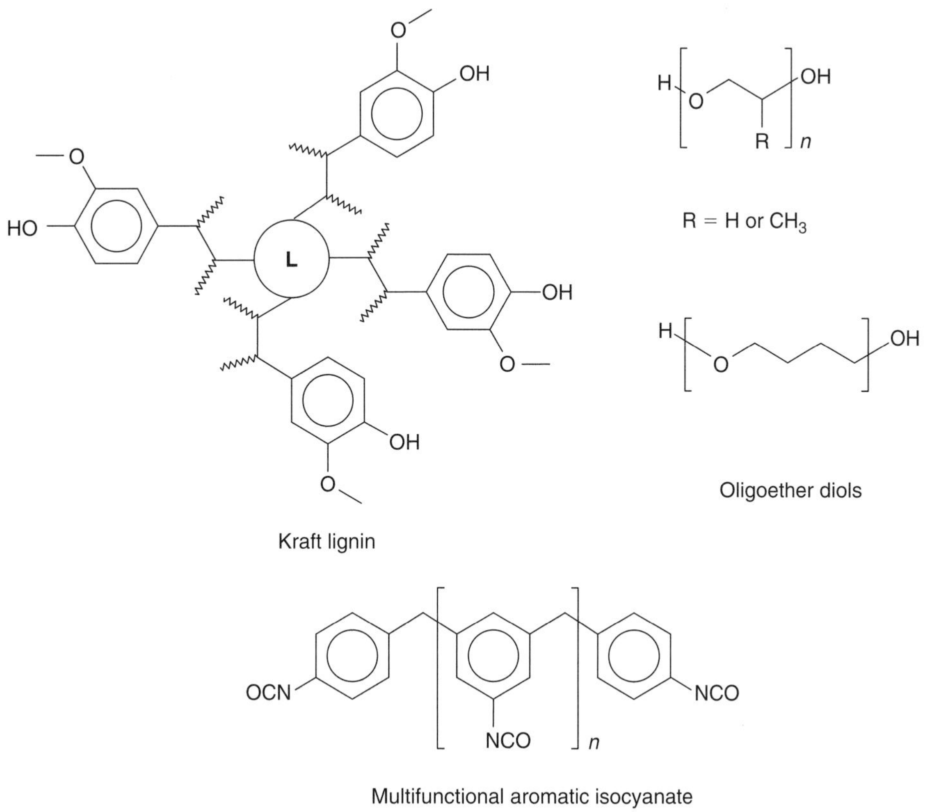

Kraft lignin

Oligoether diols

R = H or CH_3

Multifunctional aromatic isocyanate

Scheme 11.3

role provided a means to reduce the stiffness of the ensuing networks by introducing flexible oligoether sequences as spacers among the rigid aromatic domains formed by the condensation of lignin macromolecules with the isocyanate. All the materials prepared in this study in the form of sheets or foams were thoroughly characterized in terms of thermal transitions and decomposition, as well as mechanical properties. Scheme 11.3 illustrates a typical polymer structure generated by this approach.

A recent report [47] on the use of steam-exploded straw lignin as macromonomer for the synthesis of PUs, describes systems in which two diisocyanates were tested and in some instances ethylene glycol was employed as comonomer. The results of this investigation are puzzling because molecular weights could be determined for all the ensuing polymers, a fact which is in stark contrast with the expected crosslinked nature of these materials, given the OH multi-functional character of lignin. The other odd aspect of this work is that the amounts of recovered polymers were systematically much lower than the sum of the monomers used, suggesting unaccountable losses, since in principle these types of systems produce a complete yield of polymer.

11.3.1.3 Polyesters

Studies related to the synthesis of lignin-based polyesters are few and far between compared with their polyurethane counterparts. This is quite surprising considering that the synthesis and characterization of lignin alkanoates have received close scrutiny [48] and, although these products cannot be considered as interesting materials as such, they certainly represent promising additives in macromolecular compositions (see below). To the best of our knowledge, the first systematic study of the use of unmodified lignins (kraft and organosolv) for the synthesis of polyesters, was carried out some 15 years ago by Guo and Gandini and reported in several publications [6, 44, 49]. In a first series of experiments, the lignins were made to react at low temperature with different proportions of an aliphatic or aromatic acid dichloride in the presence of a proton scavenger. The FTIR spectra of the ensuing polyesters showed unambiguously that both aliphatic and phenolic hydroxyl groups had indeed

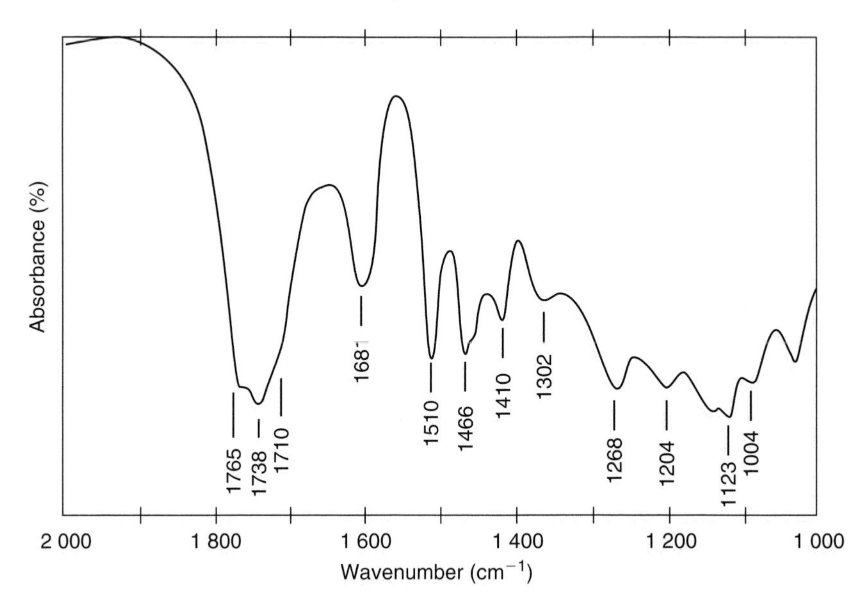

Figure 11.11 FTIR spectra of a typical KL–sebacate (Reproduced by permission of Kluwer Academic. Copyright 2002. Reprinted from Reference [1f]).

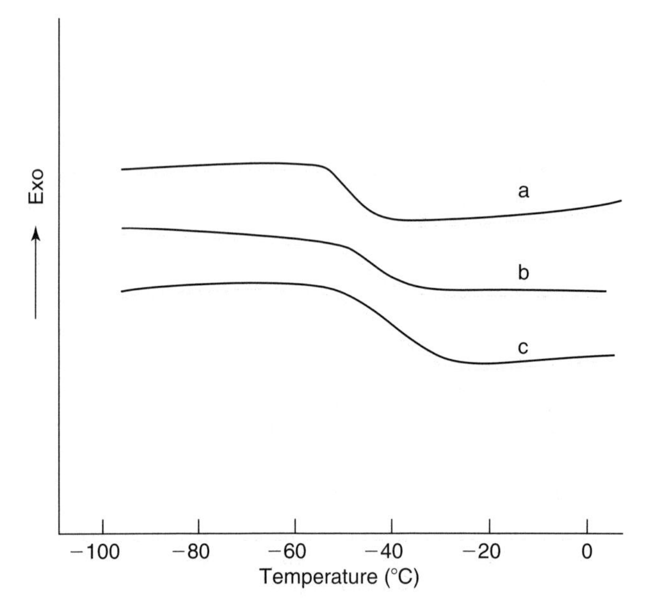

Figure 11.12 DSC thermograms of three lignin–sebacates with different KL contents. (Reproduced by permission of Pergamon Press. Copyright 1991. Reprinted from Reference [6]).

participated in the polycondensation reactions, as clearly shown in Fig. 11.11. The fact that the highest yield of insoluble polymers coincided with an OH/COCl ratio of unity indicated moreover, that all the OH groups had been involved in the construction of the networks (*i.e.* steric impediments had not played any appreciable role in these processes).

The major difference between the two sets of polyesters had to do of course with their glass transition temperature, which was obviously lower when either the aliphatic diacid had been used or lower proportions of lignin were employed, as shown in Figs 11.12 and 11.13.

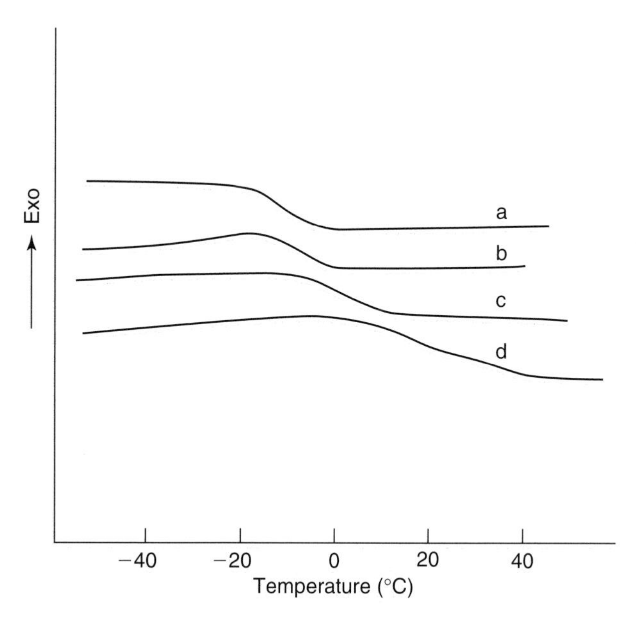

Figure 11.13 DSC thermograms of four lignin–therephthalates with different KL contents. (Reproduced by permission of Pergamon Press. Copyright 1991. Reprinted from Reference [6]).

In order to extend the range of properties and to simplify the synthetic procedures, a second series of experiments was then carried out in which, on the one hand, a third comonomer *viz.* a flexible macrodiol (PEGs of different molecular weights), was added to the systems, and, on the other hand, no solvent was employed because the lignins were thoroughly soluble in the added diol. These bulk copolyesterifications increased the degrees of freedom related to the experimental parameters since it was now possible to vary the proportion of lignin with respect to the diol (degree of chain extension), and the DP of the diol (length of the chain extensions). In this way, the range of T_g and viscoelastic properties could be amply modulated.

Typical structures of these two families of polyesters are shown in Schemes 11.4 and 11.5.

Figure 11.14 shows some of these effects through typical DSC tracings for a series of polyesters. Figure 11.15 highlights the fact that the higher the KL content in the polyesters, the higher the graphitized residue after their thermal degradation. This is an important observation because it suggests that these polymers should not propagate a flame because of their intumescent character.

Finally, a morphological analysis of some of these materials showed that it was possible to prepare single-phase networks with more than 50 per cent of KL, since a clear-cut phase separation only became visible in SEM micrographs of polyesters prepared with 65 per cent of lignin.

No substantial difference in behaviour was noticed with either the polymerization systems or the polyester properties between the uses of kraft or oxygen-organosolv lignins. The only quantitative feature favouring the latter was the fact that its reactivity was higher than that of the former as indeed expected because of its lower molecular weight [50].

Among the possible applications of these polyesters, one seems to us particularly attractive because it might replace advantageously existing materials and at the same time eliminate a health hazard. The idea is to replace styrene monomer in typical commercial unsaturated polyester compositions by prepolymers synthesized using lignin and an excess of diacid, which could be premixed with the traditional low molecular weight polyesters bearing terminal hydroxyl groups. Given the very low reactivity of carboxylic acids with OH groups, these mixtures would have an adequate pot life at room temperature and could therefore be commercialized as such. Their processing, typically with reinforcing fibres, would be carried out at high temperatures and maybe with an appropriate latent catalyst, thus producing the desired networks because of the numerous —COOH groups borne by the lignin prepolyesters. Presently, the crosslinking of the unsaturated polyesters is ensured by the free radical polymerization of styrene monomer, added to them at the processing stage, which involves the unsaturations by a grafting-through mechanism. In most of the workshops and small industrial concerns, the handling of styrene monomer constitutes

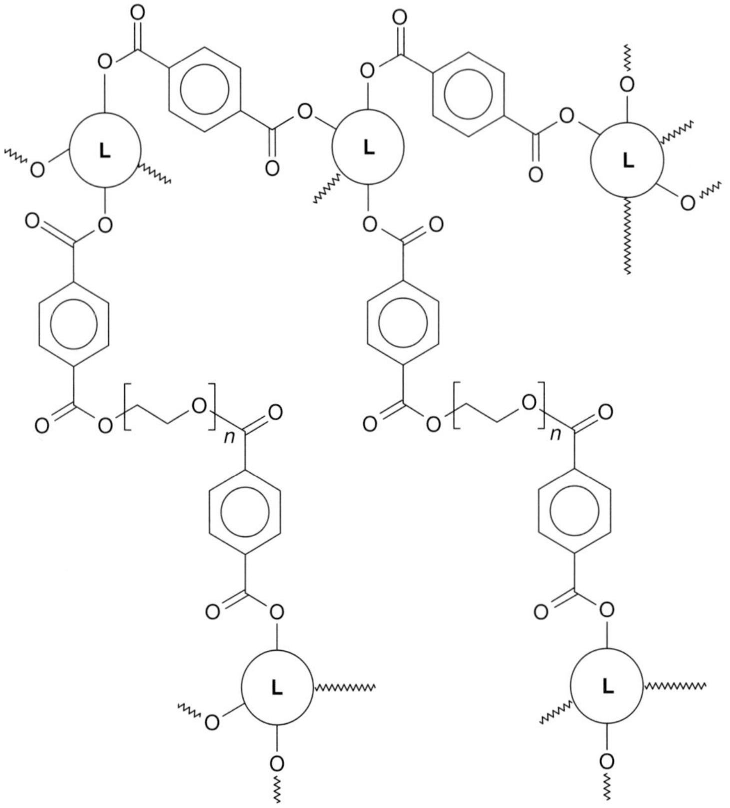

Scheme 11.4

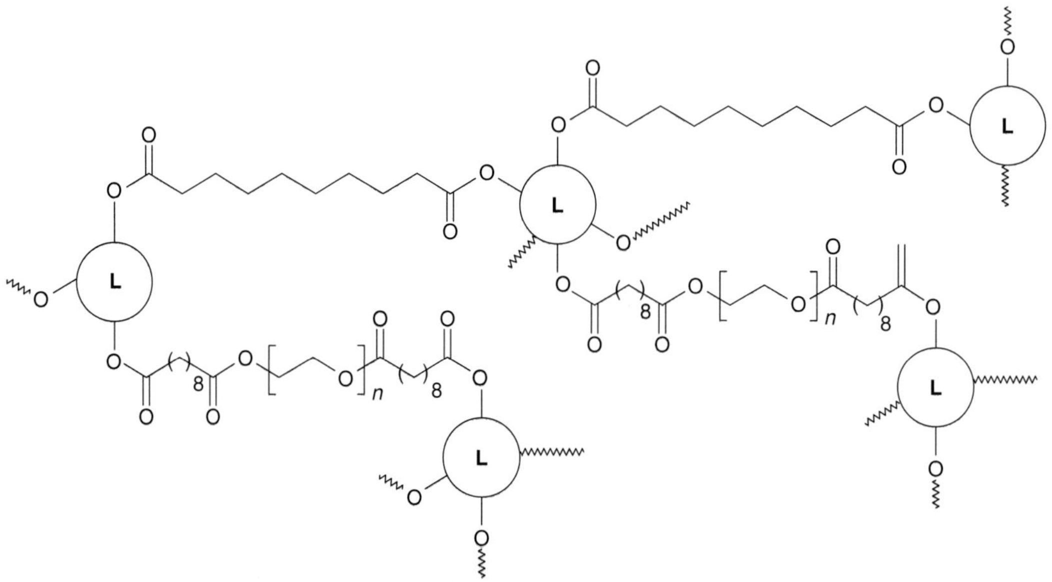

Scheme 11.5

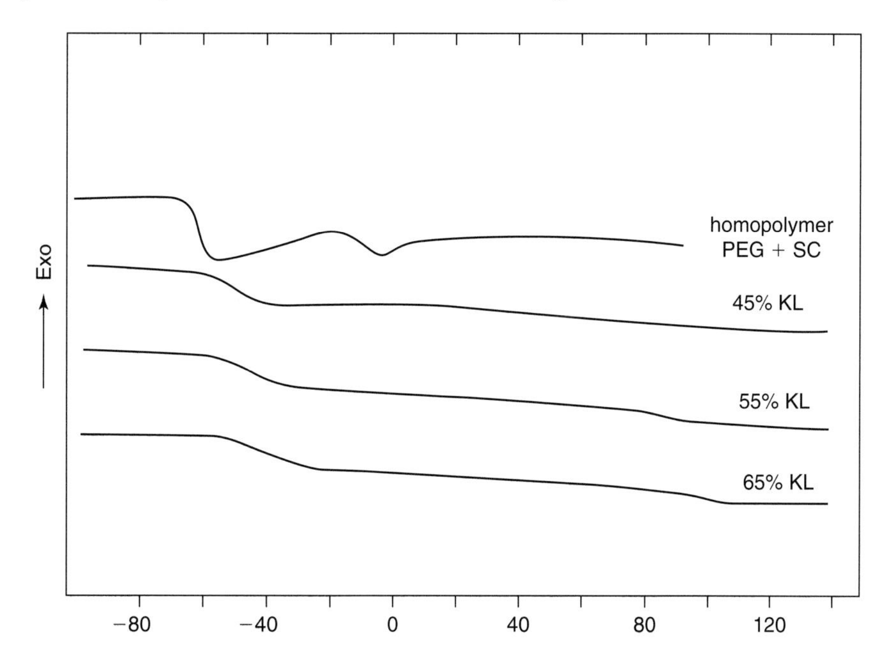

Figure 11.14 DSC tracings of four KL-PEG_{300}-sebacates with different KL contents. (Reproduced by permission of Kluwer Academic/Plenum Publishers. Copyright 2002. Reprinted from Reference [1f]).

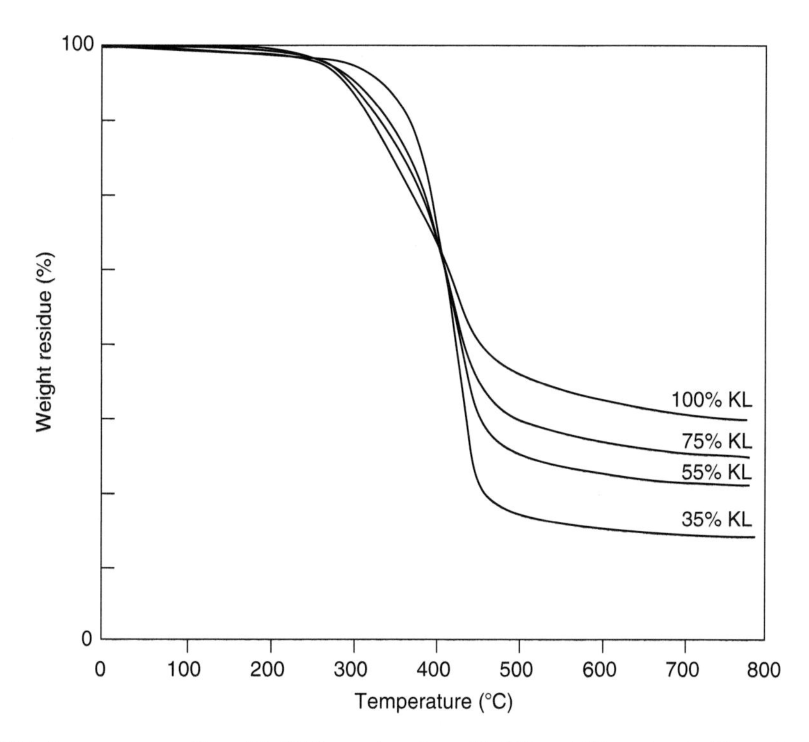

Figure 11.15 TGA thermograms of four KL-PEG_{300}-sebacates with different KL contents. (Reproduced by permission of Kluwer Academic/Plenum Publishers. Copyright 2002. Reprinted from Reference [1f]).

an obvious health risk, which would be eliminated by this approach, capable moreover of giving viable materials in which one of the components is a cheap renewable resource.

The only other study of polyesterification involving an unmodified lignin was recently published together with the already discussed study on polyurethanes [47]. Steam-explosion straw lignin was made to react with dodecandioyl dichloride in different monomer proportions in THF using triethylamine as the acid scavenger. As in the case of polyurethanes, the authors again reported quite surprisingly that the ensuing polyesters were soluble products with quite modest molecular weights. These observations are in contradiction with the normally expected crosslinking associated with systems involving a difunctional monomer (here the acid dichloride) reacting with a partner bearing numerous complementary functions (here the lignin with its multiple OH groups).

11.3.1.4 Epoxy resins

The only published report on the use of unmodified lignins for the preparation of epoxy resins was announced by IBM in the late 1990s and dealt with the solution interaction between Kraft or organosolv lignins and various epoxy monomers, followed by a crosslinking operation induced by aromatic diamines [51]. Although details were scanty, it was announced that these materials were suitable for application in the electronic industry because of their excellent adhesion on copper, very good dielectric properties and temperature stability up to more than 175°C.

11.3.1.5 Grafted lignins

The idea of using lignins as structures capable of generating polymer grafts by a free radical mechanism and the corresponding publications which claim its successful application [52] leave us somewhat perplex, as already explained in detail in previous reviews, [1b,1e] particularly with homogeneous systems using exotic initiators in dimethylsulphoxide (DMSO) (and even radioactive ^{14}C-labelled styrene [53]). It is in fact difficult to rationalize that styrene or acrylamide can undergo a normal chain growth in the presence of polyphenolic structures, which are excellent radical traps, as indeed emphasized by the numerous studies discussed above on the use of lignins as protecting additives for various commercial polymers. The lack of structural characterization to corroborate the claim of successful grafting and the absence of a study of the materials' properties contribute to generate a puzzled reaction.

11.3.1.6 Resins based on lignins and furans

A recent patent describes the use of adhesive compositions based on mixtures of lignin, furfuryl alcohol and maleic anhydride, whose setting is insured by Lewis acid catalysts [54]. Its text contains a thorough patent survey of other lignin-based resins.

11.3.2 Chemically modified lignins

Glasser and co-workers have contributed most significantly to the formulation of two families of lignin-based materials, namely polyurethanes and epoxy resins, in which lignin was modified in order to enhance its reactivity or bear novel functions.

11.3.2.1 Polyurethanes

The modifications of kraft and organosolv lignins were carried out here in order to 'bring out' their OH groups and thus make them more available for condensation with diisocyanates. The techniques used to achieve this goal consisted in blocking first of all part of the OH groups by ethylation with diethyl sulphate and then treating the ensuing lignins with propylene oxide (PO) in order to chain extend the remaining hydroxyl functions with oligomeric PO moieties terminated by primary or secondary OH groups. A variety of these lignin-based polyols (Scheme 11.6) was thus prepared and used as macromonomers in the synthesis of numerous polyurethanes, using aliphatic or aromatic diisocyanates and, in some instances, a macrodiol comonomer, mostly in the form of PEGs of different molecular weights [55].

This systematic study, carried out throughout the 1980s, constituted a very thorough approach to both the synthetic aspects, with their corollary structural characterization, and a whole set of physical and technological properties. Elastomers as well as rigid polymers were prepared and their possible applications carefully examined on the basis of the specific performances determined for each one. Further additions to the basic investigation included the use

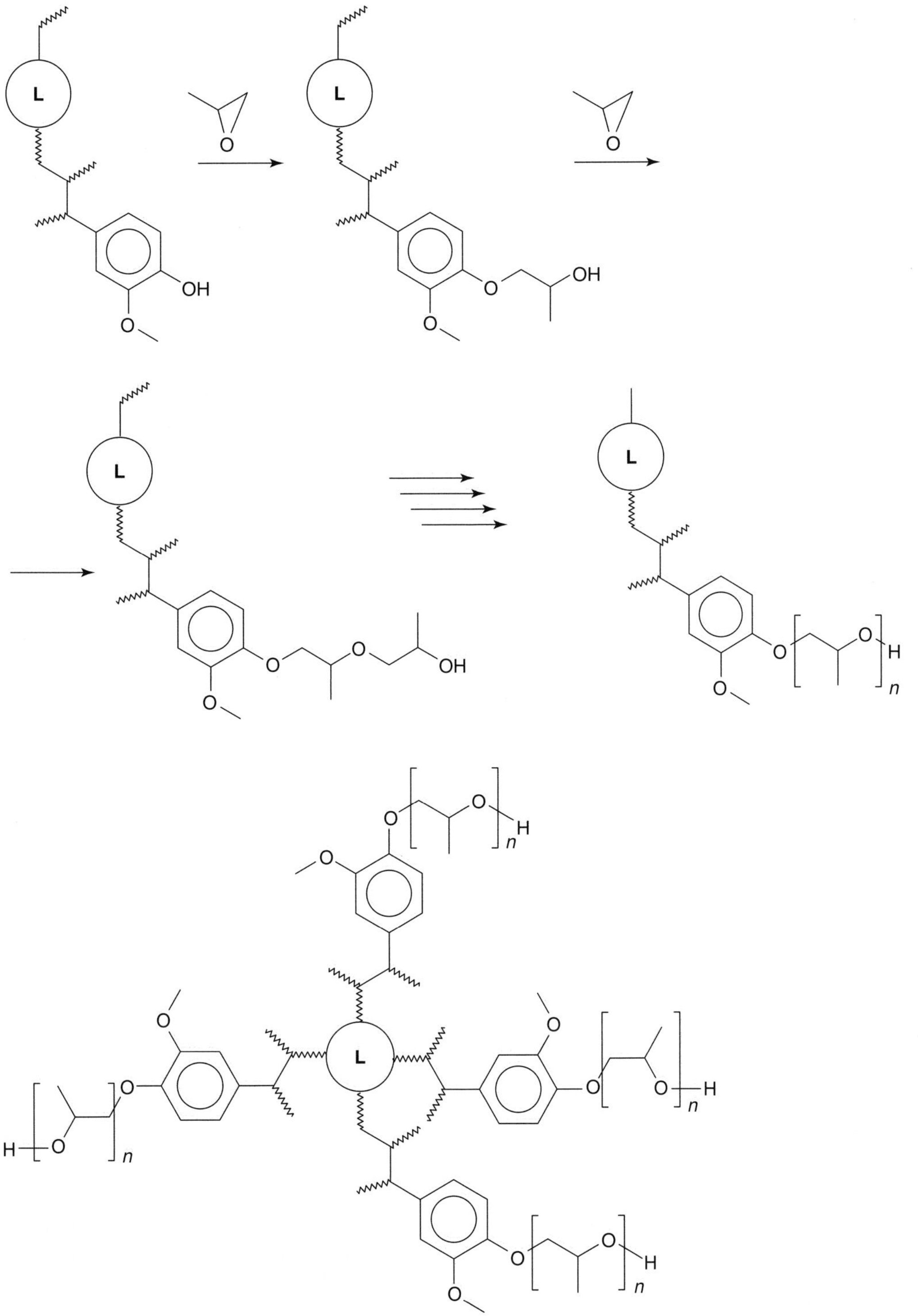

L
O
OH
O
L
O
O
O
OH
L
O
O
OH
O
O
L
O
O
O
H
n

O
O
H
n
O
L
O
O
O
O
H
n
H
O
n
O
O
O
O
H
n

Scheme 11.6

of DMTA to follow the crosslinking reactions [56] and the synthesis of interpenetrating networks with acrylic structures [57].

11.3.2.2 Epoxy resins

In this context, Glasser's group used the partially oxypropylated lignins described in the preceding section as the substrate for the synthesis of macromonomers bearing multiple oxirane groups by reacting them with epichloridin (Scheme 11.7) [58].

The crosslinking of these polyoxiranes with aromatic diamines using the standard procedures applied to epoxy resins was then studied by DMTA. Reactions proceeded normally, except for the fact that early vitrification, arising from the high aromatic content of these compositions, induced a considerable slowdown of the late curing stages.

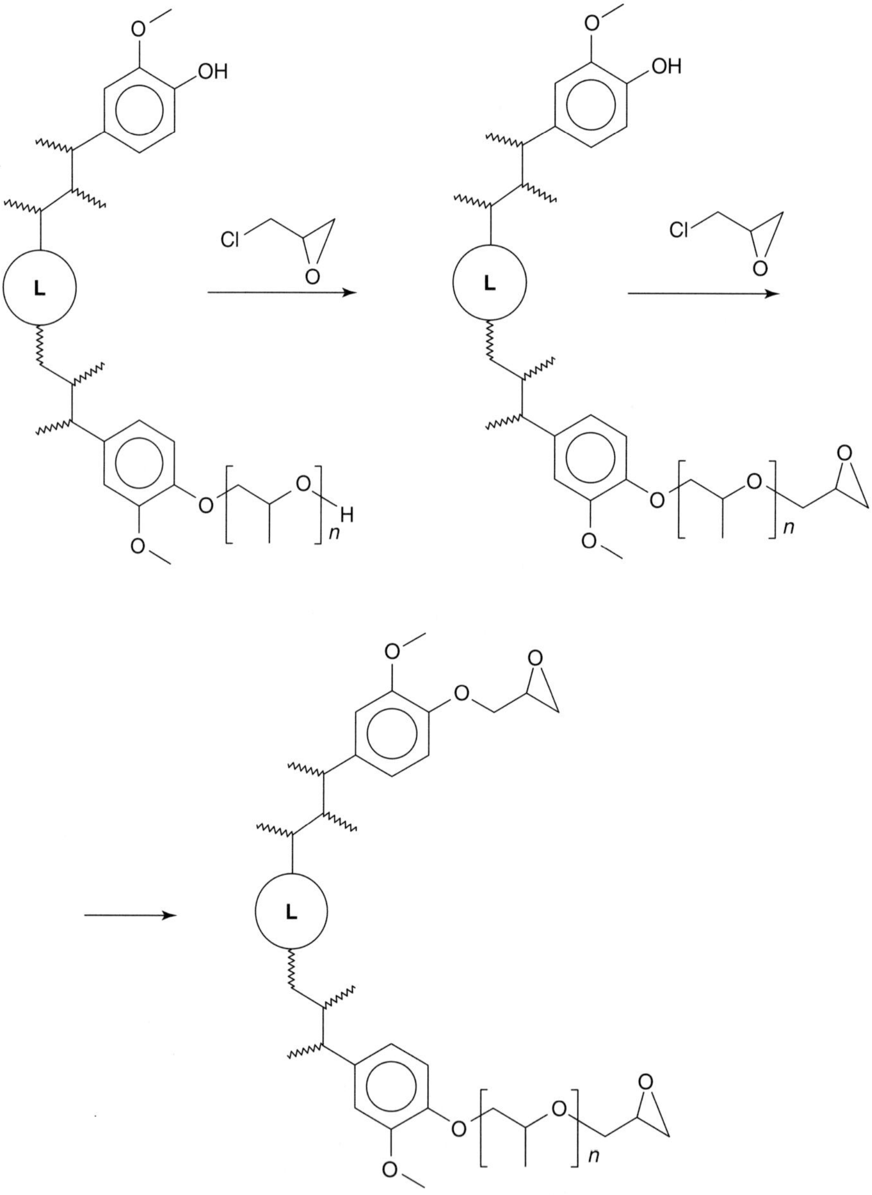

Scheme 11.7

In a recent addition to the elaboration of lignin-based epoxy resins, calcium LS was treated with epichloridrin to introduce oxirane functions in their structure [59]. Thereafter, these modified lignins were treated with maleic anhydride and the curing reaction leading to these novel epoxy resins was followed by FTIR spectroscopy [60]. More attention was devoted in this work to the kinetics and mechanism of the network formation than to any characterization of the properties of the ensuing materials, but this will probably be followed up.

11.3.2.3 Alkenyl lignins

The idea of attaching alkenyl moieties to lignins and use the modified products as macromonomers suitable for chain polymerizations has been put to test on few occasions with unconvincing results in terms of the interest of the materials obtained. The problem intrinsic to this approach is that it is impossible to prepare lignin macromonomers bearing a unique polymerizable moiety, because even when using 1:1 reaction conditions, the distribution of the alkenyl groups cannot be made to reflect that stoichiometry (*i.e.* some lignin macromonomers will contain more than one and others none). It follows that the polymerization of these derivatives will inevitably generate networks, without any straightforward means of controlling their growth.

In both reported syntheses of lignins bearing acrylic moieties, prepared respectively by the reaction of acryloyl chloride [61] and methacrylic anhydride, [62] a high degree of substitution was accompanied by free radical polymerizations or copolymerizations with ill-defined products. In the case of the reaction of lignin with chloroethyl vinyl ether, [63] the ensuing lignins bore an average of five vinylether groups per macromonomer and, as could have been easily anticipated, their cationic polymerization produced insoluble crosslinked materials.

11.3.2.4 Bulk lignin oxypropylation

We have carried out an extensive investigation of the oxypropylation of a whole variety of natural polymers using a simple bulk technique which only involves the substrate, PO and a nucleophilic catalyst. Given the general interest of this procedure and of the ensuing viscous polyols, a specific chapter of this book (Chapter 12) is devoted to the issue and therefore the conditions related to the use of lignins as substrate, the results obtained and the exploitation of the corresponding polyols is to be found there. It is important to emphasize that this process differs substantially from that adopted previously by Glasser's group and already discussed above in the context of both polyurethanes and epoxy resins.

11.4 CARBON FIBRES AND OTHER GRAPHITIC MATERIALS

It is reasonable to expect that lignins should be good precursors to graphitic structures, given their high contents of aromatic moieties. Two types of such materials have been pursued in this field, as briefly reviewed below, with particular emphasis on recent advances.

11.4.1 Activated carbon

The essence of this research is the result of a systematic study conducted over the last couple of decades by a Spanish laboratory, in which the controlled pyrolysis of KL was optimized to give activated carbons of high quality [64]. Essentially, a precarbonization treatment at 350°C under nitrogen was followed by further carbonization at 700–900°C. Finally, the product was activated either physically by a flow of carbon dioxide, or chemically with $ZnCl_2$. Depending on the specific treatment, materials with different properties were obtained, namely with specific surfaces of 1 000–2 000 $m^2 g^{-1}$ and a micro/meso/macro porosity distribution which could be readily adapted to the envisaged utilization. These materials possess characteristics which are ideally suited for applications like gas and water purification, liquid–liquid separation, selective adsorption and heterogeneous catalysis [64,65]. The lignin used as a starting material to prepare carbon was isolated from *Eucalyptus Grandis* kraft black liquors by acid precipitation. After purification and drying, it was impregnated by incipient wetting with 85 per cent (w/w) aqueous phosphoric acid at room temperature and dried for 24 h at 333 K in a vacuum dryer. The impregnated lignin was thermally treated under nitrogen at different activation temperatures (623–873 K). For example, the sample P2/698 corresponds to the activated carbon prepared at a phosphoric acid/lignin ratio of 2 and an activation temperature of 698 K. Figures 11.16 and 11.17 show the adsorption isotherms of phenol and Cr (VI), respectively, using carbonaceous substrates obtained under different conditions.

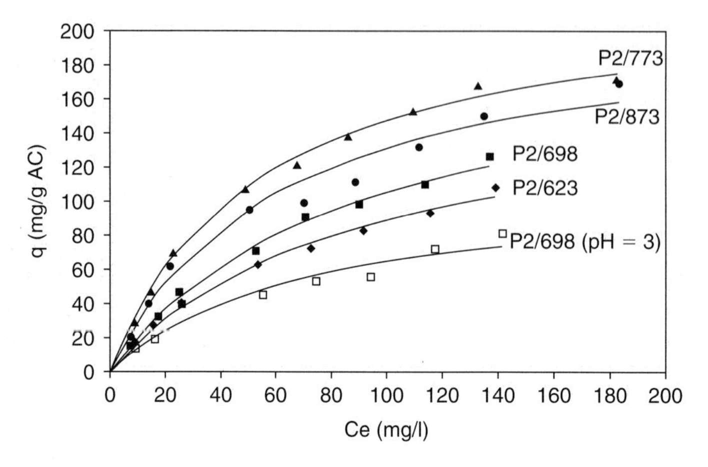

Figure 11.16 Adsorption isotherms of phenol (298 K) on the activated carbons obtained at different activation temperatures (pH = 6, unless indicated). (Reproduced by permission of Elsevier. Copyright 2004. Reprinted from Reference [65a]).

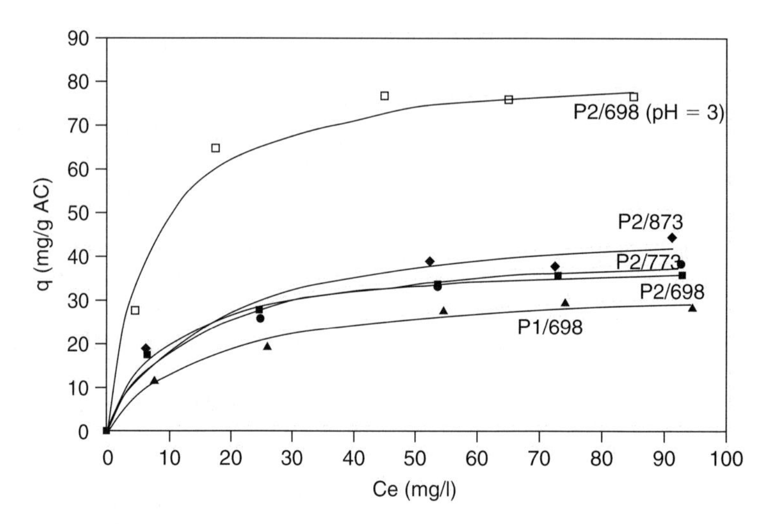

Figure 11.17 Adsorption isotherms of Cr (VI) (298 K) on activated carbons (pH = 7, unless indicated). (Reproduced by permission of Elsevier. Copyright 2004. Reprinted from Reference [65a]).

11.4.2 Carbon fibres

Research on the use of lignins as precursors to carbon fibres started some 40 years ago, but it is only from the 1990s that rational approaches have been pursued. Kadla *et al.* reviewed this topic until the beginning of the millennium [66] and have contributed significantly since then to its progress with valuable investigations.

Some of the lignins they used could be softened and spun into fibres as such, but better results were obtained from their homogeneous blends with PEO [67]. Subsequent carbonization at 1 000°C, following thermal stabilization at 250°C, gave carbon fibres with typical diameter of 30–60 μm, tensile strengths of 350–450 MPa, moduli of 30–60 GPa and elongations at break of about 1 per cent. Figure 11.18 shows the effect of increasing thermal stabilization temperature rate on the fibre stability.

Hollow core carbon fibres were also prepared from heterogeneous lignin–poly(propylene) blends, whose morphology was generated by the selective thermal ablation of the polyolefin [68].

Recent additions to these investigations deal with the use of poly(ethylene terephthalate) as the lignin blend component [69] and with the oxidative thermal stabilization of KL fibres [70].

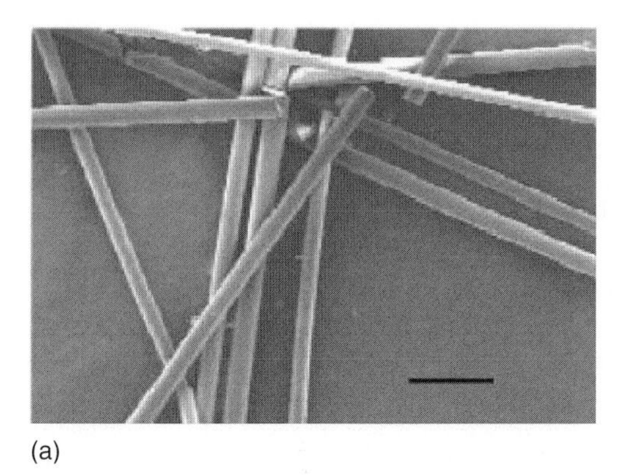

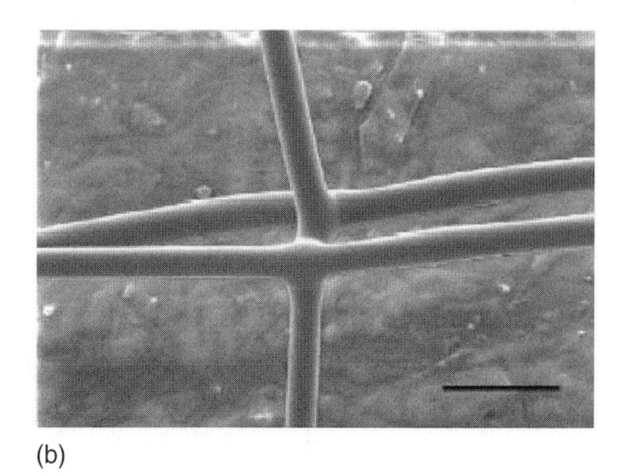

(a) (b)

Figure 11.18 Micrographs of 99 per cent hardwood kraft lignin (HKL)/1 per cent PEO (100 K) fibres after thermal stabilization at temperature rates of (a) 30°C/h and (b) 90°C/h (in both cases the bar represents 100 μm). (Reproduced by permission of Pergamon. Copyright 2002. Reprinted from Reference [67]).

11.5 LIGNIN SURFACES AND INTERFACES

Monolayers of lignin have been examined with a Langmuir troth and the observed intermolecular associations, already well-known in solution systems, point to the possibility of fairly regular extended aggregates in these two-dimensional domains [71]. Figure 11.19 displays the AFM image for a five-layer Langmuir–Blodgett (LB) film of lignin deposited onto mica. The roughness value is 3.8 nm. Figure 11.19(b) shows a lower roughness for the cast film, compared to the LB film (Fig. 11.19(a)), which may be attributed to the absence of molecular orientation in the cast film.

Very recent studies have been devoted to the preparation and characterization of smooth lignin surfaces of extremely thin films spun from solutions [72]. The interaction of these surfaces with a cellulose sphere has been quantified [73] and the corresponding forces found to behave according to the DLVO theory.

These studies open the way to a better understanding of the physical and chemical molecular behaviour of lignins faced with different structures through the construction of LB morphologies and by further studies of interfaces.

11.6 AROMATIC MONOMERS FROM LIGNIN AND MODEL POLYMERS

The possibility of degrading lignins in a controlled manner has interested wood chemists for a long time particularly because of their interest in establishing clear-cut lignin structures starting from the isolated fragments. For

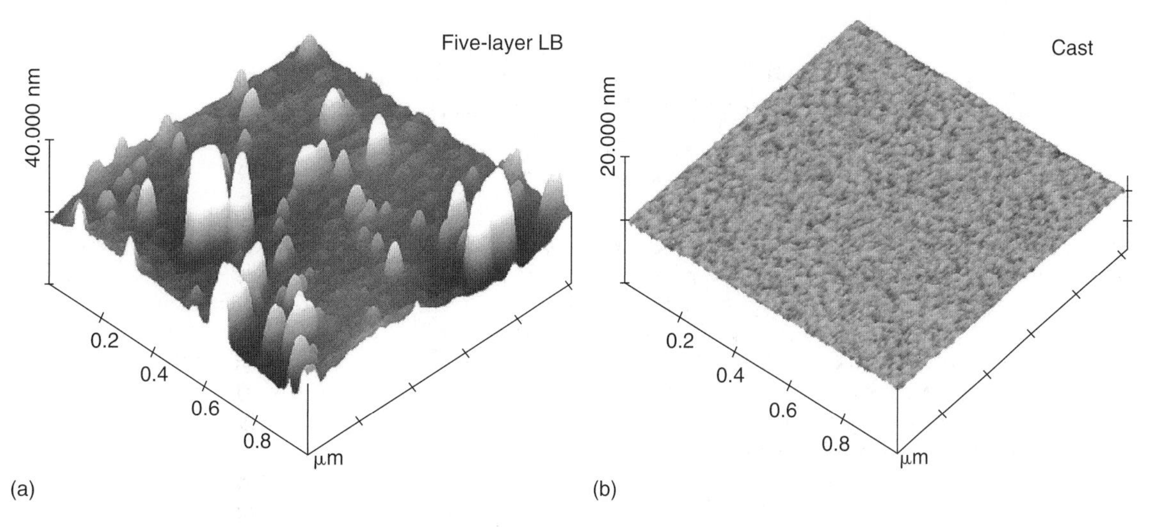

Figure 11.19 AFM images of lignin deposited on mica: (a) LB film, (b) cast film. (Reproduced by permission of the American Chemical Society. Copyright 2002. Reprinted from Reference [71b]).

a polymer chemist, this strategy must be seen in the very different context related to the possibility of preparing well-defined monomers or at least interesting precursors to monomers. This approach is obviously restricted to aromatic structures bearing either functions suitable for polycondensation (*e.g.* OH and COOH groups) or moieties like aldehyde groups that can be readily converted into polymerizable unsaturations.

To the best of our knowledge, this goal has never been attained in terms of reasonable yields of one monomer or a group of homologous structures. This is not surprising considering, on the one hand, the multiplicity of 'monomer units' present in the macromolecular structure of a given lignin and its variety as a function of species and pulping processes and, on the other hand, the complexity associated with their fine-tuned chopping.

The only viable conversion of lignin into a single aromatic compound is the synthesis of vanillin, which is a very useful commodity, but not in the realm of polymer chemistry. A recent investigation reports its oxidation to vanillic acid [74] and another, the use of this compound for leather tanning [75].

Okuda *et al.* [76] used a solvothermal treatment of organosolv lignin with a water-*p*-cresol mixture to generate a single product, *viz.* 4-hydroxyphenyl-(2-hydroxytolyl)-methane (HPHTM), in yields as high as 80 per cent without any charring. The proposed mechanism for the formation HPHTM, is shown in Scheme 11.8. According to the authors, the HPHTM is produced by the elimination of glycolaldehyde via the formation of an intermediate 'hydroxylphenyl-propane derivative (HPHTP)'.

This bis-phenol could be an interesting monomer for the synthesis of fully or partly aromatic polyesters and polyethers.

In a different vein, Bonini and D'Auria [77] examined the products arising from the degradation of steam-exploded straw lignin induced by a singlet oxygen treatment. As opposed to the solvothermal study mentioned above, this process generated a host of aromatic compounds with various moieties appended to them, such as phenolic and aliphatic OH functions, methoxy groups and α, βC=C unsaturations. This lack of selectivity seemed intrinsic to the multiple reaction pathways induced by singlet oxygen and therefore of little use for the purpose of preparing viable monomers.

If that quest has up to now eluded researchers, considerable attention has been placed on the synthesis of lignin-like polymers using model monomers. Recent results in this context include the work of Kishimoto's group [78]

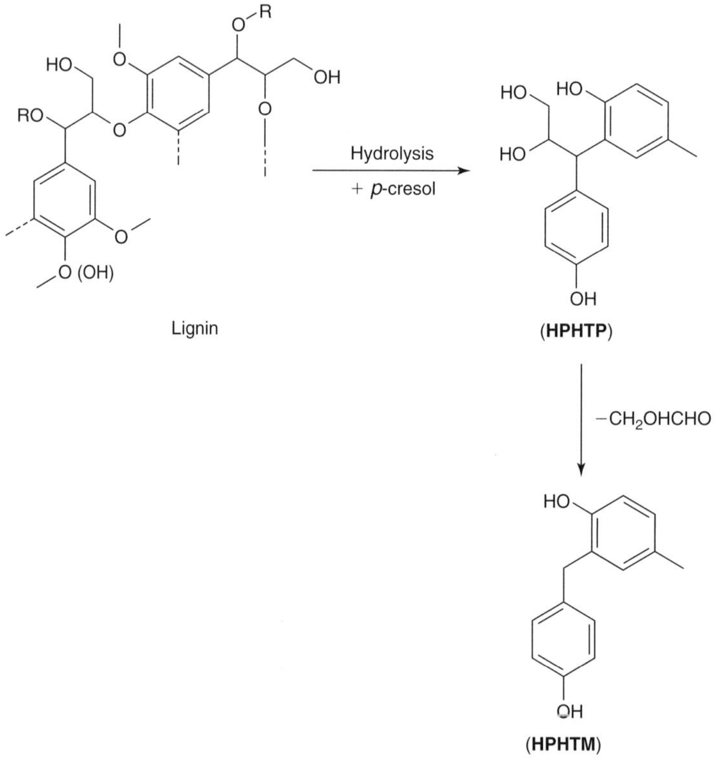

Lignin

Hydrolysis
+ *p*-cresol

(**HPHTP**)

$-CH_2OHCHO$

(**HPHTM**)

Scheme 11.8

based on the synthesis of 'β-*O*-4 type artificial lignin' through the linear polyetherification of differently substituted phenols bearing a 4-carbonyl-CH_2Br moiety using the Williamson reaction, as shown in the example of Scheme 11.9. Figure 11.20 shows the NMR spectra of both artificial and natural lignins.

Kaneko *et al.* [79] have prepared hyperbranched structures from the copolymerization of 4-hydroxycinnamic acid with 3,4-dihydroxycinnamic acid (*i.e.* monomers bearing some resemblance to lignin structural units). These

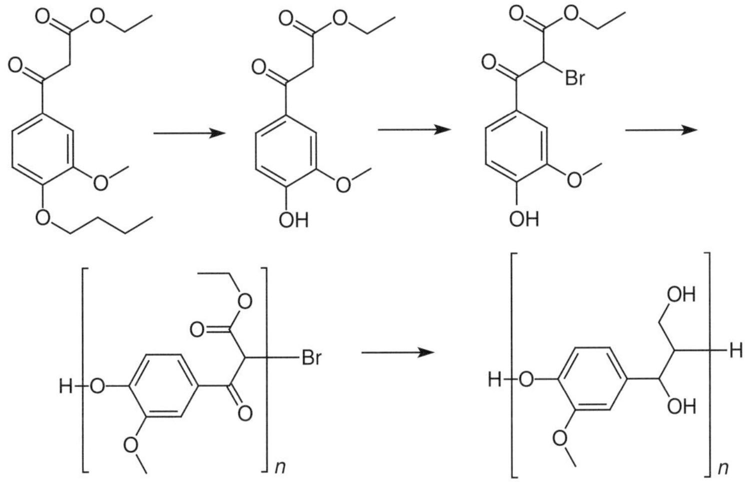

Scheme 11.9

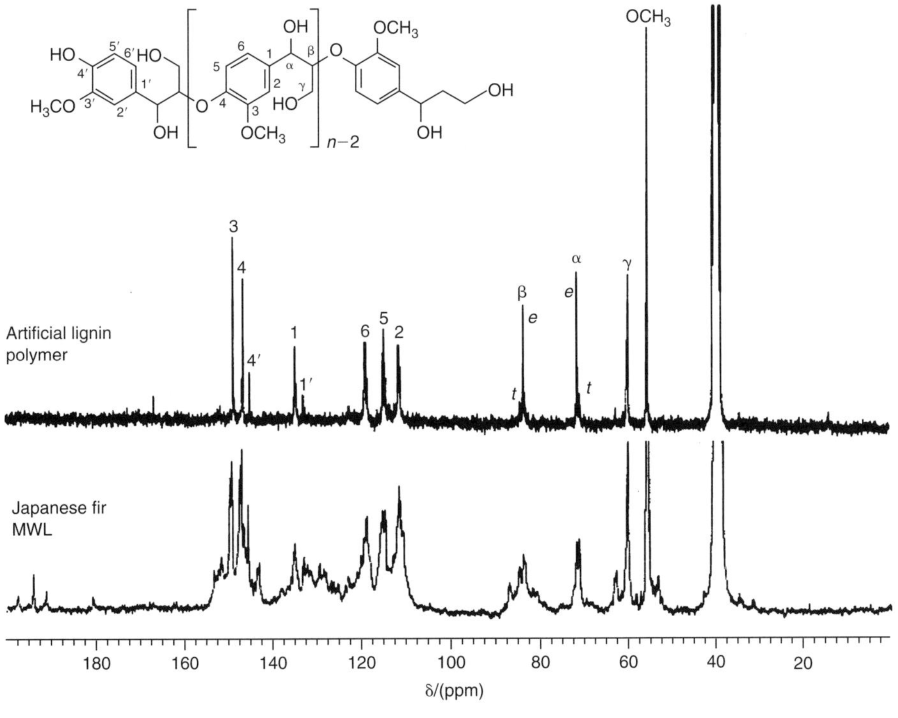

Figure 11.20 ^{13}C-NMR spectra of artificial and natural milled wood lignins (MWL). (Reproduced by permission of the Royal Society of Chemistry. Copyright 2006. Reprinted from Reference [78b]).

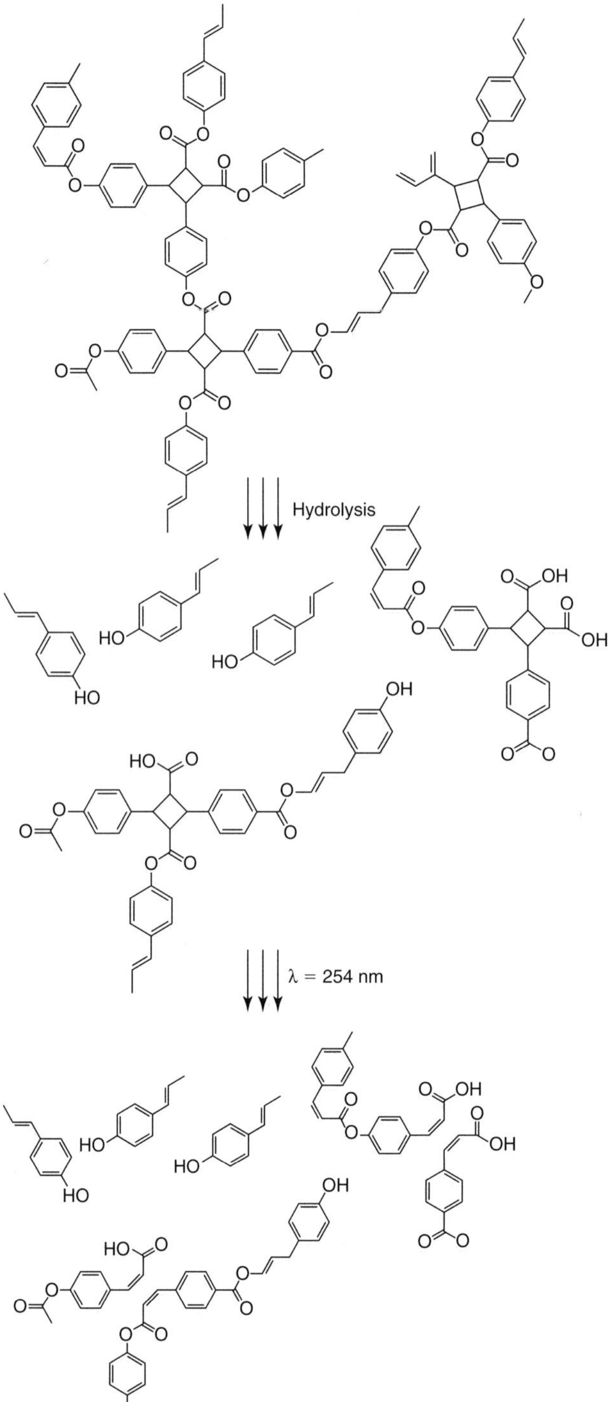

Hydrolysis

λ = 254 nm

Scheme 11.10

materials displayed liquid crystal features and were moreover readily crosslinked by UV irradiation, giving intermolecular coupling between C=C unsaturations. The degradation of these polyester networks could be carried out in two stages, *viz.* hydrolysis to split the ester group and far-UV irradiation to undo the cyclobutane moieties, as shown in Scheme 11.10.

The use of terms like 'lignin related polymers', 'artificial lignin' and 'phytomonomers' in the two latter studies are clearly overstating the issues at stake, because both the monomers used and the structure of the ensuing polymers have only faint connections with the real counterparts. In other words, the materials obtained are interesting, but their relation with lignins is quite tenuous. This assessment does not imply that such investigations are not relevant, but simply that hopefully further work will come steadily closer to the lignin paradigm.

11.7 CONCLUSIONS

Given the vast array of materials and other chemical commodities that can be prepared using lignins and the potential of industrial applications that some of them possess [80], it appears that the meagre figure of some 2 per cent as the proportion of lignins isolated from pulping liquors which is presently exploited for those purposes reflects a very poor state of affairs. A more vigorous interaction between academia and industry is required to make lignins real actors in the biorefinery scene and move from a scientific to a technological status, thus approaching the relevance that other natural polymers like cellulose, chitosan and starch have acquired over the years.

REFERENCES

1. (a) *ACS Symp. Ser.*, 1989, 397. (b) Gandini A., in Comprehensive Polymer Science, Suppl., Eds.: Aggrawal S.L. and Russo S., Pergamon Press, Oxford, 1992, **Volume 1**. p. 527. (c) Wang J., St John Manley R., Feldman D., *Prog. Polym. Sci.*, **17**, 1992, 611. (d) *ACS Symp. Ser.*, 2000, 742. (e) Gandini A., in *Les Polymères Naturels: Structure, Modifications, Applications*, Chapter VI, Editions GFP, Paris, 2000. (f) Chemical Modification, Properties and Usage of Lignin, Ed. Hu T.Q., Kluwer, New York, 2002. (g) *ACS Symp. Ser.*, 2007, 954.
2. www.ili-lignin.com.
3. Kubo S., Kadla J.F., *Biomacromolecules*, **6**, 2005, 2815.
4. Bouajila J., Dole P., Joly C., Limare A., *J. Appl. Polym. Sci.*, **102**, 2006, 1445.
5. (a) Kubo S., Kadla J.F., *Macromolecules*, **36**, 2003, 7803. (b) Kubo S., Kadla J.F., *J. Appl. Poly. Sci.*, **98**, 2005, 1437. (c) Kubo S., Kadla J.F., *Macromolecules*, **37**, 2004, 6904. (d) Kubo S., Kadla J.F., *Holzforschung*, **60**, 2006, 245.
6. Guo Z.X., Gandini A., *Eur. Polym. J.*, **27**, 1991, 1177.
7. Li Y., Mlinár J., Sarkanen S., *J. Polym. Sci. Polym. Phys. Ed.*, **35**, 1997, 1899.
8. Lee S.B., Luner P., *Tappi J.*, **55**, 1972, 116.
9. Belgacem M.N., Blayo A., Gandini A., *J. Colloid Interf. Sci.*, **182**, 1996, 431.
10. Kubo S., Kadla J.F., *Biomacromolecules*, **4**, 2003, 561.
11. Cobianu C., Ungureanu M., Ignat L., Ungureanu D., Popa V.I., *Ind. Crops Prods.*, **20**, 2004, 231.
12. Kadla J.F., Kubo S., *Composites A*, **35**, 2004, 395.
13. Zhang X., Benavente J., Garcia-Valls , *J. Power Sources*, **145**, 2005, 292.
14. Gregorová A., Cibulková Z., Kosíková B., Simon P., *Polym. Degrad. Stab.*, **89**, 2005, 553. and references therein.
15. Canetti M., Bertini F., De Chirico A., Audisio G., *Polym. Degrad. Stab.*, **91**, 2006, 494.
16. Toh K., Nakano S., Yokoyama H., Ebe K., Gotoh K., Noda H., *Polym. J.*, **37**, 2005, 633.
17. Gosselink R.J.A., Snijder M.H.B., Kranenbarg A., Keijsers E.R.P., de Jong E., Stigsson L.L., *Ind. Crops Prod.*, **20**, 2004, 191.
18. Gregorová A., Kosíková B., Moravcik R., *Polym. Degrad. Stab.*, **91**, 2006, 229.
19. Kosíková B., Gregorová A., *J. Appl. Polym. Sci.*, **97**, 2005, 924.
20. Iqbal A., Frormnn L., Saleem A., Ishaq M., *Polym. Comp.*, **28**, 2007, 186.
21. Belgacem M.N., Blayo A., Gandini A., *Ind. Crops Prod.*, **18**, 2003, 145.
22. Gosh I, Jain RK, Glasser WG, *Reference 1d*, p. 331.
23. (a) Baumberger S., *Reference 1f*, p.1. (b) Lepifre S., Baumberger S., Pollet B., Cazaux F., Coqueret X., Lapierre C., *Ind. Crops Prod.*, **20**, 2004, 219. (c) Lepitre S., Froment M., Cazaux F., Houot S., Lourdin D., Coqueret X., Lapierre C., Baumberger S., *Biomacromolecules*, **5**, 2004, 1678.
24. Berry, B.C., Viswanathan T., *Reference 1f*, p. 21.
25. Nägele, H., Pfitzer J., Nägele E., Inone R.E., Eisenreich N., Eckl W., Eyerer P., *Reference 1f*, p. 101.

26. Li Y., Sarkanen S, *Reference 1d*, p. 351.
27. Li Y., Sarkanen S., *Macromolecules*, **35**, 2002, 9707.
28. Li Y., Sarkanen S., *Macromolecules*, **38**, 2005, 2296.
29. Chen Y., Sarkanen S., *Reference* 1 g, p. 229.
30. Cazacu G., Vasile C., Stoleriu A., Constantinescu G., Pohoata V., *Proceedings of the 7th ILI Forum,* Barcelona 2005, p. 175.
31. Cazacu G., Pascu M.C., Profire L., Kowarski A.I., Mihaes M., Vasile C., *Ind. Crops Prod.*, **20**, 2004, 261.
32. Raschip I.E., Dumitriu R.P., Cazacu G., Vasile C., *Proceedings of the 7th ILI Forum*, Barcelona 2005, p. 167.
33. Cazacu G., Vasile C., Agafitei G.E., Pascu M.C., *Proceedings of the 7th ILI Forum*, Barcelona 2005, p. 171.
34. Fernandes D.M., Winkler Hechenleitner A.A., Job A.E., Radovanocic E., Gómez Pineda E.A., *Polym. Degrad. Stab.*, **91**, 2006, 1192.
35. Chen P., Zhang L., Peng S., Liao B., *J. Appl. Polym. Sci.*, **101**, 2006, 334.
36. Toriz G., Denes F., Young R.A., *Reference 1d*, p. 367.
37. Raghuraman A., Tiwari V., Thakkar J.N., Gunnarsson G.T., Shukla D., Hindle M., Desai U.R., *Biomacromolecules*, **6**, 2005, 2822.
38. Pizzi A., in *Handbook of Adhesive Technology*, Eds.: Pizzi A. and Mittal K.L., Marcel Dekker, New York, 2003, p. 589.
39. Turunen M., Alvila L., Pakkanen T.T., Rainio J., *J. Appl. Polym. Sci.*, **88**, 2003, 582.
40. Matsushita Y., Wada S., Fukushima K., Yasuda S., *Ind. Crops Prod.*, **23**, 2006, 115.
41. El. Mansouri N., Pizzi A., Salvadó J., *Holz Roh. Werkst*, **65**, 2007, 65.
42. Feldman D., Lacasse M., St. John Manley R., *J. Appl. Polym. Sci.*, **35**, 1988, 247. Natansohn A., Lacasse M., Ban D., Feldman D., *J. Appl. Polym. Sci.*, **40**, 1990, 899.
43. Detoisien M., Pla F., Gandini A., Cheradame H., *British Polym. J.*, **17**, 1985, 260. Cheradame H., Destoisien M., Gandini A., Pla F., Roux G., *British Polym. J.*, **21**, 1989, 269.
44. Gandini A., Belgacem M.N., Guo Z.-X., Montanari S., *Reference 1f,* p. 57.
45. Evtuguin D.V., Andreolety J-P., Gandini A., *Eur. Polym. J.*, **34**, 1998, 1163.
46. Hatakeyama H., *Reference 1f,* p. 41.
47. Bonini C., D'Auria M., Emanuele L., Ferri R., Pucciarello R., Sabia A.R., *J. Appl. Polym. Sci.*, **98**, 2005, 1451.
48. Glasser W.G., Jain R.K., *Holzforschung*, **47**, 1993, 225. Li S.M., Lundquist K., *J. Wood Chem. Technol.*, **17**, 1997, 391. Ralph J., Lu F.C., *J. Agric. Food Chem.*, **46**, 1998, 4616.
49. Guo Z.X., Gandini A., Pla F., *Polym. Int.*, **27**, 1992, 17.
50. Evtuguin D., Gandini A., *Acta Polym.*, **47**, 1996, 344.
51. Shaw J.M., *Print. Circ. Fabr.*, **19**(11), 1996, 38; *New Scientist*, November 29 issue, 1997.
52. Meister J.J., Chen M.J., *Macromolecules*, **24**, 1991, 6843; *J. Appl. Polym. Sci.*, **49**, 1993, 935.
53. Chen M.J., Meister J.J., *Reference 1d*, p. 321.
54. Schneider M.H., Phillips J.G., *US Patent* 6747076, 2004.
55. Kelley S.S., Ward T.C., Rials T.G., Glasser W.G., *J. Appl. Poly. Sci.*, **37**, 1989, 2961. Kelley S.S., Glasser W.G., Ward T.C., *Reference 1a*, p. 402. de Oliveira W., Glasser W.G., *Reference 1a*, p. 414.
56. Toffey A., Glasser W.G., Holzforschung, **51**, 1997, 71.
57. (a) Kelley S.S., Ward T.C., Glasser W.G., *J. Appl. Polym. Sci.*, **41**, 1990, 2813. (b) Kelley S.S., Glasser W.G., Ward T.C., *Polymer*, **30**, 1989, 2265.
58. (a) Hofman K., Glasser W.G., *J. Wood Chem. Technol.*, **13**, 1993, 73. (b) *J. Adhesion* , **40**, 1993, 229 c) Macromol. Chem. Phys., **195**, 1994, 65.
59. Zhao B., Chen G., Liu Y., Hu K., Wu R., *J. Mater. Sci. Lett.*, **20**, 2001, 859.
60. Sun G., Sun H., Liu Y., Zhao B., Zhu N., Hu K., *Polymer*, **48**, 2007, 330.
61. Naveau H.P., *Cell. Chem. Technol.*, **9**, 1975, 71.
62. Thielemans W., Wool R.P., *Biomacromolecules*, **6**, 2005, 1895.
63. Dourel P., Randrianalimanana E., Deffieux A., Fontanille M., *Eur. Polym. J.*, **24**, 1988, 843.
64. (a) Cordero T., Rodriguez-Maroo J.M., Rodriguez-Mirasol J., Rodriguez J.J., *Thermochim. Acta.*, **164**, 1990, 135. (b) Rodriguez-Mirasol J., Rodriguez J.J., Cordero T., *Carbon*, **31**, 1993, 53. (c) 87. (d) Gonzalez-Serrano E., Cordero T., Rodriguez-Mirasol J., Rodriguez J.J., *Ind. Eng. Chem. Res.*, **36**, 1997, 4832.
65. (a) Gonzalez-Serrano E., Cordero T., Rodriguez-Mirasol J., Cotoruelo L., Rodriguez J.J., *Water Res.*, **38**, 2004, 3043. (b) Rodriguez-Mirasol J., Bedia J., Cordero T., Rodriguez J.J., *Sep. Sci. Technol.*, **40**, 2005, 3113. (c) Cotoruelo L.M., Marques M.D., Rodriguez J.J., Cordero T., Rodriguez J.J., *Ind. Engin. Chem. Res.*, **46**, 2007, 4982. (d) Suhas P.J.M., Carott P.J.M., Ribeiro-Carott M.M.L., *Biores. Technol.*, **98**, 2007, 2301.
66. Kadla J.F., Kubo S., Gilbert R.D., Venditti R.A., *Reference 1f,* p. 121.
67. Kadla J.F., Kubo S., Venditti R.A., Gilbert R.D., Compere A.L., Griffith W., *Carbon*, **40**, 2002, 2913.

68. Kadla J.F., Kubo S., Venditti R.A., Gilbert R.D., *J. Appl. Polym. Sci.*, **85**, 2002, 1353.
69. Kubo S., Kadla J.F., *J. Polym. Environ.*, **13**, 2005, 97.
70. Braun J.L., Holtman K.M., Kadla J.F., *Carbon*, **43**, 2005, 385.
71. Barros, A.M., Dhanabalan A., Constantino C.J.L., Balogh D.T., Oliveira O.N., *Thin Solid Films*, , **354**, 1999, 215. Pasquini D., Balogh D.T., Antunes P.A., Constantino C.J.L., Curvelo A.A.S., Aroca R.F., Oliveira O.N., *Langmuir*, **18**, 2002, 6593. Pasquini D., Balogh D.T., Oliveira O.N., Curvelo A.A.S., *Coll. Surf. A* **252**, 2005, 193.
72. (a) Norgren M., Notley S.M., Majtnerova G.G., *Langmuir*, **22**, 2006, 1209. (b) Norgren L., Gärdlund L., Notley S.M., Htun M., Wågberg L., *Langmuir*, **23**, 2007, 3737.
73. Notley S.M., Norgren M., *Langmuir*, **22**, 2006, 11199.
74. Munavalli D.S., Chimatadar A., Nandibewoor S.T., *Ind. Eng Chem. Res.*, **46**, 2007, 1459.
75. Suparno O., Covington A.D., Phillips P.S., Evans C.S., *Resour. Conserv. Recycl.*, **45**, 2005, 114.
76. Okuda K., Man X., Umetsu M., Takami S., Adschiri T., *J. Phys.: Condens. Mat.*, **16**, 2004, S1325.
77. Bonini C., D'Auria M., *Ind. Crops Prod.*, **20**, 2004, 243.
78. (a) Kishimoto T., Uraki Y., Ubukata M., *Org. Biomol. Chem.*, **3**, 2005, 1067. (b) **4**, 2006, 1343.
79. Kaneko T., Thi T.H., Shi D.J., Akashi M., *Nat. Mater*, **5**, 2006, 966.
80. Lora J.H., Glasser W.G., *J. Polym. Environ.*, **10**, 2002, 39.

– 12 –

Partial or Total Oxypropylation of Natural Polymers and the Use of the Ensuing Materials as Composites or Polyol Macromonomers

Alessandro Gandini and Mohamed Naceur Belgacem

ABSTRACT

This chapter describes the partial and total oxypropylation of different natural polymers and more complex biomass substrates and the use of the ensuing materials as composites or polyol macromonomers, respectively. The total oxypropylation of different lignins, sugar beet pulp, cork, chitin and chitosan and olive pits is then reviewed in terms of its success in converting all these substrates into viscous polyol mixtures, which can be used to prepare polyurethane foams with thermal insulating properties similar to those of conventional industrial counterparts. The chapter finally deals with the *superficial* oxypropylation of cellulose fibres and starch granules (*i.e.* reactions limited to a certain thickness of the substrates) so that the ensuing materials keep their original morphology and properties, but now possess a thermoplastic shell which enables their conversion to single-component composites by simple thermal processing.

Keywords

Sugar beet pulp, Cork, Chitin, Chitosan, Olive pits, Polyurethane foams, Partial oxypropylation, Total oxypropylation, Cellulose fibres, Starch granules, Single-component composites

12.1 INTRODUCTION

The extensive research on the ring-opening anionic polymerization of oxiranes has provided an extremely simple approach to the grafting and chain-extension of macromolecular structures bearing hydroxyl functions [1]. Their activation by a strong Brønsted base generates the corresponding oxianions which function as initiating site for the oxirane propagation reaction. Both ethylene and propylene oxide (**PO**) respond efficiently to this type of polymerization, but the use of the former requires homologated equipment capable of ensuring safety against explosions provoked by sudden uncontrolled exothermic polymerizations. It is therefore much more common to practice oxypropylation reactions.

In the realm of polymer materials, oxypropylation has received considerable attention when applied to both small polyol molecules, like sugars and to polymers bearing multiple OH groups, like cellulose. These chain-extension reactions do not produce any increase in the OH-functionality of the starting substrate, since their role is to transport these groups as chain ends of the inserted polyether segments. The average length of these segments can be modulated as a function of the reaction parameters, but within the limitation imposed by the inevitable transfer reactions that accompany all these systems and give rise to homopolymers of the monomer used, generally in the form of α,ω-oligoether diols.

A large variety of functional polymers derived from this approach has been synthesized including, for example, block copolymers in which the polyether sequences stem from a central polymer of a very different nature, like polystyrene [1]. In the same vein, a multitude of non-ionic surfactants are today commercially available thanks to the application of these principles to oligomeric structures bearing a hydrophobic segment, like an aliphatic chain or even a poly(propylene oxide) (**PPO**) sequence, joined to a hydrophilic counterpart, which is mostly an OH-terminated oligo-ethylene oxide [2]. A third thoroughly exploited approach consists in synthesizing star-shaped macro-molecular materials using, for example, glycerol or pentaerithritol as core molecules which are subsequently decorated with polyether branches. All these applications testify to the degree of sophistication and control reached by the ring-opening polymerization of oxiranes, since the sequence length of the ensuing polyether segments can be predetermined and precisely achieved [1].

The extension of these principles to molecules and macromolecules based on renewable resources has been applied to polymer chemistry and technology, mostly in two areas requiring viscous materials, namely: (i) the preparation of liquid polyols to be used in the manufacture of polyurethanes, and (ii) the synthesis of macromolecular rheology modifiers. A typical polyol for the synthesis of polyurethane foams is obtained industrially from the oxypropylation of sorbitol, [3] whereas hydroxypropyl cellulose represents a major commodity for the rheological control of paints, foodstuff, cosmetics, etc. [4].

This chapter is devoted to a brief survey of recent work dealing with the oxypropylation of various natural polymeric substrates, all of which, of course, bear OH groups, sometimes accompanied by NH_2 functions. Whereas the oxyethylation and oxypropylation of cellulose, as well as the oxypropylation of sorbitol, are industrial processes with well-defined parameters and product properties, the studies described here are still at a research stage, although pilot plant development has been reached in some instances. The other relevant difference with respect to the two commercial commodities is that the investigations discussed below deal essentially with substrates which are in fact by-products of major processes (*i.e.* raw materials of little economic impact), but available in very large quantities. In other words, the issue here is two-fold, *viz.* the exploitation of renewable resources and their upgrading through their conversion into value-added materials.

Two distinct processes are used in the oxypropylation of molecular or macromolecular substrates *viz.* systems using a basic aqueous medium and moderate temperatures and bulk reactions in which both high temperatures and pressures are involved. The former approach is particularly suited for the attachment of short branches and has been successfully applied particularly to cellulose and starch. It does not however function satisfactorily with less hydrophilic substrates which tend to be very passive towards activation. For these situations, and indeed in a more general context, the bulk reaction provides an excellent alternative, as emphasized in all the examples discussed in this chapter. Quite apart from the fact that this process can be applied to virtually any OH bearing substrate, it has the added advantage of eliminating any workup at the end of the reaction, since all the **PO** is used up and the ensuing oxypropylated material, usually in the form of a viscous liquid, is the only product present in the reactor, apart from the catalyst which can be left as such or readily neutralized by contact with the atmosphere. Furthermore, this type of process bears green connotations because of the lack of solvent manipulation and of generation of VOC.

This chapter only deals with the studies that called upon the bulk process because, apart from the positive aspects enumerated above, they produce polyols which are ideally suited for the elaboration of different macromolecular materials, as opposed to the waterborne approach which is more often indicated for modifications associated with rheology modifiers in food and cosmetic technologies.

Most of the investigated systems which are discussed here were based on the idea of producing liquid polyols through the radical transformation of the initial morphology of the corresponding substrates, all of which are rather intractable solids. More recently, a different approach has been envisaged and successfully applied, in which the oxypropylation is limited to a certain thickness of the substrate particles so that the ensuing materials are characterized by an inner pristine core, preserving the original morphology and properties of the substrate, and a thermoplastic shell which enables the processing of a single-component composite by a simple thermal treatment like hot-pressing to produce films. This chapter is thus divided into two sections, following the corresponding different approaches.

In both instances however, the chemistry is the same and its workings can be illustrated by the simplified Scheme 12.1, for which the only difference is the extent of penetration of the oxypropylation reaction within the substrate particles. The number of OH groups involved in this chain-extension reaction, the length of the ensuing grafts, as well as the incidence of homopolymer formation, depend in both strategies on the actual reaction parameters. However, the one common prerequisite, essential for the optimization of this heterogeneous process (except in the case of fibres) is of course the maximum reduction in the substrate particle size compatible with its grinding aptitude.

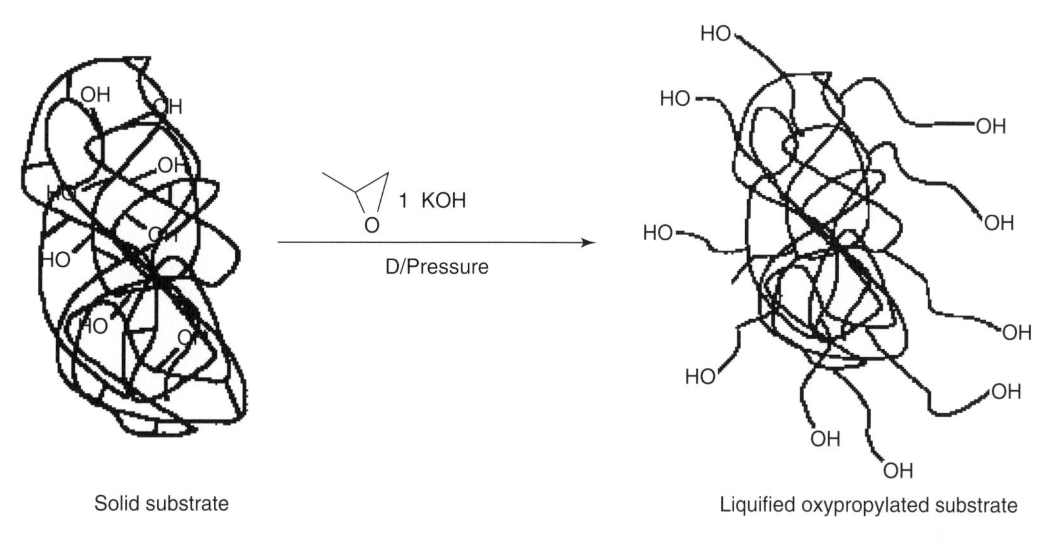

Scheme 12.1

12.2 'TOTAL' OXYPROPYLATION

This section deals with the transformation of different natural polymeric substrates, whose common feature is a high content of OH groups, into viscous polyols through their near-complete oxypropylation. Given the very pronounced similarities associated with all the systems examined below, a detailed description of the process will be provided for the first one and only relevant differences and peculiarities discussed for the others.

12.2.1 Lignin

The general structures of this fundamental vegetable polymer, as well as the applications of its various industrial forms, have already been dealt with in detail in the three preceding chapters. As already pointed out, most of the utilizations of lignin as a macromonomer are qualitatively independent of its source and subsequent isolation processes only exploit the omnipresence and relative reactivity of both aliphatic and phenolic hydroxyl groups. Oxypropylation is obviously no exception to this rule and lignin should therefore be considered here as a macromolecular polyol, potentially capable of being grafted with multiple **PPO** chains.

Glasser's pioneering work on the use of lignins as precursors for the synthesis of novel polyurethanes involved, among other approaches already discussed in Chapter 11, their oxypropylation [5]. These studies were carried out with kraft and sulphite lignins (*i.e.* high molecular weight structures) which showed some resilience to a thorough oxypropylation and hence the need of high temperatures and pressures. The authors also found that these rather drastic conditions induced some lignin self-condensation processes and the consequent formation of partly insoluble products.

More recently, this process has been revisited by us in a broader context, particularly in terms of the variety of substrates examined [6, 7]. In this comprehensive study, we called upon kraft (**KL**), ALCELL organosolv (**OL**), soda (**SL**) and an oxygen organosolv lignin (**OOL**). The relevant properties of these substrates are given in Table 12.1.

Reactions were conducted in small stainless steel autoclaves equipped with a stirrer and followed by the temperature increase associated with the onset of polymerization that produced a corresponding increase in pressure which then progressively decreased to zero, indicating the total consumption of **PO**. The parameters which were varied in order to optimize the efficiency of each lignin treatment were the **PO**/lignin ratio (1–10 w/w), the type and amount of catalyst (mostly KOH, but also NaOH and various tertiary amines, used in ratios relative to lignin of 1–10 wt%) and the temperature (140–200°C, associated with pressures of 6–20 bar). Reaction times (*i.e.* the time required for the pressure to return to zero) were found to vary very considerably, ranging from a few minutes

Table 12.1

Characterization of the lignins used in our oxypropylation studies [6, 7]

	Lignins			
	KL	OL	SL	OOL
M_n	1 220	700	1 100	1 400
M_w	11 880	1 700	6 800	4 100
M_w/M_n	9.7	2.4	6.2	2.9
Primary aliphatic OH/C9 unit	0.39	0.23	–	0.55
Secondary aliphatic OH/C9 unit	0.27	0.14	–	0.40
Aromatic OH/C9 unit	0.78	0.63	–	0.41
Total OH/C9 unit	1.44	1.00	–	1.36
Average OH/lignin molecule	10	6	9	12

to several hours, particularly as a function of the nature of the catalyst (only KOH displayed a high activity), its concentration, the temperature and the type of lignin.

The characterization of all these products comprised their structural assessment through FTIR and NMR spectroscopy, the determination of the amount of solid residue and **PO** homopolymer and the measurement of their viscosity and hydroxyl index. Figures 12.1 and 12.2 show the FTIR and ^{1}H-NMR spectra of the oxypropylated **SL**.

Although all the reactions did in fact proceed to completion in terms of **PO** consumption, not all the corresponding products were made up of 100 per cent liquid polyols, since some solid residue was obtained in several experiments, namely those involving **KL** (20–40 wt%) and **OOL** (20–30 wt%), in contrast to **OL** and **SL** which gave rise to solid-free viscous liquids. These findings confirmed Glasser's observations about the modest reactivity of **KL** [5] and provided a wider correlation between lignin structure and molecular weight and its aptitude to be oxypropylated. Thus, the high average molecular weight of **KL**, coupled with a significant steric hindrance around its OH groups, made this lignin the least reactive in the present context. **OOL** followed and its relative sluggishness

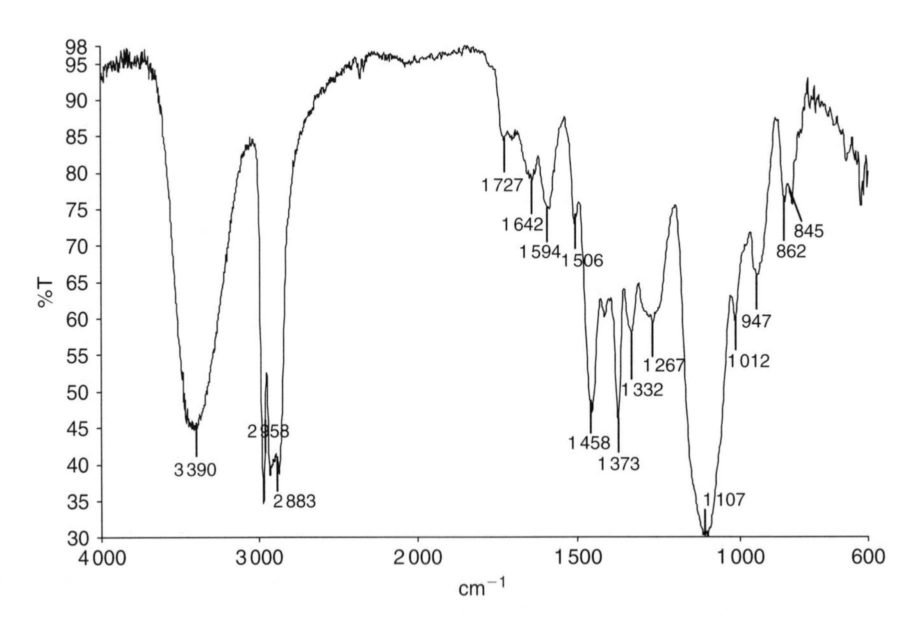

Figure 12.1 FTIR spectrum of oxypropylated **SL** (30/90 **SL/PO** wt%) (Reproduced by permission of Wiley-VCH. Copyright 2005. Reprinted from Reference [7]).

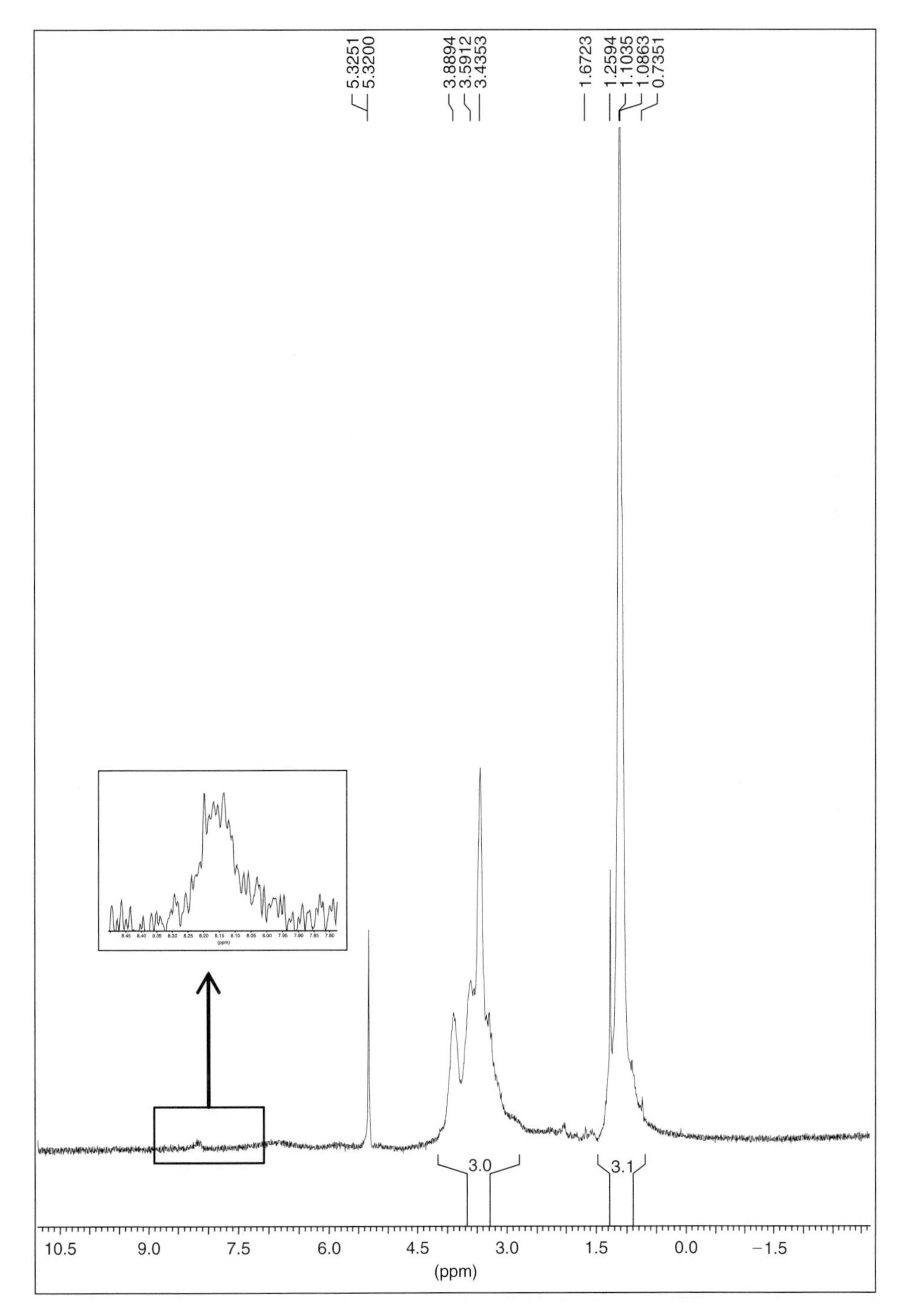

Figure 12.2 ^{1}H-NMR spectrum of oxypropylated **SL** (30/90 **SL/PO** wt%), in CD_2Cl_2 (Reproduced by permission of Wiley-VCH. Copyright 2005. Reprinted from Reference [7]).

was attributed to the presence of carbonyl and carboxylic groups in its structure. **OL** and **SL** gave the most satisfactory results in tune with their modest molecular weight.

The extent of **PO** homopolymerization with respect to lignin grafting, was also determined and found to vary between 5 and 75 wt%, the higher values being associated with the less reactive substrates. Thus, for example, with **PO/OL** = 1 and 5 wt% of KOH, a reaction conducted at 180°C lasted about half an hour and gave a product containing about 20 per cent of homopolymer. It is important to emphasize that this low molecular weight **POP** diol does not constitute a problem in terms of the use of the whole oxypropylation product as a macromonomer mixture for the synthesis of polyurethanes, or for that matter of any other polycondensate, because it must be considered as a difunctional reagent which will participate as a chain extender in the non-linear polymer growth.

The two most relevant properties of these polyols with regard to their use as polyurethane precursors are their viscosity and hydroxyl index. Values which simulate those associated with commercial counterparts were readily obtained by adjusting the key process parameters, notably the **PO/L** ratio.

The most appropriate polyols arising from this systematic study were employed for the synthesis of polyurethane foams using standard polyfunctional methylenediphenyl diisocyante (**MDI**) and all the conventional additives required for this technological processing [6, 7]. Additionally, foams were also prepared in which the lignin component was not only the corresponding oxypropylated material but also the lignin itself used as filler. The characterization of these foams, carried out before and after 3 months aging at room temperature and after 10 days accelerated aging at 70°C, included the determination of the foam density, its thermal conductivity and dimensional stability. Compared with the properties of foams prepared with conventional polyols, these novel homologous materials gave very satisfactory performances, particularly when oxypropylated **OL** and **SL** were used as polyols. Both **OL** and **KL** were found to be adequate fillers in proportions of up to about 30 wt%, although the term 'filler' is not entirely appropriate in this context since some surface OH groups must have participated in the polycondensation reaction, thus generating covalent links between the polymer matrix and the solid lignin particles, which are hence better defined as 'reactive fillers'.

In conclusion, it has been shown that it is possible to carry out the straightforward transformation of a cheap solid industrial by-product like lignin into a liquid polyol mixture that can be employed directly as a macromonomer for the fabrication of polyurethane foams, and indeed of any other macromolecular material based on polycondensation reactions involving OH groups.

12.2.2 Sugar beet pulp

Most of the European sugar production is based on the exploitation of sugar beet (**SB**) and the corresponding industrial process generates enormous amounts of pulp as a by-product. This rather intractable fibrous solid, mostly made up of polysaccharides and pectin, has found some uses as fertilizer and animal feed, but the possibility of converting it into a source of polymeric materials had eluded researchers until it was shown that this remarkably inert natural product could in fact be readily converted into a viscous liquid polyol by a simple bulk oxypropylation treatment [8, 9].

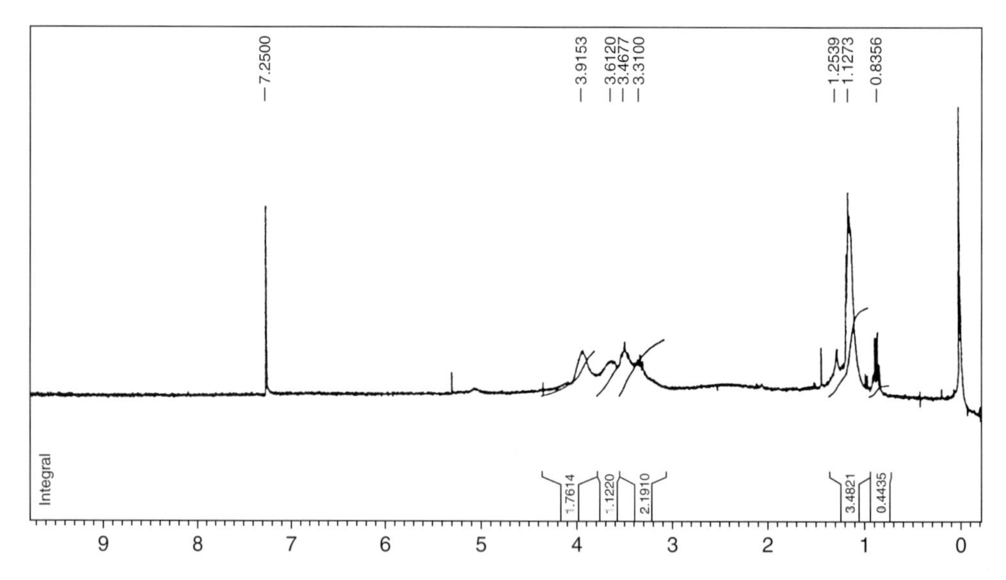

Figure 12.3 ^{1}H-NMR spectrum of oxypropylated **SB** (25/75 **SB/PO** wt%) after the extraction of the homopolymer. (Reproduced by permission of Elsevier. Copyright 2000. Reprinted from Reference [9]).

The study of this system in our laboratory preceded that related to lignins and should be considered the second application of bulk oxypropylation after Glasser's work on lignin discussed above. The systematic investigation conducted with **SB** pulp was therefore the first of its kind in terms of the optimization of the oxypropylation reaction aimed at producing a polyol with a minimum amount of solid residue and with properties adequate for its use as a macromonomer mixture in the synthesis of polyurethanes and other crosslinked macromolecular materials. Figures 12.3 and 12.4 show the ^{1}H-NMR spectra of the oxypropylated **SB** and of the accompanying **PO** homopolymer.

The relevant parameters that were varied in order to achieve this optimization [8] were the same as those described above for the oxypropylation of lignins, albeit here with only one substrate, namely the **SB** sample provided by the Italian sugar manufacturer Eridania. Likewise, the best results [8] corresponded to conditions quite similar to those established with lignin. All these aspects will therefore not be reported here for the sake of brevity, although it seems appropriate to summarize the work carried out to establish the optimal condition for the oxypropylation of **SB** (Fig. 12.5).

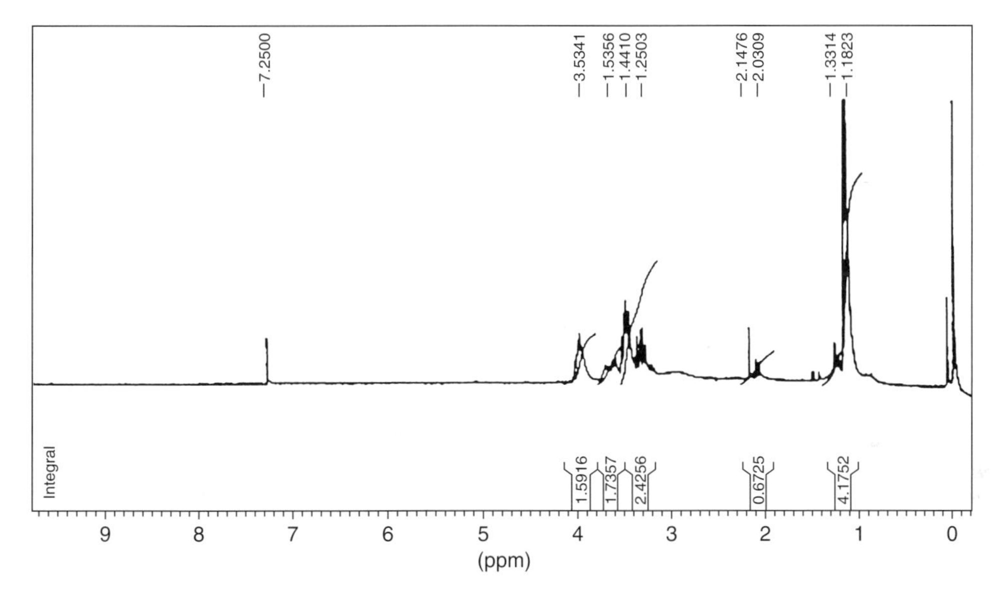

Figure 12.4 ^{1}H-NMR spectrum of the hexane-extracted **PO** homopolymer related to Figure 12.3. (Reproduced by permission of Elsevier. Copyright 2000. Reprinted from Reference [9]).

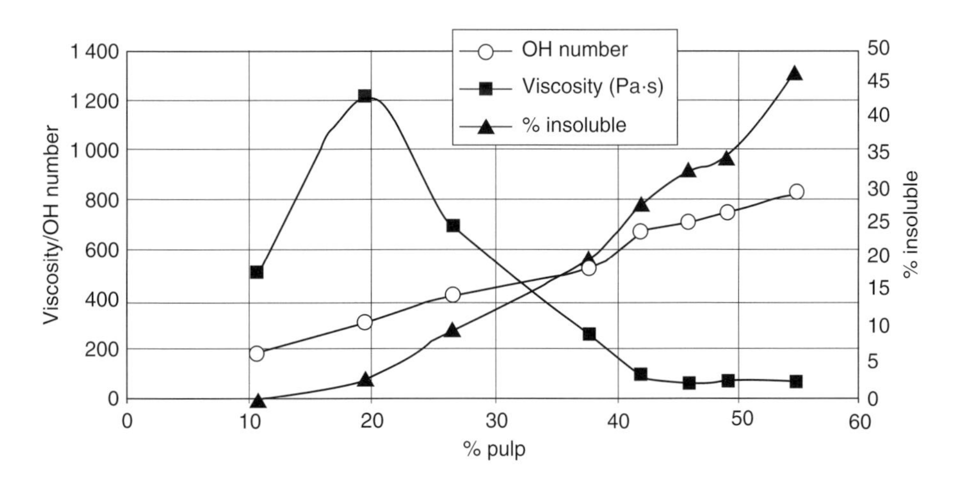

Figure 12.5 Variation of the viscosity and the OH number of the CH_2Cl_2-soluble fraction and of the percentage of the CH_2Cl_2-insoluble fraction as a function of the initial percentage of **SB** in the oxypropylation reaction mixture (Reproduced by permission of Elsevier. Copyright 2000. Reprinted from Reference [8]).

The success of this study led to the scale up of the laboratory process to a pilot plant production of tens of kilograms of polyol which were then used in the processing of a number of polyurethane products, [10] including reinforced materials prepared by the RRIM technology [11].

12.2.3 Cork

The outer bark of *Quercus suber* known as cork constitutes a dramatic example of excessive Darwinian adaptation in a vegetable species predominant in the Mediterranean basin. The composition properties and applications of cork as a material or of its major component, suberin, are discussed in Chapter 14. When cork is processed (*e.g.* for the manufacture of stop corks) a considerable amount of rejects is produced in the form of small par-ticles and larger unusable pieces. Presently, these by-products are burned to generate energy but, as in the case of lignin, this exploitation is far from satisfactory considering the interesting macromolecular structures associated with cork. Various laboratories have been actively pursuing more rational ways of exploiting this beautiful natural resource and the isolation and use of suberin as a macromolecule mixture, as discussed in Chapter 14. The oxypropylation of cork powder was undoubtedly another promising avenue worth exploring and research on this topic has already given some useful conclusions [12, 13].

Given the fact that polysaccharides and to a minor extent lignin-like structures are present in cork alongside suberin, OH groups are available in this material, even though their contribution to the overall chemical composition is smaller than in other natural polymers like cellulose, starch or lignin. Numerous experiments have shown that this relative paucity of hydroxyl functions does not preclude or indeed reduce the possibility of turning solid cork particles into a viscous liquid polyol by an oxypropylation process [12, 13] whose parameters are very similar to those optimized in the case of previous studies on lignin and **SB** pulp, as detailed above. Interestingly, both the problems and the advantages encountered in those previous studies were again met with cork and an additional improvement in the optimization of the oxypropylation reaction introduced at this point was the pre-impregnation of the substrate with an alcoholic solution of KOH, followed by the evaporation of the ethanol. This pre-treatment enhanced the substrate reactivity by activating a higher number of OH groups.

Both the polyol mixtures [12] and some polyurethane materials obtained from them [13] were thoroughly characterized.

The next step in this venture is the scaling up of the oxypropylation of cork and the processing of a variety of polymers derived from the ensuing polyols.

12.2.4 Chitin and chitosan

The increasing relevance of chitin and chitosan as sources of a variety of remarkable materials is highlighted in Chapter 25, dealing with these animal polysaccharides. The possibility of their transformation into liquid polyols arose a decade ago in the context of the search for a film forming polymer electrolyte [14]. Figure 12.6 gives the FTIR spectra of chitosan before and after oxypropylation.

Given the well-known aptitude of chitosan to generate very thin films with good mechanical properties, it was speculated that its oxypropylation might maintain the film-forming behaviour and, at the same time, generate a material capable of solvating high concentrations of lithium ions thanks to the presence of oligoether moieties. This was indeed attained by coupling the viscous oxypropylated chitosan with further ethylene- and propylene-oxide chains.

More recently, the study of the oxypropylation of both chitin and chitosan has been revived within the broader context of, on the one hand, optimizing both processes and, on the other hand, of extending the search for applications of the ensuing polyols [15].

As in the case of other natural substrates, this novel investigation is aimed at exploiting what one would define as the lower grades of both chitin and chitosan (*i.e.* the by-products resulting from the processing of both materials) compared with the high grade of the refined counterparts.

12.2.5 Olive pits

Another inedible agricultural by-product whose possible exploitation has been thus far limited to combustion for energy recovery is constituted by the olive pits separated from the pomace (which also includes the skin and pulp

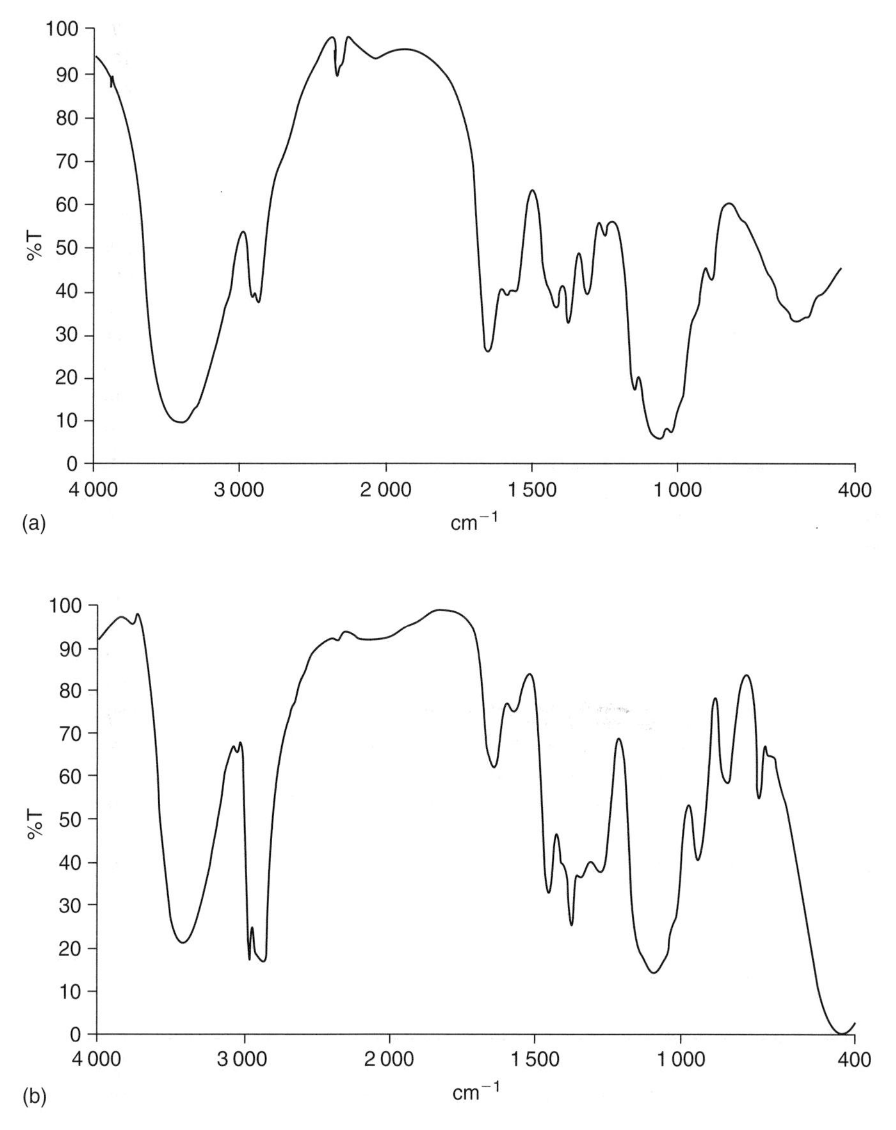

Figure 12.6 FTIR spectra of chitosan (a) before, and (b) after oxypropylation. (Reproduced by permission of Pergamon. Copyright 1998. Reprinted from Reference [14]).

solids) remaining after the oil-pressing process. The hard texture of the pits is made up of a mixture of natural polymers including lignin and polysaccharides (*i.e.* components rich in hydroxyl groups), thus potentially prone to oxypropylation.

Experiments were recently carried out [16] to assess the feasibility of transforming this otherwise intractable material into a liquid polyol following the very same approach described above for all the other studied natural polymers. Finely ground olive pits were thus pre-soaked with a KOH–ethanol solution and then, after the alcohol removal, oxypropylated in bulk at different temperatures within the range of 140–180°C. Preliminary results indicated a very efficient transformation of the substrate since it was turned into a viscous polyol leaving a minimal solid residue. As with all other natural substrates, this liquid product was a mixture of the oxypropylated pit polymers and **POP**.

The complete characterization of these novel polyol mixtures and their use as macromonomers in the synthesis of polymeric materials like polyurethanes and polyesters is in progress.

Table 12.2

Experimental conditions for the oxypropylation of different solid substrates

Solid substrate	Experiment number	Solid/PO (w/w)	Catalyst (C)	(C)/solid (%w/w)	T (°C)	Time (mn)	P[a] (bar)	Reference
KL	1	20/80	KOH	5	185	900	10.5	[6, 7]
	2	15/85	KOH	10	195	800	14.2	[6, 7]
	3	10/90	KOH	10	179	900	15.4	[6, 7]
OL	4	20/80	KOH	5	180	19	11.6	[6, 7]
	5	20/80	KOH	7	178	14	11.4	[6, 7]
	6	20/80	KOH	10	170	13	11.1	[6, 7]
	7		NaOH	5	177	21	17.3	[6, 7]
	8		KOH	5	175	20	8.5	[6, 7]
	9	30/70	KOH	5	178	18	8.9	[6, 7]
	10		NaOH	5	177	25	8.4	[6, 7]
	11	40/60	KOH	5	179	30	11.7	[6, 7]
	12		NaOH	5	178	44	18.2	[6, 7]
	13	50/50	KOH	5	183	45	16.2	[6, 7]
OOL	14	20/80	KOH	5	175	500	7.1	[6, 7]
	15	30/70	KOH	5	160	475	9.1	[6, 7]
	16	30/70	KOH	5	180	505	12.7	[6, 7]
	17	40/60	KOH	5	190	330	11.9	[6, 7]
SL	18	25/75	KOH	7.5	174	170	6.5	[7]
	19	25/75	KOH	10	140	180	14	[7]
	20	17/83	KOH	10	156	200	17.5	[7]
	21	30/70	KOH					[7]
SBP	22	30/100	NaOH	3	150	373	10	[8]
	23	30/100	KOH	3	150	330	10	[8]
	24	10/90	KOH	20	155	225	9	[8]
	25	15/85	KOH	20	145	500	8	[8]
	26	20/80	KOH	20	151	360	6	[8]
	27	26/74	KOH	20	140	315	5	[8]
Cork	28	10/90	KOH	0.5–1	236	102	15.5	[12]
	29	10/90	KOH	0.5–1	217	130	14.9	[12]
	30	10/90	KOH	0.5–1	226	145	15.5	[12]
	31	10/90	KOH	0.5–1	217	125	15.0	[12]
	32	10/90	KOH	0.5–1	258	99	15.0	[12]

[a] Maximum pressure.

Tables 12.2 and 12.3 summarize the experimental conditions used for the oxypropylation of some of the substrates discussed above, as well as the properties of the ensuing liquid polyols.

12.3 POLYURETHANE FOAMS

This section deals with the use of the polyols derived from the oxypropylation of different renewable resources, as macromonomers for the synthesis of rigid polyurethane foams (RPU), calling upon otherwise conventional formulations [6, 7].

Table 12.3

Properties of the oxypropylated substrates (see Table 12.2)

Solid substrate	Experiment number	I_{OH}	η (Pa·s)	Homopolymer content (%)	Solid residue (%)	Reference
KL	1	180	4.2	75	41	[6, 7]
	2	165	3.2	80	36	[6, 7]
	3	148	2.3	83	33	[6, 7]
OL	4	198	6.0	42	0	[6, 7]
	8	203	93	22	0	[6, 7]
	11	250	2860	20	0	[6, 7]
	13	305	–	2	0	[6, 7]
OOL	14	170	4.8	52	25	[6, 7]
	16	178	5.2	39	29	[6, 7]
	17	165	4.1	nd	39	[6,7]
SL	18	15.2	82	25	0	[6, 7]
	19	16.1	85	nd	0	[7]
	20	15.8	nd	30	0	[7]
	21	165	120	18	0	[7]
SB	14	418	710	nd	10	[7]
	15	436	700	nd	9	[8]
	16	260	49	nd	0	[8]
	17	350	305	nd	2	[8]
	18	380	450	nd	6	[8]
	19	390	520	nd	12	[8]
Cork	28	≈130	6	73.2	0.16	[12]
	29	≈130	11.9	69.8	2.50	[12]
	30	≈130	12.4	67.8	4.87	[12]
	31	≈130	13.0	68.7	2.24	[12]
	32	≈130	2.9	75.8	0.22	[12]

nd, not determined.

The procedure concerning the foam preparation consisted in mixing vigorously the polyol (or the mixture of the polyols), the surfactant, the catalysts and the water for about 30 s, followed by the addition of the blowing agent. The commercial polyisocyanate (**MDI**) was then added and the emulsified mixture finally poured into a special mould for free expansion. The RPU foams thus obtained were characterized through their T_g (DSC) and thermal conductivity, as shown in Table 12.4 [6, 7]. These data are extremely promising, since a large number of these foams gave insulating performances similar to those obtained from petroleum-based counterparts.

12.4 PARTIAL OXYPROPYLATION

The strategy behind the idea of limiting the oxypropylation reaction of a natural polymer to the outer shell of its morphology is radically different from that motivating all the bulk oxypropylation reactions described above. Although, as already pointed out, the mechanisms involved in both approaches are identical in nature, the purpose of this limited transformation is not to obtain a polyol to be exploited as a reactive component in polymer synthesis, but, instead, to generate a novel material in which the pristine core and the oxypropylated shell both play an essential role in defining its overall properties.

Two substrates have been tackled in this way up to now within the context of a research project which was started only recently. As outlined below, the aims of these partial transformations were different, but followed the same working hypothesis.

Table 12.4

Characteristics of RPU foams before and after ageing

Polyol	T_g (°C)	Before ageing		After ageing			Reference
		Density (kg/m^3)	λ (mW/m K)	λ[a] (mW/m K)	λ[b] (mW/m K)	DS[c] (%)	
CP1	58	24.8	15	17	19	+17	[6,7]
CP2	59	26.2	17	18	20	+17	[6,7]
OLOP	60	22.2	16	21	25	+11	[6,7]
	60	24.7	17	21	27	+10	[6,7]
	61	25.1	16	22	29	+15	[6,7]
SLOP	65	30.3	24	29	31	+2	[6,7]
KLOP	66	28.2	29	nd	nd	+29	[6,7]
OOLOP	66	27.9	31	nd	nd	+27	[6,7]
SBPOP	–	37.5	28	nd	nd	−1.8	[11]
	–	46.0	30	nd	nd	−1.9	[11]
	–	37.0	28	nd	nd	−3.4	[11]
Cork-OP	–	28.4	38	nd	nd	–	[12b]
	–	27.4	31	nd	nd	–	[12b]
	–	32.2	24	nd	nd	–	[12b]

[a]After ageing at room temperature for 3 months.
[b]After accelerated ageing at 70°C for 10 days.
[c]Dimensional stability.

12.4.1 Cellulose fibres

The use of cellulose fibres as reinforcing elements in composite materials with polymeric matrices has become a fast-growing topic, as discussed in Chapters 18 and 19. In most instances the appropriate chemical modification of the fibres' surface is an essential requisite imposed by the need of enhancing the interfacial compatibility between fibres and matrix, that is, of ensuring good adhesion and hence optimal mechanical properties of the composite. Of course, the ultimate optimization in this search consists in creating continuous covalent bridges between fibres and matrix which then become, by definition, a single chemical suprastructure in which the two components cannot be disjoined.

The idea of generating a composite material based exclusively on cellulose fibres' precursors arose in recent years through the concept of the partial chemical modification of the fibres, not limited this time to their surface OH groups, but, instead, to an in-depth process, itself limited to a modest thickness. In this way, a material could be generated in which the inner core maintains the original mechanical properties of the unmodified parts of the cellulose fibres, while its outer sleeve could now possess thermoplastic features given a suitable choice of chemical modification. Examples of this approach include esterification with carboxylic acids [17] and etherification with benzyl chloride [18].

Another approach has called upon the coagulation of amorphous cellulose around semi-crystalline fibres (i.e. the preparation of an all-cellulose composite) [19].

All the above systems require a sequence of steps and the use of solvents to attain the final goal, which complicates the process both in terms of economical and ecological soundness. The direct transformation of cellulose fibres into self-reinforced composites by partial oxypropylation was considered a promising alternative to this working hypothesis since it eliminated the use of solvents and could be conducted in a straightforward one-step operation. The study of this system confirmed its viability and produced interesting composite materials [20].

The bulk oxypropylation of cellulose fibres, activated by a strong Brønsted base, was investigated through a systematic study of the role of such variables as the nature of the fibres (different natural types and full cylindrical regenerated filaments), the type and amount of catalyst, the pre-activation conditions, the **PO**/cellulose ratio, the reaction temperature and its duration. In contrast with all the intensive oxypropylation processes described in the preceding section, the amount of **PO** used here was obviously much lower because the purpose of this procedure was to limit the oxypropylation reaction to the outer layers of the fibres. The other major difference with respect to the synthesis of liquid polyols was the use of more moderate temperatures and hence longer reaction times, in order to minimize the occurrence of reactions deep inside the fibres.

The most relevant results, associated with an adequate thickness of the thermoplastic sleeve surrounding the unmodified fibre cores, gave mass gains comprised between 40 and 90 per cent after the extraction of 5–10 per cent of the **PO** oligomers. These materials were flexible and could be readily converted into films by pressing them at 135°C and 11 ton. A visual and optical microscopy inspection of these semi-transparent films, showed the presence of opaque regions associated with the residual cellulose fibres. In these reactions, the oxypropylation front progressing through the fibre wall (in the case of natural fibres) or the fibre cylinder (in the case of regenerated filaments) obviously attacked preferentially the amorphous cellulose domains.

The extent of oxypropylation was assessed by FTIR spectroscopy, as shown in Fig. 12.7, and its corresponding plasticizing effect from glass transition temperature measurements by DSC. The surface of the modified fibres acquired, as expected, a hydrophobic character, compared with the pristine substrates, as revealed by contact angles of 70–80° and 25–40°, respectively. Figure 12.8 shows the changes in morphology associated with the partial oxypropylation of the fibres and their subsequent pressing to form a composite film arising from their mutual assembly through the plastic flow of their modified sleeves.

The advantages of this novel system include the extreme simplicity of the process which makes use of all the intervening components and does not require any separation or purification steps as well as the fact of being

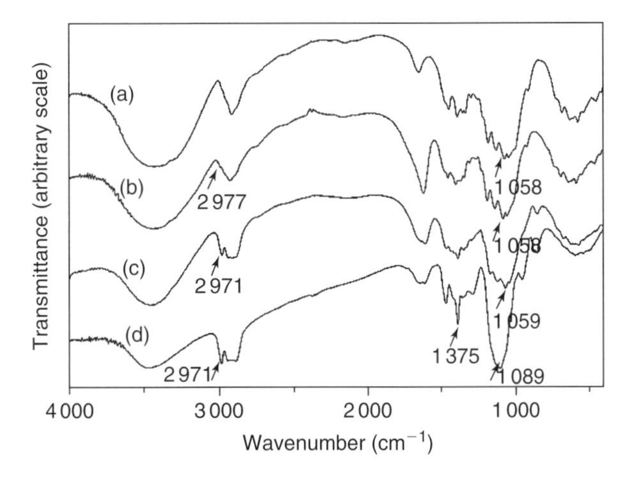

Figure 12.7 FTIR spectra of Avicell powder before (a) and after oxypropylation using [**PO**]/[cellulose OH] of 1 (b), 3 (c) and 5 (d), and after extraction of the **PO** homopolymer. (Reproduced by permission of Elsevier. Copyright 2006. Reprinted from Reference [20]).

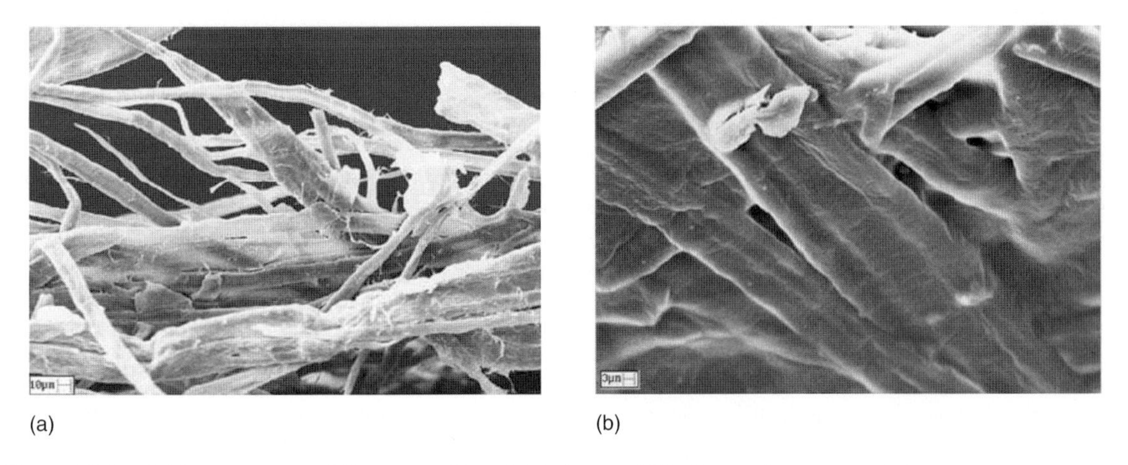

Figure 12.8 SEM micrographs of filter paper fibres (a) before, and (b) after partial oxypropylation.

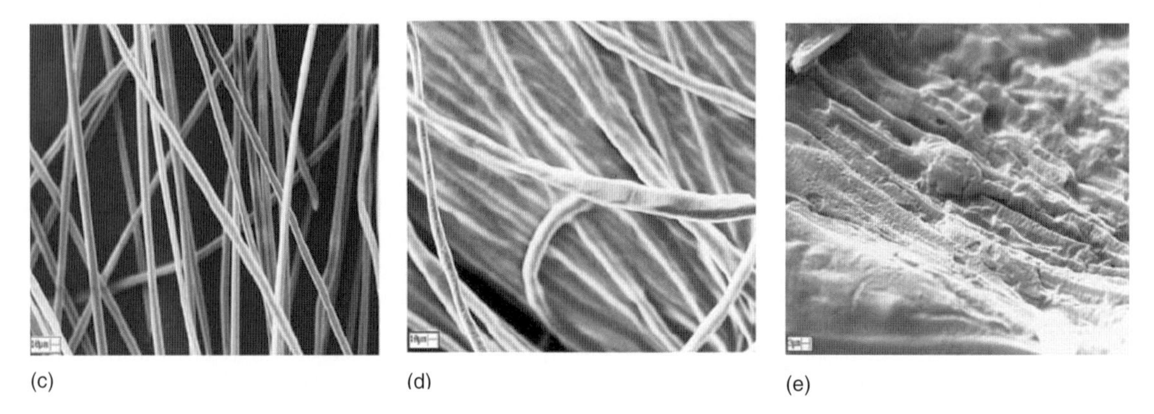

Figure 12.8 SEM micrographs of rayon filaments (c) before and (d) after partial oxypropylation and (e) hot-pressing into a film. (Reproduced by permission of Elsevier. Copyright 2006. Reprinted from Reference [20]).

economically and ecologically sound. In addition to these positive engineering aspects, it is important to add the originality and interest associated with the preparation of a composite material in which the dominant role is played by a major and ubiquitous renewable resource.

12.4.2 Starch granules

The transformation of starch granules into thermoplastic materials is one of the major issues within the broad field of novel polymers from renewable resources, as highlighted in Chapter 15, because of the ready availability of low cost starch from a variety of agricultural activities throughout the world, albeit in different macromolecular compositions.

That chapter deals comprehensively with the problems facing researchers who are actively engaged in this area which have to do, on the one hand, with the choice of a treatment capable of converting the granules into a processable material (including the actual thermo-mechanical operation and the selection of a suitable plasticizer) and, on the other hand, with ways to reduce the highly hydrophilic character of the ensuing materials. Several interesting solutions are critically discussed there together with an assessment of the state of the art and its short-term prospects for the actual production of viable materials.

The experience acquired during the research on the partial oxypropylation of cellulose fibres discussed in the previous sub-section suggested that a similar approach could be applied to starch granules with the somewhat different purpose of preparing a new form of thermoplastic starch, as such, or as a composite material in which some smaller grains would still be present. The oxypropylation of starch was not a novelty when this project was conceived but, to the best of our knowledge, had only been carried out in aqueous or organic solvent media with purposes related to food and cosmetic applications, rather than the elaboration of novel materials.

The partial oxypropylation of corn starch conducted in bulk was therefore approached [21] by studying the role of the same variables as already mentioned above in the homologous investigation on the partial oxypropylation of cellulose fibres. On the whole, the two systems behaved very similarly as clearly indicated by the characterization of the corresponding products. In particular, FTIR spectroscopy clearly showed a correlation between the amount of **PO** used in the reaction and the extent of grafting of **POP** chains from both amylose and amylopectin. Additionally, the T_g and the extent of crystallinity were lower, the higher the oxypropylation level, and the corresponding morphologies of the modified granules became increasingly akin to a thermoplastic appearance, as shown in Fig. 12.9.

This ongoing investigation has already pointed to a simple and original way of transforming starch granules by a process which, once again, does not require any separation or purification stage since the single reaction step produces the desired material as such. This new family of thermoplastic starches does not require the addition of a plasticizer since the plasticization is automatically provided by the chemical transformation. Moreover, the grafted PO chains bring about a considerable reduction in the hydrophilic character of the original substrate. Whether all these positive aspects are sufficient to justify possible applications of the materials remains to be verified by further research.

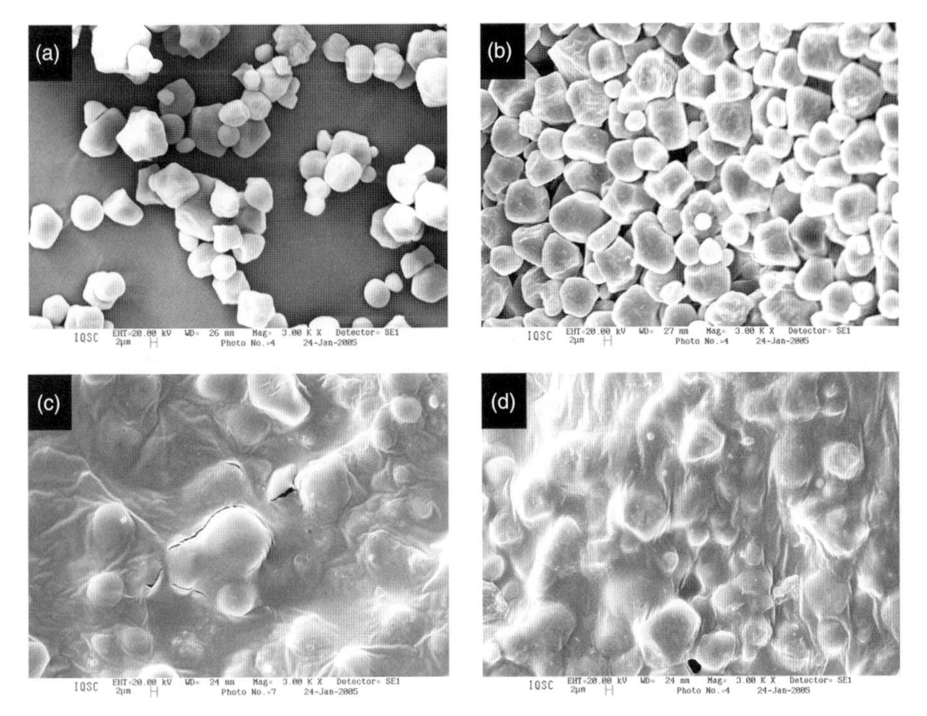

Figure 12.9 SEM micrographs of corn-starch samples: (a) before, and (b), (c), (d) after oxypropylation with increasing **PO**/starch ratios. (Reproduced by permission of the American Chemical Society. Copyright 2007. Reprinted from Reference [21]).

12.5 CONCLUSION

The general applicability of the bulk heterogeneous reaction of **PO** with natural polymers bearing hydroxyl groups in the presence of a nucleophilic catalyst has become an undisputable fact, as testified by the numerous examples given in this chapter. Whether the purpose of this operation is the total conversion of the substrate into a viscous polyol mixture or its partial modification for the preparation of composite or other materials, the only questions that require an investigation in the case of future studies, principally involving unexplored substrates, are confined to quantitative aspects, since it can be safely anticipated that the actual reaction will occur if OH groups are present in the substrate. Whereas the applications of the polyols generated by the bulk processes are clearly oriented towards the synthesis of polyurethanes and polyesters, the partial oxypropilation generates a different type of material, whose direct utilizations are obviously in the realm of composites derived entirely from renewable resources.

REFERENCES

1. Bailey F.E., Koleske J.V., *Alkylene Oxides and their Polymers*, Marcel Dekker, New York, 1990.
2. Edens M.W., *Surfactant Sci. Ser.*, **60**, 1996, 185.
3. Fis J., *Actualité Chim. (France)*, 2002, n.11–12, 54.
4. Asandei N., Perju N., Nicolescu R., Civica S., *Cell. Chem. Technol.*, **29**, 1995, 261.
5. Wu L.C.F., Glasser W.G., *J. Appl. Polym. Sci.*, **29**, 1984, 1111.
6. Gandini A., Belgacem M.N., Guo Z.X., Montanari S., in *Chemical Modification, Properties and Usage of Lignin*, Ed.: Hu T.Q., Kluwer, 2002, p. 57.
7. Nadji H., Bruzzèse C., Belgacem M.N., Benaboura A., Gandini A., *Macromol. Mater. Eng.*, **290**, 2005, 1009.
8. Pavier C., Gandini A., *Ind. Crops Prod.*, **12**, 2000, 1.
9. Pavier C., Gandini A., *Carbohydr. Polym.*, **42**, 2000, 13.
10. Pavier C., Gandini A., *Eur. Polym. J.*, **36**, 2000, 1653.
11. Pavier C., Doctorate Thesis, Grenoble National Polytechnic Institute, France, 1998.

12. (a) Evtiouguina M., Barros A.M., Cruz-Pinto J.J., Pascoal Neto C., Belgacem M.N., Pavier C., Gandini A., *Bioresour. Technol.*, **73**, 2000, 187. (b) Evtiouguina M., Barros-Timmons A., Cruz-Pinto J.J., Pacoal Neto C., Belgacem M.N., Gandini A., *Biomacromolecules*, **3**, 2002, 57.
13. Evtiouguina M., Gandini A., Pascoal Neto C., Belgacem M.N., *Polym. Intern.*, **50**, 2001, 1150.
14. Velazquez-Morales P., Le Nest J.F., Gandini A., *Electrochim. Acta*, **43**, 1998, 1275.
15. Fernandes S., Freire C.S.R., Pascoal Neto C., Gandini A., *Green Chemistry*, **10**, 2008, 93.
16. Barreiro F., Fernandes S., Gandini A., unpublished results.
17. (a) Mutsumura H., Sugyiama J., Glasser W.G., *J. Appl. Polym. Sci.*, **78**, 2000, 2242. (b) Mutsumura H., Glasser W.G., *J. Appl. Polym. Sci.*, **78**, 2000, 2254.
18. (a) Lu X., Zhang M.Q., Rong M.Z., Shi G., Yang G.C., *Compos. Sci. Technol.*, **63**, 2003, 177. (b) Lu X., Zhang M.Q., Romg M.Z., Yue D.L., Yang G.C., *Compos. Sci. Technol.*, **64**, 2004, 1301.
19. Nishino T., Matsuda I., Hirao K., *Macromolecules*, **37**, 2004, 7683.
20. Gandini A., Curvelo A.A.S., Pasquini D., de Menezes A.J., *Polymer*, **46**, 2005, 10611.
21. Gandini A., Curvelo A.A.S., Pasquini D., de Menezes A.J., *Biomacromolecules*, **8**, 2007, 2047.

– 13 –

Hemicelluloses: Major Sources, Properties and Applications

Iuliana Spiridon and Valentin I. Popa

ABSTRACT

This chapter deals with the major representatives of hemicelluloses, their properties and application. Hemicelluloses are plant cell wall polysaccharides that are not soluble in water, but they can be separated by aqueous alkali, or hydrolyzed by diluted acids. The main hemicelluloses include the following polysaccharides: xylan, glucuronoxylan, arabinoxylan, mannan, glucomannan and galactoglucomannan, and their representatives are different as a function of the plant species. The general structure of hemicelluloses is based on a sugar backbone substituted with side chains. The nature of the monosaccharide, as well as the linkages between structural units, determines some properties, such as solubility and the three-dimensional conformation of hemicelluloses. At present, there is an increasing interest to develop new applications of hemicelluloses as raw materials for chemical industry and also in the fields of food and pharmaceutical industries.

Keywords

Hemicelluloses, Polysaccharides, Xylan, Glucuronoxylan, Arabinoxylan, Mannan, Glucomannan and galactoglucomannan, Polymer, Biosynthesis, Biodegradation, Properties, Applications

13.1 STRUCTURE, SOURCES AND PROPERTIES

Hemicelluloses are the second most abundant polysaccharides in nature after cellulose. They occur in close association with cellulose and lignin and contribute to the rigidity of plant cell walls in lignified tissues. Hemicelluloses constitute about 20–30 per cent of the total mass of annual and perennial plants and have a heterogeneous composition of various sugar units, depending on the type of plant and extraction process, being classified as xylans (β-1,4-linked D-xylose units), mannans (β-1,4-linked D-mannose units), arabinans (α-1,5-linked L-arabinose units) and galactans (β-1,3-linked D-galactose units).

In Table 13.1 gives the main hemicelluloses of hardwood and softwood.

Xylan [1] is one of the major constituents (25–35 per cent) of lignocellulosic materials. Its structure has a linear backbone consisting of β-1,4-linked D-xylopyranose residues. These may be substituted with branches containing acetyl, arabinosyl and glucuronosyl residues, depending on the botanic source and method of extraction.

Xylans are the main hemicelluloses in hardwood and they also predominate in annual plants and cereals making up to 30 per cent of the cell wall material. Hardwood xylan (*O*-acetyl-4 methyl-glucuronoxylan) is substituted at irregular intervals (Fig. 13.1) with 4-*O*-methyl-α-D-glucuronic acid groups joined to xylose by α-1,2-glycosidic linkages. On an average, every tenth xylose unit has an uronic acid group attached at C2 or C3 of the xylopyranose.

Barley arabinoxylans are composed of D-xylopyranosyl (Xyl) units linked by β-(1,4) bonds with single L-arabinofuranosyl residues connected to the backbone of one or both of the α-(1,2) or α-(1,3) bonds [2, 3] Another unique

Table 13.1

The main hemicelluloses in wood

	Hardwood	Softwood
Methylglucuronoxylan	80–90	5–15
Arabinomethylglucuronoxylan	0.1–1	15–30
Glucomannan	1–5	1–5
Galactoglucomannan	0.1–1	60–70
Arabinogalactan	0.1–1	1–15
Other galactans	0.1–1	0.1–1
Pectins	1–5	1–5

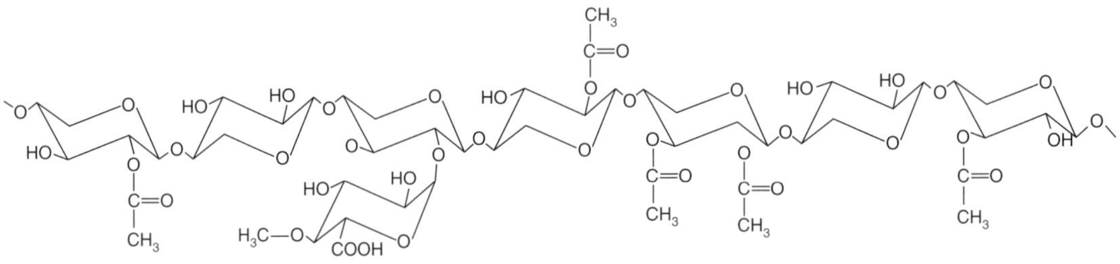

Figure 13.1 Structure of *O*-acetyl-(4-OMe-glucurono)xylan from hardwood.

feature of barley, wheat, and rye arabinoxylan, is the presence of ferulic acid that is covalently linked to the arabinose residue *via* an ester bond [4].

The physicochemical characteristics and functional properties of carbohydrate macromolecules depend on their molecular mass, degree of polymerization, branching and macroscopic structure.

Arabinoxylan or β-(1,4)-xylan polymers in an unsubstituted form tend to aggregate into insoluble complexes that are stabilized by intermolecular hydrogen bonds [5]. The polymer forms a three-fold, left-handed helix, which in the solid state appears as an extended twisted ribbon [6]. Unsubstituted polymer regions exhibit a tendency for interchain aggregation through H-bonding, which is the most likely reason why these polymers are less soluble in water. The solubility of arabinoxylan macromolecules is improved by the presence of arabinose substituents that prevent intermolecular aggregation of unsubstituted xylose residues [7]. The removal of the arabinofuranosyl residues is induced by the action of α-L-arabinofuranoside arabinofuranohydrolases (EC 3.2.1.55), enzymes which have broad specificities for the various (1 → 2)-, (1 → 3)- and (1 → 5)-α-L-arabinofuranosyl linkages of arabinan and arabinoxylans [8].

The molecular structure of sycamore xyloglucan was characterized by methylation analysis of the oligosaccharides obtained by endoglucanase treatment of the polymer. It was found that polymer structures were based on a repeating heptasaccharide unit, which consists of four residues of beta-1–4-linked glucose and three residues of terminal xylose, a single xylose residue being glycosidically linked to carbon 6 of three of the glucosyl residues [9].

Appreciable amounts of xylan, the main component of plant hemicelluloses, are currently not exploited to their full extent. Xylan and xylooligosaccharides can be converted to xylose by chemical or enzymatic hydrolysis. Subsequently, xylose can be converted biologically to single cell proteins and to a whole range of fuels and chemicals, and can also be used as a carbon source for the production of glucose isomerase. Thus, the most promising approach of economically feasible processes of xylan bioconversion is the use of microorganisms able to transform xylan directly to single cell proteins, fuels and chemicals. Microorganisms such as *Fusarium oxysporum, Aureobasidium pullulans* and *Candida shehatae* are able to transform xylose to ethanol. For the direct xylan bioconversion only anaerobic thermophilic bacteria and a few representative fungi and yeasts combine the two metabolic pathways necessary for direct xylan bioconversion [10, 11].

In the softwood, galactoglucomannan (Fig. 13.2) is the most abundant hemicellulose. Its backbone is constituted by β-1,4-linked D-glucopyranose D-mannopyranose units randomly distributed in the main chain. Partial substitutions are

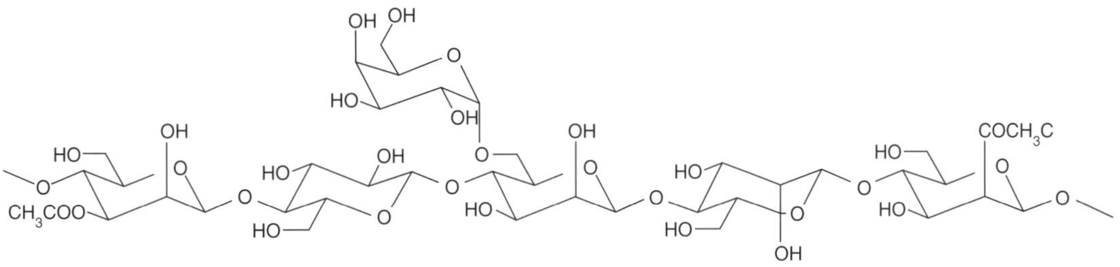

Figure 13.2 Structure of *O*-acetyl-galactoglucomannan.

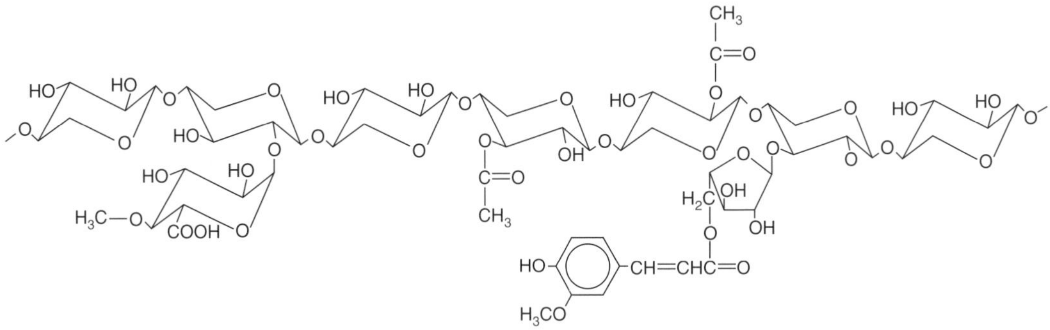

Figure 13.3 Structure of xylan from annual plants.

possible with α-D-galactose side groups, which can be attached to both mannose and glucose units by an α-1,6-linkage [12, 13]. Depending on the source of the polysaccharide, mannose/galactose ratio can vary from 1.0 to 5.3 [14, 15].

Larch arabinogalactan has a backbone of (1–3)-linked β-D-galactopyranosyl units each of which contains a side chain at position C-6. Most of these side chains are galactobiosyl units containing a (1–6)-β-D-linkage. Another side chain type that occurs is a single L-arabinofuranose unit or 3-O-(β-L-arabinopyranosyl)-α-L-arabinofuranosyl units. The preliminary X-ray fibre diffraction data [16] proved that western larch arabinogalactan can adopt a triple helix conformation.

In annual plants, the main hemicelluloses are represented by xylans (Fig. 13.3), which are more heterogeneous than the xylans from wood tissues [17]. They contain both glucuronic acid and/or its 4-*O*-methyl-ether and arabinose attached to C2 or C3 of the xylose units. Both xylan and glucomannans can be partly acetylated [18].

Arabinoxylans consist of α-L-arabinofuranose residues attached as branch-points to β-(1 → 4)-linked D-xylopyranose polymeric backbone chains. These may be 2- or 3-substituted or 2- and 3-disubstituted. In wheat flour, the distribution of the type of substitution is not random, but that of the substituted residues along the chain occurs randomly [19]. The arabinose residues may also be linked to other groups, such as glucuronic acid residues and ferulic acid crosslinks [20]. It was found that the molecules take up a twisted ribbon conformation with a three-fold symmetry. The free molecules in solution may, however, take up a wide variety of conformations, with only moderately extended structures. Although the backbone xylan structure is similar to that occurring in cellulose, there is little driving force to produce crystalline type structures as the intra- and inter-molecular hydrogen bonds associated with the 6-hydroxyl groups are necessarily absent. The side chains of arabinose reduce the interactions between chains, because of their inherently more flexible water-hungry furanose conformations. Arabinose and xylose can usually be removed by enzymatic hydrolysis with cellulases and hemicellulases [21].

The addition of the β-xylosidase and α-L-arabinofuranosidase enzymes to purified xylanases, more than doubled the degradation of xylan from 28 to 58 per cent of the total substrate, with xylose and arabinose being the major sugars produced [22]. The presence of all three purified enzymes resulted in the production of xylose, arabinose and substituted xylotetrose. β-xylosidase and α-L-arabinofuranosidase display a synergysm with endoxylanase with respect to xylan degradation. Endoxylanases generate free chain ends upon which the β-xylosidase can act, while the debranching activity of α-L-arabinofuranosidase removes the substitute arabinose that might otherwise block the enzyme's progress [22].

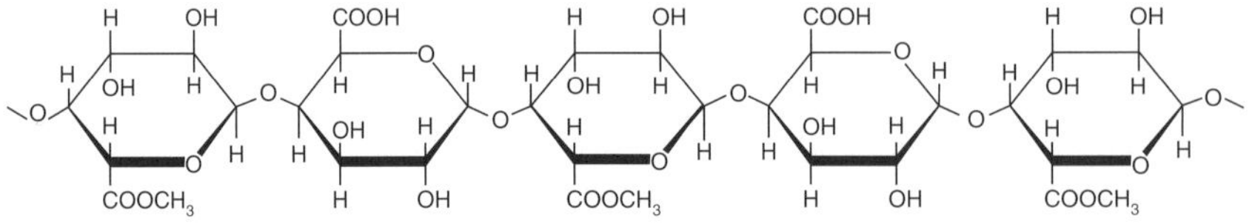

Figure 13.4 The main component of pectin.

The arabinoxylan from *Plantago ovata* is resistant against enzymatic attack. The unusual physicochemical and physiological properties of arabinoxylan may be explained directly from the polysaccharide structure which involves a β-(1 → 3)-linked xylose disaccharide attached at the 2-position to the xylan backbone and has been found to interact strongly with neighbouring arabinose residues and to introduce a conformational lock (or 'kink') into the structure, which may be released with a rise temperature [23].

Other hemicelluloses such as arabinans, galactans, arabinogalactans, rhamnogalactans, etc are regarded as pectic substances of plant cell walls. They are a complex mixture of colloidal polysaccharides, more abundant in the soft tissues of some fruit, as well as in sugar beet pulp, but less frequent in wood tissues.

Pectins (Fig. 13.4) form a hydrated crosslinked three-dimensional network in the matrix of the primary cell walls and play diverse functions in the cell physiology, growth, adhesion and separation [24]. The gel-like property of the cell wall is derived in part from pectins.

The biogenesis of pectins proceeds during deposition on the P-wall. At the same time, labelled representatives of this series, namely, glucose, glucuronic acid and xylose, appear indicating that the epimerase in the Golgi vesicles no longer quantitatively transforms the resorbed glucose into galactose. Later on, the deposition of the S-wall takes place and the activity of the epimerase is stopped, so that the pectic substances, indispensable for the extension growth of the P-wall, are replaced by hemicellulosic substances such as polyglucuronic acids and xylans.

The conversion of hexose into uronic acid and pentose units of cell wall polysaccharides in higher plants may occur, not only by the sugar nucleotide oxidation pathway, which leads to UDP-uronate, but also by the myoinositol oxidation pathway. In this case, the homocyclic inositol ring is split, oxidized at C(6) and concurrently replaced by a heterocycle with an —O—bridge between C(1) and C(5), as shown by label experiments. Glucuronic acid can be generated in this way.

Pectic substances are complex colloidal acid polysaccharides with a backbone of galacturonic acid residues linked by α-1,4-glycosidic linkages. The side chains of the pectin molecule consist of L-rhamnose, arabinose, galactose and xylose [25]. D-Galacturonic acid is the principal monosaccharide unit of pectin. The D-galacturonic acid residues are linked together by α-1,4-glycosidic linkages. The molecular mass of pectin ranges from 50 000 to 150 000. The carboxyl groups of galacturonic acid are partially esterified by methyl groups and are partially or completely neutralized by sodium, potassium or ammonium ions. Based on the type of modification, different pectic substances exist in nature. Thus, a pectin in which more than 50 per cent of the galacturonic acid residues are esterified is called high methoxyl or HM pectin, whereas a pectin in which this extent of esterification is less than 50 per cent, is called low methoxyl or LM pectin. Homogalacturonans with a linear structure make up the 'smooth region' of pectin, while the branched polysaccharides – rhamnogalacturonans and other pectin structures such as xylogalacturonan [26] and α-(1,4)-linked D-galacturonic acid substituted with β-D-xylose at the C3 position [27] – make up its 'hairy region'.

Because the predominant cation in the cell wall is the calcium ion, its middle lamella consists largely of calcium pectate. Calcium ions are involved in the stabilization of the cell wall structure by crosslinking pectin chains, the ion exchange processes and the control of the activities of the wall enzymes. An 'egg-box' model was proposed for the calcium coordination in the middle lamella, in which two helical 'polygala' fragments are mutually attracted via ionic interactions and coordination of calcium ions between chains. In these dimers, the calcium ions are sandwiched between the inner faces of both monomeric components on the specific sites [28]. Calcium egg boxes, crosslinking smooth regions in pectin chains, probably play an important structural role in keeping middle lamellas together in many non-lignified tissues.

Agar is obtained from the family of red seaweeds (*Rhodophycae*) as the carrageenans. Carrageenan compounds differ from agar in that they have sulphate groups ($—OSO_3^-$) in the place of some hydroxyl groups. Commercially, agar is produced from species of *Gelidium* and *Gracilariae*[29, 30]. Agar (Fig. 13.5) is insoluble in cold water but dissolves in boiling water to give random coils. Gelation is reported to follow a phase separation process including association on cooling (about 35°C), forming gels with up to 99.5 per cent water and remaining

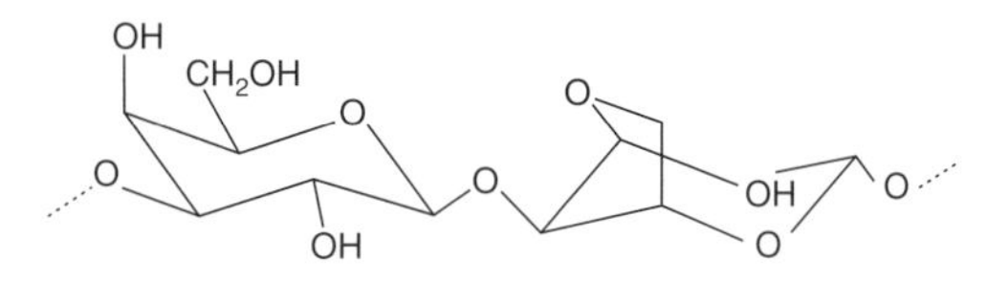

Figure 13.5 Structure of agar repeat unit.

solid up to about 85°C, although these findings have been disputed. Agar consists of a mixture of agarose and agaropectin [31, 32].

Agarose is a linear polymer with a molecular mass of about 120 000. The agarose gel contains double helices, stabilized by the presence of water molecules bound inside the double-helical cavity. It seems that this network model is excessive and that only a small proportion of double-helical junction zones would be sufficient to realize the gelation process. This is why some authors consider that network structure of agarose gel consists of single chain [33, 34].

Agaropectin is a heterogeneous mixture of smaller molecules that occur in lesser amounts. Their structures are similar, but slightly branched and sulphated, and may have methyl and pyruvic acid ketal substituents. They gel poorly and may be easily removed from the agarose molecules using their charge. The quality of agar is improved by an alkaline treatment that converts L-galactose-6-sulphate into 3,6-anhydro-L-galactose.

Another polysaccharide extracted from seaweeds is carrageenan. Carrageenans are linear, water-soluble polymers that form solutions, whose viscosity depends mainly on the concentration of the polymer, the temperature of the solution and the type of carrageenan. Chemically, they are highly sulphated galactans and, because of their half-ester sulphate moieties they are strongly anionic polyelectrolytes. In this respect, they differ from agars and alginates, the other two classes of commercially exploited seaweed hydrocolloids. Carrageenans are susceptible to depolymerization through acid-catalyzed hydrolysis. High temperatures and low pHs lead to a complete loss of functionality. Their rate of hydrolysis at a given pH and temperature is markedly lower if they are in the gel, rather than the sol state. This can be achieved by ensuring that gel-promoting cations are present in a sufficient concentration to raise the gel melting temperature above the temperature at which they will be treated. Carrageenans can be used in acidic media, provided, however, that they are not subjected to prolonged heating.

Carrageenan is frequently preferred over other thickening, suspending and binding ingredients, because it is natural, economical, readily available and functional in an extremely broad application base. Carrageenans can be produced *via* a variety of process techniques such as alcohol extraction, potassium chloride gel press or extraction with various alkalis. The process technique is important, because it influences the gel characteristics. Likewise, different seaweeds also influence the gel characteristics.

The three preferred carrageenan types are *kappa*, *iota* and *lambda*. The *kappa* and *iota* types only dissolve in a heated water medium, whereas the *lambda* type is soluble in cold water. Gels created from different carrageenan types may be fluid, elastic or rigid and are heat-reversible. Gelling temperatures and gel strength are also influenced by added ingredients such as salts and proteins.

Alginate is a gelling polysaccharide extracted from seaweeds. Most of the large brown seaweeds are potential sources of alginate, their properties being different from one species to another. The main commercial sources are species of *Ascophyllum, Durvillaea, Ecklonia, Laminaria, Lessonia, Macrocystis, Sargassum* and *Turbinaria*. The monomeric units of alginates, α L-guluronic acid residues (Fig. 13.6) and β D-mannuronic acid (Fig. 13.7) are C5 epimers of each other.

Alginates are not random copolymers, consisting instead of blocks of similar and strictly alternating residues, each of which has different conformational preferences and behaviour. Thus, poly β-(1 → 4)-linked D-mannuronate prefers forming a three-fold left-handed helix with (weak) intramolecular hydrogen bonding between the hydroxyl group in the 3-position and the subsequent ring oxygen, while poly α-(1 → 4)-linked L-guluronate forms stiffer (and more acid-stable)

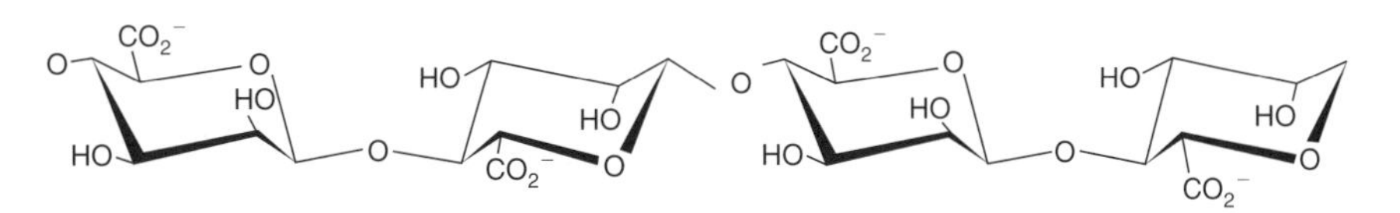

Figure 13.6 Structure of β-D-mannuronic acid.

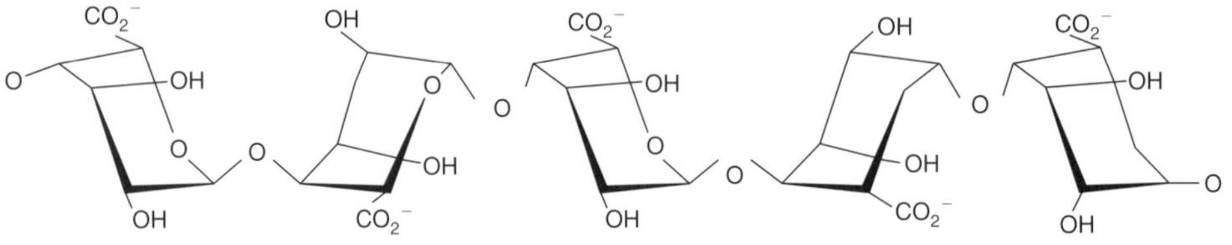

Figure 13.7 Structure of α-L-guluronic acid residues.

two fold screw helical chains, preferring intramolecular hydrogen bonding between the carboxyl group and the 2-OH group of the prior residues and the 3-OH group of the subsequent residues, the latter interaction being weaker.

It has been shown that the polymer chain is made up of three kinds of regions or blocks. The G blocks contain only units derived from L-guluronic acid, the M blocks are based entirely on D-mannuronic acid, while the MG blocks consist of alternating units of the two acids. In practice, an alginate molecule can be regarded as a block copolymer containing M, G and MG blocks, whose proportion varies as a function of the seaweed source and influences the physical properties of the corresponding alginates [35, 36]. Thus, the gel formation by addition of calcium ions, involves the G blocks so that the higher their proportion, the greater the gel strength, whereas the solubility of alginates in an acid medium, depends on the proportion of MG blocks. The monovalent cation salts [Na^+, K^+, NH_4^+, $(CH_2OH)_3NH^+$] of alginic acid, and its propylene glycol ester, dissolve in water, but alginic acid and its calcium salt do not. Neutral alginate solutions of low-to-medium viscosity can be kept at 25°C for several years, without any appreciable viscosity loss, as long as a suitable microbial preservative is added. Solutions of highly polymerized alginates will lose viscosity at room temperature within a year and in order to achieve high, stable viscosities, it is better to add calcium ions to a solution of an alginate with a moderate molecular weight. All alginate solutions depolymerize more rapidly as the temperature is raised. Alginates are most stable in the PH range of 5–9.

β-glucans were identified as major component in the bran of *Gramineae* (such as barley, oats, rye and wheat, consisting of linear unbranched polysaccharides of linked β-(1 → 3)- and β-(1 → 4)-D-glucopyranose units [37]. β-Glucans form 'worm'-like cylindrical molecules containing up to about 250 000 glucose residues that may produce crosslinks between regular areas containing consecutive cellotriose units. They form thermoreversible infinite network gels. Ninety per cent of the β-(1 → 4)-links are in cellotriosyl and cellotetraosyl units joined by single β-(1 → 3)-links, with no single β-(1 → 4)- or double β-(1 → 3)-links. The ratio of cellotriosyl/cellotetraosyl is between 2.0 and 2.4 in oats, about 3.0 in barley and about 3.5 in wheat. The high molecular mass β-glucans from barley and oats are viscous due to labile cooperative associations, whereas lower molecular mass β-glucans can form soft gels as the chains are easier to rearrange to maximize linkages. Barley β-glucan is highly viscous and pseudoplastic, both properties decreasing with increasing temperature.

The related hydrocolloid curdlan has only β-(1 → 3)-D-glucopyranose units which form thermoreversible triple-helical structures on heating and become irreversibly linked as the concentration or temperature is increased. Curdlan is a microbial fermentation extracellular polymer produced by a mutant strain of *Alcaligenes faecalis* var. *myxogenes* [38]. Curdlan was also produced by pure culture fermentation using *Agrobacterium radiobacter* NCIM 2443 using glucose, sucrose, maltose as carbon sources, sucrose being the most efficient [39, 40]. It is insoluble in water, which limits its biological applications. Water insolubility is generally attributed to the existence of extensive intra/intermolecular hydrogen bonds.

Curdlan has a moderate molecular mass (DP ~ 450) being constituted from unbranched linear 1 → 3 β-D-glucan with no side chains (Fig. 13.8). It has junction zones consisting of parallel in-phase triple right-handed six-fold helices (fibre repeat 18.78 Å) forming an uncharged rigid rod-like conformation [41]. The chains are held by intrahelix hydrogen bondings between the 2-OH groups, each formed by donation to one chain and acceptance from the other on the inside of the helix axis. As single stranded curdlan forms a six-fold helix stabilized by a chain of intramolecular hydrogen bonds between neighbouring 2-OH groups, the change from single to triple helices involves these 2-OH groups changing their hydrogen bonding allegiance from intramolecular to intermolecular.

Okuyama *et al.* [42] proposed the single right-handled six-fold helical conformation in the highly hydrated crystal, in which the water content was more than 70 wt per cent in the unit cell. This single helix is stabilized by weak intramolecular hydrogen bonds between O(5) and O(4) of the next glucose residue (distance 0.314 nm).

Curdlan gum is tasteless and produces retortable freezable food elastic gels. It is insoluble in cold water, but aqueous suspensions plasticize and briefly dissolve before producing irreversible gels (*i.e.* curdling, hence its

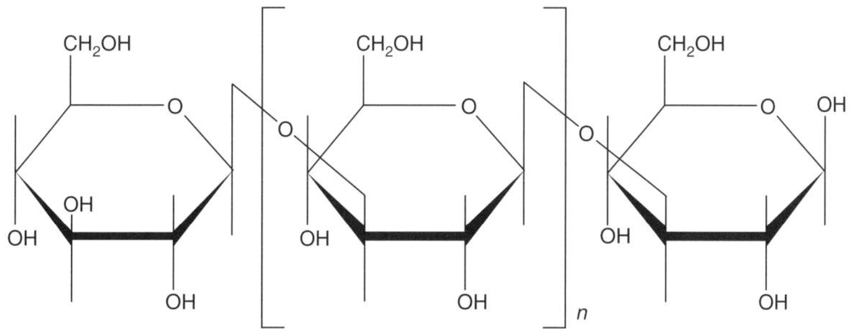

Figure 13.8 Structure of curdlan.

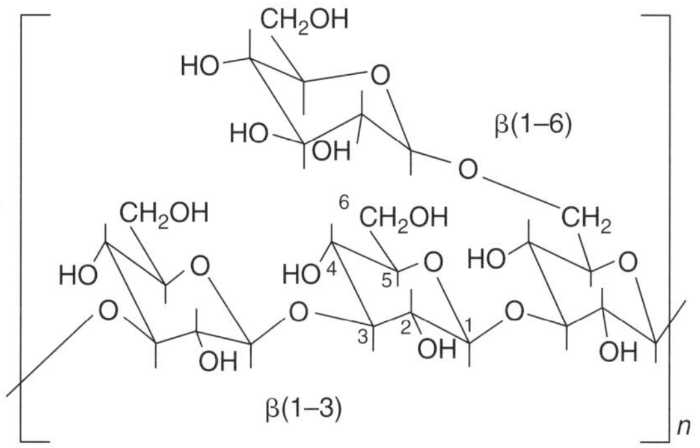

Figure 13.9 Structure of scleroglucan.

name) on heating to around 55°C, which then remain on cooling. Heating at higher temperatures produces more resilient gels by the aggregation of the triple-helical structures and synaeresis. The 'curds' consist of mixtures of single and triple helices [43–45].

Scleroglucan (from *Sclerotinia sclerotiorum*) is a 1 → 3 β-D-glucan with additional 1 → 6 β-links (Fig. 13.9) that confer water solubility to it under ambient conditions, but do not significantly interfere with a triple helix gelling process, similar to that of curdlan [46]. Similar polysaccharides can also be extracted from other sources such as waste yeast.

Scleroglucan presents a triple-helical backbone conformation [47, 48] and in solution the chains assume a rod-like triple-helical structure [49] in which the D-glucosidic side groups are on the outside and prevent the helices from coming close to each other and aggregating.

Pullulan (Fig. 13.10) is one of the few neutral water-soluble microbial polysaccharides that can be obtained in large quantities by fermentation [50, 51]. Pullulan biosynthesis is accomplished through mediation of sugar nucleotide–lipid carrier intermediates associated with the cell membrane fraction [52]. It is an extracellular, unbranched homopolysaccharide consisting of maltotriose and maltotetraose units with both α-(1 → 6) and α-(1 → 4) linkages,

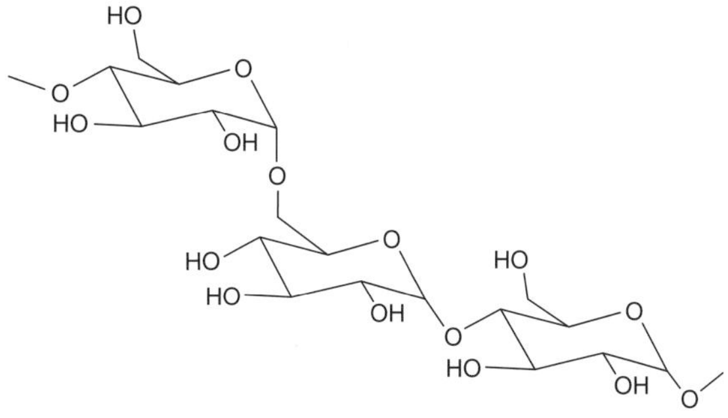

Figure 13.10 Structure of pullulan.

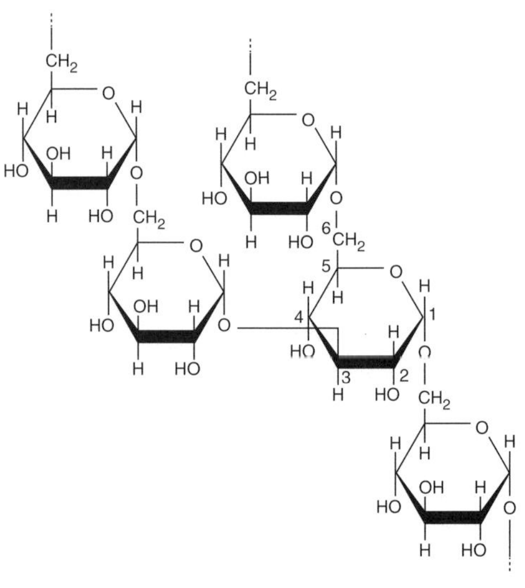

Figure 13.11 Structure of dextran.

whose molecular mass depends on the culture conditions and strain [53–55]. It adopts a random coil-type structure in aqueous solution, generates a Raman optical spectrum that closely resembles that of D-maltotriose [56]. The alternation of α-1 → 4 and α 1 → 4 is regular, resulting in structural flexibility and enhanced solubility [57].

Pullulan produces high-viscosity solutions at relatively low concentration and can be utilized to form oxygen-impermeable films, thickening or extending agents and adhesives. Pullulan has many uses as an industrial plastic. It can be processed into compression mouldings that resemble polystyrene or polyvinyl chloride in transparency, gloss, hardness, strength and toughness, but is far more elastic. It decomposes above 200°C without the formation of toxic gases.

Dextran is a microbial biopolymer [58] whose molecular structure is composed exclusively of monomeric 2-D-glucopyranosil units, linked mainly by (β-1,6) glucosidic bonds (Fig. 13.11). Its applications depend on its molecular mass [59]. Two dextran products are available in most countries for clinical purposes: dextran 70, with a molecular mass of about 70 000 and dextran 40 with a molecular mass 40 000. Dextran fractions are readily soluble in water and electrolyte media to form clear, stable solutions significantly insensitive to pH. They are also soluble in some solvents such as methyl sulphide, formamide and ethylene glycol [60].

Xanthan, a microbial polysaccharide produced by the *Xanthomonas* bacterium, represents a great scientific and industrial interest due to its properties. It is a heteropolysaccharide with a high molecular mass, between 0.9 and 1.6×10^6, as a function of the microbial source and fermentation conditions [61–63]. The pentasaccharide repeating unit (Fig. 13.12) is assembled on an isoprenoid lipid carrier by sequential addition of individual sugar residues that are donated by sugar nucleotide diphosphate precursors. Each sugar addition is catalyzed by a specific glycosyltransferase enzyme. The mannose residues of the repeating units are specifically acetylated and pyruvylated. The repeating unit is polymerized, and the polymer is subsequently secreted [64, 65].

In its native state, xanthan is a right extended chain possessing a helical conformation [66], in the solid state this polysaccharide has a S1 helix conformation [67], while the xanthan crystal chain takes an antiparallel right-handed five-fold (5/1) double helix that is stabilized by four intramolecular and one intermolecular hydrogen bonds.

In solutions of low ionic strength or at high temperature, the xanthan gum chains adopt a random coil configuration, since the anionic side chains repel each other. The addition of electrolytes reduces the electrostatic repulsion among the side chains, allowing them to wrap around and hydrogen bond to the backbone. The polymer chain straightens into a relatively rigid helical rod. This shape tends to revert to the random coil if the gum solution is highly diluted or heated. With increasing electrolyte concentration, however, the rod shape is maintained at higher temperatures and greater dilutions [68].

Gellan is an extracellular heteropolysaccharide produced by *Sphingomonas paucimobilis*, previously known as *Pseudomonas elodea*. It is composed of linear tetrasaccharide repeating units of β-D-glucose, β-D-glucuronic acid, β,-D-glucose and α-L-rhamnose (Fig. 13.13). Occasionally, it carries acetyl and glyceryl groups on the glucose units [69, 70].

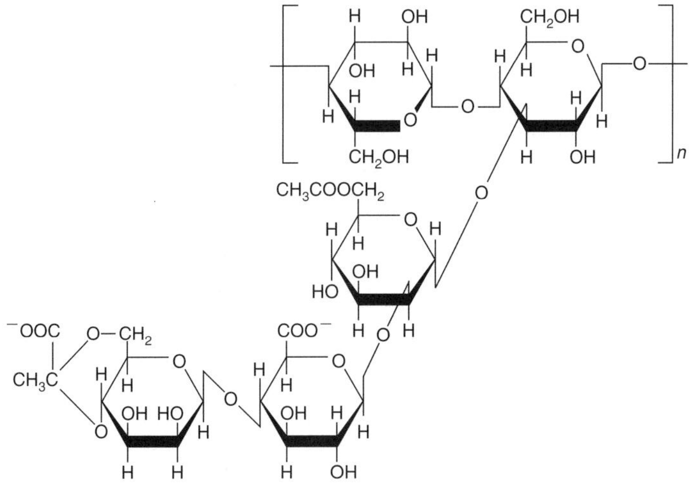

Figure 13.12 Structure of xanthan.

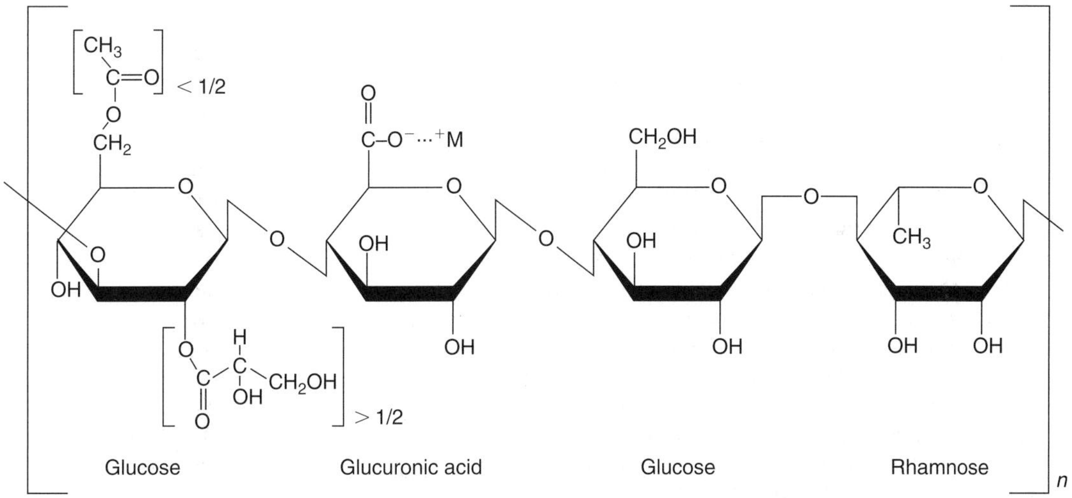

Figure 13.13 The repeat units of gellan.

The degree of ester substitution directly influences the gellan properties in solution and gel [71]. Aqueous solutions of gellan are highly viscous and show high thermal stability. The gellan gum exhibits good stability over a wide pH range of 3.5–8.0. Fibre diffraction analysis [72] showed that the two left-handed, three-fold helical chains are organized in parallel fashion in an intertwined double helix and that the duplex is stabilized by interchain hydrogen bonds at each carboxylate group.

The properties of gellan can be easily modified. A hot caustic treatment yields a polymer that has the desirable characteristic of low viscosity at high temperature. Cooling gellan in the presence of various cations (*e.g.* calcium) results in the formation of strong gels. In a heated gellan solution, the polymer chains exist as disordered coils. Cooling without the presence of metal cations leads to the formation of weak, thermoreversible interactions, such as hydrogen bonding and van der Waals attractions between adjacent polymer chains. However, with the introduction of metal cations, the interactions formed with the carboxyl groups are much stronger and more stable. Monovalent cations form stiff gels, perhaps allowing for more hydrogen bonding to occur between the polymer helices, also leading to aggregation and gel formation. Divalent cations may bridge two adjoining carboxylic acid sites, thus causing local aggregation of the helices to make a more rigid gel.

13.2 APPLICATIONS

Arabinogalactans are used as emulsifiers, stabilizers and binders in the food, pharmaceutical and cosmetic industries. Arabinogalactan has properties that make it suitable as a carrier for delivering diagnostic or therapeutic agents to hepatocytes via the asialoglycoprotein receptor [73]. Experimental studies indicated that larch arabinogalactan can stimulate natural killer cell cytotoxicity, enhance other functional aspects of the immune system, and inhibit the metastasis of tumour cells to the liver. These immune-enhancing properties also suggest an array of clinical uses, both in preventive medicine, due to its ability to build a more responsive immune system, and in clinical medicine, as a therapeutic agent in conditions associated with a lowered immune function, decreased NK activity or chronic viral infection [74].

The xylans have numerous medical applications [75, 76]. The films based on glucuronoxylan (isolated from aspen wood) and xylitol or sorbitol show low oxygen permeability and thus have a potential application in food packaging [77].

Pectin is widely used in the food industry as a gelling agent to impart a gelled texture to foods, mainly fruit-based foods such as jams and jellies. Thus, LM pectins (<50 per cent esterified), form thermoreversible gels in the presence of calcium ions and at low pH (3–4.5) whereas HM pectins rapidly form thermally irreversible gels in the presence of sufficient (*e.g.* 65 per cent by mass) sugars, such as sucrose and at low pH (<3.5); the lower the methoxyl content, the slower the set. Pectin is also used for the stabilization of low-pH drinks, including fermented drinks and mixtures of fruit juice and milk. In the medical field, pectin is used in combination with the clay kaolin (hydrated aluminium silicate) against diarrhoea [78]. It is also used as a component in the adhesive part of ostomy rings. Pectin is moreover marketed as a nutritional supplement for the management of elevated cholesterol [79, 80]. Finally, pectin has been found to alter the characteristics of the fibrin-network architecture, suggesting that it may have some antithrombotic effects [81].

Agar is used in the food industry in icings, glazes, processed cheese, jelly sweets and marshmallows and sometimes as a substitute for gelatin. Another interesting application of agar comes from the fact that it constitutes a very good microbiological medium [82].

Carrageenans and semi-synthetic sulphated polysaccharides, like laminarin sulphate, were shown to be potent angiogenesis inhibitors [83].

Alginates generally show high water absorption and may be used as low viscosity emulsifiers and shear-thinning thickeners. They can be used to stabilize phase separation in low-fat fat-substitutes, for example as alginate/caseinate blends in starch three-phase systems. Alginate is used in a wide variety of foodstuff such as pet food chunks, onion rings, stuffed olives, low-fat spreads, sauces and pie fillings [84]. Propylene glycol alginates have widespread use as acid-stable stabilizers for, for example, preserving the head on beers [85, 86]. Alginates are suitable for various potential clinical applications, such as cell encapsulation, drug delivery and tissue engineering [87, 88]. Thus, numerous studies [89–91] analyzed the biocompatibility of alginates/sodium alginate and drugs, which are the most important factors affecting the drug release from matrix tablets, or their effects on cell proliferation, cell migration etc. [92]. Because the properties of alginate hydrogels are readily controlled with the chemical structure of the sugar residues and different crosslinked molecules, and these materials interact with cells [93], they are used to engineer bones [94], cartilage [95], muscles, blood vessels [96], liver and nerve tissues [97, 98]. Finally, alginate was used as a scaffold to transplant subcultured human dental pulp cells subcutaneously into the backs of nude mice, and it was proved that subcultured dental pulp cells actively differentiated into odontoblast-like cells and induced calcification in an alginate scaffold [99].

β-glucans have important health benefits, especially in coronary heart disease, cholesterol lowering and reducing the glycemic response. These positive effects are linked to their high viscosity, although it may be that some of them arise from appetite suppression. Almost all carboxymethyl glucans possess properties useful in cosmetic and dermatological applications [100]. β-glucans were identified in different species of mushrooms and when isolated from the mature fruiting bodies of *Agaricus* species they displayed their potential therapeutical benefits [101–103].

Another β-glucan from yeast (*Saccharomyces cereviseae*) is recognized for its ability to activate nonspecific immune response [104–106].

Curdlan, at concentration between 0.05 per cent and 3 per cent, improves the texture, palatability, stability, water binding and holding of food. It can be used in fat replacement because consumers want healthier products with fewer calories, but they are unwilling to give up the mouth feel and flavour characteristics of full-fat products. When hydrated and heated, curdlan's particles mimic the mouth feel of fat containing products. Curdlan also binds the additional water in the system, so that it becomes unavailable for ice crystallization and moisture migration upon storage, freeze-thawing, or deep-fat frying. In addition, aqueous suspensions of curdlan exhibit thixotropic properties that are very important in mimicking the rheological behaviour of fats in processed liquid foods such as low-fat dressings, sauces and gravies [107, 108]. Curdlan of *Alcaligenes faecalis* was reported to display an anti-tumour activity [109, 110], while the sulphated curdlan derivative showed *in vitro* a beneficial effect for HIV-1 treatment [111, 112]. Curdlan is also effective in the prevention and treatment of ischaemic cerebrovascular disease [113]. It has been proved that laminarin from *Eisenia bicyclis* stimulates monocytes to inhibit the proliferation of U 937 cells [114, 115]. Its elicitors may be good candidates as fungicide disease control, because they are natural products and recyclable in the ecosystem [116, 117].

In the cosmetic industry, scleroglucan may be used in various skin care preparations, creams and protective lotions, while in pharmaceutical products, it may be used as a laxative, in tablet coatings and in general to stabilize suspensions [118]. Scleroglucan also proved to have an immunomodulating effect, with numerous medical applications [119–121].

Pullulan can be modified hydrophobically by the reaction with dodecanoic acid, by modulating the polymer-to-acid ratio. It was found that the addition of sodium dodecyl sulphate, alone or in mixtures with other surfactants, resulted in the formation of micelles. The long foam lifetime of these hydrophobically derivatized pullulans is suitable for applications in mineral flotations and in biomedicine (vesicle coating) [122]. Films formed from pullulan are suitable for coating foods and pharmaceuticals [123–125] and prevent oxidation, especially when the exclusion of oxygen is desirable. Pullulan also finds medical applications in vaccine production and as a plasma extender [126, 127]. Ester and ether derivatives of pullulan have adhesive qualities similar to those of gum arabic, their viscosity and adhesive properties being dependent on the degree of polymerization of the polysaccharide. Finally, modified pullulan can be used as a stationary paste that gelatinizes upon moistening.

Dextran–hemoglobin compounds may be used as blood substitutes that have oxygen delivery potential and can also function as plasma expanders. Dextran hydrogels [128] can be used to engineer tissues or to control drug delivery. Blends, based on dextran [129, 130] as well as dextran derivatives [131], have medical applications. Thus, diethylaminoethyldextran, a polycationic derivative of dextran, is recommended for numerous applications in molecular biology and the health-care sector. Applications of dextran derivatives in gene therapy have also been reported [132, 133]. Dextran sulphate is an inhibitor of the human immunodeficiency virus (HIV), binding to T-lymphocytes, but because of its low oral bioavailability, it has not been used therapeutically [134–136]. Other modified dextrans such as Sephadex@ are used extensively in the separation of biological compounds. Finally, iron dextran solutions are used for the treatment of human and veterinary anaemic iron deficiency.

Xanthan can form gels either by crosslinking with certain metal ions or by the interaction with other polysaccharides. These gels are used in the oil industry to modify the permeability profiles of heterogeneous oil reservoirs. The high solution viscosity, compatibility with salts and stability to shear and heat makes xanthan solutions well suited to meet the requirements of polymer flooding systems. In the food industry, xanthan is used in products such as beverages, desserts and dressings, but also as a gelling agent for cheese spreads, ice creams and puddings. Moreover, it can stabilize emulsified creams with pharmaceutical and cosmetic applications [137, 138]. More recently, xanthan has been used in the new clear-gel toothpastes.

Gellan has applications in the food industry as a gelling agent in frostings, glazes, icings, jam and jellies. *Sphingomonas* gellan (S-gellan), modified with diethylaminoethyl chloride-HCl (DEAE-HCl) is a polyelectrolyte that contains both positive and negative charges. The solubility of native S-gellan was improved, increased by nearly a factor of two, from 40 per cent to 75 per cent , after DEAE derivatization, while its water-holding and oil-binding capacities were drastically decreased. These improved properties and stronger bile acid binding capacity suggest that the DEAE-derivatized gellan has more advantages than gellan itself for functional food applications [139]. Films of gellan were implanted for insulin delivery in diabetic rats. It was found that their blood glucose levels were about half of those of rats implanted with blank films and that the therapeutic effect of insulin could last for 1 week. Both *in vitro* and *in vivo* studies indicated that the gellan film could be an ideal candidate for the development of protein delivery systems [140].

13.3 CONCLUDING REMARKS

The purpose of this chapter was to emphasize the interest of hemicelluloses as renewable resources capable of fulfilling a remarkable array of roles in today's science and technology. A better understanding of the physiological functions, chemistry and functionality of hemicelluloses constitutes a formidable task for the near future which should provide wider and more profound applications in areas such as materials science, medicine and biology.

REFERENCES

1. Ebringerova A., Heinze T., Xylan and xylan derivatives – Biopolymers with valuable properties, 1. Naturally occurring xylans structures, isolation procedures and properties, *Macromol. Rapid Comm.*, **21**(9), 2000, 542–556.
2. Viëtor R.J., Angelino S.A.G.F., Vosdragen A.G.J., Structural features of arabinoxylans from barley and malt cell wall material, *J. Cereal Sci.*, **15**, 1992, 213–222.
3. Carpita N.C., Structure and biogenesis of the cell walls grasses, *Annu. Rev. Plant Physiol. Plant Mol.*, **47**, 1996, 445–476.
4. Izydorczyk M.S., Biliaderis C.G., Cereal arabinoxylans: Advances in structure and physicochemical properties, *Carbohyd. Polym.*, **28**, 1995, 33–48.
5. Andrewartha K.A., Phillips D.R., Stone B.A., Solution properties of wheat-flour arabinoxylans and enzymically modified arabinoxylans, *Carbohyd. Res.*, **77**, 1979, 191–204.
6. Fincher G.B., Stone B.A., Cell walls and their components in cereal grain technology, in *Advances in Cereal Science and Technology*, Ed.: Pomeranz Y., American Association of Cereal Chemists, St. Paul, MN, 1986, **Vol. 8**, pp. 207–295.
7. Vinkx C.J.A., Delcour A., Rye (*Secale cereale* L.) arabinoxylans: A critical review, *J. Cereal Sci.*, **24**, 1996, 1–14.
8. Ferre H., Broberg A., Duus J.O., Thomsen K.K., A novel type of arabinoxylan arabinofuranosylhydrolase from germinated barley, *Eur. J. Biochem.*, **267**(22), 2000, 6633–6641.
9. Bauer W.D., Talmadge K.W., Keegstra K., Albersheim P., The structure of plant cell walls: II. The hemicellulose of the walls of suspension-cultured sycamore cells, *Plant Physiol.*, **51**(1), 1973, 174–187.
10. Kennedy J.F., Paterson M., Bioconversion of wood and cellulosic materials, in *Cellulosic Utilization [Proceedings of International Conference on cellulosics utilization in the new future]*, Eds.: Inagaki H. and Phillips G.O., Elsevier Applied Science, 1988, pp. 203–212.
11. Popa V.I., Enzymatic hydrolysis of hemicelluloses and cellulose, in *Polysaccharides. Structural Diversity and Functional Versatility*, Ed.: Dumitriu S., Marcel Dekker, 1998, pp. 969–1007.
12. Sjöström E., *Wood Chemistry. Fundamentals and Applications*, 2nd Edition, Academic Press, San Diego, CA, USA, 1993. p. 293.
13. Shimizu K., Chemistry of hemicelluloses, in *Wood and Cellulosic Chemistry,* Eds.: Hon D.N.-S., Shiraishi N., 2nd Edition, Marcel Dekker Inc., New York, USA, 2001, pp. 177–214.
14. Dea I.C.M., Morrison A., Chemistry and interactions of seed galactomannans, *Adv. Carbohyd. Chem. Biochem.*, **31**, 1975, 241–312.
15. Dey P.M., Biochemistry of plant galactomannans, *Adv. Carbohyd. Chem. Biochem.*, **35**, 1978, 341–376.
16. Chandrasekaran R., Janaswamy S., Morphology of western larch arabinogalactan, *Carbohyd. Res.*, **337**(21), 2002, 2211–2222.
17. Aspinall G.O., Chemistry of cell wall polysaccharides, in *The Biochemistry of Plants*, Ed.: Preiss J., Academic Press, London, 1980, **Vol. 3**, pp. 473–500.
18. Wilkie K.C.B., The hemicelluloses of grasses and cereals, *Adv. Carbohyd. Chem. Biochem.*, **36**, 1979, 215–264.
19. Dervilly-Pinel G., Tran V., Saulnier L., Investigation of the distribution of arabinose residues on the xylan backbone of water-soluble arabinoxylans from wheat flour, *Carbohyd. Polym.*, **55**(3), 2004, 171–177.
20. Smith M.M., Hartley R.D., Occurrence and nature of ferulic acid substitution of cell wall polysaccharides in graminaceous plants, *Carbohyd. Res.*, **118**, 1983, 65–80.
21. Sorensen H.R., Meyer A.S., Pedersen S., Enzymatic hydrolysis of water-soluble wheat arabinoxylan. 1. Synergy between alpha-L-arabinofuranosidases, endo-1, 4-beta-xylanases, and beta-xylosidase activities, *Biotechnol. Bioeng.*, **81**(6), 2003, 726–731.
22. Tuncer M., Ball A.S., Co-operative actions and degradation analysis of purified xylan-degrading enzymes from *Thermomonospora fusca* BD25 on oat-spelt xylan, *J. Appl. Microbiol.*, **94**(6), 2003, 1030–1036.
23. Chaplin M.F., The structure of *Plantago ovata* arabinoxylan, in *Gums and stabilisers for the food industry*, Eds.: Williams P.A., Phillips G.O., Royal Society of Chemistry, 2004, pp. 509–516.
24. Jarvis M.C., Briggs S.P.H., Knox J.P., Intercellular adhesion and cell separation in plant, *Plant Cell Environ.*, **26**(7), 2003, 977–989.
25. Brent L.R., Molcolm A.O., Debra M., Pectins: Structure, biosynthesis and oligosaccharide-related signaling, *Phytochemistry*, **57**, 2001, 929–967.
26. Vincker J.-P., Schols H.A., Oomen R.J.F.J., Beldman G., Visser R.G.F., Voragen A.G.J., Pectin – The hairy thing, in *Advances in Pectin and Pectinase Research,* Eds.: Voragen A.G.J., Schols H. and Visser R., Kluwer Academic Publishers, 2003, pp. 47–59.
27. Aspinall G.O., Baillie J., Gum tragacanth. I., Fractionation of the gum and structure of tragacanthic acid, *J. Chem. Soc.*, **318**, 1963, 1702–1714.
28. Grant G.T., Morris E.R., Rees D.A., Smith P.C.J., Thom D., Biological interaction between polysaccharides and divalent cations: The egg-box model, *FEBS Lett.*, **32**, 1963, 195–198.

29. Takano R., Shimoto K., Kamei K., Hara S., Hirase S., Occurrence of carrageenan structure in an agar from the red seaweed *Digenea simplex* (Wulfen) *C. Agardh* (Rhodomelaceae, Ceramiales) with a short review of carrageenan–agarocolloid hybrid in the *Florideophycidae*, *Bot. Mar.*, **46**(2), 2003, 142–150.
30. Tako M., Higa M., Medoruma K., Nakasone Y., A highly methylated agar from red seaweed, *Gracilaria arcuata*, *Bot. Mar.*, **42**(6), 1999, 513–517.
31. Mitsuiki M., Mizuno A., Motoki M., Determination of molecular weight of agars and effect of the molecular weight on the glass transition, *J. Agric. Food Chem.*, **47**(2), 1999, 473–478.
32. Lahaye M., Rochas C., Chemical structure and physico-chemical properties of agar, *Hydrobiologia*, **221**(1), 1991, 137–148.
33. Norton I.T., Goodall D.M., Austin K.R.J., Morris E.R., Rees D.A., Dynamics of molecular organization in agarose sulphate, *Biopolymers*, **25**(6), 1986, 1009–1029.
34. Arnott S., Fulmer A., Scott W.E., Dea I.C.M., Moorhouse R., Rees D.A., The agarose double helix and its function in agarose gel structure, *J. Mol. Biol.*, **90**(2), 1974, 269–284.
35. Haug A., Larsen B., Smidsrod O., Studies on the sequence of uronic acid residues in alginic acid, *Acta Chem. Scand.*, **21**, 1967, 691–704.
36. Wang, Z.-Y., Zhang Q.-Z., Konno M., Saito S., Sol–gel transition of alginate solution by addition of calcium ions: Alginate concentration dependence of gel point, *J. Phys. II France*, **3**, 1993, 1–7.
37. Kim Y.T., Kim E.H., Cheong C., Williams D.L., Kim C.W., Lim S.T., Structural characterization of β-(1 → 3, 1 → 6)-linked glucans using NMR spectroscopy, *Carbohyd. Res.*, **328**(3), 2000, 331–341.
38. Harada T., Amemura A., Saito H., Kanamaru S., Misaki A., Formation of succinoglucan and curdlan by parent and mutant strains of *Alcaligenes faecalis* var. *myxogenes* 10C3, *J. Ferment. Technol.*, **46**, 1968, 679–684.
39. Saudagar P.S., Singhal R.S., Fermentative production of curdlan, *Appl. Biochem. Biotechnol.*, **118**(1–3), 2004, 21–31.
40. McIntosh M., Stone B.A., Stanisich V.A., Curdlan and other bacterial (1 → 3)-β-D-glucans, *Appl. Microbiol. Biotechnol.*, **68**(2), 2005, 163–173.
41. Chan T.W., Tang K.Y., Analysis of a bioactive β-(1 → 3) polysaccharide (curdlan) using matrix-assisted laser-desorption/ionization time-of-flight mass spectrometry, *Rapid Commun. Mass Spectrom.*, **17**(9), 2003, 887–896.
42. Okuyama K., Otsubo A., Fukuzawa Y., Ozawa M., Harada T., Kasai N., Single-helical structure of native curdlan and its aggregation state, *J. Carbohyd. Chem.*, **10**(4), 1991, 645–656.
43. Zhang H., Nishinari K., Williams M.A., Foster T.J., Norton I.T., A molecular description of the gelation mechanism of curdlan, *Int. J. Biol. Macromol.*, **30**(1), 2002, 7–16.
44. Ikeda S., Shishido Y., Atomic force microscopy studies on heat-induced gelation of curdlan, *J. Agric. Food Chem.*, **53**(3), 2005, 786–791.
45. Zhang H., Huang L., Nishinari K., Watase M., Konno A., Thermal measurements of curdlan in aqueous suspension during gelation, *Food Hydrocolloid.*, **14**(2), 2000, 121–124.
46. Palleschi A., Bocchinfuso G., Coviello T., Alhaique F., Molecular dynamics investigations of the polysaccharide scleroglucan: First study on the triple helix structure, *Carbohyd. Res.*, **340**(13), 2005, 2154–2162.
47. Deslandes Y., Marchessault R.H., Sarko A., Triple-helical structure of (1,3)-β-D-glucan, *Macromolecules*, **13**, 1980, 1466–1471.
48. Blum T.L., Deslandes Y., Marchessault R.H., Perez S., Rinaudo M., Solid-state and solution conformations of scleroglucan, *Carbohyd. Res.*, **100**, 1982, 117–130.
49. Yanaki T., Norisuye T., Triple helix and random coil scleroglucan in dilute solution, *Polymer J.*, **15**(5), 1983, 389–396.
50. Leathers T.D., Biotechnological production and applications of pullulan, *Appl. Microbiol. Biotechnol.*, **62**(5–6), 2003, 468–473.
51. Auer D.P.F., Seviour R.J., Influence of varying nitrogen sources on polysaccharide production by *Aureobasidium pullulans* in batch culture, *Appl. Microbiol. Biotechnol.*, **32**, 1990, 637–644.
52. Badr-Eldin S.M., El-Tayeb O.M., El-Masry E.G., Mohamad O.A., El-Rahman O.A.A., Polysaccharide production by *Aureobasidium pullulans*: Factors affecting polysaccharide formation, *World J. Microbiol. Biotechnol.*, **10**(4), 1994, 423–426.
53. Wiley B.J., Ball D.H., Arcidiacono S.M., Mayer J.M., Kaplan D.L., Control of molecular weight distribution of the biopolymer pullulan produced by *Aureobasidium pullulans*, *J. Environ. Polym. Degrad.*, **1**, 1993, 3–9.
54. Roukas T., Biliaderis C.G., Evaluation of carob pod as a substrate for pullulan production by *Aureobasidium pullulans*, *Appl. Biochem. Biotechnol.*, **55**(1), 1995, 27–44.
55. Kato I., Okamoto T., Tokuya T., Takahashi A., Solution properties and chain flexibility of pullulan in aqueous solution, *Biopolymers*, **21**, 1982, 1623–1633.
56. Alasdair F.B., Lutz H., Laurence D.B., Polysaccharides vibrational Raman optical activity: Laminarin and pullulan, *J. Raman Spectrosc.*, **26**(12), 1995, 1071–1074.
57. Lee J.W., Yeomans W.G., Allen A.L., Deng F., Gross R.A., Kaplan D.L., Biosynthesis of novel exopolymers by *Aureobasidium pullulans*, *Appl. Environ. Microbiol.*, **65**(12), 1999, 5265–5271.
58. Behravan J., Bazzaz B.S., Salimi Z., Optimization of dextran production by *Leuconostoc mesenteroides* NRRL B-512 using cheap and local sources of carbohydrates and nitrogen, *Biotechnol. Appl. Biochem.*, **38**(3), 2003, 267–269.

59. Chicoine L.M., Suppiramaniam V., Vaithianathan T., Gianutsos G., Bahr B.A., Sulfate- and size-dependent polysaccharide modulation of AMPA receptor properties, *J. Neurosci. Res.*, **75**(3), 2004, 408–416.
60. Antonini E., Bellelli L., Bruzzesi M.R., Caputo A., Chiancone E., Rossi-Fanelli A., Studies on dextran and dextran derivatives. I. Properties of native dextran in different solvents, *Biopolymers*, **2**(1), 2004, 27–34.
61. Jansson P.E., Kenne L., Lindberg B., Structure of the extracellular polysccharide from *Xanthomonas campestris*, *Carbohyd. Res.*, **45**, 1975, 275–282.
62. Papagianni M., Psomas S.K., Batsilas L., Paras S.V., Kyriakidis D.A., Liakopoulou-Kyriakides M., Xanthan production by *Xanthomonas campestris* in batch cultures, *Process Biochem.*, **37**(1), 2001, 73–80.
63. Kennedy J.F., Jones P., Barker S.A., Factors affecting microbial growth and polysaccharide production during the fermentation of *Xanthomonas campestris* culture, *Enz. Microb. Technol.*, **4**, 1982, 39–43.
64. Becker A., Katzen F., Pühler A., Ielpi L., Xanthan biosynthesis and function: A biochemical-genetic perspective, *Appl. Microbiol. Biotechnol.*, **50**(2), 1998, 145–152.
65. Ielpi L., Couso R.O., Dankert M.A., Sequential assembly and polymerization of the polyprenol-linked pentasaccharide repeating unit of the xanthan polysaccharide in *Xanthomonas campestris*, *J. Bacteriol.*, **175**(9), 1993, 2490–2500.
66. Mila M., Rinaudo M., Tinland B., de Murcia G., Evidence for a single stranded xanthan by electron microscopy, *Polym. Bull*, **19**, 1988, 567–572.
67. Moorhouse R., Colegrove G.T., Sandford P.A., Baird J.K., Kang K.S., PS-60: A new gel-forming polysaccharide, *Am. Chem. Soc. Symp. Ser.*, **150**, 1981, 111–124.
68. Rinaudo M., Milas M., Xanthan properties in aqueous solution, *Carbohyd. Polym.*, **2**(4), 1982, 264–269.
69. Sanderson G.R., Gellan gum, in *Food Gels,* Ed.: Harries P., Elsevier Science Publishing, New York, 1990, pp. 201–233.
70. Kuo M.S., Dell A., Mort A.J., Identification and location of L-glycerate, unusual acyl substitution in gellan gum, *Carbohyd. Res.*, **156**, 1986, 173–187.
71. Fialho A.M., Martins L.O., Donval M.L., Leitão J.H., Ridout M.J., Jay A.J., Morris V.J., Correia I.Sá., Structures and properties of gellan polymers produced by *Sphingomonas paucimobilis* ATCC 31461 from lactose compared with those produced from glucose and from cheese whey, *Appl. Environ. Microbiol.*, **65**(6), 1999, 2485–2491.
72. Chandrasekaran R., Radha A., Thailambal V.G., Roles of potassium ions, acetyl and L-glyceryl groups in native gellan double helix: An X-ray study, *Carbohyd. Res.*, **224**, 1992, 1–17.
73. Groman E.V., Enriquez P.M., Jung C., Josephson L., Arabinogalactan for hepatic drug delivery, *Bioconjugate Chem.*, **5**(6), 1994, 547–556.
74. Gregory S., Kelly N.D., Larch arabinogalactan. Clinical relevance of a novel immune-enhancing polysaccharide, *Altern. Med. Rev.*, **4**(2), 1999, 96–103.
75. Lu Z.X., Walker K.Z., Muir J.G., O'Dea K., Arabinoxylan fibre improves metabolic control in people with Type II diabetes, *Eur. J. Clin. Nutr.*, **58**(4), 2004, 621–628.
76. Govers M.J., Gannon M.J., Dunshea F.R., Gibson P.R., Muir J.G., Wheat bran affects the site of fermentation of resistant starch and luminal indexes related to colon cancer risk: A study in pigs, *Gut*, **45**(6), 1999, 840–847.
77. Grodhal M., Eriksson L., Gatenholm P., Material properties of plasticized hardwood xylans for potential application as oxygen barrier films, *Biomacromolecules*, **5**(4), 2004, 1528–1535.
78. Fukunaga T., Sasaki M., Araki Y., Okamoto T., Yasuoka T., Tsujikawa T., Fujiyama Y., Bamba T., Effects of the soluble fibre pectin on intestinal cell proliferation, fecal short chain fatty acid production and microbial population, *Digestion*, **67**(1–2), 2003, 42–49.
79. Yamaguchi F., Uchida S., Watabe S., Kojima H., Shimizu N., Hatanaka C., Relationship between molecular weights of pectin and hypocholesterolemic effects in rats, *Biosci. Biotechnol. Biochem.*, **59**(11), 1995, 2130–2131.
80. Judd P.A., Truswell A.S., The hypocholesterolemic effects of pectins in rats, *Br. J. Nutr.*, **53**(3), 1985, 409–425.
81. Anderson J.W., Hanna T.J., Impact of nondigestible carbohydrates on serum lipoproteins and risk for cardiovascular disease, *J. Nutr.*, **129**(7), 1999, 1457–1466.
82. Chiellini F., Perspectives on *in vitro* evaluation of biomedical polymers, *J. Bioact. Compat. Pol.*, **21**(3), 2006, 257–271.
83. Paper D.H., Vogl H., Franz G., Hoffman R., Defined carrageenan derivatives as angiogenesis inhibitors, *Macromol. Symp.*, **99**, 1995, 219–225.
84. Lai L.S., Lin P.H., Application of decolourised Hsian-tsao leaf gum to low-fat salad dressing model emulsions: A rheological study, *J. Sci. Food Agric.*, **84**(11), 2004, 1307–1314.
85. Jackson G., Roberts R.T., Wainwright T., Mechanism of beer foam stabilization by propylene glycol alginate, *J. Inst. Brew.*, **86**, 1980, 34–37.
86. O'Reilly J., The role of enhanced solubility PGA in beer head retention, *Brew. Guard.*, **125**(7), 1996, 22–24.
87. Orive G., Tam S.K., Pedraz J.L., Halle J.P., Biocompatibility of alginate-poly-L-lysine microcapsules for cell therapy, *Biomaterials*, **27**(20), 2006, 3691–3700.
88. Dufrane D., Steenberghe M., Goebbels R.M., Saliez A., Guiot Y., Gianello P., The influence of implantation site on the biocompatibility and survival of alginate encapsulated pig islets in rats, *Biomaterials*, **27**(17), 2006, 3201–3208.

89. Liew C.V., Chan L.W., Ching A.L., Heng P.W., Evaluation of sodium alginate as drug release modifier in matrix tablets, *Int. J. Pharm.*, **309**(1–2), 2006, 25–37.
90. Thanos C.G., Bintz B.E., Bell W.J., Qian H., Schneider P.A., MacArthur D.H., Emerich D.F., Intraperitoneal stability of alginate-polyornithine microcapsules in rats: An FTIR and SEM analysis, *Biomaterials*, **27**(19), 2006, 3570–3579.
91. Sankalia M.G., Mashru R.C., Sankalia J.M., Sutariya V.B., Papain entrapment in alginate beads for stability improvements and site-specifica delivery: Physicochemical chracaterisation and factorial optimization using neural network modeling, *AAPS Pharm. Sci. Tech.*, **6**(2), 2005, 209–222.
92. Nagakura T., Hirata H., Tsujii M., Sugimoto T., Miyamoto K., Horiuchi T., Nagao M., Nakashima T., Uchida A., Effect of viscous injectable pure alginate sol on cultured fibroblasts, *Plast. Reconstr. Surg.*, **116**(3), 2005, 831–838.
93. Rowley J.A., Madlambayan G., Mooney D.J., Alginate hydrogels as synthetic extracellular materials, *Biomaterials*, **20**(1), 1999, 45–53.
94. Suzuki Y., Tanihara M., Suzuki K., Saitou A., Sufan W., Nishimura Y., Alginate, hydrogel linked with synthetic oligopeptide derived from BMP-2 allows ectopic osteoinduction *in vivo*, *J. Biomed. Mater. Res.*, **50**(3), 2000, 405–409.
95. Atala A., Kim W., Paige K.T., Vacanti C.A., Retik A.B., Endoscopic treatment of vesicoureteral reflux with a chondrocyte-alginate suspension, *J. Urol.*, **152**(2), 1994, 641–643.
96. Elcin Y.M., Dixit V., Gitnick T., Extensive *in vivo* angiogenesis following controlled release of human vascular endothelial cell growth factor: Implications for tissue engineering and wound healing, *Artif. Organs.*, **25**(7), 2001, 558–565.
97. Soon-Shiong P., Otterlie M., Skjak-Bræek S., Smidsrød O., Heintz R., Lanza R.P., Espevik T., An immunologic basis for the fibrotic reaction to implanted micro-capsules, *Transplant. Proc.*, **23**(1), 1991, 758–759.
98. Suzuki K., Suzuki Y., Tanihara M., Ohnishi K., Hashimoto T., Endo K., Nishimura Y.Y., Reconstruction of rat peripheral nerve gap without sutures using freeze-dried alginate gel, *J. Biomed. Mater. Res.*, **49**(4), 1999, 528–533.
99. Edwards P.C., Mason G.M., Gene-enhanced tissue engineering for dental hard tissue regeneration: (1) Overview and practical considerations, *Head Face Med.*, **2**, 2006, 12–16.
100. Zülli F., Suter F., Biltz H., Nissen H.P., Birman M., Carboxymethylated β(1,3)-glucan, a beta glucan from baker's yeast helps protect skin, *Cosmet. Toiletries*, **111**(12), 1996, 91–98.
101. Ohno N., Furukawa M., Miura N.N., Adachi Y., Motol M., Yadoma T., Antitumor β-glucan from the cultured fruit body of *Agaricus blazei*, *Biol. Pharm. Bull.*, **24**(7), 2001, 820–828.
102. Camelini C.M., Maraschin M., de Mendoca M.M., Zucco C., Ferreira A.G., Tavares L.A., Structural characterization of beta-glucans of *Agaricus brasiliensis* in different stages of fruiting body maturity and their use in nutraceutical products, *Biotechnol. Lett.*, **27**(17), 2005, 1295–1299.
103. Kaneno R., Fontanari L.M., Santos S.A., Di Stasi L.C., Rodrigues Filho E., Eira A.F., Effects of extracts from Brazilian sun-mushroom (*Agaricus blazei*) on the NK activity and lymphoproliferative responsiveness of Ehrlich tumor-bearing mice, *Food Chem. Toxicol.*, **42**(6), 2004, 909–916.
104. Di Luzio N.R., Immunopharmacology of glucan: A broad spectrum enhancer of host defense mechanisms, *Trends Pharmacol. Sci.*, **4**(8), 1983, 344–348.
105. Sakurai T., Ohno N., Yadomae T., Changes in immune mediators in mouse lung produced by administration of soluble (1,3)-beta-D-glucan, *Biol. Pharm. Bull.*, **17**(5), 1994, 617–622.
106. Ha C.H., Lim K.H., Jang S.H., Yun C.W., Paik H.D., Kim S.W., Kang C.W., Chang H.I., Immune-enhancing alkali-soluble glucans produced by wild-type and mutant *Saccharomyces cerevisiae*, *J. Microbiol. Biotechnol.*, **16**(4), 2006, 576–583.
107. Nakao Y., Konno A., Taguchi T., Tawada T., Kasai H., Toda J., Terasaki M., Curdlan: Properties and applications to foods, *J. Food Sci.*, **56**, 1991, 769–776.
108. Nakao Y., Suzuki K., Curdlan: Properties and applications to foods, *Int. Food Ingredients*, **5**, 1994.
109. Funami T., Yada H., Nakao Y., Curdlan properties for application in fat mimetics for meat products, *J. Food Sci.*, **63**(22), 1998, 283–287.
110. Sasaki T., Tanaka M., Uchida H., Effect of serum from mice treated with antitumor polysaccharides on expression of cytotoxicity by mouse peritoneal macrophages, *J. Pharmacobio-dynam.*, **5**(12), 1982, 1012–1016.
111. Morikawa K., Takeda S., Yamazaki M., Mizuno D., Induction of tumoricid activity of polymorphonuclear leukocytes by a linear β-1,3-D-glucan and other immunomodulators in murine cells, *Cancer Res.*, **45**(4), 1985, 1496–1501.
112. Aoki T., Kaneko Y., Stefanski M.S., Nguyen T., Ting R.C., Curdlan sulfate and HIV-1: *In vitro* inhibitory effects of curdlan sulfate on HIV-1 infection, *AIDS Res. Hum. Retrov.*, **7**(4), 1991, 409–415.
113. Osawa Z., Morota T., Hatanaka K., Akaike T., Matsuzaki K., Nakashima H., Yamamoto N., Suzuki E., Miyano H., Mimura T., Kaneko Y., Synthesis of sulfated derivatives of curdlan and their antiHIV activity, *Carbohyd. Polym.*, **21**, 1993, 283–288.
114. Miao H.Q., Ishai-Michaeli R., Peretz T., Vlodavsky I., Laminarin sulfate mimics the effects of heparin on smooth muscle cell proliferation and basic fibroblast growth factor-receptor binding and mitogenic activity, *J. Cell Physiol.*, **164**(3), 1995, 482–490.
115. Pang Z., Otaka K., Maoka T., Hidaka K., Ishijima S., Oda M., Ohnishi M., Structure of beta-glucan oligomer from laminarin and its effect on human monocytes to inhibit the proliferation of receptor binding in the CR3- human promonocytic cell line U937, *Biosci. Biotechnol. Biochem.*, **69**(3), 2005, 553–558.

116. Cho C.C.M., Liu E.S.L., Shin V.Y., Cho C.H., Polysaccharides: A new role in gastrointestinal protection, in *Gastrointestinal Mucosal Repair and Experimental Therapeutics*. Eds.: Cho C.-H., Wang J.-Y., Front. Gastrointest. Res., Basel, Karger, 2002, **Vol. 25**, p. 180–189.
117. Aziz A., Poinssot B., Daire X., Adrian M., Bezier A., Lambert B., Joubert J.M., Pugin A., Laminarin elicits defense responses in grapevine and induces protection against *Botrytis cinerea and Plasmopara viticola, Mol. Plant Microbe. In.*, **16**(12), 2003, 1118–1128.
118. Hoffman R., Paper D.H., Donaldson J., Vogl H., Inhibition of angiogenesis and murine tumour growth by laminarin sulphate, *Br. J. Cancer*, **73**(10), 1996, 1183–1185.
119. Coviello T., Palleschi A., Grassi M., Matricardi P., Bocchinfuso G., Alhaique F., Scleroglucan: A versatile polysaccharide for modified drug delivery, *Molecules*, **10**(1), 2005, 6–33.
120. Bousquet M., Escoula L., Peuriere S., Pipy B., Roubinet F., Chavant C., Immunopharmacologic study in mice of 2 β-1,3, beta-1, 6 polysaccharides (scleroglucan and PSAT) on the activation of macrophages and T lymphocytes, *Ann. Rech. Vet.*, **20**(2), 1989, 165–173.
121. Pretus H.A., Ensley H.E., McNamee R.B., Jones E.L., Browder I.W., Williams D.L., Isolation, physicochemical characterization and preclinical efficiency evaluation of soluble scleroglucan, *J. Pharmacol. Exp. Ther.*, **257**(1), 1991, 500–510.
122. Matsuyama H., Mangindaan R.E.P., Yano T., Protective effect of schizophyllan and scleroglucan against *Streptococcus* sp. infection in yellowtail (*Seriola quinqueradiata*), *Aquaculture*, **101**, 1992, 197–203.
123. Sallustio S., Galantini L., Gente G., Masci G., La Mesa C., Hydrophobically modified pullulans: Characterization and physicochemical properties, *J. Phys. Chem. B*, **108**(49), 2004, 18876–18883.
124. Reza M., Recent trends in the use of polysaccharides for improved delivery of therapeutic agents: Pharmacokinetic and pharmacodynamic perspectives, *Curr. Pharm. Biotechnol.*, **4**(5), 2003, 283–302.
125. Alban S., Schauerte A., Franz G., Anticoagulant sulfated polysaccharides: Part I. Synthesis and structure–activity relationships of new pullulan sulfates, *Carbohyd. Polym.*, **47**, 2002, 267–276.
126. Kandemir N., Yemenicioglu A., Mecitoglu Ç., Elmaci Z.S., Arslanogwlu A., Göksungur Y., Baysal T., Production of antimicrobial films by incorporation of partially purified lysozyme into biodegradable films of crude exopolysaccharides obtained from *Aureobasidium pullulans* fermentation, *Food Technol. Biotechnol.*, **43**(4), 2005, 343–350.
127. Uchida T., Ikegami H., Ando S., Kurimoto M., Mitsuhashi M., Naito S., Usui M., Matuhasi T., Suppression of murine IgE response with ovalbumin–pullulan conjugates: Comparison of the suppressive effect of different conjugation methods and different molecular weights of pullulan, *Int. Arch. Allergy Immunol.*, **102**(3), 1993, 276–278.
128. Ferreira L., Gil M.H., Cabrita A.M., Dordich J.S., Biocatalytic synthesis of highly ordered degradable dextran-based hydrogels, *Biomaterials*, **26**(23), 2005, 4707–4716.
129. Cascone M.G., Polacco G., Lazzeri L., Barbani N., Dextran/poly(acrylic acid) mixtures as miscible blends, *J. Appl. Polym. Sci.*, **66**(11), 1997, 2089–2094.
130. Ciardelli G., Chiono V., Vozzi G., Pracella M., Ahluwalia A., Barbani N., Cristallini C., Giusti P., Blends of poly-ε-(caprolactone) and polysaccharides in tissue engineering applications, *Biomacromolecules*, **6**(4), 2005, 1961–1976.
131. Lundblad R., Bradshaw R.A., Applications of site-specific chemical modification in the manufacture of biopharmaceuticals: I. An overview, *Biotechnol. Appl. Biochem.*, **26**(3), 1997, 143–151.
132. Kaplan J.M., Pennington S.E., George J.A., Woodworth L.A., Fasbender A., Marshall J., Cheng S.H., Wadsworth S.C., Gregory R.J., Smith A.E., Potentiation of gene transfer to the mouse lung by complexes of adenovirus vector and polycations improves therapeutic potential, *Hum. Gene Ther.*, **9**(10), 1998, 1469–1479.
133. Liptay S., Weidenbach H., Adler G., Schmid R.M., Colon epithelium can be transiently transfected with liposomes, calcium phosphate precipitation and DEAE dextran *in vivo*, *Digestion*, **59**(2), 1998, 142–147.
134. Flexner C., Barditch-Crovo P.A., Kornhauser D.M., Farzadegan H., Nerhood L.J., Chaisson R.E., Bell B.E., Lorentsen M.J., Hendrix C.W., Petty B.G., Pharmacokinetics, toxicity, and activity of intravenous dextran sulfate in human immunodeficiency virus infection, *Antimicrob. Agents Chemother.*, **35**(12), 1991, 2544–2550.
135. Dyer A.P., Banfield B.W., Martindale B., Spannier D.M., Tufaro E., Dextran sulfate can act as an artificial receptor to mediate a type-specific herpes simplex virus infection via glycoprotein B, *J. Virol.*, **71**(1), 1997, 191–198.
136. Krumbiegel M., Dimitrov D.S., Puri A., Blumenthal R., Dextran sulfate inhibits fusion of *influenza* virus and cells expressing influenza hemagglutinin with red blood cells, *Biochim. Biophys. Acta*, **1110**(2), 1992, 158–164.
137. Ruissen L.A., Vander Reijden W.A., van't Hof W., Veerman E.C.I., Nieuw Amerongen A.V., Evaluation of the use of xanthan as vehicle for cationic antifungal peptides, *J. Control Release*, **60**(1), 1999, 49–56.
138. Brookshier K.A., Tarbell J.M., Evaluation of a transparent blood analog fluid: Aqueous xanthan gum/glycerin, *Biorheology*, **30**(2), 1993, 107–116.
139. Yoo S-H., Kyung H.L., Lee J.S., Cha J., Cheon S.P., Hyeon G.L., Physicochemical properties and biological activities of DEAE-derivatized *Sphingomonas* gellan, *J. Agric. Food Chem.*, **53**(16), 2005, 6235–6239.
140. Li J., Kamath K., Dwivedi C., Gellan film as an implant for insulin delivery, *J. Biomater. Appl.*, **15**(4), 2001, 321–343.

– 14 –

Cork and Suberins: Major Sources, Properties and Applications

Armando J.D. Silvestre, Carlos Pascoal Neto and Alessandro Gandini

ABSTRACT

The main purpose of this chapter is to discuss the potential of suberin (an aromatic–aliphatic crosslinked polyester widespread in the plant kingdom) as a renewable source of chemicals and, in particular, of macromonomers. Despite being widespread in plants, only two species produce suberin-rich biomass residues in amounts that justify their exploitation as renewable sources of chemicals and monomers, namely *Quercus suber* with cork and *Betula pendula* (birch) with its outer bark. *Quercus suber* cork is a material with unique properties and applications, whereas to the best of our knowledge, the outer bark of birch finds no direct applications. Hence, this chapter first provides a general overview of the properties and applications of cork, as well as of its utilization as a starting material for the synthesis of liquid polyols, before dealing with the macromolecular structure of suberin (the major cork component), its depolymerization methods, and the composition and applications of the ensuing fragment mixtures.

Keywords

Quercus suber, Betula pendula, Cork, Suberin, Chemical composition, Aliphatic macromonomers, Polyurethanes, Polyesters

14.1 INTRODUCTION

The present chapter aims essentially at presenting the potentials of suberin as a renewable source of chemicals and more specifically of macromonomers. Suberin is a naturally occurring aromatic–aliphatic crosslinked polyester widespread in the plant kingdom, where it plays a key role as a protective barrier between the plant and the surrounding environment. It is found mainly in the cell walls of normal and wounded external tissues of aerial and/or subterranean parts of many plants, mainly in the outer bark of higher plants and in tuber periderms [1].

Despite being widespread in plants [1], only two species produce suberin-rich biomass residues in amounts that justify their exploitation as renewable sources of chemicals for polymer synthesis, namely *Quercus suber* with cork and *Betula pendula* (birch) with its outer bark. *Quercus suber* cork is a material with unique properties and applications, whereas to the best of our knowledge, the outer bark of birch finds no direct applications.

The present chapter will, therefore, first give a general overview of the properties and applications of cork, as well as of its utilization as a starting material for the synthesis of liquid polyols, before dealing with the macromolecular structure of suberin, its depolymerization methods, and the composition and applications of the ensuing fragment mixtures.

14.2 CORK

Cork is the common designation of the suberized bark of cork oak (*Quercus suber* L.), an evergreen tree that grows in the western Mediterranean region [2, 3]. Although other species contain suberized tissues, no other tree in the world gives rise to the thick layers of suberized bark (cork) generated by *Quercus suber*. Cork is therefore a unique suberized tissue and an important non-wood forest product which finds a wide domain of utilizations, such as wine and champaign stoppers, insulating materials, marine floats and household and office indoor panels, some of which date back to Egyptian, Greek and Roman times [3].

Among the countries involved in the cultivation of *Quercus suber* and the production of cork, Portugal is largely the leader with an annual production of 185 000 ton, which constitutes about 50 per cent of the world output [4]. Although *Quercus suber* is a protected species and a key element in the southern Portugal ecosystem and landscape, cork exploitation is a fully sustainable activity, if adequately carried out by professionals. Indeed, cork stripping which is carried out once every 9–12 years seems to have a positive effect on the tree's health and growth.

Cork is composed of suberin, the main component, lignin, polysaccharides and extractives, including mostly aliphatic, phenolic and triterpenenic compounds [2, 5–13]. The relative abundance of these elements is extremely variable, being influenced by the geographical origin and quality of the tree, and/or even the different parts of the tree from which the cork is harvested [5, 11, 12]. Suberin represents 38–62 per cent of the cork weight [14] and is by far the most original macromolecule present. The suberized cells, assembled in cork's peculiar hollow structure (Fig. 14.1), are responsible for the unique properties of this natural material [2, 3, 15–24], such as its remarkable elasticity, low density, impermeability to liquids and gases, low heat and sound conductivity, thermal and combustion resistance and resistance to rotting.

Apart from the major traditional cork applications mentioned above, these unique properties justify the fact that a wide variety of *high-tech* materials have been developed and are being sought in such diverse areas as industrial, household and sport [25] insulation, sealing and anti-vibration and even in the NASA Space Shuttle [26, 27]. Further applications are being developed for commercial jet liners [28]. Another interesting recent development related to a new family of composite materials, in which cork plays a key role, relates to its inclusion in cementitious materials in order to reduce their density [29].

In addition to both traditional and new applications of cork as a whole, whether used in its pristine form or after specific physical or chemical modifications, it is particularly relevant to emphasize that the interest of cork must also lie in the exploitation of some of its components, notably suberin extracted from cork processing by-products. Indeed, industrial cork processing generates substantial amounts of residues, such as the so-called *cork powder*, *black condensates* and *cooking waste waters*. Cork powder is generated during the production of granulated cork for agglomerated materials. Cork residues rejected during cork stopper production, are one of the main sources of granulated cork materials. The *cork powder* particles have an inadequate size distribution for their use in the manufacture of agglomerates and, at present, are mainly burned to produce energy [30, 31]. This by-product amounts, in Portugal, to about 40 000 ton per year (*i.e.* some 22 per cent of the national cork production). *Black condensates* are a residue of the production of black agglomerates, which involves the treatment of cork particles without any adhesive, at temperatures in the range of 250–500°C. During such thermal treatment, vapours are generated and

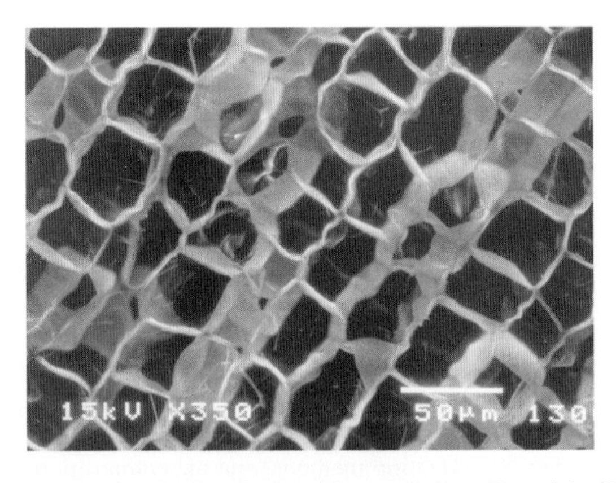

Figure 14.1 Electron microscopy of cork cell. (Reprinted with permission, Copyright 2002 by APCOR.)

later condensed in autoclave pipes. Periodically, this by-product is removed (2500 ton per year in Portugal) and again burned to produce energy [30, 31]. Finally, *cooking waste waters* are the liquid effluents released during the treatment of cork planks with boiling water, a key stage in wine stopper production. The composition of this effluent is still poorly understood, however, it is known to be rich in phenolic compounds and not to contain suberin components; therefore its exploitation is out of the scope of this chapter.

Huge amounts of cork-based by-products are therefore potentially available for more noble uses than burning and hence they must be considered as particularly attractive renewable sources of chemical commodities, which fall neatly within the scope of the biorefinery concept in forest-based industries, which calls upon the upgrading of all the by-products [32].

Suberin is the privileged by-product in the context of this chapter, although its exploitation could well be integrated with the separation and use of high added-value low molecular weight components, such as friedeline, cerine, betulin and betulinic acid [33].

Given the specific purpose of this book, the present chapter will only deal with the macromolecular materials that can be obtained from cork and suberin and disregard all aspects related to the exploitation of other cork components.

14.2.1 The oxypropylation of cork

The oxypropylation of natural polymers has been applied successfully to a host of OH-bearing macromolecules, like cellulose, starch, chitin, chitosan, lignin and more complex agricultural by-products, such as sugar-beet pulp and olive stone powder, as discussed in detail in Chapter 12. In all instances, a nucleophilic catalyst (strong Brønsted bases like KOH work best) is used to deprotonate some of the substrate's OH groups and thus generate oxianions, which initiate the anionic polymerization of propylene oxide (PO), thereby inserting polyPO grafts onto the starting macromolecules. This reaction typically transforms the solid powder of the natural polymeric material into a viscous liquid polyol bearing as many OH groups as the initial substrate, since the oxypropylation is simply a 'chain extension' process. This branching mechanism is always accompanied by some PO homopolymerization, which produces oligomeric diols. Figure 14.2 provides a schematic view of the process, which requires typically temperatures above 130°C and PO pressures above 10 bar.

Cork powder was oxypropylated under these conditions [34,35] and the ensuing polyols fully characterized in terms of structure, homopolymer content, solubility, OH index and viscosity. The latter two parameters proved to be entirely comparable with those of commercial counterparts used in the manufacture of polyurethane materials. A study was therefore conducted [36] on the reactivity of the polyol mixtures, as obtained from the oxypropylation process, towards various diisocyanates and on the structure and properties of the ensuing polyurethanes.

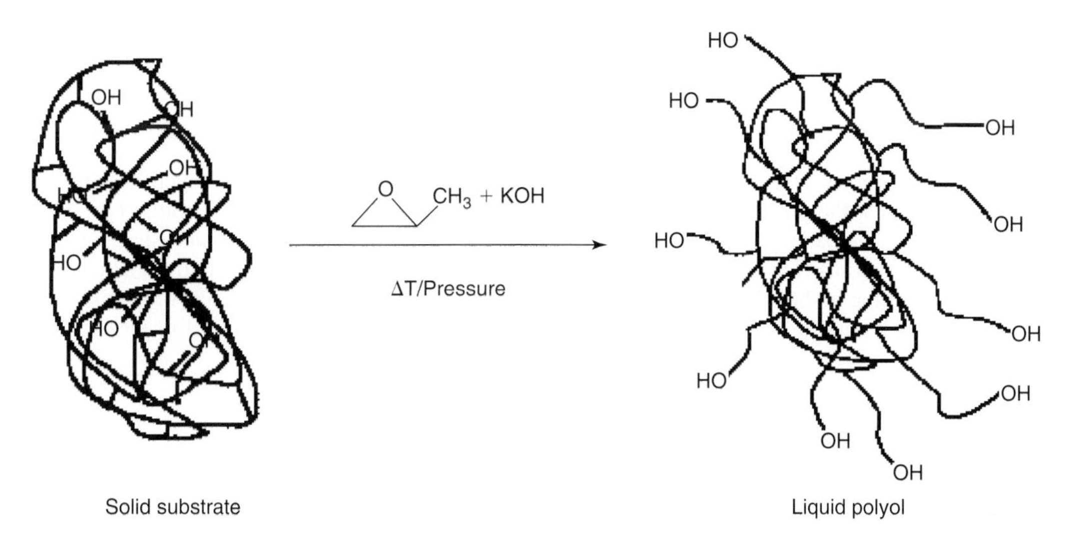

Figure 14.2 Schematic view of the oxypropylation of OH-bearing macromolecular materials.

This ongoing investigation is a good example of the interest in exploiting renewable resources, as emphasized in the broader context of Chapter 1, and the same strategy could equally be applied to other suberin-rich tree barks, such as that of birch.

14.3 SUBERIN

As mentioned above, and despite the fact that suberin is widespread in plants [1], only two species produce suberin-rich biomass residues in amounts that justify the exploitation of this natural polyester as a renewable source of chemicals for polymer synthesis, namely, *Quercus suber* and *Betula pendula. Betula pendula* is one of the most important industrial hardwood species in Northern Europe, where it is mainly used as raw material in the pulp and paper industry, which generates considerable amounts of bark. Typically, a mill with a pulp production of 400000 ton per year, leaves about 28000 ton per year of outer bark [37]. Considering a suberin content ranging from 32 to 59 per cent, [14], birch's outer bark has, like cork, an enormous potential as a source of suberin and suberin components.

Suberin contents ranging from 12 to 60 per cent have been reported for several other higher plant species namely *Acer griseum, Acer pseudoplatanus, Castanea sativa, Cupressus leylandii, Euonymus alatus, Fagus sylvatica, Fraxinus excelsior, Laburnum anagyroides, Populus tremula, Quercus ilex, Quercus robur, Ribes nigrum, Sambucus nigra* [38], *Pseudotuga menziesii* [39]) and for potato (*Solanum tuberosum*) periderm [40]. However, to the best of our knowledge, there is no industrial exploitation of these species in terms of suberin valorization. The only foreseeable exception could be, in the future, the exploitation of potato periderm as a by-product of the agro-food industry.

The interest in suberin lies in the fact that it constitutes an abundant source of ω-hydroxyfatty acids, α,ω-dicarboxylic acids and homologous mid-chain dihydroxy or epoxy derivatives [14]. Apart from this source, these compounds are not very abundant in nature. Thus, hydroxyfatty acids can only be additionally found in exploitable amounts in the seed oils of *Ricinus communis* (castor oil) and *Lesquerella spp.*, as well as in the extracellular aliphatic polyester covering most of the aerial surfaces of plants, known as cutin [41].

14.3.1 Suberin native structure

The suberin polymeric structure cannot be defined in terms of a monomer repeating unit, since the spatial arrangement of these moieties cannot be accurately defined, even when their relative abundance is known. Additionally, the latter aspect depends on the depolymerization methods used to isolate the aliphatic fraction and, moreover, as far as the aromatic domain is concerned, its identification/quantification is also quite complex because of its macromolecular nature and structural similarity with lignin [1, 42–45]. Notwithstanding these difficulties, several models have been proposed to illustrate the suberin macromolecular structure in suberized cell walls [46, 47]. In 2002, Bernards proposed a model for suberin from potato periderm, which summarized the state of the art on the structural data related to this macromolecular component [1].

The aliphatic domain of suberin (Fig. 14.3) is composed of branched polyester moieties mainly made up of long-chain ω-hydroxyfatty acids and α,ω-dicarboxylic acids, brought together through glycerol units [1, 48]. Glycerol has long been detected in suberin depolymerization extracts [43–45], but its role as a key structural building block was only demonstrated in more recent investigations [1, 39, 40, 49–52].

Solid-state NMR studies of cork [7, 53] and potato cell wall components [54–56], together with chemical analysis results, suggest the presence of two distinct aromatic domains in suberized cell walls (Fig. 14.3). The first, embedded in the aliphatic domains, is mainly composed of hydroxycinnamic acid units esterified with glycerol or ω-hydroxyfatty acids (Fig. 14.3). whereas the second is a lignin-like polymer, spatially separated from the aliphatic domains, sitting in the primary cell wall (Fig. 14.3) and composed of crosslinked hydroxycinnamic acid-based units, including amides [7, 57, 58]. The presence of polysaccharide moieties bound to the lignin-type or aliphatic domains, has also been suggested [7, 57, 58].

14.3.2 Suberin depolymerization and monomer composition

14.3.2.1 Suberin depolymerization methods

The depolymerization of suberin through ester cleavage is in general a key step both for composition analysis (usually carried out by GC–MS), and to isolate monomers for further chemical transformation. Depending either

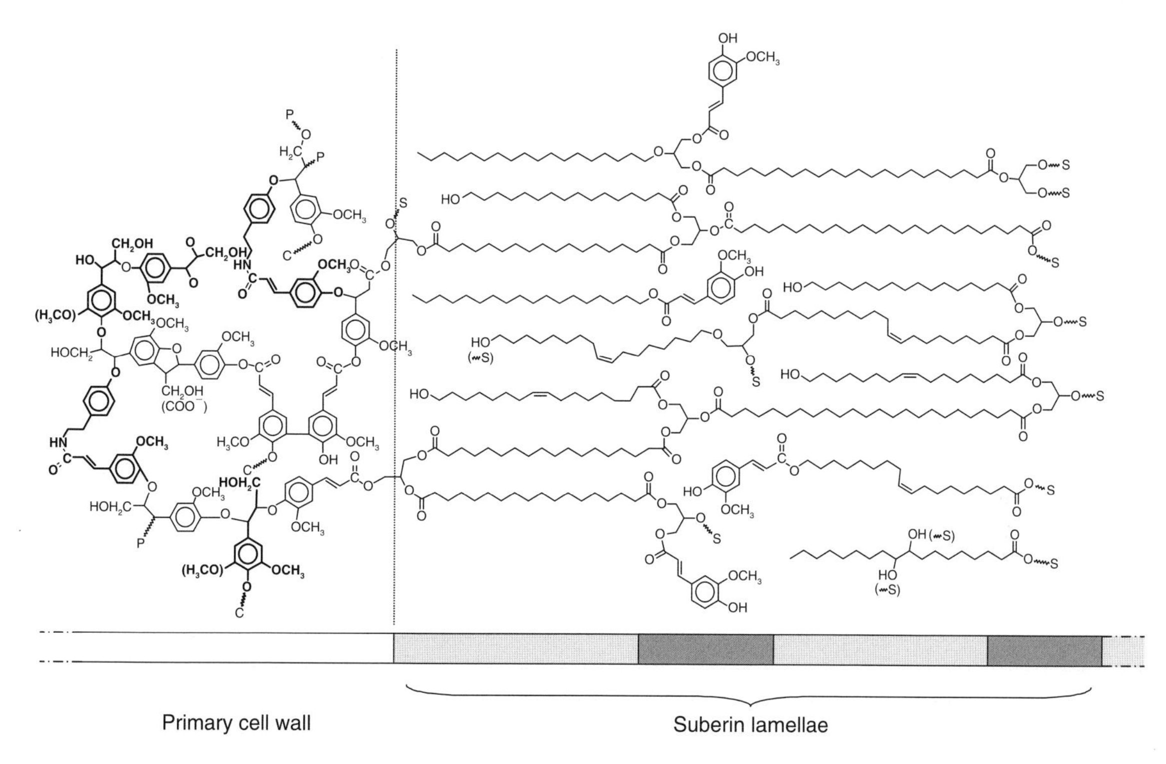

Figure 14.3 Structure proposed for suberin from potato (*Solanum tuberosum*) periderm. C: carbohydrate; P: phenolic; S: suberin. (Reprinted from Reference [1], with permission from NRC Research Press.)

on the analytical methods, or on the types of functionalities envisaged for further chemical modification, different depolymerization approaches have been implemented, involving either alkaline hydrolysis or alcoholysis (mainly methanolysis).

Alkaline methanolysis with anhydrous sodium methoxide is probably the most common depolymerization method [5, 7, 12, 14, 39, 40, 49–51, 59–66], which yields the aliphatic acid components in the form of methyl esters. Complete suberin methanolysis is normally achieved by treating it with refluxing methanol containing 3 per cent of NaOMe for about 3 h [7, 60]. However, under such conditions, epoxy moieties are, at least in part, cleaved, leading to the corresponding methoxyhydrins.

Under alkaline hydrolysis conditions, the ensuing free ω-hydroxyfatty acids and α,ω-dicarboxylic acids are obtained, with extensive cleavage of epoxy functionalities to the corresponding *vic*-diols. It has been claimed, however, that when the reaction is carried out for 15 min using KOH in ethanol:water 9:1 v:v, total depolymerization occurs with preservation of the epoxy functionalities [63], a feature that can be interesting for some polymer applications.

Methanolysis under mild conditions has been studied mainly for structural elucidation purposes [7, 51, 60]. However, such conditions can also be considered for the preferential isolation of specific groups of components, such as the alkanoic and α,ω-dicarboxylic acids generally involved in more labile structures [7,51,60]. Finally, methanolysis in the presence of calcium oxide was shown to yield oligomers [39, 40, 48–50, 59, 67], which, apart from their interest for structural elucidation, might also be considered as macromonomers for other applications. However, these mild methanolysis conditions are marred by an important drawback in terms of their use as sources of monomers/macromonomers for polymer synthesis, because they generally result in quite low yields.

14.3.2.2 Suberin monomer composition

The detailed monomer distribution of suberin from several species has been recently reviewed [14]. Table 14.1 summarizes the composition of the two most relevant suberin sources in the present context (*i.e. Quercus suber* cork and *Betula pendula* outer bark) and the structures of representative elements of each group are shown in Fig. 14.4.

Table 14.1

Relative abundance of aliphatic suberin monomers from the extractive-free *Quercus suber* cork and *Betula pendula* outer bark (adapted from [14])

	Quercus suber	*Betula pendula*
References	[7, 8, 12, 51, 60, 61, 68]	[37, 63]
Aliphatic alcohols	0.4–4.7	–
Fatty acids	**2.5–14.9**	**7.4–12.3**
C(16:0)	0–0.5	–
di(OH)-C(16:0)	0–0.9	–
C(18:1)	0–1.8	–
9,10-di(OH)-C(18:0)	0–6.6	–
9,10-epoxi-C(18:0)	0–2.2	–
C(20:0)	0–0.3	–
di-OH-C(20:0)	0–10.1	–
C(22:0)	0–2.5	–
C(24:0)		–
C(26:0)	0–2	–
ω-Hydroxyfatty acids	**36.0–61.7**	**76.7–79.7**
C(16:0)	0–1.2	
9,16-di-OH(C16:0)	0	3.2–3.7
C(18:0)	0–0.6	
C(18:1)	0–18.2	11.1–12.2
9,10-epoxy-C(18:0)	0–5.5	37.0–39.2
9,10-di(OH)-C(18:0)	0–12.7	8.4–8.6
9,10-(OH,OMe)-C(18:0)	0–7.5	
C(20:0)	0–2.2	2.8
C(20:1)	0–1.2	
C(22:0)	0–28.6	13.6–13.9
C(24:0)	0–4.6	
C(26:0)	0–4.4	
α,ω-Dicarboxylic acids	**6.1–53.3**	**10.4–12.9**
C(16:0)	0–3.1	
C(18:0)	0–0.5	0–0.9
C(18:1)	0–9.1	3.4–4.7
9,10-epoxy-(C18:0)	0–37.8	
9,10-di(OH)-C(18:0)	0–7.7	
9,10-(OH,OMe)-C(18:0)	0–20	
C(20:0)	0–4.9	
C(20:1)	0–0.3	
C(22:0)	0–7.1	6.1–8.2
C(24:0)	0–1.1	
Aromatic compounds	**0.1–7.9**	–
Ferulic acid	0.1–7.9	–

As in the case of total suberin contents in its natural precursors, the monomer composition and abundance are also highly variable, but in general the most abundant families of compounds are ω-hydroxyfatty acids, followed by α,ω-dicarboxylic acids and by smaller amounts of fatty acids, aliphatic alcohols and aromatic compounds.

ω-Hydroxyalkanoic acids are generally the most important group of components, representing between 36.0–61.7 per cent and 76.7–79.7 per cent of suberin monomers from cork and birch bark, respectively. Even C-numbered chains between C16 and C26 are frequently found, and among them, the C22 and C18 are clearly dominant. In the birch outer bark suberin, the 9,10-epoxy C18 derivative is particularly abundant, together with smaller amounts of the 9,10-dihydroxy derivative. C(18:1) and C(22:0) ω-hydroxy fatty acids are also abundant in this suberin. In cork suberin, C(18:1) and C(22:0) are in general the most abundant ω-hydroxyacids, whereas the

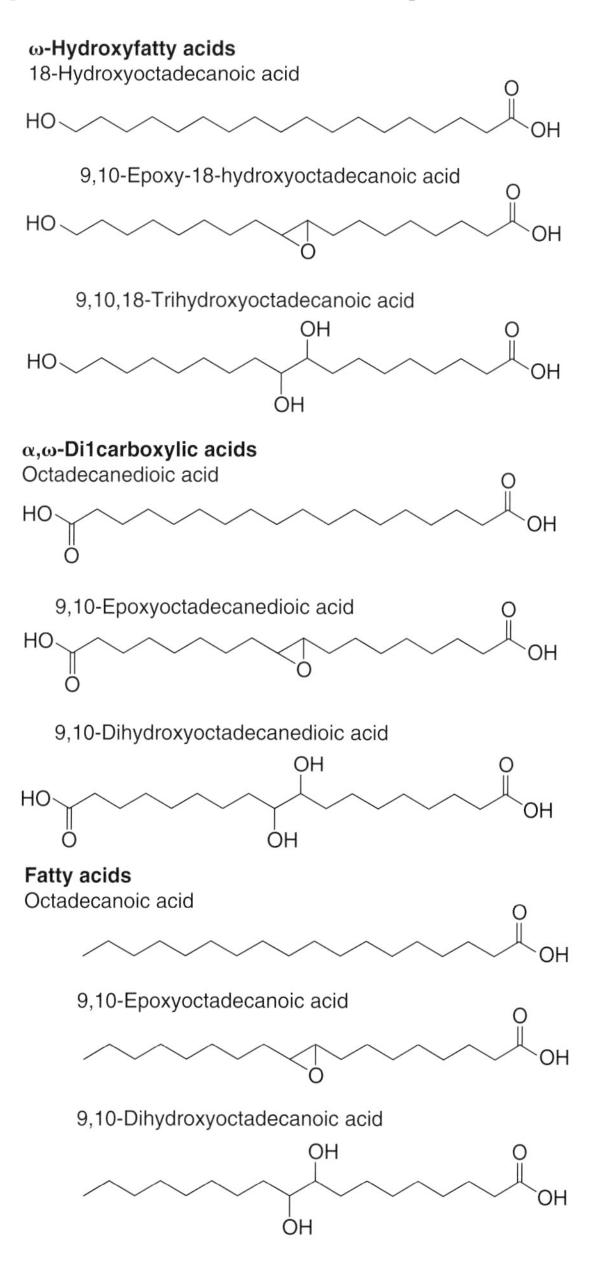

Figure 14.4 Structures of monomeric components of the most representative families of compounds formed in suberin depolymerization.

9,10-oxygenated derivatives are significantly less important than in the birch counterpart. The predominance of 9,10-dihydroxy derivative, sometimes together with the mid-chain 9,10-hydroxymethoxy derivative of the epoxide form, is certainly due to the degradation of the later under the reaction conditions used for suberin depolymerization.

α,ω-Dicarboxylic acids are generally the second most abundant group of components, representing between 6.1–53.3 per cent and 10.4–12.9 per cent of suberin monomers from cork and birch bark, respectively. Once more, the C18 and C22 structures are the prevalent components.

In the case of birch outer bark suberin, only the C(18:1) and C(22:0) homologues were found in considerable amounts, whereas in the case of cork suberin, the 9,10-epoxy and the corresponding 9,10-dihydroxy C18 α,ω-dicarboxylic acids are dominant.

Alkanoic acids represent a smaller fraction of suberin monomers, *viz.* 2.5–14.9 per cent and 7.4–12.3 per cent in cork and birch bark suberin, respectively. In cork, this fraction is mainly composed of saturated even-numbered homologues, ranging from C16 to C26, with the C18 (including the 9,10-oxygenated derivatives) and C22 as the dominant components.

Aliphatic alcohols represent less than 5 per cent of cork suberin and were not reported in the birch bark counterpart. C22 and C24 homologues are the most abundant components of this fraction.

Finally, a small fraction of aromatic compounds was also detected in cork suberin, with ferulic acid as the dominant component.

As a general overview, crucial for tailoring applications of suberin monomers in polymer synthesis, it can be concluded that C18, followed by C22 monomers, are the dominant aliphatic compounds and that most of them are carboxylic acids, bearing at least one aliphatic-OH functionality. Additionally, the 9,10-epoxide functionality is also very common.

Another important issue that should be addressed considering future applications of suberin momomers, is their absolute abundance in the hydrolysis/methanolysis mixtures. In fact, most studies have only looked at their relative abundance, and the total amount of monomers detected, relative to the mass of depolymerized suberin, is seldom provided. However, values between 27 and 74 per cent have been reported for *Quercus suber* cork [7, 8, 51], clearly showing that a non-negligible percentage of suberin is frequently *not detected* by GC–MS analysis. The exact nature of such an undetected fraction is still unknown, but it should obviously be composed by non-volatile high molecular weight components resistant to alkaline hydrolysis/methanolysis [7, 8]. It might, however, be speculated that this fraction could be composed of suberan-type oligomers. Suberan is a non-hydrolyzable highly aliphatic macromolecule found in the periderm tissue of some angiosperm species [69], whose inertness explains its detection in forest soils and fossilized samples [70, 71].

To the best of our knowledge, no such approach has been carried out for birch outer bark suberin.

In conclusion, well-established methodologies [72–74] for the selective isolation/purification of the most abundant suberin components are obviously extremely valuable for the commercial exploitation of suberin as a source of new macromolecular materials.

14.4 APPLICATIONS OF SUBERIN AND SUBERIN COMPONENTS

Although the composition of the aliphatic suberin has been thoroughly studied for many plant species [14], the physical properties of the ensuing mixtures of monomeric components, as well as their use as raw materials for polymer synthesis, are still poorly studied topics. The following sections provide a general overview of the published data on the physical properties of depolymerized suberin mixtures and on their applications in the synthesis of polymeric materials.

The only suberin depolymerization mixtures which were reported to have been used for functional characterization and for the synthesis of polymeric materials were obtained from *Quercus suber* cork, with the exception of a recent study where the reactive monomer was isolated from birch outer bark suberin [75].

14.4.1 Physical properties of depolymerized suberin

The only thorough characterization of mixtures of depolymerized suberin components (*dep-suberin*) was carried out by us in a comprehensive investigation of this material. The opaque pasty samples were obtained by the alkaline methanolysis of *Quercus suber* L. cork [8,76]. Under the conditions used for the depolymerization, most of the carboxylic acid functions were therefore converted into the corresponding methyl esters.

Given the predominance of long aliphatic chains in most of its components, which indeed impart to cork its well-known and largely exploited hydrophobic character, it seemed interesting to assess the surface properties of *dep-suberin*. A detailed study was therefore carried out using various techniques [76]. The surface energy of the solid (pasty) *dep-suberin* at 25°C, determined from contact angle measurements with liquids of different polarity, applying the Owens–Wendt approach, was $42\,mJ\,m^{-2}$, with a polar component of about $4\,mJ\,m^{-2}$. Measurements of the surface tension of the liquid samples at 50–110°C, gave a linear variation of γ with temperature, with an extrapolated value of $37\,mJ\,m^{-2}$ at 25°C. This difference was attributed to the microcrystalline character of the solid sample (see below), associated with a higher cohesive energy and, hence, a higher surface energy. Since a

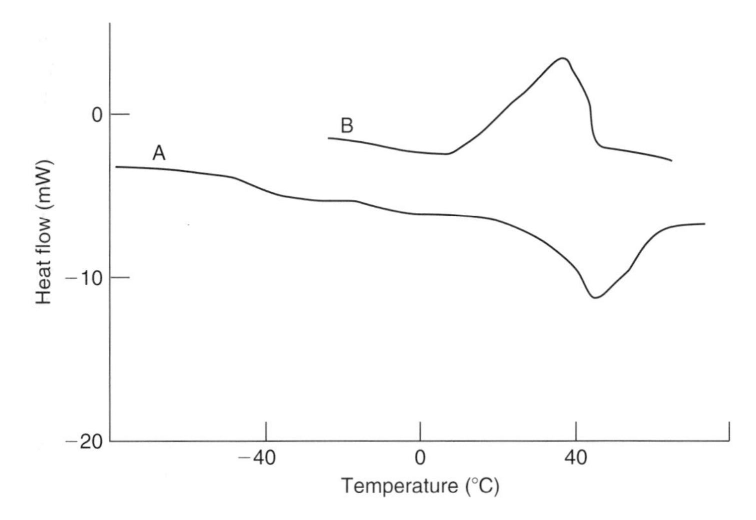

Figure 14.5 DSC thermograms of suberin. A: heating; B: cooling. (Reprinted from Reference [78], with permission from Elsevier.)

mixture of alkanes with the same range of chain lengths as the *dep-suberin* components would display a surface energy close to 28 mJ m^{-2}, it follows that (i) some of the polar groups in those components were present on the *dep-suberin* surface, as confirmed by the modest, but non-negligible, polar contribution to the surface energy, and (ii) some intermolecular interactions, mostly through hydrogen bonding, induced an increase in cohesive energy, compared with the purely dispersive alkane structures, as suggested by the correspondingly higher γ_d values obtained by both contact angle and inverse chromatography measurements [76]. Notwithstanding these fine-tuned considerations, *dep-suberin* must be considered as a rather non-polar material with surface properties resembling those of its cork precursor, whose reported values of surface energy range between 30 and 40 mJ m^{-2}[77].

The differential scanning calorimetry (DSC) tracings of *dep-suberin* (see Fig. 14.5) showed that annealing a molten sample in liquid nitrogen, produced an amorphous material with a glass transition temperature of about −50°C, which crystallized when brought to about 30°C [78]. The melting temperature of the microcrystalline phase was centred at about 40°C (broad endothermic peak). These observations were confirmed by optical microscopy carried out with reproducible temperature cycles between −20°C and 80°C [79]. A quantitative assessment of the birefringence (Fig. 14.6) showed a constant maximum value (heating cycle) up to about 0ºC, followed by a gradual decrease to zero birefringence at about 50°C. The cooling cycle reproduced the same features in reverse. The images captured in this context showed dense microcrystalline domains within an amorphous matrix [79].

Given the broad temperature range associated with the melting or forming of these crystalline phases, and the very small size of the crystals, it seems likely that the *dep-suberin* components more apt to crystallize because of their suitable structures, do so on an individual basis at their respective freezing temperature, when the liquid mixture is slowly cooled down. The result is therefore a set of microcrystals, each member belonging to a given *dep-suberin* component. Interestingly, the fact of having a rather complex mixture of compounds does not hinder the individual crystallization of some of them, most probably because the major driving force is associated with the ease of self-assembly among their *long and linear* aliphatic sequences.

The characteristic whitish and pasty appearance of these *dep-suberin* samples at room temperature reflects therefore the combination of a viscous liquid containing a substantial proportion of microcrystals.

The densities of these *dep-suberin* samples were surprisingly high, *viz. ca.* 1.08 at room temperature and above unity even up to 55°C [78], compared with those of alkanes of similar chain length, which are about 0.8 at room temperature. This clearly confirmed the existence of additional intermolecular interactions through hydrogen bonding from the OH groups borne by the different monomeric structures. Indeed, fatty acid esters, as well as fatty alcohols and diols, have densities close to those measured for *dep-suberin* in that work [78].

The thermogravimetry (TGA) of *dep-suberin* in a nitrogen atmosphere [78] showed a good thermal stability up to about 280°C, followed by a progressive weight loss, reaching a plateau at about 80 per cent volatilization at 470°C and leaving a carbonaceous residue.

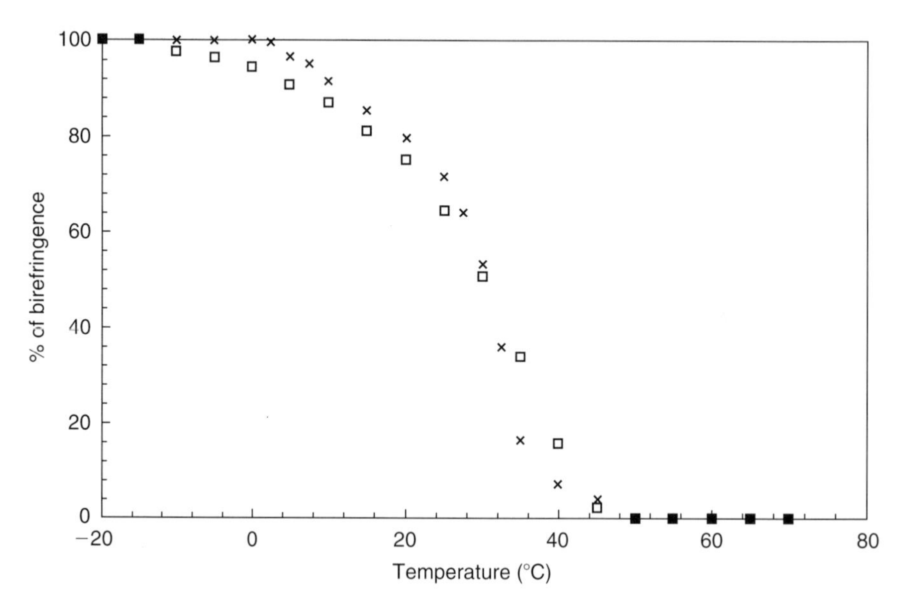

Figure 14.6 Melting (x) and recrystallization [□] of suberin, as observed by the change in birefringence intensity, as a function of temperature (the 100 per cent birefringence refers to the *maximum extent* of crystallization and *not* to the actual percentage of the crystalline phase). (Reprinted from Reference [79], with permission from Elsevier.)

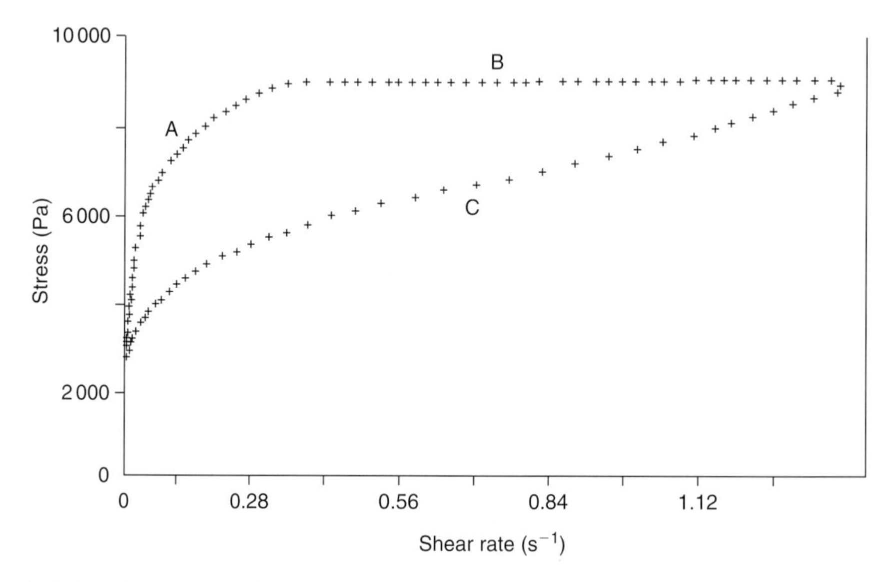

Figure 14.7 A Typical rheogram of suberin at 20°C. A: increasing stress; B: constant stress; C: decreasing stress. (Reprinted from Reference [78], with permission from Elsevier.)

The rheological properties of *dep-suberin* at room temperature were typical of a plastic response, with an important yield-stress value and a thixotropic behaviour, as shown in Fig. 14.7 [78, 80]. These features are usually associated with either intermolecular or interphase shear-induced destructuration (or both), followed by a time-dependent restructuration at rest. Since *dep-suberin* was characterized by both intermolecular association through hydrogen bonding and the existence of a liquid/crystal interphase at room temperature, its rheological study was extended to higher temperatures. The extent of yield stress decreased drastically as the temperature was raised and indeed vanished at about 50°C (*i.e.* when all the microcrystals had melted). Moreover, the rheogram at this temperature became linear, *viz*. liquid *dep-suberin* displayed a Newtonian behaviour. These observations, shown in

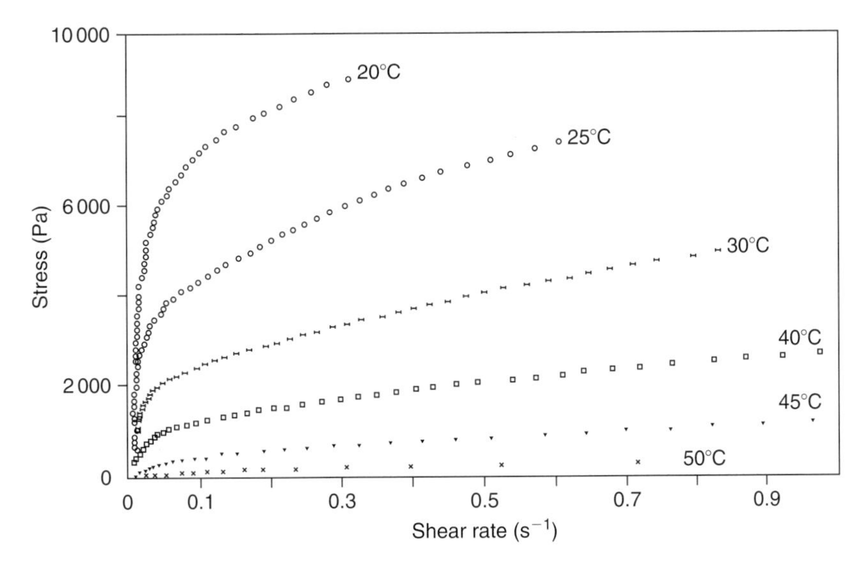

Figure 14.8 Rheograms of suberin at different temperatures (increasing stress mode). (Reprinted from Reference [78], with permission from Elsevier.)

Fig. 14.8, revealed that the major cause of its plastic behaviour at room temperature was its heterogeneous nature and the consequent strong interfacial interactions between the liquid and the microcrystals [78].

The actual values of the viscosity varied dramatically with temperature, going from 14 000 to 0.18 Pa s between 20°C and 65°C [80]. The corresponding Eyring plot [80] showed three distinct regimes (Fig. 14.9): (i) below 37°C, the presence of the microcrystalline phase induced a high value of the flow activation energy (E_a = 88 kJ mol$-^1$); (ii) above 55°C, where the sample was a homogeneous liquid, E_a dropped to 34 kJ mol$-^1$; (iii) a transition zone between these two temperatures, reflected the progressive melting of the microcrystals, which gave rise to a continuous change in the substrate solid–liquid contents and physical consistency.

Tack measurements [80] showed that the dynamic resistance of *dep-suberin* to film splitting decreased, as expected, with both increasing temperature and increasing shear rate. The temperature effect reflected mostly the melting of the crystalline phase, since the drop in tack was quite drastic between 30°C and 50°C (the melting range). The tack value registered in each test was constant with respect to time during up to 20 min.

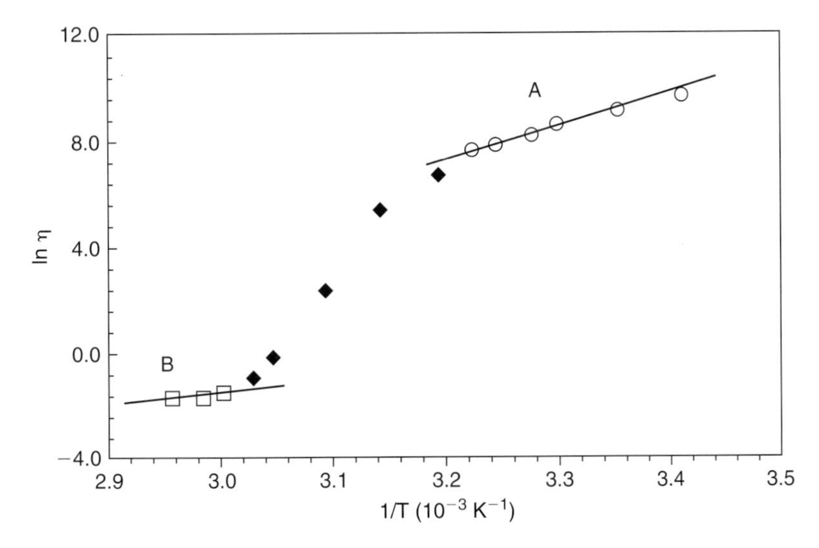

Figure 14.9 Eyring plot related to the viscous flow of suberin. (Reprinted from Reference [80], with permission from Elsevier.)

14.4.2 Depolymerized suberin mixtures as a functional additive

The microcrystalline character of *dep-suberin* mixtures described above, prompted us to examine the possible role of this material as an additive to offset printing inks, in replacement of other waxy materials like polytetrafluoroethylene (PTFE) oligomers [80]. Two reference inks were employed for this study, namely a typical vegetable-oil-based commercial ink and a 'waterless' ink containing petroleum-based diluents, to both of which *dep-suberin* was added in proportions of 2 to 10 per cent w/w. The characterization of these formulations included the determination of tack and viscosity, as well as printing tests. The presence of *dep-suberin* in the waterless ink only affected its bulk properties, by stabilizing the tack value with time and inducing a modest decrease in viscosity (with 10 per cent *dep-suberin*), without any detectable modification of the surface properties. This suggests that the hydrocarbon diluent of that ink acted as a good solvent for the *dep-suberin*, which therefore did not migrate to the surface of the printed film. With the more conventional vegetable-oil ink, *dep-suberin* induced a significant decrease in tack, small changes in viscosity and a two-fold decrease in the gloss of a printed film containing 10 per cent of it. The latter result clearly showed that at least part of the crystalline components of *dep-suberin* were not dissolved in the ink medium and could therefore migrate to the surface to produce the desired change in its optical properties.

14.4.3 Polymers from suberin monomers

Little has been published on the use of the suberin depolymerization products as monomers for the synthesis of novel macromolecular materials, which has so far concentrated on polyurethanes and polyesters using the mixture of aliphatic monomers.

14.4.3.1 Synthesis of polyurethanes

In a preliminary study also involving model compounds [81], the kinetics of urethane formation was followed by FTIR spectroscopy using an aliphatic and an aromatic monoisocyanate and their homologous diisocyanates. Both the model reactions and the polymer synthesis gave clear cut second-order behaviour, indicating that the hydroxyl groups borne by the suberin monomers displayed conventional aliphatic-OH reactivity.

The subsequent investigation [79] concentrated on the polymerization conditions and the thorough characterization of the ensuing polyurethanes, prepared using both aliphatic and aromatic diisocyanates. When the initial [NCO]/[OH] molar ratio was unity, all the polymers gave about 30 per cent of soluble material, the rest being a crosslinked product. This systematic result suggested that, on the one hand, some of the suberin monomers had a functionality higher than two, thus promoting a non-linear polycondensation leading to about 70 per cent of gel, and, on the other hand, monofunctional components must have been present in the monomer mixture, which played the role of chain-growth terminators, giving rise to the sol fraction. This conclusion was corroborated by the fact that the FTIR spectra of both fractions were practically identical, as shown in Fig. 14.10, suggesting that the solubility/insolubility factor was not based on differences in the polymer chemical structure, but instead on its macromolecular architecture.

The T_g of these polyurethanes [81] followed classical trends in that, for the networks, the use of aromatic diisocyanates resulted in high values (about 100°C) associated with the stiffness of their moieties, whereas with the aliphatic counterparts, values around room temperature indicated much higher chain flexibility. The T_gs of the soluble fractions were much lower than those of their corresponding crosslinked materials, which is in tune with the presence of very mobile open-ended branches, generated by the insertion of monofunctional monomers into the polymer structure.

14.4.3.2 Synthesis of polyesters

Benitez *et al.* [82], recently reported the synthesis of a polyester resembling cutin, a natural polymer whose structure is close to that of aliphatic suberin [83], by a circular approach, which consisted in depolymerizing cutin through ester cleavage and then submitting the ensuing monomer mixture to a chemical polyesterification process.

The crosslinked material they obtained displayed, as one would indeed expect, very similar spectroscopic features compared with those of the starting cutin. In a subsequent study in the same vein [84], glycerol derivatives of mono- and dicarboxylic acids, whose structure simulated those present in both suberin and cutin, were prepared and characterized in an effort to simulate the biological synthesis of those natural polymers and exploit their peculiar properties, particularly their tendency to form supramolecular assemblies.

In another recent study [75], 9,10-epoxy-18-hydroxyoctadecanoic acid, isolated from birch outer bark suberin was used as starting material for the lipase-catalyzed synthesis of an epoxy-functionalized polyester. Immobilized

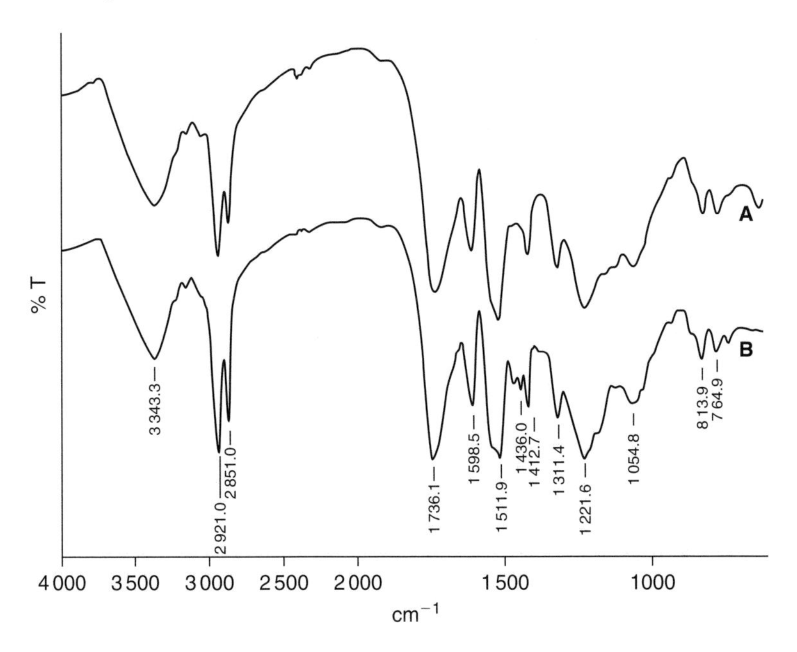

Figure 14.10 FTIR spectra of a polyurethane prepared from suberin and MDI-2.0 with $[NCO]_0/[OH]_0 = 1$. A: insoluble fraction; B: soluble fraction. (Reprinted from Reference [79], with permission from Elsevier.)

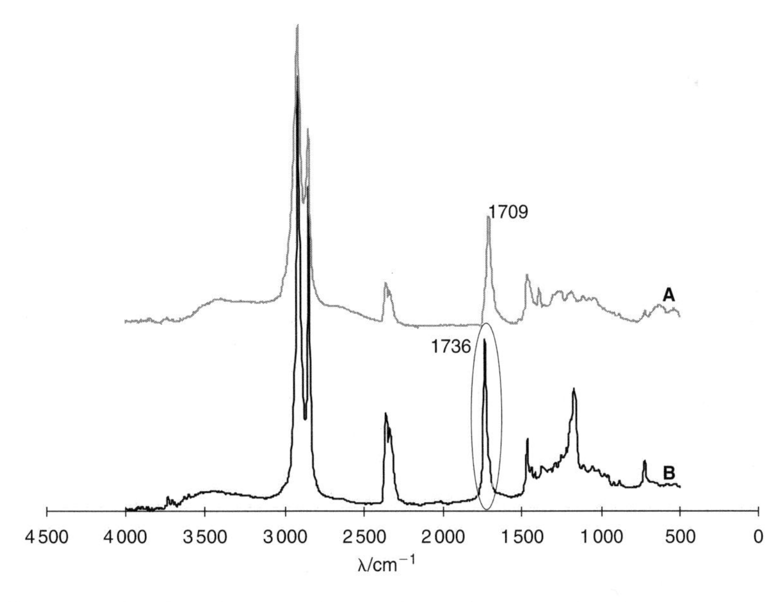

Figure 14.11 FTIR spectra of A: *dep-suberin* and B: the polyester formed by its polycondensation.

Candida antarctica lipase B was shown to catalyse efficiently the polycondensation of *cis*-9,10-epoxy-18-hydroxyoctadecanoic acid, yielding epoxy-functionalized polyesters with Mw values ranging from 15 000 to 20 000.

Another promising approach, in the sense that it avoids expensive fractionation procedures, is to promote the direct polymerization of *dep-suberin* mixtures under polycondensation or polytransesterification conditions, using *dep-suberin* mixtures issued from alkaline hydrolysis or methanolysis, respectively. Ongoing experiments prove that both types of systems are viable [85, 86], as shown by the FTIR spectrum of the product obtained from a polycondensation of a *dep-suberin* (Fig. 14.11), which clearly shows the feature of a polyester. This evidence was further corroborated by NMR spectroscopy and molecular weight measurements [85].

14.5 CONCLUDING REMARKS

The main purpose of this chapter was to emphasize the interest in considering cork and suberin, both cheap renewable resources, potentially available in very large amounts, as valuable precursors to novel macromolecular materials, particularly when exploited as by-products. Undoubtedly, the predominance of long aliphatic moieties in both these precursors, points to applications of the corresponding polymers associated with a relative softness and a highly hydrophobic character, but these obvious features should not be considered as exclusive since much more needs to be explored in this still young domain.

REFERENCES

1. Bernards M.A., Demystifying suberin, *Can. J. Bot.*, **80**(3), 2002, 227–240.
2. Silva S.P., Sabino M.A., Fernandes E.M., Correlo V.M., Boesel L.F., Reis R.L., Cork: Properties, capabilities and applications, *Int. Mater. Rev.*, **50**(6), 2005, 345–365.
3. Ciesla W.M., *Non-Wood Forest Products from Temperate Broad-Leaved Trees*, FAO, Rome, 2002.
4. CorkMasters, www.corkmasters.com (browsed May, 2007).
5. Pereira H., Chemical-Composition and Variability of Cork from *Quercus-suber* L., *Wood Sci. Technol.*, **22**(3), 1988, 211–218.
6. Lopes M., Neto C.P., Evtuguin D., Silvestre A.J.D., Gil A., Cordeiro N., Gandini A., Products of the permanganate oxidation of cork, desuberized cork, suberin and lignin from *Quercus suber* L., *Holzforschung*, **52**(2), 1998, 146–148.
7. Lopes M.H., Gil A.M., Silvestre A.J.D., Neto C.P., Composition of suberin extracted upon gradual alkaline methanolysis of *Quercus suber* L. cork, *J. Agric. Food Chem.*, **48**(2), 2000, 383–391.
8. Cordeiro N., Belgacem M.N., Silvestre A.J.D., Neto C.P., Gandini A., Cork suberin as a new source of chemicals. 1. Isolation and chemical characterization of its composition, *Int. J. Biol. Macromol.*, **22**(2), 1998, 71–80.
9. Conde E., Cadahia E., Garcia-Vallejo M.C., de Simon B.F., Polyphenolic composition of *Quercus suber* cork from different Spanish provenances, *J. Agric. Food Chem.*, **46**(8), 1998, 3166–3171.
10. Conde E., Cadahia F., GarciaVallejo M.C., deSimon B.F., Adrados J.R.G., Low molecular weight polyphenols in cork of *Quercus suber*, *J. Agric. Food Chem.*, **45**(7), 1997, 2695–2700.
11. Conde E., Cadahia E., Garcia-Vallejo M.C., Gonzalez-Adrados J.R., Chemical characterization of reproduction cork from Spanish *Quercus suber*, *J. Wood Chem. Technol.*, **18**(4), 1998, 447–469.
12. Conde E., Garcia-Vallejo M.C., Cadahia E., Variability of suberin composition of reproduction cork from *Quercus suber* throughout industrial processing, *Holzforschung*, **53**(1), 1999, 56–62.
13. Cadahia E., Conde E., de Simon B.F., Garcia-Vallejo M.C., Changes in tannic composition of reproduction cork *Quercus suber* throughout industrial processing, *J. Agric. Food Chem.*, **46**(6), 1998, 2332–2336.
14. Gandini A., Pascoal C., Silvestre A.J.D., Suberin: A promising renewable resource for novel macromolecular materials, *Progr. Polym. Sci.*, **31**(10), 2006, 878–892.
15. Rosa M.E., Osorio J., Green V., Fortes M.A., Torsion of cork under compression. in *Advanced Materials Forum Ii*, Ed. 2004, pp. 235–238.
16. Parralejo A.D., Guiberteau F., Fortes M.A., Rosa M.F., Mechanical properties of cork under contact stresses, *Rev. de Metalurgia*, **37**(2), 2001, 330–335.
17. Vaz M.F., Fortes M.A., Friction properties of cork, *J. Mater. Sci.*, **33**(8), 1998, 2087–2093.
18. Pina P., Fortes M.A., Characterization of cells in cork, *J. Phys. D: Appl. Phys.*, **29**(9), 1996, 2507–2514.
19. Rosa M.E., Fortes M.A., Water-absorption by cork, *Wood Fiber Sci.*, **25**(4), 1993, 339–348.
20. Rosa M.E., Fortes M.A., Deformation and fracture of cork in tension, *J. Mater. Sci.*, **26**(2), 1991, 341–348.
21. Rosa M.E., Fortes M.A., Temperature-induced alterations of the structure and mechanical-properties of cork, *Mater. Sci. Eng.*, **100**, 1988, 69–78.
22. Rosa M.E., Fortes M.A., Stress-relaxation and creep of cork, *J. Mater. Sci.*, **23**(1), 1988, 35–42.
23. Gibson L.J., Easterling K.E., Ashby M.F., The structure and mechanics of cork, *Proc. Royal Soc. London Series A-Math. Phys. Eng. Sci.*, **377**(1769), 1981, 99–117.
24. Gameiro C.P., Cirne J., Gary G., Experimental study of the quasi-static and dynamic behaviour of cork under compressive loading, *J. Mater. Sci.*, **42**, 2007, 4316–4324.
25. Amorim Industrial solutions, http://www.amorimsolutions.com/ (browsed May, 2007).
26. Shuttle's Cork From Portugal, *New York Times*, November, 3, 1981.
27. Thermal protection systems for space vehicles, www.p2pays.org/ref/34/33161.pdf (browsed May, 2007).
28. Airbus fait appel au groupe portugais Amorim pour fournir du liège aggloméré. http://www.bulletins-electroniques.com/actualites/43153.htm. (browsed June, 2007).

29. Karade S.R., Irle M., Maher K., Influence of granule properties and concentration on cork-cement compatibility, *Holz Als Roh-Und Werkstoff*, **64**(4), 2006, 281–286.
30. Gil L., *Cortiça, Produção Tecnologia e Aplicação*, INETI, Lisbon, 1988.
31. Gil L., Cork powder waste: An overview, *Biomass & Bioenergy*, **13**(1–2), 1997, 59–61.
32. Kamm B., Gruber P.R., Kamm M., *Biorefineries – Industrial Processes and Products*, Wiley-VCH, Weinheim, 2006.
33. Sousa A.F., Pinto P., Silvestre A.J.D., Neto C.P., Triterpenic and other lipophilic components from industrial cork byproducts, *J. Agric. Food Chem.*, **54**(18), 2006, 6888–6893.
34. Evtiouguina M., Barros-Timmons A., Cruz-Pinto J.J., Neto C.P., Belgacem M.N., Gandini A., Oxypropylation of cork and the use of the ensuing polyols in polyurethane formulations, *Biomacromolecules*, **3**(1), 2002, 57–62.
35. Evtiouguina M., Barros A.M., Cruz-Pinto J.J., Neto C.P., Belgacem N., Pavier C., Gandini A., The oxypropylation of cork residues: Preliminary results, *Bioresource Technol.*, **73**(2), 2000, 187–189.
36. Evtiouguina M., Gandini A., Neto C.P., Belgacem N.M., Urethanes and polyurethanes based on oxypropylated cork: 1. Appraisal and reactivity of products, *Polym. Int.*, **50**(10), 2001, 1150–1155.
37. Ekman R., The suberin monomers and triterpenoids from the outer bark of *betula-verrucosa* ehrh, *Holzforschung*, **37**(4), 1983, 205–211.
38. Holloway P.J., Some variations in the composition of suberin from the cork layers of higher-plants, *Phytochemistry*, **22**(2), 1983, 495–502.
39. Graca J., Pereira H., Glyceryl–acyl and aryl–acyl dimers in Pseudotsuga menziesii bark suberin, *Holzforschung*, **53**(4), 1999, 397–402.
40. Graca J., Pereira H., Suberin structure in potato periderm: Glycerol, long-chain monomers, and glyceryl and feruloyl dimers, *J. Agric. Food Chem.*, **48**(11), 2000, 5476–5483.
41. Christie W.W., The lipid library. http://www.lipidlibrary.co.uk/ (browsed January, 2006).
42. Bernards M.A., Lewis N.G., The macromolecular aromatic domain in suberized tissue: A changing paradigm, *Phytochemistry*, **47**(6), 1998, 915.
43. Kolattukudy P.E., Bio-polyester membranes of plants – cutin and suberin, *Science*, **208**(4447), 1980, 990–1000.
44. Kolattukudy P.E., Polyesters in higher plants, in *Advances in Biochemical Engineering/Biotechnology: Biopolyesters,* Eds.: Babel W. and Steinbüchel A., Springer, Berlin, Heidelberg, 2001.
45. Kolattukudy P.E., Espelie K.E., Chemistry, biochemistry, and function of suberin and associated waxes, in *Natural products of woody plants, chemical extraneous to the lignocellulosic cell wall,* Ed.: Rowe J., Springer, Berlin, Heidelberg, 1989.
46. Gil A.M., Lopes M., Rocha J., Neto C.P., A C-13 solid state nuclear magnetic resonance spectroscopic study of cork cell wall structure: The effect of suberin removal, *Int. J. Biol. Macromol.*, **20**(4), 1997, 293–305.
47. Lopes M.H., Sarychev A., Neto C.P., Gil A.M., Spectral editing of C-13 CP/MAS NMR spectra of complex systems: Application to the structural characterisation of cork cell walls, *Solid State Nuc. Magn. Reson.*, **16**(3), 2000, 109–121.
48. Graca J., Santos S., Suberin: A biopolyester of plants' skin, *Macromol. Biosc.*, **7**(2), 2007, 128–135.
49. Graca J., Pereira H., Cork suberin: A glyceryl based polyester, *Holzforschung*, **51**(3), 1997, 225–234.
50. Graca J., Pereira H., Diglycerol alkenedioates in suberin: Building units of a poly(acylglycerol) polyester, *Biomacromolecules*, **1**(4), 2000, 519–522.
51. Graca J., Pereira H., Methanolysis of bark suberins: Analysis of glycerol and acid monomers, *Phytochem. Anal.*, **11**(1), 2000, 45–51.
52. Graca J., Schreiber L., Rodrigues J., Pereira H., Glycerol and glyceryl esters of omega-hydroxyacids in cutins, *Phytochemistry*, **61**(2), 2002, 205–215.
53. Gil A.M., Lopes M.H., Neto C.P., Callaghan P.T., An NMR microscopy study of water absorption in cork, *J. Mater. Sci.*, **35**(8), 2000, 1891–1900.
54. Garbow J.R., Ferrantello L.M., Stark R.E., 13C Nuclear magnetic resonance study of suberized potato cell wall, *Plant Physiol.*, **90**, 1989, 783–787.
55. Stark R.E., Garbow J.R., Nuclear magnetic resonance relaxation studies of plant polyester dynamics. 2. Suberized potato cell walls, *Macromolecules*, **25**, 1992, 149–154.
56. Yan B., Stark R.E., Biosynthesis, molecular structure, and domain architecture of potato suberin: A C-13 NMR study using isotopically labeled precursors, *J. Agric. Food Chem.*, **48**(8), 2000, 3298–3304.
57. Neto C.P., Cordeiro N., Seca A., Domingues F., Gandini A., Robert D., Isolation and characterization of a lignin-like polymer of the cork of *Quercus suber* L., *Holzforschung*, **50**(6), 1996, 563–568.
58. Sitte P., Zum Feinbau Der Suberinschichten Im Flaschenkork, *Protoplasma*, **54**(4), 1962, 555–559.
59. Graca J., Pereira H., Feruloyl esters of omega-hydroxyacids in cork suberin, *J. Wood Chem. Technol.*, **18**(2), 1998, 207–217.
60. Bento M.F., Pereira H., Cunha M.A., Moutinho A.M.C., van den Berg K.J., Boon J.J., van den Brink O., Heeren R.M.A., Fragmentation of suberin and composition of aliphatic monomers released by methanolysis of cork from *Quercus suber* L., analysed by GC–MS, SEC and MALDI–MS, *Holzforschung*, **55**(5), 2001, 487–493.

61. GarciaVallejo M.C., Conde E., Cadahia E., deSimon B.F., Suberin composition of reproduction cork from *Quercus suber*, *Holzforschung*, **51**(3), 1997, 219–224.
62. Arno M., Serra M.C., Seoane E., Methanolysis of cork suberin – Identification and estimation of their acidic components as methyl-esters, *Anales De Quimica Serie C-Quimica Organica Y Bioquimica*, **77**(1), 1981, 82–86.
63. Ekman R., Eckerman C., Aliphatic carboxylic-acids from suberin in birch outer bark by hydrolysis, methanolysis, and alkali fusion, *Paperi Ja Puu-Paper and Timber*, **67**(4), 1985, 255.
64. Holloway P.J., Deas A.H.B., Epoxyoctadecanoic acids in plant cutins and suberins, *Phytochemistry*, **12**(7), 1973, 1721–1735.
65. Holloway P.J., Martin J.T., Baker E.A., Chemistry of plant cutins and suberins, *Anales De Quimica-International Edition*, **68**(5–6), 1972, 905.
66. Seoane E., Serra M.C., Agullo C., 2 New epoxy-acids from cork of *Quercus-suber*, *Chem. Ind.*, **15**, 1977, 662–663.
67. Graca J., Santos S., Linear aliphatic dimeric esters from cork suberin, *Biomacromolecules*, **7**(6), 2006, 2003–2010.
68. Bento M.F., Pereira H., Cunha M.A., Moutinho A.M.C., van den Berg K.J., Boon J.J., Thermally assisted transmethylation gas chromatography mass spectrometry of suberin components in cork from *Quercus suber* L., *Phytochem. Anal.*, **9**(2), 1998, 75–87.
69. Tegelaar E.W., Hollman G., Vandervegt P., Deleeuw J.W., Holloway P.J., Chemical characterization of the periderm tissue of some angiosperm species – Recognition of an insoluble, nonhydrolyzable, aliphatic biomacromolecule (uberan), *Org. Geochem.*, **23**(3), 1995, 239–251.
70. Nierop K.G.J., Origin of aliphatic compounds in a forest soil, *Org. Geochem.*, **29**(4), 1998, 1009–1016.
71. Augris N., Balesdent J., Mariotti A., Derenne S., Largeau C., Structure and origin of insoluble and non-hydrolyzable, aliphatic organic matter in a forest soil, *Org. Geochem.*, **28**(1–2), 1998, 119–124.
72. Ekman R., Eckerman C., Mattila T., Suokas E. in *Method for converting vegetable material into chemicals*, Ed. 4732708 U.S.P., 1988.
73. Krasutsky P.A., Carlson R.M., Kolomitsyn I.V., in *Isolation of natural products from birch bark*, Ed. 20030109727 U.S.P., 2003.
74. Krasutsky P.A., Carlson R.M., Nesterenko V.V., Kolomitsyn I.V., Edwardson C.F., in *Birch bark processing and the isolation of natural products from birch bark* Ed. 6815553 U.S.P., 2004.
75. Olsson A., Lindstrom M., Iversen T., Lipase-catalyzed synthesis of an epoxy-functionalized polyester from the suberin monomer *cis*-9,10-epoxy-18-hydroxyoctadecanoic acid, *Biomacromolecules*, **8**(2), 2007, 757–760.
76. Cordeiro N., Aurenty P., Belgacem M.N., Gandini A., Neto C.P., Surface properties of suberin, *J. Colloid Interface Sci.*, **187**(2), 1997, 498–508.
77. Cordeiro N., Neto C.P., Gandini A., Belgacem M.N., Characterization of the cork surface by inverse gas-chromatography, *J. Coll. Interface Sci.*, **174**(1), 1995, 246–249.
78. Cordeiro N., Belgacem N.M., Gandini A., Neto C.P., Cork suberin as a new source of chemicals: 2. Crystallinity, thermal and rheological properties, *Bioresource Technol.*, **63**(2), 1998, 153–158.
79. Cordeiro N., Belgacem M.N., Gandini A., Neto C.P., Urethanes and polyurethanes from suberin 2: Synthesis and characterization, *Ind. Crops Prod.*, **10**(1), 1999, 1–10.
80. Cordeiro N., Blayo A., Belgacem N.M., Gandini A., Neto C.P., LeNest J.F., Cork suberin as an additive in offset lithographic printing inks, *Ind. Crops Prod.*, **11**(1), 2000, 63–71.
81. Cordeiro N., Belgacem M.N., Gandini A., Neto C.P., Urethanes and polyurethanes from suberin. 1. Kinetic study, *Ind. Crops Prod.*, **6**(2), 1997, 163–167.
82. Beniitez J.J., Garcia-Segura R., Heredia A., Plant biopolyester cutin: A tough way to its chemical synthesis, *Biochim. Biophys. Acta-Gen. Subj.*, **1674**(1), 2004, 1–3.
83. Schreiber L., Franke R., Hartmann K.D., Ranathunge K., Steudle E., The chemical composition of suberin in apoplastic barriers affects radial hydraulic conductivity differently in the roots of rice (Oryza sativa L. cv. IR64) and corn (Zea mays L. cv. Helix), *J. Exp.. Bot.*, **56**(415), 2005, 1427–1436.
84. Douliez J.P., Barrault J., Jerome F., Heredia A., Navailles L., Nallet F., Glycerol derivatives of cutin and suberin monomers: Synthesis and self-assembly, *Biomacromolecules*, **6**(1), 2005, 30–34.
85. Sousa A.F., Pinto P.C.R.O., Silvestre A.J.D., Gandini A., Pascoal Neto C., *BIOPOL – 2007: 1st International conference on biodegradable polymers and sustainable composites*, Alicante, Spain, October 2007.

– 15 –

Starch: Major Sources, Properties and Applications as Thermoplastic Materials

Antonio J.F. Carvalho

ABSTRACT

This chapter reviews the general context of starch as a material. After a survey of the major sources of starch and their characteristic compositions in terms of amylase and amylopectin, the morphology of the granules and the techniques applied to disrupt them are critically examined. The use of starch for the production of polymeric materials covers the bulk of the chapter, including the major aspect of starch plasticization, the preparation and assessment of blends, the processing of thermoplastic starch (TPS), the problems associated with its degradation and the preparation of TPS composites and nanocomposites. The present and perspective applications of these biodegradable materials and the problems associated with their moisture sensitivity conclude this manuscript.

Keywords

Starch, Main sources, Macromolecular composition, Polymer structure and crystallinity, Granule disruption, Thermoplastic starch, Plasticization, Moisture uptake, Degradation, Blends, Composites

15.1 INTRODUCTION

Starch is the major carbohydrate reserve in higher plants. In contrast with cellulose that is present in dietary fibres, starch is digested by humans and represents one of the main sources of energy to sustain life. Bread, potato, rice and pasta are examples of the importance of starch in our society. Starch has also been extremely important for centuries in numerous non-food applications, for example, as glue for paper and wood [1] and as gum for the textile industry [2, 3]. Together with wood, natural fibres and leather, starch has been one of the choice materials since the inception of human technology.

Polysaccharides represent by far the most abundant biopolymers on earth, with cellulose, chitin and starch dominating. The first two are reviewed as sources of novel materials in Chapters 16–19 and 25, respectively, of this book. Starch is certainly one of the most versatile materials for potential use in polymer technology. It can be converted, on the one hand, into chemicals like ethanol, acetone and organic acids, used in the production of synthetic polymers and, on the other hand, it can produce biopolymer through fermentative processes or be hydrolyzed and employed as a monomer or oligomer. Finally, it can be grafted with a variety of reagents to produce new polymeric materials, used as such or as fillers for other polymers.

The conversion into small molecules is chemically easier for starch than for cellulose, making it an economic option to produce hydroxyl-containing compounds which can be exploited as monomers or as raw material in the production of other biopolymers like polylactic acid (PLA). This approach is in competition with other renewable resources *viz.* saccharose from sugar cane, used for the production of ethanol and biopolymers such as poly-β-hydroxybutyrate (PHB) [4, 5], and lactic acid [6] as a source of its polymer [7]. These two polymers are reviewed herein in Chapters 21 and 22, respectively. Despite the importance of starch as a raw material for the production of chemicals and other polymers, its direct use as a renewable resource commodity is undoubtedly more economical and has been a major area of research in material science over the last few decades.

The chemical modification of starch can provide tailor-made materials and has been reviewed recently [8].

The literature concerning starch is extremely vast and its chemistry and technology have been comprehensively reviewed recently [2, 3, 8, 9]. More specifically, a renewed interest has arisen in the last decade, to convert starch into a plastic material capable of replacing petroleum-based counterparts. The main aim of this chapter is to review the applications of starch in the development of new polymeric materials in which its main macromolecular structure is preserved. The preparation of monomers and oligomers will also be briefly discussed.

15.2 MAIN SOURCES OF STARCH

Several plants are commercially used for the production of starch. The choice plant depends mainly on geographic and climatic factors and on the desired functional properties of the corresponding starch [10]. It is always possible to find a highly productive plant to produce starch whatever the climate and agricultural conditions: maize in tempered and subtropical zones, cassava (manioc or tapioca) and banana in tropical environments, rice in inundated areas and potatoes in cold climates. The main plant sources are listed in Table 15.1, together with their production for 2005 [11].

Apart from the traditional crops, cassava shows a great potential because it adapts to tropical zones and constitutes therefore an important crop in developing areas of the world. New sources of starch are also arising, such as banana [12], which yields a starch of excellent quality.

The development of new uses for starch, and for materials based on starch, within the broader context of the increasing demand for materials based on renewable resources, will certainly increase the demand for starch production and hence the development of new commercial sources of starch.

15.3 STRUCTURE OF STARCH GRANULES

Starch can be found in various parts of a plant, such as the endosperm, the root, the leaf and the fruit pulp. It is deposited in the form of semi-crystalline granules which are insoluble in cold water and resemble spherulites [13] alternating amorphous and crystalline (or semi-crystalline) lamellae. Native starch is composed of two main macromolecular components, namely amylose and amylopectin [14–21]. The monomer units of these natural polymers are α-D-glucopyranosyl moieties linked by (1 → 4) and (1 → 6) bonds [22]. Amylose is a predominantly α-(1 → 4)-D-glucopyranosyl linear macromolecule. Amylopectin is a highly branched and high molecular weight macromolecule composed mostly of α-(1 → 4)-D-glucopyranose units, with α-(1 → 6)-linkages at intervals of approximately 20 units (Fig. 15.1) [15, 23].

Table 15.1

World production of the main starch crops in 2005 (1 × 1000 metric tons), (Source FAO, 2005 [11])

Crops	World production in 2005 (1 × 1000 metric tons)
Maize	711 762.87
Rice (Paddy)	621 588.53
Wheat	630 556.61
Potatoes	324 491.14
Cassava	213 024.81
Bananas	74 236.88
Yams	48 891.21
Sorghum	59 722.09

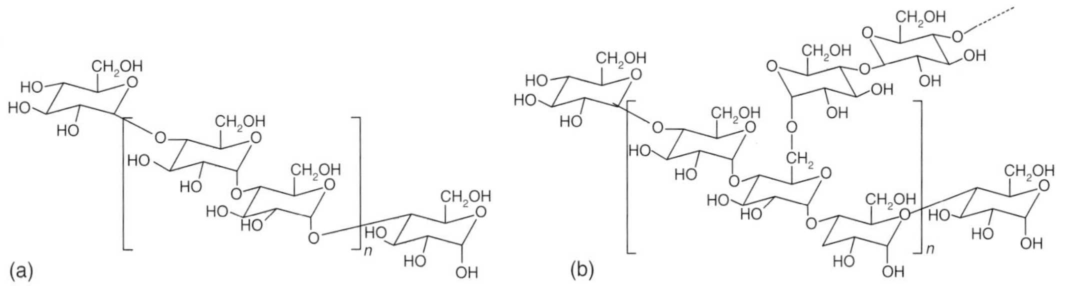

Figure 15.1 Amylose (a) and amylopectin (b) structures.

15.3.1 Granule structure

The starch granule morphology, as well as the structure of its main macromolecular constituents, have been the focus of intense research which is still ongoing because of the complexity of the problems involved. The granules have been examined using several techniques, like light and electron microscopy, X-ray and neutron scattering and more recently, atomic force microscopy [24–28]. Starch granules from different plant species are significantly different and can be, in the majority of cases, identified by light microscopy inspection. The most obvious differences in starch granules are in their shape and size which can vary considerably [1, 29, 30], as reported in Table 15.2 for some common granules.

The morphology of the starch granule varies not only according to the source plant, but also to the different parts of the same plant [16]. Other important factors influencing these aspects are the degree of polymerization (DP) of amylose and amylopectin and the possible presence of other components in the granule such as lipids, proteins and inorganic compounds [18].

The semi-crystalline nature of starch is well known and its most important feature is the alternation of long-range molecular order and amorphous regions, defining the corresponding alternation of crystalline and amorphous lamellae [14, 16, 20, 31].

An idealized structure for the starch granule was proposed recently by Gallant *et al.* [21]. This model describes the crystalline and amorphous amylopectin lamellae into effectively spherical blocklets whose diameter ranges from 20 to 500 nm. These authors also propose the existence of short radial channels of amorphous material. The granule grows alternating radial amorphous and semi-crystalline rings from the *hilum*, forming a lamellar structure. The amylopectin, which comprises around 75 per cent of the granule, is mainly responsible for the granule crystallinity. The crystalline and semi-crystalline lamellae are composed of amylopectin blocklets, forming a crystalline hard shell composed of large blocklets and a semi-crystalline soft shell composed of small blocklets. The crystalline lamellae are around 9–10 nm thick on an average and consist of ordered double-helical amylopectin side chain clusters, interfacing with the more amorphous lamellae of amylopectin branching regions. The size of the semi-crystalline soft-shell blocklets ranges from 20 to 50 nm. Figure 15.2 depicts this structure as proposed by Gallant *et al.*

Table 15.2

Size, shape and amylose content of some starch granules [2, 12, 17, 29, 30]

Source	Diameter (μm)	Amylose content (wt%)	Shape
Maize	5–25	28	Polyhedric
Waxy maize	5–25	~0	Polyhedric
High-amylose	5–35	55–85	Varied smooth spherical to elongated
Cassava	5–35	16	Semi-spherical
Potato	15–100	20	Ellipsoidal
Wheat	20–22	30	Lenticular, polyhedric
Rice normal	5/3–8	20–30	Polyhedric
Banana	26–35	9–13	Elongated oval with ridges

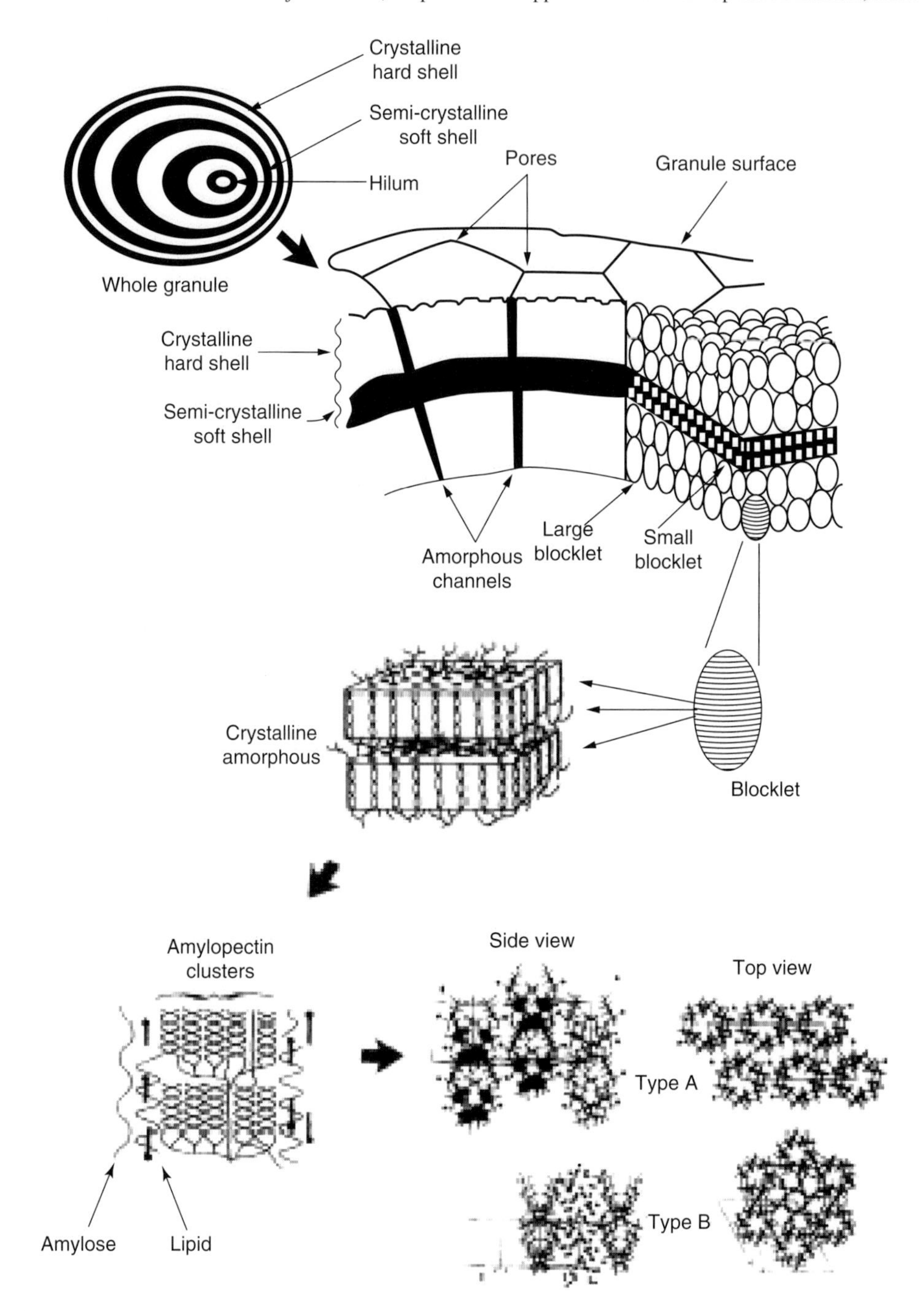

Figure 15.2 Starch granule structure. Reproduced with permission from Reference [21].

15.3.2 Molecular structure and crystallinity

The main technique used to study starch crystallinity is X-ray diffraction, from which starch can be classified to A, B and C crystallites or polymorph forms[17, 32–34]. Starches with these polymorphisms are called A-, B- and C-type, each type presenting its characteristic diffraction patterns. The most commonly observed structures in native starch are A and B, the former being associated mainly with cereal starches, while the latter dominates generally

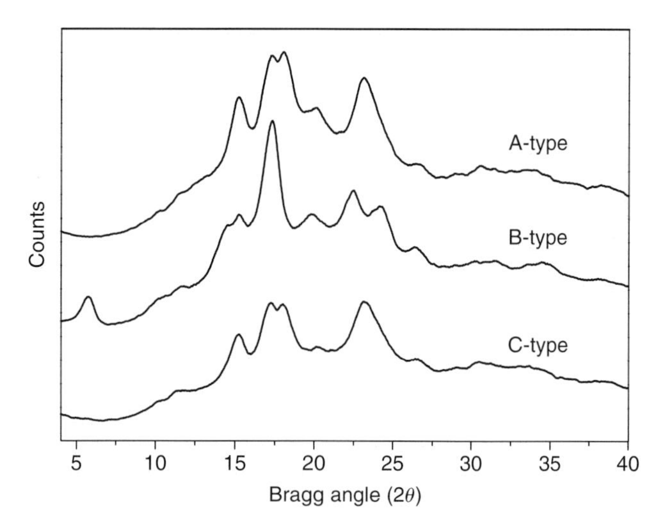

Figure 15.3 X-ray diffraction patterns of maize (A-type), potato (B-type) and cassava (C-type) starches.

in tuber starches but also occurs in maize starches with more than 30–40 per cent amylase [17]. The C pattern is a form intermediate between A and B and is usually associated with pea and various bean starches together with other root starches [34, 35]. A-type crystallites are denser and less hydrated [33] than B[32]-type counterparts, whereas the C-types arise from the joint presence of the other two homologues [17, 19, 34–40]. Figure 15.3 shows the X-ray diffraction patterns of maize (A-type), potato (B-type) and cassava (C-type) starches.

Although amylose is predominantly made up of linear macromolecules, it has been suggested that amyloses from some starch varieties may contain very long branches [15]. Amylose molecular weights [41] range from 1.5×10^5 to 2.6×10^6, with some variability as a function of the plant variety in the case of cassava [42]. Linear amylose molecules form double helices and can crystallize in a similar way as in starch granule, rendering the A and B structures [15]. Single crystals of low DP amylose in the polymorphisms A and B were also prepared and characterized [43].

Amylose also form complexes with other materials called V-structures (Verkleisterung) [2] which can be isolated in single crystals [43]. The amylose V-complexes are commonly formed by a left-handed amylose helix with six residues per turn enclosing the aliphatic tail of a lipid in its centre [18, 44, 45], each revolution around the lipid taking 8 Å [46, 47]. Depending on the condition of the V-complex formation, they can be crystalline or amorphous [48]. V-complexes are insoluble in water, even in drastic pressure and temperature conditions, a fact that makes them very important in polymer blending when water resistance is a desired property [49, 50].

Amylopectin is a highly branched molecule which is responsible for the main crystalline character of the starch granule. Its structure was modelled as a hyperbranched molecule [17, 51, 52], as proposed initially by Nikuni [53] and French [16] and later improved by Robin [37] (Fig. 15.4). In this model, short chains with 15 D-glucopyranosyl units branch out at almost regular intervals of 25 units to form either external A-chains or internal chains of amylopectin [37]. Starch crystallites are thus formed in compact areas made up of A-chains with DP 15. Less compact areas mainly occupied by B-chains, where the (1,6)-α-D-branching points are located, are placed between these compact areas.

A parallel with synthetic polymers can be made in which amylopectin is a clear cut example of a natural dendrimer [51] with a high degree of branching and a spherical shape, each generation being fully generated from branching sites with a minimum functionality of three.

15.4 DISRUPTION OF STARCH GRANULES

When starch granules are heated in an excess of water (>90 per cent w/w) or of another solvent able to form hydrogen bonding (*e.g.* liquid ammonia, formamide, formic acid, chloroacetic acids and dimethyl sulphoxide) starch undergoes an irreversible order–disorder transition known as gelatinization or destructuration. This

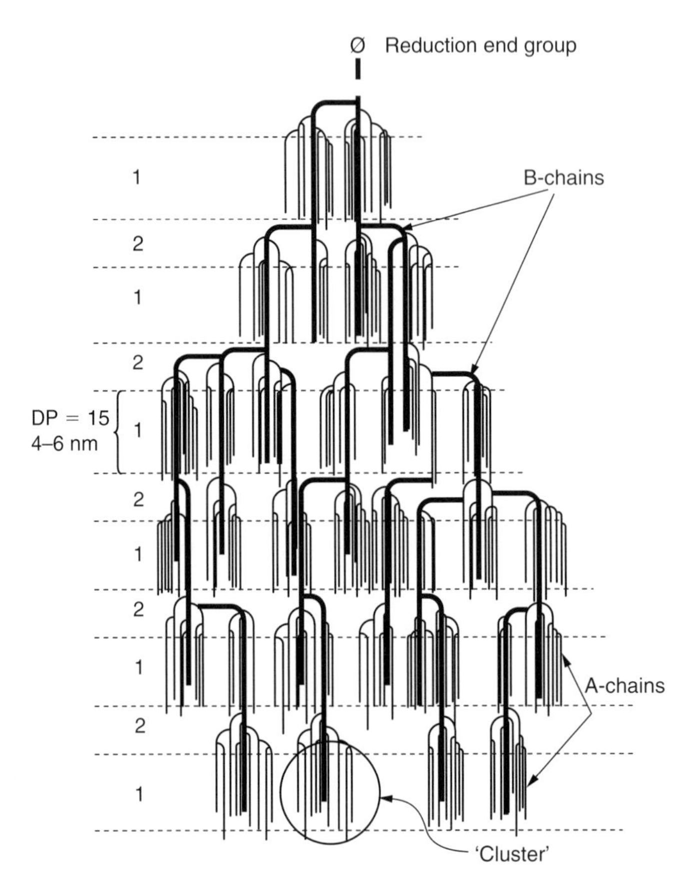

Figure 15.4 Amylopectin model, as proposed in Reference [37]. Reproduced with permission 1 = compact area; 2 = less compact area where the branch points are located, and Ø = reducing moieties.

phenomenon occurs above a characteristic temperature known as the gelatinization temperature. During gelatinization, the amylose is dissolved and progressively leached from the granules. The process can be decomposed into two steps: hydration or diffusion of the solvent through the granule takes place during the first, followed by the melting of the starch crystallites in the second. This process can be studied by a calorimetric technique such as differential scanning calorimetry (DSC) or differential thermal analysis (DTA) [13, 54, 55]. The melting temperature depends on the starch/water ratio. According to Donovan [13], with a large excess of water, only one endotherm appears in the DSC curve, corresponding to the gelatinization temperature, whereas with a modest water content, only one endotherm again appears but at a higher temperature. With intermediate water concentration, two endothermic transitions are observed. This complex behaviour has been treated quantitatively using Flory's relationship between the melting point of the crystalline phases and the quantity of added water [13, 56, 57].

There are several explanations for the multiple peaks in the DSC or DTA curves. One of them is that water is not homogenously distributed in the granule and diffusion can play an important role in that process [13]. This lack of homogeneity leads to partial melting, followed by recrystallization and re-melting, implying that this is a non-reversible process [56] and that the Flory-Huggins theory is not fully applicable to starch gelatinization [13, 56].

Besides DSC and DTA, many other techniques can also be employed to study the order–disorder transition occurring during gelatinization, including X-ray scattering, light scattering, optical microscopy (birefringence using crossed polarizers), thermomechanical analysis and NMR spectroscopy and, more recently, small-angle X-ray scattering (SAXS), wide-angle X-ray scattering (WAXS) and small-angle neutron scattering (SANS) [55]. On the bases of evidence gathered using the latter techniques, Jenkins and Donald [55] concluded that the final loss of crystallinity only occurred when the gelatinization was almost complete.

In an extrusion process, where shear forces and high pressures are applied, the entire process is obviously much more complicated. However, both with limited amounts and excess quantities of water, the main step, associated with the melting of crystallites, is the same and the final mass is an amorphous entanglement of amylose and amylopectin macromolecules.

15.5 APPLICATIONS OF STARCH AS A RAW MATERIAL FOR PLASTIC PRODUCTION

15.5.1 Strategies for the use of starch as a source of polymers

The exploitation of starch as a precursor to macromolecular materials can follow two strategies, *viz.* as a raw material for the production of chemicals used in the synthesis of other polymers or used directly as a high molecular weight polymer by keeping its molecular structure as unchanged as possible.

The first strategy, based on the use of starch for the production of other chemicals was recently reviewed by Robertson *et al.* [58], Koutinas *et al.* [59], Kennedy *et al.* [60] and Otey and Doane [61]. Three different approaches are applied in this context: (i) starch as a raw material for the production of monomers used in the synthesis of polymers which can be non-biodegradable, such as polyethylene, or biodegradable, such as PLA (the main biodegradable commercial polymer whose monomer, lactic acid, can be obtained from the fermentation of starch [62]); (ii) as a raw material for the production of biopolymers like polyhydroxyalkanoates (of which PHB is the main member); (iii) as a raw material for the production of glucose, dextrin and other hydroxyl-containing monomers used in the production of mixed compositions based on starch and other monomers.

In all the above processes, the macromolecular structure of starch is destroyed and the polymers derived from the ensuing monomers are totally different from it. It is important to emphasize that the same monomers (*e.g.* ethylene, sugars and dextrin) can be produced from other sources, both renewable, such as cellulose and sugar cane, and non-renewable such as oil.

The second strategy calls upon the use of starch as such or in combination with other materials and is therefore more interesting than the first one, if anything, in terms of cost and yield. Considering, for example, the conversion of starch into polyethylene through its fermentation to ethyl alcohol and dehydration of the latter, the maximum yield of ethylene produced from starch is close to 35 per cent [61].

In order to adjust the properties of these starch-based materials to the desired application, it is necessary to combine starch with other polymers, as frequently done in the plastic industry. The need for tuneable properties may also require starch modifications, such as esterification or etherification, grafting and reactive or melting extrusion of thermoplastic starch (TPS). The main forms of starch utilization as a polymer are (i) starch grafted with vinyl monomers, (ii) starch as a filler of other polymers and (iii) plasticized starch (PLS), commonly known as TPS.

Of the two major strategies outlined above, the second constitutes the main subject of this chapter.

15.5.2 Use of starch in plastic production

In the 1960s and 1970s, oxidized starch was used successfully in rubber and other polymer formulations [63, 64] and several technologies were developed to optimize its combination with plasticized polyvinyl chloride (PVC) [61, 65, 66].

In 1972, Griffin [66] described the use of starch as particulate filler for low density polyethylene (LDPE) with the aim of giving a paper-like texture and appearance to extrusion-blown LDPE films. The necessity for highly dried starch to avoid defects caused by water volatilization was a financial limitation to the process since it required appropriate storage of the dried starch prior to its use. Another problem was the poor adhesion of the hydrophilic starch granules to the highly hydrophobic polyethylene, which was improved by treating starch with reactive silanes, but this added an extra cost to the process.

In the 1970s and 1980s, the pollution caused by plastic packages considered 'indestructible' become a serious issue and discussions about possible solutions started based on the search for materials which can degrade in landfills, hence research on biodegradable polymers became an important topic.

LDPE–starch blends seemed an interesting approach, but starch granules were encapsulated into the LDPE matrix and thus became, in principle, inaccessible to biodegradation. Later studies carried out by Griffin demonstrated that even when encapsulated, starch could be degraded in an appropriate environment [66]. A further development called upon the use of a pro-oxidant ($Fe3^{+}$, $Mn3^{+}$) in the LDPE matrix [67].

Otey *et al.* [61, 68–71] also described blends of starch with synthetic polymers, in which gelatinized starch was used instead of starch granules. The initial films, composed of 90 per cent of starch blended with poly(ethylene-*co*-acrylic acid) (EAA), were obtained by casting from aqueous dispersions or by dry milling in a rubber mill, followed by a hot roll treatment, but only the cast films attained acceptable characteristics [71]. These materials were intended for application as mulch films for agriculture.

The next generation consisted in compositions of starch and poly(EAA) films, produced by extrusion-blowing [70]. The starch concentration varied between 10 and 40 per cent, that of EAA between 10 and 90 per cent, with some composition also containing between 0 and 50 per cent of PE. Sorbitol and glycerol were also added as plasticizers to some of these mixtures. In a further investigation, Otey *et al.* described a similar composition to which urea, starch-based polyols or glycerol, or mixtures of these materials, were added [72]. Urea was used to improve the gelatinization of starch at lower levels of moisture and the polyols were added to increase the levels of biodegradable materials in the final mixture. A typical composition comprised 40 per cent of starch, 20 per cent of EAA, 15 per cent of urea and 25 per cent of LDPE. These materials were extrusion-blown into films for mulch applications. The critical point for this application is the balance between biodegradability and resistance to it, since the film can neither disintegrate before its estimated lifetime, nor can it offer excessive resistance to biodegradation.

In the 1990s, compositions of starch processed directly in melting equipment such as extruders, were described as a new material named destructurized starch or TPS [57, 73–75]. This material was patented by the Warner-Lambert Company [76] and was described as being prepared with starch that had been heated to a high enough temperature and for enough time for the melting to occur prior to starch degradation. In this process, the starch processed in extruders contained between 5 and 40 per cent of water. It was also claimed that when starch was heated in a closed volume in appropriate moisture and temperature conditions, it became substantially compatible with hydrophobic thermoplastic synthetic polymers.

Lay *et al.* [76] also described destructurized starch compositions with one or two other polymers which included a whole variety of both natural and synthetic macromolecules.

In another patented process, starch was destructurized in the presence of low molecular weight polymers such as polyethylene–vinyl alcohol (EVOH), EAA, poly-ε-caprolactone and small amounts of moisture and of a high boiling point plasticizer, using a high shear equipment like a twin-screw extruder [77]. From these materials emerged one of the most successful commercial thermoplastic materials based on starch, which took the name of Mater-Bi®.

It is also important to mention materials with high starch content, or based exclusively on starch. The main examples of this family of products are expanded TPS compositions such as those patented by the National Starch & Chemical Investment Corp [78] used in replacement of expanded polystyrene. This was one of the first commercial TPSs placed on the market and is still sold today in growing quantities. Its success is due, not only to the replacement of a synthetic polymer by a biodegradable one, but also to its good performance and because its production cost is lower than that of expanded polystyrene.

The grafting of starch with more than 60 per cent of vinyl or acrylic monomers gave materials which showed excellent mechanical and processing properties, being virtually insoluble in water [61]. However, this process was deemed too expensive and thus limited the use of the ensuing grafted starch.

15.6 THERMOPLASTIC STARCH

15.6.1 Definition and properties

The term TPS describes an amorphous or semi-crystalline material composed of gelatinized or destructurized starch containing one or a mixture of plasticizers. TPS can be repeatedly softened and hardened so that it can be moulded/shaped by the action of heat and shear forces, allowing its processing to be conducted with the techniques commonly used in the plastic industry. The following sections are devoted to a brief description of the basics of starch extrusion and processing and to the more relevant applications of TPS [50, 57, 73–75, 79–81]. TPS or destructurized starch are also known as PLS [82], because of the inevitable presence of non-volatile plasticizers in their composition. TPS is however the predominant term used for these materials.

TPS is generally produced by processing a starch/ plasticizer(s) mixture in an extruder at temperatures between 140°C and 160°C at high pressure and high shear. Batch mixers operating in conditions similar to those of extrusion can also be used.

If the final composition contains only water as the plasticizer, in levels greater than 15–20 per cent, it keeps its thermoplastic properties. However if the processing temperature is higher than 100°C, water volatilizes and the melted material expands. If controlled, this is a desired effect, exploited in the production of expanded starch used in packaging as a shock absorber. A common feature of these processing techniques is the limited amount of plasticizer and the consequent high viscosity of the melt, but TPS can also be processed in the presence of large amounts of water, as in the technology developed by Otey *et al.* [71, 72].

The destructuration temperature profile as demonstrated by Donovan [13] in his classical paper depends on the water content of the sample. Hence, the process in the presence of limited amounts of plasticizer appears to be different from that associated with an excess of it. In general, TPS is produced in the former conditions and the shear forces play a fundamental role in its processing [57].

15.6.2 Plasticizers for TPS

The type and quantity of plasticizer employed determine the preparation/processing conditions and the mechanical and thermal properties of the final material, as discussed in several studies [83–93]. Various authors [83–85] extended existing theories related to the glass transition and the effect of plasticizers on it to the glass transition temperature of TPSs. Thus, Kalichevsy and Blanshard [85] applied Couchman and Karasz's approach despite the difficulties associated with a reliable determination of the glass transition temperature of TPSs. Orford *et al.* [83], on the other hand, estimated the T_g of pure amylose and amylopectin at 500 ± 10 K, by measuring the T_g of a series of monodisperse oligomers of increasing DP. This value is above the thermal degradation temperature of these polymers.

Numerous laboratories have tackled the effect of a series of non-volatile plasticizers, such as glycerol [84, 87, 88, 89, 90, 93, 94], urea [84, 87], fructose [85], xylitol, sorbitol, maltitol [87, 90, 94], glycols (EG, TEG, PG, PEG) [84, 87, 89, 94], ethanolamine [92] and formamide [91]. Several criteria concerning the most appropriate structures for this key role have been put forward [94], although a rough first principle simply predicts that any substance capable of forming hydrogen bonds would be able to plasticize starch [95].

The partial or total replacement of water by non-volatile organic solvents (plasticizer) such as glycerol and sugars leads to an increase in the gelatinization temperature, as showed by Perry and Donald [96] a feature which needs to be considered when processing TPS. The reason for this effect is not completely understood, but Perry and Donald attributed it to two main causes, namely, the higher molecular weight plasticizers are less able to penetrate the starch granule and less able to increase the free volume of the amorphous regions, thus being less effective than water as plasticizers. This effect has alternatively been attributed to a reduction in the activity and the volume fraction of water [97]. However, Perry and Donald [96] showed that even glycerol alone can completely gelatinize starch and induce an increase in the gelatinization temperature of approximately 60°C, compared with the water plasticized counterpart. Tan *et al.* [95] recently studied this effect and concluded that the parameters determining solvent transport through the granule, such as viscosity, diffusion and ingress rate, play an important role in determining the gelatinization temperature. Other solvent properties were also considered relevant, such as molecular size and number of possible hydrogen bonds.

15.6.3 Crystallinity in TPS

The crystalline order observed in the native starch granules is completely destroyed in TPS but, because of the mobility of the starch chains, recrystallization occurs, leading generally to the formation of B-, V- and E-type crystalline structures. B-type crystallinity appears after TPS is stored above its glass transition temperature or at high plasticizer contents [81]. V- and E-types are observed just after extrusion and are generated during processing [34, 98, 99]. V-type structures can be observed in two forms, *viz.* V_a-type (anhydrous) for materials containing low moisture contents and V_h-type (hydrated) for materials containing higher moisture contents. E-type occurs only in low moisture compositions, is unstable and rearranges to V-type when the sample is conditioned at ambient temperature with more than 30 per cent relative humidity [34].

As a consequence of their semi-crystalline character, the mechanical properties of TPSs are characteristic of partially crystalline polymers [81] with B-type crystallinity being the major factor influencing the mechanical behaviour of glycerol-PLS [100]. This recrystallization can also be a problem because after long-term storage TPSs can become rigid and brittle.

Despite the fact that TPS is considered a new material in technological terms, its basic features and processes are in fact the same as those relative to extrusion-cooking starch used in the food industry since the 1960s. This kind of processing is therefore briefly described briefly in the heading, given its importance in the development of TPS.

15.6.4 Extrusion-cooking as the basis for TPS

Extrusion has been applied to pasta processing in the food industry since the mid-1930s, but the low temperatures used (<40°C) were insufficient to cook the extruded material. In the 1960s, extrusions at higher temperatures, sufficient to cook the materials, started, and extrusion-cooking became a process of great importance [34]. This process is conducted in the presence of a limited quantity of water (10–25 per cent) at temperatures that can reach 200°C but with very short residence times, so that starch decomposition is minimized. This treatment is known as high temperature short-time (HTST) [34, 101, 102]. The molten starch shows higher viscosities than common plastics during extrusion, and under isothermal conditions it can be described as a pseudoplastic material [34]. The main structural modification associated with extrusion-cooking is the destruction of the starch granule morphology (destructurized starch). However, the process is much more complex and chemical reactions that lead to depolymerization and/or other degradation also take place. The expansion of the melted starch occurs at the extrusion head because of the fast evaporation of the moisture present in the melted starch. This expansion is also an important factor in determining the properties of the extruded material. The overall process was extensively studied by Mercier and Feillet [101] who investigated such variables as the temperature of extrusion (170–200°C), the moisture content of the product before extrusion, and its amylose content.

Extrusion-cooking can be considered the precursor of the modern technology of TPS, the main principles and the changes occurring in starch being the same [63, 75]. Wiedmann and Strobel [80] proposed that the compounding of TPS is a combination of extrusion-cooking and plastic compounding.

15.6.5 Macromolecular scission and starch degradation during destructuring/plasticization

The changes in molecular weight and its distribution play an important role on the rheological and mechanical properties of starch and have therefore received considerable attention [103–106]. The main factors affecting the molecular weight degradation during TPS preparation and processing are the specific mechanical energy applied [103], the temperature and plasticizer content [101, 104].

Gomez and Aguilera [73] studied the effects of water concentration during the extrusion of maize starch on the properties of the ensuing materials and proposed a model for starch degradation during extrusion. In this model, starch granules are converted into gelatinized starch, then into free polymer chains that, depending on the extrusion conditions, can degrade into dextrinized starch and/or oligosaccharides and sugars. One important conclusion of this work was that, when extruding starch with a moisture level below 20 per cent, a product distinct from gelatinized starch is obtained, since it has been partially dextrinized. Dextrins are dextrorotatory products of partially hydrolyzed starch that can be precipitated with alcohol from their aqueous solution [15].

Willett *et al.* [103] studied the melt rheology and molecular weight degradation of waxy maize (amylopectin) in a co-rotating twin-screw extruder by processing the native starch and re-extruding the destructurized material. The moisture content of the first extrusion was 35 per cent and the product was re-extruded with 18 or 23 per cent of moisture. Starch degradation was evaluated by multi-angle light scattering in dimethyl sulphoxide/water. The weight-average molecular weight decreased moderately with increasing specific mechanical energy, which was considered an important parameter for the prediction of the molecular weight degradation during extrusion.

Myllymäki *et al.* studied the depolymerization of barley starch during its extrusion in the presence of a mixture of water and glycerol [107]. They observed a correlation between the depolymerization of the starch chains and both the water/glycerol concentration and, to a minor extent, the screw speed.

Carvalho *et al.* [108] studied chain degradation in TPS composites of maize starch plasticized with glycerol and reinforced with wood pulp. The product was characterized by high performance size-exclusion chromatography (HPSEC). The matrix (starch/glycerol) and the composites were prepared in an intensive batch mixer at 150–160°C, with glycerol and fibre contents in the range of 30–50 per cent and 5–15 per cent, respectively. The HPSEC curves obtained for different levels of plasticization without fibre are shown in Fig. 15.5.

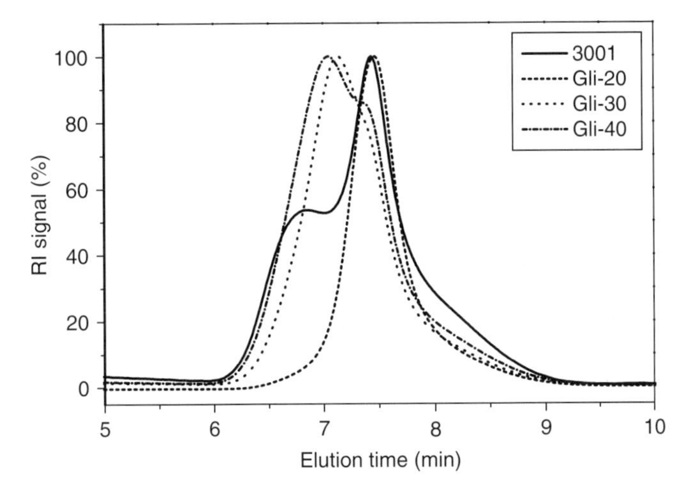

Figure 15.5 High performance size-exclusion chromatography (HPSEC) of native corn starch, and TPS with 20 per cent (Gli-20), 30 per cent (Gli-30) and 40 per cent glycerol (Gli-40). Reproduced with permission from Reference [108].

The results showed that an increase in the glycerol content reduced the starch degradation, whereas an increase in fibre content lead to its increase. The changes in the chromatographic profiles were more pronounced in the high molecular weight fraction, corresponding to amylopectin. The polydispersity index of the matrix was lower than that of native starch due to the selective breakage by shear-induced fragmentation of large amylopectin molecules. The effects of both glycerol and fibres on the molecular weight was of a similar magnitude, in the studied range, for both weight-average molecular weight, M_w, and *z*-weight-average molecular weight, M_z, and these results were expressed in terms of two equations:

$$M_W = 222\ 833 + 13\ 500G - 13\ 000\,F \tag{15.1}$$

$$M_Z = 267\ 500 + 19\ 250G - 16\ 250\,F \tag{15.2}$$

where G and F are normalized glycerol and fibre contents, respectively (30 per cent of glycerol is equal to -1 and 50 per cent of glycerol is equal to $+1$ and likewise for fibres). Glycerol and fibres showed opposite effects, as can be observed in Eq. (15.1 and 15.2). This behaviour is related to the shear-induced fragmentation, *viz.* a process in which the largest molecules are the most susceptible. As this process is highly dependent on the melt viscosity, higher viscosities induce correspondingly higher extents of degradation.

Carvalho *et al.* [109] used ascorbic and citric acid as catalysts for the controlled hydrolytic cleavage of starch macromolecules carried out by melt processing in the presence of glycerol and residual moisture. The process proved effective in providing a means for the controlled tuning of the molecular weight of starch in TPS compositions.

15.6.6 TPS blends

Blending is an important operation for the modification of polymer properties at low cost and without the need of special equipment and techniques. The possibility of blending different polymers increases considerably their range of applications and can also be used to decrease their costs. From a technological point of view, a compatible blend is achieved when the desired properties are improved as a consequence of mixing two or more polymers. In the majority of cases, this mixture is not thermodynamically compatible and constitutes hence a multiphase system [82, 110, 111].

TPS is blended mainly for two purposes. The first, and probably the most important, is to improve such properties as its water resistance and mechanical performances, whereas the second is to use it as a modifier for other polymers with the purpose of increasing the biodegradability and/or decreasing the cost of the ensuing blends. In fact, starch is cheaper than any other polymer, is readily available and renewable, so that major efforts are employed to maximize the starch content in a blend.

Blends of starch with polar polymers containing hydroxyl groups, such as poly(vinyl alcohol), copolymers of ethylene and partially hydrolyzed vinyl acetate have been prepared since the 1970s, as described by Otey *et al.* [61, 68–72]. Since starch and other natural polymers are hydrophilic, water has been commonly used as a plasticizer for these materials. The possibility of using water as plasticizer makes it possible to add the polymer to be blended as an aqueous emulsion, as for example, in the case of natural rubber latex [112], poly(vinyl acetate) and other synthetic polymer lateces [71, 113, 114]. Blends of starch and biodegradable polymers and polymers from renewable resources have been reviewed recently due to their growing importance [82, 110, 111, 115, 116]. Table 15.3 gives some polymers commonly used in blends with starch.

Starch blends can be divided into two main categories according to (1) the source and biodegradation properties of the polymer to be blended with starch and (2) the process used for its preparation. As for the first category, the sources can be obtained directly from renewable resources (biodegradable biopolymers), can be synthetic polymers from either oil or renewable resources, and in this latter case they can be biodegradable or not depending on their structure.

As for the second category, two main processing techniques are used for blending starch, namely melting and solution/dispersion. In *melting processing*, starch blends are obtained during the plasticization of starch granules in an extruder or in a batch mixer. Alternatively, two extruders are used, starch being gelatinized in a first single-screw extruder and then fed into a twin-screw extruder which processes the other component [128, 130]. In *solution or dispersion processing*, the product is often obtained by casting. The first commercial materials produced by solution or dispersion were cast films of starch/EAA blends [61, 71]. A large number of starch blends have been obtained using this procedure, especially when other natural polymers are involved. As in the case of starch, many natural polymers and also biodegradable synthetic polymers are soluble or dispersible in water. Therefore solution/dispersion blending is an interesting option for the production of these mixed materials. In some cases, as for example, when using chitosan, melting is not possible because this polymer decomposes before melting and solution blending is the only viable alternative.

Blending is one of the most promising alternatives to make starch useful as a polymer in replacement of other plastics, and the fast progress occurring in this field is attested by several reviews published recently [82, 110, 111, 115]. Indeed, the commercial plastics based on starch presently available are in the form of blends with other polymers [50, 116].

In TPS blends, starch can be the continuous or the dispersed phase, depending on the starch/second-polymer ratio and on the processing conditions. As a consequence, the interfacial interactions between starch and the other

Table 15.3

Most common polymers used in blending with thermoplastic starch

Polymer	Reference
Poly(vinyl acetate) PVAc	[71, 113]
Poly(methacrylic acid-*co*-methyl methacrylate) MAA/MMA	[114]
Poly(vinyl alcohol) PVA	[119, 120]
Poly(acrylic acid) PAA	[121, 122]
Poly(ethylene-*co*-acrylic acid) EAA	[61, 71]
Poly(ethylene-*co*-vinyl alcohol) EVOH	[49, 50, 123]
Poly(ε-caprolactone) PCL	[50, 124, 125, 123, 109]
Poly(ethylene–vinyl acetate)	[126]
Polyethylene	[127, 128, 129, 126, 130]
Poly(ester–urethanes)	[131]
Poly(D,L-lactic acid) PLA	[117, 132, 133]
Poly(3-hydroxybutyrate) PHB	[115]
Poly(3-hydroxybutyrate-*co*-3-hydroxyvalerate-) PHBV	[125, 134]
Poly(butylene succinate adipate) PBSA	[135]
Polyesteramide PEA	[124, 136]
Zein	[137, 138]
Lignin	[139, 140]
Cellulose and its derivatives	[123]
Natural rubber	[118, 112, 121]

components will determine the properties of the blend. Several approaches have been investigated in order to enhance the compatibility among the components of these blends:

(1) The use of polymers bearing polar groups, particularly those able to form hydrogen bonds (*e.g.* PVA, EAA, EVOH and natural polymers like cellulose and its derivatives, gelatin and zein [49, 50, 110, 115]).
(2) The use of mixtures of polymers where one of them acts as a compatibilizer between starch and less hydrophilic components (*e.g.* PVA in TPS/polyethylene blends or a low molecular weight polymer like poly(ethylene glycol) in TPS/PLA blends [50, 110, 115, 119]).
(3) The use of reactive compatibilizers which can promote a better interface by polymer–polymer chemical interlinking (*e.g.* methylenediphenyl diisocyanate (MDI), pyromellitic anhydride or glycidyl methacrylate [110, 115, 117, 123, 125, 126, 129]).
(4) The formation of complexes between starch and other polymers (*e.g.* V-type complexes [50, 122]).

Arvanitoyannis *et al.* [118] studied blends of starch with 1,4-transpolyisoprene (*gutta percha*) compatibilized with EAA by melt processing and observed that they were biodegradable because of the presence of starch. Their mechanical properties were improved by the addition of glycerol as plasticizer. Rouilly *et al.* [121] also prepared blends of starch and natural rubber by casting mixtures of aqueous starch with glycerol and latex. Carvalho *et al.* [112] blended native starch granules and a natural rubber latex by melt processing calling upon water as a plasticizer for starch. The stable dispersion and the good adhesion between the two natural polymers were attributed in part to the natural non-rubber constituents present in the latex. As little as 2.5 per cent w/w of rubber was sufficient to decrease drastically the brittle character of TPS. Figure 15.6 shows SEM pictures of starch/rubber blends fractured in liquid nitrogen depicting the good dispersion of rubber in the starch matrix.

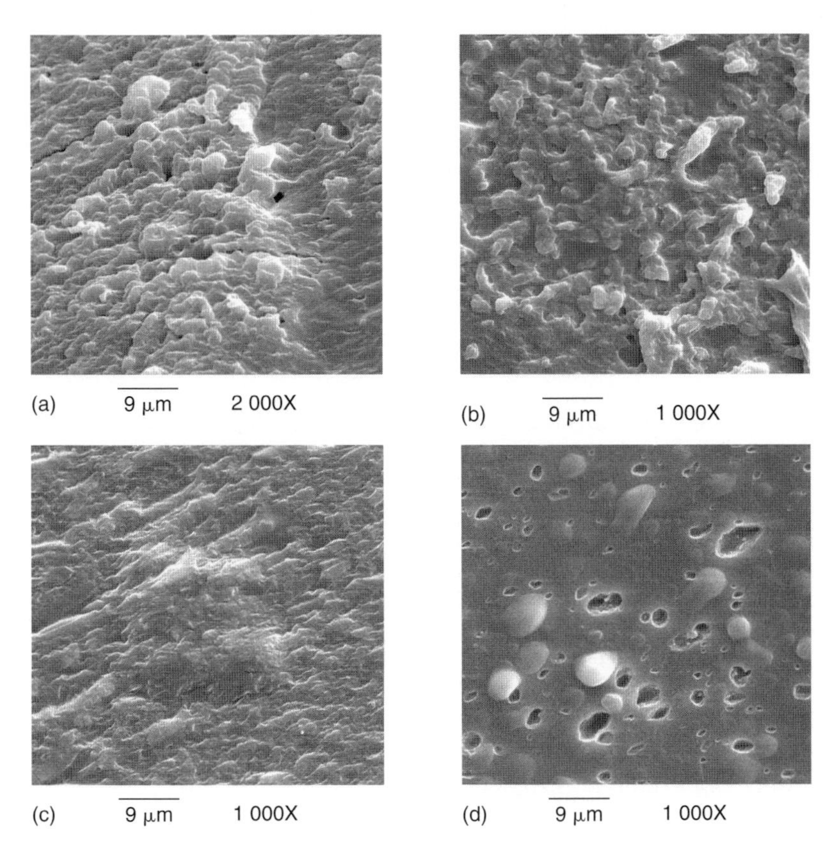

Figure 15.6 Scanning electron micrographs of fragile fractures of starch/natural rubber blends. (a) 20 per cent glycerol and 5 per cent rubber, (b) 30 per cent glycerol and 20 per cent rubber, (c) 40 per cent glycerol and 5 per cent rubber and (d) 40 per cent glycerol and 20 per cent rubber. All quantities are in w/w based on dry matter. Reproduced with permission from Reference [112].

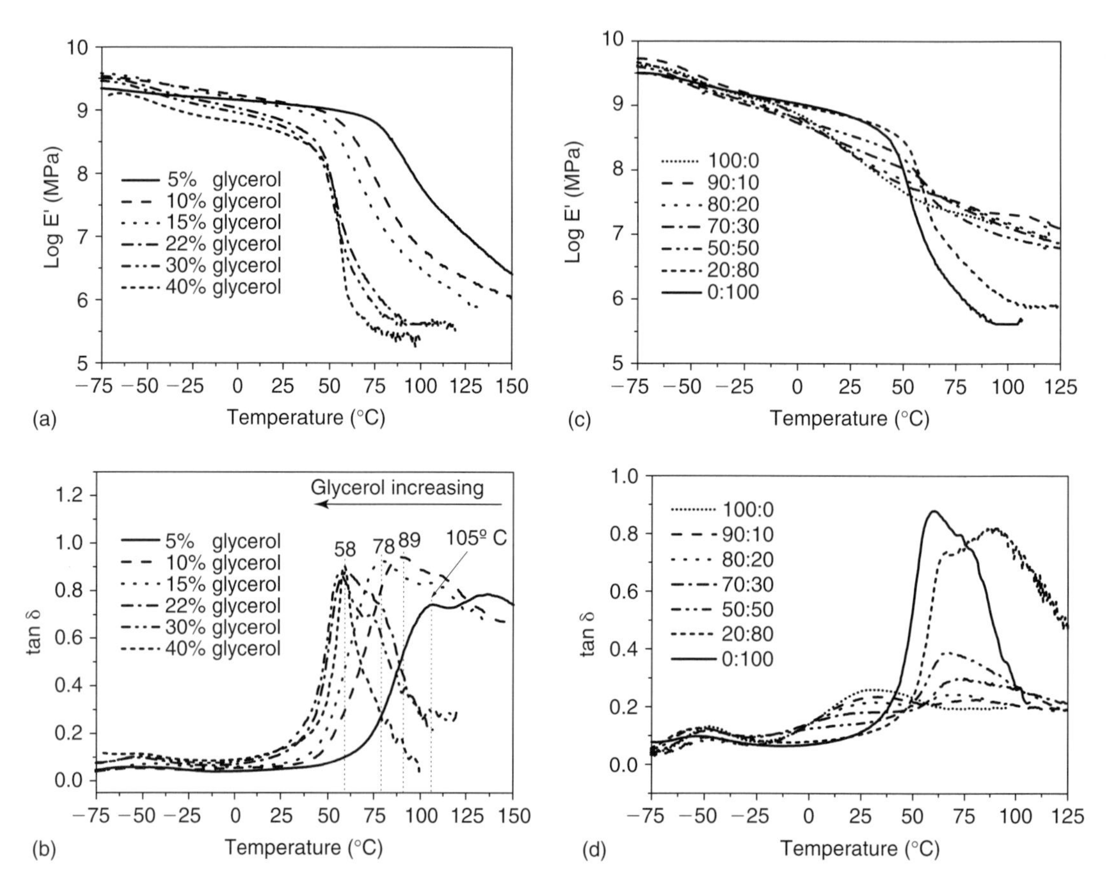

Figure 15.7 Storage modulus (E') and tan δ as a function of temperature for (a–b) plasticized zein with 5, 10, 15, 22, 30 and 40 per cent w/w of glycerol and (c–d) starch/zein blends with 22 per cent w/w of glycerol; the proportions in the legend are starch:zein. Reproduced with permission from Reference [138].

Zein is a protein obtained from maize as a by-product of maize starch production. It is completely amorphous and, despite the fact of being more hydrophobic than starch, it can also be plasticized by glycerol. It is, therefore, an interesting material for use in blends with starch because it shows some compatibility with starch while conferring to the blends a more hydrophobic character [137, 138]. Corradini *et al.* [138] described starch/zein blends prepared by melt processing. The plasticization of zein by glycerol was studied and Figs 15.7(a) and 15.7(b) show the DMA plots of the storage modulus and tan δ as a function of temperature. Above 22 per cent w/w of glycerol, its influence on the glass–rubber transition temperature seized. Consequently, the exceeding plasticizer phase separated and, in the presence of starch, migrated into the TPS-phase. Starch/zein blends are immiscible and, depending on the ratio of these two natural polymers, one or the other generated the continuous phase. Figures 15.7(c) and 15.7(d) show the dynamic mechanical properties of these blends with the two characteristic tan δ peaks corresponding to the starch/glycerol and zein/glycerol phases.

15.6.7 Composites and nanocomposites of TPS

As discussed earlier, TPS has two main drawbacks, namely it is mostly water-soluble and has poor mechanical properties; its reinforcement is one of the available options to overcome these weaknesses. Composite materials in which TPS plays the role of the matrix represent a relatively new topic and hence the literature available is rather modest. Table 15.4 lists some of these reported materials.

Table 15.4

Composites and nanocomposites based on thermoplastic starch

Matrix			Filler/Reinforcement	Reference
Commercial or common name	Description	Source		
Bioplast GS902	Potato starch blends with cellulose derivatives and synthetic polymers	Biotec Emmerich, Germany	Flax, jute, ramie, oil palm fibre	[141]
TPS/PCL-Tone P-787	Wheat starch/40% poly-ε-caprolactone – PCL	Union carbide Antwerp, Belgium	Flax and ramie	[141]
Mater-Bi ZI01U	Blends of corn starch and (poly-ε- caprolactone)	Montedison, Deutchland	Flax and ramie	[141]
TPS/TPU	Blends of TPS/ Thermoplastic polyurethane	Research[a]	Flax	[142]
TPS/PCL	Blends TPS/Poly-ε-caprolactone – PCL	Research[a]	Flax	[142]
Mater Bi	TPS blends with PCL, EVOH, etc. with at least 85% of starch	Novamont/Montedison	Non-woven of flax, hemp, ramie fibres	[143]
SCONACELL A	Modified starch blends	BSL		
Mater-Bi	TPS blends with PCL, etc.	Novamont	Flax cellulose pulp (10–40%)	[144]
TPS/PVA	TPS–polyvinyl alcohol blends	Research	Softowood fiber	[120]
TPS	Starch/Glycerol/Sorbitol–TPS	Research[a]	Regenerated cellulose fibres: Cellunier F and Temming 500	[145]
TPS	Wheat–starch/Glycerol/ Sorbitol–TPS	Research[a]	Flax and ramie	[141]
TPS	Maize starch–TPS	Research[a]	10–20% of flax fibre	[142]
TPS	Starch/Glycerol/ Formamide/Urea	Research[a]	Micro winceyette fiber	[146]
TPS	Starch/Glycerol (30–50% glycerol)	Research[a]	Kraft bleached and thermomechanical wood pulps from *Eucalyptus*	[147, 148]
TPS	Potato starch/Glycerol cast film-TPS	Research[a]	Potatoes microfibrils[b]	[149]
TPS	Waxy maize/Glycerol cast film	Research[a]	Cellulose whiskers (from tunicate)[b]	[150]
TPS	Waxy-maize/Glycerol/ Water cast film	Reseach[a]	starch nanocrystals[b]	[151]
TPS	Wheat starch/Glycerol-TPS	Research[a]	Leafwood fibres	[152]
Mater-Bi	Starch/EVOH	Novamont	Hydroxylapatite-reinforced	[153]
TPS	Maize starch/30% Glycerol	Research[a]	kaolin	[154]
TPS	Potato and wheat starch/~36.5% Glycerol	Research[a]	Up to 10 wt% of montmorillonite (Closite Na+, Closite 30B-amonium)[b]	[155, 156]

[a] Described in research papers.

[b] Filler or reinforcement with nanometric dimensions.

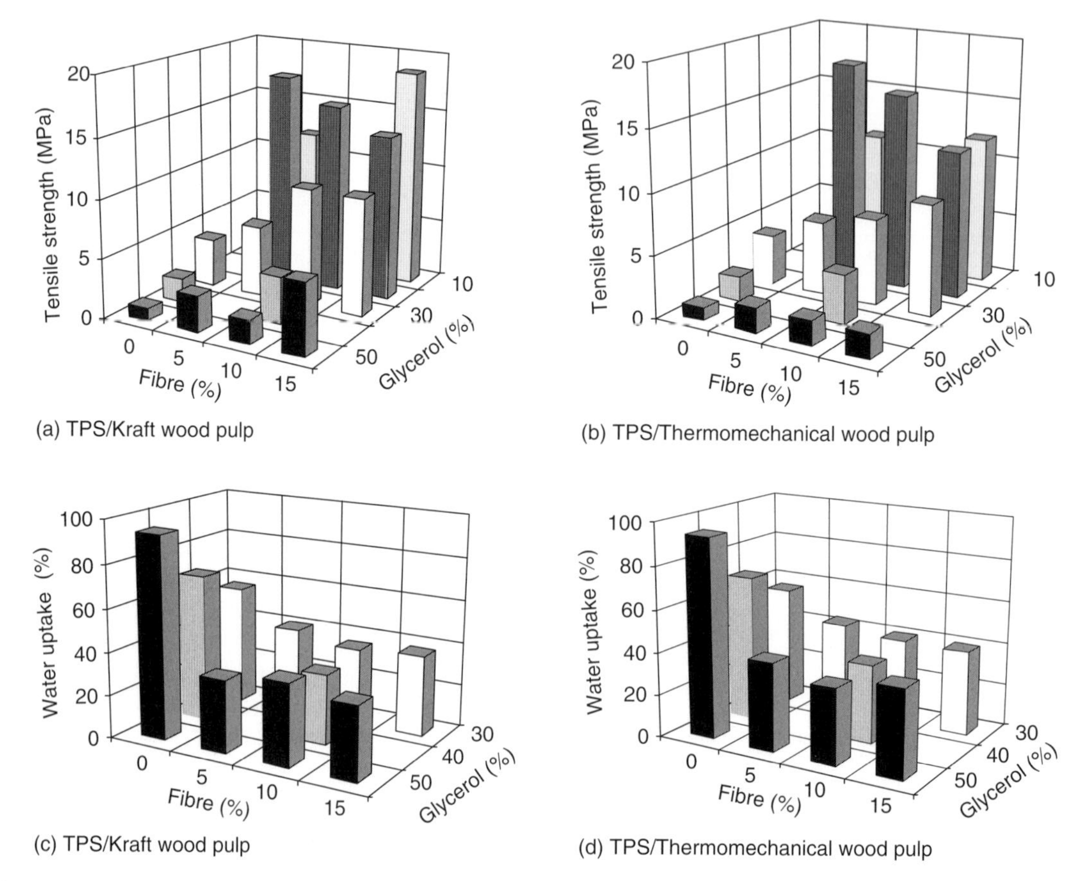

Figure 15.8 Tensile strength and equilibrium water uptake of TPS reinforced with Eucalyptus wood pulps (a–c) Kraft pulp; (b–d) thermomechanical pulp. Reproduced with permission from Reference [147].

Reinforcement has proved to be very effective in improving water resistance, and particularly the mechanical properties of TPS. Increases higher than 100 per cent in tensile strength and higher than 50 per cent in modulus were measured when TPS was reinforced with wood fibres [147–149, 152]. Figure 15.8 shows the notable increase in tensile strength and water uptake from TPS to its composites containing 10–50 wt% of glycerol and 0–15 wt% of Kraft and thermomechanical *Eucalyptus* wood pulp [147, 148]. With the addition of wood fibres, the water uptake at equilibrium decreased considerably, becoming almost independent of both fibre and glycerol contents. This behaviour was attributed mainly to the constraining effect of the reinforcement to the material expansion as water is being absorbed. The effect of lignin, present in thermomechanical fibres, and virtually absent in Kraft counterparts, was negligible.

Kaolin also played a very good role as a reinforcement for TPS matrices as shown by Carvalho *et al.* [154]. The compositions studied in this work were based on TPS containing 30 per cent of glycerol and kaolin in proportions of 10, 20, 30, 40, 50 and 60 phr. Figure 15.9 shows the variation of modulus and tensile strength of these composites as a function of kaolin content with a maximum for both at 50 phr and an increase of 135 and 50 per cent, respectively, relative to the sample prepared without kaolin. The water uptake was reduced considerably for all these kaolin-reinforced TPS composites [154].

A different approach was recently applied to cellulose fibres and starch granules to prepare single-component composites by the partial oxypropylation of these substrates. These novel materials are described in Chapter 12.

15.7 CONCLUSIONS

The necessity to replace materials based on dwindling fossil resources by homologues prepared from renewable counterparts has become an urgent matter for environmental and economical reasons. Within this context, starch

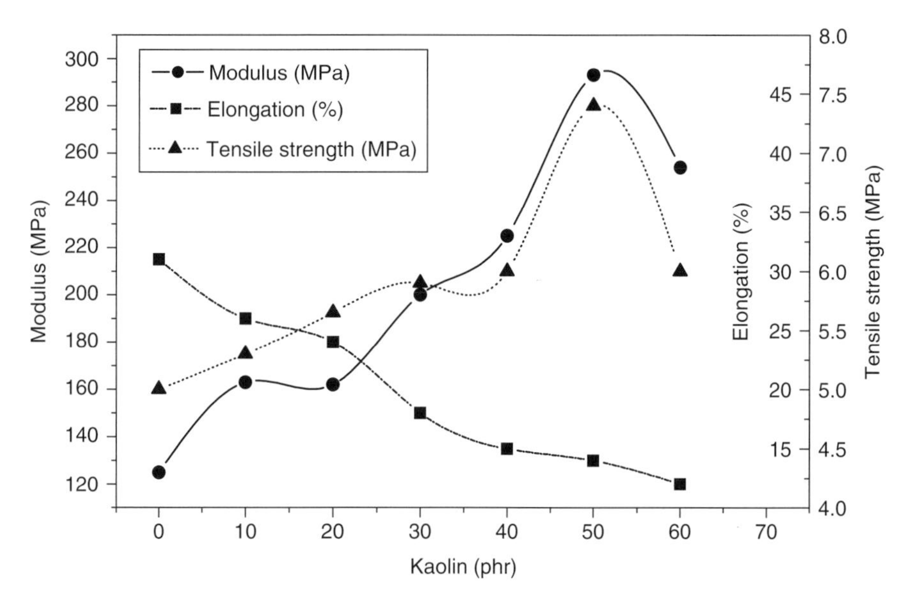

Figure 15.9 Modulus, tensile strength and elongation of TPS/kaolin composites. Reproduced with permission from Reference [154].

will certainly play a very important role as a source of viable alternatives. Its exploitation in non-food applications is not new, but had declined considerably after the Second World War because of the boom of petrochemistry and the development of polymer materials associated with it. The revival of interest in starch-based plastics began at the end of the last millennium with the emergence of new successful commercial products, and is witnessing a vigorous pursuit. The reasons for this success are to be found in the fact that starch is a cheap raw material, readily available ubiquitously (albeit from different species) and very versatile in terms of chemical and physical modifications.

ACKNOWLEDGEMENTS

I would like to thank all those who read the manuscript and provided suggestions and comments for its improvement: Dr. Debora T. Balogh, Mrs. Joan Gandini and Prof. Alessandro Gandini.

REFERENCES

1. Jarowenko W., Starch based adhesives, in *Handbook of Adhesives*, Ed.: Skeist I., 2nd Edition, Van Nostrand Reinhold Co., New York, 1977, pp. 192–211, Chapter 12.
2. Daniel J., Whistler R.L., Voragen A.C.J., Pilnik W., Starch and other polysaccharides, in *Ullmann's Encyclopedia of Industrial Chemistry*, Eds.: Elvers B., Hawkins S. and Russey W., 5th Edition, VCH Verlagsgesellschaft mbH, Weinheim, 1994, **Volume A25**, pp. 1–62.
3. Whistler R.L., Bemiller J.N., Paschall E.F., *Starch: Chemistry and Technology*, 2nd Edition, Academic Press, San Diego, CA, 1984, p. 718.
4. Baptist J.N., Process for preparing poly-β-hydroxybutyric acid, US Patent # US 3036959,1962; Baptist J.N., Process for preparing poly-β-hydroxybutyric acid, US Patent # US 3044942, 1962.
5. Baptist J.N., Werber F.X., Plasticized poly-β-hydroxybutyric acid and process, US Patent # US 3182036, 1965.
6. Prescott S.C., Dunn C.G., *Industrial Microbiology*, McGraw-Hill, New York, 1959.
7. Drumright R.E., Gruber P.R., Henton D.E., Polylactic acid technology, *Adv. Mater.*, **12**, 2000, 1841–1846.
8. Tomasik P., Schilling C.H., Chemical Modification of Starch, *Adv. Carbohydr. Chem. Biochem.*, **59**, 2004, 175–403.
9. Galliard T., *Starch: Properties and Potential*, 1st Edition, John Wiley & Sons, New York, 1987. p. 151.
10. Galliard T., Starch availability and utilization, in *Starch: Properties and Potential,* Ed.: Galliard T., 1st Edition, John Wiley & Sons, New York, 1987, pp. 1–15, Chapter 1.

11. FAO (Food and Agriculture Organization of the United Nations), 2007. *FAOSTAT* Statistical database Agriculture, Rome, Italy, data collected on January 2007.
12. Zhang P., Whistler R.L., BeMiller J.N., Hamaker B.R., Banana starch: Production, physicochemical properties, and digestibility – A review, *Carbohydr. Polym.*, **59**, 2005, 443–458.
13. Donovan J.W., Phase transitions of the starch–water system, *Biopolymers*, **18**, 1979, 263–275.
14. Jenkins P.J., Cameron R.E., Donald A.M., A universal feature in the structure of starch granules from different botanical sources, *Starch/Stärke*, **45**, 1993, 417–420.
15. Whistler R.L., Daniel J.R., Molecular structure of starch, in *Starch: Chemistry and Technology*, Eds.: Whistler R.L., Bemiller J.N. and Paschall E.F., 2nd Edition, Academic Press, San Diego, CA, 1984, pp. 153–182, Chapter 6.
16. French D., Organization of starch granules, in *Starch: Chemistry and Technology,* Eds.: Whistler R.L., Bemiller J.N. and Paschall E.F., 2nd Edition, Academic Press, San Diego, CA, 1984, pp. 183–247, Chapter 7.
17. Blanshard J.M.V., Starch granule structure and function: A physicochemical approach, in *Starch: Properties and Potential,* Ed.: Galliard T., 1st Edition, John Wiley & Sons, New York, 1987, pp. 14–54, Chapter 2.
18. Galliard T., Bowler P., Morphology and composition of starch, in *Starch: Properties and Potential*, Ed.: Galliard T., 1st Edition, John Wiley & Sons, New York, 1987, pp. 55–78, Chapter 3.
19. Gidley M.J., Bociek S.M., Molecular organization in starches: A ^{13}C CP/MAS NMR study, *J. Am. Chem. Soc.*, **107**, 1985, 7040–7044.
20. Tang H.R., Godward J., Hills B., The distribution of water in native starch granules – A multinuclear NMR study, *Carbohydr. Polym.*, **43**, 2000, 375–387.
21. Gallant D.J., Bouchet B., Baldwin P.M., Microscopy of starch: Evidence of a new level of granule organization, *Carbohydr. Polym.*, **32**, 1997, 177–191.
22. Kainuma K., Starch oligosaccharides: Linear, branched, and cyclic, in *Starch: Chemistry and Technology*, Eds.: Whistler R.L., Bemiller J.N. and Paschall E.F., 2nd Edition, Academic Press, San Diego, CA, 1984, pp. 125–152, Chapter 5.
23. Cheetham N.W.H., Tao L., Variation in crystalline type with amylose content in maize starch granules: An X-ray powder diffraction study, *Carbohydr. Polym.*, **36**, 1998, 277–284.
24. Ayoub A., Ohtani T., Sugiyama S., Atomic force microscopy investigation of disorder process in rice starch granule surface, *Starch/Stärke*, **58**, 2006, 475–479.
25. Baldwin P.M., Adler J., Davies M.C., Melia C.D., High-resolution imaging of starch granule surface by atomic force microscopy, *J. Cereal Sci.*, **27**, 1998, 255–265.
26. Ohtani T., Yoshino T., Hagiwara S., Maekawa T., High-resolution imaging of starch granule structure using atomic force microscopy, *Starch/Stärke*, **52**, 2000, 153–155.
27. Ridout M.J., Gunning A.P., Parker M.L., Wilson R.H., Morris V.J., Using atomic force microscopy to image the internal structure of starch granules, *Carbohydr. Polym.*, **50**, 2002, 123–132.
28. Ridout M.J., Parker M.L., Hedley C.L., Bogracheva T.Y., Morris V.J., Atomic force microscopy of pea starch: Granule architecture of the rug3-a, rug4-b, rug5-a and lam-c mutants, *Carbohydr. Polym.*, **65**, 2006, 64–74.
29. Snyder E.M., Industrial microscopy of starches, in *Starch: Chemistry and Technology,* Eds.: Whistler R.L., Bemiller J.N. and Paschall E.F., 2nd Edition, Academic Press, San Diego, CA, 1984, pp. 661–673, Chapter 22.
30. Fitt L.E., Snyder E.M., Photomicrographs of starches, in *Starch: Chemistry and Technology*, Eds.: Whistler R.L., Bemiller J.N. and Paschall E.F., 2nd Edition, Academic Press, San Diego, CA, 1984, pp. 675–689, Chapter 23.
31. Cameron R.E., Donald A.M., A small-angle X-ray scattering study of the annealing and gelatinization of starch, *Polymer*, **33**, 1992, 2628–2635.
32. Wu H.C.H., Sarko A., The double-helical molecular structure of crystalline B-amylose, *Carbohydr. Res.*, **61**, 1978, 7–25.
33. Wu H.C.H., Sarko A., The double-helical molecular structure of crystalline A-amylose, *Carbohydr. Res.*, **61**, 1978, 27–40.
34. Colonna P., Buleon A., Mercier C., Physically modified starches, in *Starch: Properties and Potential*, Ed.: Galliard T., 1st Edition, John Wiley & Sons, New York, 1987, pp. 79–115, Chapter 4.
35. Buléon A., Gérard C., Riekel C., Vuong R., Chanzy H., Details of the crystalline ultrastructure of C-starch granules revealed by synchrotron microfocus mapping, *Macromolecules*, **31**, 1998, 6605–6610.
36. Vermeylen R., Goderis B., Reynaers H., Delcour J.A., Amylopectin molecular structure reflected in macromolecular organization of granular starch, *Biomacromolecules*, **5**, 2004, 1775–1786.
37. Robin J.P., Mercier C., Charbonniere R., Guilbot A., Lintnerized starches. Gel filtration and enzymatic studies of soluble residues from prolonged acid treatment if potato starch, *Cereal Chem.*, **51**, 1974, 389–406.
38. Bogracheva T.Y., Morris V.J., Ring S.G., Hedley C.L., The granular structure of C-type pea starch and its role in gelatinization, *Biopolymers*, **45**, 1998, 323–332.
39. Zobel H.F., Starch crystal transformation and their industrial importance, *Starch/Stärke*, **40**, 1988, 1–7.
40. Cairns P., Bogracheva T.Y., Ring S.G., Hedley C.L., Morris V.J., Determination of the polymorphic composition of smooth pea starch, *Carbohydr. Polym.*, **32**, 1997, 275–282.

41. Young A.H., Fractionation of starch, in *Starch: Chemistry and Technology*, Eds.: Whistler R.L., Bemiller J.N. and Paschall E.F., 2nd Edition, Academic Press, San Diego, CA, 1984, pp. 249–283, Chapter 8.
42. Charoenkul N., Uttapap D., Pathipanawat W., Takeda Y., Molecular structure of starches form cassava varieties having different cooked root textures, *Starch/Stärke*, **58**, 2006, 443–452.
43. Buléon A., Duprat F., Single crystals of amylose with low degree of polymerization, *Carbohydr. Polym.*, **4**, 1984, 161–173.
44. Becker A., Hill S.E., Mitchell J.R., Relevance of amylose–lipid complexes to the behaviour of thermally processed starches, *Starch/Stärke*, **53**, 2001, 121–130.
45. Karkalas J., Ma S., Morrison W.R., Pethrick R.A., Some factors determining the thermal properties of amylose inclusion complexes with fatty acids, *Carbohydr. Res.*, **268**, 1995, 233–247.
46. Takeo K., Kuge T., Complexes of starchy materials with organic compounds .3. X-ray studies on amylose and cyclodextrin complexes, *Agric. Biol. Chem.*, **33**, 1969, 1174–1180.
47. Gidley M.J., Bociek S.M., ^{13}C CP/MAS NMR studies of amylose inclusion complexes, cyclodextrins, and the amorphous phase of starch granules: Relationships between glycosidic linkage conformation and solid-state ^{13}C chemical shifts, *J. Am. Chem. Soc*, **110**, 1988, 3820–3829.
48. Godet M.C., Bouchet B., Colonna P., Gallant D.J., Buléon A., Crystalline amylose–fatty acid complexes: Morphology and crystal thickness, *J. Food Sci.*, 1996, 1196–2101.
49. Simmons S., Thomas E.L., Structural characteristics of biodegradable thermoplastic starch/poly(ethylene vinyl alcohol) blends, *J. Appl. Polym. Sci.*, **58**, 1995, 2259–2285.
50. Bastioli C., Properties and applications of Mater-Bi starch-based materials, *Polym. Degrad. Stab.*, **59**, 1998, 263–272.
51. Matheson N.K., Caldwell R.A., α(1–4) Glucan chain disposition in models of α(1–4)(1–6) glucans: Comparison with structural data for mammalian glycogen and waxy amylopectin, *Carbohydr. Polym.*, **40**, 1999, 191–209.
52. Thompson D.B., On the non-random nature of amylopectin branching, *Carbohydr. Polym.*, **43**, 2000, 223–239.
53. Nikuni Z., Studies on starch granules, *Starch/Stärke*, **30**, 1978, 105–111.
54. Liu H., Lelievre J., Ayoung-Chee W.A., A study of starch gelatinization using differential scanning calorimetry, X-ray, and birefringence measurements, *Carbohydr. Res.*, **210**, 1991, 79–87.
55. Jenkins P.J., Donald A.M., Gelatinization of starch: A combined SAXS/WAXS/DSC and SANS study, *Carbohydr. Res.*, **308**, 1998, 133–147.
56. Biliaderis C.G., Page C.M., Maurice T.J., Juliano B.O., Thermal characterization of rice starches: A polymeric approach to phase transitions of granular starch, *J. Agric. Food Chem.*, **34**, 1986, 6–14.
57. Kokini J.L., Lai L.S., Chedid L.L., Effects of starch structure on starch rheological properties, *Food Technol.*, **46**, 1992, 124–139.
58. Robertson G.H., Wong D.W.S., Lee C.C., Wagschal K.W., Smith M.R., Orts W.J., Native or raw starch digestion: A key step in energy efficient biorefining of grain, *J. Agric. and Food Chem.*, **54**, 2006, 353–365.
59. Koutinas A.A., Wang R., Webb C., Evaluation of wheat as generic feedstock for chemical production, *Ind. Crops and Prod.*, **20**, 2004, 75–88.
60. Kennedy J.F., Cabral J.M.S., Sá-Correia I., White C.A., Starch biomass: A chemical feedstock for enzyme and fermentation processes, in *Starch: Properties and Potential*, Ed.: Galliard T., 1st Edition, John Wiley & Sons, New York, 1987, pp. 115–148, Chapter 5.
61. Otey F.H., Doane W.M., Chemicals from starch, in *Starch: Chemistry and Technology*, Eds.: Whistler R.L., Bemiller J.N. and Paschall E.F., 2nd Edition, Academic Press, San Diego, CA, 1984, pp. 389–416, Chapter 11.
62. Garlotta D., A literature review of poly(lactic acid), *J. Polym. Environ.*, **9**, 2001, 63–84.
63. Griffin G.J.L., Gelatinized starch products, in *Chemistry and Technology of Biodegradable Polymers*, Ed.: Griffin G.J.L., 1st Edition, Blackie Academic & Professional, London, 1984, pp. 135–150, Chapter 7.
64. Pfeifer V.F., Sohns V.E., Conway H.F., Lancaster E.B., Dabic S., Griffin E.L., 2-Stage process for dialdehyde starch using electrolytic regeneration of periodic acid, *Ind. Eng. Chem.*, **52**, 1960, 201–206.
65. Westhoff R.P., Otey F.H., Mehltretter C.L., Russell C.R., Starch-filled polyvinyl chloride plastics – Preparation and evaluation, *Ind. Eng. Chem. Prod. Res. Dev.*, **13**, 1974, 123–125.
66. Griffin G.J.L., Particulate starch based products, in *Chemistry and Technology of Biodegradable Polymers*, Ed.: Griffin G. J.L., 1st Edition, Blackie Academic & Professional, London, 1984, pp. 18–47, Chapter 3.
67. Arnaud R., Dabin P., Lemaire J., Al-Malaika S., Chohan S., Coke M., Scott G., Fauve A., Maaroufi A., Photooxidation and biodegradation of commercial photodegradable polyethylenes, *Polym. Degrad. Stab.*, **46**, 1994, 211–224.
68. Otey F.H., Westhoff R.P., Biodegradable film compositions prepared from starch and copolymers of ethylene and acrylic acid, US Patent # US 4133784, January 9, 1979.
69. Otey F.H., Westhoff R.P., Biodegradable starch-based blown films, US Patent # US 4337181, June 29, 1982.
70. Otey F.H., Westhoff R.P., Doane W.M., Title: Starch-based blown films, *Ind. Eng. Chem. Prod. Res. Dev.*, **19**, 1980, 592–595.

71. Otey F.H., Westhoff R.P., Russell C.R., Biodegradable films from starch and ethylene–acrylic acid copolymer, *Ind. Eng. Chem. Prod. Res. Dev.*, **16**, 1977, 305–308.
72. Otey F.H., Westhoff R.P., Doane W.M., Starch-based blown films .2., *Ind. Eng. Chem. Res.*, **26**, 1987, 1659–1663.
73. Gomez M.H., Aguilera J.M., A physicochemical model for extrusion of corn starch, *J. Food. Sci.*, **49**, 1984, 40–49.
74. Röper H., Koch H., The role of starch in biodegradable thermoplastic materials, *Starch/Stärke*, **42**, 1990, 123–130.
75. Shogren R.L., Fanta G.F., Doane W.M., Development of starch based plastics – A reexamination of selected polymer systems in historical perspective, *Starch/Stärke*, **45**, 1993, 276–280.
76. Gustav L., Rehm J., Stepto R.F., Thoma R., Sachetto J-P., Lentz D.J., Silbiger J., Polymer composition containing destructurized starch, US Patent # US 5095054, 1992.
77. Bastioli C., Lombi R., Deltredici G., Guanella I., De-structuring starch for use in biodegradable plastics articles – By heating non-dried, non-water-added starch with plasticizer and enzyme in an extruder, Eur. Pat. # EP400531-A1, 1991.
78. Lacourse N.L., Altieri P.A., Biodegradable shaped products and method of preparation, US Patent # US 5035930, 1991.
79. Lörcks J., Properties and applications of compostable starch-based plastic materials, *Polym. Degrad. Stab.*, **59**, 1998, 245–249.
80. Wiedmann W., Strobel E., Compounding of thermoplastic starch with twin-screw extruders, *Starch/Stärke*, **43**, 1991, 138–145.
81. van Soest J.J.G., de Wit D., Vliegenthart F.G., Mechanical properties of thermoplastic waxy maize starch, *J. Appl. Polm. Sci.*, **61**, 1996, 1927–1937.
82. Avérous L., Biodegradable multiphase systems based on plasticized starch: A review, *J. Macromol. Sci., Part-C.*, **C44**, 2004, 231–274.
83. Orford P.D., Parker R., Ring S.G., Smith A.C., Effect of water as a diluent on the glass transition behavior of malto-oligosaccharides, amylase and amylopectin, *Int. J. Biol. Macromol.*, **11**, 1989, 91–96.
84. Shogren R.L., Swanson C.L., Thompson A.R., Extrudates of cornstarch with urea and glycols: Structure/mechanical property relations, *Starch/Stärke*, **44**, 1992, 335–338.
85. Kalichevsky M.T., Blanshard J.M.V., The effect of fructose and water on the glass transition of amylopectin, *Carbohydr. Polym.*, **20**, 1993, 107–113.
86. Poutanen K., Forsell P., Modification of starch properties with plasticizers, *Trends Polym. Sci.*, **4**, 1996, 128–132.
87. Lourdin D., Bizot H., Colonna P., "Antiplasticization" in starch–glycerol films?, *J. Appl. Polym. Sci.*, **64**, 1997, 1047–1053.
88. Lourdin D., Coignard L., Bizot H., Colonna P., *Polymer*, **38**, 1997, 5401–5406.
89. Hulleman S.H.D., Janssen F.H.P., Feil H., The role of water during plasticization of native starches, *Polymer*, **39**, 1998, 2043–2048.
90. Mathew A.P., Dufresne A., Plasticized waxy maize starch: Effect of polyols and relative humidity on material properties, *Biomacromolecules*, **3**, 2002, 1101–1108.
91. Ma X., Yu J., Formamide as the plasticizer for thermoplastic starch, *J. Appl. Polym Sci.*, **93**, 2004, 1769–1773.
92. Huang M., Yu J., Ma X., Ethanolamine as a novel plasticizer for thermoplastic starch, *Polym. Degrad. Stab.*, **90**, 2005, 501–507.
93. Teixeira E.D., Da Roz A.L., de Carvalho A.J.F., Curvelo A.A.S., Preparation and characterisation of thermoplastic starches from cassava starch cassava root and cassava bagasse, *Macromol. Symposia.*, **229**, 2005, 266–275.
94. Da Roz A.L., Carvalho A.J.F., Gandini A., Curvelo A.A.S., The effect of plasticizers on thermoplastic starch compositions obtained by melt processing, *Carbohydr. Poly.*, **63**, 2006, 417–424.
95. Tan I., Wee C.C., Sopade P.A., Halley P.J., Investigation of the starch gelatinization phenomena in water–glycerol system: Application of modulate temperature differential scanning calorimetry, *Carbohyd. Polym.*, **58**, 2004, 191–204.
96. Perry P.A., Donald A.M., The role of plasticization in starch granule assembly, *Biomacromolecules*, **1**, 2000, 424–432.
97. Derby R.I., Miller B.S., Miller B.H., Trimbo H.B., Visual observation of wheat starch gelatinization in limited water systems, *Cereal Chem.*, **52**, 1975, 702–713.
98. Mercier C., Charbonniere R., Grebaut J. *et al*, Formation of amylose–lipid complexes by twin-screw extrusion cooking of manioc starch, *Cereal Chem.*, **57**, 1980, 4–9.
99. van Soest J.J.G., Essers P., Influence of amylose–amylopectin ratio on the properties of extrude starch plastics shets, *Pure Appl. Chem.*, **A34**, 1997, 1665–1689.
100. Hulleman .S.H.D., Kalisvaart M.G., Janssen F.H.P., Feil H., Vliegenthart J.F.G., Origins of B-type crystallinity in glycerol-plasticized, compression moulded potato starches, *Carbohydr. Polym.*, **39**, 1999, 351–360.
101. Mercier C., Feillet P., Modification of carbohydrate components by extrusion-cooking of cereal products, *Cereal Chem.*, **52**, 1975, 283–297.
102. Lawton B.T., Henderso G.A., Derlatka E.J., The effect of extruder variables on the gelatinization of corn starch, *Can. J. Chem. Eng.*, **50**, 1972, 168–172.
103. Willett J.L., Millard M.M., Jasberg B.K., Extrusion of waxy maize starch: Melt rheology and molecular weight degradation of amylopectin, *Polymer*, **38**, 1997, 5983–5989.

104. Mercier C., Effect of extrusion-cooking on potato starch using a twin-screw French extruder, *Starke*, **29**, 1977, 48–52.
105. van Soest J.J.G., Benes K., de Wit D., Vliegenthart J.F.G., The influence of starch molecular mass on the properties of extruded thermoplastic starch, *Polymer*, **37**, 1996, 3543–3552.
106. Bastioli C., Bellotti V., Rallis A., Microstructure and melt flow behavior of a starch-based polymer, *Rheol. Acta*, **33**, 1994, 307–316.
107. Myllymäki O., Eerikainen T., Suortti T., Forssele P., Linko P., Poutanen K., Depolymerization of barley starch during extrusion in water glycerol mixtures, *Food Sci. Technol., Lebensm. Wiss. Technol.*, **30**, 1997, 351–358.
108. Carvalho A.J.F., Zambon M.D., Curvelo A.A.S., Gandini A., Size exclusion chromatography characterization of thermoplastic starch composites 1. Influence of plasticizer and fibre content, *Polym. Degrad. Stab.*, **79**, 2003, 133–138.
109. Carvalho A.J.F., Zambon M.D., Curvelo A.A.S., Gandini A., Thermoplastic starch modification during melting processing: Hydrolysis catalyzed by carboxylic acids, *Carbohydr. Polym.*, **62**, 2005, 387–390.
110. Wang X.L., Yang K.K., Wang Y.Z., Properties of starch blends with biodegradable polymers, *J. Macromol. Sci., Part C*, **C43**, 2003, 385–409.
111. Amass W., Amass A., Tighe B., A review of biodegradable polymers: Uses, current development in the synthesis and characterization of biodegradable polyesters, blends of biodegradable polyesters and recent advances in biodegradable studies, *Polym. Int.*, **47**, 1998, 89–144.
112. Carvalho A.J.F., Job A.E., Alves N., Curvelo A.A.S., Gandini A., Thermoplastic starch/natural rubber blends, *Carbohydr. Polym.*, **53**, 2003, 95–99.
113. Ritter H.W., Bergner D.R., Kempf K.W., Starch-based materials and/or molded parts modified by synthetic polymer compounds and process for production the same. US Patent # US 5439953, 1995.
114. Bortnick N.M., Graham R.K., LaFleur E.E., Work W.J., Wu J.C., Melt-processed polymer blends, US Patent # US 5447669, 1995.
115. Yu L., Dean K., Li L., Polymer blends and composites from renewable resources, *Prog. Polym. Sci.*, **31**, 2006, 502–576.
116. Bastioli C., Global status of the production of biobased packaging materials, *Starch/Stärke*, **53**, 2001, 351–355.
117. Wang H., Sun X.Z., Seib P., Mechanical properties of poly(lactic acid) and wheat starch blends with methylenediphenyl diisocyanate, *J. Appl. Polym. Sci.*, **84**, 2002, 1257–1262.
118. Arvanitoyannis I., Kolokuris I., Nakayama A., Aiba S., Preparation and study of novel biodegradable blends based on gelatinized starch and 1,4-trans-polyisoprene (gutta percha) for food packaging or biomedical applications, *Carbohydr. Polym.*, **34**, 1997, 291–302.
119. Follain N., Joly C., Dole P., Roge B., Mathlouthi M., Quaternary starch based blends: Influence of a fourth component addition to the starch/water/glycerol system, *Carbohydr. Polym.*, **63**, 2006, 400–407.
120. Shogren R.L., Lawton J.W., Tiefenbacher K.F., Baked starch foams: Starch modifications and additives improve process parameters, structure and properties, *Ind. Crops. Prod.*, **16**, 2002, 69–79.
121. Rouilly A., Rigal L., Gilbert R.G., Synthesis and properties of composites of starch and chemically modified natural rubber, *Polymer*, **45**, 2004, 7813–7820.
122. Biswas A., Willet J.L., Gordon S.H., Finkenstadt V.L., Cheng H.N., Complexation and blending of starch, poly(acrylic. acid), and poly(N-vinyl pyrrolidone), *Carbohydr. Polym.*, **65**, 2006, 397–403.
123. Demirgöz D., Elvira C., Mano J.F., Cunha A.M., Piskin E., Reis R.L., Chemical modification of starch based biodegradable polymeric blends: Effects on water uptake, degradation behaviour and mechanical properties, *Polym Degrad. Stab.*, **70**, 2000, 161–170.
124. Avérous L., Moro L., Dole P., Fringant C., Properties of thermoplastic blends: Starch–polycaprolactone, *Polymer*, **41**, 2000, 4157–4167.
125. Avella M., Errico M.E., Laurienzo P., Martuscelli E., Raimo M., Rimedio R., Preparation and characterization of compatibilized polycaprolactone/starch composites, *Polymer*, **41**, 2000, 3875–3881.
126. Mani R., Bhattacharya M., Properties of injection moulded starch/synthetic polymers blends-III. Effect of amylopectin to amylose ratio in starch, *Eur. Polym. J.*, **34**, 1998, 1467–1475.
127. Griffin G.J.L., Starch polymer blends, *Polym. Degrad. Stab.*, **45**, 1994, 241–247.
128. St-Pierre N., Favis B.D., Ramsay B.A., Ramsay J.A., Verhoogt H., Processing and characterization of thermoplastic starch/polyethylene blends, *Polymer*, **38**, 1997, 647–655.
129. Bikiaris D., Prinos J., Koutsopoulos K., Vouroutzis N., Pavlidou E., Frangis N., Panayiotou C., LDPE/plasticized starch blends containing PE–g–MA copolymer as compatibilizer, *Polym. Degrad Stab.*, **59**, 1998, 287–291.
130. Rodriguez-Gonzalez F.J., Ramsay B.A., Favis B.D., High performance LDPE/thermoplastic starch blends: A sustainable alternative to pure polyethylene, *Polymer*, **44**, 2003, 1517–1526.
131. Seidenstucker T., Fritz H.G., Innovative biodegradable materials based upon starch and thermoplastic poly(ester-urethane) (TPU), *Polym. Degrad. Stab.*, **59**, 1998, 279–285.
132. Zhang J.F., Sun X.Z., Mechanical properties of poly(lactic acid)/starch composites compatibilized by maleic anhydride, *Biomacromolecules*, **5**, 2004, 1446–1451.

133. Martin O., Avérous L., Poly(lactic acid): Plasticization and properties of biodegradable multiphase systems, *Polymer*, **42**, 2001, 6209–6219.
134. Kotnis M.A., Obrien G.S., Willett J.L., Processing and mechanical-properties of biodegradable poly(hydroxybutyrate-co-valerate)–starch compositions, *J. Env. Polym. Degrad.*, **3**, 1995, 97–105.
135. Avérous L., Fringant C., Association between plasticized starch and polyesters: Processing and performances of injected biodegradable systems, *Polym. Eng. Sci.*, **41**, 2001, 727–734.
136. Schwach E., Avérous L., Starch-based biodegradable blends: Morphology and interface properties, *Polym. Int.*, **53**, 2004, 2115–2124.
137. Chanvrier H., Colonna P., Della Valle G., Lourdin D., Structure and mechanical behaviour of corn flour and starch–zein based materials in the glassy state, *Carbohydr. Polym.*, **59**, 2005, 109–119.
138. Corradini E., de Medeiros E.S., Carvalho A.J.F., Curvelo A.A.S., Mattoso L.H.C., Mechanical and morphological characterization of starch/zein blends plasticized with glycerol, *J. Appl. Polym. Sci.*, **101**, 2006, 4133–4139.
139. Baumberger S., Lapierre C., Monties B., Della Valle G., Use of kraft lignin as filler for starch films, *Polym Degrad. Stab.*, **59**, 1998, 273–277.
140. Morais L.C., Curvelo A.A.S., Zambon M.D., Thermoplastic starch–lignosulfonate blends. 1. Factorial planning as a tool for elucidating new data from high performance size-exclusion chromatography and mechanical tests, *Carbohydr. Polym.*, **62**, 2005, 104–112.
141. Wollerdorfer M., Bader H., Influence of natural fibres on the mechanical properties of biodegradable polymers, *Ind. Crops Prod.*, **8**, 1998, 105–112.
142. Mittenzwey R., Seidenstücker T., Fritz H., Süβmuth R., Prüfung der umweltverträglichkeit neu entwickelter polymerwerkstoffe auf der basis nachwachsender rohstoffe durch ein einfaches testsystem, *Starch/Stärke*, **10**, 1998, 438–443.
143. Herrmann A.S., Nickel J., Riedel U., Construction materials based upon biologically renewable resources – from components to finished parts, *Polym. Degrad. Stab.*, **59**, 1998, 251–261.
144. Puglia D., Tomassucci A., Kenny J.M., Processing, properties and stability of biodegradable composites based on Mater-Bi-(R) and cellulose fibres, *Polym. Adv. Technol.*, **14**, 2003, 749–756.
145. Funke U., Bergthaller W., Lindhauer M.G., Processing and characterization of biodegradable products based on starch, *Polym. Degrad. Stab.*, **59**, 1998, 293–296.
146. Ma X.F., Yu J.G., Kennedy J.F., Studies on the properties of natural fibers-reinforced thermoplastic starch composites, *Carbohydr. Polym.*, **62**, 2005, 19–24.
147. Carvalho A.J.F., Curvelo A.A.S., Agnelli J.A.M., Wood pulp reinforced thermoplastic starch composites, *Int. J. Polym. Mater.*, **51**, 2002, 647–660.
148. Curvelo A.A.S., Carvalho A.J.F., Agnelli J.A.M., Thermoplastic starch-cellulosic fibers composites: Preliminary results, *Carbohydr. Polym.*, **45**, 2001, 183–188.
149. Dufresne A., Vignon M.R., Improvement of starch film performance using cellulose microfibrils, *Macromolecules*, **31**, 1998, 2693–2696.
150. Anglès M.N., Dufresne A., Plasticized starch/tunicin whiskers nanocomposites materials. 2. Mechanical behavior, *Macromolecules*, **34**, 2001, 2921–2931.
151. Angellier H., Molina-Boisseau S., Dole P., Dufresne A., Thermoplastic starch–waxy maize starch nanocrystals nanocomposites, *Biomacromolecules*, **7**, 2006, 531–539.
152. Avérous A., Frigant C., Moro L., Plasticized starch–cellulose interactions in polysaccharide composites, *Polymer*, **42**, 2001, 6565–6572.
153. Reis R.L., Cunha A.M., Allan P.S., Bevis M.J., Structure development and control of injection-molded hydroxylapatite-reinforced starch/EVOH composites, *Adv. Polym. Technol.*, **16**, 1997, 263–277.
154. Carvalho A.J.F., Curvelo A.A.S., Agnelli J.A.M., A first insight on composites of thermoplastic starch and kaolin, *Carbohydr. Polym.*, **45**, 2001, 189–194.
155. Park H.M., Lee W.K., Park C.Y., Cho W.J., Ha C.S., Environmentally friendly polymer hybrids, *J Mater. Sci.*, **38**, 2003, 909–915.
156. Bagdi K., Müller P., Pukánszky B., Thermoplastic starch/layered silicate composites: Structure, interactions, properties, *Compos. Interfaces*, **13**, 2006, 1–17.

– 16 –

Cellulose Chemistry: Novel Products and Synthesis Paths

Thomas Heinze and Katrin Petzold

ABSTRACT

Novel paths for homogeneous and regioselective functionalization of cellulose are discussed. The acylation of cellulose can be efficiently carried out by homogeneous phase chemistry applying solvents based on polar aprotic media or in ionic liquids and *in situ* activation of the carboxylic acid. Some unconventional cellulose derivatives are described. The regioselective derivatization of protected cellulosics leading to 3-*O*-, 2,3-*O*-, and 6-*O*-functionalized products is of recent interest showing remarkable differences in properties compared with common cellulose derivatives. The nucleophilic displacement reactions with cellulose tosylates provide a further tool for the design of biopolymer-based structures and properties.

Keywords

Cellulose, Polysaccharide, Activation, Cellulose solvents, Esterification, Etherification, Ionic liquids, NMR spectroscopy, Nucleophile displacement, Protecting groups, Regioselective functionalization

16.1 INTRODUCTION

The chemical modification of polysaccharides is still underestimated regarding the structure and hence property design of biopolymer-based materials. At present, the cellulose derivatives commercially produced on a large scale are limited to some esters with C_2 to C_4 carboxylic acids, including mixed esters and phthalic acid half esters and some ethers with methyl-, hydroxyalkyl-, and carboxymethyl functions. In general, the organic chemistry of cellulose opens the way to a broad variety of products by esterification and etherification. In addition, novel products may be obtained by nucleophilic displacement reactions, unconventional chemistry like 'click reactions' and controlled oxidation. The aim of this chapter is to highlight selected recent advances in the chemical modification of cellulose for the synthesis of new products and alternative synthetic paths, in particular under homogeneous conditions, that is, starting with the dissolved polymer, with emphasis on the research from the authors' laboratory. Some issues related to cellulose solvents are also discussed.

16.2 CARBOXYLIC ACID ESTERS

The esterification of cellulose with carboxylic acids is among the most versatile transformations leading to a wide variety of valuable products. The commercial production of cellulose esters is carried out exclusively under heterogeneous reaction conditions, at least at the beginning of the conversion, due to the high cost of solvents and the ease of work-up procedures in the case of multiphase conversions. Completely functionalized derivatives are generally sought, because partial functionalization mainly leads to insoluble or partly soluble products. For a better control of the reaction temperature and a reduction in the amount of catalyst, acetylation is carried out preferably in methylene chloride, which is associated with the dissolution of the products formed in the final phase of the reaction. It should be pointed out that the dissolution of the cellulose acetate formed in the reaction medium must not be assimilated with the homogeneous reactions where the cellulose is initially dissolved in a solvent like *N,N*-dimethyl acetamide (DMA)/LiCl.

A variety of solvent systems have been established for the homogeneous acylation of cellulose at the laboratory scale. These homogeneous reactions permit the synthesis of highly soluble polymers, partially derivatized polymers and are the prerequisite for the application of the 'state of the art' organic reagents yielding a broad structural diversity [1]. The scope of this section is to focus on homogeneous reactions starting with the dissolved cellulose by applying novel activation procedures for the carboxylic acids.

16.2.1 Cellulose solvents

The basic requirement for cellulose dissolution is that the solvent is capable of interacting with the hydroxyl groups of the AGU, so as to eliminate, at least partially, the strong intermolecular hydrogen bonding between the polymer chains. There are two basic schemes for cellulose dissolution:

(1) where it results from physical interactions between cellulose and the solvent;
(2) where it is achieved via chemical reaction leading to covalent bond formation, these solvents usually being called 'derivatizing solvents'.

Route (1) is addressed in detail below because the 'real' solvents are mostly used as the reaction medium for the introduction of unconventional functional groups onto cellulose.

16.2.1.1 Derivatizing solvents

Certain solvents react with cellulose leading to a disruption of the hydrogen bonding by a combination of steric interactions and decrease of the number of hydroxyl groups available for hydrogen bonding. The covalent bond formed should be easily cleavable, for example, by hydrolysis or a change in the pH value of the system. Examples of derivatizing solvents, including the intermediate formed, are summarized in Fig. 16.1 [2]. One problem of this approach of cellulose dissolution is its poor reproducibility due to side reactions and the formation of undefined structures. Nevertheless, various esterification reactions were successfully carried out in these media [2].

The importance of this scheme is that it may lead to an *inverse pattern of functionalization*, provided that special reaction conditions are employed, namely, the involvement of the secondary hydroxyl groups, due to the fact that the primary counterparts that is the more reactive sites, are blocked by the moiety (nitrite, methylol, or reactive acyl function) introduced during dissolution. In some instances, the intermediate formed prior to the onset of conversion (under non-aqueous conditions) into the final product was isolated and dissolved in a common organic solvent, for example dimethyl sulfoxide (DMSO) and *N,N*-dimethyl formamide (DMF), to achieve the actual functionalization. At present, this synthesis path is not considered as a viable tool for esterification, although it gives access to novel products.

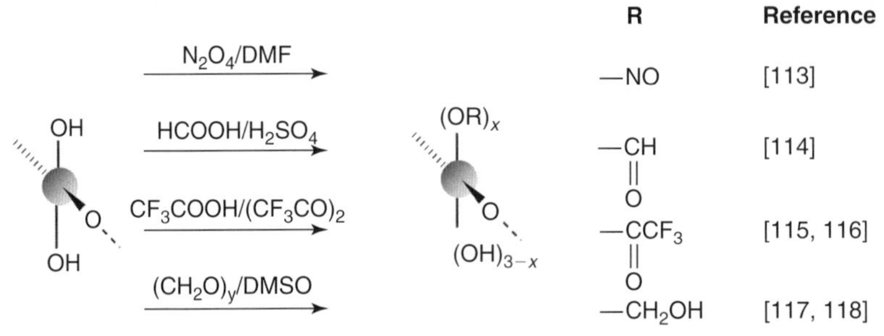

Reagent	R	Reference
N_2O_4/DMF	—NO	[113]
$HCOOH/H_2SO_4$	—CH(=O)	[114]
$CF_3COOH/(CF_3CO)_2$	—C(=O)CF_3	[115, 116]
$(CH_2O)_y$/DMSO	—CH_2OH	[117, 118]

AGU (OH)$_3$ → AGU $(OR)_x(OH)_{3-x}$

Figure 16.1 Typical derivating solvents for cellulose.

16.2.1.2 Non-derivatizing solvents

The treatment of cellulose with polar solvents capable of interacting with the hydroxyl groups of the polymer may achieve activation/dissolution in a single step. Ideally, the solvent, either mono- or poly-component, should be able to dissolve the cellulose of varying degree of polymerization (DP) and crystallinity without extensive degradation. In addition to the solution stability, the solvent should be dipolar in order to stabilize the highly polar complex of the acyl-transfer, but should not compete with cellulose for the derivatizing agent. To guarantee a complete homogeneous reaction, cellulose and all intermediate states, as well as the final product, should be soluble.

Alkali hydroxide solutions possess some of these requirements and dissolve the biopolymer of rather low DP after swelling in aqueous 8–9 per cent NaOH, subsequently freezing, thawing, and then diluting to a 5 per cent alkali concentration. Urea enhances the solubility due to the fact that it is able to weaken intermolecular hydrogen bonds [3]. However, the base will consume most of the esterifying agent before it reacts with the cellulose and therefore these systems cannot be employed in this context.

The aqueous systems Ni(tren) and Cd(tren) [tren = tris(2-aminoethyl)amine] dissolve cellulose by coordinating with the secondary hydroxyl groups at C2 and C3 [4]. Several molten salt hydrates, for example $LiClO_4 \times 3H_2O$, $LiX \times nH_2O$ ($X = I^-, NO_3^-, CH_3CO_2^-, ClO_4^-$), dissolve even high DP cellulose. Because these solvents contain water, which will compete with the cellulose for the derivatizing agent, they have also found limited interest in esterification reactions.

From the synthetic point of view, more flexibility is achieved by employing non-aqueous non-derivatizing solvents. Extensive work has been carried out on binary or ternary mixtures like inorganic or organic electrolytes in strongly dipolar aprotic solvents. The best known example is LiCl in DMA, in *N*-methyl-2-pyrrolidone (NMP), or in 1,3-dimethyl-2-imidazolidinone (DMI) [5]. The structure of these cellulose/solvent system complexes has been described by several authors, differing essentially in the role played by the Li^+ and Cl^- ions, as comprehensively discussed in a specific review [6].

Quite recent work by one of the authors showed that the combination of tetra-*n*-butylammonium fluoride trihydrate ($TBAF \times 3H_2O$) and DMSO was an efficient system capable of dissolving cellulose with a DP as high as 650 within some minutes, without any pretreatment [7]. Cellulose of higher DP was also solubilized, although, in this case, a pretreatment was necessary. A mixture of benzyltrimethylammonium fluoride monohydrate ($BTMAF \times H_2O$) and DMSO displayed a lower dissolving power [8]. Using these new solvents based on DMSO/F^- for the homogeneous acylation of cellulose, the values of the degree of substitution (DS) of the products are influenced by the medium water content [9]. Consequently, the use of water-free TBAF is of interest to understand the solubility and reactivity of cellulose dissolved in DMSO/TBAF. The dehydration of $TBAF \times 3H_2O$, resulting in the water-free salt, is impossible because anhydrous TBAF is unstable, undergoing a rapid E2 elimination giving the formation of bifluoride ions [10]. It was recently reported that anhydrous TBAF can be prepared *in situ* by reacting tetrabutylammonium cyanide and hexafluorobenzene in dry DMSO [11]. This freshly prepared DMSO/TBAF solution readily dissolved cellulose, even in the presence of the hexacyanobenzene by-product, for example, the dissolution of bleached cotton fibres (DP = 3 743) occurred within 1 min, as visualized by optical microscopy [8].

There are few single-component solvents for cellulose. Among them, *N*-ethylpyridinium chloride has a melting point of 118–120°C, which can be decreased by the addition of DMSO, DMF, pyridine, and NMP, leading to a value of 75°C [12]. *N*-methylmorpholine-*N*-oxide $\times H_2O$ (NMNO) is an efficient solvent employed in the production of Lyocell fibres [13, 14]. Its use, however, has various shortcomings with regard to cellulose chemistry namely (1) it is rather thermally unstable, becoming dangerous with a water content below the monohydrate, (2) it tolerates only some reagents yielding mostly a highly viscose gel-like state, and (3) its water content will partly consume the acylating reagent [15, 16].

More recently, some ionic liquids (IL) were shown to be good cellulose solvents over a wide range of DP values without covalent interaction, especially those based on substituted imidazolium ions (Table 16.1) [17]. Even bacterial cellulose with a DP of 6 500 was found to dissolve in one of these solvents [18].

16.2.2 Homogeneous acylation

One of the first examples of a completely homogeneous acetylation was the reaction of cellulose with acetic anhydride in *N*-ethylpyridinium chloride in the presence of pyridine, leading to a product with a DS 2.65 within the short reaction time of 44 min [19]. Although the preparation of cellulose triacetate, which is completed within 1 h, needs

Table 16.1

Results of the homogeneous acetylation of cellulose in ionic liquids

Cellulose[a]		Ionic liquid[b]	Reagent		DS[d]	Reference
Type	DP		Type	Mol/mol		
MC	290	$[C_4mim]^+Cl^-$	Acetic anhydride	3	1.87	[22]
MC	290	$[C_4mim]^+Cl^-$	Acetic anhydride	5	2.72	[22]
MC	290	$[C_4mim]^+Cl^-$	Acetic anhydride	10[c]	3.0	[22]
MC	290	$[C_4mim]^+Cl^-$	Acetyl chloride	3	2.81	[22]
MC	290	$[C_4mim]^+Cl^-$	Acetyl chloride	5	3.0	[22]
MC	290	$[C_2mim]^+Cl^-$	Acetic anhydride	3	3.0	[23]
MC	290	$[C_4dmim]^+Cl^-$	Acetic anhydride	3	2.92	[23]
MC	290	$[Admim]^+Br^-$	Acetic anhydride	3	2.67	[23]
SSP	590	$[C_4mim]^+Cl^-$	Acetyl chloride	3	3.0	[22]
SSP	590	$[C_4mim]^+Cl^-$	Acetyl chloride	5	3.0	[22]
CL	1 200	$[C_4mim]^+Cl^-$	Acetyl chloride	3	2.85	[22]
CL	1 200	$[C_4mim]^+Cl^-$	Acetyl chloride	5	3.0	[22]
BC	6 500	$[C_4mim]^+Cl^-$	Acetic anhydride	3	2.25	[18]
BC	6 500	$[C_4mim]^+Cl^-$	Acetic anhydride	5	2.50	[3]
BC	6 500	$[C_4mim]^+Cl^-$	Acetic anhydride	10	3.0	[18]
DSP	650	$[Amim]^+Cl^-$	Acetic anhydride	3	1.99	[21]
DSP	650	$[Amim]^+Cl^-$	Acetic anhydride	5	2.30	[21]

[a] MC, microcrystalline cellulose; SSP, spruce sulphite pulp; CL, cotton linters; BC, bacterial cellulose; DSP, dissolving pulp.

[b] $[C_4mim]^+Cl^-$ (1-butyl-3-methylimidazolium chloride: H_3C–N, N–$CH_2(CH_2)_2CH_3$, Cl^-), $[C_2mim]^+Cl^-$ (H_3C–N, N–CH_2CH_3, Cl^-), $[Admim]^+Br^-$ (H_3C–N, N–allyl, 2-CH_3, Br^-), $[Amim]^+Cl^-$ (H_3C–N, N–allyl, Cl^-).

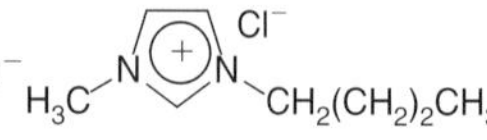

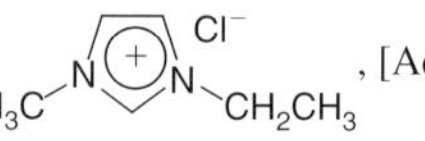

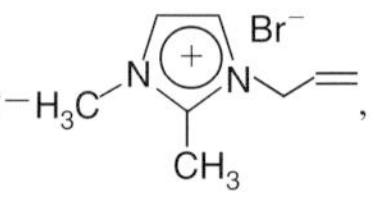

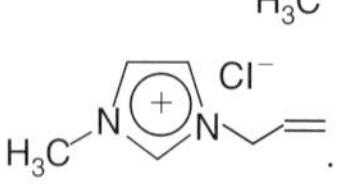

[c] Additionally 2.5 mol pyridine per mol anhydroglucose unit (AGU).

[d] Degree of substitution determined by ^{1}H-NMR spectroscopy after perpropionylation [39].

to be carried out at 85°C, it proceeds for cellulose samples with DP values below 1 000 without degradation, that is strictly polymeranalogous. Cellulose acetate samples with a predefined solubility, for example in water, acetone, or chloroform, respectively, are accessible in one step, in contrast to the heterogeneous conversion. A correlation between solubility and distribution of substituents has been attempted by means of ^{1}H NMR spectroscopy [20].

Ionic liquids are promising new solvents in the field of cellulose shaping and functionalization. The acylation of cellulose dissolved in an ionic liquid can be carried out with acetic anhydride. The reaction succeeds without an additional catalyst. Starting from DS 1.86, the cellulose acetates obtained are acetone soluble [21]. The control of the DS by the prolongation of the reaction time is possible. When acetyl chloride is used, complete acetylation is achieved in 20 min (Table 16.1) [18, 22, 23]. This method may lead to a widely applicable acylation procedure for polysaccharides, if the regeneration of the solvent becomes possible.

DMA/LiCl, widely used in peptide and polyamide chemistry, is among the best studied solvents, because it dissolves cellulose of various DP quite readily [24] and shows almost no interaction with acylating reagents, even being able to act as an acylation catalyst. The dissolution can be achieved by two routes, viz. (1) by solvent exchange, meaning the cellulose is first suspended in water and subsequently transferred into organic liquids with decreasing polarity and finally DMA/LiCl [25], and (2) after heating a suspension of cellulose in DMA, vacuum evaporating about 1/5 of the liquid (containing most of the water) and addition of the LiCl after which, during cooling to room temperature, a clear solution is obtained. The amount of LiCl is in the range of 5–15 per cent (w/w). The thermal cellulose activation under reduced pressure is far superior to the costly and time-consuming activation by solvent exchange.

In recent years, the cellulose/DMA/LiCl system has been studied intensively to develop efficient methods appropriate even for industrial application [26, 27]. The dissolution procedure and acetylation conditions allow

Table 16.2

Acetylation of different cellulose types in *N,N*-dimethyl acetamide (DMA)/LiCl with acetic anhydride (18 h at 60°C, adapted from [26, 27])

Starting materials				Molar ratio		DS
Cellulose from	M_w (g mol^{-1})	α-Cellulose content (%)	I_c (%)	AGU	Acetic anhydride	
Bagasse	116 000	89	67	1	1.5	1.0
Bagasse	116 000	89	67	1	3.0	2.1
Bagasse	116 000	89	67	1	4.5	2.9
Cotton	66 000	92	75	1	1.5	0.9
Sisal	105 000	86	77	1	1.5	1.0

Table 16.3

Preparation of aliphatic esters of cellulose in *N,N*-dimethyl acetamide (DMA)/LiCl

	Reaction conditions					Reaction product		Reference
Acid chloride	Molar ratio		Base	Time (h)	Temperature (°C)	DS	Solubility[a]	
	AGU	Agent						
Hexanoyl	1	1.0	Py	0.5	60	0.89	DMSO, NMP, Py	[119]
Hexanoyl	1	2.0	Py	0.5	60	1.70	Acetone, MEK, $CHCl_3$, AcOH, THF, DMSO, NMP, Py	
Lauroyl	1	2.0	Py	0.5	60	1.83	Py	
Stearoyl	1	1.0	Py	1	105	0.79	Acetone, MEK, $CHCl_3$, AcOH, THF, DMSO, NMP, Py	
Hexanoyl	1	8.0	TEA	8	25	2.8	DMF	[26]
Heptanoyl	1	8.0	TEA	8	25	2.4	Toluene	
Octanoyl	1	8.0	TEA	8	25	2.2	Toluene	
Phenylacetyl	1	15.0	Py	3/1.5	80/120	1.90	CH_2Cl_2	[120]
4-Methoxy-phenyl-acetyl	1	15.0	Py	3/1.5	80/120	1.8	CH_2Cl_2	
4-Tolyl-acetyl	1	15.0	Py	3/1.5	80/120	1.8	CH_2Cl_2	

[a] DMSO, dimethyl sulfoxide; NMP, *N*-methyl-2-pyrrolidone; Py, pyridine; MEK, methyl ethyl ketone; AcOH, acetic acid; THF, tetrahydrofurane; DMF, *N,N*-dimethyl formamide.

excellent control of the DS in the range 1–3. The reaction at 110°C for 4 h without additional base or catalyst gives products with almost no degradation of the starting polymer. The distribution of substituents in the order C-6 > C-2 > C-3 is determined by ^{13}C NMR spectroscopy. In addition to microcrystalline cellulose, cotton, sisal, and bagasse-based cellulose may serve as starting materials (Table 16.2). The crystallinity of the starting polymer has little effect on the homogeneous acetylation.

The acylation of cellulose with acid chlorides in DMA/LiCl is most suitable for the homogeneous synthesis of readily soluble partially functionalized long-chain aliphatic esters and substituted acetic acid esters (Table 16.3). In contrast to the anhydrides, the fatty acid chlorides are soluble in the reaction mixture and soluble polysaccharide esters may be formed with a very high efficiency. Even in the case of stearoyl chloride, 79 per cent of the reagent is consumed for the esterification of cellulose.

Table 16.4

Influence of the amount tetra-*n*-butylammonium fluoride (TBAF) trihydrate on the efficiency of the acetylation of sisal cellulose with acetic anhydride in dimethyl sulfoxide (DMSO)/TBAF (adapted from [28])

Per cent TBAF in DMSO	Cellulose acetate	
	DS	Solubility[a]
11	0.30	Insoluble
8	0.96	DMSO, Py
7	1.07	DMSO, Py
6	1.29	DMSO, DMF, Py

[a] Py, pyridine; DMF, *N,N*-dimethyl formamide.

DMSO/TBAF is highly efficient as a reaction medium for the homogeneous esterification of cellulose by transesterification and after the *in situ* activation (see below) of complex carboxylic acids. The acylation using acid chlorides and anhydrides is limited because the solution contains a certain amount of water caused by the use of the commercially available TBAF trihydrate and the residual moisture in the air-dried polysaccharides. Nevertheless, this system has shown a remarkable capacity for the esterification of lignocellulosic materials, for example, sisal fibres, which contain about 14 per cent hemicellulose [28]. The DS values of cellulose acetate prepared from these fibres with acetic anhydride in mixtures of DMSO/TBAF were found to decrease with increasing TBAF concentration from 6 to 11 per cent (Table 16.4), due to the increased rate of hydrolysis both of the anhydride and the ester moieties.

Partial dehydration of DMSO/TBAF is possible by vacuum distillation and reactions in the ensuing solvent lead to products comparable to those obtained in reactions of cellulose dissolved in anhydrous DMA/LiCl. In addition to these basic studies, the conversion of cellulose in DMSO/TBAF with more complex carboxylic acids (*e.g.* furan-2-carboxylic acid) via *in situ* activation with *N,N′*-carbonyldiimidazole (CDI) was studied (see Section 16.2.2.1.3).

16.2.2.1 Homogeneous acylation with in situ activated carboxylic acids

The synthetic approach of *in situ* activation of carboxylic acids is based on the preliminary reaction of the carboxylic acid with a specific reagent to give an intermediate reactive derivative which can be prepared prior to the reaction with cellulose or converted directly in a one-pot process. This approach opens the way to a broad variety of new esters, because for numerous acids, for example unsaturated or hydrolytically unstable ones, reactive derivatives such as anhydrides or chlorides simply cannot be synthesized. The mild reaction conditions applied for the *in situ* activation prevent common side reactions like pericyclic reactions, hydrolysis, and oxidation. Moreover, due to their hydrophobic character, numerous anhydrides are not soluble in organic media used for cellulose modification, resulting in unsatisfactory yields and insoluble products. In addition, the conversion of an anhydride is combined with the loss of half of the acid during the reaction. Consequently, *in situ* activation is much more efficient.

16.2.2.1.1 Activation with sulphonic acid chlorides

One of the early attempts of *in situ* activation was the reaction of carboxylic acids with sulphonic acid chlorides that was adopted for the homogeneous modification of cellulose. The exclusion of the base simplified the reaction medium and the isolation procedure [29]. There is an ongoing discussion about the mechanism which initiates esterification of cellulose with the carboxylic acids in the presence of *p*-toluenesulphonic acid chloride (TosCl). The mixed anhydride of *p*-toluenesulphonic acid (TosOH) and the carboxylic acid is favoured [30]. However, from ^{1}H NMR experiments with acetic acid/TosCl, it was concluded that a mixture of acetic anhydride (2.21 ppm) and acetyl chloride (2.73 ppm) was responsible for the high reactivity of this system (Figs 16.2 and 16.3).

Cellulose esters having alkyl substituents in the range of C_{12} (laurylic acid) to C_{20} (eicosanoic acid), can be obtained with almost complete functionalization (DS 2.8–2.9) within 24 h at 50°C in the presence of sulphonic acid chlorides in DMA/LiCl, using pyridine as a base [31]. This is also a general method for the *in situ* activation

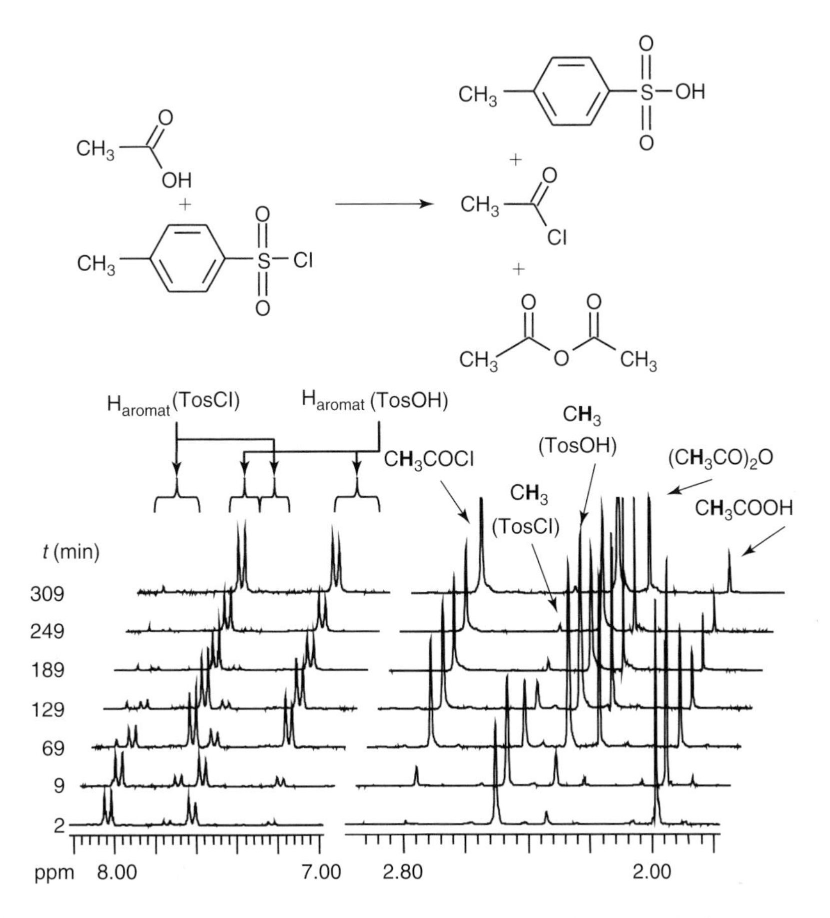

Figure 16.2 ^{1}H NMR spectroscopic investigation of the *in situ* activation of acetic acid with *p*-toluenesulphonic acid chloride showing the preferred formation of acetic anhydride and acetyl chloride (with permission from Springer, [1]).

Recycling may be possible

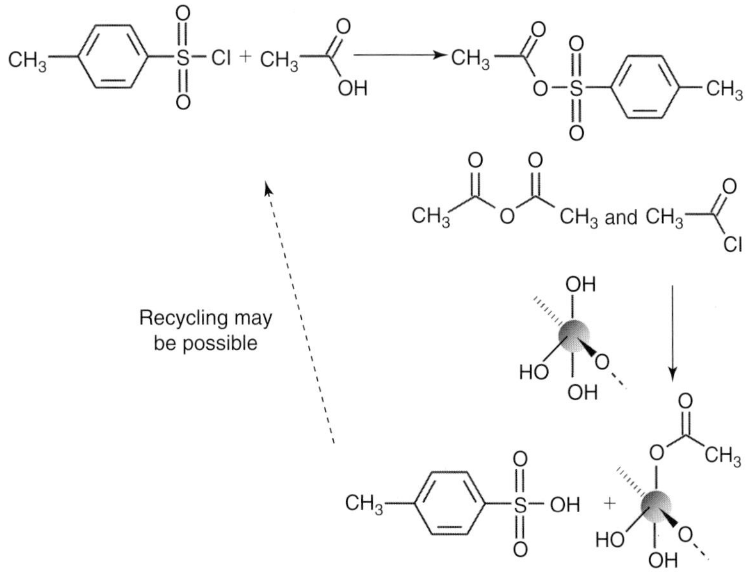

Figure 16.3 Schematic plot of the reaction by *in situ* activation of carboxylic acids with *p*-toluenesulphonic acid chloride (with permission from Springer, [1]).

Table 16.5

Esterification of cellulose dissolved in *N,N*-dimethyl acetamide (DMA)/LiCl mediated with TosCl with different carboxylic acids (adapted from [32])

Entry	Reaction conditions						Product	
	Carboxylic acid	Molar ratio					DS	Solubility in $CHCl_3$
		AGU[a]	Acid	TosCl[b]	Py[c]	Time (h)		
1	Lauric	1	2	2	0	24	1.55	+
2	Palmitic	1	2	2	0	24	1.60	+
3	Stearic	1	2	2	0	24	1.76	+
4	Lauric	1	2	2	4	24	1.79	+
5	Palmitic	1	2	2	4	24	1.71	+
6	Stearic	1	2	2	4	24	1.92	+
7	Lauric	1	2	2	0	4	1.55	+
8	Palmitic	1	2	2	0	4	1.50	+
9	Lauric	1	2	2	0	1	1.36	−
10	Palmitic	1	2	2	0	1	1.36	+

[a] Anhydroglucose unit.
[b] *p*-Toluenesulphonic acid chloride.
[c] Pyridine.

of waxy carboxylic acids. The reaction proceeds in 4 h to give partially functionalized fatty acid esters of a maximum DS (see entries 1, 7, and 9 in Table 16.5). The addition of an extra base further increases the DS by 0.1–0.2 units (see entries 1–3 and 4–6 in Table 16.5 [32]).

The *in situ* activation with TosCl is also useful for the introduction of fluorine-containing substituents, for example, 2,2-difluoroethoxy-, 2,2,2-trifluoroethoxy-, and 2,2,3,3,4,4,5,5-octafluoropentoxy functions with DS values mainly in the range of 1.0–1.5, leading to a stepwise increase in the lipophilicity of the products and an increase in their thermal stability. Structure analysis is possible by ^{19}F NMR spectroscopy [30, 33, 34]. Moreover, the *in situ* activation with TosCl enables the synthesis of water soluble cellulose esters by derivatization of cellulose with oxacarboxylic acids in DMA/LiCl [29]. The conversion of the polysaccharide with 3,6,9-trioxadecanoic acid or 3,6-dioxaheptanoic acid, in the presence of TosCl, yields non-ionic cellulose esters with DS values in the range of 0.4–3.0. In this case, the esterification is carried out without an additional base. The cellulose derivatives start to dissolve in water at a DS value as low as 0.4 and they are also soluble in common organic solvents such as acetone or ethanol.

16.2.2.1.2 Activation with dialkylcarbodiimide

Coupling reagents of the dialkylcarbodiimide type, in particular dicyclohexylcarbodiimide (DCC), are most frequently utilized for the esterification of polysaccharides with complex carboxylic acids. These reagents have a number of drawbacks: (1) they are toxic (the LD_{50} dermal, rat, of DCC is 71 mg/kg); (2) the *N,N′*-dialkylurea formed during the reaction is hard to remove from the polymer, except if it is formed in DMF and DMSO where it can be filtered off; (3) in the case of esterification in DMSO in the presence of these reagents, the oxidation of the hydroxyl functions, via the Moffatt type, may occur.

Another strategy for *in situ* activation, used for the synthesis of aliphatic cellulose esters, is the exploitation of DCC in combination with 4-pyrrolidinopyridine (PP) [35]. Among the advantages of this method are the high reactivity of the intermediately formed mixed anhydride with PP and a completely homogeneous reaction in DMA/LiCl up to hexanoic acid. If the reaction is carried out with the anhydrides of carboxylic acids, the carboxylic acid liberated is recycled by forming the mixed anhydride with PP, which is applied only in a catalytic amount. The toxic DCC can be recycled from the reaction mixture (Fig. 16.4). In addition, the method is utilized to obtain unsaturated esters (*e.g.* methacrylic-, cinnamic-, and vinyl acetic acid esters) and esters of aromatic

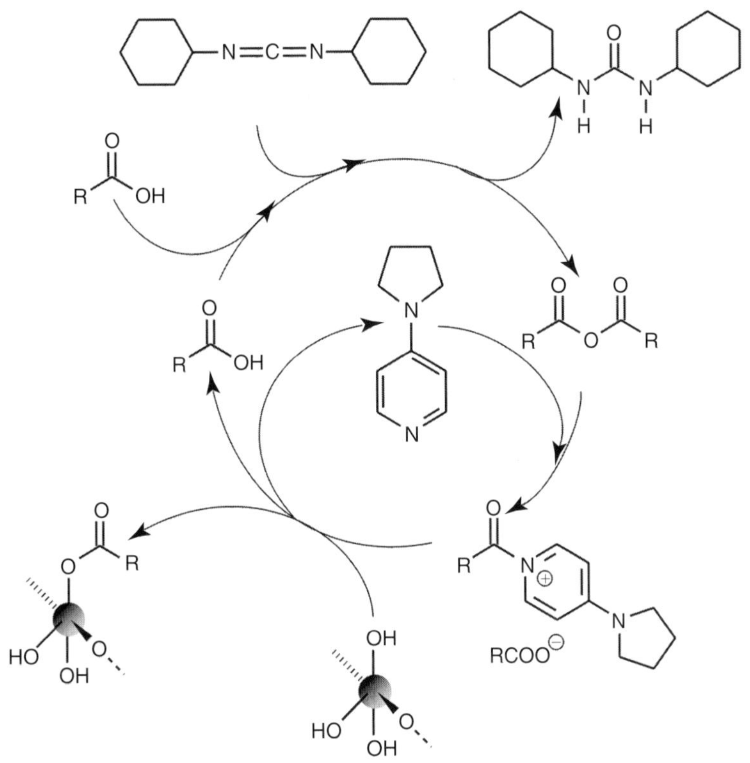

Figure 16.4 Scheme of the esterification of cellulose using PP/DCC (with permission from Springer, [1]).

carboxylic acids including cellulose (*p-N,N*-dimethylamino)benzoate. Here, the amino group containing ester can be converted to a quaternary ammonium derivative, which imparts water solubility to the material [36, 37].

16.2.2.1.3 Activation with *N,N'*-carbonyldiimidazole

A method with an enormous potential for polysaccharide modification is the homogeneous one-pot synthesis after *in situ* activation of the carboxylic acids with CDI, well known from organic chemistry [38]. It is especially suitable for the functionalization of the biopolymers because during the conversion the reactive imidazolide of the acid is generated and non-toxic by-products formed are CO_2 and imidazole only (Fig. 16.5). The imidazole is freely soluble in a broad variety of solvents including water, alcohol, ether, chloroform, and pyridine and can be easily removed. In addition, the pH is not drastically changed during the reaction resulting in negligible chain degradation.

The synthesis is generally carried out as a one-pot reaction in two steps. The acid is transformed with the CDI to give the imidazolide. The conversion of the alcohol in the first step is also possible for the esterification but yields undesired crosslinking by carbonate formation in case of a polyol like cellulose (see Fig. 16.5). Model reactions and NMR spectroscopy with acetic acid confirm that during a treatment at room temperature CDI is completely consumed within 6 h (see Fig. 16.5). Thereby, the tendency of crosslinking of remaining CDI leading to insoluble products is avoided. For instance, the reaction with (-)-menthyloxyacetic acid *in situ* activated with CDI can be carried out simply by mixing the solution of the imidazolide prepared in DMA and the cellulose in DMA/LiCl and increasing the temperature to 60°C. Pure (-)-menthyloxyacetic acid esters of cellulose with DS as high as 2.53 are obtained by precipitation in methanol and filtration (Table 16.6). The cellulose esters are characterized regarding structure and DS by means of FTIR spectroscopy, elemental analysis, ^{1}H- and ^{13}C NMR spectroscopy, and additionally by ^{1}H NMR spectroscopy after peracetylation [39]. A wide variety of carboxylic acids with, for example, unsaturated, 3-(2-furyl)-acrylcarboxylic acid, heterocyclic, furan-2-carboxylic acid, and crown ether, 4'-carboxybenzo-18-crown-6, are available by the conversion of cellulose via this path (Table 16.6).

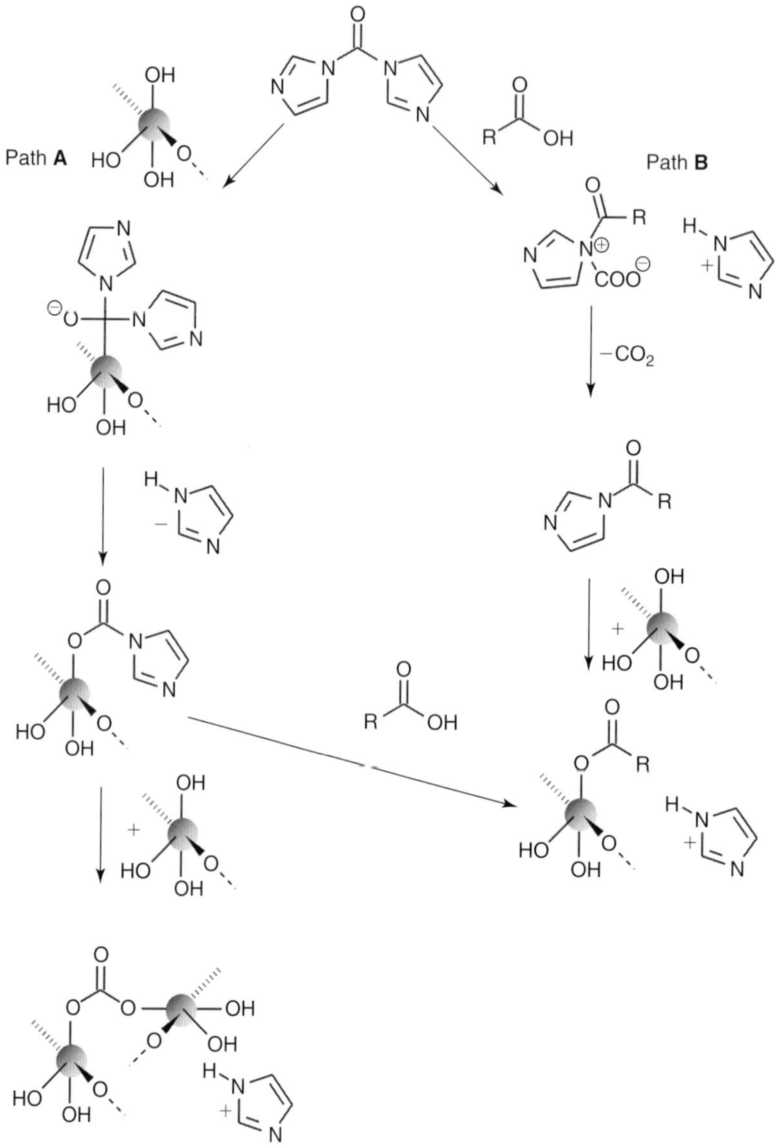

Figure 16.5 Reaction paths leading exclusively to esterification (path B) or crosslinking (path A) if the polysaccharide is treated with CDI in the first step (with permission from Springer, [1]).

In contrast, the conversion of cellulose with camphor-10-sulphonic acid via *in situ* activation with CDI is not efficient to obtain a chiral sulphonic acid ester of cellulose. Only very small amount of sulphonic acid ester functions can be introduced in agreement with results of the chemistry of low molecular weight alcohols regarding a much lower efficiency of CDI for the preparation of sulphonic acid esters [38].

In addition to DMA/LiCl, DMSO/TBAF is an appropriate reaction medium for homogeneous acylation of cellulose applying *in situ* activation with CDI. Results of reactions of cellulose with acetic-, stearic-, adamantane-1-carboxylic-, and furan-2-carboxylic acid imidazolides are summarized in Table 16.7.

16.2.2.1.4 Activation with iminium chlorides

A mild and efficient method is the *in situ* activation of carboxylic acids via iminium chlorides that are simply formed by reaction of DMF with a variety of chlorinating agents including phosphoryl chloride, phosphorus

Table 16.6

Esterification of cellulose with (-)-menthyloxyacetic acid, furan-2-carboxylic acid, 3-(2-furyl)-acrylcarboxylic acid, and 4′-carboxybenzo-18-crown-6 via *in situ* activation with *N,N′*-carbonyldiimidazole (CDI) (adapted from [39])

Conditions					Product		
Entry	Carboxylic acid	Molar ratio			DS	Solubility[b]	
		AGU[a]	Acid	CDI		DMSO	DMF
1	(-)-Menthyloxyacetic	1	2.5	2.5	0.20	+	+
2		1	5.0	5.0	1.66	–	+
3		1	7.5	7.5	2.53	–	+
4	3-(2-Furyl)-acryl-	1	2.5	2.5	0.52	+	–
5		1	5.0	5.0	1.14	+	+
6		1	7.5	7.5	1.52	+	–
7	Furan-2-	1	2.5	2.5	0.80	+	+
8		1	5.0	5.0	1.49	+	+
9		1	7.5	7.5	1.97	+	+
10	(Benzo-18-crown-6)-4′-	1	2.3	2.3	0.40	+	–

[a] Anhydroglucose unit.
[b] DMSO, dimethyl sulfoxide; DMF, *N,N*-dimethyl formamide.

Table 16.7

Homogeneous acylation of cellulose dissolved in dimethyl sulfoxide (DMSO)/tetra-*n*-butylammonium fluoride (TBAF) with different carboxylic acids, mediated by *N,N′*-carbonyldiimidazole (CDI)

Entry	Conditions				Product	
	Carboxylic acid	Molar ratio			DS	Solubility[b]
		AGU[a]	Acid	CDI		
1	Acetic	1	3	3	0.51	DMSO, DMA
2	Stearic	1	2	2	0.47	DMSO
3	Stearic	1	3	3	1.35	DMSO
4	Adamantane-1-carboxylic	1	2	2	0.50	DMA/LiCl
5	Adamantane-1-carboxylic	1	3	3	0.68	DMSO, DMA
6	Furan-2-carboxylic	1	3	3	1.91	DMSO, DMA, Py

[a] Anhydroglucose unit.
[b] DMA, *N,N*-dimethylacetamide; Py, pyridine.

trichloride and, most frequently, oxalyl chloride and subsequent reaction with the acid. During the reaction of acid iminium chlorides with alcohols mostly gaseous side products are formed and the solvent is regenerated (Fig. 16.6, Table 16.8 [40]). In a 'one-pot reaction', acylation of cellulose with the long-chain aliphatic acids (stearic acid and palmitic acid), the aromatic acid 4-nitrobenzoic acid and adamantane-1-carboxylic acid are easily achieved. NMR spectroscopy reveals that the conversion succeeds with quantitative yield in case of acetic acid.

16.2.3 Regioselectively functionalized cellulose esters

Conversion of 6-*O*-trityl cellulose (see Chapter 4) with acetic acid anhydride or propionic acid anhydride and subsequent detritylation by treating the completely functionalized polymers with HBr in acetic acid yields the

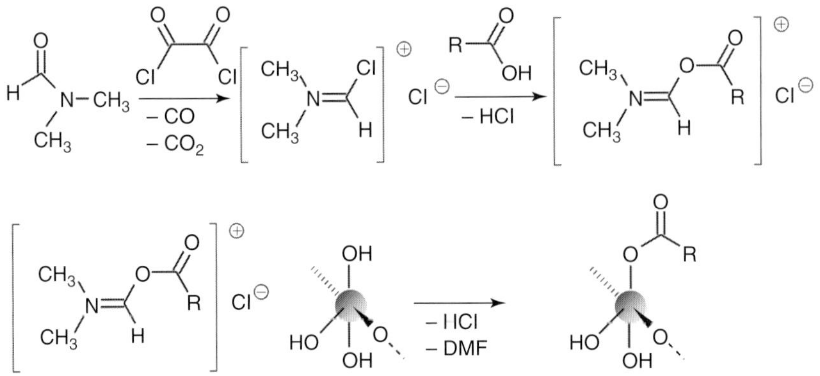

Figure 16.6 Preparation of cellulose esters via *in situ* activation by iminium chlorides (with permission from Springer, [1]).

Table 16.8

Esterification of cellulose dissolved in *N,N*-dimethyl acetamide (DMA)/LiCl via the iminium chlorides of different carboxylic acids (adapted from [40])

Entry	Carboxylic acid	Molar ratio			Product	
		AGU[a]	Acid	Oxalyl chloride	DS	Solubility[b]
1	Stearic	1	3	3	0.63	DMSO/LiCl
2	Stearic	1	5	5	1.84	THF, $CHCl_3$
3	Palmitic	1	6	6	1.89	DMSO, DMA, THF
4	Adamantane-1-carboxylic	1	1	1	0.47	DMSO/LiCl
5	Adamantane-1-carboxylic	1	3	3	1.20	DMA, DMSO, DMF
6	4-Nitrobenzoic	1	1	1	0.30	DMSO/LiCl
7	4-Nitrobenzoic	1	2	2	0.52	DMSO
8	4-Nitrobenzoic	1	3	3	0.94	DMSO

[a] Anhydroglucose unit.

[b] DMSO, dimethyl sulfoxide; THF, tetrahydrofurane; DMF, *N,N*-dimethyl formamide.

corresponding 2,3-*O*-functionalized esters [41]. Subsequent acylation of the hydroxyl groups at position 6 leads to the peracylated cellulose esters with inverse functionalization pattern useful for the peak assignment of NMR spectra. The cellulose esters form single crystals visualized by means of AFM [42]. Investigations of the structure in solution revealed large differences to cellulose esters with random distribution of substituents, for example, the chain conformation, solubility, and clustering mechanism is different [43, 44].

16.3 NUCLEOPHILIC DISPLACEMENT REACTIONS WITH CELLULOSE

It is well known that hydroxyl functions are converted to a good leaving group by the formation of the corresponding sulphonic acid esters and hence nucleophilic displacement reactions are possible. Advantageously, the formation of sulphonic acid esters is carried out homogeneously by conversion with sulphonic acid chlorides in the presence of a tertiary organic base in DMA/LiCl. In particular, homogeneous tosylation applying TosCl in the presence of triethylamine is very efficient [25, 45, 46].

Table 16.9

Selected applications of deoxycellulose (adapted from [52] and references herein, see details there)

Application field	Functionalization
Biological	
Bacteriostatic	Different cellulosics including deoxycellulose treated with 5-nitrofuroyl chloride
Immobilization of enzymes	Aminodeoxy cellulose
Anticoagulants	Sulfated aminodeox cellulose
Enzyme purification	Aminodeoxy cellulose
Chemical	
Removal of heavy metals, for example,	
Hg	Chlorodeoxy cellulose + ethylenediamine, thiourea, thiosemicarbazide, thioacetamide
Cu, Mn, Co, Ni	6-Chlorodeoxy cellulose + diamines
Hg, Ag	S-substituted 6-deoxy-6-mercapto cellulose
Preconcentration of trace elements	Tosyl cellulose treated with piperazine and CS_2 (cellulose piperazinedithiocarboxylate)
Sorption of uranium	Cellulose piperazinedithiocarboxylate
Chelating and complexing agents	Chlorodeoxy cellulose + iminodiacetic acid
	Schiff base formation from aminodeoxy cellulose
Physical	
Flame retardants	Chlorodeoxy cellulose
Propellants	Azidodeoxy cellulose
Electrostatic printing	Different cellulosics including aminodeoxy derivatives
Electron transfer catalysts	Viologens from ethyl cellulose derivatives

Applying S_N reactions, deoxycellulose products are available where the hydroxyl groups of the AGU are partially or completely replaced by other functional groups, that is, the reaction takes place at the carbon atoms. Considerable interest has found halodeoxycelluloses as starting material for S_N reactions yielding cellulose derivatives with interesting properties (Table 16.9) [47–52]. Recent studies comprise the introduction of amine/ammonium functions either by introduction of azide moieties and subsequent reduction or by using various di- and tri-amines for the S_N reaction.

The S_N reaction of tosyl cellulose with NaN_3 was comprehensively studied. It appears that not only the primary but also the secondary tosylates may be substituted. Consequently, the DS of azide moieties exceeds a value of 1, which is not possible with most of the nucleophils studied [53]. The reduction of the azide moiety is possible either with $LiAlH_4$ [54] or $NaBH_4/CoBr_2$/2,2′-bipyridine [55].

Moreover, a number of aminodeoxycellulose were accessible. Water soluble 6-trialkylammonium-6-deoxycellulose could be prepared [56]. Conversion of cellulose tosylate with diamines like 1,4-phenylenediamine yields polymers that can be applied for the immobilization of enzymes by diazo coupling, for example [57].

16.4 ETHERIFICATION OF CELLULOSE

Etherification is a very important branch of commercial cellulose derivatization. Cellulose ethers are prepared in technical scale by conversion of alkali cellulose with alkylating reagents, for example epoxides, alkyl- and carboxymethyl halides [58, 59].

In the case of heterogeneous reactions, the reactivity and the accessibility of the hydroxyl groups are determined by hydrogen bond-breaking activation steps and by interaction with the reaction media [59, 60]. Additionally, the reaction of cellulose with reagents of low steric demand leads to a random distribution of substituents within the AGU and along the polymer chain. It is well known that not only the DS but also the substitution pattern may influence the properties of cellulose ethers [61]. However, to gain detailed information about the influence of the structures on properties not only a comprehensive structure characterization but also cellulose ethers with a defined distribution of the functional groups are indispensable for the establishment of the structure–property relationships.

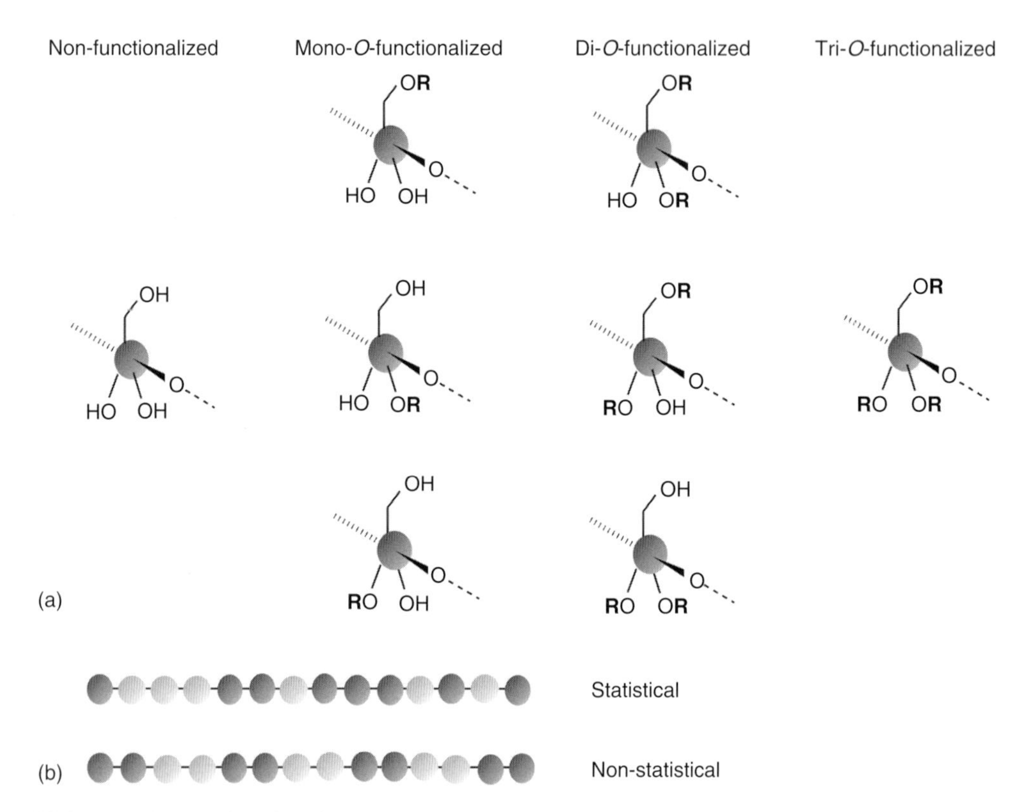

Figure 16.7 Distribution of the functional groups by regiocontrolled synthesis of cellulose derivatives: (a) within the anhydroglucose units and (b) along the polymer chain.

16.4.1 Regioselectively functionalized cellulose ether

'Regioselectivity' in cellulose chemistry means an exclusive or significant preferential reaction at one or two of the three positions 2, 3, and 6 of the AGU as well as along the polymer chain (Fig. 16.7).

In general, by chemical modification reactions of cellulose no equal reactivity of positions 2, 3, and 6 appears. For instance, homogeneous acylation in DMA/LiCl results in products with a preferred functionalization at position 6, while the carboxymethylation in 2-propanol/aqueous NaOH proceeds faster at position 2 compared to 6 and 3.

An important area in cellulose chemistry is the definite control of the substitution within the repeating unit. Up to now, the most important approach in this regard is the application of protecting groups (Fig. 16.8(a)). Other methods comprising, for example, selective cleavage of primary substituents by chemical or enzymatic treatment [20, 62, 63] play a minor role. An example is the deacetylation of cellulose acetate under aqueous acidic or alkaline conditions or in the presence of amines (Fig. 16.8(b)). In addition, activating groups like the tosyl moiety may also be disposed for selective reactions (see Section 16.3).

The most widely used protecting group is the triphenylmethyl (trityl) moiety (Fig. 16.9). Heterogeneous introduction of the trityl groups starts with an activated polymer obtained either by the deacetylation of cellulose acetate [64, 65] or by mercerization of cellulose [66] followed by a conversion in anhydrous pyridine. Furthermore, tritylation of cellulose yielding polymers with DS values of 1.0 proceeds in DMA/LiCl and DMSO/SO_2/diethylamine (DEA) [67, 68]. Methoxysubstituted triphenylmethyl compounds are more effective protecting groups for the primary hydroxyl group of cellulose [69]. The reaction of cellulose dissolved in DMA/LiCl with 4-monomethoxytriphenylmethyl chloride is 10 times faster than the conversion with unsubstituted trityl chloride. Complete functionalization of the primary hydroxyl groups is possible within 4 h and 70°C. Even after long reaction times, excess of the reagent, and elevated temperatures, alkylation of the positions 2 and 3 was less than 11 per cent, which is the same range as for the unsubstituted trityl function. Moreover, the detritylation occurs 20 times faster [61].

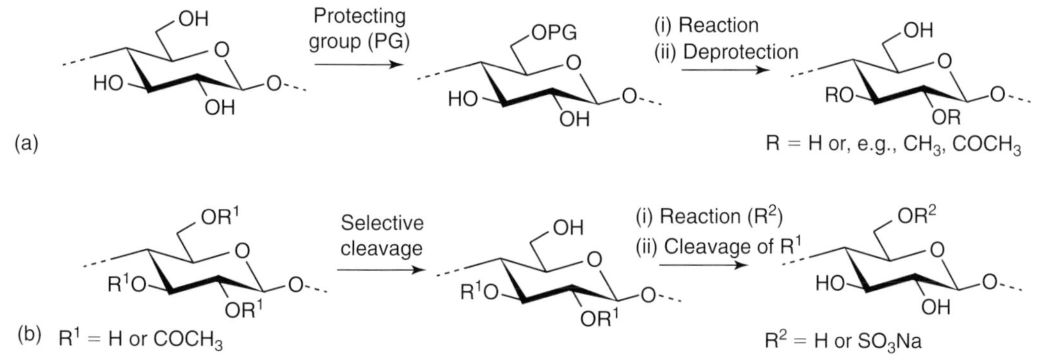

Figure 16.8 Pathways for the regioselective functionalization: (a) protecting group technique, (b) selective cleavage of primary substituents.

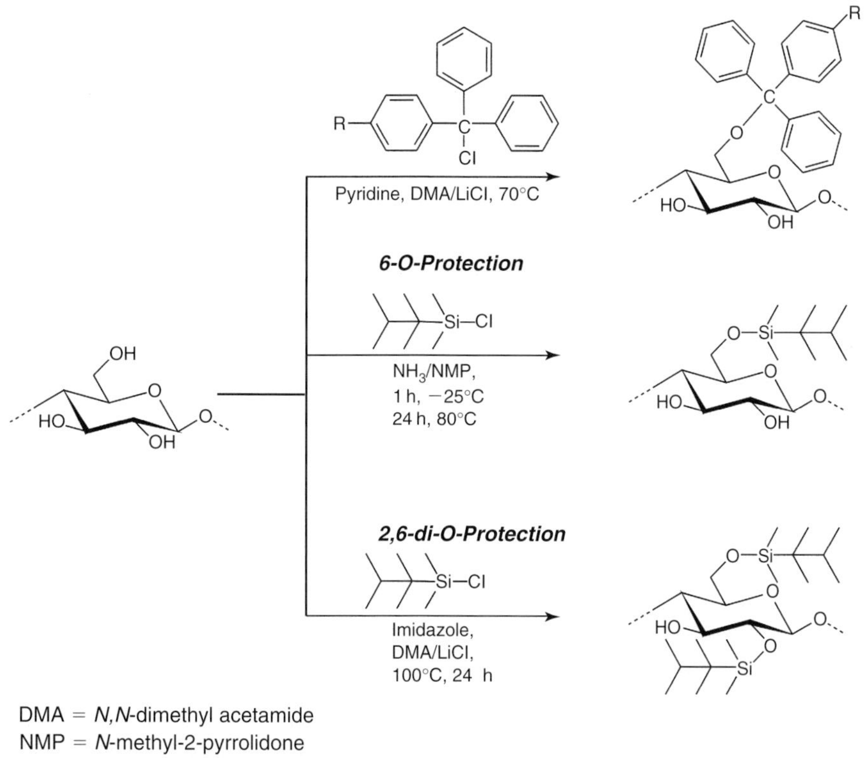

DMA = *N,N*-dimethyl acetamide
NMP = *N*-methyl-2-pyrrolidone

Figure 16.9 Protecting group technique: tritylation with trityl chloride in DMA/LiCl or silylation with thexyldimethylchlorosilane in NMP/ammonia for the regioselective blocking of the primary OH group and silylation in DMA/LiCl to protect the 6 and 2 positions simultaneously.

Trialkyl- (with at least one bulky alkyl moiety) and triaryl-silyl groups were investigated to protect the primary groups of cellulose. Pawlowski synthesized tert-butyldimethylsilyl cellulose with a DS of 0.68 in DMA/LiCl with mostly functionalization of position 6 [70]. Among this type of derivatives, 6-*O*-thexyldimethylsilyl (TDS) cellulose is most suitable (Fig. 16.9), [71, 72]. A synthesis starting with a heterogeneous phase reaction in ammonia-saturated polar aprotic liquids at −15°C by conversion of the cellulose with TDS chloride leads to a specific state of dispersion after evaporation of the ammonia at about 40°C, which does not permit any further reaction of the secondary hydroxyl groups, even with a large reagent excess, increased temperature, or long reaction time [71].

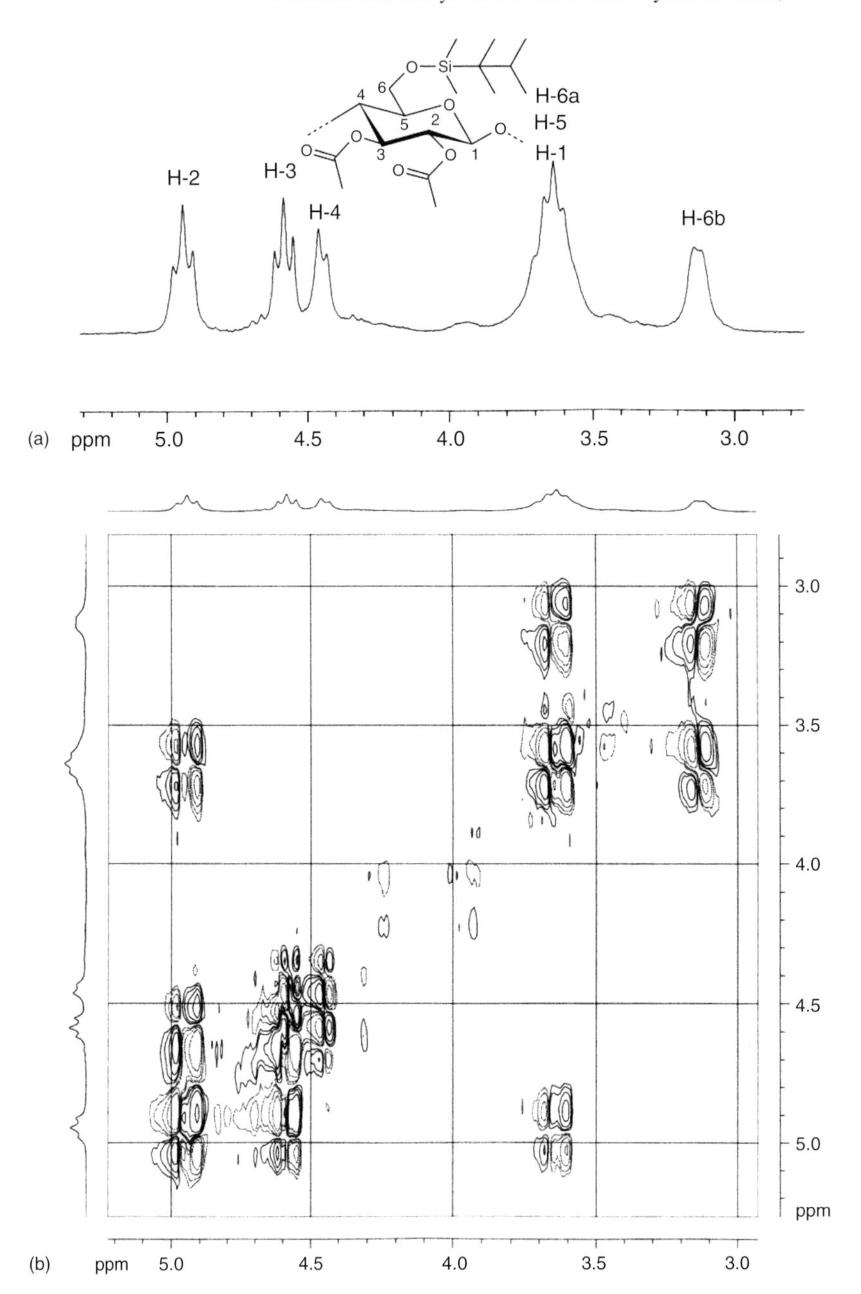

Figure 16.10 ^{1}H NMR spectrum (a) and ^{1}H/^{1}H correlated NMR spectrum (b) of peracetylated TDS cellulose in $CDCl_3$: assigned cross-peaks of the anhydroglucose unit [2,3-*O*-Ac-6-*O*-TDS] (Ac = acetyl, with permission from Springer, [71]).

The degree of TDS groups introduced by homogeneous silylation in DMA/LiCl to a total value of 0.99 was determined to be 85 per cent at position 6 only (GC/MS analysis). However, the homogeneous reaction in DMA/LiCl can be used to synthesize a 2,6-di-*O*-TDS cellulose [61, 72].

The structural uniformity and regioselectivity of the silylated cellulose products were demonstrated by means of one- and two-dimensional NMR techniques (Fig. 16.10) after subsequent acetylation of the remaining hydroxyl groups [68, 71] or after permethylation of the residual OH groups and chain degradation by means of HPLC and GC-MS [66, 69, 72, 73].

Table 16.10

Examples for regioselectively functionalized 2,3-di-*O*-cellulose ethers

(structure: OH at C6; RO, OR at positions 2,3)	R	DS	Via[a]	Reference
2,3-Di-*O*-methyl cellulose	$—CH_3$	Up to 1.77	Trt	[65]
		2.0	Trt	[66, 83]
		2.0	TDS	[72]
2,3-Di-*O*-ethyl cellulose	$—CH_2—CH_3$	Up to 1.93	Trt	[65]
2,3-Di-*O*-CM cellulose	$—CH_2—C(=O)ONa$	0.054–1.07	Trt	[75]
		Up to 1.91	MMTr	[74]
2,3-Di-*O*-hydroxyethyl cellulose	$—(CH_2—CH_2)_n—OH$	Up to MS 2.0	MMTrt	[76b]
2,3-Di-*O*-hydroxypropyl cellulose	$—CH_2—CH(OH)—CH_3$	Up to MS 2.0	MMTrt	[76b]
2,3-Di-*O*-allyl cellulose	$—CH_2CH{=}CH_2$	2.0	Trt	[78, 82]
2,3-Di-*O*-benzyl cellulose	$—CH_2—C_6H_5$	2.0	Trt	[78, 82]
2,3-Di-*O*-octadecyl cellulose	$—(CH_2)_{17}CH_3$	2.0	Trt	[84]

[a] Trt, trityl; TDS, thexyldimethylsilyl; MMTrt, 4-monomethoxytrityl.

16.4.1.1 2,3-Di-O-ethers of cellulose

The 6-mono-*O*-trityl cellulose or the more efficient 6-*O*-mono-*O*-(4-monomethyoxytrityl) derivative, and 6-mono-*O*-TDS cellulose were used to synthesize regioselectively functionalized cellulose ethers at positions 2 and 3 after the exclusive cleavage of the protecting groups (Table 16.10). In case of the trityl derivatives, the deprotection is carried out most efficiently with HCl in a suitable solvent. For TDS protected derivatives, tetrabutylammonium fluoride in THF is most successful for the cleavage of the silyl groups.

The alkylation of the 6-*O*-trityl cellulose was achieved in DMSO with solid NaOH as base and the corresponding alkyl halides at 70°C within several hours. Interestingly, a small amount of water in the mixture (about 1 ml per 60 ml DMSO) increases the conversion up to a nearly complete functionalization of the secondary hydroxyl groups [65]. The synthesis of ionic 2,3-*O*-carboxymethyl cellulose (CMC) was carried out with sodium monochloroacetate as etherifying reagent in the presence of solid NaOH in DMSO and detritylation with gaseous HCl in dichloromethane for 45 min at 0°C or alternatively with aqueous hydrochloric acid in an ethanol slurry [74]. The 2,3-*O*-CMC synthesized possess DS values up to 1.91, which shows water solubility at a DS of 0.3 [75]. 2,3-Di-*O*-hydroxyethyl- (HEC) and 2,3-di-*O*-hydroxypropyl celluloses (HPC) were synthesized by heterogeneous etherification of 6-*O*-(4-monomethyoxytrityl) cellulose (MMTC) with alkylene oxide in a 2-propanol/10 per cent NaOH–water mixture. Due to the very hydrophobic character of MMTC, the reaction was successful in the presence of anionic and non-ionic detergent in the reaction mixtures [76]. The polymers become water soluble starting with a molecular degree of substitution (MS) 0.25 (HEC) and 0.50 (HPC), while a conventional HPC is water soluble with MS > 4. ^{13}C NMR spectroscopy revealed the etherification of the secondary hydroxyl groups of the AGU. Only one signal can be observed for the CH_2 group of position 6 (Fig. 16.11). In addition, the peaks of the etherified positions 2 and 3 appear in the range of 80–83 ppm.

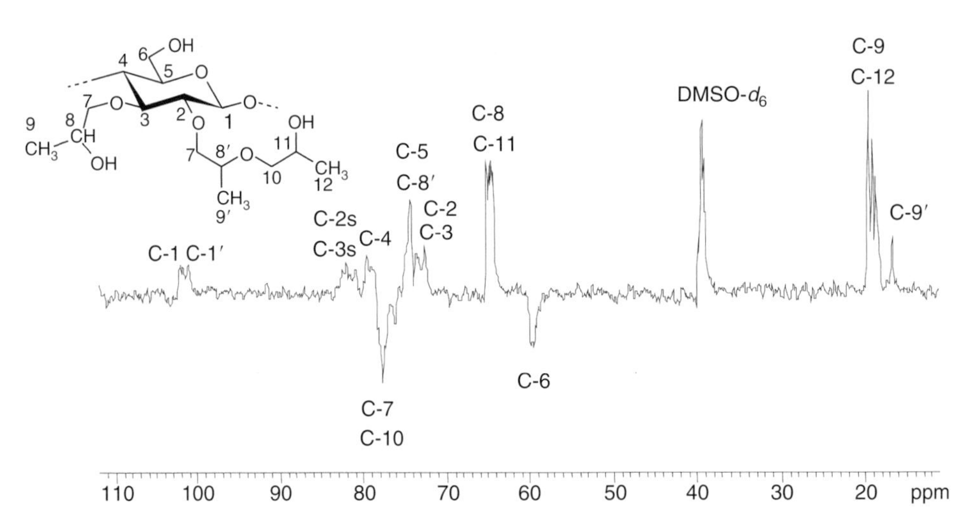

Figure 16.11 ^{13}C DEPT 135 NMR spectrum of 2,3-*O*-hydroxypropyl cellulose (molar substitution 0.33) recorded in DMSO-d_6 at 40°C (with permission from Wiley-VCH, [76b]).

Cellulose ethers may exhibit the phenomenon of thermoreversible gelation that strongly depends on the functionalization pattern as demonstrated for selectively methylated cellulose [77]. A series of 2,3-*O*-methyl celluloses was prepared starting from 6-*O*-trityl- and 6-*O*-monomethoxytrityl cellulose [66]. The total DS and the composition of the repeating units are slightly different according to varied reaction conditions. The thermal events are in strong correlation with the polymer composition concerning the different functionalization pattern of the repeating units. In case of a polymer containing tri-*O*-methylated glucose units in combination with monomethylated ones, a distinct thermal behaviour was found, that is, the methyl cellulose shows thermoreversible gelation. In contrast, a uniform 2,3-*O*-methyl cellulose shows no thermal gelation. It becomes obvious that the 2,3-*O*-methyl glucose units do not affect significantly intermolecular interactions that are necessary for the gelation.

The methylation of the 6-*O*-TDS cellulose with methyl iodide and NaH in THF yields a water insoluble product that dissolves only in NMP and methanol/chloroform in contrast to a 2,3-*O*-methyl cellulose prepared via 6-*O*-trityl cellulose [72]. Thus, the structure is more uniform compared with samples prepared via tritylation.

16.4.1.2 6-O-ethers of cellulose

Up to now, the only path to 6-*O*-cellulose ethers is the time-consuming synthesis described by Kondo [78], which comprises two different protecting groups: 6-*O*-trityl cellulose is converted with allyl chloride in the presence of NaOH resulting in a complete functionalization of positions 2 and 3; subsequent detritylation and isomerization of the allyl groups to 2,3-*O*-(1-propenyl) substituents with potassium tert-butoxide, alkylation of position 6 followed by the cleavage of the 1-propenyl groups at positions 2 and 3 with HCl in methanol. The 6-*O*-alkyl cellulose samples were included in investigations about gelation [79], about blends with synthetic polymers [80, 81], about hydrogen bond system [82, 83], about the behaviour of Langmuir–Blottgett monolayers [84], and about enzymatic degradation [85].

16.4.1.3 3-O-ethers of cellulose

The conversion of 2,6-di-*O*-TDS cellulose with an excess of alkyl halides in THF in the presence of NaH afforded the fully etherified polymers that can be desilylated with fluoride ions yielding 3-mono-*O*-functionalized cellulose ethers (Table 16.11) [86–89].

3-Mono-*O*-methyl cellulose swells in polar media like DMSO indicating a strong network of hydrogen bonds. It becomes soluble by addition of LiCl that destroys intermolecular interactions. An increase of the length of the alkyl chains changes the solubility of the 3-mono-*O*-alkyl celluloses from water (C_2) via aprotic dipolar solvents (up to C_5 alkyl chains) to non-polar solvents like THF for C_5–C_{12} alkyl chains. Light scattering investigations of 3-*O*-*n*-pentyl-, 3-*O*-*iso*-pentyl-, and 3-*O*-dodecyl cellulose in THF disclosed a different aggregation behaviour.

Table 16.11

3-*O*-mono-alkyl celluloses and their solubility

R	Solubility[a]					Reference
	Ethanol	DMSO	DMA	THF	H_2O	
$-CH_3$	−	−	−	−	−	[86]
$-CH_2-CH_3$	−	+	+	−	+	[88]
$-CH_2-CH{=}CH_2$	−	+	+	−	−	[86]
$-CH_2-CH_2-O-CH_3$	−	+	+	−	+	[89]
$-CH_2-CH_2-CH_3$	+	+	+	−	−	[89]
$-CH_2-CH_2-CH_2-CH_3$	−	+	+	−	−	[89]
$-CH_2-CH_2-CH_2-CH_2-CH_3$	+	+	+	+	−	[87]
$-CH_2-CH_2-CH-(CH_3)_2$	+	+	+	+	−	[87]
$-(CH_2)_{11}-CH_3$	−	−	−	+	−	[87]

[a] *N,N*-dimethyl acetamide (DMA), dimethyl sulfoxide (DMSO), tetrahydrofuran (THF), soluble (+), insoluble (−).

Whereas the 3-*O*-dodecyl cellulose forms molecularly dispersed solutions (at concentration less than 2 mg/l), the C_5 ethers show aggregation numbers of 6.5 (*iso*-pentyl) and 83 (*n*-pentyl) [87].

Ethyl cellulose of different functionalization pattern shows interesting thermoreversible gelation. While ethyl cellulose with DS 0.7–1.7 becomes water insoluble already at about 30°C [90], ethyl cellulose, which was prepared via induced phase separation (conversion of cellulose dissolved in DMA/LiCl with solid NaOH and ethyl iodide), possesses a distinct higher cloud point temperature of 56°C. A block-like distribution of substituents along the polymer chain is assumed for this polymer as described for CMC and MC [91, 92]. A similar value (63°C) was determined for the structural uniform 3-mono-*O*-ethyl cellulose independent of the DP [89].

One- and two-dimensional NMR spectroscopy demonstrates the uniform structure of the 3-*O*-alkyl ethers after peracetylation of the remaining OH groups as shown in Fig. 16.12 for 3-*O*-methoxyethyl-2,6-di-*O*-acetyl cellulose as a typical example [89].

16.4.2 Cellulose ethers with non-statistical distribution of functional groups

Different routes for the synthesis of cellulose ethers with a non-statistical distribution of the ether groups along the polymer chain are described. The concept of induced phase separation proceeds with the activation of cellulose by dissolution (*e.g.* in DMA/LiCl) and subsequent addition of solid water-free NaOH particles [92]. This process leads to reactive microstructures due to the regeneration of cellulose II on the interface solid particle/solution, which is shown by means of FTIR and polarizing light microscopy [91]. In this state, the reaction of cellulose with sodium monochloroacetate yields CMC with DS values up to 2.2 in a one-step procedure. The analysis of the CMC products can be realized by an HPLC separation of the repeating units after a complete depolymerization of the polymer backbone (preferably with $HClO_4$) [93] and a comparison of the mol fractions measured with values calculated by a simple statistics first described by [94], which simulated the conversion along the polymer chain without preference of any OH groups and without the influence of the DS. While the statistic data meet the mol fractions of conventionally obtained CMC completely, the mol fractions of CMCs synthesized via phase separation show significant differences from the statistical prediction. The concept of the induced phase separation can be extended to other solvent systems like DMSO/TBAF [7, 95, 96] and NMNO [4], various cellulose intermediates, for example, CA and TMSC as well as other cellulose ether syntheses [2] (Fig. 16.13).

CMC with a non-statistic functionalization pattern may be obtained by the concept of reactive structure fractions, which deals with the selective reaction in the activated non-crystalline areas of cellulose by treatment with low-concentrated aqueous NaOH [97, 98].

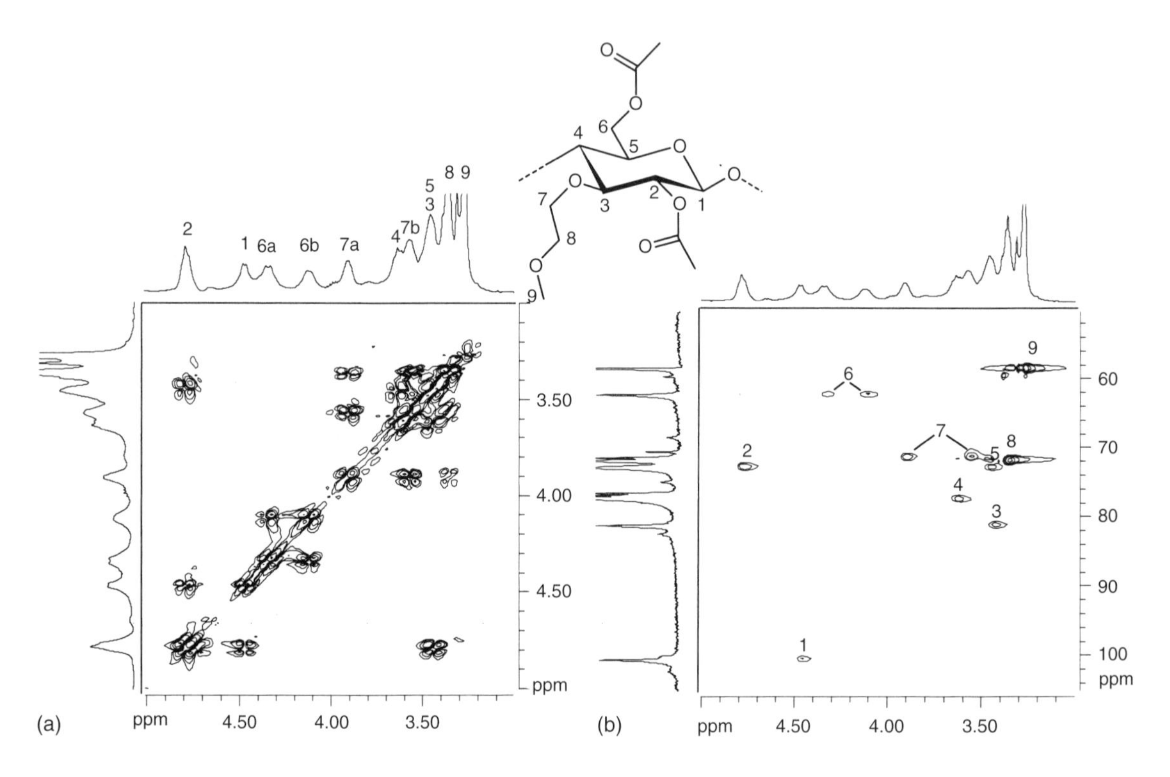

Figure 16.12 COSY (a) and HMQC NMR spectra (b) of peracetylated 3-*O*-methoxyethyl cellulose ($CDCl_3$) [89].

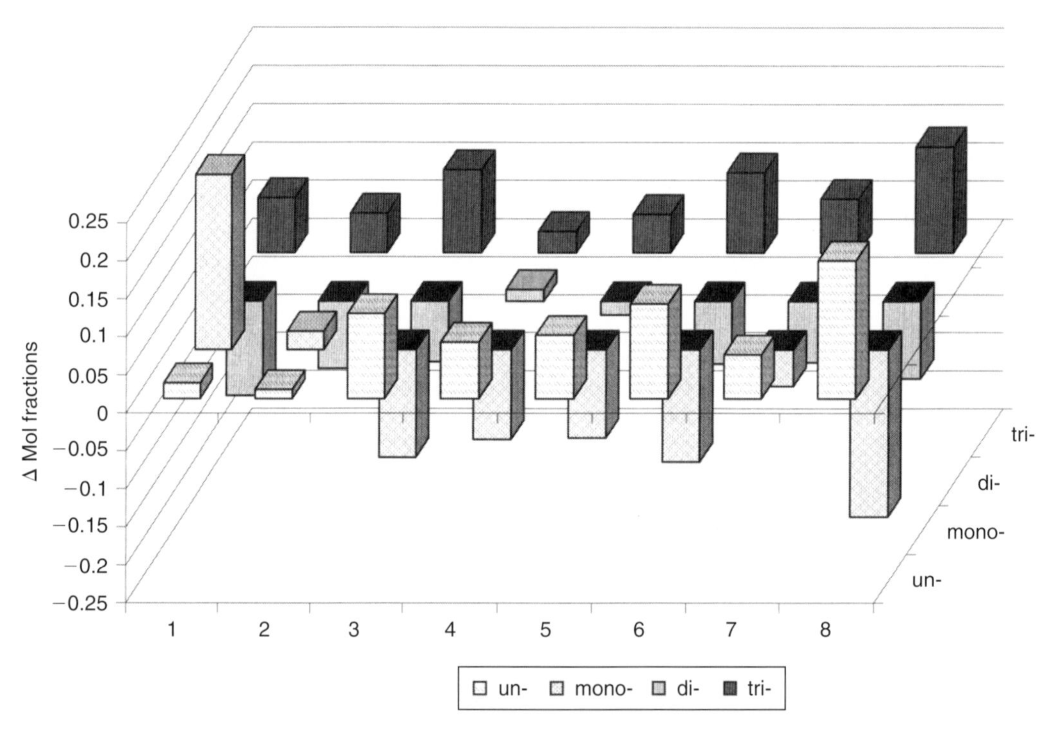

Figure 16.13 Absolute deviation of the mol fractions of un-, mono-*O*-, di-*O*-, and 2,3,6-tri-*O*-functionalized glucose from the binomial distribution of CMC and methyl cellulose (MC) synthesized via induced phase separation: **1** CMC synthesized in DMA/LiCl (DS 2.07); **2** CMC synthesized in DMSO/TBAF (DS 1.89), **3** CMC synthesized in *N*-methylmorpholine-*N*-oxide (DS 1.26), **4** CMC synthesized from cellulose acetate, **5** CMC synthesized from trimethylsilyl cellulose, **6** CMC synthesized from cellulose triflouroacetate, **7** synthesized from cellulose formiate, and **8** MC.

16.4.3 Completely functionalized cellulose ethers

Attempts to synthesize completely functionalized cellulose ethers applying both heterogeneous and homogeneous paths are known. A homogeneous synthesis path is the conversion of cellulose with alkyl halides in the presence of NaOH in the solvent system SO_2/DEA/DMSO [99]. Cellulose could be dissolved in DMSO/LiCl and alkylation is carried out in the presence of sodium dimsyl, resulting from the reaction of NaH and DMSO, with alkyl halides [100]. A facile method for the preparation of tri-*O*-alkyl cellulose deals with the conversion of cellulose acetate in DMSO using the alkyl halide in the presence of NaOH [101]. The complete methylation was described applying homogeneous etherification of cellulose in DMI/LiCl [102]. An interesting procedure for preparation of tri-*O*-allyl- and tri-*O*-crotyl cellulose uses solutions of cellulose in DMA/LiCl, the corresponding alkyl chlorides, and powdered NaOH [103]. The products were investigated regarding their thermal and mechanical properties [99, 103] and their structure in solution [87]. The tri-*O*-alkyl celluloses were characterized by means of different NMR methods involving DQF COSY, HSQC, TOCSY, and HMBC [103].

Fully functionalized trimethylsilyl cellulose (TMSC) was obtained with hexamethyldisilazane (HMDS) in liquid ammonia in a stainless steel autoclave at elevated temperature [104]. The completely silylated polymer exhibits unusual properties; it shows partial insolubility in THF, toluene, DMA, or cyclohexane compared to TMCS with a DS of 2.9 as a result of aggregation of the homopolymer structure. No glass transition was observed by DSC.

Highly functionalized TMSC was exploited for the continuous polymer fractionation [105]. Because of their stiff backbone with flexible side chains (so-called hair rod molecules) ultrathin films by Langmuir–Blodgett (LB) technique could be formed [106] yielding cellulose films after regeneration with HCl gas [107, 108].

16.4.4 Complete homogeneous etherification of cellulose

The first example for a totally homogeneous carboxymethylation is the conversion of cellulose in $Ni(tren)(OH)_2$, surprisingly leading to a CMC with a statistic functionalization pattern [4, 109]. Molten salts hydrates like $LiClO_4 \times H_2O$ were studied for homogeneous cellulose carboxymethylation [110]. Both synthesis paths result in CMCs with the same distribution of functions and properties, for example solubility, as products obtained in the slurry reaction. No complete functionalization was possible. A totally homogeneous hydroxyethylation is carried out using the solvent 6 wt per cent NaOH/4 wt per cent urea aqueous solution [111], which yields products with the analogous functionalization pattern as HEC obtained by the heterogeneous slurry process. An alternative path for cellulose etherification is the use of ionic liquids as reaction medium [22].

16.5 CONCLUSIONS

It appears that the application of dissolved cellulose opens new avenues for the design of cellulose-based products. Limitations of the application of solvents may result from high toxicity, high reactivity of the solvents leading to undesired side reactions, and the loss of solubility during reactions yielding inhomogeneous mixtures by formation of gels and pastes, which can be hardly mixed, and even by formation of de-swollen particles of low reactivity.

The combination with new solvents and esterification methodologies has driven the new synthetic paths for carboxylic ester formation of cellulose. The examples discussed illustrate the enormous structural diversity accessible by these new and efficient methods. Very promising solvents for homogeneous path chemistry of cellulose are ionic liquids even in a commercial scale that is intensively studied world-wide at present. In any case, the recycling of the solvents has to be addressed. In the field of cellulose ethers the regioselective introduction of the functional groups is still a synthetic challenge. These products will give new insights in the interaction of cellulose derivatives with each other, with other polymers and surfaces, and hence to improve common applications and to introduce cellulose-based materials in new application fields. Last but not least, it appears cellulose chemistry applying novel organic chemistry will lead to a variety of sophisticated products; impressive examples are S_N reactions. Moreover, Click chemistry with cellulose was already successfully realized [112].

ACKNOWLEDGEMENTS

Dr. Andreas Koschella, Susann Dorn, Sarah Köhler, and Stephan Daus are gratefully acknowledged for preparing tables and figures as well as for proofreading the manuscript.

REFERENCES

1. Heinze T., Liebert T., Koschella A., *Esterification of Polysaccharides*, Springer Verlag, Berlin, Heidelberg, 2006.
2. Heinze T., Liebert T., Unconventional methods in cellulose functionalization, *Prog. Polym. Sci.*, **26**, 2001, 1689–1762.
3. Zhou J., Zhang L., Solubility of cellulose in sodium hydroxide/urea aqueous solution, *Polymer J.*, **32**, 2000, 866–870.
4. Heinze T., Liebert T., Klüfers F., Meister F., Carboxymethylation of cellulose in unconventional media, *Cellulose*, **6**, 1999, 153–165.
5. Dawsey T.R., Application and limitations of lithium chloride/*N*,*N*-dimethylacetamide in the homogeneous derivatization of cellulose, in *Cellulosic Polymers, Blends and Composite*, Ed.: Gilbert R.D., Hanser, Munich, 1994, pp. 157–171.
6. El Seoud O.A., Heinze T., Organic esters of cellulose: New perspectives for old polymers, in *Polysaccharides I, Structure, Characterization and Use, Advances in Polymer Science* **186**, Ed.: Heinze T., Springer Verlag, Berlin, Heidelberg, 2005, pp. 103–149.
7. Heinze T., Dicke R., Koschella A., Kull A.H., Klohr E.-A., Koch W., Effective preparation of cellulose derivatives in a new simple cellulose solvent, *Macromol. Chem. Phys.*, **201**, 2000, 627–631.
8. Köhler S., Heinze T., New solvents for cellulose: Dimethyl sulfoxide/ammonium fluorides, *Macromol. Biosci.*, **7**, 2007, 307–314.
9. Ass B.A.P., Frollini E., Heinze T., Studies on the homogeneous acetylation of cellulose in the novel solvent dimethyl sulfoxide/tetrabutylammonium fluoride trihydrate, *Macromol. Biosci.*, **4**, 2004, 1008–1013.
10. Sharma R.K., Fry J.L., Instability of anhydrous tetra-*n*-alkylammonium fluorides, *J. Org. Chem.*, **48**, 1983, 2112–2114.
11. Sun H., DiMagno S.G., Anhydrous tetrabutylammonium fluoride, *J. Am. Chem. Soc.*, **127**, 2005, 2050–2051.
12. Philipp B., Organic solvents for cellulose, *Polym. News*, **15**, 1990, 170–175.
13. Woodings C.R., The development of advanced cellulosic fibers, *Int. J. Biol. Macromol.*, **17**, 1995, 305–309.
14. O'Driscoll C., Spinning a stronger yarn, *Chem. Brit.*, **32**, 1996, 27–29.
15. Wendler F., Graneß G., Heinze T., Characterization of autocatalytic reactions in modified cellulose/NMMO solutions by thermal analysis and UV/VIS spectroscopy, *Cellulose*, **12**, 2005, 411–422.
16. Wendler F., Fortschritte bei der Characterisierung modifizierter Cellulose/*N*-Methylmorpholin-*N*-oxid-Lösungen. Ein Beitrag zur Herstellung funktionalisierter Celluloseformkörper, Ph.D. Thesis, Friedrich Schiller University of Jena, 2006.
17. Swatloski R.P., Spear S.K., Holbrey J.D., Rogers R.D., Dissolution of cellulose with ionic liquids, *J. Am. Chem. Soc.*, **124**, 2002, 4974–4975.
18. Schlufter K., Schmauder H.-P., Dorn S., Heinze T., Efficient homogeneous chemical modification of bacterial cellulose in the ionic liquid 1-*N*-butyl-3-methylimidazolium chloride, *Macromol. Rapid Comm.*, **27**, 2006, 1670–1676.
19. Husemann E., Siefert S., *N*-ethyl-pyridinium-chlorid als Lösungsmittel und Reaktionsmedium für Cellulose, *Makromol. Chem.*, **128**, 1969, 288–291.
20. Deus C., Friebolin H., Siefert E., Partially acetylated cellulose: Synthesis and determination of the substituent distribution via proton NMR spectroscopy, *Makromol. Chem.*, **192**, 1991, 75–83.
21. Wu J., Zhang J., Zhang H., He J., Ren Q., Guo M., Homogeneous acetylation of cellulose in a new ionic liquid, *Biomacromolecules*, **5**, 2004, 266–268.
22. Heinze T., Schwikal K., Barthel S., Ionic liquids as reaction medium in cellulose functionalization, *Macromol. Biosci.*, **5**, 2005, 520–525.
23. Barthel S., Heinze T., Acylation and carbanilation of cellulose in ionic liquids, *Green Chem.*, **8**, 2006, 301–306.
24. McCormick C.L., Lichatowich D.K., Pelezo J.A., Anderson K.W., Homogeneous solution reactions of cellulose, chitin, and other polysaccharides, *Modification of Natural Polymers*, ACS Symposium Series No. 121, 1980, pp. 371–380.
25. McCormick C.L., Callais P.A., Derivatization of cellulose in lithium chloride and *N*,*N*-dimethylacetamide solutions, *Polymer*, **28**, 1987, 2317–2323.
26. El Seoud O.A., Marson G.A., Ciacco G.T., Frollini E., An efficient, one-pot acylation of cellulose under homogeneous reaction conditions, *Macromol. Chem. Phys.*, **201**, 2000, 882–889.
27. Regiani A.M., Frollini E., Marson G.A., Arantes G.M., El Seoud O.A., Some aspects of acylation of cellulose under homogeneous solution conditions, *J. Polym. Sci. A: Polym. Chem.*, **37**, 1999, 1357–1363.
28. Ciacco G.T., Liebert T.F., Frollini E., Heinze T.J., Application of the solvent dimethyl sulfoxide/tetrabutyl-ammonium fluoride trihydrate as reaction medium for the homogeneous acylation of sisal cellulose, *Cellulose*, **10**, 2003, 125–132.
29. Heinze T., Schaller J., New water soluble cellulose esters synthesized by an effective acylation procedure, *Macromol. Chem. Phys.*, **201**, 2000, 1214–1218.
30. Sealey J.E., Frazier C.E., Samaranayake G., Glasser W.G., Novel cellulose derivatives. V. Synthesis and thermal properties of esters with trifluoroethoxy acetic acid, *J. Polym. Sci. B: Polym. Phys.*, **38**, 2000, 486–494.
31. Sealey J.E., Samaranayake G., Todd J.G., Glasser W.G., Novel cellulose derivatives. IV. Preparation and thermal analysis of waxy esters of cellulose, *J. Polym. Sci. B: Polym. Phys.*, **34**, 1996, 1613–1620.

32. Heinze T., Liebert T., Pfeiffer K., Hussain M.A., Unconventional cellulose esters: Synthesis, characterization and structure–property relations, *Cellulose*, **10**, 2003, 283–296.
33. Glasser W.G., Samaranayake G., Dumay M., Dave V., Novel cellulose derivatives. III. Thermal analysis of mixed esters with butyric and hexanoic acid, *J. Polym. Sci. B: Polym. Phys.*, **33**, 1995, 2045–2054.
34. Glasser W.G., Becker U., Todd J.G., Novel cellulose derivatives. Part VI. Preparation and thermal analysis of two novel cellulose esters with fluorine-containing substituents, *Carbohyd. Polym.*, **42**, 2000, 393–400.
35. Samaranayake G., Glasser W.G., Cellulose derivatives with low DS, 1. A novel acylation system, *Carbohyd. Polym.*, **22**, 1993, 1–7.
36. Zhang Z.B., McCormick C.L., Structopendant unsaturated cellulose esters via acylation in homogeneous lithium chloride/*N,N*-dimethylacetamide solutions, *J. Appl. Polym. Sci.*, **66**, 1997, 293–305.
37. Williamson S.L., McCormick C.L., Cellulose derivatives synthesized via isocyanate and activated ester pathways in homogeneous solutions of lithium chloride/*N,N*-dimethylacetamide, *J. Macromol. Sci. Pure Appl. Chem.*, **A35**, 1998, 1915–1927.
38. Staab H.A., Syntheses with heterocyclic amides (azolides), *Angew. Chem.*, **74**, 1962, 407–423.
39. Liebert T., Heinze T., Tailored cellulose esters: Synthesis and structure determination, *Biomacromolecules*, **6**, 2005, 333–340.
40. Hussain M.A., Liebert T., Heinze T., First report on a new esterification method for cellulose, *Polym. News*, **29**, 2004, 14–16.
41. Iwata T., Azuma J., Okamura K., Muramoto M., Chun B., Preparation and NMR assignments of cellulose mixed esters regioselectively substituted by acetyl and propanoyl groups, *Carbohyd. Res.*, **224**, 1992, 277–283.
42. Iwata T., Doi Y., Azuma J., Direct imaging of single crystals of regioselectively substituted cellulose esters by atomic force microscopy, *Macromolecules*, **30**, 1997, 6683–6684.
43. Tsunashima Y., Hattori K., Substituent distribution in cellulose acetates: Its control and the effect on structure formation in solution, *J. Colloid Interf. Sci.*, **228**, 2000, 279–286.
44. Tsunashima Y., Hattori K., Kawanishi H., Horii F., Regioselectively substituted 6-*O*- and 2,3-di-*O*-acetyl-6-*O*-triphenylmethylcellulose: Its chain dynamics and hydrophobic association in polar solvents, *Biomacromolecules*, **2**, 2001, 991–1000.
45. Rahn K., Diamantoglou M., Klemm D., Berghmans H., Heinze T., Homogeneous synthesis of cellulose *p*-toluenesulfonates in *N,N*-dimethylacetamide/LiCl solvent system, *Angew. Makromol. Chem.*, **238**, 1996, 143–163.
46. Koschella A., Leermann T., Brackhagen M., Heinze T., Study of sulfonic acid esters from 1 → 4-, 1 → 3-, and 1 → 6-linked polysaccharides, *J. Appl. Polym. Sci.*, **100**, 2006, 2142–2150.
47. Ishii T., Ishizu A., Nakano J., Chlorination of cellulose with methanesulfonyl chloride in *N,N*-dimethylformamide and chloral, *Carbohyd. Res.*, **59**, 1977, 155–163.
48. Sato T., Koizumi J., Ohno Y., Endo T., An improved procedure for the preparation of chlorinated cellulose with methanesulfonyl chloride in a dimethylformamide–chloral–pyridine mixture, *J. Polym. Sci. A: Polym. Chem.*, **28**, 1990, 2223–2227.
49. Furuhata K., Aoki N., Suzuki S., Sakamoto M., Saegusa Y., Nakamura S., Bromination of cellulose with tribromoimidazole, triphenylphosphine and imidazole under homogeneous conditions in lithium bromide–dimethylacetamide, *Carbohyd. Polym.*, **26**, 1995, 25–29.
50. Heinze T., Rahn K., New polymers from cellulose sulfonates, *J. Pulp Pap. Sci.*, **25**, 1999, 136–140.
51. Kasuya N., Iiyama K., Meshitsuka G., Ishizu A., Preparation of 6-deoxy-6-fluorocellulose, *Carbohyd. Res.*, **260**, 1994, 251–257.
52. Vigo T.L., Sachinvala N., Deoxycelluloses and related structures, *Polym. Adv. Technol.*, **10**, 1999, 311–320.
53. Siegmund G., Klemm D., Cellulose: Cellulose sulfonates: preparation, properties, subsequent reactions, *Polym. News*, **27**, 2002, 84–90.
54. Liu C., Baumann H., Exclusive and complete introduction of amino groups and their *N*-sulfo and *N*-carboxymethyl groups into the 6-position of cellulose without the use of protecting groups, *Carbohyd. Res.*, **337**, 2002, 1297–1307.
55. Heinze T., Koschella A., Brackhagen M., Engelhardt J., Nachtkamp K., Studies on non-natural deoxyammonium cellulose, *Macromol. Symp.*, **244**, 2006, 74–82.
56. Koschella A., Heinze T., Novel regioselectively 6-functionalized cationic cellulose polyelectrolytes prepared via cellulose sulfonates, *Macromol. Biosci.*, **1**, 2001, 178–184.
57. Tiller J., Berlin P., Klemm D., Novel matrices for biosensor application by structural design of redox-chromogenic aminocellulose esters, *J. Appl. Polym. Sci.*, **75**, 2000, 904–915.
58. Brandt L., Cellulose ethers, in *Ullmann's Encyclopedia of Industrial Chemistry*, Eds.: Gerhartz W., Yamamoto Y.S., Campbell F.T., Pfefferfcorn R., and Rounsaville J.F., VCH, Weinheim, 1986, **Volume A5**, 461ff.
59. Klemm D., Philipp B., Heinze T., Heinze U., *Comprehensive Cellulose Chemistry*, 1st Edition, Volumes 1 and 2, Wiley-VCH, Weinheim, 1998.
60. Klemm D., Heublein B., Fink H.-P., Bohn A., Cellulose: Fascinating biopolymer and sustainable raw material, *Angew. Chem. Int. Ed.*, **44**, 2005, 3358–3393.

61. Heinze T., Chemical functionalization of cellulose, in *Polysaccharides: Structural diversity and functional versatility*, Ed.: Dumitriu S., 2nd Edition, Marcel Dekker, New York, Basel, Hong Kong, 2004, pp. 551–590, Chapter 23.
62. Wagenknecht W., Regioselectively substituted cellulose derivatives by modification of commercial cellulose acetates, *Papier* (Darmstadt, Germany), **50**, 1996, 712–720.
63. Altaner C., Saake B., Puls J., Specificity of an *Aspergillus niger* esterase deacetylating cellulose acetate, *Cellulose*, **10**, 2003, 85–95.
64. Harkness B.R., Gray D.G., Preparation and chiroptical properties of tritylated cellulose derivatives, *Macromolecules*, **23**, 1990, 1452–1457.
65. Kondo T., Gray D.G., The preparation of *O*-methyl- and *O*-ethyl-cellulose having controlled distribution of substituents, *Carbohyd. Res.*, **220**, 1991, 173–183.
66. Kern H., Choi S.W., Wenz G., Heinrich J., Ehrhardt L., Mischnik P., Garidel P., Blume A., Synthesis, control of substitution pattern and phase transitions of 2,3-di-*O*-methylcellulose, *Carbohyd. Res.*, **326**, 2000, 67–79.
67. Kasuya N., Sawatari A., A simple and facile method for triphenylmethylation of cellulose in a solution of lithium chloride in *N,N*-dimethylacetamide, *Sen'i Gakkaishi*, **56**, 2000, 249–253.
68. Hagiwara L., Shiraishi N., Yokota T., Homogeneous tritylation of cellulose in a sulphur dioxide–diethylamine–dimethylsulfoxide medium, *J. Wood Chem. Technol.*, **1**, 1981, 93–109.
69. Camacho Gómez J.A., Erler U.W., Klemm D.O., 4-Methoxy substituted trityl groups in 6-*O* protection of cellulose: Homogeneous synthesis, characterization, detritylation, *Macromol. Chem. Phys.*, **197**, 1996, 953–964.
70. Pawlowski W.P., Gilbert R.D., Fornes R.E., Purrington S.T., The liquid-crystalline properties of selected cellulose derivatives, *J. Polym. Sci. B: Polym. Phys.*, **26**, 1988, 1101–1110.
71. Petzold K., Koschella A., Klemm D., Heublein B., Silylation of cellulose and starch – Selectivity, structure analysis, and subsequent reactions, *Cellulose*, **10**, 2003, 251–269.
72. Koschella A., Klemm D., Silylation of cellulose regiocontrolled by bulky reagents and dispersity in the reaction media, *Macromol. Symp.*, **120**, 1997, 115–125.
73. Mischnik P., Lange M., Gohdes M., Stein A., Petzold K., Trialkylsilyl derivatives of cyclomaltoheptaose, cellulose, and amylose: Rearrangement during methylation analysis, *Carbohyd. Res.*, **277**, 1995, 179–187.
74. Heinze T., Röttig K., Nehls I., Synthesis of 2,3-*O*-carboxymethylcellulose, *Macromol. Rapid Comm.*, **15**, 1994, 311–317.
75. Liu H.-Q., Zhang L.-N., Takaragi A., Miyamoto T., Water solubility of regioselectively 2,3-*O*-substituted carboxymethylcellulose, *Macromol. Rapid Comm.*, **18**, 1997, 921–925.
76. (a) Yue Z., Cowie J.M.G., Preparation and chiroptical properties of a regioselectively substituted cellulose ether with PEO side chains, *Macromolecules*, **35**, 2002, 6572–6577. (b) Schaller J., Heinze T., Studies on the synthesis of 2,3-*O*-hydroxyalkyl ethers of cellulose, *Macromol. Biosci.*, **5**, 2005, 58–63.
77. Hirrien M., Chevillard C., Desbrieres J., Axelos M.A.V., Rinaudo M., Thermal gelation of methyl celluloses: New evidence for understanding the gelation mechanism, *Polymer*, **39**, 1998, 6251.
78. Kondo T., Preparation of 6-*O*-alkylcelluloses, *Carbohyd. Res.*, **238**, 1993, 231–240.
79. Itagaki H., Tokai M., Kondo T., Physical gelation process for cellulose whose hydroxyl groups are regioselectively substituted by fluorescent groups, *Polymer*, **38**, 1997, 4201–4205.
80. Kondo T., Sawatari C., Manley R.S., Gray D.G., Characterization of hydrogen-bonding in cellulose synthetic polymer blend systems with regioselectively substituted methylcellulose, *Macromolecules*, **27**, 1994, 210–215.
81. Shin J.-H., Kondo T., Cellulosic blends with poly(acrylonitrile): Characterization of hydrogen bonds using regioselectively methylated cellulose derivatives, *Polymer*, **39**, 1998, 6899–6904.
82. Kondo T., Hydrogen bonds in regioselectively substituted cellulose derivatives, *J. Polym. Sci. B: Polym. Phys.*, **32**, 1994, 1229–1236.
83. Kondo T., The relationship between intramolecular hydrogen bonds and certain physical properties of regioselectively substituted cellulose derivatives, *J. Polym. Sci. B: Polym. Phys.*, **35**, 1997, 717–723.
84. Kasai W., Kuga S., Magoshi J., Kondo T., Compression behaviour of Langmuir–Blodgett monolayers of regioselectively substituted cellulose ethers with long alkyl side chains, *Langmuir*, **21**, 2005, 2323–2329.
85. Nojiri M., Kondo T., Application of regioselectively substituted methylcelluloses to characterize the reaction mechanism of cellulase, *Macromolecules*, **29**, 1996, 2392–2395.
86. Koschella A., Heinze T., Klemm D., First synthesis of 3-*O*-functionalized cellulose ethers via 2,6-di-*O*-protected silyl cellulose, *Macromol. Biosci.*, **1**, 2001, 49–54.
87. Petzold K., Klemm D., Heublein B., Burchard W., Savin G., Investigations on structure of regioselectively functionalized celluloses in solution exemplified by using 3-*O*-alkyl ethers and light scattering, *Cellulose*, **11**, 2004, 177–193.
88. Koschella A., Fenn D., Heinze T., Water soluble 3-mono-*O*-ethyl cellulose: Synthesis and characterization, *Polym. Bull.*, **57**, 2006, 33–41.
89. Illy N.G., Regioselektive Cellulosefunktionalisierung: Herstellung, Charakterisierung und Eigenschaftsuntersuchung von 3-*O*-Celluloseethern, *Diploma Thesis*, Friedrich Schiller University of Jena, Germany, 2006.

90. Dönges R., Nonionic cellulose ethers, *Br. Polym. J.*, **23**, 1990, 315–326.
91. Liebert T., Heinze T., Induced phase separation: A new synthesis concept in cellulose chemistry, in *Cellulose Derivatives: Modification, Characterisation, and Nanostructures*, Eds.: Heinze T.J. and Glasser W.G., ACS Symposium Series No. 688, American Chemical Society, Washington/DC, 1998, pp. 61–72.
92. Heinze T., Ionische Funktionspolymere aus Cellulose: Neue Synthesekonzepte, Strukturaufklärung und Eigenschaften, *Habilitation Thesis*, Friedrich Schiller University of Jena, Germany, 1997, Shaker Verlag, Aachen, 1998.
93. Heinze T., Erler U., Nehls I., Klemm D., Determination of the substituent pattern of heterogeneously and homogeneously synthesized carboxymethyl cellulose by using high-performance liquid chromatography, *Angew. Makromol. Chem.*, **215**, 1994, 93–106.
94. Reuben J., Conner H.T., Analysis of the ^{13}C NMR spectrum of hydrolyzed *O*-(carboxymethyl)cellulose: Monomer composition and substitution patterns, *Carbohyd. Res.*, **115**, 1983, 1–13.
95. Koschella A., Neue Funktionspolymere durch regioselektiv gesteuerte Synthesen aus Cellulose und NMR-spektroskopische Charakterisierung der unkonventionellen Substituentenmuster, *Ph.D. Thesis*, Friedrich Schiller University of Jena, Germany, 2000.
96. Ramos L.A., Frollini E., Heinze T., Carboxymethylation of cellulose in the new solvent dimethyl sulfoxide/tetrabutylammonium fluoride, *Carbohyd. Polym.*, **60**, 2005, 259–267.
97. Mann G., Kunze J., Loth F., Fink H.-P., Cellulose ethers with a block-like distribution of substituents by structure-selective derivatization of cellulose, *Polymer*, **39**, 1998, 3155–3165.
98. Fink H.-P., Dautzenberg G.H., Kunze J., Philipp B., The composition of alkali cellulose – A new concept, *Polymer*, **27**, 1986, 944–948.
99. Isogai A., Ishizu A., Nakano J., Preparation of tri-*O*-alkylcelluloses by the use of a nonaqueous cellulose solvent and their physical characteristics, *J. Appl. Polym.Sci.*, **31**, 1986, 341–352.
100. Petrus L., Gray D.G., BeMiller J.N., Homogeneous alkylation of cellulose in lithium chloride/dimethyl sulfoxide solvent with dimsyl sodium activation. A proposal for the mechanism of cellulose dissolution in lithium chloride/DMSO, *Carbohyd. Res.*, **268**, 1995, 319–323.
101. Kondo T., Gray D.G., Facile method for the preparation of tri-*O*-(alkyl)cellulose, *J. Appl. Polym. Sci.*, **45**, 1992, 417–423.
102. Takaragi A., Minoda M., Miyamoto T., Liu H.Q., Zhang L.N., Reaction characteristics of cellulose in the LiCl/1,3-dimethyl-2-imidazolidinone solvent system, *Cellulose*, **6**, 1999, 93–109.
103. Sachinvala N.D., Winsor D.L., Hamed O.A., Maskos K., Niemczura W.P., Tregre G.J., Glasser W.G., Bertoniere N.R., The physical and NMR characterizations of allyl- and crotylcelluloses, *J. Polym. Sci. A: Polym. Chem.*, **38**, 2000, 1889–1902.
104. Mormann W., Demeter J., Silylation of cellulose with hexamethyldisilazane in liquid ammonia – First examples of completely trimethylsilylated cellulose, *Macromolecules*, **32**, 1999, 1706–1710.
105. Stoehr T., Petzold K., Wolf B.A., Klemm D., Continuous polymer fractionation of polysaccharides using highly substituted trimethylsilylcellulose, *Macromol. Chem. Phys.*, **199**, 1998, 1895–1900.
106. Schaub M., Wenz G., Wegner G., Stein A., Klemm D., Ultrathin films of cellulose on silicon wafers, *Adv. Mater.*, **5**, 1993, 919–922.
107. Buchholz V., Wegner G., Stemme S., Oedberg L., Regeneration, derivatization, and utilization of cellulose in ultrathin films, *Adv. Mater.*, **8**, 1996, 399–402.
108. Löscher F., Ruckstuhl T., Seeger S., Ultrathin cellulose-based layers for detection of single antigen molecules, *Adv. Mater.*, **10**, 1998, 1005–1009.
109. Burger J., Kettenbach G., Klüfers P., Coordination equilibria in transition-metal based cellulose solvents, *Macromol. Symp.*, **99**, 1995, 113–126.
110. Fischer S., Thümmler K., Pfeiffer K., Liebert T., Heinze T., Evaluation of molten inorganic hydrates as reaction medium for the derivatization of cellulose, *Cellulose*, **9**, 2002, 293–300.
111. Zhou Q., Zhang L., Li M., Wu X., Gheng G., Homogeneous hydroxyethylation of cellulose in NaOH/urea aqueous solution, *Polym. Bull.*, **53**, 2005, 243–248.
112. Liebert T., Hänsch C., Heinze T., Click chemistry with polysaccharides, *Macromol. Rapid Comm.*, **27**, 2006, 208–213.
113. Philipp B., Nehls I., Wagenknecht W., Schnabelrauch M., Carbon-^{13} NMR spectroscopic study of the homogeneous sulfation of cellulose and xylan in the dinitrogen tetroxide-DMF system, *Carbohyd. Res.*, **164**, 1987, 107–116.
114. Schnabelrauch M., Vogt S., Klemm D., Nehls I., Philipp B., Readily hydrolyzable cellulose esters as intermediates for the regioselective derivatization of cellulose. 1. Synthesis and characterization of soluble, low-substituted cellulose formats, *Angew. Makromol. Chem.*, **198**, 1992, 155–164.
115. Liebert T., Schnabelrauch M., Klemm D., Erler U., Readily hydrolyzable cellulose esters as intermediates for the regioselective derivatization of cellulose. Part II. Soluble, highly substituted cellulose trifluoroacetates, *Cellulose*, **1**, 1994, 249–258.
116. Salin B.N., Cemeris M., Mironov D.P., Zatsepin A.G., Trifluoroacetic acid as solvent for the synthesis of cellulose esters. 1. Synthesis of triesters of cellulose and aliphatic carboxylic acids, *Koksnes Kimija*, **3**, 1991, 65–69.

117. Johnson D.C., Nicholson M.D., Haigh F.C., Dimethyl sulfoxide/paraformaldehyde: A nondegrading solvent for cellulose, *Appl. Polym. Symp.*, **28**, 1976, 931–994.
118. Saikia C.N., Dutta N.N., Borah M., Thermal behavior of some homogeneously esterified products of high α-cellulose pulps of fast growing plant species, *Thermochim. Acta*, **219**, 1993, 191–203.
119. Edgar K.J., Pecorini T.J., Glasser W.G., Long-chain cellulose esters: Preparation, properties, and perspective, in *Cellulose Derivatives: Modification, Characterisation, and Nanostructures*, Eds.: Heinze T.J. and Glasser W.G., ACS Symposium Series No. 688, American Chemical Society, Washington/DC, 1998, pp. 38–60.
120. Pawlowski W.P., Sankar S.S., Gilbert R.D., Fornes R.D., Synthesis and solid state carbon-13 NMR studies of some cellulose derivatives, *J. Polym. Sci. A: Polym. Chem.*, **25**, 1987, 3355–3362.

– 17 –

Bacterial Cellulose from *Glucanacetobacter xylinus*: Preparation, Properties and Applications

Édison Pecoraro, Danilo Manzani, Younes Messaddeq and Sidney J.L. Ribeiro

ABSTRACT

This chapter deals with the cellulose produced by the *Glucanacetobacter xylinus* strain, called bacterial cellulose, which is a remarkably versatile biomaterial usable in wide variety of domains, such as paper-making, optics, electronics, acoustics and biomedical devices. Its unique structure shows entangled ultrafine fibres, which provide excellent mechanical strength, besides biodegradability, biocompatibility, high water-holding capacity and high crystallinity. Some of its applications are described, such as complementary nutrition (*nata de coco*), artificial temporary skin for wounds and burns, dental aid, artificial blood vessels and micronerve surgery, DNA separation, composite reinforcement, electronic paper, light emitting diodes and fuel cell membranes.

Keywords

Bacterial cellulose, *Glucanacetobacter xylinus*, Biosynthesis, Structure, Properties, Applications, Biocompatibility, Biomaterial, DNA separation, Biodegradability

17.1 INTRODUCTION

Cellulose is the most abundant biopolymer on earth with an estimated output of over 10^{11} tons per year. Most of its biosynthesis takes place in the cellular walls of plants, but four other sources are known, animal, bacterial, chemical and enzymatic.

The cellulose thus produced exists naturally in two native forms (native cellulose): The first is called *pure cellulose,* and includes celluloses produced in their natural state, such as cotton, bacterial cellulose and those present in some algae and some marine animals like tunicates. The second is called *complex cellulose*, and includes most of the celluloses present in nature, as the fundamental component of the cellular wall of higher plants.

Notwithstanding its specific origin in nature, cellulose is a linear polymer in which β-1,4 glycosidic moieties are joined to form cellobiose repeat units. What vary from one form of cellulose to another are, on the one hand, its degree of polymerization, which can span hundreds to thousands and, on the other hand, the supramolecular organization of its chains, which can give rise to amorphous and several types of crystalline structures. The cellulose hydroxyl groups can form intra and intermolecular hydrogen bonds which are responsible for its chemical stability, structure rigidity and tensile strength. All these fundamental aspects related to cellulose are thoroughly documented, as pointed out in Chapter 1.

The present chapter is devoted to a specific form of cellulose, bacterial cellulose, which has been gaining growing attention because of the property of the actual material associated with it.

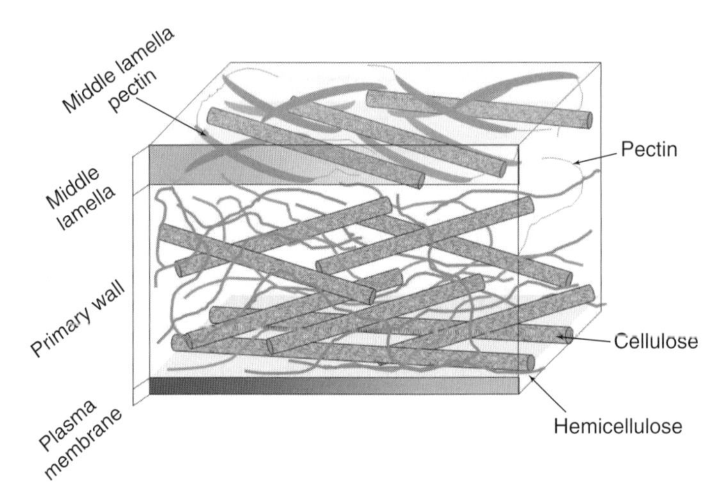

Figure 17.1 Schematic representation of the components in higher-plant cellulose.

Table 17.1

Properties of plant (PC) and bacterial (BC) cellulose

Properties	PC	BC
Fibre width	1.4–4.0 × 10^{-2} mm	70–80 nm
Crystallinity	56–65%	65–79%
Degree of polymerization	13 000–14 000	2 000–6 000
Young's modulus	5.5–12.6 GPa	15–30 GPa
Water content	60%	98.5%

Bacterial cellulose is obtained through the biosynthesis induced by bacteria belonging to genera *Glucanacetobacter*, *Rhizobium*, *Sarcina*, *Agrobacterium*, *Alcaligenes*, etc. Among them, the one possessing the best efficiency to produce cellulose, and for that reason the most studied, is *Glucanacetobacter xylinus* (also denominated as *Acetobacter xylinum*) [1]. These are gram-negative bacteria, strictly aerobic and non-photosynthetic, capable to convert glucose, glycerol and other organic substrates into cellulose within a period of a few days. They are usually found in fruits, vegetables, vinegar and alcoholic beverages.

The cellulose biosynthesized by *G. xylinus* is identical to that produced by plants, regarding its molecular formula and polymeric structure, but presents a higher crystallinity. The other fundamental difference between bacterial cellulose and its widespread plant-based counterpart stems from the fact that the former is chemically pure, free of lignin, hemicelluloses and the other natural components usually associated with the latter (Fig. 17.1).

Table 17.1 compares some properties of vegetal and bacterial cellulose.

The specific properties of bacterial cellulose make it interesting for important applications, such as a nutritional component (additive of low caloric contents, stabilizer, texture modifier, *nata de coco*), a pharmacological agent (temporary dressing, excipient, cosmetics, drug carriers), as well in telecommunications and papermaking, as summarized in Table 17.2.

17.2 CELLULOSE SYNTHESIS BY *G. XYLINUS*

G. xylinus was first described by Brown in 1886 [2, 3] who identified a jelly-like film formed over the surface of a vinegar broth fermentation. A microscopic analysis revealed the presence of bacteria distributed within the whole film. Still today, *G. xylinus* is taken as the model microorganism in the research on the biosynthesis, crystallization and structural properties of bacterial cellulose [4]. The cellulose produced by *G. xylinus* contains approximately

Table 17.2

Examples of applications of bacterial cellulose

Area	Application
Cosmetics	Stabilizer of emulsions; component of artificial nails
Textile industry	Artificial textiles; highly absorbent materials
Sports and tourism	Sporting clothes; tents; camping material
Mining and refinery	Sponges for recovery spilled oil; material for toxin adsorption
Wastes treatment	Recycling of minerals and oils
Sewage purification	Urban sewage purification; water ultrafiltration
Broadcasting	Sensitive diaphragms for microphones and stereo headphones
Forestry	Artificial wood replacer, multi-layer plywood, heavy-duty containers
Paper industry	Specialty papers, archival documents repairing, more durable banknotes, diapers, napkins
Machine industry	Car bodies, airplane parts, sealing of cracks in rocket casings
Food production	Edible cellulose (*nata de coco*)
Medicine	Temporary artificial skin for therapy of burns and ulcers, component of dental implants
Laboratory	Immobilization of proteins, cells; chromatographic techniques; medium for tissue cultures
New applications	Cellulose thin films for documents and book recovery; fibres (including optical); biodegradable plastics; oriented templates; liquid crystal displays; luminescent materials; fuel cell membranes; drug delivery; stents covering; ophthalmic, cardiovascular and neurological prostheses; bulletproof materials

98 per cent (w/w) of water. Visually, the difference between bacterial and plant cellulose relates both to appearance and water content viz. the latter has a fibrous aspect, while the former resembles a gel.

In the laboratory, the cultivation is conducted in a liquid Hestrin–Schramm medium (glucose, yeast extract, peptone, citric acid and sodium phosphate) at pH 5 and 28°C. The cultures are grown in static containers or air flow (airlift bioreactors). The growing time depends on the desired thickness of the ensuing cellulose membrane. Several studies can be found on the growth kinetics of those microorganisms, based on fermentative processes [5].

The cells of *G. xylinus* possess 100 or more pores in their membrane for cellulose extrusion. Each pore produces a cellulose chain that groups with another 36 chains to form an elementary fibril, which has a diameter of approximately 3.5 nm. About 46 adjacent fibrils join through hydrogen bonds to form a ribbon, which has a width ranging from 40 to 60 nm (Fig. 17.2(a)). The ribbons roll up to form the fibre (Fig. 17.2(b)), which gets tangled with the other fibres dispersed in the culture medium (Fig. 17.2(c)) [6]. The entangled fibres form a jelly-like film on the surface of the liquid culture medium. The film containing the bacteria is called zooglea (Fig. 17.2(d)). Its thickness depends on the cultivation time and can usually reach 1 or 2 cm (Fig. 17.2(e)).

Each extrusion pore bears a site for cellulose chains production called terminal complex (TC) and each TC accomplishes several different processes, all controlled genetically within the biosynthetic route. Many studies have been undertaken for both plant and bacterial cellulose, in order to find the proteins, enzymes and to map the genes that are involved in each synthesis step [4, 7], but, until now, none is conclusive. According to Iguchi *et al.* [8], the cellulose membrane is a shelter for the bacteria. It protects them from the lethal effect of solar ultraviolet radiation, keeps them moist in an aerobic environment, allows the diffusion of nutrients from the culture medium and protects them against predators and contamination by heavy metals. According to Borzani and de Souza [9], in a static culture, the cellulose zooglea grows towards the atmosphere, leaning on the first film formed in the liquid–gas interface. In an agitated culture, bubble columns or airlift reactors, the growing also follows to the liquid-gas interface [10].

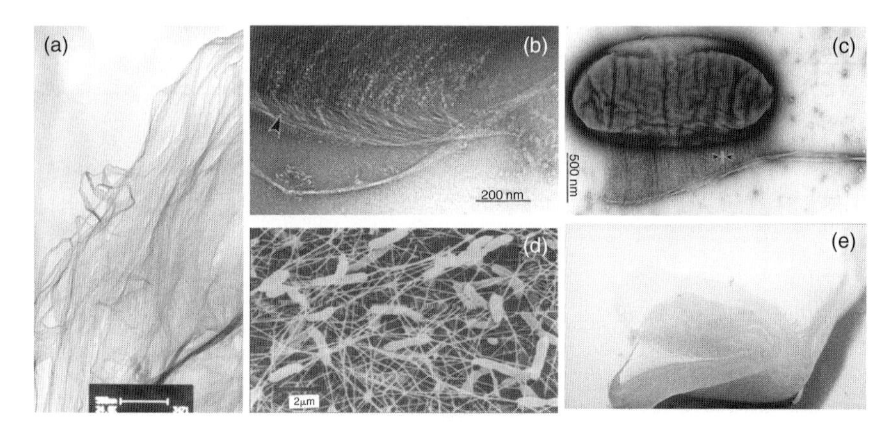

Figure 17.2 (a) Transmission electron micrograph showing the ribbons of approximately 60 nm widths; (b) on the left, the arrow indicates the place where the fibrils are rolling up; (c) SEM of *G. xylinus* producing a cellulose fibre. The suitable place among two arrows display the dimension of the fibrils: (d) SEM of the zooglea; (e) recently harvested zooglea of static culture container. (Figure 2(b) and (c) reprinted from Reference [6], with kind permission of Springer Science and Business Media. Figure 2(d) reprinted from Reference [8], with kind permission of Springer Science and Business Media.)

The more suitable purification method for the bacterial cellulose calls upon distilled water, sodium hydroxide and sodium hypochlorite solutions, since it guarantees the elimination of bacteria cells and culture medium residues from the membrane. For clinical applications, the suitable sterilization processes are gamma radiation and ethylene oxide treatment, but the latter should be avoided for bacterial cellulose dry or wet membrane, because it can cause allergic reactions, mostly when used internally.

In the bacterial cellulose biosynthesis, the cellulose chains (formed by glucose units linked through β-1,4-glycosidics bonds) interact through hydrogen bonds, assuming a parallel orientation among them. The structure and rigidity of bacterial cellulose is provided by the OH intra and intermolecular hydrogen bonds, as shown in Fig. 17.3.

The cellulose microcrystallinity results from isolated areas with orderly chains within the microfibril structures. The arrangement of those chains in an ordered structure is driven by hydrogen bonding. The remaining areas show chains distributed in a paracrystalline or amorphous phases [11]. Two crystalline allomorph phases, Iα and Iβ, were proposed for cellulose I (bacterial cellulose) [12]. The Iα phase is found predominantly in cellulose produced by algae and bacteria, while the Iβ phase is present in higher percentages in higher-plant cellulose. Atalla

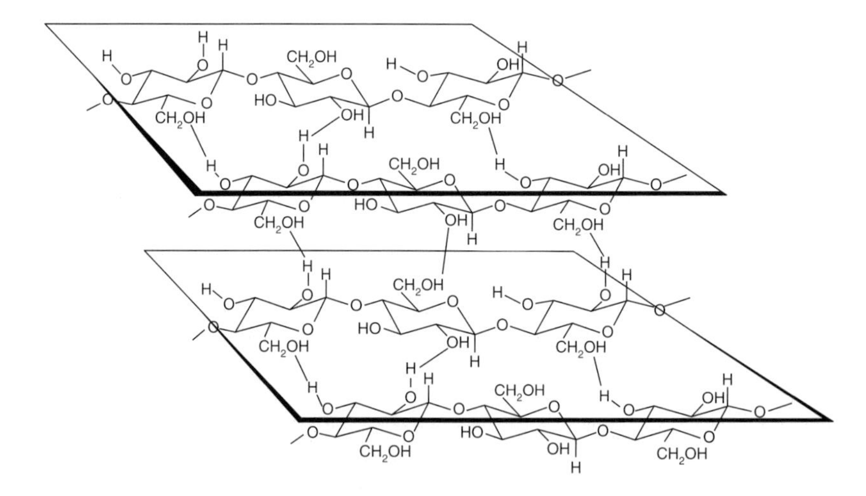

Figure 17.3 Outline of intra- and inter-molecular hydrogen bonds among cellulose chains.

Table 17.3

Iα/Iβ ratio for different cellulose sources

Type	Class	Ratio Iα/Iβ(%)
Glucanoacetobacter	Bacterial	64/36
Valonia	Vegetal (algae)	60/40
Halocynthia	Animal	10/90

and Van der Hart showed in 1984 [12] that phase Iβ is formed by cellulose chains arranged in a monoclinic unit cell, whereas phase Iα displays a triclinic arrangement. The ratio between the Iα and Iβ phases depends on the cellulose source, as shown in Table 17.3.

The structural characteristics of bacterial cellulose are directly related to two factors, namely (1) the origin of the strain, which determines the Iα/Iβ ratio and (2) the culture medium composition that influences the chain size. Such characteristics determine the degree of crystallinity of bacterial cellulose and consequently, their physico-chemical properties. Structural modifications can be accomplished in a post-production step, since it is possible to functionalize the hydroxyl groups (—OH) by methylation [13], esterification [14], sulphonation [13], nitration [13], deoxyamination [15], etc.

17.3 PROPERTIES

The results presented here refer specifically to membranes synthesized by Bionext®'s strain. Bionext is a Brazilian company that produces bacterial cellulose for medical applications. At this company, bacterial cellulose wet membranes are dried by compression and heating, a process that confers unique texture characteristics.

17.3.1 Morphology

Figure 17.4 shows the surface map of a Bionext dry membrane recorded in a Taylor Hobson Form Talysurf Series 2 profilometer.

The dry membrane can be obtained in different thicknesses, from 20 to 500 μm, depending on the need. Figure 17.5 compares the electron scanning photomicrograph of the membrane shown in Fig. 17.4 with that of one obtained by freeze drying. In both cases the fibres are made of ribbons, which are composed of a group of about 36 fibrils

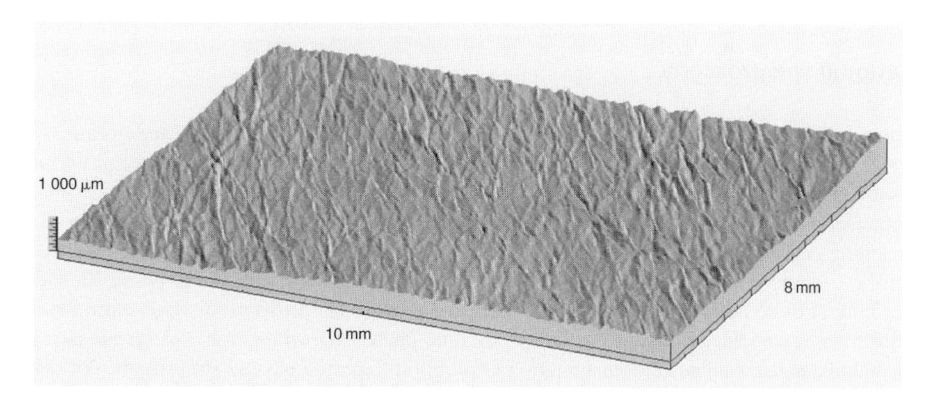

Figure 17.4 Surface map of a bacterial cellulose membrane dried by compression and heating. The scanned area is 10 mm × 8 mm. The average thickness is 200 μm.

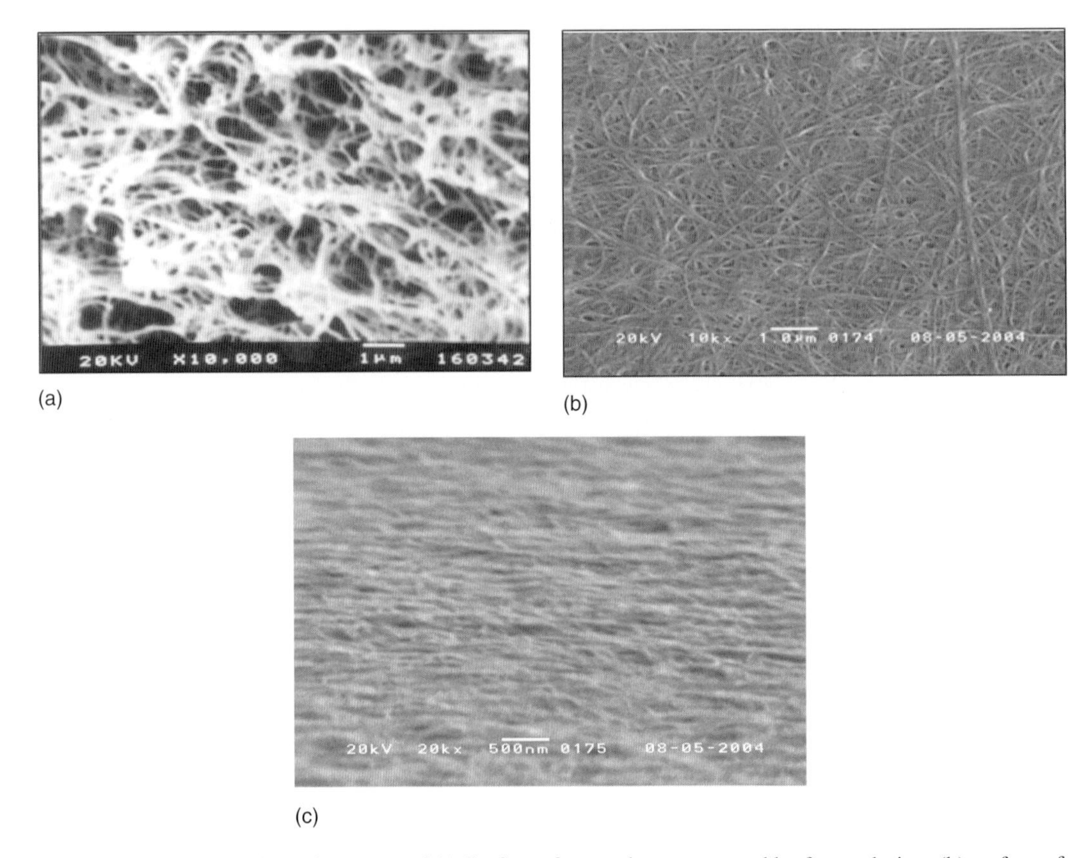

Figure 17.5 Electron scanning microscopy of (a) Surface of a membrane prepared by freeze drying; (b) surface of a membrane dried in a laboratory oven; (c) cross section of the membrane in (b).

interacting through hydrogen bonds. Figure 17.6 shows the transmission electron microscopy microphotographs of ribbons.

Under compression, the original open network, shown in Fig. 17.5(a), collapses in a fibre entanglement and this new network is so dense that it can be used as a filter for microorganisms or blood components, as shown in Fig. 17.7. Specific surface area and porosity can also be modified by a dry process. Thus, the specific area for samples pressed and heated is *ca.* $25\,m^2\,g^{-1}$, while for those freeze dried is about $45\,m^2\,g^{-1}$.

17.3.2 Vibrational spectroscopy

Figure 17.8 shows a typical FTIR spectrum of a pressed and heated bacterial cellulose membrane. Of course, the functional groups that characterize bacterial cellulose are the same as those of vegetable cellulose and the corresponding peaks are assigned in Table 17.4 [16].

Infrared spectroscopy can also be used for the determination of the Iα/Iβ ratio. The method is based on the difference in the stretching energies, and the peaks, assigned to the OH groups present in each phase. The peak at $750\,cm^{-1}$ corresponds to the triclinic phase Iα, while that at $710\,cm^{-1}$ is attributed to monoclinic phase Iβ. Hence, the ratio between the area under these peaks gives the ratio between the phases. The Iα/Iβ ratio shows variations related to the origin of cellulose. Algae and bacteria forms are rich in triclinic phase (Iα), while higher plants are rich in monoclinic phase (Iβ). For Valonia algae, Atalla [12] found a ratio of 6/4. For different *G. xylinus* strains, ratios of 6/4 and 7/3 have been reported [17, 18]. However for the Bionext strain, the ratio is 2/8, that is these bacteria produce cellulose similar to that found in higher plants, in terms of the proportion between the crystalline phases. As this strain is an over producer, its cellulose synthesis seems to be directed towards the thermodynamically most stable phase, viz. Iβ.

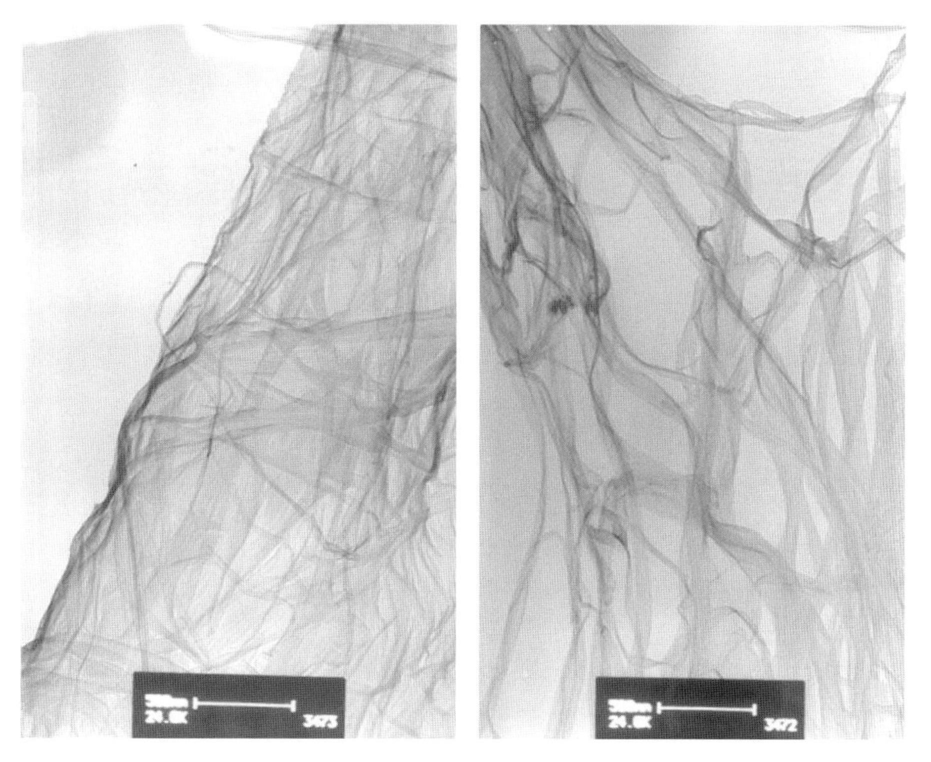

Figure 17.6 Transmission electron microscopy photomicrographs showing ribbons of 60 nm width. In this case, the ribbons compose the structure of the fibres in a membrane prepared by freeze drying (the bars correspond to 500 nm).

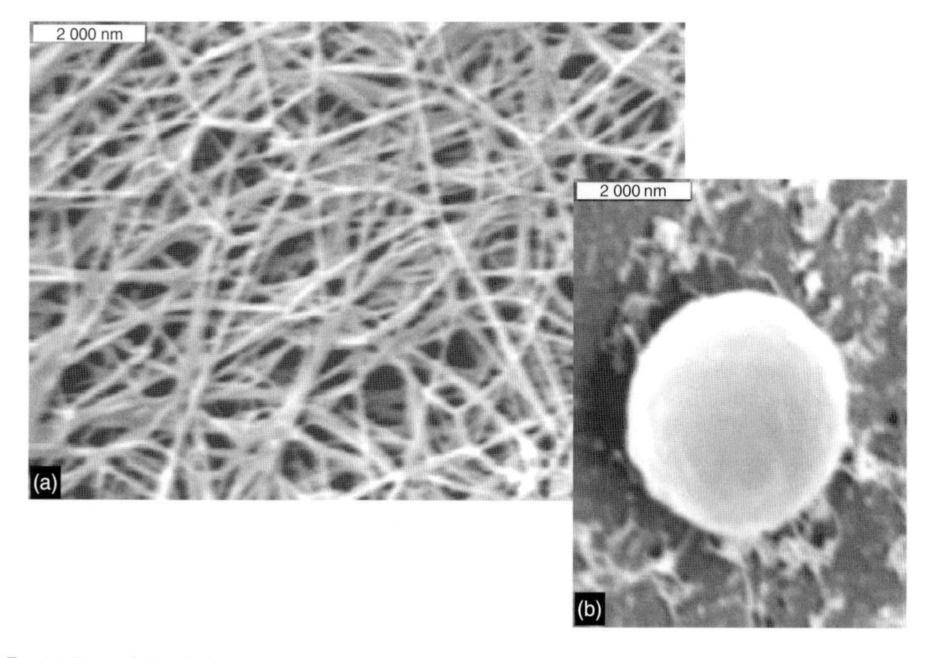

Figure 17.7 (a) Bacterial cellulose dry membrane mesh and (b) a thrombocyte (blood component). (Reprinted from Reference [5] with permission from Elsevier Science Ltd.)

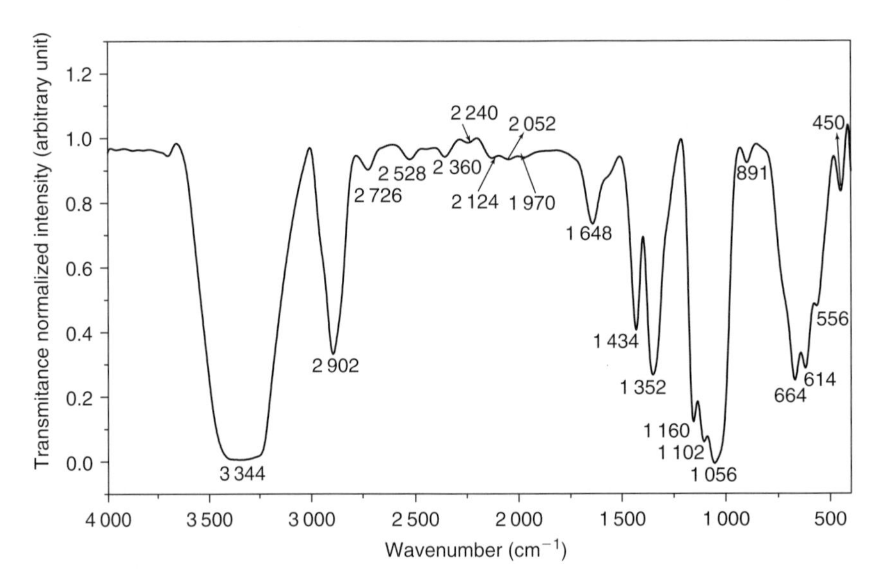

Figure 17.8 FTIR spectrum of a 200-μm thick bacterial cellulose membrane.

Table 17.4

Typical infrared absorption frequencies for bacterial cellulose [16]

Range (cm^{-1})	Assignment[a]
3 500–3 300	ν OH
3 000–2 870	ν CH and CH_2 (CHOH; CH_2OH)
1 645	δ_s HOH
1 430–1 330	δ C—OH e CH
1 200–1 000	ν C—O (—C—O—H)
1 150–1 000	ν C—O (C—O—C)
900–700	δ_{as} in plane CH_2 δ C—H
700–400	δ out of the plane OH

[a] ν = stretching; δ = angular bending; s = symmetric; as = asymmetric.

17.3.3 X-ray diffraction

The X-ray diffractogram of bacterial cellulose, Fig. 17.9, shows the presence of crystalline peaks and amorphous haloes. Due to the overlapping of the reflections from Iα and Iβ phases, it is not possible to separate the contribution of each one on the peaks at approximately $2\theta = 15°$ and $23°$. The peak at 15° corresponds to the diffractions from triclinic (100) and monoclinic (110) plans, whereas the peak at 23° corresponds to the reflections from triclinic (110) and monoclinic (200) plans [19]. Through the *Ruland* method [20], it is possible to quantify the degree of crystallinity of cellulose. The method relates the contribution of the amorphous phase on the reflexions of the crystalline phases through the expression $\%C = [I_c/(I_c + K \times I_a)] \times 100$, where $\%C$ is the percentage of the crystalline phase, I_c the integral under all the crystalline reflexion peaks, I_a the integral under the haloe that corresponds to the amorphous phase, and K is a proportionality constant, which in the case of bacterial cellulose is taken as unity.

The crystallinity of bacterial cellulose has an influence on its mechanical properties. In 1990, Brown *et al.* [21] showed that under an intense magnetic field (1.8 T), their *G. xylinus* strain produced cellulose with no crystalline regions. One characteristic of that cellulose was a lower tensile strength, compared to that of a sample obtained without magnetic field. Therefore, an increase in the amount of crystalline phase can lead to an improvement of the mechanical properties of bacterial cellulose membranes.

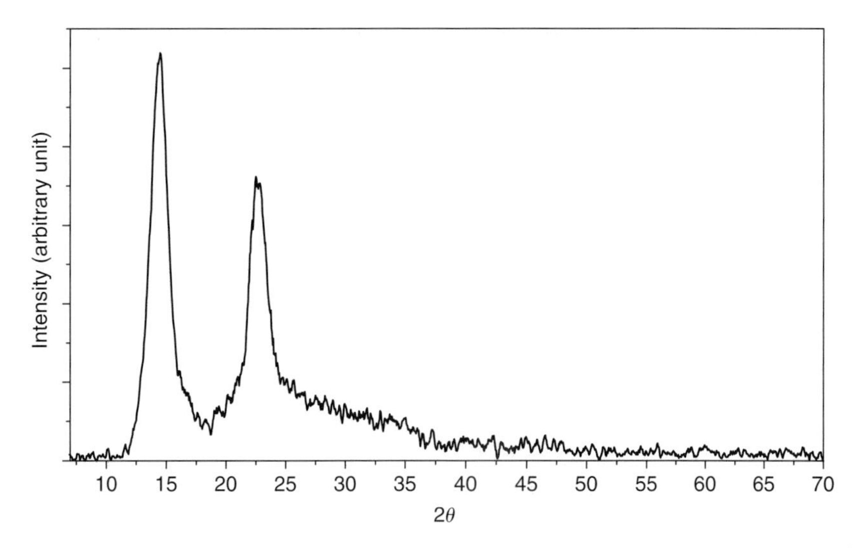

Figure 17.9 X-ray diffractogram of a bacterial cellulose membrane. The sample was supported on a borosilicate glass holder.

17.3.4 UV–visible and near-infrared absorption

Another important characteristic of bacterial cellulose is its capacity to block most of the ultraviolet radiation from sunlight. The UV spectral region is subdivided into UV-A (320–400 nm), UV-B (280–320 nm) and UV-C (below 280 nm). UV-C is not considered here, since it is absorbed by the atmosphere. Figure 17.10 shows the transmittance spectrum of a bacterial cellulose membrane from the UV (250–400 nm) to the near-infrared

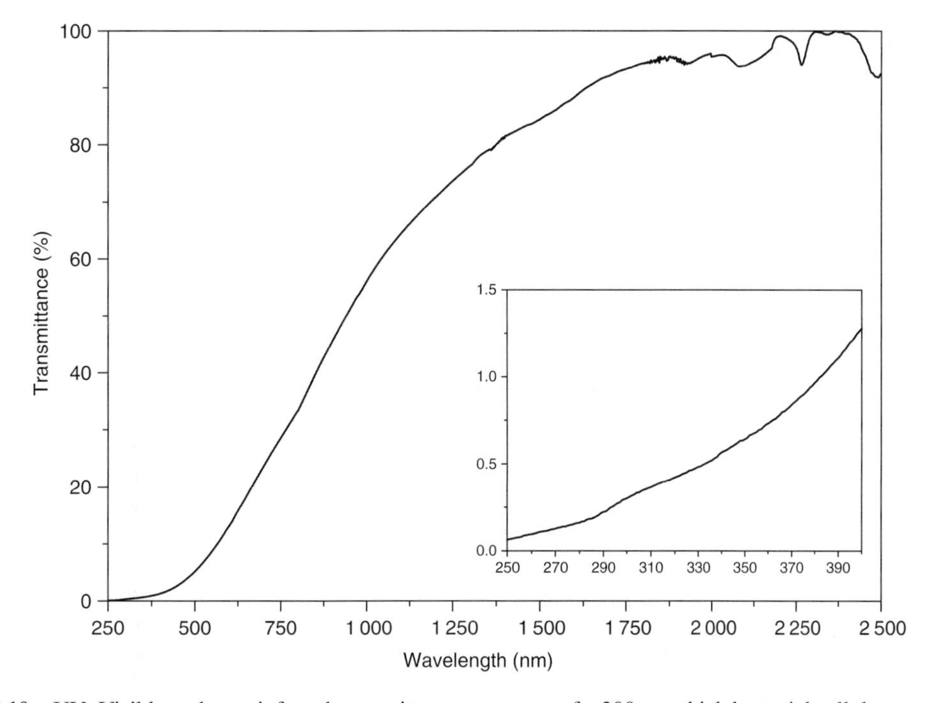

Figure 17.10 UV–Visible and near-infrared transmittance spectrum of a 200-μm thick bacterial cellulose membrane. The insert shows the magnification of the 250–400 nm range.

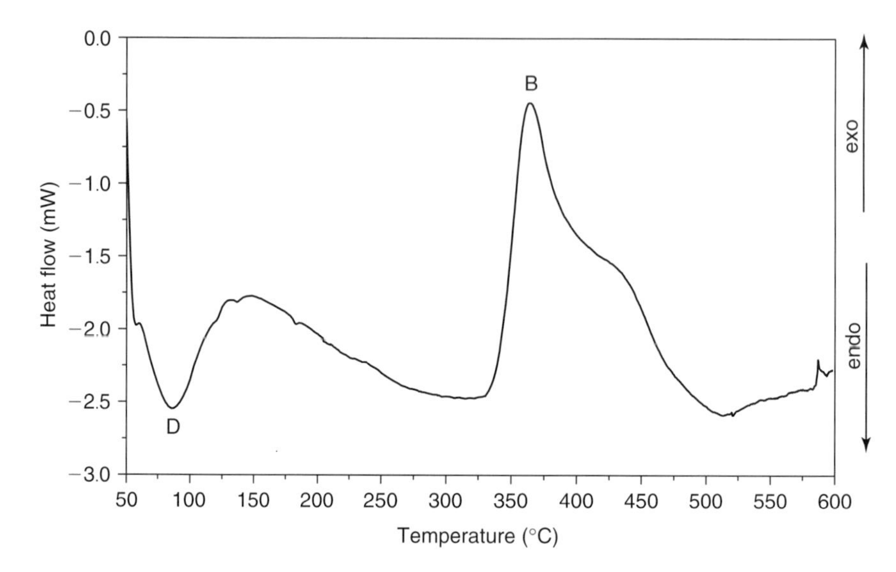

Figure 17.11 DSC thermogram of a bacterial cellulose membrane. The temperatures at D and B refer to the dehydration (87°C) and burning (363°C) peak temperatures, respectively.

(750–2 500 nm), including the visible region of the electromagnetic spectrum. The insert shows that for a 200-μm thick membrane, only 0.4 per cent of all UV-B and 1.3 per cent of UV-A radiation pass through it. For medical applications, such as temporary artificial skin, that feature is an important issue. Besides its transparency in the visible region, another characteristic of bacterial cellulose is its transparency at 1 550 nm, suggesting that it can be used for telecommunication applications.

17.3.5 Thermal behaviour

As shown in Fig. 17.11, a bacterial cellulose dry membrane can be heated up to 325°C before it starts to burn, which is at least 75°C above the burning temperature of conventional office paper. This higher thermal resistance can is attributed to the absence of additives in its composition, which are common in papermaking.

Not only by its thermal, but also by its mechanical properties (tensile strength of 200–300 MPa [22]), bacterial cellulose could also be an alternative material for industrial seals and connections.

17.4 APPLICATIONS

The applications of bacterial cellulose dry and wet membranes have increased considerably in last couple of decades, mainly in the biomedical field, as recently thoroughly reviewed [23], and high technology areas, and also in other sectors, as summarized below.

17.4.1 Food (*nata de coco*)

Probably the most popular use of bacterial cellulose is in the form of a dessert called *nata de coco*. It is a glycidic food obtained by surface fermentation with *G. xylinus*. It is well known in some Asiatic countries, mainly the Philippines. Coconut milk or water, low economical value by-products of coconut processing, are employed as culture media. Other agricultural residues such as milk, whey and even fruit juices, may also be used. The recipe to prepare *nata de coco* from bacterial cellulose is [24] as follows:

Ingredients

- 1 kg of bacterial cellulose wet membrane (bacteria free).
- Three cups white sugar.
- Three cups water.

Preparation

- Cover the wet membrane with water and boil for 10 min.
- Repeat 3 times and rinse every time.
- Boil sugar and water until syrupy. Put in membrane and cook for 10 min.
- Let stand overnight. Next day, boil again in its syrup, reduce heat and continue cooking until syrup is absorbed by membrane.
- Make another syrup (1 part sugar to 1 part water).
- Cook the membrane in second syrup for 10 min.

To preserve pack the *nata de coco* in sterilized bottle, then pour hot syrup (second syrup) and seal tightly.

17.4.2 Artificial temporary skin for wounds and burns

The idea to use bacterial cellulose for wound care started in the early 1980s with the exploratory investigation at Johnson & Johnson [25]. After that, a Brazilian company called *Biofill Industrias* (now *Fibrocel*) [26], continued to investigate the properties of microbial cellulose and nowadays *Fibrocel* commercializes *Nexfill*®, a bacterial cellulose dry bandage for burns and wounds. When the bandage made of dry or wet [27] bacterial cellulose membrane is applied on top of a burn area or wound, it relieves the pain almost completely in a few seconds and shows close adhesion to the wound bed. This effect will only be observed if the epidermis is exposed, that is it works for second or third degree burns. It is a one-time healing, since everyday clinical procedures are not necessary any longer, as in conventional treatments. The same is observed for chronic wounds, like those caused by diabetes or in bedridden patients. During cicatrization, the membrane does not develop rips or perforations, keeping the area as a closed wound, thus preventing contamination by bacteria or other infectious microorganisms. It also blocks ultraviolet radiation, besides restricting the loss of fluids by maintaining a constant evaporation rate. Furthermore, due to its transparency, it allows a straightforward wound inspection. The time of re-epithelialization is shorter than that associated with a traditional bandage. When cicatrization takes place, the membrane loses its adherence on the scarred area, breaking into small pieces and falling off. Its analgesic action mechanism is not elucidated yet. The likely mechanism involves ion capture by the cellulose hydrogen bonds. In conclusion, the treatment with bacterial cellulose membranes reduces the hospital residence time of the patient while improving his quality of life during recovery.

17.4.3 Dental wounds

Based on the results with artificial skin in burns or skin tissue loss, bacterial cellulose was tested in dental tissue regeneration. The bandage, called Gengiflex®, consists of two layers: the inner layer is composed of bacterial cellulose, which offers rigidity to the membrane, and the outer alkali-cellulose layer is chemically modified [28]. Salata *et al.* [29], in 1995, compared the biological performance of Gengiflex® and Gore-Tex® membranes using the *in-vivo* non-healing bone-defect model proposed by Dahlin *et al.* [30]. The study showed that Gore-Tex membranes (a composite with polytetrafluoroethylene, urethane and nylon) were associated with significantly less inflammation and both membranes promoted the same amount of bone formation during the same period of time. A greater amount of bone formation was present in bone defects protected by either Gore-Tex or bacterial cellulose membrane, when compared to the control sites. Gore-Tex is better tolerated by the tissues than Gengiflex.

Recently, in a similar vein, Macedo *et al.* [31] also compared bacterial cellulose and polytetrafluoroethylene (PTFE) as physical barriers used to treat bone defects in guided tissue regeneration. In this study, two osseous defects (8 mm in diameter) were performed in each hind-foot of four adult rabbits, using surgical burs with constant sterile saline solution irrigation. The effects obtained on the right hind-feet were protected with PTFE barriers, while Gengiflex membranes were used over wounds created in the left hind-feet. After 3 months, the histological evaluation of the treatments revealed that the defects covered with PTFE barriers were completely repaired with bone tissue, whereas incomplete lamellar bone formation was detected in defects treated with Gengiflex membranes, resulting in voids and lack of continuity of bone deposition. This demonstrated that the non-porous PTFE barrier is a more effective alternative to treat osseous defects than a bacterial cellulose membrane.

17.4.4 Artificial blood vessels and micronerve surgery

This particular use for bacterial cellulose was proposed by Klemm *et al.* [5] in 2001. They used BASYC® (BActerial SYnthesized Cellulose), which consists of tubes of bacterial cellulose with no sewing. The tubes have an inner diameter of 1 mm, length of about 5 mm and wall thickness of 0.7 mm. For their use in microsurgery, they must resist both mechanical strains during microsurgical preparation and anastomozing and blood pressure of the living body. Bacterial cellulose has favourable mechanical properties, including shape retention and tear resistance and a better mechanical strength than organic sheets, like polypropylene, polyethylene terephthalate or cellophane. The BASYC tubes can resist to the blood pressure of the test animal (white rat) of 0.02 MPa (150 mmHg). Beside blood vessels, the tubes can also be used as a protective cover for micronerve sutures and in the form of a practice model for the application of microsurgical suture techniques in medical training.

The results for treated blood vessels showed that after 4 weeks, the BASYC®-prosthesis was wrapped up with connective tissue, pervaded with small vessels like *vasa vasorum*. The BASYC®-interposition was completely incorporated in the body without any rejection reaction. Results from micronerve surgery (white rat sciatic nerve) showed similar results. From 4 to 26 weeks following the intervention, the bacterial cellulose tube was covered with connective tissue and contained small vessels. No inflammation reaction or capsulation of the implant was observed. The regeneration nerve was improved after 10 weeks, compared to an uncovered anastomosed nerve.

The use of bacterial cellulose in medical prosthesis is exciting, considering the perspectives of applications in cardiac bypass surgery and self-transplanted derivation vessels [23].

17.4.5 DNA separation

DNA analysis or replication for forensic investigation, clinical diagnosis or even for molecular biology procedures, demands the separation of molecular fragments from ten to thousands of base pairs (bp). Electrophoresis is one of the most powerful techniques to provide DNA separation. This method takes advantage of the sieving effect of gels or polymer solutions. The design of the optimal mesh size for DNA separation has been a major issue [32]. Recently, a new concept for separation has been developed that takes advantage of the nanospaces in nanofluid structures, for example pillars [33–35], magnetic structures [36] or nanospheres [37].

Tabuchi *et al.* in 2004 [37], proposed the use of bacterial cellulose fibrils suspension for DNA separation. A solution of 0.49 per cent hydroxypropylmethyl cellulose containing 0.3 per cent of bacterial cellulose fragments allowed excellent separation for a wide range of DNA sizes (10 bp–15 kbp). Two different types of mesh structure coexisted in the medium, viz. the mesh derived from the conventional polymer solution and that associated with the bacterial cellulose fibrils. Thus, the structure is composed of $\sim$10 μm fragments containing 10 nm to 1 μm mesh of bacterial cellulose rigid fibrils and the several 10 nm meshes of the flexible polymer network.

The system also allowed a high resolution of single-nucleotide polymorphisms, even though the viscosity was less than 5 cP. According to the authors [37], the results suggest that the ability of this medium to separate DNA is due to its double-mesh concept, combined with a stereo (obstacle) effect and that this is therefore a powerful medium currently available for the separation of DNA fragments.

17.4.6 Separation of mixtures

Bacterial cellulose can be used as a selective pervaporation membrane in water–organic mixtures. An easy experiment can be set up if a bacterial cellulose membrane is available. The membrane should be tightly fixed between two transparent containers, like a partition. Both containers should have an aperture for the introduction of the feed mixture (*e.g.*, water and olive oil) or the individual liquids. For mixtures, just fill up one container and close it, while leaving the opposite container open. For individual liquids, fill each container with the respective liquid and close both. In the case of individual liquids, the water will be pumped through the membrane for the oil container, even with the containers closed. In the case of mixtures, water will be pumped through the membrane, if at least part of the liquid is in contact with it.

Pandey *et al.* [38], in 2005, described the use of a bacterial cellulose membrane to separate water–organic mixtures (acetone, formalin, ethanol, ethylene glycol and glycerol) by a pervaporation process. The membrane showed a high sorption affinity for compounds, like glycerol and ethylene glycol, which form extensive hydrogen bonding with the cellulose chains in the membrane. The presence of water in the aqueous binary mixtures of organic chemicals, in general, increases the overall sorption via the plasticization of the membrane, which preferentially sorbs water from the binary mixtures. The total flux for the pervaporation of the aqueous (40 per cent (v/v)) binary mixtures ranged from $92\,g\,m^{-2}h^{-1}$ for glycerol/H_2O to $614\,g\,m^{-2}h^{-1}$ for ethanol/H_2O at 35°C, which further increased to $1\,429\,g\,m^{-2}h^{-1}$ at 75°C. The permeation selectivity was highest for glycerol/H_2O and lowest for ethanol/H_2O and decreased with increasing temperature. The comparatively high pervaporative separation index of $103–104\,g\,m^{-2}h^{-1}$ and the relatively low activation energy of $10\,kJ\,mol^{-1}$ emphasized the potential of bacterial cellulose membranes in the pervaporative separation of aqueous binary mixtures with organic compounds.

17.4.7 Fibre reinforcement

Because of their dimensions, ribbons and microfibrils found in the bacterial cellulose morphology, are a subject of interest for composites and the mechanical enhancement of materials. The development of new cellulose-reinforced composite materials is based on both industrial and techno-economical criteria and is therefore dominated by the search for adequate mechanical performance at the lowest cost, as thoroughly discussed in Chapter 19, which also highlights the facts that cellulose nanocrystals have the potential for excellent mechanical performances, due to a low density combined with high *moduli* and tensile strength, and are biodegradable and inexpensive, making them very attractive for industrial purposes. Grunert and Winter [39] were the first in 2000 to develop new materials containing cellulose nanocrystals from bacterial cellulose topochemically modified by trimethylsilylation. The unreacted and the surface trimethylsilylated crystals were exploited as the particulate phase in nanocomposites with crosslinked polydimethylsiloxane as the matrix. This study aimed at demonstrating that cellulose nanocrystals can improve the mechanical properties of polymers and that the surface modification of the cellulose provides an enhanced adhesion between the two components of the composite, a fundamental point which is systematically treated in Chapter 18. The dynamic mechanical analysis of the nanocomposites revealed reinforcement with regard to the pure polymer for both the composite with unmodified and that with surface trimethylsilylated cellulose fibres. The extent of reinforcement depended strongly on the temperature and on the surface chemistry of the particulate phase [39].

Nogi *et al.* [40], in 2005, used bacterial cellulose nanofibres to reinforce transparent polymers. The composites exhibited a highly luminous transmittance at a fibre content as high as 60 wt%, and a low sensitivity to matrices with a variety of refractive indices. The optical transparency was also insensitive to temperature increases up to 80°C.

In a recent original contribution to this topic, Brown and Laborie [41] prepared finely dispersed nanocomposites of bacterial cellulose in poly(ethylene oxide) by introducing the latter polymer in the former growth medium. This integrated manufacturing approach opens a novel promising route to fibre-reinforced nanocomposites based on bacterial cellulose.

17.4.8 Electronic paper

The transparency, paper-like appearance and unique microfibrillar nanostructure of bacterial cellulose sheets are characteristics that were taken into account by Shah and Brown Jr. [42], to propose the use of this material as electronic paper in 2005. The technique involved first making the cellulose sheet electrically conducting (or semiconducting) by depositing ions around the microfibrils to provide conducting pathways and then immobilizing electrochromic dyes within the microstructure. The whole system was then cased between transparent electrodes and upon the application of switching potentials (2–5 V), a reversible colour change was activated, down to a standard pixel-sized area of *ca.* $100\,\mu m^2$. Using a standard backplane or an in-plane drive circuit, a high-resolution dynamic display device using bacterial cellulose as a substrate could thus be constructed. The major advantages of such a device are its high paper-like reflectivity, flexibility, contrast and biodegradability and its potential applications could extend to e-book tablets, e-newspapers, dynamic wall papers, rewritable maps and learning tools.

17.4.9 Fuel cell membranes

Chemically modified bacterial cellulose also finds applications in electric power generation. Bacterial cellulose has the ability to catalyze the precipitation of palladium within its structure to generate a high surface area with catalytic potential [43]. Palladium-bacterial cellulose could have applications in the manufacture of both electrical and electronic devices. Since bacterial cellulose fibrils are extruded by the bacterium and then self-assemble to form a three-dimensional network structure that resembles a sponge, its physical morphology is therefore compatible with polyelectrolyte membrane technology and fuel cell development. The palladium-cellulose can be combined with enzymes immobilized in cellulose membranes for the design of biosensors and biofuel cells. The work of Evans *et al.* [43] suggests that bacterial cellulose possesses reducing groups capable of initiating the precipitation of palladium, gold and silver from aqueous solutions. Since the bacterial cellulose contained water equivalent to at least 200 times its dry weight, they dried it to a thin foil, suitable for the construction of membrane electrode assemblies. Results with palladium-cellulose showed that it was capable of catalyzing the generation of hydrogen, when incubated with sodium dithionite and generated an electrical current from hydrogen in a membrane electrode assembly containing native cellulose as the polyelectrolyte membrane. The advantages of using native and metallized bacterial cellulose membranes over other polyelectrolyte membranes such as Nafion 117®, include their higher thermal stability and lower gas crossover.

17.5 CONCLUSIONS

Although it is a reality for medical applications and exotic desserts, bacterial cellulose is still only a potential promise for other applications. This is partly due to the absence of overproducer strains, which can supply the demand for this raw material. The global production is considered small and it is made up of producing companies in Brazil, farmers in the Asian southeast and some research laboratories in Europe and the USA. Cheng *et al.* [10] published in 2002 a study for the production of bacterial cellulose pellets in airlift bioreactors. Based on their data, it would be possible to produce 1 ton of pure bacterial cellulose in a facility comprising 130 airlift bioreactors of 1 000 L capacity each, within 72 h. That pure cellulose could be used to substitute plant cellulose for papermaking. The environmental implications would be tremendous in terms of both the decrease in CO_2 emissions and the preservation of reforestry. That may be possible in the future, where genetically modified strains will produce a white and gelatinous forest.

ACKNOWLEDGEMENTS

We thank *Bionext® Produtos Biotecnológicos* and the Brazilian research agencies FAPESP and CNPq for financial support. E.P. acknowledges *Bionext* for a research grant.

REFERENCES

1. Steinbüchel A., Doi Y., *Biotechnology of Biopolymers*, Wiley-VCH Verlag, Weinheim, 2005, p. 381.
2. Brown A.J., On an acetic ferment that forms cellulose, *J. Chem. Soc.*, **49**, 1886, 172.
3. Brown A.J., The chemical action of pure cultivation of bacterium aceti, *J. Chem. Soc.*, **49**, 1886, 232.
4. Delmer D.P., Cellulose biosynthesis: Exciting times for a difficult field of study, *Annu. Rev. Plant Physiol. Plant Mol. Biol.*, **50**, 1999, 245.
5. Klemm D., Schumann D., Udhardt U., Marsch S., Bacterial synthesized cellulose – Artificial blood vessels for microsurgery, *Prog. Polym. Sci.*, **26**, 2001, 1561.
6. Hirai A., Tsuji M., Horii F., TEM study of band-like cellulose assemblies produced by *Acetobacter xylinum* at 4°C, *Cellulose*, **9**, 2002, 105.
7. Saxena I.M., Brown Jr. M., Cellulose biosynthesis: Current views and evolving concepts, *Ann. Bot.*, **96**, 2005, 9.
8. Iguchi M., Yamanaka S., Budhiono A., Bacterial Cellulose – A masterpiece of nature's arts, *J. Mater. Sci.*, **35**, 2000, 261.
9. Borzani W., de Souza S., Mechanism of the film thickness increasing during the bacterial producing of cellulose on non-agitated liquid media, *Biotechnol. Lett.*, **17**(11), 1995, 1271.

10. Cheng H.P., Wang P.M., Chen J.W., Wu W.T., Cultivation of *Acetobacter xylinum* for bacterial cellulose production in a modified airlift reactor, *Biotechnol. Appl. Biochem.*, **35**, 2002, 125.
11. Shawn D., Meder R., Cellulose hydrolysis – The role of monocomponent cellulases in crystalline cellulose degradation, *Cellulose*, **10**, 2003, 159.
12. Atalla R.H., van der Hart D.L., Native cellulose – A composite of two distinct crystalline forms, *Science*, **223**, 1984, 283.
13. Heinze T., Liebert T., Unconventional methods in cellulose functionalization, *Prog. Polym. Sci.*, **26**, 2001, 1689.
14. Greiner A., Hou H., Reuning A., Thomas A., Wendorff J.H., Zimmermann S., Synthesis and opto-electronic properties of cholesteric cellulose esters, *Cellulose*, **10**, 2003, 37.
15. Klemm D., Heublein B., Fink H-P., Bohn A., Cellulose: Fascinating biopolymer and sustainable raw material, *Angew. Chem. Int. Ed.*, **44**, 2005, 3358.
16. Zhbankov R.G., *Infrared Spectra of Cellulose and Its Derivates*, Consultants Bureau, New York, 1966, p. 304.
17. Wada M., Okano T., Sugiyama J., Allomorphs of native crystalline cellulose I evaluated by two equatorial d-spacings, *J. Wood Sci.*, **47**, 2001, 124.
18. Brown R.M., Cellulose structure and biosynthesis: What is in store for the 21st century?, *J. Polym. Sci.*, **A42**, 2004, 487.
19. Wada M., Sugiyama J., Okano T., Native celluloses on the basis of two crystalline phase (Ia/Ib) system, *J. Appl. Polym. Sci.*, **49**, 1993, 1491.
20. Ruland W., X-ray determination of crystallinity and diffuse disorder scattering, *Acta Crystallogr.*, **14**, 1961, 1180.
21. US Patent, 4,891,317, 1990.
22. Yamanaka S., Watanabe K., Applications of bacterial cellulose, in *Cellulosic Polymers – Blends and Composites*, Ed.: Gilbert R., Hanser Gardner, München, 1994, p. 207.
23. Czaja W.K., Young D.J., Kawecki M., Brown Jr. R.M., The future prospects of microbial cellulose in biomedical applications, *Biomacromolecule*, **8**, 2007, 1.
24. http://www.filipinovegetarianrecipe.com/fruit_preserves/nata_de_coco.htm (accessed 11-2006).
25. US Patent 4,655,758, 1987; 4,588,400, 1986.
26. http://www.fibrocel.com.br/ (accessed 11/2006).
27. Legeza V.I., Galenko-Yaroshevskii V.P., Zinov'ev E.V., Paramonov B.A., Kreichman G.S., Turkovskii I.I., Gumenyuk E.S., Karnovich A.G., Khripunov A.K., Effects of new wound dressings on healing of thermal burns of the skin in acute radiation disease, *Bull. Exp. Biol. Med.*, **138**, 2004, 331.
28. Novaes Jr. A.B., Novaes A.B., IMZ implants placed into extraction sockets in association with membrane therapy (Gengiflex) and porous hydroxyapatite: A case report, *Int. J. Oral Maxillofac. Implants*, **7**, 1992, 536.
29. Salata L.A., Craig G.T., Brook I.M., *In vivo* evaluation of a new membrane (gengiflex) for guided bone regeneration (GBR), in *Meeting of the British Society for Dental Research*, 1995, Manchester. *J. Dent. Res.*, **74**, 1995, 825.
30. Dahlin C., Linde A., Gottlow J., Nyman S., Healing of bone defects by guided tissue regeneration, *Plast. Reconstr. Surg.*, **81**, 1988, 672.
31. Macedo N.L., Matuda F.S., Macedo L.G.S., Monteiro A.S.F., Valera M.C., Carvalho Y.R., Evaluation of two membranes in guided bone tissue regeneration: Histological study in rabbits, *Braz. J. Oral Sci.*, **3**, 2004, 395.
32. Tabuchi M., Baba Y., Design for DNA separation medium using bacterial cellulose fibrils, *Anal. Chem.*, **77**, 2005, 7090.
33. Volkmuth W.D., Austin R.H., DNA electrophoresis in microlithographic arrays, *Nature*, **358**, 1992, 600.
34. Han J., Craighead H.G., Separation of long DNA molecules in a microfabricated entropic trap array, *Science*, **288**, 2000, 1026.
35. Hung L.R., Tegenfeldt J.O., Kraeft J.J., Strum J.C., Austin R.H., Cox E.C., A DNA prism for high-speed continuous fractionation of large DNA molecules, *Nat. Biotechnol.*, **20**, 2002, 1048.
36. Doyle P.S., Bibette J., Bancaud A.J.-L., Self-assembled magnetic matrices for DNA separation chips, *Science*, **295**, 2002, 2237.
37. Tabuchi M., Ueda M., Kaji N., Yamasaki Y., Nagasaki Y., Yoshikawa K., Kataoka K., Baba Y., Nanospheres for DNA separation chips, *Nat. Biotechnol.*, **22**, 2004, 337.
38. Pandey L.K., Saxena C., Dubey V., Studies on pervaporative characteristics of bacterial cellulose membrane, *Separation and Purification Technology*, **42**, 2005, 213.
39. Grunert M., Winter W.T., Progress in the development of cellulose-reinforced nanocomposites, *Abstr. Pap. Am. Chem. Soc.*, **219**, 2000. 126-PMSE Part 2, MAR 26.
40. Nogi M., Handa K., Nakagaito A.N., Yano H., Optically transparent bionanofiber composites with low sensitivity to refractive index of the polymer matrix, *Appl. Phys. Lett.*, **87**, 2005, 243110.
41. Brown E.E., Laborie M.-P.G., Bioengineering bacterial cellulose/poly(ethylene oxide) nanocomposites, *Biomacromolecules*, **8**, 2007, 3074.
42. Shah J., Brown Jr. R.M., Towards electronic paper displays made from microbial cellulose, *Appl. Microbiol. Biotechnol.*, **66**, 2005, 352.
43. Evans B.R., O'Neill H.M., Malyvanh V.P., Lee I., Woodward J., Palladium-bacterial cellulose membranes for fuel cells, *Biosens. Bioelectron.*, **18**, 2003, 917.

– 18 –

Surface Modification of Cellulose Fibres

Mohamed Naceur Belgacem and Alessandro Gandini

ABSTRACT

This chapter describes the most recent contributions to the realm of cellulose fibre modification, starting with a brief description of the materials and characterization techniques used in this context. The surface modification strategies are then reviewed, including the coupling with acids and anhydrides, the grafting with siloxanes, isocyanates and the grafting-from *via* free-radical initiation or ring opening polymerization, as well as the surface activation of the fibres by physical agents. The modification by admicellar configurations, the preparation of cellulose–inorganic particle hybrid materials and self-reinforced composites, are also discussed. All systems are assessed in terms of the specific type of application envisaged for the modified fibres, namely (i) as reinforcing elements in macromolecular matrices (composite materials); (ii) in textiles and (iii) for trapping organic pollutants.

Keywords

Surface modification, Cellulose fibres, Coupling by esterification, Grafting with siloxanes, Coupling by isocyanates, Grafting-from *via* free-radical initiation, Grafting-from *via* ring opening polymerization, Treatment with high energy sources, Modification by admicellar sleeving, Cellulose–inorganic hybrids, Self-reinforced composites

18.1 INTRODUCTION

The use of cellulose fibres in innovative areas of materials science has recently gained considerable attention because of three potential advantages they possess, namely: (i) their biorenewable character, (ii) their ubiquitous availability in a variety of forms, and (iii) their low cost. Cellulose fibres have been used for centuries in traditional industries, such as papermaking and textile in which they are used as a source of materials and medicine and analytical applications where they are employed as aids, excipients and column fillers among other applications. A recent addition to these numerous realms has concerned their surface modification aimed at extending their use to such novel fields as (i) reinforcing elements in macromolecular composite materials [1–5], replacing glass fibres; (ii) pollutant traps for organic molecules in a water medium [6–11]; (iii) metal-coated and magnetically active materials for microwave technologies [12, 13], (iv) conducting and photo-luminescent materials for electronic and optoelectronic devices [14–17], etc.

The driving forces related to the use of cellulose fibres in these new fields of applications resides additionally in the ease with which they can be recycled at the end of their life cycle, whether through their actual re-employment, or through combustion (energy recovery). Finally, cellulose fibres possess additional advantages like low density and modest abrasive impact.

The term 'functionalization of cellulose fibres', as used in this chapter, describes the grafting of new chemical groups at the surface or within a limited depth, in order to generate novel specific properties compared with those of the pristine material. The chemical entities involved in these transformations are predominantly the cellulose OH groups, which have also been exploited for their bulk modifications, not only in traditional processes like the synthesis of esters and ethers, but also for the preparation of new derivatives, as described in Chapter 16. The fundamental difference between bulk and surface treatments is not to be found in the underlying chemistry, but instead in the fact that the former leads to a radical transformation of the entire fibre, which almost always destroys its morphology and semi-crystalline phase, whereas the latter maintains these features virtually intact, except for a very thin outer layer whose thickness can vary from a few nanometres to a few micrometres. An extended meaning of this term includes the attachment of nanoparticles or macromolecules at the surface of the fibres through physico-chemical interactions.

The purpose of this chapter is to deal with the surface modification of cellulose fibres in order to provide them with specific functionalities, so that they can play determining roles in such applications as reinforcing elements for composite materials, self-contained composite structures, anti-pollution aids, hybrid materials, superhydrophobic surfaces and conductive and magnetic materials. Other types of surface modifications, such as those associated with dying or the manufacture of chromosorb, enzymatic and ion-exchange supports, fall outside the scope of this review. Within the structure of this book, this chapter constitutes in many ways a bridge between the chemistry associated with bulk modification treated in Chapter 16 and the processing and properties of composite materials in Chapter 19, with the addition of more specific aspects.

The surface modification of cellulose fibres, in view of their incorporation into macromolecular matrices in order to produce composite materials, has been extensively reviewed by us in recent years [18,19]. The corresponding materials have also been the subject of recent monographs [1, 5, 20–23] and are thoroughly updated in Chapter 19.

18.2 SUBSTRATES AND METHODS OF CHARACTERIZATION

A wide variety of cellulose fibres, typically going from unmodified wood fibres to very pure cellulose, have been studied in this context, which implies, on the one hand, that the modification might also involve other natural polymers, like lignin and hemicelluloses and, on the other hand, that the actual morphology can influence the kinetics and the extent of the surface modification. More specifically, the choice of materials used includes wood and annual plant fibres, as such or as obtained from delignification and bleaching processes, regular diameter regenerated filled filaments, cellophane films, semi-crystalline powders, commercial Whatman filter paper made of 'pure' cellulose and laboratory-made additive-free tracing paper.

A whole host of characterization techniques have been employed to assess the occurrence and the extent of the modification. These tools include FTIR and XPS spectroscopy, elemental analysis, contact angle measurements, inverse gas chromatography (IGC) and scanning electron microscopy (SEM). New emerging techniques, such as the take-off angle photoelectron spectroscopy, secondary ion mass spectrometry (SIMS), solid state NMR, confocal fluorescence microscopy and atomic force microscopy (AFM) have recently started to be used in this field.

18.3 SURFACE MODIFICATION STRATEGIES

Since this topic was recently reviewed [18, 19], only those past contributions characterized by a qualitative impact will be discussed again, together with the relevant publications which appeared since then.

18.3.1 Coupling with acids and anhydrides

The hydrophobization of cellulose surface was often achieved thanks to the well-known aptitude of cellulose macromolecules to undergo esterification reactions. The main coupling agents used to esterify cellulose using non-swelling solvents, in order to limit the reaction to the fibre's surface, are summarized in Scheme 18.1, namely: the classical acetic anhydride [24] (**I**), alkyl ketene dimer (AKD) **II**, alkenyl succinic anhydride (ASA) **III**, fatty acids (**IV**, with *n* varying from 6 to 22) or their chlorides, poly-(propylene-*graft*-maleic anhydride) (**V**), trifluoroethoxyacetic acid (**VI**) and its anhydride, as well as several other perfluoro-derivatives, pyromellitic anhydride (**VII**), styrene–maleic anhydride copolymer (**VIII**) and methacrylic anhydride (**IX**).

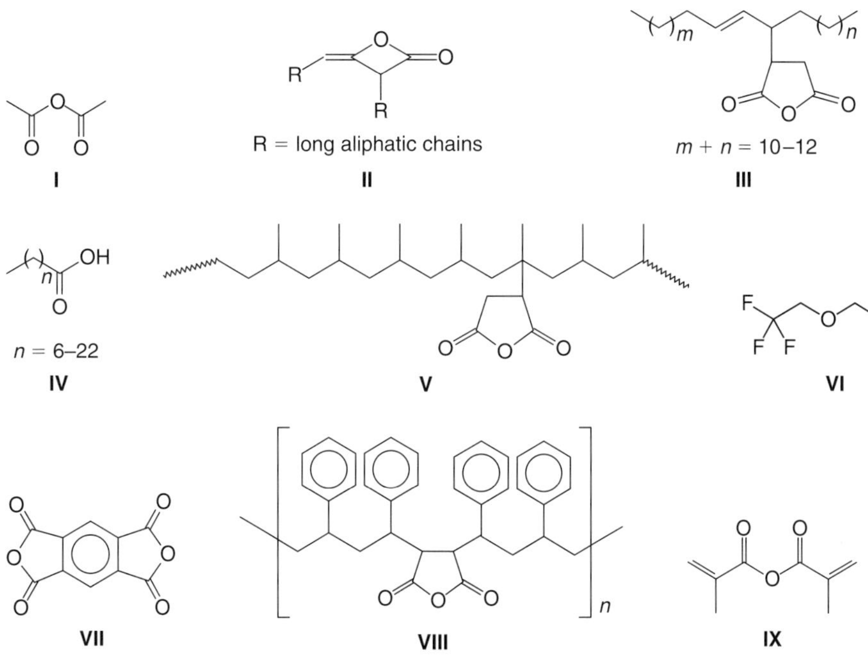

R = long aliphatic chains

$m + n = 10–12$

$n = 6–22$

I II III IV V VI VII VIII IX

Scheme 18.1

Treating cellulose fibres with coupling agents **II**, **III** and **IV** (with $n = 16$) lowered the dispersive and polar components of the surface energy when the modifications were conducted with emulsions of the coupling agents [25]. This treatment was carried out in order to compatibilize the surface of cellulose fibres with a polypropylene matrix. The same strategy was used to graft cellulose fibres with **III** or **V** [26–30]. The grafting efficiency was assessed by ESCA, FTIR, IGC, SEM and contact angle measurements. The first technique showed that the peaks at 285 and 290 eV, relative to the C—H and O—C=O moieties, respectively, increased after the treatment, particularly with **V**. FTIR showed the appearance of a peak at $1740\ cm^{-1}$ associated with the ester function. Finally, contact angle measurements showed that a drop of water formed an angle of 140° when deposited onto the surface of **V**-modified cellulose fibres and that the polar contribution to the surface energy decreased significantly, *i.e.* from 43 to $4.9\ mJ\,m^{-2}$ for the most efficient grafting agent, namely **V** with a molecular weight of 39 000 [28]. IGC and SEM corroborated the importance of the modification.

The esterification of cellulose was also carried out using lauric acid (**IV** with $n = 10$) and trifluoroethoxyacetic acid [31]. The occurrence of grafting with the latter reagent was proven by XPS, which showed the presence of a peak at 292.4 eV, assigned to CF_3 moieties and confirmed by IGC data.

Among the more recent contributions to this field, some [32–35] call upon the use of new characterization techniques, like diffuse reflectance infra red fourier transform (DRIFT) and cross polarized magic angle spinning (CP-MAS) ^{13}C-NMR spectroscopy.

The dianhydride **VII** was used to graft cellulose samples in the presence of 4-dimethylaminopyridine as the catalyst [36, 37]. FTIR spectroscopy revealed the presence of peaks at 1 852, 1 785, 1 730 and $1645\ cm^{-1}$ associated, respectively, with anhydride, ester and carboxylic acid carbonyl functions, proving that only one anhydride group had reacted with the substrate, leaving the other available for further modification (*e.g.* its participation in the growth of a polyester matrix).

Cellulose fibres modified by **I** were characterized by FTIR, SEM and WAXS to prove that the modification had preserved the substrate from structural changes and that the biodegradability of the modified fibres was maintained, albeit at a lower rate [38]. Interestingly, even bacterial cellulose (*Acetobacter xylinum*), an original novel material discussed in Chapter 17, has been submitted to surface modification with **I** [39].

A very recent investigation called upon trifluoroacetic anhydride as the esterification reagent for the surface of cellulose fibres. The reaction was carried out both in the gas [40, 41] and the liquid phase [42]. The presence of the CF_3 groups on the cellulose surface gave rise to very hydrophobic and lipophobic properties [42]. Given the high sensitivity to hydrolysis of the trifluoroacetate moiety, this modification was shown to be readily reversed by liquid water and even by a moist atmosphere. This peculiar feature has an obvious bearing in the context of applications in which the biphobic character of the fibres' surface only constitutes a temporary requirement, after which the fibres can be readily recycled following rapid hydrolysis.

Nyström *et al.* [43] recently reported the surface modification of cellulose fibres with a long perfluoroalkyl chain, which was appended either directly or through an intermediate layer of OH-bearing polymers. Again, a highly hydrophobic character was obtained, but the authors failed to recognize the hydrolytic sensitivity of the grafted perfluoroester moieties.

18.3.2 Grafting with silane coupling agents

Scheme 18.2 regroups the silane coupling molecules tested to graft cellulose fibres, namely: vinyldimethylethoxyethylsilane (**X**), vinyltrimethoxysilane ($n = 0$) (**XIa**) and vinyltriethoxysilane ($n = 1$) (**XIb**), γ-glycidylpropyltrimethoxysilane (**XII**), phenyltrimethoxysilane (**XIII**), γ-mercaptopropyltriethoxysilane (**XIV**), hexadecyltrimethoxysilane (**XV**), γ-methacrylopropyltrimethoxysilane (**XVI**), γ-aminopropyltriethoxysilane (**XVII**), (2-aminoethylamino)-propyltriethoxysilane (**XVIII**), 3-(phenylamino)-propyl-triethoxy silane (**XIX**), 3-(2-(2-aminoethylamino)-ethylamino)-propyl-triethoxysilane (**XX**) and cyanoethyltrimethoxysilane (**XXI**).

Although several papers related to the use of silane coupling agents are available in the literature, only few of them deal with a clear-cut proof of the actual occurrence of grafting and with a serious study of the reaction mechanisms involved [18, 19, 44–54]. In fact, the majority of the other studies deal with the use of the addition of a silane coupling agent to the cellulose fibres and their subsequent incorporation into polymeric matrices and evaluate the supposed grafting indirectly by monitoring the changes in the mechanical properties of the ensuing composites.

The more thorough studies showed that, contrary to glass fibre chemistry, (*i.e.*≡Si—OR + ≡Si—OH) the direct condensation reaction between the siloxane alkoxy groups and the cellulose OH functions (≡Si—OR + ≡C—OH, cellulose) *does not occur*, unless traces of water are present to generate Si—OH groups [46, 47, 53]. The occurrence of the condensation of silanol groups with cellulose (≡Si—OH + ≡C—OH, cellulose) was proven by FTIR spectroscopy, (the presence of bands at 1 134 and 1 038 cm^{-1} associated with Si—O—C), XPS spectroscopy (peaks at 102, 150 and 160 eV attributed to Si-signals) and elemental analyses (detection of silicon), but only about above 90°C. Of course, some other typical features, specific to the use of a given siloxane, were

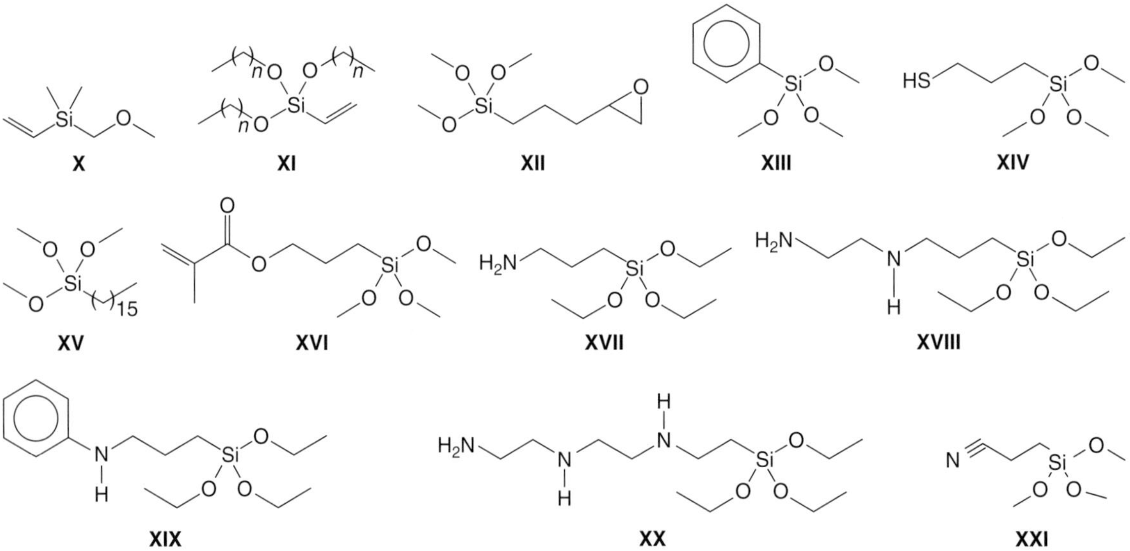

X XI XII XIII XIV

XV XVI XVII XVIII

XIX XX XXI

Scheme 18.2

detected in the FTIR spectra, namely peaks at 1712 and 1637 cm^{-1}, associated with the presence of C=O and C=C groups, respectively, for methacrylic and vinylic siloxanes [49, 52, 53] and a band at 2250 cm^{-1} for **XXI**-treated fibres revealed the presence of C≡N groups [53]. The contact angles of a drop of water were also measured in the case of the use of **XV**. The fact that their value increased from 40°, for the pristine cellulose surface to more than a 100°, for the modified surface clearly indicated the efficiency of the coupling [52, 53].

A double-grafting strategy was also applied in this context, consisting in the use of siloxanes bearing a polymerizable function [52, 53]. In this study, **XVI**-modified fibres were 'copolymerized' with styrene (**ST**) or methyl methacrylate (**MMA**) and the ensuing surface energy displayed a negligible polar component, whereas the dispersive component of the fibres, before and after the single and/or double-grafting procedures, remained practically unchanged [52]. With similar systems, **XII**- and **XVII**-modified substrates were subsequently treated with an aliphatic amine and an aliphatic oxirane, respectively, to simulate the formation of an epoxy resin. The ensuing doubly grafted fibres showed the same strong decrease of the polar component of the surface energy, which indicated that the surface had become highly hydrophobic [52].

18.3.3 Coupling with isocyanates

The chemical reaction of isocyanate-bearing molecules with cellulose fibres has been the subject of numerous studies, because urethane formation provides many advantages, namely: (i) relatively high reaction rates, (ii) the absence of elimination products and (iii) the chemical stability of the urethane moiety [18, 19, 36, 55]. The most relevant reagents used in this context are summarized in Scheme 18.3, namely: poly(styrene-*co*-isopropenyl-α-α′-dimethylbenzyl isocyanate) (**XXII**), *O*-(2-isocyanatopropyl-*O*′(methoxy)polyoxypropylene (**XXIII**), **3**-isopropenyl-α-α′-dimethylbenzyl isocyanate (**XXIV**), 2-isocyanatoethyl methacrylate (**XXV**), 1,4-butylenediisocyanate (**XXVI**), 1,4-phenyldiisocyanate (**XXVII**) and 2,4-toluene diisocyanate (**XXVIII**). More reagents were tested, as we recently reported in detail elsewhere [18, 19].

Typically, the reaction of the isocyanate functions with the OH groups of cellulose was assessed by weight gain and FTIR (bands at 3343 and 1690 cm^{-1} associated with NH, 1705 cm^{-1}, attributed to C=O and around 2900 cm^{-1} relative to aliphatic CH moieties, in the case of long chain aliphatic isocyanate). This evidence was often corroborated by XPS, contact angle measurement and IGC. When the reaction is carried out with stiff molecules bearing two NCO functions [36] or two different functions (an NCO at one end and insaturations at the other) [55], it leaves the second moieties for further exploitation. Thus, the use of reagents **XXIV** and **XXV** was followed by the 'copolymerization' with **MMA** or **ST** which confirmed the occurrence of this double-grafting through the dramatic increase in the water contact angle, the drastic reduction of the polar component of the surface energy and the notable weight gain, after the extraction of the unbound polymer [18, 19]. When the

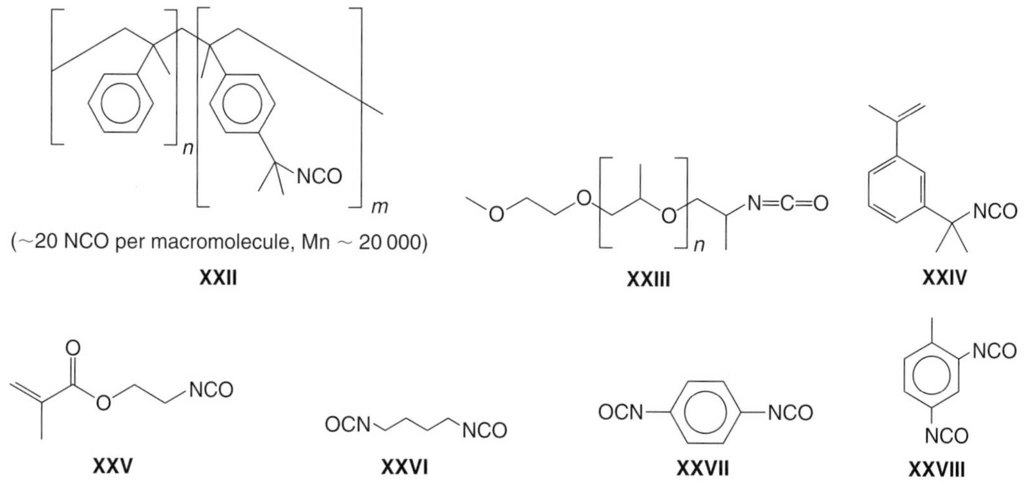

(~20 NCO per macromolecule, Mn ~ 20 000)

XXII **XXIII** **XXIV**

XXV **XXVI** **XXVII** **XXVIII**

Scheme 18.3

diisocyanate **XXVII** was grafted onto the fibre surface, the remaining NCO group was shown to participate as a comonomer in the growth of a polyurethane matrix around the fibres.

18.3.4 Grafting-from *via* free-radical initiation

For several decades after the Second World War, the direct activation of the surface of cellulose fibres aimed at generating free radicals capable of initiating the polymerization of various monomers in an aqueous medium was the subject of a very large number of studies [56–58]. This 'grafting-from' strategy gave interesting results with several optimized systems, but was never upgraded to an industrial process mainly because of the inevitable formation of appreciable quantities of homopolymers and side reactions which tended to degrade the cellulose macromolecules and discolour the ensuing materials. These classical systems were critically reviewed recently [18].

The recent discovery and rapid progress of controlled radical polymerizations [59], notably atom transfer radical polymerization (ATRP) and reversible addition-fragmentation chain transfer (RAFT), has revived interest in the possibility of grafting different macromolecular structures from the cellulose backbone [43, 60–65].

Most of these recent studies have called upon the ATRP strategy and successfully grafted various polymers on the cellulose surface. An example of this approach is shown in Scheme 18.4 [60, 61], in which 2-bromoisobutyl bromide was made to react with the cellulose surface OH groups before soaking the modified fibres into neat methyl acrylate (**MA**), in the presence of tris(2-dimethylamino)ethyl-amine and CuBr, sacrificial initiator and ethyl acetate. The grafting efficiency was evaluated by weight gain, FTIR and AFM, as shown in Fig. 18.1, which displays the images of filter paper at different grafting stages. The possibility of building grafted block copolymers was also explored and achieved through a second monomer addition, as shown in the last step of Scheme 18.4.

The use of xyloglucan as a molecular anchor constitutes an alternative approach to cellulose grafting through ATRP, as sketched in Scheme 18.5 [65].

The application of the RAFT strategy to induce a grafting-from process applied to cellulose surface has also been recently reported [66]. In this system, the cellulose surface OH groups were converted into thiocarbonyl–thio

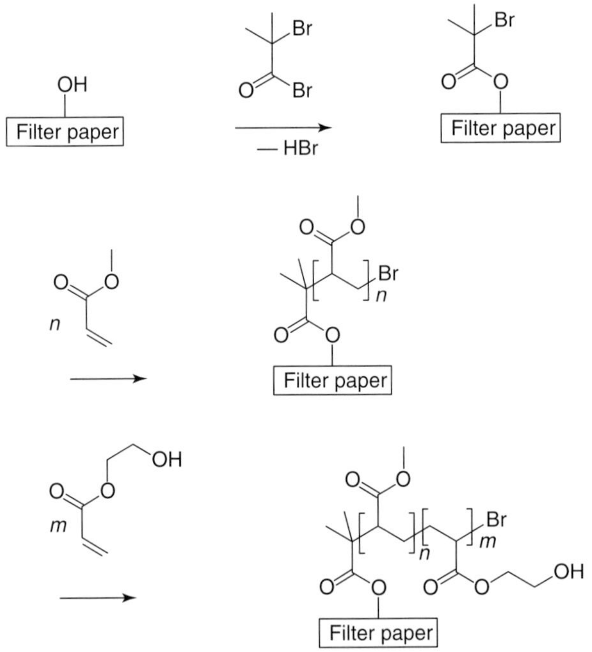

OH
Filter paper
Br
O Br
— HBr
Br
O O
Filter paper
O O
n
O O
Br
n
O O
Filter paper
OH
O O
m
O O
Br
n m
O O O O OH
Filter paper

Scheme 18.4

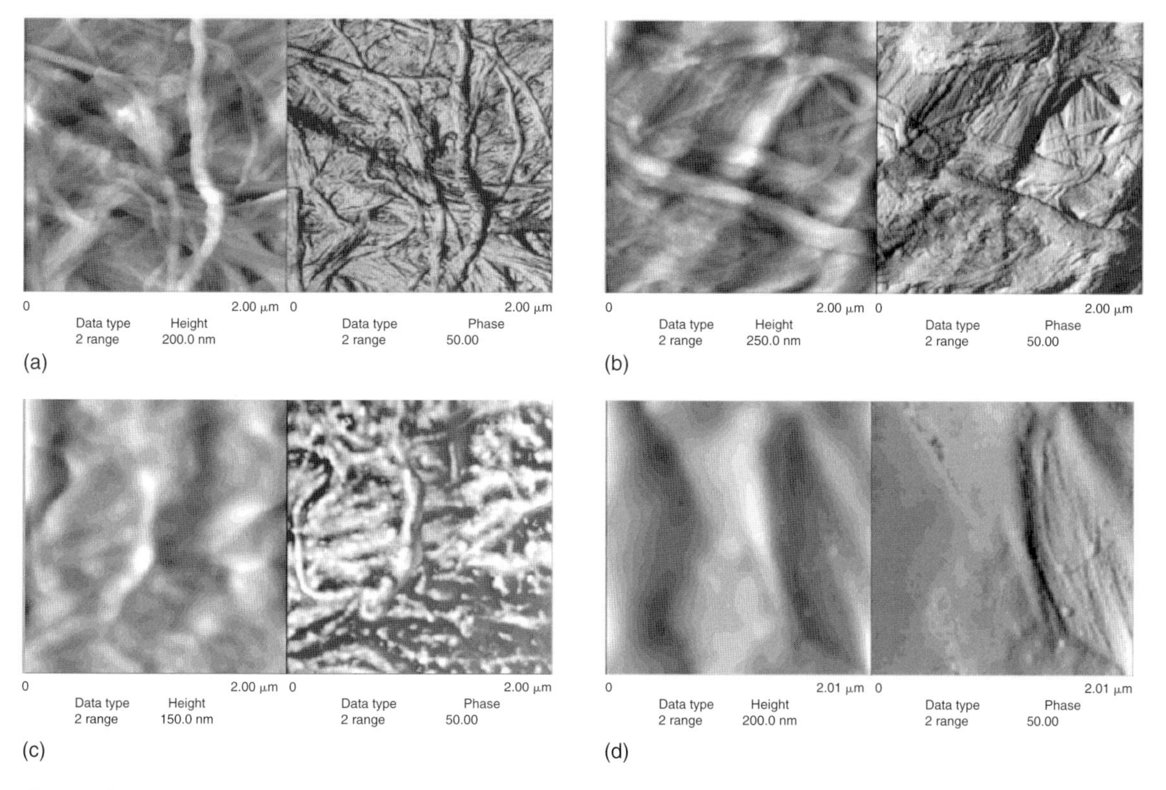

Figure 18.1 AFM-images of (a) virgin filter paper, (b) paper with the grafted initiating moieties, (c) paper grafted with **PMA** and (d) paper grafted with **PMA**-block-**PHEMA**. (Reproduced by permission of the American Chemical Society. Copyright 2003. Reprinted from Reference [61].)

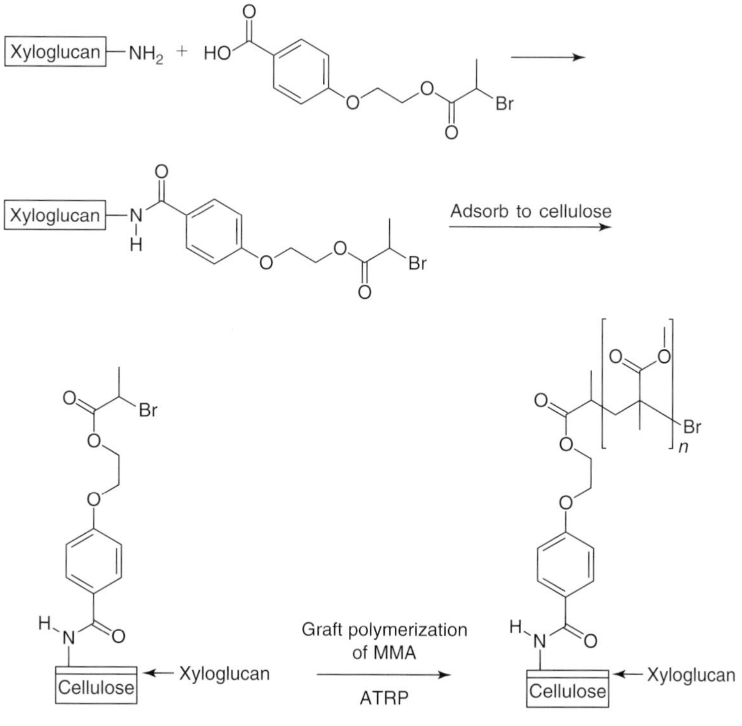

Xyloglucan—NH_2 + HO ... Br →

Xyloglucan—N(H) ... Br —Adsorb to cellulose→

Cellulose ← Xyloglucan —Graft polymerization of MMA / ATRP→ Cellulose ← Xyloglucan

Scheme 18.5

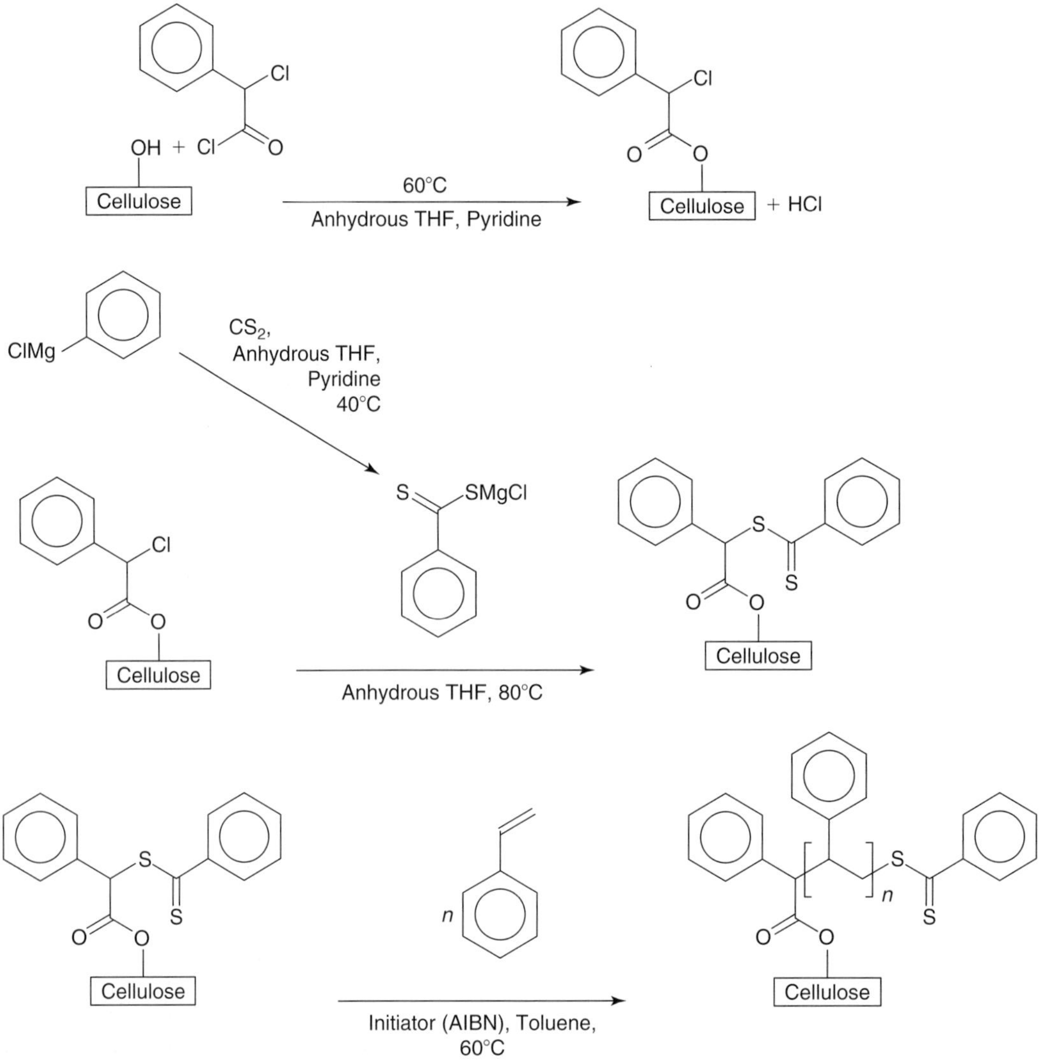

Scheme 18.6

moieties, to be used as mediators for the RAFT polymerization of **ST**, as seen in Scheme 18.6. Clear-cut proof of the successful grafting was provided by several pieces of evidence.

An alternative way to achieve this goal has recently been proposed [67–69]. It calls upon the oxidation of the fibre surface with sodium periodate and the subsequent UV photolysis of the ensuing moieties in the presence of conventional monomers, so that the free radicals generated by the cleavage of the oxidized structures can initiate their polymerization. This approach gave interesting results although the oxidation procedure was found to degrade considerably the cellulose macromolecules.

18.3.5 Grafting-from *via* ring opening polymerization

Chain extension reactions based on the activation of the cellulose surface aimed at provoking the grafting-from through the ring opening polymerization of heterocyclic monomers have been practiced industrially in the specific instance of the synthesis of hydroxypropyl and hydroxyethyl cellulose. The aim of these processes is, however, to

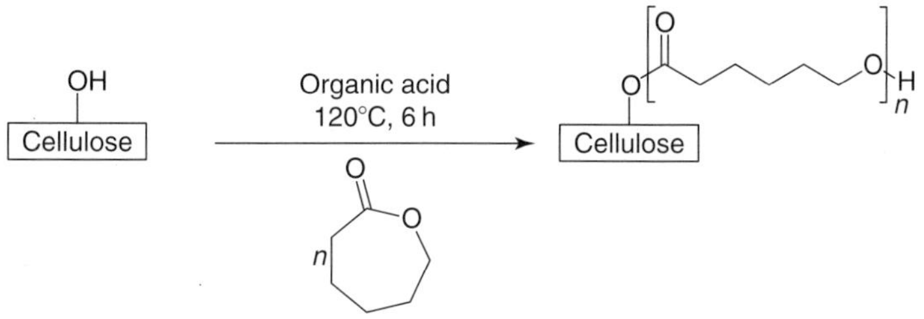

Scheme 18.7

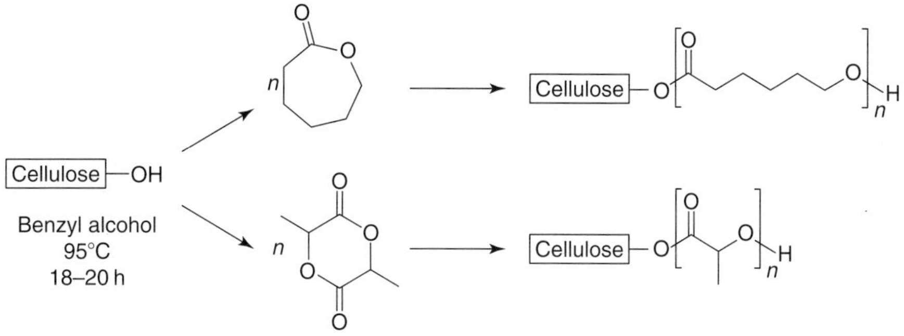

Scheme 18.8

convert the bulk of the fibres and produce novel thermoplastic polymers. A recent modification of this approach describes the partial oxypropylation of cellulose fibres in order to limit their modification to a minimum depth, thus generating a soft sleeve around them [70]. Hot-pressing of these novel materials gives rise to composites in which the matrix is made up of the oligo-(propylene oxide) grafts, and the reinforcing elements are the unmodified inner cores of the fibres. Details of this study are given in Chapter 12, together with similar approaches applied to other natural polymers.

In another vein, Haren and Cordova investigated the grafting of poly-(caprolactone) using an organic acid to activate the cellulose surface, as shown in Scheme 18.7 [71]. Ample evidence was provided for the occurrence of this reaction, which was shown to generate a weight gain of 10 per cent. A similar strategy was adopted in a later study [72] by another group in which, however, benzyl alcohol was the activating species and L-lactic acid (**LA**) was also grafted, as shown in Scheme 18.8.

Surprisingly, a water contact angle of about 110° was measured at the surface of these modified fibres, *viz.* a value which seems exceedingly high considering the structure of the grafts with their terminal OH groups. In the same publication, two other approaches were tackled in order to graft the same monomers, but through different initiation mechanisms, which provided a multiplication of the OH groups of cellulose.

Henequen microfibres were treated with two different epoxides (1,2-epoxy-5-hexene and 1,2-epoxy-5-octene), in order to append insaturations at their surface, which were then used as potential comonomer units when acrylic acid was polymerized in their presence [73]. The ensuing materials displayed a substantial increase in water absorption given the well-known hydrophilic character of acrylic acid.

18.3.6 Grafting by electrical discharges and irradiation techniques

Corona, dielectric barrier and plasma discharges, and, more recently, laser, γ-ray and vacuum UV irradiations, have been used to modify the surface of cellulose fibres. This topic was recently reviewed [18] and here we will limit our coverage to the most significant results and to an update of recent contributions.

XPS analysis showed that corona treatment oxidizes the surface of cellulose fibres [74, 75]. Thus, for example, for filter paper, regenerated cellulose films and pulps, a corona discharge induced the formation of 6.8, 1.5 and 2.1

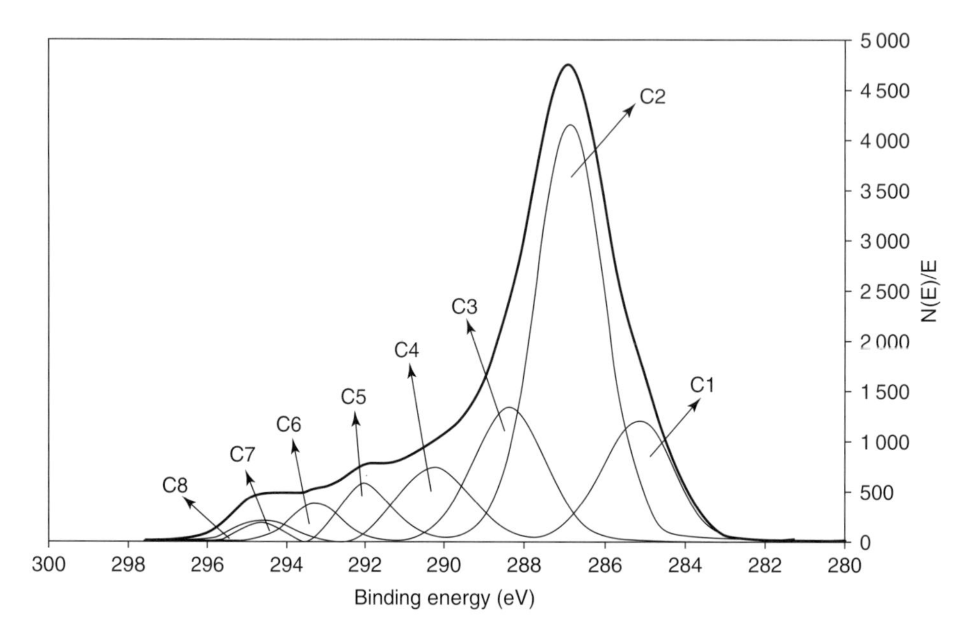

Figure 18.2 C_{1s} XPS spectra of paper (wire side) treated with RF–CF_4 plasma (P = 300 W, t = 10 min, p = 300 mTorr). (Reproduced by permission of Elsevier. Copyright 2007. Reprinted from Reference [80].)

carboxyl groups, respectively, per hundred sugar residues. An increase in both the acidic character of the surface and the dispersive energy was also established, as determined from contact angle, capillarity and IGC measurements. The increase of the dispersive energy indicated that a sort of purification of the surface took place, since there was a removal of low molecular weight substances (extractives and lignin fragments), as confirmed by XPS analysis. The electrical conductance and pH of the water suspensions of the treated fibres confirmed the formation of COOH groups.

The use of cold-plasma can be applied to modify the fibres' surface properties in two opposite directions, namely an increase in the hydrophilic character when the treatment is carried out in Argon or air [18, 76] and, on the contrary, the formation of a super-hydrophobic surface when fluorinated gases are used [18, 77–80]. A very recent publication illustrates clearly the latter approach [80]. In this study, paper sheets were plasma irradiated in the presence of CF_4 and the surface modification characterized by XPS, ATR–FTIR as a function of the power parameters and the CF_4 pressure. Interestingly, no significant difference was observed in the extent of modification to either side of the sheet. Figure 18.2 shows a typical XPS spectrum of one of these modified surfaces.

The laser treatment of cellulose surfaces is a very recent topic [81–84]. Kolar *et al.* [81] were the first to irradiate cellulose fibres and papers with both an excimer laser at 308 nm and a Nd:YAG laser at 532 nm, with the purpose of cleaning their surfaces. The use of the shorter wavelength, however, also induced some cellulose degradation. In a later study, the use of a Nd:YAG pulsed laser at 1 064 nm showed that pure microcrystalline cellulose was unaffected by the treatment [82]. This observation was challenged by a later investigation [83] which showed that the same irradiation conditions led to the surface chemical modifications of various cellulose samples, similar to those associated with thermal degradation of this natural polymer.

The effect of other excimer lasers (ArF, KrF and XeCl with wavelengths of 193, 248 and 308 nm, respectively) has also been studied on rayon [84]. The treatment at 193 nm induced significant changes in the cloth surface, as shown by cracks detected by SEM micrographs and an increase in oxidized moieties detected by XPS.

To the best of our knowledge, only one study has been published on the effect of vacuum UV irradiation on cellulose surfaces [85]. It was found that significant oxidation reactions took place (Fig. 18.3), whose effect was similar to that of strong chemical attacks, (*e.g.* chromic or nitric acid).

The γ-irradiation of cellulose in the atmosphere associates oxidation and degradation mechanisms, as one would expect from the particularly high energy involved. A recent investigation has also shown that this type of treatment favours some cellulose crosslinking [86, 87], as well as a decrease in crystallinity accompanied by thermal degradation, whose extent increases with increasing irradiation doses [88].

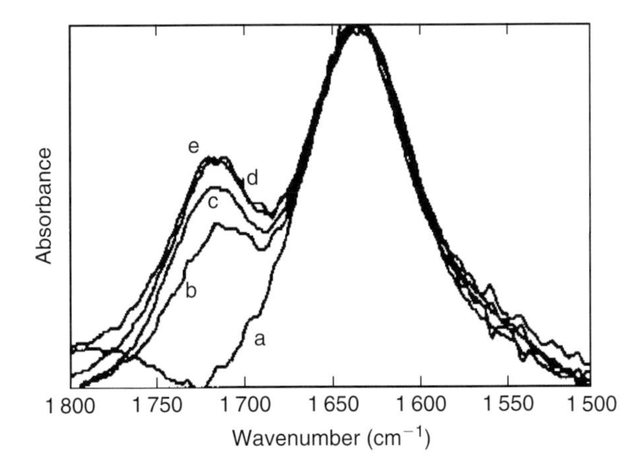

Figure 18.3 IR spectra of rayon fabrics irradiated with Vacuum UV for (a) 0, (b) 5, (c) 15, (d) 40 and (e) 60 min. (Reproduced by permission of Wiley and Sons, Inc. Copyright 1999. Reprinted from Reference [85].)

18.3.7 Modification by admicellar configurations

Boufi *et al.* have recently tackled in a systematic fashion the formation of admicelles around cellulose fibres using cationic surfactants [6–10] or fatty-esterified surfaces [11]. The purpose of this strategy is to exploit these admicellar sleeves as hydrophobic reservoirs for water-insoluble organic molecules. The first study of this series [6] made use of the admicelles to adsorb alkenyl monomers, which were then polymerized, thus generating a hydrophobic polymer sleeve around the fibres.

Subsequent investigations [7–10] were aimed, on the one hand, at enhancing the extent of admicellization by increasing the density of carboxylate groups along the fibres through TEMPO-mediated oxidations and, on the other hand, at studying quantitatively the adsorption of various organic compounds inside the admicelles. Table 18.1 gives the maximum adsorbed quantity of water-insoluble organic compounds as a function of the nature of the surfactant or the presence of the fatty acid graft.

The most relevant conclusions gathered from this ongoing investigation can be summarized as follows:

- The amount of adsorbed cationic surfactants increases with increasing cellulose surface charge, as shown in Fig. 18.4.
- Different water-insoluble organic pollutants can be successfully sucked into the admicelles by adsolubilization.

Table 18.1

Adsolubilization of organic compounds on cellulose fibres bearing surface admicelles or grafted hydrophobic chains

Organic solute	$C_{max.adsol}$ (μmol/g)		
	Cationic surfactant C18 [7, 8]	Anionic surfactant C18 [9]	Chemically grafted C12 [11]
Benzene	390–396	–	–
Naphthalene	304–295	–	–
Chlorobenzene	310–336	300	190
1,3,5-Trichlorobenzene	185–187	–	470
Nitrobenzene	194–223	340	150
2-Naphthol	290–295	660	320
Quinoline	202–374	–	330
2-Chlorophenol	169	–	–
1,4-Dichlorobenzene	223	296	250
Aniline	368	–	–
Diphenylmethane	232	410	–

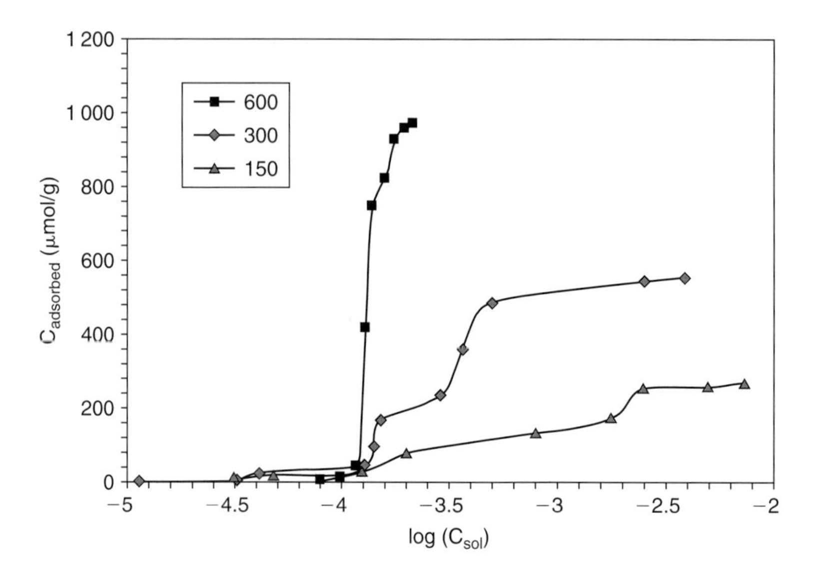

Figure 18.4 Adsorption isotherm of the C16 cationic surfactant onto oxidized cellulose fibres, as a function of the cellulose charge density, expressed in μmol/g. (Reproduced by permission of the American Chemical Society. Copyright 2005. Reprinted from Reference [10].)

- The longer the hydrophobic tail of the surfactant, the better the adsolubilization capacity of the modified surface. Thus, for example, the maximum adsolubilized quantities of 2-naphtol were 290, 180, 140 and 120 μmol/g for C18-, C16-, C14- and C12-based admicelles, respectively.
- The preliminary adsorption of cationic polyelectrolytes at the cellulose surface allowed anionic surfactants to be adsorbed at the fibre surface and form the corresponding admicelles [9].

The adsorbed surfactant molecules were found to desorb from the cellulose surface during the recycling of the substrate [7–10]. This drawback was overcome by joining the aliphatic chains chemically to the cellulose surface by esterification with fatty acid chlorides [11]. Thus, the substrates could be recycled easily and indefinitely, without any loss of their adsorption capacity.

18.3.8 Modification for advanced technologies

Beneventi *et al.* recently reported the preparation of a cellulose poly(pyrrole) composite generated by the impregnation of the fibres with $FeCl_3$ solution, followed by the *in situ* polymerization of pyrrole [14]. An alternative way of preparing conductive cellulose, based on the incorporation of carbon nanotubes into a cellulose sheet, has also been proposed recently [15, 16].

A different approach aimed at elaborating composite materials for microwave technology called upon the coating of cellulose fibres with a metal like copper [12]. Finally, magnetically active cellulose-based composites have also been described [13] by the incorporation of barium ferrite into a regenerated fibre structure.

In all these investigations, physical interactions between the active charge and the cellulose matrix dominate, rather than actual chemical bonds associated with most of the systems described previously.

18.3.9 Cellulose–inorganic particle hybrid materials

Another new field related to the use of cellulose for high value-added materials concerns the incorporation of inorganic nanoparticles into a cellulose fibre assembly. Three interesting studies are mentioned here in this context, *viz.* (i) the growth of cadmium sulphide (CdS) semi-conducting nanocrystals on the surface of regenerated cellulose fibres [89] and the study of the photoluminescence of the ensuing composite; (ii) the precipitation of titanium dioxide

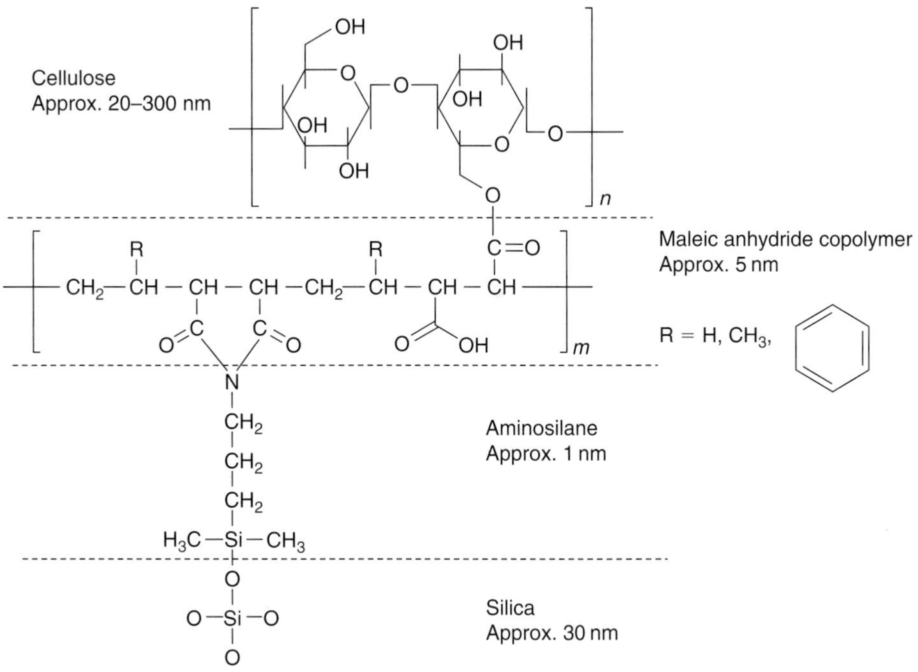

Cellulose
Approx. 20–300 nm

Maleic anhydride copolymer
Approx. 5 nm

R = H, CH_3,

Aminosilane
Approx. 1 nm

Silica
Approx. 30 nm

Scheme 18.9

nanoparticles on the surface of Eucalyptus pulp fibres [90], which resulted in a remarkable enhancement of the cellulose optical properties and (iii) the immobilization of cellulose macromolecules onto a silica surface mediated by maleic anhydride copolymers and silane coupling agent 3-aminopropyl-dimethylethoxysilane, as illustrated in Scheme 18.9 [91].

18.3.10 Self-reinforced composites

We already mentioned in Section 3.5 [70] the partial oxypropylation of cellulose fibres and the interest of the ensuing composite materials in which the unmodified fibre cores represent the reinforcing elements and their thermoplastic sleeves the source of a matrix. Other interesting approaches have been recently put forward to prepare composite materials in which cellulose or one of its derivatives prepared *in situ* are the only component. Glasser was the first to tackle this problem through the combination of cellulose esters and fibres by two distinct approaches, *viz.* (i) the incorporation of lyocell fibres into a cellulose acetate matrix [92] and (ii) the partial esterification of wood pulp fibres with *n*-hexanoic anhydride in an organic medium [93] that produced thermally deformable materials in which the thermoplastic cellulose ester constituted the matrix and the unmodified fibres the reinforcing elements.

The latter approach was also applied to the partial benzylation of sisal fibres in an aqueous medium [94]. The isolation, drying and hot-pressing of these benzylated fibres gave materials whose morphology showed features resembling those of a composite material in which the modified cellulose played the role of the thermoplastic matrix.

In another vein, a recent report [95] describes the preparation of an all-cellulose composite by dissolving pretreated ramie fibres in DMAc/LiCl and introducing untreated fibres into the ensuing solution. The composites were then isolated by coagulation with methanol and dried.

18.4 CONCLUSIONS

Together with the other three chapters devoted to cellulose-based materials, this survey has been conceived to emphasize that after a relatively dormant period, the interest in the use of cellulose fibres to participate in the technological revolution associated with a search of functional materials for highly specific and efficient purposes has became a lively reality. This is true of both the need to replace conventional reinforcing elements for composite materials for ecological and energy reasons, *i.e.* an aspect involving massive quantities of fibres, and the need to prepare high-tech devices in which cellulose-based components, even if only present in modest proportions, play a decisive role.

REFERENCES

1. Eichhorn S.J., Baillie C.A., Zafeiropoulos N., Mwaikambo L.Y., Ansell M.P., Dufresne A., Entwistle K.M., Herraro-Franco P.J., Escamilla G.C., Groom L., Hughes M., Hill C., Rials T.G., Wild P.M., *J. Mater. Sci.*, **36**, 2001, 2107.
2. *Compos. Sci. Technol.*, **63**, 2004. Eco-Composites: A special issue of 14 publications devoted to cellulose-based composite materials.
3. Mohanty A.K., Misra M., Hinrichsen G., *Macromolecular Mater. Eng.*, **1**, 2000, 276–277.
4. Gassan J., Bledzki A.K., *Prog. Polym. Sci.*, **24**, 1999, 221.
5. *Compos. Interface*, **12**, 2005. A special issue devoted to cellulose-based composites.
6. Boufi S., Gandini A., *Cellulose*, **8**, 2001, 303.
7. Aloulou F., Boufi S., Belgacem M.N., Gandini A., *Colloid Polym. Sci.*, **283**, 2004, 344.
8. Aloulou F., Boufi S., Chakchouk M., *Colloid Polym. Sci.*, **282**, 2004, 699.
9. Aloulou F., Boufi S., Beneventi D., *J. Colloid Interface Sci.*, **280**, 2004, 350.
10. Alila S., Boufi S., Belgacem M.N., Beneventi D., *Langmuir*, **21**, 2005, 8106.
11. Boufi S., Belgacem M.N., *Cellulose*, **13**, 2006, 81.
12. Zabetakis D., Dinderman M., Schoen P., *Adv. Mater.*, **17**, 2005, 734.
13. Rubacha M., *J. Appl. Polym. Sci.*, **101**, 2006, 1529.
14. Beneventi D., Alila S., Boufi S., Chaussy D., Nortier P., *Cellulose*, **13**, 2006, 725.
15. Yoon S.H., Jin H.J., Kook M.C., Pyun Y.R., *Biomacromolecules*, **7**, 2006, 1280.
16. Shah J., Brown Jr. R.M., *Appl. Microbiol. Biotechnol.*, **66**, 2005, 352.
17. Ruang D., Huang Q., Zhang L., *Macromol. Mater. Engin.*, **290**, 2005, 1017.
18. Belgacem M.N., Gandini A., *Compos. Interface*, **12**, 2005, 41.
19. Belgacem M.N., Gandini A., Natural Fibre-Surface Modification and Characterisation, in *Cellulose Fibre Reinforced Polymer Composites* (Eds. Sabu T. and Pothan L.), Old City Publishing, 2007. Chapter 3
20. Sabu T., Pothan L., *Cellulose Fibre Reinforced Polymer Composites*, Old City Publishing, 2007.
21. Pandey J.K., Kumar A.P., Misra M., Mohatny A.K., Drzal L.T., Singh R.P., *J. Nanosci. Nanotechnol.*, **5**, 2005, 497.
22. Wool R.P., Sun X.S., *Bio-Based Polymers and Composites*, Elsevier, Amesterdam, 2005. pp. 620.
23. Stevens C.V., Verhé R.G., *Renewable Bioresources: Scope and Modification for Non-Food Applications*, John Wiley & Sons, Ltd, Chichester, 2004. pp. 310.
24. Rowell R.M., *ASC Symp. Ser.*, **476**, 1990, 242.
25. Quilin D.T., Caulfield D.F., Koutsky J.A., *Mat. Res. Soc. Symp. Proc.*, **266**, 1992, 113.
26. Felix J., Gatenholm P., *J. Appl. Polym. Sci.*, **42**, 1991, 609.
27. Klason C., Kubat J., Gatenholm P., *ACS Symp. Ser.*, **489**, 1992, 82.
28. Felix J., Gatenholm P., *J. Appl. Polym. Sci.*, **50**, 1993, 699.
29. Felix J., Gatenholm P., *Nordic Pulp Paper Res. J.*, **2**, 1993, 200.
30. Felix J., Gatenholm P., *Polym. Compos.*, **14**, 1993, 449.
31. Garnier G., Glasser W.G., *Polym. Eng. Sci.*, **36**, 1996, 885.
32. Freire C.S.R., Silvestre A.J.D., Pascoal Neto C., Belgacem M.N., Gandini A., *J. Appl. Polym. Sci.*, **100**, 2006, 1093.
33. Pasquini D., Belgacem M.N., Gandini A., Curvelo A.A.S., *J. Colloid Interface. Sci.*, **295**, 2006, 79.
34. Jandura P., Kokta B.V., Riedl B., *J. Appl. Polymer Sci.*, **78**, 2000, 1354.
35. Freire C.S.R., Silvestre A.J.D., Pascoal Neto C., Gandini A., Fardim P., Holbom B., *J. Colloid Interface Sci.*, **301**, 2006, 205.
36. Gandini A., Botaro V.R., Zeno E., Bach S., *Polym. Int.*, **50**, 2001, 7.
37. Belgacem M.N., Gandini A., *ACS Symp. Ser.*, **945**, 2007, 93.
38. Frisoni G., Baiardo M., Scandola M., Lednicka D., Cnockaert M.C., Mergaert J., Swings J., *Biomacromolecules*, **2**, 2001, 476.
39. Kim D.Y., Nishiyama Y., Kuga S., *Cellulose*, **9**, 2002, 361.
40. Yuan H., Nishiyama Y., Kuga S., *Cellulose*, **12**, 2005, 543.
41. Ostenson M., Järund H., Toriz G., Gatenholm P., *Cellulose*, **13**, 2006, 157.

42. Freire C.S.R., Gandini A., Silvestre A.J.D., Pascoal Neto C., Cunha A.G., *J. Colloid Interface Sci.*, **301**, 2006, 333.
43. Nyström D., Lindqvist J., Ostmark E., Hult A., Malström E., *Chem. Commun.*, **3594**, 2006.
44. Pothan L.A., Bellman C., Kailas L., Sabu T., *J. Adhes. Sci. Technol.*, **16**, 2002, 157.
45. Sreekala M.S., Thomas S., *Compos. Sci. Technol.*, **63**, 2003, 861.
46. Valdez-Gonzalez A., Cervantes-Uc J.M., Olayo R., Herrera-Franco P.J., *Composites, Part B: Engineering*, **30**, 1999, 309.
47. Valdez-Gonzalez A., Cervantes-Uc J.M., Olayo R., Herrera-Franco P.J., *Composites, Part B: Engineering*, **30**, 1999, 321.
48. Matuana L.M., Balatinecz J.J., Park C.B., Sodhi R.N.S., *Wood Sci. Technol.*, **33**, 1999, 259.
49. Redondo S.U.A., Radovanovic E., Gonçalves M.E., Yoshida I.V.P., *J. Appl. Polym. Sci.*, **85**, 2002, 2573.
50. Ahdelmouleh M., Boufi S., Ben Salah A., Belgacem M.N., Gandini A., *Langmuir*, **18**, 2002, 3203.
51. Pickering K.L., Abdalla A., Ji C., McDonald A.G., Franich R.A., *Composites, Part A*, **34**, 2003, 915.
52. Ahdelmouleh M., Boufi S., Ben Salah A., Belgacem M.N., Gandini A., *Int. J. Adhes. Adhes.*, **24**, 2004, 43.
53. Castellano M., Fabbri P., Gandini A., Belgacem M.N., *J. Colloid Interface Sci.*, **273**, 2004, 505.
54. Park B.D., Wi S.G., Lo K.H., Singh A.P., Yoon T.H., Kim Y.S., *Biomass Bioenergy*, **27**, 2004, 353.
55. Botaro V., Gandini A., Belgacem M.N., *J. Thermoplast. Comp. Mater.*, **18**, 2005, 107.
56. Hebeish A., Guthrie J.T., *The Chemistry and Technology of Cellulosic Copolymers*, Springer-Verlag, Berrlin, 1981.
57. Samal R.K., Sahoo P.K., Samantaray H.S., *J. Macromol. Sci., Rev. Macromol. Chem.*, **C26**, 1986, 81.
58. Stannett V.T., Some recent developments in cellulose science and technology in North America, in *Cellulose: Structure and Functional Aspects* (Eds. Kennedy J.P., Phillips G.O. and Williams P.A.), Ellis Horwood, Chichester, UK, 1989, pp. 19–31.
59. Braunecker W.A., Matyjaszewski K., *Prog. Polym. Sci.* **32**, 2007, 93.
60. Carlmark A., Malmström E., *J. Am. Chem. Soc.*, **124**, 2002, 900.
61. Carlmark A., Malmström E., *Biomacromolecules*, **4**, 2003, 1740.
62. Plackett D., Jankova K., Egsgaard H., Hvilsted S., *Biomacromolecules*, **6**, 2005, 2474.
63. Coskun M., Temüz M.M., *Polym. Int.*, **54**, 2005, 342.
64. Bontempo D., Masci G., De Leonardis P., Mannina L., Capitani D., Crescenzi V., *Biomacromolecules*, **54**, 2006, 2474–2784.
65. Zhou Q., Greffe L., Baumann M.J., Malmström E., Teeri T.T., Baumer H., *Macromolecules*, **38**, 2005, 3547.
66. Roy D., Guthrie J.T., Perrier S., *Macromolecules*, **38**, 2005, 10363.
67. Margutti S., Vicini S., Proietti N., Capitani D., Conio G., Pedemonte E., Segre A.L., *Polymer*, **43**, 2002, 6183.
68. Princi E., Vicini S., Proietti N., Capitani D., *Eur. Polym. J.*, **41**, 2005, 1196.
69. Princi E., Vicini S., Pedemonte E., Gentile G., Cocca M., Mastuscelli E., *Eur. Polym. J.*, **42**, 2006, 51.
70. Gandini A., Curvelo A.A.S., Pasquini D., de Menezes A.J., *Polymer*, **46**, 2005, 10611.
71. Harén J., Cordova A., *Macromol. Rapid Commun.*, **26**, 2005, 82.
72. Lonnberg H., Zhou Q., Brumer H., Teeri T.T., Malmström E., Hult A., *Biomacromolecules*, **7**, 2006, 2178.
73. Thompson T.T., Lorià-Bastarrachea M.I., Aguilar-Vega M.J., *Carbohydrate Polym.*, **62**, 2005, 65.
74. Belgacem M.N., Czeremuszkin G., Sapieha S., Gandini A., *Cellulose*, **2**, 1995, 145.
75. Gassan J., Gutowski V.S., Bledzki A.K., *Macromolecular Mater. Eng.*, **283**, 2000, 132.
76. Yuan X., Jayaraman K., Bhattacharyya D., *J. Adhes. Sci. Technol.*, **18**, 2004, 1027.
77. Vander Wielen L.C., Ostenson M., Gatenholm P., Ragauskas A.J., *Carbohydr. Polym.*, **65**, 2006, 179.
78. Sahin H.T., Manolache S., Young R.A., Denes F., *Cellulose*, **9**, 2002, 171.
79. Navarro F., Dávalos F., Denes F., Cruz L.E., Young R.A., Ramos J., *Cellulose*, **10**, 2003, 411.
80. Sahin H.T., *Appl. Surface Sci.*, **253**, 2007, 4367.
81. Kolar J., Strlic M., Müller-Hess D., Gruber A., Troschke K., Pentzien S., Kautek W., *J. Cult. Heritage*, **1**, 2000, S221.
82. Botaro V.R., Santos C.G., Arantes Jr. G., Da Costa A.R., *Appl. Surface Sci.*, **183**, 2001, 120.
83. Strlic M., Kolar J., Vid-Simon S., Marincek M., *Appl. Surf. Sci.*, **207**, 2003, 236.
84. Kensaku M., Masatsugu I., Shunsuke O., Akihiro Y., Akihiko O., Masako S., Takahiro I., Osamu S., *Compos. Interfaces*, **7**, 2001, 497.
85. Koichi K., Victor V.N., Mikhail F.N., Masashi M., Yoshito I., Katsuhiko N., *J. Polym. Sci., Part A: Polym. Chem.*, **37**, 1999, 357.
86. Földváry C., Takács E., Wojnárovits L., *Radiat. Phys. Chem.*, **67**, 2003, 505.
87. Tóth T., Borsa J., Takács E., *Radiat. Phys. Chem.*, **67**, 2003, 513.
88. Khan F., Ahmad S.R., Kronfli E., *Biomacromolecules*, **7**, 2006, 2303.
89. Ruang D., Huang Q., Zhang L., *Macromol. Mater. Engin.*, **290**, 2005, 1017.
90. Marques P.A.A.A., Trindade T., Neto C.P., *Compos. Sci. Technol.*, **66**, 2006, 1038.
91. Freudenberg U., Zschoche S., Simon F., Janke A., Schmidt K., Behrens S.H., AUweter H., Werner C., *Biomacromolecules*, **6**, 2005, 1628.

92. Seavy K.C., Gosh I., Davis R.M., Glasser W.G., *Cellulose*, **8**(161), 2001; Franko A., Seavy K.C., Gumaer J., Glasser W.G., *Cellulose*, **8**(171) 2001.
93. Mutsumura M., Sugiyama J., Glasser W.G., *J. Appl. Polym. Sci.*, **78**(2241) 2000; Mutsumura H., Glasser W.G., *J. Appl. Polym. Sci.*, **78**(2254) 2000.
94. Lu X., Zhang M.Q., Rong M.Z., Shi G., Yang G.C., *Compos. Sci. Technol.*, **63**, 2003, 177.
95. Lu X., Zhang M.Q., Rong M.Z., Yue D.L., Yang G.C., *Compos. Sci. Technol.*, **64**, 2004, 1301.

– 19 –

Cellulose-Based Composites and Nanocomposites

Alain Dufresne

ABSTRACT

Environmental issues have recently generated considerable interest in the development of composite materials based on renewable resources such as natural fibres as low cost alternatives for glass fibres. A large number of interesting applications are emerging for these materials. In Europe, the emphasis is on the automotive industry which is looking into the use of plant fibre-based composites as a way to serve the environment and at the same time save weight (and therefore fuel). In North America, wood fibre-based composite building materials have been developing for some time. In India and South America, jute and sugar cane fibres are used in low cost housing. Wherever the industry, there will be a locally grown lignocellulosic fibre to suit the application. In addition, natural fibres present a multi-level organization and consist of several cells formed out of semi-crystalline oriented cellulose microfibrils that open opportunities in the field of nanoscience and nanotechnology.

Keywords

Cellulose, Composites, Nanocomposites, Natural fibres, Lignocellulosic fibres, Cellulose whiskers, Chitin Whiskers, Starch nanocrystals

19.1 INTRODUCTION

Composite materials (or composites for short) are engineered materials made from two or more constituents with significantly different mechanical properties, which remain separate and distinct within the finished structure. There are two categories of constituent materials: matrix and reinforcement. At least one portion of each type is required. The matrix surrounds and supports the reinforcements by maintaining their relative positions. The reinforcements impart special physical (mechanical and electrical) properties to enhance the matrix properties. A synergism produces material properties unavailable from naturally occurring materials. Due to the wide variety of matrixes and reinforcements available, the design potential for composite is huge.

The so-called natural composites like bones and woods are constructed by biological processes. The most primitive man-made composite materials comprised straw and mud in the form of bricks for building constructions. The most advanced examples are used on spacecrafts in highly demanding environments. The most visible applications are pave roadways in the form of either steel and Portland cement concrete or asphalt concrete. Engineered composite materials must be formed to shape. This involves strategically placing the reinforcements while manipulating the matrix properties. A variety of methods are used according to the end-item design requirements. These fabrication methods are commonly named moulding or casting processes, as appropriate, and both have numerous variations. The principle factors impacting the methodology are the nature of the chosen matrix and reinforcement materials. Another important factor is the gross quantity of material to be produced. Large quantities can be used to justify high capital expenditures for rapid and automated manufacturing technology. Small production quantities are accommodated with lower capital expenditures, but higher labour costs at a correspondingly lower rate.

Many commercially produced composites use a polymer matrix often called a resin. The reinforcements are often fibres but also commonly ground minerals. Strong fibres such as fibreglass, quartz, Kevlar or carbon fibres give the composite its tensile strength, while the matrix binds the fibres together, transferring the load from broken fibres to unbroken ones and between fibres that are not oriented along the tension lines. Also, unless the matrix chosen is especially flexible, it prevents the fibres from buckling in compression. In terms of stress, any fibre will provide resistance to tension, the matrix will resist shear, and all materials present will resist compression. Composite materials can be divided into two main categories, normally referred to as short-fibre reinforced materials and continuous-fibre reinforced materials, the latter often constituting a layered or laminated structure. Shocks, impacts, loadings or repeated cyclic stresses can cause the laminate to separate at the interface between two layers, a condition known as delamination. Individual fibres can separate from the matrix through a mechanism called fibre pull-out.

Nanoscience and nanotechnology correspond to science and technology that extend from about 100 nm down to atomic orders of magnitude around 0.2 nm, and to the physical phenomena and material properties observed when operating in this size range. Conceptually, nanocomposites refer to multiphase materials where at least one of the constituent phases has one dimension less than 100 nm. This field has attracted the attention, scrutiny and imagination of both scientific and industrial communities in recent years, and has opened a large window of opportunities to overcome the limitations of traditional micrometre-scale composites. Research in this area is literally exploding, because of the intellectual appeal of building blocks on the nanometre scale and because the technical innovations permit the designing and creation of new materials and structures with unprecedented flexibility, improvements in physical properties and significant industrial impact.

The large interest in the nanoscale range originates from outstanding properties. Enhanced properties can often be reached for low filler volume fraction, without a detrimental effect on other properties such as impact resistance or plastic deformation capability. Though industrial exploitation of nanocomposites is still in its infancy, the rate of technology implementation is increasing.

19.2 NATURAL FIBRES

Agro-based resources, also referred to as lignocellulosics, are resources that contain cellulose, hemicelluloses and lignin. When considering lignocellulosics as possible engineering materials, there are several basic concepts that must be taken into account [1]. First, lignocellulosics are hygroscopic resources that were designed to perform, in nature, in a wet environment. Second, nature is programmed to recycle lignocellulosics in a timely way through biological, thermal, aqueous, photochemical, chemical and mechanical degradation. In simple terms, nature builds a lignocellulosic structure from carbon dioxide and water and has all the tools to recycle it back to the starting chemicals.

There is a wide variety of agro-based or natural fibres to consider for utilization. They can be subdivided based on their origin, viz. vegetable, animal or minerals. Cellulose as a material is used by the natural world in the construction of plants and trees, and by man to make shipping sails, ropes and clothes, to give but a few examples. It is also the major constituent of paper and further processing can be performed to make cellophane and rayon. Depending on the part of the plant from which they are taken, cellulose fibres can be classified as:

- *Grasses and reeds*: The fibres come from the stem of plants, such as bamboo or sugar cane.
- *Leaf or hard fibres*: These fibres are most commonly used as reinforcing agents in polymers. They can be extracted for instance from sisal, henequen, abaca or pineapple.
- *Bast or stem fibres*: These fibres come from the inner bark of the stem of plants. Common examples are jute, flax, hemp, kenaf and ramie.
- *Straw fibres*: Examples include rice, wheat and corn straws.
- *Seed and fruit hairs*: These fibres come from seed-hairs and flosses and are primarily represented by cotton and coconut.
- *Wood fibres*: Examples include maple, yellow poplar and spruce.

In any commercial development, there must be a long-term guaranteed supply of resources. The growing of natural fibres is spread across all five continents. Quality and yield depend on the kind of plant, the grown variety, the soil and the climatic conditions. Tanzania and Brazil are the two largest producers of sisal. Henequen is produced in Mexico, kenaf is grown commercially in the United States and flax is a commodity crop grown in the European Union, as well as in many diverse agricultural systems and environment throughout the world, including Canada,

Table 19.1

Inventory of major potential bast fibre sources for the year 2000/2001 [2]

Fibre source	World production (metric tons)
Jute	2 900 000
Linseed	942 240
Kenaf	470 000
Flax	464 650
Sisal	380 000
Ramie	170 000
Hemp	157 800
Abaca	98 000

Argentina, India and Russia. Hemp originated in Central Asia, from where it spread to China, the Philippines and many other countries. Ramie fibres, mostly available and used in China, Japan and Malaysia, are the longest and one of the strongest textile fibres. The largest producers of jute are India, China and Bangladesh and coir is produced in tropical countries. The price for natural fibre varies depending on the economy of the countries where such fibres are produced. Table 19.1 shows the inventory of some of the larger sources of agricultural bast fibre that could be utilized for fibre–polymer composites. However, only a small part of these fibres has been used for industrial applications up to now, which shows that the potential of the existing bast plants has not yet been exhausted and that huge natural resources are still available.

The traditional source of natural fibres has been wood and for many countries, this will continue to be the case. Other large sources come from recycling agro-fibre-based products such as paper, waste wood and point source agricultural residues. Recycling paper products back into paper requires a wet processing and the removal of inks, salts and adhesives. Recycling the same products into composites can be done by using dry processing whereby all components are incorporated into the composite, eliminating the need of costly separation procedures. The major point source fibres are rice hulls from a rice processing plant, sun flower seed hulls from an oil processing unit and bagasse from a sugar mill.

In order to maintain the high quality of the fibres, their separation from the original plant is best done by retting, scrapping or pulping. Basically, there are two working principles to separate the bast fibres from the wood [3]. The conventional method uses breaking rollers, which alternatingly bend, buckle and soften the stalks. This method requires an intensive retting before processing which is induced by microorganisms that dissolve the lignin and pectins of the stalk. Modern technologies use swing hammer mills in most cases. The fibre decortication is provoked by the impact stress of the hammers directly on the surface of the stalks. This working principle ensures a complete separation of the fibres from the wood, even when processing freshly harvested, non-retted plants. The effective mechanical separation of the fibres and the wood inside the decorticator simplifies the subsequent fibre cleaning. The availability of large quantities of lignocellulosic fibres with well-defined mechanical properties is a general prerequisite for the successful subsequent use of these materials.

Plant fibres are bundles of elongated thick-walled dead plant cells. They are like microscopic tubes, that is cell walls surrounding the centre lumen that contributes to their water uptake behaviour (Fig. 19.1). Natural fibres display

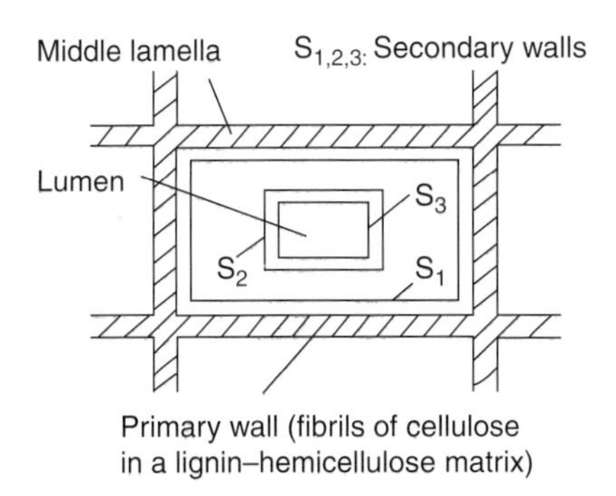

Figure 19.1 Schematic structure of a natural fibre cell. Reproduced with permission from Reference [4].

Table 19.2

Mean chemical composition of some natural fibres [7]

	Cotton	Jute	Flax	Ramie	Sisal
Cellulose	82.7	64.4	64.1	68.6	65.8
Hemicelluloses	5.7	12.0	16.7	13.1	12.0
Pectin	5.7	0.2	1.8	1.9	0.8
Lignin	–	11.8	2.0	0.6	9.9
Water soluble	1.0	1.1	3.9	5.5	1.2
Wax	0.6	0.5	1.5	0.3	0.3
Water	10.0	10.0	10.0	10.0	10.0

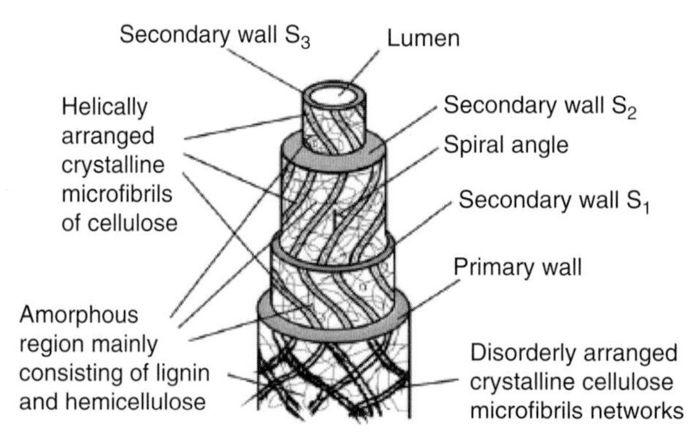

Figure 19.2 Schematic structure of an elementary plant fibre (cell). The secondary cell wall, S2, makes up about 80 per cent of the total thickness. Reproduced with permission from Reference [5].

a multi-level organization and consist of several cells formed out of semi-crystalline oriented cellulose microfibrils connected to a complete layer by lignin, hemicelluloses and in some cases pectins. Climatic conditions, age and digestion process influence not only the structure of the fibres but also their chemical composition. Table 19.2 reports the mean chemical composition of some natural fibres. With the exception of cotton, the components of natural fibres are cellulose, hemicelluloses and lignin, which determine their physical properties. Several of such cellulose–lignin/hemicellulose layers in one primary and three secondary cell walls stick together to form a multiple-layer composite. Such microfibrils have typically a diameter of about 2–20 nm, are made up of 30–100 cellulose macromolecules in extended chain conformation and provide the mechanical strength to the fibre.

The cell walls differ among themselves in their composition and orientation of the cellulose microfibrils. In most plant fibres, these microfibrils are oriented at an angle to the normal axis called the microfibrillar angle (Fig. 19.2). The characteristic value for this structural parameter varies from one plant fibre to another.

The outer cell wall is porous and contains almost all of the non-cellulose compounds, except proteins, inorganic salts and colouring matters and it is this outer cell wall that creates poor absorbency, poor wettability and other undesirable textile properties. In most applications, fibre bundles or strands are used rather than individual fibres. Within each bundle, the fibre cells overlap and are bonded together by pectins that give strength to the bundle as a whole. However, the strength of the bundle structure is significantly lower than that of the individual fibre cell and thus the potential of the individual fibres is not fully exploited.

The properties of natural fibres are strongly influenced by many factors, particularly chemical composition, internal fibre structure, microfibrillar angle, cell dimensions and defects, which differ between different parts of a plant, as well as between different plants. A weak correlation between strength and cellulose content and microfibril or spiral angle is found for different plant fibres. In general, the fibre strength increases with increasing cellulose content and decreasing spiral angle with respect to the fibre axis. This means that the most efficient cellulose fibres are those that have a high cellulose content, coupled with a low microfibril angle. Other factors that may affect the fibre properties are the maturity, the separating process, the microscopic and molecular defects, such as

Table 19.3

Physical properties of some natural fibres. Properties of some synthetic organic and inorganic fibres are added for comparison

Fibre	Density ($g\,cm^{-3}$)	Young's modulus (GPa)	Tensile strength (MPa)	Elongation (%)	Microfibrillar angle (°)
Cotton	1.5	5.5–27.6	300–1 500	3–8	–
Jute	1.3–1.5	13–26.5	393–800	1.2–1.8	8
Flax	1.5	27.6	345–1 500	2.7–3.2	5–10
Hemp	1.5	70	690	1.6	2–6.2
Ramie	1.55	61.4–128	400–938	1.2–3.8	7.5
Sisal	1.45	9.4–22	468–700	2–7	10–22
Coir	1.15–1.46	4–6	130–220	15–40	30–49
Viscose	–	11	593	11.4	–
Soft wood kraft	1.5	40	1 000	–	–
E-glass	2.5	70	2 000–3 500	2.5	–
S-glass	2.5	86	4 570	2.8	–
Aramide	1.4	63–67	3 000–3 150	3.3–3.7	–
Carbon	1.4	230–240	4 000	1.4–1.8	–

pits and knots, the type of soil and the weather conditions under which the vegetable was grown. Differences in fibre structure due to the environmental conditions during growth result in a broad range of characteristics. The mechanical properties of plant fibres are in general much lower when compared to those of the most widely used reinforcing glass fibres (Table 19.3). However, because of their low density, the specific properties which are property-to-density ratio dependent, viz. strength and stiffness, are comparable to those of glass fibres. Thus, natural fibres are in general suitable to reinforce polymer matrices, both thermoplastics and thermosets.

19.3 COMPOSITES

The use of additives in polymers is likely to grow with the introduction of improved compounding technologies and new coupling agents that permit the use of high filler/reinforcement contents. Fillings up to 75 pph could be common in the future and this would have a tremendous impact in lowering the use of petroleum-based polymers [6]. Since the price of plastics has risen sharply over the past few years, adding a natural powder or fibre to them provides a cost reduction to industry (and in some instances increases performance as well). To the agro-based industry, this represents an increased value for the agro-based component. Ideally, of course, a bio-based renewable polymer reinforced with agro-based fibres would be the most environment-friendly material.

Over the past decade there has been a growing interest in the use of lignocellulosic fibres as reinforcing elements in polymeric matrices [7, 8]. A number of researchers have been involved in investigating the exploitation of cellulosic fibres as load bearing constituents in composite materials. Prior work on lignocellulosic fibres in thermoplastics has concentrated on wood-based flour or fibres [9–13]. The majority of these studies has been on polyolefins, mainly polypropylene (PP). Compared to inorganic fillers, the main advantages of lignocellulosics are listed below:

(1) Low density: Their density, around 1.5 $g\,cm^{-3}$, is much lower than that of glass fibres, around 2.5 $g\,cm^{-3}$.
(2) Low cost and low energy consumption.
(3) High specific strength.
(4) Renewability and biodegradability.
(5) Abundant availability in a variety of forms throughout the world.
(6) Flexibility: Unlike brittle fibres, lignocellulosic fibres will not be fractured during processing.
(7) Non-abrasive nature to processing equipment, which allows high filling levels, resulting in significant cost savings and high stiffness properties.
(8) Non-toxicity.
(9) Ease of handling.

(10) Reactive surface, facilitating its chemical modification.
(11) Organic nature, resulting in the possibility to generate energy without residue after incineration at the end of their life-cycle.
(12) Economic development opportunity for non-food farm products in rural areas.

Despite these attractive aspects, lignocellulosic fibres are used only to a limited extent in industrial practice due to difficulties associated with surface interactions. It is important to keep these limitations in perspective when developing end-use applications. The primary drawback of agro-based fibres is associated with their inherent polar nature and the non-polar characteristics of most thermoplastics, which causes difficulties in compounding the filler and the matrix and, therefore, in achieving acceptable dispersion levels, which in turn generates inefficient composites. Another drawback of lignocellulosic fillers is their hydrophilic character which favours moisture absorption with a consequent swelling of the fibres and the decrease in their mechanical properties. Moisture absorption and the corresponding dimensional changes can be largely prevented if the hydrophilic filler is thoroughly encapsulated by the hydrophobic polymer matrix and there is a good adhesion between both components. However, if the adhesion level between the filler and the matrix is not good enough, diffusion pathways for moisture can pre-exist or can be created under mechanical solicitation. The existence of such pathways is also related to the filler connection and therefore to its percolation threshold.

The various approaches to the surface modification of these fibres aimed at minimizing the above drawbacks are thoroughly discussed in Chapter 18.

Yet another limitation associated with the use of lignocellulosic fillers is the fact that the processing temperature of composites must be restricted to just above 200°C (although higher temperatures can be used for short periods of time), because of their susceptibility to degradation and/or the possibility of volatile emissions that could affect the composite properties. This limits the types of thermoplastics that can be used to polymers like polyethylene, PP, poly-vinyl chloride and polystyrene, which constitute, however, about 70 per cent of all industrial thermoplastics. Nevertheless, technical thermoplastics like polyamides, polyesters and polycarbonates, which are usually processed at temperatures higher than 250°C, cannot be envisaged as matrices for these types of composite.

19.4 COMPOSITE PROCESSING

Drying the fibres is an essential prerequisite that must be applied before processing, because water on the fibre surface acts as a separating agent at the fibre–matrix interface. In addition, because of the water evaporation during processing at temperatures higher than 100°C, voids appear in the matrix. Both phenomena obviously lead to a decrease in the mechanical properties of the ensuing composites. Fibre drying can be done under different conditions, which results in different degrees of their residual moisture.

Extrusion and injection-moulding are the economically most attractive processing methods of thermoplastic-based composites. The extrusion press processing (express-processing) has been developed for the production of flax fibre reinforced PP at the research centre of Daimler Benz [7]. In this process, flax fibre non-wovens and PP melt films are alternatively deposited and moulded. A production process for PP semi-products reinforced with lignocellulosic fibres in the form of mats has been developed by BASF AG [7]. Fibre mats are produced by stitching together layers of fibres which have previously been crushed.

Beginning with bakelite in the early 1900s, engineers and scientists have continued to work to improve the various attributes of thermosets through the addition of natural fibres. Unsaturated polyester, epoxy, and vinylester resins are commonly used for preparing such composites. Fabrication techniques suitable for manufacturing natural fibre reinforced thermoset composites include the hand lay-up technique for unidirectional fibres/mats/fabric, and filament winding and pultrusion for continuous fibres. Resin transfer moulding (RTM) and prepegs can also be used. Semi-products, such as sheet moulding compounds (SMC) and bulk moulding compounds (BMC), can be obtained with short and chopped fibres.

19.5 COMPOSITE PROPERTIES

The major factors that govern the properties of short-fibre thermoplastic composites are fibre volume fraction, fibre dispersion, fibre aspect ratio and length distribution, fibre orientation and fibre–matrix adhesion. Each of these parameters is briefly discussed below.

19.5.1 Fibre volume fraction

Like other composite systems, the properties of short-fibre composites are strongly determined by the fibre concentration. The variation of the composite properties with fibre content can be predicted using the rule of mixtures, which involves the extrapolation of both matrix and fibre properties to a fibre volume fraction of 0 and 1. The following criteria must be taken into account:

(1) The composite fracture has to be fibre-controlled.
(2) The modulus of elasticity of the fibre should be greater than that of the matrix.
(3) The strain to failure of the matrix must be greater than that of the fibre.

In the case of unidirectional (or longitudinal) fibre reinforced composites, the stress is transferred from the matrix to the fibre by shear. When stressed in tension, both the fibre and the matrix elongate equally according to the principle of combined action [14]. Hence, the mechanical properties of the composite can be evaluated on the basis of the properties of the individual constituents. For a given elongation of the composite, both constituents, fibre and matrix, may be in elastic deformation; the fibre may be in elastic deformation whereas the matrix may be in plastic deformation, or both the fibre and the matrix may be in plastic deformation (Fig. 19.3).

At low fibre volume fraction, a decrease in the tensile strength is usually observed (Fig. 19.4). This is ascribed to the dilution of the matrix and the introduction of flaws at the fibre ends where a high stress concentration occurs, causing the bond between fibre and matrix to break. At high volume fraction, the stress is more evenly distributed and a reinforcement effect is observed. For all values of strain, the stress value in the composite is given by a simple mixing rule balanced by the volume fraction of each constituent, viz.:

$$\sigma_c' = \sigma_f' V_f + \sigma_m' V_m \tag{19.1}$$

where σ represents the stress value of each component at a particular strain value and V the volume fraction of each component of the composite. The subscripts c, f and m correspond to the composite, the fibre and the matrix, respectively.

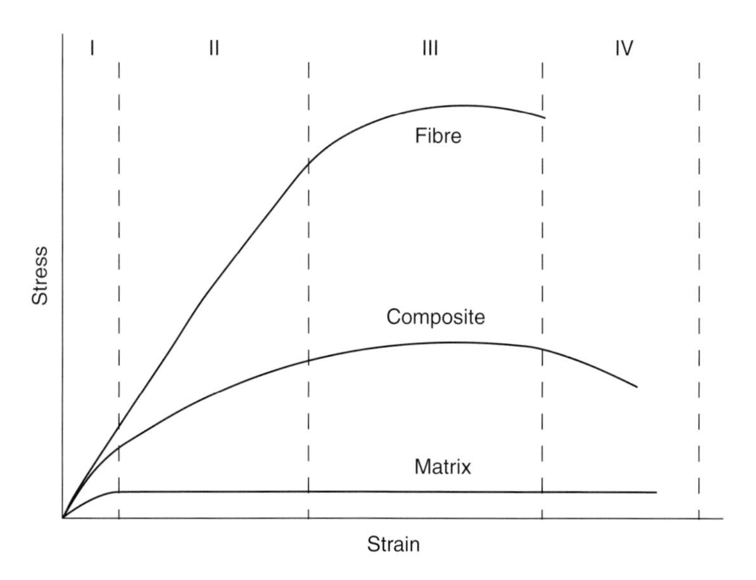

Figure 19.3 Illustration of four stages of deformation of fibres, matrix and composite. Stage I: elastic deformation of both fibres and matrix; stage II: elastic deformation of fibres and plastic deformation of matrix; stage III: plastic deformation of both fibres and matrix; stage IV: failure of both fibres and matrix. Reproduced with permission from Reference [14].

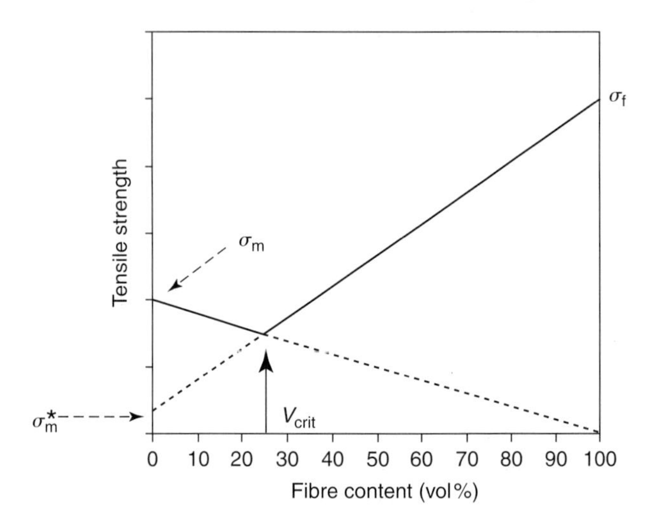

Figure 19.4 Model for the prediction of the ultimate tensile strength of unidirectional fibre-reinforced composites for which the fracture is fibre-controlled.

The fibre volume fraction for which the strength ceases to decrease and begins to increase is called the critical fibre volume fraction, V_{crit}. Below this value, the behaviour of the composite is only governed by the matrix:

$$\begin{aligned} &\text{For } V_f < V_{crit} \qquad \sigma_c = \sigma_m V_m \\ &\text{For } V_f > V_{crit} \qquad \sigma_c = \sigma_f V_f + \sigma_m^* V_m \end{aligned} \tag{19.2}$$

where σ_m is the ultimate tensile strength of the matrix, σ_f is that of the ultimate tensile strength of the fibre and σ_m^* is the stress on the matrix at a strain value where σ_f is reached. V_{crit} is an important parameter because it corresponds to the volume fraction of the fibres above which they begin to strengthen, rather than weaken the matrix. It can be calculated from the following equation:

$$V_{crit} = \frac{\sigma_m - \sigma_m^*}{\sigma_f - \sigma_m^*} \tag{19.3}$$

For a given matrix, the critical fibre volume fraction decreases with the increasing strength of the fibres. This means that for fibres which are much stiffer than the matrix, V_{crit} is very low.

The modulus of elasticity is also an important factor. Within strain limits for which both the fibre and the matrix are in elastic deformation, the modulus of the composite can be calculated using the rule of mixture:

$$E_c = E_f V_f + E_m V_m \tag{19.4}$$

where (Fig. 19.4) E_c, E_f and E_m are the modulus of elasticity of composite, fibre and matrix, respectively. When the fibre is in elastic and the matrix is in plastic deformation, the equation becomes:

$$E_c = E_f V_f + \left(\frac{\sigma_m^*}{\varepsilon}\right) V_m \tag{19.5}$$

The ratio σ_m^*/ε is the slope of the stress–strain curve of the matrix at a given strain beyond the proportional limit of the matrix.

The length of some individual natural fibres can reach up to 4 m, and when bundled with other fibres, this maximum length will be even higher. However, lignocellulosic materials are mainly used as discontinuous short fibres and are ground into fine particles with relatively low aspect ratios. These fillers generally increase the stiffness of the composites, but the strength is generally lower than that of the pristine matrix [10]. For instance, residual softwood sawdust was used as a reinforcing material in PP [15] and it was found that the tensile strength decreased regularly from 35 MPa for the unfilled matrix down to 10 MPa for the 60 wt% filled system. On the contrary, the addition of henequen fibres to a low density polyethylene matrix increased the tensile strength by 50 per cent (from 9.2 to 14 MPa) at a fibre loading of 30 vol% [16]. At the same time, the modulus increased from 275 to 860 MPa and the strain at break decreased from 42 to 5 per cent. The increase in stiffness results from the fact that lignocellulosic fillers or fibres have a higher Young's modulus, as compared to commodity thermoplastics, thereby contributing to the higher stiffness of the composites. However, an anchoring effect of the lignocellulosic filler acting as nucleating agents for the polymeric chains has been reported [17] resulting in an increase in the degree of crystallinity of the matrix. This effect seems to be strongly influenced by the lignin content and the surface aspect of the fibre [18, 19]. This transcrystallization phenomenon at the fibre–matrix interphase participates in the reinforcing effect of the filler.

In order to use models to estimate composite properties, it is necessary to know the properties of the fibres, which vary widely depending on the source, age, separating techniques, moisture content, speed of testing, history of the fibre, etc. The properties of the individual fibres are therefore very difficult to measure. Moreover, in a natural fibre–polymer composite, the lignocellulosic phase is present in a wide range of diameters and lengths, some in the form of short filaments and others in forms that seem closer to the individual fibre.

Continuous regenerated cellulose fibres are extensively used as reinforcements in composites such as tyres. However, very few studies are available on their use as reinforcement for polymer composites. Because of the strong hydrogen bonds that occur between cellulose chains, cellulose does not melt or dissolve in common solvents. Thus, it is difficult to convert the short fibres from wood pulp into continuous filaments. Regenerated cellulose fibres are produced on a commercial scale under the generic name 'Lyocell' by a spinning process from a cellulose *N*-methylmorpholine-*N*-oxide/water solution. The mechanical properties of these fibres were found to depend on the draw ratio [20, 21]. The low mechanical properties reported for unidirectional composites composed of Lyocell fibres embedded in a poly(3-hydroxybutyrate-co-3-hydroxyvalerate) matrix were ascribed to weak interfacial adhesion due to both the smooth topography of the fibres and the hydrophobic properties of the matrix [22].

19.5.2 Fibre dispersion

The primary requirement for obtaining good performances from short-fibre composites is a good dispersion level in the host polymer matrix, which is obtained if the fibres are separated from each other and each fibre is surrounded by the matrix. Clumping and agglomeration must therefore be avoided. Insufficient fibre dispersion results in an inhomogeneous mixture composed of matrix-rich and fibre-rich domains. Mixing the polar and hydrophilic fibres with a non-polar and hydrophobic matrix, can result in dispersion difficulties.

There are two major factors affecting the extent of fibre dispersion: fibre–fibre interaction, such as hydrogen bonding between the fibres, and fibre length, because of the possibility of entanglements. As mentioned above, one of the specificity of cellulose fibres as reinforcement is their poor dispersion characteristics in many thermoplastic melts, due to their hydrophilic nature. Several methods have been suggested and described in the literature to overcome this problem and improve the dispersion. Among them are:

(1) Fibre surface modification. The surface energy is closely related to the hydrophilicity of the lignocellulosic fibres.
(2) Use of dispersing agents, such as stearic acid or a mineral oil. The dispersion of lignocellulosic fibres can be improved by pretreatment with lubricants or thermoplastic polymers. An addition of 1–3 per cent stearic acid is sufficient to achieve a maximum reduction in size and number of aggregates in PP and polyethylene [7]. The use of stearic acid in HDPE/wood fibres was reported to improve the fibre dispersion and the wetting between the fibre and the matrix [9].
(3) Fibre pre-treatments, such as acetylation, or use of a coupling agent.
(4) Increased shear force and mixing time. The best processing method involves twin-screw extruder.

Some physical methods have also been suggested to improve the dispersion of short fibres within the matrix. Treatments such as stretching, calandering, thermotreatment and the production of hybrid yarns do not change the chemical composition of the fibre, but modify their structural and surface properties and thus influence their mechanical bonding with polymers.

As already mentioned, the surface modification of lignocellulosic fibres is comprehensively dealt with in Chapter 18.

19.5.3 Fibre aspect ratio and length distribution

The efficiency of a composite also depends on the amount of stress transferred from the matrix to the fibres. This can be maximized by improving the interaction and adhesion between both phases and also by maximizing the length of the fibres retained in the final composite. However, long fibres sometimes increase the amount of clumping resulting in poor dispersion of the reinforcing phase within the host matrix. The ultimate fibre length present in the composite depends on the type of compounding and moulding equipment used and the processing conditions. Several factors contribute to the fibre attrition, such as the shearing forces generated in the compounding equipment, the residence time, the temperature and the viscosity of the compound. Using a polystyrene matrix, it was shown that the extent of breakage was most severe and rapid for glass fibres, less extensive for Kevlar fibres and the least for cellulose fibres [23]. The effect of twin-screw blending of wood fibres and polyethylene was also reported [12] and it was shown that the level of fibre attrition depended on the configuration and the processing temperature.

The fibre aspect ratio, which is its length to diameter ratio is a critical parameter in a composite. A relationship has been proposed by Cox to relate the critical fibre aspect ratio, l_c/d, to the interfacial shear stress, τ_y, viz.:

$$\frac{l_c}{d} = \frac{\sigma_{fu}}{2\tau_y} \tag{19.6}$$

where, σ_{fu} is the fibre ultimate strength in tension. At controlled fibre ultimate strength in tension, this equation shows an inverse relationship between the critical aspect ratio and the interfacial shear stress, where the former decreases as the latter increases, because of efficient transfer. This means that, for each short-fibre composite system, there is a critical fibre aspect ratio that corresponds to its minimum value for which the maximum allowable stress can be achieved for a given load. This parameter is determined by the fibre properties, the matrix properties and the quality of the fibre–matrix interface.

The condition for maximum reinforcement, that is the condition ensuring maximum stress transfer to the fibres, before the composite fails, is to have a length higher than the critical length l_c (Fig. 19.5). If the fibre aspect ratio is lower than its critical value, the fibres are not loaded to their maximum stress value. A specificity of cellulose fibres is their flexibility compared to glass fibres which allows a desirable fibre aspect ratio to be maintained after processing, which is around 100 or 200 for high performance short-fibre composites.

19.5.4 Fibre orientation

Fibre orientation is another important parameter that influences the mechanical behaviour of short-fibre composites. This is because the fibres in such composites are rarely oriented in a single direction, which is necessary to obtain the maximum reinforcement effects. During the processing of short-fibre composites, a continuous and

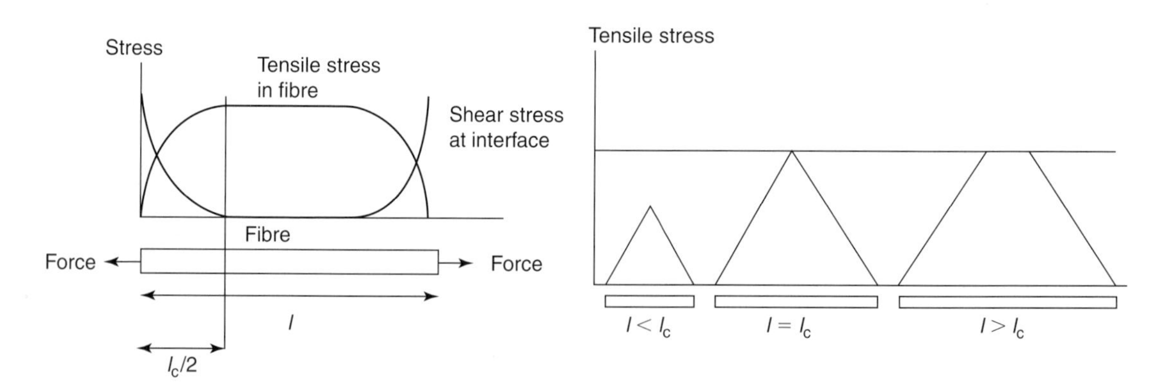

Figure 19.5 Variation of tensile stress in fibre and shear stress at interface occurring along the fibre length. If the fibre aspect ratio is lower than its critical value, l_c, the fibres are not loaded to their maximum stress value.

progressive orientation of individual fibres occurs (Fig. 19.6). This change is related to the geometrical properties of the fibres, the viscoelastic properties of the matrix and the change in shape produced by the processing. In these operations, the polymer melt undergoes both elongational and shear flow.

19.5.5 Fibre–matrix adhesion

Fibre to matrix adhesion plays a very important role in the reinforcement of composites with short fibres. During loading, loads are not applied directly to the fibres but to the matrix. It is necessary to have an effective load transfer from the matrix to the fibres for the ensuing composites to have good mechanical properties. This requires good interaction as well as adhesion between the fibres and the matrix, that is strong and efficient fibre–matrix interface.

As already pointed out, strongly hydrophilic cellulose fibres are inherently incompatible with hydrophobic polymers. When two materials are incompatible, it is often possible to introduce a third material having intermediate properties capable of reducing their interfacial energy. One way of applying this concept to the present context, is to impregnate the fibres with a polymer compatible with the matrix and, in general, this is achieved using low viscosity polymer solutions or dispersion. For a number of interesting polymers, however, the lack of solvents limits the use of this method. The compatibilization of the two components by specific chemical or physical treatments involving either is the most common approach to this problem as systematically reviewed in Chapter 18, with particular emphasis on the surface modification of the fibres. The following example illustrates the less frequent approach, based on the use of a surface modifier that bears a structure very close to that of the matrix, but which has been appropriately modified so that its macromolecules can react at the fibres' surface. Figure 19.7 shows SEM micrographs from the fractured surface of PP reinforced with cellulose fibres [15]. With the untreated matrix (Fig. 19.7(a)), a poor interfacial adhesion is clearly observed because of the absence of any physical contact between the fibre and the matrix. The micrograph in Fig. 19.7(b) corresponds to fibres in contact with a maleic

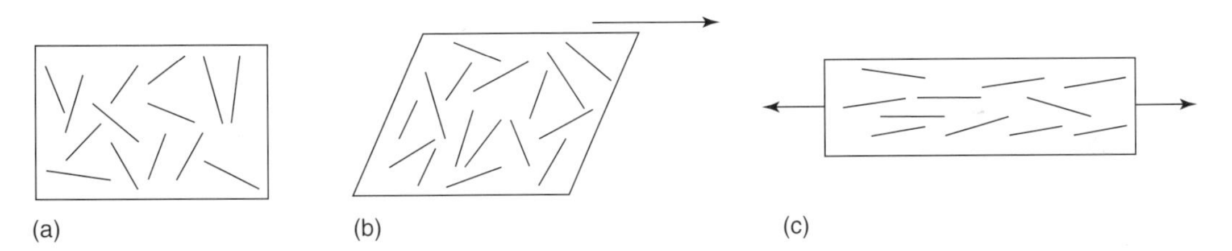

Figure 19.6 Orientation of individual fibres during processing: (a) initial random distribution, (b) rotation during shear flow and (c) alignment during elongational flow.

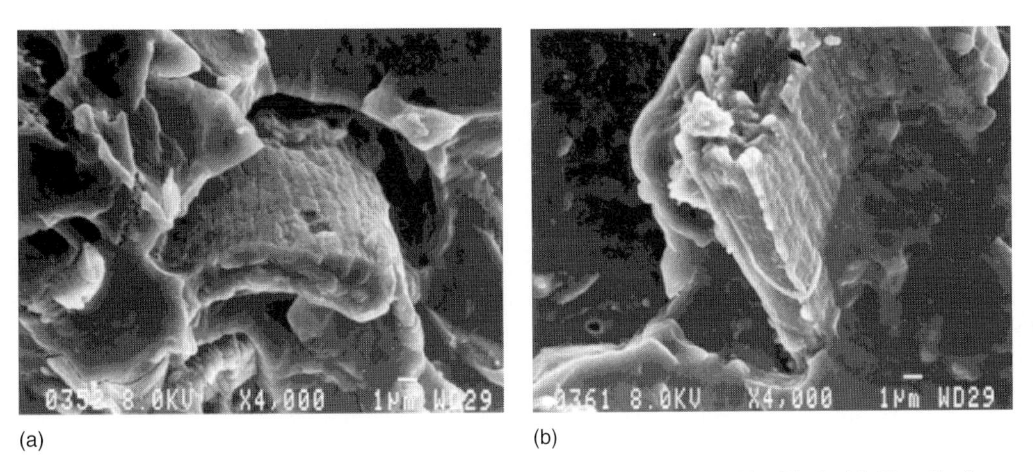

Figure 19.7 Scanning electron micrographs of a freshly fractured surface of a PP film filled with 20 wt% of raw untreated (a) and (b) MAPP coated softwood fibres. Reproduced with permission from Reference [24].

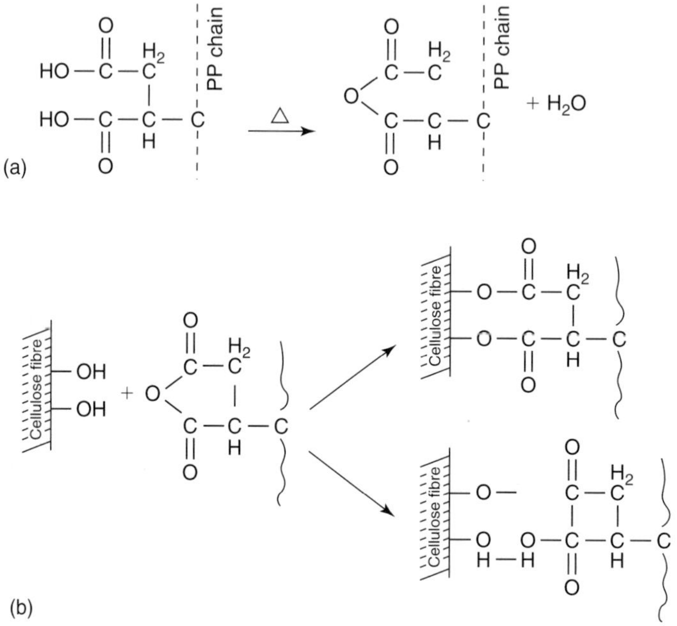

Figure 19.8 Reaction mechanism involved during the treatment of cellulose fibres with PP maleic anhydride copolymer (MAPP): (a) activation of MAPP ($T = 170°C$) before fibre treatment and (b) esterification of cellulose.

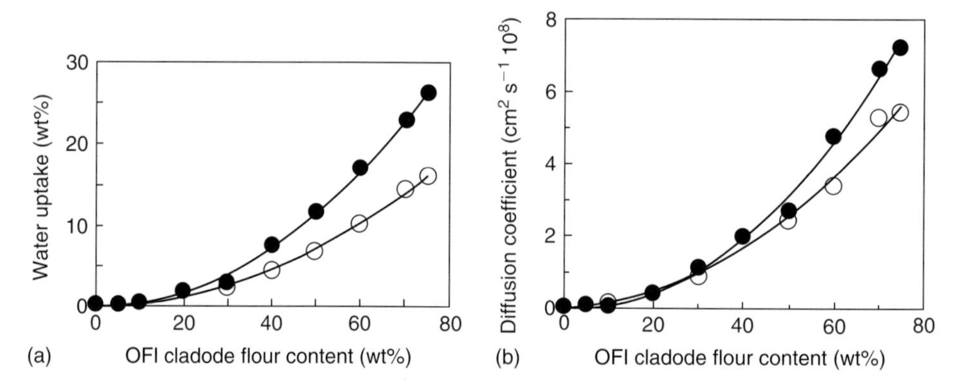

Figure 19.9 (a) Water uptake at equilibrium and (b) water diffusion coefficient of PP/*Opuntia ficus indica* cladode flour composites conditioned at 98 per cent RH versus filler loading: untreated filler (●) and MAPP coated filler (○) (the solid line serves to guide the eye). Reproduced with permission from Reference [25].

anhydride polypropylene (MAPP) graft copolymer (PP chains with pendant succinic acid moieties), which shows a good wetting, with absence of holes around the fibres. The mechanism of the reaction of MAPP with cellulose fibres can be divided into two steps (Fig. 19.8), the first being the activation of the copolymer by heat before the fibre treatment and the second, the esterification of cellulose. The fact of generating covalent bonds across the interface improved the adhesion between the matrix and the fibres, and both the Young modulus and the tensile strength were found to be higher than those obtained with the untreated fibres [15].

It has also been found that moisture absorbance of the natural fibre–polymer composite can be prevented if the fibre–matrix adhesion is optimized [15, 24]. Indeed, whereas composites based on standard PP and cellulosic fibres displayed high water content at the interphase, due to the presence of microcavities, the encapsulation of the fibres with MAPP decreased the water sensitivity of the composites in terms of both the water uptake and its diffusion coefficient [25], as shown in Fig. 19.9.

19.6 NANOCOMPOSITES

As previously mentioned, natural fibres present a multi-level organization and consist of several cells formed out of semi-crystalline oriented cellulose microfibrils. Each microfibril can be considered as a string of cellulose crystallites, linked along the chain axis by amorphous domains (Fig. 19.10) and having a modulus close to the theoretical limit for cellulose. They are biosynthesized by enzymes and deposited in a continuous fashion. A similar structure is reported for chitin, as discussed in Chapter 25. Nanoscale dimensions and impressive mechanical properties make polysaccharide nanocrystals, particularly when occurring as high aspect ratio rod-like nanoparticles, ideal candidates to improve the mechanical properties of the host material. These properties are profitably exploited by Mother Nature.

The promise behind cellulose-derived nanocomposites lies in the fact that the axial Young's modulus of the basic cellulose crystalline nanocrystal, derived from theoretical chemistry, is potentially higher than that of steel and similar to that of Kevlar. It was first experimentally studied in 1962 from the crystal deformation of cellulose I, using highly oriented fibres of bleached ramie [26]. A value of 137 GPa was reported, which differed from the theoretical estimate of 167.5 GPa calculated by Tashiro and Kobayashi [27]. The latter value is thought to be higher because the calculations had been carried out for low temperature. Force deflection data from the compression of cubes of potato tissues were fed into a model containing two structural levels, the cell structure and the cell wall structure [28], giving a maximum modulus value of 130 GPa. Eichhorn and Young [29] observed a decrease of cellulose crystallites when their crystallinity decreased. Recently, Raman spectroscopy was used to measure the elastic modulus of native cellulose crystals [30] and a value around 143 GPa was reported. However, it is worth noting that these measurements were made on epoxy/tunicin whiskers composites.

Stable aqueous suspensions of polysaccharide nanocrystals can be prepared by the acid hydrolysis of vegetable biomass. Different descriptors of the resulting colloidal suspended particles are used, including whiskers, monocrystals and nanocrystals. The designation 'whiskers' is used to describe elongated rod-like nanoparticles. These crystallites have also often been referred in the literature as microfibrils, microcrystals or microcrystallites, despite their nanoscale dimensions. Most of the studies reported in the literature refer to cellulose nanocrystals. A recent review described the properties and applications of cellulose whiskers in nanocomposites [31].

Figure 19.10 Schematic diagram showing the hierarchical structure of a semi-crystalline cellulose fibre.

The procedure for the preparation of such colloidal aqueous suspensions is described in detail in the literature for cellulose and chitin [32, 33]. The biomass is generally first submitted to a bleaching treatment with NaOH in order to purify cellulose or chitin by removing other constituents. The bleached material is then disintegrated in water, and the resulting suspension submitted to acid hydrolysis. The amorphous regions of cellulose or chitin act as structural defects and are responsible for the transverse cleavage of the microfibrils into short monocrystals by acid hydrolysis. Under controlled conditions, this transformation consists in the disruption of the amorphous regions surrounding and embedded within the cellulose or chitin microfibrils, while leaving the microcrystalline segments intact, because of the very large difference in the rate of hydrolysis between the amorphous and the crystalline domains, the latter obviously being much more resistant. The resulting suspension is subsequently diluted with water and washed by successive centrifugations. Dialysis against distilled water is then performed to remove the free acid in the dispersion. Complete dispersion of the whiskers is obtained by a sonication step. The dispersions are stored in a refrigerator after filtration to remove residual aggregates and addition of several drops of chloroform. This general procedure has to be adapted in terms of the acid hydrolysis conditions, such as time, temperature and purity of materials depending on the nature of the substrate and the geometrical characteristics of the nanocrystals.

The constitutive cellulose or chitin nanocrystals occur as elongated rod-like particles or whiskers. The length is generally of the order of a few hundreds nanometres and the width is of the order of a few nanometres. The aspect ratio of these whiskers is defined as the ratio of the length to the width. The high axial ratio of the rods is important for the determination of anisotropic phase formation and reinforcing properties. Figure 19.11 shows a transmission electron micrograph (TEM) obtained from a dilute suspension of tunicin whiskers, that is cellulose nanocrystals obtained from tunicate, a sea animal. Their average length and diameter are around 1 μm and 15 nm, respectively, and their aspect ratio was estimated to be around 67 [34].

Aqueous suspensions of starch nanocrystals can also be prepared by the acid hydrolysis of starch granules in aqueous medium using hydrochloric acid or sulphuric acid at 35°C (see also Chapter 15). Residues from the hydrolysis are called 'lintners' and 'nägeli' or amylodextrin. The degradation of native starch granules by acid hydrolysis depends on many parameters, which include the botanical origin of starch, namely crystalline type, granule morphology (shape, size, surface state) and the relative proportion of amylose and amylopectin. It also depends on the acid hydrolysis conditions, namely acid type, acid concentration, starch concentration, temperature, hydrolysis duration and stirring. A response surface methodology was used by Angellier *et al.* [35] to investigate the effect of five chosen factors on the selective sulphuric acid hydrolysis of waxy maize starch granules in order to optimize the preparation of aqueous suspensions of starch nanocrystals. These predictors were temperature, acid concentration, starch concentration, hydrolysis duration and stirring speed. The preparation of aqueous suspensions of starch nanocrystals with a yield of 15.7 wt%, was achieved after 5 days using 3.16 M H_2SO_4 at 40°C, 100 rpm and with a starch concentration of 14.7 wt%.

Compared to cellulose or chitin, the morphology of constitutive nanocrystals obtained from starch is completely different. Figure 19.12 shows a TEM obtained from a dilute suspension of waxy maize starch nanocrystals. They consist of 5–7 nm thick platelet-like particles with a length ranging 20–40 nm and a width in the range of 15–30 nm. The detailed investigation on the structure of these platelet-like nanoparticles was reported [36].

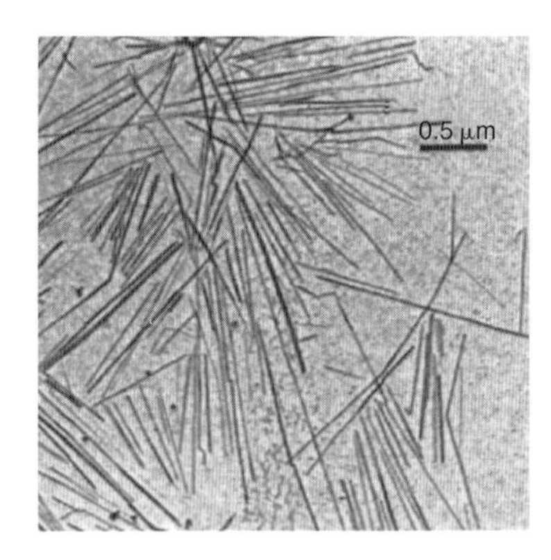

Figure 19.11 TEM of a dilute suspension of tunicin. Reproduced with permission from Reference [34].

Because of the high stability of aqueous polysaccharide nanocrystals dispersions, water is the preferred processing medium. High level of dispersion of the filler within the host matrix in the resulting composite is expected when processing nanocomposites in an aqueous medium. Therefore, this restricts the choice of the matrix to hydrosoluble polymers. The use of aqueous dispersed polymers, that is latexes, is a first alternative, which allows to employ hydrophobic polymers as matrices and ensure a good dispersion level of the filler, indispensable for homogenous composite processing. The possibility of dispersing polysaccharide nanocrystals in non-aqueous media is a second alternative which opens other possibilities for nanocomposite processing.

The first demonstration of the reinforcing effect of cellulose whiskers in a poly(St-co-BuA) matrix was reported by Favier *et al.* [37]. The authors measured, using DMA in the shear mode, a spectacular improvement in the storage modulus after adding tunicin whiskers, even at a low content, into the host polymer. This increase was especially significant above the glass–rubber transition temperature of the thermoplastic matrix, because of its poor mechanical properties in this temperature range. Figure 19.13 shows the isochronal evolution of the logarithm of the relative storage shear modulus ($\log G'_T/G'_{200}$, where G'_{200} corresponds to the experimental value measured at 200 K) at 1 Hz as a function of temperature for such composites prepared by water evaporation. In the rubbery state of the thermoplastic matrix, the modulus of the composite with a loading level as low as 6 wt%, is more than two orders of magnitude higher than that of the unfilled matrix. Moreover, the introduction of 3 wt% or more

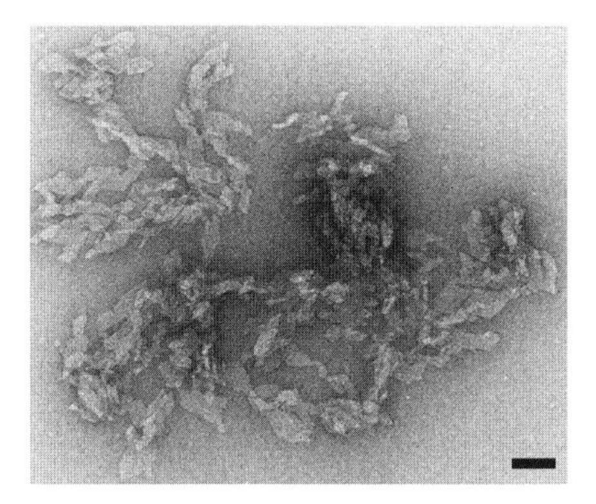

Figure 19.12 Transmission electron micrograph of a dilute suspension of hydrolyzed waxy maize starch (scale bar 50 nm). Reproduced with permission from Reference [35].

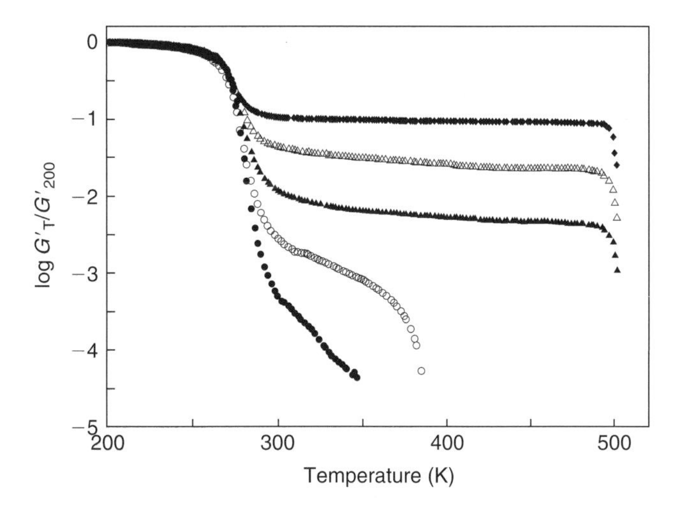

Figure 19.13 Logarithm of the normalized storage shear modulus ($\log G'_T/G'_{200}$, where G'_{200} corresponds to the experimental value measured at 200 K) versus temperature at 1 Hz for tunicin whiskers reinforced poly(St-co-BuA) nanocomposite films, obtained by water evaporation and filled with 0 (●), 1 (○), 3 (▲), 6 (△) and 14 wt% (◆) of cellulose whiskers. Reproduced with permission from Reference [31].

cellulosic whiskers provides an outstanding thermal stability to the matrix modulus up to the temperature at which cellulose starts to degrade (500 K).

The macroscopic behaviour of polysaccharide nanocrystals-based nanocomposites depends, as for any heterogeneous materials, on the specific behaviour of each phase, the composition (volume fraction of each phase), the morphology (spatial arrangement of the phases) and the interfacial properties. The outstanding properties observed for these systems were ascribed to a mechanical percolation phenomenon [37]. A good agreement between experimental and predicted data was reported when using the series–parallel model of Takayanagi, modified to include a percolation approach. Therefore, the mechanical performance of these systems was not only the result of the high mechanical properties of the reinforcing nanoparticles. It was suspected that the stiffness of the material was due to infinite aggregates of cellulose whiskers. Above the percolation threshold, the cellulose nanoparticles can connect to form a three-dimensional continuous pathway through the nanocomposite film. For rod-like particles such as tunicin whiskers with an aspect ratio of 67, the percolation threshold is close to 1 vol%. The formation of this cellulose network was supposed to result from strong interactions, like hydrogen bonds, between whiskers. This phenomenon is similar to the high mechanical properties observed for a paper sheet, which result from the hydrogen-bonding forces that hold the percolating network of fibres. This mechanical percolation effect explains both the high reinforcing effect and the thermal stabilization of the composite modulus for evaporated composite films.

Any factor that affects the formation of the percolating whisker network, or interferes with it, changes the mechanical performances of the composite [38]. Three main parameters were reported to affect the mechanical properties of such materials, viz. the morphology and dimensions of the nanoparticles, the processing method, and the microstructure of the matrix and matrix–filler interactions.

Apart from the mechanical performances, some other properties are interesting and can be improved by adding polysaccharide nanocrystals, for instance swelling properties. It was shown that the water uptake of tunicin whiskers/thermoplastic starch nanocomposites decreased as a function of the filler content [34]. For starch nanocrystals/natural rubber nanocomposites, it was shown that both the toluene uptake at equilibrium and its diffusion coefficient decreased when adding starch nanocrystals [39]. The evolution of the diffusion coefficient of toluene displayed a discontinuity around 10 per cent, suggesting a possible percolation effect of the starch nanocrystals.

The barrier properties of starch nanocrystals/natural rubber nanocomposites were also investigated [39]. For these systems, the water vapour transmission rate, the diffusion coefficient of oxygen, the permeability coefficient of oxygen and its solubility, were measured. It was observed that the permeability to water vapour, as well as to oxygen, decreased when starch nanocrystals were added. These effects were ascribed to the platelet-like morphology of the nanocrystals.

19.7 CONCLUSIONS

There is a growing trend to use lignocellulosic fibres in applications for which synthetic fibres were traditionally employed, which is ascribed as their numerous well-known advantage. Present applications of natural fibre filled composites are in the field of energy and impact absorption, such as car fenders and bicycle helmets. They also include markets that target cheaper, renewable and non-recyclable, or biodegradable materials, such as packaging and structural elements. Other uses of natural fibre-based composites are deck surface boards, picnic tables, industrial flooring, etc. In cars, about 10–15 kg of these composites, typically made up of 50 per cent natural fibres and 50 per cent PP, along with other additives, are presently being used. Examples are door panels, roof headliners, seat backs, rear decks and trunkliners.

Another interesting property of natural fibres is their hierarchical structure and the possibility to choose the scale linked to the application. Polysaccharide nanocrystals are building blocks biosynthesized to provide structural properties to living organisms. They can be isolated from cellulose-containing materials under strictly controlled conditions. Polysaccharide nanocrystals are inherently low cost materials, available from a variety of natural sources in a wide range of aspect ratios. The corresponding polymer nanocomposites display outstanding mechanical properties and can be used to process high modulus thin films. Practical applications of such fillers and their transition into industrial technology require a favourable ratio between the expected performances of the composite material and its cost.

In conclusion, this area is moving fast towards novel outstanding composite materials based on renewable resources in the form of both traditional natural fibres and their nanomorphologies, but there are still significant scientific and technological challenges to be met.

REFERENCES

1. Rowell R.M., Property enhanced natural fiber composite materials based on chemical modification, in *Science and Technology of Polymers and Advanced Materials: Emerging Technologies and Business Opportunities*, (*Proceedings of the Fourth International Conference on Frontiers of Polymers and Advanced Materials*, Cairo, Egypt, 4–9 January 1997), Plenum Press, New York N.Y., 1998, pp. 717–732.
2. Anonym., *Information Bulletin of the FAO European Cooperative Research Network on Flax and other Bast Plants*, **2**(16), 2001; Institute of Natural Fibres Coordination Centre, Poznan, Poland, Dec. 2001.
3. Munder F., Fürll C., Hempel H., Processing of bast fiber plants for industrial application, in *Natural Fibers, Biopolymers and Biocomposites*, Eds.: Mohanty A.K., Misra M. and Drzal L.T., CRC Press Taylor & Francis Group, Boca Raton, 2005, pp. 109–140, Chapter 3.
4. Bismarck A., Aranberri-Askargorta I., Springer J., Lampke T., Wielage B., Stamboulis A., Shenderovich I., Limbach H.-H., Surface characterization of flax, hemp and cellulose fibers; surface properties and the water uptake behavior, *Polym. Compos.*, **23**(5), 2002, 872–894.
5. Rong M.Z., Zhang M.Q., Liu Y., Yang G.C., Zeng H.M., The effect of fiber treatment on the mechanical properties of unidirectional sisal-reinforced epoxy composites, *Compos. Sci. Technol.*, **61**(10), 2001, 1437–1447.
6. Katz H.S., Milewski J.V., in *Handbook of Fillers for Plastics*, Van Nostrand Reinhold, New York, 1987. p. 512
7. Bledzki A.K., Gassan J., Composites reinforced with cellulose based fibres, *Prog. Polym. Sci.*, **24**, 1999, 221–274.
8. Eichhorn S.J., Baillie C.A., Zafeiropoulos N., Mwaikambo L.Y., Ansell M.P., Dufresne A., Entwistle K.M., Herrera-Franco P.J., Escamilla G.C., Groom L., Hugues M., Hill C., Rials T.G., Wild P.M., Review: Current international research into cellulosic fibres and composites, *J. Mater. Sci.*, **36**, 2001, 2107–2131.
9. Woodhams R.T., Thomas G., Rodges D.K., Wood fibers as reinforcing fillers for polyolefins, *Polym. Eng. Sci.*, **24**(15), 1984, 1166–1171.
10. Kokta B.V., Raj R.G., Daneault C., Use of wood flour as filler in polypropylene: Studies on mechanical properties, *Polym. Plast. Technol. Eng.*, **28**(3), 1989, 247–259.
11. Bataille P., Ricard L., Sapieha S., Effect of cellulose in polypropylene composites, *Polym. Compos.*, **10**(2), 1989, 103–108.
12. Yam K.L., Gogoi B.K., Lai C.C., Selke S.E., Composites from compounding wood fibers with recycled high-density polyethylene, *Polym. Eng. Sci.*, **30**(11), 1990, 693–699.
13. Sanadi A.R., Young R.A., Clemons C., Rowell R.M., Recycled newspaper fibers as reinforcing fillers in thermoplastics: Part I. Analysis of tensile and impact properties in polypropylene, *J. Reinf. Plast. Comp.*, **13**(1), 1994, 54–67.
14. Weeton J.W., Peters D.M., Thomas K.L., in *Engineers' Guide to Composite Materials*, American Society for Metals, Metals Park, Ohio, 1987.
15. Anglès M.N., Salvadó J., Dufresne A., Mechanical behavior of steam exploded residual softwood filled polypropylene composites, *J. Appl. Polym. Sci.*, **74**, 1999, 1962–1977.
16. Herrera-Franco P.J., Aguilar-Vega M.J., Effect of fiber treatment on mechanical properties of LDPE-henequen cellulosic fiber composites, *J. Appl. Polym. Sci.*, **65**, 1997, 197–207.
17. Dufresne A., Dupeyre D., Paillet M., Lignocellulosic flour reinforced poly(hydroxybutyrate-co-valerate) composites, *J. Appl. Polym. Sci.*, **87**(8), 2003, 1302–1315.
18. Luo S., Netravali A.N., Mechanical and thermal properties of environment-friendly 'green' composites made from pineapple leaf fibers and poly(hydroxybutyrate-co-valerate) resin, *Polym. Compos.*, **20**(3), 1999, 367–378.
19. Reinsch V.E., Kelley S.S., Crystallization of poly(hydroxybutyrate-co-hydroxyvalerate) in wood fiber-reinforced composites, *J. Appl. Polym. Sci.*, **64**(9), 1997, 1785–1796.
20. Mortimer S.A., Peguy A.A., Ball R.C., Influence of the physical process parameters on the structure formation of Lyocell fibers, *Cell. Chem. Technol.*, **30**(3–4), 1996a, 251–266.
21. Mortimer S.A., Peguy A.A., The formation of structure in the spinning and coagulation of Lyocell fibers, *Cell. Chem. Technol.*, **30**(1–2), 1996b, 117–132.
22. Bourban C., Karamuk E., de Fondaumiere M.J., Rufieux K., Mayer J., Wintermantel E., Processing and characterization of a new biodegradable composite made of a PHB/V matrix and regenerated cellulosic fibers, *J. Environ. Polym. Degr.*, **5**(3), 1997, 159–166.
23. Czarnecki L., White J.L., Shear flow rheological properties, fiber damage and mastication characteristics of aramid-, glass-, and cellulose fiber-reinforced polystyrene melts, *J. Appl. Polym. Sci.*, **25**(6), 1980, 1217–1244.
24. Faria H., Cordeiro N., Belgacem M.N., Dufresne A., Dwarf cavendish as a source of natural fibers in polypropylene-based composites, *Macromol. Mater. Eng.*, **291**(1), 2006, 16–26.
25. Malainine M.E., Mahrouz M., Dufresne A., Lignocellulosic flour from cladodes of *Opuntia ficus-indica* reinforced polypropylene composites, *Macromol. Mater. Eng.*, **289**(10), 2004, 855–863.
26. Sakurada I., Nukushina Y., Ito T., Experimental determination of the elastic modulus of crystalline regions oriented polymers, *J. Polym. Sci.*, **57**(165), 1962, 651–660.

27. Tashiro K., Kobayashi M., Theoretical evaluation of three-dimensional elastic constants of native and regenerated celluloses: Role of hydrogen bonds, *Polymer*, **32**(8), 1991, 1516–1526.
28. Hepworth D.G., Bruce D.M., A method of calculating the mechanical properties of nanoscopic plant cell wall components from tissue properties, *J. Mater. Sci.*, **35**(23), 2000, 5861–5865.
29. Eichhorn S.J., Young R.J., The Young's modulus of a microcrystalline cellulose, *Cellulose*, **8**(3), 2001, 197–207.
30. Šturcová A., Davies G.R., Eichhorn S.J., Elastic modulus and stress-transfer properties of tunicate cellulose whiskers, *Biomacromolecules*, **6**(2), 2005, 1055–1061.
31. Azizi Samir M.A.S., Alloin F., Dufresne A., Review of recent research into cellulosic whiskers, their properties and their application in nanocomposite field, *Biomacromolecules*, **6**(2), 2005, 612–626.
32. Wise L.E., Murphy M., D'Addiecco A.A., Chlorite holocellulose, its fractionation and bearing on summative wood analysis and on studies on hemicelluloses, *Pap. Trade J.*, **122**, 1946, 35–43.
33. Marchessault R.H., Morehead F.F., Walter N.M., Liquid crystal systems from fibrillar polysaccharides, *Nature*, **184**, 1959, 632–633.
34. Anglès M.N., Dufresne A., Plasticized starch/tunicin whiskers nanocomposites: 1. Structural analysis, *Macromolecules*, **33**(22), 2000, 8344–8353.
35. Angellier H., Choisnard L., Molina-Boisseau S., Ozil P., Dufresne A., Optimization of the preparation of aqueous suspensions of waxy maize starch nanocrystals using a response surface methodology, *Biomacromolecules*, **5**(4), 2004, 1545–1551.
36. Putaux J.L., Molina-Boisseau S., Momaur T., Dufresne A., Platelet nanocrystals resulting from the disruption of waxy maize starch granules by acid hydrolysis, *Biomacromolecules*, **4**(5), 2003, 1198–1202.
37. Favier V., Canova G.R., Cavaillé J.Y., Chanzy H., Dufresne A., Gauthier C., Nanocomposites materials from latex and cellulose whiskers, *Polym. Adv. Technol.*, **6**, 1995, 351–355.
38. Dufresne A., Comparing the mechanical properties of high performances polymer nanocomposites from biological sources, *J. Nanosci. Nanotechnol.*, **6**(2), 2006, 322–330.
39. Angellier H., Molina-Boisseau S., Lebrun L., Dufresne A., Processing and structural properties of waxy maize starch nanocrystals reinforced natural rubber, *Macromolecules*, **38**(9), 2005, 3783–3792.

– 20 –

Chemical Modification of Wood

Mohamed Naceur Belgacem and Alessandro Gandini

ABSTRACT

The main emphasis of this chapter has to do with the shift from the traditional ways of treating woods in order to improve their resistance to atmospheric and biological degradation, to chemical modifications which eliminate the problem of leaching of toxic materials into the environment. Hence, recent contributions on the chemical modification of wood by different physico-chemical treatments, such as corona and plasma discharges, or its chemical grafting through esterification, etherification, as well as urethane and siloxane formation and the reaction of wood with furfuryl alcohol, are reviewed and discussed. The chapter also covers the topic of the preparation and characterization of composite materials made of wood fibres and polymeric matrices. These surface or bulk treatments point to the relevance of a novel green approach and show that very promising results can be obtained.

Keywords

Chemical modification of wood, Corona and plasma treatment, Wood esterification, Wood etherification, Reactions between wood and isocyanates, Reactions of wood with siloxanes, Reactions of wood with furfuryl alcohol, Wood-based composites

20.1 INTRODUCTION

Wood is one of the oldest renewable resources exploited by human activity in the form of timber, tools, source of energy and of shelter. Wood is a complex natural material whose structure represents the very paradigm of a composite assembly, as briefly outlined in Chapter 1, together with general references to the traditional uses of this multifarious vegetable manifestation. Its main components (see Chapter 1) are cellulose (**A**), hemicelluloses (**B**) and lignin (**C**), accompanied by minor contributions of low molecular weight compounds and mineral salts. This book includes several chapters devoted to the utilization of both the major and some of the minor components of wood considered individually as a source of polymers.

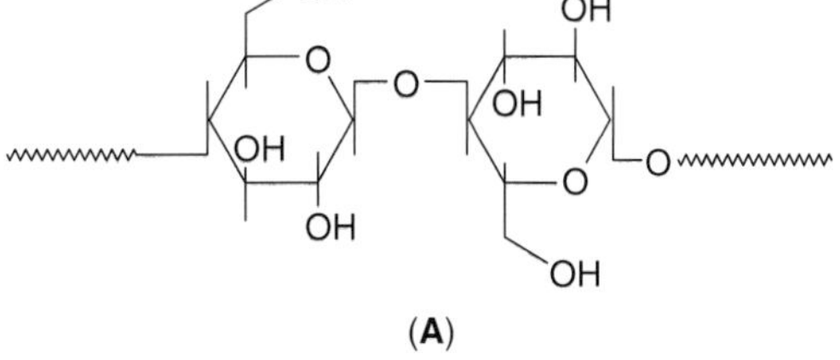

OH OH O O OH OH O O OH OH

(**A**)

Wood, as such, finds numerous applications, often depending on the specific species, because it displays remarkable properties like good mechanical strength, easy processing and an attractive and warm appearance. None of these aspects bear a direct relationship with the purposes of this book, because they are essentially related to technological issues. The same considerations apply to the use of wood in papermaking, as already emphasized in Chapter 1.

The hydrophilic character of wood, intrinsically connected with the structure of its three main macromolecular components, has been a longstanding source of problems, mostly arising from its lack of dimensional stability in moist environments. An additional major drawback is its susceptibility to photolytic and biological degradation. The interest of both scientists and technologists in modifying wood in order to overcome these drawbacks has called upon specific treatments like chemical, thermal, enzymatic or purely physical modifications. This chapter examines recent contributions to these issues, based on approaches which only involve the chemical modification of wood, carried out in bulk or at its surface. The other treatments have been aptly covered in a recent book [1] and a thorough review [2].

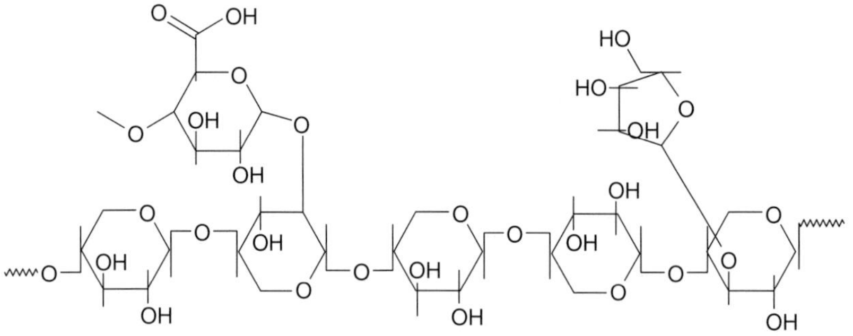

(B) 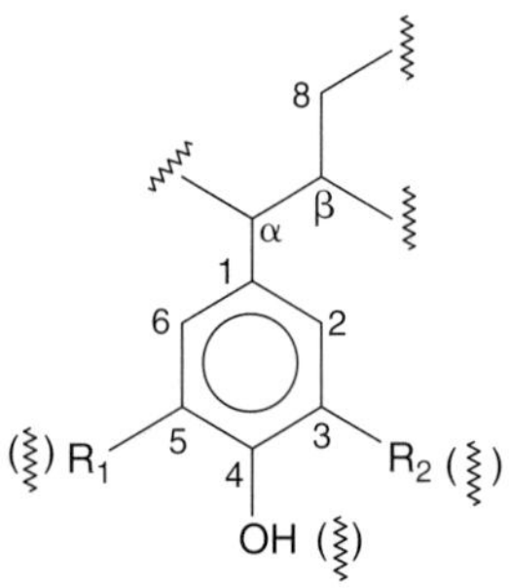 **(C)**

20.2 CHEMICAL MODIFICATION

The wood moieties exploited for chemical grafting are almost exclusively the hydroxyl functions present in the structure of the three polymers which constitute its essential composite morphology. Depending on the specific experimental conditions applied to achieve these modifications, they can be limited to the wood surface thereby preserving all its pristine bulk properties, or they can be extended inside the wood structure and, in this case, novel materials are therefore produced.

Both aspects are discussed here with a higher emphasis on surface modification.

20.2.1 Corona and plasma treatments

Corona and plasma treatments have been used to treat wood surfaces with the aim of increasing its wettability towards water and organic liquids, its adhesion with different coating lacquers and binders and its compatibility with polyolefins.

A corona treatment was applied to different wood tablets (Teak, Birch and Pine) and the wettability of the modified surfaces was measured by water contact angles which decreased from 90° to less than 40°, after a $24\,kJ\,m^{-2}$ discharge [3]. The adhesion of a water-based lacquer was also improved.

Birch was treated with an oxygen plasma in the presence or absence of hexamethyldisiloxane (HMDS) [4] and the ensuing surface characterized by atomic force microscopy (AFM), contact angle measurements and adhesion strength with polypropylene. When the treatment was by oxygen alone, the water contact angle decreased from about 90° to 55°, at an optimal treatment time (*i.e.*, 60 s) and the polar component of the surface energy increased from 12 to $40\,mJ\,m^{-2}$. A 5 min plasma treatment with HMDS yielded a better adhesion between the modified surface and polypropylene. AFM showed that the deposited polymer followed the tortuosity of the birch surface without forming a continuous film.

Yellow pine was also submitted to a plasma oxygen treatment in order to reduce its degradation by weathering [5]. Different coatings were plasma-deposited, namely poly-(dimethylsiloxane) (PDMS), benzotriazole/PDMS, ZnO/PDMS, hydroxybenzophenone/PDMS, phthalocyanine/PDMS and graphite/PDMS. ATR-FTIR and XPS analyses showed the occurrence of the grafting and the latter coating was found to be the most efficient in terms of the stabilization of weight loss during weathering tests.

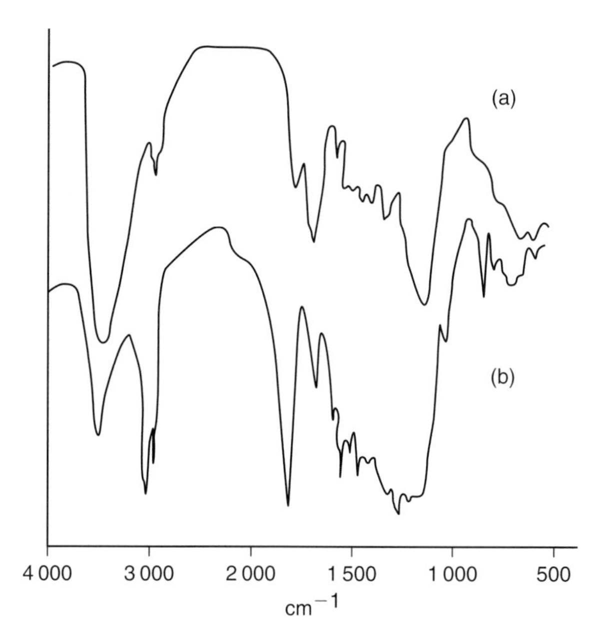

Figure 20.1 FTIR spectra of Oakwood sawdust before (a) and after (b) esterification with octanoyl chloride. (Reproduced by permission of Elsevier. Copyright 1995. Reprinted from Reference [10].)

Fir is another species whose surface was treated with oxygen plasma and corona in order to increase its adhesion with outdoor-coating systems and consequently the durability of the wood exposed to atmospheric conditions [6]. Different parameters were varied in this study and the wettability of the modified substrate tested which allowed the following optimal conditions to be established: a treatment time of 5 min, a plasma power of 600 W and a distance between the samples and the plasma source of 28 cm using oxygen or air as the plasma gas. The optimal corona voltage was around 15 kV, which yielded a drop in the water contact angle from 115° to less than 15°. Rehn and Viöl [7] showed that hydrophobic or hydrophilic wood surfaces can be generated by cold plasma treatment, depending on the gas used.

20.2.2 Esterification reactions

The esterification of wood has been investigated very extensively, as recently reviewed [1, 2]. Most of these investigations were carried out in a heterogeneous solid–liquid medium and only a few called upon the use of gaseous reagents. Only the very recent and scientifically relevant studies related to this topic will be reviewed here, since a much more detailed coverage of its historical development is available [8, 9].

Oakwood has been submitted to a solvent-free esterification procedure using fatty-acids of different chain lengths and their chlorides [10, 11]. These bulk reactions were carried out at high temperature in a nitrogen atmosphere and the modified wood samples characterized by weight gain, FTIR, SEM, moisture absorption, contact angle measurements and thermo-mechanical analyses. The weight gain reached 6 per cent for the highest grafting yields and the water contact angle was about 95°. Figure 20.1 shows the FTIR spectra of the wood sawdust before and after treatment with octanoyl chloride.

Pinus sylvestris was esterified using propionic anhydride and the FTIR spectroscopy of the ensuing sample proved that the grafting had indeed occurred [12]. *Pinus sylvestris* and mechanically pulped spruce fibres were also esterified with crotonic (**I**) and methacrylic anhydrides (**II**) and then copolymerized *in situ* with styrene (**ST**) [13].

I **II**

The fibres modified with **I** and copolymerized with **ST** using AIBN as the free-radical initiator, gave FTIR spectra with peaks which revealed the presence of C=O and C=C moieties after the first step, and the disappearance of the latter peak after the copolymerization with **ST**. The presence of grafted polystyrene was proven by the appearance of the characteristic double peak at 760 and 700 cm^{-1}, typical of monosubstituted benzene rings. The weight gain associated with these reactions was as high as 50 per cent. After the removal of wood, the FTIR and CP-MAS ^{13}C-NMR spectra of the residue were found to be similar to those of poly **ST** [13]. Figure 20.2 shows the ^{13}C-NMR spectra of the pine samples after different modifications.

Our group developed a novel strategy consisting in the use of planar stiff molecules bearing two identical reactive functions at the opposite end of their structure [14–17]. The working hypothesis was that *only one* of the functions can react with the wood OH group, whereas the other should be left to copolymerize with a subsequently

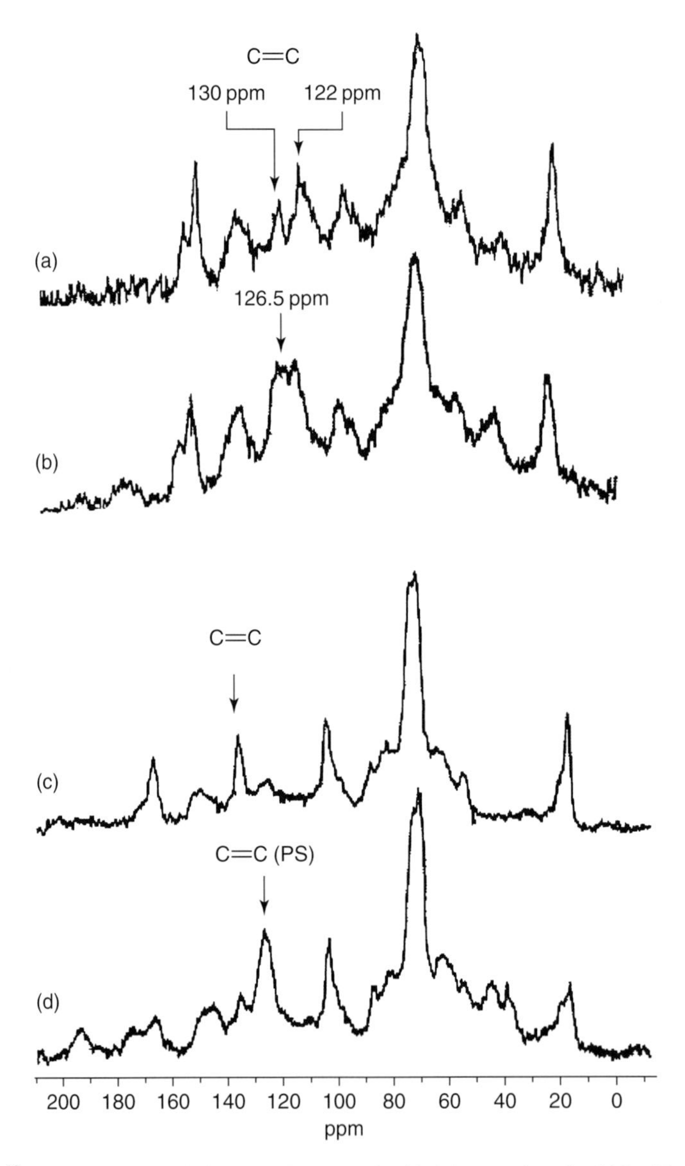

Figure 20.2 CP-MAS ^{13}C-NMR spectra of pine samples treated with (a) crotonic anhydride, (b) crotonic anhydride followed by copolymerization with styrene, (c) methacrylic anhydride and (d) methacrylic anhydride followed by copolymerization with styrene. (Reproduced by permission of Elsevier. Copyright 2000. Reprinted from Reference [13].)

added monomer. Thus, wood was treated with pyromellitic anhydride (**III**) in a dry non-swelling medium. The recovered fibres showed clearly, after extraction of the excess reagent, the relevant infrared bands arising from anhydride carbonyl groups.

III

In another study, the treatment of pine wood with **III** (see Fig. 20.3) was followed by copolymerizations with **ST** and methyl methacrylate (**MMA**). After extraction of the ungrafted poly**ST**, the samples were characterized by FTIR, IGC, anti-swelling efficiency (ASE) and weight gain. The latter measurement showed a weight gain increase of 55 per cent. The water contact angle of the initial substrate surface was about 40° just after the drop deposition and decreased to zero within a few seconds, whereas that of a sample treated with **III**, followed by **MMA** polymerization and extraction, was about 100°. The ASE was improved after both copolymerizations. Finally, IGC showed that the surface energy of wood and its acid–base character had been lowered by this type of treatment [14–17].

III-treated wood tablets were further reacted with different biocide molecules, namely 2-phenylphenol (**IV**) and tebuconazole (**V**), chosen because they bore the hydroxy group necessary for their coupling with the unreacted anhydride function remaining after the single grafting of **III** [18]. All these reactions, carried out under nitrogen, were optimized after testing solvents of different polarity, various catalysts, temperatures and durations. After extracting all unbound species and drying, these modified wood samples were characterized by weight gain, FTIR spectroscopy (multiple reflections on the block surface at different depths and transmission on wood powders mixed into KBr pellets), elemental analysis and biological tests. The occurrence of grafting was proven with most systems, albeit to different extents depending on the actual conditions. More importantly, the presence of

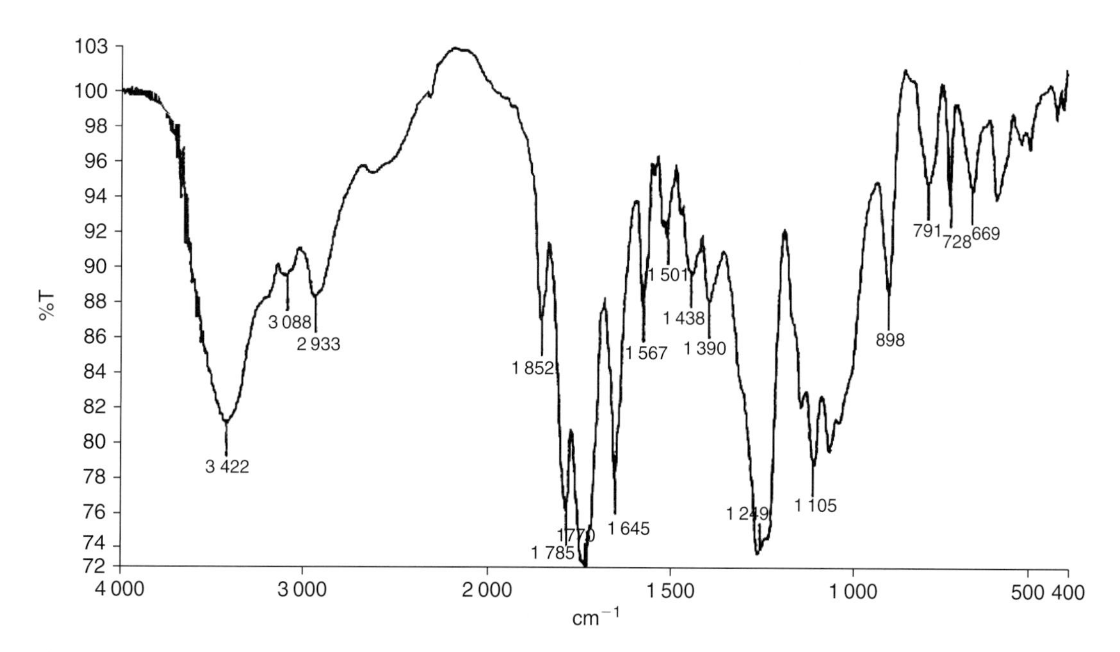

Figure 20.3 FTIR spectrum of **III**-modified pine wood [18].

the grafted biocide moieties arising from **IV** and **V** gave the correspondingly modified wood samples an excellent resistance to a wide selection of *fungi* [18].

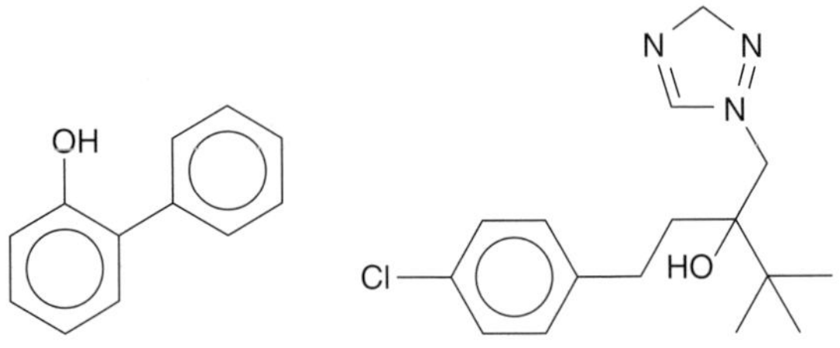

OH

Cl HO

IV **V**

The esterification of sapwood using maleic anhydride (**MA**) in the vapour phase was recently reported [19, 20]. The weight gain varied as a function of the **MA**/wood weight ratio and reached about 70 per cent for the highest ratio studied, that is, 1.6. The dimensional and antifungal stability, as well as the ASE of the treated samples, improved considerably.

Poly-(*N*-acryloyl dopamine) (**PAD**) was developed as a new wood adhesive and used to bind maple specimens [21]. Scheme 20.1 shows the different steps involved in the **PAD** synthesis. Since **PAD** undergoes substantial oxidation and crosslinking reactions at 80°C, a **PAD**–polyethylenimine (**PAD–PEI**) mixture was used to bind wood. The glued samples were characterized in terms of adhesion strength and water resistance. The **PAD–PEI** system gave a much better shear strength and a higher water resistance, compared with the corresponding system in which

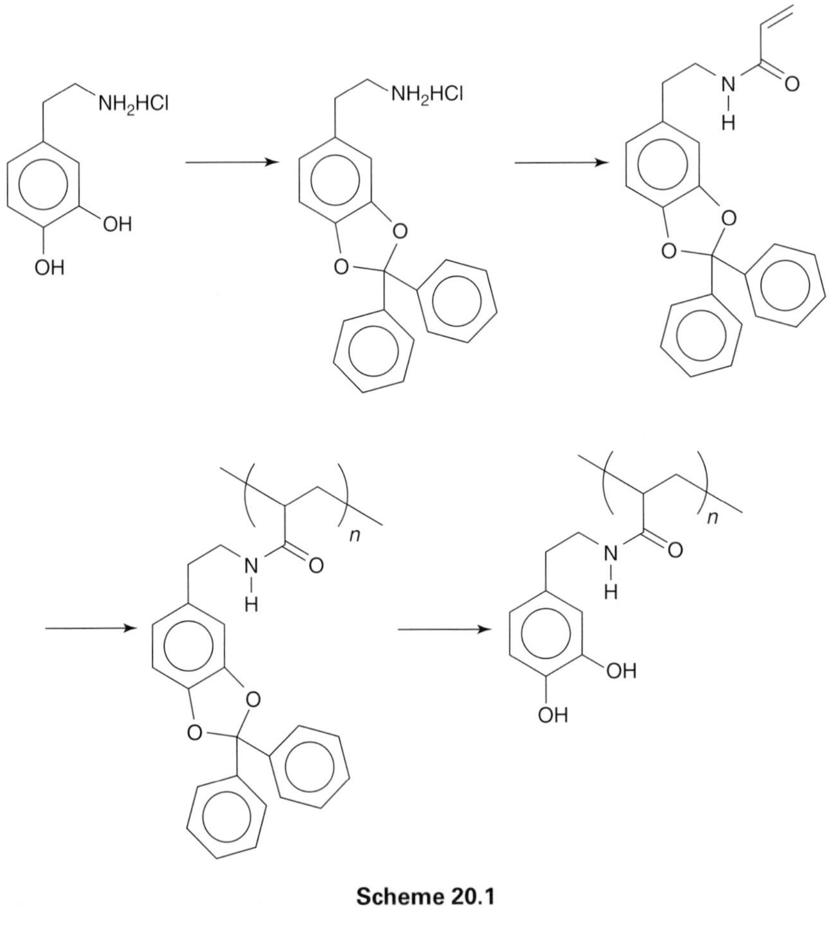

NH_2HCl NH_2HCl

OH OH O O

n *n*

Scheme 20.1

the **PAD** homopolymer was used. However, this study did not provide any evidence about the occurrence of chemical bonding between wood and adhesive.

More recently [22], a poly-(aminoamide-epichlorohydrin)/stearic anhydride compatibilizer system was developed to optimize the interface between pine flour and polyethylene in the corresponding composites. FTIR spectra showed that the compatibilizer had indeed been grafted to the wood particles and the decrease in water uptake by the composite corroborated the interest in using this novel system.

Wood specimens from *P. Roxburgii* (Chir pine) were esterified with benzoyl chloride in order to enhance their photostability [23, 24]. The samples gained about 20 per cent in weight and, after their exposure for 500 h to a 100 W xenon arc light at 30°C and 65 per cent relative humidity, showed an improved resistance to photoyellowing, thanks to a reduced extent of lignin degradation. A review dealing with the photodegradation and photostabilization of wood was recently reported [25].

20.2.3 Etherification reactions

Allyl glycidyl ether (**VI**) and glycidyl methacrylate (**VII**) were used to etherify wood blocks of *Pinus sylvestris* and mechanically pulped spruce fibres. The mechanism of the reaction consisted in the chain extension of the wood OH functions by the epoxy groups of **VI** and **VII**. The modified samples were characterized by FTIR and ^{13}C-NMR spectroscopy and by weight gain, which amounted to 7 and 20 per cent for **VI** and **VII**, respectively [12].

VI **VII**

Much more recently, **VII** was used to graft pine wood using different solvents, temperatures, catalysts and reaction times [14]. After the removal of the excess of **VII** by extraction, a weight gain of ~20 per cent was measured. These modified wood samples were then copolymerized with **MMA** or **ST**. The ensuing solvent-extracted materials (wood + **VII** + **MMA** or wood + **VII** + **ST**) had gained a further 35 per cent in weight. FTIR spectra showed unequivocally the occurrence of grafting at each step. The water contact angles on these modified wood surfaces were higher than 110° after the second grafting reactions, as shown in Fig. 20.4.

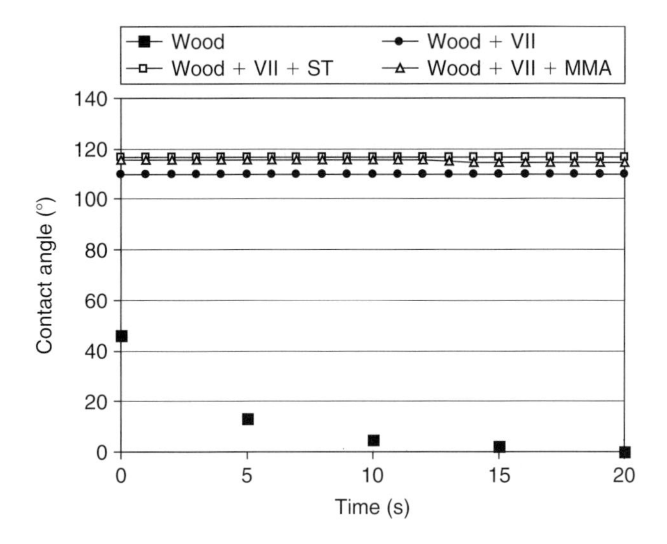

Figure 20.4 Water contact angles measured on the surface of wood before and after different modifications [18].

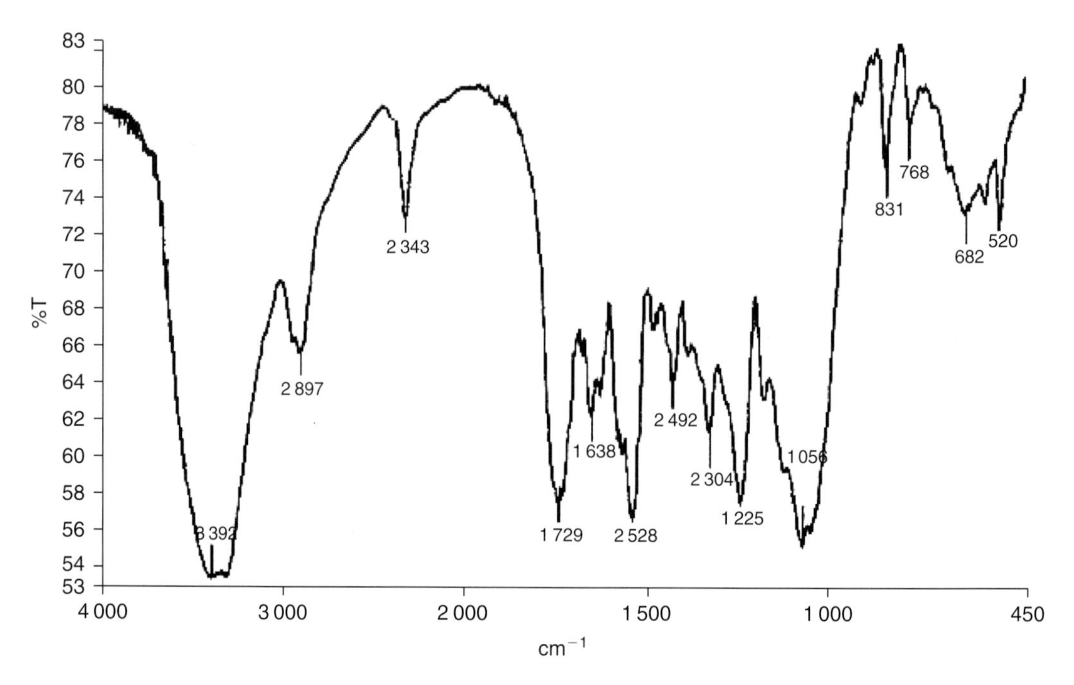

Figure 20.5 FTIR spectrum of **VIII**-modified wood [18].

Pine tablets were treated successively with **VII** and **ST** [26]. The ensuing materials were characterized by FTIR, weight gain, ASE, thermogravimetry and mechanical testing. Surprisingly, the authors claimed that wood can be grafted directly with **ST**, just by heating **ST**-impregnated wood samples in the presence of AIBN. They succeeded to graft up to 25 per cent w/w with respect to wood. Even more surprising was the reported lower weight gain associated with the double grafting strategy (wood + **VII** + **ST**), which however yielded much better results in terms of mechanical, ASE and thermal properties.

20.2.4 Reactions with isocyanates

The reaction of wood with isocyanate coupling agents was first reported in 1957 and concerned the use of phenyl isocyanate at 100–120°C [27]. Since then, numerous other reagents and wood substrates have been used, as recently reviewed [8, 9]. Hence only very recent investigations will be reported here.

1,4-Phenyl diisocyanate (**VIII**) was used as a grafting molecule following the same strategy discussed above in the case of the double anhydride **III** [14–17]. As expected, the FTIR spectrum of the modified wood surface, after extraction of the unreacted **VIII**, displayed a strong peak at 2 250 cm^{-1}, characteristic of the presence of isocyanate functions (see Fig. 20.5). These modified wood samples were treated with the biocides **IV** and **V**, which provided an excellent resistance to fungi [18].

OCN—C_6H_4—NCO

VIII

20.2.5 Reactions with siloxanes

A recent series of articles [28–30] describes the treatment of maritime pine sapwood with three different siloxanes, namely 3-isocyanatopropyl trimethoxysilane (**IX**), 3-glycidoxypropyltriethoxysilane (**X**) and *n*-propyltriethoxysilane (**XI**).

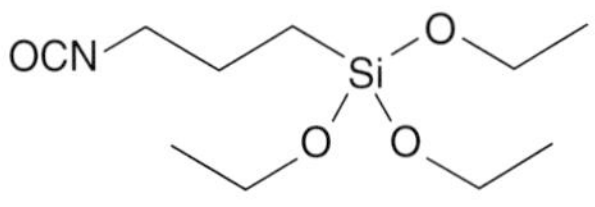

OCN Si O O O

IX

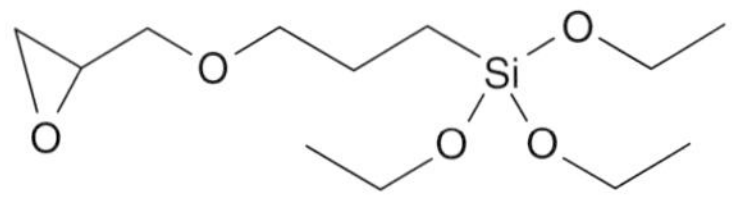

O O Si O O O

X

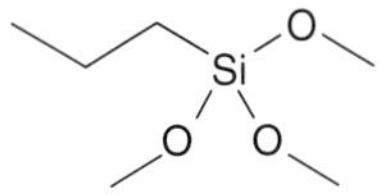

Si O O O

XI

Si O O O

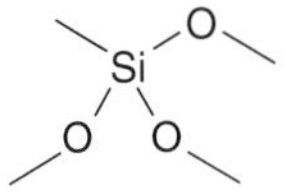

XII

These wood samples were solvent extracted and dried before grafting by impregnating them with a reagent solution under nitrogen using dibutyltin dilaurate as a catalyst and pyridine or dimethyl formamide as solvents. After these reactions, the treated samples were extracted with methylene chloride, before being characterized by weight gain, FTIR and CP-MAS NMR spectroscopy, SEM–EDX and ASE. Considerable weight gains were obtained after the optimization of the treatment. The atomic cartography showed that **IX** was mostly located in the primary wall and middle lamella regions, that is, in the lignin-rich zones. ^{29}Si-CP-MAS-NMR spectroscopy showed that the triethoxysilane moieties were not hydrolyzed during the different modification and characterization steps. The **IX**-modified wood was hydrolyzed and cocondensed with methyltrimethoxysilane (**XII**), as sketched in Scheme 20.2.

20.2.6 Reactions with furfuryl alcohol

Pinus sylvestris L. was treated with an aqueous solution of furfuryl alcohol (**XIII**) and catalyst at 80–100°C, using a process developed by Wood Polymer Technologies (Oslo, Norway) [31, 32]. The acid-catalyzed polycondensation of **XIII** has been thoroughly described in the literature [33–35].

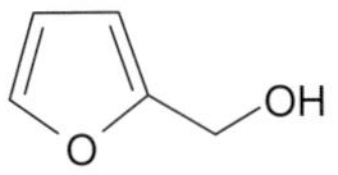

O OH

XIII

Wood—OH + OCN Si–O O O → Wood—O O N H Si–O O O

H_2O → Wood—O O N H Si–OH OH OH + —Si—O— O O →

O–Si—O—Si—O—Si—O O O O O–Si—O—Si—O—Si—O H N O O H N O O H N O O

Wood

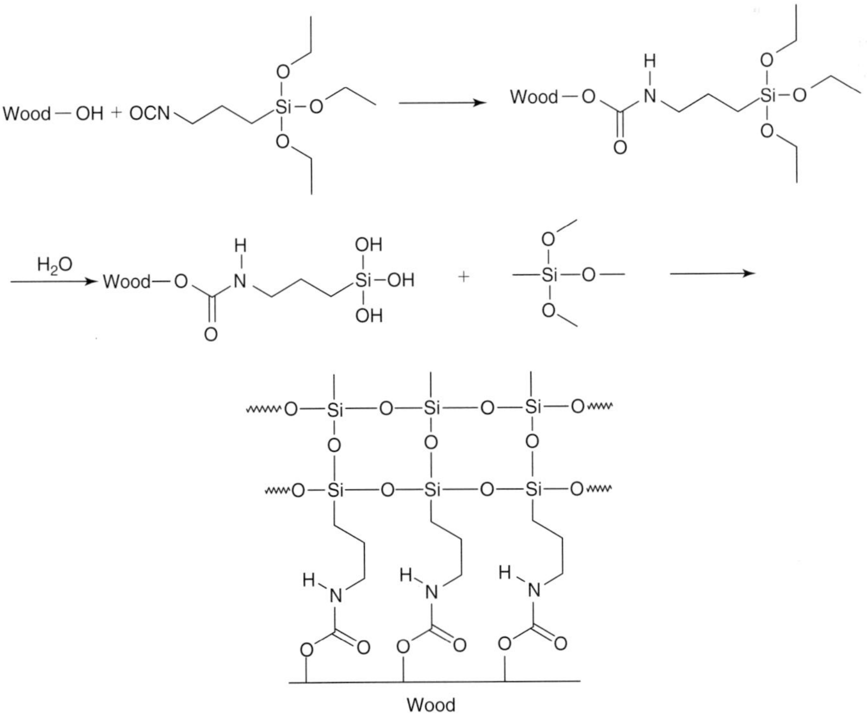

Scheme 20.2

The application of this Norwegian process has been shown to produce modified wood samples with exceptional resistance to both microbial decay and insect attacks, improved mechanical properties and remarkable dimensional stability.

Another study dealing with the treatment of wood with **XIII** involved the use of boric acid and ammonium borate as catalysts [36] and produced modified woods with improved ASE and water repellence efficiency and reduced water absorption. It is important to mention, however, that leaching tests showed that boron was slowly released into the washing water.

20.2.7 Reactions with other molecules

Ketene has been used to esterify wood (Scheme 20.3) because this reaction does not produce any condensation product, as in the case of anhydrides. However, the use of this reagent is associated with serious handling problems, since it is both toxic and explosive. Recently, Morozovs *et al.* [37] reported that, in the context of the reaction between wood and ketene gas, hardwoods are more reactive than softwoods and that the optimal reaction temperature is about 50°. The reaction with diketene (Scheme 20.3) gave a weight gain of 35 per cent, when carried out at 52°C for 3 h.

Reactions on wood with more exotic compounds have been reviewed systematically in Hill's book [1] and since then no other relevant study has appeared to the best of our knowledge.

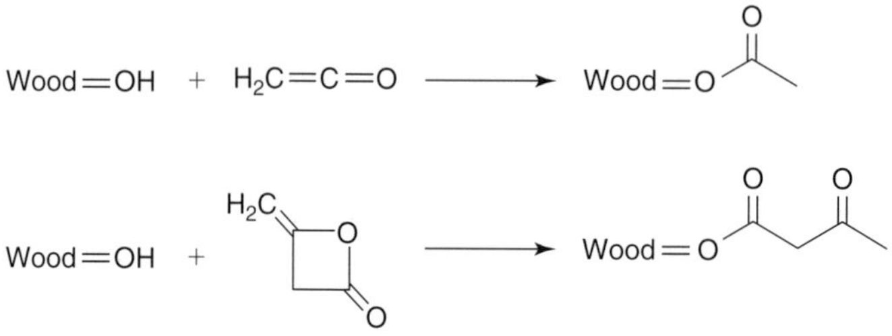

Wood=OH + $H_2C{=}C{=}O$ ⟶ Wood=O–C(=O)–CH_3

Wood=OH + diketene ⟶ Wood=O–C(=O)–CH_2–C(=O)–CH_3

Scheme 20.3

20.2.8 Wood-based composites

The use of wood fibres as reinforcing elements in macromolecular composite materials has recently gained considerable attention, as witnessed by the numerous reviews on the topic [38–42]. The main driving force related to this new strategy in composite materials is the fact that glass fibre based composites cannot be recycled at the end of their life cycle, because their burning to recover energy is accompanied by the formation of a glass residue which is particularly difficult to handle. This problem is obviously not encountered with wood fibre based counterparts, which are fully organic and therefore totally combustible. The additional advantages related to the use of wood fibres are the same as those already emphasized in Chapters 18 and 19 in the context of cellulose fibres, namely their low density, renewable character and ubiquitous availability at low cost and in a variety of forms.

Very recently, siloxane-crosslinked wood plastic composites based on the use of spruce and pine wood fibres and high density polyethylene in the presence of varying amounts of vinyltrimethoxysilane (**XIV**) have been reported [43].

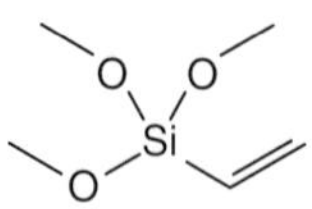

XIV

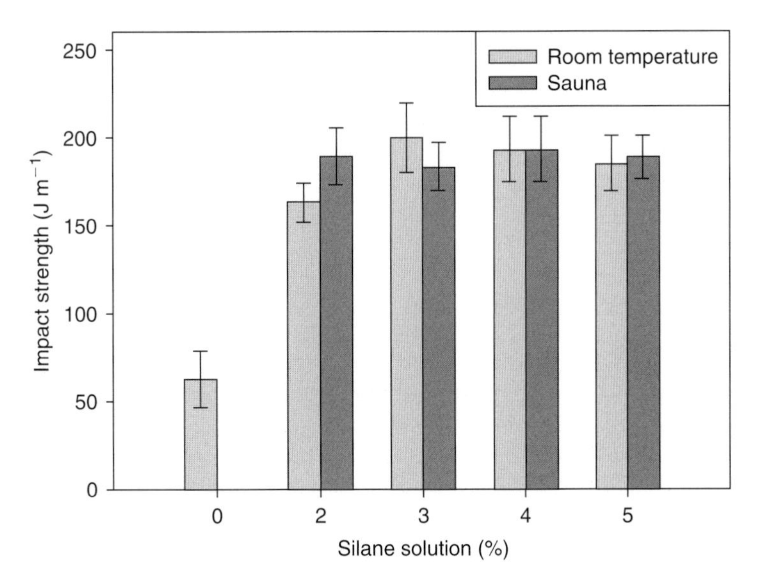

Figure 20.6 Average values of impact strength as a function of added **XIV**. (Reproduced by permission of Elsevier. Copyright 2006. Reprinted from Reference [43].)

These novel materials showed improved toughness, impact strength and creep properties compared with those of the siloxane-free counterpart, as shown in Fig. 20.6 for impact strength measurements.

The treatment of wood fibres with isocyanate-bearing molecules and their incorporation into polyethylene were also recently studied [44]. In particular, the use of poly-(diphenylmethane diisocyanate) (**XV**) increased both the modulus of rupture (MOR) and the modulus of elasticity (MOE) of the ensuing composites. The use of stearic anhydride (**XVI**) as a novel compatibilizer further improved both MOR and MOE and enhanced the water resistance of the composites.

OCN NCO NCO n

n = 0, 1, 2,...

XV

O O O 16 16

XVI

The incorporation of sisal fibres into composites containing wood particles has also attracted some attention [45]. This system involved an unsaturated polyester/styrene matrix and pine wood flour. A composite with 12 per cent (v/v) of wood particles displayed a 10-fold increase in the work of fracture upon the addition of 7 per cent (v/v) of sisal fibres.

Wood-based composites with a phenolic resin matrix have been transformed into silicon-infiltrated silicon carbide ceramics by carbonization at about 1 650°C [46]. The bending strength and the elastic modulus of these original ceramics were better than those of conventionally manufactured counterparts, whereas the fracture toughness was lower.

The performance of wood–polyethylene composites was found to improve by an appropriate compatibilization with different **MA**s and the additional incorporation of organo-clay particles [47] consisting of natural montmorillonite modified with quaternary ammonium salt. The thermal expansion coefficient and the heat of deflection values indicated that these materials displayed an improved interfacial adhesion.

The processing and processes of microcellular-foamed wood plastic composites have been the subject of a recent review [48].

20.3 CONCLUSIONS

The main emphasis of this chapter has to do with the shift from the traditional ways of treating woods involving the physical incorporation of stabilizers in order to enhance some of their physical properties but, more importantly, to improve their resistance to atmospheric and biological degradation through chemical modifications which eliminate the problem of leaching of toxic materials into the environment. The variety of these surface or bulk treatments published in the recent literature points to the relevance of this novel green approach and shows that very promising results can be obtained.

REFERENCES

1. Hill C.A.S., *Wood Modification: Chemical, Thermal and Other Processes*, John Wiley & Sons, Ltd, Chichester, 2006.
2. Lu J.Z., Wu Q., McNabb, Jr. H.S., *Wood Fibre Sci.*, **32**, 2000, 88.
3. Back E.L., Danielsson S., *Nord. Pulp Pap. Res. J.*, **2**, 1987, 53.
4. Mahlberg R., Niemi H.E.M., Denes F., Rowell R.M., *Int. J. Adhes. Adhes.*, **18**, 1998, 283.
5. Denes A.R., Young R.A., *Holzforschung*, **53**, 1999, 632.
6. Podgorski L., Chevet B., Onic L., Marlin A., *Int. J. Adhes. Adhes.*, **20**, 2000, 103.
7. Rehn P., Vlöl W., *Holz. Roh. Werkst.*, **61**, 2003, 145.
8. Ref. 1, Chapters 3 and 4.
9. Belgacem M.N., Gandini A., *Compos. Interface*, **12**, 2005, 41.
10. Thiebaud S., Borredon M.E., Baziard G., Senocq F., *Bioresource Technol.*, **59**, 1997, 103.
11. Thiebaud D., Borredon M.E., *Bioresource Technol.*, **52**, 1995, 169.
12. Cetin N.S., Hill C.A.S., *J. Wood Chem. Technol.*, **19**, 1999, 247.
13. Hill C.A.S., Cetin N.S., *Int. J. Adhes. Adhes.*, **20**, 2000, 71.
14. Bach S., Belgacem M.N., Gandini A., *Holzforschung*, **59**, 2005, 389.
15. Gandini A., Belgacem M.N., *Macromol. Symp.*, **221**, 2005, 257.
16. Gandini A., Botaro V.R., Zeno E., Bach S., *Polym. Intern.*, **50**, 2000, 7.
17. Gandini A., Belgacem M.N., *ACS Symp. Ser.*, **954**, 2007, 93.
18. Bach S., Le Sage L., Belgacem M.N., Gandini A., unpublished results. Bach, S., Doctorate Thesis, Grenoble National Polytechnic Institute, 2000.
19. Iwamoto Y., Ito T., *J. Wood Sci.*, **51**, 2005, 595.
20. Iwamoto Y., Ito T., Minato K., *J. Wood Sci.*, **51**, 2005, 601.
21. Zhang C., Li K., Simonsen J., *J. Appl. Polym. Sci.*, **89**, 2003, 1078.
22. Geng Y., Li K., Simonsen J., *J. Appl. Polym. Sci.*, **99**, 2006, 712.
23. Pandey K.K., *Polym. Degrad. Stabil.*, **90**, 2005, 9.
24. Pandey K.K., Chandrashekar N., *J. Appl. Polym. Sci.*, **99**, 2006, 2367.
25. George B., Suttie E., Merlin A., Deglise X., *Polym. Degrad. Stabil.*, **88**, 2005, 268.
26. Devi R.R., Maji T.K., *Polym. Composite*, **28**, 2007, 1.
27. Clermont L.P., Bender F., *Forest Prod. J.*, **7**, 1957, 167.
28. Sèbe G., Tingaut P., Safou-Tchiama R., Pétraut M., Grelier S., De Jéso B., *Holzforschung*, **58**, 2004, 511.
29. Tingaut P., Weigenand O., Militz H., De Jéso B., Sèbe G., *Holzforschung*, **59**, 2005, 397.
30. Tingaut P., Weigenand O., Mai C., Militz H., Sèbe G., *Holzforschung*, **60**, 2006, 271.
31. Lande S., Eikenes M., Westin M., *Scand. J. Forest Res.*, **19**, 2004, 14.
32. Lande S., Westin M., Schneider M., *Scand. J. Forest Res.*, **19**, 2004, 22.
33. Choura M., Belgacem M.N., Gandini A., *Macromolecules*, **29**, 1996, 3839.
34. Choura M., Belgacem M.N., Gandini A., *Macromol. Symp.*, **122**, 1997, 263.
35. Gandini A., Belgacem M.N., *Prog. Polym. Sci.*, **22**, 1997, 1203.
36. Baysal E., Ozaki S.K., Yalinkilic M.K., *Wood Sci. Technol.*, **38**, 2004, 405.
37. Morozovs A., Aboltins A., Zoldners J., Akerfelds I., in *Proceedings of the first European Conference on Wood Modification*, Eds.: Van Acker J. and Hill C.A.S., Ghent, Belgium, 2003, pp. 351–362.
38. Eichhorn S.J., Baillie C.A., Zafeiropoulos N., Mwaikambo L.Y., Ansell M.P., Dufresne A., Entwistle K.M., Herraro-Franco P.J., Escamilla G.C., Groom L., Hughes M., Hill C., Rials T.G., Wild P.M., *J. Mater. Sci.*, **36**, 2001, 2107.
39. Eco-Composites: A special issue of *Composite Sci. Technol.* **63**, 2004, 112, collection of 14 publications all dedicated to cellulose and wood fibres-based composite materials.
40. Mohanty A.K., Misra M., Hinrichsen G., *Macromol. Mater. Eng.*, **276/277**, 2000, 1.

41. Gassan J., Bledzki A.K., *Prog. Polym. Sci.*, **24**, 1999, 221.
42. Lu J.Z., Wu Q., McNabb, Jr. H.S., *Wood Fibre Sci.*, **32**, 2000, 88.
43. Bentsson M., Oksman K., *Comp. Sci. Technol.*, **66**, 2006, 2177.
44. Geng Y., Li K., Simonsen J., *J. Adhes. Sci. Technol.*, **19**, 2005, 987.
45. Nuñez A.J., Aranguren M.I., Berglund L.A., *J. Appl. Polym. Sci.*, **101**, 2006, 1982.
46. Hofenauer A., Treusch O., Tröger F., Wegener G., Fromm J., *Holz. Roh. Werkst.*, **64**, 2006, 165.
47. Zhong Y., Poloso T., Hetzer M., De Kee D., *Polym. Eng. Sci.*, **47**, 2007, 797.
48. Faruk O., Bledzki A.K., Matuana L.M., *Macromol. Mater. Eng.*, **292**, 2007, 113.

– 21 –

Polylactic Acid: Synthesis, Properties and Applications

L. Avérous

ABSTRACT

Polylactic acid (PLA) is at present one of the most promising biodegradable polymers (biopolymers) and has been the subject of abundant literature over the last decade. PLA can be processed with a large number of techniques and is commercially available (large-scale production) in a wide range of grades. It is relatively cheap and has some remarkable properties, which make it suitable for different applications. This chapter deals with the different syntheses to produce this biopolymer, its diverse properties and various applications. Its biodegradability is adapted to short-term packaging, and its biocompatibility in contact with living tissues is exploited for biomedical applications (implants, sutures, drug encapsulation …).

Keywords

Polylactic acid, Biopolymer, Biodegradable, Properties, Synthesis, Process, Application, Packaging, Biomedical

21.1 INTRODUCTION

Tailoring new materials within a perspective of eco-design or sustainable development is a philosophy that is applied to more and more materials. It is the reason why material components such as biodegradable polymers can be considered as 'interesting' – environmentally safe – alternatives. Besides, ecological concerns have resulted in a resumed interest in renewable resources-based products.

Figure 21.1 shows an attempt to classify the biodegradable polymers into two groups and four different families. The main groups are (i) the agro-polymers (polysaccharides, proteins, etc.) and (ii) the biopolyesters (biodegradable polyesters) such as polylactic acid (PLA), polyhydroxyalkanoate (PHA), aromatic and aliphatic copolyesters [1]. Biodegradable polymers show a large range of properties and can now compete with non-biodegradable thermoplastics in different fields (packaging, textile, biomedical, etc.). Among these biopolyesters, PLA is at present one of the most promising biopolymer. PLA has been the subject of an abundant literature with several reviews and book chapters [2–8], mainly during the last decade. PLA can be processed with a large number of techniques. PLA is commercially and largely available (large-scale production) in a wide range of grades. It has a reasonable price and some remarkable properties to fulfil different applications. For instance, the PLA production capacity of Cargill (USA) in 2006 was 140kT per year at 2–5 Euros per kg [9]. Other companies, such as Mitsui Chemical (Lacea-Japan), Treofan (Netherland), Galactic (Belgium), Shimadzu Corporation (Japan), produce smaller quantities. Some of them are only focused on the biomedical market like Boeringher Ingelheim (Germany), Purac (Netherland) or Phusis (France), because the constraints of this market are very specific. However, according to different sources, PLA consumption in 2006 was only about 60000 tons per year and, at present, only ~30 per cent of lactic acid is used for PLA production. Thus, this biopolymer presents a high potential for development.

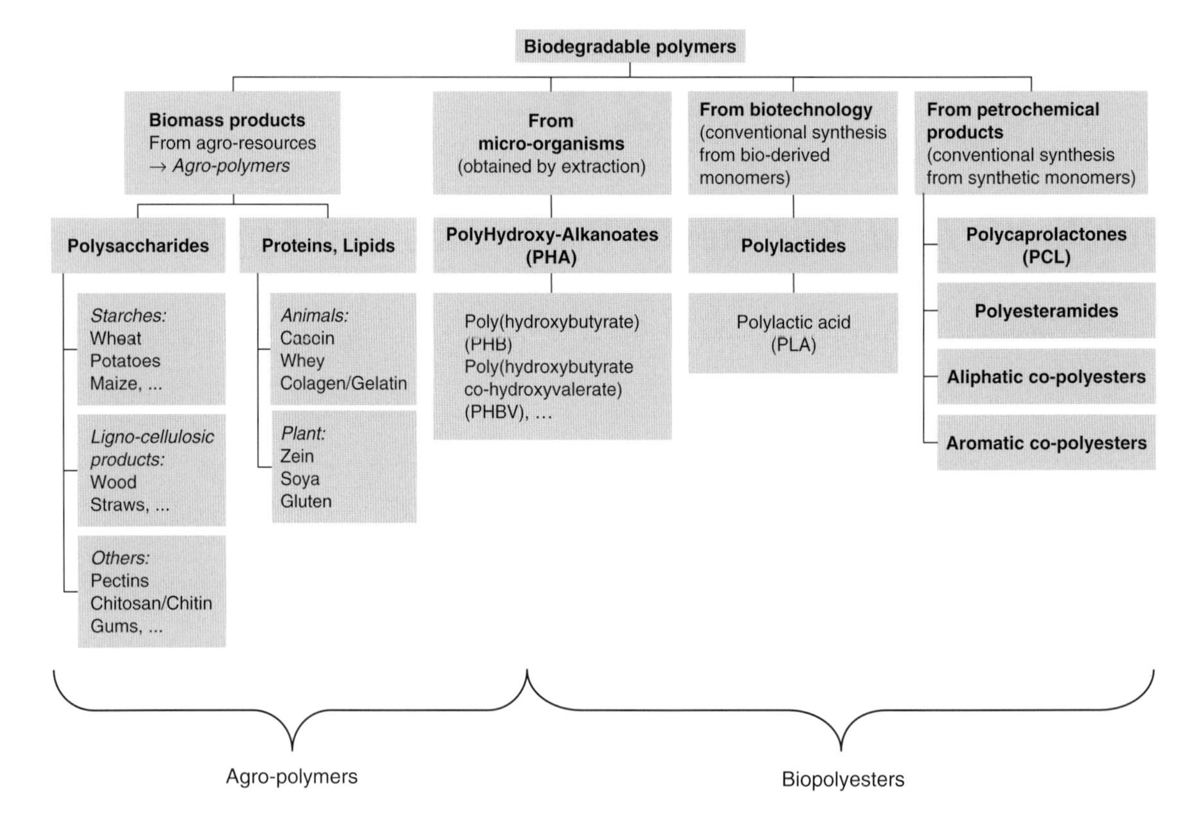

Figure 21.1 Classification of the biodegradable polymers. (Adapted from Reference [1].)

PLA belongs to the family of aliphatic polyesters commonly made from α-hydroxy acids, which also includes, for example, polyglycolic acid (PGA). It is one of the few polymers in which the stereochemical structure can easily be modified by polymerizing a controlled mixture ofland D isomers (Fig. 21.2) to yield high molecular weight and amorphous or semi-crystalline polymers. Properties can be both modified through the variation of isomers (l/d ratio) and the homo and (d,l)copolymers relative contents. Besides, PLA can be tailored by formulation involving adding plasticizers, other biopolymers, fillers, etc.

PLA is considered both as biodegradable (*e.g.* adapted for short-term packaging) and as biocompatible in contact with living tissues (*e.g.* for biomedical applications such as implants, sutures, drug encapsulation, etc.). PLA can be degraded by abiotic degradation (*i.e.* simple hydrolysis of the ester bond without requiring the presence of enzymes to catalyze it). During the biodegradation process, and only in a second step, the enzymes degrade the residual oligomers till final mineralization (biotic degradation).

As long as the basic monomers (lactic acid) are produced from renewable resources (carbohydrates) by fermentation, PLA complies with the rising worldwide concept of sustainable development and is classified as an environmentally friendly material.

21.2 SYNTHESIS OF PLA

The synthesis of PLA is a multistep process which starts from the production of lactic acid and ends with its polymerization [2–4, 6–7]. An intermediate step is often the formation of the lactide. Figure 21.2 shows that the synthesis of PLA can follow three main routes. Lactic acid is condensation polymerized to yield a low molecular weight, brittle polymer, which, for the most part, is unusable, unless external coupling agents are employed to increase its chains length. Second route is the azeotropic dehydrative condensation of lactic acid. It can yield high

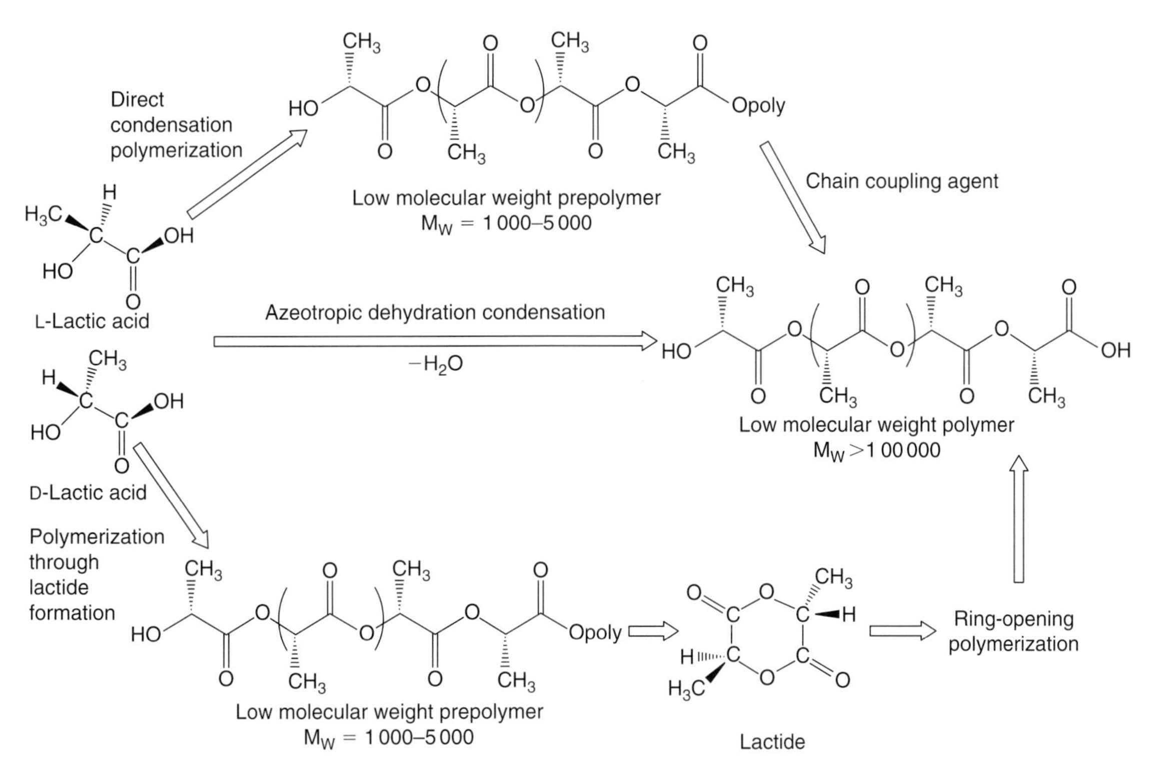

Figure 21.2 Synthesis methods for obtaining high molecular weight. (Adapted from Reference [3].)

molecular weight PLA without the use of chain extenders or special adjuvents [3]. The third and main process is ring-opening polymerization (ROP) of lactide to obtain high molecular weight PLA, patented by Cargill (US) in 1992 [8–9]. Finally, lactic acid units can be part of a more complex macromolecular architecture as in copolymers.

21.2.1 Precursors

21.2.1.1 Lactic acid

Lactic acid is a compound that plays a key role in several biochemical processes. For instance, lactate is constantly produced and eliminated during normal metabolism and physical exercise. Lactic acid has been produced on an industrial scale since the end of the nineteenth century and is mainly used in the food industry to act, for example, as an acidity regulator, but also in cosmetics, pharmaceuticals and animal feed. It is, additionally, the monomeric precursor of PLA. It can be obtained either by carbohydrate fermentation or by common chemical synthesis. Also known as 'milk acid', lactic acid is the simplest hydroxyl acid with an asymmetric carbon atom and two optically active configurations, namely the L and D isomers (Fig. 21.2), which can be produced in bacterial systems, whereas mammalian organisms only produce the L isomer, which is easily assimilated during metabolism.

Lactic acid is mainly prepared in large quantities (around 200 kT per year) by the bacterial fermentation of carbohydrates. These fermentation processes can be classified according to the type of bacteria used: (i) the hetero-fermentative method, which produces less than 1.8 mol of lactic acid per mole of hexose, with other metabolites in significant quantities, such as acetic acid, ethanol, glycerol, mannitol and carbon dioxide; (ii) the homo-fermentative method, which leads to greater yields of lactic acid and lower levels of by-products, and is mainly used in industrial processes [3]. The conversion yield from glucose to lactic acid is more than 90 per cent.

The majority of the fermentation processes use species of *Lactobacilli* which give high yields of lactic acid. Some organisms predominantly produce the L isomer, such as *Lactobacilli amylophilus*, *L. bavaricus*, *L. casei* and

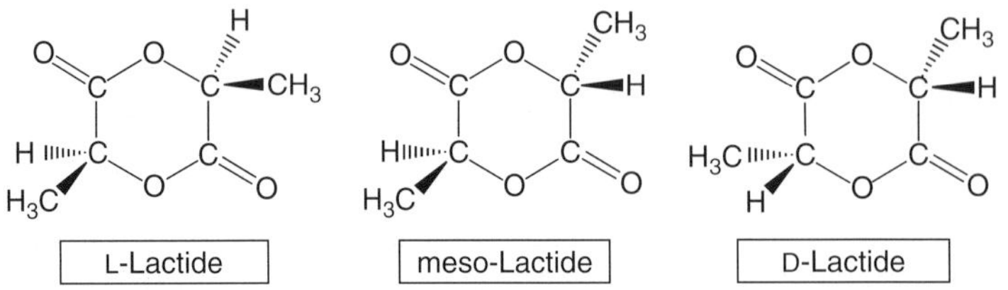

L-Lactide meso-Lactide D-Lactide

Figure 21.3 Chemical structures of L-, meso- and D-lactides.

L. maltaromicus, whereas, *L. delbrueckii*, *L. jensenii* or *L. acidophilus* produce the D isomer or a mixture of L and D [3.4]. These different bacteria are homo-fermentative. In general, the sources of basic sugars are glucose and maltose from corn or potato, sucrose from cane or beet sugar, etc. In addition to carbohydrates, other products, such as B vitamins, amino acids and different nucleotides, are formed. The processing conditions are an acid pH close to 6, a temperature around 40°C and a low oxygen concentration. The major method of separation consists in adding $CaCO_3$, $Ca(OH)_2$, $Mg(OH)_2$, NaOH, or NH_4OH to neutralize the fermentation acid and to give soluble lactate solutions, which are filtered to remove both the cells (biomass) and the insoluble products. The product is then evaporated, crystallized, and acidified with sulphuric acid to obtain the crude lactic acid. If the lactic acid is used in pharmaceutical and food applications, it is further purified to remove the residual by-products. If it is to be polymerized, it is purified by separation techniques including ultra-filtration, nano-filtration, electro-dialysis and ion-exchange processes.

21.2.1.2 Lactide

Figure 21.3 shows the different stereoforms of lactide. The cyclic dimer of lactic acid combines two of its molecules and gives rise to L-lactide or LL-lactide, D-lactide or DD-lactide, and meso-lactide or LD-lactide (a molecule of L-lactic acid associated with another one of D-lactic acid). A mixture of L- and D-lactides is a racemic lactide (rac-lactide). Lactide is usually obtained by the depolymerization of low molecular weight PLA under reduced pressure to give a mixture of L-, D- and meso-lactides. The different percentages of the lactide isomers formed depend on the lactic acid isomer feedstock, temperature and the catalyst's nature and content [3,4]. A key point in most of the processes is the separation between each stereoisomer to control the final PLA structure (*e.g.* by vacuum distillation) which is based on the boiling point differences between the meso- and the L- or D-lactide.

21.2.2 PLA polymerization

21.2.2.1 Lactic acid condensation and coupling

The condensation polymerization is the least expensive route, but it is difficult to obtain high molecular weights by this method. The use of coupling or esterification-promoting agents is required to increase the chains length [3, 4], but at the expense of an increase in both cost and complexity (multistep process). The role of chain coupling agents is to react with either the hydroxyl (OH) or the carboxyl end-groups of the PLA [3, 4, 7] thus giving telechelic polymers [10]. The nature of the chain end-groups should be fully controlled [2, 3]. The use of chain-extending agents brings some advantages, because reactions involving small amounts of them are economical and can be carried out in the melt without the need of separating the different process steps. The tunability to design copolymers with various functional groups is also greatly expanded. The disadvantages are that the final polymer may contain unreacted chain-extending agents, oligomers and residual metallic impurities from the catalyst. Moreover, some extending agents could be associated with a lack of biodegradability [2]. Examples of chain-extending agents are anhydrides, epoxides and isocyanates [11]. Similar products are used to develop compatibilization for PLA-based blends. The disadvantages of using isocyanates as chain extenders are their (eco)toxicity [3].

The advantages of esterification-promoting adjuvents are that the final product is highly purified and free from residual catalysts and/or oligomers. The disadvantages are higher costs due to the number of steps involved and the additional purification of the residual by-products [3], since these additives produce by-products that must be neutralized or removed.

21.2.2.2 Azeotropic dehydration and condensation

The azeotropic condensation polymerization is a method used to obtain high chain lengths without the use of chain extenders or adjuvents and their associated drawbacks. Mitsui Chemicals (Japan) has commercialized a process wherein lactic acid and a catalyst are azeotropically dehydrated in a refluxing, high boiling, aprotic solvent under reduced pressures to obtain high molecular weight PLA (Mw $\geq$ 300000) [2, 3]. A general procedure consists in the reduced pressure distillation of lactic acid for 2–3 h at 130°C to remove most of the condensation water. The catalyst and diphenyl ether are then added and a tube packed with molecular sieves is attached to the reaction vessel. The refluxing solvent is returned to the vessel by way of the molecular sieves during 30–40 h at 130°C. Finally, the ensuing PLA is purified [12].

This polymerization gives considerable catalyst residues because of its high concentration needed to reach an adequate reaction rate. This can cause many drawbacks during processing, such as degradation and hydrolysis. For most biomedical applications, the catalyst toxicity is a highly sensitive issue. The catalyst can be deactivated by the adding of phosphoric acid or can be precipitated and filtered out by the addition of strong acids such as sulphuric acid. Thus, residual catalyst contents can be reduced to some ppm [3].

21.2.2.3 ROP of lactide

The lactide method is the only method for producing pure high molecular weight PLA (Mw $\geq$ 100 000) [4, 6, 7, 13]. The ROP of lactide was first demonstrated by Carothers in 1932 [14], but high molecular weights were not obtained until improved lactide purification techniques were developed by DuPont in 1954 [2]. This polymerization has been successfully carried out calling upon various methods, such as solution, bulk, melt or suspension process. The mechanism involved in ROP can be ionic (anionic or cationic) or coordination–insertion, depending on the catalytic system [4, 6, 7, 13]. The role of the racemization and the extent of transesterification in the homo or copolymerization, are also decisive for the enantiomeric purity and chain architecture of the resulting macromolecules.

It has been found that trifluoromethane sulphonic acid and its methyl ester are the only cationic initiators known to polymerize lactide [15], and the mechanism of this process has been outlined in different papers [2, 3, 15].

Lactide anionic polymerizations proceed by the nucleophilic reaction of the anion with the carbonyl group and the subsequent acyl–oxygen bond cleavage, which produces an alkoxide end-group, which continues to propagate. The general mechanism for this anionic polymerization has been discussed in various publications [2, 3, 15, 16]. Some authors [16] have shown that the use of alkoxides, such as potassium methoxide, can yield well-defined polymers with negligible racemization.

Both the anionic and cationic ROPs are usually carried out in highly purified solvents, and although they show a high reactivity, they are susceptible to give racemization, transesterification and high impurity levels. For industrial and large commercial use, it is preferable to do bulk and melt polymerization with low levels of non-toxic catalysts. The use of less-reactive metal carboxylates, oxides and alkoxides has been extensively studied in this context, and it has been found that high molecular weight PLA can readily be obtained in the presence of transition metal compounds of tin [6, 7, 13], zinc [17, 18], iron [19] and aluminium [20], among others. A systematic investigation has led to the wide use of tin compounds, namely tin(II) bis-2-ethylhexanoic acid (stannous octoate) as a catalyst in PLA synthesis. This is mainly due to its high catalytic efficiency, low toxicity, food and drug contact approval and ability to give high molecular weights with low racemization [15]. The mechanisms of the polymerization with stannous octoate have been studied in detail, and it is now widely accepted that this ROP is actually initiated from compounds containing hydroxyl groups, such as water and alcohols, which are either present in the lactide feed or can be added upon demand. Figure 21.4 shows that the global mechanism is of the 'coordination–insertion' type [21], occurring in two steps: First, a complex between monomer and initiator is formed followed by a rearrangement of the covalent bonds; then, the monomer is inserted within the oxygen–metal bond of the initiator, and its cyclic structure is thus opened through the cleavage of the acyl–oxygen link, thus the metal is incorporated with an alkoxide bond into the propagating chain. It was found that the polymerization yield and the transesterification effect are affected by different parameters, such as the polymerization temperature and time, the monomer/catalyst ratio and the type of catalyst. The interaction between the time and temperature is very significant in terms of limiting the degradation reactions, which affect the molecular weight and the reaction kinetics [22]. It has also been shown that the chain length is directly controlled by the amount of OH impurities [23].

To make an economically viable PLA, Jacobsen *et al.* [21] developed a continuous one-stage process based on reactive extrusion with a twin-screw extruder. This technique requires that the bulk polymerization be close to completion within a very short time (5–7 min), which is predetermined by the residence time in the extruder.

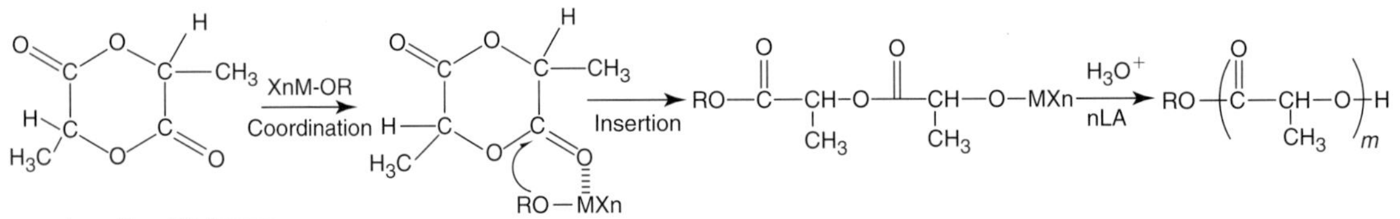

where R = Alkyl group

Figure 21.4 Coordination–insertion polymerization mechanism.

These authors showed that the addition of an equimolar content of a Lewis base, particularly triphenyl-phosphine, to stannous octoate increased the lactide polymerization rate.

21.2.3 Copolymers based on lactic acid units

A large number of macromolecular architectures of copolymers based on lactic acid have been investigated [7, 13]. Most of them are biodegradable or/and biocompatible. These copolymers can be prepared by using units containing a specific functionalized structure, thus giving rise to complex structure with unique properties. Examples of these materials are branched polyesters and graft copolymers (star, hyper-branched polymers) which involve different macromolecular architectures associated with novel materials properties and applications.

21.2.3.1 Ring-opening copolymerization

Several heterocyclic monomers can be used as comonomers with lactic acid in ring-opening copolymerizations, the most commonly used being glycolide (GA) for biomedical applications [24], caprolactone (CL) and valerolactone. The comonomer units can be inserted randomly or in block sequences.

21.2.3.2 Modification by high energy radiations and peroxides

Radical reactions applied to PLA to modify its structure have been generated by peroxides or high energy radiation [7]. Branching has been suggested to be the dominant structural change in poly(L-lactide) (PLLA) with peroxide concentrations in the range of 0.1–0.25 wt% and crosslinking above 0.25 wt% [7]. The peroxide melt-reaction with PLA has been found to cause strong modifications of the original PLA properties. A similar approach was recently developed with starch-based blends without any major improvement in their mechanical properties [25]. Irradiation of PLA causes mainly chain-scissions or crosslinking reactions, depending on the radiation intensity [26].

21.2.3.3 Graft copolymerization

Graft copolymers are often used as compatibilizers to improve the interfacial properties of blends or multiphase systems. Grafting reactions on a trunk polymer can be induced chemically, by plasma discharge, or by radiation (UV, X-rays or accelerated electrons), the latter approach giving purer products at high conversions. Plasma-induced grafting is performed by introducing an organic vapour into a plasma of inorganic gases to modify the surface properties of a substrate. Depending on the penetration depth of the irradiation, grafting can be performed either at the surface, or both on the skin and in the bulk [7].

The chemical modification of lactic acid-based polymers by graft copolymerization has been reported for the homopolymer of L-lactide and for copolymers with different L-lactide/CL contents [7, 13]. Carbohydrate polymers (*e.g.* amylose) can be modified by grafting lactic acid chains on their OH groups. A recent study [25] showed the interest of such a copolymer as a compatibilizer to improve the properties of starch/PLA blends to a better extent than the addition of peroxides or coupling agents (*e.g.* di-isocyanate) into the melt blend during the processing. Figure 21.5 shows the different steps involved in this grafting operation. After amylose purification to eliminate residual butanol and water, amylose-graft-PLA is obtained by the ROP of purified lactide with tin(II) bis (2-ethylhexanoate) in toluene at 100°C for 20 h.

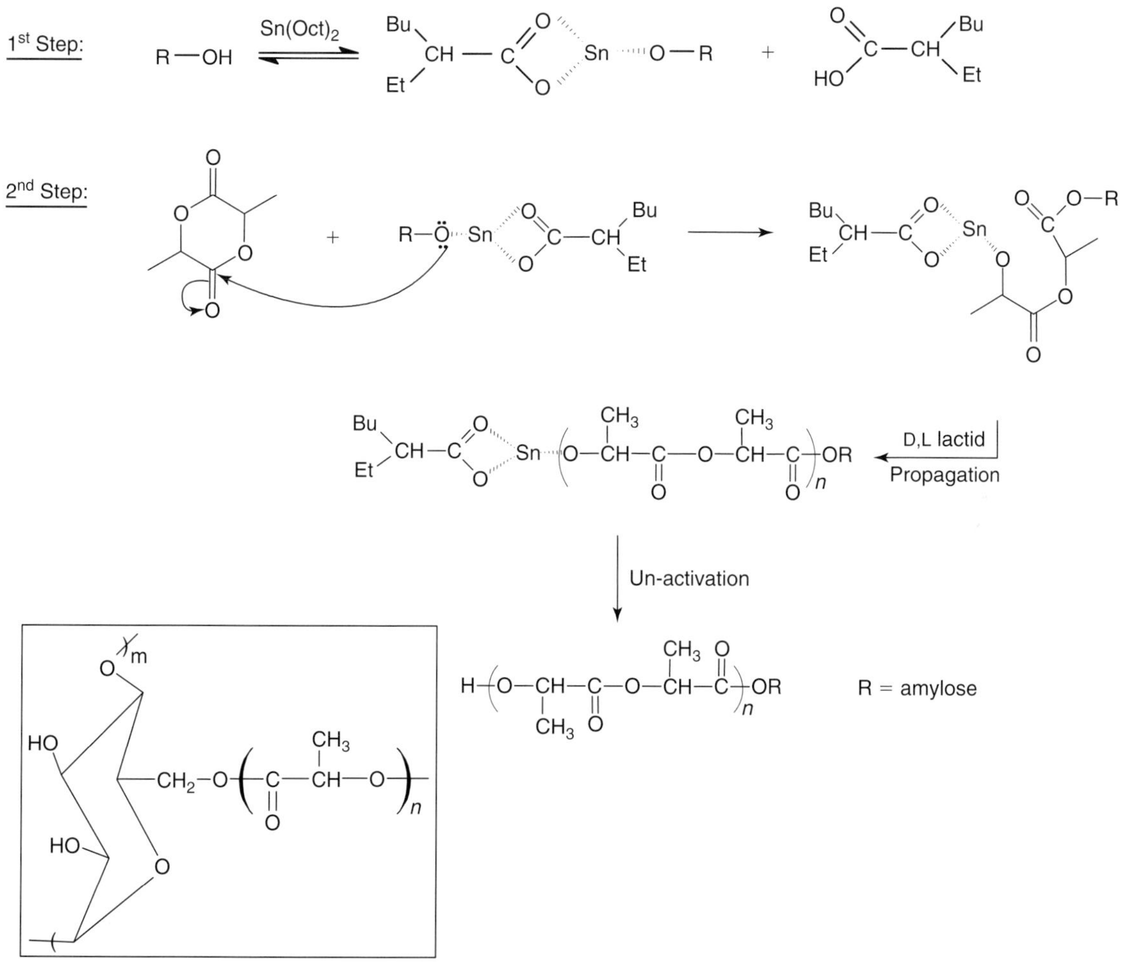

1st Step:

2nd Step:

D,L lactid

Propagation

Un-activation

R = amylose

Figure 21.5 Mechanism of the ROP synthesis of amylose-graft PLA.

21.3 PLA PROPERTIES

21.3.1 Crystallinity and thermal properties

The properties of PLA, as indeed those of other polymers, depend on its molecular characteristics, as well as on the presence of ordered structures, such as crystalline thickness, crystallinity, spherulite size, morphology and degree of chain orientation. The physical properties of polylactide are related to the enantiomeric purity of the lactic acid stereo-copolymers. Homo-PLA is a linear macromolecule with a molecular architecture that is determined by its stereochemical composition. PLA can be produced in a totally amorphous or with up to 40 per cent crystalline. PLA resins containing more than 93 per cent of L-lactic acid are semi-crystalline, but, when it contains 50–93 per cent of it, it is entirely amorphous. Both meso- and D-lactides induce twists in the very regular PLLA architecture. Macromolecular imperfections are responsible for the decrease in both the rate and the extent of PLLA crystallization. In practise, most PLAs are made up of L-and D,L-lactide copolymers, since the reaction media often contain some meso-lactide impurities.

Table 21.1 gives the details of the different crystalline structures for neat PLA. Depending on the preparation conditions, PLLA crystallizes in different forms. The α-form exhibits a well-defined diffraction pattern [27]. This structure, with a melting temperature of 185°C, is more stable than its β-counterpart, which melts at 175°C [27].

Table 21.1

PLA crystalline structures. Unit cell parameters for non-blended PLLA and stereocomplex crystals.
(Adapted from Reference [8])

		Space group	Chain orientation	Number helices/ unit cell	Helical conformation	a (nm)	b (nm)	c (nm)	α (degrees)	β (degrees)	γ (degrees)
PLLA form	α	Pseudo-orthorhombic	–	2	103	1.07	0.645	2.78	90	90	90
PLLA form	α	Pseudo-orthorhombic	–	2	103	1.07	0.62	2.88	90	90	90
PLLA form	α	Orthorhombic	Parallel	2	103	1.05	0.61	–	90	90	90
PLLA form	β	Ort horhombic	–	6	31	1.031	1.821	0.90	90	90	90
PLLA form	β	Trigonal	Random up-down	3	31	1.052	1.052	0.88	90	90	120
PLLA form	γ	Orthorombic	Antiparallel	2	31	0.995	0.625	0.88	90	90	90
Stereo complex		Triclinic	Parallel	2	31	0.916	0.916	0.870	109.2	109.2	109.8

The latter form can be prepared at a high draw ratio and a high drawing temperature [28]. The γ-form is formed by epitaxial crystallization [29]. It has been observed that a blend with equivalent poly(L-lactide) PLLA and poly(D-lactide) PDLA contents gives stereo-complexation (racemic crystallite) of both polymers. This stereocomplex has higher mechanical properties than those of both PLAs, and a higher melting temperature of 230°C. The literature reports different density data [4] for PLA, with most values for the crystalline polymer around 1.29 compared with 1.25 for the amorphous material.

The crystallization kinetics of PLA have been extensively studied and found to be rather slow, as in the case of poly(ethylene terephthalate) PET. The rate of crystallization increases with a decrease in the molecular weight and is strongly dependent on the (co)polymer composition [4]. PLLA can crystallize in the presence of D-lactide [30], however, as the structure becomes more disordered, the rate of crystallization decreases. It has been reported that the crystallization rate is essentially determined by the decrease in the melting point of the different copolymers. PDLA/PLLA stereocomplexes are very efficient nucleating agents for PLLA, with increases in both the crystallization rate and the crystallinity, the latter of up to 60 per cent [31]. Quenching decreases the crystallization time [30]. As PET, PLA can be oriented by processing and chain orientation increases the mechanical strength of the polymer. If orientation is performed at low temperature, the resulting PLLA has a higher modulus without any significant increase in crystallinity. To determine the crystallinity levels by differential scanning calorimetry (DSC), the value most often referred to in the literature concerning the PLA melt enthalpy at 100 per cent crystallinity, is $93\,J\,g^{-1}$ [7, 8, 32] The crystallization of the thermally crystallizable, but amorphous PLA, can be initiated by annealing it at temperatures between 75°C and the melting point. Annealing crystallizable PLA copolymers often produces two melting peaks [32] and different hypotheses have been put forward to explain this feature. Yasuniwa *et al.* [33] found a double melting point in PLLA polymers and attributed them to slow rates of crystallization and recrystallization.

The typical PLA glass transition temperature (T_g) ranges from 50°C to 80°C, whereas its melting temperature ranges from 130°C to 180°C. For instance, enantiomerically pure PLA is a semi-crystalline polymer with a T_g of 55°C and a T_m of 180°C. For semi-crystalline PLA, the T_m is a function of the different processing parameters and the initial PLA structure. According to Ikada and Tsuji [30], T_m increases with increasing molecular weight (Mw) to an asymptotic value, but the actual crystallinity decreases with increasing Mw. T_m, moreover, decreases with the presence of meso-lactide units in its structure [4]. Both, the degree of crystallinity and the melting temperature of PLA-based materials can be reduced by random copolymerization with different comonomers (*e.g.* GA, CL or valerolactone).

The T_g of PLA is also determined by the proportion of the different types of lactide in its macromolecular chain.

21.3.2 Surface energy

Surface energy is critically important to many processes (printing, multilayering, etc.) and it influences the interfacial tension. The surface energy of a PLA made up of 92 per cent L-lactide and 8 per cent meson-lactide was found to be $49\,mJ\,m^{-2}$, with dispersive and polar components of 37 and $11\,mJ\,m^{-2}$, respectively [34], which suggests a relatively hydrophobic structure compared with that of other biopolyesters.

21.3.3 Solubility

A good solvent for PLA and for most of the corresponding copolymers is chloroform. Other solvents are chlorinated or fluorinated organic compounds, dioxane, dioxolane and furan. Poly(rac-lactide) and poly(meso-lactide) are soluble in many other organic solvents like acetone, pyridine, ethyl lactate, tetrahydrofuran, xylene, ethyl acetate, dimethylformamide, methyl ethyl ketone. Among non-solvents, the most relative compounds are water, alcohols (*e.g.* methanol, and ethanol) and alkanes (*e.g.* hexane and heptane) [7].

21.3.4 Barrier properties

Because PLA finds a lot of applications in food packaging, its barrier properties (mainly to carbon dioxide, oxygen and water vapour) have been largely investigated [4]. The CO_2 permeability coefficients for PLA polymers are lower than those reported for crystalline polystyrene at 25°C and 0 per cent relative humidity (RH) and higher than

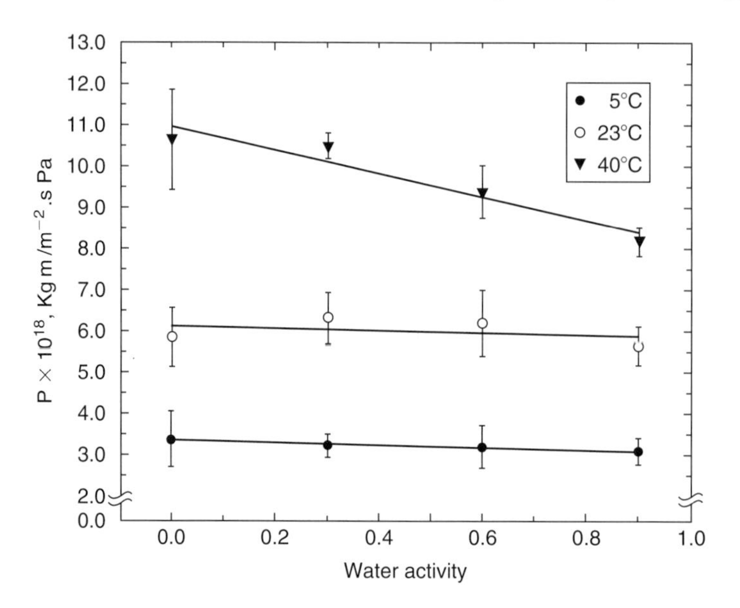

Figure 21.6 Oxygen permeability versus water activity at different temperatures, for poly(98 per cent L-lactide) films. (Source: Reference [4].)

those for PET. Since diffusion takes place through the amorphous regions of a polymer, an increase in the extent of crystallization will inevitably result in a decrease in permeability. Figure 21.6 shows the oxygen permeability for poly(98 per cent L-lactide) films as a function of the water activity. A significant increase in the oxygen permeability coefficient is shown as the temperature is increased, but, its decrease with water activity at temperatures close to T_g and its stabilization at temperatures well below T_g, are clearly visible. PET and PLA are both hydrophobic and the corresponding films absorb very low amounts of water, showing similar barrier properties, as indicated by the values of their water vapour permeability coefficient determined from 10°C to 37.8°C in the range of 40–90 per cent RH. Auras *et al.* [4] have shown that the permeability for 98 per cent L-lactide polymers is almost constant over the range studied, despite PLA being a rather polar polymer [4].

21.3.5 Mechanical properties

21.3.5.1 Solid state

The mechanical properties of PLA can vary to a large extent, ranging from soft and elastic materials to stiff and high strength materials, according to different parameters, such as crystallinity, polymer structure and molecular weight, material formulation (plasticizers, blend, composites, etc.) and processing (*e.g.* orientation). For instance, commercial PLA, such as poly(92 per cent L-lactide, 8 per cent meso-lactide), has a modulus of 2.1 GPa and an elongation at break of 9 per cent. After plasticization, its Young's modulus decreases to 0.7 MPa and the elongation at break rises to 200 per cent, with a corresponding T_g shift from 58°C to 18°C [32]. This example indicates that mechanical properties can be readily tuned to satisfy different applications.

The mechanical properties of PLA-related polymers were recently reviewed by Sodergard and Stolt [7], who showed, among other features, that the PLLA fibre modulus can be increased from 7–9 GPa to 10–16 GPa by going from melt to solution spinning. The mechanical behaviour can also be modified by preparing suitable copolymers, as in the case of the use of CL, which, with its soft segments, induces a decrease in modulus and an increase in the elongation at break, respectively.

21.3.5.2 Molten behaviour

For processing and for the corresponding applications, the knowledge of PLA melt rheology is of particular interest. A power law equation has been applied successfully by, for example, Schwach and Averous [34]. The pseudoplastic

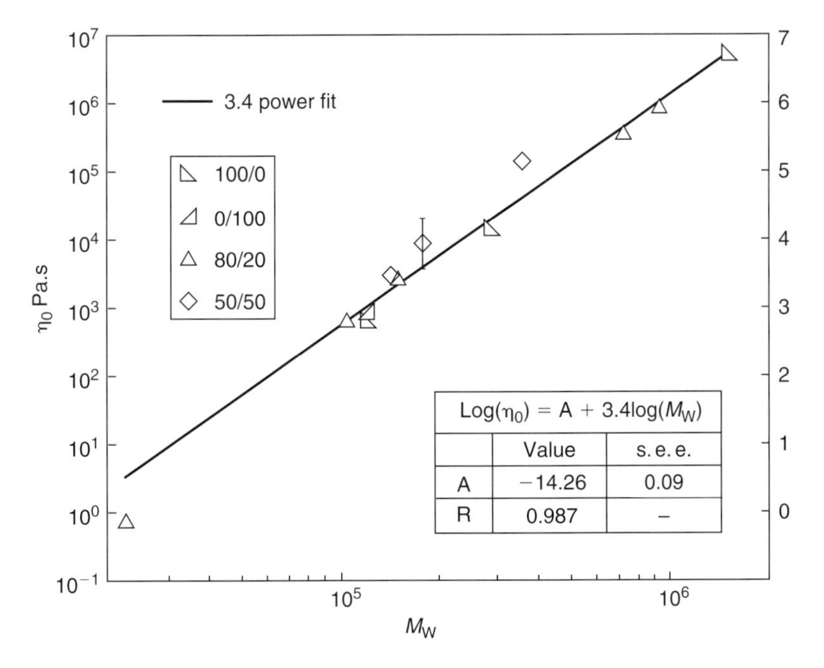

Figure 21.7 Zero-shear viscosity versus molecular weight for different L/D ratios (%). (Adapted from Reference [35].)

index is in the range 0.2–0.3, depending on the PLA structure. For instance, poly(92 per cent L-lactide, 8 per cent meso-lactide) displays a pseudoplastic index of 0.23. Figure 21.7, based on data published by Dorgan *et al.* [35], shows the evolution of the zero-shear viscosity versus molecular weight (Mw) for a wide range of L/D ratios (%), the latter parameter having virtually no effect. Static and dynamic characterizations have shown that the molecular weight between entanglements is around 10^4. Some other studies suggested that chain branching and molecular weight distribution have a significant effect on the melt viscosity of PLA [5].

21.4 DEGRADATION

21.4.1 Abiotic degradation

The main abiotic phenomena involve thermal and hydrolysis degradations during the life cycle of the material.

21.4.1.1 Thermal degradation

The thermal stability of biopolyesters is not significantly high, a fact that inevitably limits their range of applications. The PLA decomposition temperature is lies between 230°C and 260°C. Gupta and Deshmukh [36] concluded that the carbonyl carbon–oxygen linkage is the most likely bond to split under isothermal heating, as suggested by the fact that a significantly larger amount of carboxylic acid end-groups were found compared with hydroxyl end-groups. The reactions involved in the thermal degradation of lactic acid-based polymers can follow different mechanisms [7], such as thermohydrolysis, zipper-like depolymerization [36] in the presence of catalyst residues, thermo-oxidative degradation [37, 38] and transesterification reactions which give simultaneous bond breaking and bond making.

21.4.1.2 Hydrolytic degradation

PLA hydrolysis is an important phenomenon since it leads to chain fragmentation [4, 7, 39], and can be associated with thermal or biotic degradation. This process can be affected by various parameters such as the PLA structure, its molecular weight and distribution, its morphology (crystallinity), the shape of its samples and its thermal and mechanical history (including processing), as well as, of course, the hydrolysis conditions. Hydrolytic

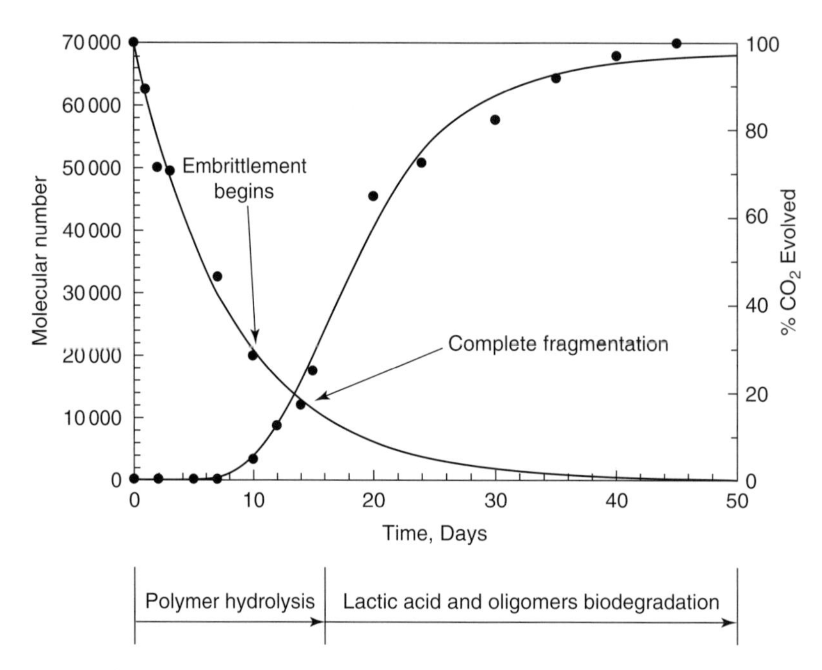

Figure 21.8 Abiotic and biotic degradations during composting stage. (Adapted from Reference [3].)

degradation is a phenomenon, which can be both desirable (*e.g.* during the composting stage) or undesirable (*e.g.* during processing or storage). The hydrolysis of aliphatic polyesters starts with a water uptake phase, followed by hydrolytic splitting of the ester bonds in a random way. The amorphous parts of the polyesters have been known to undergo hydrolysis before their crystalline regions because of a higher rate of water uptake. The initial stage is therefore located at the amorphous regions, giving the remaining non-degraded chains more space and mobility, which leads to their reorganization and hence an increased crystallinity. In the second stage, the hydrolytic degradation of the crystalline regions of the polyester leads to an increased rate of mass loss and finally to complete resorbtion [40]. The PLA degradation in an aqueous medium has been reported by Li *et al.* [40] to proceed more rapidly in the core of the sample. The explanation for this specific behaviour is an autocatalytic effect due to the increasing amount of compounds containing carboxylic end-groups. These low molar mass compounds are not able to permeate the outer shell. The degradation products in the surface layer are instead continuously dissolved in the surrounding buffer solution [40]. As expected, temperature plays a significant role in accelerating this type of degradation.

21.4.2 Biotic degradation

The biodegradation of aliphatic biopolyesters has been widely reported in the literature [5, 7, 39]. The biodegradation of lactic acid-based polymers for medical applications has been investigated in a number of studies *in vivo* [41] and some reports can also be found on their degradation in other biological systems [42]. The *in vivo* and *in vitro* degradations have been evaluated for PLA-based surgical implants [41]. *In vitro* studies have shown that the pH of the solution plays a key role in the degradation and that this analysis can be a useful predicting tool for *in vivo* PLA degradation [4]. Enzymes, such as proteinase K and pronase, have been used to bring about the *in vivo* PLA hydrolysis, although, enzymes are unable to diffuse through the crystalline parts. As expected, little enzymatic degradation occurs at the beginning of the process, but pores and fragmentation are produced, widening the accessible area to the different enzymes.

Figure 21.8 shows that during the composting stage, PLA degrades in a multistep process with different mechanisms [39]. Primarily, after exposure to moisture by abiotic mechanisms, PLA degrades by hydrolysis. First, random non-enzymatic chain-scissions of the ester groups lead to a reduction in molecular weight, with the consequent embrittlement of the polymer. This step can be accelerated by acids or bases and is affected by both temperature and moisture levels [3]. Then, the ensuing PLA oligomers can diffuse out of the bulk polymer and be attacked by micro-organisms. The biotic degradation of these residues produces carbon dioxide, water and humus (mineralization).

Studies on PLA-based multiphase materials have been carried out. Gattin *et al.* [43] have found that the physical and morphological properties of the blend play an important role in its degradation behaviour, as in the case of their comparative study of the degradation of PLA with and without plasticized starch materials [43]. These authors reported that the nature of the degradation strongly depends on the experimental biodegradation conditions. Sinha Ray *et al.* [44] prepared PLA nano-biocomposites filled with montmorillonite and studied and characterized their biodegradability.

21.5 PROCESSING

21.5.1 Multiphase materials

The extrusion of PLA-based materials is generally linked with another processing step such as thermoforming, injection moulding, fibre drawing, film blowing, bottle blowing and extrusion coating. The properties of the polymer will therefore depend on the specific conditions during the processing steps (*e.g.* the thermomechanical input). The main parameters during the melt processing are temperature, residence time, moisture content and atmosphere [1]. But, the major problem in the manufacturing of PLA-based products is the limited thermal stability during the melt processing. To overcome such a drawback or to give PLA new properties, a large number of multiphase materials have been developed, mainly by mixing PLA with others products.

21.5.1.1 Plasticization

The brittleness and stiffness of PLA can be major drawbacks for some application. According to Ljungberg *et al.* [45], any factor influencing PLA crystallinity, such as the isomer ratio, could disturb the distribution and compatibility of plasticizers with PLA and induce low efficiency and phase separation.

Lactide monomer is an effective plasticizer for PLA, but presents high migration due to its small molecular size. Oligomeric lactic acid (OLA) seems to be a better answer, since it shows low migration and high efficiency [32]. For instance, adding 20 wt% of OLA into poly(92 per cent L-lactide, 8 per cent meso-lactide) induces T_g and modulus decreases of 20°C and 63 per cent, respectively. A significant improvement of PLA (mainly PLLA) flexibility is accomplished by the incorporation of different types of citrates [45, 48] or maleates [49] whose efficiency was evaluated in terms of T_g shift and mechanical properties improvement [32]. These plasticizers are miscible with PLA up to ~25 wt%, but increasing the plasticizer content can raise the PLA crystallinity by enhancing chain mobility [32]. Low molecular weight polyethylene glycol (PEG) [32], polypropylene glycol and fatty acid are also compatible with PLA and can act as plasticizers [5].

21.5.1.2 Blends and compatibilization

A great number of articles has been published during the last few decades on PLA-based blends [4, 5, 8, 32], including starch/PLA blends, which allow reducing the material cost without sacrificing its biodegradability and maintaining certain mechanical and thermal properties. Native starch, which is composed of semi-crystalline granules (see Chapter 15), can be physically blended with PLA, but remains in a separate conglomerate form in the PLA matrix [50]. Thus, starch is typically characterized as a solid filler with poor adhesion with PLA. Such biocomposites are used as a model to test (*e.g.* carbohydrate–PLA compatibilization [51]). Most of the studies which are focused on the production of starchy blends are based on plasticized starch, the so-called thermoplastic starch. Such a processable material is obtained by the disruption of the granular starch and the transformation of its semi-crystalline granules into a homogeneous, rather amorphous material with the destruction of hydrogen bonds between the macromolecules (for more details see Chapter 15). Disruption can be accomplished by casting (*e.g.* with dry drums) or by applying thermomechanical energy in a continuous process. The combination of thermal and mechanical inputs can be obtained by extrusion. After the processing, a homogeneous material is obtained [1, 32]. A dependence of the PLA glass transition temperature on the blend composition was observed by DSC and DMA, indicating a small degree of compatibility between the blend components [32]. However, the mechanical characteristics of the blends were modest. The blend morphology (discontinuous versus co-continuous) has been investigated by Schwach and Averous [34] by microscopic observations. The full co-continuity is obtained in the domain of 60–80 per cent in volume of PLA. Despite the interest in developing plasticized starch/PLA materials, some limitations, due to the lack of affinity between the respective constituents, seem difficult to overcome. This low compatibility is mainly due to the PLA hydrophobic character.

To improve the affinity between the phases, compatibilization strategies are generally developed. This implies the addition of a compound, the compatibilizer, which can be obtained by the modification of at least one of the polymers initially present in the blend. For PLA/starch compatibilization, the literature proposes different approaches which can be classified in four groups [1, 25]: (i) the functionalization of PLA with, for example, maleic anhydride [51]; (ii) the functionalization of starch with, for example, urethane functions [25]; (iii) the starch–polyester crosslinking with a coupling agent such as a peroxide [25] and (iv) the use of copolymers, for example, starch-graft PLA [25], following the mechanism discussed above and illustrated in Fig. 21.5, for which the length of the grafts can be controlled to obtain a comb structure [52].

It is known that PLA forms miscible blends with polymers such as PEG [53]. PLA and PEG are miscible with each other when the PLA fraction is below 50 per cent [53]. The PLA/PEG blend consists of two semi-miscible crystalline phases dispersed in an amorphous PLA matrix. PHB/PLA blends are miscible over the whole range of composition. The elastic modulus, stress at yield, and stress at break decrease, whereas the elongation at break increases, with increasing polyhydroxybutyrate (PHB) content [54]. Both PLA/PGA and PLA/PCL blends give immiscible components [55], the latter being susceptible to compatibilization with P(LA-*co*-CL) copolymers or other coupling agents.

21.5.1.3 Multilayers

Developing compostable and low cost multilayer materials based for instance on plasticized starch and PLA is interesting in more than one sense. Martin *et al.* [56] carried out several studies on such a system and showed that the basic requisites for the preparation of multilayered products are to obtain sufficient adhesion between the layers, good moisture barrier properties and a uniform layer thickness distribution. Two different techniques were used to prepare the multilayers, namely, coextrusion and compression moulding. Peel strength was controlled by the compatibility between plasticized starch and PLA, which stayed low without compatibilizer. It was possible to increase the adhesion properties of the film by up to 50 per cent (*e.g.* by blending low polyester contents into the starchy core layer). There exist some inherent problems due to the multilayer flow conditions encountered in coextrusion, such as encapsulation and interfacial instability phenomena [57]. Addressing these problems is a crucial issue, since they can be detrimental to the product, affecting its quality and functionality.

21.5.1.4 Biocomposites and nano-biocomposites

Different types of fillers have been tested with PLA, such as calcium phosphate or talc [58], which show an increase in its mechanical properties. Concerning inorganic fillers, the greatest reinforcing effect is obtained with whiskers of potassium titanate and aluminium borate with a high aspect ratio. Carbon or glass fibres [59] improve the mechanical properties, particularly with fibre surface treatments capable of inducing strong interactions with PLA matrix. Different organic fillers can be associated with PLA. Biocomposites with improved mechanical properties are obtained by the association of ligno-cellulose fillers, such as paper-waste fibres and wood flour, with PLA by extrusion and compression moulding.

A significant and increasing number of papers have been published during the last 5 years on nano-biocomposites (*i.e.* nanocomposites based on a biodegradable matrix). Polylactide/layered silicate nanocomposites were largely investigated by Sinha Ray *et al.* [60, 61] and other authors [62–63]. They successfully prepared a series of biodegradable PLA nano-biocomposites using mainly melt extrusion of PLA, principally with modified montmorillonites (O-MMT), targeting nanofillers exfoliation into the matrix. Because of the interactions between the organo-clay particles which present large surface area (several hundreds m^2g^{-1}) and the PLA matrix, the nano-biocomposites dislayed improved properties, such as mechanical moduli, thermal stability, crystallization behaviour, gas barrier and biodegradability. The preparation of biodegradable nanocellular polymeric foams via nanocomposites technology based on PLA and layered silicate has been reported by different authors [61, 64] who used supercritical carbon dioxide as a foaming agent, with the silicate acting as nucleating site for cell formation. Cellular PLA structures can also be obtained by producing a co-continuous structure and extracting the co-products [65].

21.6 APPLICATIONS

At present, PLA-based materials are mainly referenced on three different markets, namely, the biomedical (initial market), the textile (mainly in Japan) and the packaging (mainly food, *i.e.* short-term applications). For instance, reported types of manufactured products are blow-moulded bottles, injection-moulded cups, spoons and forks, thermoformed cups and trays, paper coatings, fibres for textile industry or sutures, films and various moulded articles [8].

21.6.1 Biomedical applications

PLA has been widely studied for use in medical applications because of its bioresorbability and biocompatible properties in the human body. The main reported examples on medical or biomedical products are fracture fixation devices like screws, sutures, delivery systems and micro-titration plates [8].

PLA-based materials are developed for the production of screws and plates. As the bone healing progresses, it is desirable that the bone is subjected to a gradual increase in stress, thus reducing the stress-shielding effect. This is possible only if the plate loses rigidity in *in vivo* environment. To meet this need, researchers introduced resorbable polymers for bone plate applications. PLA resorbs or degrades upon implantation into the body, but most of its mechanical properties are lost within a few weeks [41]. Tormala *et al.* [66] proposed fully resorbable composites by reinforcing matrices with resorbable PLLA fibres and calcium phosphate-based glass fibres. One of the advantages often quoted for resorbable composite prostheses is that they do not need to be removed with a second operative procedure, as with metallic or non-resorbable composite implants. To improve the mechanical properties, PLA is reinforced with variety of non-resorbable materials, including carbon and polyamide fibres. Carbon fibres/PLA composites possess very high mechanical properties before their implantation, but they lose them too rapidly *in vivo* because of delamination. The long-term effects of resorbed products, and biostable or slowly eroding fibres in the living tissues are not fully known, and are concerns yet to be resolved [41].

Although PLA fibres are used in different textile applications as, for example, non-woven textile for clothes, they achieved their first commercial success as resorbable sutures. One of the first commercially available fibre-formed bioresorbable medical products is based on copolymers of GA in combination with L-lactide (Vicryl) [67]. Fibres can be produced both by solvent and by melt-spinning processes and drawn under different conditions to orient the macromolecules [7].

Micro- and nanoparticles are an important category of delivery systems used in medicine, and the use of PLA is interesting due to its hydrolytic degradability and low toxicity. The most important properties of the micro- and nanoparticles are the drug release rate and the matrix degradation rate which are affected by the particle design and the material properties [7]. Copolymers of GA and rac-lactide [5] seem to be the most suitable combinations for use as drug delivery matrices.

Porous PLA scaffolds have been found to be potential reconstruction matrices for damaged tissues and organs. There are several techniques reported for the manufacturing of such materials [7].

21.6.2 Packaging applications

Commercially available PLA packaging can provide better mechanical properties than polystyrene and have properties more or less comparable to those of PET [4, 8]. Market studies show that PLA is an economically feasible material for packaging. With its current consumption, it is at the present the most important market in volume for biodegradable packaging [4, 8]. Due to its high cost, the initial use of PLA as a packaging material has been in high value films, rigid thermoforms, food and beverage containers and coated papers. One of the first companies to use PLA as a packaging material was Danone (France) in yoghurt cups for the German market. During the last decade, the use of PLA as a packaging material has increased all across Europe, Japan and the US, mainly in the area of fresh products, where PLA is being used as a food packaging for short shelf-life products, such as fruit and vegetables. Package applications include containers, drinking cups, sundae and salad cups, wrappings for sweets, lamination films, blister packages and water bottles [9, 68]. Currently, PLA is used in compostable yard bags to promote national or regional composting programs. In addition, new applications such as cardboard or paper coatings are being pursued, for example, the fast-food market (cups, plates and the like) [9, 68]. However, to cater for a larger market, some PLA drawbacks must be overcome, such as its limited mechanical and barrier properties and heat resistance, and, in order to meet market expectations, the world production of PLA must be substantially increased.

ACKNOWLEDGEMENTS

The author wants to thank Dr. Seguinaud (Erstein-France), Dr. Pollet (ECPM-France) and Dr. Leclerc (ECPM-France) for their contribution.

REFERENCES

1. Avérous L., Biodegradable multiphase systems based on plasticized starch: a review, *J. Macromol. Sci., Polym. Rev.*, **C4**(3), 2004, 231–274.
2. Garlotta D., A literature review of poly(lactic acid), *J. Polym. Environ.*, **9**(2), 2002, 63–84.
3. Hartmann H., High molecular weight polylactic acid polymers, in *Biopolymers from Renewable Resources,* Ed.: Kaplan D.L., 1st edition, Springer-Verlag, Berlin, 1998, pp. 367–411.
4. Auras R., Harte B., Selke S., An overview of polylactides as packaging materials, *Macromol. Biosci.*, **4**, 2004, 835–864.
5. Zhang J.F., Sun X., Poly(lactic acid)based bioplastics, in *Biodegradable polymers for industrial applications,* Ed.: Smith R., CRC, 2005, Woodhead Publishing Limited, Cambridge -England, 2005, pp. 251–288, Chapter 10.
6. Mehta R., Kumar V., Bhunia H., Upahyay S.N., Synthesis of poly(lactic acid): A review, *J. Macromol. Sci., Polym. Rev.*, **45**, 2005, 325–349.
7. Sodergard A., Stolt M., Properties of lactic acid based polymers and their correlation with composition, *Prog. Polym. Sci.*, **27**, 2002, 1123–1163.
8. Doi Y., Steinbüchel A., *Biopolymers, Applications and Commercial Products – Polyesters III*, Wiley-VCH, Weiheim – Germany, 2002, p. 410.
9. http://www.natureworksllc.com.
10. Hiltunen K., Harkonen M., Seppala J.V., Vaananen T., Synthesis and characterization of lactic acid based telechelic prepolymers, *Macromolecules*, **29**(27), 1996, 8677–8682.
11. Hiltunen K., Seppala J.V., Harkonen M., Lactic acid based poly(ester-urethanes): Use of hydroxyl terminated prepolymer in urethane synthesis, *J. Appl. Polym. Sci.*, **63**, 1997, 1091–1100.
12. Ajioka M., Enomoto K., Suzuki K., Yamaguchi A., The basic properties of poly lactic acid produced by the direct condensation polymerisation of lactic acid, *J. Environ. Polym. Degrad.*, **3**(8), 1995, 225–234.
13. Stridsberg K.M., Ryner M., Albertsson A.C., Controlled ring-opening polymerization: Polymers with designed macromolecular architecture, *Adv. Polym. Sci.*, **157**, 2001, 41–65.
14. Carothers H., Dorough G.L., Van Natta F.J., The reversible polymerization of six membered cyclic esters, *J. Am. Chem. Soc.*, **54**, 1932, 761–772.
15. Kricheldorf H.R., Sumbel M., Polymerization of L, L-lactide with tin(II) and tin(IV) halogenides, *Eur. Polym. J.*, **25**(6), 1989, 585–591.
16. Kurcok P., Matuszowicz A., Jedlinski Z., Kricheldorf H.R., Dubois P., Jerome R., Substituent effect in anionic polymerization of b-lactones initiated by alkali metal alkoxides, *Macromol. Rapid Commun.*, **16**, 1995, 513–519.
17. Williams C.K., Breyfogie L.E., Choi S.K., Nam W., Young V.G., A highly active zinc catalyst for the controlled polymerization of lactide, *J. Am. Chem. Soc.*, **125**(37), 2003, 11350–11359.
18. Chabot F., Vert M., Chapelle S., Granger P., Configurational structures of lactic acid stereocopolymers as determined by 13C–1H N.M.R, *Polymer*, **24**, 1983, 53–59.
19. Stolt M., Sodergard A., Use of monocarboxylic iron derivatives in the ring-opening polymerization of L-lactide, *Macromolecules*, **32**(20), 1999, 6412–6417.
20. Dubois P., Jacobs C., Jerome R., Teyssie P., Macromolecular engineering of polylactones and polylactides. 4. Mechanism and kinetics of lactide homopolymerization by aluminum isopropoxide, *Macromolecules*, **24**, 1991, 2266–2270.
21. Jacobsen S., Fritz H., Degée P., Dubois P., Jérôme R., New developments on the ring opening polymerisation of polylactide, *Ind. Crop. Prod.*, **11**(2–3), 2000, 265–275.
22. Schwach G., Coudane J., Engel R., Vert M., Stannous octoate-versus zinc-initiated polymerization of racemic lactide, *Polym. Bull.*, **32**, 1994, 617–623.
23. Du Y.J., Lemstra P.J., Nijenhuis A.J., Van Aert H.A.M., Bastiaansen C., ABA type copolymers of lactide with poly(ethylene glycol). Kinetic, mechanistic, and model studies, *Macromolecules*, **28**(7), 1995, 2124–2132.
24. Gilding D.K., Reed A.M., Biodegradable polymers for use in surgery – Polyglycolic/poly(lactic acid) homo-and copolymers, *Polymer*, **20**, 1979, 1459–1464.
25. Schwach E., Etude de systèmes multiphasés biodegradables à base d'amidon de blé plastifié. Relations structure – propriétés. Approche de la compatibilisation, Ph.D Thesis URCA, Reims- France, 2004.
26. Gupta M.C., Deshmukh V.G., Radiation effects on poly(lactic acid), *Polymer*, **24**, 1983, 827–830.

27. Hoogsten W., Postema A.R., Pennings A.J., Brinke G., Zugenmair P., Crystal structure, conformation and morphology of solution-spun poly(L-lactide) fibres, *Macromolecules*, **23**, 1990, 634–642.
28. Puiggali J., Ikada Y., Tsuji H., Cartier L., Okihara T., Lotz B., The frustrated structure of poly(L-lactide), *Polymer*, **41**, 2000, 8921–8930.
29. Cartier L., Okihara T., Ikada Y., Tsuji H., Puiggali J., Lotz B., Epitaxial crystallization and crystalline polymorphism of polylactides, *Polymer*, **41**, 2000, 8909–8919.
30. Tsuji H., Ikada Y., Properties and morphologies of poly(L-lactide): 1. Annealing effects on properties and morphologies of poly(L-lactide), *Polymer*, **36**, 1995, 2709–2716.
31. Anderson K.S., Hillmyer M.A., Melt preparation and nucleation efficiency of polylactide stereocomplex crystallites, *Polymer*, **47**, 2006, 2030–2035.
32. Martin O., Avérous L., Poly(lactic acid): Plasticization and properties of biodegradable multiphase systems, *Polymer*, **42**(14), 2001, 6237–6247.
33. Yasuniwa M., Tsubakihara S., Sugimoto Y., Nakafuku C., Thermal analysis of the double-melting behavior of poly(L-lactic acid), *J. Polym. Sci., Polym. Phys.*, **42**, 2004, 25–32.
34. Schwach E., Averous L., Starch-based biodegradable blends: Morphology and interface properties, *Polym. Int.*, **53**(12), 2004, 2115–2124.
35. Dorgan J.R., Williams J.S., Lewis D.N., Melt rheology of poly(lactic cid): Entanglement and chain architecture effects, *J. Rheol.*, **43**(5), 1999, 1141–1155.
36. Gupta M.C., Deshmukh V.G., Thermal oxidative degradation of poly-lactic acid. Part II: Molecular weight and electronic spectra during isothermal heating, *Colloid Polym. Sci.*, **260**, 1982, 514–517.
37. Zhang X., Wyss U.P., Pichora D., Goosen M.F.A., An investigation of the synthesis and thermal stability of poly(DL-lactide), *Polym. Bull.*, **27**, 1992, 623–629.
38. McNeill I.C., Leiper H.A., Degradation studies of some polyesters and polycarbonates. 2. Polylactide: degradation under isothermal conditions, thermal degradation mechanism and photolysis of the polymer, *Polym. Degrad. Stab.*, **11**, 1985, 309–326.
39. Amass W., Amass A., Tighe B., A review of biodegradable polymers: Uses, current developments in the synthesis and characterization of biodegradable polyesters, blends of biodegradable polymers and recent advances in biodegradation studies, *Polym. Int.*, **47**, 1998, 89–144.
40. Li S.M., Garreau H., Vert M., Structure-property relationships in the case of the degradation of massive aliphatic poly-(a-hydroxy acids) in aqeous media. Part 1. Poly(D,L-lactic acid), *J. Mater. Sci. Mater. Med.*, **1**, 1990, 123–130.
41. Ramakrishna S., Mayer J., Wintermantel E., Leong K.W., Biomedical applications of polymer-composite materials: A review, *Compos. Sci. Technol.*, **61**, 2001, 1189–1224.
42. Hakkarainen M., Karlsson S., Albertsson A.C., Rapid (bio)degradation of polylactide by mixed culture of compost microorganisms low molecular weight products and matrix changes, *Polymer*, **41**, 2000, 2331–2338.
43. Gattin R., Copinet A., Bertrand C., Couturier Y., Biodegradation study of a coextruded starch and poly(lactic acid) material in various media, *J. Appl. Polym. Sci.*, **88**, 2003, 825–831.
44. Sinha Ray S., Kazunobu Y., Okamoto M., Ueda K., Control of biodegradability of polylactide via nanocomposite technology, *Macromol. Mater. Eng.*, **288**(3), 2003, 203–208.
45. Ljungberg N., Andersson T., Wesslen B., Film extrusion and film weldability of poly(lactic acid) plasticized with triacetine and tributyl citrate, *J. Appl. Polym. Sci.*, **88**, 2003, 3239–3247.
46. Labrecque L.V., Kumar R.A., Dave V., Gross R.A., McCarthy S.P., Citrate esters as plasticizer for poly (lactic acid), *J. Appl. Polym. Sci.*, **66**(18), 1997, 1507–1513.
47. Zhang J.F., Sun X., Physical characterization of coupled poly (lactic acid)/starch/maleic anhydride blends by triethyl citrate, *Macromol. Biosci.*, **4**, 2004, 1053–1060.
48. Ljungberg N., Wesslen B., The effects of plasticizers on the dynamic mechanical and thermal properties of poly(lactic acid), *J. Appl. Polym. Sci.*, **86**(5), 2002, 1227–1234.
49. Zhang J.F., Sun X., Mechanical and thermal properties of poly (lactic acid)/starch blends with dioctyl maleate, *J. Appl. Polym. Sci.*, **94**, 2004, 1697–1704.
50. Graaf R.A.D., Janssen L.P.B.M., Properties and manufacturing of a new starch plastic, *Polym. Eng. Sci.*, **41**(3), 2001, 584–594.
51. Carlson D., Nie L., Narayan R., Dubois P., Maleation of polylactide (PLA) by reactive extrusion maleation of polylactide (PLA) by reactive extrusion, *J. Appl. Polym. Sci.*, **72**(4), 1999, 477–485.
52. Nouvel C., Dubois P., Dellacherie E., Six J.L., Controlled synthesis of amphiphilic biodegradable polylactide-grafted dextran copolymers, *J. Polym. Sci., Polym. Chem.*, **42**(11), 2004, 2577–2588.
53. Tsuji H., Muramatsu H., Blends of aliphatic polyesters. IV. Morphology, swelling behavior, and surface and bulk properties of blends from hydrophobic poly(L-lactide) and hydrophilic poly(vinyl alcohol), *J. Appl. Polym. Sci.*, **81**, 2001, 2151–2160.
54. Focarete M.L., Scandola M., Dobrzynski P., Kowalczuk M., Miscibility and mechanical properties of blends of (L)-lactide copolymers with atactic poly(3-hydroxybutyrate), *Macromolecules*, **35**, 2002, 8472–8477.

55. Dell'Erba M., Groeninckx G., Maglio G., Malinconico M., Migliozzi A., Immiscible polymer blends of semicrystalline biocompatible components: Thermal properties and phase morphology analysis of PLLA/PCL blends, *Polymer*, **42**(18), 2001, 7831–7840.
56. Martin O., Schwach E., Avérous L., Couturier Y., Properties of biodegradable multilayer films based on plasticized wheat starch, *Starch/Starke*, **53**(8), 2001, 372–380.
57. Martin O., Avérous L., Comprehensive experimental study of a starch/polyesteramide coextrusion, *J. Appl. Polym. Sci.*, **86**(10), 2002, 2586–2600.
58. Kolstad J.J., Crystallization kinetics of poly(L-lactide-co-meso-lactide), *J. Appl. Polym. Sci.*, **62**, 1996, 1079–1091.
59. Wan Y.Z., Wang Y.L., Li Q.Y., Dong X.H., Influence of surface treatment of carbon fibers on interfacial adhesion strength and mechanical properties of PLA-based composites, *J. Appl. Polym. Sci.*, **80**, 2001, 367–376.
60. Sinha Ray S., Okamoto M., Polymer/layered silicate nanocomposites: A review from preparation to processing, *Prog. Polym. Sci.*, **28**(11), 2003, 1539–1641.
61. Sinha Ray S., Okamoto M., New polylactide/layered silicate nanocomposites part 6, *Macromol. Mater. Eng.*, **288**, 2003, 936–944.
62. Krikorian V., Pochan D.J., Poly(L-lactic acid)/layered silicate nanocomposite: Fabrication, characterization and properties, *Chem. Mater.*, **15**, 2003, 4317–4324.
63. Maiti P., Yamada K., Okamoto M., Ueda K., Okamoto K., New polylactide/layered silicate nanocomposites: Role of organoclays, *Chem. Mater.*, **14**, 2002, 4654–4661.
64. Fujimoto Y., Sinha Ray S., Okamoto M., Ogami A., Yamada K., Ueda K., Well-controlled biodegradable nanocomposite foams: From microcellular to nanocellular, *Macromol. Rapid. Commun.*, **24**, 2003, 457–461.
65. Sarazin P., Roy X., Favis B., Controlled preparation and properties of porous poly(image-lactide) obtained from a co-continuous blend of two biodegradable polymers, *Biomaterials*, **25**, 2004, 5965–5978.
66. Tormala P., Vasenius J., Vainionpaa S., Laiho J., Pohjonen T., Rokkanen P., Ultra-high-strength absorbable self-reinforced polyglycolide (SR-PGA) composite rods for internal fixation of bone fractures: *In vitro* and *in vivo* study, *J. Biomed. Mater. Res.*, **25**, 1991, 1–22.
67. Albertsson A.C., Varma I.K., Recent developments in ring opening polymerization of lactones for biomedical applications, *Biomacromolecules*, **4**, 2003, 1466–1486.
68. http://www.european-bioplastics.org/.

– 22 –

Polyhydroxyalkanoates: Origin, Properties and Applications

Ivan Chodak

ABSTRACT

This chapter deals with polyhydroxyalkanoates (PHAs), a class of biodegradable polyesters produced by various bacteria, but which can also be synthesized chemically. After a historical introduction, the bacterial synthesis and the structure of polyhydroxybutyrate and its copolymers with various homologous monomers are reviewed, before tackling various aspects related to the properties and behaviour of these PHAs. In particular, the mechanical and thermal properties, as well as the physical ageing, are placed in the context of possible applications. Finally, different processing technologies are examined together with the possibility of preparing plasticized or blended materials. Emphasis is also placed on the problem of economic competitivity of these polymers, compared with petroleum-based counterparts.

Keywords

Polyhydroxyalkanoates, Polyhydroxybutyrate, History, Bacterial synthesis, Chemical synthesis, Genetic engineering, Mechanical properties, Thermal transitions, Crystallization, Plasticizers, Thermal degradation, Processing, Applications

22.1 INTRODUCTION

Polyhydroxyalkanoates (PHAs) represent an important group of biodegradable plastics. They are produced by various bacteria in many grades, differing in composition, molecular weight and other parameters [1, 2]. The formation of a particular material, either homo or copolymer, depends on the type of bacteria, but even more important are the conditions of polymer formation, mainly the substrate used for feeding the bacteria and the conditions of their growth.

PHAs are generated and stored in bacteria in the form of fine powder particles. The bacteria produce the PHAs as an energy reserve, that is when overfed, the polymer is formed, which is consumed when the bacteria is short of a feeding substrate.

The simplest PHA is polyhydroxybutyrate (PHB), whose formula is shown in Fig. 22.1. This polymer and its copolymer with polyhydroxyvalerate seem to be, at present, the only PHAs relevant for practical applications.

PHB can be characterized as a rather controversial polymer. It is produced from renewable resources via a classical biotechnological process. The polymer is completely biodegradable, highly hydrophobic and thermoplastic, with high crystallinity, high melting temperature, good resistance to organic solvents and possesses excellent mechanical strength and modulus, resembling that of polypropylene [3]. In spite of these excellent properties, especially the strength parameters, the application of the polymer is limited to small volumes for rather special purposes. High volume applications are hindered by several serious drawbacks, especially pronounced brittleness, very low deformability, high susceptibility to a rapid thermal degradation, difficult processing by conventional

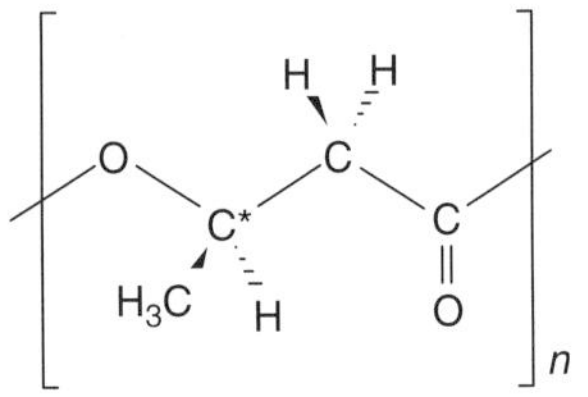

Figure 22.1 The formula of simplest PHA, polyhydroxybutyrate (PHB).

thermoplastic technologies (mainly due to fast thermal degradation) and rather high price compared to other high volume plastics. Additional problems related to processing are connected mainly with low shear strength of the melt, which needs to be addressed when considering certain applications. Its copolymer with valerate (PHBV) is much more acceptable, especially regarding improved toughness with an acceptable loss of strength and modulus. However, until now, the production process of PHBV has been more demanding, and hence its price is even higher than that of PHB. In spite of this, PHAs are considered to be promising materials worthy of being broadly investigated, and their production on a large scale is being considered or prepared in several regions of the world.

22.2 HISTORICAL REVIEW

The occurrence of PHA in bacterial cells was first described by Beijerinck in 1888 [4]. However, PHAs were studied mainly by biochemists who referred to them as 'lipids' [5], until the determination of the PHA composition by Lemoigne in 1925 [6, 7], who showed that the unknown material produced by *Bacillus megaterium* was a homopolyester of hydroxyacid, poly-3-hydroxybutyrate. That bacteria could produce polyesters was unknown to polymer chemists before 1960 and even to most biochemists and microbiologists before 1958, although their presence in bacterial cells in isolable amounts, their chemical composition, and even the fact that they were polymers, were reported in the literature as early as 1926. Generally, the scientific community did not realize the existence of these natural polyesters for so long because their discoverer, Maurice Lemoigne, published his results in little-read French journals and referred to the new material as ether-insoluble lipids [8]. At that time, many organic chemists refused to believe that there were such things as polymers. Lemoigne was a bacteriologist with training in analytical chemistry, and his series of papers, published over the 5-year period from 1923 to 1927, are remarkable for their breadth of research [9–13]. In 1923, Lemoigne reported that the acid produced by the bacteria was 3-hydroxybutyric acid [9], and in 1927, he described the procedure of obtaining the material from the cell which he characterized as a polymer of 3-hydroxybutyric acid. He named the source of the acid *lipide-β-hydroxybutyrique*, and also determined its molecular weight and melting temperature.

As pointed out by Lenz and Marchessault in an excellent review on PHAs [8], Lemoigne published these observations and interpretations at the time when Herman Staudinger was proposing the existence of high molecular weight molecules or polymers, which he termed 'macromolecules'. Lemoigne was probably familiar with the work of Emil Fischer, who had demonstrated as early as 1906 that proteins are large molecules of 'enchained' amino acid units or 'polypeptides', as he was the first to call them [14]. In 1953, Staudinger was awarded the Nobel Prize in Chemistry for his work on polymer synthesis and the concept of macromolecules [15]. There is no indication that Staudinger was aware of Lemoigne's discovery of nature's polyesters, which remained hidden from organic and polymer chemists for over 30 years, even though PHB was described in biochemistry textbooks, where, however, it was referred to as a 'lipid', and not polyester.

From 1923 until 1951, Lemoigne and co-workers published 27 papers. Lemoigne worked out a method for the quantitative analysis of PHB, and showed that it could be cast into a transparent film [16]. It was also reported that a variety of bacteria could produce PHB, and Lemoigne labelled it a 'reserve material'. It was only in the late 1950s that the important role of PHB in the overall metabolism of bacterial cells was discovered, understood and the significance of Lemoigne's earlier discoveries realized [8] by Stanier and Wilkinson and their co-workers, who showed that the PHB granules in bacteria serve as an intracellular food and energy reserve and that the polymer is produced by the cell in response to a nutrient limitation in the environment in order to prevent starvation if an essential element becomes unavailable. The nutrient limitation activates a metabolic mechanism, in which acetyl units from the tricarboxylic cycle are involved in the production of PHB. This can be reconverted into acetic acid by a series of enzymatic reactions inside the cell [17].

In 1958, Macrae and Wilkinson observed that the production of the homopolymer by *B. megaterium* was rather fast if the glucose-to-nitrogen source ratio of the substrate was high [18] and the ensuing degradation by *B. Cereus* or *B. Megaterium* occurred rapidly in the absence of any energy source [19, 20]. The authors concluded that PHB was a carbon- and energy-reserve material and suggested the pathway of PHB synthesis [19]. In the same year, it was also demonstrated that the occurrence of this reserve polymer is a widespread phenomenon in Gram-negative bacteria [21]. During the same period, Stanier, Doudoroff and co-workers found that PHB was the primary product of the oxidative and photosynthetic assimilation of organic compounds by phototropic bacteria, and they attempted to detail the biosynthesis and breakdown mechanism of PHB in the cells [22]. The reactions involved in the metabolic pathway responsible for the biosynthesis of PHB from acetic acid were first identified by Stanier and co-workers in 1959 in their studies on PHB formation in *R. rubrum*. However, only in 1973, Schlegel at the University of Gottingen, and Dawes at the University of Hull, working independently, succeeded to isolate and characterize the specific enzymes which catalyze the reactions for the synthesis of 3-hydroxybutyric acid, the PHB monomer [23–26]. Schlegel *et al.* carried out their investigations on PHB production in *Alcaligenes eutrophus*, while Dawes and co-workers studied the cycle in *Azotobacter beijerinckii*. In the same 1973 issue of *Biochemical Journal*, Schlegel and Dawes published simultaneously their discoveries on the identification of the two enzymes involved in the reactions for converting acetic acid into 3-hydroxybutyric acid in the two different bacteria [23, 24]. For both bacteria, the enzymes were a ketothiolase, which catalyzes the dimerization of the Coenzyme A derivative of acetic acid, acetyl-CoA, to acetoacetyl-CoA, and a reductase, which catalyzes the hydrogenation of the latter to [*R*]-3-hydroxybutyryl-CoA, the monomer that is polymerized to PHB by a synthase [8].

A detailed mechanism for the polymerization reaction, which was based on Merrick's suggestion, was proposed by Ballard and co-workers at ICI in 1987 and further elaborated by Doi and co-workers at the RIKEN Institute in Japan in 1992. They put forward a mechanism in which two thiol groups are involved in the active site for both the initiation and propagation reactions of the polymerization. For initiation, the two thiol groups form thioesters with two molecules of monomer, which then undergo a thioester–oxyester interchange reaction at the active site to form a dimer and release one of the thiol groups for the propagation reaction [8].

3-Hydroxybutyrate (3HB) was, for a long time, considered to be the only PHA material produced by bacteria as the reserve energy supply until, in 1974, Wallen and Rohwedder identified 3-hydroxyvalerate and 3-hydroxyhexanoate in chloroform extracts of activated sewage sludge [27]. The possibility of preparing various PHAs as a function of the substrate was first revealed by DeSmith *et al.* [28] when they cultivated *Pseudomonas oleovorans* in *n*-octane. The polymer formed was found to consist principally of 3-hydroxyoctanoate units.

In the 1970s, the interest in PHAs was rising, as reflected in a growing number of papers dealing with various aspects of these polymers. On the one hand, many different microorganisms were found to be able to produce PHAs, as described in an early review on PHAs by Daves and Senior in 1973 [29]. On the other hand, the widening research of polymer scientists on PHAs resulted in a better characterization of the materials regarding their macromolecular parameters, such as molecular weight [30], crystalline structure [31], granule morphology and other properties [32–34]. Continuing work on biochemical aspects led to a detailed knowledge on the PHB metabolism and its regulation [29, 35], the enzymology of PHB synthesis [29], the PHB intra- and extra-cellular degradation [4, 36] and the physiological function of PHB. The recognition of the relationship between PHB biosynthesis and extra-cellular environment was of particular importance, especially since it made it possible to optimize the preparation process of PHB via defining the conditions that favour PHB accumulation [24, 37]. From a practical point of view, the identification of HA units, other than 3HB, and especially the possibility to produce copolymers, revealed a way to envisage practical applications for these polymers, since it solved the serious problem of PHB brittleness. Thus, the first industrial production of PHB-*co*-valerate took place [38], although the patents on 3PHB as a potentially biodegradable plastic had already been filed in 1962 [39]. In the 1980s, extensive work was done to identify all potential HA units resulting, not only in the description of a wide range of 3HA units, but also in the discovery of 4HA [40] and 5HA [41] units in the polymer chain.

To extend the vast flexibility of nature regarding PHA production, it was found that these materials are synthesized, not only by Gram-negative bacteria, but also in a number of Gram-positive counterparts, bacteria, aerobic, anaerobic, photosynthetic bacteria, as well as in some archaebacteria [42, 43].

In 1982, Imperial Chemical Industries Ltd. (ICI) in England started to produce a thermoplastic polyester that was totally biodegradable and could be melt processed including the preparation of films and fibres [42]. The technology consisted in a large-scale fermentation process, resulting in the production of the polymer inside the bacterial cells, which at the end of the process contained as much as 90 per cent of the polymer. The bacterium employed in the process was *Alcaligenes eutrophus*, since then renamed *Ralstonia eutropha* (and more recently changed again to *Wautersia eutropha*) and the commercial polyester product with the trade name 'Biopol', was a copolyester containing randomly arranged units of [*R*]-3-hydroxybutyrate (HB), and [*R*]-3-hydroxyvalerate (HV) [44].

The fast development of molecular biology, starting in the late 1970s led to an investigation of the cloning and characterization of genes involved in the PHA biosynthesis. At the end of the 1980s, the genes coding for the enzymes involved in the PHA synthesis were cloned from *Ralstonia eutropha* and shown to be functionally active in *Escherichia coli* [45–47]. Thus, what was identified at the beginning of the twentieth century as a sudanophilic bacterial inclusion is now apparently going to the most advanced stage of development, that is, protein engineering [3]. This development may lead in the near future to an efficient production of tailor-made PHAs with a high potential for environment-friendly applications.

22.3 PREPARATION, SYNTHESIS

Bacterial synthesis is considered to be the most important process for PHA preparation at present. This approach will be discussed in this section in more detail. Besides the bacterial synthesis, the chemical synthesis of PHB via the ring-opening polymerization of butyrolactone, is also possible and will be briefly discussed at the end of this section. Recently, increased attention has been paid to the biosynthesis using genetically modified plants. This procedure has not yet reached an industrially acceptable scale up, but may be developed to an even more efficient and economical process than the bacterial fermentation technology, as discussed below.

22.3.1 Bacterial synthesis

As described above, the bacterial synthesis was the first, natural way to prepare and produce PHAs. A lot of work was devoted to unravel principles, mechanisms and possibilities to modify the process [1, 3], so that now the process is basically understood and the principles are known, enabling its modification to produce the desired products.

The first step of the biochemical pathway of PHB synthesis consists in the conversion of a selected carbon source to acetate. Then, an enzyme cofactor is attached via the formation of a thioester bond. The enzyme, called coenzyme A (CoA), is a universal carrier of acyl groups in biosynthesis and acetyl-CoA is a basic metabolic molecule found in all PHA-producing organisms. A dimer acetoacetyl-CoA is formed via reversible condensation and subsequently reduced to a monomer unit (R)-3-hydroxybutyryl-CoA. PHB is formed via the polymerization of the latter, maintaining the asymmetric centre [5]. The basic simplified process is shown in Scheme 22.1.

Propagation proceeds by bonding another monomer to the free thiol group of the active site, followed by another thioester–oxyester exchange reaction to form the trimer, and so on. The higher bond energy of the oxyester, compared to that of the thioester, is the reason for polymerization proceeding spontaneously as a thermodynamically favourable process. The synthase, therefore, acts as both initiator and catalyst for the polymerization process, which proceeds by a continuous series of insertion reactions. The enzyme is specific for monomers with the [*R*] configuration and will not polymerize identical compounds having the [*S*] configuration. Hence, all natural PHAs are totally isotactic.

It has been demonstrated that PHB is accumulated by a wide variety of bacteria under unbalanced growth, when the cells become limited for an essential nutrient (such as ammonium, sulphate, phosphate, magnesium, iron) but are exposed to an excess of carbon. Under conditions of balanced growth, CoA amounts are relatively high, so that the synthesis of PHB is inhibited and acetyl-CoA is just metabolized in the tricarboxylic acid cycle. In nutrient limitation without carbon excess, proteins cannot be synthesized, and the buildup of NADH, an universal electron carrier in biosynthesis, occurs. NADH inhibits citrate synthesis and consequently acetyl-CoA is no longer able to be oxidized rapidly enough via the tricarboxylic acid cycle and therefore accumulates. Although the equilibrium constant of the reversible condensation reaction does not favour acetoacetyl-CoA formation, at high concentrations of NADH and acetyl-CoA at low concentrations of CoA, the equilibrium is shifted in favour of the PHB biosynthesis [5].

As listed in Table 22.1, a number of bacterial genera produce PHAs, mainly PHB, under various conditions with different yields. Various microorganisms produce PHAs with different levels of efficiency and quality. However, the

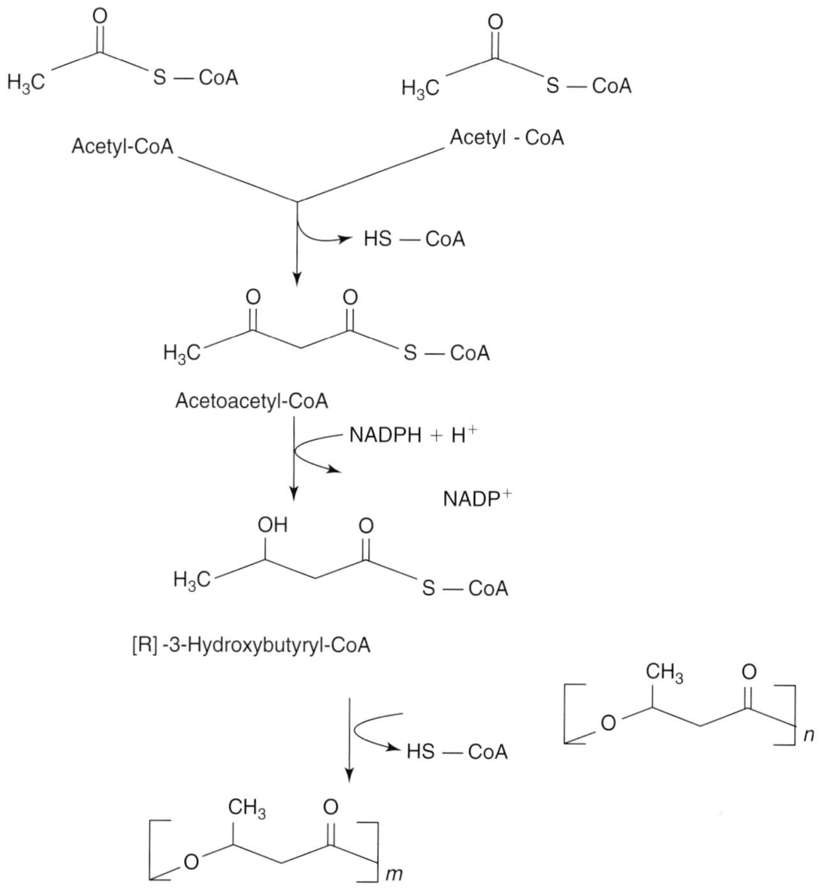

Acetyl-CoA Acetyl - CoA

HS — CoA

Acetoacetyl-CoA

NADPH + H^+

$NADP^+$

[R] -3-Hydroxybutyryl-CoA

HS — CoA

Scheme 22.1 Bacterial synthesis of PHB.

number of strains which are able to synthesize polymers intracellularly is surprising. A limited number of bacteria are able to produce random copolymers based on the usual 3HA and 3HV combination. Initially, *A. eutrophus* was grown on a variety of substrates, including natural sugar, ethanol and even gaseous mixtures of carbon dioxide and hydrogen [48] to synthesize PHB. Later, it was found that the HV comonomer could be incorporated if propionate or valerate were present in the medium [49]. The HV mole fractions were limited to the range 0–47 per cent, because of the relatively fast metabolic pathway of propionyl-CoA to acetonyl-CoA in the cell [50, 51]. Later on, a series of copolymers with 0–95 per cent of HV were produced from mixtures of butyric and valeric acids [52]. The direct incorporation of 4-hydroxybutyric acid [41, 53, 54] and 3-hydroxypropionic acid was also found to be possible, for example, a random copolyester of 3-HB and 4-HB units containing 37 per cent of 4-HB was produced when 4-hydroxybutyric acid was used as the sole carbon source [41].

The large number of bacteria able to produce PHAs is impressive. However, even more surprising is the variety of substrates which the bacteria are able to consume, producing, besides PHB and PHB-*co*-HV, a broad range of polymers, copolymers, terpolymers, including polymers containing various functional groups, such as double bonds, halogens, phenyl or epoxy moieties, depending on the type of the strain and substrate the bacteria are grown on. From this point of view, *Pseudomonas genus* has been reported as the most versatile accumulator of PHAs [37]. A brief overview of syntheses using bacteria of this type is shown in Table 22.2. The examples of some more or less special syntheses with other bacteria are listed in Table 22.3.

Among those, besides the PHB homopolymer, poly-3-hydroxybutyrate copolymers with either 3-hydroxyvalerate or 4-hydroxybutyrate are of special interest, mainly because these copolymers, while having a lower crystallinity

Table 22.1

Bacterium species accumulating PHA. Data taken from [5] and [37]

1. Acinetobacter	28. Gloeothece	55. Protomonas
2. Actinomycetes	29. Haemophilus	56. Pseudomonas
3. Alcaligenes[a]	30. Halobacterium	57. Ralstonia[a]
4. Aphanocapsa	31. Haloferax	58. Rhizobium
5. Aphanothece	32. Hyphomicrobium	59. Rhodobacter
6. Aquaspirillum	33. Lamprocystis	60. Rhodococcus
7. Asticcaulus	34. Leptothrix	61. Rhodopseudomonas
8. Azomonas	35. Lampropedia	62. Rhodospirillum
9. Azospirillum	36. Methanomonas	63. Sphaerotilus
10. Azotobacter	37. Methylobacterium	64. Spirillum
11. Bacillus	38. Methylocystis	65. Spirulina
12. Beggiatoa	39. Methylomicrobium	66. Staphylococcus
13. Beijerinckia	40. Methylomonas	67. Stella
14. Beneckea	41. Methylosinus	68. Streptomyces
15. Caryophanon	42. Methylovibrio	69. Syntrophomonas
16. Caulobacter	43. Micrococcus	70. Tetrahymena
17. Chlorofrexeus	44. Microcoleus	71. Thiobacillus
18. Chlorogloea	45. Microcystis	72. Thiocapsa
19. Chromatium	46. Moraxella	73. Thiocystis
20. Chromobacterium	47. Mycoplana	74. Thiodictyon
21. Clostridium	48. Nitrobacter	75. Thiopedia
22. Corynebacterium	49. Nitrococcus	76. Thiosphaera
23. Derxia	50. Nocardia	77. Vibrio
24. Ectothiorhodospira	51. Oceanospirillum	78. Wautersia[a]
25. Escherichia	52. Paracoccus	79. Xanthobacter
26. Ferrobacillus	53. Pedomicrobium	80. Zoogloea
27. Gamphospheria	54. Photobacterium	

[a] Different names used for the same bacteria. Wautersia has been recently used.

which could be important regarding toughness, might be produced by technologies acceptable from an economical point of view. *Ralstonia eutropha*, *Alcaligenes latus* or *Delftia acidovorans* were shown to produce random copolymers of 3-hydroxybutyric acid (3HB) and 4-hydroxybutyric acid (4HB), P(3HB-*co*-4HB), in a broad composition range using various carbon sources [41, 63, 67, 71]. When 4-hydroxybutyric acid, or 1,4-butanediol, was used as the sole carbon source for *D. acidovorans*, the homopolymer of 4-hydroxybutyric acid (P(4HB)) was synthesized [67, 82].

The isolation of these polymers from the biomass proceeds through the destruction of the cell membrane mechanically, chemically or eznymatically [42, 83], followed by the dissolution of the polymer in a suitable solvent, for example chloroform, methylene chloride, 1,2-dichloro ethane or pyridine. The remnants of the cell walls are removed by filtration and/or centrifugation. Extraction, using mixed solvents, for example water/organic solvent, is the last step applied for the final purification.

22.3.2 Chemical synthesis

It should be pointed out that all of the PHAs can be synthesized chemically from the relevant substituted propiolactones, generally using aluminium or zinc alkyl catalysts with water as the cocatalyst. Several alternative routes have been explored, for example PHBV can be produced from butyrolactone and valerolactone with an oligomeric aluminoxane catalyst [85]. Structures with partially stereoregular blocks can also be obtained. An interesting and instructive review on the chemical synthesis was published by Muller and Seebach in 1993 [84].

Table 22.2

Overview of PHAs produced by *Pseudomonas genus* on various substrates. Data taken from [37]

Bacteria	PHA produced	Substrates	Special feature	Reference
Pseudomonas oleovorans	4 to 12 C monomers (R=CH_3 to $(CH_2)_8CH_3$)	*n*-Alkanes, *n*-alkanoates, *n*-alcohols, *n*-alkenes – unsaturated monomers	Branched side chains from branched substrates	[42]
	Units contain 2 C less than the C source		Composition related to substrate	[55]
	Unsaturated medium-side-chain	*n*-Octane + *n*-octene		[55]
	PHAs containing phenyl units	Mixture of 5-phenylvaleric acid + *n*-nonanionic acid or *n*-octanionic acid	Low yield	[56]
	Copolymer containing 32% of cyano containing monomers (9-cyano-3-hydroxynonanoate and 7-cyano-3-hydroxyheptanoate	11-cyanoundecanoic acid + *n*-nonanionic acid		[57]
	Chlorinated and fluorinated PHAs			[58, 59]
	Brominated units	Nonanionic and octanionic acids + 6-bromohexanoic, 8-bromo-octanoic acid or 11-bromo-undecanoic acid		[59]
	Copolyester containing up to 37% of terminal epoxy groups	Mixture of 10-epoxyundecanoic acid and sodium octanoate		[60]
Pseudomonas putida	Seven different comonomers including 3-hydroxy decanoate, 3-hydroxybexanoate, octanoate + saturated and monounsaturated monomers of C12 and C14	Glucose		[61]
Pseudomonas sp. A33	Unsaturated medium-side-chains			[62]
	3HB + nine other components including 3-hydroxyhexadecanoate + unsaturated 3-hydroxydodecenoate, 3-hydroxytetradecenoate, 3-hydroxyhexadecenoate	1,3-Butanediol		[63]
	Polyester of hydroxyalkanoic acids with even carbon number C4, C6, C8, C10, C12	Sodium gluconate		[64]
Pseudomonas sp. GPo1	Unsaturated random copolymer poly(3-hydroxyoctanoate-*co*-3-hydroxyundecenoate)	Mixture of sodium octanoate and undecenoic acid	Fraction of unsaturated units related to the ratio of carbon sources in the medium	[60]
Burkholderia cepacia (formerly *Pseudomonas cepacia*)	3HB-*co*-3H4PE	Gluconate or sucrose		[65]
	Terpolyester 3-HB-*co*-3HV-*co*-2-methyl-3-hydroxybutyric acid	Tiglic acid with or without gluconic acid		[65]

Table 22.3

Overview of PHAs produced by selected bacteria. Data taken from [37]

Bacteria	PHA produced	Substrates	Special feature	Reference
Ralstonia eutropha	PHB, PHBV			[66]
	P3HB-*co*-4HB			[53, 67]
	P3HB-*co*-3HV-*co*-5HV			[68]
	P3HB-*co*-4HB-*co*-3HV			[69]
	P3HB-*co*-3WHV-*co*-4HV			[70]
				[70]
	P4HB pure homopolymer			[71, 72]
Alcaligenes latus	PHB			[73]
	PHBV			[74, 75]
	P3HB-*co*-3Hproprionate			[62]
	P3HB-*co*-4HB			[63]
Alcaligenes eutrophus	P3HB-*co*-3HV			[76]
Rhodospirillum rubrum	Copolymer 3HB+3HV+3H4-pentenoate	4-Pentenoic acid or pentanoic acid	Same product using both substrates	[77]
Chromobacterium violaceum	poly3HV homopolymer	Sodium valerate		[78]
Delftia acidovorans	P(3HB-*co*-4HB) copolymer	1,4-Butanediol		[65]
Methylobacterium	P(3HB-*co*-4HB) copolymer	Methanol + 3-hydroxypropionate		[79]
Comomonas testosterone	P3HB-*co*-Hcaproate			[80]
	P3HB-*co*-Hcaproate-*co*-Hoctanoate		2–4% of comonomer	[80]
Sphaerotilus natans	P3HB-*co*-3HV	Glucose + sodium propionate		[81]

The stereochemistry is very important with respect to biodegradability. In terms of stereoregularity, synthetic PHAs can be made almost identical to the corresponding biopolymers [86]. This resemblance results even in the excellent biodegradability of the material [50]. However, although academically interesting, these synthetic homologues of bacterial PHAs are unlikely ever to be competitive with PHAs produced by fermentation owing to the high lactone monomer costs [5].

22.3.3 Genetic engineering

A novel potential way of synthesizing PHAs calls upon transgenic plants prepared by cloned biosynthetic genes [87]. If this genetic engineering approach were successful and economically feasible, the basic idea would be to use the procedure for the genetic modification of plants to produce polymers through more or less standard agriculture processes, while introducing the same mechanisms as used by bacteria.

The research can be divided into two different routes. The first approach consists in cloning the appropriate genes and transferring them into other organisms. Thus, in 1988, Dennis and co-workers cloned the entire set of genes in *R. eutropha* for the three enzymes involved in the synthesis of PHB from acetyl-CoA [45]. The three genes were clustered in one operon, which was successfully introduced into *E. coli*, and the ensuing association was able to synthesize PHB in large quantities from a wide range of organic compounds. Some recombinant strains of *E. coli* can also produce the HB/HV copolymer [88]. Alternatively, strains containing only the synthase gene can express this protein in sufficiently large quantities for isolation and purification [89]. The purified enzyme can be used for *in vitro* polymerization reactions of various 3- and 4-hydroxyalkanoate-CoA monomers [90], which can have a living character, meaning that the propagating end groups remain active indefinitely and very high molecular weights can be reached [91].

The second approach consists in the successful application of genetic engineering directly to the plants, as reported by Somerville and co-workers in 1992. The reductase and synthase genes of *A. eutrophus* were inserted into a plant, *Arabidopsis thaliana*, which can also produce acetoacetyl-CoA. In this way, the transgenic plant can accumulate PHB granules to approximately 14 per cent of its dry weight [92]. A number of other reports appeared recently, dealing with various aspects of PHA (mainly PHB) production in genetically modified plants [87, 93–95]. Monsanto has recently developed a procedure for obtaining genetically modified plants able to produce small quantities of PHB [96].

To date, however, few reports on the successful production of PHAs have been published, with the exception of a number of studies with transgenic Arabidopsis. Using a variety of chimeric constructs, Steinbuchel and co-workers have determined [97] that the constitutive, chloroplast-localized expression of the β-ketothiolase (*phb*A) gene involved in PHB production, is detrimental to the efficient production of transgenic PHB. The use of either inducible or somatically activated promoters allowed the formation of transgenic PHB-producing potato (*Solanum tuberosum*) and tobacco (*Nicotiana tabacum*) plants, although the amount of PHB formed was still rather low, namely $0.09\,mg\,g^{-1}$ dry weight for potato and $3.2\,mg\,g^{-1}$ dry weight for tobacco [97]. Although very interesting from a scientific point of view, these yields are too low to be considered as a basis for a high volume production of PHA. Nevertheless, this type of research continues based on the concept that the proven genetic engineering way can be developed to an economically applicable biotechnological process, thus contributing, not only to the production of environment-favourable plastics, but also to new perspectives for agricultural development.

22.4 STRUCTURE

The structure of PHAs is based on polyester macromolecules bearing optically active carbon atoms. Two modes are generally described in biological papers, namely medium chain-length PHB (mcl PHB) as the energy reserve supply and hydroxybutyrate cyclic oligomers able to form complexes with other biomacromolecules, for example calcium polyphosphate or proteins acting as ion transfer agents [98].

Poly(3-hydroxybutyrate) is a linear polyester with helical macromolecules. The secondary structure of PHB is specified as left-hand 2_1 helix in a g^+g^+tt conformation, while the structure of oligolides consists of right-hand 3_1 helices [99]. The surface of the $3_1(+)$ helix is covered by methyl groups, leading to the lipophilic nature of the macromolecule. The carbonyl bonds in the $2_1(-)$ helix are placed perpendicularly, while in the $3_1(+)$ counterpart, they are parallel to the helix axis. The latter is the reason for ability to form ionic complexes.

The 3-hydroxyalkanoic acids are all in the [*R*] configuration resulting from the stereospecificity of the enzyme involved in the polymerization step [3]. However, certain portions of *S* units have also been reported [100]. The molecular weight of the simplest poly-3-hydroxybutyrates depends on the bacteria used and the conditions of the synthesis [101], and the way of separation and purification may also have a certain influence. Using some sophisticated procedures, higher molecular weights can be attained, for example an Mn as high as 20 kD was measured for a polymer produced by a recombinant strain of *E. coli* [102].

The formation of hydrogen bonds contributes significantly to the good properties of PHAs. The number of hydrogen bonds formed in semicrystalline polymers may increase with increasing temperature as the crystalline phase progressively melts. For example, when studying isotactic(i) PHB/catechin mixtures, the hydrogen bonding between iPHB and catechin could not be detected by FTIR at room temperature. However, strong hydrogen bonds were confirmed at 190°C, that is above the melting point of iPHB [103]. The strength of the hydrogen bonds is influenced by the tacticity of the polymer chain. An FTIR study, performed by Iriondo *et al.* on blends of poly(4-vynilphenol) (PVPh) and PHB with different tacticity [104, 105], indicated that the hydrogen bonds between PVPh and iPHB were weaker than those between PVPh and atactic(a) PHB [104].

Inoue *et al.* studied by FTIR the hydrogen bonds between catechin and PHB as a function of the latter's tacticity [106] and showed that strong inter-associated hydrogen bonds formed in aPHB/catechin and syndiotactic(s) PHB (sPHB)/catechin blends, while no evidence was found to confirm the existence of such bonds in iPHB/catechin blends [106]. Since the crystallinity of PHB is significantly affected by the tacticity, and the crystalline phase restrains the formation of hydrogen bonds, they also carried out FTIR measurements on the melt at 190°C to separate the effect of tacticity from that of crystallinity. In this case, strong inter-associated hydrogen bonds were detected in iPHB/catechin blends as well as in aPHB/catechin and sPHB/catechin blends [106]. Furthermore, it was found that in the melt, a higher PHB tacticity facilitates the access of the OH group of catechin to the carbonyl group of PHB, and hence the formation of hydrogen bonds.

22.5 ULTIMATE PROPERTIES

22.5.1 General features

Biodegradability is arguably the major advantage of PHAs in terms of their perspective uses and applications. This favourable feature can compete to a certain extent against disadvantages such as higher prices, when compared with conventional plastics. However, a more comprehensive assessment of ultimate properties is crucial when considering any application aimed at everyday commercial products. Table 22.4 gives a comparison of the properties of both PHB and a few PHA copolymers with those of the two most common polyolefins, namely polypropylene and low density polyethylene.

It can be seen that the properties of PHB are rather close to those of polypropylene, outperforming polyethylene in most parameters. The factor of primary negative importance is the low deformation at break, related to low film toughness and unacceptable rigidity and brittleness. The reason for the PHB brittleness arises mainly from the presence of large crystals in the form of spherulites. On the other hand, the high strength and E modulus represent a suitable starting point for possible polymer modifications, since there is a large margin in these strength parameters to increase deformability and toughness.

Copolymers exhibit properties much closer to those of LDPE, but their availability and price still represent a hindrance for these materials to be considered as serious competitors against commodity polyolefins.

Interesting information can be drawn from a comparison of the properties of various commercial biodegradable plastics, as seen in Table 22.5.

Table 22.4

A comparison of the physical properties of PHB and some of its copolymers with those of polypropylene (PP), and low density polyethylene (LDPE). Most data taken from [3]

	PHB	20V[a]	6HA[b]	PP	LDPE
Melting temperature (°C)	175	145	133	176	110
Glass transition temperature (°C)	4	−1	−8	−10	−30
Crystallinity (%)	60	ng	ng	50	50
Density ($g\,cm^{-3}$)	1.25	ng	ng	0.91	0.92
E modulus (Mpa)	3.5	0.8	0.2	1.5	0.2
Tensile strength (Mpa)	0	20	17	38	10
Elongation at break (%)	5	50	680	400	600

[a] Poly(3-hydroxybutyrate-*co*-20 mol% hydroxyvalerate).
[b] Poly(3-hydroxybutyrate-*co*-6 mol% Has), HAs (hydroxyalkanoates) = 3% 3-hydroxydecano-ate, 3% 3-hydroxydodecanoate, <1% 3-hydroxyoctanoate, <1% 5-hydroxy-dodecanoate, ng = negligible.

Table 22.5

A comparison of the physical properties of PHBV with those of other biodegradable plastic

Polymer	PHBV	PLA	PCL	PEA	PBSA	PBAT
Density	1.25	1.25	1.11	107	1.23	1.21
Melting temperature (ºC)	153	152	65	112	114	110–115
Tg (ºC)	5	58	−61	−29	−45	−30
Crystallinity (%)	51	0–1	67	33	41	20–35
Modulus (MPa)	900	2000	190	260	250	52
Elongation at break (%)	15	9	>500	420	>500	>500
Water permeability $g\,m^{-2}$ per day	21	172	177	680	330	550

PHBV – poly-(3-hydroxybutyrate-*co*-3-valerate), Monsanto (Biopol D400G, HV 7%); PLA – poly(lactic acid), Dow-Cargill (Nature Works); PCL – polycaprolactone, Solvay (CAPA 680); PEA – polyesteramide, Bayer (BAK 1095); PBSA – poly(butylene succinate-*co*-adipate), Showa (Bionolle 3000); PBAT – aromatic copolyester, Eastman (Eastar bio 14766).

The density of all these materials is above unity, which may be considered as a certain disadvantage. The melting temperatures are within a reasonable range, except that of PCL, which is a little low for practical applications, considering its possible exposure to sunshine in summer. In terms of the glass transition temperature, these materials may be separated into two groups, one containing just PLA with a T_g well above RT, and the other with T_gs below the temperatures commonly reached in winter in the majority of inhabited countries. Thus, PLA can be considered for applications where common polystyrene is used at present, whereas the other materials can compete mainly with packaging foils made from polyethylene. PHAs are in a special position from this point of view, since, on the one hand, their T_g lies somewhat below, but still close to common RT, so that the material is brittle. On the other hand, it is possible to lower their T_g and thus make the materials ductile by plasticizing them, blending them with miscible additives or by the formation of two-phase systems by blending them with suitable immiscible polymers. The crystallinity of these biodegradable materials lies within a reasonable range, typical of conventional thermoplastics, going from amorphous PLA up to PCL with crystallinity slightly higher than that of LDPE. The modulus is more or less reciprocal to the glass transition temperature with its highest value for brittle PLA and its lowest for PBAT. The values of elongation at break correspond to the T_g being rather high for polymers in a plastic state. From the point of view of PHAs, its low water permeability is certainly an interesting property, suggesting possible applications for packaging.

The material properties of PHAs can be adjusted by varying the HV content, its increase resulting in an increase in the impact strength and a decrease in melting temperature and glass transition [107] (see Figs 22.2 and 22.3), tensile strength [108], crystallinity and water permeability [109]. Different results have been published concerning the mechanical properties of P3HB-*co*-P3HV copolymers [110]. A rather unusual behaviour was reported suggesting that the 3HB and 3HV units are isodimorphous, that is because of their similarity in shape and size, the 3HV units are incorporated into the P(3HB) crystal-lattice [111]. It follows that the properties of P(3HB-*co*-3HV) are not significantly improved in comparison with the P(3HB) homopolymer. Copolymers with improved properties can be prepared via the copolymerization of 3HB with longer chain hydroxyalkanoic acids, which form a separate crystalline lattice or do not crystallize at all. Examples of this type of copolymers, showing exclusion of the longer chain comonomer units, are the copolymers of 3HB and 3-hydroxyhexanoate (3HH) [112].

22.5.2 Glass transition, melting and crystallization

PHB synthesized by bacteria and separated by standard procedures is a semicrystalline polymer with a rather high crystallinity, which can reach 80 per cent. It was found surprising that bacteria could produce crystalline polymers and even more surprising that they were able to use it as a feedstock. Until the mid-1880s, the prevailing belief was that PHA *in vivo* was indeed a crystalline solid [113, 114]. The first doubts about this morphology of P3HB were raised because of the observation that a short centrifugation resulted in an irreversible loss of intracellular degradability. The final solution regarding these inconsistencies was obtained from solution-state NMR spectra which showed [115] that poly(3-hydroxybutyrate) *in vivo* is not crystalline, but rather a completely amorphous material. It was later demonstrated that centrifugation gave rise to completely different features in the NMR spectra and the simultaneous appearance of X-ray powder diffraction patterns, typical of crystalline P3HB [116]. The amorphous character of *in vivo* P3HB is an obvious and fully acceptable idea, but the question is immediately raised regarding the mechanism triggering its crystallization when the inclusions are isolated. The existence of *in vivo* plasticizers or nucleation inhibitors was put forward as an explanation [3], but no proof has yet been found for this proposal.

The crystalline structure of PHB is orthorhombic with dimensions of the basic crystalline cell $a = 5.76\,\text{Å}$, $b = 13.20\,\text{Å}$, $c = 5.96\,\text{Å}$. A partially planar zig-zag structure can be formed as a result of the mechanical uniaxial load from the amorphous phase between lamellae [117, 118]. PHB forms extremely thin lamellar crystals, with a thickness comprised between 4 and 7 nm and a prevailing size of 5 nm [98].

A number of data are available of the glass transition and melting temperatures of P3HB and its copolymers with P3HV. The equilibrium melting point $T_m{}^o$ is 470 ± 2 K, according to Barham *et al.* [119], although the former author in a later paper reported a broad range of $T_m{}^o$, changing with molecular weight, from 352 K (Mw $\sim$ 20 000) up to the previously reported 471 K (Mw $\sim$ 300 000) [120]. The melting enthalpy for the 100 per cent crystalline PHB is 146 J g^{-1} [119]. The real melting temperature, as measured by optical observation, does vary significantly with molecular weight, as seen in Fig. 22.2, which shows that the T_m of oligomeric PHB increases almost linearly in the Mn range of 350 to around 700. The curve becomes less steep and the linearity disappears. Finally, the T_m

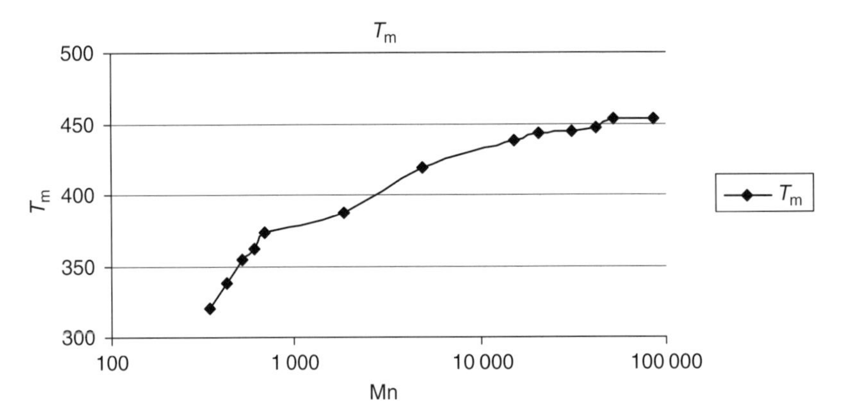

Figure 22.2 Melting temperature of P3HB as a function of its molecular weight. (Data taken from [121], original source [122].)

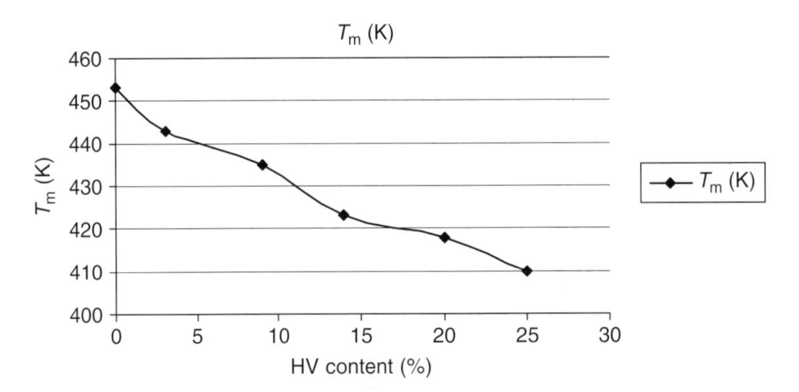

Figure 22.3 Melting temperature variation as a function of the comonomer content for poly-3-hydroxybutyrate-*co*-3-hydroxyvalerate copolymers. (Data from [121], original source [44].)

of polymers with molecular weights above 50000 levels off at a value of 453 K. A practically linear decrease in T_m was reported for a set of P3HB-3HV copolymers with rising HV contents, as seen in Fig. 22.3.

Due to its low nucleation density, a small number of large, rather imperfect, crystallites grow in PHB, affecting its mechanical properties, especially the elongation at break and the brittleness of moulded products and films. This effect can be eliminated to a certain extent by selecting the crystallization conditions, especially the crystallization temperature, since the morphology of the crystalline region depends significantly on this parameter [119]. The addition of nucleating agents and suitable post-treatments after extrusion or casting can also lead to much improved properties [40].

PHB copolymers with higher alkanoates usually exhibit lower T_g values than that of the corresponding homopolymer. The poly([*R*]-3-hydroxybutyrate-*co*-5%-[*R*]-3-hydroxyhexanoate) (P(3HB-*co*-5%-3HH)) with $M_w = 0.82 \times 10^6$ and DPI = 2.2, and poly([*R*]-3-hydroxybutyrate-*co*-12%-[*R*]-3-hydroxyhexanoate) (P(3HB-*co*-12 per cent-3HH)) with $M_w = 0.93 \times 10^6$ and DPI = 2.2 were reported to have T_gs of 277 and 270 K, respectively, with the corresponding T_m at 417 and 385 K [110]. It is interesting that the homopolymer of 4-hydroxybutyric acid (P(4HB)) showed much lower values for both its glass transition (~−50°C) and melting (~60°C) temperatures, as measured by calorimetric analysis [67, 82].

22.5.3 Ageing of PHB by physical processes

It is known that PHB ages upon storing with an increase in its Young's modulus accompanied by a decrease in its elongation at break which translates into a pronounced increase in brittleness of the material within a few months after thermal processing [123]. This phenomenon is only partially understood and has been rationalized by two alternative explanations. The first is termed physical ageing and is based on the relaxation process usually observed in amorphous polymers. When the material is cooled from the melt to glassy state, the amorphous domains do not reach their equilibrium free volume. If the material is stored close to its T_g, the chains reorganize slowly and the corresponding free volume decrease occurs. Thus, the reason for the extended effect in PHB arises from the fact that its T_g is only a little below room temperature. Reorganization proceeds extremely slowly (within few months) compared to other polymers, but the movements are not completely frozen as they would be if the storing temperature were well below T_g, as is the case for most amorphous polymers.

It is believed that physical ageing is responsible for the pronounced brittleness of PHB [124–126]. A decrease in the maximal intensity of the loss factor $tg\delta$ around T_g with storage confirmed the occurrence of a relaxation process [124]. Hurrell and Cameron provided an identical conclusion by investigating the process by SAXS and proved that the microstructural changes observed do not fully correspond with the changes in mechanical properties [126].

The other explanation of the process consists in suggesting a secondary crystallization proceeding very slowly because of the low chain mobility associated with the temperature being close to T_g. In this process, the lamellae become progressively more perfect, leading to a reduction of the interphase between amorphous and crystalline domains. Some of the chains in the amorphous phase are more stretched and this introduces a tension within it [127]. This progressive crystallization produces a growth of 7 per cent crystallinity within 8 days from the initial value 56 per cent, measured immediately after the thermal processing. Moreover, storing at temperatures higher than RT leads to an increase in the ultimate crystallinity, a process which is more or less completed after a month with a crystallinity of about 65 per cent and an end to the evolution of the mechanical properties [125]. Other authors, however, have observed that the crystallinity only increases during the first few hours after cooling the melt [128].

A detailed investigation of the ageing behaviour of PHB copolymers containing 8 or 12 per cent of valerate units has been reported [129]. Both extruded and compression moulded samples were investigated by DSC, DMTA, dielectric spectroscopy, and thermally stimulated discharge. Besides changes in the amorphous and crystalline phases, the authors considered the importance of the interphase region on the ageing process, consisting in relaxation above T_g due to morphological reorganization in the interphase [129]. This opinion is supported by fracture mechanics data indicating that the ageing process does not consist of a simple embrittlement, but rather of a reduction in the energy dissipating properties of the material, accompanied by the ability to survive high stress levels [130]. From this point of view, the changes in FTIR spectra of PHB during storage may be of interest. Two IR peaks have been found [131] to be sensitive to storage, namely at $1\,685\,cm^{-1}$ and $3\,435\,cm^{-1}$, corresponding to carbonyl and hydroxyl stretching, respectively. Both peaks grow during storage with no levelling off after 14 days [131], indicating that some interactions between the functional groups of PHB may result from storage. It is not clear to what extent these interactions are related to the physical ageing. Nevertheless, it is worth considering the importance of these effects in addition to the changes in the crystalline structure, especially in the interphase region.

22.5.4 Solubility and plasticization

The solubility of PHB in various solvents (see Table 22.6) is of importance, not only from a scientific point of view, but also in terms of the selection of a suitable solvent for the preparation and purification of the polymer from bacteria.

Although generally PHB is considered to be a hydrophobic material, it is, in fact, slightly hydrophilic [132, 133], that is the ratio between its hydrophobic (dispersive) and hydrophilic (polar and electrostatic) interactions depends, to a certain extent, on the morphology and degree of orientation of the anisotropic units. While for isotropic structures, the presence of imperfect crystallites may result in the formation of sites for water absorption (especially accessible carbonyl moieties), in anisotropic materials (*e.g.* fibres), the sorption capacity is substantially lower. On the other hand, the sorption of some organic solvents, such as acetone, does not depend on the PHB morphology [134]. Thus, the small molecules of water can act as a kind of structural probe, while the larger acetone molecules do not recognize the difference between the iso- and aniso-structures of PHB. The linear dependence of the diffusion coefficient on the sorbed water concentration in PHB indicates that water acts as a plasticizer for this matrix. Similar effects have been obtained for various organic solvents. Limiting diffusion coefficient values increase in the order benzene $=$ ethanol $<$ hexane $<$ acetone.

Table 22.6

Solubility of PHB in various solvents [83, 121]

Good	Partial	Not soluble
Chloroform	Dioxane	Water
Dichloromethane1	Toluene	Methanol
1,2,2-Tetrachloroethane	Pyridine	Ethanol
Ethylene carbonate	Benzene	Isopropanol
Propylene carbonate	Xylene	Cyclohexanol
Acetic anhydride		Diethyl ether
N,N-dimethylformamide		*n*-Hexane
Ethylacetoacetate		Cyclohexanone
Acetic acid		Ethyl acetate
Higher alcohols (C > 3)		Ethyl methyl ketone
2,2,2-Trifluoro ethanol		Tetrahydrofurane
		Butyl acetate
		Valeric acid
		Diluted mineral acids

Butyl chloride is reported as a solvent for the synthetic PHB at 13°C [83].

From a practical point of view, chloroform or dichloromethane is commonly used for the extraction of PHB or its copolymers from the bacteria, for example in [124], a preparation of a bacterial P(3HB-*co*-4HB) copolymer was described based on the fermentation of *D. acidovorans* from 1,4-butanediol [82], where the product was extracted from the lyophilized cells with hot chloroform and purified by reprecipitation with hexane. Then, the precipitate was fractionated in boiling acetone, and the soluble part taken as the P(3HB-*co*-4HB) sample containing 7 mol% of 3HB units.

The solubility of PHAs differs depending on their composition, for example after a PHA synthesis by *Pseudomonas sp*., the acetone-insoluble fraction contained the poly-3-hydroxybutyrate homopolymer, while the acetone-soluble fraction was reported to consist of seven different 3HA units ranging from C4 to C12 [64].

Another reason for investigating solubility and/or miscibility with various liquids of low molecular size and relatively high boiling points, is their possible application as plasticizers, as reported in some studies [135, 136]. Soybean oil (SO), epoxidized soybean oil (ESO), dibutyl phthalate (DBP) and triethyl citrate (TEC) have also been described as plasticizers [137]. PHBV blends containing 20 wt% of plasticizer were prepared by evaporating the solvent from the blend solutions. The compatibility of the plasticizers was then examined using DSC and SEM and their efficiency estimated according to the T_g decrease. DPB and TEC were found to be more effective than SO and ESO. The extent of the decrease in the T_g induced by plasticizers varies, leading, in some cases, to an increase in the elongation at break, as in the case of the addition of around 32 wt% of oxypropyl glycerol, dibutyl sebacate or dioctyl sebacate to PHB, which resulted in an elongation at break of more than 250 per cent [138].

It has to be stressed that some plasticizers which are commonly and successfully used with other polymers, may act as prodegradants for the thermal degradation of PHAs. Glycerol has been reported to act in such a way, while glycerol triacetate (triacetin), although effective as a PHB plasticizer, was found to be inert as a prodegradant [139]. Acetyl tributyl citrate was also examined in this context [140].

Obviously, the biodegradability of a given plasticizer is a fundamental prerequisite for its use in biodegradable plastics such as PHAs, as are any possible health hazards associated with it or with its decomposition products. Both aspects have been taken into account in studies involving additives such as citrate-based plasticizers [141, 142], or the new benzoate plasticizer, Benzoflex® 2888, a blend of diethylene-, triethylene-, and dipropylene-glycol dibenzoate [143].

22.6 THERMAL AND HYDROLYTIC DEGRADATION

The very high susceptibility of PHB and also of other PHAs, to thermal degradation, represents a serious problem, especially with respect to processing. Considering its melting point, the processing temperature of PHB should be at least 190°C. At this temperature, thermal degradation proceeds rapidly, so that it is impossible to avoid certain, sometimes substantial, decomposition consisting of a decrease in molecular weight and, consequently, a

Table 22.7

GPC molecular weights after the thermal treatment (1, 10 and 80 min) of PHB and PHB with 20 wt% of glycerol triacetate (TAC) and glycerol plasticizers, as a function of the annealing temperature. Data taken from [139]

Sample	Annealing		MW $g\,mol^{-1}$
	°C	Minutes	
PHB	170	1	270 000
	180	10	97 000
	200	10	29 000
	220	10	4 500
PHB/TAC	170	1	240 000
	200	10	30 000
	220	10	6 000
PHB/glycerol	170	1	170 000
	200	10	9 000
PHB	190	80	6 000
PHB/TAC	190	80	9 000
PHB/glycerol	190	80	<3 000

Scheme 22.2 Thermal *cis*-elimination in polyesters.

pronounced effect on the mechanical and other ultimate properties. The extent of this degradation is illustrated by the changes in molecular weight after annealing at various temperatures shown in Table 22.7.

The thermal degradation of polyesters is generally described either by a *cis*-elimination or a transesterification. The former process dominates with ester moieties with acidic C═H bonds due to activation via the carboxyl groups, as shown in Scheme 22.2, with the formation of polymer fragments [144]. *Cis*-elimination is a non-catalyzed process, almost uninfluenced by the presence of additives or impurities, and its rate is similar in the liquid or gas phase.

P(3HB) with three carbon atoms in the main-chain monomer unit, degrades by the *cis*-elimination reaction, releasing crotonic acid and linear oligomers of 3HB containing crotonyl chain ends as volatile products [145]. However, the changes in molecular weight indicate that more complex side reactions occur. Gel permeation chromatography (GPC) measurements of PHB heated at various temperatures, as a function of time, revealed that the molecular weight initially decreased, but then increased slightly before continuing to decline [146]. This effect was attributed to the polycondensation of the initial hydroxyl and carboxyl groups formed by the elimination process. Severe degradation conditions (*e.g.* up to 300°C under vacuum) lead to the formation of crotonic and isocrotonic acids and the dimer, trimer and tetramer of PHB [147].

Transesterification must also be considered when blends of PHB with other polymers are investigated, especially if they contain hydroxyl, carboxyl or other reactive moieties. A significant decrease in the activation energy of the thermal decomposition of PHB was observed if PMMA was present as a second component in the blend, although the changes were not attributed to any particular PHB/PMMA interaction, whereas only marginal changes were registered if PHB was mixed with polypropylene, as revealed by thermogravimetry [148].

The presence of various additives alters the thermal degradation kinetics with many species acting as prodegradants. Aluminium compounds and fumed silica have been reported to have a slight stabilizing effect, followed by a prodegradant activity [2]. Other inorganic compounds have been shown to be prodegradants, as revealed by dynamic TG experiments [149] and the same effect was observed for impurities present in technical PHB when compared with a carefully purified sample [150].

As expected, degradation has a negative influence on the mechanical properties of PHB [151]. From this point of view, it is important to estimate the effect of plasticizers on the thermal degradation. As shown in Table 22.7, glycerol triacetate does not play any significant role, while glycerol has a detrimental effect on the molecular weight during processing [139], which is attributed to a transesterification reaction in which the hydroxyl groups of glycerol play a crucial role.

Given the high rate of thermal degradation of PHB, its processing by injection moulding or extrusion must be carried out with the lowest possible temperature and residence time [154, 155].

Stabilization of PHB by conventional methods is almost impossible. Antioxidants are ineffective, so are other additives, such as acid acceptors, moisture adsorbers, chelates, blockers of hydroxyl and carboxyl groups (acetic anhydride or diazomethane, respectively). Some improvement was achieved by the addition of sulphur and additives releasing sulphur dioxide. The presence of certain inorganic additives (CaO, PbO, PbO_2, Al_2O_3, ZnO) leads to destabilization according to the acidity of the additive. On the other hand, more stable products are formed by the reaction with MgO, so that the decomposition reaction increases to about 670 K, that is about 100 K higher compared with PHB alone [149]. The presence of aluminium complexes results in crosslinking, which to a certain extent, heals the effect of the thermal main-chain cleavage [2].

Since the thermal stabilization of PHB has not been satisfactorily solved yet, it is necessary to look for processing procedures operating at lower temperatures. From this point of view, the addition of plasticizers is definitely an option, since it leads to a decrease in the melting temperature of 15–20°C, although this requires rather high amounts of the plasticizer, up to 30 wt%.

It is interesting that the presence of a second monomer in PHA copolymers has a certain stabilizing effect in their thermal degradation, as revealed by thermogravimetry for PHB and its copolymers with hydroxyvalerate and hydroxyhexanoate [156]. Moreover, in the case of the PHBV copolymer, two peaks were observed in the thermograms, indicating that the two different monomers evaporated at different temperatures [157]. The thermal degradation of various P(3HB-*co*-4HB) copolymers containing between 11 and 82 mol% of 4HB units, was reported to occur by a random chain scission process involving *cis*-elimination below 200°C [158]. On the other hand, Abate *et al.* reported that the main degradation mechanism of P(3HB-*co*-4HB) containing 97 mol% of 4HB units is an unzipping reaction starting at the chain ends [227]. They also suggested that the existence of 3HB unit would cause a competition between their *cis*-elimination and the unzipping reaction [gi]. Two processes were also identified during the non-isothermal degradation of P(4HB) at temperatures below and above 350°C and the major volatile product was γ-butyrolactone, regardless of the degradation temperature, indicating that its selective formation via an unzipping reaction predominated at temperatures below 300°C. At temperatures above 300°C, both the *cis*-elimination reaction of 4HB units and the formation of cyclic macromolecules of P(4HB) via intramolecular transesterification took place in addition to the unzipping reaction [158, 159].

As mentioned above, PHB takes up minor quantities of water upon storage [152], that is, about 0.2 wt%. Despite this modest moisture content, the possibility of polymer degradation by hydrolysis must be taken into account since PHAs degrade in water at room temperature, albeit at a very low rate. This hydrolytic degradation is strongly accelerated at higher temperatures or in an alkaline medium [153].

22.7 MODIFICATION

The modification of PHAs is aimed at improving certain ultimate properties. Since the price of copolymers is still higher compared to P3HB, the most interesting approach seems to be to modify the simplest homologue, that is P3HB, to prepare PHA-based biodegradable materials with good mechanical and other properties and acceptable prices. An alternative route consists in synthesizing various copolymers, especially P3HB-3HV, with the aim of developing a process producing them at a lower price. However, the modification to change properties is interesting even in the case of copolymers.

Generally, the methods of modification are based on chemical (changes of chemical structure, *e.g.* grafting, crosslinking, transesterification) or physical procedures (mixing with low or high molecular additives, optimization of crystallization) and these two approaches are discussed below.

22.7.1 Chemical methods

As briefly discussed above in the context of ultimate properties, the introduction of comonomer, especially hydroxyvalerate or hydroxyhexanoate, leads to higher elongation and toughness [5] and a modest improvement of thermal stability [156], accompanied by a decrease in modulus and, in many cases, of strength.

Transesterification with glycol is frequently used to prepare telechelic diols of PHB and P(HB-*co*-HV) in the presence of dibutyl tin dilaurate as catalyst [160]. The ensuing oligomers with well-defined end groups are suitable for the synthesis of block copolymers [161, 162], whose mechanical properties depend on the ratio of hard and soft segments. The synthesis of poly(H3B-*co*-caprolactone, CL) uses this principle, where the transesterification of the homopolymers, catalyzed by acids, results in the formation of random or microblock copolymers [163]. With rising CL content, a decrease in T_g is observed from 2 down to −42°C, for CL contents of 0 and 72 per cent, respectively.

Crosslinking is a basic method for preparing thermosets, but it is also commonly used to modify a number of elastomers or thermoplasts. Polyhydroxyalkanoates can be crosslinked if the process is initiated by the thermal decomposition of peroxides, although the crosslinking efficiency is not too high, so that a crosslinking coagent is recommended together with an initiator. An increase in viscosity by a factor of 20 was observed when PHB was treated with a mixture of dicumyl peroxide and triallyl cyanurate as a coagent [2]. Crosslinking does not affect the chain cleavage resulting from thermal degradation, but it eliminates, and sometimes overrules, the decrease of molecular weight due to degradation. During the thermal degradation of PHB via *cis*-elimination, crotonyl and carboxyl groups are formed, and the former can be expected to react with free radicals formed by the thermal decomposition of the peroxide to form crossbonds. Initiation by irradiation is claimed to be ineffective, since it leads to a pronounced degradation of the PHA chains [164]. However, this depends on the actual composition of the specific PHA, since successful crosslinking was reported using electron beam (EB) irradiation of unsaturated PHAs [165], and the ensuing network had elastomeric properties and was still completely biodegradable. A rather successful crosslinking was also reported using initiation by gamma irradiation, especially if the process was carried out under nitrogen instead of air [166]. In this case, the copolymer was poly(3-hydroxyoctanoate-*co*-undecenoate), for which a much higher efficiency can be expected, compared to PHB or PHBV, not only because of the presence of unsaturations, but also as a result of the long saturated carbon chains, which can be sites of hydrogen abstraction forming free radicals ready to recombine and crosslink. Crosslinking results in a modest increase in T_g and an improved shape stability at temperatures above T_m.

Crosslinking has been suggested to be a reason for the improved thermal stability of PHB/poly (glycidyl methacrylate) blends. In this case crossbonds are formed via the reaction of the epoxy groups of PGMA with the carboxyls of PHB [167].

Free-radical initiated reactions can also be used, besides crosslinking, for other modifications. The thermal decomposition of benzoyl peroxide was successfully used to initiate the grafting of maleic anhydride on PHB which was reported to produce a material with increased thermal stability and biodegradability, the latter attributed to the appended maleic moieties [168, 169].

PHAs containing specific functional groups (halogen, nitrile, double bond) can be modified, not only by crosslinking via peroxides or sulphur, but also by more sophisticated procedures. Thus double bonds can be transformed into epoxy moieties, which can be further used for a number of modifications, including crosslinking [170]. The grafting of poly(3-hydroxyoctanoate-*co*-3-hydroxyundecenoate) was shown to proceed also via the oxidation of the pendant double bonds into carboxyls and the esterification of these side moieties with poly(ethylene glycol). This process enhanced the hydrophilic character of the modified polymers, making them soluble in polar solvents, such as alcohols and water/acetone mixtures [171].

22.7.2 Physical modification

As already discussed above, the use of plasticizers produces a decrease in the processing temperature and hence a more convenient processing regime, resulting from the decrease in the melting temperature [2]. Some losses in strength and modulus must be accepted in this context, depending on the amount of plasticizer added.

A hot rolling treatment has been shown to substantially improve the ductility of PHB. It is proposed that rolling results in the healing of cracks present inside the spherulites leading to more ductile films [172]. A more sophisticated approach seems to be the solid-state processing of PHB powders, which prevents thermal degradation, since the extrusion proceeds well below the polymer melting temperature. The improved mechanical properties were attributed in this case to structural differences at the molecular and supramolecular levels [173]. The changes are apparently similar to those induced by hot rolling; but much higher due to the higher stress.

An interesting and rather simple option for toughness improvement consists in annealing the material at or above 120°C [125]. The elongation at break was found to increase up to 30 per cent [127], with values of up to 60 per cent in individual cases [125]. A change in lamellar morphology during annealing was suggested as an explanation of this effect, resulting in a substantial increase in the mobile phase, as indicated by the area under the $t_g\delta$ peak measured by DMTA. The reason for the morphology changes can be rationalized by the fact that the annealing could be considered as if crystallization proceeded from the melt at a much lower rate.

In conclusion, these types of physical modifications can improve significantly the mechanical properties of PHB, although the extent of these changes is insufficient in terms of its long-term properties.

Blending with other polymers and mixing with solid isotropic or anisotropic particles are efficient ways to produce new materials with highly modified properties using physical modification principles. This topic falls somewhat outside the scope of this chapter, but enough data can be found in the recent literature [128, 174]. Hence, only a few remarks will be given here.

Blends of various PHAs or PHAs with other biodegradable polymers are commonly used for adjusting the properties of the final materials, while maintaining their biodegradability. The blends of two different PHAs are usually immiscible [175], at least in the crystalline state [176], even in the case of a blend of isotactic bacterial PHB and synthetic atactic PHB, which is miscible only in a limited concentration range [177].

PHA blends with poly(vinyl alcohol) (PVA) are interesting materials whose miscibility depends on the composition and the PVA tacticity [178]. PHB is claimed to be miscible with polyethylene glycol [179] and with poly D-lactide [180]; its miscibility with polyethylene oxide depends on the blend composition [181]. Immiscible, but well-compatible blends of PHB, were prepared with poly(butylene succinate-*co*-butylene adipate) and poly(butylene succinate-*co*-caprolactone) [182].

Important materials have been developed based on blends of PHAs with PCL. These blends are immiscible and mainly incompatible [183], but compatibilization can be reached by the addition of poly(HB-*co*-CL) copolymers [184], or by crosslinking [185].

Unlike blends with biodegradable polymers, where the main purpose is to prepare biodegradable materials with modified properties, blends with non-biodegradable polymers are prepared as materials containing a polymer from renewable resources, aiming at attaining certain unique properties, for example an improved barrier behaviour. Examples of such PHB blends include the use of polyvinyl acetate [186, 187], polymethyl methacrylate [188], polyvinyl phenol (miscible blends due to hydrogen-specific interactions) [105] and various rubbers [189], including epoxidized natural rubber [190]. Miscible blends of PHB with poly(vinylidene chloride-*co*-acrylonitrile) [191] and epichlorhydrin [192] have also been reported.

22.8 PROCESSING

The processing of PHAs depends very much on the particular grade of the polymer used. These considerations apply essentially to PHB and perhaps PHBV, since the other PHAs are still in the development stage and hence their processing industrial technology is not yet a concrete issue.

The melt processing of PHB can be performed using conventional technologies; however, it is complicated by two factors. The first problem is the low viscosity of its melt and the low melt elasticity, requiring rather precise conditions for extrusion or injection moulding. The second problem is the low thermal stability of PHAs in general and of PHB in particular, as discussed above, which gives rise, during melt processing, to a substantial decrease in molecular weight leading to both worse ultimate properties and a further decrease in melt viscosity. The extrusion of PHB, recommended for Biomer, Germany, consists in melting the material in the first section of the equipment and a gradual temperature decrease in the next zones. The temperatures for Biomer grade P226 decrease from 185°C in zone 1, down to 150°C in zone 4 and 145°C in the die. For other grades (Biomer P209 and P249), temperatures are lower by 5–10°C.

Processing at lower temperatures can be achieved by the addition of plasticizers or by copolymerization with higher PHAs, as already discussed. In both cases, the ultimate properties change, so that a compromise must be found between an acceptable extent of thermal degradation and the desired properties.

22.8.1 Special modes of processing

Special procedures for PHB processing are applied both to avoid or decrease thermal degradation, and to produce unconventional products. In some cases, the two targets overlap, so that for certain special products unconventional processing is used. The three most relevant applications of these special processes are briefly discussed.

22.8.1.1 Solid-state processing

This technique was recently described in detail by Radusch in his review [193], and its principles are also applicable to other PHAs. The process consists in sintering powders at very high pressures and elevated temperatures, but below their melting point, and involves a two-step solid-state transformation [194, 195]. Deformation degrees expressed as draw ratios (DRs) 1.5 and 2.75 were achieved, depending on the processing conditions. Together with this increase in DR, the density rose as well with an increase in the processing temperature, as seen in Table 22.8, in which data for melt-processed PHB are shown for the sake of comparison.

After solid-state processing, a very different crystalline structure was found, compared to melt-processed PHB, as revealed by WAXS and, in spite of a certain orientation due to solid-state processing, a slight decrease of crystallinity was also reported. The mechanical properties of the solid-state extruded PHB depends substantially on the initial molecular weight of the polymer, as well as on the processing temperature, as seen in Tables 22.9 and 22.10.

Other properties are also significantly affected by solid-state processing, compared to the melt counterpart, viz. dynamic mechanical properties, as well as relaxation behaviour, for example creep [193].

Table 22.8

Changes in draw ratio and density of both solid-state extruded and melt-processed PHB with temperature. Data taken from [193]

Temperature (°C)	25	70	100	120	Melt
Draw ratio, λ/λ_0	1.6	1.8	2.1	2.7	~1.0
Density	1.11	1.24	1.25	1.25	1.25

Table 22.9

Dependence of tensile stress and elongation at break on the initial molecular weight of solid-state extruded PHB. Data taken from [193]

MW (kg mol^{-1})	171	350	500	1 000
Tensile strength at break (MPa)	15	38	48	54
Elongation at break (%)	1.0	6.6	7.2	9.1

Table 22.10

Dependence of tensile stress and elongation at break on processing temperature of solid-state extruded PHB. Data taken from [193]

Temperature (°C)	70	100	120	185	185[a]
Tensile strength at break (MPa)	35	48	57	10	39
Elongation at break (%)	2.1	9.2	18.0	0.5	2.4

[a] Melt processed material.

22.8.1.2 Foaming

Biodegradable foams represent interesting products. Depending on their structure and properties, they can be used for simple impact-damping fillings of packages, for transport of brittle products, as well as for more sophisticated applications, mostly based on large surface areas, such as filtering, absorption of various species or drug release. Biodegradability is required for easy waste disposal in the former case, or for performance improvement in the latter.

The present industrial applications of PHB-, or more generally, PHA-based foams, are modest because their preparation is not easy, mainly due to problems of thermal degradation, which are particularly serious in this context, since they make foaming difficult and produce cell collapse after the application of the foaming agents because of the low viscosity of the material.

Few examples of PHB foaming are to be found in the literature. In one of these, PHB was mixed with polyvinyl alcohol and starch, to prepare a composite, for which an optimal viscosity could be attained. Azodicarbon amide was used as the foaming agent. A biodegradability study revealed a positive effect of the larger surface on the rate of biodegradation in foamed products, compared with bulk materials with the same composition [196].

The development of foams for medical purposes was described by Radusch [193]. To eliminate thermal degradation, as well as to avoid the problems associated with the low melt viscosity, a process was developed based on solvent casting and subsequent particle leaching. Well-defined porous structures can be formed in such a way [197–200], for example, PHB was dissolved in chloroform, the ensuing solution was mixed with water-soluble particles like NaCl, films were then prepared by solvent casting and evaporation and, subsequently, the salt particles were washed out with water. Various foamed structures were thus prepared, depending on the particle size and amount, as well as other parameters, such as the PHB type and its concentration in the chloroform solution.

22.8.1.3 Fibres

The preparation of fibres based on PHA is not easy. Usually, exact procedures have to be developed and small deviations from the optimal process may be detrimental to spinning, drawing or to satisfactory final properties of the fibres. Melt spinning seems to be the simplest, although not easiest, way for PHA fibre preparation. For a delicate procedure such as spinning and subsequent drawing, the general processing problems are quite pronounced. Gel spinning was also described in the literature and reported to give fibres with a higher DR and improved mechanical properties, compared to PHB fibres made by melt spinning.

Several methods for both melt and gel spinning have been developed, differing in certain details of the process. A schematic description of the basic steps of these various procedures is given in Table 22.11.

The basic requirements for melt spinning/drawing seem to be the presence of nucleating agents, for example boron nitride [201], on the one hand and the drawing, or at least pre-drawing, while the material has not yet developed a fully crystalline structure, on the other hand. The drawability depends on the drawing temperature to a certain extent, producing the highest DR (about 5) at 120°C. Surprisingly, the drawn materials exhibit, in some cases, a rubber-like elastic behaviour, indicating that drawing leads to changes similar to cold rolling.

The surface of melt-spun fibres consists of many large spherulites, as indicated by SEM observations [210]. After drawing to DR = 6, the fibres have a fibrillar structure and their surface is fairly smooth. Annealing under tension leads to the formation of a more perfect fibrillar structure in the core, and this effect seems to be increasing with a rising tension load during annealing [210]. A study aimed at improving high speed melt spinning and spin drawing was reported by Schmack *et al.* [211]. A spinning line, consisting of an extruder, spinning pump, heated godets and two winders, enabled reaching speeds in the range 2 000–6 000 $m\,min^{-1}$.

22.9 APPLICATIONS

No high volume applications of PHAs exist; at present there are only trials and small pilot productions. Two perspective areas of applications are considered here, namely health care and packaging. In the field of medical and pharmaceutical applications, the hydrolytic degradation of PHAs is required as the main degradation mechanism, while in environmental applications, enzymatic degradation mechanisms (biodegradation) are relevant.

Isotropic PHB foils possess excellent barrier properties against gas permeation. This is the rationale for potential applications in packaging, especially for food. The substitution of polyethylene terephthalate (PET) for bottle production seems to be especially attractive, considering the volume of PET waste. Another option consists in the modification of paper since paper/poly(HB-*co*-HV) foils are completely biodegradable, unlike paper coated with conventional foils [212]. The simplest application is in containers, plastic bags and foils, commonly produced

Table 22.11

Procedures for the preparation of PHB fibres

Process steps	Remarks	Reference
Melt spinning		[201]
Melt spinning followed by a pre-orientation	Gordeyev *et al.*, 1977	
Melt quenching below T_g Subsequent drawing		[202, 203]
Melt spinning		
Drawing of melt spun PHB immediately afterwards	Pre-oriented material	
Drawing to high draw ratios	Pre-oriented fibres can be drawn after a few weeks of storing at RT	[201]
Dissolve PHB in chloroform	Molecular weight about 300 000 and T_m = 180°C	[204]
Filtrate the solution before spinning to remove impurities, as well as high molecular weight portions		
Spinning and pre-drawing (DR = 2)	Extruder heated in four regions between 170°C and 182°C	
Extruding	Filaments about 0.3 mm in diameter	[117]
Drawn at 110°C immediately after melt spinning	Maximum DR achieved = 6	
Gel spinning		
Dissolving the PHB in 1,2-dichloroethane	PHB concentration as high as possible	[205]
Solid gel prepared by evaporation of part of the solvent		
Gel extrudable at about 170°C		
Consequently processed in three stages		
Winding the fibre on a speed-controlled drum	Preconditioning stage	
Continuous hot drawing between two rollers at 120°C	Total DR around 10	
Fibres stretched at RT to 180%Fixed and annealed at 150°C for 1 h	DR = 1.8 of the length after second step	
Other procedures		
Solution-cast high molecular weight films stretching to DR higher than 6	Silicon oil bath at 160°C	[206]
Annealing	To avoid brittleness increase due to storing	
Centrifugation – spinning	Pores with diameter in the range 1–15 μm	[207]
PHB-co-higher PHA copolymers		
Solvent – cast films	PHB-hydroxyhexanoate copolymer	[110]
Melted in a hot press and subsequently quenched in ice water	More or less amorphous films	
Oriented by cold drawing	DR 2–5	
Annealed	Various temperatures, 23–140°C	
Blends		
Solvent casting	Poly(L-lactic acid) and PHB	[208]
Uniaxial drawing	At 2°C for PHB-rich blends (close to PHB's T_g) or 60°C for PLLA-rich blends	
Solvent casting	PHB-*co*-hydroxyvalerate and polyalcohols	[209]

from polyolefins. The first commercial product of this type was biodegradable bottles for shampoo produced by the German company Wella AG [213]. However, its volume was marginal, main reason being the high price of PHAs. Thus, the competition with polyolefins in the near future does not look too optimistic for PHAs. Perhaps a better chance can be seen in the partial substitution of PET bottles, where the difference in the price of the basic materials is less dramatic, although in this case, many technological details have to be solved, in particular concerning the low thermal stability and melt viscosity of PHAs.

The application in human medicine is more realistic, since a combination of biodegradability, hydrophobicity and biocompatibility, together with other interesting specific properties seems to be relevant [25, 214, 216]. PHB composites with apatite can be used as biodegradable bone fracture fixations or even as bone repair materials [217, 218], taking into account the piezoelectric properties of PHB, which are similar to those of bones. Another potential area is controlled drug release, where the advantages of PHAs are far from being fully exploited [215].

PHB fibres were considered to be mainly used for the production of scaffolds [219]. From the point of view of medical applications, an interesting paper by Schmack *et al.* deals with the effect of electron beam irradiation on the properties and degradation of PHB fibres with the aim of estimating the consequences of sterilization of medical devices by that technique. The application as scaffolds was also suggested for centrifugally spun fibres [207], since the fibres, treated with acid or alkali, had rather good adhesion towards cells and could be potentially used as wound scaffold. In this context, an interesting procedure [220] called upon PHA foam prepared by solution casting and subsequent solvent leaching of water-soluble particles, to be applied as scaffold in tissue regeneration. A very well-controlled open pore structure was purported to be suitable for this particular purpose. Other applications of PHAs have been reviewed [128, 212].

22.10 PRODUCERS

Several companies offer PHAs on a commercial basis, as seen in Table 22.12, although their production involves marginal amounts. Among these, Procter & Gamble is the only company producing higher PHAs, namely poly(hydroxyl-3-butyrate-*co*-hexanoate) under the trade name Nodax. The other companies offer PHB or PHBV. Monsanto produces around 800 tons per year of various grades of Biopol, using *Ralstonia eutropha*, having taken over the process developed by ICI and used for years by Zeneca Bio Products, a company formed by the partial splitting of ICI in 1993 [1]. Little is known about the production of Enmat made in China. Biomer in Munich produces the PHB and PHBV of the same trademark, which are reported to be top quality and of excellent purity, but expensive.

A range of PHBV copolymers with valerate comonomer contents varying between 0 and 24 per cent are also produced and sold in the US [221] and in Japan [222]. PHB was also produced on a commercial scale by Chemie Linz for a few years starting in the late 1980s. Unlike the ICI/Zeneca process, the fast-growing bacteria *Alcaligenes latus* were used, which synthesized the homopolymer at a much higher rate [223]. Somewhat misleading information relates to the Brazilian production. On the one hand, the web site [224] claims that the company is called Copersucar and the product Biocycle. On the other hand, a presentation authored by Braunegg, Bona, Kutschera and Ortega, which can be found on another web site [225] presents Biocycle as part of PHB Industrial S/A. Moreover, the production is quoted as taking place in Serrana in the state of Sao Paolo, whereas PHB Industrial S/A is owned by Irmaos Biagi S/A and the Balbo Group, which are large Brazilian producers of sugar and ethanol, with an equal 50 per cent share. Although the data on the company itself are not quite clear in the source cited, the process is well described, including its economy. *Wautersia eutropha* (formerly *Ralstonia eutropha* or *Alcaligenes eutrophus*) are used as the polymer-producing bacteria. The substrate is sucrose made from sugar cane, providing 0.33 kg of PHB per kg of sucrose. Since the yield of sucrose from sugar cane is estimated at 10.8 tons per hectare, the production of PHB per hectare is 3.6 tons. A power plant is also a part of the sucrose production, with bagasse, a waste of sugar cane, being used as fuel. The overall energy balance shows a

Table 22.12

Companies offering PHAs on a commercial basis. Information taken from www.biodegrad.net/biopolymer

Company	Product	Country	Web site
Monsanto–Metabolix	Biopol	USA	www.monsanto.com
Metabolix/ADM	Metabolix	USA	www.metabolix.com
Tianan	Enmat	China	www.fs-tianan.com
Copersucar	Biocycle	Brazil	www.copersucar.com.br
Biomer	Biomer L	Germany	www.biomer.de
Procter & Gamble	Nodax	USA	www.pg.com

total energy production of 12.61 MWh ha^{-1}, while the consumption is 0.95 for the sugar mill and 3.60 MHh ha^{-1} for the PHB production. Thus, a surplus of 8.06 MWh ha^{-1} is left, which can substantially improve the economy of the process. The pilot plant started in 1995 and in 2000 the scaling up raised the production to 50 t PHB per year in 2000. The construction of an industrial large-scale plant should start in 2007, and its production is planned to begin in 2008 with a starting annual output of 4 000 tons.

Currently, the P(4HB) homopolymer has been commercially produced by Tepha Inc., USA, in a large-scale fermentation facility of genetically engineered *E. coli* and developed as a new absorbable material for implantable medical applications [226].

22.11 OUTLOOK

PHAs, and especially PHB/PHBV, promise to be interesting materials for the future development of the plastic industry. However, in spite of a long history and growing interest, these materials seem to be far from reaching a large-scale production. The problems are associated with, on the one hand, some incompletely solved features regarding their properties, especially the design of appropriate processing technologies in relation to their low thermal stability and, on the other hand, the search for proper thermal stabilization, improving the toughness of PHB or, alternatively, lowering the price of all these materials.

It seems that although a number of medical applications are becoming a reality, their required amount of PHAs is not sufficient for running a plant at an economically feasible capacity, estimated to be at least 20 000 tons per year.

The market is ready for biodegradable plastics in many different areas of application, if they prove to have acceptable processing parameters, ultimate properties and viable prices. Since other biodegradable plastics are developing rapidly, for example polylactic acid (see Chapter 21) and starch (see Chapter 15), if PHAs should aspire to be among them, their competitive industrial production should start soon.

REFERRENCES

1. Braunegg G., Lefebvre G., Genser K.F., *J. Biotechnol.*, **65**, 1998, 127–161.
2. Billingham N.C., Henman T.J., Holmes P.A., in *Development in Polymer Degradation 7*, Ed.: Grassie N., Elsevier Science Publishers, Boston/London, 1987, p. 81. Chapter 3.
3. Sudesh K., Abe H., Doi Y., *Progr. Polym. Sci.*, **25**, 2000, 1503–1555.
4. Chowdhury A.A., *Archiv. Mikrobiol.*, **47**, 1963, 167–200.
5. de Konings, G.J.M., Prospects of bacterial poly[(R)-3-hydroxyalkanoates], thesis, TUE Eindhoven, 1993
6. Lemoigne M., *Ann. Inst. Pasteur Paris*, **39**, 1925, 144.
7. Lemoigne M., *Bull. Soc. Chim. Biol.*, **8**, 1926, 770–782.
8. Lenz R., Marchessault R.H., Bacterial polyesters: Biosynthesis, biodegradable plastics and biotechnology, *Biomacromolecules*, **6**, 2005, 1–8.
9. Lemoigne M., *C. R. Acad. Sci.*, **176**, 1923, 1761.
10. Lemoigne M., *C. R. Acad. Sci.*, **178**, 1924, 1093.
11. Lemoigne M., *C. R. Acad. Sci.*, **179**, 1924, 253.
12. Lemoigne M., *C. R. Acad. Sci.*, **180**, 1925, 1539.
13. Lemoigne M., *Bull. Soc. Chim. Biol.*, **9**, 1927, 446.
14. Morawetz H., *In Polymers: The origin and growth of a science*, John Wiley, New York, 1985. pp. 41–416.
15. Furukawa Y., *In Inventing Polymer Science*, University of Pennsylvania Press, Philadelphia, 1998.
16. Hocking P.J., Marchessault R.H., Biopolyesters, in *Chemistry and Technology of Biodegradable Polymers* (Ed. Griffin G.J.L.), Blackie Academic & Professional, London, Glasgow, NY, Tokio, Melbourne, Madras, 1994.
17. Stanier R.Y., Doudoroff M., Kunisawa R., Contopoulou R., *Proc. NAS*, **45**, 1959, 1246.
18. Macrae R.M., Wilkinson J.F., *Proc. R. Phys. Soc. Edin.*, **27**, 1958, 73–78.
19. Macrae R.M., Wilkinson J.F., *J. Gen. Microbiol.*, **19**, 1958, 210–222.
20. Williamson D.H., Wilkinson J.F., *J. Gen. Microbiol.*, **19**, 1958, 198.
21. Forsyth W.G.C., Hayward A.C., Roberts J.B., *Nature*, **182**, 1958, 800–801.
22. Doudoroff M., Stanier R.Y., *Nature*, **183**, 1959, 1440.
23. Oeding V., Schlegel H.G., *Biochem. J.*, **134**, 1973, 239.

24. Schlegel H.G., Gottschalk G., von Bartha R., *Nature*, **191**, 1971, 463–465.
25. Senior P.J., Dawes E.A., *Biochem. J.*, **134**, 1973, 225.
26. Stockdale H., Ribbons D.W., Dawes E.A., *J. Bacteriol.*, **95**, 1968, 1798.
27. Wallen L.L., Rohwedder W.K., *Environ. Sci. Technol.*, **8**, 1974, 576–579.
28. De Smet M.J., Eggink G., Witholt B., Kingma J., Wynberg H., *J. Bacteriol.*, **154**, 1983, 870–878.
29. Daves E.A., Senior P.J., *Adv. Microbiol. Physiol.*, **10**, 1973, 135–266.
30. Lundgren D.G.R., Alper C., Schnaitman C., Marchessault R.H., *J. Bacteriol.*, **89**, 1965, 245–251.
31. Cornibert J., Marchessault R.H., *J. Mol. Biol.*, **71**, 1972, 735–756.
32. Boatman E.S., *J. Cell. Biol.*, **20**, 1964, 297–311.
33. Merick J.M., Dourdoroff M., *Nature*, **189**, 1961, 890–892.
34. Ellar D., Lendgren D.G., Okamura K., Marchessault R.H., *J. Mol. Biol.*, **35**, 1968, 489–502.
35. Kominek L.A., Halvorsen H.O., *J. Bacteriol.*, **90**, 1965, 1251–1259.
36. Merrick J.M., Yu C.I., *Biochemistry*, **5**, 1966, 3563–3568.
37. Braunegg G., in *Sustainable Poly(hydroxyalkanoate) (PHA) Production Degradable Polymers: Principles and Application*, Ed.: Scott G., 2nd Edition, Kluwer Academic Publisher, Dordrecht/Boston/London, 2002, Chapter 8.
38. Holmes P.A., Wright L.F. Collins S.H. European Patent, 52,459, 1981.
39. Baptist J.N., US Patent, 3,044,942, 1962.
40. Doi Y., *Microbial Polyesters*, VCH Publishers, NY, 1990.
41. Kunioka M., Nakamura Y., Doi Y., *Polym. Commun.*, **29**, 1988, 174–176.
42. Anderson A.J., Dawes E.A., *Microbiol. Rev.*, **54**, 1990, 450–472.
43. Steinbüchel A., Polyhydroxyalkanoic acids, in *Biomaterials*, Ed.: Byrom D., MacMillan, Baingstoke, 1991, pp. 125–213.
44. Holmes P.A., in *Developments in Crystalline Polymers*, Ed.: Basset D.C., Vol. 2, Elsevier Applied Science, London, 1988, pp. 1–65.
45. Slater S.C., Voige W.H., Dennis D.E., *J. Bacteriol.*, **170**, 1988, 4431–4436.
46. Schubert P., Steinbuchel A., Schlegel H.G., *J. Bacteriol.*, **170**, 1988, 5837–5847.
47. Peoples O.P., Sinskey A.J., *J. Biol. Chem.*, **264**, 1989, 293–297.
48. Byrom D., *Trends Biotechnol.*, **5**, 1987, 246.
49. Holmes P.A., *Phys. Technol.*, **16**, 1985, 32.
50. Araki T., Hayase S., *J. Polym. Sci. Pol. Chem. Ed.*, **17**, 1979, 1877.
51. Doi Y., Kunioka M., Nakamura Y., Soga K., *J. Chem. Soc. Chem. Commun.*, **23**, 1986, 1696.
52. Doi Y., Tanaki A., Kunioka M., Soga K., *Appl. Microbiol. Biotechnol.*, **28**, 1988, 330.
53. Doi Y., Kunioka M., Nakamura Y., Soga K., *Macromolecules*, **21**, 1988, 2722–2727.
54. Kunioka M., Kawaguchi Y., Doi Y., *Appl. Microbiol. Biotechnol.*, **30**, 1989, 569.
55. Huisman G.W., de Leeuw O., Eggink G., Witholt B., *Appl. Environ. Microbiol.*, **55**, 1989, 1949–1954.
56. Kim Y.B., Lenz R.W., Fuller R.C., *Macromolecules*, **24**, 1991, 5256–5260.
57. Lenz R.W., Kim Y.B., Fuller R.C., *FEMS Microbiol. Rev.*, **103**, 1992, 207–208.
58. Abe C., Taima Y., Nakamura Y., Doi Y., *Polym. Commun.*, **31**, 1990, 44–406.
59. Doi Y., Abe C., *Macromolecules*, **23**, 1990, 3377–3705.
60. Bear M.M., Leboucherdurand M.A., Langlois V., Lenz R.W., Goodwin S., Guerin P., *React. Funct. Polym.*, **34**, 1997, 65–77.
61. Huijberts G.N.M., Eggink G., de Waard P., Huisman G.W., Witholt B., *Appl. Environ. Microbiol.*, **58**, 1992, 536–544.
62. Hiramitsu M., Doi Y., *Polymer*, **34**, 1993, 4782–4786.
63. Hiramitsu M., Koyama N., Doi Y., *Biotechnol. Lett.*, **15**, 1993, 461–464.
64. Abe H., Doi Y., Fukushima T., Eya H., *Int. J. Biol. Macromol.*, **16**, 1994, 115–119.
65. Fuchtenbusch B., Fabritius D., Steinbüchel A., *FEMS Microbiol. Lett.*, **138**, 1996, 153–160.
66. Stageman J.F., European Patent No. 124, 1984, 309.
67. Saito Y., Nakamura S., Hiramitsu M., Doi Y., *Polym. Int.*, **39**, 1996, 169–174.
68. Doi Y., Kunioka M., Nakamura Y., Soga K., *Macromol. Chem. Rapid Commun.*, **8**, 1987, 631–635.
69. Kunioka M., Nakamura Y., Doi Y., *Polym. Commun.*, **29**, 1988, 174–176.
70. Valentin H.E., Schönebaum A., Steinbüchel A., *Appl. Microbiol. Biotechnol.*, **36**, 1992, 507–514.
71. Nakamura S., Doi Y., Scandola M., *Macromolecules*, **25**, 1992, 4237–4241.
72. Steinbüchel A., Valentin H.E., Schönebaum A., *J. Environ. Polym. Degrad.*, **2**, 1994, 67–74.
73. Palleroni N.J., Palleroni A.V., *Int. J. Syst. Bacteriol.*, **28**, 1978, 416–424.
74. Chen G.-Q., König K.H., Lafferty R.M., *Antonie van Leeuwenhoek*, **60**, 1991, 61–66.
75. Ramsay B.A., Lomaliza K., Chavarie C., Dubé B., Bataille P., Ramsay J.A., *Appl. Environ. Microb.*, **56**, 1990, 2093–2098.
76. Nakamura K., Goto Y., Yoshie N., Inoue Y., Chujo R., *Int. J. Biol. Macromol.*, **14**(2), 1992, 117–118.

77. Ulmer H.W., Gross R.A., Posada M., Weisbach P., Fuller R.C., and Lenz, R.W., *Macromolecules*, **27**, 1994, 1675–1679.
78. Steinbüchel A., Debsi E.-M., Marchessault R.H., Timm A., *Appl. Microbiol. Biotechnol.*, **39**, 1993, 443–449.
79. Kang C.K., Lee H.S., Kim J.H., *Biotechnol. Lett.*, **15**, 1993, 1017–1020.
80. Caballero K.P., Karel S.F., Register R.A., *Int. J. Biol. Macromol.*, **17**, 1995, 86–92.
81. Takeda M., Matsuoka H., Ban H., Ohashi Y., Hikuma M., Koizumi J., *Appl. Microbiol. Biotechnol.*, **44**, 1995, 37–42.
82. Saito Y., Doi Y., *Int. J. Biol. Macromol.*, **16**, 1994, 99.
83. Lapčík, L'., Raab, M., *Nauka o materiálech*, Univerzita Tomáše Bati, Zlín, 2001.
84. Muller H.M., Seebach D., *Angew. Chem*, **105**, 1993, 483.
85. Bastioli C., *Polym. Degrad. Stabil.*, **59**, 1998, 263–272.
86. Agostini D.E., Lando J.B., Shelton J.R., *J. Polym. Sci. Partr A1*, **9**, 1971, 1789.
87. Poirier Y., Dennis D.E., Klomparens K., Somerville C., *Science*, **256**, 1992, 520–523.
88. Slater S., Gallaher T., Dennis D., *Appl. Environ. Microb.*, **58**, 1992, 1089.
89. Gerngross T.V., Snell K.D., Peoples O.P., Sinskey A.J., Cauhai E., Masamune S., Stubbe J., *Biochemistry*, **33**, 1994, 9311.
90. Zhang S., Lenz R.W., Goodwin S., in *Biopolymers: Polyesters I*, Eds.: Doi Y. and Steinbüchel A., Wiley-VCH, Weinheim, 2002, pp. 353–372.
91. Su L., Lenz R.W., Takagi Y., Zhang S., Goodwin S., Zhong L., Martin D.P., *Macromolecules*, **33**, 2000, 229.
92. Poirier Y., Dennis D., Klompareus K., Nawrath C., *FEMS Microbiol. Rev.*, **103**, 1992, 237.
93. Allenbach L., Poirier Y., *Plant Physiol.*, **124**, 2000, 1159–1168.
94. Valentin H.E., Mitsky T.A., Mahadeo D.A., Tran M., Gruys K.J., *Appl. Environ. Microb.*, **66**, 2000, 5253–5258.
95. ec Saruul P., Srienc F., Somers D.A., Samac D.A., *Crop Sci.*, **42**, 2002, 919–927.
96. Steinbüchel A., Doi Y., Biopolymers, Volume 4: Polyesters III-Applications and Commercial products., Wiley-VCH, Weinheim (Germany), 2002. 398 pp.
97. Bohmert K., Balbo J., Steinbüchel A., Tischendorf G., Willmitzer L., *Plant Physiol.*, **128**, April 2002, 1282–1290.
98. Seebach D., Fritz M.G., *Int. J. Biol. Macromol.*, **25**, 1999, 217.
99. Kyles R.E., Tonelli A.E., *Macromolecules*, **36**, 2003, 1125.
100. Haywood G.W., Anderson A.J., Williams G.A., Dawes E.A., Ewing D.F., *Int. J. Biol. Macromol.*, **13**, 1991, 83–88.
101. Byrom D., in *Plastics from Microbes: Microbial Synthesis of Polymers and Polymer Precursors*, Ed.: Mobley D.P., Hanser, Munich, 1994, pp. 5–33.
102. Kusaka S., Abe H., Lee S.Y., Doi Y., *Appl. Microbiol. Biotechnol.*, **47**, 1997, 140–143.
103. He Y., Zhu B., Inoue Y., *Prog. Polym. Sci.*, **29**, 2004, 1021–1051.
104. Iriondo P., Iruin J.J., Fernandez-Berridi M.J., *Macromolecules*, **29**, 1996, 5605–5610.
105. Iriondo P., Iruin J.J., Fernandez-Berridi M.J., *Polymer*, **36**, 1995, 3235–3237.
106. Zhu B., Li J., He Y., Osanai Y., Matsumura S., Inoue Y., *Green Chem.*, **5**, 2003, 580–586.
107. Amass W., Amass A., Tighe B., *Polym. Int.*, **47**, 1998, 89–144.
108. Shogren R.L., *J. Environ. Polym. Degr.*, **5**(2), 1997, 91–95.
109. Kotnis M.A., O'Brien G.S., Willett J.L., *J. Environ. Polym. Degr.*, **3**(2), 1995, 97–103.
110. Fischer J.J., Aoyagi Y., Enoki M., Doi Y., Iwata T., *Polym. Degrad. Stabil.*, **83**, 2004, 453.
111. Yoshie N., Saito Y., Inoue Y., *Macromolecules*, **34**, 2001, 8953–8960.
112. Doi Y., Kitamura S., Abe H., *Macromolecules*, **28**, 1995, 4822–4828.
113. Alper R., Lundgren D.G., Marchessault R.H., *Biopolymers*, **1**, 1963, 545–546.
114. Lundgren D.G., Pfister R.M., Merrick J.M., *J. Gen. Microbiol.*, **34**, 1964, 441–446.
115. Barnard G.N., Sanders J.K.M., *J. Biol. Chem.*, **264**, 1989, 3286–3291.
116. Kawaguchi Y., Doi Y., *FEMS Microbiol. Lett.*, **79**, 1990, 151–156.
117. Yamane H., Terao K., Hiki S., Kimura Y., *Polymer*, **42**, 2001, 3241.
118. Orts W.J., Bluhm T.L., Hamer G.K., Marchessault R.H., *Macromolecules*, **23**, 1990, 5368.
119. Barham P.J., Keller A., Otun E.L., Holmes P.A., *J. Mater. Sci.*, **19**, 1984, 2781.
120. Organ S.J., Barham P.J., *Polymer*, **34**, 1993, 2169.
121. Noda I., Marchessault R.H., Terada M., *Polymer Data Handbook*, Oxford University Press, Oxford (UK), 1999.
122. Marchessault R.H. et al, *Can. J. Chem.*, **59**, 1981, 38.
123. Abate R., Ballistreri A., Montaudo G., Giuffrida M., Impallomeni G., *Macromolecules*, **28**, 1995, 7911.
124. Scandola M., Cerrorulli G., Pizzoli M., *Macromol. Chem. Rapid Commun*, **10**, 1989, 47.
125. Koning G.J.M., Lemstra P.J., *Polymer*, **34**, 1993, 4089.
126. Hurrell B.L., Cameron R.E., *Polym. Int.*, **45**, 1998, 308.
127. Koning G.J.M., Scheeren A.H.C., Lemstra P.J., Peeters M., Reynaers H., *Polymer*, **35**, 1994, 4598.

128. Chodák I., Polyhydroxyalkanoates: Properties and modification for high volume applications, in *Degradable Polymers, Principles and Applications*, Ed.: Scott G., 2nd Edition, Kluwer Academic Publishers, Dordrecht/Boston/London, 2002, p. 295.
129. Chambers R., Daly J.H., Hayward D., Liggat J.J., *J. Mater. Sci.*, **36**, 2001, 3785–3792.
130. Hobbs J.K., *J. Mater. Sci.*, **33**, 1998, 2509–2514.
131. Karpátyová A., Chodák I., unpublished results.
132. Iordanskii A.L., Razumovskii L.P., Krivandin A.V., Lebedeva T.L., *Desalination*, **104**, 1996, 27.
133. Iordanskii A.L., Kamaev P.P., Ol'khov A.A., Wasserman A.M., *Desalination*, **126**, 1999, 139.
134. Kamaev P.P., Aliev I.I., Iordanskii A.L., Wassermanm A.M., *Polymer*, **42**, 2001, 515.
135. Innocenti-Mei L.H., Bartoli J., Baltieri R.C., *Macromol. Symp.*, **197**, 2003, 77–87.
136. Baltieri R.C., Innocenti-Mei L.H., Bartoli J., *Macromol. Symp.*, **197**, 2003, 33–44.
137. Choi J.S., Park W.H., *Polym. Test.*, **23**, 2004, 447–453.
138. Savenkova L., Gercberga Z., Nikolaeva V., Dzene A., Bibers I., Kalnin M., *Process Biochem.*, **35**, 2000, 573.
139. Janigová I., Lacík I., Chodák I., *Polym. Degrad. Stabil.*, **77**, 2002, 35–451.
140. Erceg M., Kovacic T., Klaric I., *Polym. Degrad. Stabil.*, **90**, 2005, 313–318.
141. Edenbaum J. (Ed.), *Plastics Additives and Modifiers Handbook*, 1992, Van Nostrand Reinhold, New York.
142. Tickner J.A., Rossi M., Haiama N., Lappe M., Hunt P., *The Use of Di(2-ethylhexyl) Phthalate in PVC Medical Devices: Exposure, Toxicity and Alternatives*, Center for Sustainable Production, Lowell, MA, 1999, http://www.sustainableproduction.org/downloads/DEHP%20Full%20Text.pdf.
143. *SpecialChem Newsletter*. Benzoate plasticizer for flexible PVC injection moulded toy applications. *Plastics Additives and Compounding*; July 23, 2001, http://www.specialchem4polymers.com/resources/articles/article.aspx?id=449.
144. Taylor R., in *Chemistry of Functional Groups*, Ed.: Patai S., Applied Science Publishers, London, 1979, Chapter 15.
145. Kim K.J., Doi Y., Abe H., *Polym. Degrad. Stabil.*, **91**, 2006, 769.
146. Grassie N., Murray E.J., Holmes P.A., *Polym. Degrad. Stabil.*, **6**, 1984, 95.
147. Grassie N., Murray E.J., Holmes P.A., *Polym. Degrad. Stabil.*, **6**, 1984, 47.
148. Rychly J., Csomorova K., Janigova I., Broska R., Bakos D., *Iran. J. Polym. Sci. Technol.*, **4**, 1995, 274–282.
149. Csomorova K., Rychly J., Bakos D., Janigova I., *Polym. Degrad. Stabil.*, **43**, 1994, 441–446.
150. Kopinke F.D., Remmler M., Mackenzie K., *Polym. Degrad. Stabil.*, **52**, 1996, 25–38.
151. Hoffmann A., Kreuzberger S., Hinrichsen G., *Polym. Bull.*, **33**, 1994, 355–362.
152. Weber E.J., Heusinger H., *Radiochim. Acta*, **4**, 1965, 92.
153. Albertsson A.-C., Karlsson S., *Acta Polym.*, **46**, 1995, 114–123.
154. Tormala P., Rokkanen P., Vainiopaa S., Laiho J., Hepponen V.-P., Pohjonen T., US Patent Application, 4,968,317, 1990.
155. Wang Y.D., Yamamoto T., Cakmak M., *J. Appl. Polym. Sci.*, **61**, 1996, 1957.
156. He J.-D., Cheung M.K., Yu P.H., Chen G.-Q., *J. Appl. Polym. Sci.*, **82**, 2001, 90–98.
157. Li S.-D., Yu P.H., Cheung M.K., *J. Appl. Polym. Sci.*, **80**, 2001, 2237–2244.
158. Kunioka M., Doi Y., *Macromolecules*, **23**, 1990, 1933.
159. Kim K.J., Doi Y., Abe H., Martin D.P., *Polym. Degrad. Stabil.*, **91**, 2006, 2333.
160. Hirt T.D., Neuenschwander P., Suter U.W., *Macromol. Chem. Phys.*, **197**, 1996, 1609.
161. Andrade A.P., Witholt B., Hany R., Egli T., Li Z., *Macromolecules*, **35**, 2002, 684.
162. Saad G.R., Seliger H., *Polym. Degrad. Stabil.*, **83**, 2004, 101.
163. Impallomeni G., Giuffrida M., Barbuzzi T., Musumarra G., Ballistreri A., *Biomacromolecules*, **3**, 2002, 835.
164. Luo S., Netravali A.N., *J. Appl. Polym. Sci.*, **73**, 1999, 1059.
165. Biresaw G., Carriere C.J., *Polym. Prepr.*, **41**, 2000, 64–65.
166. Dufresne A., Reche L., Marchessault R.H., Lacroix M., *Int. J. Biol. Macromol.*, **29**, 2001, 73–82.
167. Lee S.N., Lee M.Y., Park W.H., *J. Appl. Polym. Sci.*, **83**, 2002, 2945.
168. Chen Ch., Fei B., Peng S., Zhuang Y., Dong L., Feng Z., *J. Appl. Polym. Sci.*, **84**, 2002, 1789.
169. Chen C., Peng S., Fei B., Zhuang Y., Dong L., Feng Z., Chen S., Xia H., *J. Appl. Polym. Sci.* **88**, 2003, 659.
170. Lee M.Y., Park W.H., *Polym. Degrad. Stabil.*, **65**, 1999, 137.
171. Domenek S., Langlois V., Renard E., *Polym. Degrad. Stabil.*, **92**, 2007, 1384–1392.
172. Barham P.I., Keller A., *J. Polym. Sci. Polym. Phys. Ed.*, **24**, 1986, 69.
173. Luepke T., Radusch H.J., Metzner K., *Polymer*, **127**, 1998, 227–240.
174. Mai Y.-W., Yu Z.-Z. (Eds.), *Polymer Nanocomposites*, Woodhead Pulbisher Ltd, Boca Raton, USA, 2006.
175. Abe H., Doi Y., Kumagai Y., *Macromolecules*, **27**, 1994, 6012.
176. Gassner F., Owen A.J., *Polym. Int.*, **39**, 1996, 215.
177. Pearce R., Jesudason J., Orts W., Marchessault R.H., Bloembergen S., *Polymer*, **33**, 1992, 4647.
178. Yoshie N., Azuma Y., Sakurai M., Inoue Y., *J. Appl. Polym. Sci.*, **56**, 1995, 17.
179. Park S.H., Yoon J.S., Lee H.S., Choi S.J., *Polym. Eng. Sci.*, **35**, 1995, 1636.
180. Koyama N., Doi Y., *Can. J. Microbiol.*, **41**, 1995, 316.

181. He Y., Asakawa N., Inoue Y., *Polym. Int.*, **49**, 2000, 609.
182. Ha Ch.S., Cho W.J., *Prog. Polym. Sci.*, **27**, 2002, 759.
183. Jones D.S., Djokic J., McCoy C.P., Gorman S.P., *Biomaterials*, **23**, 2002, 4449.
184. Kim B.O., Woo S.I., *Polym. Bull.*, **41**, 1998, 707.
185. Chodak I., Mikova G., Slovak Patent Application 5082-2005-22, 2006.
186. Shafee E.E., *Eur. Polym. J.*, **37**, 2001, 451.
187. Xing P., Ai X., Dong L., Feng Z., *Macromolecules*, **31**, 1998, 6898.
188. Ceccorulli G., Scandola M., Adamus G., *J. Polym. Sci. Part B: Pol. Phys.*, **40**, 2002, 1390.
189. Abbate M., Martuscelli E., Ragosta G., Scarinzi G., *J. Mater. Sci.*, **26**, 1991, 1119.
190. Han C.C., Ismail J., Kammer H.W., *Polym. Degrad. Stabil.*, **85**, 2004, 947–955.
191. Lee J.C., Nakajima K., Ikehara T., Nishi T., *J. Polym. Sci. Part B: Pol. Phys.*, **35**, 1997, 2645.
192. Paglia E.D., Beltrame P.L., Cannetti M., Seves A., Marcanall B., Martuscelli E., *Polymer*, **34**, 1993, 996.
193. Radusch H.-J., Unconventional processing methods for poly(hydroxybutyrate), in *Handbook of Engineering Polymers*, Eds.: Fakirov S. and Bhattacharryya D., Hanser Publishing, Munich; Hanser Gardener Publications, Cincinnati, 2007, Chapter 24, pp. 717–746.
194. Lüpke Th., Radusch H.-J., Metzner K., Solid state processing of PHB-powders, *Macromol. Symp.*, **127**, 1998, 227–240.
195. Lüpke Th., Metzner K., Radusch H-J., Verfahren zur Forngebung von pulverformigen, thermisch instabilen Thermoplastformmassen, Patent DE 19637904.0, 18.09.1996, 1996.
196. Grosu E., Nemes E., Rapa M., Scheau A., Cornea P.C., Lupescu I., *Proceedings of the International Conference on Biofoams*, Capri, September 2007.
197. Radusch H-J., Lüpke Th., Grunz A., Le H.H., Macroporous foams from PHB for biomedical use – Manufacturing and properties, in *Conference Proceedings ICCE/9*, Ed.: Huy D., San Diego, University of New Orleans, pp. 647–648.
198. Radusch H-J., Lüpke Th., Grunz A. and Le H.H., Non-thermal route for preparation of biodegradable PHB foams, in *Proceedings of the Regional Meeting PPS 2003*, Eds.: Papaspyrides C.D. and Mitsoulls E., Athens, p. 61.
199. Radusch H-J., Lüpke Th., Grunz A., Le H.H., Unconventional methods for processing of thermal sensitive poly(hydroxyalkanontes) – Solid state processing and SC/PL foaming, *Proceedings of the 20th Annual Meeting of the Polymer Processing Society, PPS-20*, Eds.: Goettler I.A. and Isayev A.I. Akron, OH, pp. CD9.1–16.
200. Grunz A., Le H.H., Lüpke Th., Radusch H-J., *Gummi Fasern Kunstoffe*, **58**, 2005, 297–302.
201. Gordeyev S.A., Nekrasov Yu. P., Ward I.M., *IVth Int Symposium on Polymer for Advanced Technologies,* Leipzig, 1977, PVII.10.
202. Yokouchi M., Chatani Y., Tadakoro K., Teranishi K., Tani H., *Polymer*, **14**, 1973, 267.
203. Nicholson T.M., Unwin P.A., Ward I.M., *J. Chem. Soc. Faraday Trans.*, **91**, 1995, 2623.
204. Gordeyev S.A., Nekrasov Yu. P., *J. Mater. Sci. Lett.*, **18**, 1999, 1691.
205. Gordeyev S.A., Nekrasov Yu. P., Shilton S.J., *J. Appl. Polym. Sci.*, **81**, 2001, 2260.
206. Kusaka S., Iwata T., Doi Y.J., *Macromol. Sci. Pure Appl. Chem.*, **A35**, 1998, 319.
207. Foster L.J.R., Davies S.M., Tighe B.J., *J. Biomater. Sci. Polym. Ed.*, **12**, 2001, 317.
208. Park J.W., Doi Y., Iwata T., *Biomacromolecules*, **5**, 2004, 1557.
209. Cyras V.P., Fernandez N.G., Vazquez A., *Polym. Int.*, **48**, 1999, 705.
210. Yamane H., Terao K., Hiki S., Kawahara Y., Kimura Y., Saito T., *Polymer*, **42**, 2001, 7873.
211. Schmack G., Jehnichen D., Vogel R., Tändler B., *J. Polym. Sci. Pol. Phys.*, **38**, 2000, 2841.
212. Hocking P.J., Marchessault R.H., in *Biopolyeters; Chemistry and Technology of Biodegradable Polymers*, Ed.: Griffin G.J.L., Blackie Academic & Professional, London, Glasgow, NY, Tokio, Melbourne, Madras, 1994, p. 48.
213. www.usask.ca/agriculture/plantsci/classes/plsc416/projects_2002/immel/applications.html.
214. Williams S.F., Martin D.P., Horowitz D.M., Peoples O.P., *Int. J. Biol. Macromol.*, **25**, 1999, 111.
215. Eligio T., Rieumont J., Sánchez R., Silva J.F.S., *Die Angew. Makromolekulare Chemie*, **270**, 1999, 69.
216. Hoffman A.S., *Adv. Drug Deliv. Rev.*, **43**, 2002, 3.
217. Ni J., Wang M., *Mater. Sci. Eng. C*, **20**, 2002, 101.
218. Chen L.J., Wang M., *Biomaterials*, **23**, 2002, 2631.
219. Roth H., Patent EP 0567 845 B1, 1998.
220. Grunz A., Le H.H., Luepke T., Radusch H.-J., Gummi Asbest Kunstoffe, **58**, 2005, 297.
221. Luzier W.D., *Proc. Natl Acad. Sci. USA*, **89**, 1995, 839–842.
222. Lenz R.W., National Technical Information Service, *Report No. PB 95-199071*, US Department of Commerce, 1995.
223. Hrabak O., *FEMS Microbiol. Rev.*, **103**, 1992, 251–256.
224. www.biodegrad.net/biopolymer.
225. rns.tugraz.at/…/home/server/httpd//htdocs
226. Martin D.P., Williams S.F., *Biochem. Eng. J.*, **16**, 2003, 97.
227. Abate R., Ballistreri A., Montaudo G., *Macromolecules*, **27**, 1994, 332.

– 23 –

Proteins as Sources of Materials

Lina Zhang and Ming Zeng

ABSTRACT

Proteins are natural, renewable, and biodegradable polymers which have attracted considerable attention in recent years in terms of advances in genetic engineering, eco-friendly materials, and novel composite materials based on renewable sources. This chapter reviews the protein structures, their physicochemical properties, their modification and their application, with particular emphasis on soy protein, zein, wheat protein, and casein. Firstly, it presents an overview of the structure, classification, hydration–dehydration, solubility, denaturation, and new concepts on proteins. Secondly, it concentrates on the physical and chemical properties of the four important kinds of proteins. Thirdly, the potential applications of proteins, including films and sheets, adhesives, plastics, blends, and composites, etc. are discussed.

Keywords

Protein, Protein structure, Physicochemical properties, Industrial application, Denaturation, Soy protein, Zein, Wheat protein, Casein, Films, Adhesives, Plastics, Blends, Composites, 'Green' materials, Biodegradability

23.1 INTRODUCTION

The term protein comes from the Greek προτειυ. Proteins, as naturally occurring polymers, are important renewable resources produced by animals, plants, and bacteria. In terms of potential sources of materials, soy protein, zein, and wheat proteins (WP) are among the main plant proteins; casein, collagen protein, and silk fibroin represent relevant animal proteins; and lactate dehydrogenase, chymotrypsin, and fumarase constitute major bacterial proteins. A number of proteins have received much attention for the production of biodegradable polymers but, few have led to actual industrial scale-up due to performance difficulties and high production costs. In this chapter, an attempt is made to review the materials based on proteins derived from renewable resources abundantly available in nature, with particular emphasis on soy protein, zein, WP, and casein [1–4].

23.2 STRUCTURES

23.2.1 Primary structure

Proteins are natural macromolecules consisting of 20 different amino acid residues arranged in a highly sophisticated three-dimensional structure. Protein structures can be described at four levels as shown in Fig. 23.1 [5], including the primary structure (amino acid sequence), the secondary structure (conformation), the tertiary structure

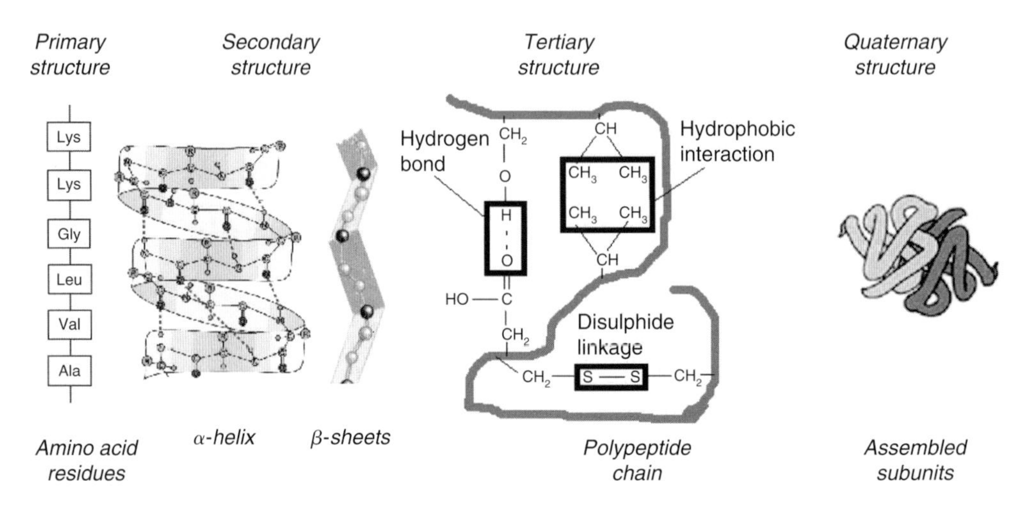

Figure 23.1 Proteins levels. (Reprinted with permission from Reference [5].)

(overall folding of the polypeptide chain), and the quaternary structure (specific association of multiple polypeptide chains). Amino acids have two functional groups, namely, an amino group ($—NH_2$) and a carboxyl group (—COOH). These groups are joined to a single (aliphatic) carbon, so the amino acids in proteins are all α-amino acids. The sequence of amino acids in a protein is termed its primary structure, and proteins are linear macromolecules formed by linking the α-carboxyl group of one amino acid to the α-amino group of another amino acid with a peptide bond (also called an amide bond). A peptide bond has several properties, such as resistance to hydrolysis, a planar geometry, a hydrogen-bond donor (the NH group) and a hydrogen-bond acceptor (the carbonyl group), the distinctive hydrogen bonding, and the uncharged peptide bond, which allows proteins to form tightly packed globular structures.

23.2.2 Secondary structure

The secondary structure of proteins includes helices, sheets, turns, loops, and random coils. The two most important secondary structures of proteins, the α helix and the β sheet, were predicted by Linus Pauling in the early 1950s. Pauling and Corey recognized that the folding of peptide chains, among other criteria, should preserve the bond angles and planar configuration of the peptide bond, as well as keep atoms from coming together so closely that they repelled each other through van der Waals interactions. Finally, Pauling predicted that hydrogen bonds must be able to stabilize the folding of the peptide backbone [6]. The α helix, a rod like structure, is formed by the hydrogen bonding of the backbone carbonyl oxygen of each residue to the backbone NH of the fourth residue along. The backbone atoms pack closely and form favourable van der Waals interactions, while the side chains project out from the helix. The amino acid residue, proline (without an NH group), interrupts the hydrogen-bonding pattern, leading to a kink in a helix. There are 3.6 residues per turn in the α helix; in other words, the helix repeats itself every 36 residues, with 10 turns of the helix in that interval. The pitch of the α helix is 1.5 Å and the number of residues per turn (5.4) is 3.6.

A β pleated sheet differs markedly from the rod like α helix. A polypeptide chain in a β sheet is almost fully extended rather than being tightly coiled as in the α helix. The β sheet involves hydrogen bonding between backbone residues in adjacent chains. In the β sheet, a single chain forms hydrogen bonds with its neighbouring chains, with the donor (amide) and acceptor (carbonyl) groups pointing sideways rather than along the chain, as in the α helix. The distance between adjacent amino acids along a β strand is approximately 3.5 Å. A β sheet is formed by linking two or more β strands by hydrogen bonds. In the β strand, the polypeptide chain is nearly fully extended.

Most proteins have compact, globular shapes, 'non-regular' structures requiring reversals in the direction of their polypeptide chains [7]. To make a spherical fold for globular proteins, the residues between regular helices and strands need to make sharp turns. Turns, reverse turns, or β-turns, were first recognized by Venkatachalam *et al* [8–10].

23.2.3 Tertiary structure

The tertiary structure of proteins describes the pattern of folding of secondary structures into a compact, more sophisticated molecule that can carry out biological functions. The tertiary structures of water-soluble proteins have the following common morphological features: (1) an interior formed of amino acids with hydrophobic side chains, and (2) a surface formed largely of hydrophilic amino acids that interact with the aqueous environment, directed by the hydrophobic interactions between the interior residues.

23.2.4 Quaternary structure

The tertiary-structured proteins may further associate into a higher degree of complexity, called quaternary structure. In most cases, the subunits are held together by non-covalent bonds. For example, haemoglobin has a molecular weight of 64 000 and is composed of four subunits, each of molecular weight 16 000. Salt bridges, hydrogen bonds, hydrophobic, and van der Waals interaction act in an additive fashion to specifically associate the subunits.

23.3 PHYSICOCHEMICAL PROPERTIES

23.3.1 Classification

Proteins can be assigned to one of three global classes, on the basis of shape and solubility: fibrous, globular, or membrane [11]. Fibrous proteins tend to have relatively simple, regular linear structures, and often serve structural roles in cells. The fibrous protein keratin forms structures such as hair and fingernails. The springy nature of wool is based on its composition of α helices that are coiled around and crosslinked to each other through cystine residues. Globular proteins are roughly spherical in shape. The polypeptide chain is compactly folded so that hydrophobic amino acid side chains are in the interior of the molecule and the hydrophilic side chains are on the outside exposed to the solvent. Consequently, globular proteins are very soluble in aqueous media. Globular proteins, such as most enzymes, usually consist of a combination of the two secondary structures, with important exceptions. For example, haemoglobin is almost entirely α-helical, and antibodies are composed almost entirely of β structures.

Membrane proteins are found in association with the various membrane systems of cells. For interaction with the non-polar phase within membranes, membrane proteins have hydrophobic amino acid side chains oriented outwards, and are insoluble in aqueous media.

23.3.2 Hydration and dehydration

Protein–water interactions play an important role in the maintenance of the three-dimensional structure of proteins, and their study has provided significant advances in our understanding of the involvement of water in protein functionality, stability, and dynamics [12]. Protein hydration is a process from its dry state to the solution state, which occurs over a very wide water activity range. For example, serine can be both a hydrogen-bond acceptor and donor, and is soluble in water as a result of the formation of hydrogen bonds. Serine on the inside of a protein, away from water, can form hydrogen bonds with other amino acids. Water molecules bind to both polar and non-polar groups in proteins via dipole–dipole, dipole-induced dipole, and charge–dipole interactions. The hydration of a protein, therefore, is related to its amino acid composition and is affected by solution conditions such as pH, temperature, and ionic strength [13]. Lyophilized proteins cannot go beyond their lowest hydration level of about 0.01 g H_2O per gram protein, namely about 8 mol of water per mole of protein [14]. When the protein is dehydrated to a certain level, its conformational flexibility decreases in order to maintain a local free-energy minimum. Therefore, the level of hydration significantly affects the biological properties (*e.g.* enzyme activity) of proteins [15]. In general, enzymes require only a small amount of water to express their catalytic activity [16].

The functional properties are illustrated by the water–protein interactions of soy protein isolate (SPI). SPI, with high solubility, or excessive thermally induced insolubilization, or compact calcium-induced aggregates, gives rise to low water-imbibing capacity (WIC) values. The highest WIC results from the balance between intermediate solubility and the formation of aggregates with good hydration properties. The hydration properties and viscosity of the SPI suspensions are strongly determined by the amount and properties of the insoluble fraction [17, 18].

23.3.3 Solubility

The solubility of proteins is an important property that affects and predicts other functional properties. The solubility of proteins is a thermodynamic manifestation of the equilibrium between protein–solvent and protein–protein interactions. Hydrophobic interactions promote protein intermolecular interactions, resulting in a decreased solubility. Polar and charged amino acid residues, on the other hand, promote protein–water interactions, resulting in increased solubility. As a rule, proteins containing more polar and charged groups, globular in shape, with relatively small molecular weights, have better solubility. The highly hydrophobic proteins, proteins with random structure, or highly aggregated protein polymers, are generally insoluble or unstable in solution. Many thermodynamic variables, such as temperature, pH, ionic strength, and other parameters of the solution system, affect protein solubility and compatibility. At constant pH and ionic strength, the solubility of most proteins increases with an increase in temperature. Intermolecular interactions significantly affect the overall solubility of the system, especially when the protein concentration is high. Processing equipment and stirring conditions also contribute to protein hydration and solubilization. The added energy for stirring gives a high degree of deformation of dispersed particles with a low (related to the dispersion medium) viscosity in flow. In the shear field of a protein solution system, the macromolecules may orient themselves, interact with each other more intensively, and self-associate or dissociate frequently. A weak association of protein molecules may break down in a shear field, thereby increasing the cosolubility of proteins in a multicomponent solution system.

23.3.4 Denaturation

Native proteins having biological functions can exist only in living organisms. With respect to the overall structure of a native protein, any conformational change could mean a certain degree of denaturation. The loss of structural order in these complex macromolecules, so-called denaturation, is accompanied by a loss of function. The nitrogen solubility index, or protein dispersibility index, expresses the percentages of the total content of nitrogen and protein, respectively, and also indicates the extent of protein denaturation. A more 'restrictive' definition of denaturation, therefore, specifies the loss of the most characteristic properties of the protein, such as enzyme activity and solubility. Theoretically, proteins have to maintain their natural conformation to perform their specific functions, hence any factor that changes the conformation of proteins is a potential source of denaturation. Therefore, the denaturation of proteins includes thermal denaturation, denaturation by changing the pH, by using high concentrations of urea, guanidinium chloride and other guanidinium salts, inorganic salts, by organic solvents, and by detergents. Heat is the most common factor that causes the denaturation of proteins. Heat treatment of proteins increases molecular motion, leading to the rupture of various intermolecular and intramolecular bonds of the native protein structure. As a physicochemical process, protein denaturation can be reversible or irreversible, depending on the process of denaturation and the conditions after denaturation. Denatured proteins may refold back to their original structure and resume their biological functions. The process of protein folding, unfolding, and refolding is still an attractive research area, although it has been extensively studied for decades [19]. Methods normally used to denature proteins include exposure to heat, acid/alkali treatment, or the addition of organic solvents, detergents and urea. Wet and dry heat, grinding, freezing, pressure, irradiation and high frequency sound waves can also be used for denaturing proteins [20].

23.4 IMPORTANT PROTEINS AS SOURCES OF MATERIALS

23.4.1 Soy protein

The soybean plant called 'The Gold that Grows', originated in China, has attracted attention in recent years because of its versatile uses [21]. Soybeans contain approximately 42 per cent protein, 20 per cent oil, 33 per cent carbohydrates, and 5 per cent ash on a dry basis. The storage soy proteins consist of a mixture of proteins (α-,β-, and γ-conglycinins, glycinin, and other globulins) ranging in molecular weight from about 140 to 300 kDa and differing in physicochemical and other properties.

Soy protein has an isoelectric point at pH 4.5. The numerical coefficient is the characteristic sedimentation constant in water at 20°C. Four such fractions are separable and are designated as 2S, 7S, 11S, and 15S. The two major fractions are 7S (35 per cent) and 11S (52 per cent). The 7S fraction is highly heterogeneous, and its principal component is β-conglycinin with a molecular mass of the order of 150–190 kDa. The 11S fraction consists of glycinin with a molecular mass of 320–360 kDa. 11S is a quaternary structure composed of three acidic and three basic subunits of 35 and 20 kDa with isoelectric points between pH 4.7–5.4 and 8.0–8.5, respectively. The other minor fractions are 2S (8 per cent) and 15S (5 per cent). The glass transition temperature of soy 7S and 11S globulin fractions, isolated from defatted soy flour (DSF) by the method of Thanh and Shibasaki, was studied as a function of moisture content, using differential scanning calorimetry (DSC). The DSC scans of 7S and 11S fractions with 10 per cent water content showed an endothermic transition at 120°C and 150°C, respectively [22]. There are two glass transitions (T_{g1} and T_{g2}) in the soy protein plasticized with glycerol, corresponding to glycerol-rich and protein-rich domains, respectively. The T_{g1} of the sheets decreases from −28.5°C to −65.2°C with an increase of glycerol content from 25 to 50 wt per cent, whereas the T_{g2} is almost invariable at about 44°C [23]. The radius of gyration of protein-rich domain ranges from 59 to 60 nm, suggesting the existence of a stable protein domain. Their endothermic peaks at about 100°C on the DSC curves are assigned to the evaporation of residual moisture in the samples. The improvement of the functional and the mechanical properties of proteins by altering their molecular structure or conformation through physical, chemical, or enzymatic agents at the secondary, tertiary, and quaternary levels, has been well documented in the literature [24, 25]. Although soybeans have been consumed as food for thousands of years, soy polymer technology can create an age of green plastic in the twenty-first century [26].

23.4.2 Zein

Zein is extracted from corn grain and has been examined as a possible raw material for polymer applications. It is one of the few cereal proteins extracted in a relatively pure form and constitutes a unique and complex material, zein comes from the alcohol soluble protein of corn, classified as a prolamin, and is the principle storage protein of corn and constituting 44–79 per cent of the endosperm protein. Biologically, zein is a mixture of proteins varying in molecular size and solubility, which can be separated by differential solubility to give four related zeins with distinct types:α, β, and γ and δ. Commercial zein is made up of α-zein, which is by far the most abundant, accounting for around 70 per cent of the total. Zein is known for its solubility in binary solvents containing a lower aliphatic alcohol and water, such as aqueous ethanol or isopropanol. The molecular weight of zein lies in the range of 9.6–44 kDa [27]. For zein to reach its full potential, research must find ways to overcome two main problems, *viz.* a prohibitive cost and a poor resistance to water. The latter drawback makes zein unacceptable for some applications for which it had shown great promise in the past, such as the processing of films and coatings. Moreover, zein's soft, ductile nature after being precipitated from a solvent, promises interesting applications as a plastic material, either alone or in blends [28].

23.4.3 Wheat gluten

Wheat gluten is composed of a mixture of complex protein molecules that can be separated into glutenins and gliadins on the basis of their extractability in aqueous ethanol. Gliadin proteins ($M_W < 50$ kDa) provide the viscous component of gluten and constitute a heterogeneous group of proteins characterized by single polypeptide chains associated by hydrogen bonding and hydrophobic interactions, having intramolecular disulphide bonds, and being soluble in a 70 per cent ethanol/water solution. Glutenins comprise a diverse number of protein molecules grouped into high molecular weight subunits from 80 to several million kDa. Hydrogen bonding between the repeat regions of high molecular weight proteins has been found to be responsible for the elasticity of gluten. Glutenins with low molecular weights between 30 and 80 kDa are partially soluble in a 70 per cent ethanol/water mixture. Gliadins and glutenins are present in almost equal quantities in wheat gluten and have similar amino acid compositions, being high in both glutamine and proline. They also have a considerable number of non-polar amino acids containing aliphatic or aromatic groups. These groups, together with a few readily ionizable amino acids, are responsible for the insolubility of gluten in water. The amount, size distribution, and molecular architecture of glutenins and gliadins greatly influence the rheological, processing, mechanical, and physicochemical properties of gluten [29]. WP are one of the most important resources with the lowest price among plant proteins. They have good viscoelastic properties, strong tensile strength (TS) and excellent barrier properties for gas and water. The investigation of wheat protein based materials has greatly attracted the attention of chemists and material scientists in recent years [30].

23.4.4 Casein

Casein is the major component (80 per cent) of milk, with molecular weights between 1 and 20 kDa and includes four distinct types: α–s1, α–s2, β, and k. Casein is the predominant phosphoprotein that precipitates at pH 4.6 (20°C) and is characterized by an open, random coil structure. By treating acid-precipitated caseins with alkali solution caseinates are produced. Both caseins and caseinates form transparent films from aqueous solutions without any treatment because of their random coil nature and numerous hydrogen bonds. Caseins have shown to be useful in adhesives, micro encapsulation, food ingredients, and pharmaceuticals [31].

23.5 POTENTIAL APPLICATIONS AS MATERIALS

23.5.1 Films and sheets

Soy protein, zein, gluten, rapeseed protein, casein, and collagen are currently being investigated to prepare edible and non-edible films. Most protein films are produced by casting from solutions, in which proteins, plasticizers, and other agents are dissolved in an appropriate solvent. Extrusion, which is widely used in the plastics industry, is an alternative method that needs to be investigated for the industrial production of protein films. This is a challenge to researchers, since few reports have been published up to now. The thermoplastic behaviour of proteins has been exploited to make films by thermal or thermomechanical processes under low moisture conditions. For soy protein films, heat-curing reduces moisture content, water vapour permeability, elongation, and total soluble matter (TSM), and increases total colour difference and TS pressure, individually and interactively with temperature, significantly affects the film moisture content, TS and TSM [32].

The term edible film is defined as a free standing thin layer of edible material which can be used as a food product or a wrapper for foods. As an edible food wrapper, it has many advantages over the conventional non-edible wrappers, including (1) its biodegradable which makes it consumable with the packaged product, (2) its protective shell which preserves the quality of the packaged food and prolongs its shelf life, and (3) the possibility to load it with additives to enhance the sensory and nutritional properties of the food. Edible films from plant proteins have been investigated extensively in recent years and new products are continuously being developed [33].

A notable feature of soy proteins is the strong pH dependence of the molecular conformation and the associated functional properties, such as surface activity, film structure, surface dilatational viscoelasticity and, especially, the rate of adsorption at a fluid interface. Optimum functionality occurs at pH < 5, which limits the application of soy globulins as food ingredients [34]. Whole soy flour and apple pectin have also been used as raw materials for producing hydrocolloid edible films. The best ratio between the two components (2:1 $mg\,cm^{-2}$, pectin:soy flour) was determined in order to obtain films which could be perfectly handled for their consistence. Films have also been prepared in the presence of transglutaminase, an enzyme able to produce isopeptide bonds among the soy polypeptide chains. The latter films showed a smoother surface and a higher homogeneity, as demonstrated by microstructural analyses, whereas studies of their mechanical properties indicated that transglutaminase increased their strength and reduced their flexibility [35]. Calcium salts crosslinking interactions with SPI and glucono-δ-lactone (GDL) gave rise to edible films with rigid three-dimensional structure. GDL contributed to the formation of a homogeneous film structure due to increased protein–solvent attraction. The TS of calcium-sulphate-treated SPI films (8.6 MPa) is higher than that of calcium-chloride counterparts (6.4 MPa) and then that of the control (5.5 MPa). The puncture strength (PS) of calcium-sulphate-treated SPI films (9.8 MPa) is higher than that of the calcium-chloride counterparts (8.5 MPa) and then that of the control (5.9 MPa). Moreover, SPI films formulated with GDL have a higher elongation at break (39.4 per cent) than that of the control (18.2 per cent). Calcium salts and GDL-treated SPI films have a lower water vapour permeability than that of the control [36].

The addition of a plasticizer to make a very good film is essential, since it modifies the three-dimensional organization of the proteins, decreases their attractive intermolecular forces, resulting in a decreased cohesion, elasticity, mechanical properties, and rigidity [37–43]. The most used plasticizers are glycerol and sorbitol, and acetamide, used to plasticize soy protein, has also been recently reported [44]. The effects of plasticizers (glycerol, sorbitol, and 1:1 mixture of glycerol and sorbitol) on the moisture sorption characteristics of hydrophilic SPI films have been investigated at three levels of plasticizer concentration (0.3, 0.5, and 0.7 g plasticizer per gram SPI). Under given relative humidity (RH) conditions, films with higher glycerol ratios absorbed more moisture with higher initial adsorption rates, and films with higher plasticizer contents exhibited higher equilibrium moisture contents. Plasticizer and absorbed water loosened the film synergistically, resulting in a higher elongation but a lower TS. Films with lower glycerol contents were more sensitive to RH variations, as compared to those richer in glycerol [45].

The addition of the anionic surfactant sodium dodecyl sulphate (SDS) into glycerin-plasticized SPI leads to changes in TS, solubility, and water vapour barrier properties of the corresponding films. This can be attributed to the disruption of hydrophobic associations among neighbouring protein molecules, as the non-polar portions of the SDS molecules attach themselves onto the hydrophobic amino acid residues within the film structure. It has been demonstrated that adding SDS to film-forming solutions prior to casting, greatly modifies the properties of the ensuing SPI films. In particular, SDS can improve the water vapour barrier ability and the extendibility of the films, both desirable attributes when assessing the potential of such films for packaging applications [46].

Heat treatment is well-known to generate crosslinks in some proteins, such as soy and whey. Indeed, heat favours soy protein crosslinking by disturbing the protein structure and exposing sulphydryl and hydrophobic groups. Sulphydryl groups have been reported to be responsible for the formation of disulphide linkages which generate a three-dimensional network [47]. To elucidate the effect of γ-irradiation on the physicochemical properties, the molecular and mechanical properties of the SPI films were examined after the γ-irradiation of the film-forming solution at various radiation doses. The γ-irradiation causes the disruption of the ordered structure of the soy protein molecules, as well as degradation, crosslinking, and aggregation of SPI in solution, leading to a decrease in its viscosity. However, the mean TS of the SPI films increased by a factor of 2 after γ-irradiation, as a result of the reduction in water vapour permeability by 13 per cent. The microstructures observed by scanning electron microscopy (SEM) showed that the irradiated SPI films had a smoother and glossier surface than the control film [48]. In addition, γ-irradiation combined with thermal treatments have been used to crosslink sterilized biofilms based on SPI (S system) and a 1:1 mixture of SPI and whey protein isolate, WPI (SW system). This double treatment improved significantly the mechanical properties, namely, the puncture strength and puncture deformation, for all types of protein films. The incorporation of carboxymethylcellulose (CMC) also showed a significant improvement in water vapour permeability for irradiated films of the S system and for non-irradiated films of the SW system.

Zein has been tested as a possible polymeric material because of its film-forming ability. Zein-based plastic sheets and films have been formed by extrusion through a slit-die or blowing head. Zein was plasticized with oleic acid and formed into a wet mouldable mass (resin) to feed the extruders. Both single- and twin-screw extruded sheets showed higher elongation at break, lower TS, and lower Young's modulus than non-extruded samples. Stress–strain plots for extruded samples gave evidence of plastic behaviour. Blown film extrusion can be affected by feed moisture content and barrel temperatures. The optimal moisture content was determined at 14–15 per cent, while temperature at the three extruder zones was maintained at 20–25°C, 20–25°C, and 35°C, respectively. Temperature at the blowing head was 45°C. Film samples blown after either single- or twin-screw extrusion indicated similar tensile properties to those of slit-die extruded samples [49].

23.5.2 Adhesives

Protein-based adhesives can be traced back to the beginning of the last century. Blends of adhesive grade soy flour with casein (ground or screened) have been prepared to obtain glues with composite performance for making panels and flush door assemblies. Composite adhesives for making plant fibre boxes for food were obtained by blending SPI with varying amounts of poly(vinyl alcohol) or poly(vinyl acetate) [50]. The formation of electrostatic/covalent complexes upon mixing SPI with sodium alginate/PGA (propylene glycol alginate) under alkaline conditions has been reported [51]. Films formed from covalent complexes had greater stability in water as compared to those obtained from protein–alginate complexes. The adhesive performance of soybean proteins is dependent on the particle size, the nature of surface, the structure of the protein, its viscosity and pH. Other factors, which can affect their performance, are the processing parameters such as the press temperature, the pressure, and the time [52]. The major advantage of soy glues is that they can be cured either hot or cold. The major disadvantages of soybean protein-based adhesives are low gluing strength and poor water resistance [53]. To improve the water performance,

SPIs have been modified using SDS and guanidine hydrochloride (GuHCl). The SDS-modified SPI containing 91 per cent protein has a water-soluble mass of 1.7 per cent. To be considered as a water-resistant adhesive, its water-soluble mass of adhesive should be less than 2 per cent. The wet shear strength test showed 100 per cent cohesive failure within the fibreboard, indicating that the modified SPI had good water resistance. Drying treatment significantly affected the final adhesion performance. Its shear strength did not change much, but the percentage of cohesive failure within the fibreboard increased markedly as the drying temperature was increased. All the unsoaked, soaked, and wet specimens glued by the adhesives treated at 70°C or 90°C had 100 per cent cohesive failure within fibreboard. The viscosity also increased greatly with an increase in the drying temperature [54, 55].

Citric acid (CA) was thermochemically reacted with food quality SPI, distillers dried grains (DDG), produced from corn dry milling, and corn gluten meal (CGM), produced from corn wet milling, to generate acid-stable products with enhanced metal-binding properties. CA dehydrates at high temperature to form the corresponding anhydride that can interact with the nucleophilic functional groups of proteins or carbohydrates to generate ester or acyl derivatives. The effects of temperature, CA concentration, pH, and reaction time were evaluated to show that SPI, DDG, and CGM, when heated in the range of 110–120°C with CA at 1:1 w/w ratio, under endogenous acidic conditions for 24 h, yielded products with reaction efficiencies >60 per cent, which possessed 4.13, 4.19, and 4.26 mmol COOH per gram, respectively. FTIR data of the original heated proteins, compared with their respective CA products demonstrated additional peaks indicative of ester and carboxyl linkages. The SPI/CA, DDG/CA, and CGM/CA products effectively bound 1.18, 1.07, and 0.98 mmol of Cu^{2+}per gram, respectively, when analysed by ion plasma spectrometry. Solid-state NMR supported the metal-binding characteristics of the CA reaction products and demonstrated that Al^{3+} was bound ionically to their carboxyl groups. Amino acid composition studies showed diminished amounts of amino acids with nucleophilic reactive groups in all three CA reaction products. The CA reaction products were highly resistant to acid hydrolysis with 6 N HCl for 4 h at 145°C. Moreover, the products generated possessed cation-exchange capabilities and a potential biodegradability that may have an outlet for industrial wastewater treatment [56].

Recently, urea and urease inhibitor *N*-(*n*-butyl) thiophosphoric triamide (*n* BTPT) were used to modify wheat straw–soy flour particleboards. Boric and citric acid, along sodium hypophosphite monohydrate, were used to modify soy carbohydrates. Particleboard bonded by urea and high concentrations of *n* BTPT-treated soy flour showed improved mechanical properties, whereas that bonded by boric acid-treated soy flour had better water resistance. The adhesive made from soy flour treated with 1.5 M urea, 0.4 per cent *n* BTPT, 7 per cent CA, 4 per cent NaH_2PO_2, 3 per cent boric acid, and 1.85 per cent NaOH, produced particleboard with the maximum mechanical strength and water resistance [57].

23.5.3 Plastics

A great deal of research of soybean plastics was conducted in the 1930s and 1940s. At that time, soybean products were incorporated into phenolic resins mainly as filler or extender to decrease the cost of the plastics, because petroleum was expensive whereas soybeans were abundantly available. Decreasing petroleum prices and better-performing petroleum-based plastics, dominated the market after World War II. At that time, Henry Ford mixed soy protein with phenol–formaldehyde resin to produce automobile body parts [58]. Brother and Mckinney reported making plastics by using soy protein and various crosslinking agents [59]. Most plastics, at present, are petroleum-based and do not degrade over many decades under normal environmental conditions. As a result, efforts towards developing environment-friendly and biodegradable 'green' plastics for various commercial applications have gained significant momentum in recent years. SPI-based 'green' plastics have been shown to suffer from high moisture sensitivity and low strength. These properties have limited their use in most commercial applications.

In addition, SPI-based plastics are also difficult to process into sheets without any plasticizer. Thermoplastic sheets prepared from SPI with ethylene glycol (EG), as the plasticizer, have been obtained by compression moulding under a pressure of 15 MPa at 150°C. With increasing EG content, the TS and Young's modulus decreased and the elongation at break increases. The water resistance of the thermoplastic sheet of SPI increased with an increase in the EG content and was much higher than that of thermoplastic starch sheets or cellulose films. Further investigation has been carried on SPI sheets containing 50 per cent EG, which display a maximum water resistance in boiling water, good mechanical properties, and a light transmittance of 82 per cent at 800 nm because of the interchain hydrogen bonds and novel crystals [60].

The effects of water, glycerol, methyl glucoside, $ZnSO_4$, epichlorohydrin, and glutaric dialdehyde on the mechanical properties of soy protein plastic sheets have been studied. The thermal transition temperatures and dynamic mechanical properties of soy protein plastics have also been investigated in terms of the effect of the moisture and glycerol contents on their properties. The glass transition temperatures of the sheets varied from *ca.* −7°C to 50°C with moisture contents ranging from 26 to 2.8 per cent and 30 parts. And the moisture range cited is right of glycerol. The soy protein plastic sheets are usually in their glassy states at room temperature, unless they contain a high level of moisture. The β-transitions of soy protein plastic sheets lie in the range of −33°C to −72°C depending on their moisture content. In the presence of two parts of $ZnSO_4$, the water absorption of the soy protein sheets decreased by 30 per cent. Therefore, soy protein sheets absorbed or lost moisture, depending on the RH of the environment [61].

The effect of storage time on the thermal and mechanical properties of SPI plastics was also investigated and showed that the glass transition temperature and the dynamic storage modulus increased and the loss tangent decreased during storage. The excess enthalpy of relaxation of SPI plastics has an exponential relationship with the storage time, indicating a fast aging rate at the beginning of the storage. The SPI plastics containing glycerol have the slowest aging rate and are fairly stable after 60 days, with about 8.8 MPa TS and 168 per cent strain at break. Urea-modified SPI plastics also displayed slow aging and became relatively stable after 60 days, with about 10 MPa TS and 72 per cent elongation [62, 63].

Processing and modification routes of proteins have been used to produce and to improve properties of biodegradable plastics. SPI, acid-treated and crosslinked by acetic acid and glyoxal have been subsequently compounded, extruded, and injection moulded. Heat treatment is also used as a possible methodology to crosslink the protein structure. The moulded specimens are tested in terms of their tensile properties and solubility at different pH, and are also evaluated for their degree of crosslinking and molecular weight distribution. The ensuing plastics are rigid and brittle with stiffness ranging from 1 436 MPa for SPI, to 1 229 MPa for glyoxal crosslinked SPI, up to 2 698 MPa for heat-treated SPI. The solubility profiles have been studied as a function of the pH of the immersion solutions and the crosslinking degree of each material. A reduction in protein solubility with decreasing pH was observed, with a minimum between pH 4 and 5 and a resolubilization of the protein at pH lower than 4 and greater than 8. Higher levels of crosslinking resulted in a decrease in the solubility and an aggregation of the protein molecules [64, 65].

A great deal of research has been recently focusing on the modification of waterborne polyurethane (WPU) with natural polymers including proteins. WPUs grafted, mixed and crosslinked with casein were prepared, respectively, by incorporating casein into the WPU and its prepolymer aqueous dispersion [66–68]. The particle size and content of casein play an important role in enhancing the mechanical properties of the composite sheets. The WPU/casein sheets possessed good mechanical properties, optical transmittance, and miscibility between the two components. Moreover, blend sheets crosslinked with ethanedial were successfully prepared. By introducing ethanedial into WPU/casein (1:1 by weight), the mechanical properties and water resistivity of the blend materials were enhanced, obviously as a result of the formation of a network. When the ethanedial content was about 2 wt%, the blend sheets showed significantly higher TS, water resistivity, thermal stability, and kept roughly the same elongation at break as those of uncrosslinked blends.

Soy dreg (SD) is an abundant by-product from the isolation process of soy protein with only a tenth of the price of SPI in China. Thus, using SD as the raw material for preparing biodegradable plastics is not only helpful in solving the environmental problems, but also in enhancing the value of agricultural by-products. Water-resistant composite plastics have been prepared from SPI or SD, poly(3-caprolactone) and toluene-2,4-diisocyanate as the compatibilizer by blending and one-step reactive extrusion, followed by compression moulding [69]. The resultant SPI and SD composite materials exhibited high water resistance and good TS (14.8 MPa for SPI-35 and 16.3 MPa for SD-35). Moreover, the SD sheets containing cellulose possessed a higher TS than those of the SPI series, when the SD content was 30–35 per cent, whereas the latter had a better biodegradability and water resistance. By burying the two materials in soil and culturing them in a mineral salt medium containing microorganisms, both of them were almost completely degraded.

23.5.4 Blend and composite materials

The blending of two or more polymers, as a simple and convenient procedure, is very important for preparing biodegradable materials based on renewable resources. The general methods include casting from solution, modification during extrusion, semi-interpenetrating polymer networks (semi-IPN), and nanocomposites [70, 71]. Microporous membranes have been prepared by blending cellulose and SPI in an NaOH/thiourea aqueous solution. Cellulose, immersed in a 6 wt% NaOH/5 wt% thiourea was medium, was kept below −8°C for 12 h, and then stirred vigorously at 20°C for 1 h to obtain a cellulose solution. SPI was dispersed in the solvent to form a slurry with an SPI content of 40–50 wt%. The mixture of the cellulose and SPI solutions was cast on a glass plate to give a gel sheet, and then coagulated with a 5 wt% H_2SO_4 aqueous solution to generate transparent membranes. The blend membranes were hydrolyzed with a 5 wt% NaOH aqueous solution, and dried in air, coded as CS2-40 and CS2-50. The resulting microporous membranes kept a high TS in both dry and wet states. Interestingly, the cellulose/SPI microporous membranes containing a small amount of SPI can be suitable for the culture of Vero cells as shown in Fig. 23.2 [72]. Therefore, the membranes are good candidates for application in separation technology and biomedical fields.

Blends of soy protein and biodegradable polyesters have been prepared using glycerol as a compatibilizing agent. Good miscibility was obtained only when the soy protein was initially combined with glycerol under high shear at elevated temperatures in an extruder. The extrusion conditions and appropriate screw configuration were the critical factors that affected the reactivity of the protein and hence, the properties of the blends. Under these conditions, partial denaturing of the soy protein leads to specific interactions between the functional groups of the protein and the glycerol. Screws with large kneading blocks that produced high shear mixing were preferred and led to thermoplastic blends with high elongation and TS. Moreover, under appropriate processing conditions, even this low grade protein could be compounded to yield low cost, biodegradable materials that could replace low density polyethylene products [73].

The aqueous dispersion of DSF containing soy protein, soy carbohydrate, and soy whey, has been blended with a styrene–butadiene latex to form elastomer composites. The inclusion of soy carbohydrate increased the tensile stress in the small strain region, but reduced the elongation at break. The inclusion of soy carbohydrate and soy whey also improved the recovery behaviour in the non-linear region. At small strain, the shear elastic modulus of 30 per cent filled composites at 140°C, is about 500 times higher than that of the unfilled elastomer, indicating a significant reinforcement effect generated by DSF. Compared with SPI, the stress softening effect and recovery behaviour under dynamic strain indicated that the addition of soy carbohydrate and soy whey had enhanced the filler–rubber interactions [74]. Blends of SPI with 10, 20, 30, 40, and 50 per cent poly(ethylene-*co*-ethyl acrylate-*co*-maleic anhydride) (PEEAMA), with or without the addition of 2.0 wt% methylene diphenyl diisocyanate (MDI), were prepared

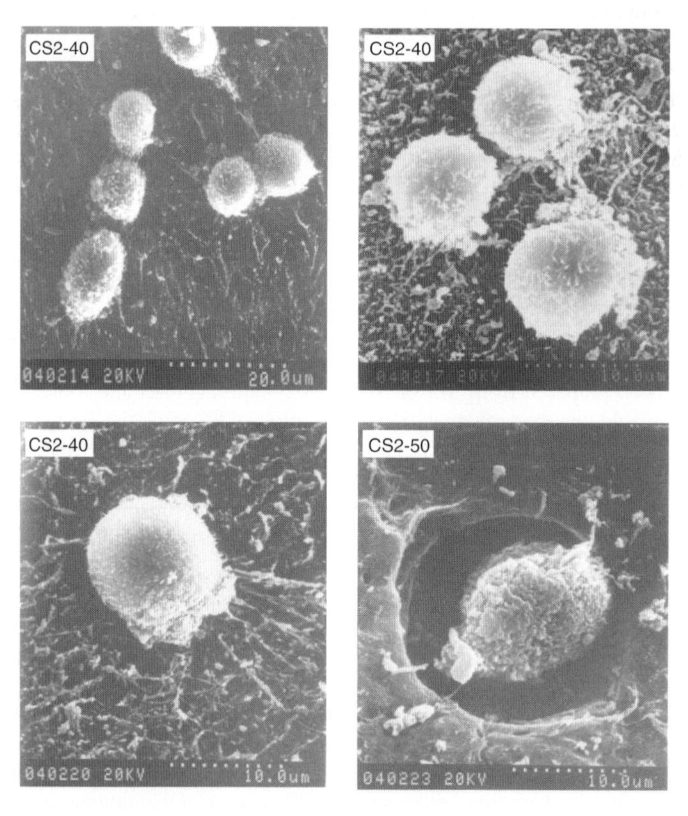

Figure 23.2 SEM photographs of the Vero cells cultured on the free surfaces of the protein based CS2-40 and CS2-50 (see text) membranes. (Reprinted with permission from Reference [72].)

in an intensive mixer at 150°C for 5 min. The blends were then compression-moulded into a tensile bar at 140°C. These materials showed two composition-dependent glass transition temperatures. Furthermore, as the SPI content increased, the melting temperature of the PEEAMA remained constant, but its heat of fusion decreased. These results indicated that SPI and PEEAMA are partially miscible. Increasing the PEEAMA content resulted in a decrease in the modulus and TS and an increase in the elongation and toughness of the blends. Water absorption of the blends also decreased with increasing PEEAMA content. Incorporating MDI further decreased the water absorption of the blends. The mechanism of water sorption of SPI was relaxation controlled, and that of the blends, diffusion controlled [75].

Composites are processed by a variety of methods including compression. Environment-friendly fibre-reinforced composites have been fabricated using ramie fibres and SPI. Based on the interfacial shear strength results and fibre strength distribution, three different fibre lengths and fibre weight contents (FWC) were used to fabricate short fibre-reinforced SPI composites. The fracture stress of the composites increased with an increase in fibre length and FWC. The addition of glycerol increased the fracture strain and reduced the resin fracture stress and modulus as a result of plasticization. The short fibres acted as flaws leading to a reduction in the tensile properties. On further increasing the fibre length and the FWC, a significant increase in the Young's modulus and fracture stress and a decrease in the fracture strain were observed as the fibres started to control the tensile properties of the composites. The ramie fibres and SPI polymers formed compatible moderate-strength composites. However, there is significant scope to improve the tensile properties of both the SPI polymer and the composites containing natural fibres by optimizing the associated processes [76]. Grass fibres were treated with an alkali solution, leading to a more homogenous dispersion of the biofibre in the matrix as well as an increase in the aspect ratio of the fibre in the composite, resulting in an improvement of the mechanical properties, including tensile and flexural properties, as well as impact strength. Additionally, the alkali solution treatment increased the concentration of hydroxyl groups on the fibre surface, resulting in a better interaction between the fibres and the matrix [77].

Nanocomposites have been prepared using a colloidal suspension of chitin and cellulose whiskers as a filler to reinforce SPI plastics with glycerol as the plasticizer. The strong interactions between fillers and between the filler and the SPI matrix played an important role in reinforcing the composites without interfering with their biodegradability. Furthermore, the incorporating of chitin whisker or cellulose whiskers into the SPI matrix led to an improvement in the mechanical properties and the water resistance of the SPI-based nanocomposites [78, 79]. The SPI/chitin whisker nanocomposites, with whiskers having a length of 500 + 50 and a diameter of 50 + 10 on average, respectively, under 43 per cent RH, exhibited a strong increase in both TS and Young's modulus, from 3.3 MPa for a GSPI sheet without chitin whiskers, to 8.4 MPa, and from 26 MPa for the GSPI sheet to 158 MPa, respectively, with increasing chitin whisker content from 0 to 25 wt%, as shown in Fig. 23.3. Further, incorporating of whiskers led to an improvement in water resistance.

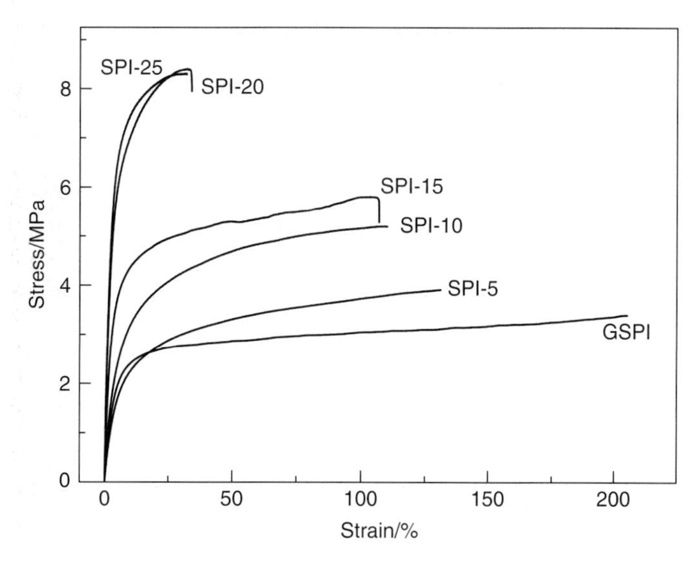

Figure 23.3 Stress–strain curves of GSPI sheet and SPI/chitin whisker nanocomposites conditioned at 43 per cent RH. (Reprinted with permission from Reference [78].

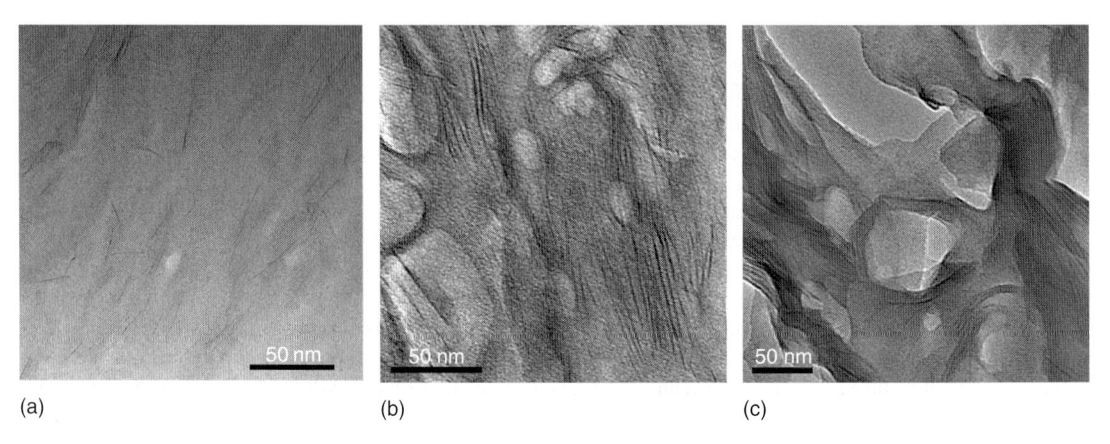

Figure 23.4 TEM images of SPI/MMT plastics, (a) MS-8, (b) MS-16, and (c) MS-24 (see text for the meaning of the symbols). (Reprinted with permission from Reference [80].)

High performance natural polymer/clay materials with an intercalated or highly exfoliated structure have been prepared by several research groups. The heterogeneous distribution of the surface positive charges of SPI provided the possibility for the negatively charged soy protein to intercalate and exfoliate the layered Na^+-montmorillonite (MMT). At least two types of interactions were involved in these composites, namely, surface electrostatic interactions and hydrogen bondings. These highly exfoliated or intercalated SPI/MMT materials were successfully prepared from the solution-intercalated SPI/MMT nanocomposties via extrusion mixing and compression moulding [80]. Using 30 wt% glycerol as a plasticizer, the SPI/MMT plastic sheets were prepared by a compression-moulding process at 140°C and 20 MPa. The X-ray diffraction (XRD) patterns of the nanocomposite plastics (MS-0 to MS-24) with an MMT content varying from 0 to 24 wt%, indicated that the high degree disordered structure of the materials had been maintained in the compression-moulding process. In addition, the diffraction peak of MS-24 shifted to a lower angle ($2\theta \approx 1.4°$), indicating a higher interlayer spacing of 6.2 nm. The microstructure of these materials was visualized by TEM, and the images are shown in Fig. 23.4. For the MS-8 plastic sheet (Fig. 4(a)), the delaminated silicate lamellas were randomized during the extrusion and compression-moulding processes. The dimension of the silicate layers were diminished to about 30 nm in length and 1 nm in thickness, indicating that the layered MMT had been highly exfoliated by the soy protein molecules. For the MS-16 plastic sheet (Fig. 4(b)), most of the layered MMT tactoids were intercalated with a *d*-spacing of about 6 nm. Simultaneously, some conglomerations of MMT occurred within the soy protein matrix. When the MMT content reached 24 wt%, the degree of conglomeration became serious in the MS-24, leading to the obvious phase separation between the components (Fig. 4(c)). Moreover, their thermal stability and mechanical properties were improved as a result of the fine dispersion of the MMT layers and the strong restriction effects on the interfaces. The values of the Young's modulus increased from 180 to 540 MPa by increasing the MMT content from 0 (MS-0) to 16 wt% (MS-16), and the TS increased from 8.8 MPa for the MS-0 to 15.4 MPa for the MS-16 sheet. Within the whole MMT content range, the elongation at break kept decreasing with increasing MMT content. All these results confirm that the strong electrostatic and hydrogen bonding interactions between the soy protein and the highly dispersed MMT layers were sufficient to restrict the segmental motions of the soy protein macromolecules, leading to the improvement of their modulus and TS. The thermal stability of these materials showed a weight loss between 0°C and 120°C, attributed to their absorbed moisture, followed by a second weight loss at 120–250°C, mainly related to the evaporation of glycerol. Thereafter, the pure SPI material is less important than that of the material with MMT.

Mixtures of SPI and native or modified (crosslinked) maize starch have been extruded in a twin-screw extruder at screw speeds of 80, 120, and 160 rpm and a moisture content of 250 g kg^{-1} (dry basis). The specific mechanical energy dissipation and the water solubility index of the ensuing blends were lower than those of native starch/SPI and their sectional expansion indices higher. The type of starch mixed with soybean also affects the expansion of the extrudates, and another factor, in addition to phase transition, could be related to the effect of soybean on that expansion. The effect of the presence of modified starch on the expansion of extrudates can result from the reduction in the specific mechanical energy dissipation, or the higher resistance of modified starch to degradation. No effect of increasing screw speed on the specific mechanical energy dissipation into the feed during extrusion was observed, as the flow rate of the feed was not held constant [81]. Extruded samples of starch–casein blends were processed by using a single-screw extruder. The independent variables of the process, namely, temperature (126–194°C), moisture content (18–29 per cent), and starch–casein composition (5–95 per cent), affected significantly the physicochemical and textural properties of the ensuing blends. The highest values for expansion and water absorption index were found when a higher starch proportion was present in the blends, at a barrel temperature of 126°C and a moisture content higher than 25 per cent. By increasing the barrel temperature from 126°C to 194°C, the water solubility index and colour parameter were increased. The compression force (CF) was found to be strongly dependent on moisture content and casein proportion in the blend, the highest CF values being found at starch concentration around 50 per cent and 25 per cent moisture content [82].

SPI–lignosulphonate (LS), alkaline lignin and SPI–LS–celloluse blends exhibited higher TS, elongation, and Young's modulus than the corresponding material containing only SPI. The improvement of these properties is attributed mainly to the existence of intermolecular interactions among the components, to a beneficial microphase separation, and to the formation of a physically crosslinked network between SPI and LS. Therefore, a physical network between SPI molecules and an active LS molecule placed at its centre is proposed as the most plausible structure [83].

Thermally processed WP and polyvinylalcohol (PVOH) blends have been studied by solid-state high resolution NMR spectroscopy. The intermolecular hydrogen bonding interactions between WP and PVOH induced some miscibility in the system at the nanometre scale, especially when the PVOH content was low. The TS and modulus of the blends were improved when compared to those of WP. However, the intermolecular interactions were relatively weak and could not be further enhanced by increasing the PVOH content, because such an increase enhanced the immiscible character of the blend composites [84].

Water-blown low density rigid polyurethane foams have been prepared with poly(ether polyol)s, polymeric isocyanates, DSF, water, a catalyst mixture, and a surfactant the immiscible character of the blend composites. Soy flour and the initial water content were varied from 0 to 40 per cent and from 4.5 to 5.5 per cent of the poly(ether polyol) content, respectively. The addition of soy flour in the rigid polyurethane foam system contributed to a higher glass transition temperature, and increasing the initial water content also resulted in an increase in the glass transition temperature [85].

SPI, soy fibre, and corn starch together with 0–40 per cent polyether polyol were also incorporated into a flexible polyurethane foam formulation. Stress–strain curves of the control foam and foams containing 10–20 per cent biomass material exhibited a considerable plateau stress region, but not for foams extended with 30–40 per cent of them. An increase in the biomass content produced an increase in the foam density, whereas an increase in the initial water content produced the opposite effect. Foams extended with 30 per cent SPI, as well as those extended with 30 per cent soy fibre, displayed considerably higher resilience values than all other extended foams. The comfort factor increased by increasing the biomass content, and foams containing 10–40 per cent biomass showed significantly lower values in compression-set than the control foam [86].

A novel series of protein composites have been prepared from 30 to 50 wt% polyurethane prepolymer (PUP) with SD, soy whole flour (SWF) and SPI, by a compression-moulding process at 120°C, without the addition of any plasticizer. The toughness, thermal stability, and water resistivity of these composites was significantly improved. By increasing the PUP content, elastomeric materials could be obtained. With an increase in cellulose content in the system, the TS and water resistivity of the ensuing composites increased [87].

It is particularly interesting to focus on the organization at the interface zein matrix/starch granule, in order to understand the physicochemical properties involved in the formation of cereal endosperm. To study the influence of the starch–zein ratio on material properties, corn flour and starch–zein (5–50 weight per cent) based samples were prepared by extrusion and thermomoulding and then analysed at a moisture content of 12.0 per cent (wet basis). Amorphous starch was ductile, whereas blends and corn flour samples were brittle. The blend morphology, observed by confocal scanning light microscopy, showed that the proteins had to undergo aggregation during the thermomechanical processing, which largely conditioned the mechanical properties of the ensuing materials [88].

Casein has been found to be an excellent candidate to produce oil-in-water emulsions that have both high physical and oxidative stability. The differences in the physical properties and oxidative stability of corn oil-in-water emulsions stabilized by casein, WPI, or SPI at pH 3.0, have been investigated. Emulsions have been prepared with 5 per cent corn oil and 0.2–1.5 per cent protein. Physically stable, monomodal emulsions have been prepared with 1.5 per cent casein, 1.0 or 1.5 per cent SPI, and $\geq$0.5 per cent WPI. The oxidation stability of the different protein-stabilized emulsions was in the order of casein $>$ WPI $>$ SPI, as determined by monitoring both lipid hydroperoxide and headspace hexanal formation. The degree of positive charge on the protein-stabilized emulsion droplets was not the only factor involved in the inhibition of lipid oxidation, because the charge of the emulsion droplets

(WPI > casein ≥ SPI) did not parallel the corresponding oxidation stability. Other potential reasons for the differences in the oxidation stability of the protein-stabilized emulsions include differences in the interfacial film thickness, protein chelating properties, and differences in the free radical scavenging amino acids [89].

REFERENCES

1. Fukushima D., *Advance of Science and Technology*, Korin Press, Tokyo, 1988. p. 21–49.
2. Pernollet J.C., Mosse J., in *Seed Proteins,* Eds.: Daussant J., Mosse J. and Vaughan J., Academic Press, London, 1983, p. 155.
3. Wall J.S., Paulis J.W., in *Advances in Cereal Science and Technology*, Ed.: Pomeranz Y., American Association of Cereal Chemists, St. Paul, Minn., 1978, **Volume 2**, pp. 135–219.
4. Osborne T.B., *The Vegetable Proteins*, Longmans, Green and Co., London, 1924.
5. Kumar R., Liu D., Zhang L., *J. Biobased Mater., Bioenergy*, **2**, 2008, 1.
6. Pauling L., Corey R.B., Branson H.R., *Proceedings of the National Academy of Science, USA*, **37**, 1951, 205.
7. Lewis P.N., Monany F.A., Scheraga H.A., *Proceedings of the National Academy of Science, USA*, **68**, 1971, 2293.
8. Venkatachalam C.M., *Biopolymers*, **6**, 1968, 1425.
9. Richardson J.S., *Protein Chem.*, **34**, 1981, 167.
10. Leszczynski J.F., Rose G.D., *Science*, **234**, 1986, 849.
11. Han X.Q., in *Protein Based Surfactants*, Eds.: Nnanna I.A. and Xia J., Marcel Dekker Inc., USA, 2001.
12. Rupley J.A., Careri G., *Adv. Prot. Chem.*, **41**, 1991, 37.
13. Bull H.B., Breese K., *Arch. Biochem. Biophys*, **128**, 1968, 488.
14. Gregory R.B., in *Protein–Solvent Interactions,* Ed.: Gregory R.B., Marcel Dekker, New York, 1993, pp. 191–264.
15. Cerofolini G.F., Cerofolini M.J., *J. Colloid Interface Sci.*, **78**, 1980, 65.
16. Zaks P., Klibanov A., *Science*, **224**, 1984, 1249.
17. Anon M.C., Sorgentini D.A., Wagner J.R., *J. Agric. Food Chem.*, **49**, 2001, 4852.
18. Jovanovich G., Puppo M.C., Giner S.A., Anon M.C., *J. Food Eng.*, **56**, 2003, 331.
19. Li M.S., Klimov D.K., Thirumalaib D., *Polymer*, **45**, 2004, 573.
20. Kumar R., Choudhary V., Mishra S., Varma I.K., Mattiason B., *Ind. Crop. Prod.*, **16**, 2002, 155.
21. Friedman M., Brandon D.L., *J. Agric. Food Chem.*, **49**(3), 2001, 1069.
22. Wang S., Sue H.J., Jane J., *J. Macromol. Sci., Pure Appl. Chem.*, **A33**, 1996, 557.
23. Chen P., Zhang L., *Macromol. Biosci.*, **5**, 2005, 237.
24. Feeney R.E., Whitaker J.R., *Adv. Chem. Ser 160*, ACS, Washington DC, p. 3, 1977.
25. Feeney R.E., Whitaker J.R., in *Chemical and Enzymatic Modification of Plant Proteins in New Protein Foods,* Eds.: Altschul A.M. and Wilcke H.L., Academic Press, New York, 1985, pp. 181–219.
26. Swain S.N., Biswal S.M., *J. Polym. Environ.*, **12**, 2004, 1.
27. Pomes A.F., Zein, in *Encyclopedia of Polymer Science and Technology*, Ed.: Mark H., Wiley, New York, 1971, **Volume 15**, pp. 125–132.
28. Lawton J.W., *Cereal Chem.*, **79**(1), 2002, 1.
29. Hernandez-munoz P., Kanavouras A. *et al.*, *J. Agric. Food Chem.*, **51**(26), 2003, 7647.
30. Zhang X., Burgar I., Do M.D., Lourbakos E., *Biomacromolecules*, **6**(3), 2005, 1661.
31. Silva G.A., Vaz C.M., Coutinho O.P., Cunha A.M., Reis R.L., *J. Mater. Sci. – Mater. Med.*, **14**, 2003, 1055.
32. Kim K.M., Weller C.L., Hanna M.A., Gennadios A., *Lebensm-Wiss. Technol.*, **35**, 2002, 140.
33. Shih F.F., *Nahrung*, **42**, 1998, 254.
34. Nino M.R.R., Sanchez C.C., Ruiz-Henestrosa V.P., Patino J.M.R., *Food Hydrocolloids*, **19**, 2005, 417–428.
35. Mariniello L., Pierro P.D., Esposito C., Sorrentino A., Masi P., Porta R., *J. Biotechnol.*, **102**, 2003, 191.
36. Park S.K., Rhee C.O., Bae D.H., Hettiarachchy N.S., *J. Agric. Food Chem.*, **49**, 2001, 2308.
37. Lieberman E.R., Gilbert S.G., *J. Polym. Sci.*, **41**, 1973, 33.
38. Donhowe .I.G., Fennema O., *J. Food Process. Preserv.*, **17**, 1993, 247.
39. Gennadios A., Weller C.L., Testia R.F., *Trans. ASAE.*, **36**, 1993, 465.
40. Gontard N., Guilbert S., Cuq J.L., *J. Food. Sci.*, **58**, 1993, 206.
41. McHugh T.H., Krochta J.M., *J. Agric. Food Chem.*, **42**, 1994, 841.
42. Park H.J., Weller J.M., Vergano P.J., Testin R.F., *Trans. ASAE.*, **37**, 1994, 1281.
43. Cuq B., Gontard N., Cuq J.L., Guilbert S., *J. Agric. Food Chem.*, **45**, 1997, 622.
44. Liu D., Zhang L., *Macromol. Mater. Eng.*, **291**, 2006, 820.
45. Cho S.Y., Rhee C., *Lebensm-Wiss. Technol.*, **35**, 2002, 151.
46. Rhim J.W., Gennadios A., Weller C.L., Hanna M.A., *Ind. Crop. Prod.*, **15**, 2002, 199.

47. Sabato S.F., Ouattara B., Lacroix M., *J. Agric. Food Chem.*, **49**(3), 2001, 1397.
48. Lee M., Lee S., Song K.B., *Radiat. Phys. Chem.*, **72**, 2005, 35.
49. Wang Y., Padua G.W., *Macromol. Mater. Eng.*, **288**, 2003, 886.
50. Brown O.E., Labelling adhesives, US Patent 4,675,351, 1987.
51. Shih F.F., *J. Am. Oil Chem. Soc.*, **71**, 1994, 1281.
52. Lambuth A.L., Soybean glues, in *Handbook of Adhesives,* Ed.: Skeist I., 2nd Edition, Van Nostrand, New York, 1977, p. 172.
53. Pizzi A., Mittal K.L., *Adv. Wood Adhes. Technol.*, Marcel Dekker Inc, New York, 1998. p. 259.
54. Zhong Z., Sun X.S., Wang D., Ratto J.A., *J. Polym. Environ.*, **11**(4), 2003, 137.
55. Zhong Z., Sun X.S., Fang X., Ratto J.A., *Int. J. Adhes. Adhes.*, **22**, 2002, 267.
56. Sessa D.J., Wing R.E., *Nahrung*, **42**, 1998, 266.
57. Chenga E., Suna X., Karrb G.S., *Compos.: Part A*, **35**, 2004, 297.
58. Anonymous, *Time*, August, 25, 1941, p. 63, 1941.
59. Brother G.H., McKinney L.L., *Ind. Eng. Chem.*, **32**, 1940, 1002.
60. Wu Q., Zhang L., *Ind. Eng. Chem. Res.*, **40**, 2001, 1879.
61. Zhang J., Mungara P., Jane J., *Polymer*, **42**(6), 2001, 2569.
62. Mo X., Sun X., *J. Polym. Environ.*, **11**(1), 2003, 15.
63. Mo X., Sun X., *J. Polym. Environ.*, **8**(4), 2000, 2000.
64. Vaz C.M., Doeveren P.F.N.M.V., Yilmaz G., Graaf L.A.d., Reis R.L., Cunhal A.M., *J. Appl. Polym. Sci.*, **97**, 2005, 604.
65. Lodha P., Netravali A.N., *Ind. Crop. Prod.*, **21**, 2005, 49.
66. Wang N., Zhang L., Lu Y., *Ind. Eng. Chem. Res.*, **43**, 2004, 336.
67. Wang N., Zhang L., Lu Y., Du Y., *J. Appl. Polym. Sci.*, **91**, 2004, 332.
68. Wang N., Zhang L., Gu J., *J. Appl. Polym. Sci.*, **95**, 2005, 465.
69. Deng R., Chen Y., Chen P., Zhang L., Liao B., *Polym. Degrad. Stabil.*, **91**, 2006, 2189.
70. Wu Q., Zhang L., *Ind. Eng. Chem. Res.*, **40**, 2001, 1879.
71. Wu Q., Zhang L., *J. Appl. Polym. Sci.*, **82**, 2001, 3373.
72. Chen Y., Zhang L., Gu J., Liu J., *J. Membrane Sci.*, **241**, 2004, 393.
73. Graiver D., Waikul L.H., Berger C., Narayan R., *J. Appl. Polym. Sci.*, **92**, 2004, 3231.
74. Jong L., *J. Appl. Polym. Sci.*, **98**, 2005, 353.
75. Zhong Z., Sun S.X., *J. Appl. Polym. Sci.*, **88**, 2003, 407.
76. Lodha P., Netravali A.N., *J. Mater. Sci.*, **37**, 2002, 3657.
77. Liua W., Mohantyb A.K., Askelanda P., Drzala L.T., Misraa M., *Polymer*, **45**, 2004, 7589.
78. Lu Y., Weng L., Zhang L., *Biomacromolecules*, **5**, 2004, 1046.
79. Wang Y., Cao X., Zhang L., *Macromol. Biosci.*, **6**, 2006, 524.
80. Chen P., Zhang L., *Biomacromolecules*, **7**, 2006, 1700.
81. Seker M., *J. Sci. Food Agric.*, **85**, 2005, 1161.
82. Fernández-Gutiérrez J.A., San Martín-Martínez E., Cruz-Orea A., *Starch/Starke*, **56**, 2004, 190.
83. Zhang L., *Modified Materials from Natural Polymers and their Applications*, Chinese chemical industry press, Beijing, 2006, Chapter 4.
84. Zhang X., Burgar I., Lourbakos E., Beh H., *Polymer*, **45**, 2004, 3305.
85. Chang L., Xue Y., Hsieh F., *J. Appl. Polym. Sci.*, **80**, 2001, 10.
86. Lin Y., Hsieh F., Huff H.E., *J. Appl. Polym. Sci.*, **65**, 1997, 695.
87. Chen Y., Zhang L., Du L., *Ind. Eng. Chem. Res.*, **42**, 2003, 6786.
88. Chanvrier H., Colonna P., Valle G.D., Lourdin D., *Carbohyd. Polym.*, **59**, 2005, 109.
89. Hu M., McClements D.J., Decker E.A., *J. Agric. Food Chem.*, **51**(6), 2003, 1696.

– 24 –

Polyelectrolytes Derived from Natural Polysaccharides

Marguerite Rinaudo

ABSTRACT

This chapter describes the main properties and methods for the characterization of polyelectrolytes derived from the biomass. The most important sources are plants, with cellulose and starch, which turn to polyelectrolytes after chemical modifications. Carboxymethylcellulose is the main cellulose derivative used in many industrial applications as good thickener and hydrophilic polymer for aqueous media. Cationic starches are mainly used in the paper industry for filler retention or paper wet-strength. Natural polyelectrolytes are produced by algae with anionic alginates and carrageenans as the major representatives, which are used in food applications and for biomedical devices. In this respect, alginates are often associated in an electrostatic complex with a pseudo-natural polyelectrolyte (chitosan), a cationic polymer extracted from crustaceous shells.

Keywords

Polyelectrolyte, Carboxymethylcellulose, Pectin, Carrageenan, Alginate, Cationic starch, Galactomannan, Lignosulphonate, Ion exchange, Conformational transition, Physical gelation, Electrostatic complex, Chitosan

24.1 INTRODUCTION

Plant biomass, including algae, is a very important source of renewable polysaccharides, some of which have original properties compared with those of synthetic polymers, as shown in Table 24.1 [1–3]. Cellulose and starch are discussed respectively in Chapters 16 and 15. The corresponding animal-derived chitin [4] is covered in Chapter 25. The microbial water soluble polysaccharides like hyaluronan, xanthan and succinoglycan [1, 2, 5] are dealt with in Chapter 13. This chapter focuses on the polyelectrolytes based on polysaccharides from plant and algae.

24.2 POLYELECTROLYTE CHARACTERIZATION

When a polymer bears ionic groups regularly appended on its chain, it is called a polyelectrolyte. The parameter which controls its thermodynamic properties in solution is the charge parameter λ proportional to the linear charge density, introduced in the polyelectrolyte theory proposed by Katchalsky [6] and later by Manning [7]. It is expressed as:

$$\lambda = (\nu/h)(e^2/D\mathrm{kT}) \tag{24.1}$$

where ν is the number of ionic charges along a chain with a contour length h, e is the electronic charge, D is the dielectric constant taken as that of water ($D = 78$) and kT is the Boltzman term; λ is also written as $\lambda = (\nu/h)Q$

Table 24.1

The most important natural polysaccharide and lignin sources

Sources	Polymer extracted	Initial polymer or derivatives	Main properties
Wood	Lignin	Alkali-lignin, Lignosulphonate (from sulphite paper process)	Clay deflocculant Additive in drilling fluids
	Cellulose	Fibres: textile, paper, composites Derivatives: methylcellulose, carboxymethylcellulose....	Fibrous substrate Gelling and/or thickening polymers
	Hemicelluloses		
Fruits	Pectins	Copolymer based on galacturonic acid repeat units with different degrees of methylation	Gelling polymer depending on the cations and temperature
Algae	Carrageenans (extracted from red algae)	Sulphated alternated copolymers (λ,κ,ι forms)	Thickening or gelling polymers depending on the cations and temperature
	Alginates (extracted from brown algae)	Block copolymers based on guluronic and mannuronic units	Gelling polymer in the presence of divalent counterions
Cereals	Starch	Cationic starch	Fibres and fillers retention in paper industry
	Cellulose	Fibres: textile, paper, composites Derivatives: methylcellulose, carboxymethylcellulose....	Fibrous substrates Gelling and/or thickening polymers

with Q, the Bjerrum length, that is 7.2 Å at 25°C in aqueous solution. Hence, λ is directly imposed by the distance b between two ionic sites projected on the axis of the chain (b is the length of a monomeric unit if each monomer has an ionic charge); in the present context, the ionic sites are mainly $—COO^-$ (in carboxymethylcelluloses, pectins and alginates), $—NH_3^+$ (for chitosan in acidic media) and $—SO_3^-$ (in carrageenan). In the case of polyelectrolytes with $—COO^-$ and $—NH_3^+$ functions, the net charge will depend strongly on the pH, in connection with the dissociation equilibrium. The electrostatic potential of the polyelectrolyte grows progressively as the degree of polymerization increases and goes to a limit as soon as the number of charges (or degree of polymerization) is larger than 15. Then, the thermodynamic properties become independent of the molecular weight.

In addition, b is directly related to the conformation of the polymer; viz. single or double chain helix formation implies a decrease of the length b. This helical conformation, which often exists in stereoregular polysaccharides, is stabilized by an intrachain and interchain H-bond network. Thus, the study of electrostatic properties will help to characterize the conformation of these polymers as a function of the experimental conditions (pH, ionic concentration, temperature) [8, 9]. One of the most useful experiments is the determination of the activity coefficient of counterions (γ) obtained by potentiometry (or conductimetry). Its theoretical value is directly related to the charge parameter and the valence of the counterion. The main relations, when λ is higher than unity, are:

$$\gamma_1 \sim (2\lambda)^{-1} \quad \text{for monovalent counterions} \tag{24.2}$$

and

$$\gamma_2 \sim (4\lambda)^{-1} \quad \text{for divalent counterions} \tag{24.3}$$

When this experiment is combined with additional determinations, such as optical rotation, which indicates that an ordered conformation is formed in a given situation and molar mass (one or two chains are associated), the determination of this parameter, inversely proportional to b, gives information on the conformation of the chain, allowing to conclude whether it is a single helix, a double helix on itself or a double helix (or a helical dimer) made of two chains. A double helix was clearly demonstrated with κ-carrageenan, as discussed later [10].

Differential scanning calorimetry (DSC), circular dichroism and NMR spectroscopy are convenient methods to demonstrate the existence of a helical conformation (or at least the existence of an ordered conformation in a given situation) and to determine the thermodynamic conditions for the helix–coil transition [11, 12].

An original behaviour in salt-free solution is observed by viscometry or light scattering and neutron scattering. A peak is observed in the reduced viscosity plotted as a function of polymer concentration, as shown in Fig. 24.1; these results were obtained with the sodium salt of a short polygalacturonate. First, each curve passes through a maximum, whose position depends on the external salt concentration: it was demonstrated that the maximum is located at $C_p = 2C_s$ or $2\lambda C_s$ (when λ is larger than 1) [13]. C_p is the polymer concentration expressed in charge equivalents per litre. On the right side of the peak, interchain electrostatic interactions are established between the chains in solution whereas, on the left side, chains are progressively diluted and controlled by the external salt concentration which remains constant and high in comparison with the polymer concentration. Under the conditions corresponding to the right side of the viscosity peak, a peak (at q_{max}) is observed in light or neutron scattering (the q vector range necessary to evidence this peak depends on the range of polymer concentrations covered) [14, 15]. q_{max} increases when the polymer concentration increases, indicating that a preferential distance ($d \sim q_{max}^{-1}$) exists in solution in the absence (or at low concentration) of external salt. The position of the peak, q_{max}, increases as a function of $C^{1/2}$ (C being the polymer concentration), as predicted for a hexagonal packing. This peak is suppressed in the presence of an excess of salt, when C_s is larger than the value corresponding to $C_p/C_s \sim 2$ or 2λ (when λ is larger than 1). The position of the viscosity peak was shown to be independent of the molar mass of the polymer [8, 16].

The original behaviour also concerns the ionic selectivity in some polysaccharides, which is related to the formation of ion pairs and which occurs in the range of charge parameters larger than 1. This feature was demonstrated from ultrasound experiments on carrageenans [17] and alginate, and also on carboxymethylcelluloses (CMC) [18]. However, for $\lambda < 1$, selectivity was shown even in the coiled conformation for the K^+ and Na^+ forms of κ-carrageenan and at the same time it was found that K^+ formed more ion pairs and favoured the double helix formation compared with Na^+.

From all these original properties, it follows that to characterize a polyelectrolyte in aqueous solution for its molar mass and/or dimensions, it is necessary to isolate the chains by screening the long-range electrostatic repulsions. This is achieved in the presence of 0.05 M, or better 0.1 M, monovalent external salt (NaCl, $NaNO_3$, etc.), allowing SEC and viscometry to be used as normally with neutral polymers.

Another very important point in this type of study is the purification technique adapted to ionic polysaccharides, which involves the exchange of multivalent counterions with monovalent ones and the prevention of aggregate formation mainly due to the large amount of —OH groups in polysaccharides forming cooperative H-bonds [19].

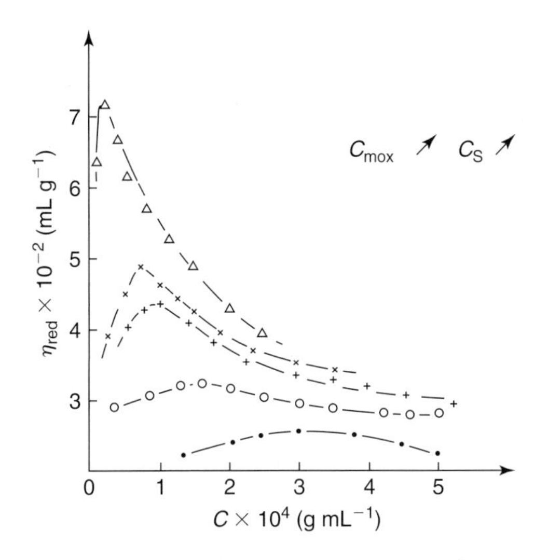

Figure 24.1 Variation of the reduced viscosity for sodium polygalacturonate (Mw = 25 500) as a function of the polymer concentration at different NaCl concentrations. •, 5×10^{-4} M; ○, 2×10^{-4} M; +, 1×10^{-4} M; ×, 5×10^{-5} M; Δ, 1×10^{-5} M. (Reproduced from Reference [13], with the permission of ACS.).

The rheology of many polysaccharides is particularly interesting compared with that of synthetic polymers because of their local stiffness which imposes a large intrinsic viscosity for a given molecular weight. The worm-like chain model was applied to analyze the behaviour of ionic polysaccharides and to characterize their local stiffness by a persistence length L_t, corresponding to the value at the ionic concentration considered: $L_t = L_p + L_e$, where L_p is the intrinsic persistence length obtained when L_t is extrapolated to infinite salt concentration to screen the electrostatic contribution (L_e), assuming that the θ-conditions are approached. Concurrently, the L_p value was calculated for a few polysaccharides by molecular modelling and found to be in good agreement with the experimental values [20–22]. For alginates rich in G units, L_p was found to be 9 nm whereas for a sample rich in M units, L_p decreased to 4 nm [23].

This treatment allows one to predict the dimensions of the chain (and particularly the radius of gyration R_g) and the intrinsic viscosity ([η]).The equation related to the radius of gyration at a given ionic concentration is:

$$R_g^{\,2} = \alpha_s^{\,2}(LL_t)/3 \quad \text{with} \quad L_t = L_p + L_e \tag{24.4}$$

where the contour length L (proportional to M) is large compared with L_t; α_s is the excluded volume which contains a large contribution from the electrostatic interactions. Otherwise, it is necessary to use the development proposed for the θ-state by Benoit and Doty, if it is accepted that with an excess of salt the θ-conditions are approached [24]. The model of worm-like chain polyelectrolyte is given in detail by Reed [25a].

The local stiffness of polysaccharides explains the high viscosity obtained at a given molecular weight, compared with flexible polymers, as well as the relatively low sensitivity to the external salt concentration (since $L_p > L_e$ in many conditions). It was shown that the specific viscosity is directly related to the overlap parameter $C[\eta]$ at zero shear rate and expressed by the following relation:

$$\eta sp = C[\eta](1 + k_1(C[\eta]) + k_2(C[\eta])^2 + k_3(C[\eta])^3) \tag{24.5}$$

in which $k_1 = 0.4$, $k_2 = k_1/2!$; $k_3 = k_1/3!$ [25b]. This relation is very important because it allows one to predict that, for a given polymer, under given thermodynamic conditions, all the viscosity values for different polymer concentrations and molecular weights are going on the same curve, when plotted as a function of $C[\eta]$. The k_1 value represents the Huggins constant and for many perfectly water soluble polysaccharides it equals 0.4. The parameter L_p (which controls [η]) is obtained experimentally (when the ionic concentration of the eluent is known) from the curve $R_g(M)$ established from steric exclusion chromatography (SEC) experiments, using a multidetection equipment in which three detectors are on line, namely, a differential refractometer, a multiangle light scattering detector and a viscometer [9].

When an anionic polymer, such as an alginate, is mixed with a cationic polymer (like chitosan in acidic conditions), an electrostatic complex is formed, whose stability depends on the pH and the salt concentration [26]. The mechanism of complex formation was established following conductimetric measurements and can be expressed by the following equilibrium:

$$\text{----COO}^-\text{Na}^+ + \text{Cl}^- \ {}^+\text{NH}_3\text{----} \leftrightarrow \text{----COO}^- \ {}^+\text{NH}_3\text{----} + \text{Na}^+ + \text{Cl}^- \tag{24.6}$$

In stoichiometric conditions, the complex is usually insoluble and can be exploited to obtain fibres, films or capsules. The chitosan–alginate complex was examined for alginates with different molecular weights and M/G ratios [27, 28] given its many applications, particularly in the biomedical field.

24.3 CELLULOSE IONIC DERIVATIVES

24.3.1 Carboxymethylcelluloses

Cellulose ethers are obtained by the substitution of hydroxyl groups with ether groups. Different derivatives are commercially available, but the most important among the water soluble products are the CMC. These compounds, depending on the substituent density (or the degree of substitution, DS, characterizing the average number of substituents per glucose unit) become soluble in water when the degree of carboxymethylation is larger than about 0.5. CMC is the most important cellulose ether, with a production of approximately 300 000 tons per year. It is produced by the reaction of alkali-cellulose with sodium chloroacetate or chloroacetic acid [29, 30]. We have produced, on a laboratory scale, soluble CMC with DS varying from 0.5 to 3. The typical raw materials for cellulose ether production are wood pulp or cotton linters, the latter being preferred for the production of high-viscosity ethers because of their higher degree of polymerization. The production of CMC can be carried out at atmospheric pressure, an advantage compared with the majority of the other cellulose ethers. In order to improve the diffusion of alkali and the etherifying reagent into the cellulose, inert solvents such as isopropyl alcohol or *t*-butyl alcohol are used. This leads to a more uniform substitution, and hence a higher water solubility. After etherification, the reaction mass or slurry may be neutralized with hydrochloric or acetic acid giving a crude CMC with a salt content (sodium glycolate or sodium chloride) of up to 40 per cent. If further purification is needed, these salts are extracted with water–alcohol mixtures (normally ethanol or methanol) before drying, grinding, screening and storage.

The CMC sodium salt is a white, odourless, hygroscopic and non-toxic solid. The DS of commercial samples may be between 0.3 and 1.2 (the majority of them have a DS between 0.65 and 0.85), although clear and fibre-free CMC solutions require a minimum DS value of about 0.5. Most CMC solutions are highly pseudoplastic and often show a thixotropic behaviour, which decreases when the macromolecules are more uniformly substituted. Recently, CMC were prepared from different non-wood fibres [31, 32]. Figure 24.2 shows the rheological behaviour of CMC from Abaca fibres after a first reaction giving a DS of 0.95 (ABE 1; $[\eta] = 601\,\mathrm{mL\,g^{-1}}$) and after a second reaction increasing it to 2.4 (ABE 2; $[\eta] = 410\,\mathrm{mL\,g^{-1}}$). At $100\,\mathrm{g\,L^{-1}}$, ABE 1 displayed a gel-like behaviour due to loose interchain interactions (which exist even at $30\,\mathrm{g\,L^{-1}}$), but ABE 2 had a viscoelastic behaviour at low frequency $G'' > G'$ [32a]. The rheological behaviour of CMC in the semi-dilute regime was also studied recently [32b].

The viscosity of aqueous solutions varies as a function of pH, showing a maximum at pH 6–7. It is sensitive to the added salt concentration which is explained by the polyelectrolyte properties of these water soluble derivatives. The role of the charge density on the solution properties (activity of counterions, pK etc.) was abundantly examined in our laboratory and will be briefly recalled here. The intrinsic pK of the water soluble CMCs in their acidic form, whatever their DS is, was found to be 3 ± 0.2 [33]. The activity coefficients of monovalent and divalent counterions depend directly on the DS and follow the theoretical prediction, as discussed in Reference [33]. Using CMC prepared in our laboratory and covering a wide range of DS up to 3, we demonstrated that an ionic selectivity among monovalent counterions occurred for DS $\geqslant 1$ with Li > Na > K > Cs > TMA. These important results were obtained by ultrasound absorption experiments, as shown in Fig. 24.3, where the acidic form was neutralized with different hydroxides. The limit for the appearance of ionic selectivity is in good agreement with the critical charge parameter $\lambda = 1$ introduced by Manning for the condensation [7]. This indicated that among the atmospheric counterions, a small fraction formed ion pairs.

CMC have good film forming properties, innocuity and an excellent behaviour as a protecting colloid and an adhesive, which make their field of applications very wide, including textiles, paper making, paints, drilling muds, detergents, foodstuffs (under the reference E466), cosmetics, pharmaceuticals and agricultural aids [3a].

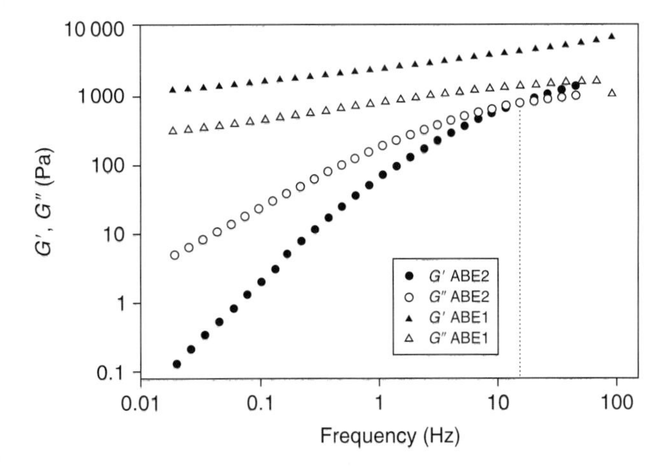

Figure 24.2 Storage and loss moduli as a function of frequency for $100\,\mathrm{g\,L^{-1}}$ abaca CMC samples isolated after one (ABE 1) and two treatments (ABE 2). (Reproduced with permission from Reference [32]).

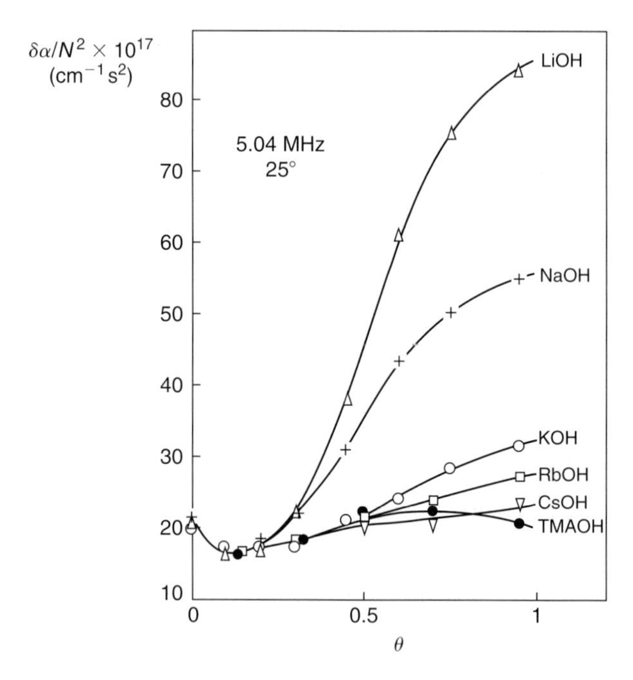

Figure 24.3 Ionic selectivity demonstrated by ultrasound absorption on CMC with a DS = 2.5, showing the role of the counterion. Tetramethylammonium counterion, TMA, is chosen as reference as it does not form ion pairs. (Reproduced with permission from Reference [18].)

CMC can be crosslinked with epichlorhydrine (in alkaline conditions) or formaldehyde (in acidic conditions) to give cation-exchange gels or ultrafiltration membranes [34, 35]. At low pH, CMC may form crosslinks through lactonization between carboxylic acid and free hydroxyl groups.

It has been shown recently that CMC adsorbs on cellulosic fibres improving the stabilization of their dispersion in an aqueous medium. This phenomenon was studied by electrokinetic measurements and rheology [36], and it was also demonstrated that these modified fibres can be dried and redispersed easily in a reversible way. Some of these results are given in Fig. 24.4 and show that the role of these additives is nearly independent of the CMC characteristics. Among different additives tested in this context, CMC were found as the most efficient.

Apart from (sodium) CMC, other mixed ethers exist in the market, such as carboxymethylhydroxyethylcellulose (CMHEC) used in drilling muds or completion fluids, and carboxymethylmethylcellulose (CMMC), used as a binder and as an adhesive for tobacco leaves.

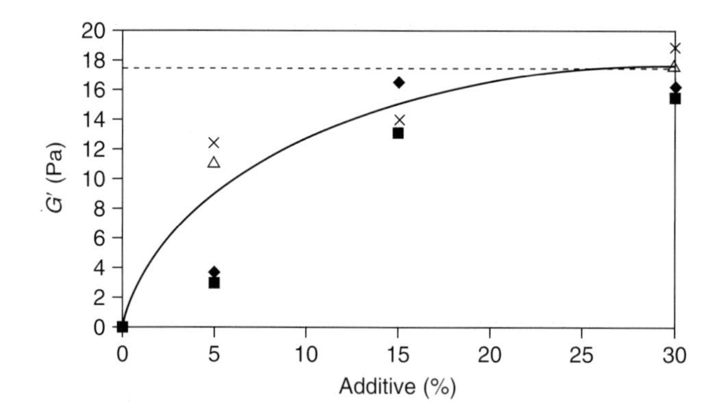

Figure 24.4 Storage modulus (G') at 0.17 Hz for the redispersion of a cellulose microfibril suspension ($4\,g\,L^{-1}$) treated with CMC with different molecular weights and different DS. The dotted line is the initial never-dried suspension. ◆ CMC DS = 0.7 Mw = 1×10^6; 7HF; ■ CMC DS = 0.7 Mw = 3×10^4; 7ULC; Δ CMC DS = 1.2 Mw = 3×10^5; 12M8P; X CMC DS = 2 Mw = 6×10^5; X8212. (Reproduced with permission from Reference [36].)

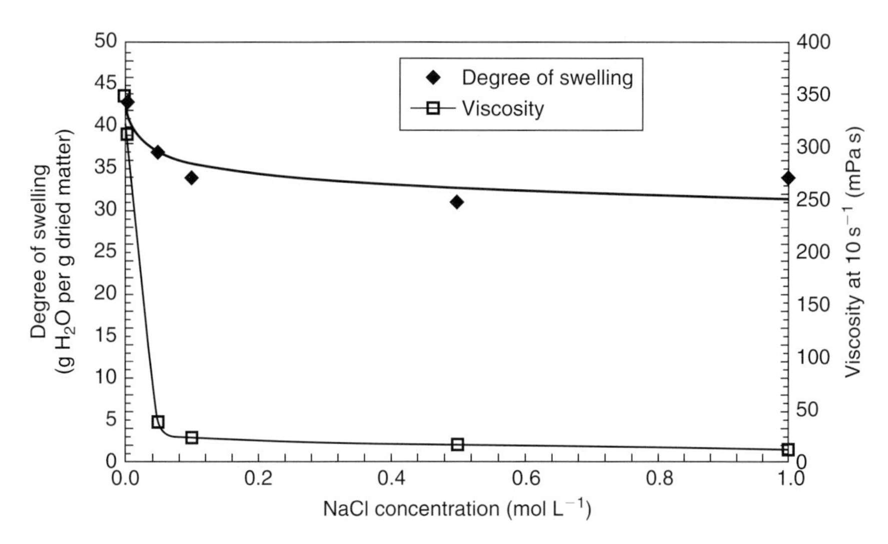

Figure 24.5 Degree of swelling of a suspension of cationic cellulose fibres (average DS = 0.45) and viscosity at $10\,s^{-1}$ in the presence of different external salt concentrations, at fibre concentration $C = 9.83\,g\,L^{-1}$. (Reproduced with permission from Reference [37].)

24.3.2 Cationic and carboxymethylated fibres

Heterogeneous polyelectrolyte systems can be prepared by mild chemical modifications of cellulose fibres; partially carboxymethylated pulps have been known to produce dispersible papers for a long time. More recently, we developed a technique to prepare cationic fibres with different degrees of substitution using cellulose from wheat bran, an agricultural residue [37]. The ensuing suspensions had a yield stress over $5\,g\,L^{-1}$, directly related to the DS and the fibre concentration; higher fibre concentrations gave a gel-like consistency. The degree of swelling and the viscosity of a suspension of cationic fibres depend directly on the salt concentration in tune with the behaviour of soluble polyelectrolytes (Fig. 24.5).

24.3.3 Other charged cellulose derivatives

Cellulose xanthate is produced as an intermediate in the viscose process (even though it is not generally used as a final product) used to prepare regenerated cellulose fibres. The progressively decreasing importance of this technology due to its high cost and impact on the environment, has reduced the interest in cellulose xanthate, which has been employed for the selective flocculation of minerals [38].

Phosphorylated derivatives are also available. Cellulose pulp or linters can react with phosphoric acid in a urea melt or with a mixture of phosphoric acid and phosphorus pentoxide in an alcoholic medium to prepare cellulose phosphate. Cellulose phosphites and phosphonates are produced via transesterification with alkyl phosphites. All these compounds have fire-retarding properties and ion-exchange properties and cellulose phosphate is used in textile and paper manufacture, as well as in the treatment of kidney stones.

Other derivatives include cellulose sulphates and borates, which however have not yet found industrial applications.

24.4 STARCH IONIC DERIVATIVES

24.4.1 Starch succinates

Succinic anhydride reacts directly with starch to form a half-ester called starch succinate. Starch can also react with alkyl or alkenyl derivatives of succinic anhydride to form the corresponding succinates, among which one of the most used is starch sodium octenylsuccinate. Due to the presence of hydrophobic and hydrophilic groups, this product has interfacial activity and emulsifying properties. Alkenylsuccinates are used in food, pharmaceutical and industrial applications [39].

24.4.2 Starch phosphates

Starch phosphates are derivatives obtained with phosphoric acid and include mono-, di- and tri-starch phosphate esters. The di- and tri-esters form crosslinked systems that contain also mono-phosphate. Monoesters can be produced by the reaction of starch with inorganic phosphates, with or without urea and with organic phosphorus-containing reagents. These products are anionic compounds that produce higher viscosity and more stable dispersions than unmodified starch. Their dispersions are very stable to freezing and because of their ionic properties, are good emulsifying agents. Starch phosphates are used as adhesives in papermaking, textiles, pharmaceuticals, foods, agriculture, flocculation and foundry [40]. Other ionic inorganic esters, such as starch sulphates, are used in foods, pharmaceuticals and petroleum recovery.

24.4.3 Cationic starch

Cationic starches are obtained by the reaction of starch with reagents containing amino, imino, ammonium or sulphonium groups. The two main types of commercial products are the tertiary amino and quaternary ammonium starch ethers. Among the reagents that can add quaternary ammonium groups to starch, probably the most popular is the 2,3-epoxypropyltrimethylammonium chloride. This reaction was examined in our laboratory where different DS were achieved on amylose and amylopectin separately and the products tested in calcium carbonate adsorption in relation with the mechanisms of dispersion and flocculation of small particles [41, 42]. The adsorption isotherm of amylopectins was investigated on water soluble polymers as a function of the DS and it was shown that the amount adsorbed decreased with increasing DS. This was related to the mechanism of adsorption: neutral amylopectin adsorbed forming loops and trains, but it adsorbed flat on the surface when its cationic DS increased; at the same time, the electrokinetic potential of the particles became highly positive and the particles were finely dispersed in the aqueous suspension [42]. Some relevant data are recalled in Fig. 24.6.

In some cases, cationization can be combined with other treatments, such as acid hydrolysis, oxidation or dextrinization to produce derivatives with a large range of viscosities. The key factor in the usefulness of these products is their affinity of negatively charged substrates. For that reason, cationic starches are used in papermaking (an example is the HI-CAT® cationic starch from Roquette, France). When they are used as a wet-end additive, the affinity between the positively charged cationic starch and the negatively charged cellulose fibres gives rise to an almost complete and irreversible absorption of starch. They are also used for filler retention in paper sheet formation and in other applications in textiles, flocculation, detergents, cosmetics and adhesives [43].

24.5 SEAWEED POLYSACCHARIDES

The cell walls of seaweeds contain polysaccharides which provide algae with their flexibility and help adapting them to the variety of water movements in which they grow. They also swell in water and thus preserve hydration. These polymers, often named phycocolloids, are usually extracted from the algae with water in different temperature conditions. The three main commercial polysaccharides are agars (including agarose, a neutral polysaccharide), alginates (carboxylic polysaccharides) and carrageenans (sulphated polysaccharides) and are used mainly for their thickening and gelling properties, depending on the thermodynamic conditions and on their molecular structures, as discussed below. Red seaweeds contain mainly agars and carrageenans, whereas brown algae produce alginates.

24.5.1 Alginates

Brown algae contain large amounts of anionic polysaccharides in their cell walls. Alginate, also named alginic acid or algin, was discovered in 1880. The quantity and quality of the alginates extracted depend on the algae and on the harvesting season. The total production of alginates, extracted mainly from the species of the orders *Laminariales* and *Fucales* is around 40 000 tons per year, of which 30 per cent is used in the food industry. They can be isolated under different ionic forms and applied in foods in the acid (E400), sodium (E401), potassium (E402), ammonium (E403) or calcium form (E404) (E is the code for food additives in the EU regulation).

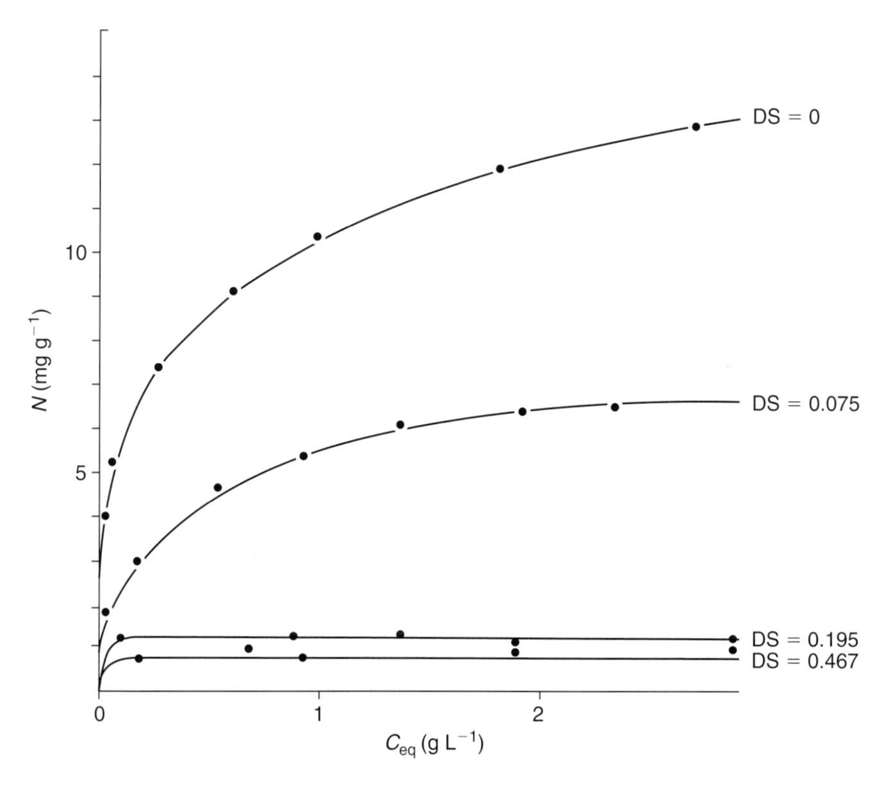

Figure 24.6 Adsorption isotherms for cationic amylopectins with different DS on calcite dispersed in water. $T = 25°C$. Amount adsorbed N expressed in milligrams per gram of solid; equilibrium polymer concentration expressed in grams per litre. (Reproduced with permission from Reference [41].)

Alginates are linear block copolymers composed of 1,4-linked β-D-mannuronic acid (M) with $^4C_1$ ring conformation and α-L-guluronic acid (G) with $^1C_4$ conformation, present in varying proportions and in their pyranosic structure. They are formed of three types of blocks: alternated M and G blocks, resistant to acid hydrolysis, the most flexible part of the chain, blocks of GG and of MM, with a DP $\geqslant$ 20. In relation to their chemical structure, it was demonstrated that the physical properties of these polymers in aqueous media depends, not only on the M/G ratio, but also on the distribution of the M and G units along the chain. The GG blocks in which an axial–axial linkage is involved, are more rigid than the diequatorially linked MM blocks; hence, the stiffness of the chain, as well as its calcium complex formation, depends on the composition and distribution of the M and G units, as discussed by Smidsrod [44]. G blocks of more than 6–10 residues each, form stable crosslinked junctions (and gels) with divalent counterions (Ca, Ba, Sr), but not with Mg, as was also found with pectins with a low degree of methylation. It was also shown that the relative viscosity increases rapidly over a critical amount of divalent counterions (with Ca, Ba, Sr, but not with Mg) when the chains interact [45].

At low pH, alginates form acidic gels stabilized by H-bonds. The homopolymeric G blocks form the junctions and the stability of the gels is determined by the relative content and length of the G blocks.

For their characterization, alginates must first be purified and isolated under their sodium form. NMR spectroscopy (^1H and ^{13}C) is the most powerful technique to characterize the chemical composition and the microstructure of alginates [46–49]. Purified alginates, isolated under the sodium salt form, were also characterized by steric exclusion chromatography (SEC)with three detectors on line. For commercial products, molecular weights may range between 32 000 and 400 000. A further means of characterization is their intrinsic viscosity using the Mark–Houwink relation:

$$[\eta](\text{mL g}^{-1}) = KM^a \tag{24.7}$$

with $K = 2 \times 10^{-3}$ and $a = 0.97$, in a 0.1 M NaCl aqueous solution at 25°C [50].

The dimensions of the alginate chains (radius of gyration and intrinsic viscosity) in aqueous solutions depend on the external salt concentration, their expansion being directly related to the thickening performance of the polymer. It seems, however, that the expansion of polysaccharides, even in external salt excess, is larger than that of a random coil in the absence of electrostatic repulsion, because of the semi-rigid character usually associated with this type of polymers [9]. The relative extension of the three types of blocks in alginates increases in the following order [51]:

$$\mathrm{MG} < \mathrm{MM} < \mathrm{GG}$$

which is in agreement with the L_p values discussed above [23].

The most important technical properties of alginates are their thickening character (increase in the solvent viscosity upon dissolution), their ionic exchange aptitude, and their gel-forming ability in the presence of multivalent counterions. These features are a direct consequence of the fact that alginates are polyelectrolytes and follow, therefore, the usual behaviour of charged polymers. Alginates, in their monovalent salt form, are water soluble whatever be the temperature and their ionic properties (electrostatic short- and long-range interactions), as well as their conformation and molecular weight, are important factors that control their rheological behaviour. The viscosity of their aqueous solutions depends on their concentration, molecular weight and external salt concentration (because of the screening effect of electrostatic interactions). The viscosity of alginate solutions is nearly constant between pH 6 and 8 but, at moderate concentrations, it increases below pH 4.5 and reaches a maximum around 3–3.5 and then decreases (alginic acid forms gels). This behaviour arises from the fact that H-bond attractions dominate over electrostatic repulsions, as was observed recently with hyaluronan [52]. In dilute solutions, the viscosity decreases when pH decreases from pH $\sim$ 6 and in the range of pH 1–4, alginic acid is hydrolyzed. These solutions are considered to be stable in the range of pH 5–10. The intrinsic pK of alginic acid is about 3, as is the case for many polyuronic acids [53].

Another characteristic of alginates, in addition to their gelling and stabilizing properties, is their ability to retain water. Because of their linear structure and high molecular weight, alginates also form strong films and good fibres in the solid state.

The mechanism of gelation is related to the chemical structure of these polymers. First, electrostatic properties are important for the selective interactions with divalent counterions. A specific cooperative Ca interaction forms on G blocks, which is the basis of the junction zones and the crosslink of an ionic network, as soon as the degree of polymerization is larger than 20 (Fig. 24.7); this interaction is not observed for M unit blocks [54, 55].

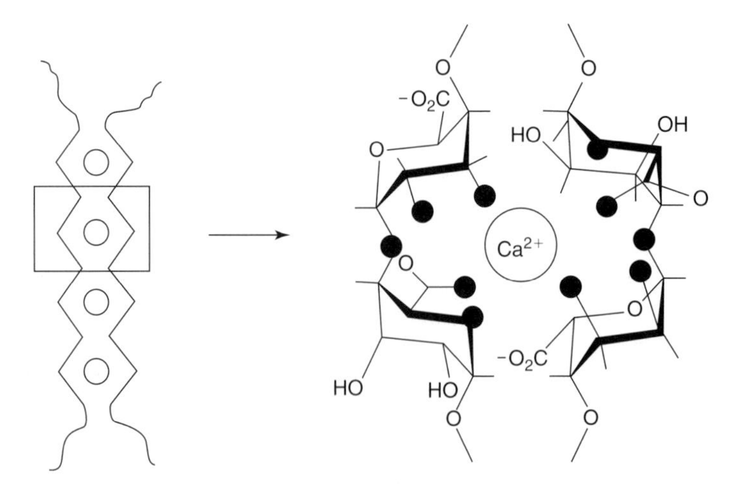

Figure 24.7 Specific interaction of calcium ion with α-L-guluronic-box block. Dark circles represent the oxygen atoms involved in the coordination of the calcium ion. (Reproduced from Reference [57] with the permission of ACS.)

The conformation of acidic polysaccharides and their interactions with calcium ions was examined by molecular simulation, and the authors demonstrated the existence of specific calcium binding with poly-α-L-guluronate [56, 57]. The mechanism of complex formation involves calcium interactions with different oxygen atoms of two adjacent guluronic acid units and with two inter-chain units, as visualized in the egg-box model (Fig. 24.7). The mechanism of gelation is a two-step process: first step is a dimer formation, followed by precipitation for small chains, or gelation for long ones formed with different types of blocks.

The properties of the ensuing gels depend on the molecular characteristics of the alginates; the stability of the gels and their physical properties depend directly on their G content and on the length of their G blocks. It is clear that the stiffness of the gels increases when the G content and the length of the G blocks increase [58]. Figure 24.8 shows that the gel strength increases for a given molar structure when the molecular weight increases up to a limit around $M = 3 \times 10^5$, a behaviour similar that of κ-carrageenan [59].

Alginates have interesting ion-exchange properties: most monovalent counterions (except Ag^+) form soluble alginate salts, whereas divalent and multivalent cations (except Mg^{2+}) form gels or precipitates. The affinity was found to follow the order [60–63]:

$$\mathrm{Mn} < \mathrm{Zn,Co,Ni} < \mathrm{Ca} < \mathrm{Sr} < \mathrm{Ba} < \mathrm{Cd} < \mathrm{Cu} < \mathrm{Pb}$$

The amount of salt required to induce gelation increased in the order [64, 65]:

$$\mathrm{Ba} < \mathrm{Pb} < \mathrm{Cu} < \mathrm{Sr} < \mathrm{Cd} < \mathrm{Ca} < \mathrm{Zn} < \mathrm{Ni} < \mathrm{Co} < \mathrm{Mn,Fe} < \mathrm{Mg}$$

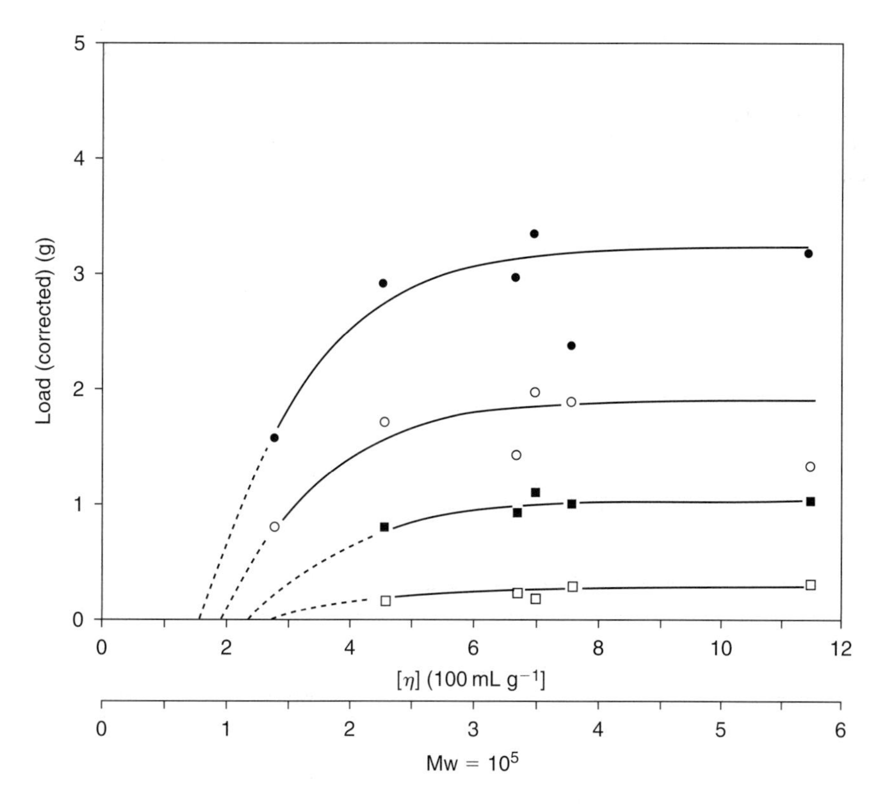

Figure 24.8 Influence of the molar mass, intrinsic viscosity and polymer concentration on the gel strength of alginate from *L. hyperborea* (outer cortex) $F_G = 0.75$, $N_{G>1} = 17.5$. Concentration: (•) 1 per cent, (○) 0.8 per cent, (■) 0.6 per cent and (□) 0.4 per cent. (Reproduced with permission from Reference [58].)

The efficiency of a divalent ion as a precipitant for alginate depends, not only on its affinity for the alginate, but also on the amount of ions which must be bound to the alginate for gel formation. This sequence is not exactly the same as the one given above.

Purified alginates are produced under different salt forms: Na, K, NH_4, Ca and/or propylene glycol derivative as food additives for thickening soups and jellies. They are used as anti-acid preparation such as Gaviscon®, in mould-making materials in dentistry in the presence of a slow release calcium salt to control the gelation delay. Alginates in the form of fibres or films are commercialized as haemostatic materials and wound dressings, like AlgiDERM® or Sorbsan® made of sterile purified calcium alginate fibres. Calcium-based gels are often used in a bead form as immobilization matrices for animal cells or plant protoplast [66–68]. Finally, polyelectrolyte complexes are often used in biomedical applications; thus, chitosan–alginate beads were evaluated for tissue engineering [69] and wound dressing [70].

24.5.2 Carrageenans

Agars and carrageenans are the most important polymers extracted from Red algae. Many Rhodophycean galactans, with a few exceptions, have a linear chain structure of β-D-galactopyranose residues linked through positions 1 and 3 (A units) and α-galactopyranose residues linked through positions 1 and 4 (B units) arranged in an alternated $(AB)_n$ sequence. Units A may carry methyl ether groups at position 6, sulphate hemiester groups at position 2, 4 or 6 and a few A units may carry pyruvic acid, linked as a cyclic ketal, bridging *O*-4 and *O*-6 (1-carboxyethylidene groups). Units B can occur in either D or L form, can carry methyl groups at position 2, 4-*O*-methyl-α-L-galactopyranosyl groups at position 6 and sulphate hemiester groups at position 2 or 6, or both; B units can be wholly or partly converted into 3,6-anhydro forms by the elimination of the sulphate moiety from position 6 (by enzymatic elimination or alkaline treatment) [71]. The ideal composition of the polysaccharides obtained depends on the species, habitat, harvesting season and on the extraction conditions.

Carrageenans belong to two families: the κ-family includes ι- and κ-carrageenans, which contain as B unit a 3,6-anhydro-D-galactose, forming gels in the presence of K^+ counterions (see below), as well as the μ- and ν-carrageenans, without the anhydrogalactose, also named the precursors; the λ-family includes λ- and ξ-carrageenans with no anhydrogalactose, nor gelling properties [71].

The main physical properties of these three carrageenans are:

- Kappa (κ) → strong and rigid gels (polysaccharides extracted from *Kappaphycus cottonii, Chondrus, Hypnea, Furcellaria*).
- Iota (ι) → soft gels (polysaccharides extracted from *Euchema spinosum, Hypnea, Gigartina*).
- Lambda (λ) → thickening polymer (polysaccharides extracted from *Gigartina pistillata, Chondrus crispus*).

All of them are soluble in hot water but form gels on cooling, except the lambda type which remains water soluble at low temperature. Emphasis will be placed below on the properties and characterization of ι- and κ-carrageenans.

^{1}H NMR is the most powerful technique available to identify the type of carrageenan, but it requires the purification of the samples, preferably under the sodium salt form, to be sure that the polymer is in the coiled conformation in dilute solution, even at ambient temperature. This technique also gives rapid access to the quantitative determination of the different substituents, if any, and to get information on the purity of the sample tested. This point is important because the chemical structure directly controls the physical properties of the polysaccharide.

The application of steric exclusion chromatography was again adopted to determine the molecular weight distribution. For κ-carrageenan at 25°C in 0.1 M NaCl, the intrinsic viscosity gives access to the molecular weight, using the following relation, obtained from fractions isolated by preparative gel chromatography [72]:

$$[\eta](\mathrm{mL\,g^{-1}}) = 3.1\times10^{-3}M^{0.95} \tag{24.8}$$

For stereoregular charged polysaccharides, the formation of a helical conformation and eventually gelation depend on the ionic concentration, the nature of electrolyte and the temperature. The conformation results from a balance between H-bonds (which stabilize the helical conformation, but which are destabilized when the temperature increases) and electrostatic repulsions between the charges on the polymer (which are screened by external salt addition). The helix–coil transition for κ-carrageenan was demonstrated by different techniques, viz. conductivity, optical rotation, NMR and DSC [73–75]. Optical rotation measurements showed that an ordered conformation is stabilized at low temperature and goes to a disordered conformation when the temperature is increased. This conformational transition is characterized by T_m, the temperature for conformational helix–coil change, determined at half transition. In fact the helix–coil transition is perfectly reversible (no hysteresis) at low ionic concentration (and low polymer concentration playing the role of electrolyte), but hysteresis appears when the ionic concentration increases [76]. The hysteresis is modified after ageing because of an increase in the degree of aggregation, which is in fact related to a synaeresis phenomenon. At the same time, at lower temperatures, the molecular weight is doubled and the activity coefficient γ of monovalent counterions decreases. At 15°C and 35°C, the activity of Na^+ is practically not influenced by neither temperature nor polymer concentration and tends to 0.71 at infinite dilution [76, 77]. This value is in good agreement with the prediction from polyelectrolyte theories and corresponds to a single linear chain. The same values are obtained for Na^+ and K^+ at 35°C, but at 15°C, a transition is observed when the polymer concentration increases and the K^+ activity coefficient goes to 0.37, corresponding to a double charge parameter λ, indicating the association of two chains to form a double helix (as corroborated by optical rotation) [76]. This behaviour also demonstrates the existence of an ionic selectivity, as discussed below. A doubling of the molecular weight of ι-carrageenan was also found in non-aggregating conditions [77].

The different values of T_m (temperature for conformational change), T_F (melting temperature) and T_G (gelling temperature) were determined from heating and cooling curves obtained by optical rotation or DSC of aqueous solutions of κ-carrageenan and plotted in a log–log representation of these characteristic temperatures (T^{-1}) as a function of the total ionic concentration C_T (taking into account the polyelectrolyte contribution, γC_p, C_p being the polyelectrolyte concentration expressed in charge equivalents per litre). The helix–coil transition is perfectly reversible (no hysteresis) at low ionic concentration (C_T lower than a critical value C_T* around 7.5×10^{-3} M in KCl and 2×10^{-1} M in NaCl), but hysteresis appears when the ionic concentration increases ($T_G \sim T_m < T_F$). Figure 24.9 shows the phase diagram of κ-carrageenan in the presence of K^+ and Na^+ counterions and salt excess [78].

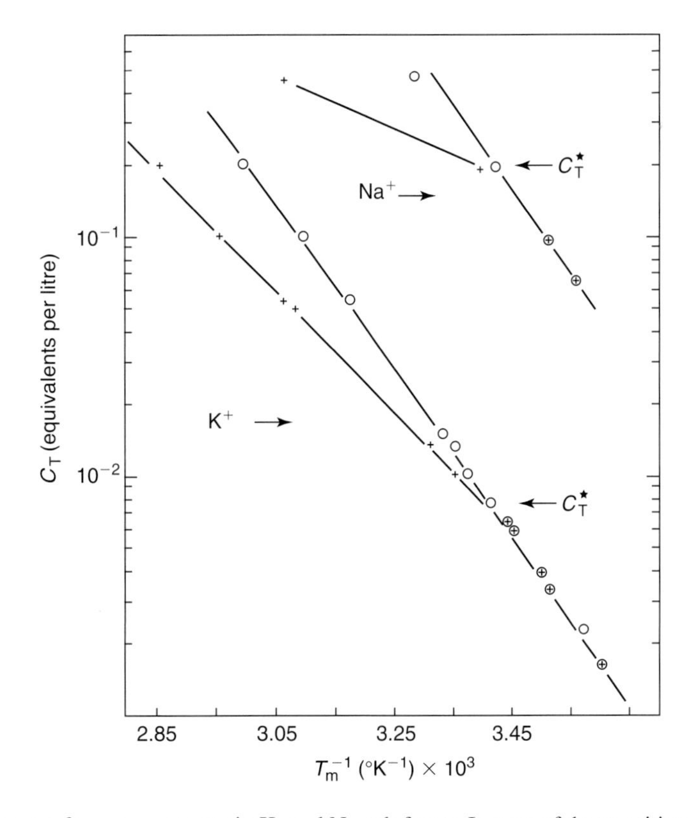

Figure 24.9 Phase diagram for κ-carrageenan in K- and Na-salt forms. Inverse of the transition temperature (T_m) as a function of the total ionic concentration (C_T). T_m (+,○) when $C_T < C_T$* and T_F (+) and T_G (○) when $C_T > C_T$* with the logarithm of the total concentration of counterions C_T (including external salt and free counterions from the polymer). (Reproduced from Reference [78] with the permission of ACS.)

This figure gives evidence of ionic selectivity, which appears in the coil conformation first and is associated with ion pair formation, as demonstrated using ultrasound absorption [17]. The ion pair formation with K^+ reduces the net charge of the polymer and favours the helical dimer formation.

For κ-carrageenan, the ionic selectivity among monovalent cations is very high whereas it is low among divalent counterions [76]. The T_m values were determined at a constant ionic concentration (0.1 M) and T_m in the presence of monovalent counterions varied according to the following sequence:

$$Rb^+ > K^+, Cs^+ > Na^+ > Li^+, NH_4^+ > R_4N^+$$

No selectivity was observed among the monovalent anions with the exception of I^-, which stabilizes the helix, but prevents gelation [76, 79].

Agarose, κ-carrageenans and ι-carrageenans are recognized as gelling polymers, but their specific properties depend on their chemical structure; the conditions of gelation, as well as its mechanism, have been abundantly discussed in the literature [80]. The physical thermoreversible gel formation is based on the establishment of H-bonds between double helices which associate in aggregates (or junction zones) giving rise to a three-dimensional network. It has been shown that gelation proceeds following a two-step process, as clearly demonstrated for thermodynamic conditions around $C_T{}^*$ and a polymer concentration larger than its overlap concentration. The width of the hysteresis in temperature is directly related to the degree of aggregation of the double helices and to the charge density of the polymers (this width decreases from agarose to κ-carrageenan to ι-carrageenan). The related rigidity of the gels formed in the presence of K^+ counterions follows the same trend and varies in the following order:

$$\text{agarose} > \kappa\text{-carrageenan} > \iota\text{-carrageenan}$$

(even if the salt effect is not the same for agarose).

The mechanical properties of the gels formed in the presence of different counterions follow the same order as that of the stability of the double helices: potassium-κ-carrageenan gel forms at lower polymer and ionic concentrations and has a higher modulus and a higher melting temperature than the corresponding sodium-κ-carrageenan gel, which forms at a much higher ionic concentration and has a lower modulus. It was shown that ι-carrageenan displays only a very low ionic selectivity [81].

The mechanical properties of the κ-carrageenan gels were investigated by compression measurements (Fig. 24.10) and it was demonstrated that the elastic modulus, when the gels were formed at a given KCl concentration, increased when the molecular weight increased up to $M \sim 250\,000$ and that the modulus was directly related to the polymer concentration. For the same polymer concentration, the modulus was found to depend not only on the molecular weight, but also on the ionic KCl concentration (causing the screening of the electrostatic repulsion) and went to a limit for the same range of molecular weights [59]. Conversely, the yield stress for gel rupture increased linearly with molecular weight in the range covered [82].

The gels from ι-carrageenan are less strong and have less synaeresis and their properties often depend on the presence of κ impurities. Their elastic modulus increases when the ionic concentration increases up to 0.25 M and decreases for higher concentrations due to a salting-out effect; thus, a $10\,g\,L^{-1}$ gel of ι-carrageenan formed in 0.25 M KCl has an elastic modulus of 0.32×10^4 Pa, while for a pure κ-carrageenan counterpart in 0.25 M KCl it is 6.6×10^4 Pa. To conclude, the stiffness of ι-carrageenan gels is low because of shorter junction zones with lower stability (larger charge density) and they shrink strongly in non-solvents [83].

A general relationship was found to apply to all these physical gels, in 0.25 M KCl, in terms of the variation of their elastic modulus, viz:

$$E\,(\text{Pa}) = K\,C^{2\pm0.1} \tag{24.9}$$

in which C is expressed in grams per litre and $K = 745$ and 43 for κ- and ι-carrageenans, respectively. Comparative data for different polysaccharide gels are also available [84].

For industrial applications, 30 000 tons per year of the three types of carrageenans are produced. When used for food products, carrageenans have the EU additive number E407. Their essential characteristics make them ideal for

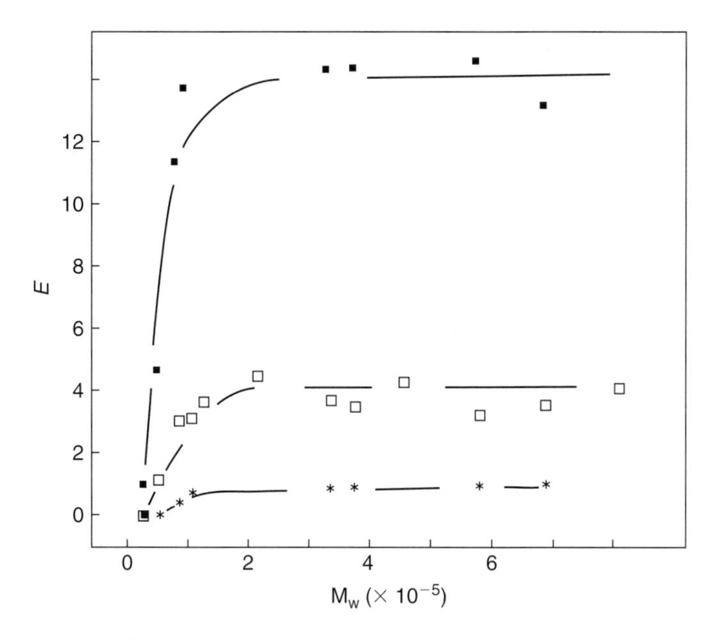

Figure 24.10 Elastic modulus (10^4 Pa) obtained in compression measurements, as a function of the molecular weight of different fractions of κ-carrageenan in 0.1 M KCl for 5 g L^{-1} (*); 10 g L^{-1} (□); 20 g L^{-1} (■). (Reproduced with permission from Reference [59].)

use as thickening and stabilizing agents. Other applications include the solidification of flans, yoghurts, chocolate milk, ice creams and milk puddings and the solidification and emulsification of solutions, the prevention of chocolate milk from creaming and sedimenting. Toothpastes and canned and frozen petfoods also contain carrageenans to promote solidification. Beers are clarified with carrageenans used to complex and precipitate proteinaceous impurities. As agarose and alginates, carrageenans are used for encapsulation. The specific features brought about by these incorporations of carrageenans depend on the composition of the systems into which they are introduced.

24.6 PECTINS

Pectins are located in the middle lamella and primary cell walls of plant tissues and consist primarily of 1,4-linked α-D-galacturonic units or their methyl esters with some interruptions by rhamnogalacturonan region kinking the linear polygalacturonic backbone [85]. They also contain branched chains composed of neutral sugars such as galactose or arabinose. These neutral sugars amount to 10–15 per cent of the pectic dried weight and are concentrated in blocks (called the hairy region) [86]. Pectins are mostly extracted from apple marks and citrus peels. Pectins with different degrees of esterification (DE) represent about 70 per cent of all natural structures, depending on the age of the source and on the presence of enzymes (causing blockwise or random distribution of free carboxylic groups in relation with the nature of the enzyme); if DE $<$ 50 per cent, they are insolubilized by calcium association in the plant and extraction will impose the use of chelating agents, such as sodium oxalate. Two categories of these pectins are recognized: HM pectins (high methoxyl; DE $>$ 50 per cent) form gels in acid conditions and in the presence of sucrose to decrease the water activity. The gelation is based on H-bond association and is thermoreversible [87]. LM pectins (low methoxyl; DE $<$ 50 per cent) form gels in the presence of calcium ions [88]. HM pectins can be extracted by water, mineral acids or bases. These differences are shown in Fig. 24.11, where the activity coefficient of calcium is determined on pectins with different DE (random distribution of the carboxylic groups). Aggregation of chains in the dilute regime occurs when DE $<$ 50 per cent [89].

LM pectins are immobilized *in situ* via metallic ions and need a sequestering agent to displace the counterions. Different treatments were compared between sugar beet and potato pulps [90, 91]. Before pectin extraction, a pretreatment of the plant material is necessary to inactivate enzymes (water at 85°C, 20 min). HM pectins naturally occur in sugar beet pulp and they were extracted in alkaline conditions (50 mM NaOH, pH 12) and precipitated with ethanol after neutralization to pH 6.5–7. On the contrary, for potato pulp containing an LM pectin, the

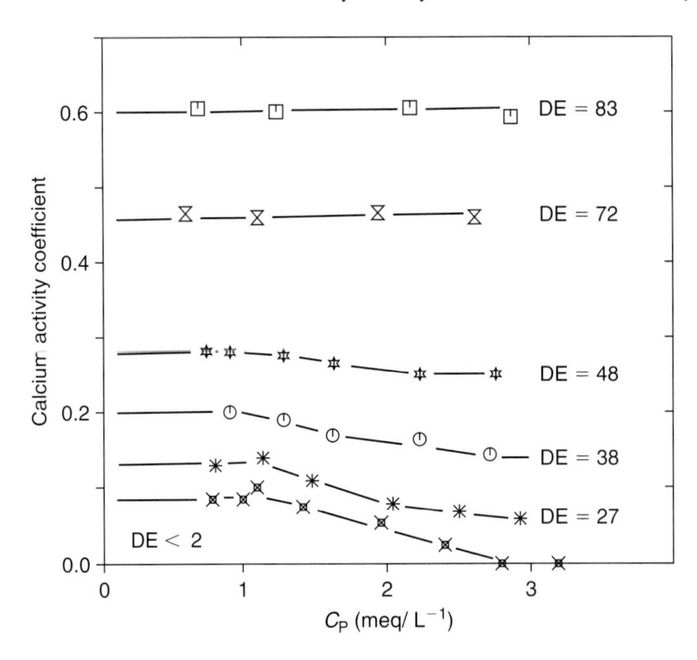

Figure 24.11 Calcium activity coefficient as a function of polymer concentration for different pectins. The DE are expressed in group per cent and vary from less than 2 to 83 per cent. (Reproduced with permission from Reference [89].)

extraction was performed in acidic conditions (pH 3.5 in the presence of 0.75 per cent hexametaphosphate, 75°C, 1 h). The pectins were precipitated at pH 2 with HCl. The precipitate was redispersed in water with NaOH up to pH 6.5–7 and reprecipitated with ethanol. Potato pulp was also extracted in alkaline conditions with hexametaphosphate, precipitated at pH 2, redispersed in water to pH 6.5–7 and precipitated with ethanol. The LM pectins obtained in alkaline conditions had a higher gelling ability because of deesterification and partial deacetylation. The alkaline treatment also has a fundamental role in the deesterification and produces a random hydrolysis of the ester groups. For both sources, pectins with good gelling properties in the presence of calcium ions were obtained.

The interaction with calcium is very cooperative due to the stereochemistry of the 1,4-linked monomeric galacturonic units, leading to the formation of polar cavity that can be occupied by calcium or related cations. The mechanism of interaction is described by the egg-box model already discussed above for alginates (see Fig. 24.7). A series of galacturonic acid oligomers were investigated to test the minimum carboxylic group concentration necessary to get a stable junction. A two-step process leads to gelation: first, a dimer formation occurs for DP > 10 (or 5 Ca^{2+}), followed by aggregation forming the junction zones [53]. The influence of the uronic acid configuration was investigated by Kohn [54], who showed that specific interactions of Ca^{2+} occur with L-guluronic and D-galacturonic acid units, but not with D-mannuronic acid counterparts.

The ability to gel can be determined from the dependence of the viscosity or light scattering on the concentration of added counterions. At a given polymer concentration, the critical amount of cations at the gel point is directly related to the degree of esterification, but also to the distribution of carboxyl groups along the chain [92]. This is shown in Fig. 24.12, where two samples with the same average DE, but one bearing a blockwise distribution of carboxylic groups (a) and the other a random distribution (b) are shown.

It was also shown that the divalent counterions form gels with the affinity sequence Ba > Sr > Ca, but Mg does not form gels nor dimers [89, 93, 94]. This association was investigated by conductimetry, from which the transport coefficient (*f*, which is close to the activity coefficient) of the divalent counterions was determined on dilute solutions, as given in Table 24.2, where the experimental values are compared with the theoretical transport coefficients calculated from the Manning theory. These data explain, on the basis of the charge parameter, why, in the presence of Mg, the pectin remains as a single chain, whereas with Ba and Sr, the values are much lower, especially when DE < 50 per cent, indicating dimer formation with some additional aggregation [89].

As mentioned above [26], LM pectins form polyelectrolyte complexes with chitosan, which were envisaged as encapsulating materials for shark liver oil [95].

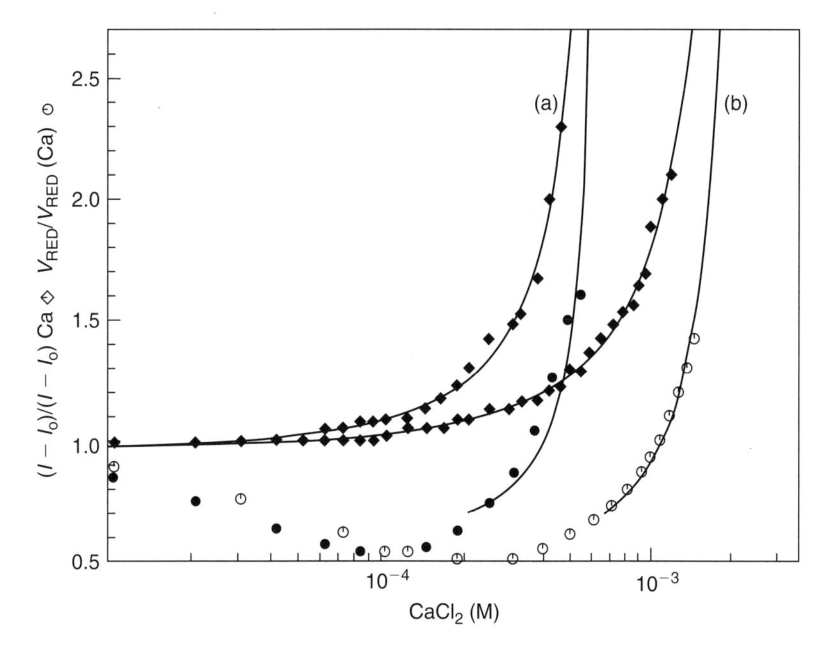

Figure 24.12 Changes in reduced viscosity (○, ●) and scattered light (◆) during the addition of $CaCl_2$ to pectin solutions (polymer concentration: $0.2\,g\,L^{-1}$; 5×10^{-3} M $CaCl_2$). Role of the carboxyl group distribution: (a) DE = 30 per cent blockwise distribution, obtained by pectin-esterase hydrolysis; (b) DE = 30 per cent random distribution, obtained by alkaline hydrolysis. (Reproduced from Reference [92], with permission from John Wiley & Sons, Inc., on behalf of the Society of Chemical Industry.)

Table 24.2

Transport parameters (*f*) of divalent counterions for pectins with different DE. Comparison of experimental values with the theoretical one calculated with Manning's theory [92].

DE (%)	λ	*f* for Ba^{+2}	*f* for Sr^{+2}	*f* for Mg^{+2}	Theoretical *f*
<2	1.58	0.095	0.110	0.280	0.276
21.4	1.26	0.140	0.171	0.290	0.344
72.1	0.45	0.475	0.529	0.630	0.889

24.7 GALACTOMANNAN IONIC DERIVATIVES

Galactomannans are neutral polysaccharides isolated from seeds (carob, guar, locust bean, tara, etc.). Their main chain is made of [(1 → 4)-β-D-Man] (Man = mannose; M units) with different degrees of substitution on *O*-6 with α-D-galactopyranosyl units (G). Their composition (or M/G ratio) is easily determined by ^1H NMR.

The solubility of galactomannans depends on the M/G ratio and on the distribution of galactose units along the mannan chain, the larger being the galactose content, the higher the solubility in water. The M/G ratio varies from 1/1 (in Mimosa scrabella) to 1/5 (in locust bean gum) and the rheological properties of galactomannan solutions depend on this ratio, because of the increase in loose interchain interactions when the G content decreases [96]. Molecular modelling was applied to the determination of the persistence length of a galactomannan model in which M/G ≈ 1 and a good agreement was obtained between experimental SEC determinations and the corresponding calculated; L_p was found in the range of 9.5 nm [97].

Although the water solubility of galactomannans is relatively low, the ensuing solutions display interesting rheological properties as thickeners. They have been modified by the specific oxidation at C-6 position with TEMPO in order to increase their solubility by transforming them into new polyelectrolytes [98, 99].

24.8 LIGNINS AND LIGNOSULPHONATES

Although lignins are not polysaccharides, a brief mention of their interest when they are in a polyelectrolyte form seems justified to conclude this chapter.

Lignins extracted from wood have modest polyelectrolyte behaviour and are usually soluble in alkaline media thanks to the presence of phenolic and a few carboxylic groups. A birch lignin extracted by 40 wt% sodium benzoate at 150°C displayed a net charge of about 3×10^{-3} equivalents per gram of lignin [100].

Lignosulphonate is a by-product of the sulphite process in the manufacturing of pulp (see Chapter 10). It is a complex mixture of small-to-moderate-size polymers and oligomers bearing sulphonate groups, which display a strong polyelectrolyte character and are often used to deflocculate clay-based muds.

24.9 CONCLUSION

The impressive variety of polysaccharide structures which display polyelectrolyte properties can be therefore exploited in a wide range of applications, those associated with human consumption and health care being of particular interest. Renewable resources play therefore a particularly useful role in providing readily available and often cheap macromolecular materials which are rarely matched by equivalent counterparts derived from fossil or inorganic sources.

REFERENCES

1. Rinaudo M., Auzely R., Mazeau K., Polysaccharides and carbohydrate polymers, in *Encyclopedia of Polymer Science and Technology*, John Wiley & Sons, New-York, 2004, pp. 200–261, Chapter 11.
2. Rinaudo M., Polysaccharides, in *Kirk-Othmer Encyclopedia of Chemical Technology*, Vol. 20, 5th Edition, John Wiley & Sons, New-York, 2006, pp. 549–586.
3. (a) Rinaudo M., Reguant J., *Polysaccharide derivatives*, in *Natural Polymers and Agrofibers Composites*, Eds.: Frollini E., Leão A.L. and Mattoso L.H.C., São Carlos, Brazil, 2000, pp. 15–39. (b) Ballerini D., Production d'éthanol à partir de biomasse, *Actual. Chim.*, **11–12**, 2002, 83–87.
4. Rinaudo M., Chitin and Chitosan: Properties and applications, *Prog. Polym. Sci.*, **31**, 2006, 603–632.
5. Geremia R., Rinaudo M., Biosynthesis, structure, and physical properties of some bacterial polysaccharides, in *Polysaccharides: Structural Diversity and Functional Versatility*, Ed.: Dimitriu S., Marcel Dekker, New-York, 2004, pp. 411–430, Chapter 15.
6. (a) Lifson S., Katchalsky A., The electrostatic free energy of polyelectrolyte solutions. II. Fully stretched macromolecules, *J. Polym. Sci.*, **13**, 1954, 43–55. (b) Katchalsky A., Problems in the physical chemistry of polyelectrolytes, *J. Polym. Sci.*, **12**, 1954, 159–182.
7. (a) Manning G.S., Limiting laws and counterion condensation in polyelectrolyte solutions. I. Colligative properties, *J. Chem. Phys.*, **51**, 1969, 924–933. (b) Manning G.S., Limiting laws and counterion condensation in polyelectrolyte solutions. III. An analysis based on the Mayer ionic solution theory, *J. Chem. Phys.*, **51**, 1969, 3249–3252.
8. (a) Milas M., Rinaudo M., On the electrostatic interactions of ionic polysaccharides in solution, *Curr. Trends Polym. Sci.*, **2**, 1997, 47–67. (b) Rinaudo M., Milas M., Ionic selectivity of polyelectrolytes in salt free solutions, in *Polyelectrolytes and their Applications*, Eds.: Rembaum A. and Sélégny E., D. Reidel Publishing Company, Dordrecht (Holland), 1975, pp. 31–49.
9. Rinaudo M., Advances in characterisation of polysaccharides in aqueous solution and gel state, in *Polysaccharides: Structural Diversity and Functional Versatility*, Ed.: Dimitriu S., Marcel Dekker, New-York, 2004, pp. 237–252, Chapter 8.
10. Rochas C., Rinaudo M., Activity coefficients of counterions and conformation in kappa-carrageenan systems, *Biopolymers*, **19**, 1980, 1675–1687.
11. Rochas C., Rinaudo M., Calorimetric determination of the conformational transition of kappa-carrageenan, *Carbohydr. Res.*, **105**, 1982, 227–236.
12. Ciencia M.M., Milas M., Rinaudo M., On the specific role of coions and counterions on kappa-carrageenan conformation, *Int. J. Biol. Macromol.*, **20**, 1997, 35–41.
13. Malovikova A., Milas M., Rinaudo M., Borsali R., Viscosimetric behavior of Na-polygalacturonate in the presence of low salt content, in *Macro-ion characterization from dilute solutions to complex fluids*, Symposium series No. 548, American Chemical Society, New-York 1994, pp. 315–321.
14. Morfin I., Reed W.F., Rinaudo M., Borsali R., Further evidence of liquid-like correlations in polyelectrolyte solutions, *J. Phys. II*, **4**, 1994, 1001–1019.

15. Milas M., Rinaudo M., Duplessix R., Borsali R., Lindner P., Small angle neutron scattering from polyelectrolyte solutions: From disordered to ordered xanthan chain conformation, *Macromolecules*, **28**, 1995, 3119–3124.
16. Roure I., Rinaudo M., Milas M., Viscometric behavior of dilute polyelectrolytes. Role of electrostatic interactions, *Ber. Bunsenges. Phys. Chem.*, **100**, 1996, 703–706.
17. Rinaudo M., Rochas C., Michels B., Etude par absorption ultrasonore de la fixation sélective du potassium sur le carraghénane, *J. Chim. Phys.*, **80**(3), 1983, 305–308.
18. Zana R., Tondre C., Rinaudo M., Milas M., Etude ultrasonore de la fixation sur site des ions alcalins de densités de charge variable, *J. Chim. Phys.*, **68**, 1971, 1258–1260.
19. (a) Rinaudo M., Polysaccharide characterization in relation with some original properties, *J. Appl. Polym. Sci.: Appl. Polym. Symp.*, **52**, 1993, 11–13. (b) Rinaudo M., Non-covalent interactions in polysaccharide systems, *Macromol. Biosci.*, **6**, 2006, 590–610.
20. Haxaire K., I. Braccini I., M. Milas M., Rinaudo M., Perez S., Conformational behavior of hyaluronan in relation to its physical properties as probed by molecular modelling, *Glycobiology*, **10**, 2000, 587–594.
21. Petkowicz C.L.O., Rinaudo M., Milas M., Mazeau K., Bresolin T., Reicher F., Ganter J.L.M.S., Conformation of galactomanan, experimental and modeling approaches, *Food Hydrocolloid.*, **13**, 1999, 263–266.
22. Mazeau K., Perez S., Rinaudo M., Predicted influence of *N*-acetyl group content on the conformational extension of chitin and chitosan chains, *J. Carbohydr. Chem.*, **19**, 2000, 1269–1284.
23. Rinaudo M., Wormlike chain behaviour of some bacterial polysaccharides, in *Macromolecules 1992*, Ed.: Kahovec J., VSP, Utrecht, 1993, pp. 207–219.
24. Benoit H., Doty P., Light scattering from non-gaussian chains, *J. Phys. Chem.*, **57**, 1953, 958–963.
25. (a) Reed W., Light-scattering results on polyelectrolyte conformations, diffusion and interparticles interactions and correlations, in *Macroion Characterization. From Dilute Solution to Complex Fluids*, Ed.: Schmitz K.S., American Chemical Symposium series 548, American Chemical Society, New-York, 1984, pp. 297–314. (b) Matsuoka S., Cowman M.K., The intrinsic viscosity of Hyaluronan, in Hyaluronan, Vol. 1: Chemical, Biochemical and Biological Aspects, Ed.: Kennedy J.F., Phillips G.O., Williams P.A. and Hascall V.C., Woodhead Publishing, Cambridge (GB), 2002, 79–88.
26. Arguelles-Monal W., Cabrera G., Peniche C., Rinaudo M., Conductometric study of the inter-polyelectrolyte reaction between chitosan and poly(galacturonic acid), *Polymer*, **41**, 2000, 2373–2378.
27. Bercheran-Maron L., Peniche C., Arguelles-Monal W., Study of the interpolyelectrolyte reaction between chitosan and alginate: Influence of alginate composition and chitosan molecular weight, *Int. J. Biol. Macromol.*, **34**, 2004, 127–133.
28. Gaserod O., Smidsrod O., Skjak-Braek G., Microcapsules of alginate-chitosan-I. A quantitative study of the interaction between alginate and chitosan, *Biomaterials*, **19**, 1998, 1815–1825.
29. Rinaudo M., Hudry-Clergeon G., Etude des O-carboxyméthylcelluloses à degré de substitution variable. I: Préparation et caractérisation des produits, *J. Chim. Phys.*, **64**, 1967, 1746–1752.
30. Milas M., Interaction avec les cations compensateurs et sélectivité ionique dans les solutions aqueuses de polyélectrolyte, Influence de la densité de charge, Ph.D. Thesis, Grenoble, 1974.
31. Barba C., Montane D., Farriol X., Rinaudo M., Synthesis and characterization of carboxymethylcelluloses (CMC) from non-wood fibers I. Accessibility of cellulose fibers and CMC synthesis, *Cellulose*, **9**, 2003, 319–326.
32. (a) Barba C., Montane D., Farriol X., Desbrieres J., Rinaudo M., Synthesis and characterization of carboxymethylcelluloses from non-wood pulps II. Rheological behavior of CMC in aqueous solution, *Cellulose*, **9**, 2003, 327–335. (b) Renaud M., Belgacem M., Rinaudo M., Rheological behaviour of polysaccharide aqueous solutions, *Polymer*, **46**, 2005, 12348–12358.
33. Rinaudo M., Comparison between experimental results obtained with hydroxylated polyacids and some theoretical models, in *Polyelectrolytes*, Eds.: Sélégny E., Mandel M. and Strauss U.P., D. Reidel Publishing Company, Dordrecht (Holland), 1974, pp. 157–193.
34. Hofer F., Synthèse et caractérisation de gels échangeurs de cations obtenus par réticulation de polysaccharides carboxyliques, Ph.D. Thesis, Grenoble, 1974.
35. Canova P., Préparation et comportement d'une membrane obtenue par réticulation d'un polysaccharide ionique, Ph.D. Thesis, Grenoble, 1975.
36. Lowys M.P., Desbrieres J., Rinaudo M., Rheological characterization of cellulosic microfibril suspensions. Role of polymeric additives, *Food Hydrocolloid.*, **15**, 2001, 25–32.
37. Rinaudo M., Lowys M.P., Desbrieres J., Characterization and properties of cationic cellulosic microfibrils, *Polymer*, **41**, 2000, 607–613.
38. Baudet G., Morio M., Rinaudo M., Nematollahi H., Synthèse et caractérisation de floculants sélectifs à base de dérivés xanthés de la cellulose et de l'amylose, *Minéralurgie*, **3**, 1978, 19–35.
39. Trubiano P.C., Succinate and substituted succinate derivatives of starch, in *Modified Starch: Properties and Uses*, Ed.: Wurzburg O.B., CRC Press, Boca Raton (USA), 1986, pp. 131–147.
40. Solarek D.B., Phosphorilated starches and miscellaneous inorganic esters, in *Modified Starch: Properties and Uses*, Ed.: Wurzburg O.K., CRC Press, Boca Raton (USA), 1986, pp. 97–112.

41. Noik C., Interaction de polymères hydrophiles à une interface solide, Ph.D. Thesis, Grenoble, 1982.
42. Rinaudo M., Noik C., Adsorption of polysaccharides on a calcite using spin labelled polymers, *Polym. Bull.*, **9**, 1983, 543–547.
43. Solarek D.B., Cationic starches, in *Modified Starch: Properties and Uses*, Ed.: Wurzburg O.B., CRC Press, Boca Raton (USA), 1986, pp. 114–129.
44. Smidsrod O., Some physical properties of alginates in solution and in gel state. Ph.D. Thesis, Trondheim (Norway), 1973.
45. Bouffar-Roupe C., Structure et propriétés gélifiantes des alginates, Ph.D. Thesis, Grenoble (France), 1989.
46. Grasdalen H., Larsen B., Smidsroed O., Carbon-13 NMR studies of monomeric composition and sequence in alginate, *Carbohydr. Res.*, **89**, 1981, 179–191.
47. Grasdalen H., Larsen B., Smidsroed O., A proton magnetic resonance study of the composition and sequence of uronate residues in alginates, *Carbohydr. Res.*, **68**, 1979, 23–31.
48. Grasdalen H., Larsen B., Smidsroed O., Carbon-13 NMR studies of alginate, *Carbohydr. Res.*, **56**, 1977, C11–C15.
49. Grasdalen H., High field ^1H NMR spectroscopy of alginate: Sequential structure and linkage conformations, *Carbohydr. Res.*, **118**, 1983, 255–260.
50. Rinaudo M., Graebling D., On the viscosity of sodium alginates in the presence of external salt, *Polym. Bull.*, **15**, 1986, 253–256.
51. Smidsrod O., Glover R.M., Whittington S.G., Relative extension of alginates having different chemical composition, *Carbohydr. Res.*, **27**, 1973, 107–118.
52. Gatej I., Popa M., Rinaudo M., Role of the pH on hyaluronan behavior in aqueous solution, *Biomacromolecules*, **6**, 2005, 61–67.
53. Ravanat G., Rinaudo M., Investigation on oligo- and polygalacturonic acids by potentiometry and circular dichroism, *Biopolymers*, **19**, 1980, 2209–2222.
54. Kohn R., Ion binding on polyuronates. Alginate and pectin, *Pure Appl. Chem.*, **30**, 1975, 371–397.
55. Grant G.T., Morris E.R., Rees D.A., Smith P.J.C., Thom D., Biological interactions between polysaccharides and divalent cations: The egg-box model, *FEBS Lett.*, **32**, 1973, 195–198.
56. Braccini I., Grasso R.P., Perez S., Conformational and configurational features of acidic polysaccharides and their interactions with calcium ions: A molecular modelling investigation, *Carbohydr. Res.*, **317**, 1999, 119–130.
57. Braccini I., Perez S., Molecular basis of Ca^{2+}-induced gelation in alginates and pectins: The egg-box model revisited, *Biomacromolecules*, **2**(4), 2001, 1089–1096.
58. Martinsen A., Skjak-Braek G., Smidsrod O., Alginate as immobilization material: I. Correlation between chemical and physical properties of alginate gel beads, *Biotech. Bioeng.*, **33**, 1989, 79–89.
59. Rochas C., Rinaudo M., Landry S., Role of the molecular weight on the mechanical properties of kappa carrageenan gels, *Carbohydr. Polym.*, **12**, 1990, 255–266.
60. Kohn R., Furda I., Haug A., Smidsroed O., Binding of calcium and potassium ions to some polyuronides and monouronates, *Acta Chem. Scand.*, **22**, 1968, 3098–3102.
61. Smidsroed O., Haug A., Dependence upon uronic acid composition of some ion-exchange properties of alginates, *Acta Chem. Scand.*, **22**, 1968, 1989–1997.
62. Takahashi T., Ishiwatari Y., Shirai H., Ion exchange of alginic acid. IV. Selective ion exchange behavior of alginate in mixed solutions of metal ions, *Kogyo Kagaku Zasshi*, **66**(10), 1963, 1458–1461.
63. Takahashi T., Emura E., Ion exchange of alginic acid. II. Selective ion-exchange properties of alginates for metallic ions, *Kogyo Kagaku Zasshi*, **63**(6), 1960, 1025–1026.
64. Smidsrod O., Haug A., Dependence upon the gel–sol state of the ion-exchange properties of alginates, *Acta Chem. Scand.*, **26**(5), 1972, 2063–2074.
65. Haug A., Smidsroed O., Effect of divalent metals on the properties of alginate solutions. II. Comparison of different metal ions, *Acta Chem. Scand.*, **19**(2), 1965, 341–351.
66. (a) Fundueanu G., Esposito E., Mihai D., Carpov A., Desbrieres J., Rinaudo M., Nastruzzi C., Preparation and characterization of Ca-alginate microspheres by a new emulsification method, *Int. J. Pharm.*, **170**, 1998, 11–21. (b) Fundueanu G., Nastruzzi C., Carpov A., Desbrieres J., Rinaudo M., Physico-chemical characterization of Ca-alginate microparticles produced with alternative strategies, *Biomaterials*, **20**, 1999, 1427–1435.
67. (a) Smidsrod O., Skjak-Braek G., Alginate as immobilization matrix for cells, *Trends Biotechnol.*, **8**, 1990, 71–78. (b) Draget K.I., Skjak-Braek G., Smidsrod O., Alginate based new materials, *Int. J. Biol. Macromol.*, **21**(1–2), 1997, 47–55.
68. Draget K.I., Oestgaard K., Smidsrod O., Alginate-based solid media for plant tissue culture, *Appl. Microbiol. Biotech.*, **31**(1), 1989, 79–83.
69. Majima T., Funakosi T., Iwasaki N., Yamane S-T., Harada K., Nonaka S., Minami A., Nishimura S-I., Alginate and chitosan polyion complex hybrid fibers for scaffolds in ligament and tendon tissue engineering, *J. Orthop. Sci.*, **10**, 2005, 302–307.

70. Wang L., Khor E., Wee A., Lim L.Y., Chitosan–alginate PEC membrane as a wound dressing: Assessment of incisional wound healing, *J. Biomed. Mater. Res.*, **63**, 2002, 610–618.
71. (a) Heyraud A., Rinaudo M., Rochas C., Physical and chemical properties of phycocolloids, in *Introduction to Applied Phycology*, Ed.: Akatsuka I., SPB Academic Publishing, Dordrecht (The Netherlands), 1989, pp. 59–84. (b) Rinaudo M., Les alginates et les carraghénanes, *Actual. Chim.*, **11–12**, 2002, 35–38.
72. Lecacheux D., Panaras R., Brigant G., Martin G., Molecular weight distribution of carrageenan by size exclusion chromatography and low angle laser light scattering, *Carbohydr. Polym.*, **5**, 1985, 423–440.
73. Rochas C., Etude de la transition sol-gel du kappa-carraghénane. Ph.D. Thesis, Grenoble (France), 1982.
74. Rochas C., Rinaudo M., Calorimetric determination of the conformational transition of kappa-carrageenan, *Carbohydr. Res.*, **105**, 1982, 227–236.
75. Morris E.R., Rees D.A., Calorimetric and chiroptical evidence of aggregate-driven helix formation in carrageenan systems, *Carbohydr. Res.*, **80**, 1980, 317–323.
76. Rochas C., Rinaudo M., Activity coefficients of counterions and conformation in kappa-carrageenan systems, *Biopolymers*, **19**, 1980, 1675–1687.
77. Viebke, Borgström J., Piculell L., Characterization of kappa- and iota-carrageenan coils and helices by MALLS/GPC, *Carbohydr. Polym.*, **27**, 1995, 145–154.
78. Rinaudo M., Rochas C., Investigations on aqueous solution properties of kappa carrageenans, in *Solution Properties of Polysaccharides*, Ed.: Brant D.A., ACS Symposium Series 150, American Chemical Society, Washington D.C., 1981, pp. 367–378.
79. Borgstroem J., Quist P.O., Piculell L., A novel chiral nematic phase in aqueous κ-carrageenan, *Macromolecules*, **29**(18), 1996, 5926–5933.
80. Rochas C., Rinaudo M., Mechanism of gel formation in kappa-carrageenan, *Biopolymers*, **23**, 1984, 735–745.
81. Rinaudo M., Karimian A., Milas M., Polyelectrolyte behaviour of carrageenans in aqueous solutions, *Biopolymers*, **18**, 1979, 1673–1683.
82. Landry S., Relation entre la structure moléculaire et les propriétés mécaniques des gels de carraghénanes, Ph.D. Thesis. Grenoble, 1987.
83. Rinaudo M., Landry S., On the volume change on non covalent gels in solvent-non solvent mixtures, *Polym. Bull.*, **17**, 1987, 563–565.
84. Rinaudo M., Gelation of polysaccharides, *J. Intel. Mat. Syst. Str.*, **4**, 1993, 210–215.
85. Visser J., Voragen A.G.J., *Pectins and Pectinases, Progress in Biotechnology*, vol. 14, Elsevier, Amsterdam, 1996.
86. Ralet M.-C., Thibault J.-F., Interchain heterogeneity of enzymatically deesterified lime pectins, *Biomacromolecules*, **3**, 2002, 917–925.
87. Rinaudo M., Physicochemical properties of pectins in solution and gel states, in *Pectins and Pectinases*, Eds.: Visser J. and Voragen A.G., Elsevier, Amsterdam, 1996, pp. 21–33.
88. Rinaudo M., Effect of chemical structure of pectins on their interactions with calcium, in *Plant Cell Wall Polymers: Biogenesis and Biodegradation*, Eds.: Lewis G. and Paice M.G., ACS Symposium Series, American Chemical Society, Washington D.C., 1989, Chapter 23, pp. 324–332.
89. Thibault J.F., Rinaudo M., Interactions of mono- and divalent counterions with alcali- and enzyme-deesterified pectins in salt-free solutions, *Biopolymers*, **24**, 1985, 2131–2134.
90. Turquois T., Rinaudo M., Taravel F.R., Heyraud A., Extraction of highly gelling pectins from sugar beet pulp, in *Hydrocolloids 1: Physical Chemistry and Industrial Application of Gels, Polysaccharides and Proteins*, Ed.: Nishinari K., Elsevier, 2000, pp. 229–235.
91. Turquois T., Rinaudo M., Taravel F.R., Heyraud A., Extraction of highly gelling pectic substances from sugar beet pulp and potato pulp: Influence of extrinsic parameters on their gelling properties, *Food Hydrocolloid.*, **13**, 1999, 255–262.
92. Thibault J.F., Rinaudo M., Gelation of pectinic acids in the presence of calcium counterions, *Br. Polym. J.*, **17**(2), 1985, 181–184.
93. Thibault J.F., Rinaudo M., Chain association of pectic molecules during calcium-induced elation, *Biopolymers*, **25**, 1986, 455–468.
94. Malovikova A., Rinaudo M., Milas M., Comparative interactions of magnesium and calcium counterions with polygalacturonic acid, *Biopolymers*, **34**, 1994, 1059–1064.
95. Diaz-Rojas E.I., Pacheco-Aguilar R., Lizardi J., Argüelles-Monal W., Valdez M.A., Rinaudo M., Goycoolea F.M., Linseed pectin: Gelling properties and performance as an encapsulation matrix for shark liver oil, *Food Hydrocolloid.*, **18**(2), 2004, 293–304.
96. Ganter J., Milas M., Rinaudo M., Study of solution properties of galactomannan from the seeds of Mimosa scabrella, *Carbohydr. Polym.*, **17**, 1992, 171–175.

97. Petkowicz C.L.O., Rinaudo M., Milas M., Mazeau K., Bresolin T., Reicher F., Ganter J.L.M.S., Conformation of galactomanan, experimental and modeling approaches, *Food Hydrocolloid.*, **13**, 1999, 263–266.
98. Frollini E., Reed W.F., Milas M., Rinaudo M., Polyelectrolytes from polysaccharides: Selective oxidation of guar gum – A revisited reaction, *Carbohydr. Polym.*, **27**, 1995, 129–135.
99. Sierakowski M.R., Milas M., Desbrieres J., Rinaudo M., Specific modifications of galactomannans, *Carbohydr. Polym.*, **42**, 2000, 51–57.
100. Rinaudo M., Pla F., Fractionnement par filtration sur gel et distribution en masses moléculaires sur un haut polymère naturel: la lignine, *Chimie Analytique*, **49**, 1967, 320–326.

– 25 –

Chitin and Chitosan: Major Sources, Properties and Applications

C. Peniche, W. Argüelles-Monal and F.M. Goycoolea

ABSTRACT

Chitin is widely distributed in nature, constituting an important renewable resource. The main sources of chitin generally used are the crustacean wastes of the fishing industry. The chapter gives a brief account of the main processes employed in chitin isolation and the preparation of chitosan by extensive deacetylation of chitin. The common methods of characterization of chitin and chitosan, in terms of degree of acetylation and molecular weight, are discussed. Their crystalline structure and their solution properties, are also described. The capacity of chitin and chitosan of forming complexes with metal ions is shown, and mention is made to some of its diverse applications. The ability of chitosan to form polyelectrolyte complexes with polyanions, the cooperativity of this reaction and the properties of chitosan-based polyelectrolyte complex membranes, are also examined. The chapter ends with a review of the applications of chitin and chitosan in medicine, pharmacy, agriculture, the food industry, cosmetics, among others.

Keywords

Chitin, Chitosan, Demineralization, Deproteinization, Deacetylation, Degree of acetylation, Metal ions complexation, Polyelectrolyte complex, Biomaterial, Flocculation, Antimicrobial activity

25.1 INTRODUCTION

Chitin is the second most abundant polysaccharide in nature after cellulose. It is widely distributed in the animal and vegetal kingdom, constituting an important renewable resource. It was first isolated from fungi by Braconnot in 1811 [1], but the name chitin – from the Greek χιτων which means tunic or cover – was given by Odier, who in 1923 isolated it from the elytrum of the cock-chafer beetle by treatment with hot alkaline solutions [2]. Because of its insolubility in the vast majority of common solvents, chitin was considered an intractable polymer and for many years it remained mainly a laboratory curiosity. However, as will be shown below, at present chitin and its derivatives have become polymers of great interest in a large variety of areas of human activity.

Chitin is generally represented as a linear polysaccharide composed of $\beta(1 \rightarrow 4)$ linked units of *N*-acetyl-2-amino-2-deoxy-D-glucose (Fig. 25.1(a)). Although it has been proposed that, depending on the source, a variable, but always small, proportion of these structural units are deacetylated in natural chitin, this has never been proved unambiguously, with recent evidence pointing to the contrary.

The great structural similarity existing between chitin and cellulose is shown in Fig. 25.1. The difference between them consists in that the hydroxyl group of carbon C2 in cellulose (Fig. 25.1(b)) is substituted by an acetamide group in chitin. Both biopolymers play similar roles, since they both act as structural support and defence materials in living organisms.

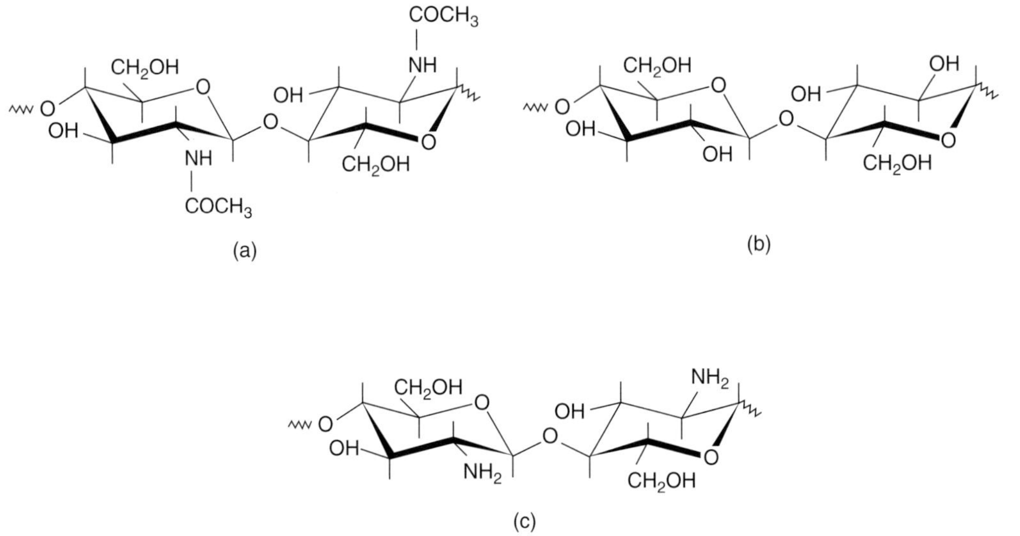

Figure 25.1 Schematic representation of (a) completely acetylated chitin; (b) cellulose and (c) completely deacetylated chitosan. The structural similarity between them becomes evident.

Chitosan is a linear polysaccharide obtained by extensive deacetylation of chitin. It is mainly composed of two kinds of $\beta(1 \rightarrow 4)$ linked structural units *viz.* 2-amino-2-deoxy-D-glucose and *N*-acetyl-2-amino-2-deoxy-D-glucose. The chemical structure of a completely deacetylated chitosan is represented in Fig. 25.1(c). However, since it is virtually impossible to completely deacetylate chitin, what is usually known as chitosan is a family of chitins with different but always low degrees of acetylation. The capacity of chitosan to dissolve in dilute aqueous solutions is the commonly accepted criterion to differentiate it from chitin.

Chitin is the most abundant organic component of the skeletal structure of many classes comprising the group of invertebrates, such as arthropods, mollusks and annelids. In animals, chitin occurs associated with other constituents, such as lipids, calcium carbonate, proteins and pigments. It has been estimated that the crustacean chitin present in the sea amounts to 1 560 million tons [3]. Chitin is also found as a major polymeric constituent of the cell wall of fungi and algae. Fungal chitin exhibits some advantages as compared with animal chitin, such as a greater uniformity in composition, a continuous availability in time and the absence of inorganic salts in its matrix. However, in fungi chitin is associated with other polysaccharides, such as cellulose, glucan, mannan and polygalactosamine, which makes its isolation difficult [4].

Chitosan is also present in significant quantities in some fungi, such as *Mucor rouxii* (30 per cent) and *Choanephora cucurbitarum* (28 per cent), although again associated with other polysaccharides.

In order to improve the properties of these unique polysaccharides and to develop new advanced materials, much attention has been paid to their chemical modification. These polymers have two reactive groups suitable for this purpose, namely, primary (C6) and secondary (C3) hydroxyl groups in the case of chitin whereas chitosan has additionally the amino (C2) group on each deacetylated unit. All these functions are susceptible to a variety of classical reactions which can be applied here in a controlled fashion to obtain a vast array of novel materials based on the two polysaccharides which can also be modified by either crosslinking or graft copolymerization. This topic has been extensively studied and thoroughly documented [5–7].

25.2 METHODS OF PREPARATION

25.2.1 Isolation of chitin

Commercial chitin is extracted from crustacean wastes of the fishing industry, the main chitin sources being the shells of shrimp, crab, lobster, prawn and krill. These crustacean wastes consist of chitin (20–30 per cent), protein

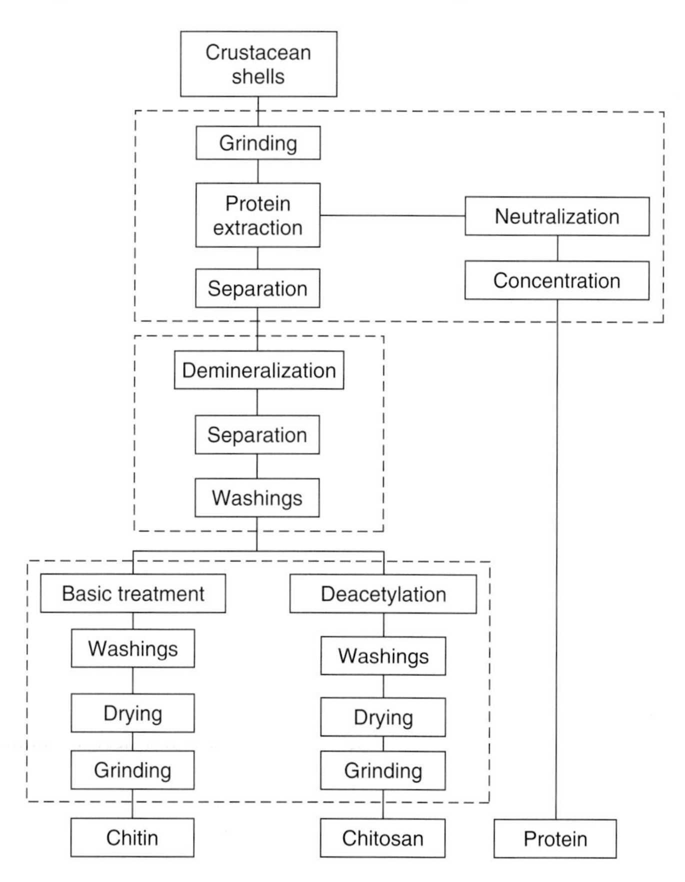

Figure 25.2 Schematic representation of the different steps involved in chitin/chitosan preparation procedures. In this diagram a second treatment with dilute NaOH is considered for chitin isolation in order to remove any residual protein.

(30–40 per cent), inorganic salts (mainly calcium carbonate and phosphate) (30–50 per cent) and lipids (0–14 per cent). These percentages vary considerably with the species and the season [8]. Consequently, the extraction techniques reported are very wide-ranging, since they depend significantly on the composition of the source. The majority of the techniques developed rely on chemical processes of hydrolysis of the protein and the removal of the inorganic material. A feasible sequence of isolation steps is schematically represented in Fig. 25.2. Some processes include a decolouration step of chitin by solvent extraction or oxidation of the remaining pigments. These isolation methods generally consume great amounts of water and energy, and frequently give rise to corrosive wastes. At present, enzymatic treatments are being investigated as a promising alternative. To this end processes have been reported that use enzymatic extracts or isolated enzymes and biological fermentations, but they still lack the effectiveness of the chemical methods, mainly with respect to the removal of the inorganic materials [9].

Chitin isolation processes are generally performed through the following consecutive steps: raw material conditioning, protein extraction (deproteinization), removal of inorganic components (demineralization) and decolouration. This sequence is preferred if the isolated protein is to be used as food additive for livestock feeding. Otherwise, demineralization can be carried out first [10]. A brief account of these processes will be given below. A more detailed description of chitin isolation (and chitosan preparation) can be found elsewhere [5, 8, 11].

25.2.1.1 Raw material conditioning

The crustacean shells are ground to the appropriate size (generally a few millimetres) and washed profusely with water to remove any organic material adhered to their surface. Acosta *et al.* have employed a pretreatment of the wastes with boiling water for 24 h followed by drying at 80°C for 24 h [12].

25.2.1.2 Deproteinization

The procedure most frequently used to separate the proteins consists in treating the crustacean shells with diluted aqueous NaOH solutions (1–10 per cent) at elevated temperatures (65–100°C). The reaction time usually varies from 0.5 to 72 h, depending on the treatment. Prolonged treatments or too high temperatures may provoke chain scission and partial deacetylation of the polymer. Sometimes, it is preferred to perform two consecutive treatments of 1–2 h. Other reagents employed for the removal of the proteins include: Na_2CO_3, $NaHCO_3$, KOH, K_2CO_3, $Ca(OH)_2$, Na_2SO_3, $NaHSO_3$, Na_3PO_4 and Na_2S [5].

The protein extracted can be recovered by lowering the pH of the solution to its isoelectric point for precipitation. The recovered protein can be used as a high grade additive for livestock starter feeds. This decreases the manufacturing costs of chitin.

The deproteinization processes using enzymatic extracts or isolated enzymes and biological fermentations have been tested with relative success, since they minimize the chemical degradation of chitin and lead to environmentally cleaner operations. However, enzymatic/microbiological treatments have the drawback of being time consuming and leaving the material with 1–7 per cent of residual protein [13]. These remnants can be diminished with the use of detergents [13] or with the application of physicomechanical techniques [14].

25.2.1.3 Demineralization

The main inorganic component of crustacean shells is calcium carbonate, which is usually eliminated with dilute HCl solutions (up to 10 wt-per cent) at room temperature, although other acids (*e.g.* HNO_3, HCOOH, HNO_3, H_2SO_4 and CH_3COOH) have also been employed. Demineralization occurs according to the following reaction:

$$CaCO_3 + 2HCl \rightarrow CO_2\uparrow + CaCl_2 + H_2O \tag{25.1}$$

It is evident that if the amount of acid employed is below the stoichiometric ratio, the demineralization reaction will not be completed. The acid concentration and the reaction time depend on the source, but these parameters must be carefully controlled in order to minimize the hydrolytic depolymerization and deacetylation of chitin. High temperature treatments must also be avoided to prevent thermal degradation [10]. An alternative treatment for demineralization makes use of the complexing agent EDTA (ethylenediamine tetra acetic acid) in basic media [15].

25.2.1.4 Decolouration

The colour of crustacean shells is mainly due to the presence of pigments such as astaxanthin, cantaxanthin, astacene, lutein and β-carotene. The above treatments are not usually capable of eliminating these pigments which are frequently extracted at room temperature with acetone, chloroform, ether, ethanol, ethyl acetate or a mixture of solvents [8]. Traditional oxidizing agents such as H_2O_2 (0.5–3 per cent) and NaClO (0.32 per cent) have also been employed, but these reagents may attack the free amino groups and introduce modifications in the polymer. For strongly coloured shells, such as those of the common lobster carapaces, treatments with mixtures of acetone and NaOCl at room temperature have been reported [12].

25.2.2 Preparation of chitosan

Chitin deacetylation is performed by the hydrolysis of the acetamide groups at high temperature in a strongly alkaline medium. The reaction is generally carried out heterogeneously using concentrated (40–50 per cent) NaOH or KOH solutions at temperatures above 100°C, preferably in an inert atmosphere or in the presence of reducing agents, such as $NaBH_4$ or thiophenol, in order to avoid depolymerization. The specific reaction conditions depend on several factors, such as the starting material, the previous treatment and the desired degree of acetylation. Nevertheless, with only one alkaline treatment, the maximum deacetylation degree attained will not surpass 75–85 per cent. Prolonged treatments cause the degradation of the polymer without resulting in an appreciable increase in the deacetylation degree [16].

Several treatments have been developed to prepare fully deacetylated chitosan [17–19]. Their common characteristic is that they involve the repetition of consecutive deacetylation–washing–drying treatments as many times as required. Lamarque *et al.* developed a multistep heterogeneous deacetylation process for the production of well-defined chitosans with high deacetylation degrees and higher molecular weights than those usually reported in the literature. The procedure was applied to α- and β-chitins in the presence of 50 per cent (w/v) NaOH, at temperatures ranging from 80°C to 110°C under an argon atmosphere [20].

Chitin, just like cellulose, is a semi-crystalline polymer, so that when deacetylation is performed in heterogeneous conditions, the reaction takes place mainly at the amorphous regions, whereas homogeneous conditions bring about a more uniform modification of the polymer. The latter reaction is carried out using alkali chitin, which is obtained by performing successive freezing–thawing cycles of an alkaline aqueous suspension of chitin until dissolution. Homogeneous deacetylation is achieved with more moderate alkali concentrations (about 13 wt-per cent), at 25–40°C for 12–24 h [21]. It has been shown that while chitosans obtained by the heterogenous process are polydispersed in terms of the acetylation degree of their chains, chitosans obtained under homogeneous conditions do not exhibit chain compositional dispersion.

The main drawback of the methods usually employed to produce chitosan is that they involve long processing times and expend large amounts of alkali. In order to overcome this, a number of unconventional deacetylation methods, such as thermomechanical processes using a cascade reactor operating under alkali conditions [22], flash treatment under saturated steam [23], microwave dielectric heating [14] and intermittent water washings [17], have been developed.

25.3 CHARACTERIZATION OF CHITIN AND CHITOSAN

25.3.1 Determination of the degree of acetylation

The degree of acetylation (or deacetylation) and the molecular mass are undoubtedly the most important parameters to establish the chemical and physical identity of chitin and chitosan. Both parameters vary with the biological source of the raw material and the preparation method. They also dictate the physicochemical, functional and biological properties of these polysaccharides, essential to fit an application or end product. Solubility, pK_o, viscosity, gelling capacity, among other properties, are all dependent on these parameters.

The degree of acetylation is defined as the fraction or percentage of *N*-acetylated glycosidic units in chitin or chitosan. It is designated indistinctly as F_A or DA (although DA is sometimes also expressed as DA per cent). It is also customary to express this parameter as degree of deacetylation ($= 1 - F_A$ or DD $= 100 - $ DA per cent). Table 25.1 shows a summary of the main spectroscopic, titration, circular dichroism, enzymatic, chromatographic, and thermal analytical methods that have been documented for the determination of DA in chitin and chitosan samples. Although the list of methods and references for each is not exhaustive, it purports to present an overview of the state-of-the-art of the various techniques, so as to serve as a guideline to help choose the best method to suit specific needs. To a great extent, the preferred method is influenced by the instrumental availability and the solubility of the sample. We have therefore separated the various analytical and instrumental methods in two categories, namely, those suitable for the analysis of samples in solution and those suitable for samples in the solid state. The former are best suited for the analysis of chitosan and the latter for both chitin and chitosan, regardless of their solubility. For routine analytical purposes of soluble chitosan, the measurement of *N*-acetyl content by the first-derivative UV-method gives very good results [24, 25], while IR spectroscopy remains the most simple and reliable method for samples of chitin or chitosan in the solid state [26–28].

It is important to recall that in addition to the net degree of acetylation, the distribution of the acetyl groups in the chitosan chain varies with the preparation protocol. Homogeneously deacetylated chitosan samples, with F_A varying in the range 0.04–0.49, showed that the ^{1}H-NMR diad frequencies distribution was close to random (Bernoullian). Conversely, chitosan samples within the same range of F_A, but produced under heterogeneous conditions, seemed to have a slightly more blockwise distribution [29]. This seems to be dependent on the degree of acetylation, as it is known that the alkaline reaction proceeds preferentially at the amorphous regions, and for chitosans with a $F_A > 0.44$, it will also proceed in the crystalline regions to give random type copolymers [30, 31]. This will result therefore, in the existence of a polydispersity with respect to DA among different chains on a chitin or chitosan sample. An important conclusion which also arises from these results is that the value of DA measured in a bulk heterogeneously prepared sample has an average character.

25.3.2 Molecular mass determination

The molecular weight and its distribution affect the physical, chemical and biological properties of chitosan, such as (i) the mechanical properties of hydrogels, (ii) the pore size of membranes, scaffolds and microcapsules [32],

Table 25.1

Characterization methods for determination of the degree of acetylation in chitin or chitosan

Method	Analytical conditions	Notes	References
I. Soluble samples			
High field ^1H NMR spectroscopy	Depolymerized chitosan samples dissolved in D_2O (pD 3–4) at 90°C	In all cases, DA is calculated from the relative areas of H-1 resonances of A and/or D or in combination with *N*-acetyl group resonances to the total area of resonances	[29]
	Native chitosan samples in DCl 2% at 70°C; pulse sequence: $\tau = 40\,s$, $\theta = 45°$		[210]
	Native chitosan samples in DCl 2% at 27°C; pulse sequence: $\tau = 13\,s$, $\theta = 90°$	Acquisition time optimized based on inversion-recovery T_1 measurements	[211]
UV spectroscopy	Chitosan dissolved in 0.01 M acetic acid; calibration curve with of *N*-acetyl glucosamine	First derivative of UV absorbance is recorded at the point of null contribution of acetic acid ($\lambda \sim 200\,nm$)	[24, 25]
Circular dichroism	Chitosan dissolved in acetic acid; calibration curve with *N*-acetyl glucosamine	Based on n–π^* transition of amide groups with a c.d. band located in the region of $\lambda \sim 211\,nm$	[212]
Potentiometry	Chitosan dissolved in excess of HCl	Titration with NaOH measures the consumption of HCl needed to protonate the amino groups	[11, 213]
Conductimetry	Chitosan dissolved in excess of HCl	Titration with NaOH needed to neutralize the free amino groups in chitosan	[214]
Enzymatic method	Chitosan dissolved in 0.2% acetic acid and incubated with 5 units of each exo-β-D-glucoseaminidase, β-*N*-acetylhexosaminidase and chitosanase for 12 h at 40°C	Liberated amounts of glucosamine and *N*-acetyl glucosamine monosaccharides are determined by a specific colorimetric or HPLC method	[215]
II. Insoluble samples			
FTIR spectroscopy	Ground samples are mixed with dry KBr (~1:100) (*NB*: thin films can also be cast from concentrated solutions of soluble samples)	Several sets of measuring and reference bands calibrated with DA determined by NMR and UV spectroscopy and titration methods	[26–28]
^{13}C CP-MAS NMR	No sample preparation	Relaxation delay optimized. DA is calculated from the relative areas of CH_3 to the total resonances	[214, 216]
^{15}N CP-MAS NMR	No sample preparation	Not suitable for DA $< 10\%$. Particularly useful for chitin/chitosan complexes associated with other polysaccharides	[217, 218]
Elemental analysis	Powder analysed directly	Calculated from the C/N ratio	[219]
Thermal analysis	Powder analysed directly	Based on the enthalpy of degradation of sample	[220]
Pyrolysis – GC	Pyrolysis at 450°C under He	DA calculated from programs of sample and of *N*-acetylglucosamine	[221]
HPLC	Sample hydrolyzed with 2.41 M H_2SO_4	Based on the determination of liberated acetyl groups	[222]

(iii) the particle size and drug release properties of nanoparticles [33], (iv) the effect of chitosan on the permeability of epithelial cells [34] and (v) its antimicrobial activity [35]. Therefore, this parameter affects directly the performance of chitosan in biotechnology, food, pharmaceutical, biomedical and biological applications.

The main methods used to determine the molecular weight of chitin and chitosan are similar to those used for any polymer, namely, viscosimetry, light scattering, gel permeation chromatography, osmometry and sedimentation equilibrium by ultracentrifugation. The first three are briefly discussed in the following subsections since they are the most employed techniques.

25.3.2.1 *Viscosimetry*

Viscosimetry is the mostly utilized method to determine the molecular weight of chitosan due to its simplicity. The method has the disadvantage of not being absolute because it relies on the correlation between the values of intrinsic viscosity with those of molecular weight, as determined by an absolute method for fractions of a given polymer. This relationship is given by the well-known Mark–Houwink–Sakurada equation.

$$[\eta] = KM_\upsilon^a \tag{25.2}$$

where M_ν is the viscosimetric-average molecular weight and K and a are constants that depend on the nature of the polymer, the solvent system and the temperature. Experimentally, the intrinsic viscosity of chitosan solutions is calculated using the Huggins and Kraemer equations [36].

The degree of acetylation and the polydispersion with respect to the size and the content of residual *N*-acetylated groups must be known for the calibration step, given their influence on the rheological behaviour of the corresponding solutions. The tendency of chitosan to form aggregates must also be taken into account. An exhaustive updated list of calibration constants obtained under different conditions and solvent systems has recently been reviewed [37]. The constants given by Rinaudo [38] are perhaps, to date, the most realistic ones, as they call upon a calibration from size exclusion chromatography and hence a large number of monodisperse fractions. This issue is further discussed in Section 25.3.4.

25.3.2.2 *Static light scattering*

Static light scattering (SLS) provides the direct determination of the polymer weight-average molecular weight, M_w, along with the *z*-average radius of gyration, $<S^2>^{1/2}$, and the second virial coefficient, Γ_2, without need of calibration [39]. The angular and concentration dependence of the excess Rayleigh ratio $R_{vv}(q)$ of dilute polymer solutions is measured and related to M_w, $<S^2>^{1/2}$ and Γ_2 by the following equation:

$$\frac{KC}{R_{vv}(q)} = \frac{1}{M_w}\left(1 + \frac{1}{3}\left\langle S^2\right\rangle^{1/2} q^2\right) + 2\Gamma_2 C \qquad C \to 0;\ \theta \to 0 \tag{25.3}$$

where, for vertically polarized incident radiation, the optical constant is given by $K = 4\pi^2 n_o^2(dn/dC)^2/(N_A\lambda_o^4)$ and $q = (4\pi n_o/\lambda_o)\sin(\theta/2)$, with N_A, n, C, λ_o and θ being Avogadro's constant, the solvent refractive index, the polymer concentration ($g\,cm^{-3}$ or $g\,g^{-1}$), the light wavelength in the vacuum, and the scattering angle, respectively. In order to calculate M_w, $<S^2>^{1/2}$ and Γ_2 using Eq. (25.3), the refractive index increment, *dn/dC* for a given polymer–solvent system, must also be determined independently (*i.e.* using an interferometer) with high precision. The *dn/dC* values of chitosan in 0.15 M ammonium acetate/0.2 M acetic acid buffer (pH 4.5) have been found to vary from 0.195 to 0.154 for chitosans with DA values of 5.2 and 70.6, respectively [40]. The simultaneous double extrapolation to both zero angle and zero concentration is conducted by constructing the well-known Zimm plot.

Despite being an absolute method that provides very important information on the fundamental properties of the polymer in solution, including its conformation, chain stiffness and its interaction with the environment, the use of SLS is limited as it requires expensive equipment and strict experimental conditions including very pure and fully soluble polymers and dust-free. In addition, Zimm plots of experimental data can be irregularly shaped and difficult to interpret when polymer aggregation or association occurs [41]. Several examples of further studies addressing the macromolecular dimensions and conformation of chitin [42] and chitosan [40, 43–48] can be found in the literature.

25.3.2.3 High performance size exclusion liquid chromatography

Size exclusion chromatography coupled with HPLC (SEC–HPLC or HPSEC) permits not only the determination of the molecular weight, but also its statistical distribution, hence the index of polydispersity [49]. Separation of the polymer species takes place according to their molecular size, or more strictly, their hydrodynamic volume. This method is claimed to be rapid, reproducible and simple compared with light scattering, osmometry or ultracentrifugation.

By coupling a refractive index detector to the high performance size exclusion liquid chromatography (HPSEC) system, it is possible to determine the amount of polymer eluting at a given retention time (t_R) or retention volume (V_R). In order to determine the molecular weight using such an instrumental set up, the system must be calibrated in terms of the expected V_R of a polymer fraction of known M_w. Since there are no chitosan standards, in order to construct a universal calibration curve, pullulan standards of narrow M_w distribution have been used, as for other polysaccharides. Isogai has reviewed the various methods for obtaining the molecular mass of chitin and chitosan by HPSEC, with emphasis on methodological aspects and solvent systems [50]. An ASTM guide describes the application of this technique to chitosan [51].

It is crucial that the separation mechanisms in HPSEC are determined by the size of the molecules and not by their charge, nor by ion exclusion effects or adsorption into the column. These remarks are particularly relevant to chitosan which is a polyelectrolyte. The addition of sodium acetate or ammonium acetate in the buffer used as the mobile phase generates enough ionic strength to overcome these unwanted features.

In order to avoid the use of standards, multidetection SEC–HPLC systems incorporate, in addition to the refractive index detector, a multiple angle laser light scattering (MALLS) detector. This enables the determination of the molecular weight and radius of gyration of the individual fractions as they elute out of the column, hence to obtain their distribution as a function of concentration. Multidetection HPSEC is undoubtedly the preferred technique for the analysis of the distribution of chitosan's molar masses and sizes [38, 52, 53].

The MALLS detector uses the same fundamental principles of light scattering discussed above (Eq. 25.3). Other detectors that can be coupled to a multidetection system include viscosity and dynamic light scattering units, which enable the determination of additional fundamental parameters providing information on the molecular shape and conformation of chitosan in solution. Thus, using the viscosity detector coupled to the MALLS counterpart, Brugnerotto *et al.* derived the Mark–Houwink–Sakurada K and a constants (Eq. 25.2) for chitosans of varying DA [54].

25.3.3 Crystallinity

Chitin is found in nature as highly crystalline ordered networks. Two major polymorphs have been firmly established for chitin in nature, namely α and β-chitin. Each can readily be distinguished on the basis of its X-ray diffraction patterns [55], solid state ^{13}C CP-MAS NMR [56], infrared [57] and Raman spectroscopy [58].

α-Chitin is the most stable and ubiquitous form of chitin occurring in the exoskeleton of arthropods and in fungi. Its chains are arranged along a two-chain orthorhombic unit cell with dimensions $a = 0.474$, $b = 1.032$ and $c = 1.886$ nm [55]. Space groups are consistent with a P_{212121} symmetry, which requires the antiparallel arrangement of the chains.

On the other hand, β-chitin occurs commonly in squid pen [59], in diatoms and in deep-sea organisms (*e.g. Tevnia jerichonana*) [60] from which it has been isolated. The early crystallographic studies established that the unit cell of β-chitin is monoclinic and has a P_{21} space group, with its chains arranged in a parallel structure. In contrast with α-chitin, β-chitin can incorporate small molecules into its crystal lattice to form various complexes, in particular anhydrous, monohydrated and dehydrated forms have been identified [61]. The inclusion of water between the dense sheet planes formed by the close stacking of chains leads to an increase in the distance between sheets along the intersheet axis (c axis). Recent X-ray crystallographic studies carried out with highly crystalline β-chitin have provided intersheet (c axis) d-spacing values of 0.92, 1.03 and 1.1 nm for anhydrous, mono- and dehydrated forms, respectively [62]. Differential scanning calorimetry (DSC) evidence showed that the incorporation of water in highly crystalline β-chitin is reversible and that the thermal transitions between the various forms can be induced by heating and cooling cycles. Yet another key feature of β-chitin is its ability to undergo conversion into its α-chitin polymorph by precipitation from a formic acid solution or by treatment with cold 6 M HCl [63].

Highly deacetylated chitosan in the solid state can also exhibit a crystalline structure. Two hydrated crystalline forms of chitosan have been identified, namely 'L-2' and 'Tendon', the latter being the most abundant hydrated polymorph of chitosan [64].

25.3.4 Solution properties

Because of its high cohesive energy related to strong intermolecular interactions through hydrogen bonds, chitin like cellulose is difficult to dissolve. Roberts has grouped the solvent systems for chitin into three categories *viz.* aqueous solutions of neutral salts, acid solvents and organic solvents [5]. The former group rarely gives good solutions, while the solvents of the second class generally provoke degradation of the polymer chain. Up to recent years, the most common solvent for chitin was the system composed by *N,N*-dimethyl acetamide (DMA) containing LiCl in amounts up to 8 per cent. It has been recently shown that chitin can be dissolved in calcium chloride dihydrate–saturated methanol and that this system is stable for long periods of time at room temperature [65].

Vincendon showed that chitin dissolves in DMA–5 per cent LiCl via strong interaction of one LiCl molecule with intermolecularly hydrogen bonded labile protons (OH or NH) of *N*-acetylglucosamine residues [66]. This interaction is assumed to destroy the intermolecular hydrogen bonds, allowing chitin to swell and then to dissolve. Terbojevich *et al.* have demonstrated that chitin chains are rather stiff in DMA–5 per cent LiCl and have a high persistence length, ranging from 150 to 400 Å [67]. A later paper by this group estimated the Mark–Houwink constants as $a = 0.88$ and $K = 0.021\,\text{mLg}^{-1}$ in this solvent system at 25°C [42].

In contrast to chitin, the presence of free amino groups along the chitosan chains allows this macromolecule to dissolve in dilute aqueous acidic solvents through the protonation of these groups and the formation of the corresponding chitosan salt. It is therefore important to realize that the polyelectrolyte character of chitosan influences its solution properties.

25.3.4.1 Acid-base properties

An important parameter that characterizes the polyelectrolyte behaviour of this weak polybase is its intrinsic pK, pK_0, which is the pK-value extrapolated to zero charge. According to Katchalsky, the apparent value of pK, pKa, is defined by [68]

$$\text{pKa} = \text{pH} + \log\frac{1-\alpha}{\alpha} = \text{pK}_0 - \frac{\varepsilon\Delta\Psi(1-\alpha)}{kT} \qquad (25.4)$$

where Ka is the dissociation constant of the conjugate acid *viz.*

$$\sim\!\sim NH_3^+ \underset{}{\overset{Ka}{\rightleftharpoons}} \sim\!\sim NH_2 + H^+ \qquad \text{(I)}$$

A study on this issue has shown that Ka depends largely on the degree of neutralization, particularly at low degrees of *N*-acetylation [69]. For a given value of α, the higher the acetyl content, the higher the pKa value. The computed value for pK_0 (6.5) was shown to be independent of the degree of acetylation of the sample up to 0.25. Nevertheless, investigations carried out by Rinaudo *et al.* for chitosan hydrochloride and acetate, combining potentiometric, conductimetric and viscosimetric experiments, gave a pK_0 value of 6.0 ± 0.1 [70, 71], that is, lower than that reported for D-glucosamine (between 7.5 and 7.8), indicating the influence of the chemical environment on the strength of this polybase.

25.3.4.2 Conductimetric behaviour

According to Manning's theory for polyelectrolytes [72], a charged macromolecule in solution is submitted to *counterion condensation* when the so-called charge parameter, ξ, is higher than unity, enabling this parameter to decrease.

Since the charge parameter is proportional to the charge density, the former depends on the degree of *N*-acetylation and the degree of dissociation of the amino groups and therefore on the solution pH. For chitosan, the charge parameter becomes [38]

$$\xi = 1.38 \times (1 - \text{DA}) \times \alpha \qquad (25.5)$$

where DA is the degree of *N*-acetylation and α the degree of protonation. The expression

$$\chi = cf(\lambda_P + \lambda_c) \tag{25.6}$$

defines how the specific conductivity, χ, is affected by the presence of the polyelectrolyte following Manning's theory [72]. Here, c is the polyelectrolyte concentration, λ_P and λ_c are the limiting equivalent conductivities for the polyelectrolyte and the corresponding counterion, respectively, and f is the transport coefficient which is directly related to the charge parameter.

The values for a fully protonated chitosan chain with DA = 0.20 in the presence of univalent counterions has been estimated as $f = 0.82$ and $\lambda_P = 31.4 \times 10^{-4} \mathrm{m}^2 \cdot \mathrm{S\ mol}^{-1}$ [73], the former being in good agreement with the theoretical value ($f = 0.79$).

25.3.4.3 *Dilute solution properties*

The most relevant information about the behaviour, size and conformation of macromolecules must come from studies of dilute solutions. Many such investigations have been published and controversies have arisen which have not yet been solved.

Concerning the chemical composition, two factors must be taken into account when examining the stiffness of chitosan chains, namely (i) the presence of acetylated units should increase the rigidity because of steric reasons and/or hydrogen bonding between two adjacent residues and (ii) the polyelectrolyte character of this biopolymer derived from the presence in solution of ammonium ions will tend to expand the chains at low or moderate ionic strength, thus increasing the excluded volume by electrostatic repulsions. It is apparent that each of these two factors will increase when the other decreases, sometimes counterbalancing each other depending on their relative strength.

The analysis of the stiffness parameter, $\boldsymbol{B}$, proposed by Smidsrød and Haug [74] for samples of chitosan with different degrees of acetylation led to an important conclusion. At contents of amino groups of about 80 per cent there is no effect of chemical composition on the chain stiffness [38, 75–77]. However, when the content of *N*-acetylglucosamine residues increases, the rigidity of the polysaccharide increases indicating that the role of acetyl groups dominates over the polyelectrolyte effect. (It should be remembered that the higher the stiffness of the polyelectrolyte chain, the lower the $\boldsymbol{B}$ value). A possible explanation for this behaviour is that if the charge parameter of a fully protonated chitosan chain is evaluated (see Eq. 25.6), the result is that for DA ≥ 0.28, $\xi \geq 1$. On the contrary, when DA < 0.28, then $\xi < 1$, making the charge density lower and the role of acetyl groups more important than the polyelectrolyte effect.

The worm-like chain has proved to be an appropriate model for studying the conformational characteristics of chitosan. Considering that the persistence length is a parameter that characterizes the stiffness of a worm-like chain, many authors have evaluated it with conflicting results [38, 76, 78], denoting that many factors influence its experimental determination.

A study carried out with chitosan samples prepared by both homogenous and heterogeneous processes helps to understand the role of the *N*-acetyl group content and its distribution on the stiffness of the polymer chains. Theoretical and experimental data have shown that the intrinsic persistence length remains constant ($L_P = 110$ Å) for heterogeneous chitosans with degrees of acetylation lower than 25 per cent. For homogeneous chitosans, L_P increases with the degree of acetylation. The stiffness of chitosan chains appears greater for homogeneously *N*-reacetylated chitosan [79].

25.3.4.4 *Rheological behaviour*

The investigation of the rheological behaviour of chitosan solutions has been carried out by studying the influence of different parameters such as chemical composition, concentration, ionic strength, pH and the effect of the type of acid [54, 80–85]. Furthermore, the role of hydrophobic association phenomena on the viscoelastic behaviour of chitosan solutions is an interesting feature [86] which requires further understanding.

25.4 INTERACTION WITH METAL IONS

The capacity of chitin and chitosan to form complexes with metal ions, particularly with transition and post-transition metal ions, has been widely documented and comprehensive reviews on the subject are available [11, 87, 88].

However, in spite of this vast literature it is generally difficult to compare the results from different authors. This is because the adsorption capacity and the rate of adsorption of metal ions by these biopolymers depend on a number of factors such as the physical state of the polymer (powder, flakes, or even if was reprecipitated), its degree of crystallinity, the degree of acetylation and the chain length. Temperature, stirring rate, contact time with the metal ion solution, pH and ionic strength, are also factors that influence adsorption.

Various authors have reported different ordering for the metal ions sorption capacity of chitosan. For instance, while Muzzarelli and Tubertini conclude that the adsorption capacity of chitosan follows the Irvin and Williams series [89], Koshijima *et al.* [90] order it in the series Cr(III) $<$ Co(II) $<$ Pb(II) $<$ Mn(II) $<$ Cd(II) $<$ Ag(I) $<$ Ni (II)$<$ Fe(II) $<$ Cu(II) $<$ Hg(II) and Masri *et al.* [91] arrange them as follows: Cr(III) $<$ Fe(II) $<$ Mn(II) $<$ Co(II) $<$ Cd(II) $<$ Cu(II) $<$ Ni(II) $<$ Ag(I) $<$ Pb(II) $<$ Hg(II).

Chitosan exhibits a superior metal ion sequestering ability than chitin. The presence of the amino groups in the structural unit of chitosan is definitely the main cause of its complexing capacity, but the binding mechanism of metal ions to chitosan is not yet completely understood. Various processes such as adsorption, ion exchange and chelation have been considered as the mechanisms responsible for complex formation between chitosan and metal ions. The type of the interaction prevailing depends on the metal ion, its chemistry and the pH [92]. Ogawa *et al.* studied the complex formation of chitosan with transition metal ions [93] by immersing stretched films of chitosan (DD $=$ 99.5 per cent) in 0.1–0.4 M solutions of Co(II), Ni(II), Cu(II), Zn(II) and Hg(II) salts for 1 h. The films were then washed, dried and inspected by X-ray diffraction. The first structure they proposed as the most probable for these complexes involved the binding of the pendant metal ion to one amino group of the dimer residue of chitosan.

The generally accepted interpretation at present is that under heterogeneous conditions at pH $<$ 6, chitosan acts as a poly(monodentate) ligand, while at higher pH, it behaves as a poly(bidentate) ligand, forming chelates. However, in solution, the formation of complexes in which two amino groups (belonging to the same chain or to two different chains) are coordinated with the same metal ion, can also be envisaged [5].

The elevated metal ions adsorption capacity of chitosan is of course particularly interesting for applications in different contexts such as metal ion recovery from solutions, decontamination of residual industrial waters, inorganic chromatography or catalyst support. To this end, a number of chitin and chitosan derivatives with improved properties have been prepared. Thus, chitosan has been crosslinked with di/polyfunctional reagents, such as glutaraldehyde for preventing its dissolution in an acid medium. Other modifications include the synthesis of polyampholytes with increased chelating ability (*N*-carboxymethyl and *N*-(*O*-carboxybenzyl) chitosan [94]); derivatives containing sulphur, such as chitosan dithiocarbamate [95] and *N*-(2-hydroxy-3-mercaptopropyl) chitosan [96], with superior sequestring capacity for Hg^{2+}, and grafting with other polymers, like polyacrylonitrile [97] and polyvinylpyrrolidone [98]. The structure of some of these chitosan derivatives are shown in Fig. 25.3. A more comprehensive account of chitin and chitosan modifications for ion binding purposes can be found elsewhere [88].

25.5 CHITOSAN IN POLYELECTROLYTE COMPLEXES

Polymers exhibit a high tendency to interact and generate interpolymer complexes and hence supramolecular structures. These interactions constitute the basis of many biological reactions occurring in living organisms. They also provide a route to the preparation of novel polymeric materials displaying physical properties different from those of the individual constituent macromolecules. As a result, in the last few years many studies have been devoted to the elucidation of the nature of these interactions as well as to the evaluation of the physical properties of the resulting materials [99–103]. Interpolymer complexes are often classified according to the nature of their interactions such as charge-transfer or coordination complexes, stereocomplexes resulting from van der Waals forces, hydrogen-bonded complexes and polyelectrolyte complexes (PEC) [99].

As a cationic biopolymer, chitosan may react with any other anionically charged polyelectrolyte, giving rise to the formation of PEC [101, 104]. There are reports of PEC between chitosan and carboxymethyl cellulose (CMC) [105–108], alginate [109–112], poly(acrylic acid) [107, 113, 114], pectin [73, 115–117], carrageenans [118–121], heparin [122] and others [123–131].

Gene therapy is currently one of the most advanced strategies to solve important health problems such as cancer, AIDS and cardiovascular diseases. The ideal gene delivery system must be capable of protecting the DNA during the transfer of genetic material until it reaches the target cells of a patient. Chitosan is a good candidate for gene delivery systems through the formation of PEC with negatively charged DNA [132–134].

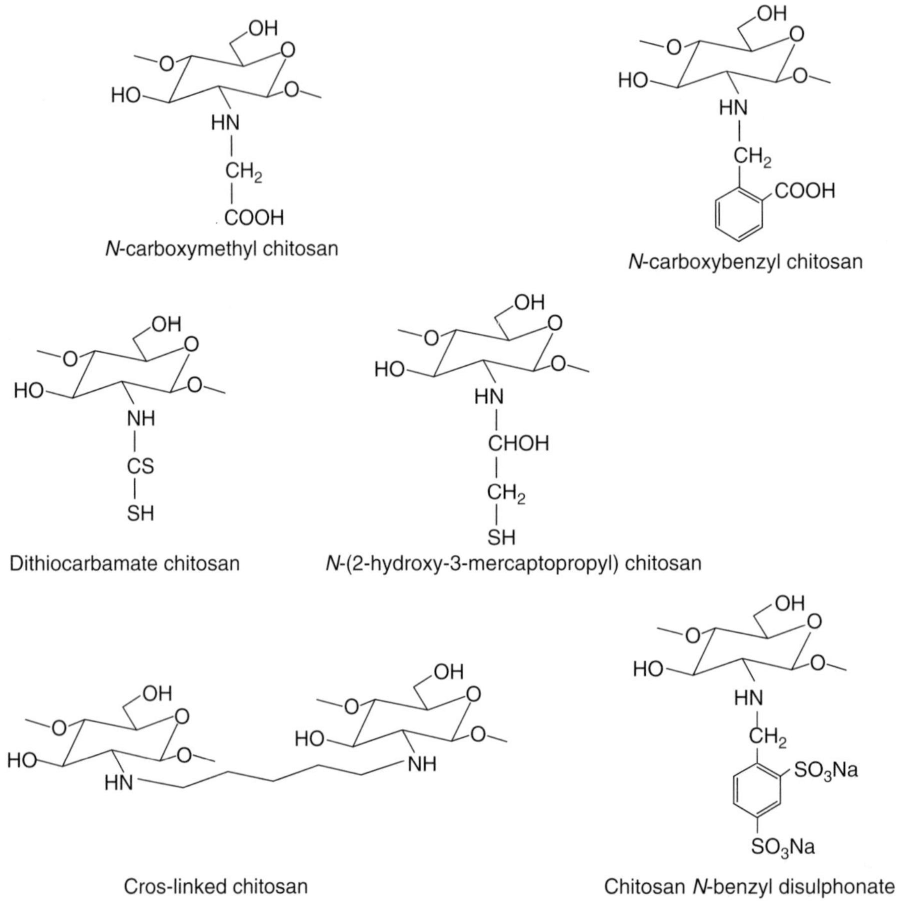

N-carboxymethyl chitosan

N-carboxybenzyl chitosan

Dithiocarbamate chitosan

N-(2-hydroxy-3-mercaptopropyl) chitosan

Cros-linked chitosan

Chitosan *N*-benzyl disulphonate

Figure 25.3 Some of the many chitosan derivatives developed for increasing its performance as metal ion scavenger.

The interaction of chitosan with carboxylic polyacids leads to the formation of an insoluble PEC. Chitosan is insoluble in neutral and basic media, therefore the stoichiometric PEC is usually obtained by mixing equimolar quantities of the polyacid and chitosan hydrochloride. The complex is formed according to the following reaction:

$$\underset{C_0(1-\theta)}{\sim\sim COOH} + \underset{C_0(1-\theta)}{{}^{+}NH_3-} \rightleftharpoons \underset{C_0\theta}{\sim\sim COO^{-}\ {}^{+}NH_3 \sim\sim} + \underset{C_0\theta}{H^{+}} \qquad \text{(II)}$$

As a result, the pH of the solution decreases. The degree of conversion of the complex, θ, expressed as the ratio of the concentration of interchain salt bonds formed, C_k, to the initial concentration of any of the polyelectrolytes, C_0, can be evaluated from potentiometric measurements by:

$$\theta = \frac{([H^+] - [H^+]_{PA})}{C_0} \qquad (25.7)$$

where $[H^+]$ and $[H^+]_{PA}$ are the concentration of hydrogen ions in the reaction mixture and in a solution of the polyacid at the same concentration, respectively. However, the equilibrium (II) can be shifted to the right by addition of a strong base. In so doing, θ can be assessed from the data of the potentiometric titration of the mixture.

Reactions between polyelectrolytes are accompanied by the release into the medium of ionic species with different mobilities, thus making conductimetry a powerful technique to study these types of processes. Taking into account the additivity of the contributions of all ionic species, the knowledge of the conductivity of the reaction medium allows the degree of conversion to be determined [73, 111].

The stoichiometry of PECs of weak polyacids and weak polybases is pH dependent, because of the variation of their dissociation degree with pH. The composition of the PEC is then given by:

$$\alpha_{PA}[PA] = \alpha_{PB}[PB]$$

where [PA] and [PB] are the molar concentrations of the polyacid and the polybase in the PEC and α_{PA} and α_{PB} their respective dissociation degrees [99, 111].

An important characteristic of interpolyelectrolyte reactions is their cooperativity, which is determined by the influence of neighbouring functional groups on the reactivity of the next one. The cooperativity of interpolyelectrolyte reactions is well-illustrated in the case of the PEC between chitosan and poly(acrylic acid) [135]. The equilibrium (II) can be decomposed into the following contributions:

$$\sim\sim COOH \underset{}{\overset{K_d}{\rightleftharpoons}} \sim\sim COO^- + H^+ \quad \text{(III)}$$

$$\sim\sim COO^- + {}^+NH_3\sim\sim \underset{}{\overset{K_2}{\rightleftharpoons}} \sim\sim COO^{-+}NH_3\sim\sim \quad \text{(IV)}$$

The existence of cooperativity implies that the equilibrium constant for the formation of bond i, K_2^i, is bigger than that for the formation of bond $i - 1$, K_2^{i-1}. The magnitude of this effect can be adequately quantified by expressing K_2 as $K_2 = K_2^0 \times \exp^{-m\theta}$, where K_2^0 is the equilibrium constant for the formation of the first interchain bond [136]. A marked increase in K_2 with θ, denotes a cooperative reaction resulting in the formation of long sequences of consecutive bonds [135].

Most of applications of PEC are in membranes. The wall of microcapsules consists of a membrane of PEC, whose characteristics govern properties that are usually decisive for their end use. The properties of PEC membranes are strongly dependent on the preparation conditions, *viz.* pH, ionic strength, temperature, concentration and molar ratio of reacting polyelectrolytes, among others [137]. These membranes adsorb water until a maximum swelling is achieved after which they slowly shrink to an equilibrium size [110, 115, 137]. The contraction experienced by the PEC membranes can be explained considering two factors: (a) the small ions trapped diffuse out of the membrane, decreasing the osmotic pressure and (b) at pH 5.5 the free carboxylic groups of the CMC chains are mostly as sodium carboxylate and the amino groups of chitosan are protonated, so that new salt bonds can be formed. The segmental mobility in the swollen state must be sufficient for this reaction to proceed [137]. As a result, θ rises with time, increasing the retractile force of the polymer network.

However, if the membrane is dried after a first swelling cycle, further swelling sequences show the typical pattern of a polymer network. This behaviour indicates that the rearrangements of the polymer chains during the first cycle, which induces an increase in the degree of complexation, no longer take place [110].

During the swelling process, the integrity of the PEC network is maintained by the crosslinking of chains produced by the interchain $—NH_3^+\ {}^-OOC—$ salt bonds. However, at low or high pH, these bonds can be broken, resulting in the disintegration of the PEC and the dissolution of the membrane. This pH instability can be modified by heating PEC membranes at 120°C under a nitrogen atmosphere, a process which forms covalent amide bonds according to the following reaction [117, 138, 139]:

$$\left|—COO^{-+}H_3N—\right| \xrightarrow{\Delta T} \left|—COO—HN—\right| + H_2O \quad \text{(V)}$$

The membrane then becomes insoluble whatever the solution pH and moreover, the degree of swelling at a given pH decreases as a result of reaction (V). The extent of amide bond formation for a given membrane depends on the heating time, and therefore this parameter can be used to regulate the swelling behaviour of these PEC membranes.

It is interesting to note that when chitosan/CMC PEC membranes in the swollen state are placed in solutions containing $CaCl_2$, a collapse in volume is observed [140, 141]. This effect is not induced by other salts, such as NaCl, KCl, $NaNO_3$ and Na_2SO_4, indicating the existence of a specific role of the Ca^{2+} ions on the free carboxylate groups of the PEC, possibly through the formation of Ca^{2+} complexes with fixed ligands of the network. This complexation can be visualized as the formation of new crosslinks, which would increase the retractile force of the PEC's macromolecular network, thus provoking its contraction. When the pH of the formation of the membranes is increased, this effect is enhanced because of the greater amount of available free carboxylate groups. The variation of the swelling degree with pH and Ca^{2+} concentration can be exploited to control the permeability of solute fluxes through the membranes [140].

25.6 APPLICATIONS IN MEDICINE AND PHARMACY

Chitosan is a very useful polymer for biomedical applications because of its biocompatibility, biodegradability and low toxicity. Chitin has also been described as biocompatible and biodegradable, and has found applications for specific purposes, such as sponges and bandages for the treatment of wounds and suture threads. However, it has generally received less attention than chitosan, because of its insolubility in water and low reactivity.

The biodegradability of chitin and chitosan is mainly due to their susceptibility to enzymatic hydrolysis by lysozyme, a non-specific proteolytic enzyme present in all tissues of the human body. Lipase, an enzyme present in the saliva and in human gastric and pancreatic fluids, can also degrade chitosan [142]. The products of the enzymatic degradation of chitosan are non-toxic. The degree of acetylation, the molecular weight, the pH and even the method of preparation of chitosan affect biodegradation.

In contact with blood, chitosan activates the formation of clots as a result of the interaction of the amino groups with the acid groups of blood cells [143]. It is therefore a good haemostatic agent. However, it has been found that water-soluble chitosan and chitosan oligomers do not present thrombogenic activity, whereas the sulphated derivatives of chitosan exhibit anticoagulant activity [143]. It has also been claimed that chitosan is hypocholesterolemic and hypolipidemic [144]. It has antimicrobial, antiviral and antitumoural activity [145]. The immunoadjuvant activity of chitosan has also been recognized [146].

All these interesting characteristics have led to the development of numerous applications of chitosan and its derivatives not only in biomedicine such as surgical sutures, biodegradable sponges and bandages [143], matrices (in microspheres, microcapsules, membranes and compressed tablets) for the delivery of drugs [147], but also in orthopaedic materials and dentistry [148].

25.6.1 Chitosan as a biomaterial

The great variety of applications of chitosan in the field of biomaterials is due to its excellent properties when interacting with the human body: bioactivity, antimicrobial activity, immunostimulation, chemotactic action, enzymatic biodegradability, mucoadhesion and epithelial permeability which supports the adhesion and proliferation of different cell types [149].

Chitosan has been tested for applications such as contact lenses, tissue adhesive, preventing bacterial adhesion, sutures and others [150]. However, this biopolymer has been thoroughly investigated mainly in two biomedical fields. On the one hand, it has been used together with chitin in the treatment of wounds, ulcers and burns, calling upon its haemostatic properties and its accelerating wound healing effect. On the other hand, given its cell affinity and biodegradability, it has been applied in tissue regeneration and restoration, including its perspective use as a structural material in tissue engineering.

25.6.1.1 Treatment of wounds and burns

This is undoubtedly one of the most promising medical applications for chitin and chitosan. The adhesive properties of chitosan, together with its antifungal and bactericidal character, and its permeability to oxygen, are very important properties associated with the treatment of wounds and burns. Different derivatives of chitin and chitosan have been patented for this purpose in the form of membranes, woven fibres, hydrogels, etc. Some of these formulations have been released on the market, like *Beschitin* in Japan (based on chitin) or *HemCon* in USA (based on chitosan).

Both polysaccharides promote granulation (with angiogenesis) and cell organization during wound healing. The reepithelization and tissue regeneration are strongly improved in open wounds, and at the same time scarring is reduced. The analgesic effect of both of chitin and chitosan has also been reported [151].

PEC of chitosan and heparin have been also studied as matrices for wound healing, because it has been demonstrated that heparin interacts and stabilizes growth factors involved in the wound healing process [152].

25.6.1.2 Tissue engineering

Tissue engineering techniques generally require the use of three-dimensional (3D) supports for initial cell attachment and subsequent tissue formation. Chitosan has similar structural characteristics as glycosaminoglycans (GAGs) found in extracellular matrix of several human tissues. It has therefore been widely employed in tissue engineering, since it facilitates cell attachment and the maintenance of differentiating functions. Gelatin/chitosan or collagen/chitosan supports (some of them crosslinked with glutaraldehyde) have been investigated with promising results in cell growth and proliferation in tracheal [153], cartilage [154], nerve [155] and bone tissue [156] repair and regeneration. In the latter context, a composite with hydroxyapatite was used. Chitosan/chondroitin sulphate membranes have been shown to support chondrogenesis and are, therefore, promising materials for cartilage repair [157].

Platelet-derived growth factor releasing chitosan/chondroitin sulphate sponges have also been evaluated in bone regeneration [158]. A human skin equivalent composed of collagen–GAG–chitosan has been reported [159]. The biological properties of chitosan–chondroitin sulphate and chitosan–hyaluronate PEC have been studied [160]. These materials were cytocompatible, but better cell attachment and proliferation was attained with pure chitosan. Chitosan–GAG materials, including chitosan–heparin, have been evaluated as modulators of vascular cells proliferation. In this same vein, antigenic collagen–hyaluronic acid mixtures have been proposed to promote fibroblasts growth for tissue repairing.

Porous 3D scaffolds of chitosan/calcium phosphate composites have also been proposed for bone regeneration [148]. Calcium phosphate greatly reinforced the chitosan matrix and also modulated the release burst effect, when loaded with gentamicin. In addition, good cellular biocompatibility was observed. The preparation of chitosan/tricalcium phosphate sponges by mixing and freeze drying, in some cases loaded with platelet-derived growth factor, has also been reported [161].

An interesting strategy in tissue engineering is the encapsulation or immunoisolation of pancreatic and hepatic cells [162] (Fig. 25.4). Langerhans islets have been enclosed in chitosan/calcium alginate capsules with the aim of developing an artificial pancreas for the treatment of diabetes mellitus [163].

25.6.2 Pharmaceutical applications

Chitosan has been widely used as matrix in drug-release systems in the form of beads and granules, as promising vehicles for oral drug sustained-release formulations [164].

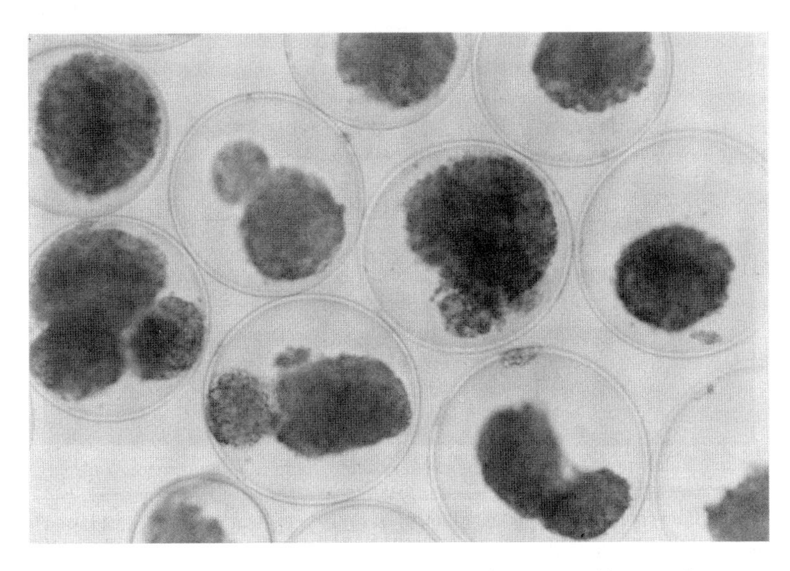

Figure 25.4 Cells enclosed in chitosan/calcium alginate microcapsules prepared by complex coacervation. Micrographs obtained by optical microscopy at 60X. Reproduced with permission from Reference [162].

Chitosan films exhibit low swelling in water, but membranes with diverse hydrophilic aptitudes can be prepared through the formation of mixtures or semi-interpenetrated and interpenetrated networks of chitosan with highly hydrophilic polymers like poly(vinyl alcohol), poly(vinylpyrrolidone) and gelatin [164]. PEC of chitosan with polyanions of natural origin like alginate, pectin or CMC or with synthetic ones, like poly(acrylic acid), have been investigated as matrices for controlled-release systems [165].

A pH-sensitive drug delivery system based on a glutaraldehyde crosslinked chitosan/gelatin hybrid network has been described. This gel swells at low pH and de-swells at high pH, thus exhibiting pH-dependent release of drugs [166].

Chitosan/polyethylene glycol/alginate miscrospheres have been proposed as good candidates for delivery of LMW heparin with antithrombotic properties [167]. Chitosan/CMC microcapsules of different compositions were prepared and tested as a protective matrix against the acid pH of the stomach for the oral administration of proteins and drugs [168]. Chitosan/xanthan microspheres are pH sensitive and biodegradable, and bestow a protective effect upon the drug in gastric and intestinal environment. They have also been proposed as a potential drug delivery system in the gastrointestinal tract [169].

Chitosan nanoparticles have been used for the nasal dosage of drugs and vaccines, since it has been demonstrated that they enhance the penetration of macromolecules through the nasal barrier [170].

Transdermal-drug delivery (TDD) devices using chitosan have been prepared for the permeation-controlled delivery of propranolol hydrochloride [171]. These chitosan-based TDD systems were composed of chitosan membranes with various crosslink densities as drug release controlling tools and chitosan gel as the drug reservoir. Drug release is highly affected by the crosslinking density of chitosan.

Chitosan has been considered a good candidate for non-viral gene delivery systems, since cationically charged chitosan can be complexed with negatively charged plasmid DNA. Self-aggregates of hydrophobically modified chitosan by deoxycholic acid (mean diameter of *ca.* 160 nm) can form charge complexes when mixed with plasmid DNA. These self-aggregate/DNA complexes are considered useful for the transfer of genes into mammalian cells *in vitro* and serve as a good delivery system [172].

25.7 APPLICATIONS IN AGRICULTURE

The utilization of chitin and chitosan in agriculture has followed four main directions: (a) protection of plants against plagues and diseases during pre-harvest and post-harvest, (b) promotion of antagonist microorganism action and biological control, (c) support of beneficial plant–microorganism symbiotic relationships and (d) plant growth regulation and development.

Chitin and its derivatives have been widely used for inducing defensive mechanisms in plants. The well-documented elicitor effect of chitin on plants confers them protection against many vegetable diseases [173]. Chitin and chitosan have fungicidal activity against many phytopathogenic fungi. The antiviral and antibacterial activity of chitosan and its derivatives have also been recognized [174].

These polysaccharides have been employed with success for controlling the parasitic nematodes in soils. The addition of chitin to the soil increases the population of chitinolytic microorganisms that destroy the eggs and cuticles of young nematodes which have chitin in their composition [175]. The chitinase and chitosanase activity in seeds protected by films of chitin and its derivatives has also been reported [176]. The antimicrobial properties of chitosan and its excellent film forming aptitude have been exploited in the post-harvest preservation of fruits and vegetables. Covering fruits and vegetables with a chitosan film imparts them antimicrobial protection and increased shelf life [177].

Chitin and chitosan in soil enhance plant–microorganism symbiotic interactions to the benefit of plants, as in the case of micorrizas. They also enhance the action of plague-controlling biological organisms such as *Tricoderma sp., Bacilus sp.* [178] and are good candidates for the encapsulation of biocides. Hence their efficiency in the control against pathogenic microorganisms and plagues is increased.

Chitosan and its derivatives induce favourable changes in the metabolism of plants and fruits. This results in an increased germination and higher crop yields [176, 179].

25.8 APLICATIONS IN THE FOOD INDUSTRY

Even though the use of chitosan as a food additive has not yet been approved by the FDA in the USA [180], nor by the Codex Committee on Food Additives in Europe, food is one of the industrial sectors in which chitosan has drawn considerable attention in recent years [181]. The main applications of chitosan in food are summarized below with reference to selected examples.

25.8.1 Flocculation

At present, the use of chitosan in processed food is almost entirely limited to the treatment of liquid wastewaters [182]. Indeed, chitosan is an effective flocculant due to its capacity to associate with proteins, polysaccharides, lipids and pigments. The polycationic character of chitosan confers it the capacity to associate electrostatically with negatively charged molecules, such as proteins, phospholipids, carboxylic acids, etc. Therefore, chitosan has long been known to be a highly effective agent for the recovery of proteins [183] and phospholipids [184] from the whey produced in the manufacture of cheese. A preferential affinity for β-lactoglobulin and bovine serum albumin whey proteins was found using particles consisting of a chitosan–alginate PEC [185]. Other studies have demonstrated that chitosan can be used successfully to remove up to 99 per cent of emulsified lipids from wastewater produced in large amounts by palm oil-processing mills [186]. Some important factors that affect the level of lipids removal are the pH and the surface area of the chitosan substrate.

Yet another important area of application is the clarification and deacidification of fruit juices, for example, pineapple [187] and apple [188], as well as the adsorption of polyphenols (catechins, flavans, proanthocyanidins, cinnamic acids, etc.) in white wine, responsible for browning and maderization [189].

25.8.2 Antimicrobial activity

The well-established antimicrobial properties of chitin and chitosan against a wide spectrum of bacteria, fungi and viruses can lead to a potentially large reduction in the amount of synthetic food preservatives currently used. Although the precise mechanisms of antimicrobial action of chitin and chitosan are yet to be elucidated [35], there is a consensus that the interaction of chitosan with the negatively charged cell membranes, leading to the leakage of proteinaceous and other intracellular constituents, is determinant. As for the role of chitosan M_w, the experimental evidence is consistent with the fact that a low M_w enhances the antibacterial activity [190]. The most probable means of application are either through packaging wraps or directly as edible coatings for the preservation of fruits, vegetables and meat and fish products [191–196].

25.8.3 Other functional properties

The role of chitosan as a stabilizer of w/o/w is another of its potential usages as an industrial food additive [197]; this property has been attributed to a presumed amphiphilic character of chitosan and/or to the fact that it can increase the viscosity of the continuous aqueous phase. Additionally, the effectiveness of chitosan treatment on oxidative stability showed that the addition of chitosan at 1 per cent produced a decrease of 70 per cent in the 2-thiobarbituric acid (TBA) contents of meat after 3 days of storage at 4°C [198].

Another potential use of chitosan is for the improvement of the functional properties of fibril proteins of meat and fish products during frozen storage. Thus, the cryoprotective effect of chitosan isolated from various sources (DA in the range 0.14–0.17) has been evaluated, and a reduction in protein denaturation during storage at −20°C was demonstrated in mixtures of lizardfish (*Saurida wanieso*) myofibrillar protein containing 5 per cent chitosan [199].

25.9 APPLICATIONS IN COSMETICS

Because of its cationic character, chitosan interacts with negatively charged biological surfaces, such as skin and hair [200]. Other relevant characteristics of chitosan for cosmetic applications are its high molecular weight, water retention and film formation capacity as well as its heavy metal ion complexing ability.

High molecular weight chitosan decreases the loss of trans-epidermic water, increasing the humidity of the skin which preserves its softness and flexibility. Chitosan is an excellent ingredient for allergic skins and formulations with high molecular weight samples were found to reduce skin irritation [200]. Therefore it is very convenient to use chitosan in formulations containing alcohol, such as aftershave lotions and deodorants, again in order to reduce skin irritation. Additionally, chitosan also inhibits inflammatory processes and promotes the regeneration of damaged tissues [201]. Sun-protecting emulsions that incorporate chitosan have a positive effect on water resistance, conferring an increased protection to the skin [202]. High molecular weight chitosan is a valuable component in deodorants because it absorbs humidity thus reducing transpiration. Besides, its antibacterial properties protect the skin [203].

When added to hair-care cosmetic preparations, chitosan interacts with keratin forming a uniform and elastic film which is more stable to high humidity than those usually formed with synthetic polymers. Moreover, it reduces the electrostatic charges, so that the hair does not stand on end, preserving the hair do. The hair treated with chitosan formulations has less tendency to adhere and is more easily brushed than with traditional fixers [204].

The bactericidal properties of chitosan prevent bad breath. Low molecular weight chitosan has also been shown to inhibit the oral adsorption of streptococci and have been proposed as a potential anticavity agent [205]. This makes chitosan a good additive in oral hygienic products such as tooth pastes, oral rinsing solutions and chewing gums.

25.10 OTHER APPLICATIONS

In addition to the already mentioned role of chitosan and some of its derivatives in the treatment of waste waters in the food industry, these substrates have also been applied for the treatment of residual waters from the mining and chemical industries, usually contaminated with heavy metals like mercury, cadmium, lead and copper. These processes can be very efficient, even if the metal content in the effluents is as low as 10–50 ppm [88].

Chitosan and chitin derivatives have been shown to increase the breaking strength and folding endurance of paper, without affecting its brightness. The water resistance of paper, as well as its sorption capacity for dyes, is also improved together with its electrical resistance [206].

Chitosan membranes have been successfully used in pervaporation processs, *e.g.* chitosan/poly(vinyl alcohol) (75:25 wt. per cent) blends showed excellent performance with good mechanical strength during the pervaporation dehydration of isopropanol [207].

Chitin and chitosan have been extensively employed for the immobilization of enzymes and cells applied in food technology and in biosensors [208]. Enzymes immobilized in these polymers possess an increased stability and are more temperature resistant. The immobilized enzyme can usually be recovered after reaction and reused. Immobilization is often achieved using glutaraldehyde as the crosslinking agent, although the enzymes have also been encapsulated by various techniques [209].

Other applications of chitin and chitosan include the spinning of fibres, preferably with the latter because of its higher solubility [223]; as additives in inks and dyes for the improvement of their rheological properties and colourfastness [224]; and in the coating of fabrics to control their shrinkage and enhance their dyefastness [223].

REFERENCES

1. Braconnot H., Sur la Nature des Champignons, *Ann. Chi. Phys.*, **79**, 1811, 265–304.
2. Odier A., Mémoire sur la Composition Chimique des Parties Cornées des Insectes, *Mém. Soc. Histoire Nat.*, **1**, 1823, 29–42.
3. Cauchie H.M., An attempt to estimate crustacean chitin production in the hydrosphere, in *Advances in Chitin Science*, Eds.: Domard A., Roberts G.A.F., and Vårum K.M., Jacques André Publisher, Lyon, 1998, **Volume II**, pp. 32–39.
4. Peter M.G., Chitin and chitosan in fungi, in *Polysaccharides II: Polysaccharides from Eukaryotes*, Ed.: Steinbüchel A., Wiley-VCH, Weinheim; *Biopolymers*, **6**, 2002, pp. 123–157.
5. Roberts G.A.F., *Chitin Chemistry*, The Macmillan Press Ltd, London, 1992. p. 350.
6. Jenkins D.W., Hudson S.M., Review of vinyl graft copolymerization featuring recent advances toward controlled radical-based reactions and illustrated with chitin/chitosan trunk polymers, *Chem. Rev.*, **101**(11), 2001, 3245–3273.
7. Lim S.-H., Hudson S.M., Review of chitosan and its derivatives as antimicrobial agents and their uses as textile chemicals, *J. Macromol. Sci.—Polym. Rev.*, **43**(2), 2003, 223–269.
8. No H.K., Meyers S.P., Preparation and characterization of chitin and chitosan – A review, *J. Aquat. Food Prod. Tech.*, **4**(2), 1995, 27–52.
9. Beaney P., Lizardi-Mendoza J., Healy M., Comparison of chitins produced by chemical and bioprocessing methods, *J. Chem. Technol. Biotechnol.*, **80**(2), 2005, 145–150.
10. Percot A., Viton C., Domard A., Optimization of chitin extraction from shrimp shells, *Biomacromolecules*, **4**, 2003, 12–18.
11. Muzzarelli R.A.A., *Chitin*, Pergamon Press, Oxford, 1977. p. 326.
12. Acosta N., Jiménez C., Borau V., Heras A., Extraction and characterization of chitin from crustaceans, *Biomass Bioenergy*, **5**, 1993, 145–153.

13. Synowiecki J., The recovery of protein hydrolyzate during enzymatic isolation of chitin from shrimp Crangon processing discard, *Food Chem.*, **68**(2), 2000, 147–152.
14. Goycoolea F.M., Higuera-Ciapara I., Hernández G., Lizardi J., García K.-D., Preparation of chitosan from squid (*Loligo spp.*) pen by a microwave-accelerated thermochemical process, *Adv. Chitin Sci.*; *Proceedings of the 7th International Conference on Chitin Chitosan and Euchis'97*, Jacques André Publisher, Lyon, France, 1997, pp. 78–83.
15. Austin P.R., Brine C.J., Castle J.E., Zikakis J.P., Chitin: New facets of research, *Science*, **212**(4496), 1981, 749–753.
16. Tsaih M.L., Chen R.H., The effect of reaction time and temperature during heterogenous alkali deacetylation on degree of deacetylation and molecular weight of resulting chitosan, *J. Appl. Polym. Sci.*, **88**, 2003, 2917–2923.
17. Mima S., Miya R., Iwamoto R., Yoshikawa S., Highly deacetylated chitosan and its properties, *J. Appl. Polym. Sci.*, **28**, 1983, 1909–1917.
18. Domard A., Rinaudo M., Properties and characterization of fully deacetylated chitosan., *Int. J. Biol. Macromol.*, **5**(1), 1983, 49–52.
19. Mima S., Miya M., Iwamoto R., Yoshikawa S., Highly deacetylated chitosan and its properties, *J. Appl. Polym. Sci.*, **28**, 1983, 1909–1917.
20. Lamarque G., Viton C., Domard A., Comparative study of the second and third heterogeneous deacetylations of α- and β-chitins in a multistep process, *Biomacromolecules*, **5**(5), 2004, 1899–1907.
21. Sannan T., Kurita K., Iwakura Y., Studies on chitin, 2. Effect of deacetylation on solubility, *Makromol. Chem.*, **177**(12), 1976, 3589–3600.
22. Pelletier A., Lemire Y., Sygusch J., Chornet E., Overend R.P., Chitin/chitosan transformation by thermo-mechano-chemical treatment including characterization by enzymatic depolymerization., *Biotechnol. Bioeng.*, **36**(3), 1990, 310–315.
23. Focher B., Beltrame P.L., Naggi A., Torri G., Alkaline *N*-deacetylation of chitin enhanced by flash treatments. Reaction kinetics and structure modifications, *Carbohyd. Polym.*, **12**, 1990, 405–418.
24. Muzzarelli R.A.A., Rocchetti R., Determination of the degree of acetylation of chitosans by first derivative ultraviolet spectrophotometry, *Carbohyd. Polym.*, **5**, 1985, 461–472.
25. Tan S.C., Khor E., Tan T.K., Wong S.M., The degree of deacetylation of chitosan: Advocating the first derivative UV spectrophotometry method of determination, *Talanta*, **45**, 1998, 713–719.
26. Baxter A., Dillon M., Taylor K.D.A., Roberts G.A.F., Improved method for IR determination of the degree of *N*-acetylation of chitosan, *Int. J. Biol. Macromol.*, **14**, 1992, 166–169.
27. Brugnerotto J., Lizardi J., Goycoolea F.M., Argüelles-Monal W., Desbriéres J., Rinaudo M., An infrared investigation in relation with chitin and chitosan characterization, *Polymer*, **42**, 2001, 3569–3580.
28. Duarte M.L., Ferreira M.C., Marvão M.R., Rocha J., An optimised method to determine the degree of acetylation of chitin and chitosan by FTIR spectroscopy, *Int. J. Biol. Macromol.*, **31**, 2002, 1–8.
29. Vårum K.M., Anthonsen M.W., Grasdalen H., Smidsrød O., Determination of the degree of *N*-acetylation and the distribution of *N*-acetyl groups in partially *N*-deacetylated chitins (chitosans) by high field n.m.r. spectroscopy, *Carbohyd. Res.*, **211**, 1991, 17–23.
30. Sashiwa H., Saimoto H., Shigemasa Y., Ogawa R., Tokura S., Distribution of the acetamide group in partially deacetylated chitins, *Carbohyd. Polym.*, **16**, 1991, 291–296.
31. Sashiwa H., Saimoto H., Shigemasa Y., Tokura S., *N*-Acetyl group distribution in partially deacetylated chitins prepared under homogeneous conditions, *Carbohyd. Res.*, **242**, 1993, 167–172.
32. Tsaih M.L., Chen R.H., Molecular weight determination of 83% degree of decetylation chitosan with non-gaussian and wide range distribution by high-performance size exclusion chromatography and capillary viscometry, *J. Appl. Polym. Sci.*, **71**, 1999, 1905–1913.
33. Janes K.A., Alonso M.J., Depolymerized chitosan nanoparticles for protein delivery: Preparation and characterization, *J. Appl. Polym. Sci.*, **88**, 2003, 2769–2776.
34. Schipper N.G., Vårum K.M., Artursson P., Chitosans as absorption enhancers for poorly absorbable drugs. 1: Influence of molecular weight and degree of acetylation on drug transport across human intestinal epithelial (Caco-2) cells, *Pharm. Res.*, **13**, 1996, 1686–1692.
35. Rabea E.I., Badawy M.E.-T., Stevens C.V., Smagghe G., Steurbaut W., Chitosan as antimicrobial agent: Applications and mode of action, *Biomacromolecules*, **4**(6), 2003, 1457–1465.
36. Sun S.F., *Physical Chemistry of Macromolecules: Basic Principles and Issues*, Wiley, New York, 1994. p. 188.
37. Argüelles-Monal W.M., Caballero A.H.H., Acosta N., Galed G., Gallardo A., Miralles B., Peniche C., Román J.S., Caracterización de Quitina y Quitosano, in *Quitina y quitosano: obtención, caracterización y aplicaciones*, Ed.: Pastor de Abram A., Pontificia Universidad Católica del Perú, Lima, 2004. Chapter 4, pp. 157–206.
38. Rinaudo R., Milas M., Dung M.L., Characterisation of chitosan. Influence of ionic strength and degree of acetylation on chain expansion, *Int. J. Biol. Macromol.*, **15**(5), 1993, 281–285.
39. Chu B., *Laser Light Scattering: Basic Principles and Practice*, 2nd Edition, Academic Press, New York, 1991. p. 352.
40. Sorlier P., Viton C., Domard A., Relation between solution properties and degree of acetylation of chitosan: Role of aging, *Biomacromolecules*, **3**, 2002, 1336–1342.

41. Anthonsen M.W., Vårum K.M., Hermansson A.M., Smidsrød O., Brant D.A., Aggregates in acidic solutions of chitosans detected by static laser light scattering, *Carbohyd. Polym.*, **25**, 1994, 13–23.
42. Terbojevich M., Cosani A., Bianchi E., Marsano E., Solution behaviour of chitin in dimethylacetamide/LiCl, in *Advances in Chitin Science*, Ed.: Domard A., Jeuriaux C., Muzzarelli R.A.A. and Roberts G.A.F., Jacques André Publisher, Lyon, 1996, **Volume 1**, pp. 333–339.
43. Berth G., Dautzenberg H., Peter M.G., Physico-chemical characterization of chitosans varying in degree of acetylation, *Carbohyd. Polym.*, **36**, 1998, 205–216.
44. Buhler E., Rinaudo M., Structural and dynamical properties of semirigid polyelectrolyte solutions: A light-scattering study, *Macromolecules*, **33**, 2000, 2098–2106.
45. Cölfen H., Berth G., Dautzenberg H., Hydrodynamic studies on chitosans in aqueous solution, *Carbohyd. Polym.*, **45**, 2001, 373–383.
46. Lamarque G., Lucas J.M., Viton C., Domard A., Physicochemical behavior of homogeneous series of acetylated chitosans in aqueous solution: Role of various structural parameters, *Biomacromolecules*, **6**, 2005, 131–142.
47. Pa J.-H., Yu T.L., Light scattering study of chitosan in acetic acid aqueous solutions, *Macromol. Chem. Phys.*, **202**, 2001, 985–991.
48. Schatz C., Viton C., Delair T., Pichot C., Domard A., Typical physicochemical behaviors of chitosan in aqueous solution, *Biomacromolecules*, **4**(3), 2003, 641.
49. Yau W.W., Kirkland J.J., Bly D.D., *Modern Size Exclusion Liquid Chromatography*, J. Wiley and Sons, New York, 1979. p. 75.
50. Isogai A., Molecular mass distribution of chitin and chitosan, in *Chitin Handbook*, Eds.: Muzzarelli R.A.A. and Peter M.G., European Chitin Society, Atec, Grottammare, Italy, 1997, pp. 103–108.
51. ASTM, Standard guide for characterization and testing of chitosan salts as starting materials intended for use in biomedical and tissue-engineered medical product applications, ASTM F 2103–01, 1–8, 2001.
52. Beri R.G., Walker J., Reese E.T., Rollings J.E., Characterization of chitosans via coupled size-exclusion chromatography and multiple-angle laser light-scattering technique, *Carbohyd. Res.*, **238**, 1993, 11–12.
53. Ottøy M.H., Vårum K.M., Christensen B.E., Anthonsen M.W., Smidsød O., Preparative and analytical size-exclusion chromatography of chitosans, *Carbohyd. Polym.*, **31**, 1996, 253–261.
54. Brugnerotto J., Desbrieres J., Heux L., Mazeau K., Rinaudo M., Overview on structural characterization of chitosan molecules in relation with their behavior in solution, in *Natural and Synthetic Polymers: Challenges and Perspectives*, Ed.: Argüelles-Monal W., Wiley-VCH, Weinheim; *Macromol. Symp.*, **168**, 2001, pp. 1–20.
55. Blackwell J., Minke R., Gardner K.H., *Proceedings of the First International Conference on Chitin/Chitosan*, Ed.: Muzzarelli R.A.A. and Pariser E.R.) MIT Sea Grant Program, Cambridge M.A., 1978, pp. 108–123.
56. Tanner S.F., Chanzy H., Vincendon M., Roux J.C., Gaill F., High-resolution solid-state carbon-13 nuclear magnetic resonance study of chitin, *Macromolecules*, **23**, 1990, 3576–3586.
57. Iwamoto R., Miya M., Mima S., in *Chitin and Chitosan and Related Enzymes*, Eds.: Hirano S. and Tokura S., The Japanese Society of Chitin and Chitosan, Tottori, 1982, pp. 82–86.
58. Focher B., Naggi A., Torri G., Cosani A., Terbojevich M., Structural differences between chitin polymorphs and their precipitates from solutions – Evidence from CP-MAS 13C-NMR, FT-IR and FT-Raman spectroscopy, *Carbohyd. Polym.*, **17**, 1992, 97–102.
59. Kurita K., Tomita K., Tada T., Ishii S., Nishimura S.-I., Shimoda K., Squid chitin as a potential alternative chitin source: deacetylation behavior and characteristic properties, *J. Polym. Sci., Polym. Chem.*, **31**(2), 1993, 485–491.
60. Chanzy H., Chitin crystals, in A*dvances in Chitin Science,* Eds.: Domard A., Roberts G.A.F. abd Vårum K.M., Jacques André Publisher, Lyon, 1997, **Volume II**, pp. 11–21.
61. Blackwell J., Structure of β-chitin or parallel chain systems of poly-β-(1 → 4)-*N*-acetyl-D-glucosamine, *Biopolymers*, **7**(3), 1969, 281–298.
62. Saito Y., Kumagai H., Wada M., Kuga S., Thermally reversible hydration of β-chitin, *Biomacromolecules*, **3**(3), 2002, 407–410.
63. Saito Y., Putaux J.-L., Okano T., Gaill F., Chanzy H., Structural aspects of the swelling of β-chitin in HCL and its conversion into α-chitin, *Macromolecules*, **30**, 1997, 3867–3873.
64. Okuyama K., Noguchi K., Miyazawa T., Yui T., Ogawa K., Molecular and crystal structure of hydrated chitosan, *Macromolecules*, **30**, 1997, 5849–5855.
65. Shirai A., Takahashi K., Rujiravanit R., Nishi N., Tokura S., Regeneration of chitin using new solvent system, *Asia-Pacific Chitin and Chitosan Symposium*, Bangi, 1994, pp. 53–60, 1995.
66. Vincendon M., Proton NMR study of the chitin dissolution mechanism, *Makromol. Chem.*, **186**(9), 1985, 1787–1795.
67. Terbojevich M., Carraro C., Cosani A., Marsano E., Solution studies of the chitin–lithium chloride-*N*,*N*-dimethylacetamide system, *Carbohyd. Res.*, **180**(1), 1988, 73–86.
68. Katchalsky A., Mazur J., Spitnik P., Polybase properties of poly(vinylamine), *J. Polym. Sci.*, **23**, 1957, 513–530.
69. Domard A., pH and CD measurements on a fully deacetylated chitosan: Application to copper(II)-polymer interactions, *Int. J. Biol. Macromol.*, **9**(2), 1987, 98–104.

70. Rinaudo M., Pavlov G., Desbrieres J., Solubilization of chitosan in strong acid medium, *Int. J. Polym. Anal. Charac.*, **5**(3), 1999, 267–276.
71. Rinaudo M., Pavlov G., Desbrieres J., Influence of acetic acid concentration on the solubilization of chitosan, *Polymer*, **40**(25), 1999, 7029–7032.
72. Manning G.S., Limiting laws and counterion condensation in polyelectrolyte solutions II. Self-diffusion of the small ions, *J. Chem. Phys.*, **2**(3), 1969, 934–938.
73. Arguelles-Monal W., Cabrera G., Peniche C., Rinaudo M., Conductometric study of the inter-polyelectrolyte reaction between chitosan and poly(galacturonic acid), *Polymer*, **41**(7), 1999, 2373–2378.
74. Smidsrød O., Haug A., Estimation of the relative stiffness of the molecular chain in polyelectrolytes from measurements of viscosity at different ionic strengths, *Biopolymers*, **10**, 1971, 1213–1227.
75. Lang E.R., Kienzle-Sterzer C.A., Rodriguez-Sanchez D., Rha C., Rheological behavior of a typical random coil polyelectrolyte: Chitosan, *Chitin Chitosan*; *Proceeding 2nd International Conference*, Japanese Society of Chitin and Chitosan, 1982, pp. 34–38, 1982.
76. Kienzle-Sterzer C., Rodriguez-Sanchez D., Rha C., Solution properties of chitosan: Chain conformation, *Chitin, Chitosan, Relat. Enzymes*; *Proceedings of the Joint US–Japan. Seminar on Advances in Chitin, Chitosan, Related Enzymes*, 1984, pp. 383–93, 1984.
77. Anthonsen M.W., Varum K.M., Smidsrod O., Solution properties of chitosans: Conformation and chain stiffness of chitosans with different degrees of *N*-acetylation, *Carbohyd. Polym.*, **22**(3), 1993, 193–201.
78. Terbojevich M., Cosani A., Conio G., Marsano E., Bianchi E., Chitosan: Chain rigidity and mesophase formation, *Carbohyd. Res.*, **209**, 1991, 251–260.
79. Brugnerotto J., Desbrieres J., Roberts G., Rinaudo M., Characterization of chitosan by steric exclusion chromatography, *Polymer*, **42**(25), 2001, 9921–9927.
80. Kienzle-Sterzer C.A., Rodriguez-Sanchez D., Rha C.K., Flow behavior of a cationic biopolymer: Chitosan, *Polym.r Bull.*, **13**(1), 1985, 1–6.
81. Mucha M., Rheological characteristics of semi-dilute chitosan solutions, *Macromol. Chem. Phys.*, **198**(2), 1997, 471–484.
82. Argüelles-Monal W., Goycoolea F.M., Peniche C., Higuera-Ciapara I., Rheological study of the chitosan/glutaraldehyde chemical gel system, *Polym. Gels Networks*, **6**, 1998, 429–440.
83. Desbrieres J., Viscosity of semiflexible chitosan solutions: Influence of concentration, temperature, and role of intermolecular interactions, *Biomacromolecules*, **3**(2), 2002, 342–349.
84. Hamdine M., Heuzey M.-C., Begin A., Effect of organic and inorganic acids on concentrated chitosan solutions and gels, *Int. J. Biol. Macromol.*, **37**(3), 2005, 134–142.
85. Cho J., Heuzey M.-C., Begin A., Carreau P.J., Viscoelastic properties of chitosan solutions: Effect of concentration and ionic strength, *J. Food Eng.*, **74**(4), 2006, 500–515.
86. Amiji M.M., Pyrene fluorescence study of chitosan selfassociation in aqueous solution, *Carbohyd. Polym.*, **26**, 1995, 211–213.
87. Muzzarelli R.A.A., Natural Chelating Polymers, Ed.: Press A., New York, 1973, pp. 134.
88. Varma A.J., Deshpande S.V., Kennedy J.F., Metal complexation by chitosan and its derivatives: A review, *Carbohyd. Polym.*, **55**, 2004, 77–93.
89. Muzzarelli R.A.A., Tubertini O., Purification of thallium(I) nitrate by column chromatography on chitosan, *Microchem. Acta*, **5**, 1970, 892–899.
90. Koshijima T., Tanaka R., Muraki E., Yamada A., Yaku F., Chelating polymers derived from cellulose and chitin. I. Formation of polymer complexes with metal ions, *Cellulose Chem. Technol.*, **7**(2), 1973, 197–208.
91. Masri M.S., Reutger F.W., Friedman M., Binding of metal cations by natural substances, *J. Appl. Polym. Sci.*, **18**(3), 1974, 675–681.
92. Vold I.M.N., Vårum K.M., Guibal E., Smidsrød O., Binding of ions to chitosan – Selectivity studies, *Carbohyd. Polym.*, **54**, 2003, 471–477.
93. Ogawa K., Oka K., Miyanishi T., Hirano S., X-ray diffraction study of chitosan–metal complexes, in *Chitin, Chitosan and Related Enzimes*, Ed.: Zikakis J.P., Academic Press, Orlando, FL, 1984, pp. 327–345.
94. Muzzarelli R.A.A., Tanfani F., *N*-(carboxymethyl) chitosans and *N*-(*O*-carboxybenzyl) chitosans: Novel chelating polyampholytes, in *Chitin and Chitosan*, Eds.: Hirano S. and Tokura S., The Japanese Society of Chitin and Chitosan, Totori, 1982, pp. 45–53.
95. Muzzarelli R.A.A., Tanfani F., *N*-(*O*-Carboxybenzyl) chitosan, *N*-carboxymethyl chitosan, and chitosan dithiocarbamate: new chelating derivatives of chitosan, *Pure Appl. Chem.*, **54**(11), 1982, 2141–2150.
96. Argüelles-Monal W., Peniche C., Preparation and characterization of a mercaptan derivative of chitosan for the removal of mercury from brines, *Die Angewandte Makromol. Chemie*, **207**, 1993, 1–8.
97. Kang D.W., Choi H.R., Kweon D.K., Stability constants of amidoximated chitosan-g-poly(acrylonitrile) copolymer for heavy metal ions, *J. Appl. Polym. Sci.*, **73**(4), 1999, 469–476.

98. Yazdani-Pedram M., Retuert J., Homogeneous grafting reaction of vinyl pyrrolidone onto chitosan, *J. Appl. Polym. Sci.*, **63**(10), 1997, 1321–1326.
99. Tsuchida E., Abe K., Interactions between macromolecules in solution and intermacromolecular complexes, *Adv. Polym. Sci.*, **45**, 1982, 1–130.
100. Philipp B., Dautzenberg H., Linow K.J., Koetz J., Dawydoff W., Polyelectrolyte complexes – Recent developments and open problems, *Progr. Polym. Sci.*, **14**(1), 1989, 91–172.
101. Peniche C., Argüelles-Monal W., Chitosan based polyelectrolyte complexes, in *Natural and Synthetic Polymers: Challenges and Perspectives*, Ed. Argüelles-Monal W., Wiley-VCH, Weinheim; *Macromol. Symp.*, **168**, 2001, pp. 103–116.
102. Dragan S., Cristea M., Polyelectrolyte complexes. Formation, characterization and applications, *Recent Res. Dev. Polym. Sci.*, **7**, 2003, 149–181.
103. Kabanov V.A., Polyelectrolyte complexes in solution and in bulk, *Russ. Chem. Rev.*, **74**(1), 2005, 3–20.
104. Kubota N., Shimoda K., Macromolecule complexes of chitosan, in *Polysaccharides*, Ed.: Dumitriu S., 2nd Edition, Marcel Dekker, Inc., New York, 2005, pp. 679–706.
105. Fukuda H., Kikuchi Y., Polyelectrolyte complexes of sodium carboxymethylcellulose with chitosan, *Makromol. Chem.*, **180**(6), 1979, 1631–1633.
106. Kikuchi Y., Oshima A., Structure and properties of a polyelectrolyte complex consisting of carboxymethyl cellulose, poly(vinyl alcohol) sulfate, and chitosan, *Nippon Kagaku Kaishi*(8), 1979, 1101–1105.
107. Arguelles-Monal W., Peniche-Covas C., Study of the interpolyelectrolyte reaction between chitosan and carboxymethyl cellulose, *Makromolekulare Chemie, Rapid Commun.*, **9**(10), 1988, 693–697.
108. Arguelles-Monal W., Garciga M., Peniche-Covas C., Study of the stoichiometric polyelectrolyte complex between chitosan and carboxymethyl cellulose, *Polym. Bull. (Berlin, Germany)*, **23**(3), 1990, 307–313.
109. Lee K.Y., Park W.H., Ha W.S., Polyelectrolyte complexes of sodium alginate with chitosan or its derivatives for microcapsules, *J. Appl. Polym. Sci.*, **63**(4), 1997, 425–432.
110. Cárdenas A., Argüelles-Monal W., Goycoolea F.M., Higuera-Ciapara I., Peniche C., Diffusion through membranes of the polyelectrolyte complex of chitosan and alginate, *Macromol. Biosci.*, **3**(10), 2003, 535–539.
111. Becheran-Maron L., Peniche C., Argüelles-Monal W., Study of the interpolyelectrolyte reaction between chitosan and alginate: Influence of alginate composition and chitosan molecular weight, *Int. J. Biol. Macromol.*, **34**(1–2), 2004, 127–133.
112. Peniche C., Howland I., Carrillo O., Zaldivar C., Argüelles-Monal W., Formation and stability of shark liver oil loaded chitosan/calcium alginate capsules, *Food Hydrocolloids*, **18**(5), 2004, 865–871.
113. Chavasit V., Kienzle-Sterzer C., Torres J.A., Formation and characterization of an insoluble polyelectrolyte complex chitosan–polyacrylic acid, *Polym. Bull.*, **19**, 1988, 223–230.
114. Skorikova E.E., Vikhoreva G.A., Kalyuzhnaya R.I., Zezin A.B., Gal'braikh L.S., Kabanov V.A., Polyelectrolyte complexes based on chitosan, *Vysokomol. Soedin., Seriya A*, **30**(1), 1988, 44–49.
115. Yao K.D., Tu H., Cheng F., Zhang J.W., Liu J., pH-Sensitivity of the swelling of a chitosan–pectin polyelectrolyte complex, *Angewandte Makromol. Chem.*, **245**, 1997, 63–72.
116. Rashidova S.S., Milusheva R.Y., Semenova L.N., Mukhamedjanova M.Y., Voropaeva N.L., Vasilyeva S., Faizieva R., Ruban I.N., Characteristics of interactions in the pectin–chitosan system, *Chromatographia*, **59**, 2004, 779–782.
117. Bernabe P., Peniche C., Argüelles-Monal W., Swelling behavior of chitosan/pectin polyelectrolyte complex membranes. Effect of thermal cross-linking, *Polym. Bull.*, **55**(5), 2005, 367–375.
118. Jiang S., Lin R., Reaction of chitosan with carrageenan in solution and the characteristics of the complex, *Zhongguo Haiyang Yaowu*, **13**(1), 1994, 19–23.
119. Hugerth A., Caram-Lelha N., Sundelof L.-O., The effect of charge density and conformation on the polyelectrolyte complex formation between carrageenan and chitosan, *Carbohyd. Polym.*, **34**(3), 1997, 149–156.
120. Goycoolea F.M., Argüelles-Monal W., Peniche C., Higuera-Ciapara I., Effect of chitosan on the gelation of κ-carrageenan under various salt conditions, *Hydrocolloids*, (*based on the Presentations at* (*the*) *Osaka City University International Symposium 98, Joint Meeting with the 4th International Conference on Hydrocolloids*), Osaka, October 4–10, 1998, 2, 211–216, 2000.
121. Arguelles-Monal W., Goycoolea F.M., Lizardi J., Peniche C., Higuera-Ciapara I., Chitin and chitosan in gel network systems, in *ACS Symposium Series*, Eds.: Bohidar H.B., Dubin P., and Osada Y., 2003, pp. 102–121.
122. Kikuchi K., Noda A., Polyelectrolyte complexes of heparin with chitosan, *J. Appl. Polym. Sci.*, **20**(9), 1976, 2561–2563.
123. Kikuchi Y., Fukuda H., Polyelectrolyte complex of sodium dextransulfate with chitosan, *Nippon Kagaku Kaishi*(9), 1976, 1505–1508.
124. Fukuda H., Kikuchi Y., Polyelectrolyte complexes of chitosan with sodium carboxymethyldextran, *Bull. Chem. Soc. Japan*, **51**(4), 1978, 1142–1144.
125. Hirano S., Mizutani C., Yamaguchi R., Miura O., Formation of the polyelectrolyte complexes of some acidic glycosaminoglycans with partially *N*-acylated chitosans, *Biopolymers*, **17**(3), 1978, 805–810.

126. Pushpa S., Srinivasan R., Polyelectrolyte complexes of glycol chitosan with some mucopolysaccharides: Dielectric properties and electric conductivity, *Biopolymers*, **23**(1), 1984, 59–69.
127. Stoilova O., Koseva N., Manolova N., Rashkov I., Polyelectrolyte complex between chitosan and poly(2-acryloylamido-2-methylpropanesulfonic acid), *Polym. Bull.*, **43**(1), 1999, 67–73.
128. Berth G., Voigt A., Dautzenberg H., Donath E., Moehwald H., Polyelectrolyte complexes and layer-by-layer capsules from chitosan/chitosan sulfate, *Biomacromolecules*, **3**(3), 2002, 579–590.
129. Gamzazade A.I., Nasibov S.M., Formation and properties of polyelectrolyte complexes of chitosan hydrochloride and sodium dextransulfate, *Carbohyd. Polym.*, **50**(4), 2002, 339–343.
130. Mincheva R., Manolova N., Paneva D., Rashkov I., Novel polyelectrolyte complexes between *N*-carboxyethylchitosan and synthetic polyelectrolytes, *Eur. Polym. J.*, **42**(4), 2006, 858–868.
131. Rusu-Balaita L., Desbrieres J., Rinaudo M., Formation of a biocompatible polyelectrolyte complex: Chitosan–hyaluronan complex stability, *Polym. Bull.*, **50**(1–2), 2003, 91–98.
132. Mansouri S., Lavigne P., Corsi K., Benderdour M., Beaumont E., Fernandes J.C., Chitosan–DNA nanoparticles as non-viral vectors in gene therapy: Strategies to improve transfection efficacy, *Eur. J. Pharm. Biopharm.*, **57**(1), 2004, 1–8.
133. Liu W., Sun S., Cao Z., Zhang X., Yao K., Lu W.W., Luk K.D.K., An investigation on the physicochemical properties of chitosan/DNA polyelectrolyte complexes, *Biomaterials*, **26**(15), 2005, 2705–2711.
134. Strand S.P., Danielsen S., Christensen B.E., Vrum K.M., Influence of chitosan structure on the formation and stability of DNA–chitosan polyelectrolyte complexes, *Biomacromolecules*, **6**(6), 2005, 3357–3366.
135. Perez-Gramatges A., Argüelles-Monal W., Peniche-Covas C., Thermodynamics of complex formation of poly(acrylic acid) with poly(*N*-vinyl-2-pyrrolidone) and chitosan, *Polym. Bull.*, **37**(1), 1996, 127–134.
136. Lutsenko V.V., Zezin A.B., Kalyuzhnaya R.I., Thermodynamics of the cooperative interaction of polyelectrolytes in aqueous solutions, *Vysokomol. Soedin., Seriya A*, **16**(11), 1974, 2411–2417.
137. Argüelles-Monal W., Hechavarria O.L., Rodriguez L., Peniche C., Swelling of membranes from the polyelectrolyte complex between chitosan and carboxymethyl cellulose, *Polym. Bull.*, **31**(4), 1993, 471–478.
138. Grishina N.V., Rogacheva V.B., Lopatina L.I., Zezin A.B., Kabanov V.A., Transformation of the structure and properties of a complex of poly(acrylic acid) and linear polyethylenimine during intracomplex amidation in aqueous solutions, *Vysokomol. Soedin., Seriya A*, **27**(6), 1985, 1154–1159.
139. Peniche C., Argüelles-Monal W., Davidenko N., Sastre R., Gallardo A., San Román J., Self-curing membranes of chitosan/PAA IPNs obtained by radical polymerization: preparation, characterization and interpolymer complexation, *Biomaterials*, **20**(20), 1999, 1869–1878.
140. Barroso F., Argüelles W., Peniche C., Swelling and permeability of chitosan/carboxymethyl cellulose polyelectrolyte complex membranes: Effect of pH and Ca^{2+} ions, *Adv. Chitin Sci.*, **2**, 1997, 573–579.
141. Barroso F., Argüelles-Monal W., Peniche-Covas C., Evaluación y permeabilidad de membranas del complejo polielectrolito de la quitosana y la carboximetil celulosa, *Revista Cubana de Quimica*, **8**(9), 1996–1997, 101–108.
142. Pantaleone D., Yalpani M., Scollar M., Unusual susceptibility of chitosan to enzymic hydrolysis, *Carbohyd. Res.*, **237**, 1992, 325–332.
143. Hirano S., Noshiki Y., Kinugawa J., Higashijima H., Hayashi T., Chitin and chitosan for use as novel biomedical materials, in *Adv. Biomed. Polym.*, Ed.: Gebelein L.G., Plenum, New York, 1987, p. 285.
144. Muzzarelli R.A.A., *Recent results in the oral administration of chitosan, EUCHIS*, Universitaet Potsdam, Potsdam, 1999. pp. 212–216, 2000.
145. Domard A., Domard M., Chitosan: Structure-properties relationship and biomedical applications, in *Polymeric Biomaterials*, Ed.: Dumitriu S., 2nd Edition, Marcel Dekker, Inc., New York, 2002, pp. 187–212.
146. Suzuki K., Tokoro A., Okawa Y., Suzuki S., Suzuki M., Effect of *N*-acetylchito-oligosaccharides on activation of phagocytes, *Microbiol. Immunol.*, **30**(8), 1986, 777–787.
147. Peniche C., Argüelles-Monal W., Peniche H., Acosta N., Chitosan: An attractive biocompatible polymer for microencapsulation, *Macromol. Biosci.*, **3**(10), 2003, 511–520.
148. Gallardo A., Aguilar M.R., Elvira C., Peniche C., Román J.S., Chitosan based microcomposites – From biodegradable microparticles to self-curing hydrogels, in *Biodegradable Systems in Tissue Engineering*, Eds.: Reis R. and Román J.S., CRC Press, Boca Ratón, 2005, pp. 145–162.
149. Pena J., Izquierdo-Barba I., Martinez A., Vallet-Regi M., New method to obtain chitosan/apatite materials at room temperature, *Solid State Sci.*, **8**, 2006, 513–519.
150. Singla A.K., Chawla M., Chitosan: Some pharmaceutical and biological aspects – An update, *J. Pharm. Pharmacol.*, **53**(8), 2001, 1047–1067.
151. Okamoto Y., Kawakami K., Miyakate K., Morimoto M., Shigemasa Y., Minami S., Analgesic effect of chitin and chitosan, *Carbohyd. Polym.*, **49**, 2002, 249–252.
152. Kweon D.K., Song S.B., Park Y.Y., Preparation of water-soluble chitosan/heparin complex and its application as wound healing accelerator, *Biomaterials*, **24**(9), 2003, 1595–1601.

153. Risbud M., Endres M., Ringe J., Bhonde R., Sittinger M., Biocompatible hydrogel supports the growth of respiratory epithelial cells: Possibilities in tracheal tissue engineering, *J. Biomed. Mater. Res. Part A*, **56**, 2001, 120–127.
154. Risbud M., Ringe J., Bhonde R., Sittinger M., *In vitro* expression of cartilage-specific markers by chondrocytes on a biocompatible hydrogel: Implications for engineering cartilage tissue, *Cell Transplant.*, **10**, 2001, 755–763.
155. Cheng M., Deng J., Yang F., Gong Y., Zhao N.Z., Zhang X., Study on physical properties and nerve cell affinity of composite films from chitosan and gelatin solutions, *Biomaterials*, **24**, 2003, 2871–2880.
156. Zhao F., Yin Y., Lu W.W., Leong J.C., Zhang W., Zhang J., Zhang M., Yao K., Preparation and histological evaluation of biomimetic three-dimensional hydroxyapatite/chitosan–gelatin network composite scaffolds, *Biomaterials*, **23**, 2002, 3227–3234.
157. Sechriest V.F., Miao Y.J., Niyibizi C., Westerhausen-Larson A., Matthew H.W., Evans C.H., Fu F.H., Suh J.K., GAG-augmented polysaccharide hydrogel: A novel biocompatible and biodegradable material to support chondrogenesis, *J. Biomed. Mater. Res. Part A*, **49**, 2000, 534–541.
158. Park Y.J., Lee Y.M., Lee J.Y., Seol Y.J., Chung C.P., Lee S.J., Controlled release of platelet-derived growth factor-BB from chondroitin sulfate–chitosan for guided bone regeneration, *J. Control. Rel.*, **67**, 2000, 385–394.
159. Shahabeddin L., Berthod F., Damour O., Collombel C., Characterization of skin reconstructed on a chitosan–cross-linked collagen–glycosaminoglycan matrix, *Skin Pharmacol.*, **3**, 1990, 107–114.
160. Denuziere A., Ferrier D., Damour O., Domard A., Chitosan–chondroitin sulfate and chitosan–hyaluronate polyelectrolyte complexes: Biological properties, *Biomaterials*, **19**(14), 1998, 1275–1285.
161. Lee Y.M., Park Y.J., Lee S.J., Ku Y., Han S.B., Klokkevold P.R., Chung C.P., The bone regenerative effect of platelet-derived growth factor-BB delivered with a chitosan/tricalcium phosphate sponge carrier, *J. Periodontol.*, **71**(3), 2000, 418–424.
162. Peniche C., Argüelles W., Gallardo A., Elvira C., Román J.S., Quitosano: un polisacárido natural biodegradable y biocompatible con aplicaciones en biotecnología y biomedicina, *Revista Plásticos Modernos*, **81**(535), 2001, 81–91.
163. Zhou D., Kintsourashvili E., Mamujee S., Vacek I., Sun A.M., Bioartificial pancreas: Alternative supply of insulin-secreting cells, in *Bioartificial organs II. Technology, medicine & materials*, Eds.: Hunkeler D., Prokop A., Cherington A.D., Rajotte R.V., and Sefton M., *Annals of the New York Academy of Sciences*, New York; *Ann. NY. Acad. Sci.* **875**, 1999, pp. 208–218.
164. Ravi-Kumar M.N.V., A review of chitin and chitosan applications, *React. Funct. Polym.*, **46**, 2000, 1–27.
165. Berger J., Reist M., Mayer J.M., Felt O., Gurny R., Structure and interactions in chitosan hydrogels formed by complexation or aggregation for biomedical applications, *Eur. J. Pharm. Biopharm.*, **57**(1), 2004, 35–52.
166. Yao K.D., Yin Y.J., Xu M.X., Wang Y.F., Investigation of pH-sensitive drug delivery system of chitosan/gelatin hybrid polymer network, *Polym. Int.*, **38**, 1995, 77–82.
167. Chandi T., Rao G.H., Wilson R.F., Das G.S., Delivery of LMW heparin via surface coated chitosan/peg–alginate microspheres prevent thrombosis, *Drug. Deliv.*, **9**, 2000, 87–96.
168. Bayoni M.A., Influence of polymers weight ratio and pH of polymers solution on the characteristics of chitosan carboxymethyl cellulose microspheres containig theophyline, *Boll. Chem. Farm.*, **142**, 2003, 336–342.
169. Chellet F., Tabrizian M., Dumitriu S., Chornet R., Rivard C.H., Yahian L., Study of biodegradation behavior of chitosan–xanthan microspheres in simulated physiological media, *J. Biomed. Mater. Res. Part B*, **53**, 2000, 592–599.
170. Hejazi R., Amiji M., Chitosan-based delivery systems: Physicochemical properties and pharmaceutical applications, in *Polymeric Biomaterials*, Ed.: Dumitriu S., 2nd Edition, Marcel Decker, Inc., New York, 2002, pp. 213–237.
171. Thacharodi D., Rao K.P., Development and *in vitro* evaluation of chitosan based transdermal drug delivery systems for controlled delivery of propranolol hydrochloride, *Biomaterials*, **16**, 1995, 145–148.
172. Lee K.Y., Kwon I.C., Kim Y.H., Jo W.H., Jeong S.Y., Preparation of chitosan self aggregates as a gene delivery system, *J. Control. Rel.*, **51**, 1998, 213–220.
173. Yamaguchi T., Ito Y., Shibuya N., Oligosaccharide elicitors and their receptors for plant defense responses, *Trends Glycosci. Glycotechnol.*, **12**(64), 2000, 113–120.
174. Struszczyk H., Pospieszny H., Kotlinski K., Some new applications of chitosan in agriculture, in *Chitin and Chitosan Sources, Chemistry, Biochemistry, Physical Properties and Applications*, Eds.: Bræk G.S., Anthonsen T. and Sandford P., Elsevier Applied Sci., London, 1989, pp. 733–742.
175. Gooday G.W., The ecology of chitin degradation, *Adv. Micro. Ecol.*, **11**, 1990, 387–419.
176. Hirano S., Hayashi M., Nishida T., Yamamoto T., Chitinase activity of some seeds during their germination process and its induction by treating with chitosan and derivatives, in *Chitin and Chitosan Sources, Chemistry, Biochemistry, Physical Properties and Applications*, Eds.: Bræk G.S., Anthonsen T. and Sandford P., Elsevier Applied Sci., London, 1989, pp. 733–742.
177. Galed G., Fernández-Valle M.E., Martínez A., Heras A., Application of MRI to monitor the process of ripening and decay in citrus treated with chitosan, *Magnet. Reson. Imaging*, **22**, 2004, 127–137.

178. Schisler D.A., Slininger P.J., Behle R.W., Jackson M.A., Formulation of *Bacilus sp.* for biological control of plant diseases, *Phytopathology*, **94**(11), 2004, 1267–1271.
179. Hadwiger L.A., Methods for treating cereal crops with chitosan, US, Patent Patent # US 5,104,437, 1992, 6 pp.
180. FDA 2002, Agency Response Letter, *GRAS Notice No. GRN 000073.* Letter to Lee B. Dexter, Lee B. Dexter and Associates, from Linda Kahl, Division of Biotech and GRAS Notice Review, Office of Food Additive Safety, Center for Food Safety and Applied Nutrition, U.S. Food and Drug Administration, February 2, 2002, http://www.cfsan.fda.gov/~rdb/opa-g073.html
181. Shahidi F., Arachi J.K., Jeon Y.J., Food applications of chitin and chitosan, *Trends Food Sci. Technol.*, **10**, 1999, 37–51.
182. Tharanathan R.N., Kittur F.S., Chitin – The undisputed biomolecule of great potential, *Crit. Rev. Food Sci.*, **43**, 2003, 61–87.
183. Bough W.A., Landes D., Recovery and nutritional evaluation of proteinaceous solids separated from whey by coagulation with chitosan, *J. Dairy Sci.*, **59**, 1976, 874–1880.
184. Hwang D.-C., Damodaran S., Selective precipitation and removal of lipids from cheese whey using chitosan, *J. Agric. Food Chem.*, **43**(1), 1995, 33–37.
185. Savant V.D., Torres J.A., Chitosan-based coagulating agents for treatment of cheddar cheese whey, *Biotechnol. Prog.*, **16**, 2000, 1091–1097.
186. Ahmad A.L., Sumathi S., Hameed B.H., Chitosan: A natural biopolymer for the adsorption of residue oil from oily wastewater, *Adsorpt. Sci. Technol.*, **22**(1), 2004, 75–88.
187. Noomhorm A., Kupongsak S., Chandrkrachang S., Deacetylated chitin used as adsorbent in production of clarified pineapple syrup, *J. Sci. Food Agric.*, **76**, 1998, 226–232.
188. Soto-Perlata N.V., Muller H., Knoor D., Effect of chitosan treatment on the clarity and color of apple juice, *J. Food Sci.*, **54**, 1999, 495–496.
189. Spagna G., Pifferi P.G., Rangoni C., Mattivi F., Nicolini G., Palmonari R., The stabilization of white wines by adsorption of phenolic compounds on chitin and chitosan, *Food Res. Int.*, **29**(3–4), 1996, 241–248.
190. Sekiguchi S., Miura Y., Kaneko H., Nishimura S.I., Nishi N., Iwase M., Tokura S., Molecular weight dependency of antimicrobial activity by chitosan oligomers, in *Food Hydrocolloids: Structures, Properties and Functions*, Eds.: Nishinari K. and Doi E., Plenum Press, New York, 1993, pp. 71–76.
191. Coma V., Martial-Gros A., Garreau S., Copinet A., Salin F., Deschamps A., Edible antimicrobial films based on chitosan matrix, *J. Food Sci.*, **67**, 2002, 1162–1169.
192. Ghaouth A.E., Arul J., Wilson C., Benhamou N., Biochemical and cytochemical aspects of chitosan and *Botrytis cinerea* in bell pepper fruit, *Postharvest Biol. Tec.*, **12**, 1997, 183–194.
193. Ghaouth A.E., Ponnampalam R., Cataigne F., Arul J., Chitosan coating to extend the storage life of tomatoes, *Hort Sci.*, **27**(9), 1992, 1016–1018.
194. Ouattar B., Simard R.E., Piett G., Begin A., Hollye R.A., Inhibition of surface spoilage bacteria in processed meats by application of antimicrobial films prepared with chitosan, *Int. J. Food Microbiol.*(62), 2000, 139–148.
195. Darmadji P., Izumimoto M., Effects of chitosan in meat preservation, *Meat Sci.*, **38**, 1994, 243–254.
196. Tsai G.J., Su W.H., Chen H.C., Pan C.L., Antimicrobial activity of shrimp chitin and chitosan from different treatments and applications of fish preservation, *Fisheries Sci.*, **68**, 2002, 170–177.
197. Schulz P.C., Rodríguez M.S., Blanco L.F.D., Pistonesi M., Agulló E., Emulsification properties of chitosan, *Colloid Polym. Sci.*, **276**, 1998, 1159–1165.
198. Xie W., Xu P., Liu Q., Antioxidant activity of water-soluble chitosan derivatives, *Bioorg. Med. Chem. Lett.*, **11**, 2001, 1699–1701.
199. Arredondo E., Yamashita Y., Ichikawa H., Goto S., Osatomi K., Nozaki Y., Effect of chitosan from shrimp, squid and crab on the state of water and denaturation of myofibriliar protein during frozen storage, in *Adv. Chitin Sci.*, Eds.: Domard A., Roberts G.A.F. and Vårum K.M., Jacques André Publisher, Lyon, 1997, **Volume II**, pp. 815–822.
200. Juneau A., Georgalas A., Kapino R., Chitosan in cosmetics: Technical aspects when formulating, *Cosmet. Toiletries*, **116**(8), 2001, 73–80.
201. Wachter R., Stenberg E., HYDAGEN CMF in Cosmetic applications. Efficacy in different *in-vitro* and *in-vivo* measurements, in *Advances in Chitin Science*, Eds.: Domard A., Roberts G.A.F., and Varum K.M., Lyon, **I**, 1997, pp. 381–388.
202. Horner V., Pittermann W., Wachter R., Efficiency of high molecular weight chitosan in skin care application, in *Adv. Chitin Sci.*, Eds.: Domard A., Roberts G.A.F., and Vårum K.M., Jacques Andre Publisher, Lyon, 1997, **Volume II**, pp. 671–677.
203. Hohle M., Griesbach V., Chitosan: A deodorizing component, *Cosmet. Toiletries*, **114**(12), 1998, 61–64.
204. Dee G.J., Rhode O., Wachter R., Chitosan multi-functional marine polymer, *Cosmet. Toiletries*, **116**(2), 2001, 39–44.
205. Sano H., Shibasaki K., Matsukubo T., Takaesu Y., Effect of chitosan rinsing on reduction of dental plaque formation., *Bull. Tokyo Dent. Coll.*, **44**(1), 2003, 9–16.
206. Muzzarelli R.A.A., Chitin and its derivatives. New trends of applied research, *Carbohyd. Polym*, **3**, 1983, 53–75.

207. Svang-Ariyaskul A., Huang R.Y.M., Douglas P.L., Pal R., Feng X., Chen P., Liu L., Blended chitosan and polyvinyl alcohol membranes for the pervaporation dehydration of isopropanol, *J. Membrane Sci.*, **280**, 2006, 815–823.
208. Wang G., Maogen Z., Waldemar G., Highly sensitive sensors based on the immobilization of tyrosinase in chitosan, *Bioelectrochemistry*, **57**(1), 2002, 33–38.
209. Taqieddin E., Amiji M., Enzyme immobilization in novel alginate–chitosan core-shell microcapsules, *Biomaterials*, **25**(10), 2004, 1937–1945.
210. Hirai A., Odani H., Nakajima A., Determination of degree of deacetylation of chitosan by ^{1}H NMR spectroscopy, *Polym. Bull.*, **26**, 1991, 87–94.
211. Fernandez-Megia E., Novoa-Carballal R., Quiñoá E., Riguera R., Optimal routine conditions for the determination of the degree of acetylation of chitosan by ^{1}H-NMR, *Carbohyd. Polym.*, **61**(2), 2005, 155–161.
212. Domard A., Determination of *N*-acetyl content in chitosan samples by c.d. measurements, *Int. J. Biol. Macromol.*, **9**(6), 1987, 333–336.
213. Broussignac P., Chitosan, a natural polymer not well known by the industry, *Chim. Ind. Genie Chim.*, **99**, 1968, 1241.
214. Raymond L., Morin F.G., Marchessault R.H., Degree of deacetylation of chitosan using conductometric titration and solid-state NMR, *Carbohyd. Res.*, **246**(1), 1993, 331–336.
215. Nanjo F., Katsumi R., Sakai K., Enzymatic method for determination of the degree of deacetylation of chitosan, *Anal. Biochem.*, **193**(2), 1991, 164–167.
216. Duarte M.L., Ferreira M.C., Marvão M.R., Rocha J., Determination of the degree of acetylation of chitin materials by ^{13}C CP/MAS spectroscopy, *Int. J. Biol. Macromol.*, **28**, 2001, 359–363.
217. Yu G., Morin F.G., Nobes G.A.R., Marchessault R.H., Degree of acetylation of chitin and extent of grafting PHB on chitosan determined by solid state ^{15}N NMR, *Macromolecules*, **32**, 1999, 518–520.
218. Heux L., Brugnerotto J., Desbriéres J., Versali M.-F., Rinaudo M., Solid state NMR for determination of degree of acetylation of chitin and chitosan, *Biomacromolecules*, **1**, 2000, 746–751.
219. Inoue Y., NMR determination of the degree of acetylation, in *Chitin Handbook*, Eds.: Muzzarelli R.A.A. and Peter M.G., European Chitin Society, Grottamare, 1997, pp. 133–136, Chapter 3.
220. García-Alonso J., Peniche-Covas C., Nieto J.M., Determination of the degree of acetylation of chitin and chitosan by thermal analysis, *J. Thermal Anal.*, **28**(1), 1983, 189–193.
221. Sato H., Mizutani S., Tsuge S., Ohtani H., Aoi K., Takasu A., Okada M., Kobayashi S., Kiyosada T., Shoda S., Determination of the degree of acetylation of chitin/chitosan by pyrolysis-gas chromatography in the presence of oxalic acid, *Anal. Chem.*, **70**(1), 1998, 7–12.
222. Niola F., Basora N., Chornet F., Vidal P.F., A rapid method for the determination of the degree of *N*-acetylation of chitin–chitosan samples by acid hydrolysis and HPLC, *Carbohyd. Res.*, **238**, 1993, 1–9.
223. Hudson, S.M., Application of chitin and chitosan as fiber and textile chemicals, in *Advances in Chitin Science*, Eds.: Domard A., Roberts G.A.F. and Vårum K.M., Jacques André Publisher, Lyon, 1997, **Volume II**, pp. 590–599.
224. Maghami G.G., Roberts G.A.F., Studies on the adsorption of anionic dyes on chitosan, *Makromol. Chem.*, **189**, 1988, 2239–2243.

Index

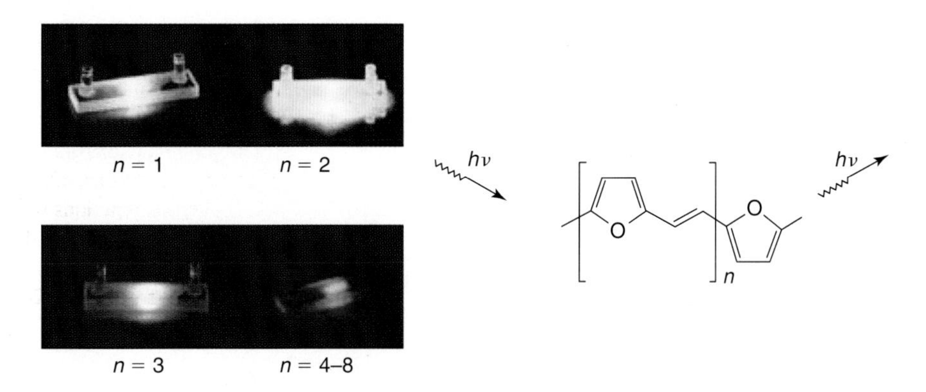

Figure 6.4 Photoluminescence of oligomers oligo(furylene vinylene)s with different lengths [57b].